CONSTRUCTION

Canadian Edition

APPLIED CALCULUS

FOR BUSINESS, ECONOMICS, AND THE SOCIAL AND LIFE SCIENCES

Laurence D. Hoffmann
Smith Barney

Gerald L. Bradley
Claremont McKenna College

Dot Miners
Brock University

McGraw-Hill
Ryerson
Connect. Learn. Succeed.

APPLIED CALCULUS FOR BUSINESS, ECONOMICS, AND THE SOCIAL AND LIFE SCIENCES
CANADIAN EDITION

ISBN-13: 978-0-07-068717-2
ISBN-10: 0-07-068717-X

1 2 3 4 5 6 7 8 9 10 CTPS 1 9 8 7 6 5 4 3 2

Printed and bound in China.

Care has been taken to trace ownership of copyright material contained in this text; however, the publisher
will welcome any information that enables them to rectify any reference or credit for subsequent editions.

Executive Sponsoring Editor: *Leanna MacLean*
Sponsoring Editor: *James Booty*
Marketing Manager: *Cathie Lefebvre*
Developmental Editors: *Amy Rydzanicz & Sarah Fulton*
Senior Editorial Associates: *Stephanie Giles & Erin Catto*
Supervising Editor: *Jessica Barnoski*
Copy Editor: *Julia Cochrane*
Proofreader: *Laurel Sparrow*
Production Coordinator: *Scott Morrison*
Cover Design: *Word & Image Design Studio Inc.*
Cover Image: *Peter Hinson/Jim Denevan*
Interior Design: *Laserwords Private Limited*
Page Layout: *Aptara, Inc.*
Printer: *China Translation & Printing Services Limited*

Library and Archives Canada Cataloguing in Publication

Hoffmann, Laurence D., 1943–
    Applied calculus for business, economics, and the social and life sciences / Laurence D. Hoffmann,
    Gerald L. Bradley, Dot Miners.—1st Canadian ed.
    Includes index.
    ISBN 978-0-07-068717-2
    1. Calculus—Textbooks. I. Bradley, Gerald L., 1940- II. Miners, Dot III. Title.

QA303.2.H64 2012
515   C2011–904043–3

CONTENTS

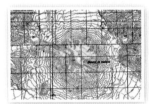

Appendix A Algebra Review, L'Hôpital's Rule, and Summation Notation 849

Answers

PREFACE

Overview of the Canadian Edition

Applied Calculus for Business, Economics, and the Social and Life Sciences, Canadian Edition, provides a sound, intuitive understanding of the basic concepts students need as they pursue careers in business, economics, and the life and social sciences. Students achieve success using this text as a result of the authors' applied and real-world orientation to concepts, problem-solving approach, straightforward and concise writing style, and comprehensive exercise sets.

Improved Exercise Sets

Many new routine and application exercises have been added to the already extensive problem sets. New applied problems help demonstrate the practicality of the material. Four hundred different applications provide a balanced representation of multiple disciplines ensuring that instructors have an application to suit each student (*see inside front and back covers for an Index of Selected Applications*). Hoffmann also maintains a balance between mathematical details on the one hand, and readability and accessibility on the other.

Graphing Calculator Introduction

The Graphing Calculator Introduction can now be found on Connect. This introduction includes instructions regarding common calculator keystrokes, terminology, and introductions to more advanced calculator applications.

Appendix: Algebra Review, L'Hôpital's Rule, and Summation Notation

The appendix reviews topics from high school algebra needed for calculus.

New to the Canadian Edition

- Learning Objectives have been included for ease of use.
- Each chapter opening page now begins with a vignette written by the Canadian author. The vignette links the chapter opening image to a concept developed within the chapter and is cross-referenced to an example or exercise in the text to highlight the real-world relevance of the application.
- Sequencing Suggestions, made by the Canadian author, offer options for alternative sequences throughout the text provided on chapter opening pages in "Option Notes."
- Metric units have been used wherever appropriate in exercises, examples, and applications.
- In response to reviewer feedback, advanced level questions are now designated by an A*.
- Graphing explorations have been modified to be applicable to any technology, such as Maple, Matlab, and Wolfram Alpha, instead of being specific to a graphing calculator.
- Concept Summaries, organized by sub-heading, have been added to aid students in synthesizing the important concepts discussed within the chapter.
- Algebra Warm-Up and Algebra Insight, located in chapters 1 to 4, provide students with additional algebraic guidance prior to chapter exercises.

KEY FEATURES OF THIS TEXT

EXAMPLE 4.1.1

Sketch the graphs of $y = 2^x$ and $y = \left(\frac{1}{2}\right)^x$.

Solution

Begin by constructing a table of values for $y = 2^x$ and $y = \left(\frac{1}{2}\right)^x$:

x	-15	-10	-1	0	1	3	5	10	15
$y = 2^x$	0.00003	0.001	0.5	1	2	8	32	1024	32 768
$y = \left(\frac{1}{2}\right)^x$	32 768	1024	2	1	0.5	0.125	0.313	0.001	0.00003

The pattern of values in this table suggests that the functions $y = 2^x$ and $y = \left(\frac{1}{2}\right)^x$ have the following features:

The function $y = 2^x$

always increasing

$$\lim_{x \to -\infty} 2^x = 0$$

$$\lim_{x \to +\infty} 2^x = +\infty$$

The function $y = \left(\frac{1}{2}\right)^x$

always decreasing

$$\lim_{x \to -\infty} \left(\frac{1}{2}\right)^x = +\infty$$

$$\lim_{x \to +\infty} \left(\frac{1}{2}\right)^x = 0$$

Applications

Throughout the text great effort is made to ensure that topics are applied to practical problems soon after their introduction, providing methods for dealing with both routine computations and applied problems. These problem-solving methods and strategies are introduced in applied examples and practised throughout in the exercise sets. Applications are clearly labelled with a subject title.

Procedural Examples and Boxes

Each new topic is approached with careful clarity by providing step-by-step problem-solving techniques through frequent procedural examples and summary boxes.

The Trapezoidal Rule

$$\int_a^b f(x)\, dx \approx \frac{\Delta x}{2}[f(x_1) + 2f(x_2) + \cdots + 2f(x_n) + f(x_{n+1})].$$

The trapezoidal rule is illustrated in Example 6.3.1.

Domain Convention ■ Unless otherwise specified, if a formula (or several formulas, as in Example 1.1.3) is used to define a function f, then we assume the domain of f to be the set of all numbers for which $f(x)$ is defined (as a real number). We refer to this as the **natural domain** of f.

Definitions

Definitions and key concepts are set off in shaded boxes to provide easy referencing for the student.

Just-In-Time Reviews

These references, located in the margins, are used to quickly remind students of important concepts from algebra or precalculus as they are being used in examples and review.

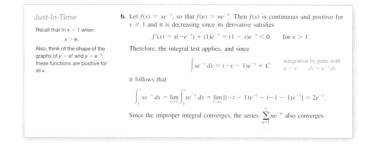

Exercise Sets

New problems have been added to increase the effectiveness of the highly praised exercise sets! Routine problems have been added where needed to ensure students have enough practice to master basic skills, and a variety of applied problems have been added to help demonstrate the practicality of the material.

Writing Exercises

These problems, designated by writing icons, challenge a student's critical thinking skills and invite students to research topics on their own.

Graphing Exercises

Graph icons designate problems in each exercise set that require graphing technology. With the availability of free online graphing programs with very user-friendly interfaces, functions can easily be graphed to answer these questions.

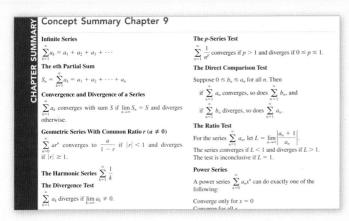

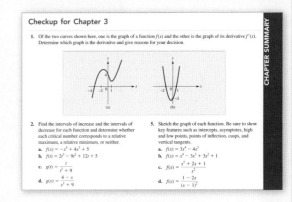

Chapter Summary

Chapter Summary material aids the student in synthesizing the important concepts discussed within the chapter, including a master list of key technical terms and formulas introduced in the chapter.

Chapter Checkup

Chapter Checkups provide a quick quiz for students to test their understanding of the concepts introduced in the chapter.

Review Exercises

A wealth of additional routine and applied problems is provided within the end-of-chapter exercise sets, offering further opportunities for practice.

Explore!

Now that Wolfram Alpha is available free online or on a smart phone, and Maple and Mathematica are available on most campuses, it is easier to reinforce concepts through graphs. These explorations are tied to specific examples to illustrate graphical interpretations of an algebraic answer, or to extend a concept further.

EXPLORE!

Refer to Example 5.1.8. Graph the position function $s(t) = -3t^2 + 66t$. Locate the stopping time, 4 seconds, and the corresponding position on the graph. Work the problem again for a car travelling at 30 m/s. Enter the new position function and compare the graphs. In this case, what is happening at 4 seconds?

EXAMPLE 5.1.8

A car is travelling along a straight, level road at 24 m/s when the driver is forced to apply the brakes to avoid an accident. If the brakes supply a constant deceleration of 6 m/s², how far does the car travel before coming to a complete stop?

Solution

Let $s(t)$ denote the distance travelled by the car in t seconds after the brakes are applied. Since the car decelerates at 6 m/s², it follows that $a(t) = -6$; that is,

$$\frac{dv}{dt} = a(t) = -6.$$

Integrating, the velocity at time t is given by

$$v(t) = \int \frac{dv}{dt}\, dt = \int -6\, dt = -6t + C_1.$$

THINK ABOUT IT

THINK ABOUT IT

MODELLING DATA—SUNFLOWER EXPORTS FROM MANITOBA

(Photo: CORBIS/Royalty Free)

Mathematical models can be used to study trends in infectious diseases, in inflation and unemployment, and in enrollment in university, and to study any statistical data that can be approximated by a continuous function. We will look at making mathematical models of some agricultural data, in particular sunflower product exports from Manitoba.*

Year	Exports (millions of dollars)
2002	44.5
2003	45.7
2004	32.0
2005	17.6
2006	28.6
2007	70.8
2008	72.6
2009	61.2
2010	30.0

The goal in modelling is to find a relatively simple function $f(t)$ that provides a close approximation to the data, in this case the value of the exports of sunflower products. One of the simplest approaches to constructing such functions is to use *polynomial regression*, a technique that produces best-fit polynomials of specified degree that are good approximations to observed data. The higher the degree of the polynomial, the better the approximation, but the more complicated the function. Care

*Janet Honey, "Crops in Manitoba 2009–2010," Department of Agribusiness and Agricultural Economics, University of Manitoba.

Think About It Essays

The modelling-based Think About It essays show students how material introduced in the chapter can be used to construct useful mathematical models while explaining the modelling process and providing an excellent starting point for projects or group discussions.

ALGEBRA WARM-UP

Make each expression ready to differentiate by changing the exponents, but do not differentiate:

1. $\dfrac{2}{x^3}$ 2. $\dfrac{2}{3\sqrt{x}}$ 3. $\dfrac{2\sqrt{x}}{5}$ 4. $\dfrac{2}{\sqrt[3]{x}}$ 5. $\dfrac{7}{x^5} - 10\sqrt{x} + 6x^4$

Algebra Insight and Algebra Warm-Up

In the first few chapters, prior to some exercise sets, there are sets of algebra questions. These questions reinforce the algebra needed for the exercises of that section. The answers to all Algebra Insight and Algebra Warm-Up exercises are included in the back of the book.

Concept Summary

At the end of each chapter, there is a Concept Summary listing the formulas used and sometimes including a relevant example.

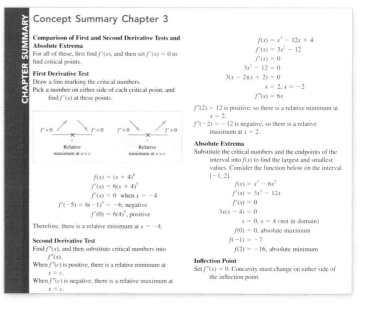

Concept Summary Chapter 3

Comparison of First and Second Derivative Tests and Absolute Extrema
For all of these, first find $f'(x)$, and then set $f'(x) = 0$ to find critical points.

First Derivative Test
Draw a line marking the critical numbers. Pick a number on either side of each critical point, and find $f'(x)$ at these points.

$f' > 0$ ⟍ $f' < 0$ Relative maximum at $x = c$
$f' < 0$ ⟋ $f' > 0$ Relative minimum at $x = c$

$$f(x) = (x + 4)^6$$
$$f'(x) = 6(x + 4)^5$$
$$f'(x) = 0 \text{ when } x = -4$$
$$f'(-5) = 6(-1)^5 = -6, \text{ negative}$$
$$f'(0) = 6(4)^5, \text{ positive}$$

Therefore, there is a relative minimum at $x = -4$.

Second Derivative Test
Find $f''(x)$, and then substitute critical numbers into $f''(x)$.
When $f''(c)$ is positive, there is a relative minimum at $x = c$.
When $f''(c)$ is negative, there is a relative maximum at $x = c$.

$$f(x) = x^3 - 12x + 4$$
$$f'(x) = 3x^2 - 12$$
$$f'(x) = 0$$
$$3x^2 - 12 = 0$$
$$3(x - 2)(x + 2) = 0$$
$$x = 2, x = -2$$
$$f''(x) = 6x$$

$f''(2) = 12$ is positive, so there is a relative minimum at $x = 2$.
$f''(-2) = -12$ is negative, so there is a relative maximum at $x = -2$.

Absolute Extrema
Substitute the critical numbers and the endpoints of the interval into $f(x)$ to find the largest and smallest values. Consider the function below on the interval $[-1, 2]$.

$$f(x) = x^3 - 6x^2$$
$$f'(x) = 3x^2 - 12x$$
$$f'(x) = 0$$
$$3x(x - 4) = 0$$
$$x = 0, x = 4 \text{ (not in domain)}$$
$$f(0) = 0, \text{ absolute maximum}$$
$$f(-1) = -7$$
$$f(2) = -16, \text{ absolute minimum}$$

Inflection Point
Set $f''(x) = 0$. Concavity must change on either side of the inflection point.

Supplements

Instructor's Solutions Manual

The *Instructor's Solutions Manual* contains comprehensive, worked-out solutions for all even-numbered problems in the text and is available on Connect.

Computerized Test Bank

Brownstone Diploma testing software, available on Connect, offers instructors a quick and easy way to create customized exams and view student results. The software utilizes an electronic test bank of short answer, multiple choice, and true/false questions tied directly to the text.

Connect

McGraw-Hill Connect™ is a web-based assignment and assessment platform that gives students the means to better connect with their coursework, with their instructors, and with the important concepts that they will need to know for success now and in the future.

With Connect, instructors can deliver assignments, quizzes, and tests online. Questions corresponding to all sections of the text are presented in an auto-gradeable format and tied to the text's learning objectives. Track individual student performance—by question, by assignment, or in relation to the class overall—with detailed grade reports. Integrate grade reports easily with Learning Management Systems (LMS) such as WebCT and Blackboard. And much more.

By choosing Connect, instructors are providing their students with a powerful tool for improving academic performance and truly mastering course material. Connect allows students to practise important skills at their own pace and on their own schedule. Importantly, students' assessment results and instructors' feedback are all saved online—so students can continually review their progress and plot their course to success.

Connect also provides 24/7 online access to an eBook—an online edition of the text—to aid students in successfully completing their work, wherever and whenever they choose.

KEY FEATURES

Smart Grading

When it comes to studying, time is precious. Connect helps students learn more efficiently by providing feedback and practice material when they need it, where they need it.

- Automatically score assignments, giving students immediate feedback on their work and side-by-side comparisons with correct answers
- Access and review each response; manually change grades or leave comments for students to review
- Reinforce classroom concepts with practice tests and instant quizzes

Instructor Library

The Connect Instructor Library is your course creation hub. It provides all the critical resources you'll need to build your course, just how you want to teach it.

- Assign eBook readings and draw from a rich collection of textbook-specific assignments

- Access instructor resources, including ready-made PowerPoint presentations and media to use in your lectures
- View assignments and resources created for past sections
- Post your own resources for students to use

eBook

Connect reinvents the textbook learning experience for the modern student. Every Connect subject area is seamlessly integrated with Connect eBooks, which are designed to keep students focused on the concepts key to their success.

- Provide students with a Connect eBook, allowing for anytime, anywhere access to the textbook
- Merge media with the text's narrative to engage students and improve learning and retention
- Pinpoint and connect key concepts in a snap using the powerful eBook search engine
- Manage notes, highlights, and bookmarks in one place for simple, comprehensive review

CourseSmart Electronic Textbook

CourseSmart is a new way for faculty to find and review eBooks. It's also a great option for students who are interested in accessing their course materials digitally and saving money. At CourseSmart, students can save up to 50% off the cost of a print book, reduce their impact on the environment, and gain access to powerful Web tools for learning, including full text search, notes and highlighting, and email tools for sharing notes between classmates. **CourseSmart.com**

Blackboard

McGraw-Hill Higher Education and Blackboard have teamed up.

Blackboard, the Web-based course-management system, has partnered with McGraw-Hill to better allow students and faculty to use online materials and activities to complement face-to-face teaching. Blackboard features exciting social learning and teaching tools that foster more logical, visually impactful, and active learning opportunities for students. You'll transform your closed-door classrooms into communities where students remain connected to their educational experience 24 hours a day.

This partnership allows you and your students access to McGraw-Hill's Connect™ and Create™ right from within your Blackboard course—all with one single sign-on.

Not only do you get single sign-on with Connect™ and Create™, you also get deep integration of McGraw-Hill content and content engines right in Blackboard. Whether you're choosing a book for your course or building Connect™ assignments, all the tools you need are right where you want them—inside of Blackboard.

Gradebooks are now seamless. When a student completes an integrated Connect™ assignment, the grade for that assignment automatically (and instantly) feeds into your Blackboard grade centre.

McGraw-Hill and Blackboard can now offer you easy access to industry-leading technology and content, whether your campus hosts it or we do. Be sure to ask your local McGraw-Hill representative for details.

Tegrity

Tegrity Campus is a service that makes class time available all the time by automatically capturing every lecture in a searchable format for students to review when they study and complete assignments. With a simple one-click start-and-stop process, you capture all computer screens and corresponding audio. Students can replay any part of any class with easy-to-use browser-based viewing on a PC or Mac. Educators know that the more students can see, hear, and experience class resources, the better they learn. With Tegrity Campus, students quickly recall key moments by using the unique search feature. This search helps students efficiently find what they need, when they need it across an entire semester of class recordings. Help turn all your students' study time into learning moments immediately supported by your lecture. To learn more about Tegrity, watch a two-minute Flash demo at http://tegritycampus.mhhe.com.

Create

McGraw-Hill's Create Online gives you access to the most abundant resource at your fingertips—literally. With a few mouse clicks, you can create customized learning tools simply and affordably. McGraw-Hill Ryerson has included many of our market-leading textbooks within Create Online for eBook and print customization as well as many licensed readings and cases. For more information, go to www.mcgrawhillcreate.com.

iLearning Services

At McGraw-Hill Ryerson, we take great pride in developing high-quality learning resources while working hard to provide you with the tools necessary to utilize them. We want to bring your teaching to life, and we do this by integrating technology, events, conferences, training, and other services. We call it iServices. For more information, contact your iLearning Slaes Specialist.

McGraw-Hill Ryerson National Teaching and Learning Conference Series

The educational environment has changed tremendously in recent years, and McGraw-Hill Ryerson continues to be committed to helping you acquire the skills you need to succeed in this new milieu. Our innovative Teaching and Learning Conference Series brings faculty together from across Canada with 3M Teaching Excellence award winners to share teaching and learning best practices in a collaborative and stimulating environment. Pre-conference workshops on general topics, such as teaching large classes and technology integration, are also offered.

Acknowledgments

Special thanks for their advice and recommendations for revisions found in the Canadian Edition go to the following reviewers:

Richard Blute, *University of Ottawa*
Elena Devdariani, *Carleton University*
Vivian Fayowski, *University of Northern British Columbia*
Barry Ferguson, *University of Waterloo*
Raymond Grinnell, *University of Toronto at Scarborough*
Veselin Jungic, *Simon Fraser University*
Merzik Kamel, *University of New Brunswick in Saint John*
David Lozinski, *McMaster University*
Paul McNicholas, *University of Guelph*
Peter Penner, *University of Manitoba*
Allyson Rozell, *Kwantlen Polytechnic University*
Manuele Santoprete, *Wilfrid Laurier University*
Jonathan P. Seldin, *University of Lethbridge*
George Stoica, *University of New Brunswick*
Ken Towson, *Capilano University*
Margaret Wyeth, *University of Victoria*

Special thanks also to the following McGraw-Hill Ryerson and freelance team members for their help in heading the Canadian Edition in the right direction: Sponsoring Editor, James Booty; Developmental Editors, Amy Rydzanicz and Sarah Fulton; Supervising Editor, Jessica Barnoski; Copy Editor, Julia Cochrane; and Proofreader, Laurel Sparrow. Thanks also go to Robert Raphael of Concordia University for his thorough technical check of the text and solutions.

Paper production in Canada produces revenues of $29.6 billion. Demand and supply fluctuate. In recent years, newsprint demand has decreased because more people are reading newspapers online. Supply can also fluctuate when unexpected warm spells in winter cause ice roads to melt so that logging trucks cannot use them. For a mathematical model of supply and demand of paper, see Example 1.4.4. (Photo: CORBIS/ Royalty Free)

FUNCTIONS, GRAPHS, AND LIMITS

CHAPTER OUTLINE

LEARNING OBJECTIVES

After completing this chapter, you should be able to

LO1 Evaluate functions at a given point. Evaluate composite functions.

LO2 Plot graphs of functions.

LO3 Find the equation of a line, given either two points or one point and a slope. Formulate equations of lines from given application data.

LO4 With mathematical modelling, express given information as a function of one or more variables.

LO5 Evaluate limits of polynomials and rational functions.

LO6 Discuss the continuity of a function. Evaluate one-sided limits.

SECTION 1.1

L01

Evaluate functions at a given point. Evaluate composite functions.

Just-In-Time

A review of algebra can be found in Appendices A1 and A2, but specific instances of algebra that are relevant to an adjacent example, or expansions of certain concepts, can be found in these Just-In-Time notes.

Functions

In many practical situations, the value of one quantity may depend on the value of a second. For example, consumer demand for beef may depend on the current market price, the amount of air pollution in a metropolitan area may depend on the number of cars on the road, or the value of a rare coin may depend on its age. Such relationships can often be represented mathematically as **functions.**

Loosely speaking, a function consists of two sets and a rule that associates elements in one set with elements in the other. For instance, suppose you want to determine the effect of price on the number of units of a particular commodity that will be sold at that price. To study this relationship, you need to know the set of admissible prices, the set of possible sales levels, and a rule for associating each price with a particular sales level. Here is the definition of function we shall use.

Function ■ A **function** is a rule that assigns to each object in a set A exactly one object in a set B. The set A is called the **domain** of the function, and the set of assigned objects in B is called the **range.**

For most functions in this book, the domain and range will be collections of real numbers and the function itself will be denoted by a letter such as f. The value that the function f assigns to the number x in the domain is then denoted by $f(x)$ (read as "f of x"), which is often given by a formula, such as $f(x) = x^2 + 4$.

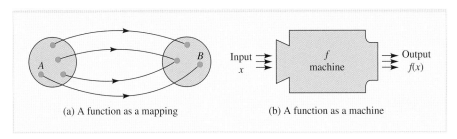

(a) A function as a mapping   (b) A function as a machine

FIGURE 1.1 Interpretations of the function $f(x)$.

It may help to think of such a function as a "mapping" from numbers in A to numbers in B (Figure 1.1a), or as a "machine" that takes a given number in A and converts it into a number in B through a process indicated by the functional rule (Figure 1.1b). For instance, the function $f(x) = x^2 + 4$ can be thought of as an "f machine" that accepts an input x and then squares it and adds 4 to produce an output $y = x^2 + 4$.

No matter how you choose to think of a functional relationship, it is important to remember that *a function assigns one and only one number in the range (output) to each number in the domain (input)*. Here is an example.

EXAMPLE 1.1.1

Find $f(3)$ if $f(x) = x^2 + 4$.

Solution

$$f(3) = 3^2 + 4 = 13.$$

Observe the convenience and simplicity of functional notation. In Example 1.1.1, the compact formula $f(x) = x^2 + 4$ completely defines the function, and you can indicate that 13 is the number the function assigns to 3 by simply writing $f(3) = 13$.

It is often convenient to represent a functional relationship by an equation $y = f(x)$, and in this context, x and y are called **variables.** In particular, since the numerical value of y is determined by that of x, we refer to y as the **dependent variable** and to x as the **independent variable.** Note that there is nothing sacred about the symbols x and y. For example, the function $y = x^2 + 4$ can just as easily be represented by $s = t^2 + 4$ or by $w = u^2 + 4$.

Functional notation can also be used to describe tabular data. For instance, Table 1.1 lists the cash receipts from Canadian fruit farms.

TABLE 1.1 Canadian Fruit Farm Cash Receipts

Year	Period, n	Value in millions of dollars
2004	1	614
2005	2	597
2006	3	724
2007	4	717
2008	5	743

SOURCE: Adapted from Statistics Canada (Farm Cash Receipts, 21-011-X). "Statistical Overview of Horticulture 2008: Canadian Horticulture Cash Receipts by Sector." © Agriculture and Agri-Food Canada. Reproduced with the permission of the Minister of Public Works and Government Services Canada, 2010. (Photo: © D. Hurst/Alamy)

We can describe these data as a function f defined by the rule

$$f(n) = [\text{annual farm receipts at the end of the } n\text{th period}].$$

Thus, $f(1) = 614, f(2) = 597, \ldots, f(5) = 743$. Note that the domain of f is the set of integers $A = \{1, 2, \ldots, 5\}$.

The use of functional notation is illustrated further in Examples 1.1.2 and 1.1.3. In Example 1.1.2, notice that letters other than f and x are used to denote the function and its independent variable.

Just-In-Time

Recall that $x^{1/2} = \sqrt{x}$ and $x^{1/3} = \sqrt[3]{x}$. These are two ways of writing the same expression. To generalize, $x^{a/b} = \sqrt[b]{x^a}$ whenever a and b are positive integers. Example 1.1.2 uses the case when $a = 1$ and $b = 2$.

EXAMPLE 1.1.2

If $g(t) = (t - 2)^{1/2}$, find (if possible) $g(27)$, $g(5)$, $g(2)$, and $g(1)$.

Solution

Rewrite the function as $g(t) = \sqrt{t - 2}$. (If you need to brush up on fractional powers, consult the discussion of exponential notation in Appendix A1.) Then

$$g(27) = \sqrt{27 - 2} = \sqrt{25} = 5$$
$$g(5) = \sqrt{5 - 2} = \sqrt{3} \approx 1.7321$$

and
$$g(2) = \sqrt{2 - 2} = \sqrt{0} = 0$$

However, $g(1)$ is undefined since

$$g(1) = \sqrt{1 - 2} = \sqrt{-1},$$

and negative numbers do not have real square roots.

EXPLORE!

Plot the graph of
$f(x) = \sqrt{x - 22}$. At first use
the default scale of your
graphing utility, or $x = -10$ to 10.
Can you see anything?
Change the scale, using $x = 0$
to a larger number of your
choice. Can you see why no
graph appeared when the
scale only went up to $x = 10$
or $x = 20$?

Functions are often defined using more than one formula, where each individual formula describes the function on a subset of the domain. A function defined in this way is sometimes called a **piecewise-defined function.** Here is an example of such a function.

EXAMPLE 1.1.3

Find $f\left(-\dfrac{1}{2}\right)$, $f(1)$, and $f(2)$ for

$$f(x) = \begin{cases} \dfrac{1}{x - 1} & \text{if } x < 1 \\ 3x^2 + 1 & \text{if } x \geq 1 \end{cases}$$

Solution

Since $x = -\dfrac{1}{2}$ satisfies $x < 1$, use the top part of the formula to find

$$f\left(-\frac{1}{2}\right) = \frac{1}{-1/2 - 1} = \frac{1}{-3/2} = -\frac{2}{3}.$$

However, $x = 1$ and $x = 2$ satisfy $x \geq 1$, so $f(1)$ and $f(2)$ are both found by using the bottom part of the formula:

$$f(1) = 3(1)^2 + 1 = 4 \qquad\qquad f(2) = 3(2)^2 + 1 = 13$$

EXPLORE!

Plot the graph of
$f(x) = \dfrac{x - 1}{x^2 - 5x - 6}$. If it is
difficult to see, change the y
scale to smaller values such
as -20 to 20. Factor the
denominator to find where a
zero in the denominator would
cause major problems with
the function. What happens
to the graph at these two
values of x?

Domain Convention ■ Unless otherwise specified, if a formula (or several formulas, as in Example 1.1.3) is used to define a function f, then we assume the domain of f to be the set of all numbers for which $f(x)$ is defined (as a real number). We refer to this as the **natural domain** of f.

Determining the natural domain of a function often amounts to excluding all numbers x that result in dividing by 0 or in taking the square root of a negative number. This procedure is illustrated in Example 1.1.4.

EXAMPLE 1.1.4

Find the domain and range of each function.

a. $f(x) = \dfrac{1}{x - 3}$ **b.** $g(t) = \sqrt{t - 2}$

Just-In-Time

Recall that when $x^2 = 25$, then
$x = \pm\sqrt{25} = \pm 5$. However,
the symbol $\sqrt{}$ means to take
the *positive* square root. So
$\sqrt{25} = 5$, $\sqrt{4} = 2$, and
$\sqrt{256} = 16$, for example. Thus,
the value of $\sqrt{a}$ for any a is
defined as a positive number.

Solution

a. Since division by any number other than 0 is possible, the domain of f is the set of all numbers x such that $x - 3 \neq 0$, that is, $x \neq 3$. The range of f is the set of all numbers y except 0, since for any $y \neq 0$, there is an x such that $y = \dfrac{1}{x - 3}$, in particular, $x = 3 + \dfrac{1}{y}$.

b. Since negative numbers do not have real square roots, $g(t)$ can be evaluated only when $t - 2 \geq 0$, so the domain of g is the set of all numbers t such that $t \geq 2$. The range of g is the set of all nonnegative numbers, for if $y \geq 0$ is any such number, there is a t such that $y = \sqrt{t - 2}$ namely, $t = y^2 + 2$.

Functions Used in Economics

Several functions are associated with the marketing of a particular commodity:

The **demand function** $D(x)$ for the commodity is the price $p = D(x)$ that must be charged for each unit of the commodity if x units are to be sold (demanded).

The **supply function** $S(x)$ for the commodity is the unit price $p = S(x)$ at which producers are willing to supply x units to the market.

The **revenue function** $R(x)$ obtained from selling x units of the commodity is given by the product

$$R(x) = (\text{number of items sold})(\text{price per item})$$
$$= xp(x)$$

The **cost function** $C(x)$ is the cost of producing x units of the commodity.

The **profit function** $P(x)$ is the profit obtained from selling x units of the commodity and is given by the difference

$$P(x) = \text{revenue} - \text{cost}$$
$$= R(x) - C(x) = xp(x) - C(x)$$

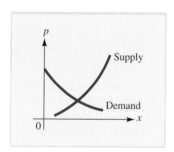

Generally speaking, the higher the unit price, the smaller the number of units demanded, and vice versa. Conversely, an increase in unit price leads to an increase in the number of units supplied. Thus, demand functions are typically decreasing ("falling" from left to right), while supply functions are increasing ("rising"), as illustrated in the margin. Here is an example that uses several of these special economic functions.

EXAMPLE 1.1.5

Market research indicates that consumers will buy x thousand units of a particular kind of coffee maker when the unit price, in dollars, is

$$p(x) = -0.27x + 51.$$

The cost, in thousands of dollars, of producing the x thousand units is

$$C(x) = 2.23x^2 + 3.5x + 85.$$

a. What are the revenue and profit functions, $R(x)$ and $P(x)$, for this production process?

b. For what values of x is production of the coffee makers profitable?

Solution

a. The revenue, in thousands of dollars, is

$$R(x) = xp(x) = -0.27x^2 + 51x,$$

Just-In-Time

When factoring, always take a common factor first. Then, if the expression is of the form $x^2 + bx + c$, find two numbers that multiply to make the last number and add to make the middle coefficient. Here the factors of 34 can be 17, 2 or $-17, -2$. Always check by multiplying out the brackets— this small step can save many errors!

and the profit, in thousands of dollars, is

$$P(x) = R(x) - C(x)$$
$$= -0.27x^2 + 51x - (2.23x^2 + 3.5x + 85)$$
$$= -2.5x^2 + 47.5x - 85$$

b. Production is profitable when $P(x) > 0$. We find that

$$P(x) = -2.5x^2 + 47.5x - 85$$
$$= -2.5(x^2 - 19x + 34)$$
$$= -2.5(x - 2)(x - 17)$$

Since the coefficient -2.5 is negative, it follows that $P(x) > 0$ only if the terms $(x - 2)$ and $(x - 17)$ have different signs; that is, when $x - 2 > 0$ and $x - 17 < 0$. Thus, production is profitable for $2 < x < 17$.

Example 1.1.6 illustrates how functional notation is used in a practical situation. Notice that to make the algebraic formula easier to interpret, letters suggesting the relevant practical quantities are used for the function and its independent variable. (In this example, the letter C stands for cost and q stands for quantity manufactured.)

EXAMPLE 1.1.6

Suppose the total cost, in dollars, of manufacturing q units of a certain commodity is given by the function $C(q) = q^3 - 30q^2 + 500q + 200$.

a. Compute the cost of manufacturing 10 units of the commodity.

b. Compute the cost of manufacturing the 10th unit of the commodity.

Solution

a. The cost, in dollars, of manufacturing 10 units is the value of the total cost function when $q = 10$. That is,

$$\text{Cost of 10 units} = C(10)$$
$$= (10)^3 - 30(10)^2 + 500(10) + 200$$
$$= 3200$$

b. The cost of manufacturing the 10th unit is the difference between the cost of manufacturing 10 units and the cost of manufacturing 9 units. First, calculate $C(9)$:

$$C(9) = (9)^3 - 30(9)^2 + 500(9) + 200 = 2999.$$

Then

$$\text{Cost of 10th unit} = C(10) - C(9) = \$3200 - \$2999 = \$201.$$

Composition of Functions

There are many situations in which a quantity is given as a function of one variable that, in turn, can be written as a function of a second variable. By combining the functions in an appropriate way, you can express the original quantity as a function of the second variable. This process is called **composition of functions** or **functional composition**.

For instance, suppose environmentalists estimate that when p thousand people live in a certain city, the average daily level of carbon monoxide in the air is $c(p)$ parts per million, and that separate demographic studies indicate that the population in t years will be $p(t)$ thousand. What level of pollution should be expected in t years? To answer this question, substitute $p(t)$ into the pollution formula $c(p)$ to express c as a composite function of t.

We shall return to the pollution problem in Example 1.1.11 with specific formulas for $c(p)$ and $p(t)$, but first you need to see a few examples of how composite functions are formed and evaluated. Here is a definition of functional composition.

> **Composition of Functions**   ■   Given functions $f(u)$ and $g(x)$, the composition $f(g(x))$ is the function of x formed by substituting $u = g(x)$ for u in the formula for $f(u)$.

Note that the composite function $f(g(x))$ makes sense only if the domain of f contains the range of g. In Figure 1.2, the definition of composite function is illustrated as an "assembly line" in which "raw" input x is first converted into a transitional product $g(x)$ that acts as input that the f machine uses to produce $f(g(x))$.

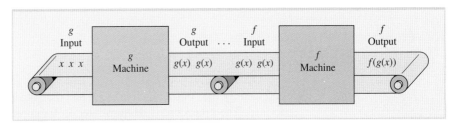

FIGURE 1.2   The composition $f(g(x))$ as an assembly line.

EXAMPLE 1.1.7

Find the composite function $f(g(x))$, where $f(u) = u^2 + 3u + 1$ and $g(x) = x + 1$.

Solution

Replace u by $x + 1$ in the formula for $f(u)$ to get

$$f(g(x)) = (x + 1)^2 + 3(x + 1) + 1$$
$$= (x^2 + 2x + 1) + (3x + 3) + 1$$
$$= x^2 + 5x + 5$$

> **NOTE**   By reversing the roles of f and g in the definition of the composite function, you can define the composition $g(f(x))$. In general, $f(g(x))$ and $g(f(x))$ will *not* be the same. For instance, with the functions in Example 1.1.7, you first write
>
> $$g(w) = w + 1 \quad \text{and} \quad f(x) = x^2 + 3x + 1$$
>
> and then replace w by $x^2 + 3x + 1$ to get
>
> $$g(f(x)) = (x^2 + 3x + 1) + 1$$
> $$= x^2 + 3x + 2$$
>
> which is equal to $f(g(x)) = x^2 + 5x + 5$ only when $x = -\dfrac{3}{2}$ (you should verify this).   ■

Example 1.1.7 could have been worded more compactly as follows: Find the composite function $f(x + 1)$, where $f(x) = x^2 + 3x + 1$. The use of this compact notation is illustrated further in Example 1.1.8.

EXAMPLE 1.1.8

Find $f(x - 1)$ if $f(x) = 3x^2 + \dfrac{1}{x} + 5$.

Solution

At first glance, this problem may look confusing because the letter x appears both as the independent variable in the formula defining f and as part of the expression $x - 1$. Because of this, you may find it helpful to begin by writing the formula for f in more neutral terms, such as

$$f(\square) = 3(\square)^2 + \frac{1}{\square} + 5.$$

To find $f(x - 1)$, you simply insert the expression $x - 1$ in each box, getting

$$f(x - 1) = 3(x - 1)^2 + \frac{1}{x - 1} + 5.$$

Occasionally, you will have to "take apart" a given composite function $g(h(x))$ and identify the "outer function" $g(u)$ and the "inner function" $h(x)$ from which it was formed. The procedure is demonstrated in Example 1.1.9.

EXAMPLE 1.1.9

If $f(x) = \dfrac{5}{x - 2} + 4(x - 2)^3$, find functions $g(u)$ and $h(x)$ such that $f(x) = g(h(x))$.

Solution

The form of the given function is

$$f(x) = \frac{5}{\square} + 4(\square)^3,$$

where each box contains the expression $x - 2$. Thus, $f(x) = g(h(x))$, where

$$\underbrace{g(u) = \frac{5}{u} + 4u^3}_{\text{outer function}} \qquad \text{and} \qquad \underbrace{h(x) = x - 2.}_{\text{inner function}}$$

Actually, in Example 1.1.9, there are infinitely many pairs of functions $g(u)$ and $h(x)$ that combine to give $g(h(x)) = f(x)$ (for example, $g(u) = \dfrac{5}{u + 1} + 4(u + 1)^3$ and $h(x) = x - 3$). The particular pair selected in the solution to this example is the most natural one and reflects most clearly the structure of the original function $f(x)$.

EXAMPLE 1.1.10

A **difference quotient** is an expression of the general form

$$\frac{f(x + h) - f(x)}{h},$$

where f is a given function of x and h is a non-zero number. Difference quotients will be used in Chapter 2 to define the *derivative,* one of the fundamental concepts of calculus. Find the difference quotient for $f(x) = x^2 - 3x$.

Solution

You find that

$$
\begin{aligned}
\frac{f(x + h) - f(x)}{h} &= \frac{[(x + h)^2 - 3(x + h)] - (x^2 - 3x)}{h} \\
&= \frac{(x^2 + 2xh + h^2 - 3x - 3h) - (x^2 - 3x)}{h} \quad \text{expand the numerator} \\
&= \frac{2xh + h^2 - 3h}{h} \quad \text{combine terms in the numerator} \\
&= \frac{h(2x + h - 3)}{h} \quad \text{factor out } h \\
&= 2x + h - 3 \quad \text{divide by } h
\end{aligned}
$$

Example 1.1.11 illustrates how a composite function may arise in an applied problem.

EXAMPLE 1.1.11

An environmental study of a certain community suggests that the average daily level of carbon monoxide in the air is $c(p) = 0.5p + 1$ parts per million when the population is p thousand. It is estimated that t years from now the population of the community will be $p(t) = 10 + 0.1t^2$ thousand.

a. Express the level of carbon monoxide in the air as a function of time.

b. When will the carbon monoxide level reach 6.8 parts per million?

Solution

a. Since the level of carbon monoxide is related to the variable p by the equation

$$c(p) = 0.5p + 1$$

and the variable p is related to the variable t by the equation

$$p(t) = 10 + 0.1t^2,$$

it follows that the composite function

$$c(p(t)) = c(10 + 0.1t^2) = 0.5(10 + 0.1t^2) + 1 = 6 + 0.05t^2$$

expresses the level of carbon monoxide in the air as a function of the variable t.

Just-In-Time

When taking the square root of both sides of an equation, take a look at the context of the question. Mathematically there are positive and negative answers, but the function is only valid t years from now, so we discard the negative answer.

b. Set $c(p(t))$ equal to 6.8 and solve for t to get

$$6 + 0.05t^2 = 6.8$$
$$0.05t^2 = 0.8$$
$$t^2 = \frac{0.8}{0.05} = 16$$
$$t = \sqrt{16} = 4 \qquad \text{discard } t = -4$$

That is, 4 years from now the level of carbon monoxide will be 6.8 parts per million.

ALGEBRA WARM-UP

Factor each expression.

1. $x^2 - 7x + 6$
2. $x^2 - 8x + 12$
3. $3x^2 + 6x - 48$
4. $x^2 - 100$
5. $36 - 4x^2$

EXERCISES ▪ 1.1

In Exercises 1 through 14, compute the indicated values of the given function.

1. $f(x) = 3x + 5; f(0), f(-1), f(2)$
2. $f(x) = -7x + 1; f(0), f(1), f(-2)$
3. $f(x) = 3x^2 + 5x - 2; f(0), f(-2), f(1)$
4. $h(t) = (2t + 1)^3; h(-1), h(0), h(1)$
5. $g(x) = x + \dfrac{1}{x}; g(-1), g(1), g(2)$
6. $f(x) = \dfrac{x}{x^2 + 1}; f(2), f(0), f(-1)$
7. $h(t) = \sqrt{t^2 + 2t + 4}; h(2), h(0), h(-4)$
8. $g(u) = (u + 1)^{3/2}; g(0), g(-1), g(8)$
9. $f(t) = (2t - 1)^{-3/2}; f(1), f(5), f(13)$
10. $f(t) = \dfrac{1}{\sqrt{3 - 2t}}; f(1), f(-3), f(0)$
11. $f(x) = x - |x - 2|; f(1), f(2), f(3)$
12. $g(x) = 4 + |x|; g(-2), g(0), g(2)$
13. $h(x) = \begin{cases} -2x + 4 & \text{if } x \le 1 \\ x^2 + 1 & \text{if } x > 1 \end{cases}; h(3), h(1), h(0), h(-3)$

14. $f(t) = \begin{cases} 3 & \text{if } t < -5 \\ t + 1 & \text{if } -5 \le t \le 5 \\ \sqrt{t} & \text{if } t > 5 \end{cases}; f(-6), f(-5), f(16)$

In Exercises 15 through 18, determine whether or not the given function has the set of all real numbers as its domain.

15. $g(x) = \dfrac{x}{1 + x^2}$
16. $f(x) = \dfrac{x + 1}{x^2 - 1}$
17. $f(t) = \sqrt{1 - t}$
18. $h(t) = \sqrt{t^2 + 1}$

In Exercises 19 through 24, determine the domain of the given function.

19. $g(x) = \dfrac{x^2 + 5}{x + 2}$
20. $f(x) = x^3 - 3x^2 + 2x + 5$
21. $f(x) = \sqrt{2x + 6}$
22. $f(t) = \dfrac{t + 1}{t^2 - t - 2}$

23. $f(t) = \dfrac{t + 2}{\sqrt{9 - t^2}}$

24. $h(s) = \sqrt{s^2 - 4}$

In Exercises 25 through 32, find the composite function $f(g(x))$.

25. $f(u) = 3u^2 + 2u - 6$, $g(x) = x + 2$

26. $f(u) = u^2 + 4$, $g(x) = x - 1$

27. $f(u) = (u - 1)^3 + 2u^2$, $g(x) = x + 1$

28. $f(u) = (2u + 10)^2$, $g(x) = x - 5$

29. $f(u) = \dfrac{1}{u^2}$, $g(x) = x - 1$

30. $f(u) = \dfrac{1}{u}$, $g(x) = x^2 + x - 2$

31. $f(u) = \sqrt{u + 1}$, $g(x) = x^2 - 1$

32. $f(u) = u^2$, $g(x) = \dfrac{1}{x - 1}$

In Exercises 33 through 38, find the difference quotient, $\dfrac{f(x + h) - f(x)}{h}$, of f.

33. $f(x) = 4 - 5x$

34. $f(x) = 2x + 3$

35. $f(x) = 4x - x^2$

36. $f(x) = x^2$

37. $f(x) = \dfrac{x}{x + 1}$

38. $f(x) = \dfrac{1}{x}$

In Exercises 39 through 42, first obtain the composite functions $f(g(x))$ and $g(f(x))$, and then find all numbers x (if any) such that $f(g(x)) = g(f(x))$.

39. $f(x) = \sqrt{x}$, $g(x) = 1 - 3x$

40. $f(x) = x^2 + 1$, $g(x) = 1 - x$

41. $f(x) = \dfrac{2x + 3}{x - 1}$, $g(x) = \dfrac{x + 3}{x - 2}$

42. $f(x) = \dfrac{1}{x}$, $g(x) = \dfrac{4 - x}{2 + x}$

In Exercises 43 through 50, find the indicated composite function.

43. $f(x - 2)$ where $f(x) = 2x^2 - 3x + 1$

44. $f(x + 1)$ where $f(x) = x^2 + 5$

45. $f(x - 1)$ where $f(x) = (x + 1)^5 - 3x^2$

46. $f(x + 3)$ where $f(x) = (2x - 6)^2$

47. $f(x^2 + 3x - 1)$ where $f(x) = \sqrt{x}$

48. $f\left(\dfrac{1}{x}\right)$ where $f(x) = 3x + \dfrac{2}{x}$

49. $f(x + 1)$ where $f(x) = \dfrac{x - 1}{x}$

50. $f(x^2 - 2x + 9)$ where $f(x) = 2x - 20$

In Exercises 51 through 56, find functions $h(x)$ and $g(u)$ such that $f(x) = g(h(x))$.

51. $f(x) = (x - 1)^2 + 2(x - 1) + 3$

52. $f(x) = (x^5 - 3x^2 + 12)^3$

53. $f(x) = \dfrac{1}{x^2 + 1}$

54. $f(x) = \sqrt{3x - 5}$

55. $f(x) = \sqrt[3]{2 - x} + \dfrac{4}{2 - x}$

56. $f(x) = \sqrt{x + 4} - \dfrac{1}{(x + 4)^3}$

CONSUMER DEMAND   *In Exercises 57 through 60, the demand function $p = D(x)$ and the total cost function $C(x)$ for a particular commodity are given in terms of the level of production x. In each case, find*

   (a) the revenue $R(x)$ and profit $P(x)$

   (b) all values of x for which production of the commodity is profitable

57. $D(x) = -0.02x + 29$
   $C(x) = 1.43x^2 + 18.3x + 15.6$

58. $D(x) = -0.37x + 47$
   $C(x) = 1.38x^2 + 15.15x + 115.5$

59. $D(x) = -0.5x + 39$
   $C(x) = 1.5x^2 + 9.2x + 67$

60. $D(x) = -0.09x + 51$
   $C(x) = 1.32x^2 + 11.7x + 101.4$

61. MANUFACTURING COST   Suppose the total cost of manufacturing q units of a certain commodity is $C(q)$ thousand dollars, where
$$C(q) + 0.01q^2 + 0.9q + 2.$$

a. Compute the cost of manufacturing 10 units.

b. Compute the cost of manufacturing the 10th unit.

62. **MANUFACTURING COST**   Suppose the total cost, in dollars, of manufacturing q units of a certain commodity is given by the function
$$C(q) = q^3 - 30q^2 + 400q + 500.$$
   a. Compute the cost of manufacturing 20 units.
   b. Compute the cost of manufacturing the 20th unit.

63. **DISTRIBUTION COST**   Suppose that the number of worker-hours required to distribute new telephone books to x percent of the households in a certain rural community is given by the function
$$W(x) = \frac{600x}{300 - x}.$$
   a. What is the domain of the function W?
   b. For what values of x does $W(x)$ have a practical interpretation in this context?
   c. How many worker-hours were required to distribute new telephone books to the first 50% of the households?
   d. How many worker-hours were required to distribute new telephone books to the entire community?
   e. What percentage of the households in the community had received new telephone books by the time 150 worker-hours had been expended?

64. **WORKER EFFICIENCY**   An efficiency study of the morning shift at a certain factory indicates that an average worker who arrives on the job at 8:00 A.M. will have assembled
$$f(x) = -x^3 + 6x^2 + 15x$$
   DVD players x hours later.
   a. How many DVD players will such a worker have assembled by 10:00 A.M.? [*Hint:* At 10:00 A.M., $x = 2$.]
   b. How many DVD players will such a worker assemble between 9:00 A.M. and 10:00 A.M.?

65. **IMMUNIZATION**   Suppose that during a nationwide program to immunize the population against a certain form of influenza, public health officials found that the cost of inoculating x percent of the population was approximately
$$C(x) = \frac{150x}{200 - x} \text{ million dollars.}$$
   a. What is the domain of the function C?
   b. For what values of x does $C(x)$ have a practical interpretation in this context?
   c. What was the cost of inoculating the first 50% of the population?

   d. What was the cost of inoculating the second 50% of the population?
   e. What percentage of the population had been inoculated by the time $37.5 million had been spent?

66. **TEMPERATURE CHANGE**   Suppose that t hours past midnight, the temperature in Toronto was
$$C(t) = -\frac{1}{6}t^2 + 2t + 10 \text{ degrees Celsius.}$$
   a. What was the temperature at 9:00 A.M.?
   b. By how much did the temperature increase or decrease between 6:00 A.M. and 9:00 A.M.?

67. **POPULATION GROWTH**   It is estimated that t years from now, the population of a certain suburban community will be $P(t) = 20 - \dfrac{6}{t + 1}$ thousand.
   a. What will the population of the community be 9 years from now?
   b. By how much will the population increase during the 9th year?
   c. What happens to $P(t)$ as t gets larger and larger? Interpret your result.

68. **EXPERIMENTAL PSYCHOLOGY**   To study the rate at which animals learn, a psychology student performs an experiment in which a rat is sent repeatedly through a laboratory maze. Suppose that the time required for the rat to traverse the maze on the nth trial is approximately
$$T(n) = 3 + \frac{12}{n}$$
   minutes.
   a. What is the domain of the function T?
   b. For what values of n does $T(n)$ have meaning in the context of the psychology experiment?
   c. How long does it take the rat to traverse the maze on the 3rd trial?
   d. On which trial does the rat first traverse the maze in 4 min or less?
   e. According to the function T, what will happen to the time required for the rat to traverse the maze as the number of trials increases? Will the rat ever be able to traverse the maze in less than 3 min?

69. **BLOOD FLOW**   Biologists have found that the speed of blood in an artery is a function of the distance of the blood from the artery's central axis. According to **Poiseuille's law,** the speed (in centimetres per second) of blood that is r centimetres from the central axis of an artery is given by the

function $S(r) = C(R^2 - r^2)$, where C is a constant and R is the radius of the artery. Suppose that for a certain artery, $C = 1.76 \times 10^5$ and $R = 1.2 \times 10^{-2}$ cm.

  a. Compute the speed of the blood at the central axis of this artery.

  b. Compute the speed of the blood midway between the artery's wall and its central axis.

70. **POSITION OF A MOVING OBJECT** A ball is dropped from the top of a building. Its height (in metres) after t seconds is given by the function $H(t) = -4.9t^2 + 80$.

  a. What is the height of the ball after 2 s?

  b. How far will the ball travel during the third second?

  c. How tall is the building?

  d. When will the ball hit the ground?

71. **ISLAND ECOLOGY** Observations show that on an island of area A square kilometres, the average number of animal species is approximately equal to $s(A) = 1.13\sqrt[3]{A}$.

  a. On average, how many animal species would you expect to find on an island of area 12 km²?

  b. If s_1 is the average number of species on an island of area A and s_2 is the average number of species on an island of area $2A$, what is the relationship between s_1 and s_2?

  c. What size must an island be to have an average of 100 animal species?

72. **POPULATION DENSITY** Observations suggest that for herbivorous mammals, the number of animals, N, per square kilometre can be estimated by the formula $N = \dfrac{91.2}{m^{0.73}}$, where m is the average mass of the animal in kilograms.

  a. Assuming that the average elk on a particular reserve has mass 300 kg, approximately how many elk would you expect to find per square kilometre in the reserve?

  b. Using this formula, it is estimated that there is fewer than one animal of a certain species per square kilometre. How large can the average animal of this species be?

  c. One species of large mammal has twice the average mass of a second species. If a particular reserve contains 100 animals of the larger species, how many animals of the smaller species would you expect to find there?

73. **CONSUMER DEMAND** An importer of Brazilian coffee estimates that local consumers will buy approximately $Q(p) = \dfrac{4374}{p^2}$ kilograms of the coffee per week when the price is p dollars per kilogram. It is estimated that t weeks from now the price of this coffee will be

$$p(t) = 0.04t^2 + 0.2t + 12$$

dollars per kilogram.

  a. Express the weekly demand (kilograms sold) for the coffee as a function of t.

  b. How many kilograms of the coffee will consumers be buying from the importer 10 weeks from now?

  c. When will the demand for the coffee be 30.375 kg?

74. **MANUFACTURING COST** At a certain factory, the total cost of manufacturing q units during the daily production run is $C(q) = q^2 + q + 900$ dollars. On a typical workday, $q(t) = 25t$ units are manufactured during the first t hours of a production run.

  a. Express the total manufacturing cost as a function of t.

  b. How much will have been spent on production by the end of the third hour?

  c. When will the total manufacturing cost reach $11 000?

75. **AIR POLLUTION** An environmental study of a certain suburban community suggests that the average daily level of carbon monoxide in the air will be $c(p) = 0.4p + 1$ parts per million when the population is p thousand. It is estimated that t years from now the population of the community will be $p(t) = 8 + 0.2t^2$ thousand.

  a. Express the level of carbon monoxide in the air as a function of time.

  b. What will the carbon monoxide level be 2 years from now?

  c. When will the carbon monoxide level reach 6.2 parts per million?

76. What is the domain of $f(x) = \dfrac{7x^2 - 4}{x^3 - 2x + 4}$?

77. What is the domain of $f(x) = \dfrac{4x^2 - 3}{2x^2 + x - 3}$?

78. For $f(x) = 2\sqrt{x - 1}$ and $g(x) = x^3 - 1.2$, find $g(f(4.8))$. Use two decimal places.

79. For $f(x) = 2\sqrt{x-1}$ and $g(x) = x^3 - 1.2$, find $f(g(2.3))$. Use two decimal places.

80. EFFECT OF MEDICATION   Amanjot has a fever and takes a fever-reducing medication. Her temperature t hours later is $T(t) = 37 + \dfrac{2}{t+1}$.

   a. What is her temperature after 1 hour?
   b. When will her temperature be 37.5°C?

81. WORKING TIME   Ben can do $10t$ questions in t hours. Adam can do $4t$ questions in t hours but he has already done 10 of them. Corey can do $6t$ questions in t hours but he has already done 9 of them. They each have 20 questions to do for an assignment.

   a. Write an equation for the number done by each student after time t.
   b. If they all sit down to work now, who will finish first?
   c. Plot the graph of each equation and a horizontal line at $y = 20$ questions. Find the intersection point of each equation with $y = 20$. Write a sentence about the meaning of the intersection point.

SECTION 1.2

L02

Plot graphs of functions.

The Graph of a Function

Graphs have visual impact. They also reveal information that may not be evident from verbal or algebraic descriptions. Two graphs depicting practical relationships are shown in Figure 1.3.

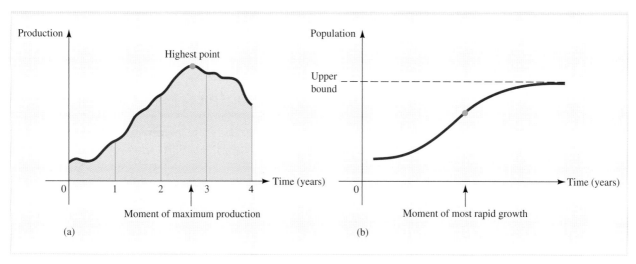

FIGURE 1.3   (a) A production function. (b) Bounded population growth.

The graph in Figure 1.3a describes the variation in total industrial production in a certain country over a 4-year period of time. Notice that the highest point on the graph occurs near the end of the third year, indicating that production was greatest at that time.

The graph in Figure 1.3b represents population growth when environmental factors impose an upper bound on the possible size of the population. It indicates that the *rate* of population growth increases at first and then decreases as the size of the population gets closer and closer to the upper bound.

**Rectangular
Coordinate System**

To represent graphs in the plane, we shall use a **rectangular (Cartesian) coordinate system.** To construct such a system, we begin by choosing two perpendicular number lines that intersect at the origin of each line. For convenience, one line is taken to be horizontal and is called the **x axis,** with positive direction to the right. The other line, called the **y axis,** is vertical with positive direction upward. Scaling on the two coordinate axes is often the same, but this is not necessary. The coordinate axes separate the plane into four parts called **quadrants,** which are numbered counterclockwise I through IV, as shown in Figure 1.4.

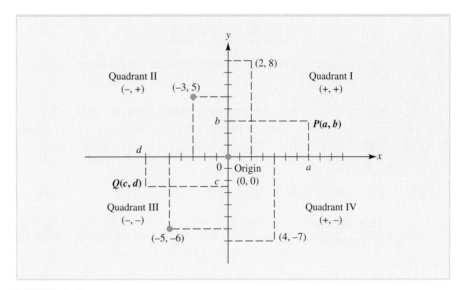

FIGURE 1.4   A rectangular coordinate system.

Any point P in the plane can be associated with a unique ordered pair of numbers (a, b) called the **coordinates** of P. Specifically, a is called the **x coordinate** (or **abscissa**) and b is called the **y coordinate** (or **ordinate**). To find a and b, draw vertical and horizontal lines through P. The vertical line intersects the x axis at a, and the horizontal line intersects the y axis at b. Conversely, if c and d are given, the vertical line through c and horizontal line through d intersect at the unique point Q with coordinates (c, d).

Several points are plotted in Figure 1.4. In particular, note that the point $(2, 8)$ is 2 units to the right of the vertical axis and 8 units above the horizontal axis, while $(-3, 5)$ is 3 units to the left of the vertical axis and 5 units above the horizontal axis. Each point P has unique coordinates (a, b), and conversely each ordered pair of numbers (c, d) uniquely determines a point in the plane.

**The Distance
Formula**

There is a simple formula for finding the distance D between two points in a coordinate plane.

Figure 1.5 shows the points $P(x_1, y_1)$ and $Q(x_2, y_2)$. Note that the difference $x_2 - x_1$ of the x coordinates and the difference $y_2 - y_1$ of the y coordinates represent the lengths of the sides of a right triangle, and the length of the hypotenuse is the required distance D between P and Q. Thus, the Pythagorean theorem gives us the **distance formula** $D = \sqrt{(x_2 - x_1)^2 + (y_2 - y_1)^2}$. To summarize:

Just-In-Time

Recall the Pythagorean theorem, $c^2 = a^2 + b^2$, where a and b are the smaller sides of the right angled triangle and c, the hypotenuse, is the largest side, opposite the right angle.

The Distance Formula   ■   The distance between the points $P(x_1, y_1)$ and $Q(x_2, y_2)$ is given by

$$D = \sqrt{(x_2 - x_1)^2 + (y_2 - y_1)^2}.$$

NOTE   The distance formula is valid for all points in the plane even though we have considered only the case in which Q is above and to the right of P.

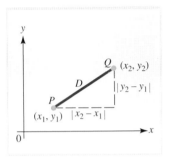

FIGURE 1.5   The distance formula.

EXAMPLE 1.2.1

Find the distance between the points $P(-2, 5)$ and $Q(4, -1)$.

Solution

In the distance formula, we have $x_1 = -2$, $y_1 = 5$, $x_2 = 4$, and $y_2 = -1$, so the distance between P and Q may be found as follows:

$$D = \sqrt{(4 - (-2))^2 + (-1 - 5)^2} = \sqrt{72} = 6\sqrt{2}.$$

The Graph of a Function

To represent a function $y = f(x)$ geometrically as a graph, we plot values of the independent variable x on the (horizontal) x axis and values of the dependent variable y on the (vertical) y axis. The graph of the function is defined as follows.

The Graph of a Function   ■   The graph of a function f consists of all points (x, y) where x is in the domain of f and $y = f(x)$, that is, all points of the form $(x, f(x))$.

In Chapter 3, you will study efficient techniques involving calculus that can be used to draw accurate graphs of functions. For many functions, however, you can make a fairly good sketch by plotting a few points, as illustrated in Example 1.2.2.

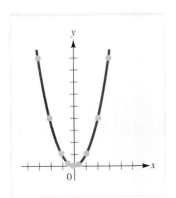

FIGURE 1.6   The graph of $y = x^2$.

EXAMPLE 1.2.2

Graph the function $f(x) = x^2$.

Solution

Begin by constructing the table of values.

x	-3	-2	-1	$-\dfrac{1}{2}$	0	$\dfrac{1}{2}$	1	2	3
$y = x^2$	9	4	1	$\dfrac{1}{4}$	0	$\dfrac{1}{4}$	1	4	9

Then plot the points (x, y) and connect them with the smooth curve shown in Figure 1.6.

NOTE Many different curves pass through the points in Example 1.2.2. Several of these curves are shown in Figure 1.7. There is no way to guarantee that the curve we draw through the plotted points is the actual graph of f. However, in general, the more points that are plotted, the more likely the graph is to be reasonably accurate. ■

EXPLORE!

Plot $f(x) = x^2$, $g(x) = x^2 + 3$, and $h(x) = x^2 - 2$ on the same graph. Can you see the relationship between these functions? What would the graph of $s(x) = x^2 + 8$ look like? Imagine it, and then plot it.

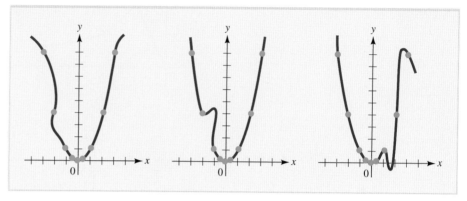

FIGURE 1.7   Other graphs through the points in Example 1.2.2.

Example 1.2.3 illustrates how to sketch the graph of a function defined by more than one formula.

EXAMPLE 1.2.3

Graph the function

$$f(x) = \begin{cases} 2x & \text{if } 0 \le x < 1 \\ \dfrac{2}{x} & \text{if } 1 \le x < 4 \\ 3 & \text{if } x \ge 4 \end{cases}$$

Solution

When making a table of values for this function, remember to use the formula that is appropriate for each particular value of x. Using the formula $f(x) = 2x$ when $0 \le x < 1$, the formula $f(x) = \dfrac{2}{x}$ when $1 \le x < 4$, and the formula $f(x) = 3$ when $x \ge 4$, you can compile this table of values:

x	0	$\dfrac{1}{2}$	1	2	3	4	5	6
$f(x)$	0	1	2	1	$\dfrac{2}{3}$	3	3	3

Now plot the corresponding points $(x, f(x))$ and draw the graph as in Figure 1.8. Notice that the pieces for $0 \le x < 1$ and $1 \le x < 4$ are connected to one another at $(1, 2)$ but that the piece for $x \ge 4$ is separated from the rest of the graph. (The open dot at $\left(4, \dfrac{1}{2}\right)$ indicates that the graph approaches this point but that the point is not actually on the graph.)

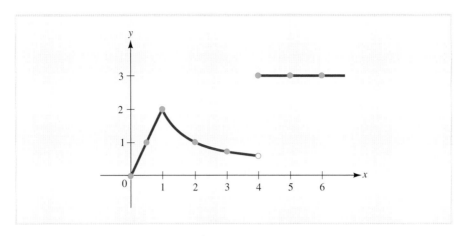

FIGURE 1.8   The graph of $f(x) = \begin{cases} 2x & 0 \le x < 1 \\ \dfrac{2}{x} & 1 \le x < 4 \\ 3 & x \ge 4 \end{cases}$

Intercepts

When the height of a graph is zero, the graph crosses the x axis. These points are called roots of the function. A function may have several roots or no roots. The points where a graph crosses the x axis are also called **x intercepts,** and, similarly, a **y intercept** is the point where a graph crosses the y axis. Intercepts are key features of a graph and can be determined using algebra or technology in conjunction with these criteria.

How to Find the x and y Intercepts ▪ To find any x intercept of a graph, set $y = 0$ and solve for x. To find any y intercept, set $x = 0$ and solve for y. For a function f, the only y intercept is $y_0 = f(0)$, but finding x intercepts may be difficult.

EXAMPLE 1.2.4

Graph the function $f(x) = -x^2 + x + 2$. Include all x and y intercepts.

Solution

The y intercept is $f(0) = 2$. To find the x intercepts, solve the equation $f(x) = 0$. Factoring, we find that

$$-x^2 + x + 2 = 0$$
$$-(x + 1)(x - 2) = 0$$
$$x = -1, x = 2$$

Thus, the x intercepts are $(-1, 0)$ and $(2, 0)$.

Next, make a table of values and plot the corresponding points $(x, f(x))$.

x	-3	-2	-1	0	1	2	3	4
$f(x)$	-10	-4	0	2	2	0	-4	-10

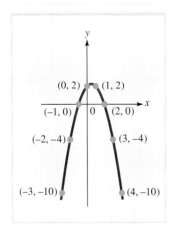

FIGURE 1.9   The graph of $f(x) = -x^2 + x + 2$.

The graph of f is shown in Figure 1.9.

> **NOTE** The factoring in Example 1.2.4 is fairly straightforward, but in other problems, you may need to review the factoring procedure provided in Appendix A2. ■

Graphing Parabolas

The graphs in Figures 1.6 and 1.9 are called **parabolas.** In general, the graph of $y = Ax^2 + Bx + C$ is a parabola as long as $A \neq 0$. All parabolas have a "U shape," and the parabola $y = Ax^2 + Bx + C$ opens up if $A > 0$ and down if $A < 0$. The "peak" or "valley" of the parabola is called its **vertex** and occurs where $x = \dfrac{-B}{2A}$ (Figure 1.10; also see Exercise 72). These features of the parabola are easily obtained by the methods of calculus developed in Chapter 3. Note that to get a reasonable sketch of the parabola $y = Ax^2 + Bx + C$, you need only determine three key features:

1. The location of the vertex $\left(\text{where } x = \dfrac{-B}{2A} \right)$.
2. Whether the parabola opens up ($A > 0$) or down ($A < 0$).
3. Any intercepts. The intercepts are the roots of the quadratic function. Real roots can be found by rearranging the function as $y = a(x - \text{root}_1)(x - \text{root}_2)$.

For instance, in Figure 1.9, the parabola $y = -x^2 + x + 2$ opens downward (since $A = -1$ is negative) and has its vertex (high point) where $x = \dfrac{-B}{2A} = \dfrac{-1}{2(-1)} = \dfrac{1}{2}$.

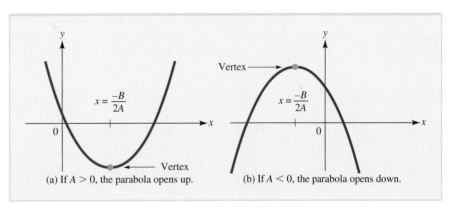

(a) If $A > 0$, the parabola opens up.   (b) If $A < 0$, the parabola opens down.

FIGURE 1.10 The graph of the parabola $y = Ax^2 + Bx + C$.

In Chapter 3, we will develop a procedure in which the graph of a function of practical interest is first obtained by calculus and then interpreted to obtain useful information about the function, such as its largest and smallest values. In Example 1.2.5 we preview this procedure by using what we know about the graph of a parabola to determine the maximum revenue obtained in a production process.

EXAMPLE 1.2.5

A manufacturer determines that when x hundred units of a particular commodity are produced, they can all be sold for a unit price given by the demand function $p = 60 - x$ dollars. At what level of production is revenue maximized? What is the maximum revenue?

Solution

The revenue derived from producing x hundred units and selling them all at $60 - x$ dollars is $R(x) = x(60 - x)$ hundred dollars. Note that $R(x) \geq 0$ only for $0 \leq x \leq 60$. The graph of the revenue function

$$R(x) = x(60 - x) = -x^2 + 60x$$

is a parabola that opens downward (since $A = -1 < 0$) and has its high point (vertex) where

$$x = \frac{-B}{2A} = \frac{-60}{2(-1)} = 30,$$

as shown in Figure 1.11. Thus, revenue is maximized when $x = 30$ hundred units are produced, and the corresponding maximum revenue is

$$R(30) = 30(60 - 30) = 900$$

hundred dollars. The manufacturer should produce 3000 units and at that level of production should expect maximum revenue of \$90 000.

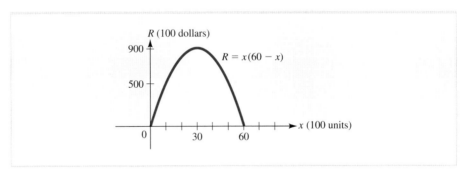

FIGURE 1.11   A revenue function.

Just-In-Time

Completing the square is reviewed in Appendix A2 and illustrated in Examples A.2.12 and A.2.13.

Note that we can also find the largest value of $R(x) = -x^2 + 60x$ by completing the square:

$$R(x) = -x^2 + 60x = -(x^2 - 60x) \qquad \text{factor out } -1, \text{ the coefficient of } x^2$$

$$= -(x^2 - 60x + 900) + 900 \qquad \text{complete the square inside}$$
$$\text{parentheses by adding}$$
$$-900 + 900 \qquad (-60/2)^2 = 900$$

$$= -(x - 30)^2 + 900$$

Thus, $R(30) = 0 + 900 = 900$, and if c is any number other than 30, then

$$R(c) = -(c - 30)^2 + 900 < 900 \qquad \text{since } -(c - 30)^2 < 0$$

So the maximum revenue is \$90 000 when $x = 30$ (3000 units).

Intersections of Graphs

Sometimes it is necessary to determine when two functions are equal. For instance, an economist may wish to compute the market price at which the consumer demand for a commodity will be equal to supply. Or a political analyst may wish to predict how long it will take for the popularity of a certain challenger to reach that of the incumbent. We shall examine some of these applications in Section 1.4.

In geometric terms, the values of x for which two functions $f(x)$ and $g(x)$ are equal are the x coordinates of the points where their graphs intersect. In Figure 1.12, the graph of $y = f(x)$ intersects the graph of $y = g(x)$ at two points, labelled P and Q. To find the points of intersection algebraically, set $f(x)$ equal to $g(x)$ and solve for x. This procedure is illustrated in Example 1.2.6.

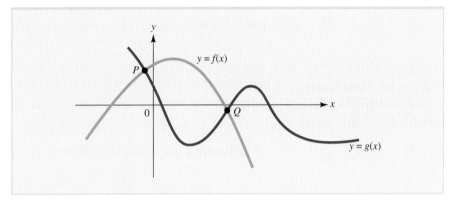

FIGURE 1.12   The graphs of $y = f(x)$ and $y = g(x)$ intersect at P and Q.

Just-In-Time

The **quadratic formula** is used in Example 1.2.6. Recall that this result says that the equation $Ax^2 + Bx + C = 0$ has real solutions if and only if $B^2 - 4AC \geq 0$, in which case the solutions are

$$r_1 = \frac{-B + \sqrt{B^2 - 4AC}}{2A}$$

and

$$r_2 = \frac{-B - \sqrt{B^2 - 4AC}}{2A}.$$

A review of the quadratic formula may be found in Appendix A2.

EXAMPLE 1.2.6

Find all points of intersection of the graphs of $f(x) = 3x + 2$ and $g(x) = x^2$.

Solution

You must solve the equation $x^2 = 3x + 2$. Rewrite the equation as $x^2 - 3x - 2 = 0$ and apply the quadratic formula to obtain

$$x = \frac{-(-3) \pm \sqrt{(-3)^2 - 4(1)(-2)}}{2(1)} = \frac{3 \pm \sqrt{17}}{2}.$$

The solutions are

$$x = \frac{3 + \sqrt{17}}{2} \approx 3.56 \quad \text{and} \quad x = \frac{3 - \sqrt{17}}{2} \approx -0.56.$$

(The computations were done on a calculator, with results rounded off to two decimal places.)

Computing the corresponding y coordinates from the equation $y = x^2$, you find that the points of intersection are approximately $(3.56, 12.67)$ and $(-0.56, 0.31)$. (As a result of round-off errors, you will get slightly different values for the y coordinates if you substitute into the equation $y = 3x + 2$.) The graphs and the intersection points are shown in Figure 1.13.

EXPLORE!

Plot graphs of $f(x) = x^2$ and $g(x) = 3x + 10$. Look at the graphs to find the intersection points. Solve the equation $x^2 - 3x - 10 = 0$. Are the answers the same as you observed in the graph?

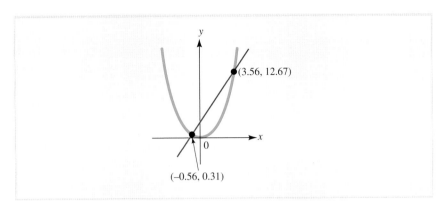

FIGURE 1.13   The intersection of the graphs of $f(x) = 3x + 2$ and $g(x) = x^2$.

Power Functions, Polynomials, and Rational Functions

A **power function** is a function of the form $f(x) = x^n$, where n is a real number. For example, $f(x) = x^2$, $f(x) = x^{-3}$, and $f(x) = x^{1/2}$ are all power functions. So are $f(x) = \dfrac{1}{x^2}$ and $f(x) = \sqrt[3]{x}$, since they can be rewritten as $f(x) = x^{-2}$ and $f(x) = x^{1/3}$, respectively.

A **polynomial** is a function of the form

$$p(x) = a_n x^n + a_{n-1} x^{n-1} + \cdots + a_1 x + a_0,$$

where n is a nonnegative integer and $a_0, a_1, \ldots, a_n$ are constants. If $a_n \neq 0$, the integer n is called the **degree** of the polynomial. For example, $f(x) = 3x^5 - 6x^2 + 7$ is a polynomial of degree 5. It can be shown that the graph of a polynomial of degree n is an unbroken curve that crosses the x axis no more than n times. To illustrate some of the possibilities, the graphs of three polynomials of degree 3 are shown in Figure 1.14.

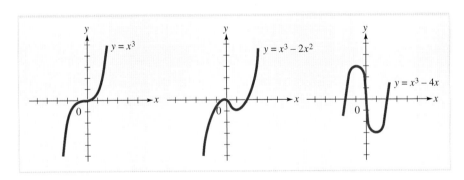

FIGURE 1.14   Three polynomials of degree 3.

A quotient $\dfrac{p(x)}{q(x)}$ of two polynomials $p(x)$ and $q(x)$ is called a **rational function.** Such functions appear throughout this text in examples and exercises. Graphs of three rational functions are shown in Figure 1.15. You will learn how to sketch such graphs in Section 3.3 of Chapter 3.

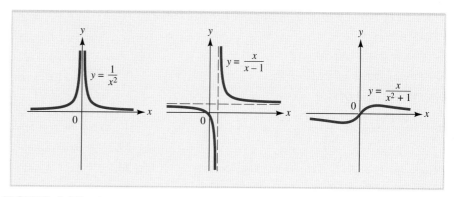

FIGURE 1.15   Graphs of three rational functions.

The Vertical Line Test   It is important to realize that not every curve is the graph of a function (Figure 1.16). For instance, the graph of the circle centred at the origin with radius $\sqrt{5}$ is represented by the relation $x^2 + y^2 = 5$. Then, since the points $(1, 2)$ and $(1, -2)$ both lie on the circle, we would have $y = 2$ and $y = -2$ when $x = 1$, contrary to the requirement that a function assigns one and *only* one value to each number in its domain. Thus, $x^2 + y^2 = 5$ is *not* a function. The **vertical line test** is a geometric rule for determining whether a curve is the graph of a function.

> **The Vertical Line Test** ■ A curve is the graph of a function if and only if no vertical line intersects the curve more than once.

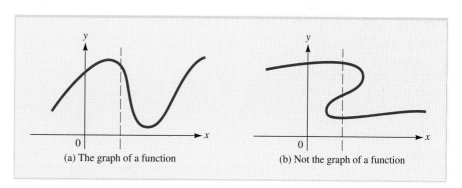

(a) The graph of a function   (b) Not the graph of a function

FIGURE 1.16   The vertical line test.

EXERCISES ■ 1.2

In Exercises 1 through 6, plot the given points in a rectangular coordinate plane.

1. $(4, 3)$

2. $(-2, 7)$

3. $(5, -1)$

4. $(-1, -8)$

5. $(0, -2)$

6. $(3, 0)$

In Exercises 7 through 10, find the distance between the given points P and Q.

7. $P(3, -1)$ and $Q(7, 1)$

8. $P(4, 5)$ and $Q(-2, -1)$

9. $P(7, -3)$ and $Q(5, 3)$

10. $P\left(0, \dfrac{1}{2}\right)$ and $Q\left(\dfrac{-1}{5}, \dfrac{3}{8}\right)$

In Exercises 11 and 12, classify each function as a polynomial, a power function, or a rational function. If the function is not one of these types, classify it as "different."

11. **a.** $f(x) = x^{1.4}$
   b. $f(x) = -2x^3 - 3x^2 + 8$
   c. $f(x) = (3x - 5)(4 - x)^2$
   d. $f(x) = \dfrac{3x^2 - x + 1}{4x + 7}$

12. **a.** $f(x) = -2 + 3x^2 + 5x^4$
   b. $f(x) = \sqrt{x} + 3x$
   c. $f(x) = \dfrac{(x - 3)(x + 7)}{-5x^3 - 2x^2 + 3}$
   d. $f(x) = \left(\dfrac{2x + 9}{x^2 - 3}\right)^3$

In Exercises 13 through 28, sketch the graph of the given function. Include all x and y intercepts.

13. $f(x) = x$

14. $f(x) = x^2$

15. $f(x) = \sqrt{x}$

16. $f(x) = \sqrt{1 - x}$

17. $f(x) = 2x - 1$

18. $f(x) = 2 - 3x$

19. $f(x) = x(2x + 5)$

20. $f(x) = (x - 1)(x + 2)$

21. $f(x) = -x^2 - 2x + 15$

22. $f(x) = x^2 + 2x - 8$

23. $f(x) = x^3$

24. $f(x) = -x^3 + 1$

25. $f(x) = \begin{cases} x - 1 & \text{if } x \le 0 \\ x + 1 & \text{if } x > 0 \end{cases}$

26. $f(x) = \begin{cases} 2x - 1 & \text{if } x < 2 \\ 3 & \text{if } x \ge 2 \end{cases}$

27. $f(x) = \begin{cases} x^2 + x - 3 & \text{if } x < 1 \\ 1 - 2x & \text{if } x \ge 1 \end{cases}$

28. $f(x) = \begin{cases} 9 - x & \text{if } x \le 2 \\ x^2 + x - 2 & \text{if } x > 2 \end{cases}$

In Exercises 29 through 34, find the points of intersection (if any) of the given pair of curves and draw the graphs.

29. $y = 3x + 5$ and $y = -x + 3$

30. $y = 3x + 8$ and $y = 3x - 2$

31. $y = x^2$ and $y = 3x - 2$

32. $y = x^2 - x$ and $y = x - 1$

33. $3y - 2x = 5$ and $y + 3x = 9$

34. $2x - 3y = -8$ and $3x - 5y = -13$

In Exercises 35 through 38, the graph of a function f(x) is given. In each case find
   (a) the y intercept
   (b) all x intercepts
   (c) the largest value of f(x) and the value(s) of x for which it occurs
   (d) the smallest value of f(x) and the value(s) of x for which it occurs

35.

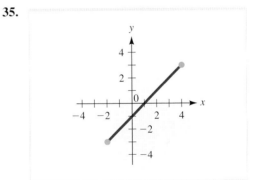

36.

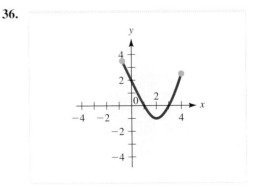

37.

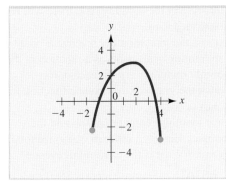

38.

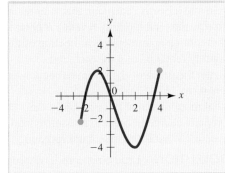

39. MANUFACTURING COST   A manufacturer can produce a backpack that can hold a laptop and books at a cost of $40 each. It is estimated that if the backpacks are sold for p dollars each, consumers will buy $120 - p$ of them a month. Express the manufacturer's monthly profit as a function of price, graph this function, and use the graph to estimate the optimal selling price.

40. MANUFACTURING COST   A manufacturer can produce bicycle tires at a cost of $20 each. It is estimated that if the tires are sold for p dollars each, consumers will buy $1560 - 12p$ of them each month. Express the manufacturer's monthly profit as a function of price, graph this function, and use the graph to determine the optimal selling price. How many tires will be sold each month at the optimal price?

41. RETAIL SALES   The owner of a toy store can obtain a popular board game at a cost of $15 per set. She estimates that if each set sells for x dollars, then $5(27 - x)$ sets will be sold each week. Express the owner's weekly profit from the sales of this game as a function of price, graph this function, and estimate the optimal selling price. How many sets will be sold each week at the optimal price?

42. RETAIL SALES   A bookstore can obtain an atlas from the publisher at a cost of $10 per copy. The manager estimates that if the store sells the atlas for x dollars per copy, approximately $20(22 - x)$ copies will be sold each month. Express the bookstore's monthly profit from the sale of the atlas as a function of price, graph this function, and use the graph to estimate the optimal selling price.

43. CONSUMER EXPENDITURE   Suppose $x = -200p + 12\ 000$ units of a particular commodity are sold each month when the market price is p dollars per unit. The total monthly consumer expenditure E is the total amount of money spent by consumers during each month.

  a. Express the total monthly consumer expenditure E as a function of the unit price p and sketch the graph of $E(p)$.

  b. Discuss the economic significance of the p intercepts of the expenditure function $E(p)$.

  c. Use the graph in part (a) to determine the market price that generates the greatest total monthly consumer expenditure.

44. CONSUMER EXPENDITURE   Suppose that x thousand units of a particular commodity are sold each month when the price is p dollars per unit, where

$$p(x) = 5(24 - x)$$

The total monthly consumer expenditure E is the total amount of money consumers spend during each month.

  a. Express the total monthly expenditure E as a function of the unit price p and sketch the graph of $E(p)$.

  b. Discuss the economic significance of the p intercepts of the expenditure function $E(p)$.

  c. Use the graph in part (a) to determine the market price that generates the greatest total monthly consumer expenditure. How many units will be sold during each month at the optimal price?

45. MOTION OF A PROJECTILE   If an object is thrown vertically upward from the ground with an initial speed of 49 m/s, its height (in metres) t seconds later is given by the function $H(t) = -4.9t^2 + 49t$.

  a. Graph the function $H(t)$.

  b. Use the graph in part (a) to determine when the object will hit the ground.

  c. Use the graph in part (a) to estimate how high the object will rise.

46. MOTION OF A PROJECTILE   A missile is projected vertically upward from an underground bunker in such a way that t seconds after launch, it is s metres above the ground, where

$$s(t) = -4.9t^2 + 245t - 5$$

   a. How deep is the bunker?
   b. Sketch the graph of $s(t)$.
   c. Use the graph in part (b) to determine when the missile is at its highest point. What is its maximum height?

47. PROFIT   Suppose that when the price of a certain commodity is p dollars per unit, then x hundred units will be purchased by consumers, where $p = -0.05x + 38$. The cost of producing x hundred units is $C(x) = 0.02x^2 + 3x + 574.77$ hundred dollars.
   a. Express the profit P obtained from the sale of x hundred units as a function of x. Sketch the graph of the profit function.
   b. Use the profit curve found in part (a) to determine the level of production x that results in maximum profit. What unit price p corresponds to maximum profit?

48. BLOOD FLOW   Recall from Exercise 69, Section 1.1, that the speed of blood located r centimetres from the central axis of an artery is given by the function $S(r) = C(R^2 - r^2)$, where C is a constant and R is the radius of the artery. What is the domain of this function? Sketch the graph of $S(r)$.

49. ROAD SAFETY   When an automobile is being driven at v kilometres per hour, the average driver requires D metres of visibility to stop safely, where $D = 0.0554v^2 + 0.789v$. Sketch the graph of $D(v)$.

50. POSTAGE RATES   Effective January 2010, Canada Post specified the cost of mailing a nonstandard or oversize letter as $1.22 for the first 100 g, $2.00 for the next 100 g, and $2.25 for the next 100 g. Let $P(w)$ be the postage required for mailing this type of letter, with a weight of w grams, for $0 \le w \le 300$.
   a. Describe $P(w)$ as a piecewise-defined function.
   b. Sketch the graph of P.

51. REAL ESTATE   A real estate company manages 150 apartments in Regina. All the apartments can be rented at a price of $1200 per month, but for each $100 increase in the monthly rent, there will be five additional vacancies.
   a. Express the total monthly revenue R obtained from renting apartments as a function of the monthly rental price p for each unit.
   b. Sketch the graph of the revenue function found in part (a).
   c. What monthly rental price p should the company charge in order to maximize total revenue? What is the maximum revenue?

52. REAL ESTATE   Suppose it costs $500 each month for the real estate company in Exercise 51 to maintain and advertise each unit that is unrented.
   a. Express the total monthly revenue R obtained from renting apartments as a function of the monthly rental price p for each unit.
   b. Sketch the graph of the revenue function found in part (a).
   c. What monthly rental price p should the company charge in order to maximize total revenue? What is the maximum revenue?

53. AIR POLLUTION   Lead emissions have been a major source of air pollution, but in the last 25 years Health Canada has regulated the amount of lead in gasoline, paints, and food cans. Using data gathered by the U.S. Environmental Protection Agency in the 1990s, it can be shown that the formula

$$N(t) = -35t^2 + 299t + 3347$$

estimates the total amount of lead emission N (in thousands of tons) occurring in the United States t years after the base year 1990.
   a. Sketch the graph of the pollution function $N(t)$.
   b. Approximately how much lead emission did the formula predict for the year 1995? (The actual amount was about 3924 thousand tons.)
   c. Based on this formula, when during the decade 1990–2000 would you expect the maximum lead emission to have occurred?
   d. Can this formula be used to accurately predict the current level of lead emission? Explain.

54. ARCHITECTURE   A train track crosses an arch-shaped bridge over a road. The arch has a parabolic shape. It is 6 m wide at the base and just tall enough to allow a truck 5 m high and 4 m wide to pass.
   a. Assuming that the arch has an equation of the form $y = ax^2 + b$, use the given information to find a and b. Explain why this assumption is reasonable.
   b. Sketch the graph of the arch equation you found in part (a).

In Exercises 55 through 58, use the vertical line test to determine whether or not the given curve is the graph of a function.

55.

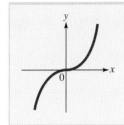

56.

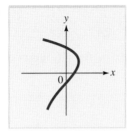

57.

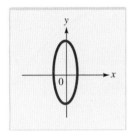

58.

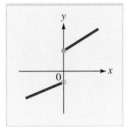

59. What viewing rectangle should be used to get an adequate graph for the quadratic function

$$f(x) = -9x^2 + 3600x - 358\,200?$$

60. What viewing rectangle should be used to get an adequate graph for the quadratic function

$$f(x) = 4x^2 - 2400x + 355\,000?$$

61. a. Graph the functions $y = x^2$ and $y = x^2 + 3$. How are the graphs related?

  b. Without further computation, graph the function $y = x^2 - 5$.

  c. Suppose $g(x) = f(x) + c$, where c is a constant. How are the graphs of f and g related? Explain.

62. a. Graph the functions $y = x^2$ and $y = -x^2$. How are the graphs related?

  b. Suppose $g(x) = -f(x)$. How are the graphs of f and g related? Explain.

63. a. Graph the functions $y = x^2$ and $y = (x - 2)^2$. How are the graphs related?

  b. Without further computation, graph the function $y = (x + 1)^2$.

  c. Suppose $g(x) = f(x - c)$, where c is a constant. How are the graphs of f and g related? Explain.

64. It costs \$90 to rent a piece of equipment plus \$21 for every day of use.

  a. Make a table showing the number of days the equipment is rented and the cost of renting for 2 days, 5 days, 7 days, and 10 days.

  b. Write an algebraic expression representing the cost, y, as a function of the number of days, x.

  c. Graph the expression in part (b).

65. A company that manufactures lawn mowers has determined that a new employee can assemble N mowers per day after t days of training, where

$$N(t) = \frac{45t^2}{5t^2 + t + 8}.$$

  a. Make a table showing the numbers of mowers assembled for training periods of lengths $t = 2$ days, 3 days, 5 days, 10 days, and 50 days.

  b. Based on the table in part (a), what do you think happens to $N(t)$ for very long training periods?

  c. Graph $N(t)$. Extend the t scale for large values to confirm your answer for part b.

66. Use a graphing utility to graph $y = x^4$, $y = x^4 - x$, $y = x^4 - 2x$, and $y = x^4 - 3x$ on the same coordinate axes, using $x = -2$ to 2. What effect does the added term involving x have on the shape of the graph? Repeat using $y = x^4$, $y = x^4 - x^3$, $y = x^4 - 2x^3$, and $y = x^4 - 3x^3$. Adjust the x scale appropriately.

67. Graph $f(x) = \dfrac{-9x^2 - 3x - 4}{4x^2 + x - 1}$. Determine the values of x for which the function is defined.

68. Graph $f(x) = \dfrac{8x^2 + 9x + 3}{x^2 + x - 1}$. Determine the values of x for which the function is defined.

69. Graph $g(x) = -3x^3 + 7x + 4$. Find the x intercepts.

70. Use the distance formula to show that the circle with centre (a, b) and radius R has the equation

$$(x - a)^2 + (y - b)^2 = R^2$$

71. Use the result in Exercise 70 to answer the following questions.

  a. Find the equation for the circle with centre $(2, -3)$ and radius 4.

  b. Find the centre and radius of the circle with the equation

$$x^2 + y^2 - 4x + 6y = 11$$

  c. Describe the set of all points (x, y) such that

$$x^2 + y^2 + 4y = 2x - 10$$

72. Show that the vertex of the parabola $y = Ax^2 + Bx + C$ $(A \neq 0)$ occurs at the point where $x = \dfrac{-B}{2A}$. [*Hint:* First verify that

$$Ax^2 + Bx + C = A\left[\left(x + \frac{B}{2A}\right)^2 + \left(\frac{C}{A} - \frac{B^2}{4A^2}\right)^2\right].$$ Then note that the largest or

smallest value of $f(x) = Ax^2 + Bx + C$ must occur where $x = \dfrac{-B}{2A}$.]

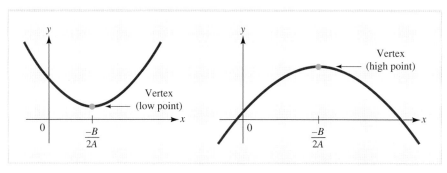

EXERCISE 72

SECTION 1.3

Linear Functions

L03

Find the equation of a line, given either two points or one point and a slope. Formulate equations of lines from given application data.

In many practical situations, the rate at which one quantity changes with respect to another is constant. Here is a simple example from economics.

EXAMPLE 1.3.1

A manufacturer's total cost consists of a fixed overhead of $200 plus production costs of $50 per unit. Express the total cost as a function of the number of units produced and draw the graph.

EXPLORE!

Plot $C_1(x) = 50x + 200$, $C_2(x) = 50x + 300$, and $C_3(x) = 50x + 400$ on the same graph. Can you see how increased overhead affects cost?

Solution

Let x denote the number of units produced and $C(x)$ the corresponding total cost. Then,

$$\text{Total cost} = (\text{cost per unit})(\text{number of units}) + \text{overhead},$$

where

$$\text{Cost per unit} = 50$$
$$\text{Number of units} = x$$
$$\text{Overhead} = 200$$

Hence,

$$C(x) = 50x + 200.$$

The graph of this cost function is sketched in Figure 1.17.

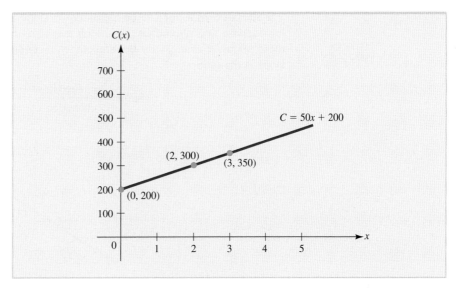

FIGURE 1.17   The cost function $C(x) = 50x + 200$.

The total cost in Example 1.3.1 increases at a constant rate of $50 per unit. As a result, its graph in Figure 1.17 is a straight line that increases in height by 50 units for each 1-unit increase in x.

In general, a function whose value changes at a constant rate with respect to its independent variable is said to be a **linear function.** Such a function has the form

$$f(x) = mx + b$$

where m and b are constants, and its graph is a straight line. For example, $f(x) = \dfrac{3}{2} + 2x$, $f(x) = -5x$, and $f(x) = 12$ are all linear functions. To summarize:

Linear Functions ■ A linear function is a function that changes at a constant rate with respect to its independent variable.

The graph of a linear function is a straight line.

The equation of a linear function can be written in the form

$$y = mx + b$$

where m and b are constants.

The Slope of a Line   A surveyor might say that a hill with a rise of 2 m for every 1 m of run has a **slope** of

$$m = \frac{\text{rise}}{\text{run}} = \frac{2}{1} = 2.$$

The steepness of a line can be measured by slope in much the same way. In particular, suppose (x_1, y_1) and (x_2, y_2) lie on a line, as indicated in Figure 1.18. Between these points, x changes by the amount $x_2 - x_1$ and y by the amount $y_2 - y_1$. The slope is the ratio

$$\text{Slope} = \frac{\text{change in } y}{\text{change in } x} = \frac{y_2 - y_1}{x_2 - x_1}.$$

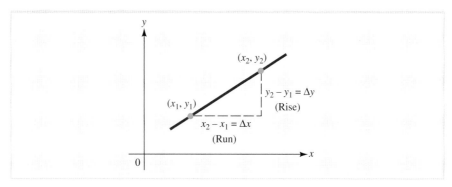

FIGURE 1.18   Slope $= \dfrac{y_2 - y_1}{x_2 - x_1} = \dfrac{\Delta y}{\Delta x}.$

It is sometimes convenient to use the symbol Δy instead of $y_2 - y_1$ to denote the change in y. The symbol Δy is read "delta y." Similarly, the symbol Δx is used to denote the change $x_2 - x_1$.

The Slope of a Line ■ The slope of the nonvertical line passing through the points (x_1, y_1) and (x_2, y_2) is given by the formula

$$\text{Slope} = \frac{\Delta y}{\Delta x} = \frac{y_2 - y_1}{x_2 - x_1}.$$

The use of this formula is illustrated in Example 1.3.2.

EXAMPLE 1.3.2

Find the slope of the line joining the points $(-2, 5)$ and $(3, -1)$.

Solution

$$\text{Slope} = \frac{\Delta y}{\Delta x} = \frac{-1 - 5}{3 - (-2)} = \frac{-6}{5}.$$

The line is shown in Figure 1.19.

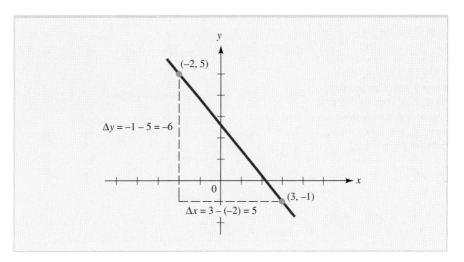

FIGURE 1.19   The line joining $(-2, 5)$ and $(3, -1)$.

The sign and magnitude of the slope of a line indicate the line's direction and steepness, respectively. The slope is positive if the height of the line increases as x increases and is negative if the height decreases as x increases. The absolute value of the slope is large if the slant of the line is severe and small if the slant of the line is gradual. The situation is illustrated in Figure 1.20.

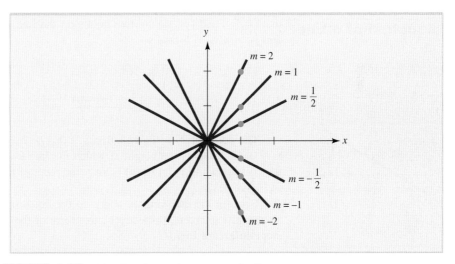

FIGURE 1.20   The direction and steepness of a line.

Horizontal and Vertical Lines

Horizontal and vertical lines (Figure 1.21a and 1.21b) have particularly simple equations. The y coordinates of all points on a horizontal line are the same. Hence, a horizontal line is the graph of a linear function of the form $y = b$, where b is a constant. The slope of a horizontal line is zero, since changes in x produce no changes in y.

EXPLORE!

Plot $y_1 = 3x + 4$, $y_2 = 8x + 4$, and $y_3 = 0.5x + 4$ on the same graph. Can you plot three more lines on the same graph, with the same slopes as y_1, y_2, and y_3 but with different y intercepts?

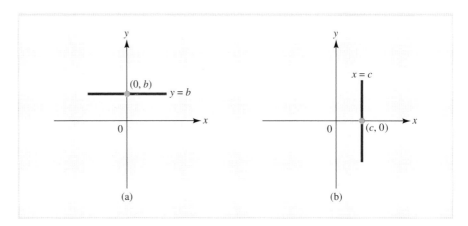

FIGURE 1.21   Horizontal and vertical lines.

The x coordinates of all points on a vertical line are equal. Hence, vertical lines are characterized by equations of the form $x = c$, where c is a constant. The slope of a vertical line is undefined because only the y coordinates of points on the line can change, so the denominator of the quotient $\dfrac{\text{change in } y}{\text{change in } x}$ is zero.

The Slope-Intercept Form of the Equation of a Line

The constants m and b in the equation $y = mx + b$ of a nonvertical line have geometric interpretations. The coefficient m is the slope of the line. To see this, suppose that (x_1, y_1) and (x_2, y_2) are two points on the line $y = mx + b$. Then $y_1 = mx_1 + b$ and $y_2 = mx_2 + b$, and so

$$
\begin{aligned}
\text{Slope} &= \frac{y_2 - y_1}{x_2 - x_1} = \frac{(mx_2 + b) - (mx_1 + b)}{x_2 - x_1} \\
&= \frac{mx_2 - mx_1}{x_2 - x_1} = \frac{m(x_2 - x_1)}{x_2 - x_1} = m
\end{aligned}
$$

The constant b in the equation $y = mx + b$ is the value of y corresponding to $x = 0$. Hence, b is the height at which the line $y = mx + b$ crosses the y axis, and the corresponding point $(0, b)$ is the y intercept of the line. The situation is illustrated in Figure 1.22.

Because the constants m and b in the equation $y = mx + b$ correspond to the slope and y intercept, respectively, this form of the equation of a line is known as the **slope-intercept form.**

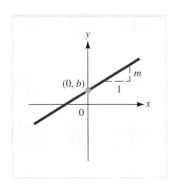

FIGURE 1.22   The slope and y intercept of the line $y = mx + b$.

The Slope-Intercept Form of the Equation of a Line   ■   The equation

$$y = mx + b$$

is the equation of the line with slope m and y intercept b.

The slope-intercept form of the equation of a line is particularly useful when geometric information about a line (such as its slope or y intercept) is to be determined from the line's algebraic representation. Here is a typical example.

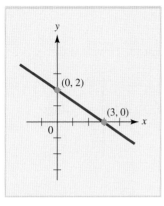

FIGURE 1.23   The line
$3y + 2x = 6$.

EXAMPLE 1.3.3

Find the slope and y intercept of the line $3y + 2x = 6$ and draw the graph.

Solution

First put the equation $3y + 2x = 6$ in slope-intercept form $y = mx + b$. To do this, solve for y to get

$$3y = -2x + 6$$

$$y = -\frac{2}{3}x + 2$$

It follows that the slope is $-\frac{2}{3}$ and the y intercept is 2.

To graph a linear function, plot two of its points and draw a straight line through them. In this case, you already know the point $(0, 2)$, at the y intercept. A convenient choice for the x coordinate of the second point is $x = 3$, since the corresponding y coordinate is $y = -\frac{2}{3}(3) + 2 = 0$. Draw a line through the points $(0, 2)$ and $(3, 0)$ to obtain the graph shown in Figure 1.23.

The Point-Slope Form of the Equation of a Line

Geometric information about a line can be obtained readily from the slope-intercept formula $y = mx + b$. Another form of the equation of a line, however, is usually more efficient for problems in which the geometric properties of a line are known and the goal is to find the equation of the line.

> **The Point-Slope Form of the Equation of a Line** ■ The equation
>
> $$y - y_0 = m(x - x_0)$$
>
> is the equation of the line that passes through the point (x_0, y_0) and has slope equal to m.

The point-slope form of the equation of a line is simply the formula for slope in disguise. To see this, suppose the point (x, y) lies on the line that passes through a given point (x_0, y_0) and has slope m. Using the points (x, y) and (x_0, y_0) to compute the slope, you get

$$\frac{y - y_0}{x - x_0} = m,$$

which you can put in point-slope form,

$$y - y_0 = m(x - x_0),$$

by simply multiplying both sides by $x - x_0$.

The use of the point-slope form of the equation of a line is illustrated in Examples 1.3.4 and 1.3.5.

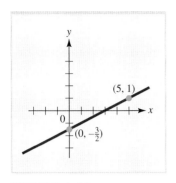

FIGURE 1.24   The line $y = \frac{1}{2}x - \frac{3}{2}$.

EXAMPLE 1.3.4

Find the equation of the line that passes through the point (5, 1) with slope $\frac{1}{2}$.

Solution

Use the formula $y - y_0 = m(x - x_0)$ with $(x_0, y_0) = (5, 1)$ and $m = \frac{1}{2}$ to get

$$y - 1 = \frac{1}{2}(x - 5)$$

$$y = \frac{1}{2}x - \frac{3}{2}$$

The graph is shown in Figure 1.24.

For practice, solve the problem in Example 1.3.4 using the slope-intercept formula. Notice that the solution based on the point-slope formula is more efficient.

In Chapter 2, the point-slope formula will be used extensively for finding the equation of the tangent line to the graph of a function at a given point. Example 1.3.5 illustrates how the point-slope formula can be used to find the equation of a line through two given points.

EXAMPLE 1.3.5

Find the equation of the line that passes through the points (3, −2) and (1, 6).

Solution

First compute the slope:

$$m = \frac{6 - (-2)}{1 - 3} = \frac{8}{-2} = -4.$$

Then use the point-slope formula with (1, 6) as the given point (x_0, y_0) to get

$$y - 6 = -4(x - 1) \quad \text{or} \quad y = -4x + 10.$$

Convince yourself that the resulting equation would have been the same if you had chosen (3, −2) to be the given point (x_0, y_0). The graph is shown in Figure 1.25.

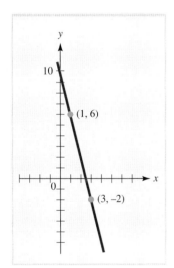

FIGURE 1.25   The line $y = -4x + 10$.

NOTE   The general form for the equation of a line is $Ax + By + C = 0$, where A, B, and C are constants, with A and B not both equal to 0. If $B = 0$, the line is vertical, and when $B \neq 0$, the equation $Ax + By + C = 0$ can be rewritten as $By = -Ax - C$. Then dividing by B gives

$$y = \left(\frac{-A}{B}\right)x + \left(\frac{-C}{B}\right).$$

Comparing this equation with the slope-intercept form $y = mx + b$, we see that the slope of the line is given by $m = -A/B$ and the y intercept by $b = -C/B$. The line is horizontal (slope 0) when $A = 0$.

Practical
Applications

If the rate of change of one quantity with respect to a second quantity is constant, the function relating the quantities must be linear. The constant rate of change is the slope of the corresponding line. Examples 1.3.6 and 1.3.7 illustrate techniques you can use to find the appropriate linear functions in such situations.

EXAMPLE 1.3.6

Since the beginning of the year, the price of a bottle of pop at a local discount super-market has been rising at a constant rate of 2 cents per month. By November 1, the price had reached $1.56 per bottle. Express the price of the pop as a function of time and determine the price at the beginning of the year.

Solution

Let x denote the number of months that have elapsed since the first of the year and y the price of a bottle of pop (in cents). Since y changes at a constant rate with respect to x, the function relating y to x must be linear, and its graph is a straight line. Since the price y increases by 2 each time x increases by 1, the slope of the line must be 2. The fact that the price was 156 cents ($1.56) on November 1, 10 months after the first day of the year, implies that the line passes through the point (10, 156). To write an equation defining y as a function of x, use the point-slope formula

$$y - y_0 = m(x - x_0)$$

with $m = 2, x_0 = 10, y_0 = 156$

to get $y - 156 = 2(x - 10)$ or $y = 2x + 136$

The corresponding line is shown in Figure 1.26. Notice that the y intercept is (0, 136), which implies that the price of pop at the beginning of the year was $1.36 per bottle.

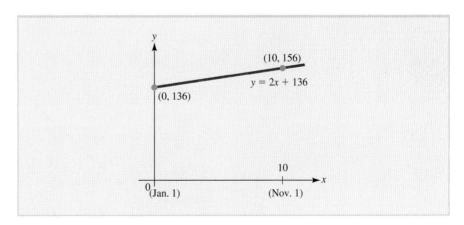

FIGURE 1.26   The rising price of pop: $y = 2x + 136$.

Sometimes it is hard to tell how two quantities, x and y, in a data set are related by simply examining the data. In such cases, it may be useful to graph the data to see if the points (x, y) follow a clear pattern (say, lie along a line). Here is an example of this procedure.

TABLE 1.2 Greenhouse Gas Emissions Edmonton 1990–2007

Year	Greenhouse Gas Emissions TCO$_2$ (tonnes/year)
1990	343 812
2002	382 218
2003	394 457
2004	402 085
2005	404 368
2006	412 730
2007	421 471

SOURCE: *EcoVision Annual Report,* 2008, Table 3. "City Operations—Greenhouse Gas Emissions Trends" and Chart 17 "Edmonton's Population," City of Edmonton.

EXAMPLE 1.3.7

The city of Edmonton published a report (available online) called the *EcoVision Annual Report*. This states goals for the environment and statistics on progress so far. The following table gives the annual greenhouse gas emissions.

Solution

The graph is shown in Figure 1.27. Note that the initial point of 1990 has a large effect on the relationship, but the pattern does suggest that we may be able to get useful information by finding a line that "best fits" the data in some meaningful way. One such procedure, called "least-squares approximation," requires the approximating line to be positioned so that the sum of squares of vertical distances from the data points to the line is minimized. The least-squares procedure, which will be developed in Section 7.4 of Chapter 7, can be carried out easily with most calculators. For the greenhouse gas data, this procedure produces the "best-fitting line" $y = 3132x + 5\ 898\ 570$. Sometimes a linear relationship is used to interpolate for unknown data. We could use this formula to determine the expected tonnes of emissions in 2000:

$$y(2000) = 3132(2000) + 5\ 898\ 570 = 365\ 430$$

Although the greenhouse gases have increased in this time period, the population of Edmonton has also increased considerably. Exercise 63 looks at the tonnes of greenhouse gas per person, which shows a very different relationship. Exercise 63 indicates that the work that the City of Edmonton is doing in decreasing greenhouse gases is on its way to success.

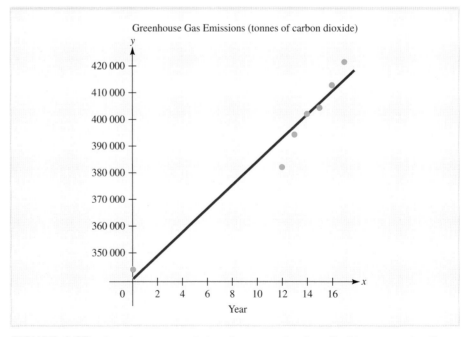

FIGURE 1.27 Greenhouse gas emissions in tonnes of carbon dioxide per year for City of Edmonton.

SOURCE: *EcoVision Annual Report,* 2008, Table 3. "City Operations—Greenhouse Gas Emissions Trends" and Chart 17 "Edmonton's Population," City of Edmonton.

Parallel and Perpendicular Lines

In applications, it is sometimes necessary or useful to know whether two given lines are parallel or perpendicular. A vertical line is parallel only to other vertical lines and is perpendicular to any horizontal line. Cases involving nonvertical lines can be handled by the following slope criteria.

> **Parallel and Perpendicular Lines** ■ Let m_1 and m_2 be the slopes of the nonvertical lines L_1 and L_2. Then
>
> L_1 and L_2 are **parallel** if and only if $m_1 = m_2$.
>
> L_1 and L_2 are **perpendicular** if and only if $m_2 = -\dfrac{1}{m_1}$.

These criteria are demonstrated in Figure 1.28. We close this section with an example illustrating one way the criteria can be used.

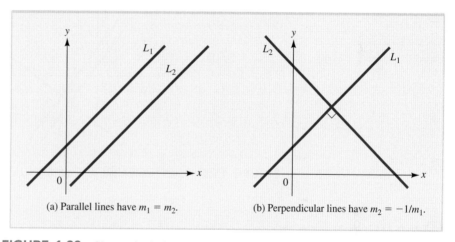

(a) Parallel lines have $m_1 = m_2$.   (b) Perpendicular lines have $m_2 = -1/m_1$.

FIGURE 1.28   Slope criteria for parallel and perpendicular lines.

EXPLORE!

Plot graphs of $y_1 = 2x - 3$ and $y_2 = 1 - 0.5x$. Are they perpendicular? Plot the graph of $y_3 = 5x - 7$. What is the slope of a line perpendicular to this? Plot your line and check it out!

EXAMPLE 1.3.8

Let L be the line $4x + 3y = 3$.

a. Find the equation of the line L_1 parallel to L through $P(-1, 4)$.
b. Find the equation of the line L_2 perpendicular to L through $Q(2, -3)$.

Solution

By rewriting the equation $4x + 3y = 3$ in the slope-intercept form $y = -\dfrac{4}{3}x + 1$, we see that L has slope $m_L = -\dfrac{4}{3}$.

a. Any line parallel to L must also have slope $m = -\dfrac{4}{3}$. The required line L_1 contains $P(-1, 4)$, so

$$y - 4 = -\frac{4}{3}(x + 1)$$

$$y = -\frac{4}{3}x + \frac{8}{3}$$

b. A line perpendicular to L must have slope $m = -\dfrac{1}{m_L} = \dfrac{3}{4}$. Since the required line L_2 contains $Q(2, -3)$, we have

$$y + 3 = \frac{3}{4}(x - 2)$$

$$y = \frac{3}{4}x - \frac{9}{2}$$

The given line L and the required lines L_1 and L_2 are shown in Figure 1.29.

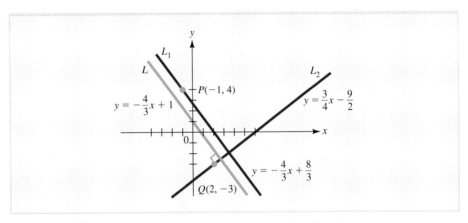

FIGURE 1.29   Lines parallel and perpendicular to a given line L.

EXERCISES ■ 1.3

In Exercises 1 through 8, find the slope (if possible) of the line that passes through the given pair of points.

1. $(2, -3)$ and $(0, 4)$

2. $(-1, 2)$ and $(2, 5)$

3. $(2, 0)$ and $(0, 2)$

4. $(5, -1)$ and $(-2, -1)$

5. $(2, 6)$ and $(2, -4)$

6. $\left(\dfrac{2}{3}, -\dfrac{1}{5}\right)$ and $\left(-\dfrac{1}{7}, \dfrac{1}{8}\right)$

7. $\left(\dfrac{1}{7}, 5\right)$ and $\left(-\dfrac{1}{11}, 5\right)$

8. $(-1.1, 3.5)$ and $(-1.1, -9)$

In Exercises 9 through 12, find the slope and intercepts of the line shown. Then find an equation for the line.

9.

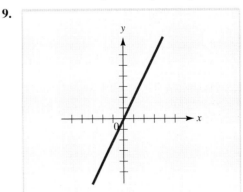

10.

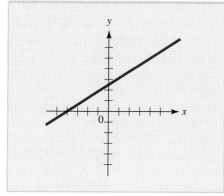

11.

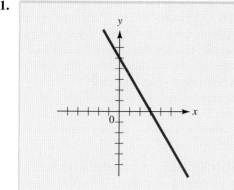

12.

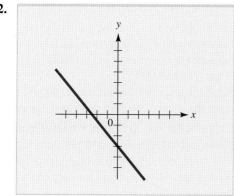

In Exercises 13 through 20, find the slope and intercepts of the line whose equation is given, and then sketch the graph of the line.

13. $x = 3$

14. $y = 5$

15. $y = 3x$

16. $y = 3x - 6$

17. $3x + 2y = 6$

18. $5y - 3x = 4$

19. $\dfrac{x}{2} + \dfrac{y}{5} = 1$

20. $\dfrac{x + 3}{-5} + \dfrac{y - 1}{2} = 1$

In Exercises 21 through 36, write the equation for the line with the given properties.

21. Through $(2, 0)$ with slope 1

22. Through $(-1, 2)$ with slope $\dfrac{2}{3}$

23. Through $(5, -2)$ with slope $-\dfrac{1}{2}$

24. Through $(0, 0)$ with slope 5

25. Through $(2, 5)$ and parallel to the x axis

26. Through $(2, 5)$ and parallel to the y axis

27. Through $(1, 0)$ and $(0, 1)$

28. Through $(2, 5)$ and $(1, -2)$

29. Through $\left(-\dfrac{1}{5}, 1\right)$ and $\left(\dfrac{2}{3}, \dfrac{1}{4}\right)$

30. Through $(-2, 3)$ and $(0, 5)$

31. Through $(1, 5)$ and $(3, 5)$

32. Through $(1, 5)$ and $(1, -4)$

33. Through $(4, 1)$ and parallel to the line $2x + y = 3$

34. Through $(-2, 3)$ and parallel to the line $x + 3y = 5$

35. Through $(3, 5)$ and perpendicular to the line $x + y = 4$

36. Through $\left(-\dfrac{1}{2}, 1\right)$ and perpendicular to the line $2x + 5y = 3$

37. **MANUFACTURING COST**   A manufacturer's total cost consists of a fixed overhead of $5000 plus production costs of $60 per unit. Express the total cost as a function of the number of units produced and draw the graph.

38. **MANUFACTURING COST**   A manufacturer estimates that it costs $75 to produce each unit of a particular commodity. The fixed overhead is $4500. Express the total cost of production as a function of the number of units produced and draw the graph.

39. **CREDIT CARD DEBT**   The Canadian Bankers Association website shows the net retail volume of credit card purchases in billions of dollars. In 2000 it was $125.18 billion and in 2009 it was

$296.21 billion. Use just these two numbers and assume a linear relationship, with retail volume V growing at a constant rate.

a. Express V as a linear function of time t, where t is the number of years after 2000. Draw the graph.

b. Use the function in part (a) to predict the total volume in 2015.

c. Why should more data be taken into account when modelling a relationship? What happened in 2009 that may have affected the data? Check out the website and write a sentence about why a linear relationship based on two pieces of data, extrapolated to the future, may not give a good prediction.

40. CAR RENTAL   A car rental agency charges $75 per day plus 70 cents per kilometre.

a. Express the cost of renting a car from this agency for 1 day as a function of the number of kilometres driven and draw the graph.

b. How much does it cost to rent a car for a l-day trip of 50 km?

c. The agency also offers a rental for a flat fee of $125 per day. How many kilometres must you drive on a l-day trip for this to be the better deal?

41. SPORTS REGISTRATION   Soccer registration takes place the first weekend in May. Players have the option of registering online before that date. At the registration desk on the first weekend in May, 16 players can be registered per hour. After 3 hours, 70 players have registered, including those who preregistered.

a. Express the number of players registered as a function of time and draw the graph.

b. How many players were registered after 2 h?

c. How many players preregistered online?

42. MEMBERSHIP FEES   Membership in a swimming club costs $250 for the 12-week summer season. If a member joins after the start of the season, the fee is prorated; that is, it is reduced linearly.

a. Express the membership fee as a function of the number of weeks that have elapsed by the time the membership is purchased and draw the graph.

b. Compute the cost of a membership that is purchased 5 weeks after the start of the season.

43. LINEAR DEPRECIATION   A doctor owns $3500 worth of medical books, which, for tax purposes, are assumed to depreciate linearly to zero over a 10-year period. That is, the value of the books decreases at a constant rate so that it is equal to

zero at the end of 10 years. Express the value of the books as a function of time and draw the graph.

44. LINEAR DEPRECIATION   A manufacturer buys $20 000 worth of machinery that depreciates linearly so that its trade-in value after 10 years will be $1000.

a. Express the value of the machinery as a function of its age and draw the graph.

b. Compute the value of the machinery after 4 years.

c. When does the machinery become worthless? The manufacturer might not wait this long to dispose of the machinery. Discuss the issues the manufacturer may consider in deciding when to sell.

45. WATER CONSUMPTION   Since the beginning of the month, a local reservoir has been losing water at a constant rate. On the 12th of the month, the reservoir held 800 million litres of water, and on the 21st it held only 647 million litres.

a. Express the amount of water in the reservoir as a linear function of time and draw the graph.

b. How much water was in the reservoir on the 8th of the month?

46. PRINTING COST   A publisher estimates that the cost of producing between 1000 and 10 000 copies of a certain textbook is $50 per copy; between 10 001 and 20 000, the cost is $40 per copy; and between 20 001 and 50 000, the cost is $35 per copy.

a. Find a function $F(N)$ that gives the total cost of producing N copies of the text for $1000 \leq N \leq 50\ 000$.

b. Sketch the graph of the function $F(N)$ you found in part (a).

47. STOCK PRICES   A certain stock had an initial public offering (IPO) price of $10 per share and is traded 24 hours per day. Sketch the graph of the share price over a 2-year period for each of the following cases.

a. The price increases steadily to $50 a share over the first 18 months and then decreases steadily to $25 per share over the next 6 months.

b. The price takes just 2 months to rise at a constant rate to $15 a share and then slowly drops to $8 over the next 9 months before steadily rising to $20.

c. The price steadily rises to $60 a share during the first year, at which time an accounting scandal is uncovered. The price jumps down to $25 a share and then steadily decreases to $5 over the next 3 months before rising at a constant rate to close at $12 at the end of the 2-year period.

48. AN ANCIENT FABLE   In Aesop's fable about the race between the tortoise and the hare, the tortoise trudges along at a slow, constant rate from start to finish. The hare starts out running steadily at a much more rapid pace, but halfway to the finish line, stops to take a nap. Finally, the hare wakes up, sees the tortoise near the finish line, desperately resumes his old rapid pace, and is nosed out by a hair. On the same coordinate plane, graph the respective distances of the tortoise and the hare from the starting line regarded as functions of time.

49. GROWTH OF A CHILD   The average height, H, in centimetres of a child of age A years can be estimated by the linear function $H = 6.5A + 50$. Use this formula to answer these questions.
  a. What is the average height of a 7-year-old child?
  b. How old must a child be to have an average height of 150 cm?
  c. What is the average height of a newborn baby? Does this answer seem reasonable?
  d. According to this formula, what is the average height of a 20-year-old? Does this answer seem reasonable?

50. CAR POOLING   To encourage motorists to form car pools, the transit authority in a certain metropolitan area has been offering a special reduced rate at toll bridges for vehicles containing four or more people. When the program began 30 days ago, 157 vehicles qualified for the reduced rate during the morning rush hour. Since then, the number of vehicles qualifying has been increasing at a constant rate, and today 247 vehicles qualified.
  a. Express the number of vehicles qualifying each morning for the reduced rate as a function of time and draw the graph.
  b. If the trend continues, how many vehicles will qualify during the morning rush hour 14 days from now?

51. TEMPERATURE CONVERSION
  a. Temperature measured in degrees Fahrenheit is a linear function of temperature measured in degrees Celsius. Use the fact that 0°C is equal to 32°F and 100°C is equal to 212°F to write an equation for this linear function.
  b. Use the function you obtained in part (a) to convert 15°C to Fahrenheit.
  c. Convert 68°F to Celsius.
  d. What temperature is the same in both the Celsius and Fahrenheit scales?

52. ENTOMOLOGY   It has been observed that the number of chirps made by a cricket each minute depends on the temperature. Crickets won't chirp if the temperature is 3°C or less, and observations yield the following data:

Number of chirps (N)	0	5	10	20	60
Temperature T (°C)	3	3.5	4	5	9

  a. Express T as a linear function of N.
  b. How many chirps would you expect to hear if the temperature is 12°C? If you hear 37 chirps in 30 seconds, what is the approximate temperature?

53. APPRECIATION OF ASSETS   The value of a certain rare book doubles every 10 years. In 1900, the book was worth $100.
  a. How much was it worth in 1930? In 1990? What about the year 2020?
  b. Is the value of the book a linear function of its age? Answer this question by interpreting an appropriate graph.

54. AIR POLLUTION   In certain parts of the world, the number of deaths, N, per week has been observed to be linearly related to the average concentration x of sulphur dioxide in the air. Suppose there are 97 deaths when $x = 100$ mg/m^3 and 110 deaths when $x = 500$ mg/m^3.
  a. What is the functional relationship between N and x?
  b. Use the function in part (a) to find the number of deaths per week when $x = 300$ mg/m^3. What concentration of sulfur dioxide corresponds to 100 deaths per week?
  c. Research data on how air pollution affects the death rate in a population. Summarize your results in a one-paragraph essay.

55. COLLEGE ADMISSIONS   At an Alternative Health College, a standardized test score is used as part of the admission criteria. The average scores of incoming students have been declining at a constant rate in recent years. In 2000, the average score was 575, while in 2010 it was 545.
  a. Express the average score as a function of time.
  b. If the trend continues, what will the average score of incoming students be in 2015?
  c. If the trend continues, when will the average score be 527?

56. NUTRITION   Each hundred grams of Food I contains 3 g of carbohydrate and 2 g of protein, and each hundred grams of Food II contains 5 g of carbohydrate and 3 g of protein. Suppose x hundred grams of Food I are mixed with y hundred grams of Food II. The foods are combined to produce a blend that contains exactly 73 g of carbohydrate and 46 g of protein.

    a. Explain why there are $3x + 5y$ grams of carbohydrate in the blend and why we must have $3x + 5y = 73$. Find a similar equation for protein. Sketch the graphs of both equations.

    b. Where do the two graphs in part (a) intersect? Interpret the significance of this point of intersection.

57. ACCOUNTING   For tax purposes, the book value of certain assets is determined by depreciating the original value of the asset linearly over a fixed period of time. Suppose an asset originally worth V dollars is linearly depreciated over a period of N years, at the end of which it has a scrap value of S dollars.

    a. Express the book value B of the asset t years into the N-year depreciation period as a linear function of t. [*Hint:* Note that $B = V$ when $t = 0$ and $B = S$ when $t = N$.]

    b. Suppose a \$50 000 piece of office equipment is depreciated linearly over a 5-year period, with a scrap value of \$18 000. What is the book value of the equipment after 3 years?

58. ALCOHOL ABUSE CONTROL   Ethyl alcohol is metabolized by the human body at a constant rate (independent of concentration). Suppose the rate is 10 mL per hour.

    a. How much time is required to eliminate the effects of 1 L of beer containing 3% ethyl alcohol?

    b. Express the time T required to metabolize the effects of drinking ethyl alcohol as a function of the amount A of ethyl alcohol consumed.

    c. Discuss how the function in part (b) can be used to determine a reasonable "cutoff" value for the amount of ethyl alcohol A that each individual may be served at a party.

59. Graph $y = \dfrac{25}{7}x + \dfrac{13}{2}$ and $y = \dfrac{144}{45}x + \dfrac{630}{229}$ on the same set of coordinate axes. Are the two lines parallel?

60. Graph $y = \dfrac{54}{270}x - \dfrac{63}{19}$ and $y = \dfrac{139}{695}x - \dfrac{346}{14}$ on the same set of coordinate axes. Adjust the range settings until both lines are displayed. Are the two lines parallel?

61. EQUIPMENT RENTAL   A rental company rents a chain saw for a \$60 flat fee plus an hourly fee of \$5 per hour.

    a. Make a chart showing the number of hours the equipment is rented and the cost for renting the equipment for 2 hours, 5 hours, 10 hours, and t hours of time.

    b. Write an algebraic expression representing the cost y as a function of the number of hours t. Assume t can be measured to any decimal portion of an hour. (In other words, assume t is any nonnegative real number.)

    c. Graph the expression from part (b).

    d. Use the graph to approximate, to two decimal places, the number of hours the equipment was rented if the bill is \$216.25 (before taxes).

62. ASTRONOMY   The following table gives the length of the year L (in Earth years) of each planet in the solar system along with the mean (average) distance D of the planet from the sun, in astronomical units (1 astronomical unit is the mean distance of Earth from the sun).

Planet	Mean Distance from Sun, D	Length of Year, L
Mercury	0.388	0.241
Venus	0.722	0.615
Earth	1.000	1.000
Mars	1.523	1.881
Jupiter	5.203	11.862
Saturn	9.545	29.457
Uranus	19.189	84.013
Neptune	30.079	164.783

    a. Plot the point (D, L) for each planet on a coordinate plane. Do these quantities appear to be linearly related?

    b. For each planet, compute the ratio $\dfrac{L^2}{D^3}$. Interpret what you find by expressing L as a function of D.

c. What you have discovered in part (b) is one of Kepler's laws, named for the German astronomer Johannes Kepler (1571–1630). Read an article about Kepler and describe his place in the history of science.

63. ENVIRONMENT Consider the data in Example 1.3.7 for Edmonton. The table below shows the tonnes of greenhouse gas per person.

Year	Tonnes of Greenhouse Gas per Year per Person
1990	0.5520
2002	0.5315
2003	0.5602
2004	0.5361
2005	0.5260
2006	0.5234
2007	0.5276

a. Graph these data. Do they form a linear relationship? Write a sentence about the differences between this graph and the graph of Example 1.3.7.

b. If you were mayor of a city, would you make a goal of a certain number of tonnes per year or a certain number of tonnes per person per year?

64. PERPENDICULAR LINES Show that if a nonvertical line L_1 with slope m_1 is perpendicular to a line L_2 with slope m_2, then $m_2 = -1/m_1$. [*Hint:* Find expressions for the slopes of the lines L_1 and L_2 in the accompanying figure. Then apply the Pythagorean theorem along with the distance formula from Section 1.2 to the right triangle OAB to obtain the required relationship between m_1 and m_2.]

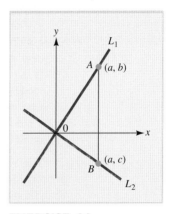

EXERCISE 64

65. JUST FOR YOU Write down your student number. Consider the number as abcdefg, adding more letters if required. Consider the point (ab, fg) on a line with slope d. If your student number has fewer or more digits, use the first and last two digits and the middle digit. For example, if your student number is 3412785, the point will be (34, 85) and the slope will be 2.

a. Find the equation of the line. Plot this line.

b. Plot three more lines, one with the same slope, one perpendicular to the line, and one with the same y intercept as the line. Print out the plot. Label each line with its equation. Label the point that uses your student number digits.

Functional Models

L04

With mathematical modelling, express given information as a function of one or more variables.

Practical problems in business, economics, and the physical and life sciences are often too complicated to be precisely described by simple formulas, and one of our basic goals is to develop mathematical methods for dealing with such problems. To this end, we shall use a procedure called **mathematical modelling,** which may be described in four stages:

Stage 1 (Formulation): Given a real-world situation (for example, the growth of Twitter, an influenza pandemic, global weather patterns), we make enough simplifying assumptions to allow a mathematical formulation. This may require gathering and analyzing data and using knowledge from a variety of different areas to identify key variables and establish equations relating those variables. This formulation is called a **mathematical model.**

Stage 2 (Analysis of the Model): We use mathematical methods to analyze or "solve" the mathematical model. Calculus will be the primary tool of analysis in this text, but in practice, a variety of tools, such as statistics, numerical analysis, and computer science, may be brought to bear on a particular model.

Stage 3 (Interpretation): After the mathematical model has been analyzed, any conclusions that may be drawn from the analysis are applied to the original real-world problem, both to gauge the accuracy of the model and to make predictions. For instance, analysis of a model of a particular business may predict that profit will be maximized by producing 200 units of a certain commodity.

Stage 4 (Testing and Adjustment): In this final stage, the model is tested by gathering new data to check the accuracy of any predictions inferred from the analysis. If the predictions are not confirmed by the new evidence, the assumptions of the model are adjusted and the modelling process is repeated. Referring to the business example described in Stage 3, it may be found that profit begins to wane at a production level significantly less than 200 units, which would indicate that the model requires modification.

The four stages of mathematical modelling are displayed in Figure 1.30. In a good model, the real-world problem is idealized just enough to allow mathematical analysis but not so much that the essence of the underlying situation has been compromised. For instance, if we assume that weather is strictly periodic, with rain occurring every 10 days, the resulting model will be relatively easy to analyze but will clearly be a poor representation of reality.

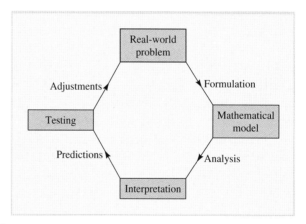

FIGURE 1.30   A diagram illustrating the mathematical modelling procedure.

In preceding sections, you have seen models representing quantities such as manufacturing cost, price and demand, air pollution levels, and population size, and you will encounter many more in subsequent chapters. Some of these models, especially those analyzed in the Think About It essays at the end of each chapter, are more detailed and illustrate how decisions are made about assumptions and predictions. Constructing and analyzing mathematical models is one of the most important skills you will learn in calculus, and the process begins with learning how to set up and solve word problems. Examples 1.4.1 through 1.4.7 illustrate a variety of techniques for addressing such problems.

Elimination of Variables

In Examples 1.4.1 and 1.4.2, the quantity we are seeking is expressed most naturally in terms of two variables. We will have to eliminate one of these variables before we can write the quantity as a function of a single variable.

EXAMPLE 1.4.1

The highway department is planning to build a picnic area for motorists along a major highway. It is to be rectangular with an area of 5000 m² and is to be fenced off on the three sides not adjacent to the highway. Express the number of metres of fencing required as a function of the length of the unfenced side.

Solution

It is natural to start by introducing two variables, say x and y, to denote the lengths of the sides of the picnic area (Figure 1.31). Expressing the number of metres, F, of fencing required in terms of these two variables, we get

$$F = x + 2y$$

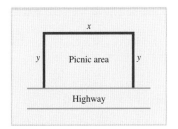

FIGURE 1.31 Rectangular picnic area.

Since the goal is to express the number of metres of fencing as a function of x alone, you must find a way to express y in terms of x. To do this, use the fact that the area is to be 5000 m² and write

$$\underbrace{xy}_{\text{area}} = 5000.$$

Solve this equation for y:

$$y = \frac{5000}{x},$$

and substitute the resulting expression for y into the formula for F to get

$$F(x) = x + 2\left(\frac{5000}{x}\right) = x + \frac{10\,000}{x}.$$

A graph of the relevant portion of this rational function is sketched in Figure 1.32. Notice that there is some length x for which the amount of required fencing is minimal. In Chapter 3, you will compute this optimal value of x using calculus.

EXPLORE!

Plot $F(x) = x + \dfrac{10\,000}{x}$.

Move along the curve with the cursor, checking the (x, y) coordinates displayed. For what value of x is the function a minimum, and what is this minimum?

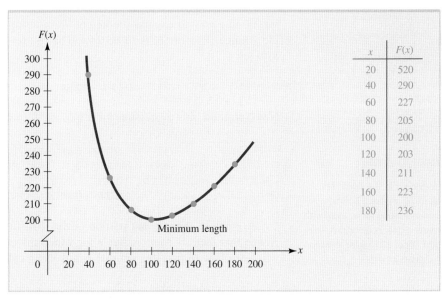

x	$F(x)$
20	520
40	290
60	227
80	205
100	200
120	203
140	211
160	223
180	236

FIGURE 1.32 The length of fencing: $F(x) = x + \dfrac{10\,000}{x}$.

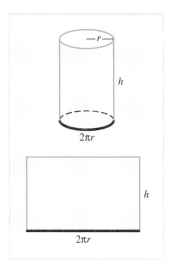

FIGURE 1.33   Cylindrical can for Example 1.4.2.

EXAMPLE 1.4.2

A cylindrical can is to have a capacity (volume) of 24π cm^3. The cost of the material used for the top and bottom of the can is 3 cents/cm^2, and the cost of the material used for the curved side is 2 cents/cm^2. Express the cost of constructing the can as a function of its radius.

Solution

Let r denote the radius of the circular top and bottom, h the height of the can, and C the cost (in cents) of constructing the can. Then

$$C = \text{cost of top} + \text{cost of bottom} + \text{cost of side},$$

where, for each component of cost,

$$\text{Cost} = (\text{cost per square centimetre})(\text{number of square centimetres}).$$

The area of the circular top (or bottom) is πr^2, and the cost per square centimetre of the top (or bottom) is 3 cents. Hence,

$$\text{Cost of top} = 3\pi r^2 \quad \text{and} \quad \text{Cost of bottom} = 3\pi r^2.$$

To find the area of the curved side, imagine the top and bottom of the can removed and the side cut and spread out to form a rectangle, as shown in Figure 1.33. The height of the rectangle is the height, h, of the can. The length of the rectangle is the circumference, $2\pi r$, of the circular top (or bottom) of the can. Hence, the area of the rectangle (or curved side) is $2\pi rh$ square centimetres. Since the cost of the side is 2 cents/cm^2, it follows that

$$\text{Cost of side} = 2(2\pi rh) = 4\pi rh.$$

Putting it all together,

$$C = 3\pi r^2 + 3\pi r^2 + 4\pi rh = 6\pi r^2 + 4\pi rh.$$

Since the goal is to express the cost as a function of the radius alone, you must find a way to express the height, h, in terms of r. To do this, use the fact that the volume, $V = \pi r^2 h$, is to be 24π. That is, set $\pi r^2 h$ equal to 24π and solve for h to get

$$\pi r^2 h = 24\pi \quad \text{or} \quad h = \frac{24}{r^2}.$$

Now substitute this expression for h into the formula for C:

$$C(r) = 6\pi r^2 + 4\pi r \left(\frac{24}{r^2} \right)$$

or

$$C(r) = 6\pi r^2 + \frac{96\pi}{r}$$

A graph of the relevant portion of this cost function is sketched in Figure 1.34. Notice that there is some radius r for which the cost is minimal. In Chapter 3, you will learn how to find this optimal radius using calculus.

Just-In-Time

Leaving π in the equation leads to an exact solution. A decimal value for π, even with many digits, is an approximation. Also, it is easier to write "π" than $3.14159. \ldots$

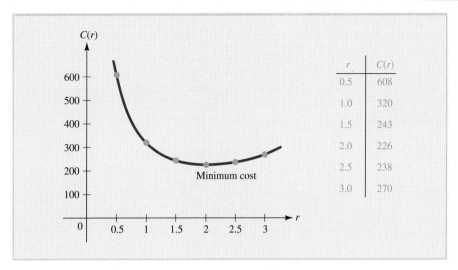

r	$C(r)$
0.5	608
1.0	320
1.5	243
2.0	226
2.5	238
3.0	270

FIGURE 1.34   The cost function: $C(r) = 6\pi r^2 + \dfrac{96\pi}{r}$.

Just-In-Time

Canada ranks 28th out of 29 in OECD countries for water consumption. Although Canada has 20% of the world's drinking water, there have been severe droughts in Alberta and Saskatchewan, in particular in 2002 and 2009.

EXAMPLE 1.4.3

During a drought, residents of a prairie town are faced with a severe water shortage. To discourage excessive use of water, one proposal is to drastically increase rates. The charges would be assessed per unit of 10 000 L. The proposed monthly rate for a family of four was $1.22 per unit of water for the first 12 units, $10 per unit for the next 12 units, and $50 per unit thereafter. Express the monthly water bill for a family of four as a function of the amount of water used.

Solution

Let x denote the number of 10 000-L units of water used by the family during the month and $C(x)$ the corresponding cost in dollars. If $0 \le x \le 12$, the cost is simply the cost per unit times the number of units used:

$$C(x) = 1.22x.$$

If $12 < x \le 24$, each of the first 12 units costs $1.22, and so the total cost of these 12 units is $1.22(12) = 14.64$ dollars. Each of the remaining $x - 12$ units costs $10, and hence the total cost of these units is $10(x - 12)$ dollars. The cost of all x units is the sum

$$C(x) = 14.64 + 10(x - 12) = 10x - 105.36.$$

If $x > 24$, the cost of the first 12 units is $1.22(12) = 14.64$ dollars, the cost of the next 12 units is $10(12) = 120$ dollars, and that of the remaining $x - 24$ units is $50(x - 24)$ dollars. The cost of all x units is the sum

$$C(x) = 14.64 + 120 + 50(x - 24) = 50x - 1065.36.$$

Combining these three formulas, we can express the total cost as the piecewise-defined function

$$C(x) = \begin{cases} 1.22x & \text{if } 0 \le x \le 12 \\ 10x - 105.36 & \text{if } 12 < x \le 24 \\ 50x - 1065.36 & \text{if } x > 24 \end{cases}$$

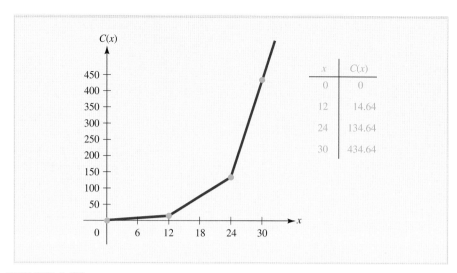

x	$C(x)$
0	0
12	14.64
24	134.64
30	434.64

FIGURE 1.35   The cost of water in a prairie town with proposed rate increases.

The graph of this function is shown in Figure 1.35. Notice that the graph consists of three line segments, each one steeper than the preceding one. What aspect of the practical situation is reflected by the increasing steepness of the lines?

Modelling in Business and Economics

Business and economic models often involve issues such as pricing, cost control, and optimization of profit. In Chapter 3, we shall examine a variety of such models. Here is an example in which profit is expressed as a function of the selling price of a particular product.

EXAMPLE 1.4.4

(Photo: © Ingram Publishing/ SuperStock)

A manufacturer can produce printer paper at a cost of $2 per packet of 500 sheets. The paper has been selling for $5 per packet, and at that price, consumers have been buying 4000 packets a month. The manufacturer is planning to raise the price of the paper and estimates that for each $1 increase in the price, 400 fewer packets will be sold each month.

a. Express the manufacturer's monthly profit as a function of the price at which the packets are sold.

b. Sketch the graph of the profit function. What price corresponds to maximum profit? What is the maximum profit?

Solution

a. Begin by stating the desired relationship in words:

Profit = (number of packets sold)(profit per packet).

Since the goal is to express profit as a function of price, the independent variable is price and the dependent variable is profit. Let p denote the price at which each packet will be sold and let $P(p)$ be the corresponding monthly profit.

Next, express the number of packets sold in terms of the variable p. You know that 4000 packets are sold each month when the price is $5 and that 400 fewer will be sold each month for each $1 increase in price. Since the number of $1 increases is the difference $p - 5$ between the new and old selling prices, you must have

$$\text{Number of packets sold} = 4000 - 400(\text{number of \$1 increases})$$
$$= 4000 - 400(p - 5)$$
$$= 6000 - 400p$$

The profit per packet is simply the difference between the selling price p and the cost $2. Thus,

$$\text{Profit per packet} = p - 2,$$

and the total profit is

$$P(p) = (\text{number of packets sold})(\text{profit per packet})$$
$$= (6000 - 400p)(p - 2)$$
$$= -400p^2 + 6800p - 12\,000$$

b. The graph of $P(p)$ is the downward-opening parabola shown in Figure 1.36. Maximum profit will occur at the value of p that corresponds to the highest point on the profit graph. This is the vertex of the parabola, which we know occurs where

$$p = \frac{-B}{2A} = \frac{-(6800)}{2(-400)} = 8.5.$$

Thus, profit is maximized when the manufacturer charges $8.50 for each packet, and the maximum monthly profit is

$$P_{\text{max}} = P(8.5) = -400(8.5)^2 + 6800(8.5) - 12\,000$$
$$= 16\,900$$

dollars. Notice that if the manufacturer tries to charge too little or too much, the production and sale of printer paper become unprofitable. In fact, if the price is less than $2 or greater than $15 per packet, the profit function $P(p)$ becomes negative and the manufacturer experiences a loss. This fact is represented by the portion of the profit curve in Figure 1.36 that lies below the p axis.

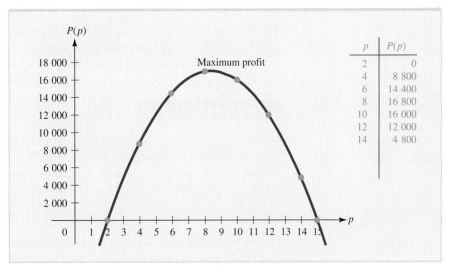

FIGURE 1.36   The profit function $P(p) = (6000 - 400p)(p - 2)$.

Market Equilibrium   Recall from Section 1.1 that the **demand function** $D(x)$ for a commodity relates the number of units x that are produced to the unit price $p = D(x)$ at which all x units are demanded (sold) in the marketplace. Similarly, the **supply function** $S(x)$ gives the corresponding price $p = S(x)$ at which producers are willing to supply x units to the marketplace. Usually, as the price of a commodity increases, more units of the commodity will be supplied and fewer will be demanded. Likewise, as the production level x increases, the supply price $p = S(x)$ also increases but the demand price $p = D(x)$ decreases. This means that a typical supply curve is rising, while a typical demand curve is falling, as indicated in Figure 1.37.

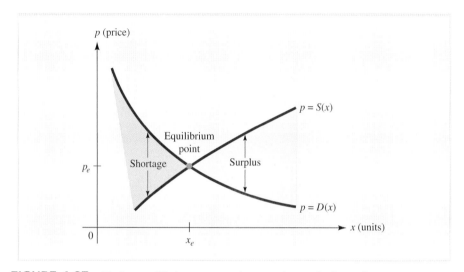

FIGURE 1.37   Market equilibrium occurs when supply equals demand.

The **law of supply and demand** says that in a competitive market environment, supply tends to equal demand, and when this occurs, the market is said to be in **equilibrium.** Thus, market equilibrium occurs precisely at the production level x_e, where $S(x_e) = D(x_e)$. The corresponding unit price p_e is called the **equilibrium price;** that is,

$$p_e = D(x_e) = S(x_e).$$

When the market is not in equilibrium, it has a **shortage** when demand exceeds supply ($D(x) > S(x)$) and a **surplus** when supply exceeds demand ($S(x) > D(x)$). This terminology is illustrated in Figure 1.38 and in Example 1.4.5.

EXAMPLE 1.4.5

Market research indicates that manufacturers will supply x units of a particular commodity to the marketplace when the price is $p = S(x)$ dollars per unit and that the same number of units will be demanded (bought) by consumers when the price is $p = D(x)$ dollars per unit, where the supply and demand functions are given by

$$S(x) = x^2 + 14 \qquad \text{and} \qquad D(x) = 174 - 6x.$$

a. At what level of production x and unit price p is market equilibrium achieved?

b. Sketch the supply and demand curves, $p = S(x)$ and $p = D(x)$, on the same graph and interpret.

Solution

a. Market equilibrium occurs when

$$S(x) = D(x)$$
$$x^2 + 14 = 174 - 6x$$
$$x^2 + 6x - 160 = 0 \qquad \text{subtract } 174 - 6x \text{ from both sides}$$
$$(x - 10)(x + 16) = 0 \qquad \text{factor}$$
$$x = 10 \quad \text{or} \quad x = -16$$

Since only positive values of the production level x are meaningful, we reject $x = -16$ and conclude that equilibrium occurs when $x_e = 10$. The corresponding equilibrium price can be obtained by substituting $x = 10$ into either the supply function or the demand function. Thus,

$$p_e = D(10) = 174 - 6(10) = 114.$$

b. The supply curve is a parabola and the demand curve is a line, as shown in Figure 1.38. Notice that no units are supplied to the market until the price reaches \$14 per unit and that 29 units are demanded when the price is 0. For $0 \le x < 10$, there is a market shortage since the supply curve is below the demand curve. The supply curve crosses the demand curve at the equilibrium point (10, 114), and for $10 < x \le 29$, there is a market surplus.

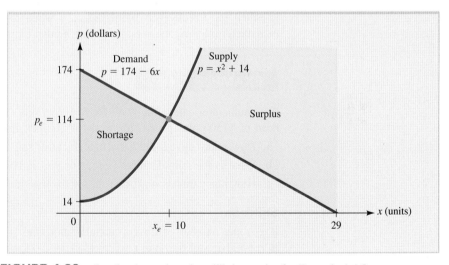

FIGURE 1.38 Supply, demand, and equilibrium point for Example 1.4.5.

Break-Even Analysis

Intersections of graphs arise in business in the context of **break-even analysis.** In a typical situation, a manufacturer wishes to determine how many units of a certain commodity have to be sold for total revenue to equal total cost. Suppose x denotes the number of units manufactured and sold, and let $C(x)$ and $R(x)$ be the corresponding total cost and total revenue, respectively. A pair of cost and revenue curves is sketched in Figure 1.39.

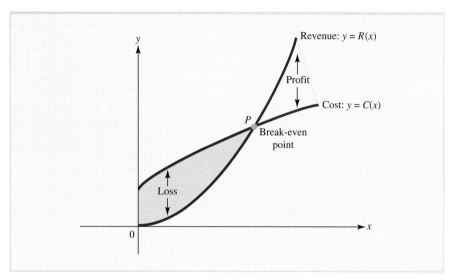

FIGURE 1.39   Cost and revenue curves, with a break-even point at P.

Because of fixed overhead costs, the total cost curve is initially higher than the total revenue curve. Hence, at low levels of production, the manufacturer suffers a loss. At higher levels of production, however, the total revenue curve is the higher one and the manufacturer realizes a profit. The point at which the two curves cross is called the **break-even point,** because when total revenue equals total cost, the manufacturer breaks even, experiencing neither a profit nor a loss. Here is an example.

EXAMPLE 1.4.6

A manufacturer can sell a duvet for $110 per unit. Total cost consists of a fixed overhead of $7500 plus production costs of $60 per unit.

  a. How many duvets must the manufacturer sell to break even?
  b. What is the manufacturer's profit or loss if 100 units are sold?
  c. How many duvets must be sold for the manufacturer to realize a profit of $1250?

Solution

If x is the number of duvets manufactured and sold, the total revenue is given by $R(x) = 110x$ and the total cost by $C(x) = 7500 + 60x$.

  a. To find the break-even point, set $R(x)$ equal to $C(x)$ and solve:

$$110x = 7500 + 60x$$
$$50x = 7500$$
$$x = 150$$

It follows that the manufacturer will have to sell 150 duvets to break even (see Figure 1.40).

  b. The profit $P(x)$ is revenue minus cost. Hence,

$$P(x) = R(x) - C(x) = 110x - (7500 + 60x) = 50x - 7500.$$

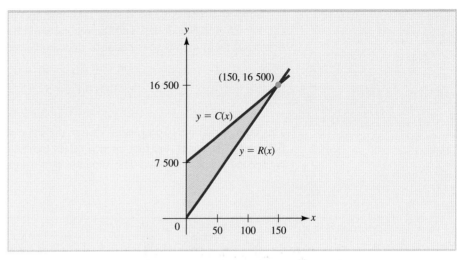

FIGURE 1.40   Revenue $R(x) = 110x$ and cost $C(x) = 7500 + 60x$.

The profit from the sale of 100 duvets is

$$P(100) = 50(100) - 7500$$
$$= -2500$$

The minus sign indicates a negative profit (that is, a loss), which was expected since 100 duvets is less than the break-even level of 150 duvets. It follows that the manufacturer will lose $2500 if 100 units are sold.

c. To determine the number of duvets that must be sold to generate a profit of $1250, set the formula for profit, $P(x)$, equal to 1250 and solve for x:

$$P(x) = 1250$$
$$50x - 7500 = 1250$$
$$50x = 8750$$
$$x = \frac{8750}{50} = 175$$

Thus, 175 duvets must be sold to generate the desired profit.

EXPLORE!

Plot lines representing these three choices of daily car rental charges:

A. $20 plus 70 cents/km
B. $30 plus 60 cents/km
C. $40 plus 50 cents/km
D. flat rate of $110

From the graphs, determine the best rate for 50 km, for 200 km, and for 300 km. What is the minimum distance travelled to make the flat rate a better choice?

Example 1.4.7 illustrates how break-even analysis can be used as a tool for decision making.

EXAMPLE 1.4.7

A certain car rental agency charges $25 plus 60 cents/km. A second agency charges $30 plus 50 cents/km. Which agency offers the better deal?

Solution

The answer depends on the number of kilometres the car is driven. For short trips, the first agency charges less than the second, but for long trips, the second agency charges less. Break-even analysis can be used to find the number of kilometres for which the two agencies charge the same.

Suppose a car is to be driven x kilometres. Then the first agency will charge $C_1(x) = 25 + 0.60x$ dollars and the second will charge $C_2(x) = 30 + 0.50x$ dollars. If you set these expressions equal to one another and solve, you get

$$25 + 0.60x = 30 + 0.50x$$
$$0.1x = 5$$
$$x = 50$$

This implies that the two agencies charge the same amount if the car is driven 50 km. For shorter distances, the first agency offers the better deal, and for longer distances, the second agency is better. The situation is illustrated in Figure 1.41.

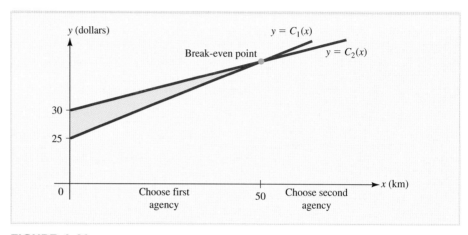

FIGURE 1.41   Car rental costs at competing agencies.

EXERCISES ■ 1.4

1. **SALES REVENUE** When x time-share condos on an island are built, they can all be sold at a price of p thousand dollars per condo, where $p = -6x + 100$.
   a. Express the revenue $R(x)$ as a function of x.
   b. How much revenue is obtained when $x = 15$ time-share condos are built and sold?

2. **MANUFACTURING PROFIT** A manufacturer estimates that it costs $14 to produce a USB key that sells for $23 per USB. There is also a fixed cost of $1200.
   a. Express the cost, $C(x)$, and revenue, $R(x)$, as functions of the number of USB keys, x, that are produced and sold.
   b. What is the profit function for this commodity?
   c. How much profit is generated when 2000 USB keys are produced?

3. **MANUFACTURING PROFIT** A manufacturer estimates that each fluorescent low-energy light bulb can be sold for $3 more than it costs to produce. A fixed cost of $17 000 is also associated with the production of the light bulbs.
   a. Express the total profit, $P(x)$, as a function of the level of production x.
   b. What is the profit (or loss) when $x = 20\,000$ light bulbs are produced? When 5000 light bulbs are produced?

4. The sum of two numbers is 18. Express the product of the numbers as a function of the smaller number.

5. The product of two numbers is 318. Express the sum of the numbers as a function of the smaller number.

6. **LANDSCAPING** A landscaper wishes to make a rectangular flower garden that is twice as long as it is wide. Express the area of the garden as a function of its width.

7. **FENCING** A farmer wishes to fence off a rectangular field with 1000 m of fencing. If the long side of the field is along a stream (and does not require fencing), express the area of the field as a function of its width.

8. **FENCING** A city recreation department plans to build a rectangular playground with an area of 3600 m^2. The playground is to be surrounded by a fence. Express the length of the fencing as a function of the length of one of the sides of the playground, draw the graph, and estimate the dimensions of the playground requiring the least amount of fencing.

9. **AREA** Express the area of a rectangular field with perimeter 320 m as a function of the length of one of its sides. Draw the graph and estimate the dimensions of the field of maximum area.

10. **PACKAGING** A closed box with a square base is to have a volume of 1500 cm^3. Express its surface area as a function of the length of its base.

11. **PACKAGING** A closed box with a square base has a surface area of 4000 cm^2. Express its volume as a function of the length of its base.

In Exercises 12 through 16, you need to know that a cylinder of radius r and height h has volume $V = \pi r^2 h$ and lateral (side) surface area $S = 2\pi rh$. A circular disk of radius r has area $A = \pi r^2$.

12. **PACKAGING** A pop can holds 370 mL, which is equivalent to 370 cm^3. Express the surface area of the can as a function of its radius.

13. **PACKAGING** A closed cylindrical can has surface area 120π cm^2. Express the volume of the can as a function of its radius.

14. **PACKAGING** A closed cylindrical can has radius r and height h.
   a. If the surface area, S, of the can is a constant, express the volume, V, of the can in terms of S and r.
   b. If the volume, V, of the can is a constant, express the surface area, S, in terms of V and r.

15. **PACKAGING** A cylindrical can is to hold 70π cm^3 of frozen orange juice. The cost per square centimetre of constructing the metal top and bottom is twice the cost per square centimetre of constructing the cardboard side. Express the cost of constructing the can as a function of its radius if the cost of the side is 0.03 cent per square centimetre.

16. **PACKAGING** A small cylindrical can with no top is made from 27π cm^2 of metal. Express the volume of the can as a function of its radius.

17. **POPULATION GROWTH** In the absence of environmental constraints, a population grows at a rate proportional to its size. Express the rate of population growth as a function of the size of the population.

PEDIATRIC DRUG DOSAGE *Several different formulas have been proposed for determining the appropriate dose of a drug for a child in terms of the adult dosage. Suppose that A milligrams is the adult dosage of a certain drug and C milligrams is the appropriate dosage for a child of age N years. Then Cowling's rule says that*

$$C = \left(\frac{N + 1}{24}\right)A,$$

while Friend's rule says that

$$C = \frac{2NA}{25}.$$

Exercises 18 through 20 require these formulas.

18. If an adult dose of ibuprofen is 300 mg, what dose does Cowling's rule suggest for an 11-year-old child? What dose does Friend's rule suggest for the same child?

19. Assume an adult dose of $A = 300$ mg, so that Cowling's rule and Friend's rule become functions of the child's age, N. Sketch the graphs of these two functions.

20. For what child's age is the dosage suggested by Cowling's rule the same as that predicted by Friend's rule? For what ages does Cowling's rule suggest a larger dosage than Friend's rule? For what ages does Friend's rule suggest the larger dosage?

21. **PEDIATRIC DRUG DOSAGE** As an alternative to Friend's rule and Cowling's rule, pediatricians sometimes use the formula

$$C = \frac{SA}{1.7}$$

to estimate an appropriate drug dosage for a child whose surface area is S square metres, when the adult dosage of the drug is A milligrams. In turn, the surface area of the child's body is estimated by the formula

$$S = 0.0072W^{0.425}H^{0.725},$$

where W and H are, respectively, the child's weight in kilograms and height in centimetres.

a. The adult dosage for a certain drug is 250 mg. How much of the drug should be given to a child who is 91 cm tall and weighs 18 kg?

b. A drug is prescribed for two children, one of whom is twice as tall and twice as heavy as the other. Show that the larger child should receive approximately 2.22 times as much of the drug as the smaller child.

22. AUCTION BUYER'S PREMIUM Usually, when you purchase a lot in an auction, you pay not only your winning bid price but also a buyer's premium. At one auction house, the buyer's premium is 17.5% of the winning bid price for purchases up to $50 000. For larger purchases, the buyer's premium is 17.5% of the first $50 000 plus 10% of the purchase price above $50 000.

a. Find the total price a buyer pays (bid price plus buyer's premium) at this auction house for purchases of $1000, $25 000, and $100 000.

b. Express the total purchase price of a lot at this auction house as a function of the final (winning) bid price. Sketch the graph of this function.

23. TRANSPORTATION COST A bus company has adopted the following pricing policy for groups that wish to charter its buses. Groups containing no more than 40 people will be charged a fixed amount of $2400 (40 times $60). In groups containing between 40 and 80 people, everyone will pay $60 minus 50 cents for each person in excess of 40. The company's lowest fare of $40 per person will be offered to groups that have 80 members or more. Express the bus company's revenue as a function of the size of the group. Draw the graph.

24. ADMISSION FEES A local natural history museum charges admission to groups according to the following policy. Groups of fewer than 50 people are charged a rate of $3.50 per person, while groups of 50 people or more are charged a reduced rate of $3 per person.

a. Express the amount a group will be charged for admission as a function of its size and draw the graph.

b. How much money will a group of 49 people save in admission costs if it can recruit 1 additional member?

25. INCOME TAX The accompanying table represents the 2010 Canada Revenue Agency federal income tax rate schedule for single taxpayers.

a. Express an individual's income tax as a function of the taxable income x for $0 \le x \le \$140\ 000$ and draw the graph.

b. The graph in part (a) should consist of four line segments. Compute the slope of each segment. What happens to these slopes as the taxable income increases? Interpret the slopes in practical terms.

If the Taxable Income Is		The Income Tax Is	
Over	But Not Over		Of the Amount Over
0	$40 726	15%	0
$40 726	$81 452	$6109 + 22%	$40 726
$81 452	$126 264	$15 069 + 26%	$81 452
$126 264		$26 720+ 29%	$126 264

26. MARKETING A company makes two shampoos, Shiny Sleek and Bodiful. The manager estimates that if x percent of the total available marketing budget is spent on marketing Shiny Sleek, then the total profit gained from both products will be P thousand dollars, where

$$P(x) = \begin{cases} 20 + 0.7x & \text{for } 0 \le x < 30 \\ 26 + 0.5x & \text{for } 30 \le x < 72 \\ 80 - 0.25x & \text{for } 72 \le x \le 100 \end{cases}$$

a. Sketch the graph of $P(x)$.

b. What is the company's profit when the marketing budget is split equally between the two products?

c. Express the total profit P in terms of the percentage y of the budget that is spent on marketing Bodiful.

27. CONSTRUCTION COST A company plans to construct a new building and parking lot on a rectangular plot of land 100 m wide and 120 m long. The building is to be 20 m high and have a rectangular footprint with perimeter 320 m, as shown in the accompanying figure.

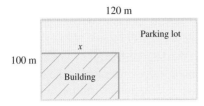

EXERCISE 27

a. Express the volume $V(x)$ of the building as a function of the length of its longer side x.
b. Graph the volume function in part (a) and determine the dimensions of the building of greatest volume that satisfies the stated requirements.
c. Suppose the company decides to construct the building of maximum volume. If it costs $75/m^3 to construct the building and $50/m^2 for the parking lot, what is the total cost of construction?

28. **VOLUME OF A TUMOUR** The shape of a cancerous tumour is roughly spherical and has volume

$$V = \frac{4}{3}\pi r^3,$$

where r is the radius in centimetres.
a. When first observed, the tumour has radius 0.73 cm, and 45 days later, the radius is 0.95 cm. By how much does the volume of the tumour increase during this period?
b. After being treated with chemotherapy, the radius of the tumour decreases by 23%. What is the corresponding percentage decrease in the volume of the tumour?

29. **POSTER DESIGN** A rectangular poster contains 25 cm^2 of print surrounded by margins of 2 cm on each side and 4 cm on the top and bottom. Express the total area of the poster (printing plus margins) as a function of the width of the printed portion.

30. **CONSTRUCTION COST** A small closed box with a square base is to have a volume of 250 cm^3. The material for the top and bottom of the box costs 2 cents/cm^2, and the material for the sides costs 1 cent/cm^2. Express the construction cost of the box as a function of the length of its base.

31. **CONSTRUCTION COST** An open box with a square base is to be built for $48. The sides of the box will cost $3/m^2, and the base will cost $4/m^2. Express the volume of the box as a function of the length of its base.

32. **CONSTRUCTION COST** An open box is to be made from a square piece of cardboard, 18 cm by 18 cm, by removing a small square from each corner and folding up the flaps to form the sides. Express the volume of the resulting box as a function of the length x of a side of the removed squares. Draw the graph and estimate the value of x for which the volume of the resulting box is greatest.

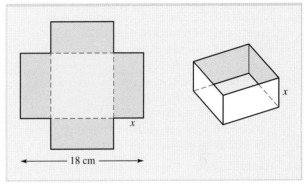

EXERCISE 32

33. **RETAIL SALES** A manufacturer has been selling lamps at a price of $50 per lamp, and at this price consumers have been buying 3000 lamps a month. The manufacturer wishes to raise the price and estimates that for each $1 increase in the price, 1000 fewer lamps will be sold each month. The manufacturer can produce the lamps at a cost of $29 per lamp. Express the manufacturer's monthly profit as a function of the price at which the lamps are sold, draw the graph, and estimate the optimal selling price.

34. **RETAIL SALES** A bookstore can obtain a certain gift book from the publisher at a cost of $3 per book. The bookstore has been offering the book at a price of $15 per copy and has been selling 200 copies a month at this price. The bookstore is planning to lower its price to stimulate sales and estimates that for each $1 reduction in the price, 20 more books will be sold each month. Express the bookstore's monthly profit from the sale of this book as a function of the selling price, draw the graph, and estimate the optimal selling price.

35. **PRODUCTION COST** A company has received an order from the city recreation department to manufacture 8000 Styrofoam kickboards for its summer swimming program. The company owns several machines, each of which can produce 30 kickboards per hour. The cost of setting up the

machines to produce these particular kickboards is $20 per machine. Once the machines have been set up, the operation is fully automated and can be overseen by a single production supervisor earning $19.20 per hour. Express the cost of producing the 8000 kickboards as a function of the number of machines used, draw the graph, and estimate the number of machines the company should use to minimize cost.

36. **AGRICULTURAL YIELD**   A citrus grower estimates that if 60 orange trees are planted, the average yield per tree will be 400 oranges. The average yield will decrease by 4 oranges per tree for each additional tree planted on the same acreage. Express the grower's total yield as a function of the number of additional trees planted, draw the graph, and estimate the total number of trees the grower should plant to maximize yield.

37. **HARVESTING**   One year a potato producer in Prince Edward Island can get $200/tonne for his potatoes on July 1. After that, the price drops by $1.25 per tonne per day. On July 1, there are 4 tonnes of potatoes in the field and it is estimated that the crop is increasing at the rate of 0.03 tonnes per day. Express the revenue from the sale of the potatoes as a function of the time at which the crop is harvested, draw the graph, and estimate when the potatoes should be harvested to maximize revenue.

MARKET EQUILIBRIUM *In Exercises 38 through 41, supply and demand functions, S(x) and D(x), are given for a particular commodity in terms of the level of production x. In each case:*
   (a) Find the value of x_e for which equilibrium occurs and the corresponding equilibrium price p_e.
   (b) Sketch the graphs of the supply and demand curves, p = S(x) and p = D(x), on the same graph.
   (c) For what values of x is there a market shortage? A market surplus?

38. $S(x) = 4x + 200$ and $D(x) = -3x + 480$

39. $S(x) = 3x + 150$ and $D(x) = -2x + 275$

40. $S(x) = x^2 + x + 3$ and $D(x) = 21 - 3x^2$

41. $S(x) = 2x + 7.43$ and $D(x) = -0.21x^2 - 0.84x + 50$

42. **SUPPLY AND DEMAND**   When blenders are sold for p dollars each, manufacturers will supply $\dfrac{p^2}{10}$ blenders to local retailers, while the local demand will be $60 - p$ blenders. At what market price will the manufacturers' supply of blenders be equal to the consumers' demand for the blenders? How many blenders will be sold at this price?

43. **SUPPLY AND DEMAND**   Producers will supply x computer surge protectors to the market when the price is $p = S(x)$ dollars per unit, and consumers will demand (buy) x units when the price is $p = D(x)$ dollars per unit, where

$$S(x) = 2x + 15 \quad \text{and} \quad D(x) = \frac{385}{x + 1}.$$

   a. Find the equilibrium production level x_e and the equilibrium price p_e of these surge protectors.
   b. Draw the supply and demand curves on the same graph.
   c. Where does the supply curve cross the y axis? Describe the economic significance of this point.

44. **SPY STORY**   The hero of a popular spy story has escaped from the headquarters of an international diamond-smuggling ring in the tiny Mediterranean country of Azusa. Our hero, driving a stolen milk truck at 72 km/h, has a 40-min head start on his pursuers, who are chasing him in a Ferrari going 168 km/h. The distance from the smugglers' headquarters to the border, and freedom, is 83.8 km. Will he make it?

45. **AIR TRAVEL**   Two flights bound for Toronto leave Calgary 30 min apart. The first travels 880 km/h, while the second goes 1040 km/h. At what time will the second plane pass the first?

46. **BREAK-EVEN ANALYSIS**   A furniture manufacturer can sell dining room tables for $500 each. The manufacturer's total cost consists of a fixed overhead of $30 000 plus production costs of $350 per table.

   a. How many tables must the manufacturer sell to break even?
   b. How many tables must the manufacturer sell to make a profit of $6000?
   c. What will the manufacturer's profit or loss be if 150 tables are sold?
   d. On the same set of axes, graph the manufacturer's total revenue and total cost functions. Explain how the overhead can be read from the graph.

47. **PUBLISHING DECISION**   An author must decide between two publishers who are vying for his new book. Publisher A offers royalties of 1% of net proceeds on the first 30 000 copies and 3.5% on all copies in excess of that figure, and expects to net $2 on each copy sold. Publisher B will pay no royalties on the first 4000 copies sold but will pay 2% on the net proceeds of all copies sold in excess of 4000 copies, and expects to net $3 on each copy sold. Suppose the author expects to sell N copies. State a simple criterion based on N for deciding how to choose between the publishers.

48. **CHEQUING ACCOUNT**   The charge for maintaining a chequing account at a certain bank is $12 per month plus 10 cents for each cheque that is written. A competing bank charges $10 per month plus 14 cents per cheque. Find a criterion for deciding which bank offers the better deal.

49. **PHYSIOLOGY**   The pupil of the human eye is roughly circular. If the intensity of light, I, entering the eye is proportional to the area of the pupil, express I as a function of the radius, r, of the pupil.

50. **RECYCLING**   To raise money for a scout group summer camp, the scouts collect aluminum pop cans. They have collected 120 kg in the last 80 days. The price they will be paid is 70 cents/ kg. Aluminum prices fluctuate and the price is falling at 0.3 cents/day. They can only make one trip to the recycling company as they only have access to a truck for one day. If they continue to collect at the same rate, write the scout group revenue as a function of the additional days the project runs. Draw the graph and estimate when the scout group should conclude the project and deliver the pop cans to maximize its revenue.

51. **LIFE EXPECTANCY**   In Canada in 1900, the life expectancy of a newborn male child was 47 years, and a female 50. By 2000, this had grown to 75.3 years for a male and 81.4 for a female. Assume these life expectancies increased linearly with time between 1900 and 2000.
    a. Find the function, $M(t)$, that represents the life expectancy of a newborn male child t years after 1900 and the function, $F(t)$, that represents the life expectancy of a newborn female child.
    b. In 2009 the life expectancies were 78.69 for males and 83.91 for females. Do these statistics agree with the linear relationship from the

previous century? Research life expectancies in Canada for the current year and investigate deviations from the trend.

52. **A* BIOCHEMISTRY**   In biochemistry, the rate R of an enzymatic reaction is found to be given by the equation

$$R = \frac{R_m[S]}{K_m + [S]},$$

where K_m is a constant (called the **Michaelis constant**), R_m is the maximum possible rate, and $[S]$ is the substrate concentration.* Rewrite this equation so that $y = \dfrac{1}{R}$ is expressed as a function of $x = \dfrac{1}{[S]}$, and sketch the graph of this function. (This graph is called the **Lineweaver-Burk double reciprocal plot**.)

53. **A* SUPPLY AND DEMAND**   Producers will supply q units of a certain commodity to the market when the price is $p = S(q)$ dollars per unit, and consumers will demand (buy) q units when the price is $p = D(q)$ dollars per unit, where

$$S(q) = aq + b \quad \text{and} \quad D(q) = cq + d$$

for constants a, b, c, and d.
    a. What can you say about the signs of the constants a, b, c, and d if the supply and demand curves are as shown in the accompanying figure?
    b. Express the equilibrium production level q_e and the equilibrium price p_e in terms of the coefficients a, b, c, and d.
    c. Use your answer in part (b) to determine what happens to the equilibrium production level q_e as a increases. What happens to q_e as d increases?

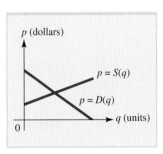

EXERCISE 53

*Mary K. Campbell, *Biochemistry,* Philadelphia, PA: Saunders College Publishing, 1991, pp. 221–226.

54. PUBLISHING PROFIT   It costs a publisher $74 200 to prepare a book for publication (typesetting, illustrating, editing, and so on); printing and binding costs are $5.50 per book. The book is sold to bookstores for $19.50 per copy.

   a. Make a table showing the cost of producing 2000, 4000, 6000, and 8000 books. Use four significant digits.

   b. Make a table showing the revenue from selling 2000, 4000, 6000, and 8000 books. Use four significant digits.

   c. Write an algebraic expression representing the cost, C, as a function of the number of books, x, that are produced.

   d. Write an algebraic expression representing the revenue, R, as a function of the number of books, x, sold.

   e. Graph both functions on the same coordinate axes.

   f. Find the point on your graph where cost equals revenue.

   g. Use the graph to determine how many books need to be made to produce revenue of at least $85 000. How much profit is made for this number of books?

SECTION 1.5   Limits

L05

Evaluate limits of polynomials and rational functions.

As you will see in subsequent chapters, calculus is an enormously powerful branch of mathematics with a wide range of applications, including curve sketching, optimization of functions, analysis of rates of change, and computation of area and probability. What gives calculus its power and distinguishes it from algebra is the concept of limit, and the purpose of this section is to introduce this important concept. Our approach will be intuitive rather than formal. The ideas outlined here form the basis for a more rigorous development of the laws and procedures of calculus and lie at the heart of much of modern mathematics.

Intuitive Introduction to the Limit

Roughly speaking, the limit process involves examining the behaviour of a function, $f(x)$, as x approaches a number, c, that may or may not be in the domain of f. Limiting behaviour occurs in a variety of practical situations. For instance, absolute zero, the temperature, T_c, at which all molecular activity ceases, can be approached but never actually attained in practice. Similarly, economists who speak of profit under ideal conditions or engineers profiling the ideal specifications of a new engine are really dealing with limiting behaviour.

To illustrate the limit process, consider a manager who determines that when x percent of her company's plant capacity is being used, the total cost of operation is C hundred thousand dollars, where

$$C(x) = \frac{8x^2 - 636x - 320}{x^2 - 68x - 960}.$$

The company has a policy of rotating maintenance in an attempt to ensure that approximately 80% of capacity is always in use. What cost should the manager expect when the plant is operating at this ideal capacity?

It may seem that we can answer this question by simply evaluating $C(80)$, but attempting this evaluation results in the meaningless fraction $\dfrac{0}{0}$. However, it is still possible to evaluate $C(x)$ for values of x that approach 80 from the right ($x > 80$, when capacity is temporarily overutilized) and from the left ($x < 80$, when capacity is underutilized). A few such calculations are summarized in the following table.

	x approaches 80 from the left →			←x approaches 80 from the right		
x	79.8	79.99	79.999	80	80.0001	80.001	80.04
C(x)	6.99782	6.99989	6.99999	✕	7.000001	7.00001	7.00043

The values of $C(x)$ displayed on the lower line of this table suggest that $C(x)$ approaches the number 7 as x gets closer and closer to 80. Thus, it is reasonable for the manager to expect a cost of $700 000 when 80% of plant capacity is utilized.

The functional behaviour in this example can be described by saying "$C(x)$ has the limiting value 7 as x approaches 80" or, equivalently, by writing

$$\lim_{x \to 80} C(x) = 7.$$

More generally, the limit of $f(x)$ as x approaches the number c can be defined informally as follows.

> **The Limit of a Function** ■ If $f(x)$ gets closer and closer to a number L as x gets closer and closer to c from both sides, then L is the limit of $f(x)$ as x approaches c. The behaviour is expressed by writing
>
> $$\lim_{x \to c} f(x) = L.$$

Geometrically, the limit statement $\lim_{x \to c} f(x) = L$ means that the height of the graph $y = f(x)$ approaches L as x approaches c, as shown in Figure 1.42. This interpretation is illustrated along with the tabular approach to computing limits in Example 1.5.1.

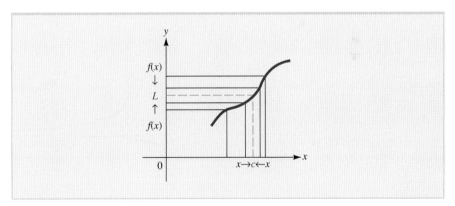

FIGURE 1.42 If $\lim_{x \to c} f(x) = L$, the height of the graph of f approaches L as x approaches c.

EXPLORE!

Plot $f(x) = \dfrac{\sqrt{x} - 1}{x - 1}$. Does there seem to be a problem at $x = 1$? Trace the x and y coordinates near $x = 1$. Substitute $x = 1$ into the function. Now does there seem to be a problem?

EXAMPLE 1.5.1

Use a table to estimate the limit

$$\lim_{x \to 1} \frac{\sqrt{x} - 1}{x - 1}.$$

Solution

Let

$$f(x) = \frac{\sqrt{x} - 1}{x - 1},$$

and compute $f(x)$ for a succession of values of x approaching 1 from the left and from the right:

$$x \rightarrow 1 \leftarrow x$$

x	0.99	0.999	0.9999	1	1.0001	1.001	1.01
$f(x)$	0.50126	0.50013	0.50001	✕	0.49999	0.49988	0.49876

The numbers on the bottom line of the table suggest that $f(x)$ approaches 0.5 as x approaches 1; that is,

$$\lim_{x \to 1} \frac{\sqrt{x} - 1}{x - 1} = 0.5.$$

The graph of $f(x)$ is shown in Figure 1.43. The limit computation says that the height of the graph of $y = f(x)$ approaches $L = 0.5$ as x approaches 1. This corresponds to the hole in the graph of $f(x)$ at $(1, 0.5)$. We will compute this same limit using an algebraic procedure in Example 1.5.6.

When holes or jumps occur in a function, the graph of the function is not continuous. Continuity will be discussed in detail in Section 1.6.

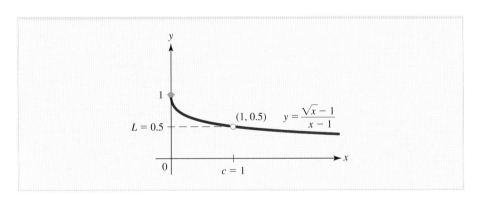

FIGURE 1.43   The function $f(x) = \dfrac{\sqrt{x} - 1}{x - 1}$ tends toward $L = 0.5$ as x approaches $c = 1$.

It is important to remember that limits describe the behaviour of a function *near* a particular point, not necessarily *at* the point itself. This is illustrated in Figure 1.44. For all three functions graphed, the limit of $f(x)$ as x approaches 3 is equal to 4. Yet the functions behave quite differently at $x = 3$ itself. In Figure 1.44a, $f(3)$ is equal to the limit 4; in Figure 1.44b, $f(3)$ is different from 4; and in Figure 1.44c, $f(3)$ is not defined at all.

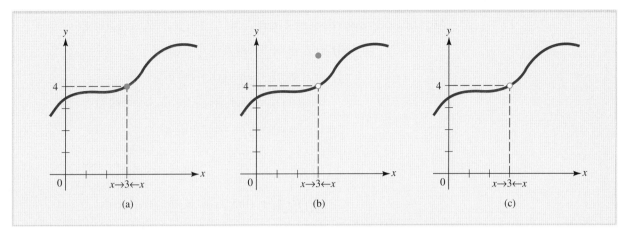

FIGURE 1.44   Three functions for which $\lim_{x \to 3} f(x) = 4$.

Figure 1.45 shows the graph of two functions that do not have a limit as x approaches 2. The limit does not exist in Figure 1.45a because $f(x)$ tends toward 5 as x approaches 2 from the right and tends toward a different value, 3, as x approaches 2 from the left. The function in Figure 1.45b has no finite limit as x approaches 2 because the values of $f(x)$ increase without bound as x tends toward 2 and hence tend to no finite number L. Such *infinite limits* will be discussed later in this section.

EXPLORE!

Graph $f(x) = \dfrac{2}{(x-2)^2}$ and

$f(x) = \begin{cases} 3, & x \leq 2 \\ 5, & x > 2 \end{cases}$.

Can you match the two functions to the two graphs on this page?

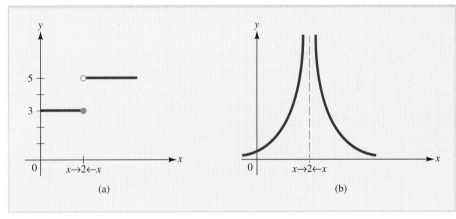

FIGURE 1.45   Two functions for which $\lim_{x \to 2} f(x)$ does not exist.

Properties of Limits   Limits obey certain algebraic rules that can be used to simplify computations. These rules, which should seem plausible on the basis of our informal definition of limit, are proven formally in more theoretical courses.

Algebraic Properties of Limits ■ If $\lim\limits_{x \to c} f(x)$ and $\lim\limits_{x \to c} g(x)$ exist, then

$$\lim_{x \to c} [f(x) + g(x)] = \lim_{x \to c} f(x) + \lim_{x \to c} g(x)$$

$$\lim_{x \to c} [f(x) - g(x)] = \lim_{x \to c} f(x) - \lim_{x \to c} g(x)$$

$$\lim_{x \to c} [kf(x)] = k \lim_{x \to c} f(x) \quad \text{for any constant } k$$

$$\lim_{x \to c} [f(x)g(x)] = [\lim_{x \to c} f(x)][\lim_{x \to c} g(x)]$$

$$\lim_{x \to c} \frac{f(x)}{g(x)} = \frac{\lim\limits_{x \to c} f(x)}{\lim\limits_{x \to c} g(x)} \quad \text{if } \lim_{x \to c} g(x) \neq 0$$

$$\lim_{x \to c} [f(x)]^P = [\lim_{x \to c} f(x)]^P \quad \text{if } [\lim_{x \to c} f(x)]^P \text{ exists}$$

That is, the limit of a sum, a difference, a multiple, a product, a quotient, or a power exists and is the sum, difference, multiple, product, quotient, or power of the individual limits, as long as all expressions involved are defined.

Here are two elementary limits that we will use along with the limit rules to compute limits involving more complex expressions.

Limits of Two Linear Functions ■ For any constant k,

$$\lim_{x \to c} k = k \quad \text{and} \quad \lim_{x \to c} x = c.$$

That is, the limit of a constant is the constant itself, and the limit of $f(x) = x$ as x approaches c is c.

In geometric terms, the limit statement $\lim\limits_{x \to c} k = k$ says that the height of the graph of the constant function $f(x) = k$ approaches k as x approaches c. Similarly, $\lim\limits_{x \to c} x = c$ says that the height of the linear function $f(x) = x$ approaches c as x approaches c. These statements are illustrated in Figure 1.46.

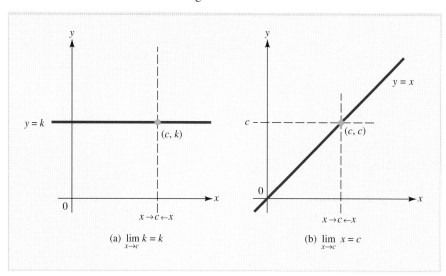

FIGURE 1.46 Limits of two linear functions.

Computation of Limits

Examples 1.5.2 through 1.5.6 illustrate how the properties of limits can be used to calculate limits of algebraic functions. In Example 1.5.2, you will see how to find the limit of a polynomial.

EXAMPLE 1.5.2

Find $\lim\limits_{x \to -1} (3x^3 - 4x + 8)$.

Solution

Apply the properties of limits to obtain

$$\lim_{x \to -1} (3x^3 - 4x + 8) = 3\left(\lim_{x \to -1} x\right)^3 - 4\left(\lim_{x \to -1} x\right) + \lim_{x \to -1} 8$$
$$= 3(-1)^3 - 4(-1) + 8$$
$$= 9$$

In Example 1.5.3, you will see how to find the limit of a rational function whose denominator does not approach zero.

EXPLORE!

Graph $f(x) = \dfrac{x^2 + x - 2}{x - 1}$.
Move the cursor along the line, watching the coordinates of the points as you move. Notice that $f(1)$ does not exist, but move in very close from both sides of $x = 1$. Can you suggest a value for $f(1)$ that would fill the hole at $x = 1$?

EXAMPLE 1.5.3

Find $\lim\limits_{x \to -1} \dfrac{3x^3 - 8}{x - 2}$.

Solution

Since $\lim\limits_{x \to -1} (x - 2) \neq 0$, you can use the quotient rule for limits to get

$$\lim_{x \to 1} \frac{3x^3 - 8}{x - 2} = \frac{\lim\limits_{x \to 1}(3x^3 - 8)}{\lim\limits_{x \to 1}(x - 2)} = \frac{3\lim\limits_{x \to 1} x^3 - \lim\limits_{x \to 1} 8}{\lim\limits_{x \to 1} x - \lim\limits_{x \to 1} 2} = \frac{3 - 8}{1 - 2} = 5.$$

In general, you can use the properties of limits to obtain the following formulas, which can be used to evaluate many limits that occur in practical problems.

> **Limits of Polynomials and Rational Functions** ■ If $p(x)$ and $q(x)$ are polynomials, then
>
> $$\lim_{x \to c} p(x) = p(c)$$
>
> and
>
> $$\lim_{x \to c} \frac{p(x)}{q(x)} = \frac{p(c)}{q(c)} \quad \text{if } q(c) \neq 0.$$

In Example 1.5.4, the denominator of the given rational function approaches zero, while the numerator does not. When this happens, you can conclude that the limit does not exist. The absolute value of such a quotient increases without bound and hence does not approach any finite number.

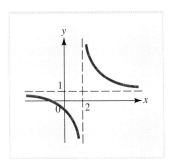

FIGURE 1.47   The graph of $f(x) = \dfrac{x + 1}{x - 2}$.

Just-In-Time

In the limit, we can cancel terms that are very close to zero but are not exactly zero. This occurs for $x - 1$ in Example 1.5.5. Then the function looks continuous at $x = 1$ on a graph, but in reality there is a hole at that point.

When we have a factor that does not cancel, such as $x - 2$ in this example, the break is very noticeable on the graph. Although we can only "see" one break in the graph, there are still two places where the curve is not continuous.

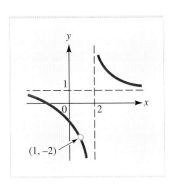

FIGURE 1.48   The graph of $f(x) = \dfrac{x^2 - 1}{x^2 - 3x + 2}$.

EXAMPLE 1.5.4

Find $\displaystyle\lim_{x \to 2} \frac{x + 1}{x - 2}$.

Solution

The quotient rule for limits does not apply in this case since the limit of the denominator is

$$\lim_{x \to 2}(x - 2) = 0.$$

Since the limit of the numerator is $\displaystyle\lim_{x \to 2}(x + 1) = 3$, which is not equal to zero, you can conclude that the limit of the quotient does not exist.

The graph of the function $f(x) = \dfrac{x + 1}{x - 2}$ in Figure 1.47 gives you a better idea of what is actually happening in this example. Note that $f(x)$ increases without bound as x approaches 2 from the right and decreases without bound as x approaches 2 from the left.

In Example 1.5.5, the numerator and the denominator of the given rational function *both* approach zero. When this happens, you should try to simplify the function algebraically to find the desired limit.

EXAMPLE 1.5.5

Find $\displaystyle\lim_{x \to 1} \frac{x^2 - 1}{x^2 - 3x + 2}$.

Solution

As x approaches 1, both the numerator and the denominator approach zero, and you can draw no conclusion about the size of the quotient. To proceed, observe that the given function is not defined when $x = 1$ but that for all other values of x, you can cancel the common factor $x - 1$ to obtain

$$\frac{x^2 - 1}{x^2 - 3x + 2} = \frac{(x - 1)(x + 1)}{(x - 1)(x - 2)}$$

$$= \frac{x + 1}{x - 2}, \quad x \neq 1$$

(Since $x \neq 1$, you are not dividing by zero.) Now take the limit as x approaches (but is not equal to) 1 to get

$$\lim_{x \to 1} \frac{x^2 - 1}{x^2 - 3x + 2} = \lim_{x \to 1} \frac{x + 1}{x - 2} = \frac{\lim_{x \to 1}(x + 1)}{\lim_{x \to 1}(x - 2)} = \frac{2}{-1} = -2.$$

The graph of the function $f(x) = \dfrac{x^2 - 1}{x^2 - 3x + 2}$ is shown in Figure 1.48. Note that it is like the graph in Figure 1.47 with a hole at the point $(1, -2)$.

In general, when both the numerator and the denominator of a quotient approach zero as x approaches c, your strategy will be to simplify the quotient algebraically (as in Example 1.5.5 by cancelling $x - 1$). In most cases, the simplified form of the quotient will be valid for all values of x except $x = c$. Since you are interested in the behaviour of the quotient *near* $x = c$ and not *at* $x = c$, you may use the simplified form of the quotient to calculate the limit. In Example 1.5.6, we use this technique to obtain the limit we estimated using a table in Example 1.5.1.

EXAMPLE 1.5.6

Find $\lim\limits_{x \to 1} \dfrac{\sqrt{x} - 1}{x - 1}$.

Just-In-Time

When factoring a difference of squares, recall that

$$(a - b)(a + b) = a^2 - b^2.$$

For example, $x^2 - 9 = (x - 3)(x + 3)$. But in Example 1.5.6, one of the squares is $(\sqrt{x})^2$. Try multiplying out $(\sqrt{x} - 1)(\sqrt{x} + 1)$ and you will see the same pattern as in the difference of squares formula.

Solution

Both the numerator and the denominator approach 0 as x approaches 1. To simplify the quotient, we rationalize the numerator (that is, multiply numerator and denominator by $\sqrt{x} + 1$) to get

$$\frac{\sqrt{x} - 1}{x - 1} = \frac{(\sqrt{x} - 1)(\sqrt{x} + 1)}{(x - 1)(\sqrt{x} + 1)}$$

$$= \frac{x - 1}{(x - 1)(\sqrt{x} + 1)}$$

$$= \frac{1}{\sqrt{x} + 1}, \quad x \neq 1$$

and then take the limit to obtain

$$\lim_{x \to 1} \frac{\sqrt{x} - 1}{x - 1} = \lim_{x \to 1} \frac{1}{\sqrt{x} + 1} = \frac{1}{2}.$$

Limits Involving Infinity

Long-term behaviour is often a matter of interest in business and economics or the physical and life sciences. For example, a biologist may wish to know the population of a bacterial colony or a population of fruit flies after an indefinite period of time, or a business manager may wish to know how the average cost of producing a particular commodity is affected as the level of production increases indefinitely.

In mathematics, the infinity symbol ∞ is used to represent either unbounded growth or the result of such growth. The following are the definitions of limits involving infinity that we will use to study long-term behaviour.

Just-In-Time

Have you ever heard someone say, "What will happen in the long run?" A limit as time approaches infinity is the "long run."

Limits at Infinity ■ If the values of the function $f(x)$ approach the number L as x increases without bound, we write

$$\lim_{x \to +\infty} f(x) = L.$$

Similarly, we write

$$\lim_{x \to -\infty} f(x) = M$$

when the functional values $f(x)$ approach the number M as x decreases without bound.

Geometrically, the limit statement $\lim\limits_{x \to +\infty} f(x) = L$ means that as x increases without bound, the graph of $f(x)$ approaches the horizontal line $y = L$, while $\lim\limits_{x \to -\infty} f(x) = M$ means that the graph of $f(x)$ approaches the line $y = M$ as x decreases without bound. The lines $y = L$ and $y = M$ that appear in this context are called **horizontal asymptotes** of the graph of $f(x)$. There are many different ways for a graph to have horizontal asymptotes, one of which is shown in Figure 1.49. We will have more to say about asymptotes in Chapter 3 as part of a general discussion of graphing with calculus.

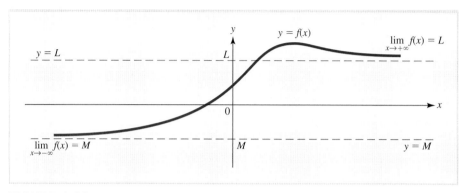

FIGURE 1.49   A graph illustrating limits at infinity and horizontal asymptotes.

The algebraic properties of limits listed earlier in this section also apply to limits at infinity. In addition, since any reciprocal power $1/x^k$ for $k > 0$ becomes smaller and smaller in absolute value as x either increases or decreases without bound, we have these useful rules:

Reciprocal Power Rules ■ If A and k are constants with $k > 0$ and x^k is defined for all x, then

$$\lim_{x \to +\infty} \frac{A}{x^k} = 0 \quad \text{and} \quad \lim_{x \to +\infty} \frac{A}{x^k} = 0.$$

The use of these rules is illustrated in Example 1.5.7.

EXAMPLE 1.5.7

Find $\lim\limits_{x \to +\infty} \dfrac{x^2}{1 + x + 2x^2}$.

Solution

To get a feeling for what happens with this limit, we evaluate the function

$$f(x) = \frac{x^2}{1 + x + 2x^2}$$

at $x = 100, 1000, 10\,000$, and $100\,000$ and display the results in a table:

				$x \to +\infty$
x	100	1000	10 000	100 000
$f(x)$	0.49749	0.49975	0.49997	0.4999975

EXPLORE!

Plot the three functions on the same graph:

$$f(x) = \frac{11x^4 + 2x}{5x^4 + 3},$$

$$g(x) = \frac{11x^3 + 2x}{5x^4 + 3},$$

$$h(x) = \frac{11x^5 + 2x}{5x^4 + 3}$$

Investigate their behaviour as x becomes large. Notice that one goes to zero, one goes to a horizontal asymptote, and one becomes very large. Can you match the function with the behaviour by finding the limits algebraically?

The functional values on the bottom line in the table suggest that $f(x)$ tends toward 0.5 as x grows larger and larger. To confirm this observation analytically, consider the dominant term in the denominator and how this term influences asymptotic behaviour. We divide each term in $f(x)$ by the dominant term that appears in the denominator $1 + x + 2x^2$, namely, by x^2. This enables us to find $\lim\limits_{x \to +\infty} f(x)$ by applying reciprocal power rules as follows:

$$\lim_{x \to +\infty} \frac{x^2}{1 + x + 2x^2} = \lim_{x \to +\infty} \frac{x^2/x^2}{1/x^2 + x/x^2 + 2x^2/x^2}$$

$$= \lim_{x \to +\infty} \frac{1}{1/x^2 + 1/x + 2}$$

$$= \frac{\lim\limits_{x \to +\infty} 1}{\lim\limits_{x \to +\infty} 1/x^2 + \lim\limits_{x \to +\infty} 1/x + \lim\limits_{x \to +\infty} 2} \quad \text{several algebraic properties of limits}$$

$$= \frac{1}{0 + 0 + 2} = 0.5 \quad \text{reciprocal power rule}$$

The graph of $f(x)$ is shown in Figure 1.50. For practice, verify that $\lim\limits_{x \to -\infty} f(x) = 0.5$ also.

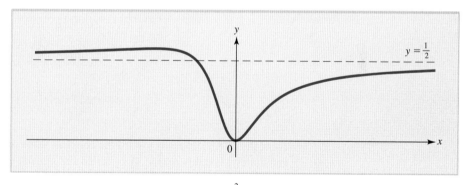

FIGURE 1.50   The graph of $f(x) = \dfrac{x^2}{1 + x + 2x^2}$.

Here is a general description of the procedure for evaluating the limit of a rational function at infinity.

Procedure for Evaluating the Limit at Infinity of $f(x) = p(x)/q(x)$

Step 1. Divide each term in $f(x)$ by the highest power x^k that appears in the denominator polynomial $q(x)$.

Step 2. Compute $\lim\limits_{x \to +\infty} f(x)$ or $\lim\limits_{x \to -\infty} f(x)$ using algebraic properties of limits and the reciprocal power rules.

EXAMPLE 1.5.8

Find $\lim\limits_{x \to +\infty} \dfrac{2x^2 + 3x + 1}{3x^2 - 5x + 2}$.

Solution

The highest power in the denominator is x^2. Divide the numerator and denominator by x^2 to get

$$\lim_{x \to +\infty} \frac{2x^2 + 3x + 1}{3x^2 - 5x + 2} = \lim_{x \to +\infty} \frac{2 + 3/x + 1/x^2}{3 - 5/x + 2/x^2} = \frac{2 + 0 + 0}{3 - 0 + 0} = \frac{2}{3}.$$

Just-In-Time

Although this example has no *x* and no numbers, the variable *N* performs in the same way as *x*. *A* and *B* act in the same way as more familiar numbers.

EXAMPLE 1.5.9

If a crop is planted in soil where the nitrogen level is N, then the crop yield Y can be modelled by the *Michaelis-Menten* function

$$Y(N) = \frac{AN}{B + N}, \quad N \geq 0,$$

where A and B are positive constants. What happens to crop yield as the nitrogen level is increased indefinitely?

Solution

We wish to compute

$$\lim_{N \to +\infty} Y(N) = \lim_{N \to +\infty} \frac{AN}{B + N}$$

$$= \lim_{N \to +\infty} \frac{AN/N}{B/N + N/N}$$

$$= \lim_{N \to +\infty} \frac{A}{B/N + 1} = \frac{A}{0 + 1}$$

$$= A$$

Thus, the crop yield tends toward the constant value A as the nitrogen level N increases indefinitely. For this reason, A is called the *maximum attainable yield*.

If the functional values $f(x)$ increase or decrease without bound as x approaches c, then technically $\lim\limits_{x \to c} f(x)$ does not exist. However, the behaviour of the function in such a case is more precisely described by using the following notation, which is illustrated in Example 1.5.10.

Infinite Limits ▪ We say that $\lim\limits_{x \to c} f(x)$ is an **infinite limit** if $f(x)$ increases or decreases without bound as $x \to c$. We write

$$\lim_{x \to c} f(x) = +\infty$$

if $f(x)$ increases without bound as $x \to c$ or

$$\lim_{x \to c} f(x) = -\infty$$

if $f(x)$ decreases without bound as $x \to c$.

EXAMPLE 1.5.10

Find $\displaystyle\lim_{x \to +\infty} \frac{-x^3 + 3x + 1}{x - 3}$.

Solution

The dominant term or highest power in the denominator is x. Divide the numerator and denominator by x to get

$$\lim_{x \to +\infty} \frac{-x^3 + 3x + 1}{x - 3} = \lim_{x \to +\infty} \frac{-x^2 + 3 + 1/x}{1 - 3/x}.$$

Since

$$\lim_{x \to +\infty}\left(-x^2 + 3 + \frac{1}{x}\right) = -\infty \quad \text{and} \quad \lim_{x \to +\infty}\left(1 - \frac{3}{x}\right) = 1,$$

it follows that

$$\lim_{x \to +\infty} \frac{-x^3 + 2x + 1}{x - 3} = -\infty.$$

EXERCISES ■ 1.5

In Exercises 1 through 6, find $\displaystyle\lim_{x \to a} f(x)$ *if it exists.*

1.   **2.**   **3.**

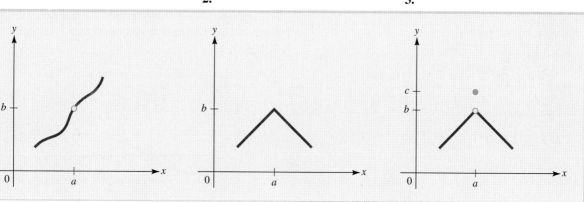

4.   **5.**   **6.**

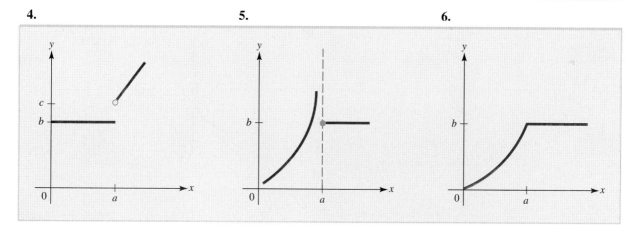

In Exercises 7 through 26, find the indicated limit if it exists.

7. $\lim_{x \to 2}(3x^2 - 5x + 2)$

8. $\lim_{x \to -1}(x^3 - 2x^2 + x - 3)$

9. $\lim_{x \to 0}(x^5 - 6x^4 + 7)$

10. $\lim_{x \to -1/2}(1 - 5x^3)$

11. $\lim_{x \to 3}(x - 1)^2(x + 1)$

12. $\lim_{x \to -1}(x^2 + 1)(1 - 2x)^2$

13. $\lim_{x \to 1/3}\dfrac{x + 1}{x + 2}$

14. $\lim_{x \to 1}\dfrac{2x + 3}{x + 1}$

15. $\lim_{x \to 5}\dfrac{x + 3}{5 - x}$

16. $\lim_{x \to 3}\dfrac{2x + 3}{x - 3}$

17. $\lim_{x \to 1}\dfrac{x^2 - 1}{x - 1}$

18. $\lim_{x \to 3}\dfrac{9 - x^2}{x - 3}$

19. $\lim_{x \to 5}\dfrac{x^2 - 3x - 10}{x - 5}$

20. $\lim_{x \to 2}\dfrac{x^2 + x - 6}{x - 2}$

21. $\lim_{x \to 4}\dfrac{(x + 1)(x - 4)}{(x - 1)(x - 4)}$

22. $\lim_{x \to 0}\dfrac{x(x^2 - 1)}{x^2}$

23. $\lim_{x \to -2}\dfrac{x^2 - x - 6}{2x^2 + 3x + 2}$

24. $\lim_{x \to 1}\dfrac{x^2 + 4x - 5}{x^2 - 1}$

25. $\lim_{x \to 4}\dfrac{\sqrt{x} - 2}{x - 4}$

26. $\lim_{x \to 9}\dfrac{\sqrt{x} - 3}{x - 9}$

For Exercises 27 through 36, find $\lim_{x \to +\infty} f(x)$ and $\lim_{x \to -\infty} f(x)$. If the limiting value is infinite, indicate whether it is $+\infty$ or $-\infty$.

27. $f(x) = x^3 - 4x^2 - 4$

28. $f(x) = 1 - x + 2x^2 - 3x^3$

29. $f(x) = (1 - 2x)(x + 5)$

30. $f(x) = (1 + x^2)^3$

31. $f(x) = \dfrac{x^2 - 2x + 3}{2x^2 + 5x + 1}$

32. $f(x) = \dfrac{1 - 3x^3}{2x^3 - 6x + 2}$

33. $f(x) = \dfrac{2x + 1}{3x^2 + 2x - 7}$

34. $f(x) = \dfrac{x^2 + x - 5}{1 - 2x - x^3}$

35. $f(x) = \dfrac{3x^2 - 6x + 2}{2x - 9}$

36. $f(x) = \dfrac{1 - 2x^3}{x + 1}$

In Exercises 37 and 38, the graph of a function $f(x)$ is given. Use the graph to determine $\lim_{x \to +\infty} f(x)$ and $\lim_{x \to -\infty} f(x)$.

37.

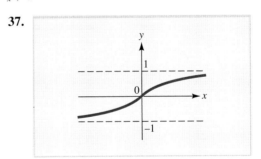

38.

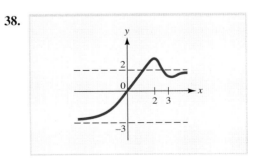

In Exercises 39 through 42, copy and complete the table by evaluating $f(x)$ at the specified values of x. Then use the table to estimate the indicated limit or show it does not exist.

39. $f(x) = x^2 - x;$ $\lim\limits_{x \to 2} f(x)$

x	1.9	1.99	1.999	2	2.001	2.01	2.1
$f(x)$							

40. $f(x) = x - \dfrac{1}{x};$ $\lim\limits_{x \to 0} f(x)$

x	−0.09	−0.009	0	0.0009	0.009	0.09
$f(x)$						

41. $f(x) = \dfrac{x^3 + 1}{x - 1};$ $\lim\limits_{x \to 1} f(x)$

x	0.99	0.999	1	1.001	1.01	1.1
$f(x)$						

42. $f(x) = \dfrac{x^3 + 1}{x + 1};$ $\lim\limits_{x \to -1} f(x)$

x	−1.1	−1.01	−1.001	−1	−0.999	−0.99	−0.9
$f(x)$							

In Exercises 43 through 50, find the indicated limit or show that it does not exist using the following facts about limits involving the functions $f(x)$ and $g(x)$:

$$\lim_{x \to c} f(x) = 5 \quad \text{and} \quad \lim_{x \to \infty} f(x) = -3$$
$$\lim_{x \to c} g(x) = -2 \quad \text{and} \quad \lim_{x \to \infty} g(x) = 4$$

43. $\lim\limits_{x \to c} [2f(x) - 3g(x)]$

44. $\lim\limits_{x \to c} f(x)g(x)$

45. $\lim\limits_{x \to c} \sqrt{f(x) + g(x)}$

46. $\lim\limits_{x \to c} f(x)[g(x) - 3]$

47. $\lim\limits_{x \to c} \dfrac{f(x)}{g(x)}$

48. $\lim\limits_{x \to c} \dfrac{2f(x) - g(x)}{5g(x) + 2f(x)}$

49. $\lim\limits_{x \to \infty} \dfrac{2f(x) - g(x)}{x + f(x)}$

50. $\lim\limits_{x \to \infty} \sqrt{g(x)}$

51. A wire is stretched horizontally, as shown in the accompanying figure. An experiment is conducted in which different weights are attached at the centre and the corresponding vertical displacements are measured. When too much weight is added, the wire snaps. Based on the data in the following table, what do you think is the maximum possible displacement for this kind of wire?

Weight W (kg)	15	16	17	18	17.5	17.9	17.99
Displacement y (cm)	1.7	1.75	1.78	Snaps	1.79	1.795	Snaps

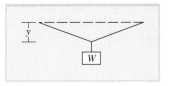

EXERCISE 51

52. If $1000 is invested at 5% compounded n times per year, the balance after 1 year will be $1000(1 + 0.05x)^{1/x}$, where $x = \dfrac{1}{n}$ is the length of the compounding period. For example, if $n = 4$, the compounding period is $\dfrac{1}{4}$ year long. For what is called *continuous compounding* of interest, the balance after 1 year is given by the limit

$$B = \lim_{x \to 0^+} 1000(1 + 0.05x)^{1/x}.$$

Estimate the value of this limit by completing the following table:

x	1	0.1	0.01	0.001	0.0001
$1000(1 + 0.05x)^{1/x}$					

53. PER CAPITA EARNINGS Studies indicate that t years from now, the population of a certain country will be $p = 0.2t + 1500$ thousand people, and that the gross earnings of the country will be E million dollars, where

$$E(t) = \sqrt{9t^2 + 0.5t + 179}.$$

  a. Express the per capita earnings of the country, $P = E/p$, as a function of time, t. (Take care with the units.)

  b. What happens to the per capita earnings in the long run (as $t \to \infty$)?

54. **PRODUCTION**   A business manager determines that t months after production begins on a new product, the number of units produced will be P thousand, where

$$P(t) = \frac{6t^2 + 5t}{(t + 1)^2}.$$

What happens to production in the long run (as $t \to \infty$)?

55. **ANIMAL BEHAVIOUR**   In some animal species, the intake of food is affected by the amount of vigilance maintained by the animal while feeding. In essence, it is hard to eat heartily while watching for predators that may eat you. In one model,* if the animal is foraging on plants that offer a bite of size S, the intake rate of food $I(S)$ is given by a function of the form

$$I(S) = \frac{aS}{S + c},$$

where a and c are positive constants.

What happens to the intake $I(S)$ as bite size S increases indefinitely? Interpret your result.

56. **EXPERIMENTAL PSYCHOLOGY**   To study the rate at which animals learn, a psychology student performs an experiment in which a rat is sent repeatedly through a laboratory maze. Suppose the time required for the rat to traverse the maze on the nth trial is approximately

$$T(n) = \frac{5n + 17}{n}$$

minutes. What happens to the time of traverse as the number of trials n increases indefinitely? Interpret your result.

57. **AVERAGE COST**   A business manager determines that the total cost of producing x units of a particular commodity may be modelled by the function

$$C(x) = 7.5x + 20\,000$$

(dollars). The average cost is $A(x) = \dfrac{C(x)}{x}$. Find $\lim\limits_{x \to +\infty} A(x)$ and interpret your result.

58. **REVENUE**   The organizer of a sports event estimates that if the event is announced x days in advance, the revenue obtained will be $R(x)$ thousand dollars, where

$$R(x) = 400 + 120x - x^2.$$

The cost of advertising the event for x days is $C(x)$ thousand dollars, where

$$C(x) = 2x^2 + 300.$$

a. Find the profit function $P(x) = R(x) - C(x)$ and sketch its graph.

b. How many days in advance should the event be announced in order to maximize profit? What is the maximum profit?

c. What is the ratio of revenue to cost

$$Q(x) = \frac{R(x)}{C(x)}$$

at the optimal announcement time found in part (b)? What happens to this ratio as $x \to 0$? Interpret these results.

59. **EXPLOSION AND EXTINCTION**   Two species coexist in the same ecosystem. Species I has population $P(t)$ in t years, while Species II has population $Q(t)$, both in thousands, where P and Q are modelled by the functions

$$P(t) = \frac{30}{3 + t} \quad \text{and} \quad Q(t) = \frac{64}{4 - t}$$

for all times $t \geq 0$ for which the respective populations are nonnegative.

a. What is the initial population of each species?

b. What happens to $P(t)$ as t increases? What hapens to $Q(t)$?

c. Sketch the graphs of $P(t)$ and $Q(t)$.

d. Species I is said to face **extinction** in the long run, while the population of Species II explodes in what is known as a **doomsday** scenario. Write a paragraph on what circumstances might result in either explosion or extinction of a species.

60. **POPULATION**   An urban planner models the population $P(t)$ (in thousands) of a certain community t years from now by the function

$$P(t) = \frac{40t}{t^2 + 10} - \frac{50}{t + 1} + 70.$$

a. What is the current population of the community?

b. By how much does the population change during the 3rd year? Is the population increasing or decreasing over this time period?

c. What happens to the population in the long run (as $t \to \infty$)?

---

*A. W. Willius and C. Fitzgibbon, "Costs of Vigilance in Foraging Ungulates," *Animal Behaviour*, Vol. 47, pt. 2 (Feb. 1994).

61. CONCENTRATION OF DRUG The concentration of drug in a patient's bloodstream t hours after an injection is $C(t)$ milligrams per millilitre, where

$$C(t) = \frac{0.4}{t^{1.2} + 1} + 0.013.$$

a. What is the concentration of drug immediately after the injection (when $t = 0$)?

b. By how much does the concentration change during the 5th hour? Does it increase or decrease over this time period?

c. What is the residual concentration of drug, that is, the concentration that remains in the long run (as $t \to \infty$)?

62. ENVIRONMENT The cost of removing x percent of pollutants from discharges into the atmosphere by a factory is $C(x) = \dfrac{35\,000x}{100 - x}$ dollars.

a. Find the cost of removing 90% of the pollutants.

b. Write a sentence to describe the meaning of the limit of this function as x approaches 100%.

63. The accompanying graph represents a function, $f(x)$, that oscillates between 1 and -1 more and more frequently as x approaches 0 from either the right or the left. Does $\lim\limits_{x \to 0} f(x)$ exist? If so, what is its value? [*Note:* For students with experience in trigonometry, the function $f(x) = \sin\!\left(\dfrac{1}{x}\right)$ behaves in this way.]

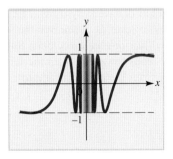

EXERCISE 63

64. BACTERIAL GROWTH The accompanying graph shows how the growth rate $R(T)$ of a bacterial colony changes with temperature, T.*

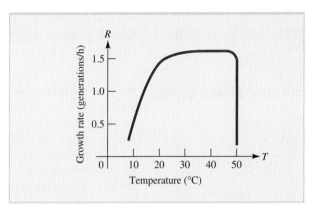

EXERCISE 64

a. Over what range of values of T does the growth rate $R(T)$ double?

b. What can be said about the growth rate for $25 < T < 45$?

c. What happens when the temperature reaches roughly 45°C? Does it make sense to compute $\lim\limits_{T \to 50} R(T)$?

d. Write a paragraph describing how temperature affects the growth rate of a species.

65. A* Evaluate the limit

$$\lim_{x \to +\infty} \frac{a_n x^n + a_{n-1}x^{n-1} + \cdots + a_1 x + a_0}{b_m x^m + b_{m-1}x^{m-1} + \cdots + b_1 x + b_0}$$

for constants $a_0, a_1, \ldots, a_n$ and $b_0, b_1, \ldots, b_m$ in each of the following cases:

a. $n < m$

b. $n = m$

c. $n > m$

[*Note:* There are two possible answers, depending on the signs of a_n and b_m.]

Source: Michael D. La Grega, Phillip L. Buckingham, and Jeffrey C. Evans, *Hazardous Waste Management.* New York: McGraw-Hill, 1994, pp. 565–566. Reprinted by permission.

SECTION 1.6

L06

Discuss the continuity of a function. Evaluate one-sided limits.

One-Sided Limits and Continuity

The dictionary defines continuity as "unbroken or uninterrupted succession." Continuous behaviour is an important part of our lives. For instance, the growth of a tree is continuous, as are the motion of a rocket and the volume of water flowing into a bathtub. In this section, we shall discuss what it means for a function to be continuous and examine a few important properties of such functions.

One-Sided Limits   Informally, a continuous function is one whose graph can be drawn without the "pencil" leaving the paper (Figure 1.51a). Not all functions have this property, but those that do play a special role in calculus. A function is *not* continuous where its graph has a hole or gap (Figure 1.51b), but what do we really mean by holes and gaps in a graph? To describe such features mathematically, we require the concept of a **one-sided limit** of a function, that is, a limit in which the approach is either from the right or from the left, rather than from both sides as required for the two-sided limit introduced in Section 1.5.

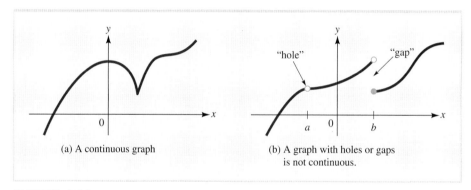

(a) A continuous graph

(b) A graph with holes or gaps is not continuous.

FIGURE 1.51   Continuity and discontinuity.

For instance, Figure 1.52 shows the graph of inventory, I, as a function of time, t, for a company that immediately restocks to level L_1 whenever the inventory falls to a certain minimum level, L_2 (this is called *just-in-time inventory*). Suppose the first restocking time occurs at $t = t_1$. Then as t tends toward t_1 from the left, the limiting value of $I(t)$ is L_2, while if the approach is from the right, the limiting value is L_1.

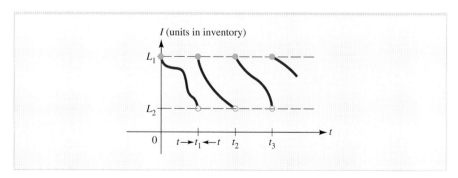

FIGURE 1.52   One-sided limits in a just-in-time inventory example.

Another example of a discontinuous function is the price of a coupon-paying bond. The value is saw-toothed with discontinuous jumps whenever a coupon is paid out.

Here is the notation we will use to describe one-sided limiting behaviour.

> **One-Sided Limits**   ■   If $f(x)$ approaches L as x tends toward c from the left $(x < c)$, we write $\lim\limits_{x \to c^-} f(x) = L$. Likewise, if $f(x)$ approaches M as x tends toward c from the right $(c < x)$, then we write $\lim\limits_{x \to c^+} f(x) = M$.

Using this notation in our inventory example, we write

$$\lim_{x \to t_1^-} I(t) = L_2 \quad \text{and} \quad \lim_{x \to t_1^+} I(t) = L_1.$$

Here are two more examples involving one-sided limits.

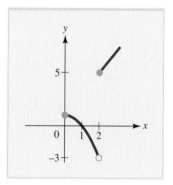

FIGURE 1.53   The graph of
$f(x) = \begin{cases} 1 - x^2 & \text{if } 0 \le x < 2 \\ 2x + 1 & \text{if } x \ge 2 \end{cases}$

EXAMPLE 1.6.1

For the function

$$f(x) = \begin{cases} 1 - x^2 & \text{if } 0 \le x < 2 \\ 2x + 1 & \text{if } x \ge 2 \end{cases}$$

evaluate the one-sided limits $\lim\limits_{x \to 2^-} f(x)$ and $\lim\limits_{x \to 2^+} f(x)$.

Solution

The graph of $f(x)$ is shown in Figure 1.53. Since $f(x) = 1 - x^2$ for $0 \le x < 2$, we have

$$\lim_{x \to 2^-} f(x) = \lim_{x \to 2^-} (1 - x^2) = -3.$$

Similarly, $f(x) = 2x + 1$ if $x \ge 2$, so

$$\lim_{x \to 2^+} f(x) = \lim_{x \to 2^+} (2x + 1) = 5.$$

EXAMPLE 1.6.2

Find $\lim \dfrac{x - 2}{x - 4}$ as x approaches 4 from the left and from the right.

Solution

First, note that for $2 < x < 4$ the quantity

$$f(x) = \frac{x - 2}{x - 4}$$

is negative, so as x approaches 4 from the left, $f(x)$ *decreases* without bound. We denote this fact by writing

$$\lim_{x \to 4^-} \frac{x-2}{x-4} = -\infty.$$

Likewise, as x approaches 4 from the right (with $x > 4$), $f(x)$ increases without bound and we write

$$\lim_{x \to 4^+} \frac{x-2}{x-4} = +\infty.$$

The graph of f is shown in Figure 1.54.

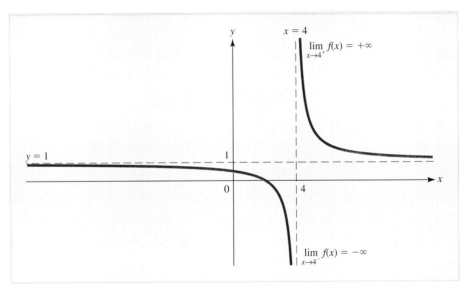

FIGURE 1.54   The graph of $f(x) = \dfrac{x-2}{x-4}$.

Notice that the two-sided limit $\lim_{x \to 4} f(x)$ does *not* exist for the function in Example 1.6.2 since the functional values $f(x)$ do not approach a single value L as x tends toward 4 from each side. In general, we have the following useful criterion for the existence of a limit.

Existence of a Limit ■ The two-sided limit $\lim_{x \to c} f(x)$ exists if and only if the two one-sided limits $\lim_{x \to c^-} f(x) = \lim_{x \to c^+} f(x)$ both exist and are equal, and then

$$\lim_{x \to c} f(x) = \lim_{x \to c^-} f(x) = \lim_{x \to c^+} f(x).$$

EXAMPLE 1.6.3

Determine whether $\lim_{x \to 1} f(x)$ exists, where

$$f(x) = \begin{cases} x+1 & \text{if } x < 1 \\ -x^2 + 4x - 1 & \text{if } x \geq 1 \end{cases}$$

Solution

Computing the one-sided limits at $x = 1$, we find

$$\lim_{x \to 1^-} f(x) = \lim_{x \to 1^-} (x + 1) = (1) + 1 = 2 \quad \text{since } f(x) = x + 1 \text{ when } x < 1$$

and

$$\lim_{x \to 1^+} f(x) = \lim_{x \to 1^+} (-x^2 + 4x - 1) \quad \text{since } f(x) = -x^2 + 4x - 1 \text{ when } x \geq 1$$
$$= -(1)^2 + 4(1) - 1 = 2$$

Since the two one-sided limits are equal, it follows that the two-sided limit of $f(x)$ at $x = 1$ exists, and we have

$$\lim_{x \to 1} f(x) = \lim_{x \to 1^-} f(x) = \lim_{x \to 1^+} f(x) = 2.$$

The graph of $f(x)$ is shown in Figure 1.55.

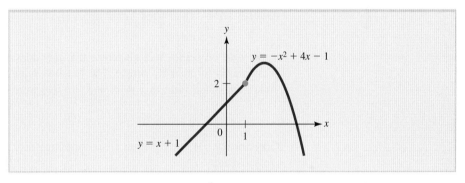

FIGURE 1.55   The graph of $f(x) = \begin{cases} x + 1 & \text{if } x < 1 \\ -x^2 + 4x - 1 & \text{if } x \geq 1 \end{cases}$

Continuity   At the beginning of this section, we observed that a continuous function is one whose graph has no holes or gaps. A hole at $x = c$ can arise in several ways, three of which are shown in Figure 1.56.

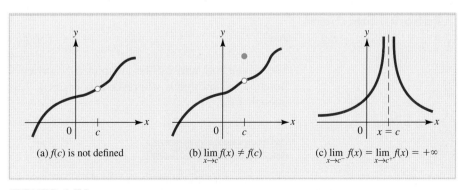

(a) $f(c)$ is not defined   (b) $\lim_{x \to c} f(x) \neq f(c)$ (c) $\lim_{x \to c^-} f(x) = \lim_{x \to c^+} f(x) = +\infty$

FIGURE 1.56   Three ways the graph of a function can have a hole at $x = c$.

The graph of $f(x)$ will have a gap at $x = c$ if the one-sided limits $\lim\limits_{x \to c^-} f(x)$ and $\lim\limits_{x \to c^+} f(x)$ are not equal. Three ways this can happen are shown in Figure 1.57.

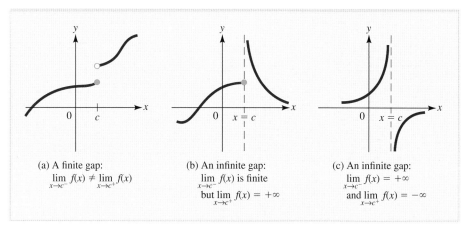

(a) A finite gap:
$\lim\limits_{x \to c^-} f(x) \neq \lim\limits_{x \to c^+} f(x)$

(b) An infinite gap:
$\lim\limits_{x \to c^-} f(x)$ is finite
but $\lim\limits_{x \to c^+} f(x) = +\infty$

(c) An infinite gap:
$\lim\limits_{x \to c^-} f(x) = +\infty$
and $\lim\limits_{x \to c^+} f(x) = -\infty$

FIGURE 1.57   Three ways for the graph of a function to have a gap at $x = c$.

So what properties will guarantee that $f(x)$ does not have a hole or gap at $x = c$? The answer is surprisingly simple. The function must be defined at $x = c$; it must have a finite, two-sided limit at $x = c$; and $\lim\limits_{x \to c} f(x)$ must equal $f(c)$. To summarize:

Continuity ■ A function f is continuous at c if all three of these conditions are satisfied:

a. $f(c)$ is defined.

b. $\lim\limits_{x \to c} f(x)$ exists.

c. $\lim\limits_{x \to c} f(x) = f(c)$.

If $f(x)$ is not continuous at c, it is said to have a **discontinuity** there.

Continuity of Polynomials and Rational Functions

Recall that if $p(x)$ and $q(x)$ are polynomials, then

$$\lim_{x \to c} p(x) = p(c)$$

and

$$\lim_{x \to c} \frac{p(x)}{q(x)} = \frac{p(c)}{q(c)} \quad \text{if } q(c) \neq 0$$

These limit formulas can be interpreted as saying that **a polynomial or a rational function is continuous wherever it is defined.** This is illustrated in Examples 1.6.4 through 1.6.7.

EXAMPLE 1.6.4

Show that the polynomial $p(x) = 3x^3 - x + 5$ is continuous at $x = 1$.

Solution

Verify that the three criteria for continuity are satisfied. Clearly $p(1)$ is defined; in fact, $p(1) = 7$. Moreover, $\lim\limits_{x \to 1} p(x)$ exists and $\lim\limits_{x \to 1} p(x) = 7$. Thus,

$$\lim\limits_{x \to 1} p(x) = 7 = p(1),$$

as required for $p(x)$ to be continuous at $x = 1$.

EXAMPLE 1.6.5

Show that the rational function $f(x) = \dfrac{x + 1}{x - 2}$ is continuous at $x = 3$.

Solution

Note that $f(3) = \dfrac{3 + 1}{3 - 2} = 4$. Since $\lim\limits_{x \to 3}(x - 2) \neq 0$,

$$\lim\limits_{x \to 3} f(x) = \lim\limits_{x \to 3}\frac{x + 1}{x - 2} = \frac{\lim\limits_{x \to 3}(x + 1)}{\lim\limits_{x \to 3}(x - 2)} = \frac{4}{1} = 4 = f(3),$$

as required for $f(x)$ to be continuous at $x = 3$.

EXPLORE!

Graph $f(x) = \dfrac{x^3 - 8}{x - 2}$. Does this graph appear continuous? Check the coordinates near $x = 2$ to investigate. Factor the numerator (with technology if you wish). Can you see now why the graph appears as it does?

EXAMPLE 1.6.6

Discuss the continuity of each of the following functions:

a. $f(x) = \dfrac{1}{x}$

b. $g(x) = \dfrac{x^2 - 1}{x + 1}$

c. $h(x) = \begin{cases} x + 1 & \text{if } x < 1 \\ 2 - x & \text{if } x \geq 1 \end{cases}$

Solution

The functions in parts (a) and (b) are rational and are therefore continuous wherever they are defined (that is, wherever their denominators are not zero).

a. $f(x) = \dfrac{1}{x}$ is defined everywhere except $x = 0$, so it is continuous for all $x \neq 0$ (Figure 1.58a).

b. Since $x = -1$ is the only value of x for which $g(x)$ is undefined, $g(x)$ is continuous except at $x = -1$ (Figure 1.58b).

c. This function is defined in two pieces. First check for continuity at $x = 1$, the value of x that separates the two pieces. You find that $\lim\limits_{x \to 1} h(x)$ does not exist, since $h(x)$ approaches 2 from the left and 1 from the right. Thus, $h(x)$ is not continuous at 1 (Figure 1.58c). However, since the polynomials $x + 1$ and $2 - x$ are each continuous for every value of x, it follows that $h(x)$ is continuous at every number x other than 1.

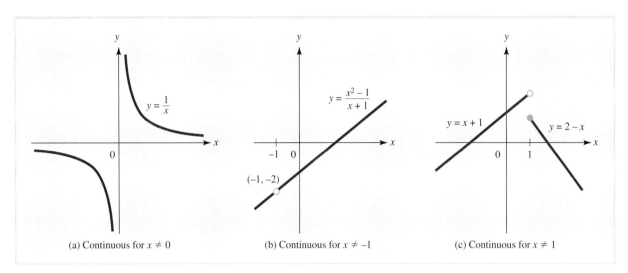

FIGURE 1.58   Functions for Example 1.6.6.

EXAMPLE 1.6.7

For what value of the constant A is the following function continuous for all real x?

$$f(x) = \begin{cases} Ax + 5 & \text{if } x < 1 \\ x^2 - 3x + 4 & \text{if } x \geq 1 \end{cases}$$

Solution

Since $Ax + 5$ and $x^2 - 3x + 4$ are both polynomials, it follows that $f(x)$ will be continuous everywhere except possibly at $x = 1$. Moreover, $f(x)$ approaches $A + 5$ as x approaches 1 from the left and 2 as x approaches 1 from the right. Thus, for $\lim\limits_{x \to 1} f(x)$ to exist, we must have $A + 5 = 2$ or $A = -3$, in which case

$$\lim_{x \to 1} f(x) = 2 = f(1).$$

This means that f is continuous for all x only when $A = -3$.

Continuity on an Interval

For many applications of calculus, it is useful to have definitions of continuity on open and closed intervals.

> **Continuity on an Interval** ■ A function $f(x)$ is said to be continuous on an open interval $a < x < b$ if it is continuous at each point $x = c$ in that interval.
>
> Moreover, f is continuous on the closed interval $a \leq x \leq b$ if it is continuous on the open interval $a < x < b$ and
>
> $$\lim_{x \to a^+} f(x) = f(a) \quad \text{and} \quad \lim_{x \to b^-} f(x) = f(b).$$

In other words, continuity on an interval means that the graph of f is "one piece" throughout the interval.

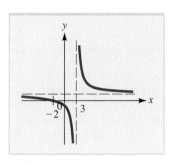

FIGURE 1.59   The graph of $f(x) = \dfrac{x + 2}{x - 3}$.

EXAMPLE 1.6.8

Discuss the continuity of the function

$$f(x) = \frac{x + 2}{x - 3}$$

on the open interval $-2 < x < 3$ and on the closed interval $-2 \le x \le 3$.

Solution

The rational function $f(x)$ is continuous for all x except $x = 3$. Therefore, it is continuous on the open interval $-2 < x < 3$ but not on the closed interval $-2 \le x \le 3$, since it is discontinuous at the endpoint 3 (where its denominator is zero). The graph of f is shown in Figure 1.59.

The Intermediate Value Property

An important feature of continuous functions is the **intermediate value property,** which says that if $f(x)$ is continuous on the interval $a \le x \le b$ and L is a number between $f(a)$ and $f(b)$, then $f(c) = L$ for some number c between a and b (see Figure 1.60). In other words, *a continuous function attains all values between any two of its values.* For instance, a tree that grew in height from 3 m to 7 m over a 5-year period must have been exactly 5 m tall at some point, since height is a continuous function of time.

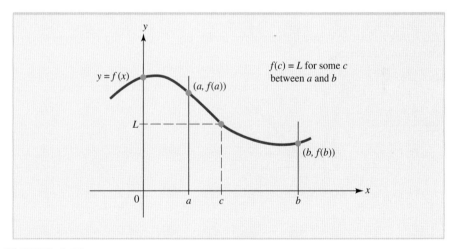

FIGURE 1.60   The intermediate value property.

The intermediate value property has a variety of applications. In Example 1.6.9, we show how it can be used to estimate a solution of a given equation.

EXAMPLE 1.6.9

Show that the equation $x^2 - x - 1 = \dfrac{1}{x + 1}$ has a solution for $1 < x < 2$.

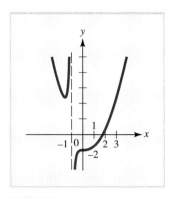

FIGURE 1.61   The graph of $y = x^2 - x - 1 - \dfrac{1}{x + 1}$.

Solution

Let $f(x) = x^2 - x - 1 - \dfrac{1}{x + 1}$. Then $f(1) = -\dfrac{3}{2}$ and $f(2) = \dfrac{2}{3}$. Since $f(x)$ is continuous for $1 \le x \le 2$ and the graph of f is below the x axis at $x = 1$ and above the x axis at $x = 2$, it follows from the intermediate value property that the graph must cross the x axis somewhere between $x = 1$ and $x = 2$ (see Figure 1.61). In other words, there is a number c such that $1 < c < 2$ and $f(c) = 0$, so

$$c^2 - c - 1 = \frac{1}{c + 1}.$$

NOTE   The root-location procedure described in Example 1.6.9 can be applied repeatedly to estimate the root c to any desired degree of accuracy. For instance, the midpoint of the interval $1 \le x \le 2$ is $d = 1.5$ and $f(1.5) = -0.65$, so the root c must lie in the interval $1.5 < x < 2$ (since $f(2) > 0$), and so on.

"That's nice," you say, "but I can solve it with my laptop to find a much more accurate estimate for c with much less effort." You are right, of course, but how do you think your computer program makes its estimation? Perhaps not by the method just described, but certainly by some similar algorithmic procedure. It is important to understand such procedures as you use the technology that utilizes them.

EXERCISES ■ 1.6

In Exercises 1 through 4, find the one-sided limits $\lim\limits_{x \to 2^-} f(x)$ *and* $\lim\limits_{x \to 2^+} f(x)$ *from the given graph of f and determine whether* $\lim\limits_{x \to 2} f(x)$ *exists.*

1.

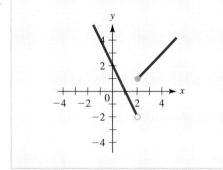

2.

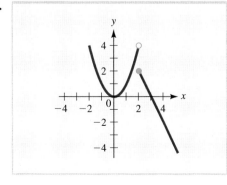

3.

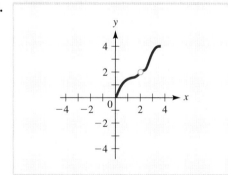

4.

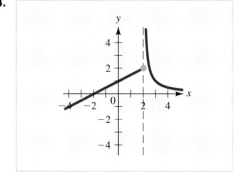

In Exercises 5 through 16, find the indicated one-sided limit. If the limiting value is infinite, indicate whether it is $+\infty$ or $-\infty$.

5. $\lim_{x \to 4^+} (3x^2 - 9)$

6. $\lim_{x \to 1^-} x(2 - x)$

7. $\lim_{x \to 3^+} \sqrt{3x - 9}$

8. $\lim_{x \to 2^-} \sqrt{4 - 2x}$

9. $\lim_{x \to 2^-} \dfrac{x + 3}{x + 2}$

10. $\lim_{x \to 2^-} \dfrac{x^2 + 4}{x - 2}$

11. $\lim_{x \to 0^+} (x - \sqrt{x})$

12. $\lim_{x \to 1^-} \dfrac{x - \sqrt{x}}{x - 1}$

13. $\lim_{x \to 3^+} \dfrac{\sqrt{x + 1} - 2}{x - 3}$

14. $\lim_{x \to 5^+} \dfrac{\sqrt{2x - 1} - 3}{x - 5}$

15. $\lim_{x \to 3^-} f(x)$ and $\lim_{x \to 3^+} f(x)$,

where $f(x) = \begin{cases} 2x^2 - x & \text{if } x < 3 \\ 3 - x & \text{if } x \ge 3 \end{cases}$

16. $\lim_{x \to -1^-} f(x)$ and $\lim_{x \to -1^+} f(x)$,

where $f(x) = \begin{cases} \dfrac{1}{x - 1} & \text{if } x < -1 \\ x^2 + 2x & \text{if } x \ge -1 \end{cases}$

In Exercises 17 through 28, decide if the given function is continuous at the specified value of x.

17. $f(x) = 5x^2 - 6x + 1$ at $x = 2$

18. $f(x) = x^3 - 2x^2 + x - 5$ at $x = 0$

19. $f(x) = \dfrac{x + 2}{x + 1}$ at $x = 1$

20. $f(x) = \dfrac{2x - 4}{3x - 2}$ at $x = 2$

21. $f(x) = \dfrac{x + 1}{x - 1}$ at $x = 1$

22. $f(x) = \dfrac{2x + 1}{3x - 6}$ at $x = 2$

23. $f(x) = \dfrac{\sqrt{x} - 2}{x - 4}$ at $x = 4$

24. $f(x) = \dfrac{\sqrt{x} - 2}{x - 4}$ at $x = 2$

25. $f(x) = \begin{cases} x + 1 & \text{if } x \le 2 \\ 2 & \text{if } x > 2 \end{cases}$ at $x = 2$

26. $f(x) = \begin{cases} x + 1 & \text{if } x < 0 \\ x - 1 & \text{if } x \ge 0 \end{cases}$ at $x = 0$

27. $f(x) = \begin{cases} x^2 + 1 & \text{if } x \le 3 \\ 2x + 4 & \text{if } x > 3 \end{cases}$ at $x = 3$

28. $f(x) = \begin{cases} \dfrac{x^2 + 1}{x + 1} & \text{if } x < -1 \\ x^2 - 3 & \text{if } x \ge -1 \end{cases}$ at $x = -1$

In Exercises 29 through 42, list all the values of x for which the given function is not continuous.

29. $f(x) = 3x^2 - 6x + 9$

30. $f(x) = x^5 - x^3$

31. $f(x) = \dfrac{x + 1}{x - 2}$

32. $f(x) = \dfrac{3x - 1}{2x - 6}$

33. $f(x) = \dfrac{3x + 3}{x + 1}$

34. $f(x) = \dfrac{x^2 - 1}{x + 1}$

35. $f(x) = \dfrac{3x - 2}{(x + 3)(x - 6)}$

36. $f(x) = \dfrac{x}{(x + 5)(x - 1)}$

37. $f(x) = \dfrac{x}{x^2 - x}$

38. $f(x) = \dfrac{x^2 - 2x + 1}{x^2 - x - 2}$

39. $f(x) = \begin{cases} 2x + 3 & \text{if } x \le 1 \\ 6x - 1 & \text{if } x > 1 \end{cases}$

40. $f(x) = \begin{cases} x^2 & \text{if } x \le 2 \\ 9 & \text{if } x > 2 \end{cases}$

41. $f(x) = \begin{cases} 3x - 2 & \text{if } x < 0 \\ x^2 + x & \text{if } x \ge 0 \end{cases}$

42. $f(x) = \begin{cases} 2 - 3x & \text{if } x \le -1 \\ x^2 - x + 3 & \text{if } x > -1 \end{cases}$

43. WEATHER Environment Canada uses the following formula to determine the wind chill factor. T_{air} is the air temperature in degrees Celsius, and V is the wind velocity in kilometres per hour. WC is the wind chill temperature:

$$WC = \begin{cases} \begin{aligned} &13.12 + 0.6215T_{air} \\ &- 11.37V^{0.16} + 0.3965T_{air}V^{0.16} \end{aligned} & \text{if } V > 4.8 \\ 0 & \text{if } V < 4.8 \end{cases}$$

a. What is the wind chill when the temperature is $-10°C$ and the wind speed is 25 km/h?

b. If the temperature is $-1°C$, is there continuity between the real outdoor temperature and the wind chill temperature as the wind increases to 4.8 km/h?

44. ELECTRIC FIELD INTENSITY If a hollow sphere of radius R is charged with one unit of static electricity, then the field intensity $E(x)$ at a point P located x units from the centre of the sphere satisfies

$$E(x) = \begin{cases} 0 & \text{if } 0 < x < R \\ \dfrac{1}{2x^2} & \text{if } x = R \\ \dfrac{1}{x^2} & \text{if } x > R \end{cases}$$

Sketch the graph of $E(x)$. Is $E(x)$ continuous for $x > 0$?

45. POSTAGE The cost of sending a parcel from Canada to England is given by the function $p(x)$:

$$p(x) = \begin{cases} 8.23 & \text{if } 0 < x \le 250 \\ 16.47 & \text{if } 250 < x \le 500 \\ 31.29 & \text{if } 500 < x \le 1000 \end{cases}$$

where x is the weight of the parcel in grams and $p(x)$ is the corresponding postage in dollars. These rates are for the Canada Post small packet category. Sketch the graph of $p(x)$ for $0 < x \le 1000$. For what values of x is $p(x)$ discontinuous for $0 < x \le 1000$?

46. WATER POLLUTION A ruptured pipe in a North Atlantic oil rig produces a circular oil slick that is y metres thick at a distance x metres from the rupture. Turbulence makes it difficult to directly measure the thickness of the slick at the source (where $x = 0$), but for $x > 0$, it is found that

$$y = \frac{0.5(x^2 + 3x)}{x^3 + x^2 + 4x}.$$

Assuming the oil slick is continuously distributed, how thick would you expect it to be at the source?

47. ENERGY CONSUMPTION The accompanying graph shows the amount of gasoline in the tank of Jan's car over a 30-day period. When is the graph discontinuous? What do you think happens at these times?

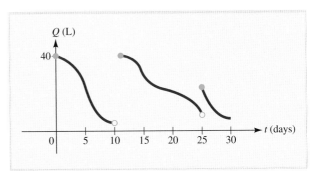

EXERCISE 47

48. INVENTORY The accompanying graph shows the number of units in inventory at a certain business over a 2-year period. When is the graph discontinuous? What do you think is happening at those times?

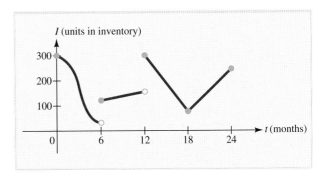

EXERCISE 48

49. COST-BENEFIT ANALYSIS In certain situations, it is necessary to weigh the benefit of pursuing a certain goal against the cost of achieving that goal. For instance, suppose that in order to remove x percent of the pollution from an oil spill, it costs C thousand dollars, where

$$C(x) = \frac{12}{100 - x}.$$

a. How much of the pollutant can be removed with $50 000?

b. Sketch the graph of the cost function.

c. What happens as $x \to 100^-$? Is it possible to remove all of the pollutant?

50. EARNINGS On January 1, 2011, Sam started working for an insurance company with an annual salary of \$48 000, paid each month on the last day of that month. On July 1, he received a commission of \$2000 for his work, and on September 1, his base salary was raised to \$54 000 per year. Finally, on December 21, he received a Christmas bonus of 1% of his base salary.

a. Sketch the graph of Sam's cumulative earnings E as a function of time t (days) during the year 2011.

b. For what values of t is the graph of $E(t)$ discontinuous?

51. COST MANAGEMENT A business manager determines that when x percent of her company's plant capacity is being used, the total cost of operation is C hundred thousand dollars, where

$$C(x) = \frac{8x^2 - 636x - 320}{x^2 - 68x - 960}.$$

a. Find $C(0)$ and $C(100)$.

b. Explain why the result of part (a) cannot be used along with the intermediate value property to show that the cost of operation is exactly \$700 000 when a certain percentage of plant capacity is being used.

52. AIR POLLUTION It is estimated that t years from now the population of a certain suburban community will be p thousand people, where

$$p(t) = 20 - \frac{7}{t + 2}.$$

An environmental study indicates that the average level of carbon monoxide in the air will be c parts per million when the population is p thousand, where

$$c(p) = 0.4\sqrt{p^2 + p + 21}.$$

What happens to the level of pollution c in the long run (as $t \to \infty$)?

In Exercises 53 and 54, find the values of the constant A so that the function f(x) is continuous for all x.

53. $f(x) = \begin{cases} Ax - 3 & \text{if } x < 2 \\ 3 - x + 2x^2 & \text{if } x \geq 2 \end{cases}$

54. $f(x) = \begin{cases} 1 - 3x & \text{if } x < 4 \\ Ax^2 + 2x - 3 & \text{if } x \geq 4 \end{cases}$

55. Discuss the continuity of the function

$$f(x) = x\left(1 + \frac{1}{x}\right) \text{ on the open interval } 0 < x < 1$$

and on the closed interval $0 \leq x \leq 1$.

56. Discuss the continuity of the function

$$f(x) = \begin{cases} x^2 - 3x & \text{if } x < 2 \\ 4 + 2x & \text{if } x \geq 2 \end{cases}$$

on the open interval $0 < x < 2$ and on the closed interval $0 \leq x \leq 2$.

57. Show that the equation $\sqrt[3]{x - 8} + 9x^{2/3} = 29$ has at least one solution for the interval $0 \leq x \leq 8$.

58. Show that the equation $\sqrt[3]{x} = x^2 + 2x - 1$ must have at least one solution on the interval $0 \leq x \leq 1$.

59. Investigate the behaviour of $f(x) = \dfrac{2x^2 - 5x + 2}{x^2 - 4}$ when x is near (a) 2 and (b) -2. Does the limit exist at these values of x? Is the function continuous at these values of x?

60. Explain why there must have been some time in your life when your weight in pounds was the same as your height in inches.

61. Explain why there is a time every hour when the hour hand and minute hand of a clock coincide.

62. At age 15, Nan is twice as tall as her 5-year-old brother Dan, but on Dan's 21st birthday, they find that he is 15 cm taller. Explain why there must have been a time when they were exactly the same height.
[*Note:* More algebraic concepts are reviewed in Appendix A.]

Concept Summary Chapter 1

Straight Line

$y = mx + b$, m is the slope, b is the y-intercept

$y - y_0 = m(x - x_0)$

Modelling

$Q = kx$ direct proportion

$Q = \dfrac{k}{x}$ indirect proportion

Limits, I

when approaching $\dfrac{0}{0}$, factor and cancel:

$$\lim_{x \to 2} \frac{x^2 - 4}{x - 2} = \lim_{x \to 2} \frac{(x - 2)(x + 2)}{x - 2} = 4$$

Limits, II

$\lim_{x \to \infty} f(x)$

Divide every term by the highest power of x in the denominator:

$$\lim_{x \to \infty} \frac{1}{x^n} = 0$$

$$\lim_{x \to \infty} \frac{5x^3 - 6x^2 + 2x - 1}{9x^4 + 7x - 2} = \lim_{x \to \infty} \frac{5/x - 6/x^2 + 2/x^3 - 1/x^4}{9 + 7/x^3 - 2/x^4} = 0$$

ALGEBRA INSIGHT 🅧

Factoring

Factoring will be needed throughout the course, especially difference of squares and quadratic factoring. Cover up the right side of the page, leaving just the column of questions. When you have completed the questions, check your answers.

   It is very important to multiply out the quadratic factors to be sure the factoring is correct. Missing something small, such as a minus sign, can lead to large errors for longer questions later on. A quick multiplying out is well worth it.

1. $x^2 + x - 2$ $(x + 2)(x - 1)$
2. $x^2 - 2x + 1$ $(x - 1)(x - 1)$
3. $x^2 + 2x - 8$ $(x + 4)(x - 2)$
4. $x^2 - 1$ $(x - 1)(x + 1)$
5. $x^2 - 6x + 9$ $(x - 3)(x - 3)$
6. $4 - x^2$ $(2 - x)(2 + x)$
7. $-x^2 + 3x + 4$ $-(x + 1)(x - 4)$
8. $x^2 + 2x - 3$ $(x + 3)(x - 1)$

Checkup for Chapter 1

1. Specify the domain of the function

$$f(x) = \frac{2x - 1}{\sqrt{4 - x^2}}.$$

2. Find the composite function $g(h(x))$, where

$$g(u) = \frac{1}{2u + 1} \quad \text{and} \quad h(x) = \frac{x + 2}{2x + 1}.$$

3. Find the equation for each line.

   a. through the point $(-1, 2)$ with slope $-\dfrac{1}{2}$

   b. with slope 2 and y intercept -3

4. Sketch the graph of each function. Be sure to show all intercepts and any high or low points.

   a. $f(x) = 3x - 5$

   b. $f(x) = -x^2 + 3x + 4$

5. Find each limit. If the limit is infinite, indicate whether it is $+\infty$ or $-\infty$.

   a. $\displaystyle\lim_{x \to -1} \frac{x^2 + 2x - 3}{x - 1}$

   b. $\displaystyle\lim_{x \to 1} \frac{x^2 + 2x - 3}{x - 1}$

   c. $\displaystyle\lim_{x \to 1} \frac{x^2 - x - 1}{x - 2}$

   d. $\displaystyle\lim_{x \to +\infty} \frac{2x^3 + 3x - 5}{-x^2 + 2x + 7}$

6. Determine whether the function $f(x)$ is continuous at $x = 1$.

$$f(x) = \begin{cases} 2x + 1 & \text{if } x \leq 1 \\ \dfrac{x^2 + 2x - 3}{x - 1} & \text{if } x > 1 \end{cases}$$

7. **PRICE OF GASOLINE** Since the beginning of the year, the price of gasoline has been increasing at a constant rate of 2 cents per litre per month. By June 1, the price had reached $1.06 per litre.

a. Express the price of gasoline as a function of time and draw the graph.

b. What was the price at the beginning of the year?

c. What will the price be on November 1?

8. **DISTANCE** A truck is 300 km due east of a car and is travelling west at a constant speed of 30 km/h. Meanwhile, the car is going north at a constant speed of 60 km/h. Express the distance between the car and the truck as a function of time.

9. **SUPPLY AND DEMAND** Suppose it is known that producers will supply x units of a certain commodity to the market when the price is $p = S(x)$ dollars per unit and that the same number of units will be demanded (bought) by consumers when the price is $p = D(x)$ dollars per unit, where

$$S(x) = x^2 + A \quad \text{and} \quad D(x) = Bx + 59$$

for constants A and B. It is also known that no units will be supplied until the unit price is $3 and that market equilibrium occurs when $x = 7$ units.

a. Use this information to find A and B and the equilibrium unit price.

b. Sketch the supply and demand curves on the same graph.

c. What is the difference between the supply price and the demand price when 5 units are produced? When 10 units are produced?

10. **BACTERIAL POPULATION** The population (in thousands) of a colony of bacteria t minutes after the introduction of a toxin is given by the function

$$f(t) = \begin{cases} t^2 + 7 & \text{if } 0 \leq t < 5 \\ -8t + 72 & \text{if } x \geq 5 \end{cases}$$

a. When does the colony die out?

b. Explain why the population must be 10 000 some time between $t = 1$ and $t = 7$.

11. **CONSUMER APPLICATIONS** The number of home-cooked meals Canadians are consuming has dropped steadily in recent years. Canadian households consumed 380 homemade meals a year in 2009 on average, and 423 in 2003.* If there is a linear relationship, express the number of homemade meals as a function of year. If this trend continued, when would only 365 meals be prepared at home, that is, one third of the number possible?

_____

*Canadian Press February, 2010.

Review Exercises

1. Specify the domain of each function.
a. $f(x) = x^2 - 2x + 6$
b. $f(x) = \dfrac{x - 3}{x^2 + x - 2}$
c. $f(x) = \sqrt{x^2 - 9}$

2. Specify the domain of each function.
a. $f(x) = 4 - (3 - x)^2$
b. $f(x) = \dfrac{x - 1}{x^2 - 2x + 1}$
c. $f(x) = \dfrac{1}{\sqrt{4 - 3x}}$

3. Find the composite function $g(h(x))$.
a. $g(u) = u^2 + 2u + 1, h(x) = 1 - x$
b. $g(u) = \dfrac{1}{2u + 1}, h(x) = x + 2$

4. Find the composite function $g(h(x))$.
a. $g(u) = (1 - 2x)^2, h(x) = \sqrt{x + 1}$
b. $g(u) = \sqrt{1 - u}, h(x) = 2x + 4$

5. a. Find $f(3 - x)$ if $f(x) = 4 - x - x^2$.
b. Find $f(x^2 - 3)$ if $f(x) = x - 1$.
c. Find $f(x + 1) - f(x)$ if $f(x) = \dfrac{1}{x - 1}$.

6. **a.** Find $f(x - 2)$ if $f(x) = x^2 - x + 4$.

    b. Find $f(x^2 + 1)$ if $f(x) = \sqrt{x} + \dfrac{2}{x - 1}$.

    c. Find $f(x + 1) - f(x)$ if $f(x) = x^2$.

7. Find functions $h(x)$ and $g(u)$ such that $f(x) = g(h(x))$.

    a. $f(x) = (x^2 + 3x + 4)^5$

    b. $f(x) = (3x + 1)^2 + \dfrac{5}{2(3x + 1)^3}$

8. Find functions $h(x)$ and $g(u)$ such that $f(x) = g(h(x))$.

    a. $f(x) = (x - 1)^2 - 3(x - 1) + 1$

    b. $f(x) = \dfrac{2(x + 4)}{2x - 3}$

9. Graph the quadratic function $f(x) = x^2 + 2x - 8$.

10. Graph the quadratic function $f(x) = 3 + 4x - 2x^2$.

11. Find the slope and y intercept of each line and draw the graph.

    a. $y = 3x + 2$

    b. $5x - 4y = 20$

12. Find the slope and y intercept of each line and draw the graph.

    a. $2y + 3x = 0$

    b. $\dfrac{x}{3} + \dfrac{y}{2} = 4$

13. Find the equation of each line.

    a. slope 5 and y intercept $(0, -4)$

    b. slope -2 and contains $(1, 3)$

    c. contains $(5, 4)$ and is parallel to $2x + y = 3$

14. Find the equation of each line.

    a. passes through the points $(-1, 3)$ and $(4, 1)$

    b. x intercept $(3, 0)$ and y intercept $\left(0, -\dfrac{2}{3}\right)$

    c. contains $(-1, 3)$ and is perpendicular to $5x - 3y = 7$

15. Find the points of intersection (if any) of the given pair of curves and draw the graphs.

    a. $y = -3x + 5$ and $y = 2x - 10$

    b. $y = x + 7$ and $y = -2 + x$

16. Find the points of intersection (if any) of the given pair of curves and draw the graphs.

    a. $y = x^2 - 1$ and $y = 1 - x^2$

    b. $y = x^2$ and $y = 15 - 2x$

17. Find c so that the curve $y = 3x^2 - 2x + c$ passes through the point $(2, 4)$.

18. Find c so that the curve $y = 4 - x - cx^2$ passes through the point $(-2, 1)$.

In Exercises 19 through 32, either find the given limit or show that it does not exist. If the limit is infinite, indicate whether it is $+\infty$ or $-\infty$.

19. $\displaystyle\lim_{x \to 1} \dfrac{x^2 + x - 2}{x^2 - 1}$

20. $\displaystyle\lim_{x \to 2} \dfrac{x^2 - 3x}{x - 1}$

21. $\displaystyle\lim_{x \to 2} \dfrac{x^3 - 8}{2 - x}$

22. $\displaystyle\lim_{x \to 1} \left(\dfrac{1}{x^2} - \dfrac{1}{x}\right)$

23. $\displaystyle\lim_{x \to 0} \left(2 - \dfrac{1}{x^3}\right)$

24. $\displaystyle\lim_{x \to -\infty} \left(2 + \dfrac{1}{x^2}\right)$

25. $\displaystyle\lim_{x \to -\infty} \dfrac{x}{x^2 + 5}$

26. $\displaystyle\lim_{x \to 0} \left(x^3 - \dfrac{1}{x^2}\right)$

27. $\displaystyle\lim_{x \to -\infty} \dfrac{x^4 + 3x^2 - 2x + 7}{x^3 + x + 1}$

28. $\displaystyle\lim_{x \to -\infty} \dfrac{x^3 - 3x + 5}{2x + 3}$

29. $\displaystyle\lim_{x \to -\infty} \dfrac{1 + \dfrac{1}{x} + \dfrac{1}{x^2}}{x^2 + 3x - 1}$

30. $\displaystyle\lim_{x \to -\infty} \dfrac{x(x - 3)}{7 - x^2}$

31. $\displaystyle\lim_{x \to 0^-} x\sqrt{1 - \dfrac{1}{x}}$

32. $\displaystyle\lim_{x \to 0^+} \sqrt{x\left(1 + \dfrac{1}{x^2}\right)}$

In Exercises 33 through 36, list all values of x for which the given function is not continuous.

33. $f(x) = \dfrac{x^2 - 1}{x + 3}$

34. $f(x) = 5x^3 - 3x + \sqrt{x}$

35. $h(x) = \begin{cases} x^3 + 2x - 33 & \text{if } x \le 3 \\ \dfrac{x^2 - 6x + 9}{x - 3} & \text{if } x > 3 \end{cases}$

36. $g(x) = \dfrac{x^3 + 5x}{(x - 2)(2x + 3)}$

37. PRICE Due to advances in technology, the price of cell phones on the market drops. Suppose that x months from now, the price of a certain model bought on a pay-as-you-go plan will be P dollars per unit, where

$$P(x) = 40 + \frac{30}{x + 1}.$$

  a. What will the price be 5 months from now?

  b. By how much will the price drop during the 5th month?

  c. When will the price be $43?

  d. What happens to the price in the long run (as x becomes very large)?

38. ENVIRONMENTAL ANALYSIS An environmental study of a certain community suggests that the average daily level of particulate matter (one of the components of smog) in the air will be $Q(p) = 10\sqrt{0.5p + 19.4}$ micrograms per cubic metre (μg/m^3) when the population is p thousand. It is estimated that t years from now, the population will be $p(t) = 8 + 0.2t^2$ thousand.

  a. Express the level of particulate matter in the air as a function of time.

  b. What will the particulate matter level be 3 years from now?

  c. When will the particulate matter level reach 50 μg/m^3 if pollution controls are not put in place?

39. EDUCATIONAL FUNDING A university has launched a fund-raising campaign for student scholarships. Suppose that university officials estimate that it will take $f(x) = \dfrac{10x}{150 - x}$ weeks to reach x percent of their goal.

  a. Sketch the relevant portion of the graph of this function.

  b. How long will it take to reach 50% of the campaign's goal?

  c. How long will it take to reach 100% of the goal?

40. CONSUMER EXPENDITURE The demand for a certain commodity is $D(x) = -50x + 800$; that is, x units of the commodity will be demanded by consumers when the price is $p = D(x)$ dollars per unit. Total consumer expenditure $E(x)$ is the amount of money consumers pay to buy x units of the commodity.

  a. Express consumer expenditure as a function of x, and sketch the graph of $E(x)$.

  b. Use the graph in part (a) to determine the level of production x at which consumer expenditure is largest. What price p corresponds to maximum consumer expenditure?

41. MICROBIOLOGY A spherical cell of radius r has volume $V = \dfrac{4}{3}\pi r^3$ and surface area $S = 4\pi r^2$. Express V as a function of S. If S is doubled, what happens to V?

42. NEWSPAPER CIRCULATION The reader hits for the Web edition of a local newspaper are increasing at a constant rate. Three months ago there were 3200 hits a day on average. Today the average number is 4400.

  a. Express the reader hits as a function of time and draw the graph.

  b. What will the average number of predicted hits be 2 months from now?

43. MANUFACTURING EFFICIENCY A manufacturing firm has received an order to make 400 000 souvenir soldier uniform buttons commemorating the 200th anniversary of the Battle of Queenston Heights, in Niagara, in the war of 1812. The firm owns several machines, each of which can produce 200 buttons per hour. The cost of setting up the machines to produce the buttons is $80 per machine, and the total operating cost is $5.76 per hour. Express the cost of producing the 400 000 buttons as a function of the number of machines used. Draw the graph and estimate the number of machines the firm should use to minimize cost.

44. OPTIMAL SELLING PRICE   A manufacturer can produce bookcases at a cost of $80 each. Sales figures indicate that if the bookcases are sold for x dollars each, approximately $150 - x$ will be sold each month. Express the manufacturer's monthly profit as a function of the selling price x, draw the graph, and estimate the optimal selling price.

45. OPTIMAL SELLING PRICE   A retailer can obtain cameras from the manufacturer at a cost of $150 each. The retailer has been selling the cameras at a price of $340 each, and at this price, consumers have been buying 40 cameras per month. The retailer is planning to lower the price to stimulate sales and estimates that for each $5 reduction in the price, 10 more cameras will be sold each month. Express the retailer's monthly profit from the sale of the cameras as a function of the selling price. Draw the graph and estimate the optimal selling price.

46. STRUCTURAL DESIGN   A cylindrical can with no top is to be constructed for 80 cents. The cost of the material used for the bottom is 3 cents/cm^2, and the cost of the material used for the curved side is 2 cents/cm^2. Express the volume of the can as a function of its radius.

47. PROPERTY TAX   A homeowner is trying to decide between two competing property tax propositions. With Proposition A, the homeowner will pay $100 plus 8% of the assessed value of her home, while Proposition B requires a payment of $1900 plus 2% of the assessed value. Assuming the homeowner's only consideration is to minimize her tax payment, develop a criterion based on the assessed value V of her home for deciding between the propositions.

48. INVENTORY ANALYSIS   A vet clinic maintains inventory of a special type of dog food over a particular 30-day month as follows:

days 1–9	30 cans
days 10–15	17 cans
days 16–23	12 cans
days 24–30	steadily decreasing from 12 units to 0 units

Sketch the graph of the inventory as a function of time t (days). At what times is the graph discontinuous?

49. BREAK-EVEN ANALYSIS   A manufacturer can sell its low-weight back-packing sleeping bags for $80 each. The total cost consists of a fixed overhead of $4500 plus production costs of $50 per sleeping bag.

   a. How many sleeping bags must the manufacturer sell to break even?

   b. What is the manufacturer's profit or loss if 200 sleeping bags are sold?

   c. How many sleeping bags must the manufacturer sell to realize a profit of $900?

50. PRODUCTION MANAGEMENT   During the summer, a group of students builds octagonal picnic tables in a converted garage. The rental for the garage is $1500 for the summer, and the materials needed to build a picnic table cost $125. The picnic tables can be sold for $275 each.

   a. How many picnic tables must the students sell to break even?

   b. How many picnic tables must the students sell to make a profit of at least $1000?

51. LEARNING   Some psychologists believe that when a person is asked to recall a set of facts, the rate at which the facts are recalled is proportional to the number of relevant facts in the subject's memory that have not yet been recalled. Express the recall rate as a function of the number of facts that have been recalled.

52. COST-EFFICIENT DESIGN   A cable is to be run from a power plant on one side of a river that is 900 m wide to a factory on the other side, 3000 m downstream. The cable will be run in a straight line from the power plant to some point P on the opposite bank and then along the bank to the factory. The cost of running the cable across the water is $5/m, while the cost over land is $4/m. Let x be the distance from P to the point directly across the river from the power plant. Express the cost of installing the cable as a function of x.

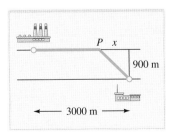

EXERCISE 52

53. **CONSTRUCTION COST** A window with a 7-m perimeter (frame) is made of a semicircular stained glass pane above a rectangular clear pane, as shown in the accompanying figure. Clear glass costs 0.9 cents/cm² and stained glass costs 3 cents/cm². Express the cost of the window as a function of the radius of the stained glass pane.

EXERCISE 53

54. **MANUFACTURING OVERHEAD** A furniture manufacturer can sell end tables for $125 each. It costs the manufacturer $85 to produce each table, and it is estimated that revenue will equal cost when 200 tables are sold. What is the overhead associated with the production of the tables? [*Note:* Overhead is the cost when 0 units are produced.]

55. **MANUFACTURING COST** A manufacturer can produce 5000 cartons of soy milk per day. There is a fixed (overhead) cost of $1500 per day and a variable cost of $2 per carton produced. Express the daily cost C as a function of the number of cartons produced and sketch the graph of $C(x)$. Is $C(x)$ continuous? If not, where do its discontinuities occur?

56. At what time between 3 P.M. and 4 P.M. will the minute hand coincide with the hour hand? [*Hint:* The hour hand moves $\dfrac{1}{12}$ as fast as the minute hand.]

57. The radius of Earth is roughly 6400 km, and an object located x kilometres from the centre of Earth weighs $w(x)$ newtons, where

$$w(x) = \begin{cases} Ax & \text{if } x \leq 6400 \\ \dfrac{B}{x^2} & \text{if } x > 6400 \end{cases}$$

and A and B are positive constants. Assuming that $w(x)$ is continuous for all x, what must be true about A and B? Sketch the graph of $w(x)$.

58. In each of these cases, find the value of the constant A that makes the given function $f(x)$ continuous for all x.

a. $f(x) = \begin{cases} 2x + 3 & \text{if } x < 1 \\ Ax - 1 & \text{if } x \geq 1 \end{cases}$

b. $f(x) = \begin{cases} \dfrac{x^2 - 1}{x + 1} & \text{if } x < -1 \\ Ax^2 + x - 3 & \text{if } x \geq -1 \end{cases}$

59. The accompanying graph represents a function $g(x)$ that oscillates more and more frequently as x approaches 0 from either the right or the left but with decreasing magnitude. Does $\lim\limits_{x \to 0} g(x)$ exist? If so, what is its value? [*Note:* For students with experience in trigonometry, the function $g(x) = |x| \sin(1/x)$ behaves in this way.]

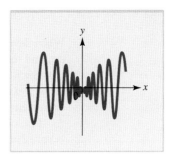

EXERCISE 59

60. Graph $f(x) = \dfrac{3x^2 - 6x + 9}{x^2 + x - 2}$. Determine the values of x where the function is undefined.

61. Graph $y = \dfrac{21}{9}x - \dfrac{84}{35}$ and $y = \dfrac{654}{279}x - \dfrac{54}{10}$ on the same set of coordinate axes. Are the two lines parallel?

62. For $f(x) = \sqrt{x + 3}$ and $g(x) = 5x^2 + 4$, find
a. $f(g(-1.28))$
b. $g(f(\sqrt{2}))$

Use three decimal place accuracy.

63. Graph $y = \begin{cases} x^2 + 1 & \text{if } x \leq 1 \\ x^2 - 1 & \text{if } x > 1 \end{cases}$. Find the points of discontinuity.

64. Graph the function $f(x) = \dfrac{x^2 - 3x - 10}{1 - x} - 2$. Find the x and y intercepts. For what values of x is this function defined?

THINK ABOUT IT

ALLOMETRIC MODELS

(Photo: © IT Stock/PunchStock)

When developing a mathematical model, the first task is to identify quantities of interest, and the next is to find equations that express relationships between these quantities. Such equations can be quite complicated, but many important relationships can be expressed in the relatively simple form $y = Cx^k$, in which one quantity y is expressed as a constant multiple of a power function of another quantity x.

In biology, the study of the relative growth rates of various parts of an organism is called **allometry,** from the Greek words *allo* (other or different) and *metry* (measure). In allometric models, equations of the form $y = Cx^k$ are often used to describe the relationship between two biological measurements.

Whenever possible, allometric models are developed using basic assumptions from biological (or other) principles. For example, it is reasonable to assume that the body volume and hence the weight of most animals is proportional to the cube of the linear dimension of the body, such as height for animals that stand up or length for four-legged animals. Thus, it is reasonable to expect the weight of a snake to be proportional to the cube of its length, and indeed, observations of the hognose snake of Kansas indicate that the weight w (grams) and length L (metres) of a typical such snake are related by the equation*

$$w = 440L^3.$$

Although a cubic relationship seems a logical model, in reality, experimental results imply a slightly different relationship. The model needs modification. For the western hognose snake, it turns out that the equations $w = 446L^{2.99}$ and $w = 429L^{2.90}$ provide better approximations for the weight of male and female western hognose snakes, respectively.

The *basal metabolic rate M* of an animal is the rate of heat produced by its body when it is resting and fasting. Basal metabolic rates have been studied since the 1830s and continue to be studied. The models continue to be modified. Allometric equations of the form $M = cw^r$, for constants c and r, have long been used to build models

---

*Edward Batschelet, *Introduction to Mathematics for Life Scientists,* 3rd ed., New York: Springer-Verlag, 1979, p. 178.

relating the basal metabolic rate to the body weight w of an animal. This model is based on the assumption that basal metabolic rate M is proportional to S, the surface area of the body, so that $M = aS$, where a is a constant. To set up an equation relating M and w, we need to relate the weight w of an animal to its surface area S. Assuming that all animals are shaped like spheres or cubes, and that the weight of an animal is proportional to its volume, we can show (see Exercises 1 and 2) that the surface area is proportional to $w^{2/3}$, so that $S = bw^{2/3}$, where b is a constant. Putting the equations $M = aS$ and $S = bw^{2/3}$ together, we obtain the allometric equation

$$M = abw^{2/3} = kw^{2/3},$$

where $k = ab$.

When more refined modelling assumptions are used, it is found that the basal metabolic rate M is better approximated if the exponent 3/4 is used in the allometric equation rather than the exponent 2/3.* The Swiss agricultural chemist Max Kleiber concluded that

$$M = 70w^{3/4},$$

where M is measured in kilocalories per day and w is measured in kilograms.

There is still work being done on the validity of this model.* A model is a way of expressing a situation mathematically, but with more data and new ways of investigating data (computers did not exist when allometry was proposed), a model can be changed and improved.

Questions

1. What weight does the allometric equation $w = 440L^3$ predict for a western hognose snake that is 0.7 m long? If this snake is male, what does the equation $w = 446L^{2.99}$ predict for its weight? If it is female, what does the equation $w = 429L^{2.90}$ predict for its weight?

2. What basal metabolic rates are predicted by the equation $M = 70w^{3/4}$ for animals with weights of 50 kg, 100 kg, and 350 kg?

3. Observations show that the brain weight b, measured in grams, of adult female primates is given by the allometric equation $b = 0.064w^{0.822}$, where w is the weight in grams of the primate. What are the predicted brain weights of female primates weighing 5 kg, 10 kg, 25 kg, 50 kg, and 100 kg?

4. If $y = Cx^k$, where C and k are constants, then y and x are said to be related by a positive allometry if $k > 1$ and a negative allometry if $k < 1$. The *weight-specific metabolic rate* of an animal is defined to be the basal metabolic rate M of the animal divided by its weight, w, that is, M/w. Show that if the basal metabolic rate is the positive allometry of the weight in Exercise 2, then the weight-specific metabolic rate is a negative allometry equation of the weight.

*Vaclav Smil, "Laying down the law, every living thing obeys the rules of scaling discovered by Max Kleiber," *Nature,* Vol. 403 (10 February 2000).

*Gary C. Packard and Geoffrey Birchard, "Traditional allometric analysis fails to provide a valid predictive model for mammalian metabolic rates," *The Journal of Experimental Biology,* 211, 3581–3587 (2008).

5. Show that if we assume that all animal bodies are shaped like cubes, then the surface area S of an animal body is proportional to $V^{2/3}$, where V is the volume of the body. By combining this fact with the assumption that the weight w of an animal is proportional to its volume, show that S is proportional to $w^{2/3}$.

6. Show that if we assume that all animal bodies are shaped like spheres, then the surface area S of an animal body is proportional to $V^{2/3}$, where V is the volume of the body. By combining this fact with the assumption that the weight w of an animal is proportional to its volume, show that S is proportional to $w^{2/3}$. [*Hint:* Recall that a sphere of radius r has surface area $4\pi r^2$ and volume $\dfrac{4}{3}\pi r^3$.]

A windsurfer flies across the water. His velocity is the rate of change of his displacement, and his acceleration is the rate of change of the velocity. This chapter shows how to find these rates of change. See what happens to the windsurfer in Example 2.1.2. (Photo: CORBIS/ Royalty Free)

DIFFERENTIATION: BASIC CONCEPTS

CHAPTER OUTLINE

1 The Derivative
2 Techniques of Differentiation
3 Product and Quotient Rules; Higher-Order Derivatives
4 The Chain Rule
5 Marginal Analysis and Approximations Using Increments
6 Implicit Differentiation and Related Rates

   Chapter Summary
      Concept Summary
      Checkup for Chapter 2
      Review Exercises
   Think about It

LEARNING OBJECTIVES

After completing this chapter, you should be able to

LO1 Distinguish between instantaneous and average rates of change both graphically and numerically. Use the limit definition of the derivative.

LO2 Calculate derivatives using derivative rules.

LO3 Use the chain, product, and quotient rules to differentiate a given function. Calculate second and higher-order derivatives.

LO4 Use the chain rule to find derivatives.

LO5 Use marginal analysis to solve business problems, and differentials to approximate by increments.

LO6 Differentiate functions defined implicitly. Solve related rate application problems.

OPTION NOTE: Chapter 4, Sections 4.1–4.3, can follow Chapter 2 as an option, with Section 4.4 following Chapter 3.

L01

Distinguish between instantaneous and average rates of change both graphically and numerically. Use the limit definition of the derivative.

The Derivative

Calculus is the mathematics of change, and the primary tool for studying change is a procedure called **differentiation.** In this section, we shall introduce this procedure and examine some of its uses, especially in computing rates of change. Here, and later in this chapter, we shall encounter rates such as velocity, acceleration, production rates with respect to labour level or capital expenditure, the rate of growth of a population, the infection rate of a susceptible population during an epidemic, and many others.

Calculus was developed in the seventeenth century by Isaac Newton (1642–1727), G. W. Leibniz (1646–1716), and others at least in part in an attempt to deal with two geometric problems:

Tangent problem: Find a tangent line at a particular point on a given curve.
Area problem: Find the area of the region under a given curve.

The area problem involves a procedure called **integration** in which quantities such as area, average value, present value of an income stream, and blood flow rate are computed as a special kind of limit of a sum. We shall study this procedure in Chapters 5 and 6. The tangent problem is closely related to the study of rates of change, and we will begin our study of calculus by examining this connection.

Slope and Rates of Change

Recall from Section 1.3 that a linear function $L(x) = mx + b$ changes at a constant rate m with respect to the independent variable x. That is, the rate of change of $L(x)$ is given by the slope or steepness of its graph, the line $y = mx + b$ (Figure 2.1a). For a function $f(x)$ that is not linear, the rate of change is not constant but varies with x. In particular, when $x = c$, the rate is given by the steepness of the graph of $f(x)$ at the point $P(c, f(c))$, which can be measured by the slope of the *tangent line* to the graph at P (Figure 2.1b). The relationship between rate of change and slope is illustrated in Example 2.1.1.

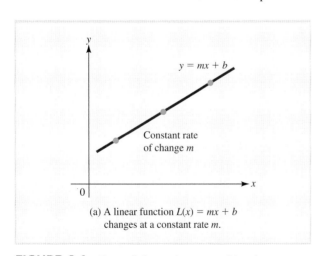

(a) A linear function $L(x) = mx + b$
changes at a constant rate m.

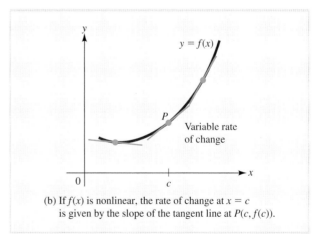

(b) If $f(x)$ is nonlinear, the rate of change at $x = c$
is given by the slope of the tangent line at $P(c, f(c))$.

FIGURE 2.1   Rate of change is measured by slope.

EXAMPLE 2.1.1

The graph in Figure 2.2 gives the relationship between the percentage of unemployment U and the corresponding percentage of inflation I. Use the graph to estimate the rate at which I changes with respect to U when the level of unemployment is 3% and again when it is 10%.

Solution

From the figure, we estimate the slope of the tangent line at the point (3, 14), corresponding to $U = 3$, to be approximately -14. That is, when unemployment is 3%, inflation I is *decreasing* at the rate of 14 percentage points for each percentage point increase in unemployment U.

At the point (10, -5), the slope of the tangent line is approximately -0.4, which means that when there is 10% unemployment, inflation is decreasing at a rate of only 0.4 percentage points for each percentage point increase in unemployment.

In Example 2.1.2, we show how slope and rate of change can be computed analytically using a limit.

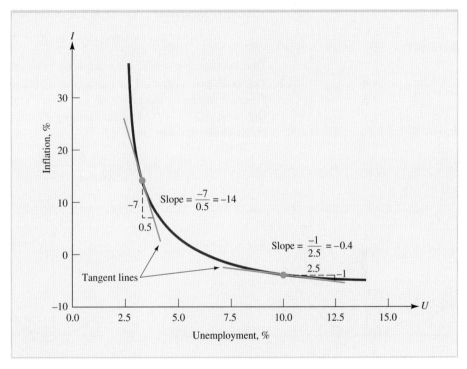

FIGURE 2.2   Inflation as a function of unemployment.

SOURCE: Adapted from Robert Eisner, *The Misunderstood Economy: What Counts and How to Count It,* Boston, MA: Harvard Business School Press, 1994, p. 173.

EXAMPLE 2.1.2

In a windsurfing competition, one windsurfer turns and then moves in a straight line. He clicks his GPS and finds that after t seconds he has moved $s(t) = 5t^2$ metres.

  a. What is the windsurfer's velocity after $t = 2$ s?

  b. How is the velocity found in part (a) related to the graph of $s(t)$?

Solution

  a. The velocity after 2 seconds is the *instantaneous* rate of change of $s(t)$ at $t = 2$. Unless the windsurfer has a speedometer, it is hard to simply "read" his velocity.

(Photo: © CORBIS/Royalty Free)

However, we can measure the distance he moves as time t changes by a small (nonzero) amount h from $t = 2$ to $t = 2 + h$ and then compute the *average* rate of change of $s(t)$ over the time period $[2, 2 + h]$ by the ratio

$$
\begin{aligned}
v_{\text{ave}} &= \frac{\text{distance travelled}}{\text{elapsed time}} = \frac{s(2 + h) - s(2)}{(2 + h) - 2} \\
&= \frac{5(2 + h)^2 - 5(2)^2}{h} = \frac{5(4 + 4h + h^2) - 5(4)}{h} \\
&= \frac{20h + 5h^2}{h} = 20 + 5h
\end{aligned}
$$

Since the elapsed time h is small, we would expect the average velocity v_{ave} to closely approximate the instantaneous ("speedometer") velocity v_{ins} at time $t = 2$. Thus, it is reasonable to compute the instantaneous velocity by the limit

$$
v_{\text{ins}} = \lim_{h \to 0} v_{\text{ave}} = \lim_{h \to 0}(20 + 5h) = 20.
$$

That is, after 2 seconds, the windsurfer is travelling at a rate of 20 m/s.

b. The procedure described in part (a) is represented geometrically in Figure 2.3. Figure 2.3a shows the graph of $s(t) = 5t^2$, along with the points $P(2, 20)$ and $Q(2 + h, 5(2 + h)^2)$. The line joining P and Q is called a *secant line* of the graph of $s(t)$ and has slope

$$
m_{\text{sec}} = \frac{5(2 + h)^2 - 20}{(2 + h) - 2} = 20 + 5h.
$$

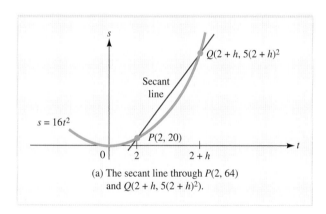

(a) The secant line through $P(2, 64)$ and $Q(2 + h, 5(2 + h)^2)$.

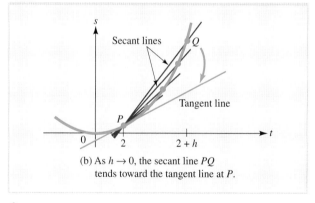

(b) As $h \to 0$, the secant line PQ tends toward the tangent line at P.

FIGURE 2.3   Computing the slope of the tangent line to $s = 5t^2$ at $P(2, 20)$.

As indicated in Figure 2.3b, when we take smaller and smaller values of h, the corresponding secant lines PQ tend toward the position of what we intuitively think of as the tangent line to the graph of $s(t)$ at P. This suggests that we can compute the slope m_{tan} of the tangent line by finding the limiting value of m_{sec} as h tends toward 0; that is,

$$
m_{\text{tan}} = \lim_{h \to 0} m_{\text{sec}} = \lim_{h \to 0}(20 + 5h) = 20.
$$

Thus, the slope of the tangent line to the graph of $s(t) = 5t^2$ at the point where $t = 2$ is the same as the instantaneous rate of change of $s(t)$ with respect to t when $t = 2$.

The procedure used in Example 2.1.2 to find the velocity of a windsurfer can be used to find other rates of change. Suppose we wish to find the rate at which the function $f(x)$ is changing with respect to x when $x = c$. We begin by finding the **average rate of change** of $f(x)$ as x varies from $x = c$ to $x = c + h$, which is given by the ratio

$$\text{Rate}_{\text{ave}} = \frac{\text{change in } f(x)}{\text{change in } x} = \frac{f(c + h) - f(c)}{(c + h) - c}$$

$$= \frac{f(c + h) - f(c)}{h}$$

This ratio can be interpreted geometrically as the slope of the secant line from the point $P(c, f(c))$ to the point $Q(c + h, f(c + h))$, as shown in Figure 2.4a.

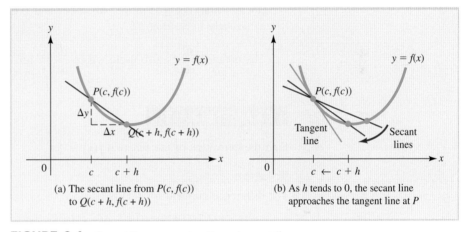

(a) The secant line from $P(c, f(c))$ to $Q(c + h, f(c + h))$

(b) As h tends to 0, the secant line approaches the tangent line at P

FIGURE 2.4 Secant lines approximating a tangent line.

We then compute the instantaneous rate of change of $f(x)$ at $x = c$ by finding the limiting value of the average rate as h tends to 0; that is,

$$\text{Rate}_{\text{ins}} = \lim_{h \to 0} \text{Rate}_{\text{ave}} = \lim_{h \to 0} \frac{f(c + h) - f(c)}{h}.$$

This limit also gives the slope of the tangent line to the curve $y = f(x)$ at the point $P(c, f(c))$, as indicated in Figure 2.4b.

The Derivative The expression

$$\frac{f(x + h) - f(x)}{h}$$

is called the **difference quotient** for the function $f(x)$, and we have just seen that rates of change and slope can both be computed by finding the limit of an appropriate difference quotient as h tends to 0. To unify the study of these and other applications that involve taking the limit of a difference quotient, we introduce the following terminology and notation.

The Derivative of a Function ■ The **derivative** of the function $f(x)$ with respect to x is the function $f'(x)$ given by

$$f'(x) = \lim_{h \to 0} \frac{f(x + h) - f(x)}{h}$$

(read $f'(x)$ as "f prime of x"). The process of computing the derivative is called **differentiation,** and we say that $f(x)$ is **differentiable** at $x = c$ if $f'(c)$ exists, that is, if the limit that defines $f'(x)$ exists when $x = c$.

NOTE We use h to represent a small change in x, the independent variable, in difference quotients to simplify algebraic computations. However, when it is important to emphasize that, say, the variable x is being incremented, we will denote the increment by Δx (read as "delta x"). Similarly, Δt and Δs denote small (incremental) changes in the variables t and s, respectively. This notation is used extensively in Section 2.5.

The notation $\dfrac{df}{dx}$ is also used, where $\dfrac{df}{dx} = \lim_{\Delta x \to 0} \dfrac{\Delta y}{\Delta x}$.

> **EXAMPLE 2.1.3**

Find the derivative of the function $f(x) = 5x^2$.

Solution

The difference quotient for $f(x)$ is

$$\frac{f(x + h) - f(x)}{h} = \frac{5(x + h)^2 - 5x^2}{h}$$

$$= \frac{5(x^2 + 2hx + h^2) - 5x^2}{h}$$

$$= \frac{10hx + 5h^2}{h} \qquad \text{combine terms}$$

$$= 10x + 5h \qquad \text{cancel common } h \text{ terms}$$

Thus, the derivative of $f(x) = 5x^2$ is the function

$$f'(x) = \lim_{h \to 0} \frac{f(x + h) - f(x)}{h} = \lim_{h \to 0}(10x + 5h)$$

$$= 10x$$

Once we have computed the derivative of a function $f(x)$, it can be used to streamline any computation involving the limit of the difference quotient of $f(x)$ as h tends to 0. For instance, notice that the function in Example 2.1.3 is essentially the same as the distance function $s = 5t^2$ that appears in the windsurfer problem in Example 2.1.2. Using the result of Example 2.1.3, the velocity of the windsurfer at time $t = 2$ can now be found by simply substituting $t = 2$ into the formula for the derivative $s'(t)$:

$$\text{Velocity} = s'(2) = 10(2) = 20.$$

Likewise, the slope of the tangent line to the graph of $s(t)$ at the point $P(2, 20)$ is given by

$$\text{Slope} = s'(2) = 20.$$

For future reference, our observations about rates of change and slope may be summarized as follows in terms of the derivative notation.

Slope as a Derivative ■ The slope of the tangent line to the curve $y = f(x)$ at the point $(c, f(c))$ is $m_{\tan} = f'(c)$.

Instantaneous Rate of Change as a Derivative ■ The rate of change of $f(x)$ with respect to x when $x = c$ is given by $f'(c)$.

In Example 2.1.4, we find the equation of a tangent line. Then, in Example 2.1.5, we consider a business application involving a rate of change.

EXAMPLE 2.1.4

First compute the derivative of $f(x) = x^3$ and then use it to find the slope of the tangent line to the curve $y = x^3$ at the point where $x = -1$. What is the equation of the tangent line at this point?

Solution

According to the definition of the derivative,

$$
\begin{aligned}
f'(x) &= \lim_{h \to 0} \frac{f(x + h) - f(x)}{h} \\
&= \lim_{h \to 0} \frac{(x + h)^3 - x^3}{h} \\
&= \lim_{h \to 0} \frac{(x^3 + 3x^2h + 3xh^2 + h^3) - x^3}{h} \\
&= \lim_{h \to 0} (3x^2 + 3xh + h^2) \\
&= 3x^2
\end{aligned}
$$

Thus, the slope of the tangent line to the curve $y = x^3$ at the point where $x = -1$ is $f'(-1) = 3(-1)^2 = 3$ (Figure 2.5). To find the equation of the tangent line, we also need the y coordinate of the point of tangency, namely, $y = (-1)^3 = -1$. Therefore, the tangent line passes through the point $(-1, -1)$ with slope 3. By applying the point-slope formula, we obtain the equation

$$
\begin{aligned}
y - (-1) &= 3[x - (-1)] \\
y &= 3x + 2
\end{aligned}
$$

Just-In-Time

Recall that $(a + b)^3 = a^3 + 3a^2b + 3ab^2 + b^3$. This is a special case of the binomial theorem for the exponent 3 and is used to expand the numerator of the difference quotient in Example 2.1.4.

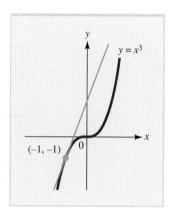

FIGURE 2.5 The graph of $y = x^3$.

Just-In-Time

The point-slope formula $y - y_0 = m(x - x_0)$ is a very important one when using derivatives to find the equations of tangent lines. Keep it handy in your mind!

EXAMPLE 2.1.5

A manufacturer determines that when x thousand units of a particular commodity are produced, the profit generated will be

$$P(x) = -400x^2 + 6800x - 12\,000$$

dollars. At what rate is profit changing with respect to the level of production x when 9000 units are produced?

Solution

We find that

$$P'(x) = \lim_{h \to 0} \frac{P(x+h) - P(x)}{h}$$

$$= \lim_{h \to 0} \frac{[-400(x+h)^2 + 6800(x+h) - 12\,000] - (-400x^2 + 6800x - 12\,000)}{h}$$

$$= \lim_{h \to 0} \frac{-400h^2 - 800hx + 6800h}{h} \quad \text{expand } (x+h)^2 \text{ and combine terms}$$

$$= \lim_{h \to 0} (-400h - 800x + 6800) \quad \text{cancel common } h \text{ terms}$$

$$= -800x + 6800$$

Thus, when the level of production is $x = 9$ (9000 units), the profit is changing at a rate of

$$P'(9) = -800(9) + 6800 = -400$$

in dollars per thousand units.

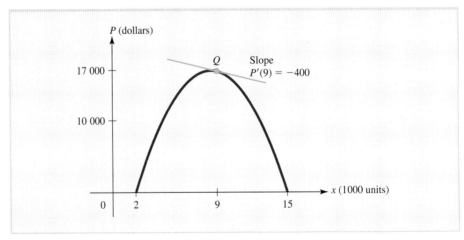

FIGURE 2.6   The graph of the profit function $P(x) = -400x^2 + 6800x - 12\,000$.

In Example 2.1.5, we found that $P'(9) = -400$, which means that the tangent line to the profit curve $y = P(x)$ is sloped *downward* at the point Q where $x = 9$, as shown in Figure 2.6. There is a decrease in profits per unit change in production. Since the tangent line at Q slopes downward, the profit curve must be falling at Q, so profit must be *decreasing* when 9000 units are being produced.

Compare the profits when 8500 units are produced and when 9000 units are produced. We have $P(8.5) = \$16\ 900$ and $P(9) = \$16\ 800$. The profit is decreasing. However, the derivative is a constantly decreasing function of x and the rate of decrease changes as the slope of the tangent changes. This can be seen in the parabola of Figure 2.6.

The significance of the sign of the derivative is summarized in the following box and in Figure 2.7. We will have much more to say about the relationship between the shape of a curve and the sign of derivatives in Chapter 3, where we develop a general procedure for curve sketching.

Significance of the Sign of the Derivative $f'(x)$ ■ If the function f is differentiable at $x = c$, then

$$f \text{ is } \textbf{increasing} \text{ at } x = c \text{ if } f'(c) > 0$$

and

$$f \text{ is } \textbf{decreasing} \text{ at } x = c \text{ if } f'(c) < 0.$$

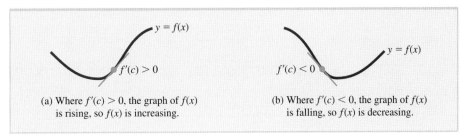

(a) Where $f'(c) > 0$, the graph of $f(x)$ is rising, so $f(x)$ is increasing.

(b) Where $f'(c) < 0$, the graph of $f(x)$ is falling, so $f(x)$ is decreasing.

FIGURE 2.7   The significance of the sign of the derivative $f'(c)$.

Derivative Notation   The derivative $f'(x)$ of $y = f(x)$ is sometimes written as $\dfrac{dy}{dx}$ or $\dfrac{df}{dx}$ (read as "dee y by dee x" or "dee f by dee x"). In this notation, the value of the derivative at $x = c$ (that is, $f'(c)$) is written as

$$\left.\frac{dy}{dx}\right|_{x=c} \quad \text{or} \quad \left.\frac{df}{dx}\right|_{x=c}.$$

For example, if $y = x^2$, then

$$\frac{dy}{dx} = 2x,$$

and the value of this derivative at $x = -3$ is

$$\left.\frac{dy}{dx}\right|_{x=-3} = 2x\big|_{x=-3} = 2(-3) = -6.$$

The $\dfrac{dy}{dx}$ notation for derivative suggests slope, $\dfrac{\Delta y}{\Delta x}$, and can also be thought of as the rate of change of y with respect to x. Sometimes it is convenient to condense a statement such as "when $y = x^2$, then $\dfrac{dy}{dx} = 2x$" by writing simply

$$\frac{d}{dx}(x^2) = 2x,$$

which reads "the derivative of x^2 with respect to x is $2x$."

Example 2.1.6 illustrates how the different notational forms for the derivative can be used.

Just-In-Time

Multiplying out,
$(\sqrt{x+h} - \sqrt{x})(\sqrt{x+h} - \sqrt{x}) = ((\sqrt{x+h})^2 - (\sqrt{x})^2)$, which can also be seen from the difference of squares formula. This simplifies to $x + h - x$, a much easier expression to work with.

EXAMPLE 2.1.6

First compute the derivative of $f(x) = \sqrt{x}$ and then use it to do the following:

a. Find the equation of the tangent line to the curve $y = \sqrt{x}$ at the point where $x = 4$.

b. Find the rate at which $y = \sqrt{x}$ is changing with respect to x when $x = 1$.

Solution

The derivative of $y = \sqrt{x}$ with respect to x is given by

$$\frac{d}{dx}(\sqrt{x}) = \lim_{h \to 0} \frac{f(x + h) - f(x)}{h}$$

$$= \lim_{h \to 0} \frac{\sqrt{x + h} - \sqrt{x}}{h}$$

$$= \lim_{h \to 0} \frac{(\sqrt{x + h} - \sqrt{x})(\sqrt{x + h} + \sqrt{x})}{h(\sqrt{x + h} + \sqrt{x})}$$

$$= \lim_{h \to 0} \frac{x + h - x}{h(\sqrt{x + h} + \sqrt{x})}$$

$$= \lim_{h \to 0} \frac{h}{h(\sqrt{x + h} + \sqrt{x})}$$

$$= \lim_{h \to 0} \frac{1}{\sqrt{x + h} + \sqrt{x}}$$

$$= \frac{1}{2\sqrt{x}} \quad \text{if } x > 0$$

a. When $x = 4$, the corresponding y coordinate on the graph of $f(x) = \sqrt{x}$ is $y = \sqrt{4} = 2$, so the point of tangency is $P(4, 2)$. Since $f'(x) = \dfrac{1}{2\sqrt{x}}$, the slope of the tangent line to the graph of $f(x)$ at the point $P(4, 2)$ is given by

$$f'(4) = \frac{1}{2\sqrt{4}} = \frac{1}{4},$$

and by substituting into the point-slope formula, we find that the equation of the tangent line at P is

$$y - 2 = \frac{1}{4}(x - 4)$$

$$y = \frac{1}{4}x + 1$$

b. The rate of change of $y = \sqrt{x}$ when $x = 1$ is

$$\left.\frac{dy}{dx}\right|_{x=1} = \frac{1}{2\sqrt{1}} = \frac{1}{2}.$$

NOTE   Notice that the function $f(x) = \sqrt{x}$ in Example 2.1.6 is defined at $x = 0$ but its derivative $f'(x) = \dfrac{1}{2\sqrt{x}}$ is not. This example shows that a function and its derivative do not always have the same domain. ▪

Differentiability and Continuity

If a function $f(x)$ is differentiable where $x = c$, then the graph of $y = f(x)$ has a non-vertical tangent line at the point $P(c, f(c))$ and at all points on the graph that are "near" P. We would expect such a function to be continuous at $x = c$ since a graph with a tangent line at the point P certainly cannot have a hole or gap at P. To summarize:

> **Continuity of a Differentiable Function** ▪ If the function $f(x)$ is differentiable at $x = c$, then it is also continuous at $x = c$.

Verification of this observation is outlined in Exercise 62. Notice that we are *not* claiming that a continuous function must be differentiable. Indeed it can be shown that a continuous function $f(x)$ will not be differentiable at $x = c$ if $f'(x)$ becomes infinite at $x = c$ or if the graph of $f(x)$ has a sharp point at $P(c, f(c))$; that is, a point where the curve makes an abrupt change in direction. If $f(x)$ is continuous at $x = c$ but $f'(c)$ is infinite, the graph of f may have a vertical tangent at the point $P(c, f(c))$ (Figure 2.8a) or a cusp at P (Figure 2.8b). The absolute value function $f(x) = |x|$ is continuous for all x but has a sharp point at the origin $(0, 0)$ (see Figure 2.8c and Exercise 61). Another graph with a sharp point is shown in Figure 2.8d.

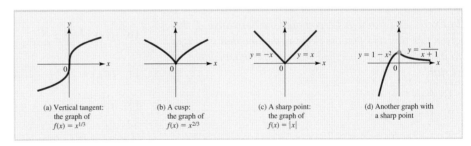

(a) Vertical tangent: the graph of $f(x) = x^{1/3}$

(b) A cusp: the graph of $f(x) = x^{2/3}$

(c) A sharp point: the graph of $f(x) = |x|$

(d) Another graph with a sharp point

FIGURE 2.8   Graphs of four continuous functions that are not differentiable at $x = 0$.

In general, the functions you encounter in this text will be differentiable at almost all points. In particular, polynomials are everywhere differentiable and rational functions are differentiable wherever they are defined.

An example of a practical situation involving a continuous function that is not always differentiable is provided by the circulation of blood in the human body.* It

---

*www.cvphysiology.com/Blood%20Pressure/BP003.htm.

is tempting to think of blood coursing smoothly through the veins and arteries in a constant flow, but in actuality, blood is ejected by the heart into the arteries in discrete bursts. This results in an *arterial pulse*, which can be used to measure the heart rate by applying pressure to an accessible artery, as at the wrist. A blood pressure curve showing a pulse is sketched in Figure 2.9. Notice how the curve quickly changes direction at the minimum (*diastolic*) pressure, as a burst of blood is injected by the heart into the arteries, causing the blood pressure to rise rapidly until the maximum (*systolic*) pressure is reached, after which the pressure gradually declines as blood is distributed to tissue by the arteries. The blood pressure function is continuous but not differentiable at $t = 0.75$ seconds, where the infusion of blood occurs.

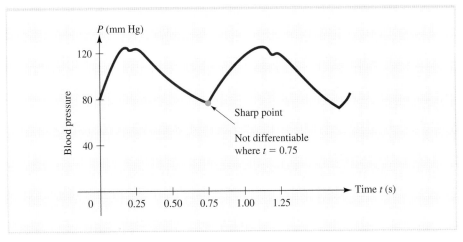

FIGURE 2.9   Graph of blood pressure showing an arterial pulse.

EXERCISES ■ 2.1

In Exercises 1 through 12, compute the derivative of the given function and find the slope of the line that is tangent to its graph for the specified value of the independent variable.

1. $f(x) = 4; x = 0$

2. $f(x) = -3; x = 1$

3. $f(x) = 5x - 3; x = 2$

4. $f(x) = 2 - 7x; x = -1$

5. $f(x) = 2x^2 - 3x - 5; x = 0$

6. $f(x) = x^2 - 1; x = -1$

7. $f(x) = x^3 - 1; x = 2$

8. $f(x) = -x^3; x = 1$

9. $g(t) = \dfrac{2}{t}; t = \dfrac{1}{2}$

10. $f(x) = \dfrac{1}{x^2}; x = 2$

11. $H(u) = \dfrac{1}{\sqrt{u}}; u = 4$

12. $f(x) = \sqrt{x}; x = 9$

In Exercises 13 through 24, compute the derivative of the given function and find the equation of the line that is tangent to its graph for the specified value $x = c$.

13. $f(x) = 2; c = 13$

14. $f(x) = 3; c = -4$

15. $f(x) = 7 - 2x; c = 5$

16. $f(x) = 3x; c = 1$

17. $f(x) = x^2; c = 1$

18. $f(x) = 2 - 3x^2; c = 1$

19. $f(x) = \dfrac{-2}{x}; c = -1$

20. $f(x) = \dfrac{3}{x^2}; c = \dfrac{1}{2}$

21. $f(x) = 2\sqrt{x}; c = 4$

22. $f(x) = \dfrac{1}{\sqrt{x}}; c = 1$

23. $f(x) = \dfrac{1}{x^3}; c = 1$

24. $f(x) = x^3 - 1; c = 1$

In Exercises 25 through 32, find the rate of change $\dfrac{dy}{dx}$ where $x = x_0$.

25. $y = 3; x_0 = 2$

26. $y = -17; x_0 = 14$

27. $y = 3x + 5; x_0 = -1$

28. $y = 6 - 2x; x_0 = 3$

29. $y = x(1 - x); x_0 = -1$

30. $y = x^2 - 2x; x_0 = 1$

31. $y = x - \dfrac{1}{x}; x_0 = 1$

32. $y = \dfrac{1}{2 - x}; x_0 = -3$

33. Let $f(x) = x^2$.
   a. Compute the slope of the secant line joining the points on the graph of f whose x coordinates are $x = -2$ and $x = -1.9$.
   b. Use calculus to compute the slope of the line that is tangent to the graph when $x = -2$ and compare with the slope found in part (a).

34. Let $f(x) = 2x - x^2$.
   a. Compute the slope of the secant line joining the points where $x = 0$ and $x = \dfrac{1}{2}$.
   b. Use calculus to compute the slope of the tangent line to the graph of $f(x)$ at $x = 0$ and compare with the slope found in part (a).

35. Let $f(x) = x^3$.
   a. Compute the slope of the secant line joining the points on the graph of f whose x coordinates are $x = 1$ and $x = 1.1$.
   b. Use calculus to compute the slope of the line that is tangent to the graph when $x = 1$ and compare with the slope found in part (a).

36. Let $f(x) = \dfrac{x}{x - 1}$.
   a. Compute the slope of the secant line joining the points where $x = -1$ and $x = -\dfrac{1}{2}$.
   b. Use calculus to compute the slope of the tangent line to the graph of $f(x)$ at $x = -1$ and compare with the slope found in part (a).

37. Let $f(x) = 3x^2 - x$.
   a. Find the average rate of change of $f(x)$ with respect to x as x changes from $x = 0$ to $x = \dfrac{1}{16}$.
   b. Use calculus to find the instantaneous rate of change of $f(x)$ at $x = 0$ and compare with the average rate found in part (a).

38. Let $f(x) = x(1 - 2x)$.
   a. Find the average rate of change of $f(x)$ with respect to x as x changes from $x = 0$ to $x = \dfrac{1}{2}$.
   b. Use calculus to find the instantaneous rate of change of $f(x)$ at $x = 0$ and compare with the average rate found in part (a).

39. Let $s(t) = \dfrac{t - 1}{t + 1}$.
   a. Find the average rate of change of $s(t)$ with respect to t as t changes from $t = -\dfrac{1}{2}$ to $t = 0$.
   b. Use calculus to find the instantaneous rate of change of $s(t)$ at $t = -\dfrac{1}{2}$ and compare with the average rate found in part (a).

40. Let $s(t) = \sqrt{t}$.
   a. Find the average rate of change of $s(t)$ with respect to t as t changes from $t = 1$ to $t = \dfrac{1}{4}$.
   b. Use calculus to find the instantaneous rate of change of $s(t)$ at $t = 1$ and compare with the average rate found in part (a).

41. Complete the table by filling in the missing interpretations using the example as a guide.

	If $f(t)$ represents . . .	then $\dfrac{f(t_0 + h) - f(t_0)}{h}$ represents . . . and	$\lim\limits_{h \to 0} \dfrac{f(t_0 + h) - f(t_0)}{h}$ represents . . .
Example	The number of bacteria in a colony at time t	The average rate of change of the bacteria population during the time interval $[t_0, t_0 + h]$	The instantaneous rate of change of the bacteria population at time $t = t_0$
	a. The temperature in Edmonton t hours after midnight on a certain day		
	b. The blood alcohol level t hours after a person consumes an ounce of alcohol		
	c. The 30-year fixed mortgage rate t years after 2005		

42. Complete the table by filling in the missing interpretations using the example as a guide.

	If $f(x)$ represents . . .	then $\dfrac{f(x_0 + h) - f(x_0)}{h}$ represents . . . and	$\lim\limits_{h \to 0} \dfrac{f(x_0 + h) - f(x_0)}{h}$ represents . . .
Example	The cost of producing x units of a particular commodity	The average rate of change of cost as production changes from x_0 to $x_0 + h$ units	The instantaneous rate of change of cost with respect to production level when x_0 units are produced
	a. The revenue obtained when x units of a particular commodity are produced		
	b. The amount of unexpended fuel (in kilograms) left in a rocket when it is x metres above the ground		
	c. The volume (in cubic centimetres) of a peach 4 weeks after watering the tree with x litres of fertilizer		

43. RENEWABLE RESOURCES

a. The accompanying graph shows how the volume V of lumber in a tree varies with time t (the age of the tree). Use the graph to estimate the rate at which V is changing with respect to time when $t = 30$ years. What seems to be happening to the rate of change of V as t increases without bound (that is, in the long run)?

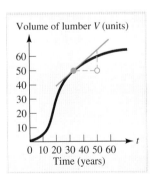

EXERCISE 43(a)   Graph showing how the volume V of lumber in a tree varies with time t.
SOURCE: Adapted from Robert H. Frank, *Microeconomics and Behavior*, 2nd ed., New York: McGraw-Hill, 1994, p. 623.

  b. Look at the picture from the British Columbia Forestry Department. Write about the reasons that trees in this situation should be cut at 80 years.

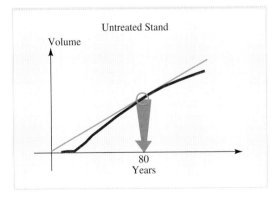

EXERCISE 43(b)
SOURCE: BC Ministry of Forests—Research Branch— Forest Productivity Section, http://www.for.gov.bc.ca/hre/ syldemo/syl2vol.htm.

44. POPULATION GROWTH   The accompanying graph shows how a population P of fruit flies (*Drosophila*) changes with time t (days) during an experiment. Use the graph to estimate the rate at which the population is growing after 20 days and also after 45 days. At what time is the population growing at the greatest rate?

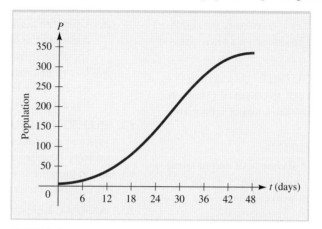

EXERCISE 44   Growth curve for a population of fruit flies.
SOURCE: Adapted from E. Batschelet, *Introduction to Mathematics for Life Scientists,* 3rd ed., New York: Springer-Verlag, 1979, p. 355.

45. THERMAL INVERSION   Air temperature usually decreases with increasing altitude. However, thanks to a phenomenon called *thermal inversion,* the temperature of air warmed by the sun may rise above the temperature at lower altitudes. This causes pollution to stay trapped at low altitudes. Use the accompanying graph to estimate the rate at which temperature T is changing with respect to altitude h at altitudes of 1000 m and 2000 m.

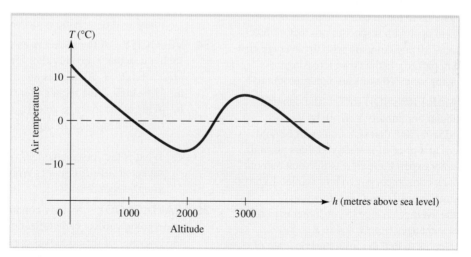

EXERCISE 45   Temperature change with altitude during a temperature inversion.
SOURCE: E. Batschelet, *Introduction to Mathematics for Life Scientists,* 3rd ed., New York: Springer-Verlag, 1979, p. 150.

46. PROFIT   A manufacturer can produce MP3 players at a cost of $50 each. It is estimated that if these are sold for p dollars each, consumers will buy $q = 120 - p$ each month.
   a.   Express the manufacturer's profit P as a function of q.
   b.   What is the average rate of profit obtained as the level of production increases from $q = 0$ to $q = 20$?
   c.   At what rate is profit changing when $q = 20$ MP3 players are produced? Is the profit increasing or decreasing at this level of production?

47. PROFIT A manufacturer determines that when x hundred units of a particular commodity are produced, the profit will be $P(x) = 4000(15 - x)(x - 2)$ dollars.
  a. Find $P'(x)$.
  b. Determine where $P'(x) = 0$. What is the significance of the level of production x_m where this occurs?

48. MANUFACTURING OUTPUT At a certain factory, it is determined that an output of Q units is to be expected when L worker-hours of labour are employed, where

$$Q(L) = 3100 \sqrt{L}.$$

  a. Find the average rate of change of output as the labour employment changes from $L = 3025$ worker-hours to 3100 worker-hours.
  b. Use calculus to find the instantaneous rate of change of output with respect to labour level when $L = 3025$.

49. COST OF PRODUCTION A business manager determines that the cost of producing x units of a particular commodity is C thousand dollars, where

$$C(x) = 0.04x^2 + 5.1x + 40.$$

  a. Find the average cost as the level of production changes from $x = 10$ to $x = 11$ units.
  b. Use calculus to find the instantaneous rate of change of cost with respect to production level when $x = 10$ and compare with the average cost found in part (a). Is the cost increasing or decreasing when 10 units are being produced?

50. CONSUMER EXPENDITURE The demand for a particular commodity is given by $D(x) = -35x + 200$; that is, x units will be sold (demanded) at a price of $p = D(x)$ dollars per unit.
  a. Consumer expenditure $E(x)$ is the total amount of money consumers pay to buy x units. Express consumer expenditure E as a function of x.
  b. Find the average change in consumer expenditure as x changes from $x = 4$ to $x = 5$.
  c. Use calculus to find the instantaneous rate of change of expenditure with respect to x when $x = 4$. Is expenditure increasing or decreasing when $x = 4$?

51. UNEMPLOYMENT In economics, the graph in Figure 2.2 on page 99 is called the **Phillips curve**, after A. W. Phillips, a New Zealander associated with the London School of Economics. Until Phillips published his ideas in the 1950s, many economists believed that unemployment and inflation were linearly related. Consider two possible linear relationships between these variables.
  a. Using the tangent line at 3% unemployment to represent the linear relationship, find the corresponding inflation rate at 5% unemployment.
  b. Using the tangent line at 10% to represent the linear relationship, find the corresponding inflation rate at 5% unemployment.

52. BLOOD PRESSURE Refer to the graph of blood pressure as a function of time shown in Figure 2.9 on page 108. Estimate the average rate of change in blood pressure over the time periods [0.7, 0.75] and [0.75, 0.8]. Interpret your results.

53. ANIMAL BEHAVIOUR Experiments indicate that when a flea jumps, its height (in metres) after t seconds is given by the function

$$H(t) = 4.4t - 4.9t^2.$$

  a. Find $H'(t)$. At what rate is $H(t)$ changing after 1 second? Is it increasing or decreasing?
  b. At what time is $H'(t) = 0$? What is the significance of this time?
  c. When does the flea "land" (return to its initial height)? At what rate is $H(t)$ changing at this time? Is it increasing or decreasing?

54. VELOCITY A toy rocket rises vertically in such a way that t seconds after lift-off, it is $h(t) = -4.9t^2 + 70t$ metres above the ground.
  a. How high is the rocket after 6 seconds?
  b. What is the average velocity of the rocket over the first 6 seconds of flight (between $t = 0$ and $t = 6$)?
  c. What is the (instantaneous) velocity of the rocket at lift-off ($t = 0$)? What is its velocity after 6 seconds?

55. CARDIOLOGY A study conducted on a patient undergoing cardiac catheterization indicated that the diameter of the aorta was approximately D millimetres when the aortic pressure was p millimetres of mercury, where

$$D(p) = -0.0009p^2 + 0.13p + 17.81,$$

for $50 \leq p \leq 120$.
  a. Find the average rate of change of the aortic diameter D as p changes from $p = 60$ to $p = 61$.
  b. Use calculus to find the instantaneous rate of change of diameter D with respect to aortic

pressure p when $p = 60$. Is the pressure increasing or decreasing when $p = 60$?

c. For what value of p is the instantaneous rate of change of D with respect to p equal to 0? What is the significance of this pressure?

56. a. Find the derivative of the linear function $f(x) = 3x - 2$.

b. Find the equation of the tangent line to the graph of this function at the point where $x = -1$.

c. Explain how the answers to parts (a) and (b) could have been obtained from geometric considerations with no calculation whatsoever.

57. a. Find the derivatives of the functions $y = x^2$ and $y = x^2 - 3$ and account geometrically for their similarity.

b. Without further computation, find the derivative of the function $y = x^2 + 5$.

58. a. Find the derivative of the function $y = x^2 + 3x$.

b. Find the derivatives of the functions $y = x^2$ and $y = 3x$ separately.

c. How is the derivative in part (a) related to those in part (b)?

d. In general, if $f(x) = g(x) + h(x)$, what would you guess is the relationship between the derivative of f and those of g and h?

59. a. Compute the derivatives of the functions $y = x^2$ and $y = x^3$.

b. Examine your answers in part (a). Can you detect a pattern? What do you think is the derivative of $y = x^4$? How about the derivative of $y = x^{27}$?

60. Use calculus to prove that if $y = mx + b$, the rate of change of y with respect to x is constant.

61. Let f be the absolute value function; that is,

$$f(x) = \begin{cases} x & \text{if } x \geq 0 \\ -x & \text{if } x < 0 \end{cases}$$

Show that

$$f'(x) = \begin{cases} 1 & \text{if } x > 0 \\ -1 & \text{if } x < 0 \end{cases}$$

and explain why f is not differentiable at $x = 0$.

62. A* Let f be a function that is differentiable at $x = c$.

a. Explain why

$$f'(c) = \lim_{x \to c} \frac{f(x) - f(c)}{x - c}.$$

b. Use the result of part (a) together with the fact that

$$f(x) - f(c) = \left[\frac{f(x) - f(c)}{x - c} \right](x - c)$$

to show that

$$\lim_{x \to c} [f(x) - f(c)] = 0.$$

c. Explain why the result obtained in part (b) shows that f is continuous at $x = c$.

63. A* Show by plotting a graph that $f(x) = \dfrac{|x^2 - 1|}{x - 1}$ is not differentiable at $x = 1$.

64. A* Find the x values at which the peaks and valleys of the graph of $y = 2x^3 - 0.8x^2 + 4$ occur on a graph of the function. Use four decimal places.

SECTION 2.2

L02

Calculate derivatives using derivative rules.

Techniques of Differentiation

If we had to use the limit definition every time we wanted to compute a derivative, it would be both tedious and difficult to use calculus in applications. Fortunately, this is not necessary, and in this section and in Section 2.3 we develop techniques that greatly simplify the process of differentiation. We begin with a rule for the derivative of a constant.

The Constant Rule ■ For any constant c,

$$\frac{d}{dx}[c] = 0.$$

That is, the derivative of a constant is zero.

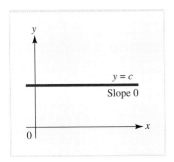

FIGURE 2.10   The graph of $f(x) = c$.

You can see this by considering the graph of a constant function $f(x) = c$, which is a horizontal line (see Figure 2.10). Since the slope of such a line is 0 at all its points, it follows that $f'(x) = 0$. Here is a proof using the limit definition:

$$f'(x) = \lim_{h \to 0} \frac{f(x + h) - f(x)}{h}$$

$$= \lim_{h \to 0} \frac{c - c}{h} = 0 \qquad \text{since } f(c + h) = c \text{ for all } x$$

EXAMPLE 2.2.1

$$\frac{d}{dx}[-15] = 0.$$

The next rule is one of the most useful because it tells you how to take the derivative of any power function $f(x) = x^n$. Note that the rule applies not only to functions like $f(x) = x^5$ but also to $g(x) = \sqrt[5]{x^4} = x^{4/5}$ and $h(x) = \dfrac{1}{x^3} = x^{-3}$. Functions such as $x^{\sqrt{2}}$ can also be differentiated using the power rule, but we do not define what is meant by such functions until Chapter 4.

The Power Rule   ■   For any real number n,

$$\frac{d}{dx}[x^n] = nx^{n-1}.$$

In words, to find the derivative of x^n, reduce the exponent of n by 1 and multiply the new power of x by the original exponent.

Just-In-Time

We will use the two forms of square roots, $\sqrt{x} = x^{1/2}$ and $\dfrac{1}{\sqrt{x}} = x^{-1/2}$, often in calculus. We use the fractional exponent form for the power rule and the square root form for calculations.

According to the power rule, the derivative of $y = x^3$ is $\dfrac{d}{dx}(x^3) = 3x^2$, which agrees with the result found directly in Example 2.1.4 of Section 2.1. You can use the power rule to differentiate radicals and reciprocals by first converting them to power functions with fractional and negative exponents, respectively. (You can find a review of exponential notation in Appendix A1 at the back of the book.) For example, recall that $\sqrt{x} = x^{1/2}$, so the derivative of $y = \sqrt{x}$ is

$$\frac{d}{dx}(\sqrt{x}) = \frac{d}{dx}(x^{1/2}) = \frac{1}{2}x^{-1/2},$$

which agrees with the result of Example 2.1.6 of Section 2.1. In Example 2.2.2, we verify the power rule for a reciprocal power function.

Just-In-Time

Simplify $\dfrac{A/B}{C}$ by putting both denominators together:

$$\frac{A/B}{C} = \frac{A}{BC}$$

For example,

$$\frac{x/3}{2} = \frac{x}{3 \times 2} = \frac{x}{6}.$$

EXAMPLE 2.2.2

Verify the power rule for the function $F(x) = \dfrac{1}{x^2} = x^{-2}$ by showing that its derivative is $F'(x) = -2x^{-3}$.

Solution

The derivative of $F(x)$ is given by

$$F'(x) = \lim_{h \to 0} \frac{F(x+h) - F(x)}{h}$$

$$= \lim_{h \to 0} \frac{\dfrac{1}{(x+h)^2} - \dfrac{1}{x^2}}{h}$$

$$= \lim_{h \to 0} \frac{\dfrac{x^2 - (x+h)^2}{x^2(x+h)^2}}{h} \qquad \text{put the numerator over a common denominator}$$

$$= \lim_{h \to 0} \frac{[x^2 - (x^2 + 2hx + h^2)]}{x^2 h(x+h)^2} \qquad \text{simplify the complex fraction}$$

$$= \lim_{h \to 0} \frac{-2xh - h^2}{x^2 h(x+h)^2} \qquad \text{combine terms in the numerator}$$

$$= \lim_{h \to 0} \frac{-2x - h}{x^2(x+h)^2} \qquad \text{cancel common } h \text{ factors}$$

$$= \frac{-2x}{x^2(x^2)}$$

$$= \frac{-2}{x^3}$$

$$= -2x^{-3}$$

as claimed by the power rule.

Just-In-Time

Recall that $x^{-n} = 1/x^n$ when n is a positive integer, and that $x^{a/b} = \sqrt[b]{x^a}$ whenever a and b are positive integers.

     Note the steps for negative exponents and roots:
1. Change form to negative or fractional exponents.
2. Differentiate.
3. Simplify.

Here are several additional applications of the power rule:

$$\frac{d}{dx}(x^7) = 7x^{7-1} = 7x^6$$

$$\frac{d}{dx}(\sqrt[3]{x^2}) = \frac{d}{dx}(x^{2/3}) = \frac{2}{3}x^{2/3-1} = \frac{2}{3}x^{-1/3}$$

$$\frac{d}{dx}\left(\frac{1}{x^5}\right) = \frac{d}{dx}(x^{-5}) = -5x^{-5-1} = -5x^{-6}$$

$$\frac{d}{dx}(x^{1.3}) = 1.3x^{1.3-1} = 1.3x^{0.3}$$

A general proof for the power rule in the case where n is a positive integer is outlined in Exercise 76. The case where n is a negative integer and the case where n is a rational number ($n = r/s$ for integers r and s with $s \neq 0$) are outlined in exercises in Section 2.3 and Section 2.6, respectively.

     The constant rule and the power rule provide simple formulas for finding derivatives of a class of important functions, but to differentiate more complicated expressions, we need to know how to manipulate derivatives algebraically. The next two rules tell us that derivatives of multiples and sums of functions are multiples and sums of the corresponding derivatives.

The Constant Multiple Rule   ■   If c is a constant and $f(x)$ is differentiable, then $cf(x)$ is differentiable and

$$\frac{d}{dx}[cf(x)] = c\frac{d}{dx}[f(x)].$$

That is, the derivative of a multiple is the multiple of the derivative.

EXAMPLE 2.2.3

$$\frac{d}{dx}(3x^4) = 3\frac{d}{dx}(x^4) = 3(4x^3) = 12x^3$$

$$\frac{d}{dx}\left(\frac{-7}{\sqrt{x}}\right) = \frac{d}{dx}(-7x^{-1/2}) = -7\left(\frac{-1}{2}x^{-3/2}\right) = \frac{7}{2}x^{-3/2}$$

The Sum Rule   ■   If $f(x)$ and $g(x)$ are differentiable, then the sum $S(x) = f(x) + g(x)$ is differentiable, and $S'(x) = f'(x) + g'(x)$; that is,

$$\frac{d}{dx}[f(x) + g(x)] = \frac{d}{dx}[f(x)] + \frac{d}{dx}[g(x)].$$

In words, the derivative of a sum is the sum of the separate derivatives.

EXAMPLE 2.2.4

$$\frac{d}{dx}[x^{-2} + 7] = \frac{d}{dx}[x^{-2}] + \frac{d}{dx}[7] \qquad \frac{d}{dx}[2x^5 - 3x^{-7}] = 2\frac{d}{dx}[x^5] - 3\frac{d}{dx}[x^{-7}]$$

$$= -2x^{-3} + 0 \qquad\qquad\qquad = 2(5x^4) - 3(-7x^{-8})$$

$$= -2x^{-3} \qquad\qquad\qquad\qquad = 10x^4 + 21x^{-8}$$

By combining the power rule, the constant multiple rule, and the sum rule, you can differentiate any polynomial. Here is an example.

EXAMPLE 2.2.5

Differentiate the polynomial $y = 5x^3 - 4x^2 + 12x - 8$.

Solution

Differentiate this sum term by term to get

$$\frac{dy}{dx} = \frac{d}{dx}[5x^3] + \frac{d}{dx}[-4x^2] + \frac{d}{dx}[12x] + \frac{d}{dx}[-8]$$

$$= 5\frac{d}{dx}[x^3] - 4\frac{d}{dx}[x^2] + 12\frac{d}{dx}[x] + \frac{d}{dx}[-8]$$

$$= 15x^2 - 8x^1 + 12x^0 + 0 \qquad \text{recall } x^0 = 1$$

$$= 15x^2 - 8x + 12$$

EXAMPLE 2.2.6

It is estimated that x months from now, the population of a certain community will be $P(x) = x^2 + 20x + 8000$.

a. At what rate will the population be changing with respect to time 15 months from now?

b. By how much will the population actually change during the 16th month?

Solution

a. The rate of change of the population with respect to time is the derivative of the population function. That is,

$$\text{Rate of change} = P'(x) = 2x + 20.$$

The rate of change of the population 15 months from now will be

$$P'(15) = 2(15) + 20 = 50 \text{ people per month.}$$

b. The actual change in the population during the 16th month is the difference between the population at the end of 16 months and the population at the end of 15 months. That is,

$$\text{Change in population} = P(16) - P(15) = 8576 - 8525$$
$$= 51 \text{ people}$$

> **NOTE** In Example 2.2.6, the actual change in population during the 16th month in part (b) differs from the monthly rate of change at the beginning of the month in part (a) because the rate varies during the month. The instantaneous rate in part (a) can be thought of as the change in population that would occur during the 16th month if the rate of change of population remained constant. ▪

Units of Derivatives

It is important to know the meaning of a derivative in terms of the units used. A derivative is the limit of the quotient of rise and run on a graph. Therefore, the units are the rise units over the run units, or the units of $f(x)$ over the units of x.

In Example 2.2.6, the units of $P(x)$ are people and the units of x are months. Therefore, the units of $P'(x)$ are people/month.

Function	Variable	Units of Derivative
displacement, metres	t, seconds	m/s
weight loss, kg	t, weeks	kg/week
annual salary, \$/year	years	\$/year/year (annual salary is \$/year, but we calculate a rate of change per year)
unemployment rate, %	month	change in %/month

Thinking of the units can help you interpret the answer to a derivative question.

Rectilinear Motion

The motion of an object along a line is called *rectilinear motion*. For example, the motion of a rocket early in its flight can be regarded as rectilinear. When studying rectilinear motion, we will assume that the object involved is moving along a coordinate axis. If the function $s(t)$ gives the *position* of the object at time t, then the rate of change of $s(t)$ with respect to t is its *velocity* $v(t)$, and the time derivative of the velocity is its *acceleration* $a(t)$. That is, $v(t) = s'(t)$ and $a(t) = v'(t)$.

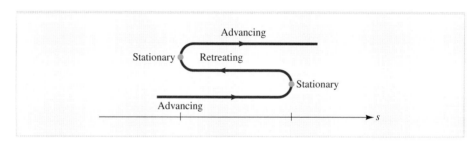

FIGURE 2.11   A diagram for rectilinear motion.

The object is said to be *advancing* (moving forward) when $v(t) > 0$ and *retreating* (moving backward) when $v(t) < 0$. When $v(t) = 0$, the object is neither advancing nor retreating and is said to be *stationary*; see Figure 2.11. Finally, the object is *accelerating* when $a(t) > 0$ and *decelerating* when $a(t) < 0$. To summarize:

Rectilinear Motion ■ If the **position** at time t of an object moving along a straight line is given by $s(t)$, then the object has

$$\textbf{velocity} \ \ v(t) = s'(t) = \frac{ds}{dt}$$

and

$$\textbf{acceleration} \ \ a(t) = v'(t) = \frac{dv}{dt}.$$

The object is **advancing** when $v(t) > 0$, **retreating** when $v(t) < 0$, and **stationary** when $v(t) = 0$. It is **accelerating** when $a(t) > 0$ and **decelerating** when $a(t) < 0$.

If position is measured in metres and time in seconds, velocity is measured in metres per second (m/s) and acceleration in metres per second per second (written as m/s²).

EXAMPLE 2.2.7

The position at time t of an insect, moving along a line searching for food particles, is given by $s(t) = t^3 - 6t^2 + 9t + 5$ centimetres after t seconds.

a. Find the velocity of the insect and discuss its motion between times $t = 0$ and $t = 4$.

b. Find the total distance travelled by the insect between times $t = 0$ and $t = 4$.

c. Find the acceleration of the insect and determine when the insect is accelerating and decelerating between times $t = 0$ and $t = 4$.

Solution

a. The velocity is $v(t) = \dfrac{ds}{dt} = 3t^2 - 12t + 9$. The insect is stationary when

$$v(t) = 3t^2 - 12t + 9 = 3(t - 1)(t - 3) = 0,$$

that is, at times $t = 1$ and $t = 3$. Otherwise, the insect is either advancing or retreating, as described in the following table.

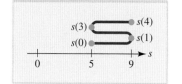

FIGURE 2.12 The motion of an insect: $s(t) = t^3 - 6t^2 + 9t + 5$.

Interval	Sign of v(t)	Description of the Motion
$0 < t < 1$	+	Advances from $s(0) = 5$ to $s(1) = 9$
$1 < t < 3$	−	Retreats from $s(1) = 9$ to $s(3) = 5$
$3 < t < 4$	+	Advances from $s(3) = 5$ to $s(4) = 9$

The motion of the insect is summarized in the diagram in Figure 2.12.

b. The insect travels from $s(0) = 5$ to $s(1) = 9$, then back to $s(3) = 5$, and finally to $s(4) = 9$. Thus, the total distance travelled by the insect is

$$D = \underbrace{|9 - 5|}_{0 < t < 1} + \underbrace{|5 - 9|}_{1 < t < 3} + \underbrace{|9 - 5|}_{3 < t < 4} = 12.$$

c. The acceleration of the insect is

$$a(t) = \frac{dv}{dt} = 6t - 12 = 6(t - 2).$$

The insect is accelerating ($a(t) > 0$) when $2 < t < 4$ and decelerating ($a(t) < 0$) when $0 < t < 2$.

NOTE An accelerating object is not necessarily "speeding up," nor is a decelerating object always "slowing down." For instance, the object in Example 2.2.9 has negative velocity and is accelerating for $2 < t < 3$. This means that the velocity is increasing over this time period; that is, becoming *less negative*. In other words, the object is actually slowing down. ∎

The Motion of a Projectile

An important example of rectilinear motion is the motion of a projectile. Suppose an object is projected (e.g., thrown, fired, or dropped) vertically in such a way that the only acceleration acting on the object is the constant downward acceleration g due to gravity. Near sea level, g is approximately 9.8 m/s^2. It can be shown that at time t, the height of the object is given by the formula

$$H(t) = -\frac{1}{2}gt^2 + V_0 t + H_0,$$

where H_0 and V_0 are the initial height and velocity, respectively, of the object. Here is an example using this formula.

EXAMPLE 2.2.8

Suppose a person standing at the top of a building 19.6 m high throws a ball vertically upward with an initial velocity of 14.7 m/s (see Figure 2.13).

a. Find the ball's height and velocity at time t.

b. When does the ball hit the ground and what is its impact velocity?

c. When is the velocity 0? What is the significance of this time?

d. How far does the ball travel during its flight?

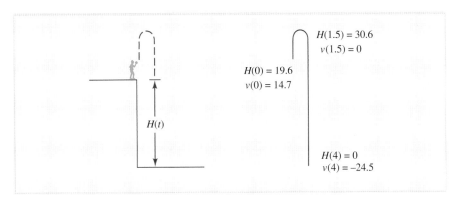

FIGURE 2.13   The motion of a ball thrown upward from the top of a building.

Solution

a. Since $g = 9.8$, $V_0 = 14.7$, and $H_0 = 19.6$, the height of the ball above the ground at time t is

$$H(t) = -4.9t^2 + 14.7t + 19.6 \text{ metres.}$$

The velocity at time t is

$$v(t) = \frac{dH}{dt} = -9.8t + 14.7.$$

b. The ball hits the ground when $H = 0$. Solve the equation $-4.9t^2 + 14.7t + 19.6 = 0$. Factor out -4.9; then $-4.9(t - 4)(t + 1) = 0$. Then $t = 4$ or $t = -1$. Disregarding the negative time $t = -1$, which is not meaningful in this context, conclude that impact occurs when $t = 4$ s and that the impact velocity is

$$v(4) = (-9.8 \text{ m/s}^2)(4 \text{ s}) + 14.7 \text{ m/s} = -24.5 \text{ m/s}.$$

The negative sign means the ball is coming down at the moment of impact.

c. The velocity is zero when $v(t) = -9.8t + 14.7 = 0$, which occurs when $t = 1.5$ s. For $t < 1.5$, the velocity is positive and the ball is rising, and for $t > 1.5$, the ball is falling (see Figure 2.13). Thus, the ball is at its highest point when $t = 1.5$ s.

d. The ball starts at $H(0) = 19.6$ m and rises to a maximum height of $H(1.5) = 30.6$ m before falling to the ground. Thus,

$$\text{Total distance travelled} = \underbrace{(30.6 - 19.6)}_{up} + \underbrace{30.6}_{down} = 41.6 \text{ m.}$$

ALGEBRA WARM UP

Make each expression ready to differentiate by changing the exponents, but do not differentiate:

1. $\dfrac{2}{x^3}$

2. $\dfrac{2}{3\sqrt{x}}$

3. $\dfrac{2\sqrt{x}}{5}$

4. $\dfrac{2}{\sqrt[3]{x}}$

5. $\dfrac{7}{x^5} - 10\sqrt{x} + 6x^4$

EXERCISES ▪ 2.2

In Exercises 1 through 28, differentiate the given function. Simplify your answers.

1. $y = -2$

2. $y = 3$

3. $y = 5x - 3$

4. $y = -2x + 7$

5. $y = x^{-4}$

6. $y = x^{7/3}$

7. $y = x^{3.7}$

8. $y = 4 - x^{-1.2}$

9. $y = \pi r^2$

10. $y = \dfrac{4}{3}\pi r^3$

11. $y = \sqrt{2x}$

12. $y = 2\sqrt[4]{x^3}$

13. $y = \dfrac{9}{\sqrt{t}}$

14. $y = \dfrac{3}{2t^2}$

15. $y = x^2 + 2x + 3$

16. $y = 3x^5 - 4x^3 + 9x - 6$

17. $f(x) = x^9 - 5x^8 + x + 12$

18. $f(x) = \dfrac{1}{4}x^8 - \dfrac{1}{2}x^6 - x + 2$

19. $f(x) = -0.02x^3 + 0.3x$

20. $f(u) = 0.07u^4 - 1.21u^3 + 3u - 5.2$

21. $y = \dfrac{1}{t} + \dfrac{1}{t} - \dfrac{2}{\sqrt{t}}$

22. $y = \dfrac{3}{x} - \dfrac{2}{x^2} + \dfrac{2}{3x^3}$

23. $f(x) = \sqrt{x^3} + \dfrac{1}{\sqrt{x^3}}$

24. $f(t) = 2\sqrt{t^3} + \dfrac{4}{\sqrt{t}} - \sqrt{2}$

25. $y = \dfrac{x^2}{16} + \dfrac{2}{x} - x^{3/2} + \dfrac{1}{3x^2} + \dfrac{x}{3}$

26. $y = \dfrac{7}{x^{1.2}} + \dfrac{5}{x^{-2.1}}$

27. $y = \dfrac{x^5 - 4x^2}{x^3}$ [*Hint:* Divide first.]

28. $y = x^2(x^3 - 6x + 7)$ [*Hint:* Multiply first.]

In Exercises 29 through 34, find the equation of the line that is tangent to the graph of the given function at the specified point.

29. $y = -x^3 - 5x^2 + 3x - 1; (-1, -8)$

30. $y = x^5 - 3x^3 - 5x + 2; (1, -5)$

31. $y = 1 - \dfrac{1}{x} + \dfrac{2}{\sqrt{x}}; \left(4, \dfrac{7}{4}\right)$

32. $y = \sqrt{x^3} - x^2 + \dfrac{16}{x^2}; (4, -7)$

33. $y = (x^2 - x)(3 + 2x); (-1, 2)$

34. $y = 2x^4 - \sqrt{x} + \dfrac{3}{x}; (1, 4)$

In Exercises 35 through 40, find the equation of the line that is tangent to the graph of the given function at the point $(c, f(c))$ for the specified value of $x = c$.

35. $f(x) = -2x^3 + \dfrac{1}{x^2}; x = -1$

36. $f(x) = x^4 - 3x^3 + 2x^2 - 6; x = 2$

37. $f(x) = x - \dfrac{1}{x^2}; x = 1$

38. $f(x) = x^3 + \sqrt{x}; x = 4$

39. $f(x) = -\dfrac{1}{3}x^3 + \sqrt{8x}; x = 2$

40. $f(x) = x(\sqrt{x} - 1); x = 4$

In Exercises 41 through 46, find the rate of change of the function $f(x)$ with respect to x for the prescribed value $x = c$.

41. $f(x) = 2x^4 + 3x + 1; x = -1$

42. $f(x) = x^3 - 3x + 5; x = 2$

43. $f(x) = x - \sqrt{x} + \dfrac{1}{x^2}; x = 1$

44. $f(x) = \sqrt{x} + 5x; x = 4$

45. $f(x) = \dfrac{x + \sqrt{x}}{\sqrt{x}}; x = 1$

46. $f(x) = \dfrac{2}{x} - x\sqrt{x}; x = 1$

In Exercises 47 through 50, find the derivative, paying special attention to the units.

47. The population is $P(t) = 2t^3 - 5t^2 + 4$ thousand people after t months. Find and interpret $P'(1)$.

48. The temperature is $T(t) = t + \dfrac{1}{t}$ degrees Celsius after t hours, $t > 0$. Find and interpret $T'(1)$.

49. The height of a plant is $L(t) = t\sqrt{t} + t^2$ centimetres after t weeks. Find and interpret $L'(1)$.

50. Revenue is $R(x) = (4 - x)x^{-1}$ hundred dollars when x thousand items are produced. Find and interpret $R'(3)$.

51. ANNUAL EARNINGS The gross annual earnings of a certain company are $A(t) = 0.1t^2 + 10t + 20$ thousand dollars t years after its formation in 2006.
   a. At what rate were the gross annual earnings of the company growing with respect to time in 2010?
   b. At what percentage rate were the gross annual earnings growing with respect to time in 2010 (using percentage rate = 100(rate/function))?

52. WORKER EFFICIENCY An efficiency study of the morning shift at a certain factory indicates that an average worker who arrives on the job at 8:00 A.M. will have assembled $f(x) = -x^3 + 6x^2 + 15x$ units x hours later.
   a. Derive a formula for the rate at which the worker will be assembling units after x hours.
   b. At what rate will the worker be assembling units at 9:00 A.M.?
   c. How many units will the worker actually assemble between 9:00 A.M. and 10:00 A.M.?

53. EDUCATIONAL TESTING It is estimated that x years from now, the average SAT mathematics score of the incoming students at a private college will be $f(x) = -6x + 582$.
   a. Derive an expression for the rate at which the average SAT score will be changing with respect to time x years from now.

b. What is the significance of the fact that the expression in part (a) is a constant? What is the significance of the fact that the constant in part (a) is negative?

54. **PUBLIC TRANSPORTATION**   After x weeks, the number of people using a new rapid transit system was approximately $N(x) = 6x^3 + 500x + 8000$.
    a. At what rate was the use of the system changing with respect to time after 8 weeks?
    b. By how much did the use of the system change during the eighth week?

55. **PROPERTY TAX**   Records indicate that x years after 2006, the average property tax on a three-bedroom house in a certain community was $T(x) = 20x^2 + 40x + 600$ dollars.
    a. At what rate was the property tax increasing with respect to time in 2006?
    b. By how much did the tax change between the years 2006 and 2010?

56. **ADVERTISING**   A manufacturer of motorcycles estimates that if x thousand dollars is spent on advertising, then

$$M(x) = 2300 + \frac{125}{x} - \frac{517}{x^2}, \quad 3 \le x \le 18,$$

motorcycles will be sold. At what rate will sales be changing when $9000 is spent on advertising? Are sales increasing or decreasing for this level of advertising expenditure?

57. **POPULATION GROWTH**   It is projected that x months from now, the population of a certain town will be $P(x) = 2x + 4x^{3/2} + 5000$.
    a. At what rate will the population be changing with respect to time 9 months from now?
    b. At what percentage rate will the population be changing with respect to time 9 months from now? (Use percentage rate = rate/function expressed as a percent.)

58. **POPULATION GROWTH**   It is estimated that t years from now, the population of a certain town will be $P(t) = t^2 + 200t + 10\,000$.
    a. Express the percentage rate of change of the population as a function of t, simplify this function algebraically, and draw its graph.
    b. What will happen to the percentage rate of change of the population in the long run (that is, as t grows very large)?

59. **SPREAD OF AN EPIDEMIC**   A medical research team determines that t days after an

epidemic begins, $N(t) = 10t^3 + 5t + \sqrt{t}$ people will be infected, for $0 \le t \le 20$. At what rate is the infected population increasing on the ninth day?

60. **SPREAD OF AN EPIDEMIC**   A disease is spreading in such a way that after t weeks, the number of people infected is $N(t) = 5175 - t^3$ $(t - 8)$, $0 \le t \le 8$. At what rate is the epidemic spreading after 3 weeks?

61. **AIR POLLUTION**   An environmental study of a certain suburban community suggests that t years from now, the average level of carbon monoxide in the air will be $Q(t) = 0.05t^2 + 0.1t + 3.4$ parts per million.
    a. At what rate will the carbon monoxide level be changing with respect to time 1 year from now?
    b. By how much will the carbon monoxide level change this year?
    c. By how much will the carbon monoxide level change over the next 2 years?

62. **WEB SITE HITS**   It is estimated that t months from now, the number of weekly hits to a tourist information website will be $C(t) = 100t^2 + 400t + 5000$.
    a. Derive an expression for the rate at which the number of hits will be changing with respect to time t months from now.
    b. At what rate will the number of hits be changing with respect to time 5 months from now? Will the number be increasing or decreasing at that time?
    c. By how much will the number of hits actually change during the 6th month?

63. **SALARY INCREASES**   Your starting salary will be $45\,000, and you will get a raise of $2000 each year.
    a. Express the rate of change of your salary as a function of time.
    b. Find the percentage rate of change of salary and sketch a graph of this function.

64. **A* ORNITHOLOGY**   An ornithologist determines that the body temperature of a certain species of bird fluctuates over roughly a 17-hour period according to the cubic formula $T(t) = -68.07t^3 + 30.98t^2 + 12.52t + 37.1$ for $0 \le t \le 0.713$, where T is the temperature in degrees Celsius measured t days from the beginning of a period.
    a. Compute and interpret the derivative $T'(t)$.
    b. At what rate is the temperature changing at the beginning of the period ($t = 0$) and at the end of the period ($t = 0.713$)? Is the temperature increasing or decreasing at each of these times?

c. At what time is the temperature not changing (neither increasing nor decreasing)? What is the bird's temperature at this time? Interpret your result.

65. A* COST MANAGEMENT A company uses a truck to deliver its products. To estimate costs, the manager models gas consumption by the function

$$G(x) = \frac{1}{100}\left(\frac{1920}{x} + \frac{5x}{8}\right)$$

litres per kilometre, assuming that the truck is driven at a constant speed of x kilometres per hour, for $x \geq 5$. The driver is paid \$30 per hour to drive the truck 400 km, and gasoline costs \$1.05/L.

a. Find an expression for the total cost $C(x)$ of the trip.

b. At what rate is the cost $C(x)$ changing with respect to x when the truck is driven at 64 km/h? Is the cost increasing or decreasing at that speed?

RECTILINEAR MOTION *In Exercises 66 through 69, $s(t)$ is the position of a particle moving along a straight line at time t.*

(a) Find the velocity and acceleration of the particle.

(b) Find all times in the given interval when the particle is stationary.

66. $s(t) = t^2 - 2t + 6$ for $0 \leq t \leq 2$

67. $s(t) = 3t^2 + 2t - 5$ for $0 \leq t \leq 1$

68. $s(t) = t^3 - 9t^2 + 15t + 25$ for $0 \leq t \leq 6$

69. $s(t) = t^4 - 4t^3 + 8t$ for $0 \leq t \leq 4$

70. MOTION OF A PROJECTILE A stone is dropped from a height of 50 m.

a. When will the stone hit the ground?

b. With what velocity does it hit the ground?

71. MOTION OF A PROJECTILE You are standing on the top of a building and throw a ball vertically upward. After 2 seconds, the ball passes you on the way down, and 2 seconds after that, it hits the ground below.

a. What is the initial velocity of the ball?

b. How high is the building?

c. What is the velocity of the ball when it passes you on the way down?

d. What is the velocity of the ball as it hits the ground?

72. SPY STORY Our friend, the spy who escaped from the diamond smugglers in Chapter 1 (Exercise 44 of Section 1.4), is on a secret mission in space. An encounter with an enemy agent leaves him with a mild concussion and temporary amnesia. Fortunately, he has a book that gives the formula for the motion of a projectile and the values of g for various celestial bodies (9.8 m/s^2 on Earth, 1.6 m/s^2 on the moon, 3.7 m/s^2 on Mars, and 8.9 m/s^2 on Venus). To deduce his whereabouts, he throws a rock vertically upward (from ground level) and notes that it reaches a maximum height of 11.6 m and hits the ground 5 seconds after it leaves his hand. Where is he?

73. A* Find numbers a, b, and c such that the graph of the function $f(x) = ax^2 + bx + c$ has x intercepts at $(0, 0)$ and $(5, 0)$ and a tangent with slope 1 when $x = 2$.

74. A* Find the equations of all the tangents to the graph of the function

$$f(x) = x^2 - 4x + 25$$

that pass through the origin $(0, 0)$.

75. A* Prove the sum rule for derivatives. *Hint:* Note that the difference quotient for $f + g$ can be written as

$$\frac{(f + g)(x + h) - (f + g)(x)}{h} = \frac{[f(x + h) + g(x + h)] - [f(x) + g(x)]}{h}.$$

76. A* a. If $f(x) = x^4$, show that $\dfrac{f(x + h) - f(x)}{h} = 4x^3 + 6x^2h + 4xh^2 + h^3$.

b. If $f(x) = x^n$ for a positive integer n, show that

$$\frac{f(x + h) - f(x)}{h} = nx^{n-1} + \frac{n(n - 1)}{2}x^{n-2}h + \cdots + nxh^{n-2} + h^{n-1}.$$

c. Use the result in part (b) in the definition of the derivative to prove the power rule:

$$\frac{d}{dx}[x^n] = nx^{n-1}.$$

77. RECYCLING   The city of Edmonton has increased the amount of residential waste recycled through blue bags and depots.
   a. For what years is the rate of change approximately constant?
   b. The data (shown by the circles) for each year were modelled using a polynomial by a computer. At what times is the rate of change zero in the polynomial model?
   c. Using the rate of change of part (a), write an equation for the tangent to the graph. If this rate of change continued, when would Edmonton have 90% of its garbage recycled?
   d. Would it be possible to have 100% of the residential waste recycled?

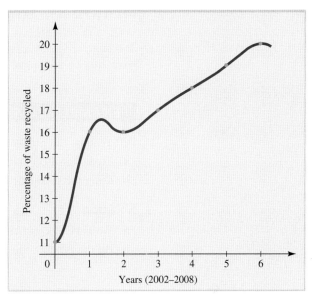

EXERCISE 77
SOURCE: *Ecovision Annual Report*, 2008, Chart 13, "% Residential Waste Collected Through Blue Bags, Bins, and Depots," City of Edmonton.

| **SECTION 2.3** | **Product and Quotient Rules; Higher-Order Derivatives** |

L03

Use the chain, product, and quotient rules to differentiate a given function. Calculate second and higher-order derivatives.

Based on your experience with the multiple and sum rules in Section 2.2, you may think that the derivative of a product of functions is the product of separate derivatives, but it is easy to see that this conjecture is false. For instance, if $f(x) = x^2$ and $g(x) = x^3$, then $f'(x) = 2x$ and $g'(x) = 3x^2$, so

$$f'(x)g'(x) = (2x)(3x^3) = 6x^3,$$

while $f(x)g(x) = x^2x^3 = x^5$ and

$$[f(x)g(x)]' = (x^5)' = 5x^4.$$

The correct formula for differentiating a product can be stated as follows.

> **The Product Rule**   ■   If $f(x)$ and $g(x)$ are differentiable at x, then so is their product $P(x) = f(x)g(x)$, and
>
> $$\frac{d}{dx}[f(x)g(x)] = f(x)\frac{d}{dx}[g(x)] + g(x)\frac{d}{dx}[f(x)],$$
>
> or, equivalently,
>
> $$(fg)' = fg' + gf'.$$
>
> In words, the derivative of the product fg is f times the derivative of g plus g times the derivative of f.

Applying the product rule to our introductory example, we find that

$$(x^2x^3)' = x^2(x^3)' + (x^3)(x^2)'$$
$$= (x^2)(3x^2) + (x^3)(2x) = 3x^4 + 2x^4 = 5x^4$$

which is the same as the result obtained by direct computation:

$$(x^2 x^3)' = (x^5)' = 5x^4.$$

Here are two additional examples.

EXAMPLE 2.3.1

Differentiate the product $P(x) = (x - 1)(3x - 2)$ by
  a. expanding $P(x)$ and using the polynomial rule
  b. using the product rule

Solution

  a. We have $P(x) = 3x^2 - 5x + 2$, so $P'(x) = 6x - 5$.
  b. By the product rule

$$P'(x) = (x - 1)(3) + (3x - 2)(1)$$
$$= 3x - 3 + 3x - 2$$
$$= 6x - 5$$

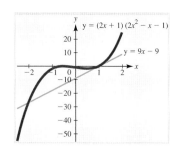

EXAMPLE 2.3.2

For the curve $y = (2x + 1)(2x^2 - x - 1)$:
  a. Find y'.
  b. Find an equation for the tangent line to the curve at the point where $x = 1$.
  c. Find all points on the curve where the tangent line is horizontal.

Solution

  a. Using the product rule, we get

$$y' = (2x + 1)[2x^2 - x - 1]' + [2x + 1]'(2x^2 - x - 1)$$
$$= (2x + 1)(4x - 1) + (2)(2x^2 - x - 1)$$

  b. When $x = 1$, the corresponding y value is

$$y(1) = [2(1) + 1][2(1)^2 - 1 - 1] = 0,$$

so the point of tangency is $(1, 0)$. The slope at $x = 1$ is

$$y'(1) = [2(1) + 1][4(1) - 1] + 2[2(1)^2 - 1 - 1] = 9.$$

Substituting into the point-slope formula, we find that the tangent line at $(1, 0)$ has the equation

$$y - 0 = 9(x - 1)$$
$$y = 9x - 9$$

(Note that this y refers to a tangent line; the other y referred to a function. Don't confuse the many uses of the letter y!)

Just-In-Time

In part (b), $x = 1$ was substituted right away, without simplifying the expression. This is the best approach when substituting $x = 1$ or $x = 0$. Errors can creep in when simplifying.

c. Horizontal tangents occur where the slope is zero, that is, where $y' = 0$. Expanding the expression for the derivative and combining terms, we obtain

$$y' = (2x + 1)(4x - 1) + (2)(2x^2 - x - 1) = 12x^2 - 3.$$

Solving $y' = 0$, we find that

$$y' = 12x^2 - 3 = 0$$

$$x^2 = \frac{3}{12} = \frac{1}{4}$$

$$x = \frac{1}{2} \quad \text{and} \quad x = -\frac{1}{2}$$

Substituting $x = \frac{1}{2}$ and $x = -\frac{1}{2}$ into the formula for y, we get $y\left(\frac{1}{2}\right) = -2$ and $y\left(-\frac{1}{2}\right) = 0$, so horizontal tangents occur at the points $\left(\frac{1}{2}, -2\right)$ and $\left(-\frac{1}{2}, 0\right)$ on the curve.

EXAMPLE 2.3.3

A manufacturer determines that t months after a new product is introduced to the market, $x(t) = t^2 + 3t$ hundred units can be produced and then sold at a price of $p(t) = -2t^{3/2} + 30$ dollars per unit.

  a. Express the revenue $R(t)$ for this product as a function of time.
  b. At what rate is revenue changing with respect to time after 4 months? Is revenue increasing or decreasing at this time?

Solution

  a. The revenue is given by

$$R(t) = x(t)p(t) = (t^2 + 3t)(-2t^{3/2} + 30)$$

hundred dollars.

  b. The rate of change of revenue $R(t)$ with respect to time is given by the derivative $R'(t)$, which we find using the product rule:

$$R'(t) = (t^2 + 3t)\frac{d}{dt}[-2t^{3/2} + 30] + (-2t^{3/2} + 30)\frac{d}{dt}[t^2 + 3t]$$

$$= (t^2 + 3t)\left[-2\left(\frac{3}{2}t^{1/2}\right)\right] + (-2t^{3/2} + 30)(2t + 3)$$

$$= (t^2 + 3t)(-3t^{1/2}) + (-2t^{3/2} + 30)(2t + 3)$$

At time $t = 4$, the revenue is changing at the rate

$$R'(4) = [(4)^2 + 3(4)][-3(4)^{1/2}] + [-2(4)^{3/2} + 30][2(4) + 3]$$

$$= -14$$

Thus, after 4 months, the revenue is changing at a rate of 14 hundred dollars ($1400) per month. It is *decreasing* at that time since $R'(4)$ is negative.

A proof of the product rule is given at the end of this section. It is also important to be able to differentiate quotients of functions, and for this purpose, we have the following rule, a proof of which is outlined in Exercise 69.

CAUTION: A common error is to assume that $\left(\dfrac{f}{g}\right)' = \dfrac{f'}{g'}$.

The Quotient Rule ■ If $f(x)$ and $g(x)$ are differentiable functions, then so is the quotient $Q(x) = f(x)/g(x)$, and

$$\frac{d}{dx}\left[\frac{f(x)}{g(x)}\right] = \frac{g(x)\dfrac{d}{dx}[f(x)] - f(x)\dfrac{d}{dx}[g(x)]}{g^2(x)} \quad \text{if } g(x) \neq 0,$$

or, equivalently,

$$\left(\frac{f}{g}\right)' = \frac{gf' - fg'}{g^2}$$

Another way of remembering the quotient rule is to think of the function as two parts, *Hi* on top and *Lo* on the bottom:

$$f(x) = \frac{Hi}{Lo}.$$

Then

$$f'(x) = \frac{Lo \; d(Hi) - Hi \; d(Lo)}{\text{square the bottom and away we go!}},$$

where $d(Hi)$ and $d(Lo)$ are those functions differentiated, as in Hi' and Lo'. Remembering *Hi* and *Lo* will keep them in the right place more easily than using f and g.

Just-In-Time

When an expression has a denominator we have to divide each of the terms by the denominator. Using numbers, we can see that

$$\frac{4 + 2 + 8}{4} = \frac{14}{4} = 3\frac{1}{2}$$

or $\dfrac{4}{4} + \dfrac{2}{4} + \dfrac{8}{4} = 3\dfrac{1}{2}$

In the last line of part (b), we use

$$\frac{A + B}{C} = \frac{A}{C} + \frac{B}{C}.$$

We can never break up the denominator:

$$\frac{A}{B + C} \neq \frac{A}{B} + \frac{A}{C}.$$

EXAMPLE 2.3.4

Differentiate the quotient $Q(x) = \dfrac{x^2 - 5x + 7}{2x}$ by

a. dividing through first

b. using the quotient rule

Solution

a. Dividing by the denominator $2x$, we get

$$Q(x) = \frac{1}{2}x - \frac{5}{2} + \frac{7}{2}x^{-1},$$

so

$$Q'(x) = \frac{1}{2} - 0 + \frac{7}{2}(-x^{-2}) = \frac{1}{2} - \frac{7}{2x^2}.$$

b. By the quotient rule,

$$Q'(x) = \frac{(2x)\dfrac{d}{dx}[x^2 - 5x + 7] - (x^2 - 5x + 7)\dfrac{d}{dx}[2x]}{(2x)^2}.$$

$$= \frac{(2x)(2x - 5) - (x^2 - 5x + 7)(2)}{4x^2}$$

$$= \frac{4x^2 - 10x - (2x^2 - 10x + 14)}{4x^2}$$

$$= \frac{2x^2 - 14}{4x^2}$$

$$= \frac{1}{2} - \frac{7}{2x^2}$$

EXAMPLE 2.3.5

A biologist models the effect of introducing a toxin to a bacterial colony by the function

$$P(t) = \frac{t + 1}{t^2 + t + 4},$$

where P is the population of the colony (in millions) t hours after the toxin is introduced.

a. At what rate is the population changing when the toxin is introduced? Is the population increasing or decreasing at this time?

b. At what time does the population begin to decrease? By how much does the population increase before it begins to decline?

Solution

a. The rate of change of the population with respect to time is given by the derivative $P'(t)$, which we compute using the quotient rule:

$$P'(t) = \frac{(t^2 + t + 4)\dfrac{d}{dt}[t + 1] - (t + 1)\dfrac{d}{dt}[t^2 + t + 4]}{(t^2 + t + 4)^2}$$

$$= \frac{(t^2 + t + 4)(1) - (t + 1)(2t + 1)}{(t^2 + t + 4)^2}$$

$$= \frac{t^2 + t + 4 - (2t^2 + 3t + 1)}{(t^2 + t + 4)^2}$$

$$= \frac{-t^2 - 2t + 3}{(t^2 + t + 4)^2}$$

The toxin is introduced when $t = 0$, and at that time the population is changing at the rate

$$P'(0) = \frac{0 + 0 + 3}{(0 + 0 + 4)^2} = \frac{3}{16} = 0.1875.$$

That is, the population is initially changing at a rate of 0.1875 million (187 500) bacteria per hour, and it is increasing since $P'(0) > 0$.

b. The population is decreasing when $P'(t) < 0$. Since the numerator of $P'(t)$ can be factored as

$$-t^2 - 2t + 3 = -(t^2 + 2t - 3) = -(t - 1)(t + 3).$$

we can write

$$P'(t) = \frac{-(t - 1)(t + 3)}{(t^2 + t + 4)^2}.$$

The denominator $(t^2 + t + 4)^2$ and the factor $t + 3$ are both positive for all $t \geq 0$, which means that

$$\text{for } 0 \leq t < 1 \quad P'(t) > 0 \text{ and } P(t) \text{ is increasing}$$
$$\text{for } t > 1 \quad P'(t) < 0 \text{ and } P(t) \text{ is decreasing}$$

Thus, the population begins to decline after 1 hour.
The initial population of the colony is

$$P(0) = \frac{0 + 1}{0 + 0 + 4} = \frac{1}{4}$$

million, and after 1 hour, the population is

$$P(1) = \frac{1 + 1}{1 + 1 + 4} = \frac{1}{3}$$

million. Therefore, before the population begins to decline, it increases by

$$P(1) - P(0) = \frac{1}{3} - \frac{1}{4} = \frac{1}{12}$$

million, that is, by approximately 83 333 bacteria.

The quotient rule is somewhat cumbersome, so don't use it unnecessarily. Consider Example 2.3.6.

EXAMPLE 2.3.6

Differentiate the function $y = \dfrac{2}{3x^2} - \dfrac{x}{3} + \dfrac{4}{5} + \dfrac{x + 1}{x}$.

Solution

Don't use the quotient rule! Instead, rewrite the function as

$$y = \frac{2}{3}x^{-2} - \frac{1}{3}x + \frac{4}{5} + 1 + x^{-1}$$

and then apply the power rule term by term to get

$$\frac{dy}{dx} = \frac{2}{3}(-2x^{-3}) - \frac{1}{3} + 0 + 0 + (-1)x^{-2}$$
$$= -\frac{4}{3}x^{-3} - \frac{1}{3} - x^{-2}$$
$$= -\frac{4}{3x^3} - \frac{1}{3} - \frac{1}{x^2}$$

The Second Derivative   In applications, it may be necessary to compute the rate of change of a function that is itself a rate of change. For example, the acceleration of a car is the rate of change with respect to time of its velocity, which in turn is the rate of change with respect to time of its position. If the position is measured in kilometres and time in hours, the velocity (rate of change of position) is measured in kilometres per hour, and the acceleration (rate of change of velocity) is measured in kilometres per hour, per hour.

Statements about the rate of change of a rate of change are used frequently in economics. During a recession, for example, you may hear that unemployment is still increasing but at a decreasing rate. That is, people are still losing jobs but not as many each month as before.

The rate of change of the function $f(x)$ with respect to x is the derivative $f'(x)$. But $f'(x)$ is itself a function and the rate of change of the function $f'(x)$ with respect to x is *its* derivative $(f'(x))'$. This notation is awkward, so we write the derivative of the derivative of $f(x)$ as $(f'(x))' = f''(x)$ and refer to it as the *second derivative* of $f(x)$ (read $f''(x)$ as "f double prime of x"). If $y = f(x)$, then the second derivative of y with respect to x is written as y'' or as $\dfrac{d^2y}{dx^2}$. Here is a summary of the terminology and notation used for second derivatives.

The Second Derivative ■ The second derivative of a function is the derivative of its derivative. If $y = f(x)$, the second derivative is denoted by

$$\frac{d^2y}{dx^2} \quad \text{or} \quad f''(x).$$

The second derivative gives the rate of change of the rate of change of the original function.

NOTE The derivative $f'(x)$ is sometimes called the **first derivative** to distinguish it from the **second derivative** $f''(x)$. ▪

You don't have to use any new rules to find the second derivative of a function. Just find the first derivative and then differentiate again.

EXAMPLE 2.3.7

Find the second derivative of the function $f(x) = 5x^4 - 3x^2 - 3x + 7$.

Solution
Compute the first derivative,

$$f'(x) = 20x^3 - 6x - 3,$$

and then differentiate again to get

$$f''(x) = 60x^2 - 6.$$

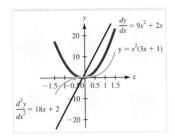

EXAMPLE 2.3.8

Find the second derivative of $y = x^2(3x + 1)$.

Solution

According to the product rule,

$$\frac{d}{dx}[x^2(3x + 1)] = x^2\frac{d}{dx}[3x + 1] + (3x + 1)\frac{d}{dx}[x^2]$$
$$= x^2(3) + (3x + 1)(2x)$$
$$= 9x^2 + 2x$$

Therefore, the second derivative is

$$\frac{d^2y}{dx^2} = \frac{d}{dx}[9x^2 + 2x]$$
$$= 18x + 2$$

Look at the graphs of the function, the first derivative, and the second derivative in the margin. Notice that the function has a positive or zero slope throughout, so the first derivative is on the x axis or above it in the positive region. The first derivative has a constantly increasing slope—it becomes less negative and then more positive—so the second derivative is a straight line. We will discuss this more later, but it is a good idea to keep in mind that derivatives are slopes of tangents to graphs.

> **NOTE**   Before computing the second derivative of a function, always take time to simplify the first derivative as much as possible. The more complicated the form of the first derivative, the more tedious the computation of the second derivative.

The second derivative will be used in Section 3.2 to obtain information about the shape of a graph, and in Sections 3.4 and 3.5 in optimization problems. Here is a more elementary application illustrating the interpretation of the second derivative as the rate of change of a slope. The derivative indicates whether the slope is increasing or decreasing and how rapidly the change in slope occurs.

EXAMPLE 2.3.9

An efficiency study of the morning shift at a certain warehouse indicates that an average worker who arrives at 8:00 A.M. will have retrieved and moved

$$Q(t) = -t^3 + 6t^2 + 24t$$

crates of produce with a forklift t hours later.

  a. Compute the worker's rate of production at 11:00 A.M.

  b. At what rate is the worker's rate of production changing with respect to time at 11:00 A.M.?

Solution

  a. The worker's rate of production is the first derivative

$$R(t) = Q'(t) = -3t^2 + 12t + 24$$

of the output $Q(t)$. The rate of production at 11:00 A.M. ($t = 3$) is

$$R(3) = Q'(3) = -3(3)^2 + 12(3) + 24$$
$$= 33 \text{ crates per hour}$$

b. The rate of change of the rate of production is the second derivative

$$R'(t) = Q''(t) = -6t + 12$$

of the output function. At 11:00 A.M., this rate is

$$R'(3) = Q''(3) = -6(3) + 12$$
$$= -6 \text{ crates per hour per hour}$$

The negative sign indicates that the worker's rate of production is decreasing; that is, the worker is slowing down. The rate of this decrease in efficiency at 11:00 A.M. is 6 units per hour per hour.

Recall from Section 2.2 that the **acceleration** $a(t)$ of an object moving along a straight line is the derivative of the velocity $v(t)$, which in turn is the derivative of the position function $s(t)$. Thus, the acceleration may be thought of as the second derivative of position; that is,

$$a(t) = \frac{d^2 s}{dt^2}.$$

This notation is used in Example 2.3.10.

EXAMPLE 2.3.10

If the position of a dog running along a sidewalk in a straight line is given by $s(t) = t^3 - 3t^2 + 4t$ metres at time t seconds, find its velocity and acceleration when $t = 2$ s.

Solution

The velocity of the dog is

$$v(t) = \frac{ds}{dt} = 3t^2 - 6t + 4.$$

After 2 seconds, the velocity is $3(4) - 6(2) + 4 = 4$, or 4 m/s.
    The acceleration of the dog is

$$a(t) = \frac{dv}{dt} = \frac{d^2 s}{dt^2} = 6t - 6.$$

After 2 seconds, the acceleration is $6(2) - 6 = 6$, or 6 m/s^2.

Higher-Order Derivatives   If you differentiate the second derivative $f''(x)$ of a function $f(x)$ one more time, you get the third derivative $f'''(x)$. Differentiate again and you get the fourth derivative, which is denoted by $f^{(4)}(x)$ since the prime notation $f''''(x)$ is cumbersome. In general, the derivative obtained from $f(x)$ after n successive differentiations is called the **nth derivative** or **derivative of order n** and is denoted by $f^{(n)}(x)$.

The *n*th Derivative ■ For any positive integer *n*, the *n*th derivative of a function is obtained from the function by differentiating successively *n* times. If the original function is $y = f(x)$, the *n*th derivative is denoted by

$$\frac{d^n y}{dx^n} \quad \text{or} \quad f^{(n)}(x).$$

Just-In-Time

Note the three steps for differentiating the reciprocal of a power:
1. Change the form.
2. Differentiate.
3. Simplify back to the original form using positive exponents.

EXAMPLE 2.3.11

Find the fifth derivative of each function:

a. $f(x) = 4x^3 + 5x^2 + 6x - 1$ **b.** $y = \dfrac{1}{x}$

Solution

a. $f'(x) = 12x^2 + 10x + 6$

$f''(x) = 24x + 10$

$f'''(x) = 24$

$f^{(4)}(x) = 0$

$f^{(5)}(x) = 0$

b. $\dfrac{dy}{dx} = \dfrac{d}{dx}(x^{-1}) = -x^{-2} = -\dfrac{1}{x^2}$

$\dfrac{d^2 y}{dx^2} = \dfrac{d}{dx}(-x^{-2}) = 2x^{-3} = \dfrac{2}{x^3}$

$\dfrac{d^3 y}{dx^3} = \dfrac{d}{dx}(2x^{-3}) = -6x^{-4} = -\dfrac{6}{x^4}$

$\dfrac{d^4 y}{dx^4} = \dfrac{d}{dx}(-6x^{-4}) = 24x^{-5} = \dfrac{24}{x^5}$

$\dfrac{d^5 y}{dx^5} = \dfrac{d}{dx}(24x^{-5}) = -120x^{-6} = -\dfrac{120}{x^6}$

Derivation of the Product Rule

The product and quotient rules are not easy to prove. In both cases, the key is to express the difference quotient of the given expression (the product *fg* or quotient *f/g*) in terms of difference quotients of *f* and *g*. Here is a proof of the product rule. The proof for the quotient rule is outlined in Exercise 69.

To show that $\dfrac{d}{dx}(fg) = f\dfrac{dg}{dx} + g\dfrac{df}{dx}$, begin with the appropriate difference quotient and rewrite the numerator by subtracting and adding the quantity $f(x + h)g(x)$ as follows:

$$\frac{d}{dx}(fg) = \lim_{h \to 0} \frac{f(x + h)g(x + h) - f(x)g(x)}{h}$$

$$= \lim_{h \to 0} \left[\frac{f(x + h)g(x + h) - f(x + h)g(x) + f(x + h)g(x) - f(x)g(x)}{h} \right]$$

$$= \lim_{h \to 0} \left[\frac{f(x + h)g(x + h) - f(x + h)g(x)}{h} + \frac{f(x + h)g(x) - f(x)g(x)}{h} \right]$$

$$= \lim_{h \to 0} \left(f(x + h) \left[\frac{g(x + h) - g(x)}{h} \right] + g(x) \left[\frac{f(x + h) - f(x)}{h} \right] \right)$$

Now let h approach zero. Since

$$\lim_{h \to 0} \frac{f(x + h) - f(x)}{h} = \frac{df}{dx},$$

$$\lim_{h \to 0} \frac{g(x + h) - g(x)}{h} = \frac{dg}{dx},$$

and

$$\lim_{h \to 0} f(x + h) = f(x),$$

by the continuity of $f(x)$, it follows that

$$\frac{d}{dx}(fg) = f\frac{dg}{dx} + g\frac{df}{dx}.$$

EXERCISES ▪ 2.3

In Exercises 1 through 20, differentiate the given function.

1. $f(x) = (2x + 1)(3x - 2)$

2. $f(x) = (x - 5)(1 - 2x)$

3. $y = 10(3u + 1)(1 - 5u)$

4. $y = 400(15 - x^2)(3x - 2)$

5. $f(x) = \frac{1}{3}(x^5 - 2x^3 + 1)\left(x - \frac{1}{x}\right)$

6. $f(x) = -3(5x^3 - 2x + 5)(\sqrt{x} + 2x)$

7. $y = \frac{x + 1}{x - 2}$

8. $y = \frac{2x - 3}{5x + 2}$

9. $f(t) = \frac{1}{t^2 - 2}$

10. $f(x) = \frac{1}{x - 2}$

11. $y = \frac{3}{x + 5}$

12. $y = \frac{t^2 + 1}{1 - t^2}$

13. $f(x) = \frac{x^2 - 3x + 2}{2x^2 + 5x - 1}$

14. $f(t) = \frac{t^2 + 2t + 1}{t^2 + 3t - 1}$

15. $f(x) = \frac{(2x - 1)(x + 3)}{(x + 1)}$

16. $g(x) = \frac{(x^2 + x + 1)(4 - x)}{2x - 1}$

17. $f(x) = (2 + 5x)^2$

18. $f(x) = \left(x + \frac{1}{x}\right)^2$

19. $g(t) = \frac{t^2 + \sqrt{t}}{2t + 5}$

20. $h(x) = \frac{x}{x^2 - 1} + \frac{4 - x}{x^2 + 1}$

In Exercises 21 through 26, find an equation for the tangent line to the given curve at the point where $x = x_0$.

21. $y = (5x - 1)(4 + 3x); x_0 = 0$

22. $y = (x^2 + 3x - 1)(2 - x); x_0 = 1$

23. $y = \frac{x}{2x + 3}; x_0 = -1$

24. $y = \frac{x + 7}{5 - 2x}; x_0 = 0$

25. $y = (3\sqrt{x} + x)(2 - x^2); x_0 = 1$

26. $y = \frac{2x - 1}{1 - x^3}; x_0 = 0$

In Exercises 27 through 31, find all points on the graph of the given function where the tangent line is horizontal.

27. $f(x) = (x + 1)(x^2 - x - 2)$

28. $f(x) = (x - 1)(x^2 - 8x + 7)$

29. $f(x) = \dfrac{x + 1}{x^2 + x + 1}$

30. $f(x) = \dfrac{x^2 + x - 1}{x^2 - x + 1}$

31. $f(x) = x^3(x - 5)^2$

In Exercises 32 through 35, find the rate of change $\dfrac{dy}{dx}$ for the prescribed value of x_0.

32. $y = (x^2 + 2)(x + \sqrt{x});\ x_0 = 4$

33. $y = (x^2 + 3)(5 - 2x^3);\ x_0 = 1$

34. $y = \dfrac{2x - 1}{3x + 5};\ x_0 = 1$

35. $y = x + \dfrac{3}{2 - 4x};\ x_0 = 0$

The normal line to the curve $y = f(x)$ at the point $P(x_0, f(x_0))$ is the line perpendicular to the tangent line at P. In Exercises 36 through 39, find the equation of the normal line to the given curve at the prescribed point.

36. $y = x^2 + 3x - 5;\ (0, -5)$

37. $y = \dfrac{2}{x} - \sqrt{x};\ (1, 1)$

38. $y = (x + 3)(1 - \sqrt{x});\ (1, 0)$

39. $y = \dfrac{5x + 7}{2 - 3x};\ (1, -12)$

40. a. Differentiate the function $y = 2x^2 - 5x - 3$.
   b. Now factor the function in part (a) as $y = (2x + 1)(x - 3)$ and differentiate using the product rule. Show that the two answers are the same.

41. a. Use the quotient rule to differentiate the function $y = \dfrac{2x - 3}{x^3}$.
   b. Rewrite the function as $y = x^{-3}(2x - 3)$ and differentiate using the product rule.
   c. Rewrite the function as $y = 2x^{-2} - 3x^{-3}$ and differentiate.
   d. Show that your answers to parts (a), (b), and (c) are the same.

In Exercises 42 through 47, find the second derivative of the given function. In each case, use the appropriate notation for the second derivative and simplify your

answer. (Don't forget to simplify the first derivative as much as possible before computing the second derivative.)

42. $f(x) = 5x^{10} - 6x^5 - 27x + 4$

43. $f(x) = \dfrac{2}{x}x^5 - 4x^3 + 9x^2 - 6x - 2$

44. $y = 5\sqrt{x} + \dfrac{3}{x^2} + \dfrac{1}{3\sqrt{x}} + \dfrac{1}{2}$

45. $y = \dfrac{2}{3x} - \sqrt{2x} + \sqrt{2}x - \dfrac{1}{6\sqrt{x}}$

46. $y = (x^2 - x)\left(2x - \dfrac{1}{x}\right)$

47. $y = (x^3 + 2x - 1)(3x + 5)$

48. DEMAND AND REVENUE   The manager of a company that produces noise-reducing music headphones determines that when x thousand are produced, they will all be sold when the price is

$$p(x) = \frac{1000}{0.3x^2 + 8}$$

dollars per headphone set.
   a. At what rate is demand $p(x)$ changing with respect to the level of production x when 3000 $(x = 3)$ of these headphones are produced?
   b. The revenue derived from the sale of x thousand headphones is $R(x) = xp(x)$ thousand dollars. At what rate is revenue changing when 3000 headphones are produced? Is revenue increasing or decreasing at this level of production?

49. SALES   The manager of the Many Facets jewellery store models total sales by the function

$$S(t) = \frac{2000t}{4 + 0.3t},$$

where t is time (years) since the year 2008 and S is measured in thousands of dollars.
   a. At what rate were sales changing in the year 2010?
   b. What happens to sales in the long run (that is, as $t \to +\infty$)?

50. PROFIT   Bea Johnson, the owner of the Bea Nice boutique, estimates that when a particular kind of perfume is priced at p dollars per bottle, she will sell

$$B(p) = \frac{500}{p + 3}, \quad p \geq 5,$$

bottles per month at a total cost of

$$C(p) = 0.2p^2 + 3p + 200 \text{ dollars.}$$

a. Express Bea's profit $P(p)$ as a function of the price p per bottle.

b. At what rate is the profit changing with respect to p when the price is \$12 per bottle? Is profit increasing or decreasing at that price?

51. **ADVERTISING** A company manufactures a laptop screen protector that protects the screen from breaking. The marketing manager determines that t weeks after an advertising campaign begins, $P(t)$ percent of the potential market is aware of this product, where

$$P(t) = 100\left[\frac{t^2 + 5t + 5}{t^2 + 10t + 30}\right].$$

a. At what rate is the market percentage $P(t)$ changing with respect to time after 5 weeks? Is the percentage increasing or decreasing at this time?

b. What happens to the percentage $P(t)$ in the long run, that is, as $t \to +\infty$? What happens to the rate of change of $P(t)$ as $t \to +\infty$?

52. **BACTERIAL POPULATION** A bacterial colony is estimated to have a population of

$$P(t) = \frac{24t + 10}{t^2 + 1}$$

million t hours after the introduction of a toxin.

a. At what rate is the population changing 1 hour after the toxin is introduced ($t = 1$)? Is the population increasing or decreasing at this time?

b. At what time does the population begin to decline?

53. **POLLUTION CONTROL** A study indicates that spending money on pollution control is effective up to a point but eventually becomes wasteful. Suppose it is known that when x million dollars is spent on controlling pollution, the percentage of pollution removed is given by

$$P(x) = \frac{100\sqrt{x}}{0.03x^2 + 9}.$$

a. At what rate is the percentage of pollution removal $P(x)$ changing when \$16 million is spent? Is the percentage increasing or decreasing at this level of expenditure?

b. For what values of x is $P(x)$ increasing? For what values of x is $P(x)$ decreasing?

54. **PHARMACOLOGY** An oral painkiller is administered to a patient, and t hours later, the concentration of drug in the patient's bloodstream is given by

$$C(t) = \frac{2t}{3t^2 + 16}.$$

a. At what rate $R(t)$ is the concentration of drug in the patient's bloodstream changing t hours after being administered? At what rate is $R(t)$ changing at time t?

b. At what rate is the concentration of drug changing after 1 hour? Is the concentration changing at an increasing or decreasing rate at this time?

c. When does the concentration of the drug begin to decline?

d. Over what time period is the concentration changing at a declining rate?

55. **WORKER EFFICIENCY** An efficiency study of the morning shift at a certain manufacturing plant indicates that an average worker arriving on the job at 8:00 A.M. will have produced $Q(t) = -t^3 + 8t^2 + 15t$ units t hours later.

a. Compute the worker's rate of production $R(t) = Q'(t)$.

b. At what rate is the worker's rate of production changing with respect to time at 9:00 A.M.?

56. **POPULATION GROWTH** It is estimated that t years from now, the population of a certain suburban community will be $P(t) = 20 - \dfrac{6}{t + 1}$ thousand.

a. Derive a formula for the rate at which the population will be changing with respect to time t years from now.

b. At what rate will the population be growing 1 year from now?

c. By how much will the population actually increase during the second year?

d. At what rate will the population be growing 9 years from now?

e. What will happen to the rate of population growth in the long run?

In Exercises 57 through 60, the position $s(t)$ of an object moving along a straight line is given. Units of position are metres and time has units of seconds. In each case:
(a) Find the object's velocity $v(t)$ and acceleration $a(t)$.
(b) Find all times t when the acceleration is 0.

57. $s(t) = 3t^5 - 5t^3 - 7$

58. $s(t) = 2t^4 - 5t^3 + t - 3$

59. $s(t) = -t^3 + 7t^2 + t + 2$

60. $s(t) = 4t^{5/2} - 15t^2 + t - 3$

61. VELOCITY   An object moves along a straight line so that after t minutes, its distance from its starting point is $D(t) = 10t + \dfrac{5}{t+1} - 5$ metres.

a. At what velocity is the object moving at the end of 4 min?

b. How far does the object actually travel during the 5th minute?

62. ACCELERATION   After t hours of an 8-hour trip, a car has gone $D(t) = 64t + \dfrac{10}{3}t^2 - \dfrac{2}{9}t^3$ kilometres.

a. Derive a formula expressing the acceleration of the car as a function of time.

b. At what rate is the velocity of the car changing with respect to time at the end of 6 hours? Is the velocity increasing or decreasing at this time?

c. By how much does the velocity of the car actually change during the 7th hour?

63. ACCELERATION   If an object is dropped or thrown vertically, its height (in metres) after t seconds is $H(t) = -4.9t^2 + S_0t + H_0$, where S_0 is the initial speed of the object and H_0 is its initial height.

a. Derive an expression for the acceleration of the object.

b. How does the acceleration vary with time?

c. What is the significance of the fact that the answer to part (a) is negative?

64. Find $f^{(4)}(x)$ for $f(x) = x^5 - 2x^4 + x^3 - 3x^2 + 5x - 6$.

65. Find $\dfrac{d^3y}{dx^3}$ if $y = \sqrt{x} - \dfrac{1}{2x} + \dfrac{x}{\sqrt{2}}$.

66. a. Show that
$$\frac{d}{dx}[fgh] = fg\frac{dh}{dx} + fh\frac{dg}{dx} + gh\frac{df}{dx}.$$
[*Hint:* Apply the product rule twice.]

b. Find $\dfrac{dy}{dx}$, where $y = (2x + 1)(x - 3)(1 - 4x)$.

67. A* a. By combining the product rule and the quotient rule, find an expression for $\dfrac{d}{dx}\left[\dfrac{fg}{h}\right]$.

b. Find $\dfrac{dy}{dx}$, where $y = \dfrac{(2x + 7)(x^2 + 3)}{3x + 5}$.

68. A*   The product rule tells you how to differentiate the product of any two functions, while the constant multiple rule tells you how to differentiate products in which one of the factors is constant. Show that the two rules are consistent. In particular, use the product rule to show that $\dfrac{d}{dx}[cf] = c\dfrac{df}{dx}$ if c is a constant.

69. A*   Derive the quotient rule. [*Hint:* Show that the difference quotient for f/g is
$$\frac{1}{h}\left[\frac{f(x+h)}{g(x+h)} - \frac{f(x)}{g(x)}\right] = \frac{g(x)f(x+h) - f(x)g(x+h)}{g(x+h)g(x)h}.$$
Before letting h approach zero, rewrite this quotient using the trick of subtracting and adding $g(x)f(x)$ in the numerator.]

70. A*   Prove the power rule $\dfrac{d}{dx}[x^n] = nx^{n-1}$ for the case where $n = -p$ is a negative integer. [*Hint:* Apply the quotient rule to $y = x^{-p} = \dfrac{1}{x^p}$.]

71.   Use a graphing utility to sketch the curve $f(x) = x^2(x - 1)$, and on the same set of coordinate axes, draw the tangent line to the graph of $f(x)$ at $x = 1$. Find where $f'(x) = 0$.

72.   Use a graphing utility to sketch the curve $f(x) = \dfrac{3x^2 - 4x + 1}{x + 1}$, and on the same set of coordinate axes, draw the tangent lines to the graph of $f(x)$ at $x = -2$ and at $x = 0$. Find where $f'(x) = 0$.

73.   Graph $f(x) = x^4 + 2x^3 - x + 1$ using a graphing utility. Find the minima and maxima of this function by checking the coordinates along the curve of the graph. Find the derivative function $f'(x)$ algebraically and graph $f(x)$ and $f'(x)$ on the same axes. Find the x-intercepts of $f'(x)$. Explain why the maximum or minimum of $f(x)$ occurs at the x-intercepts of $f'(x)$.

74.   Repeat Exercise 73 for the product function $f(x) = x^3(x - 3)^2$.

SECTION 2.4

L04

Use the chain rule to find derivatives.

The Chain Rule

In many practical situations, the rate at which one quantity is changing can be expressed as a product of other rates. For example, suppose a car is travelling at 80 km/h at a particular time when gasoline is being consumed at a rate of 0.25 L/km. Then, to find out how much gasoline is being used each hour, multiply the rates:

$$(0.25 \text{ L/km})(80 \text{ km/h}) = 20 \text{ L/h}.$$

Or, suppose the total manufacturing cost at a certain factory is a function of the number of units produced, which in turn is a function of the number of hours the factory has been operating. If C, q, and t denote the cost, units produced, and time, respectively, then

$$\frac{dC}{dq} = \begin{bmatrix} \text{rate of change of cost} \\ \text{with respect to output} \end{bmatrix} \quad \text{(dollars per unit)}$$

and

$$\frac{dq}{dt} = \begin{bmatrix} \text{rate of change of output} \\ \text{with respect to time} \end{bmatrix} \quad \text{(units per hour).}$$

The product of these two rates is the rate of change of cost with respect to time; that is,

$$\frac{dC}{dt} = \frac{dC}{dq}\frac{dq}{dt} \quad \text{(dollars per hour).}$$

This formula is a special case of an important result in calculus called the **chain rule.**

The Chain Rule ■ If $y = f(u)$ is a differentiable function of u and $u = g(x)$ is in turn a differentiable function of x, then the composite function $y = f(g(x))$ is a differentiable function of x with derivative given by the product

$$\frac{dy}{dx} = \frac{dy}{du}\frac{du}{dx}$$

or, equivalently, by

$$\frac{dy}{dx} = f'(g(x))g'(x).$$

NOTE One way to remember the chain rule is to pretend that the derivatives $\dfrac{dy}{du}$ and $\dfrac{du}{dx}$ are quotients and to "cancel" du; that is,

$$\frac{dy}{dx} = \frac{dy}{d\bcancel{u}}\frac{d\bcancel{u}}{dx}.$$

To illustrate the use of the chain rule, suppose you wish to differentiate the function $y = (3x + 1)^2$. Your first instinct may be to "guess" that the derivative is

$$2(3x + 1) = 6x + 2.$$

But this guess cannot be correct since expanding $(3x + 1)^2$ and differentiating each term yields

$$\frac{dy}{dx} = \frac{d}{dx}[(3x + 1)^2] = \frac{d}{dx}[9x^2 + 6x + 1] = 18x + 6,$$

which is 3 times our "guess" of $6x + 2$. However, writing $y = (3x + 1)^2$ as $y = u^2$, where $u = 3x + 1$, gives

$$\frac{dy}{du} = \frac{d}{du}[u^2] = 2u \quad \text{and} \quad \frac{du}{dx} = \frac{d}{dx}[3x + 1] = 3,$$

and the chain rule tells you that

$$\frac{dy}{dx} = \frac{dy}{du}\frac{du}{dx}$$
$$= 2u(3)$$
$$= 6u$$
$$= 6(3x + 1)$$
$$= 18x + 6$$

which coincides with the correct answer found earlier by expanding $(3x + 1)^2$. Examples 2.4.1 and 2.4.2 illustrate various ways of using the chain rule.

EXAMPLE 2.4.1

Find $\dfrac{dy}{dx}$ if $y = (x^2 + 2)^3 - 3(x^2 + 2)^2 + 1$.

Solution

Note that $y = u^3 - 3u^2 + 1$, where $u = x^2 + 2$. Thus,

$$\frac{dy}{du} = 3u^2 - 6u \quad \text{and} \quad \frac{du}{dx} = 2x,$$

and, according to the chain rule,

$$\frac{dy}{dx} = \frac{dy}{du}\frac{du}{dx} = (3u^2 - 6u)(2x).$$

In Sections 2.1 through 2.3, you have seen several applications (e.g., slope, rates of change) that require the evaluation of a derivative at a particular value of the independent variable. There are two basic ways of doing this when the derivative is computed using the chain rule.

For instance, suppose in Example 2.4.1 we wish to evaluate $\dfrac{dy}{dx}$ when $x = -1$. One way to proceed would be to first express the derivative in terms of x alone by substituting $x^2 + 2$ for u as follows:

$$\frac{dy}{dx} = (3u^2 - 6u)(2x)$$
$$= [3(x^2 + 2)^2 - 6(x^2 + 2)](2x) \quad \text{replace } u \text{ with } x^2 + 2$$
$$= 6x(x^2 + 2)[(x^2 + 2) - 2] \quad \text{factor out } 6x(x^2 + 2)$$
$$= 6x(x^2 + 2)(x^2) \quad \text{combine terms in the brackets}$$
$$= 6x^3(x^2 + 2)$$

Then, substituting $x = -1$ into this expression, we would get

$$\left.\frac{dy}{dx}\right|_{x=-1} = 6(-1)^3[(-1)^2 + 2] = -18.$$

Alternatively, we could compute $u(-1) = (-1)^2 + 2 = 3$ and then substitute directly into the formula $\dfrac{dy}{dx} = (3u^2 - 6u)(2x)$ to obtain

$$\begin{aligned}
\left.\frac{dy}{dx}\right|_{x=-1} &= (3u^2 - 6u)(2x)\big|_{x=-1,\, u=3} \\
&= [3(3)^2 - 6(3)][2(-1)] \\
&= (9)(-2) \\
&= -18
\end{aligned}$$

Both methods yield the correct result, but since it is easier to substitute numbers than algebraic expressions, the second (numerical) method is often preferable, unless for some reason you need to have the derivative function $\dfrac{dy}{dx}$ expressed in terms of x alone. In Example 2.4.2, the numerical method for evaluating a derivative computed with the chain rule is used to find the slope of a tangent line.

EXAMPLE 2.4.2

Consider the function $y = \dfrac{u}{u + 1}$, where $u = 3x^2 - 1$.

a. Use the chain rule to find $\dfrac{dy}{dx}$.

b. Find an equation for the tangent line to the graph of $y(x)$ at the point where $x = 1$.

Solution

a. We use the quotient rule to find that

$$\begin{aligned}
\frac{dy}{du} &= \frac{(u + 1)(1) - u(1)}{(u + 1)^2} \\
&= \frac{1}{(u + 1)^2}
\end{aligned}$$

and then find

$$\frac{du}{dx} = 6x$$

According to the chain rule, it follows that

$$\frac{dy}{dx} = \frac{dy}{du}\frac{du}{dx} = \left[\frac{1}{(u + 1)^2}\right](6x) = \frac{6x}{(u + 1)^2}.$$

b. To find an equation for the tangent line to the graph of $y(x)$ at $x = 1$, we need to know the value of y and the slope at the point of tangency. Since

$$u(1) = 3(1)^2 - 1 = 2,$$

the value of y when $x = 1$ is

$$y(1) = \frac{(2)}{(2) + 1} = \frac{2}{3}$$

and the slope is

$$\frac{dy}{dx}\Big|_{x=1, u=2} = \frac{6(1)}{(2 + 1)^2} = \frac{6}{9} = \frac{2}{3}.$$

Therefore, by applying the point-slope formula for the equation of a line, we find that the tangent line to the graph of $y(x)$ at the point where $x = 1$ has the equation

$$\frac{y - \frac{2}{3}}{x - 1} = \frac{2}{3}$$

$$y = \frac{2}{3}x$$

In many practical problems, a quantity is given as a function of one variable, which, in turn, can be written as a function of a second variable, and the goal is to find the rate of change of the original quantity with respect to the second variable. Such problems can be solved by means of the chain rule. Here is an example.

EXAMPLE 2.4.3

The cost of producing x units of a particular commodity is $C(x) = \frac{1}{3}x^2 + 4x + 53$ dollars, and the production level t hours into a particular production run is $x(t) = 0.2t^2 + 0.03t$ units. At what rate is cost changing with respect to time after 4 hours?

Solution
We find that

$$\frac{dC}{dx} = \frac{2}{3}x + 4 \quad \text{and} \quad \frac{dx}{dt} = 0.4t + 0.03,$$

so, according to the chain rule,

$$\frac{dC}{dt} = \frac{dC}{dx}\frac{dx}{dt} = \left(\frac{2}{3}x + 4\right)(0.4t + 0.03).$$

When $t = 4$, the level of production is

$$x(4) = 0.2(4)^2 + 0.03(4) = 3.32 \text{ units,}$$

and by substituting $t = 4$ and $x = 3.32$ into the formula for $\frac{dC}{dt}$, we get

$$\frac{dC}{dt}\Big|_{t=4, x=3.32} = \left[\frac{2}{3}(3.32) + 4\right][0.4(4) + 0.03] = 10.1277.$$

Thus, after 4 hours, cost is increasing at a rate of approximately $10.13 per hour.

Just-In-Time

Think of the chain rule like a Boston cream doughnut. First do the derivative of the whole doughnut, then open it up and do the derivative of the cream inside. Don't forget the cream inside! (Photo: Dot Miners)

Sometimes when dealing with a composite function $y = f(g(x))$ it may help to think of f as the "outer" function and g as the "inner" function, as indicated here:

"outer" function ⟶

$$y = f(g(x))$$

⟵ "inner" function

Then the chain rule

$$\frac{dy}{dx} = f'(g(x))g'(x)$$

says that *the derivative of $y = f(g(x))$ with respect to x is given by the derivative of the outer function evaluated at the inner function times the derivative of the inner function.* In Example 2.4.4, we emphasize this interpretation by using a box ($\square$) to indicate the location and role of the inner function in computing a derivative with the chain rule.

EXAMPLE 2.4.4

Differentiate the function $f(x) = \sqrt{x^2 + 3x + 2}$.

Solution

The form of the function is

$$f(x) = (\square)^{1/2},$$

where the box $\square$ contains the expression $x^2 + 3x + 2$. Then

$$(\square)' = (x^2 + 3x + 2)' = 2x + 3,$$

and, according to the chain rule, the derivative of the composite function $f(x)$ is

$$f'(x) = \frac{1}{2}(\square)^{-1/2}(\square)'$$

$$= \frac{1}{2}(\square)^{-1/2}(2x + 3)$$

$$= \frac{1}{2}(x^2 + 3x + 2)^{-1/2}(2x + 3) = \frac{2x + 3}{2\sqrt{x^2 + 3x + 2}}$$

The General Power Rule

In Section 2.2, you learned the rule

$$\frac{d}{dx}[x^n] = nx^{n-1}$$

for differentiating power functions. Combining this rule with the chain rule results in the following rule for differentiating functions of the general form $[h(x)]^n$.

The General Power Rule ■ For any real number n and differentiable function h,

$$\frac{d}{dx}[h(x)]^n = n[h(x)]^{n-1}\frac{d}{dx}[h(x)].$$

To derive the general power rule, think of $[h(x)]^n$ as the composite function

$$[h(x)]^n = g[h(x)], \quad \text{where } g(u) = u^n.$$

Then $g'(u) = nu^{n-1}$ and $h'(x) = \dfrac{d}{dx}[h(x)]$, and, by the chain rule,

$$\frac{d}{dx}[h(x)]^n = \frac{d}{dx}g[h(x)] = g'[h(x)]h'(x) = n[h(x)]^{n-1}\frac{d}{dx}[h(x)].$$

The use of the general power rule is illustrated in Examples 2.4.5 through 2.4.7.

EXAMPLE 2.4.5

Differentiate the function $f(x) = (2x^4 - x)^3$.

Solution

One way to solve this problem is to expand the function and rewrite it as

$$f(x) = 8x^{12} - 12x^9 + 6x^6 - x^3$$

and then differentiate this polynomial term by term to get

$$f'(x) = 96x^{11} - 108x^8 + 36x^5 - 3x^2.$$

But see how much easier it is to use the general power rule. According to this rule,

$$f'(x) = 3(2x^4 - x)^2 \frac{d}{dx}[2x^4 - x] = 3(2x^4 - x)^2(8x^3 - 1).$$

Not only is this method easier, but the answer even comes out in factored form!

In Example 2.4.6, the solution to Example 2.4.4 is written more compactly with the aid of the general power rule.

EXAMPLE 2.4.6

Differentiate the function $f(x) = \sqrt{x^2 + 3x + 2}$.

Solution

Rewrite the function as $f(x) = (x^2 + 3x + 2)^{1/2}$ and apply the general power rule:

$$f'(x) = \left[\frac{1}{2}(x^2 + 3x + 2)^{-1/2}\right]\frac{d}{dx}[x^2 + 3x + 2]$$

$$= \frac{1}{2}(x^2 + 3x + 2)^{-1/2}(2x + 3)$$

$$= \frac{2x + 3}{2\sqrt{x^2 + 3x + 2}}$$

EXAMPLE 2.4.7

Differentiate the function $f(x) = \dfrac{1}{(2x + 3)^5}$.

Solution

Although you can use the quotient rule, it is easier to rewrite the function as

$$f(x) = (2x + 3)^{-5}$$

and apply the general power rule to get

$$f'(x) = -5(2x + 3)^{-6}\frac{d}{dx}[2x + 3] = -5(2x + 3)^{-6}(2) = -\frac{10}{(2x + 3)^6}.$$

The chain rule is often used in combination with the other rules you learned in Sections 2.2 and 2.3. Example 2.4.8 involves the product rule.

Just-In-Time

When simplifying by factoring, look at the bracketed parts and find the largest exponent of each that can come out as a factor. Then check out what is left and put it in square brackets to make it easier to follow. Remember to multiply out in your head to make sure you factored correctly.

Then simplify the terms in the square brackets.

To put this mathematically, if A and B are the bracketed factors, then if $m > n$ and $q > p$, $rA^mB^p + sA^nB^q = A^nB^p(rA^{m-n} + sB^{q-p})$ for any constants r and s.

EXAMPLE 2.4.8

Differentiate the function $f(x) = (3x + 1)^4(2x - 1)^5$ and simplify your answer. Then find all values of $x = c$ for which the tangent line to the graph of $f(x)$ at $(c, f(c))$ is horizontal.

Solution

First apply the product rule to get

$$f'(x) = (3x + 1)^4\frac{d}{dx}[(2x - 1)^5] + (2x - 1)^5\frac{d}{dx}[(3x + 1)^4].$$

Continue by applying the general power rule to each term:

$$f'(x) = (3x + 1)^4[5(2x - 1)^4(2)] + (2x - 1)^5[4(3x + 1)^3(3)]$$
$$= 10(3x + 1)^4(2x - 1)^4 + 12(2x - 1)^5(3x + 1)^3$$

Finally, simplify your answer by factoring:

$$f'(x) = 2(3x + 1)^3(2x - 1)^4[5(3x + 1) + 6(2x - 1)]$$
$$= 2(3x + 1)^3(2x - 1)^4[15x + 5 + 12x - 6]$$
$$= 2(3x + 1)^3(2x - 1)^4(27x - 1)$$

The tangent line to the graph of $f(x)$ is horizontal at points $(c, f(c))$ where $f'(c) = 0$. By solving

$$f'(x) = 2(3x + 1)^3(2x - 1)^4(27x - 1) = 0,$$

we see that $f'(c) = 0$ where

$$3c + 1 = 0 \quad \text{and} \quad 2c - 1 = 0 \quad \text{and} \quad 27c - 1 = 0,$$

that is, at $c = -\dfrac{1}{3}$, $c = \dfrac{1}{2}$, and $c = \dfrac{1}{27}$.

EXAMPLE 2.4.9

Find the second derivative of the function $f(x) = \dfrac{3x - 2}{(x - 1)^2}$.

Solution

Using the quotient rule, along with the general power rule (applied to $(x - 1)^2$), we get

$$f'(x) = \frac{(x - 1)^2(3) - (3x - 2)[2(x - 1)(1)]}{(x - 1)^4}$$

$$= \frac{(x - 1)[3(x - 1) - 2(3x - 2)]}{(x - 1)^4}$$

$$= \frac{3x - 3 - 6x + 4}{(x - 1)^3}$$

$$= \frac{1 - 3x}{(x - 1)^3}$$

Using the quotient rule again, this time applying the general power rule to $(x - 1)^3$, we find that

$$f''(x) = \frac{(x - 1)^3(-3) - (1 - 3x)[3(x - 1)^2(1)]}{(x - 1)^6}$$

$$= \frac{-3(x - 1)^2[(x - 1) + (1 - 3x)]}{(x - 1)^6}$$

$$= \frac{-3(-2x)}{(x - 1)^4}$$

$$= \frac{6x}{(x - 1)^4}$$

EXAMPLE 2.4.10

An environmental study of a certain suburban community suggests that the average daily level of carbon monoxide in the air will be $c(p) = \sqrt{0.5p^2 + 17}$ parts per million when the population is p thousand. It is estimated that t years from now, the population of the community will be $p(t) = 3.1 + 0.1t^2$ thousand. At what rate will the carbon monoxide level be changing with respect to time 3 years from now?

Solution

The goal is to find $\dfrac{dc}{dt}$ when $t = 3$. Since

$$\frac{dc}{dp} = \frac{1}{2}(0.5p^2 + 17)^{-1/2}[0.5(2p)] = \frac{1}{2}p(0.5p^2 + 17)^{-1/2}$$

and

$$\frac{dp}{dt} = 0.2t,$$

it follows from the chain rule that

$$\frac{dc}{dt} = \frac{dc}{dp}\frac{dp}{dt} = \frac{1}{2}p(0.5p^2 + 17)^{-1/2}(0.2t) = \frac{0.1pt}{\sqrt{0.5p^2 + 17}}.$$

When $t = 3$,

$$p(3) = 3.1 + 0.1(3)^2 = 4,$$

and by substituting $t = 3$ and $p = 4$ into the formula for $\dfrac{dc}{dt}$, we get

$$\frac{dc}{dt} = \frac{0.1(4)(3)}{\sqrt{0.5(4)^2 + 17}}$$

$$= \frac{1.2}{\sqrt{25}} = \frac{1.2}{5} = 0.24 \text{ parts per million per year.}$$

EXAMPLE 2.4.11

The manager of an appliance manufacturing firm determines that when rice cookers are priced at p dollars each, the number sold each month can be modelled by

$$D(p) = \frac{8000}{p}.$$

The manager estimates that t months from now, the unit price of the rice cookers will be $p(t) = 0.06t^{3/2} + 22.5$ dollars. At what rate will the monthly demand for rice cookers $D(p)$ be changing 25 months from now? Will it be increasing or decreasing at this time?

Solution

We want to find $\dfrac{dD}{dt}$ when $t = 25$. We have

$$\frac{dD}{dp} = \frac{d}{dp}\left[\frac{8000}{p}\right] = -\frac{8000}{p^2}$$

and

$$\frac{dp}{dt} = \frac{d}{dt}[0.06t^{3/2} + 22.5] = 0.06\left(\frac{3}{2}t^{1/2}\right) = 0.09t^{1/2},$$

so it follows from the chain rule that

$$\frac{dD}{dp} = \frac{dD}{dp}\frac{dp}{dt} = \left[-\frac{8000}{p^2}\right](0.09t^{1/2}).$$

When $t = 25$, the unit price is

$$p(25) = 0.06(25)^{3/2} + 22.5 = 30$$

dollars, and we have

$$\left.\frac{dD}{dt}\right|_{t=25,\, p=30} = \left[-\frac{8000}{30^2}\right][0.09(25)^{1/2}] = -4.$$

That is, 25 months from now, the demand for rice cookers will be changing at a rate of 4 units per month and will be decreasing since $\dfrac{dD}{dt}$ is negative.

ALGEBRA WARM-UP

Simplify by factoring.

1. $10x^3(2x - 1)^4 + 3x^2(2x - 1)^5$

2. $12x^2(3x + 1)^3 + 2x(2x - 1)^4$

3. $15(2 - 5x)^4(3x - 1)^4 - 20(2 - 5x)^3(3x - 1)^5$

4. $-8(7 - 4x)^2(3 - 2x)^3 - 12(7 - 4x)^3(3 - 2x)^4$

EXERCISES ■ 2.4

In Exercises 1 through 12, use the chain rule to compute the derivative $\dfrac{dy}{dx}$, and simplify your answer.

1. $y = u^2 + 1; u = 3x - 2$

2. $y = 1 - 3u^2; u = 3 - 2x$

3. $y = \sqrt{u}; u = x^2 + 2x - 3$

4. $y = 2u^2 - u + 5; u = 1 - x^2$

5. $y = \dfrac{1}{u^2}; u = x^2 + 1$

6. $y = \dfrac{1}{u^2}; u = 3x^2 + 5$

7. $y = \dfrac{1}{u - 1}; u = x^2$

8. $y = \dfrac{1}{\sqrt{u}}; u = x^2 - 9$

9. $y = u^2 + 2u - 3; u = \sqrt{x}$

10. $y = u^3 + u; u = \dfrac{1}{\sqrt{x}}$

11. $y = u^2 + u - 2; u = \dfrac{1}{x}$

12. $y = u^2; u = \dfrac{1}{x - 1}$

In Exercises 13 through 20, use the chain rule to compute the derivative $\dfrac{dy}{dx}$ for the given value of x.

13. $y = u^2 - u; u = 4x + 3$ for $x = 0$

14. $y = u + \dfrac{1}{u}; u = 5 - 2x$ for $x = 0$

15. $y = 3u^4 - 4u + 5; u = x^3 - 2x - 5$ for $x = 2$

16. $y = u^5 - 3u^2 + 6u - 5; u = x^2 - 1$ for $x = 1$

17. $y = \sqrt{u}; u = x^2 - 2x + 6$ for $x = 3$

18. $y = 3u^2 - 6u + 2; u = \dfrac{1}{x^2}$ for $x = \dfrac{1}{3}$

19. $y = \dfrac{1}{u}; u = 3 - \dfrac{1}{x^2}$ for $x = \dfrac{1}{2}$

20. $y = \dfrac{1}{u + 1}; u = x^3 - 2x + 5$ for $x = 0$

In Exercises 21 through 42, differentiate the given function and simplify your answer.

21. $f(x) = (2x + 3)^{1.4}$

22. $f(x) = \dfrac{1}{\sqrt{5 - 3x}}$

23. $f(x) = (2x + 1)^4$

24. $f(x) = \sqrt{5x^6 - 12}$

25. $f(x) = (x^5 - 4x^3 - 7)^8$

26. $f(t) = (3t^4 - 7t^2 + 9)^5$

27. $f(t) = \dfrac{2}{5t^2 - 6t + 2}$

28. $f(x) = \dfrac{2}{(6x^2 + 5x + 1)^2}$

29. $g(x) = \dfrac{1}{\sqrt{4x^2 + 1}}$

30. $f(s) = \dfrac{1}{\sqrt{5s^3 + 2}}$

31. $f(x) = \dfrac{3}{(1 - x^2)^4}$

32. $f(x) = \dfrac{2}{3(5x^4 + 1)^2}$

33. $h(s) = (1 + \sqrt{3s})^5$

34. $g(x) = \sqrt{1 + \dfrac{1}{3x}}$

35. $f(x) = (x + 2)^3(2x - 1)^5$

36. $f(x) = 2(3x + 1)^4(5x - 3)^2$

37. $G(x) = \sqrt{\dfrac{3x + 1}{2x - 1}}$

38. $f(y) = \left(\dfrac{y + 2}{2 - y}\right)^3$

39. $f(x) = \dfrac{(x + 1)^5}{(1 - x)^4}$

40. $F(x) = \dfrac{(1 - 2x)^2}{(3x + 1)^3}$

41. $f(y) = \dfrac{3y + 1}{\sqrt{1 - 4y}}$

42. $f(x) = \dfrac{1 - 5x^2}{\sqrt[3]{3 + 2x}}$

In Exercises 43 through 50, find the equation of the line that is tangent to the graph of f for the given value of x.

43. $f(x) = \sqrt{3x + 4}$; $x = 0$

44. $f(x) = (9x - 1)^{-1/3}$; $x = 1$

45. $f(x) = (3x^2 + 1)^2$; $x = -1$

46. $f(x) = (x^2 - 3)^5(2x - 1)^3$; $x = 2$

47. $f(x) = \dfrac{1}{(2x - 1)^6}$; $x = 1$

48. $f(x) = \left(\dfrac{x + 1}{x - 1}\right)^3$; $x = 3$

49. $f(x) = \sqrt[3]{\dfrac{x}{x + 2}}$; $x = -1$

50. $f(x) = x^2\sqrt{2x + 3}$; $x = -1$

In Exercises 51 through 56, find all values of x = c so that the tangent line to the graph of f(x) at (c, f(c)) is horizontal.

51. $f(x) = (x^2 + x)^2$

52. $f(x) = x^3(2x^2 + x - 3)^2$

53. $f(x) = \dfrac{x}{(3x - 2)^2}$

54. $f(x) = \dfrac{2x + 5}{(1 - 2x)^3}$

55. $f(x) = \sqrt{x^2 - 4x + 5}$

56. $f(x) = (x - 1)^2(2x + 3)^3$

In Exercises 57 and 58, differentiate the given function f(x) by two different methods, first by using the general power rule and then by using the product rule. Show that the two answers are the same.

57. $f(x) = (3x + 5)^2$

58. $f(x) = (7 - 4x)^2$

In Exercises 59 through 64, find the second derivative of the given function.

59. $f(x) = (3x + 1)^5$

60. $f(t) = \dfrac{2}{5t + 1}$

61. $h(t) = (t^2 + 5)^8$

62. $y = (1 - 2x^3)^4$

63. $f(x) = \sqrt{1 + x^2}$

64. $f(u) = \dfrac{1}{(3u^2 - 1)^2}$

65. ANNUAL EARNINGS   The gross annual earnings of a certain company are $f(t) = \sqrt{10t^2 + t + 229}$ thousand dollars t years after its formation in January 2007.
   a. At what rate will the gross annual earnings of the company be growing in January 2012?
   b. At what percentage rate will the gross annual earnings be growing in January 2012?

66. MANUFACTURING COST   At a certain factory, the total cost of manufacturing q units is $C(q) = 0.2q^2 + q + 900$ dollars. It has been determined that approximately $q(t) = t^2 + 100t$ units are manufactured during the first t hours of a production run. Compute the rate at which the total manufacturing cost is changing with respect to time 1 h after production starts.

67. CONSUMER DEMAND   An importer of Brazilian coffee estimates that local consumers will buy approximately $D(p) = \dfrac{4374}{p^2}$ kilograms of the coffee per week when the price is p dollars per kilogram. It is also estimated that t weeks from now, the price of Brazilian coffee will be $p(t) = 0.02t^2 + 0.1t + 6$ dollars per kilogram.

a. At what rate will the demand for coffee be changing with respect to price when the price is $9?

b. At what rate will the demand for coffee be changing with respect to time 10 weeks from now? Will the demand be increasing or decreasing at this time?

68. **CONSUMER DEMAND**   When printer paper is sold for p dollars per packet, consumers will buy $D(p) = \dfrac{40\,000}{p}$ units per month. It is estimated that t months from now, the price of the paper will be $p(t) = 0.4t^{3/2} + 6.8$ dollars per unit. At what rate will the monthly demand for the paper be changing with respect to time 4 months from now?

69. **AIR POLLUTION**   It is estimated that t years from now, the population of a certain suburban community will be $p(t) = 20 - \dfrac{6}{t+1}$ thousand. An environmental study indicates that the average daily level of carbon monoxide in the air will be $c(p) = 0.5\sqrt{p^2 + p + 58}$ parts per million when the population is p thousand.

a. At what rate will the level of carbon monoxide be changing with respect to population when the population is 18 thousand people?

b. At what rate will the carbon monoxide level be changing with respect to time 2 years from now? Will the level be increasing or decreasing at this time?

70. **JUST FOR YOU**   Plot the function $I(x) = (ax^2 - bx + c)^2$ with technology, where a, b, c, and d are numbers that come from your age as follows. The month of your birthday is a, the last digit of the year is b, and the day is c. By hand, calculate the derivative and find the equation of the tangent at (d, e), where d and e are the middle two digits of your birth year. Plot the function and the tangent together on the same graph. You will know if you are correct by the appearance and position of the tangent line.

71. **MAMMALIAN GROWTH**   Observations show that the length L in millimetres from nose to tip of tail of a Siberian tiger that is less than 6 months old, can be estimated using the function $L = 0.25w^{2.6}$,

where w is the weight of the tiger in kilograms. Furthermore, when a tiger is less than 6 months old, its weight can be estimated in terms of its age A in days by the function $w = 3 + 0.21A$.

a. At what rate is the length of a Siberian tiger increasing with respect to its weight when it weighs 60 kg?

b. How long is a Siberian tiger when it is 100 days old? At what rate is its length increasing with respect to time at this age?

72. **QUALITY OF LIFE**   A demographic study models the population p (in thousands) of a community by the function

$$p(Q) = 3Q^2 + 4Q + 200,$$

where Q is a quality-of-life index that ranges from $Q = 0$ (extremely poor quality) to $Q = 10$ (excellent quality). Suppose the index varies with time in such a way that t years from now,

$$Q(t) = \dfrac{t^2 + 2t + 3}{2t + 1}$$

for $0 \le t \le 10$.

a. What value of the quality-of-life index should be expected 4 years from now? What will the corresponding population be at this time?

b. At what rate is the population changing with respect to time 4 years from now? Is the population increasing or decreasing at this time?

73. **WATER POLLUTION**   When organic matter is introduced into a body of water, the oxygen content of the water is temporarily reduced by oxidation. Suppose that t days after untreated sewage is dumped into a particular lake, the proportion of the usual oxygen content in the water of the lake that remains is given by the function

$$P(t) = 1 - \dfrac{12}{t+12} + \dfrac{144}{(t+12)^2}.$$

a. At what rate is the oxygen proportion $P(t)$ changing after 10 days? Is the proportion increasing or decreasing at this time?

b. Is the oxygen proportion increasing or decreasing after 15 days?

c. If there is no new dumping, what would you expect to eventually happen to the proportion of oxygen? Use a limit to verify your conjecture.

74. **PRODUCTION**   The number of units Q of a particular commodity that will be produced

when L worker-hours of labour are employed is modelled by

$$Q(L) = 300L^{1/3}.$$

Suppose that the labour level varies with time in such a way that t months from now, $L(t)$ worker-hours will be employed, where

$$L(t) = \sqrt{739 + 3t - t^2}$$

for $0 \le t \le 12$.

a. How many worker-hours will be employed in producing the commodity 5 months from now? How many units will be produced at this time?

b. At what rate will production be changing with respect to time 5 months from now? Will production be increasing or decreasing at this time?

75. PRODUCTION The number of units Q of a particular commodity that will be produced with K thousand dollars of capital expenditure is modelled by

$$Q(K) = 500K^{2/3}.$$

Suppose that capital expenditure varies with time in such a way that t months from now there will be $K(t)$ thousand dollars of capital expenditure, where

$$K(t) = \frac{2t^4 + 3t + 149}{t + 2}.$$

a. What will be the capital expenditure 3 months from now? How many units will be produced at this time?

b. At what rate will production be changing with respect to time 5 months from now? Will production be increasing or decreasing at this time?

76. A* DEPRECIATION The value V (in thousands of dollars) of an industrial machine is modelled by

$$V(N) = \left(\frac{3N + 430}{N + 1} \right)^{2/3},$$

where N is the number of hours the machine is used each day. Suppose further that usage varies with time in such a way that

$$N(t) = \sqrt{t^2 - 10t + 45},$$

where t is the number of months the machine has been in operation.

a. How many hours per day will the machine be used 9 months from now? What will be the value of the machine at this time?

b. At what rate is the value of the machine changing with respect to time 9 months from now? Will the value be increasing or decreasing at this time?

77. A* INSECT GROWTH The growth of certain insects varies with temperature. Suppose a particular species of insect grows in such a way that the volume of an individual in cubic centimetres is

$$V(T) = 0.41(-0.01T^2 + 0.4T + 3.52)$$

when the temperature is T degrees Celsius, and that its mass is m grams, where

$$m(V) = \frac{0.39V}{1 + 0.09V}.$$

a. Find the rate of change of the insect's volume with respect to temperature.

b. Find the rate of change of the insect's mass with respect to volume.

c. When $T = 10°C$, what is the insect's volume? At what rate is the insect's mass changing with respect to temperature when $T = 10°C$?

78. COMPOUND INTEREST If \$10 000 is invested at an annual rate r (expressed as a decimal) compounded weekly, the total amount (principal P and interest) accumulated after 10 years is given by the formula

$$A = 10\,000 \left(1 + \frac{r}{52} \right)^{520}.$$

Find the rate of change of A with respect to r.

79. A* LEARNING When you first begin to study a topic or practise a skill, you may not be very good at it, but in time, you will approach the limits of your ability. One model for describing this behaviour involves the function

$$T = aL\sqrt{L - b},$$

where T is the time required for a particular person to learn the items on a list of L items and a and b are positive constants. Find the derivative $\dfrac{dT}{dL}$ and interpret it in terms of the learning model.

80. A man walks along a straight road with velocity $v(t)$ metres per hour after t hours, where

$$v(t) = (2t + 9)^2(8 - t)^3 \text{ for } 0 \le t \le 5.$$

a. Find the acceleration $a(t)$ of the man at time t.

b. When is the man stationary for $0 \le t \le 5$? Find the acceleration at each time.

c. When is the acceleration zero for $0 \le t \le 5$? Find the velocity at each time.

d. Use a graphing utility to plot the graphs of the velocity $v(t)$ and acceleration $a(t)$ on the same screen.

e. The man is said to be *speeding up* when $v(t)$ and $a(t)$ have the same sign (both positive or both negative). Use a graphing utility to determine when (if ever) this occurs for $0 \le t \le 5$.

81. A leaf is blown by the wind in a straight line in both directions in such a way that its position at time t is given by

$$s(t) = (3 + t - t^2)^{3/2} \quad \text{for } 0 \le t \le 2.$$

a. What are the leaf's velocity $v(t)$ and acceleration $a(t)$ at time t?

b. When is the leaf stationary for $0 \le t \le 2$? Where is the leaf and what is its acceleration at each time?

c. When is the acceleration zero for $0 \le t \le 2$? What are the leaf's position and velocity at each time?

d. Use a graphing utility to draw the graphs of the leaf's position $s(t)$, velocity $v(t)$, and acceleration $a(t)$ on the same screen for $0 \le t \le 2$.

e. The leaf is said to be *slowing down* when $v(t)$ and $a(t)$ have opposite signs (one positive, the other negative). Use your graphs to determine when (if ever) this occurs for $0 \le t \le 2$.

82. **A*** Suppose $L(x)$ is a function with the property that $L'(x) = \dfrac{1}{x}$. Use the chain rule to find the derivatives of the following functions and simplify your answers.

a. $f(x) = L(x^2)$

b. $f(x) = L\left(\dfrac{1}{x}\right)$

c. $f(x) = L\left(\dfrac{2}{3\sqrt{x}}\right)$

d. $f(x) = L\left(\dfrac{2x + 1}{1 - x}\right)$

83. **A*** Prove the general power rule for $n = 2$ by using the product rule to compute $\dfrac{dy}{dx}$ if $y = [h(x)]^2$.

84. **A*** Prove the general power rule for $n = 3$ by using the product rule and the result of Exercise 83 to compute $\dfrac{dy}{dx}$ if $y = [h(x)]^2$. [*Hint:* Begin by writing y as $h(x)[h(x)]^2$.]

85. Find the derivative of $f(x) = \sqrt[3]{3.1x^2 + 19.4}$ with a graphing utility. Explore the graph of $f(x)$. How many horizontal tangents does the graph have?

86. Find the derivative of $f(x) = (2.7x^3 - 3\sqrt{x} + 5)^{2/3}$ with a graphing utility. Explore the graph of $f(x)$. How many horizontal tangents does it have?

SECTION 2.5

L05

Use marginal analysis to solve business problems, and differentials to approximate by increments.

Marginal Analysis and Approximations Using Increments

Calculus is an important tool in economics. We briefly discussed sales and production in Chapter 1, where we introduced economic concepts such as cost, revenue, profit, supply, demand, and market equilibrium. In this section, we will use the derivative to explore rates of change involving economic quantities.

Marginal Analysis

In economics, the use of the derivative to approximate the change in a quantity that results from a 1-unit increase in production is called **marginal analysis.** This depends on the assumption that 1 unit is a small number, so the choice of units must also be considered. For instance, suppose $C(x)$ is the total cost of producing x units of a particular commodity. If x_0 units are currently being produced, then the derivative

$$C'(x_0) = \lim_{h \to 0} \frac{C(x_0 + h) - C(x_0)}{h}$$

is called the **marginal cost** of producing x_0 units. The limiting value that defines this derivative is approximately equal to the difference quotient of $C(x)$ when $h = 1$; that is,

$$C'(x_0) \approx \frac{C(x_0 + 1) - C(x_0)}{1} = C(x_0 + 1) - C(x_0),$$

where the symbol $\approx$ is used to indicate that this is an approximation, not an equality. But $C(x_0 + 1) - C(x_0)$ is just the cost of increasing the level of production by 1 unit, from x_0 to $x_0 + 1$. To summarize:

> **Marginal Cost** ■ If $C(x)$ is the total cost of producing x units of a commodity, then the **marginal cost** of producing x_0 units is the derivative $C'(x_0)$, which approximates the additional cost $C(x_0 + 1) - C(x_0)$ incurred when the level of production is increased by 1 unit, from x_0 to $x_0 + 1$.

The geometric relationship between the marginal cost $C'(x_0)$ and the additional cost $C(x_0 + 1) - C(x_0)$ is shown in Figure 2.14.

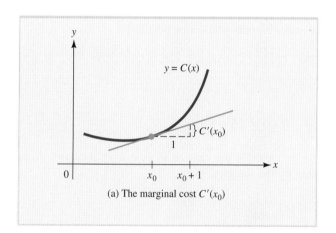

(a) The marginal cost $C'(x_0)$

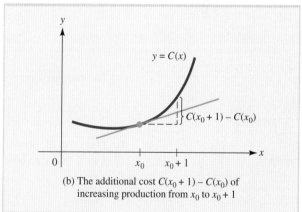

(b) The additional cost $C(x_0 + 1) - C(x_0)$ of increasing production from x_0 to $x_0 + 1$

FIGURE 2.14 Marginal cost $C'(x_0)$ approximates $C(x_0 + 1) - C(x_0)$.

The preceding discussion applies not only to cost, but also to other economic quantities. Here is a summary of what is meant by marginal revenue and marginal profit and how these marginal quantities can be used to estimate 1-unit changes in revenue and profit.

> **Marginal Revenue and Marginal Profit** ■ Suppose $R(x)$ is the revenue generated when x units of a particular commodity are produced, and $P(x)$ is the corresponding profit. Suppose $x = x_0$ units are being produced.
>
> The **marginal revenue** is $R'(x_0)$. It approximates $R(x_0 + 1) - R(x_0)$, the additional revenue generated by producing one more unit.
>
> The **marginal profit** is $P'(x_0)$. It approximates $P(x_0 + 1) - P(x_0)$, the additional profit obtained by producing one more unit.

Marginal analysis is illustrated in Example 2.5.1.

EXAMPLE 2.5.1

A manufacturer estimates that when x units of a particular commodity are produced, the total cost will be $C(x) = \frac{1}{8}x^2 + 3x + 98$ dollars and, furthermore, that all x units will be sold when the price is $p(x) = \frac{1}{3}(75 - x)$ dollars per unit.

a. Find the marginal cost and the marginal revenue.
b. Use marginal cost to estimate the cost of producing the 9th unit.
c. What is the actual cost of producing the ninth unit?
d. Use marginal revenue to estimate the revenue derived from the sale of the 9th unit.
e. What is the actual revenue derived from the sale of the 9th unit?

Solution

a. The marginal cost is $C'(x) = \frac{1}{4}x + 3$. Since x units of the commodity are sold at a price of $p(x) = \frac{1}{3}(75 - x)$ dollars per unit, the total revenue is

$$R(x) = (\text{number of units sold})(\text{price per unit})$$
$$= xp(x) = x\left[\frac{1}{3}(75 - x)\right] = 25x - \frac{1}{3}x^2$$

The marginal revenue is

$$R'(x) = 25 - \frac{2}{3}x.$$

b. The cost of producing the 9th unit is the change in cost as x increases from 8 to 9 and can be estimated by the marginal cost

$$C'(8) = \frac{1}{4}(8) + 3 = \$5.$$

c. The actual cost of producing the 9th unit is

$$C(9) - C(8) = \$5.12,$$

which is reasonably well approximated by the marginal cost $C'(8) = \$5$.

d. The revenue obtained from the sale of the 9th unit is approximated by the marginal revenue

$$R'(8) = 25 - \frac{2}{3}(8) = \$19.67.$$

e. The actual revenue obtained from the sale of the ninth unit is

$$R(9) - R(8) = \$19.33.$$

In Example 2.5.2, a marginal economic quantity is used to analyze a production process.

EXAMPLE 2.5.2

A manufacturer of cameras estimates that when x hundred cameras are produced, the total profit will be

$$P(x) = -0.0035x^3 + 0.07x^2 + 25x - 200$$

thousand dollars.

a. Find the marginal profit function.

b. What are the marginal profits when the levels of production are $x = 10$, $x = 50$, and $x = 80$?

c. Interpret these results.

Solution

a. The marginal profit is given by the derivative

$$P'(x) = -0.0035(3x^2) + 0.07(2x) + 25$$
$$= -0.0105x^2 + 0.14x + 25$$

b. The marginal profits at $x = 10$, $x = 50$, and $x = 80$ are

$$P'(10) = -0.0105(10)^2 + 0.14(10) + 25 = 25.35$$
$$P'(50) = -0.0105(50)^2 + 0.14(50) + 25 = 5.75$$
$$P'(80) = -0.0105(80)^2 + 0.14(80) + 25 = -31$$

Graph labeled $P(x) = -0.0035x^3 + 0.07x^2 + 25x - 200$, vertical axis Profit $P(x)$, horizontal axis x.

c. The fact that $P'(10) = 25.35$ means that a 1-unit increase in production from 10 to 11 hundred cameras increases profit by approximately 25.35 thousand dollars ($25 350), so the manager may be inclined to increase production at this level. However, since $P'(50) = 5.75$, increasing the level of production from 50 units to 51 units increases the profit by only about $5750, thus providing relatively little incentive for the manager to make a change. Finally, since $P'(80) = -31$ is negative, the profit will actually *decrease* by approximately $31 000 if the production level is raised from 80 to 81 units. The manager may wish to consider decreasing the level of production in this case. Look at the graph of $P(x)$ in the margin and visualize these answers on the slope of the graph.

Approximation by Increments

Marginal analysis is an important example of a general approximation procedure based on the fact that since

$$f'(x_0) = \lim_{h \to 0} \frac{f(x_0 + h) - f(x_0)}{h},$$

then for small h, the derivative $f'(x_0)$ is approximately equal to the difference quotient

$$\frac{f(x_0 + h) - f(x_0)}{h}.$$

We indicate this approximation by writing

$$f'(x_0) \approx \frac{f(x_0 + h) - f(x_0)}{h}$$

or, equivalently,

$$f(x_0 + h) - f(x_0) \approx f'(x_0)h.$$

To emphasize that the incremental change is in the variable x, we write $h = \Delta x$ (read Δx as "delta x") and summarize the incremental approximation formula as follows.

> **Approximation by Increments** ■ If $f(x)$ is differentiable at $x = x_0$ and Δx is a small change in x, then
>
> $$f(x_0 + \Delta x) \approx f(x_0) + f'(x_0)\Delta x$$
>
> or, equivalently, if
>
> $$\Delta f = f(x_0 + \Delta x) - f(x_0),$$
>
> then
>
> $$\Delta f \approx f'(x_0)\Delta x.$$

For example, consider the sensitivity of an investment portfolio of value P. If $P' = 30\,000$, for instance, we say it describes how our portfolio value increases by \$30 000 per dollar increase in stock price. What happens when the stock price goes up \$0.20? Intuitively we think the gain is $0.20(\$30\,000) = \$6000$.

Looking at the formula, $P(x + 0.2) - P(x) = P'(x)(0.2)$, which gives the same result.

Here is an example of how this approximation formula can be used in economics.

EXAMPLE 2.5.3

Suppose the total cost in dollars of manufacturing q units of a certain commodity is $C(q) = 3q^2 + 5q + 10$. If the current level of production is 40 units, estimate how the total cost will change if 40.5 units are produced.

Solution

In this problem, the current value of production is $q = 40$ and the change in production is $\Delta q = 0.5$. By the approximation formula, the corresponding change in cost is

$$\Delta C = C(40.5) - C(40) \approx C'(40)\Delta q = [C'(40)](0.5).$$

Since

$$C'(q) = 6q + 5 \text{ and } C'(40) = 6(40) + 5 = 245,$$

it follows that

$$\Delta C \approx [C'(40)](0.5) = 245(0.5) = \$122.50.$$

For practice, compute the actual change in cost caused by the increase in the level of production from 40 to 40.5 and compare your answer with the approximation. Is the approximation a good one?

Suppose you wish to compute a quantity Q using a formula $Q(x)$. If the value of x used in the computation is not precise, then its inaccuracy is passed on or *propagated* to the computed value of Q. Example 2.5.4 shows how the **propagated error** can be estimated.

EXAMPLE 2.5.4

In an analysis comparing yellow plums grown under different conditions, the size of a roughly spherical plum is estimated by measuring its diameter and using the formula $V = \frac{4}{3}\pi R^3$ to compute its volume. If the diameter of a plum is measured as 2.5 cm with a maximum error of 2%, how accurate is the volume measurement?

Solution

A sphere of radius R and diameter $x = 2R$ has volume

$$V = \frac{4}{3}\pi R^3 = \frac{4}{3}\pi\left(\frac{x}{2}\right)^3 = \frac{1}{6}\pi x^3,$$

so the volume using the estimated diameter $x = 2.5$ cm is

$$V = \frac{1}{6}\pi(2.5)^3 \approx 8.181 \text{ cm}^3.$$

The error made in computing this volume using the diameter 2.5 when the actual diameter is $2.5 + \Delta x$ is

$$\Delta V = V(2.5 + \Delta x) - V(2.5) \approx V'(2.5)\Delta x.$$

The measurement of the diameter can be off by as much as 2%; that is, by as much as $0.02(2.5) = 0.05$ cm in either direction. Hence, the maximum error in the measurement of the diameter is $\Delta x = \pm 0.05$, and the corresponding maximum error in the calculation of volume is

$$\text{Maximum error in volume} = \Delta V \approx [V'(2.5)](\pm 0.05).$$

Since

$$V'(x) = \frac{1}{6}\pi(3x^2) = \frac{1}{2}\pi x^2 \quad \text{and} \quad V'(2.5) = \frac{1}{2}\pi(2.5)^2 \approx 9.817,$$

it follows that

$$\text{Maximum error in volume} = (9.817)(\pm 0.05) \approx \pm 0.491.$$

Thus, at worst, the calculation of the volume as 8.181 cm^3 is off by 0.491 cm^3, so the actual volume V must satisfy

$$7.690 \le V \le 8.672 \text{ cm}^3.$$

In Example 2.5.5, the desired change in the function is given, and the goal is to estimate the necessary corresponding change in the variable.

EXAMPLE 2.5.5

The daily output at a certain factory is $Q(L) = 900L^{1/3}$ units, where L denotes the size of the labour force measured in worker-hours. Currently, 1000 worker-hours of labour are used each day. Use calculus to estimate the number of additional worker-hours of labour that will be needed to increase daily output by 15 units.

Solution
Solve for ΔL using the approximation formula

$$\Delta Q \approx Q'(L)\Delta L$$

with $\Delta Q = 15$, $L = 1000$, and $Q'(L) = 300L^{-2/3}$ to get

$$15 \approx 300(1000)^{-2/3}\Delta L$$

$$\Delta L \approx \frac{15}{300}(1000)^{2/3}$$

$$= \frac{15}{300}(10)^2$$

$$= 5 \text{ worker-hours.}$$

Approximation of Percentage Change

We defined the **percentage rate of change** of a quantity as the change in the quantity as a percentage of its size prior to the change; that is,

$$\text{Percentage change} = \frac{\text{change in quantity}}{\text{size of quantity}}.$$

The formula can include multiplying by 100 to express the fraction as a percent. This formula can be combined with the approximation formula and written in functional notation as follows.

Approximation Formula for Percentage Change ■ If Δx is a (small) change in x, the corresponding percentage change in the function $f(x)$ is

$$\text{Percentage change in } f = \frac{\Delta f}{f(x)} \approx \frac{f'(x)\Delta x}{f(x)}.$$

EXAMPLE 2.5.6

The GDP of a certain country was $N(t) = t^2 + 5t + 200$ billion dollars t years after 2004. Use calculus to estimate the percentage change in the GDP during the first quarter of 2012.

Solution

Use the formula

$$\text{Percentage change in } N \approx \frac{N'(t)\Delta t}{N(t)}$$

with $t = 8$, $\Delta t = 0.25$, and $N'(t) = 2t + 5$ to get

$$\text{Percentage change in } N \approx \frac{N'(8)0.25}{N(8)}$$

$$= \frac{[2(8) + 5](0.25)}{(8)^2 + 5(8) + 200}$$

$$\approx 1.73\%$$

Differentials   Sometimes the increment Δx is referred to as the *differential of x* and is denoted by dx, and then our approximation formula can be written as $df \approx f'(x)dx$. If $y = f(x)$, the *differential of y* is defined to be $dy = f'(x)dx$. To summarize:

> **Differentials** ■ The **differential of x** is $dx = \Delta x$, and if $y = f(x)$ is a differentiable function of x, then $dy = f'(x)\, dx$ is the **differential of y.**

EXAMPLE 2.5.7

In each case, find the differential of $y = f(x)$.
a. $f(x) = x^3 - 7x^2 + 2$
b. $f(x) = (x^2 + 5)(3 - x - 2x^2)$

Solution

a. $dy = f'(x)\, dx = (3x^2 - 14x)\, dx$
b. By the product rule,

$$dy = f'(x)\, dx = [(x^2 + 5)(-1 - 4x) + (2x)(3 - x - 2x^2)]\, dx$$

A geometric interpretation of the approximation of Δy, the actual change in y, by the differential dy is shown in Figure 2.15. Note that since the slope of the tangent line at $P(x, f(x))$ is $f'(x)$, the differential $dy = f'(x)dx$ is the change in the height of the tangent that corresponds to a change from x to $x + \Delta x$. Also, Δy is the change in the height of the curve corresponding to this change in x. Hence, approximating Δy by the differential dy is the same as approximating the change in the height of a curve by the change in the height of the tangent line. If Δx is small, it is reasonable to expect this to be a good approximation.

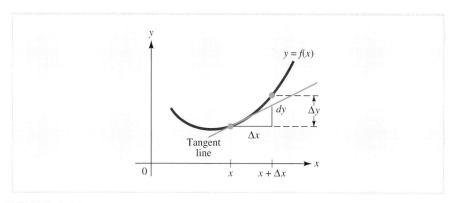

FIGURE 2.15   Approximation of Δy by the differential dy.

EXERCISES ■ 2.5

MARGINAL ANALYSIS   *In Exercises 1 through 6,*
C(x) is the total cost of producing x units of a particular
commodity and p(x) is the price at which all x units will
be sold. Assume p(x) and C(x) are in dollars.
  (a) Find the marginal cost and the marginal revenue.
  (b) Use the marginal cost to estimate the cost of
      producing the 4th unit.
  (c) Find the actual cost of producing the 4th unit.
  (d) Use the marginal revenue to estimate the
      revenue derived from the sale of the 4th unit.
  (e) Find the actual revenue derived from the sale
      of the 4th unit.

1. $C(x) = \dfrac{1}{5}x^2 + 4x + 57;\ p(x) = \dfrac{1}{4}(36 - x)$

2. $C(x) = \dfrac{1}{4}x^2 + 3x + 67;\ p(x) = \dfrac{1}{5}(45 - x)$

3. $C(x) = \dfrac{1}{3}x^2 + 2x + 39;\ p(x) = -x^2 - 4x + 80$

4. $C(x) = \dfrac{5}{9}x^2 + 5x + 73;\ p(x) = -x^2 - 2x + 33$

5. $C(x) = \dfrac{1}{4}x^2 + 43;\ p(x) = \dfrac{3 + 2x}{1 + x}$

6. $C(x) = \dfrac{2}{7}x^2 + 65;\ p(x) = \dfrac{12 + 2x}{3 + x}$

In Exercises 7 through 32, use increments to make the
required estimate.

7. Estimate how much the function $f(x) = x^2 - 3x + 5$
     will change as x increases from 5 to 5.3.

8. Estimate how much the function $f(x) = \dfrac{x}{x + 1} - 3$
     will change as x decreases from 4 to 3.8.

9. Estimate the percentage change in the function
     $f(x) = x^2 + 2x - 9$ as x increases from 4 to 4.3.

10. Estimate the percentage change in the function
     $f(x) = 3x + \dfrac{2}{x}$ as x decreases from 5 to 4.6.

11. MARGINAL ANALYSIS   A manufacturer's total
     cost is $C(q) = 0.1q^3 - 0.5q^2 + 500q + 200$ dollars,
     where q is the number of units produced.
     a. Use marginal analysis to estimate the cost of
          manufacturing the 4th unit.
     b. Compute the actual cost of manufacturing the
          4th unit.

12. MARGINAL ANALYSIS   A manufacturer's total
     monthly revenue is $R(q) = 240q - 0.05q^2$ dollars
     when q units are produced and sold during the
     month. Currently, the manufacturer is producing
     80 units per month and is planning to increase the
     monthly output by 1 unit.

a. Use marginal analysis to estimate the additional revenue that will be generated by the production and sale of the 81st unit.

b. Use the revenue function to compute the actual additional revenue that will be generated by the production and sale of the 81st unit.

13. **MARGINAL ANALYSIS**   Suppose the total cost in dollars of manufacturing q units is $C(q) = 3q^2 + q + 500$.

a. Use marginal analysis to estimate the cost of manufacturing the 41st unit.

b. Compute the actual cost of manufacturing the 41st unit.

14. **AIR POLLUTION**   An environmental study of a certain community suggests that t years from now, the average level of carbon monoxide in the air will be $Q(t) = 0.05t^2 + 0.1t + 3.4$ parts per million. By approximately how much will the carbon monoxide level change during the coming 6 months?

15. **PHONE APPS**   It is projected that t weeks after a new graphing calculator application is available for a smart phone, the number of people who have downloaded it will be $C(t) = 100t^2 + 400t + 5000$. Estimate the amount by which the number of people downloading it will increase during the next 6 months.

16. **MANUFACTURING**   A manufacturer's total cost is $C(q) = 0.1q^3 + 0.5q^2 + 500q + 200$ dollars when the level of production is q units. The current level of production is 4 units, and the manufacturer is planning to increase this to 4.1 units. Estimate how the total cost will change as a result.

17. **MANUFACTURING**   A manufacturer's total monthly revenue is $R(q) = 240q - 0.05q^2$ dollars when q units are produced during the month. Currently, the manufacturer is producing 80 units per month and is planning to decrease the monthly output by 0.65 units. Estimate how the total monthly revenue will change as a result.

18. **EFFICIENCY**   An efficiency study of the morning shift at a certain factory indicates that an average worker arriving on the job at 8:00 A.M. will have assembled $f(x) = -x^3 + 6x^2 + 15x$ units x hours later. Approximately how many units will the worker assemble between 9:00 A.M. and 9:15 A.M.?

19. **PRODUCTION**   At a certain factory, the daily output is $Q(K) = 600K^{1/2}$ units, where K denotes the capital investment measured in units of $1000. The current capital investment is $900\,000. Estimate the effect that an additional capital investment of $800 will have on the daily output.

20. **PRODUCTION**   At a certain factory, the daily output is $Q(L) = 60\,000L^{1/3}$ units, where L denotes the size of the labour force measured in worker-hours. Currently, 1000 worker-hours of labour are used each day. Estimate the effect on output that will be produced if the labour force is cut to 940 worker-hours.

21. **PROPERTY TAX**   A projection made in January of 2004 determined that x years later, the average property tax on a three-bedroom house in a certain community will be $T(x) = 60x^{3/2} + 40x + 1200$ dollars. Estimate the percentage change by which this projection predicts that the property tax will increase during the first half of the year 2012.

22. **POPULATION GROWTH**   A 5-year projection of population trends suggests that t years from now, the population of a certain community will be $P(t) = -t^3 + 9t^2 + 48t + 200$ thousand.

a. Find the rate of change of population $R(t) = P'(t)$ with respect to time t.

b. At what rate does the population growth rate $R(t)$ change with respect to time?

c. Use increments to estimate how much $R(t)$ changes during the first month of the 4th year. What is the actual change in $R(t)$ during this time period?

23. **PRODUCTION**   At a certain factory, the daily output is $Q = 3000K^{1/2}L^{1/3}$ units, where K denotes the firm's capital investment measured in units of $1000 and L denotes the size of the labour force measured in worker-hours. Suppose that the current capital investment is $400\,000 and that 1331 worker-hours of labour are used each day. Use marginal analysis to estimate the effect that an additional capital investment of $1000 will have on the daily output if the size of the labour force is not changed.

24. **PRODUCTION**   The daily output at a certain factory is $Q(L) = 300L^{2/3}$ units, where L denotes the size of the labour force measured in worker-hours. Currently, 512 worker-hours of labour are used each day. Estimate the number of additional worker-hours of labour that will be needed to increase daily output by 12.5 units.

25. **MANUFACTURING**   A manufacturer's total cost is $C(q) = \frac{1}{6}q^3 + 642q + 400$ dollars when q units are produced. The current level of production is 4 units. Estimate the amount by which the manufacturer should decrease production to reduce the total cost by $130.

26. **A* GROWTH OF A CELL**   A certain cell has the shape of a sphere. The formulas $S = 4\pi r^2$ and $V = \frac{4}{3}\pi r^3$ are used to compute the surface area and volume of the cell, respectively. Estimate the effect on S and V produced by a 1% increase in the radius r.

27. **A* CARDIAC OUTPUT**   *Cardiac output* is the volume (cubic centimetres) of blood pumped by a person's heart each minute. One way of measuring cardiac output C is by Fick's formula

    $$C = \frac{a}{x - b},$$

    where x is the concentration of carbon dioxide in the blood entering the lungs from the right side of the heart and a and b are positive constants. If x is measured as $x = c$ with a maximum error of 3%, what is the maximum percentage error that can be incurred by measuring cardiac output with Fick's formula? (Your answer will be in terms of a, b, and c.)

28. **A* MEDICINE**   A tiny spherical balloon is inserted into a clogged artery. If the balloon has an inner diameter of 0.01 mm and is made from material that is 0.0005 mm thick, approximately how much material is inserted into the artery? [*Hint:* Think of the amount of material as a change in volume ΔV, where $V = \frac{4}{3}\pi r^3$.]

29. **A* ARTERIOSCLEROSIS**   In *arteriosclerosis*, fatty material called plaque gradually builds up on the walls of arteries, impeding the flow of blood, which in turn can lead to stroke and heart attacks. Consider a model in which the carotid artery is represented as a circular cylinder with cross-sectional radius $R = 0.3$ cm and length L. Suppose it is discovered that plaque that is 0.07 cm thick is distributed uniformly over the inner wall of the carotid artery of a particular patient. Use increments to estimate the percentage of the total volume of the artery that is blocked by plaque. [*Hint:* The volume of a cylinder of radius R and length L is $V = \pi R^2 L$. Does it matter that we have not specified the length L of the artery?]

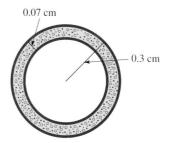

EXERCISE 29

30. **BLOOD CIRCULATION**   In Exercise 69, Section 1.1, we introduced an important law attributed to the French physician Jean Poiseuille. Another law discovered by Poiseuille says that the volume of the fluid flowing through a small tube in unit time under fixed pressure is given by the formula $V = kR^4$, where k is a positive constant and R is the radius of the tube. This formula is used in medicine to determine how wide a clogged artery must be opened to restore a healthy flow of blood. Suppose the radius of a certain artery is increased by 5%. Approximately what effect does this have on the volume of blood flowing through the artery?

31. **A* EXPANSION OF MATERIAL**   The (linear) **thermal expansion coefficient** of an object is defined to be

    $$\sigma = \frac{L'(T)}{L(T)},$$

    where $L(T)$ is the length of the object when the temperature is T. (The Greek letter σ is pronounced "sigma.") Suppose the 50-m span of a bridge is built of steel with $\sigma = 1.4 \times 10^{-5}$ per degree Celsius. Approximately how much will the length change during a year when the temperature varies from $-20°C$ (winter) to $35°C$ (summer)?

32. **RADIATION**   Stefan's law in physics states that a body emits radiant energy according to the formula $R(T) = kT^4$, where R is the amount of energy emitted from a surface with temperature T (in degrees Kelvin) and k is a positive constant. Estimate the percentage change in R that results from a 2% increase in T.

Implicit Differentiation and Related Rates

The functions you have worked with so far have all been given by equations of the form $y = f(x)$, in which the dependent variable y on the left is given explicitly by an expression on the right involving the independent variable x. A function in this form is said to be in **explicit form.** For example,

$$y = x^2 + 3x + 1, \quad y = \frac{x^3 + 1}{2x - 3}, \quad \text{and} \quad y = \sqrt{1 - x^2}$$

are all functions in explicit form.

Sometimes practical problems will lead to equations in which the function y is not written explicitly in terms of the independent variable x, for example, equations such as

$$x^2 y^3 - 6 = 5y^3 + x \quad \text{and} \quad x^2 y + 2y^3 = 3x + 2y.$$

Since it has not been solved for y, such an equation is said to **define y implicitly as a function of x** and the function y is said to be in **implicit form.**

Differentiation of Functions in Implicit Form

Suppose you have an equation that defines y implicitly as a function of x and you want to find the derivative $\dfrac{dy}{dx}$. For instance, you may be interested in the slope of a line that is tangent to the graph of the equation at a particular point. One approach might be to solve the equation for y explicitly and then differentiate using the techniques you already know. Unfortunately, it is not always possible to find y explicitly. For example, there is no obvious way to solve for y in the equation $x^2 y + 2y^3 = 3x + 2y$.

Moreover, even when you can solve for y explicitly, the resulting formula is often complicated and unpleasant to differentiate. For example, the equation $x^2 y^3 - 6 = 5y^3 + x$ can be solved for y to give

$$x^2 y^3 - 5y^3 = x + 6$$
$$y^3(x^2 - 5) = x + 6$$
$$y = \left(\frac{x + 6}{x^2 - 5}\right)^{1/3}$$

The computation of $\dfrac{dy}{dx}$ for this function in explicit form would be tedious, involving both the chain rule and the quotient rule. Fortunately, there is a simple technique based on the chain rule that you can use to find $\dfrac{dy}{dx}$ without first solving for y explicitly.

This technique, known as **implicit differentiation,** consists of differentiating both sides of the given (defining) equation with respect to x and then solving algebraically for $\dfrac{dy}{dx}$. Here is an example illustrating the technique.

EXAMPLE 2.6.1

Find $\dfrac{dy}{dx}$ if $x^2 y + y^2 = x^3$.

Solution

You are going to differentiate both sides of the given equation with respect to x. So that you won't forget that y is actually a function of x, temporarily replace y by $f(x)$ and begin by rewriting the equation as

$$x^2 f(x) + (f(x))^2 = x^3.$$

Now differentiate both sides of this equation term by term with respect to x:

$$\frac{d}{dx}[x^2 f(x) + (f(x))^2] = \frac{d}{dx}[x^3]$$

$$\underbrace{\left[x^2 \frac{df}{dx} + f(x) \frac{d}{dx}(x^2) \right]}_{\frac{d}{dx}[x^2 f(x)]} + \underbrace{2 f(x) \frac{df}{dx}}_{\frac{d}{dx}[(f(x))^2]} + \underbrace{3x^2}_{\frac{d}{dx}[x^3]}$$

Thus, we have

$$x^2 \frac{df}{dx} + f(x)(2x) + 2 f(x) \frac{df}{dx} = 3x^2$$

$$x^2 \frac{df}{dx} + 2 f(x) \frac{df}{dx} = 3x^2 - 2x f(x) \qquad \text{gather all } \frac{df}{dx} \text{ terms on}$$
$$\text{one side of the equation}$$

$$[x^2 + 2 f(x)] \frac{df}{dx} = 3x^2 - 2x f(x) \qquad \text{combine terms}$$

$$\frac{df}{dx} = \frac{3x^2 - 2x f(x)}{x^2 + 2 f(x)} \qquad \text{solve for } \frac{df}{dx}$$

Finally, replace $f(x)$ by y to get

$$\frac{dy}{dx} = \frac{3x^2 - 2xy}{x^2 + 2y}.$$

NOTE Temporarily replacing y by $f(x)$ as in Example 2.6.1 is a useful device for illustrating the implicit differentiation process, but as soon as you feel comfortable with the technique, try to leave out this unnecessary step and differentiate the given equation directly. Just keep in mind that y is really a function of x and remember to use the chain rule when it is appropriate.

Here is an outline of the procedure.

Implicit Differentiation ■ Suppose an equation defines y implicitly as a differentiable function of x. To find $\frac{dy}{dx}$:

1. Differentiate both sides of the equation with respect to x. Remember that y is really a function of x and use the chain rule when differentiating terms containing y.

2. Solve the differentiated equation algebraically for $\frac{dy}{dx}$ in terms of x and y.

Computing the Slope of a Tangent Line by Implicit Differentiation

In Examples 2.6.2 and 2.6.3, you will see how to use implicit differentiation to find the slope of a tangent line.

Just-In-Time

Remember that the equation of a circle with centre (a, b) and radius c is
$(x - a)^2 + (y - b)^2 = c^2$.
In Example 2.6.2, the centre of the circle is at the origin $(0, 0)$.

EXAMPLE 2.6.2

Find the slope of the tangent line to the circle $x^2 + y^2 = 25$ at the point $(3, 4)$. What is the slope at the point $(3, -4)$?

Solution

Differentiating both sides of the equation $x^2 + y^2 = 25$ with respect to x gives

$$2x + 2y\frac{dy}{dx} = 0$$

$$\frac{dy}{dx} = -\frac{x}{y}$$

The slope at $(3, 4)$ is the value of the derivative when $x = 3$ and $y = 4$:

$$\left.\frac{dy}{dx}\right|_{(3,4)} = \left.-\frac{x}{y}\right|_{x=3,\, y=4} = -\frac{3}{4}.$$

Similarly, the slope at $(3, -4)$ is the value of $\frac{dy}{dx}$ when $x = 3$ and $y = -4$:

$$\left.\frac{dy}{dx}\right|_{(3,4)} = \left.-\frac{x}{y}\right|_{x=3,\, y=4} = -\left(\frac{3}{-4}\right) = \frac{3}{4}.$$

The graph of the circle is shown in Figure 2.16 together with the tangent lines at $(3, 4)$ and $(3, -4)$.

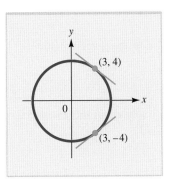

FIGURE 2.16 The graph of the circle $x^2 + y^2 = 25$.

EXAMPLE 2.6.3

Find all points on the graph of the equation $x^2 - y^2 = 2x + 4y$ where the tangent line is horizontal. (This time we will use y' instead of $\frac{dy}{dx}$.) Does the graph have any vertical tangents?

Solution

Differentiate both sides of the given equation with respect to x to get

$$2x - 2yy' = 2 + 4y'$$
$$2x - 2 = 4y' + 2yy'$$
$$2x - 2 = y'(4 + 2y)$$
$$y' = \frac{2x - 2}{4 + 2y}$$

There is a horizontal tangent at each point on the graph where the slope is zero, that is, where the *numerator* $2x - 2$ of y' is zero:

$$2x - 2 = 0$$
$$x = 1$$

To find the corresponding value of y, substitute $x = 1$ into the given equation and solve using the quadratic formula (or technology):

$$1 - y^2 = 2(1) + 4y$$
$$y^2 + 4y + 1 = 0$$
$$y = -0.27, -3.73$$

Thus, the given graph has horizontal tangents at the points $(1, -0.27)$ and $(1, -3.73)$.

Since the slope of a vertical line is undefined, the given graph can have a vertical tangent only where the *denominator* $4 + 2y$ of y' is zero:

$$4 + 2y = 0$$
$$y = -2$$

To find the corresponding value of x, substitute $y = -2$ into the given equation:

$$x^2 - (-2)^2 = 2x + 4(-2)$$
$$x^2 - 2x + 4 = 0$$

But this quadratic equation has no real solutions, which implies that the given graph has no vertical tangents. The graph is shown in Figure 2.17.

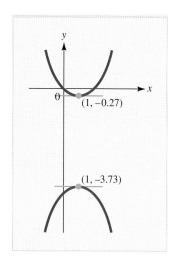

FIGURE 2.17 The graph of the equation $x^2 - y^2 = 2x + 4y$.

Application to Economics

Implicit differentiation is used in economics in both practical and theoretical work. In Section 4.3, it will be used to derive certain theoretical relationships. A more practical application of implicit differentiation is given in Example 2.6.4, which is a preview of the discussion of level curves of a function of two variables given in Section 7.1.

EXAMPLE 2.6.4

Suppose the output at a certain factory is $Q = 2x^3 + x^2y + y^3$ units, where x is the number of hours of skilled labour used and y is the number of hours of unskilled labour. The current labour force consists of 30 h of skilled labour and 20 h of unskilled labour. Use calculus to estimate the change in unskilled labour y that should be made to offset a 1-h increase in skilled labour x so that output will be maintained at its current level.

Solution

The current level of output is the value of Q when $x = 30$ and $y = 20$. That is,

$$Q = 2(30)^3 + (30)^2(20) + (20)^3 = 80\ 000 \text{ units.}$$

If output is to be maintained at this level, the relationship between skilled labour x and unskilled labour y is given by the equation

$$80\ 000 = 2x^3 + x^2y + y^3,$$

which defines y implicitly as a function of x.

The goal is to estimate the change in y that corresponds to a 1-unit increase in x when x and y are related by this equation. As you saw in Section 2.5, the change in y caused by a 1-unit increase in x can be approximated by the derivative $\dfrac{dy}{dx}$ or y'. To find this derivative, use implicit differentiation. (Remember that the derivative of the constant 80 000 on the left-hand side is zero.)

$$0 = 6x^2 + x^2y' + 2xy + 3y^2y'$$
$$-x^2y' - 3y^2y' = 6x^2 + 2xy$$
$$-(x^2 + 3y^2)y' = 6x^2 + 2xy$$
$$y' = -\frac{6x^2 + 2xy}{x^2 + 3y^2}$$

Just-In-Time

If you need to find the derivative at a certain point, it is often easier to substitute the coordinates (x, y) for the point before solving the equation for y'. If your algebra skills are rusty, this is the way to go!

Now evaluate this derivative when $x = 30$ and $y = 20$ to conclude that

$$\text{Change in } y \approx \frac{dy}{dx}\bigg|_{x=30,\, y=20} = -\frac{6(30)^2 + 2(30)(20)}{(30)^2 + 3(20)^2} \approx -3.14$$

That is, to maintain the current level of output, unskilled labour should be decreased by approximately 3.14 hours to offset a 1-hour increase in skilled labour.

Related Rates

In certain practical problems, x and y are related by an equation and can be regarded as functions of a third variable t, which often represents time. Then implicit differentiation can be used to relate $\dfrac{dx}{dt}$ to $\dfrac{dy}{dt}$. This kind of problem is said to involve **related rates.** Here is a general procedure for analyzing related rates problems.

> **A Procedure for Solving Related Rates Problems**
> **1.** Draw a diagram (if appropriate) and assign variables.
> **2.** Find a formula relating the variables.
> **3.** Use implicit differentiation to find how the rates are related.
> **4.** Substitute any given numerical information into the equation in step 3 to find the desired rate of change.

Here are four applied problems involving related rates.

EXAMPLE 2.6.5

The manager of a bicycle company determines that when q hundred bicycles are produced, the total cost of production is C thousand dollars, where $C^2 - 3q^3 = 4275$. When 1500 bicycles are being produced, the level of production is increasing at the rate of 20 bicycles per week. What is the total cost at this time and at what rate is it changing?

Solution

We want to find $\dfrac{dC}{dt}$ when $q = 15$ (1500 bicycles) and $\dfrac{dq}{dt} = 0.2$ (20 bicycles/week with q measured in hundreds of units). Differentiating the equation $C^2 - 3q^3 = 4275$ implicitly with respect to time, we get

$$2C\frac{dC}{dt} - 9q^2\frac{dq}{dt} = 0$$

so that

$$2C\frac{dC}{dt} = 9q^2\frac{dq}{dt}$$

and

$$\frac{dC}{dt} = \frac{9q^2}{2C}\frac{dq}{dt}.$$

When $q = 15$, the cost C satisfies

$$C^2 - 3(15)^3 = 4275$$
$$C^2 = 4275 + 3(15)^3 = 14\ 400$$
$$C = 120$$

and by substituting $q = 15$, $C = 120$, and $\dfrac{dq}{dt} = 0.2$ into the formula for $\dfrac{dC}{dt}$, we obtain

$$\frac{dC}{dt} = \left[\frac{9(15)^2}{2(120)}\right](0.2) = 1.6875$$

thousand dollars ($1687.50) per week. To summarize, the cost of producing 1500 bicycles is $120\ 000$ ($C = 120$), and at this level of production, total cost is increasing at a rate of $1687.50 per week.

EXAMPLE 2.6.6

A storm at sea has damaged an oil rig. Oil is spilling from the rupture at a constant rate of 2 m³/min, forming a slick that is roughly circular in shape and 1.5 cm thick.

   a. How fast is the radius of the slick increasing when the radius is 25 m?
   b. Suppose the rupture is repaired in such a way that the flow is shut off instantaneously. If the radius of the slick is increasing at a rate of 0.4 m/min when the flow stops, what is the total volume of oil that spilled onto the sea?

Solution

We can think of the slick as a cylinder of oil of radius r metres and thickness $h = 0.015$ m. Such a cylinder will have volume $V = \pi r^2 h = 0.015\pi r^2$ cubic metres, because h is constant.

Differentiating implicitly in this equation with respect to time t, we get

$$\frac{dV}{dt} = 0.015\left(2\pi r\frac{dr}{dt}\right) = 0.03\pi r\frac{dr}{dt},$$

and since $\dfrac{dV}{dt} = 2$ at all times, we obtain the rate relationship

$$2 = 0.03\pi r\frac{dr}{dt}.$$

a. We want to find $\dfrac{dr}{dt}$ when $r = 25$. Substituting into the rate relationship we have just obtained, we find that

$$2 = (0.03)\pi(25)\frac{dr}{dt},$$

so that

$$\frac{dr}{dt} = \frac{2}{(0.03)\pi(25)} \approx 0.849.$$

Thus, when the radius is 25 m, it is increasing at about 0.849 m/min.

b. We can compute the total volume of oil in the spill if we know the radius of the slick at the instant the flow stops. Since $\dfrac{dr}{dt} = 0.4$ at that instant, we have

$$2 = 0.03\pi r(0.4)$$

and the radius is

$$r = \frac{2}{0.03\pi(0.4)} \approx 53.1.$$

Therefore,

$$V = 0.015\pi(53.1)^2 \approx 133.$$

The total amount of oil spilled is 133 m^3.

EXAMPLE 2.6.7

A lake is polluted by waste from a plant located on its shore. Ecologists determine that when the level of pollutant is x parts per million (ppm), there will be F fish of a certain species in the lake, where

$$F = \frac{32\,000}{3 + \sqrt{x}}.$$

When there are 4000 fish left in the lake, the pollution is increasing at a rate of 1.4 ppm/year. At what rate is the fish population changing at this time?

Solution

We want to find $\dfrac{dF}{dt}$ when $F = 4000$ and $\dfrac{dx}{dt} = 1.4$. When there are 4000 fish in the lake, the level of pollution x satisfies

$$4000 = \frac{32\,000}{3 + \sqrt{x}}$$

$$4000(3 + \sqrt{x}) = 32\,000$$

$$3 + \sqrt{x} = 8$$

$$\sqrt{x} = 5$$

$$x = 25$$

We find that

$$\frac{dF}{dx} = \frac{32\,000(-1)}{(3 + \sqrt{x})^2}\left(\frac{1}{2}\frac{1}{\sqrt{x}}\right) = \frac{-16\,000}{\sqrt{x}\,(3 + \sqrt{x})^2},$$

and according to the chain rule

$$\frac{dF}{dt} = \frac{dF}{dx}\frac{dx}{dt} = \left[\frac{-16\,000}{\sqrt{x}(3 + \sqrt{x})^2}\right]\frac{dx}{dt}.$$

Substituting $x = 25$ and $\dfrac{dx}{dt} = 1.4$, we find that

$$\frac{dF}{dt} = \left[\frac{-16\,000}{\sqrt{25}(3 + \sqrt{25})^2}\right](1.4) = -70,$$

so the fish population is decreasing by 70 fish per year.

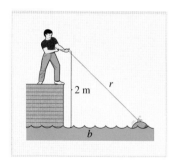

EXAMPLE 2.6.8

A person stands at the end of a pier that is 2 m above the water and pulls in a rope attached to a buoy. If the rope is hauled in at a rate of 0.8 m/s, how fast is the buoy moving when it is 1.5 m from the pier?

Solution

Consider the triangle formed by the pier, the buoy, and the rope, using r for the rope length and b for the distance from the pier to the buoy.

Use the Pythagorean theorem and implicit differentiation to find

$$r^2 = b^2 + 2^2$$

$$2rr' = 2bb'$$

$$b' = \frac{rr'}{b}$$

We know that when $b = 1.5$, $r' = 0.8$. Also, $r^2 = 1.5^2 + 2^2$, giving $r = 2.5$.

Then

$$b' = \frac{rr'}{b}$$

$$= \frac{(2.5)(0.8)}{1.5}$$

$$\approx 1.33$$

The buoy approaches the dock at a rate of 1.33 m/s.

ALGEBRA WARM-UP

The derivative, $\frac{dy}{dx}$, can be written as y'. Keep this in mind for the first three questions. Solve for y'.

1. $x^2y' + 2xy + 4y' = 3x - 1$

2. $3x^2y^3 + 3x^3y^2y' + 2 = 7xy' + 7y$

3. $5y' - 2xy - x^2y' = 6xyy' - 3y^2 - 9$

Solve for $\frac{dy}{dx}$.

4. $2\frac{dy}{dx} + 6xy + 3x^2\frac{dy}{dx} = 5\frac{dy}{dx}$

5. $2y - 2x\frac{dy}{dx} - 5x^3\frac{dy}{dx} - 15x^2y = 12$

EXERCISES ■ 2.6

In Exercises 1 through 8, find $\frac{dy}{dx}$ in two ways:

    (a) by implicit differentiation
    (b) by differentiating an explicit formula for y
In each case, show that the two answers are the same.

1. $2x + 3y = 7$

2. $5x - 7y = 3$

3. $x^3 - y^2 = 5$

4. $x^2 + y^3 = 12$

5. $xy = 4$

6. $x + \dfrac{1}{y} = 5$

7. $xy + 2y = 3$

8. $xy + 2y = x^2$

In Exercises 9 through 22, find $\frac{dy}{dx}$ by implicit differentiation.

9. $x^2 + y^2 = 25$

10. $x^2 + y = x^3 + y^2$

11. $x^3 + y^3 = xy$

12. $5x - x^2y^3 = 2y$

13. $y^2 + 2xy^2 - 3x + 1 = 0$

14. $\dfrac{1}{x} + \dfrac{1}{y} = 1$

15. $\sqrt{x} + \sqrt{y} = 1$

16. $\sqrt{2x} + y^2 = 4$

17. $xy - x = y + 2$

18. $y^2 + 3xy - 4x^2 = 9$

19. $(2x + y)^3 = x$

20. $(x + 2y)^2 = y$

21. $(x^2 + 3y^2)^5 = 2xy$

22. $(3xy^2 + 1)^4 = 2x - 3y$

In Exercises 23 through 30, find the equation of the tangent line to the given curve at the specified point.

23. $x^2 = y^3$; $(8, 4)$

24. $x^2 - y^3 = 2x$; $(1, -1)$

25. $xy = 2$; $(2, 1)$

26. $\dfrac{1}{x} - \dfrac{1}{y} = 2$; $\left(\dfrac{1}{4}, \dfrac{1}{2}\right)$

27. $xy^2 - x^2y = 6$; $(2, -1)$

28. $x^2y^3 - 2xy = 6x + y + 1$; $(0, -1)$

29. $(1 - x + y)^3 = x + 7; (1, 2)$

30. $(x^2 + 2y)^3 = 2xy^2 + 64; (0, 2)$

In Exercises 31 through 36, find all points (both coordinates) on the given curve where the tangent line is (a) horizontal and (b) vertical.

31. $x + y^2 = 9$

32. $x^2 + xy + y = 3$

33. $xy = 16y^2 + x$

34. $\dfrac{y}{x} - \dfrac{x}{y} = 5$

35. $x^2 + xy + y^2 = 3$

36. $x^2 - xy + y^2 = 3$

In Exercises 37 and 38, use implicit differentiation to find the second derivative $\dfrac{d^2y}{dx^2}$.

37. $x^2 + 3y^2 = 5$

38. $xy + y^2 = 1$

39. MANUFACTURING   The output at a certain plant is $Q = 0.08x^2 + 0.12xy + 0.03y^2$ units per day, where x is the number of hours of skilled labour used and y is the number of hours of unskilled labour used. Currently, 80 hours of skilled labour and 200 hours of unskilled labour are used each day. Use calculus to estimate the change in unskilled labour that should be made to offset a 1-hour increase in skilled labour so that output will be maintained at its current level.

40. MANUFACTURING   The output of a certain plant is $Q = 0.06x^2 + 0.14xy + 0.05y^2$ units per day, where x is the number of hours of skilled labour used and y is the number of hours of unskilled labour used. Currently, 60 hours of skilled labour and 300 hours of unskilled labour are used each day. Use calculus to estimate the change in unskilled labour that should be made to offset a 1-hour increase in skilled labour so that output will be maintained at its current level.

41. SUPPLY RATE   When the price of a ruled notebook is p dollars per notebook, the manufacturer is willing to supply x hundred notebooks, where
$$3p^2 - x^2 = 12.$$
At what rate is the supply changing when the price is \$4 per notebook and increasing at a rate of 87 cents per month?

42. DEMAND RATE   When the price of a 400-g bag of almonds is p dollars per bag, customers demand x hundred bags of almonds, where
$$x^2 + 3px + p^2 = 79.$$
At what rate is the demand x changing with respect to time when the price is \$5 per bag and decreasing at a rate of 30 cents per month?

43. DEMAND RATE   When the price of a certain commodity is p dollars per unit, consumers demand x hundred units of the commodity, where
$$75x^2 + 17p^2 = 5300.$$
At what rate is the demand x changing with respect to time when the price is \$7 and decreasing at a rate of 75 cents per month? (That is, $\dfrac{dp}{dt} = -0.75$.)

44. REFRIGERATION   An ice block used in a cooler at a campsite is modelled as a cube of side s. The block has volume 125 000 cm³ and is melting at a rate of 1000 cm³/h.

   a. What is the current length s of each side of the cube? At what rate is s currently changing with respect to time t?

   b. What is the current rate of change of the surface area S of the block with respect to time? [*Note:* A cube of side s has volume $V = s^3$ and surface area $S = 6s^2$.]

45. MEDICINE   A tiny spherical balloon is inserted into a clogged artery and is inflated at a rate of 0.002π mm³/min. At what rate is the radius of the balloon growing when the radius is $R = 0.005$ mm?

   [*Note:* A sphere of radius R has volume $V = \dfrac{4}{3}\pi R^3$.]

46. POLLUTION CONTROL   An environmental study for a certain community indicates that there will be $Q(p) = p^2 + 4p + 900$ units of a harmful pollutant in the air when the population is p thousand people. If the population is currently 50 000 and is increasing at a rate of 1500 per year, at what rate is the level of pollution increasing?

47. GROWTH OF A MUSHROOM   A type of mushroom is modelled as being roughly spherical, with radius R. If the radius of the mushroom is currently $R = 0.54$ cm and is increasing at a rate of 0.13 cm/h overnight, what is the corresponding rate of change of the volume $V = \dfrac{4}{3}\pi R^3$?

48. BOYLE'S LAW   Boyle's law states that when gas is compressed at constant temperature, the pressure P and volume V of a given sample satisfy the equation $PV = C$, where C is constant. Suppose that at a certain time the volume is 40 cm^3, the pressure is 70 kPa, and the volume is increasing at a rate of 12 cm^3/s. At what rate is the pressure changing at this instant? Is it increasing or decreasing?

49. METABOLIC RATE   The *basal metabolic rate* is the rate of heat produced by an animal per unit time. Observations indicate that the basal metabolic rate in kilocalories per day of a warm-blooded animal of mass m kilograms is given by

$$M = 70m^{3/4}.$$

a. Find the rate of change of the metabolic rate of an 80-kg cougar that is gaining mass at a rate of 0.8 kg per day.

b. Find the rate of change of the metabolic rate of a 50-kg ostrich that is losing mass at a rate of 0.5 kg per day.

50. SPEED OF A LIZARD   Herpetologists have proposed using the formula $s = 1.1m^{0.2}$ to estimate the maximum sprinting speed s (m/s) of a lizard of mass m (g). At what rate is the maximum sprinting speed of an 11-g lizard increasing if the lizard is growing at a rate of 0.02 g per day?

51. PRODUCTION   At a certain factory, output is given by $Q = 60\,K^{1/3}L^{2/3}$ units, where K is the capital investment (in thousands of dollars) and L is the size of the labour force, measured in worker-hours. If output is kept constant, at what rate is capital investment changing at a time when $K = 8$, $L = 1000$, and L is increasing at a rate of 25 worker-hours per week?
[*Note:* Output functions of the general form $Q = AK^{\alpha}L^{1-\alpha}$, where A and α are constants with $0 \le \alpha \le 1$, are called **Cobb-Douglas production functions.** Such functions appear in examples and exercises throughout this text, especially in Chapter 7.]

52. WATER POLLUTION

A circular oil slick spreads in such a way that its radius is increasing at a rate of 7 m/h. How fast is the area of the slick changing when the radius is 50 m?

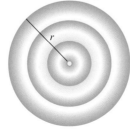

EXERCISE 52

53. A* A 1.8-m-tall man walks at a rate of 1.5 m/s away from the base of a street light that is 3.6 m above the ground. At what rate is the length of his shadow changing when he is 6 m away from the base of the light?

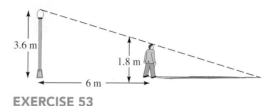

3.6 m

1.8 m

6 m

EXERCISE 53

54. CHEMISTRY   In an *adiabatic* chemical process, there is no net change (gain or loss) of heat. Suppose a container of oxygen is subjected to such a process. Then if the pressure on the oxygen is P and its volume is V, it can be shown that $PV^{1.4} = C$, where C is a constant. At a certain time, $V = 5$ m^3, $P = 0.6$ kPa, and P is increasing at 0.23 kPa/s. What is the rate of change of V? Is V increasing or decreasing?

55. MANUFACTURING   At a certain factory, output Q is related to inputs x and y by the equation

$$Q = 2x^3 + 3x^2y^2 + (1 + y)^3.$$

If the current levels of input are $x = 30$ and $y = 20$, use calculus to estimate the change in input y that should be made to offset a decrease of 0.8 units in input x so that output will be maintained at its current level.

56. A* LUMBER PRODUCTION   To estimate the amount of wood in the trunk of a tree, it is reasonable to assume that the trunk is a cut-off cone (see the figure).

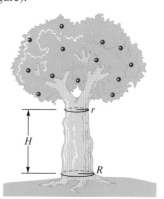

r

H

R

EXERCISE 56

If the upper radius of the trunk is r, the lower radius is R, and the height is H, the volume of wood is given by

$$V = \frac{\pi}{3}H(R^2 + rR + r^2).$$

Suppose r, R, and H are increasing at the respective rates of 10 cm/year, 8 cm/year, and 24 cm/year. At what rate is V increasing at a time when $r = 60$ cm, $R = 90$ cm, and $H = 450$ cm?

57. A* BLOOD FLOW   One of Poiseuille's laws says that the speed of blood flowing under constant pressure in a blood vessel at a distance r from the centre of the vessel is given by

$$v = \frac{K}{L}(R^2 - r^2),$$

where K is a positive constant, R is the radius of the vessel, and L is the length of the vessel. Suppose the radius R and length L of the vessel change with time in such a way that the speed of blood flowing at the centre is unaffected; that is, v does not change with time. Show that in this case, the relative rate of change of L with respect to time must be twice the relative rate of change of R.

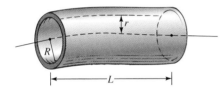

EXERCISE 57

58. A* MANUFACTURING   At a certain factory, output Q is related to inputs u and v by the equation

$$Q = 3u^2\frac{2u + 3v}{(u + v)^2}.$$

If the current levels of input are $u = 10$ and $v = 25$, use calculus to estimate the change in input v that should be made to offset a decrease of 0.7 units in input u so that output will be maintained at its current level.

59. A* Show that the tangent line to the curve

$$\frac{x^2}{a^2} + \frac{y^2}{b^2} = 1$$

at the point (x_0, y_0) is

$$\frac{x_0 x}{a^2} + \frac{y_0 y}{b^2} = 1.$$

60. A* Consider the equation $x^2 + y^2 = 6y - 10$.
   a. Show that there are no points (x, y) that satisfy this equation. [*Hint:* Complete the square or use the quadratic formula.]
   b. Show that by applying implicit differentiation to the given equation, you obtain

$$\frac{dy}{dx} = \frac{x}{3 - y}.$$

The point of this exercise is to show that one must be careful when applying implicit differentiation. Just because it is possible to find a derivative formally by implicit differentiation does not mean that the derivative has any meaning.

61. A* Prove the power rule $\dfrac{d}{dx}[x^n] = nx^{n-1}$ for the case where $n = r/s$ is a rational number. [*Hint:* Note that if $y = x^{r/s}$, then $y^s = x^r$, and use implicit differentiation.]

62. Use a graphing utility to graph the curve $5x^2 - 2xy + 5y^2 = 8$. Draw the tangent line to the curve at $(1, 1)$. How many horizontal tangent lines does the curve have? Find the equation of each horizontal tangent.

63. Use a graphing utility to graph the curve $11x^2 + 4xy + 14y^2 = 21$. Draw the tangent line to the curve at $(-1, 1)$. How many horizontal tangent lines does the curve have? Find the equation of each horizontal tangent.

64. Answer the following questions about the curve $x^3 + y^3 = 3xy$ (called the **folium of Descartes**).
   a. Find the equation of each horizontal tangent to the curve.
   b. The curve intersects the line $y = x$ at exactly one point other than the origin. What is the equation of the tangent line at this point?
   c. Graph the curve using technology.

65. Use a graphing utility to graph the curve $x^2 + y^2 = \sqrt{x^2 + y^2} + x$. Find the equation of each horizontal tangent to the curve. (The curve is called a **cardioid**.)

Concept Summary Chapter 2

Rule

Example

Power Rule

$f(x) = x^n$

$f'(x) = nx^{n-1}$

$f(x) = x^3$

$f'(x) = 3x^2$

Product Rule

$f(x) = uv$

$f'(x) = uv' + vu'$

$f(x) = (x^2 + 3)(2x^5 - 7)$

$f'(x) = (x^2 + 3)(10x^4) + (2x^5 - 7)(2x)$

Quotient Rule

$f(x) = \dfrac{u}{v}$

$f'(x) = \dfrac{vu' - uv'}{v^2}$

or when $f(x) = \dfrac{Hi}{Lo}$,

$f'(x) = \dfrac{Lo\ d(Hi) - Hi\ d(Lo)}{\text{square the bottom and away we go!}}$

where $d(Hi)$ and $d(Lo)$ are those functions differentiated, as in Hi' and Lo'.

$f(x) = \dfrac{x^2 + 3}{2x^3 - 7}$

$f'(x) = \dfrac{(2x^3 - 7)(2x) - (x^2 + 3)(6x^2)}{(2x^3 - 7)^2}$

$= \dfrac{2x(2x^3 - 7 - 3x(x^2 + 3))}{(2x^3 - 7)^2}$

$= \dfrac{2x(2x^3 - 7 - 3x^3 - 9x)}{(2x^3 - 7)^2}$

$= \dfrac{-2x(x^3 + 9x + 7)}{(2x^3 - 7)^2}$

Chain Rule + Power Rule

$f(x) = u^n$

$f'(x) = nu^n \dfrac{du}{dx}$

$f(x) = (x^3 - 3x^2 - 4)^5$

$f'(x) = 5(x^3 - 3x^2 - 4)^4(3x^2 - 6x)$

Chain + Product + Power Rules

$f(x) = (x^2 - 2)(x^3 + 4)^6$

$f'(x) = (x^2 - 2)(6)(x^3 + 4)^5(3x^2) + (x^3 + 4)^6(2x)$

Implicit Differentiation

$\dfrac{d}{dx}(y^n) = ny^{n-1}y'$

using product rule:

$\dfrac{d}{dx}(x^3y^2) = x^3(2yy') + y^2(3x^2)$

solving for y':

$$y^3 + x^2 = 7x^4y^2 - 2y + 3$$

$$3y^2y' + 2x = 7x^4(2yy') + 28x^3y^2 - 2y'$$

$$3y^2y' - 14x^4yy' + 2y' = 28x^3y^2 - 2x$$

$$y' = \dfrac{28x^3y^2 - 2x}{3y^2 - 14x^4y + 2}$$

ALGEBRA INSIGHT

These questions model the algebra needed in some of the Review questions (13, 27, 29, and 37).

1. Simplify $(3x + 2)(-2) - (1 - 2x)(3)$.

2. Simplify by factoring $2(x + 4)^3 + 6x(x + 4)^2$.

3. Simplify by factoring $(x + 1)^2 - (x - 1)(2)(x + 1)$.

4. Solve for y' in $5(1 - 2xy^3)^4(-2y^3 - 6xy^2y') = 1 + 4y'$.

5. Simplify $(t^2 + 12)(2) - (2t + 1)(2t)$.

Checkup for Chapter 2

1. In each case, find the derivative $\dfrac{dy}{dx}$.

   a. $y = 3x^4 - 4\sqrt{x} + \dfrac{5}{x^2} - 7$

   b. $y = (3x^3 - x + 1)(4 - x^2)$

   c. $y = \dfrac{5x^2 - 3x + 2}{1 - 2x}$

   d. $y = (3 - 4x + 3x^2)^{3/2}$

2. Find the second derivative of the function $f(t) = t(2t + 1)^2$.

3. Find an equation for the tangent line to the curve $y = x^2 - 2x + 1$ at the point where $x = -1$.

4. Find the rate of change of the function $f(x) = \dfrac{x + 1}{1 - 5x}$ with respect to x when $x = 1$.

5. **PROPERTY TAX**   Records indicate that x years after the year 2008, the average property tax on a four-bedroom house in a suburb of a major city was $T(x) = 3x^2 + 40x + 1800$ dollars.

   a. At what rate was the property tax increasing with respect to time in 2011?

   b. At what percentage rate was the property tax increasing in 2011?

6. **MOTION ON A LINE**   A car moves along a row between cars in a parking lot as the driver looks for a space. Its position at time t is given by $s(t) = 2t^3 - 3t^2 + 2$ metres for $t \geq 0$.

   a. Find the velocity $v(t)$ and acceleration $a(t)$ of the car after 2 seconds.

   b. When is the car stationary? When is it advancing? Retreating?

   c. What is the change in the odometer of the car for $0 \leq t \leq 2$?

7. **PRODUCTION COST**   Suppose the cost of producing x units of a particular commodity is $C(x) = 0.04x^2 + 5x + 73$ hundred dollars.

   a. Use marginal cost to estimate the cost of producing the sixth unit.

   b. What is the actual cost of producing the sixth unit?

8. **INDUSTRIAL OUTPUT**   At a certain factory, the daily output is $Q = 500L^{3/4}$ units, where L denotes the size of the labour force in worker-hours. Currently, 2401 worker-hours of labour are used each day. Use calculus (increments) to estimate the effect on output of increasing the size of the labour force by 200 worker-hours from its current level.

9. **PEDIATRIC MEASUREMENT**   Pediatricians use the formula $S = 0.2029w^{0.425}$ to estimate the surface area S (m^2) of a child 1 m tall who weighs w kilograms. A particular child weighs 30 kg and is gaining weight at a rate of 0.13 kg per week while remaining 1 m tall. At what rate is this child's surface area changing?

10. **BREAD BAKING**   The winner of a baking competition is to be chosen according to the risen volume of a dinner roll. The rolls are modelled as spheres of radius r centimetres.

   a. At what rate is the volume $V = \dfrac{4}{3}\pi r^3$ changing with respect to r when $r = 3$?

   b. Estimate the percentage error that can be allowed in the measurement of the radius r to ensure that there will be no more than an 8% error in the calculation of volume.

Review Exercises

In Exercises 1 and 2, use the definition of the derivative to find $f'(x)$.

1. $f(x) = x^2 - 3x + 1$

2. $f(x) = \dfrac{1}{x - 2}$

In Exercises 3 through 13, find the derivative of the given function.

3. $f(x) = 6x^4 - 7x^3 + 2x + \sqrt{2}$

4. $f(x) = x^3 = \dfrac{1}{3x^5} + 2\sqrt{x} - \dfrac{3}{x} + \dfrac{1 - 2x}{x^3}$

5. $y = \dfrac{2 - x^2}{3x^2 + 1}$

6. $y = (x^3 + 2x - 7)(3 + x - x^2)$

7. $f(x) = (5x^4 - 3x^2 + 2x + 1)^{10}$

8. $f(x) = \sqrt{x^2 + 1}$

9. $y = \left(x + \dfrac{1}{x}\right)^2 - \dfrac{5}{\sqrt{3x}}$

10. $y = \left(\dfrac{x + 1}{1 - x}\right)^2$

11. $f(x) = (3x + 1)\sqrt{6x + 5}$

12. $f(x) = \dfrac{(3x + 1)^3}{(1 - 3x)^4}$

13. $y = \sqrt{\dfrac{1 - 2x}{3x + 2}}$

In Exercises 14 through 17, find the equation of the tangent line to the graph of the given function at the specified point.

14. $f(x) = x^2 - 3x + 2; x = 1$

15. $f(x) = \dfrac{4}{x - 3}; x = 1$

16. $f(x) = \dfrac{x}{x^2 + 1}; x = 0$

17. $f(x) = \sqrt{x^2 + 5}; x = -2$

18. In each case, find the rate of change of $f(t)$ with respect to t at the given value of t.
    a. $f(t) = t^3 - 4t^2 + 5r\sqrt{t} - 5$ at $t = 4$
    b. $f(t) = \dfrac{2t^2 - 5}{1 - 3t}$ at $t = -1$

19. In each case, find the rate of change of $f(t)$ with respect to t at the given value of t.
    a. $f(t) = t^3(t^2 - 1)$ at $t = 0$
    b. $f(t) = (t^2 - 3t + 6)^{1/2}$ at $t = 1$

20. In each case, find the percentage rate of change of the function $f(t)$ with respect to t at the given value of t.
    a. $f(t) = t^2 - 3t + \sqrt{t}$ at $t = 4$
    b. $f(t) = \dfrac{t}{t - 3}$ at $t = 4$

21. In each case, find the percentage rate of change of the function $f(t)$ with respect to t at the given value of t.
    a. $f(t) = t^2(3 - 2t)^3$ at $t = 1$
    b. $f(t) = \dfrac{1}{t + 1}$ at $t = 0$

22. Use the chain rule to find $\dfrac{dy}{dx}$.
    a. $y = 5u^2 + u - 1; u = 3x + 1$
    b. $y = \dfrac{1}{u^2}; u = 2x + 3$

23. Use the chain rule to find $\dfrac{dy}{dx}$.
    a. $y = (u + 1)^2; u = 1 - x$
    b. $y = \dfrac{1}{\sqrt{u}}; u = 2x + 1$

24. Use the chain rule to find $\dfrac{dy}{dx}$ for the given value of x.
    a. $y = u - u^2; u = x - 3$ for $x = 0$
    b. $y = \left(\dfrac{u - 1}{u + 1}\right)^{1/2}, u = \sqrt{x} - 1$ for $x = \dfrac{34}{9}$

25. Use the chain rule to find $\dfrac{dy}{dx}$ for the given value of x.
    a. $y = u^3 - 4u^2 + 5u + 2; u = x^2 + 1$ for $x = 1$
    b. $y = \sqrt{u}, u = x^2 + 2x - 4$ for $x = 2$

26. Find the second derivative of each function:
    a. $f(x) = 6x^5 - 4x^3 + 5x^2 - 2x + \dfrac{1}{x}$
    b. $z = \dfrac{2}{1 + x^2}$
    c. $y = (3x^2 + 2)^4$

CHAPTER SUMMARY

27. Find the second derivative of each function.
    a. $f(x) = 4x^3 - 3x$
    b. $f(x) = 2x(x + 4)^3$
    c. $f(x) = \dfrac{x - 1}{(x + 1)^2}$

28. Find $\dfrac{dy}{dx}$ by implicit differentiation.

    a. $5x + 3y = 12$
    b. $(2x + 3y)^5 = x + 1$

29. Find $\dfrac{dy}{dx}$ by implicit differentiation.

    a. $x^2 y = 1$
    b. $(1 - 2xy^3)^5 = x + 4y$

30. Use implicit differentiation to find the slope of the line that is tangent to the given curve at the specified point.
    a. $xy^3 = 8$; $(1, 2)$
    b. $x^2 y - 2xy^3 + 6 = 2x + 2y$; $(0, 3)$

31. Use implicit differentiation to find the slope of the line that is tangent to the given curve at the specified point.

    a. $x^2 + 2y^3 = \dfrac{3}{xy}$; $(1, 1)$

    b. $y = \dfrac{x + y}{x - y}$; $(6, 2)$

32. Use implicit differentiation to find $\dfrac{d^2 y}{dx^2}$ if $4x^2 + y^2 = 1$.

33. Use implicit differentiation to find $\dfrac{d^2 y}{dx^2}$ if $3x^2 - 2y^2 = 6$.

34. A baseball is hit vertically upward from ground level with an initial velocity of 49 m/s.
    a. When will the ball pass the bat?
    b. What is the impact velocity?
    c. When will the ball reach its maximum height? What is the maximum height?

35. **POPULATION GROWTH**   Suppose that a 5-year projection of population trends suggests that t years from now, the population of a certain community will be P thousand, where
    $$P(t) = t^3 - 9t^2 + 48t + 200.$$
    a. At what rate will the population be growing 3 years from now?
    b. At what rate will the rate of population growth be changing with respect to time 3 years from now?

In Exercises 36 and 37, s(t) denotes the position, in metres, of an object moving along a line after t seconds.
    (a) Find the velocity and acceleration of the object and describe its motion during the indicated time interval.
    (b) Compute the total distance travelled by the object during the indicated time interval.

36. $s(t) = 2t^3 - 21t^2 + 60t - 25$; $1 \le t \le 6$

37. $s(t) = \dfrac{2t + 1}{t^2 + 12}$; $0 \le t \le 4$

38. **RAPID TRANSIT**   After x weeks, the number of people using a new rapid transit system is approximately $N(x) = 6x^3 + 500x + 8000$.
    a. At what rate is the use of the system changing with respect to time after 8 weeks?
    b. By how much does the use of the system change during the eighth week?

39. **PRODUCTION**   It is estimated that the weekly output at a certain plant is $Q(x) = 50x^2 + 9000x$ units, where x is the number of workers employed at the plant. Currently 30 workers are employed at the plant.
    a. Use calculus to estimate the change in the weekly output that will result from the addition of 1 worker to the force.
    b. Compute the actual change in output that will result from the addition of 1 worker.

40. **MOSQUITO NUMBERS**   It is projected that t hours from now, the number of mosquitoes at a campground near a marsh will be $P(t) = 3t + 5t^{3/2} + 6000$. At what percentage rate will the number be changing with respect to time 4 h from now?

41. **PRODUCTION**   At a certain factory, the daily output is $Q(L) = 20\,000L^{1/2}$ units, where L denotes the size of the labour force measured in worker-hours. Currently 900 worker-hours of labour are used each day. Use calculus to estimate the effect on output that will be produced if the labour force is cut to 885 worker-hours.

42. **GROSS DOMESTIC PRODUCT**   The gross domestic product of a certain country was $N(t) = t^2 + 6t + 300$ billion dollars t years after 2004. Use calculus to predict the percentage change in the GDP during the second quarter of 2012.

43. **POLLUTION**   The level of air pollution in a certain city is proportional to the square of the population. Use calculus to estimate the percentage

by which the air pollution level will increase if the population increases by 5%.

44. **WIND CHILL**   Consider the wind chill formula from Environment Canada used in Section 1.6,

$$\text{Wind Chill} = 13.12 + 0.6215T_{air} - 11.37V^{0.16} + 0.3965T_{air}V^{0.16}$$

degrees Celsius, where V is the wind velocity in kilometres per hour.
   a. Find the derivative of the wind chill with respect to the velocity of the wind when the air temperature is $-10°C$. Interpret your answer.
   b. Find the derivative of the wind chill with respect to the air temperature when the wind blows at 30 km/h. Interpret your answer.

45. **POPULATION DENSITY**   The formula $D = 36m^{-1.14}$ is sometimes used to determine the ideal population density D (individuals per square kilometre) for a large animal of mass m kilograms.
   a. What is the ideal population density for humans, assuming that a typical human weighs about 70 kg?
   b. Consider an island of area 3000 km². Two hundred animals of mass $m = 30$ kg are brought to the island, and t years later, the population is given by
$$P(t) = 0.43t^2 + 13.37t + 200$$
   How long does it take for the ideal population density to be reached? At what rate is the population changing when the ideal density is attained?

46. **BACTERIAL GROWTH**   The population P of a bacterial colony t days after observation begins is modelled by the cubic function
$$P(t) = 1.035t^3 + 103.5t^2 + 6900t + 230\ 000.$$
   a. Compute and interpret the derivative $P'(t)$.
   b. At what rate is the population changing after 1 day? After 10 days?
   c. What is the initial population of the colony? How long does it take for the population to double? At what rate is the population growing at the time it doubles?

47. **PRODUCTION**   The output at a certain factory is $Q(L) = 600L^{2/3}$ units, where L is the size of the labour force. The manufacturer wishes to increase output by 1%. Use calculus to estimate the percentage increase in labour that will be required.

48. **PRODUCTION**   The output Q at a certain factory is related to inputs x and y by the equation
$$Q = x^3 + 2xy^2 + 2y^3.$$

If the current levels of input are $x = 10$ and $y = 20$, use calculus to estimate the change in input y that should be made to offset an increase of 0.5 in input x so that output will be maintained at its current level.

49. **A* BLOOD FLOW**   Physiologists have observed that the flow of blood from an artery into a small capillary is given by the formula
$$F = kD^2\sqrt{A - C}\ (\text{cm}^3/\text{s}),$$
   where D is the diameter of the capillary, A is the pressure in the artery, C is the pressure in the capillary, and k is a positive constant.
   a. By how much is the flow of blood F changing with respect to pressure C in the capillary if A and D are kept constant? Does the flow increase or decrease with increasing C?
   b. What is the percentage rate of change of flow F with respect to A if C and D are kept constant?

50. **A* POLLUTION CONTROL**   It is estimated that t years from now, the population of a certain suburban community will be $p(t) = 10 - \dfrac{20}{(t + 1)^2}$ thousand. An environmental study indicates that the average daily level of carbon monoxide in the air will be $c(p) = 0.8\sqrt{p^2 + p + 139}$ units when the population is p thousand. At what percentage rate will the level of carbon monoxide be changing with respect to time 1 year from now?

51. You measure the radius of a circle to be 12 cm and use the formula $A = \pi r^2$ to calculate the area. If your measurement of the radius is accurate to within 3%, how accurate is your calculation of the area?

52. **A*** Estimate what will happen to the volume of a cube if the length of each side is decreased by 2%. Express your answer as a percentage.

53. **A* PRODUCTION**   The output at a certain factory is $Q = 600K^{1/2}L^{1/3}$ units, where K denotes the capital investment and L is the size of the labour force. Estimate the percentage increase in output that will result from a 2% increase in the size of the labour force if capital investment is not changed.

54. **BLOOD FLOW**   The speed of blood flowing along the central axis of a certain artery is $S(R) = 1.8 \times 10^5 R^2$ centimetres per second, where R is the radius of the artery. A medical researcher measures the radius of the artery to be

1.2×10^{-2} cm and makes an error of 5×10^{-4} cm. Estimate the amount by which the calculated value of the speed of the blood will differ from the true speed if the incorrect value of the radius is used in the formula.

55. **AREA OF A TOMATO**   You measure the radius of a spherical cherry tomato to be 1.2 cm and use the formula $S = 4\pi r^2$ to calculate the surface area. If your measurement of the radius r is accurate to within 3%, how accurate is your calculation of the surface area?

56. **CARDIOVASCULAR SYSTEM**   One model of the cardiovascular system relates $V(t)$, the stroke volume of blood in the aorta at a time t during systole (the contraction phase), to the pressure $P(t)$ in the aorta during systole by the equation

$$V(t) = [C_1 + C_2 P(t)]\left(\frac{3t^2}{T^2} - \frac{2t^3}{T^3}\right),$$

where C_1 and C_2 are positive constants and T is the (fixed) time length of the systole phase.* Find a relationship between the rates $\dfrac{dV}{dt}$ and $\dfrac{dP}{dt}$.

57. **CONSUMER DEMAND**   When toasters are sold for p dollars each, local consumers will buy $D(p) = \dfrac{32\,670}{2p+1}$ toasters. It is estimated that t months from now, the unit price of the toasters will be $p(t) = 0.04t^{3/2} + 44$ dollars. Compute the rate of change of the monthly demand for toasters with respect to time 25 months from now. Will the demand be increasing or decreasing?

58. At noon, a truck is at the intersection of two roads and is moving north at 70 km/h. An hour later, a car passes through the same intersection, travelling east at 105 km/h. How fast is the distance between the car and the truck changing at 2 P.M.?

59. **FISH IN A LAKE**   It is projected that t years from now, the number of a certain type of fish in a lake will be $P(t) = 200 - \dfrac{60}{t+1}$. Calculate the number of fish in the lake 3 years from now and the rate of change of the number of fish at that time.

60. **WORKER EFFICIENCY**   At a mail-order business, it is found that an average worker arriving

on the job at 8:00 A.M. will have processed $Q(t) = -t^3 + 9t^2 + 12t$ orders t hours later.
   a. Compute the worker's rate of processing orders $R(t) = Q'(t)$.
   b. At what rate is the worker's rate of processing changing with respect to time at 9:00 A.M.?
   c. Use calculus to estimate the change in the worker's rate of processing orders between 9:00 and 9:06 A.M.
   d. Compute the actual change in the worker's rate of processing between 9:00 and 9:06 A.M.

61. **TRAFFIC SAFETY**   A car is travelling at 30 m/s when the driver applies the brakes to avoid hitting an obstruction in the road. After t seconds, the car is $s = 30t - 3t^2$ metres from the point where the brakes were applied. How long does it take for the car to come to a stop, and how far does it travel before stopping?

62. **CONSTRUCTION MATERIAL**   Sand is leaking from a bag in such a way that after t seconds, there are

$$S(t) = 50\left(1 - \frac{t^2}{15}\right)^3$$

kilograms of sand left in the bag.
   a. How much sand was originally in the bag?
   b. At what rate is sand leaking from the bag after 1 second?
   c. How long does it take for all of the sand to leak from the bag? At what rate is the sand leaking from the bag at the time it empties?

63. **OIL PRICES**   It is projected that t months from now, due to social unrest in an oil-producing country, the price per barrel of oil will be P dollars, where

$$P(t) = -t^3 + 2t^2 + 2t + 72.$$

   a. At what rate will the price per barrel be increasing with respect to time 4 months from now?
   b. At what rate will the rate of price increase be changing with respect to time 4 months from now?

64. **PRODUCTION COST**   At a certain factory, approximately $q(t) = t^2 + 50t$ units are manufactured during the first t hours of a production run, and the total cost of manufacturing q units is $C(q) = 0.1q^2 + 10q + 400$ dollars. Find the rate at which the manufacturing cost is changing with respect to time 2 hours after production starts.

*J. G. Defares, J. J. Osborn, and H. H. Hura, *Acta Physiol. Pharm. Neerl.*, Vol. 12, 1963, pp. 189–265.

65. **PRODUCTION COST**   It is estimated that the monthly cost of producing x units of a particular commodity is $C(x) = 0.06x + 3x^{1/2} + 20$ hundred dollars. Suppose production is decreasing at a rate of 11 units per month when the monthly production is 2500 units. At what rate is the cost changing at this level of production?

66. Estimate the largest percentage error you can allow in the measurement of the radius of a sphere if you want the error in the calculation of its surface area using the formula $S = 4\pi r^2$ to be no greater than 8%.

67. A soccer ball made of leather that is 0.3 cm thick has an inner diameter of 22 cm. Estimate the volume of its leather shell. [*Hint:* Think of the volume of the shell as a certain change ΔV in volume.]

68. A car travelling north at 60 km/h and a truck travelling east at 45 km/h leave an intersection at the same time. At what rate is the distance between them changing 2 hours later?

69. A child is flying a kite at a height of 25 m above her hand. If the kite moves horizontally at a constant speed of 2 m/s, at what rate is string being released when the kite is 30 m away from the child?

70. When the price of a 400-g bag of frozen shrimp is p dollars per bag, the manufacturer is willing to supply x thousand bags, where
$$x^2 - 2x\sqrt{p} - p^2 = 31.$$
How fast is the supply changing when the price is $9 per bag and is increasing at a rate of 20 cents/week?

71. A 3-m-long ladder leans against the side of a wall. The top of the ladder is sliding down the wall at a rate of 0.9 m/s. How fast is the foot of the ladder moving away from the building when the top is 2 m above the ground?

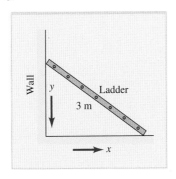

EXERCISE 71

72. **A*** A lantern falls from the top of a building in such a way that after t seconds, it is $h(t) = 50 - 4t^2$ metres above ground. A woman 1.6 m tall originally standing directly under the lantern sees it start to fall and walks away at a constant rate of 2 m/s. At what rate is the length of the woman's shadow changing when the lantern is 3 m above the ground?

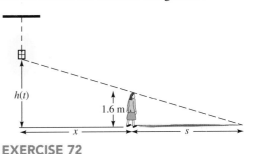

EXERCISE 72

73. A baseball diamond is a square that is 27.4 m on each side. A runner runs from second base to third at 7 m/s. How fast is the distance s between the runner and home base changing when he is 5 m from third base?

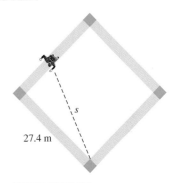

EXERCISE 73

74. **MANUFACTURING COST**   Suppose the total manufacturing cost C at a certain factory is a function of the number q of units produced, which in turn is a function of the number t of hours during which the factory has been operating.

   a. What quantity is represented by the derivative $\dfrac{dC}{dq}$? In what units is this quantity measured?

   b. What quantity is represented by the derivative $\dfrac{dq}{dt}$? In what units is this quantity measured?

   c. What quantity is represented by the product $\dfrac{dC}{dq}\dfrac{dq}{dt}$? In what units is this quantity measured?

75. **A*** A new type of engine for a remote-controlled car is being investigated. The car moves along a straight line from point P. The velocity of the car is directly proportional to the product of the time the car has been moving and the distance it has moved from P. It is also known that at the end of 5 seconds, the car is 20 m from P and is moving at a rate of 4 m/s. Find the acceleration of the car at this time (when $t = 5$).

76. **A*** Find all the points (x, y) on the graph of the function $y = 4x^2$ with the property that the tangent to the graph at (x, y) passes through the point $(2, 0)$.

77. **A*** Suppose y is a linear function of x; that is, $y = mx + b$. What will happen to the percentage rate of change of y with respect to x as x increases without bound? Explain.

78. **A*** Find an equation for the tangent line to the curve
$$\frac{x^2}{a^2} - \frac{y^2}{b^2} = 1$$
at the point (x_0, y_0).

79. Let $f(x) = (3x + 5)(2x^3 - 5x + 4)$. Use a graphing utility to graph $f(x)$ and $f'(x)$ on the same set of coordinate axes. Compare the graphs and find where $f'(x) = 0$.

80. Use a graphing utility to graph $f(x) = \dfrac{(2x + 3)}{(1 - 3x)}$ and $f'(x)$ on the same set of coordinate axes. Compare the graphs and find where $f'(x) = 0$.

81. The curve $y^2(2 - x) = x^3$ is called a **cissoid**.
    a. Use a graphing utility to sketch the curve.
    b. Find equations for the tangent lines to the curve at all points where $x = 1$.
    c. What happens to the curve as x approaches 2 from the left?
    d. Does the curve have a tangent line at the origin? If so, what is its equation?

82. An object moves along a straight line in such a way that its position at time t is given by
$$s(t) = t^{5/2}(0.73t^2 - 3.1t + 2.7) \quad \text{for } 0 \le t \le 2.$$
    a. With technology, find the velocity $v(t)$ and the acceleration $a(t)$ and then graph $s(t)$, $v(t)$, and $a(t)$ on separate graphs using the same scale $0 \le t \le 2$. Line the graphs up under each other.
    b. With technology, solve $v(t) = 0$ in the time interval of the plots. What is the object's position at this time?
    c. When does the smallest value of $a(t)$ occur? Where is the object at this time and what is its velocity?

THINK ABOUT IT

MODELLING THE BEHAVIOUR OF THE CODLING MOTH

(Photos: Left: © Nigel Cattlin/Photo Researchers, Inc.; Right: © Runk/Schoenberger/Grant Heilman)

In the middle of the Niagara Fruit Belt, the Ontario Ministry of Agriculture research station at Vineland researches the codling moth. The larvae cause serious damage to apples by eating the fruit and seeds. With organic farming on the rise, research has been done into pheromones emitted by the cocoon-spinning larvae. These pheromones attract other larvae to pupate in the same area. When many pupae are in one area, they can be found and destroyed, preventing a female from hatching up to 130 eggs on the leaves of the apple tree.*

After the codling moth larva, also known as the common apple worm, hatches from its egg, it goes looking for an apple. The time between hatching and finding an apple is called the **searching period.** Once a codling moth finds an apple, it squirms into the apple and eats the fruit and seeds of the apple, thereby ruining it. After about 4 weeks, the codling moth backs out of the apple and forms a cocoon.

Observations on the behaviour of codling moths indicate that the length of the searching period, $S(T)$, and the percentage of larvae that survive the searching period, $N(T)$, depend on the air temperature, T. Methods of data analysis (polynomial regression) applied to data recorded from observations suggest that if T is measured in degrees Celsius with $20 \leq T \leq 30$, then $S(T)$ and $N(T)$ may be modelled** by

$$S(T) = (-0.03T^2 + 1.6T - 13.65)^{-1} \text{ days}$$

and

$$N(T) = -0.85T^2 + 45.4T - 547$$

Use these formulas to answer the following questions.

---

*H. Fraser, "'Cuddlemone'—Pheromone-based attraction in codling moth larvae," www.omafra.gov.on.ca, 18 February 2009.
**P. L. Shaffer and H. J. Gold, "A simulation model of population dynamics of the codling moth *Cydia Pomonella,*" *Ecological Modeling,* Vol. 30, 1985, pp. 247–274.

(Photo: © IT Stock Free/Alamy)

Questions

1. What do these formulas for $S(T)$ and $N(T)$ predict for the length of the searching period and the percentage of larvae surviving the searching period when the air temperature is 25°C?

2. Sketch the graph of $N(T)$, and determine the temperature at which the largest percentage of codling moth larvae survive. Then determine the temperature at which the smallest percentage of larvae survive. (Remember, $20 \leq T \leq 30$.)

3. Find $\dfrac{dS}{dT}$, the rate of change of the searching period with respect to temperature T. When does this rate equal zero? What (if anything) occurs when $\dfrac{dS}{dT} = 0$?

4. Find $\dfrac{dN}{dS}$, the rate of change of the percentage of larvae surviving the searching period with respect to the length of the searching period using the chain rule,

$$\frac{dN}{dT} = \frac{dN}{dS}\frac{dS}{dT}.$$

What (if any) information does this rate provide?

Determining when an assembly line is working at peak efficiency is one application of the derivative. Peak efficiency is not the same as producing the maximum number of items. How is it different? See Example 3.2.6 for the answer! (Photo: © Getty/Royalty Free)

Chapter 3

ADDITIONAL APPLICATIONS OF THE DERIVATIVE

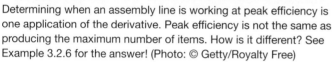

CHAPTER OUTLINE

LEARNING OBJECTIVES

After completing this chapter, you should be able to

L01 Determine intervals where a function is increasing or decreasing. Find relative extrema of a function using the first derivative test.

L02 Determine intervals where a function is concave up or concave down. Use the second derivative test. Find inflection points of a function.

L03 Sketch curves using the first and second derivatives.

L04 Calculate the absolute extrema of a function. Apply these methods to solve problems in elasticity of demand.

L05 Construct equations for applied optimization problems and solve to optimize one variable.

OPTION NOTE: Chapter 4, Sections 4.1, 4.2, and 4.3, can be done before Chapter 3 if exponential and logarithmic functions are required earlier in the course. Omit Example 4.3.12 and Section 4.3 Exercises 39 to 46.

Then Section 4.4 can be taught after completing Chapter 3.

Optional exercises pertaining to this chapter, but involving exponential and logarithmic functions, are noted at the beginning of Sections 3.3 and 3.4 Exercises.

SECTION 3.1

L01

Determine intervals where a function is increasing or decreasing. Find relative extrema of a function using the first derivative test.

Increasing and Decreasing Functions; Relative Extrema

Intuitively, we regard a function $f(x)$ as *increasing* where the graph of f is rising and *decreasing* where the graph is falling. For instance, the graph in Figure 3.1 shows the tourism expenditures in Canada during the period 2000–2009. The scale shows quarterly data seasonally adjusted. The shape of the graph suggests that the spending increased for most of this time, but there were some noticeable decreases where expenditures were falling. In 2009 the recession reduced expenditures, and the exchange rate with the United States dollar and travel restrictions influenced tourism spending. There are peaks and valleys where the expenditures change from decreasing to increasing and vice versa.

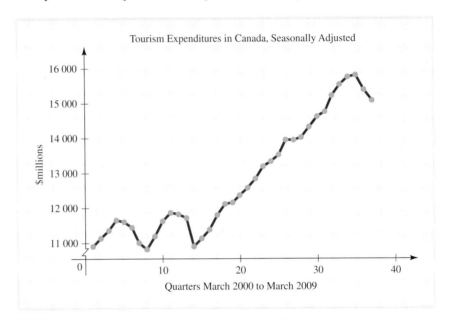

FIGURE 3.1 Tourism expenditures in Canada 2000–2009, seasonally adjusted and in constant 2002 dollars.*

*SOURCE: Adapted from the Statistics Canada CANSIM database, http://cansim2.gc.ca, table 387-0001, retrieved April 27, 2010.

Here is a more formal definition of increasing and decreasing functions, which are illustrated graphically in Figure 3.2.

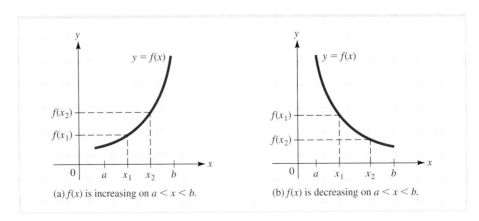

(a) $f(x)$ is increasing on $a < x < b$.

(b) $f(x)$ is decreasing on $a < x < b$.

FIGURE 3.2 Intervals of increase and decrease.

Increasing and Decreasing Functions ■ Let $f(x)$ be a function defined on the interval $a < x < b$, and let x_1 and x_2 be two numbers in the interval. Then

$f(x)$ is **increasing** on the interval if $f(x_2) > f(x_1)$ whenever $x_2 > x_1$.

$f(x)$ is **decreasing** on the interval if $f(x_2) < f(x_1)$ whenever $x_2 > x_1$.

As demonstrated in Figure 3.3a, if the graph of a function $f(x)$ has tangent lines with only positive slopes on the interval $a < x < b$, then the graph is rising and $f(x)$ is increasing on the interval. Since the slope of each such tangent line is given by the derivative $f'(x)$, it follows that $f(x)$ is increasing (graph rising) on intervals where $f'(x) > 0$. Similarly, $f(x)$ is decreasing (graph falling) on intervals where $f'(x) < 0$ (Figure 3.3b).

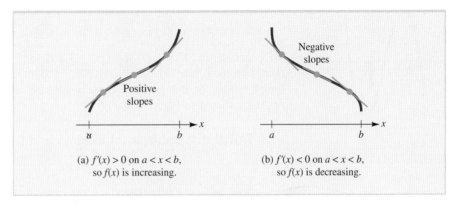

(a) $f'(x) > 0$ on $a < x < b$,
so $f(x)$ is increasing.

(b) $f'(x) < 0$ on $a < x < b$,
so $f(x)$ is decreasing.

FIGURE 3.3   Derivative criteria for increasing and decreasing functions.

Thanks to the intermediate value property (Section 1.6), we know that a continuous function cannot change sign without first becoming 0. This means that if we mark on a number line all numbers x where $f'(x)$ is discontinuous or $f'(x) = 0$, the line will be divided into intervals where $f'(x)$ does not change sign. Therefore, if we pick a test number c from each such interval and find that $f'(c) > 0$, we know that $f'(x) > 0$ for *all* numbers x in the interval and $f(x)$ must be increasing (graph rising) throughout the interval. Similarly, if $f'(c) < 0$, it follows that $f(x)$ is decreasing (graph falling) throughout the interval. These observations may be summarized as follows.

Procedure for Using the Derivative to Determine Intervals of Increase and Decrease for a Function f

Step 1.   Find all values of x for which $f'(x) = 0$ or $f'(x)$ does not exist, and mark these numbers on a number line. This divides the line into a number of open intervals.

Step 2.   Choose a test number c from each interval $a < x < b$ determined in step 1 and evaluate $f'(c)$. Then,

If $f'(c) > 0$, the function $f(x)$ is increasing (graph rising) on $a < x < b$.
If $f'(c) < 0$, the function $f(x)$ is decreasing (graph falling) on $a < x < b$.

This procedure is illustrated in Examples 3.1.1 and 3.1.2.

EXPLORE!

Find and graph the derivative of $f(x)$ in Example 3.1.1. Compare the graph with Figure 3.4. Can you identify the intervals where $f(x)$ is increasing or decreasing by examining the behaviour of $f'(x)$?

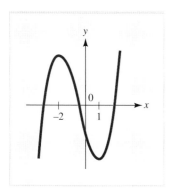

FIGURE 3.4   The graph of $f(x) = 2x^3 + 3x^2 - 12x - 7$.

EXAMPLE 3.1.1

Find the intervals of increase and decrease for the function

$$f(x) = 2x^3 + 3x^2 - 12x - 7$$

Solution

The derivative of $f(x)$ is

$$f'(x) = 6x^2 + 6x - 12 = 6(x + 2)(x - 1),$$

which is continuous everywhere, with $f'(x) = 0$ where $x = 1$ and $x = -2$. The numbers -2 and 1 divide the x axis into three open intervals, namely, $x < -2$, $-2 < x < 1$, and $x > 1$. Choose a test number c from each of these intervals, say, $c = -3$ from $x < -2$, $c = 0$ from $-2 < x < 1$, and $c = 2$ from $x > 1$. Then evaluate $f'(c)$ for each test number:

$$f'(-3) = 24 > 0 \qquad f'(0) = -12 < 0 \qquad f'(2) = 24 > 0$$

We conclude that $f'(x) > 0$ for $x < -2$ and for $x > 1$, so $f(x)$ is increasing (graph rising) on these intervals. Similarly, $f'(x) < 0$ on $-2 < x < 1$, so $f(x)$ is decreasing (graph falling) on this interval. These results are summarized in Table 3.1. The graph of $f(x)$ is shown in Figure 3.4.

TABLE 3.1   Intervals of Increase and Decrease for
$f(x) = 2x^3 + 3x^2 - 12x - 7$

Interval	Test Number c	$f'(c)$	Conclusion	Direction of Graph
$x < -2$	-3	$f'(-3) > 0$	f is increasing	Rising
$-2 < x < 1$	0	$f'(0) < 0$	f is decreasing	Falling
$x > 1$	2	$f'(2) > 0$	f is increasing	Rising

NOTATION   Henceforth, we shall indicate an interval where $f(x)$ is increasing by an "up arrow" ($\nearrow$) and an interval where $f(x)$ is decreasing by a "down arrow" ($\searrow$). Thus, the results in Example 3.1.1 can be represented by this arrow diagram:

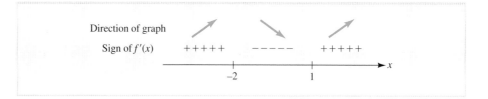

EXAMPLE 3.1.2

Find the intervals of increase and decrease for the function

$$f(x) = \frac{x^2}{x - 2}.$$

Solution

The function is defined for $x \neq 2$, and its derivative is

$$f'(x) = \frac{(x-2)(2x) - x^2(1)}{(x-2)^2} = \frac{x(x-4)}{(x-2)^2}$$

which is discontinuous at $x = 2$ and has $f'(x) = 0$ at $x = 0$ and $x = 4$. Thus, there are four intervals on which the sign of $f'(x)$ does not change: namely, $x < 0$, $0 < x < 2$, $2 < x < 4$, and $x > 4$. Choosing test numbers in these intervals (say, -2, 1, 3, and 5, respectively), we find that

$$f'(-2) = \frac{3}{4} > 0 \qquad f'(1) = -3 < 0 \qquad f'(3) = -3 < 0 \qquad f'(5) = \frac{5}{9} > 0$$

We conclude that $f(x)$ is increasing (graph rising) for $x < 0$ and for $x > 4$ and that it is decreasing (graph falling) for $0 < x < 2$ and for $2 < x < 4$. These results are summarized in the arrow diagram displayed next (the dashed vertical line indicates that $f(x)$ is not defined at $x = 2$).

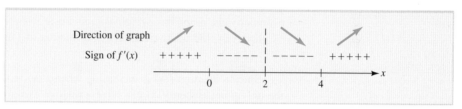

Intervals of increase and decrease for $f(x) = \dfrac{x^2}{x-2}$.

The graph of $f(x)$ is shown in Figure 3.5. Notice how the graph approaches the vertical line $x = 2$ as x tends toward 2. This behaviour identifies $x = 2$ as a *vertical asymptote* of the graph of $f(x)$. We will discuss asymptotes in Section 3.3. (Graphs made with technology often include an asymptote line; remember that this is not part of the function.)

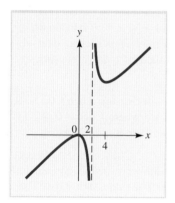

FIGURE 3.5   The graph of $f(x) = \dfrac{x^2}{x-2}$.

Relative Extrema

The simplicity of the graphs in Figures 3.4 and 3.5 may be misleading. A more general graph is shown in Figure 3.6. Note that peaks occur at C and E and valleys occur at

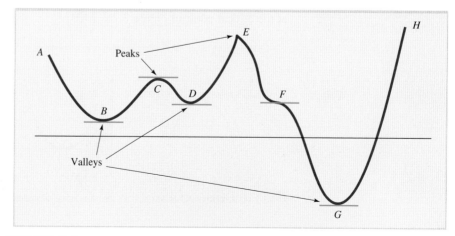

FIGURE 3.6   A graph with various kinds of peaks and valleys.

B, D, and G, and while there are horizontal tangents at B, C, D, and G, no tangent can be drawn at the sharp point E. Moreover, there is a horizontal tangent at F that is neither a peak nor a valley. In this section and the next, you will see how the methods of calculus can be used to locate and identify the peaks and valleys of a graph. This, in turn, provides the basis for a curve-sketching procedure and for methods of optimization.

Less informally, a peak on the graph of a function f is known as a **relative maximum** of f, and a valley is a **relative minimum.** Thus, a relative maximum is a point on the graph of f that is at least as high as any nearby point on the graph, while a relative minimum is at least as low as any nearby point. Collectively, relative maxima and minima are called **relative extrema.** In Figure 3.6, the relative maxima are located at C and E, and the relative minima are at B, D, and G. Note that a relative extremum does not have to be the highest or lowest point on the entire graph. For instance, in Figure 3.6, the lowest point on the graph is at the relative minimum G, but the highest point occurs at the right endpoint H. Here is a summary of this terminology.

Relative Extrema ■ The graph of the function $f(x)$ is said to have a *relative maximum* at $x = c$ if $f(c) \geq f(x)$ for all x in an interval $a < x < b$ containing c. Similarly, the graph has a *relative minimum* at $x = c$ if $f(c) \leq f(x)$ on such an interval. Collectively, the relative maxima and minima of f are called its *relative extrema.*

Since a function $f(x)$ is increasing when $f'(x) > 0$ and decreasing when $f'(x) < 0$, the only points where $f(x)$ can have a relative extremum are where $f'(x) = 0$ or $f'(x)$ does not exist. Such points are so important that we give them a special name.

Critical Numbers and Critical Points ■ A number c in the domain of $f(x)$ is called a **critical number** if either $f'(c) = 0$ or $f'(c)$ does not exist. The corresponding point $(c, f(c))$ on the graph of $f(x)$ is called a **critical point** for $f(x)$. *Relative extrema can occur only at critical points.*

It is important to note that while relative extrema occur at critical points, *not all critical points correspond to relative extrema.* For example, Figure 3.7 shows three different situations where $f'(c) = 0$, so a horizontal tangent line occurs at the critical point $(c, f(c))$. The critical point corresponds to a relative maximum in Figure 3.7a and to a relative minimum in Figure 3.7b, but neither kind of relative extremum occurs at the critical point in Figure 3.7c.

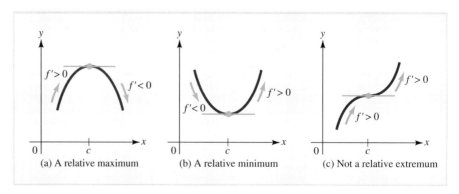

(a) A relative maximum   (b) A relative minimum   (c) Not a relative extremum

FIGURE 3.7   Three critical points $(c, f(c))$ where $f'(c) = 0$.

Three functions with critical points at which the derivative is undefined are shown in Figure 3.8. In Figure 3.8c, the tangent line is vertical at $(c, f(c))$, so the slope $f'(c)$ is undefined. In Figures 3.8a and 3.8b, no (unique) tangent line can be drawn at the sharp point $(c, f(c))$.

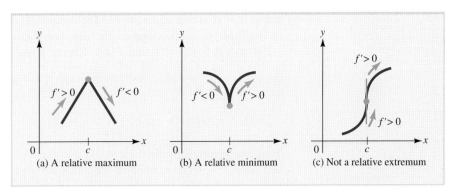

(a) A relative maximum   (b) A relative minimum   (c) Not a relative extremum

FIGURE 3.8   Three critical points $(c, f(c))$ where $f'(c)$ is undefined.

The First Derivative Test for Relative Extrema

Figures 3.7 and 3.8 also suggest a method for using the sign of the derivative to classify critical points as relative maxima, relative minima, or neither. Suppose c is a critical number of f and that $f'(x) > 0$ to the left of c, while $f'(x) < 0$ to the right. Geometrically, this means the graph of f goes up before the critical point $P(c, f(c))$ and then comes down, which implies that P is a relative maximum. Similarly, if $f'(x) < 0$ to the left of c and $f'(x) > 0$ to the right, the graph goes down before $P(c, f(c))$ and up afterward, so there must be a relative minimum at P. On the other hand, if the derivative has the same sign on both sides of c, then the graph either rises through P or falls through P, and no relative extremum occurs there. These observations can be summarized as follows.

The First Derivative Test for Relative Extrema ■ Let c be a critical number for $f(x)$ (that is, $f(c)$ is defined and either $f'(c) = 0$ or $f'(c)$ does not exist). Then the critical point $P(c, f(c))$ is

a **relative maximum** if $f'(x) > 0$ to the left of c and $f'(x) < 0$ to the right of c

a **relative minimum** if $f'(x) < 0$ to the left of c and $f'(x) > 0$ to the right of c

not a relative extremum if $f'(x)$ has the same sign on both sides of c

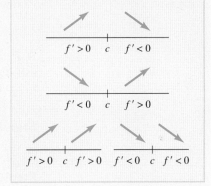

EXAMPLE 3.1.3

Find all critical numbers of the function

$$f(x) = 2x^4 - 4x^2 + 3,$$

and classify each critical point as a relative maximum, a relative minimum, or neither.

Solution

The polynomial $f(x)$ is defined for all x, and its derivative is

$$f'(x) = 8x^3 - 8x = 8x(x^2 - 1) = 8x(x - 1)(x + 1)$$

Since the derivative exists for all x, the only critical numbers are where $f'(x) = 0$; that is, $x = 0$, $x = 1$, and $x = -1$. These numbers divide the x axis into four intervals, on each of which the sign of the derivative does not change, namely, $x < -1$, $-1 < x < 0$, $0 < x < 1$, and $x > 1$. Choose a test number c in each of these intervals (say, -5, $-\dfrac{1}{2}$, $\dfrac{1}{4}$, and 2, respectively) and evaluate $f'(c)$ in each case:

$$f'(-5) = -960 < 0 \quad f'\left(-\frac{1}{2}\right) = 3 > 0 \quad f'\left(\frac{1}{4}\right) = -\frac{15}{8} < 0 \quad f'(2) = 48 > 0$$

Thus, the graph of f falls for $x < -1$ and for $0 < x < 1$, and rises for $-1 < x < 0$ and for $x > 1$, so there must be a relative maximum at $x = 0$ and relative minima at $x = -1$ and $x = 1$, as indicated in this arrow diagram.

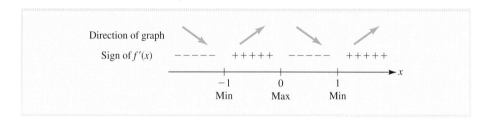

Applications

Once you determine the intervals of increase and decrease of a function f and find its relative extrema, you can obtain a rough sketch of the graph of the function. Here is a step-by-step description of the procedure for sketching the graph of a continuous function $f(x)$ using the derivative $f'(x)$. In Section 3.3, we will extend this procedure to cover the situation where $f(x)$ is discontinuous.

A Procedure for Sketching the Graph of a Continuous Function f(x) Using the Derivative f'(x)

Step 1. Determine the domain of $f(x)$. Set up a number line restricted to include only those numbers in the domain of $f(x)$.

Step 2. Find $f'(x)$ and mark each critical number on the restricted number line obtained in step 1. Then analyze the sign of the derivative to determine intervals of increase and decrease for $f(x)$ on the restricted number line.

Step 3. For each critical number c, find $f(c)$ and plot the critical point $P(c, f(c))$ on a coordinate plane, with a "cap" ⌢ at P if it is a relative maximum (↗↘), or a "cup" ⌣ if P is a relative minimum (↘↗). Plot intercepts and other key points that can easily be found.

Step 4. Sketch the graph of f as a smooth curve joining the critical points in such a way that it rises where $f'(x) > 0$, falls where $f'(x) < 0$, and has a horizontal tangent where $f'(x) = 0$.

EXPLORE!

Graph f(x) from Example 3.1.4. On the same window, graph $g(x) = x^4 + 8x^3 + 18x^2 + 2$, which is f(x) with a change in the constant term from -8 to $+2$.
How does this affect the critical values?
How does this affect the derivative?
How does this affect the value of the function at the relative maximum and minimum?

EXAMPLE 3.1.4

Sketch the graph of the function $f(x) = x^4 + 8x^3 + 18x^2 - 8$.

Solution

Since $f(x)$ is a polynomial, it is defined for all x. Its derivative is

$$f'(x) = 4x^3 + 24x^2 + 36x = 4x(x^2 + 6x + 9) = 4x(x + 3)^2.$$

Since the derivative exists for all x, the only critical numbers are where $f'(x) = 0$, namely, at $x = 0$ and $x = -3$. These numbers divide the x axis into three intervals, on each of which the sign of the derivative $f'(x)$ does not change, namely, $x < -3$, $-3 < x < 0$, and $x > 0$. Choose a test number c in each interval (say, -5, -1, and 1, respectively), and determine the sign of $f'(c)$:

$$f'(-5) = -80 < 0 \qquad f'(-1) = -16 < 0 \qquad f'(1) = 64 > 0.$$

Thus, the graph of f has horizontal tangents where x is -3 and 0, and the graph is falling (f decreasing) in the intervals $x < -3$ and $-3 < x < 0$ and is rising (f increasing) for $x > 0$, as indicated in this arrow diagram:

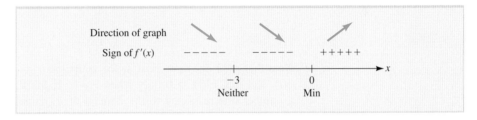

Interpreting the diagram, we see that the graph falls to a horizontal tangent at $x = -3$, then continues falling to the relative minimum at $x = 0$, after which it rises indefinitely. We find that $f(-3) = 19$ and $f(0) = -8$. To begin the sketch, plot a "cup" $\smile$ at the critical point $(0, -8)$ to indicate that a relative minimum occurs there (if it had been a relative maximum, you would have used a "cap" $\frown$), and a "twist" $\curvearrowright$ at $(-3, 19)$ to indicate a falling graph with a horizontal tangent at this point. This is shown in the preliminary graph in Figure 3.9a. Finally, complete the sketch by passing a smooth curve through the critical points in the directions indicated by the arrows, as shown in Figure 3.9b.

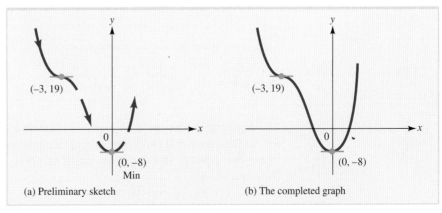

(a) Preliminary sketch

(b) The completed graph

FIGURE 3.9   The graph of $f(x) = x^4 + 8x^3 + 18x^2 - 8$.

EXAMPLE 3.1.5

Find the intervals of increase and decrease and the relative extrema of the function $g(t) = \sqrt{3 - 2t - t^2}$. Sketch the graph.

Solution

Since $\sqrt{u}$ is defined only for $u \geq 0$, the domain of g is the set of all t such that $3 - 2t - t^2 \geq 0$. Factoring the expression $3 - 2t - t^2$, we get

$$3 - 2t - t^2 = (3 + t)(1 - t)$$

Note that $3 + t \geq 0$ when $t \geq -3$ and that $1 - t \geq 0$ when $t \leq 1$. We have $(3 + t)(1 - t) \geq 0$ when *both* terms are nonnegative, that is, when $t \geq -3$ *and* $t \leq 1$, or, equivalently, when $-3 \leq t \leq 1$. We would also have $(3 + t)(1 - t) \geq 0$ if $3 + t \leq 0$ and $1 - t \leq 0$, but this is not possible. (Do you see why?) Thus, $g(t)$ is defined only for $-3 \leq t \leq 1$.

Next, using the chain rule, we compute the derivative of $g(t)$:

$$g'(t) = \frac{1}{2} \frac{1}{\sqrt{3 - 2t - t^2}}(-2 - 2t)$$

$$= \frac{-1 - t}{\sqrt{3 - 2t - t^2}}$$

Note that $g'(t)$ does not exist at the endpoints $t = -3$ and $t = 1$ of the domain of $g(t)$, and that $g'(t) = 0$ only when $t = -1$. Mark these three critical numbers on a number line restricted to the domain of g (namely, $-3 \leq t \leq 1$), and determine the sign of the derivative $g'(t)$ on the subintervals $-3 < t < -1$ and $-1 < t < 1$ to obtain the arrow diagram shown in Figure 3.10a. Finally, compute $g(-3) = g(1) = 0$ and $g(-1) = 2$ and observe that the arrow diagram suggests that there is a relative maximum at $(-1, 2)$. The completed graph is shown in Figure 3.10b.

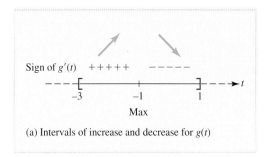

(a) Intervals of increase and decrease for $g(t)$

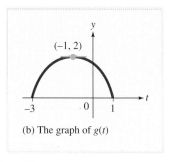

(b) The graph of $g(t)$

FIGURE 3.10   Sketching the graph of $g(t) = \sqrt{3 - 2t - t^2}$.

Sometimes, the graph of $f(x)$ is known and the relationship between the sign of the derivative $f'(x)$ and intervals of increase and decrease can be used to determine the general shape of the graph of $f'(x)$. The procedure is illustrated in Example 3.1.6.

EXPLORE!

Plot $f(x) = x^3 - x^2 - 4x + 4$ and $f'(x) = 3x^2 - 2x - 4$ on the same graph. How do the relative extrema of $f(x)$ relate to the features of the graph of $f'(x)$? What are the largest and smallest values of $f(x)$ over the interval $[-2, 1]$?

EXAMPLE 3.1.6

The graph of a function $f(x)$ is shown. Sketch a possible graph for the derivative $f'(x)$.

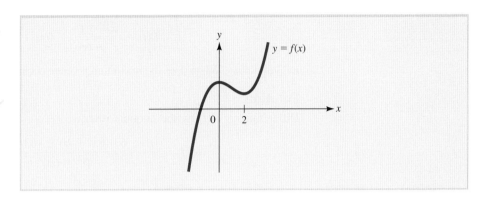

Solution

Since the graph of $f(x)$ is falling for $0 < x < 2$, we have $f'(x) < 0$ and the graph of $f'(x)$ is below the x axis on this interval. Similarly, for $x < 0$ and for $x > 2$, the graph of $f(x)$ is rising, so $f'(x) > 0$ and the graph of $f'(x)$ is above the x axis on both these intervals. The graph of $f(x)$ is flat (horizontal tangent line) at $x = 0$ and $x = 2$, so $f'(0) = f'(2) = 0$, and $x = 0$ and $x = 2$ are the x intercepts of the graph of $f'(x)$. Here is one possible graph that satisfies these conditions:

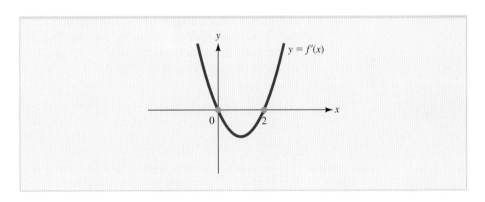

Curve sketching using the derivative $f'(x)$ will be refined by adding features using the second derivative $f''(x)$ in Section 3.2, and a general curve-sketching procedure involving derivatives and limits will be presented in Section 3.3. The same reasoning used to analyze graphs can be applied to determining optimal values, such as the minimum cost of a production process or the maximal sustainable yield for a salmon fishery. Optimization is illustrated in Example 3.1.7 and is discussed in more detail in Sections 3.4 and 3.5.

EXAMPLE 3.1.7

The revenue derived from the sale of a new kind of motorized skateboard t weeks after its introduction is given by

$$R(t) = \frac{63t - t^2}{t^2 + 63}, \quad 0 \le t \le 63,$$

million dollars. When does the maximum revenue occur? What is the maximum revenue?

Solution

Differentiating $R(t)$ by the quotient rule, we get

$$
\begin{aligned}
R'(t) &= \frac{(t^2 + 63)(63 - 2t) - (63t - t^2)(2t)}{(t^2 + 63)^2} \\
&= \frac{63t^2 - 2t^3 + 63^2 - 126t - 126t^2 + 2t^3}{(t^2 + 63)^2} \\
&= \frac{-63t^2 - 126t + 63^2}{(t^2 + 63)^2} \\
&= \frac{-63(t^2 + 2t - 63)}{(t^2 + 63)^2} \\
&= \frac{-63(t - 7)(t + 9)}{(t^2 + 63)^2}
\end{aligned}
$$

By setting the numerator in this expression for $R'(t)$ equal to 0, we find that $t = 7$ is the only solution of $R'(t) = 0$ in the interval $0 \le t \le 63$, and hence it is the only critical number of $R(t)$ in its domain. The critical number divides the domain $0 \le t \le 63$ into two intervals, $0 \le t < 7$ and $7 < t \le 63$. Evaluating $R'(t)$ at test numbers in each interval (say, at $t = 1$ and $t = 9$), we obtain the arrow diagram shown here.

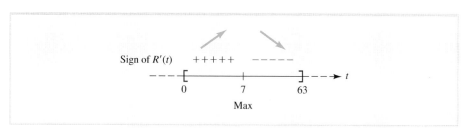

The arrow pattern indicates that the revenue increases to a maximum at $t = 7$, after which it decreases. At the optimal time $t = 7$, the revenue produced is

$$R(7) = \frac{63(7) - (7)^2}{(7)^2 + 63} = 3.5 \text{ million dollars.}$$

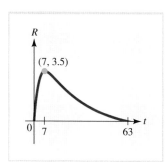

FIGURE 3.11   The graph of $R(t) = \dfrac{63t - t^2}{t^2 + 63}$ for $0 \le t \le 63$.

The graph of the revenue function $R(t)$ is shown in Figure 3.11. It suggests that immediately after its introduction, the motorized skateboard is a very popular product, producing peak revenue of \$3.5 million after only 7 weeks. However, its popularity, as measured by revenue, then begins to wane. After 63 weeks, revenue ceases altogether as presumably the skateboards are taken off the shelves and replaced by something new.

EXERCISES ■ 3.1

In Exercises 1 through 4, specify the intervals on which the derivative of the given function is positive and the intervals on which it is negative.

1.

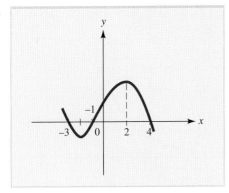

3.

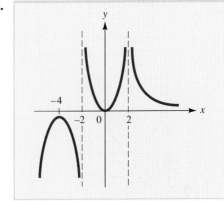

2.

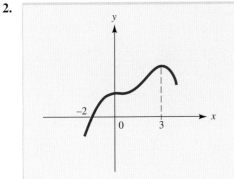

4.

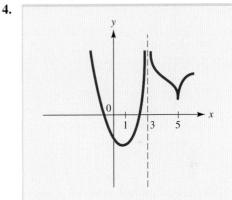

The curves A, B, C, and D shown here are the graphs of the derivatives of the functions whose graphs are shown in Exercises 5 through 8. Match each graph with the graph of its derivative.

5.

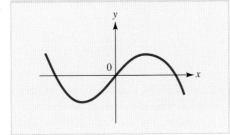

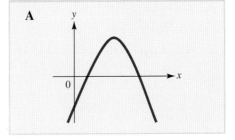

6.

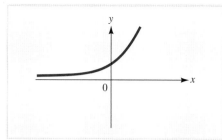

B

7.

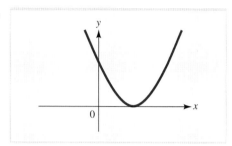

C

8.

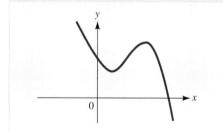

D

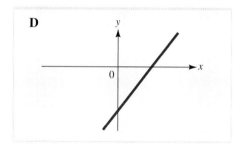

In Exercises 9 through 22, find the intervals of increase and decrease for the given function.

9. $f(x) = x^2 - 4x + 5$

10. $f(t) = t^3 + 3t^2 + 1$

11. $f(x) = x^3 - 3x - 4$

12. $f(x) = \dfrac{1}{3}x^3 - 9x + 2$

13. $g(t) = t^5 - 5t^4 + 100$

14. $f(x) = 3x^5 - 5x^3$

15. $f(t) = \dfrac{1}{4 - t^2}$

16. $g(t) = \dfrac{1}{t^2 + 1} - \dfrac{1}{(t^2 + 1)^2}$

17. $h(u) = \sqrt{9 - u^2}$

18. $f(x) = \sqrt{6 - x - x^2}$

19. $F(x) = x + \dfrac{9}{x}$

20. $f(t) = \dfrac{t}{(t + 3)^2}$

21. $f(x) = \sqrt{x} + \dfrac{1}{\sqrt{x}}$

22. $G(x) = x^2 - \dfrac{1}{x^2}$

In Exercises 23 through 34, determine the critical numbers of the given function and classify each critical point as a relative maximum, a relative minimum, or neither.

23. $f(x) = 3x^4 - 8x^3 + 6x^2 + 2$

24. $f(x) = 324x - 72x^2 + 4x^3$

25. $f(t) = 2t^3 + 6t^2 + 6t + 5$

26. $f(t) = 10t^6 + 24t^5 + 15t^4 + 3$

27. $g(x) = (x - 1)^5$

28. $F(x) = 3 - (x + 1)^3$

29. $f(t) = \dfrac{t}{t^2 + 3}$

30. $f(t) = t\sqrt{9 - t}$

31. $h(t) = \dfrac{t^2}{t^2 + t - 2}$

32. $g(x) = 4 - \dfrac{2}{x} + \dfrac{3}{x^2}$

33. $S(t) = (t^2 - 1)^4$

34. $F(x) = \dfrac{x^2}{x - 1}$

In Exercises 35 through 44, use calculus to sketch the graph of the given function.

35. $f(x) = x^3 - 3x^2$

36. $f(x) = 3x^4 - 4x^3$

37. $f(x) = 3x^4 - 8x^3 + 6x^2 + 2$

38. $g(x) = 3 - (x + 1)^3$

39. $f(t) = 2t^3 + 6t^2 + 6t + 5$

40. $f(x) = x^3(x + 5)^2$

41. $g(t) = \dfrac{t}{t^2 + 3}$

42. $f(x) = \dfrac{x + 1}{x^2 + x + 1}$

43. $f(x) = 3x^5 - 5x^3 + 4$

44. $H(x) = \dfrac{1}{50}(3x^4 - 8x^3 - 90x^2 + 70)$

In Exercises 45 through 48, the derivative of a function f(x) is given. In each case, find the critical numbers of f(x) and classify each as corresponding to a relative maximum, a relative minimum, or neither.

45. $f'(x) = x^2(4 - x^2)$

46. $f'(x) = \dfrac{x(2 - x)}{x^2 + x + 1}$

47. $f'(x) = \dfrac{(x + 1)^2(4 - 3x)^3}{(x^2 + 1)^2}$

48. $f'(x) = x^3(2x - 7)^2(x + 5)$

In Exercises 49 through 52, the graph of a function f is given. In each case, sketch a possible graph for f'.

49.

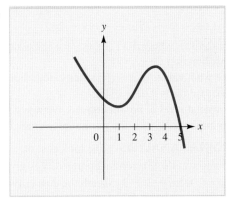

50.

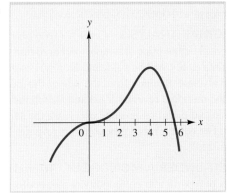

51.

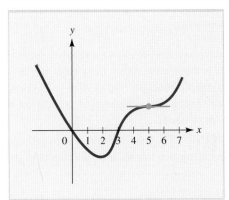

52.

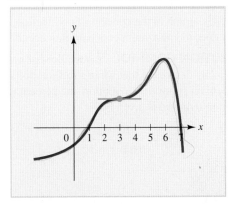

53. AVERAGE COST   The total cost of producing x units of a certain commodity is $C(x)$ thousand dollars, where

$$C(x) = x^3 - 20x^2 + 179x + 242$$

a. Find $A'(x)$, where $A(x) = C(x)/x$ is the average cost function.
b. For what values of x is $A(x)$ increasing? For what values is it decreasing?
c. For what level of production x is average cost minimized? What is the minimum average cost?

54. MARGINAL ANALYSIS   The total cost of producing x units of a certain commodity is given by $C(x) = \sqrt{5x + 2} + 3$. Sketch the cost curve and find the marginal cost. Does marginal cost increase or decrease with increasing production?

55. MARGINAL ANALYSIS   Let $p = (10 - 3x)^2$ for $0 \le x \le 3$ be the price at which x hundred units of a chocolate candy will be sold, and let $R(x) = xp(x)$ be the revenue obtained from the sale of the x units. Find the marginal revenue $R'(x)$ and sketch the revenue and marginal revenue curves on the same graph. For what level of production is revenue maximized?

56. PROFIT UNDER A MONOPOLY   To produce x units of a particular commodity, a monopolist has total cost of

$$C(x) = 2x^2 + 3x + 5$$

and total revenue of $R(x) = xp(x)$, where $p(x) = 5 - 2x$ is the price at which the x units will be sold. Find the profit function $P(x) = R(x) - C(x)$ and sketch its graph. For what level of production is profit maximized?

57. MEDICINE   The concentration of a drug t hours after being injected into the arm of a patient is given by

$$C(t) = \frac{0.15t}{t^2 + 0.81}.$$

Sketch the graph of the concentration function. When does the maximum concentration occur?

58. POLLUTION CONTROL   Councillors of a certain city determine that when x million dollars are spent on controlling pollution, the percentage of pollution removed is given by

$$P(x) = \frac{100\sqrt{x}}{0.04x^2 + 12}.$$

a. Sketch the graph of $P(x)$.
b. What expenditure results in the largest percentage of pollution removal?

59. ADVERTISING   A travel company determines that if x thousand dollars are spent on advertising a vacation spot, then $S(x)$ people will buy vacations there, where

$$S(x) = -2x^3 + 27x^2 + 132x + 207, \quad 0 \le x \le 17.$$

a. Sketch the graph of $S(x)$.
b. How many vacations will be sold if nothing is spent on advertising?
c. How much should be spent on advertising to maximize sales? What is the maximum sales level?

60. ADVERTISING   Answer the questions in Exercise 59 for the sales function

$$S(x) = \frac{200x + 1500}{0.02x^2 + 5}.$$

61. MORTGAGE REFINANCING   In a certain country, when interest rates are low and refinancing costs are also low, many homeowners take the opportunity to refinance their mortgages. As rates start to rise, more people rush in to refinance while they can still do so profitably. Eventually, however, rates reach a level where refinancing begins to wane.

Suppose in a certain community $M(r)$ thousand mortgages will be refinanced when the 30-year fixed mortgage rate is r percent, where

$$M(r) = \frac{1 + 0.05r}{1 + 0.004r^2} \quad \text{for } 1 \le r \le 8.$$

a. For what values of r is $M(r)$ increasing?
b. For what interest rate r is the number of refinanced mortgages maximized? What is this maximum number?

62. POPULATION DISTRIBUTION   A demographic study of a certain city indicates that $P(r)$ hundred people live r kilometres from the civic centre, where

$$P(r) = \frac{5(3r + 1)}{r^2 + r + 2}.$$

a. What is the population at the civic centre?
b. For what values of r is $P(r)$ increasing? For what values is it decreasing?
c. At what distance from the civic centre is the population largest? What is this largest population?

63. **BUILDING PERMITS**   The graph shows the number of building permits issued in Canada from August 2005 to March 2009.*

    a. Relative minima occur regularly. The last plotted point on the curve is March 2009. What is the most frequent month for a minimum number of building permits? What happened in 2009 that caused the lower minimum?

    b. Between which two months is the rate of change greatest?

    c. Between October and December 2008, what was the approximate rate of decrease?

    d. Between which two months is the rate of change nearest to zero?

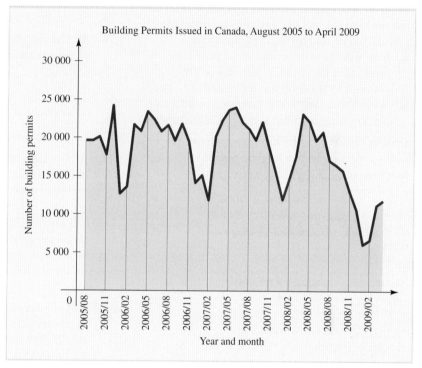

Building Permits Issued in Canada, August 2005 to April 2009

EXERCISE 63

64. **A* DEPRECIATION**   The value V (in thousands of dollars) of an industrial machine is modelled by

$$V(N) = \left(\frac{3N + 430}{N + 1}\right)^{2/3},$$

where N is the number of hours the machine is used each day. Suppose further that usage varies with time in such a way that

$$N(t) = \sqrt{t^2 - 10t + 61},$$

where t is the number of months the machine has been in operation.

    a. Over what time interval is the value of the machine increasing? When is it decreasing?

    b. At what time t is the value of the machine the largest? What is this maximum value?

65. **A* FISHERY MANAGEMENT**   The manager of a fishery determines that t weeks after 300 fish of a particular species are released into a pond, the average weight of an individual fish (in kilograms) for the first 10 weeks will be

$$w(t) = 3 + t - 0.05t^2.$$

He further determines that the proportion of the fish that are still alive after t weeks is given by

$$p(t) = \frac{31}{31 + t}.$$

* Adapted from the Statistics Canada CANSIM database, http://cansim2.statcan.gc.ca, Table 026-0002, retrieved April 27, 2010.

a. The expected yield $Y(t)$ of the fish after t weeks is the total weight of the fish that are still alive. Express $Y(t)$ in terms of $w(t)$ and $p(t)$ and sketch the graph of $Y(t)$ for $0 \leq t \leq 10$.

b. When is the expected yield $Y(t)$ the largest? What is the maximum yield?

66. A* FISHERY MANAGEMENT   Suppose, for the situation described in Exercise 65, that it costs the fishery $C(t) = 50 + 1.2t$ hundred dollars to maintain and monitor the pond for t weeks after the fish are released, and that each fish harvested after t weeks can be sold for $2.75 per kilogram.

a. If all fish that remain alive in the pond after t weeks are harvested, express the profit obtained by the fishery as a function of t.

 b. When should the fish be harvested in order to maximize profit? What is the maximum profit?

67. GROWTH OF A SPECIES   The percentage of codling moth eggs* that hatch at a given temperature T (degrees Celsius) is given by

$$H(T) = -0.53T^2 + 25T - 209 \qquad \text{for } 15 \leq T \leq 30$$

Sketch the graph of the hatching function $H(T)$. At what temperature T ($15 \leq T \leq 30$) does the maximum percentage of eggs hatch? What is the maximum percentage?

68. Sketch the graph of a function that has all of the following properties:
- $f'(0) = f'(1) = f'(2) = 0$
- $f'(x) < 0$ when $x < 0$ and $x > 2$
- $f'(x) > 0$ when $0 < x < 1$ and $1 < x < 2$

69. Sketch the graph of a function that has all of the following properties:
- $f'(0) = f'(1) = f'(2) = 0$
- $f'(x) < 0$ when $0 < x < 1$
- $f'(x) > 0$ when $x < 0$, $1 < x < 2$, and $x > 2$

70. Sketch the graph of a function that has all of the following properties:
- $f'(x) > 0$ when $x < -5$ and when $x > 1$
- $f'(x) < 0$ when $-5 < x < 1$
- $f(-5) = 4$ and $f(1) = -1$

71. Sketch the graph of a function that has all of the following properties:
- $f'(x) < 0$ when $x < -1$
- $f'(x) > 0$ when $-1 < x < 3$ and when $x > 3$
- $f'(-1) = 0$ and $f'(3) = 0$

72. A* Find constants a, b, and c so that the graph of the function $f(x) = ax^2 + bx + c$ has a relative maximum at $(5, 12)$ and crosses the y axis at $(0, 3)$.

73. A* Find constants a, b, c, and d so that the graph of the function $f(x) = ax^3 + bx^2 + cx + d$ has a relative maximum at $(-2, 8)$ and a relative minimum at $(1, -19)$.

74. Sketch the graph of $f(x) = (x - 1)^{2/5}$. Explain why $f'(x)$ is not defined at $x = 1$.

75. Sketch the graph of $f(x) = 1 - x^{3/5}$.

76. A* Use calculus to prove that the relative extremum of the quadratic function

$$f(x) = ax^2 + bx + c$$

occurs when $x = -\dfrac{b}{2a}$. Where is $f(x)$ increasing, and where is it decreasing?

77. A* Use calculus to prove that the relative extremum of the quadratic function $y = (x - p)(x - q)$ occurs midway between its x intercepts.

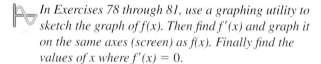 *In Exercises 78 through 81, use a graphing utility to sketch the graph of $f(x)$. Then find $f'(x)$ and graph it on the same axes (screen) as $f(x)$. Finally find the values of x where $f'(x) = 0$.*

78. Make a polynomial function based on the number part of your street address. Repeat the digits if necessary until you have a four-digit number, $ABCD$. Your function will be $m(x) = Ax^4 + Bx^3 - Cx^2 + D$. Plot this function using a scale that shows the relative extrema. Find the derivative, solving for critical points using technology if necessary. Change the scale of the original plot if necessary. Plot the derivative with the same scale. Print out both plots, labelling the critical points of $m(x)$ on both graphs. Label intervals where the function is increasing and intervals where it is decreasing, and label any relative extrema.

79. $f(x) = (x^2 + x - 1)^3(x + 3)^2$

80. $f(x) = x^5 - 7.6x^3 + 2.1x^2 - 5$

81. $f(x) = (1 - x^{1/2})^{1/2}$

---

*See Think About It at the end of Chapter 2.

SECTION 3.2

L02

Determine intervals where a function is concave up or concave down. Use the second derivative test. Find inflection points of a function.

Concavity and Points of Inflection

In Section 3.1, you saw how to use the sign of the derivative $f'(x)$ to determine where $f(x)$ is increasing or decreasing and where its graph has relative extrema. In this section, you will see that the second derivative $f''(x)$ also provides useful information about the graph of $f(x)$. By way of introduction, here is a brief description of a situation from industry that can be analyzed using the second derivative.

The number of units that a factory worker can produce in t hours after arriving at work is often given by a function $Q(t)$ like the one whose graph is shown in Figure 3.12. Notice that at first the graph is not very steep. The steepness increases, however, until the graph reaches a point of maximum steepness, after which the steepness begins to decrease. This reflects the fact that at first the worker's rate of production is low. The rate of production increases, however, as the worker settles into a routine and continues to increase until the worker is performing at maximum efficiency. Then fatigue sets in and the rate of production begins to decrease. The point P on the output curve where maximum efficiency occurs is known in economics as the **point of diminishing returns.**

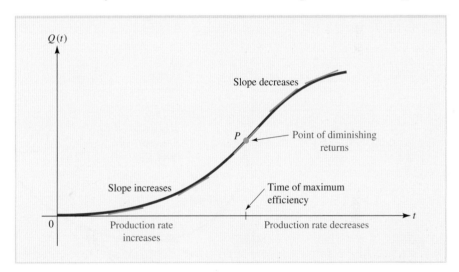

FIGURE 3.12   The output $Q(t)$ of a factory worker t hours after coming to work.

The behaviour of the graph of this production function on either side of the point of diminishing returns can be described in terms of its tangent lines. To the left of this point, the slope of the tangent increases as t increases. To the right of this point, the slope of the tangent decreases as t increases. It is this increase and decrease of slopes that we shall examine in this section with the aid of the second derivative. (We shall return to the questions of worker efficiency and diminishing returns in Example 3.2.6, later in this section.)

Concavity

The increase and decrease of tangential slope will be described in terms of a graphical feature called **concavity.** Here is the definition that we will use.

> **Concavity** ■ If the function $f(x)$ is differentiable on the interval $a < x < b$, then the graph of f is
>
> **concave upward** on $a < x < b$ if f' is increasing on the interval
>
> **concave downward** on $a < x < b$ if f' is decreasing on the interval

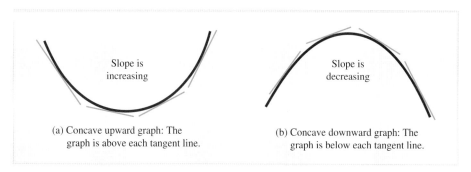

FIGURE 3.13   Concavity.

Equivalently, a graph is concave upward on an interval if it lies above all of its tangent lines on the interval (Figure 3.13a), and concave downward on an interval where it lies below all of its tangent lines (Figure 3.13b). Informally, a concave upward portion of a graph "holds water," while a concave downward portion "spills water," as illustrated in Figure 3.14.

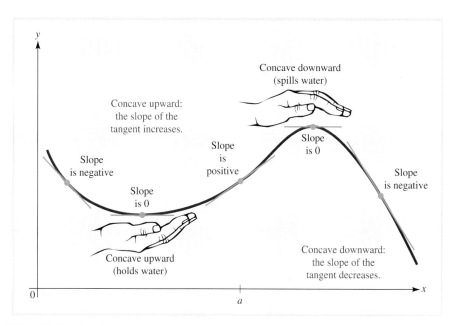

FIGURE 3.14   Concavity and the slope of the tangent.

Determining Intervals of Concavity Using the Sign of f''

There is a simple characterization of the concavity of the graph of a function $f(x)$ in terms of the second derivative $f''(x)$. In particular, in Section 3.1, we observed that a function $f(x)$ is increasing where its derivative is positive. Thus, the derivative function $f'(x)$ must be increasing where *its* derivative $f''(x)$ is positive. Suppose $f''(x) > 0$ on an interval $a < x < b$. Then $f'(x)$ is increasing, which in turn means that the graph of $f(x)$ is concave upward on this interval. Similarly, on an interval $a < x < b$ where $f''(x) < 0$, the derivative $f'(x)$ is decreasing and the graph of $f(x)$ is concave downward. Using these observations, we can modify the procedure for determining intervals of increase and decrease developed in Section 3.1 to obtain this procedure for determining intervals of concavity.

Just-In-Time

A way to remember this is to think the graph is a smile ☺ when concave up, "smiling positive," and the graph is a frown ☹ when concave down, "feeling negative." So a positive $f''(x)$ is smiling concave up, and a negative $f''(x)$ is frowning concave down.

Second Derivative Procedure for Determining Intervals of Concavity for a Function f

Step 1. Find all values of x for which $f''(x) = 0$ or $f''(x)$ does not exist, and mark these numbers on a number line. This divides the line into a number of open intervals.

Step 2. Choose a test number c from each interval $a < x < b$ determined in step 1 and evaluate $f''(c)$. Then,

    If $f''(c) > 0$, the graph of $f(x)$ is concave upward on $a < x < b$.

    If $f''(c) < 0$, the graph of $f(x)$ is concave downward on $a < x < b$.

EXAMPLE 3.2.1

Determine the intervals of concavity for the function

$$f(x) = 2x^6 - 5x^4 + 7x - 3.$$

Solution

Differentiate to find that

$$f'(x) = 12x^5 - 20x^3 + 7$$

and

$$f''(x) = 60x^4 - 60x^2 = 60x^2(x^2 - 1) = 60x^2(x - 1)(x + 1)$$

The second derivative $f''(x)$ is continuous for all x and $f''(x) = 0$ for $x = 0$, $x = 1$, and $x = -1$. These numbers divide the x axis into four intervals on which $f''(x)$ does not change sign, namely, $x < -1$, $-1 < x < 0$, $0 < x < 1$, and $x > 1$. Evaluating $f''(x)$ at test numbers in each of these intervals (say, at $x = -2$, $x = -\dfrac{1}{2}$, $x = \dfrac{1}{2}$, and $x = 5$, respectively), we find that

$$f''(-2) = 720 > 0 \qquad f''\left(-\frac{1}{2}\right) = -\frac{45}{4} < 0$$

$$f''\left(\frac{1}{2}\right) = -\frac{45}{4} < 0 \qquad f''(5) = 36\,000 > 0$$

EXPLORE!

Following Example 3.2.1, plot $f(x) = 2x^6 - 5x^4 + 7x - 3$, $f'(x) = 12x^5 - 20x^3 + 7$, and $f''(x) = 60x^4 - 60x^2$ on the same graph or on separate graphs directly under each other so that all can be seen. How does the behaviour of $f''(x)$ relate to the concavity of $f(x)$, specifically at $x = -1$, 0, and 1?

Thus, the graph of $f(x)$ is concave up for $x < -1$ and for $x > 1$ and concave down for $-1 < x < 0$ and for $0 < x < 1$, as indicated in this concavity diagram.

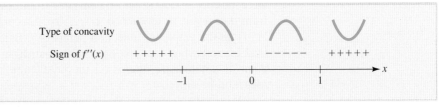

Type of concavity   Sign of $f''(x)$ $+++++$ $-----$ $-----$ $+++++$ -1 0 1 x

Intervals of concavity for $f(x) = 2x^6 - 5x^4 + 7x - 3$.

The graph of $f(x)$ is shown in Figure 3.15.

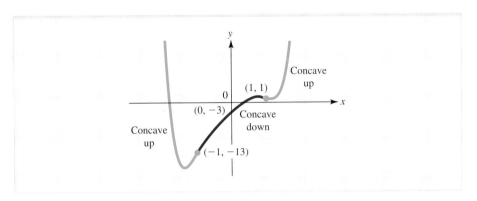

FIGURE 3.15 The graph of $f(x) = 2x^6 - 5x^4 + 7x - 3$.

NOTE Do not confuse the concavity of a graph with its direction (rising or falling). A function f may be increasing or decreasing on an interval regardless of whether its graph is concave upward or concave downward on that interval. The four possibilities are illustrated in Figure 3.16.

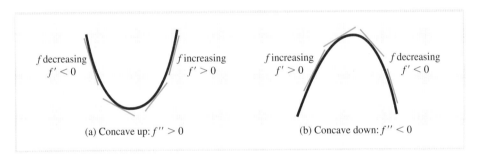

FIGURE 3.16 Possible combinations of increase, decrease, and concavity.

Inflection Points

A point $P(c, f(c))$ on the graph of a function f is called an *inflection point* of f if the concavity of the graph changes at P; that is, the graph of f is concave upward on one side of P and concave downward on the other side. Such transition points provide useful information about the graph of f. For instance, the concavity diagram for the function $f(x) = 2x^6 - 5x^4 + 7x - 3$ analyzed in Example 3.2.1 shows that the concavity of the graph of f changes from upward to downward at $x = -1$ and from downward to upward at $x = 1$, so the corresponding points $(-1, -13)$ and $(1, 1)$ on the graph are inflection points of f. Inflection points may also be of practical interest when interpreting a mathematical model based on f. For example, the point of diminishing returns on the production curve shown in Figure 3.12 is an inflection point.

At an inflection point $P(c, f(c))$, the graph of f can be neither concave upward ($f''(c) > 0$) nor concave downward ($f''(c) < 0$). Therefore, if $f''(c)$ exists, we must have $f''(c) = 0$. To summarize:

Inflection Points ■ An inflection point (or point of inflection) is a point $(c, f(c))$ on the graph of a function f where the concavity changes. At such a point, either $f''(c) = 0$ or $f''(c)$ does not exist.

Procedure for Finding the Inflection Points for a Function f

Step 1. Compute $f''(x)$ and determine all points in the domain of f where either $f''(c) = 0$ or $f''(c)$ does not exist.

Step 2. For each number c found in step 1, determine the sign of $f''(x)$ to the left and to the right of $x = c$, that is, for $x < c$ and for $x > c$. If $f''(x) > 0$ on one side of $x = c$ and $f''(x) < 0$ on the other side, then $(c, f(c))$ is an inflection point for f.

EXAMPLE 3.2.2

In each case, find all inflection points of the given function.

   a. $f(x) = 3x^5 - 5x^4 - 1$ **b.** $g(x) = x^{1/3}$

Solution

a. Note that $f(x)$ exists for all x and that

$$f'(x) = 15x^4 - 20x^3$$
$$f''(x) = 60x^3 - 60x^2 = 60x^2(x - 1)$$

Thus, $f''(x)$ is continuous for all x and $f''(x) = 0$ when $x = 0$ and $x = 1$. Testing the sign of $f''(x)$ on each side of $x = 0$ and $x = 1$ (say, at $x = -1$, $x = \dfrac{1}{2}$, and $x = 2$) gives

$$f''(-1) = -120 < 0 \qquad f''\left(\frac{1}{2}\right) = -\frac{15}{2} < 0 \qquad f''(2) = 240 > 0,$$

which leads to the concavity pattern shown in this diagram:

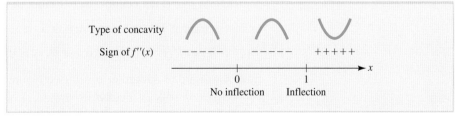

Thus, the concavity does not change at $x = 0$ but changes from downward to upward at $x = 1$. Since $f(1) = -3$, it follows that $(1, -3)$ is an inflection point of f. The graph of f is shown in Figure 3.17a.

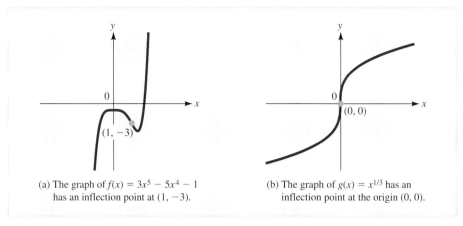

(a) The graph of $f(x) = 3x^5 - 5x^4 - 1$ has an inflection point at $(1, -3)$.

(b) The graph of $g(x) = x^{1/3}$ has an inflection point at the origin $(0, 0)$.

FIGURE 3.17 Two graphs with inflection points.

b. The function $g(x)$ is continuous for all x and since

$$g'(x) = \frac{1}{3}x^{-2/3} \quad \text{and} \quad g''(x) = -\frac{2}{9}x^{-5/3}$$

it follows that $g''(x)$ is never 0 but does not exist when $x = 0$. Testing the sign of $g''(x)$ on each side of $x = 0$, we obtain the results displayed in this concavity diagram:

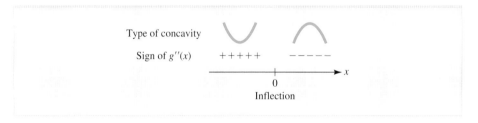

Since the concavity of the graph changes at $x = 0$ and $g(0) = 0$, there is an inflection point at the origin, $(0, 0)$. The graph of g is shown in Figure 3.17b.

NOTE A function can have an inflection point only where it is continuous. In particular, if $f(c)$ is not defined, there cannot be an inflection point corresponding to $x = c$ even if $f''(x)$ changes sign at $x = c$. For example, if $f(x) = \frac{1}{x}$, then $f''(x) = \frac{2}{x^3}$, so $f''(x) < 0$ if $x < 0$ and $f''(x) > 0$ if $x > 0$. The concavity changes from downward to upward at $x = 0$ (see Figure 3.18a), but there is no inflection point at $x = 0$ since $f(0)$ is not defined.

Moreover, just knowing that $f(c)$ is defined and that $f''(c) = 0$ does not guarantee that $(c, f(c))$ is an inflection point. For instance, if $f(x) = x^4$, then $f(0) = 0$ and $f''(x) = 12x^2$, so $f''(0) = 0$. However, $f''(x) > 0$ for any number $x \neq 0$, so the graph of f is always concave upward, and there is no inflection point at $(0, 0)$ (see Figure 3.18b).

Do you think that if $f(c)$ is defined and $f''(c) = 0$, then you can at least conclude that either an inflection point or a relative extremum occurs at $x = c$? For the answer, see Exercise 70.

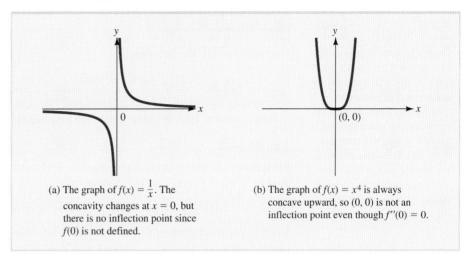

(a) The graph of $f(x) = \frac{1}{x}$. The concavity changes at $x = 0$, but there is no inflection point since $f(0)$ is not defined.

(b) The graph of $f(x) = x^4$ is always concave upward, so $(0, 0)$ is not an inflection point even though $f''(0) = 0$.

FIGURE 3.18　A graph need not have an inflection point where $f'' = 0$ or f'' does not exist.

Curve Sketching with the Second Derivative

Geometrically, inflection points occur at "twists" on a graph. Here is a summary of the graphical possibilities.

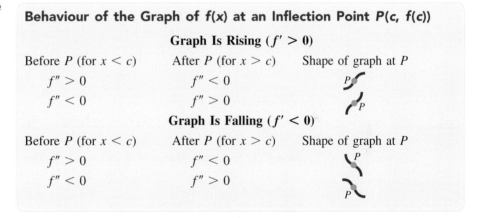

Behaviour of the Graph of $f(x)$ at an Inflection Point $P(c, f(c))$

Graph Is Rising ($f' > 0$)

Before P (for $x < c$)	After P (for $x > c$)	Shape of graph at P
$f'' > 0$	$f'' < 0$	
$f'' < 0$	$f'' > 0$	

Graph Is Falling ($f' < 0$)

Before P (for $x < c$)	After P (for $x > c$)	Shape of graph at P
$f'' > 0$	$f'' < 0$	
$f'' < 0$	$f'' > 0$	

Figure 3.19 shows several ways in which inflection points can occur on a graph.

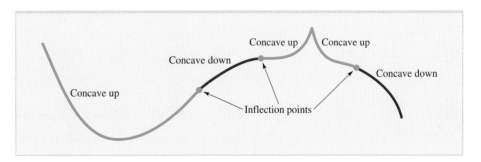

FIGURE 3.19　A graph showing concavity and inflection points.

By adding the criteria for concavity and inflection points to the first derivative methods discussed in Section 3.1, you can sketch a variety of graphs with considerable detail. Here is an example.

EXPLORE!

Graphically confirm the results of Example 3.2.3 by plotting $f(x)$, $f'(x)$, and $f''(x)$ either on the same graph or under each other on the same page. Look at the relative extrema of $f(x)$ in the first graph, and compare them with the values of $f'(x)$ at the same x values. Look at the concavity of $f(x)$ and how it relates to the sign of $f''(x)$ in the third graph.

EXAMPLE 3.2.3

Determine where the function

$$f(x) = 3x^4 - 2x^3 - 12x^2 + 18x + 15$$

is increasing and where it is decreasing, and where its graph is concave up and where it is concave down. Find all relative extrema and points of inflection and sketch the graph.

Solution

First, note that since $f(x)$ is a polynomial, it is continuous for all x, as are the derivatives $f'(x)$ and $f''(x)$. The first derivative of $f(x)$ is

$$f'(x) = 12x^3 - 6x^2 - 24x + 18 = 6(x - 1)^2(2x + 3),$$

and $f'(x) = 0$ only when $x = 1$ and $x = -1.5$. The sign of $f'(x)$ does not change in each of the intervals $x < -1.5$, $-1.5 < x < 1$, and $x > 1$. Evaluating $f'(x)$ at test numbers in each interval (say, at -2, 0, and 3) gives the arrow diagram shown. Note that there is a relative minimum at $x = -1.5$ but no extremum at $x = 1$.

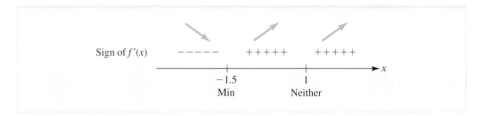

The second derivative is

$$f''(x) = 36x^2 - 12x - 24 = 12(x - 1)(3x + 2),$$

and $f''(x) = 0$ only when $x = 1$ and $x = -\dfrac{2}{3}$. The sign of $f''(x)$ does not change on each of the intervals $x < -\dfrac{2}{3}$, $-\dfrac{2}{3} < x < 1$, and $x > 1$. Evaluating $f''(x)$ at test numbers in each interval gives the concavity diagram shown.

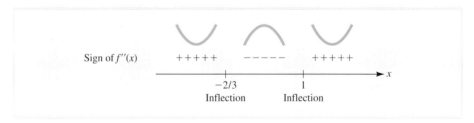

The patterns in these two diagrams suggest that there is a relative minimum at $x = -1.5$ and inflection points at $x = -\dfrac{2}{3}$ and $x = 1$ (since the concavity changes at both points).

To sketch the graph, evaluate $f(x)$ at these values of x:

$$f(-1.5) \approx -17.06 \quad f\left(-\dfrac{2}{3}\right) = -1.15 \quad f(1) = 22$$

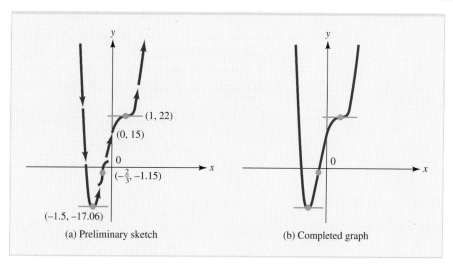

FIGURE 3.20   The graph of $f(x) = 3x^4 - 2x^3 - 12x^2 + 18x + 15$.

and plot a "cup" $\smile$ at $(-1.5, -17.06)$ to mark the relative minimum located there. Likewise, plot twists $\int$ at $\left(-\dfrac{2}{3}, -1.15\right)$ and at $(1, 22)$ to mark the shape of the graph near the inflection points. Use the arrow and concavity diagrams to get the preliminary diagram shown in Figure 3.20a. Finally, complete the sketch as shown in Figure 3.20b by passing a smooth curve through the minimum point, the inflection points, and the y intercept $(0, 15)$.

Sometimes you are given the graph of a derivative function $f'(x)$ and asked to analyze the graph of $f(x)$ itself. For instance, it would be quite reasonable for a manufacturer who knows the marginal cost $C'(x)$ associated with producing x units of a particular commodity to want to know as much as possible about the total cost $C(x)$. The following example illustrates a procedure for carrying out this kind of analysis.

EXAMPLE 3.2.4

The graph of the derivative $f'(x)$ of a function $f(x)$ is shown in the figure. Find intervals of increase and decrease and concavity for $f(x)$ and locate all relative extrema and inflection points. Then sketch a curve that has all these features.

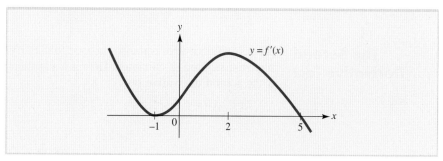

Solution

First, note that for $x < -1$, the graph of $f'(x)$ is above the x axis, so $f'(x) > 0$ and the graph of $f(x)$ is rising. The graph of $f'(x)$ is falling for $x < -1$, which means that $f''(x) < 0$ and the graph of $f(x)$ is concave down. Other intervals can be analyzed in a similar fashion, and the results are summarized in the table.

x	Feature of $y = f'(x)$	Feature of $y = f(x)$
$x < -1$	f' is positive; decreasing	f is increasing; concave down
$x = -1$	x intercept; horizontal tangent	Horizontal tangent; possible inflection point ($f'' = 0$)
$-1 < x < 2$	f' is positive; increasing	f is increasing; concave up
$x = 2$	Horizontal tangent	Possible inflection point
$2 < x < 5$	f' is positive; decreasing	f is increasing; concave down
$x = 5$	x intercept	Horizontal tangent
$x > 5$	f' is negative; decreasing	f is decreasing; concave down

Since the concavity changes at $x = -1$ (down to up), an inflection point occurs there, along with a horizontal tangent. At $x = 2$, there is also an inflection point (concavity changes from up to down), but no horizontal tangent. The graph of $f(x)$ is rising to the left of $x = 5$ and falling to the right, so there must be a relative maximum at $x = 5$.

One possible graph that has all the features required for $y = f(x)$ is shown in Figure 3.21. Note, however, that since you are not given the values of $f(-1)$, $f(2)$, and $f(5)$, many other graphs will also satisfy the requirements.

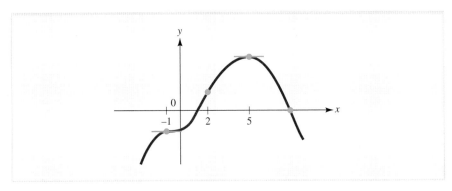

FIGURE 3.21   A possible graph for $y = f(x)$.

The Second Derivative Test

The second derivative can also be used to classify a critical point of a function as a relative maximum or minimum. Here is a statement of the procedure, which is known as the **second derivative test.**

The Second Derivative Test ■ Suppose $f''(x)$ exists on an open interval containing $x = c$ and that $f'(c) = 0$.

If $f''(c) > 0$, then f has a relative minimum at $x = c$.

If $f''(c) < 0$, then f has a relative maximum at $x = c$.

However, if $f''(c) = 0$ or if $f''(c)$ does not exist, the test is inconclusive and f may have a relative maximum, a relative minimum, or no relative extremum at all at $x = c$.

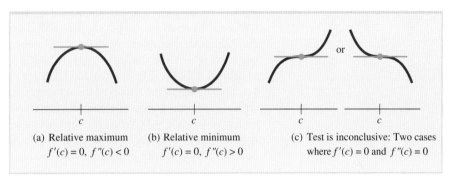

(a) Relative maximum $f'(c) = 0,\ f''(c) < 0$

(b) Relative minimum $f'(c) = 0,\ f''(c) > 0$

(c) Test is inconclusive: Two cases where $f'(c) = 0$ and $f''(c) = 0$

FIGURE 3.22 The second derivative test.

To see why the second derivative test works, look at Figure 3.22, which shows four possibilities that can occur when $f'(c) = 0$. Figure 3.22a suggests that at a relative maximum, the graph of f must be concave downward, so $f''(c) < 0$. Likewise, at a relative minimum (Figure 3.22b), the graph of f must be concave upward, so $f''(c) > 0$. On the other hand, if $f'(c) = 0$ and $f''(c)$ is neither positive nor negative, the test is inconclusive. Figure 3.22c suggests that if $f'(c) = 0$ and $f''(c) = 0$, there may be an inflection point at $x = c$. There may also be a relative extremum. For example, $f(x) = x^4$ has a relative minimum at $x = 0$ and $g(x) = -x^4$ has a relative maximum (see Figure 3.24).

The second derivative test is illustrated in Example 3.2.5.

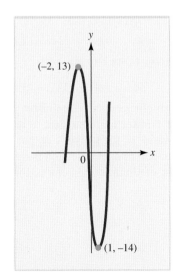

FIGURE 3.23 The graph of $f(x) = 2x^3 + 3x^2 - 12x - 7$.

EXAMPLE 3.2.5

Find the critical points of $f(x) = 2x^3 + 3x^2 - 12x - 7$ and use the second derivative test to classify each critical point as a relative maximum or minimum.

Solution

Since the first derivative

$$f'(x) = 6x^2 + 6x - 12 = 6(x + 2)(x - 1)$$

is zero when $x = -2$ and $x = 1$, the corresponding points $(-2, 13)$ and $(1, -14)$ are the critical points of f. To test these points, compute the second derivative

$$f''(x) = 12x + 6$$

and evaluate it at $x = -2$ and $x = 1$. Since

$$f''(-2) = -18 < 0,$$

it follows that the critical point $(-2, 13)$ is a relative maximum, and since

$$f''(1) = 18 > 0,$$

it follows that the critical point $(1, -14)$ is a relative minimum. For reference, the graph of f is sketched in Figure 3.23.

NOTE   Although it was easy to use the second derivative test to classify critical points in Example 3.2.5, the test does have some limitations. For some functions, the work involved in computing the second derivative is time-consuming, which may diminish the appeal of the test. Moreover, the test applies only to critical points at which the derivative is zero and not to those where the derivative is undefined. Finally, if both $f'(c)$ and $f''(c)$ are zero, the second derivative test tells you nothing about the nature of the critical point. This is illustrated in Figure 3.24, which shows the graphs of three functions whose first and second derivatives are both zero when $x = 0$. When it is inconvenient or impossible to apply the second derivative test, you may still be able to use the first derivative test described in Section 3.1 to classify critical points.

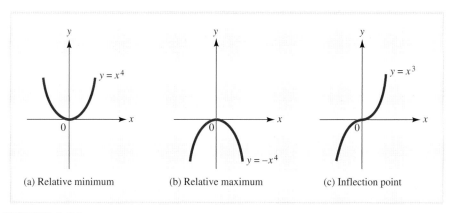

(a) Relative minimum   (b) Relative maximum   (c) Inflection point

FIGURE 3.24   Three functions whose first and second derivatives are zero at $x = 0$.

In Example 3.2.6, we return to the questions of worker efficiency and diminishing returns examined in the illustration at the beginning of this section. Our goal is to maximize a worker's *rate* of production, that is, the derivative of the worker's output. Hence, we will set to zero the *second* derivative of output and find an inflection point of the output function, which we interpret as the point of diminishing returns for production.

EXAMPLE 3.2.6

An efficiency study of the morning shift at a factory indicates that an average worker who starts at 8:00 A.M. will have produced $Q(t) = -t^3 + 9t^2 + 12t$ units t hours later. At what time during the morning is the worker performing most efficiently?

Solution

The worker's rate of production is the derivative of the output $Q(t)$; that is,

$$R(t) = Q'(t) = -3t^2 + 18t + 12$$

Assuming that the morning shift runs from 8:00 A.M. until noon, the goal is to find the largest rate $R(t)$ for $0 \le t \le 4$. The derivative of the rate function is

$$R(t) = Q''(t) = -6t + 18$$

which is zero when $t = 3$, positive for $0 < t < 3$, and negative for $3 < t < 4$, as indicated in the arrow diagram shown.

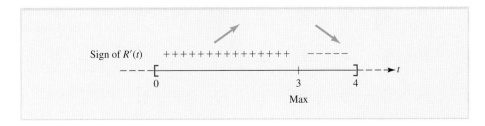

Thus, the rate of production $R(t)$ increases for $0 < t < 3$, decreases for $3 < t < 4$, and has its maximum value when $t = 3$ (11:00 A.M.). This means that the output function $Q(t)$ has an inflection point at $t = 3$, since $Q''(t) = R'(t)$ changes sign at that time. The graph of the production function $Q(t)$ is shown in Figure 3.25a, while that of the production rate function $R(t)$ is shown in Figure 3.25b.

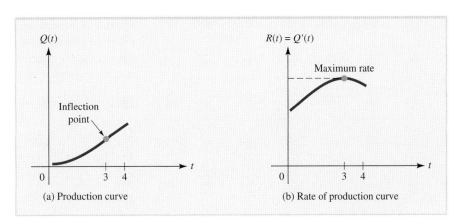

FIGURE 3.25   The production of an average worker.

EXERCISES ▪ 3.2

In Exercises 1 through 4, determine where the second derivative of the function is positive and where it is negative.

1.

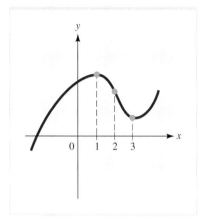

2.

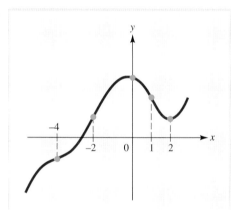

3.

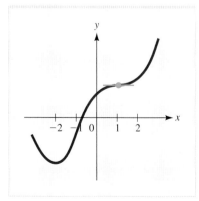

4.

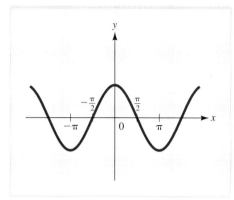

In Exercises 5 through 12, determine where the graph of the given function is concave upward and where it is concave downward. Find the coordinates of all inflection points.

5. $f(x) = x^3 + 3x^2 + x + 1$

6. $f(x) = x^4 - 4x^3 + 10x - 9$

7. $f(x) = x(2x + 1)^2$

8. $f(s) = s(s + 3)^2$

9. $g(t) = t^2 - \dfrac{1}{t}$

10. $F(x) = (x - 4)^{7/3}$

11. $f(x) = x^4 - 6x^3 + 7x - 5$

12. $g(x) = 3x^5 - 25x^4 + 11x - 17$

In Exercises 13 through 26, determine where the given function is increasing and where it is decreasing, and where its graph is concave up and where it is concave down. Find the relative extrema and inflection points and sketch the graph of the function.

13. $f(x) = \dfrac{1}{3}x^3 - 9x + 2$

14. $f(x) = x^3 + 3x^2 + 1$

15. $f(x) = x^4 - 4x^3 + 10$

16. $f(x) = x^3 - 3x^2 + 3x + 1$

17. $f(x) = (x - 2)^3$

18. $f(x) = x^5 - 5x$

19. $f(x) = (x^2 - 5)^3$

20. $f(x) = (x - 2)^4$

21. $f(s) = 2s(s + 4)^3$

22. $f(x) = (x^2 - 3)^2$

23. $g(x) = \sqrt{x^2 + 1}$

24. $f(x) = \dfrac{x^2}{x^2 + 3}$

25. $f(x) = \dfrac{1}{x^2 + x + 1}$

26. $f(x) = x^4 + 6x^3 - 24x^2 + 24$

In Exercises 27 through 38, use the second derivative test to find the relative maxima and minima of the given function.

27. $f(x) = x^3 + 3x^2 + 1$

28. $f(x) = x^4 - 2x^2 + 3$

29. $f(x) = (x^2 - 9)^2$

30. $f(x) = x + \dfrac{1}{x}$

31. $f(x) = 2x + 1 + \dfrac{18}{x}$

32. $f(x) = \dfrac{x^2}{x - 2}$

33. $f(x) = x^2(x - 5)^2$

34. $f(x) = \left(\dfrac{x}{x + 1}\right)^2$

35. $h(t) = \dfrac{2}{1 + t^2}$

36. $f(s) = \dfrac{s + 1}{(s - 1)^2}$

37. $f(x) = \dfrac{(x - 2)^3}{x^2}$

38. $h(t) = \dfrac{(t + 3)^3}{(t - 1)^2}$

In Exercises 39 through 42, the second derivative $f''(x)$ of a function is given. In each case, use this information to determine where the graph of $f(x)$ is concave upward and where it is concave downward and find all values of x for which an inflection point occurs. (You are not required to find $f(x)$ or the y coordinates of the inflection points.)

39. $f''(x) = x^2(x - 3)(x - 1)$

40. $f''(x) = x^3(x^2 + 2x - 3)$

41. $f''(x) = (x - 1)^{1/3}$

42. $f''(x) = \dfrac{x^2 + x - 2}{x^2 + 4}$

In Exercises 43 through 46, the first derivative $f'(x)$ of a certain function $f(x)$ is given. In each case,

    (a) Find the intervals on which f is increasing and the intervals on which it is decreasing.

    (b) Find the intervals on which the graph of f is concave up and the intervals on which it is concave down.

    (c) Find the x coordinates of the relative extrema and inflection points of f.

    (d) Sketch a possible graph for $f(x)$.

43. $f'(x) = x^2 - 4x$ **44.** $f'(x) = x^2 - 2x - 8$

45. $f'(x) = 5 - x^2$ **46.** $f'(x) = x(1 - x)$

47. Sketch the graph of a function that has all of the following properties:

- $f'(x) > 0$ when $x < -1$ and when $x > 3$
- $f'(x) < 0$ when $-1 < x < 3$
- $f''(x) < 0$ when $x < 2$
- $f''(x) > 0$ when $x > 2$

48. Sketch the graph of a function f that has all of the following properties:

- The graph has discontinuities at $x = -1$ and at $x = 3$.
- $f'(x) > 0$ for $x < 1$, and $x \neq -1$.
- $f'(x) < 0$ for $x > 1$, and $x \neq 3$.
- $f''(x) > 0$ for $x < -1$ and $x > 3$ and $f''(x) < 0$ for $-1 < x < 3$.
- $f(0) = 0 = f(2)$ and $f(1) = 3$.

In Exercises 49 through 52, the graph of a derivative function $y = f'(x)$ is given. Describe the function $f(x)$ and sketch a possible graph of $y = f(x)$.

49.

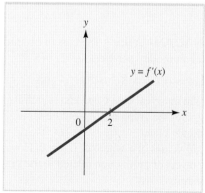

50.

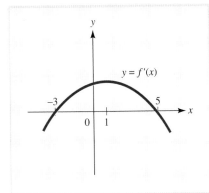

51.

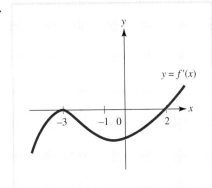

52.

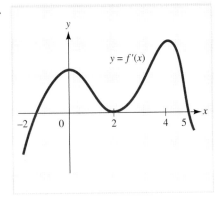

53. MARGINAL ANALYSIS The cost of producing x units of a commodity per week is

$$C(x) = 0.3x^3 - 5x^2 + 28x + 200$$

a. Find the marginal cost $C'(x)$ and sketch its graph along with the graph of $C(x)$ on the same coordinate plane.

b. Find all values of x where $C''(x) = 0$. How are these levels of production related to the graph of the marginal cost?

54. MARGINAL ANALYSIS The profit obtained from producing x thousand units of a particular commodity each year is $P(x)$ dollars, where

$$P(x) = -x^{9/2} + 90x^{7/2} - 5000.$$

a. Find the marginal profit $P'(x)$, and determine all values of x such that $P'(x) = 0$.

b. Sketch the graph of marginal profit along with the graph of $P(x)$ on the same coordinate plane.

c. Find $P''(x)$, and determine all values of x such that $P''(x) = 0$. How are these levels of production related to the graph of marginal profit?

55. SALES A company estimates that if x thousand dollars are spent on marketing a certain product, then $S(x)$ units of the product will be sold each month, where

$$S(x) = -x^3 + 33x^2 + 60x + 1000.$$

a. How many units will be sold if no money is spent on marketing?

b. Sketch the graph of $S(x)$. For what value of x does the graph have an inflection point? What is the significance of this marketing expenditure?

56. SALES An auto sales company estimates that when x thousand dollars are spent on the marketing of a certain model, $Q(x)$ cars will be sold, where

$$Q(x) = -4x^3 + 252x^2 - 3200x + 17\ 000$$

for $10 \le x \le 40$. Sketch the graph of $Q(x)$. Where does the graph have an inflection point? What is the significance of the marketing expenditure that corresponds to this point?

57. WORKER EFFICIENCY An efficiency study of the morning shift (from 8:00 A.M. to 12:00 noon) at a large bakery for specialty baked goods indicates that an average worker who arrives on the job at 8:00 A.M. will have taken Q trays of baked goods from the ovens t hours later, where

$$Q(t) = -t^3 + \frac{9}{2}t^2 + 15t.$$

a. At what time during the morning is the worker performing most efficiently?

b. At what time during the morning is the worker performing least efficiently?

58. WORKER EFFICIENCY An efficiency study of the evening shift (from 5:00 P.M. to 9:00 P.M.) at a certain factory indicates that an average worker who arrives on the job at 5:00 P.M. will have assembled $Q(t) = -t^3 + 6t^2 + 15t$ units t hours later.

a. At what time during the evening is the worker performing most efficiently?
b. At what time during the evening is the worker performing least efficiently?

59. **POPULATION GROWTH**   A 5-year projection of population trends suggests that t years from now, the population of a certain community will be $P(t) = -t^3 + 9t^2 + 48t + 50$ thousand.
   a. At what time during the 5-year period will the population be growing most rapidly?
   b. At what time during the 5-year period will the population be growing least rapidly?
   c. At what time is the rate of population growth changing most rapidly?

60. **ADVERTISING**   The manager of the Footloose sandal company determines that t months after initiating an advertising campaign, $S(t)$ hundred pairs of sandals will be sold, where

$$S(t) = \frac{3}{t+2} - \frac{12}{(t+2)^2} + 5.$$

   a. Find $S'(t)$ and $S''(t)$.
   b. At what time will sales be maximized? What is the maximum level of sales?
   c. The manager plans to end the advertising campaign when the sales rate is minimized. When does this occur? What are the sales level and sales rate at this time?

61. **HOUSING STARTS**   Suppose that in a certain community, there will be $M(r)$ thousand new houses built when the 30-year fixed mortgage rate is r percent, where

$$M(r) = \frac{1 + 0.02r}{1 + 0.009r^2}.$$

   a. Find $M'(r)$ and $M''(r)$.
   b. Sketch the graph of the construction function $M(r)$.
   c. At what rate of interest r is the rate of construction of new houses minimized?

62. **GOVERNMENT SPENDING**   During a recession in a certain country, the government decides to stimulate the economy by providing funds to hire unemployed workers for infrastructure projects. Suppose that t months after the stimulus program begins, $N(t)$ thousand people are unemployed, where

$$N(t) = \frac{1}{20}(-t^3 + 45t^2 + 408t + 3078).$$

a. What is the maximum number of unemployed workers? When does the maximum level of unemployment occur?
b. In order to avoid overstimulating the economy (and inducing inflation), a decision is made to end the stimulus program as soon as the rate of unemployment begins to decline. When does this occur? At this time, how many people are unemployed?

63. **SPREAD OF A DISEASE**   An epidemiologist determines that a particular epidemic spreads in such a way that t weeks after the outbreak, N hundred new cases will be reported, where

$$N(t) = \frac{5t}{12 + t^2},$$

   a. Find $N'(t)$ and $N''(t)$.
   b. At what time is the epidemic at its worst? What is the maximum number of reported new cases?
   c. Health officials declare the epidemic to be under control when the rate of reported new cases is minimized. When does this occur? What number of new cases will be reported at that time?

64. **A* THE SPREAD OF AN EPIDEMIC**   Let $Q(t)$ denote the number of people in a city of population N_0 who have been infected with a certain disease t days after the beginning of an epidemic. Studies indicate that the rate $R(Q)$ at which an epidemic spreads is jointly proportional to the number of people who have contracted the disease and the number who have not, so $R(Q) = kQ(N_0 - Q)$. Sketch the graph of the rate function, and interpret your graph. In particular, what is the significance of the highest point on the graph of $R(Q)$?

65. **A* SPREAD OF A RUMOUR**   The rate at which a rumour spreads through a community of P people is jointly proportional to the number of people N who have heard the rumour and the number who have not. Show that the rumour is spreading most rapidly when half the people have heard it.

66. **A* POPULATION GROWTH**   Studies show that when environmental factors impose an upper bound on the possible size of a population $P(t)$, the population often tends to grow in such a way that the percentage rate of change of $P(t)$ satisfies

$$\frac{100P'(t)}{P(t)} = A - BP(t)$$

where A and B are positive constants. Where does the graph of $P(t)$ have an inflection point? What is the significance of this point? (Your answer will be in terms of A and B.)

67. **A* TISSUE GROWTH**   Suppose a particular tissue culture has area $A(t)$ at time t and a potential maximum area M. Based on properties of cell division, it is reasonable to assume that the area A grows at a rate jointly proportional to $\sqrt{A(t)}$ and $M - A(t)$; that is,

$$\frac{dA}{dt} = k\sqrt{A(t)}\,[M - A(t)]$$

where k is a positive constant.

a. Let $R(t) = A'(t)$ be the rate of tissue growth. Show that $R'(t) = 0$ when $A(t) = M/3$.

b. Is the rate of tissue growth greatest or least when $A(t) = M/3$? [*Hint:* Use the first derivative test or the second derivative test.]

c. Based on the given information and what you discovered in part (a), what can you say about the graph of $A(t)$?

68. **A*** Water is poured at a constant rate into the vase shown in the accompanying figure. Let $h(t)$ be the height of the water in the vase at time t (assume the vase is empty when $t = 0$). Sketch a rough graph of the function $h(t)$. In particular, what happens when the water level reaches the neck of the vase?

EXERCISE 68

69. The position of a boy with a kite, running back and forth in a straight line, is $s(t) = t^4 - 2t^2$ metres t seconds after passing his mother.

a. Find the times of any relative maxima and minima of his position.

b. Find the times of any relative maxima and minima of his velocity.

c. Find the times of any relative maxima and minima of his acceleration.

  d. Plot graphs of the position, velocity, and acceleration, using the same scale for each. Label the relative extrema of each, noting that the relative extrema of one of the functions correspond to the zero of the derivative of that function.

70. Let $f(x) = x^4 + x$. Show that even though $f''(0) = 0$, the graph of f has neither a relative extremum nor an inflection point where $x = 0$. Sketch the graph of $f(x)$.

71. **A*** Use calculus to show that the graph of the quadratic function $y = ax^2 + bx + c$ is concave upward if a is positive and concave downward if a is negative.

72. **A*** If $f(x)$ and $g(x)$ are continuous functions that both have an inflection point at $x = c$, is it true that the sum $h(x) = f(x) + g(x)$ must also have an inflection point at $x = c$? Either explain why this must always be true or find functions $f(x)$ and $g(x)$ for which it is false.

73. **A*** Suppose $f(x)$ and $g(x)$ are continuous functions with $f'(c) = 0$. If both f and g have an inflection point at $x = c$, does $P(x) = f(x)g(x)$ have an inflection point at $x = c$? Either explain why this must always be true or find functions $f(x)$ and $g(x)$ for which it is false.

SECTION 3.3   Curve Sketching

L03

Sketch curves using the first and second derivatives.

So far in this chapter, you have seen how to use the derivative $f'(x)$ to determine where the graph of $f(x)$ is rising and falling and to use the second derivative $f''(x)$ to determine the concavity of the graph. While these tools are adequate for locating the high and low points of a graph and for sculpting its twists and turns, other graphical features are best described using limits.

Recall from Section 1.5 that a limit of the form $\lim\limits_{x \to +\infty} f(x)$ or $\lim\limits_{x \to -\infty} f(x)$, in which the independent variable x either increases or decreases without bound, is called a *limit at infinity.* On the other hand, if the functional values $f(x)$ themselves grow without bound as x approaches a number c, we say that $f(x)$ has an *infinite limit* at $x = c$ and write $\lim\limits_{x \to c} f(x) = +\infty$ if $f(x)$ increases indefinitely as x approaches c or $\lim\limits_{x \to c} f(x) = -\infty$ if $f(x)$ decreases indefinitely. Collectively, limits at infinity and infinite limits are referred to as **limits involving infinity.** Our first goal in this section is to see how limits involving infinity may be interpreted as graphical features. We will then combine this information with the derivative methods of Sections 3.1 and 3.2 to form a general procedure for sketching graphs.

Vertical Asymptotes

Limits involving infinity can be used to describe graphical features called *asymptotes.* In particular, the graph of a function $f(x)$ is said to have a **vertical asymptote** at $x = c$ if $f(x)$ increases or decreases without bound as x tends toward c, from either the right or the left.

For instance, consider the rational function

$$f(x) = \frac{x + 1}{x - 2}.$$

As x approaches 2 from the left ($x < 2$), the functional values decrease without bound, but they increase without bound if the approach is from the right ($x > 2$). This behaviour is illustrated in the table and demonstrated graphically in Figure 3.26.

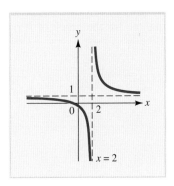

FIGURE 3.26   The graph of $f(x) = \dfrac{x + 1}{x - 2}$.

x	1.95	1.97	1.99	1.999	2	2.001	2.005	2.01
$f(x) = \dfrac{x + 1}{x - 2}$	−59	−99	−299	−2999	Undefined	3001	601	301

The behaviour in this example can be summarized as follows using the one-sided limit notation introduced in Section 1.6:

$$\lim_{x \to 2^-} \frac{x + 1}{x - 2} = -\infty \quad \text{and} \quad \lim_{x \to 2^+} \frac{x + 1}{x - 2} = +\infty.$$

In a similar fashion, we use the limit notation to define the concept of a vertical asymptote.

Vertical Asymptotes   ■   The line $x = c$ is a *vertical asymptote* of the graph of $f(x)$ if either

$$\lim_{x \to c^-} f(x) = +\infty \ (\text{or} -\infty)$$

or

$$\lim_{x \to c^+} f(x) = +\infty \ (\text{or} -\infty)$$

In general, a rational function $R(x) = \dfrac{p(x)}{q(x)}$ has a vertical asymptote $x = c$ whenever $q(c) = 0$ but $p(c) \neq 0$. Here is an example of a function with a vertical asymptote.

EXAMPLE 3.3.1

Determine all vertical asymptotes of the graph of

$$f(x) = \frac{x^2 - 9}{x^2 + 3x}$$

Solution

Let $p(x) = x^2 - 9$ and $q(x) = x^2 + 3x$ be the numerator and denominator, respectively, of $f(x)$. Then $q(x) = 0$ when $x = -3$ and when $x = 0$. However, for $x = -3$, we also have $p(-3) = 0$ and

$$\lim_{x \to -3} \frac{x^2 - 9}{x^2 + 3x} = \lim_{x \to -3} \frac{(x - 3)(x + 3)}{x(x + 3)} = \lim_{x \to -3} \frac{x - 3}{x} = 2$$

This means that the graph of $f(x)$ has a hole at the point $(-3, 2)$ and $x = -3$ is *not* a vertical asymptote of the graph.

On the other hand, for $x = 0$ we have $q(0) = 0$ but $p(0) \ne 0$, which suggests that the y axis (the vertical line $x = 0$) is a vertical asymptote for the graph of $f(x)$. This asymptotic behaviour is verified by noting that

$$\lim_{x \to 0^-} \frac{x^2 - 9}{x^2 + 3x} = +\infty \quad \text{and} \quad \lim_{x \to 0^+} \frac{x^2 - 9}{x^2 + 3x} = -\infty$$

The graph of $f(x)$ is shown in Figure 3.27.

EXPLORE!

Plot $f(x) = \dfrac{x^2 - 9}{x^2 + 3x}$. Move the cursor along $f(x)$ to confirm that it is not defined at $x = -3$ and $x = 0$. How does the graphing program indicate these undefined points?

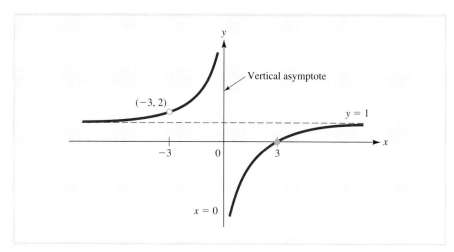

FIGURE 3.27   The graph of $f(x) = \dfrac{x^2 - 9}{x^2 + 3x}$

Horizontal Asymptotes

In Figure 3.27, note that the graph approaches the horizontal line $y = 1$ as x increases or decreases without bound; that is,

$$\lim_{x \to -\infty} \frac{x^2 - 9}{x^2 + 3x} = 1 \quad \text{and} \quad \lim_{x \to +\infty} \frac{x^2 - 9}{x^2 + 3x} = 1$$

In general, when a function $f(x)$ tends toward a finite value b as x either increases or decreases without bound (or both as in our example), then the horizontal line $y = b$ is called a **horizontal asymptote** of the graph of $f(x)$. A horizontal asymptote gives the value of the function's long-term behaviour, or its behaviour in the long run. Here is a definition.

> **Horizontal Asymptotes** ■ The horizontal line $y = b$ is called a *horizontal asymptote* of the graph of $y = f(x)$ if
> $$\lim_{x \to -\infty} f(x) = b \quad \text{or} \quad \lim_{x \to \infty} f(x) = b$$

Recall the reciprocal power rules for limits (Section 1.5):

$$\lim_{x \to +\infty} \frac{A}{x^k} = 0$$

and

$$\lim_{x \to -\infty} \frac{A}{x^k} = 0$$

for constants A and k, with $k > 0$ and x^k defined for all x.

EXAMPLE 3.3.2

Determine all horizontal asymptotes of the graph of

$$f(x) = \frac{x^2}{x^2 + x + 1}$$

Solution

Dividing each term in the rational function $f(x)$ by x^2 (the highest power of x in the denominator) gives

$$\lim_{x \to +\infty} f(x) = \lim_{x \to +\infty} \frac{x^2}{x^2 + x + 1} = \lim_{x \to +\infty} \frac{x^2/x^2}{x^2/x^2 + x/x^2 + 1/x^2}$$

$$= \lim_{x \to +\infty} \frac{1}{1 + 1/x + 1/x^2} = \frac{1}{1 + 0 + 0} = 1 \qquad \text{reciprocal power rule}$$

and, similarly,

$$\lim_{x \to -\infty} f(x) = \lim_{x \to -\infty} \frac{x^2}{x^2 + x + 1} = 1$$

Thus, the graph of $f(x)$ has $y = 1$ as a horizontal asymptote. The graph is shown in Figure 3.28.

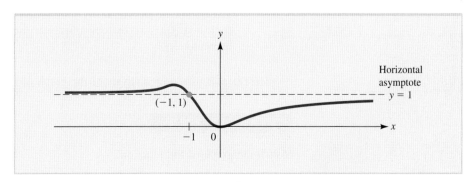

FIGURE 3.28   The graph of $f(x) = \dfrac{x^2}{x^2 + x + 1}$.

NOTE   The graph of a function $f(x)$ can never cross a vertical asymptote $x = c$ because at least one of the one-sided limits $\lim\limits_{x \to c^-} f(x)$ and $\lim\limits_{x \to c^+} f(x)$ must be infinite. However, it is quite possible for a graph to cross its horizontal asymptotes. For instance, in Example 3.3.2, the graph of $y = \dfrac{x^2}{x^2 + x + 1}$ crosses the horizontal asymptote $y = 1$ at the point where

$$\frac{x^2}{x^2 + x + 1} = 1$$
$$x^2 = x^2 + x + 1$$
$$x = -1$$

that is, at the point $(-1, 1)$.

A General Graphing Procedure

We now have the tools we need to describe a general procedure for sketching a variety of graphs.

A General Procedure for Sketching the Graph of $f(x)$

Step 1. Find the domain of $f(x)$ (that is, where $f(x)$ is defined).

Step 2. Find and plot all intercepts. The y intercept (where $x = 0$) is usually easy to find, but the x intercepts (where $f(x) = 0$) may require a graphing utility.

Step 3. Determine all vertical and horizontal asymptotes of the graph. Draw the asymptotes in a coordinate plane.

Step 4. Find $f'(x)$ and use it to determine the critical numbers of $f(x)$ and its intervals of increase and decrease.

Step 5. Determine the x and y coordinates of all relative extrema. Plot each relative maximum with a cap ($\frown$) and each relative minimum with a cup ($\smile$).

Step 6. Find $f''(x)$ and use it to determine intervals of concavity and points of inflection. Plot each inflection point with a twist ($\sim$) to suggest the shape of the graph near the point.

Step 7. You now have a preliminary graph, with asymptotes in place, intercepts plotted, arrows indicating the direction of the graph, and caps, cups, and twists suggesting the shape at key points. Plot additional points if needed, and complete the sketch by joining the plotted points in the directions indicated. Be sure to remember that the graph cannot cross a vertical asymptote.

Here is a step-by-step analysis of the graph of a rational function.

EXAMPLE 3.3.3

Sketch the graph of the function

$$f(x) = \frac{x}{(x + 1)^2}$$

Solution

Steps 1 and 2. The function is defined for all x except $x = -1$, and the only intercept is the origin $(0, 0)$.

Step 3. The line $x = -1$ is a vertical asymptote of the graph of $f(x)$ since $f(x)$ decreases indefinitely as x approaches -1 from either side; that is,

$$\lim_{x \to -1^-} \frac{x}{(x + 1)^2} = \lim_{x \to -1^+} \frac{x}{(x + 1)^2} = -\infty$$

Moreover, since

$$\lim_{x \to -\infty} \frac{x}{(x + 1)^2} = \lim_{x \to +\infty} \frac{x}{(x + 1)^2} = 0$$

the line $y = 0$ (the x axis) is a horizontal asymptote. Draw the dashed lines $x = -1$ and $y = 0$ on a coordinate plane.

Step 4. Applying the quotient rule, compute the derivative of $f(x)$:

$$f'(x) = \frac{(x + 1)^2(1) - x[2(x + 1)(1)]}{(x + 1)^4} = \frac{1 - x}{(x + 1)^3}$$

Since $f'(1) = 0$, it follows that $x = 1$ is a critical number. Note that even though $f'(-1)$ does not exist, $x = -1$ is not a critical number since it is not in the domain of $f(x)$. Place $x = 1$ and $x = -1$ on a number line with a dashed vertical line at $x = -1$ to indicate the vertical asymptote there. Then evaluate $f'(x)$ at appropriate test numbers (say, at -2, 0, and 3) to obtain the arrow diagram shown.

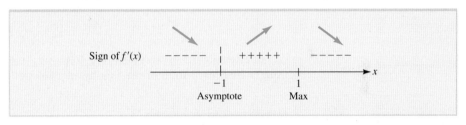

Step 5. The arrow pattern in the diagram obtained in step 4 indicates that there is a relative maximum at $x = 1$. Since $f(1) = \dfrac{1}{4}$, plot a cap at $\left(1, \dfrac{1}{4}\right)$.

Step 6. Apply the quotient rule again to get

$$f''(x) = \frac{2(x - 2)}{(x + 1)^4}$$

Since $f''(x) = 0$ at $x = 2$ and $f''(x)$ does not exist at $x = -1$, plot -1 and 2 on a number line and check the sign of $f''(x)$ on the intervals $x < -1$, $-1 < x < 2$, and $x > 2$ to obtain the concavity diagram shown.

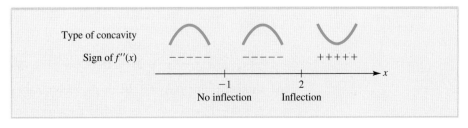

Note that the concavity changes at $x = 2$. Since $f(2) = \dfrac{2}{9}$, plot a twist $\searrow$ at $\left(2, \dfrac{2}{9}\right)$ to indicate the inflection point there.

Step 7. The preliminary graph is shown in Figure 3.29a. Note that the vertical asymptote (dashed line) breaks the graph into two parts. Join the features in each separate part by a smooth curve to obtain the completed graph shown in Figure 3.29b.

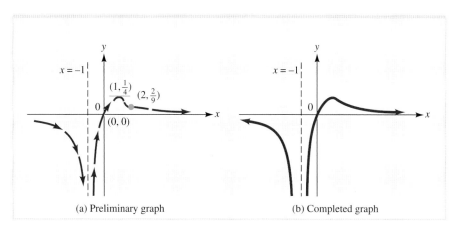

(a) Preliminary graph   (b) Completed graph

FIGURE 3.29   The graph of $f(x) = \dfrac{x}{(x + 1)^2}$.

In Example 3.3.4, we sketch the graph of a more complicated rational function using a condensed form of the step-by-step solution featured in Example 3.3.3.

EXAMPLE 3.3.4

Sketch the graph of

$$f(x) = \frac{3x^2}{x^2 + 2x - 15}$$

Solution

Since $x^2 + 2x - 15 = (x + 5)(x - 3)$, the function $f(x)$ is defined for all x except $x = -5$ and $x = 3$. The only intercept is the origin $(0, 0)$.

We see that $x = 3$ and $x = -5$ are vertical asymptotes because if we write $f(x) = p(x)/q(x)$, where $p(x) = 3x^2$ and $q(x) = x^2 + 2x - 15$, then $q(3) = 0$ and $q(-5) = 0$, while $p(3) \neq 0$ and $p(-5) \neq 0$. Moreover, $y = 3$ is a horizontal asymptote since

$$\lim_{x \to +\infty} f(x) = \lim_{x \to +\infty} \frac{3x^2}{x^2 + 2x - 15} = \lim_{x \to +\infty} \frac{3}{1 + 2/x - 15/x^2} = \frac{3}{1 + 0 - 0} = 3$$

and, similarly, $\lim_{x \to -\infty} f(x) = 3$. Begin the preliminary sketch by drawing the asymptotes $x = 3$, $x = -5$, and $y = 3$ as dashed lines on a coordinate plane.

Next, use the quotient rule to obtain

$$f'(x) = \frac{(x^2 + 2x - 15)(6x) - (2x + 2)(3x^2)}{(x^2 + 2x - 15)^2} = \frac{6x(x - 15)}{(x^2 + 2x - 15)^2}$$

We see that $f'(x) = 0$ when $x = 0$ and $x = 15$ and that $f'(x)$ does not exist when $x = -5$ and $x = 3$. Put $x = -5, 0, 3,$ and 15 on a number line and obtain the arrow diagram shown in Figure 3.30a by determining the sign of $f'(x)$ at appropriate test numbers (say, at $-7, -1, 2, 5,$ and 20). Interpreting the arrow diagram, we see that there is a relative maximum at $x = 0$ and a relative minimum at $x = 15$. Since $f(0) = 0$ and $f(15) \approx 2.81$, we plot a cap at $(0, 0)$ on our preliminary graph and a cup at $(15, 2.81)$.

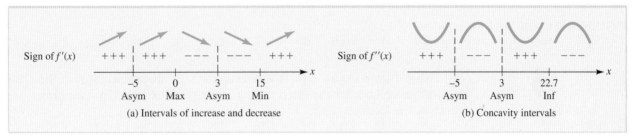

FIGURE 3.30   Arrow and concavity diagrams for $f(x) = \dfrac{3x^2}{x^2 + 2x - 15}$.

Apply the quotient rule again to get

$$f''(x) = \frac{-6(2x^3 - 45x^2 - 225)}{(x^2 + 2x - 15)^3}$$

We see that $f''(x)$ does not exist when $x = -5$ and $x = 3$, and that $f''(x) = 0$ when

$$2x^3 - 45x^2 - 225 = 0$$

$$x \approx 22.7 \qquad \text{found by technology}$$

Put $x = -5, 3,$ and 22.7 on a number line and obtain the concavity diagram shown in Figure 3.30b by determining the sign of $f''(x)$ at appropriate test numbers (say, at $-6, 0, 4,$ and 25). The concavity changes at all three subdivision numbers, but only $x = 22.7$ corresponds to an inflection point since $x = -5$ and $x = 3$ are not in the domain of $f(x)$. We find that $f(22.7) \approx 2.83$ and plot a twist ($\int$) at $(22.7, 2.83)$.

The preliminary graph is shown in Figure 3.31a. Note that the two vertical asymptotes break the graph into three parts. Join the features in each separate part by a smooth curve to obtain the completed graph shown in Figure 3.31b.

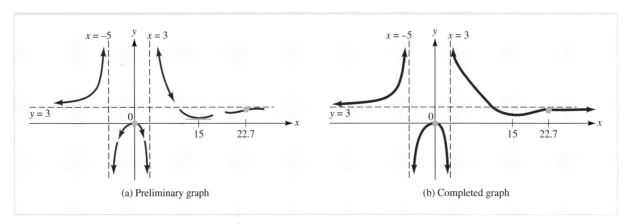

FIGURE 3.31   The graph of $f(x) = \dfrac{3x^2}{x^2 + 2x - 15}$.

When $f'(c)$ does not exist at a number $x = c$ in the domain of $f(x)$, there are several possibilities for the graph of $f(x)$ at the point $(c, f(c))$. Two such cases are examined in Example 3.3.5.

EXAMPLE 3.3.5

Sketch the graphs of $f(x) = x^{2/3}$ and $g(x) = (x - 1)^{1/3}$.

Solution

Both functions are defined for all x. For $f(x) = x^{2/3}$, we have

$$f'(x) = \frac{2}{3}x^{-1/3} \quad \text{and} \quad f''(x) = -\frac{2}{9}x^{-4/3}, \quad x \neq 0.$$

The only critical point is $(0, 0)$, and the intervals of increase and decrease and concavity are as shown:

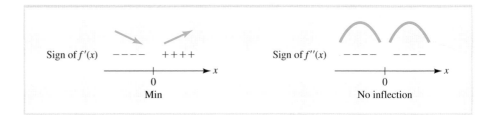

Interpreting these diagrams, we conclude that the graph of $f(x)$ is concave downward for all $x \neq 0$ and is falling for $x < 0$ and rising for $x > 0$. Thus, the graph of $f(x)$ has a relative minimum at the origin $(0, 0)$ and its shape there is $\curlyvee$, called a cusp. The graph of $f(x)$ is shown in Figure 3.32a.

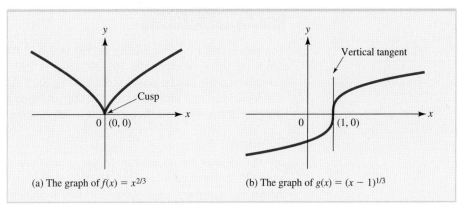

(a) The graph of $f(x) = x^{2/3}$ (b) The graph of $g(x) = (x - 1)^{1/3}$

FIGURE 3.32   A graph with a cusp and another with a vertical tangent.

The derivatives of $g(x) = (x - 1)^{1/3}$ are given by

$$g'(x) = \frac{1}{3}(x - 1)^{-2/3} \quad \text{and} \quad g''(x) = -\frac{2}{9}(x - 1)^{-5/3}, \quad x \ne 1.$$

The only critical point is $(1, 0)$, and we obtain the intervals of increase and decrease and concavity shown:

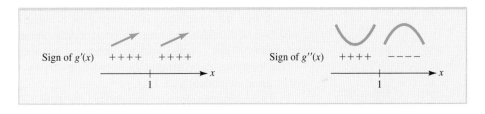

Thus, the graph of $g(x)$ is rising for all $x \ne 1$ and is concave upward for $x < 1$ and concave downward for $x > 1$. This means that $(1, 0)$ is an inflection point; in addition, note that

$$\lim_{x \to 1^-} g'(x) = \lim_{x \to 1^+} g'(x) = +\infty$$

Geometrically, this means that as x approaches 0 from either side, the tangent line at $(x, g(x))$ becomes steeper and steeper (with positive slope). This can be interpreted as saying that the graph of $g(x)$ has a tangent line at $(1, 0)$ with infinite slope, that is, a **vertical tangent** line. The graph of $g(x)$ is shown in Figure 3.32b.

   OPTION NOTE: Examples 4.4.1 and 4.4.2 may be done at this time.

Sometimes, it is useful to represent observations about a quantity in graphical form. This procedure is illustrated in Example 3.3.6.

EXAMPLE 3.3.6

The population of a community is 230 000 in 1990 and increases at an increasing rate for 5 years, reaching the 300 000 level in 1995. It then continues to rise, but at a decreasing rate until it peaks at 350 000 in 2002. After that, the population decreases at a decreasing rate for 3 years to 320 000 and then at an increasing rate, approaching 280 000 in the long run. Represent this information in graphical form.

Solution

Let $P(t)$ denote the population of the community t years after the base year 1990, where P is measured in units of 10 000 people. Since the population increases at an increasing rate for 5 years from 230 000 to 300 000, the graph of $P(t)$ rises from $(0, 23)$ to $(5, 30)$ and is concave upward for $0 < t < 5$. The population then continues to increase until 2002, but at a decreasing rate, until it reaches a maximum value of 350 000. That is, the graph continues to rise from $(5, 30)$ to a high point at $(12, 35)$ but is now concave downward. Since the graph changes concavity at $x = 5$ (from up to down), it has an inflection point at $(5, 30)$.

For the next 3 years, the population decreases at a decreasing rate, so the graph of $P(t)$ is falling and concave downward for $12 < t < 15$. Since the population continues to decrease from the 320 000 level reached in 2005 but at an increasing rate, the graph falls from the point $(15, 32)$ for $t > 15$ but is concave upward. The change in concavity at $x = 15$ (from down to up) means that $(15, 32)$ is another inflection point.

The statement that "the population decreases at an increasing rate" for $t > 15$ means that the population is changing at a negative rate, which is becoming less negative with increasing time. In other words, the decline in population slows down after 2005. This, coupled with the statement that the population "approaches 280 000 in the long run," suggests that the population curve $y = P(t)$ flattens out and approaches $y = 28$ asymptotically as $t \to +\infty$.

These observations are summarized in Table 3.2 and are represented in graphical form in Figure 3.33.

TABLE 3.2   Behaviour of a Population $P(t)$

Time Period	The function $P(t)$ is . . .	and the graph of $P(t)$ is . . .
$t = 0$	$P(0) = 23$	at the point $(0, 23)$
$0 < t < 5$	increasing at an increasing rate	rising and concave upward
$t = 5$	$P(5) = 30$	at the inflection point $(5, 30)$
$5 < t < 12$	increasing at a decreasing rate	rising and concave downward
$t = 12$	$P(12) = 35$	at the high point $(12, 35)$
$12 < t < 15$	decreasing at a decreasing rate	falling and concave downward
$t = 15$	$P(15) = 32$	at the inflection point $(15, 32)$
$t > 15$	decreasing at an increasing rate and gradually tending toward 28	falling and concave upward asymptotically approaching $y = 28$

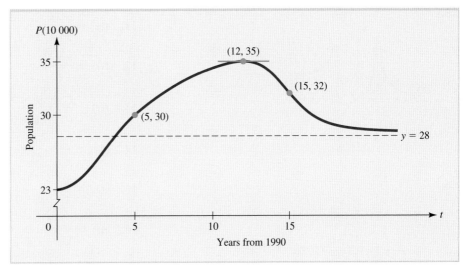

FIGURE 3.33   Graph of a population function.

EXERCISES ■ 3.3

> **OPTION NOTE:**   Section 4.4 Exercises 5–20 may be done at this time if exponential and logarithmic functions are to be included here.

In Exercises 1 through 8, name the vertical and horizontal asymptotes of the given graph.

1.

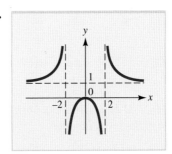

2.

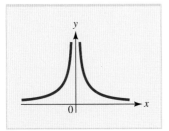

3.

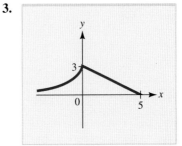

4.

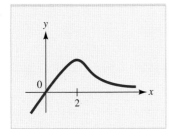

5.

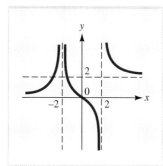

6.

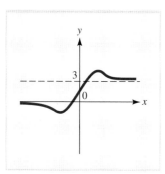

7.

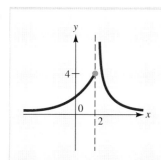

8.

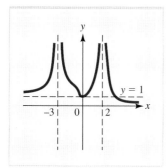

In Exercises 9 through 16, find all vertical and horizontal asymptotes of the graph of the given function.

9. $f(x) = \dfrac{3x - 1}{x + 2}$

10. $f(x) = \dfrac{x}{2 - x}$

11. $f(x) = \dfrac{x^2 + 2}{x^2 + 1}$

12. $f(t) = \dfrac{t + 2}{t^2}$

13. $f(t) = \dfrac{t^2 + 3t - 5}{t^2 - 5t + 6}$

14. $g(x) = \dfrac{5x^2}{x^2 - 3x - 4}$

15. $h(x) = \dfrac{1}{x} - \dfrac{1}{x - 1}$

16. $g(t) = \dfrac{t}{\sqrt{t^2 - 4}}$

In Exercises 17 through 32, sketch the graph of the given function.

17. $f(x) = x^3 + 3x^2 - 2$

18. $f(x) = x^5 - 5x^4 + 93$

19. $f(x) = x^4 + 4x^3 + 4x^2$

20. $f(x) = 3x^4 - 4x^2 + 3$

21. $f(x) = (2x - 1)^2(x^2 - 9)$

22. $f(x) = x^3 - 3x^4$

23. $f(x) = \dfrac{1}{2x + 3}$

24. $f(x) = \dfrac{x + 3}{x - 5}$

25. $f(x) = x - \dfrac{1}{x}$

26. $f(x) = \dfrac{x^2}{x + 2}$

27. $f(x) = \dfrac{1}{x^2 - 9}$

28. $f(x) = \dfrac{1}{\sqrt{1 - x^2}}$

29. $f(x) = \dfrac{x^2 - 9}{x^2 + 1}$

30. $f(x) = \dfrac{1}{\sqrt{x}} - \dfrac{1}{x}$

31. $f(x) = x^{3/2}$

32. $f(x) = x^{4/3}$

In Exercises 33 through 38, diagrams indicating intervals of increase or decrease and concavity are given. Sketch a possible graph for a function with these characteristics.

33.

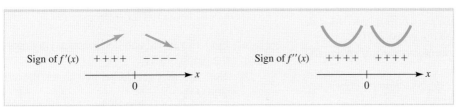

34.

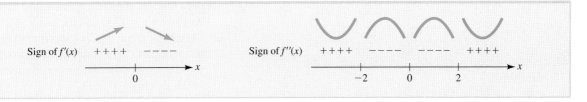

35.

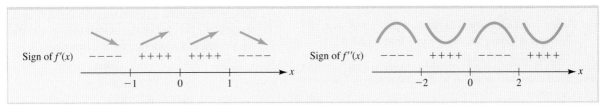

36.

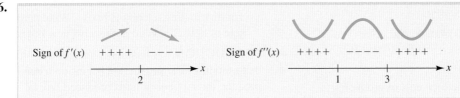

37.

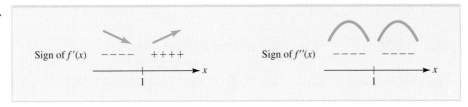

38.

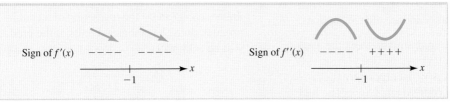

In Exercises 39 through 42, the derivative f '(x) of a differentiable function f(x) is given. In each case,
   (a) Find intervals of increase and decrease for f(x).
   (b) Determine values of x for which relative maxima and minima occur on the graph of f(x).
   (c) Find f"(x) and determine intervals of concavity for the graph of f(x).
   (d) For what values of x do inflection points occur on the graph of f(x)?

39. $f'(x) = x^3(x - 2)^2$ **40.** $f'(x) = x^2(x + 1)^3$

41. $f'(x) = \dfrac{x + 3}{(x - 2)^2}$ **42.** $f'(x) = \dfrac{x + 2}{(x - 1)^2}$

43. Find constants A and B so that the graph of the function

$$f(x) = \frac{Ax - 3}{5 + Bx}$$

has $x = 2$ as a vertical asymptote and $y = 4$ as a horizontal asymptote. Once you find A and B, sketch the graph of $f(x)$.

44. Find constants A and B so that the graph of the function

$$f(x) = \frac{Ax + 2}{8 - Bx}$$

has $x = 4$ as a vertical asymptote and $y = -1$ as a horizontal asymptote. Once you find A and B, sketch the graph of $f(x)$.

45. AVERAGE COST   The total cost of producing x units of a particular commodity is C thousand dollars, where $C(x) = 3x^2 + x + 48$, and the average cost is

$$A(x) = \frac{C(x)}{x} = 3x + 1 + \frac{48}{x}$$

   a. Find all vertical and horizontal asymptotes of the graph of $A(x)$.
   b. Note that as x gets larger and larger, the term $\dfrac{48}{x}$ in $A(x)$ gets smaller and smaller. What does this say about the relationship between the average cost curve $y = A(x)$ and the line $y = 3x + 1$?
   c. Sketch the graph of $A(x)$, incorporating the result of part (b) in your sketch. [*Note:* The line $y = 3x + 1$ is called an *oblique* (or *slant*) *asymptote* of the graph.]

46. INVENTORY COST   A manufacturer estimates that if each shipment of raw materials contains

x units, the total cost in dollars of obtaining and storing the year's supply of raw materials will be

$$C(x) = 2x + \frac{80\ 000}{x}$$

   a. Find all vertical and horizontal asymptotes of the graph of $C(x)$.
   b. Note that as x gets larger and larger, the term $\dfrac{80\ 000}{x}$ in $C(x)$ gets smaller and smaller. What does this say about the relationship between the cost curve $y = C(x)$ and the line $y = 2x$?
   c. Sketch the graph of $C(x)$, incorporating the result of part (b) in your sketch. [*Note:* The line $y = 2x$ is called an *oblique* (or *slant*) *asymptote* of the graph.]

47. DISTRIBUTION COST   The number of worker-hours W required to distribute new telephone books to x percent of the households in a certain community is modelled by the function

$$W(x) = \frac{200x}{100 - x}$$

   a. Sketch the graph of $W(x)$.
   b. Suppose only 1500 worker-hours are available for distributing telephone books. What percentage of households do not receive new books?

48. PRODUCTION   A business manager determines that t months after production begins on a new product, the number of units produced will be P million per month, where

$$P(t) = \frac{t}{(t + 1)^2}$$

   a. Find $P'(t)$ and $P''(t)$.
   b. Sketch the graph of $P(t)$.
   c. What happens to production in the long run (as $t \to \infty$)?

49. SALES   A company estimates that if x thousand dollars are spent on the marketing of a certain product, then $Q(x)$ thousand units of the product will be sold, where

$$Q(x) = \frac{7x}{27 + x^2}$$

   a. Sketch the graph of the sales function $Q(x)$.
   b. For what marketing expenditure x are sales maximized? What is the maximum sales level?
   c. For what value of x is the sales rate minimized?

50. CONCENTRATION OF DRUG   A patient is given an injection of a particular drug at noon, and samples of blood are taken at regular intervals to determine the concentration of drug in the patient's system. It is found that the concentration increases at an increasing rate with respect to time until 1 P.M., and for the next 3 hours continues to increase but at a decreasing rate until the peak concentration is reached at 4 P.M. The concentration then decreases at a decreasing rate until 5 P.M., after which it decreases at an increasing rate toward zero. Sketch a possible graph for the concentration of drug $C(t)$ as a function of time.

51. **BACTERIAL POPULATION**   The population of a bacterial colony increases at an increasing rate for 1 hour, after which it continues to increase but at a rate that gradually decreases toward zero. Sketch a possible graph for the population $P(t)$ as a function of time t.

52. **EPIDEMIOLOGY**   Epidemiologists studying a contagious disease observe that the number of newly infected people increases at an increasing rate during the first 3 years of the epidemic. At that time, a new drug is introduced, and the number of infected people declines at a decreasing rate. Two years after its introduction, the drug begins to lose effectiveness. The number of new cases continues to decline for 1 more year but at an increasing rate before rising again at an increasing rate. Draw a possible graph for the number of new cases $N(t)$ as a function of time.

53. **ADOPTION OF TECHNOLOGY**   In a certain area a new cell phone rate plan becomes available. Draw a possible graph for the percentage of households cancelling their landlines and using a cell phone instead if the percentage grows at an increasing rate for the first 2 years, after which the rate of increase declines, with the market penetration of the technology eventually approaching 90%.

54. **EXPERIMENTAL PSYCHOLOGY**   To study the rate at which animals learn, a psychology student performed an experiment in which a rat was sent repeatedly through a laboratory maze. Suppose the time required for the rat to traverse the maze on the nth trial was approximately $f(n) = 3 + \dfrac{12}{n}$ minutes.
 a. Graph the function $f(n)$.
 b. What portion of the graph is relevant to the practical situation under consideration?
 c. What happens to the graph as n increases without bound? Interpret your answer in practical terms.

55. **AVERAGE TEMPERATURE**   A researcher models the temperature T (in degrees Celsius) during the time period from 6 A.M. to 6 P.M. in a certain city by the function

$$T(t) = \frac{-1}{36}t^3 + \frac{1}{8}t^2 + \frac{7}{3}t - 2 \quad \text{for } 0 \leq t \leq 12,$$

where t is the number of hours after 6 A.M.
 a. Sketch the graph of $T(t)$.
 b. At what time is the temperature the greatest? What is the highest temperature of the day?

56. **IMMUNIZATION**   During a nationwide program to immunize the population against a new strain of influenza, public health officials determined that the cost of inoculating x percent of the susceptible population would be approximately

$$C(x) = \frac{1.7x}{100 - x}$$

million dollars.
 a. Sketch the graph of the cost function $C(x)$.
 b. Suppose $40 million is available for providing immunization. What percentage of the susceptible population will not be inoculated?

57. **POLITICAL POLLING**   A poll commissioned by a politician estimates that t days after she comes out in favour of a controversial bill, the percentage of her constituency (those who support her at the time she declares her position on the bill) that still supports her is given by

$$S(t) = \frac{100(t^2 - 3t + 25)}{t^2 + 7t + 25}$$

The vote is to be taken 10 days after she announces her position.
 a. Sketch the graph of $S(t)$ for $0 \leq t \leq 10$.
 b. When is her support at its lowest level? What is her minimum support level?
 c. The derivative $S'(t)$ may be thought of as an approval rate. Is her approval rate positive or negative when the vote is taken? Is the approval rate increasing or decreasing at this time? Interpret your results.

58. **ADVERTISING**   A manufacturer of motorcycles estimates that if x thousand dollars are spent on advertising, then for $x > 0$,

$$M(x) = 2300 + \frac{125}{x} - \frac{500}{x^2}$$

motorcycles will be sold.
 a. Sketch the graph of the sales function $M(x)$.
 b. What level of advertising expenditure results in maximum sales? What is the maximum sales level?

59. COST MANAGEMENT   A company uses a truck to deliver its products. To estimate costs, the manager models gas consumption by the function

$$G(x) = \frac{1}{160}\left(\frac{256}{x} + \frac{5}{8}x\right)$$

litres per kilometre, assuming that the truck is driven at a constant speed of x kilometres per hour for $x \geq 5$. The driver is paid \$35 per hour to drive the truck 600 km, and gasoline costs \$1.10 per litre. Highway regulations require 50 km/h $\leq x \leq$ 100 km/h.

a. Find an expression for the total cost $C(x)$ of the trip. Sketch the graph of $C(x)$ for the legal speed interval 50 km/h $\leq x \leq$ 100 km/h.

b. What legal speed will minimize the total cost of the trip? What is the minimal total cost?

60. Let $f(x) = x^{1/3}(x - 4)$.

a. Find $f'(x)$ and determine the intervals of increase and decrease for $f(x)$. Locate all relative extrema on the graph of $f(x)$.

b. Find $f''(x)$ and determine the intervals of concavity for $f(x)$. Find all inflection points on the graph of $f(x)$.

c. Find all intercepts for the graph of $f(x)$. Does the graph have any asymptotes?

d. Sketch the graph of $f(x)$.

61. Repeat Exercise 60 for the function

$$f(x) = x^{2/3}(2x - 5)$$

62. Repeat Exercise 60 for the function

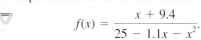

$$f(x) = \frac{x + 9.4}{25 - 1.1x - x^2}.$$

63. Let $f(x) = \dfrac{x - 1}{x^2 - 1}$ and $g(x) = \dfrac{x - 1.01}{x^2 - 1}$.

a. Use a graphing utility to sketch the graph of $f(x)$. What happens at $x = 1$?

b. Sketch the graph of $g(x)$. Now what happens at $x = 1$?

64. **A*** Find constants A, B, and C so that the function $f(x) = Ax^3 + Bx^2 + C$ has a relative extremum at $(2, 11)$ and an inflection point at $(1, 5)$. Sketch the graph of f.

SECTION 3.4	# Optimization; Elasticity of Demand

L04

Calculate the absolute extrema of a function. Apply these methods to solve problems in elasticity of demand.

You have already seen several situations where the methods of calculus were used to determine the largest or smallest value of a function of interest (for example, maximum profit or minimum cost). In most such optimization problems, the goal is to find the absolute maximum or absolute minimum of a particular function on some relevant interval. The absolute maximum of a function on an interval is the largest value of the function on that interval, and the absolute minimum is the smallest value. Here is a definition of absolute extrema.

> **Absolute Maxima and Minima of a Function** ■ Let f be a function defined on an interval I that contains the number c. Then,
>
> $f(c)$ is the *absolute maximum* of f on I if $f(c) \geq f(x)$ for all x in I
>
> $f(c)$ is the *absolute minimum* of f on I if $f(c) \leq f(x)$ for all x in I
>
> Collectively, absolute maxima and minima are called *absolute extrema.*

Absolute extrema often, but not always, coincide with relative extrema. For example, in Figure 3.34 the absolute maximum and relative maximum on the interval $a \leq x \leq b$ are the same, but the absolute minimum occurs at the left endpoint, $x = a$.

(It is important to note that infinity is not considered an absolute maximum and negative infinity is not considered an absolute minimum.)

In this section, you will learn how to find absolute extrema of functions on intervals. We begin by considering intervals $a \leq x \leq b$ that are "closed," which means that they include both endpoints, a and b. It can be shown that a function that is continuous on such an interval has both an absolute maximum and an absolute minimum on the interval. Moreover, each absolute extremum must occur either at an

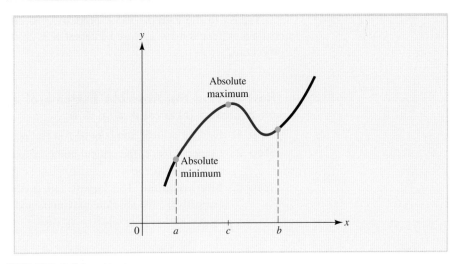

FIGURE 3.34   Absolute extrema.

endpoint of the interval (at a or b) or at a critical number c between a and b (Figure 3.35). To summarize:

The Extreme Value Property ■ A function $f(x)$ that is continuous on the closed interval $a \leq x \leq b$ attains its absolute extrema on the interval either at an endpoint of the interval (a or b) or at a critical number c such that $a < c < b$.

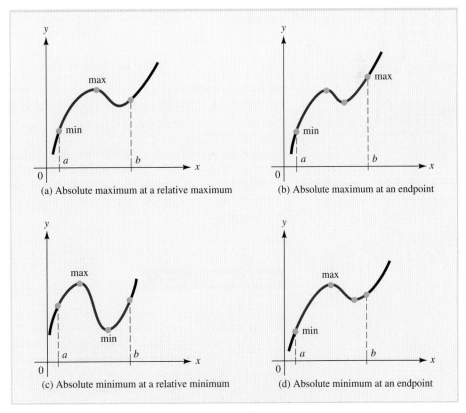

FIGURE 3.35   Absolute extrema of a continuous function on $a \leq x \leq b$.

Thanks to the extreme value property, you can find the absolute extrema of a continuous function on a closed interval $a \le x \le b$ by using this straightforward procedure.

How to Find the Absolute Extrema of a Continuous Function f on a Closed Interval $a \le x \le b$

Step 1. Find all critical numbers of f in the open interval $a < x < b$.

Step 2. Compute $f(x)$ at the critical numbers found in step 1 and at the endpoints $x = a$ and $x = b$.

Step 3. **Interpretation:** The largest and smallest values found in step 2 are, respectively, the absolute maximum and absolute minimum values of $f(x)$ on the closed interval $a \le x \le b$.

The procedure is illustrated in Examples 3.4.1 through 3.4.3.

EXAMPLE 3.4.1

Find the absolute maximum and absolute minimum of the function

$$f(x) = 2x^3 + 3x^2 - 12x - 7$$

on the interval $-3 \le x \le 0$.

Solution

From the derivative

$$f'(x) = 6x^2 + 6x - 12 = 6(x + 2)(x - 1)$$

the critical numbers are $x = -2$ and $x = 1$. Of these, only $x = -2$ lies in the interval $-3 \le x \le 0$. Compute $f(x)$ at $x = -2$ and at the endpoints $x = -3$ and $x = 0$.

$$f(-2) = 13 \quad f(-3) = 2 \quad f(0) = -7$$

Compare these values to conclude that the absolute maximum of f on the interval $-3 \le x \le 0$ is $f(-2) = 13$ and the absolute minimum is $f(0) = -7$.

Notice that we did not have to classify the critical points or draw the graph to locate the absolute extrema. The sketch in Figure 3.36 is presented only for the sake of illustration.

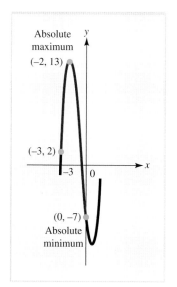

FIGURE 3.36 The absolute extrema on $-3 \le x \le 0$ for $y = 2x^3 + 3x^2 - 12x - 7$.

EXAMPLE 3.4.2

For several weeks, the highway department has been recording the speed of traffic flowing past a certain Toronto street corner. The data suggest that between 1:00 P.M. and 6:00 P.M. on a normal weekday, the speed of the traffic is approximately $V(t) = t^3 - 10.5t^2 + 30t + 20$ kilometres per hour, where t is the number of hours past noon. At what time between 1:00 P.M. and 6:00 P.M. is the traffic moving the fastest, and at what time is it moving the slowest?

EXPLORE!

Refer to Example 3.4.2.
Because of a change to make
a one-way street, the speed
past the same corner is now
given by

$V_1(t) = t^3 - 10.5t^2 + 30t + 25.$

Graph $V(t)$ and $V_1(t)$ from 1 P.M.
to 6 P.M. At what time between
1 P.M. and 6 P.M. is the
maximum speed achieved
using $V_1(t)$? At what time is the
minimum speed achieved?

Solution

The goal is to find the absolute maximum and absolute minimum of the function $V(t)$ on the interval $1 \le t \le 6$. From the derivative

$$V'(t) = 3t^2 - 21t + 30 = 3(t^2 - 7t + 10) = 3(t - 2)(t - 5),$$

we get the critical numbers $t = 2$ and $t = 5$, both of which lie in the interval $1 \le t \le 6$.

Compute $V(t)$ for these values of t and at the endpoints $t = 1$ and $t = 6$ to get

$$V(1) = 40.5 \qquad V(2) = 46 \qquad V(5) = 32.5 \qquad V(6) = 38$$

Since the largest of these values is $V(2) = 46$ and the smallest is $V(5) = 32.5$, the traffic is moving fastest at 2:00 P.M., when its speed is 46 km/h, and slowest at 5:00 P.M., when its speed is 32.5 km/h. For reference, the graph of V is sketched in Figure 3.37.

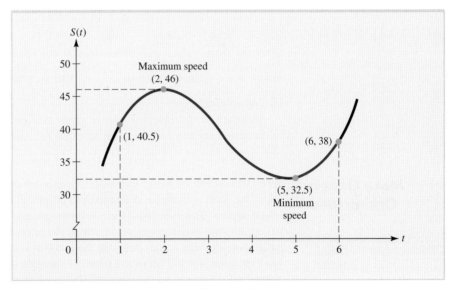

FIGURE 3.37   Traffic speed $V(t) = t^3 - 10.5t^2 + 30t + 20$.

EXAMPLE 3.4.3

When you cough, the radius of your trachea (windpipe) decreases, affecting the speed of the air in the trachea. If r_0 is the normal radius of the trachea, the relationship between the speed V of the air and the radius r of the trachea during a cough is given by a function of the form $V(r) = ar^2(r_0 - r)$, where a is a positive constant.* Find the radius r for which the speed of the air is greatest.

---

*Philip M. Tuchinsky, "The Human Cough," *UMAP Modules 1976: Tools for Teaching,* Lexington, MA: Consortium for Mathematics and Its Application, Inc., 1977.

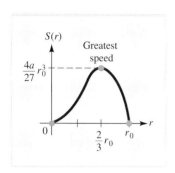

FIGURE 3.38 The speed of air during a cough,

$$V(r) = ar^2(r_0 - r).$$

Solution

The radius r of the contracted trachea cannot be greater than the normal radius r_0 or less than zero. Hence, the goal is to find the absolute maximum of $V(r)$ on the interval $0 \le r \le r_0$.

First differentiate $V(r)$ with respect to r using the product rule and factor the derivative as follows (note that a and r_0 are constants):

$$V'(r) = -ar^2 + (r_0 - r)(2ar) = ar[-r + 2(r_0 - r)] = ar(2r_0 - 3r).$$

Then set the factored derivative equal to zero and solve to get the critical numbers:

$$ar(2r_0 - 3r) = 0$$

$$r = 0 \quad \text{or} \quad r = \frac{2}{3}r_0$$

Both of these values of r lie in the interval $0 \le r \le r_0$, and one is actually an endpoint of the interval. Compute $V(r)$ for these two values of r and for the other endpoint $r = r_0$ to get

$$V(0) = 0 \quad V\left(\frac{2}{3}r_0\right) = \frac{4a}{27}r_0^3 \quad V(r_0) = 0$$

Compare these values and conclude that the speed of the air is greatest when the radius of the contracted trachea is $\frac{2}{3}r_0$, that is, when it is two-thirds the radius of the uncontracted trachea.

A graph of the function $V(r)$ is given in Figure 3.38. Note that the r intercepts of the graph are obvious from the factored function $V(r) = ar^2(r_0 - r)$.

More General Optimization

When the interval on which you wish to maximize or minimize a continuous function is not of the form $a \le x \le b$, the procedure illustrated in Examples 3.4.1 through 3.4.3 no longer applies. This is because there is no longer any guarantee that the function actually has an absolute maximum or minimum on the interval in question. On the other hand, if an absolute extremum does exist and the function is continuous on the interval, the absolute extremum will still occur at a relative extremum or endpoint contained in the interval. Several possibilities for functions on unbounded intervals are illustrated in Figure 3.39.

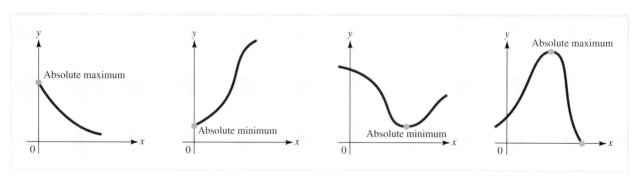

FIGURE 3.39 Extrema for functions defined on unbounded intervals.

To find the absolute extrema of a continuous function on an interval that is not of the form $a \le x \le b$, you still evaluate the function at all the critical points and endpoints that are contained in the interval. However, before you can draw any final conclusions, you must find out if the function actually has relative extrema on the interval. One way to do this is to use the first derivative to determine where the function is increasing and where it is decreasing and then to sketch the graph. The technique is illustrated in Example 3.4.4.

EXAMPLE 3.4.4

If they exist, find the absolute maximum and absolute minimum of the function $f(x) = x^2 + \dfrac{16}{x}$ on the interval $x > 0$.

Solution
The function is continuous on the interval $x > 0$ since its only discontinuity occurs at $x = 0$. The derivative is

$$f'(x) = 2x - \frac{16}{x^2} = \frac{2x^3 - 16}{x^2} = \frac{2(x^3 - 8)}{x^2}$$

which is zero when

$$x^3 - 8 = 0$$
$$x^3 = 8$$
$$x = 2$$

Since $f'(x) < 0$ for $0 < x < 2$ and $f'(x) > 0$ for $x > 2$, the graph of f is decreasing for $0 < x < 2$ and increasing for $x > 2$, as shown in Figure 3.40. It follows that

$$f(2) = 2^2 + \frac{16}{2} = 12$$

is the absolute minimum of f on the interval $x > 0$ and that there is no absolute maximum.

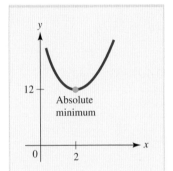

FIGURE 3.40 The function $f(x) = x^2 + \dfrac{16}{x}$ on the interval $x > 0$.

The procedure illustrated in Example 3.4.4 can be used whenever we wish to find the largest or smallest value of a function f that is continuous on an interval I on which it has *exactly one* critical number c. In particular, if this condition is satisfied and $f(x)$ has a *relative* maximum (minimum) at $x = c$, it also has an *absolute* maximum (minimum) there. To see why, suppose the graph has a relative minimum at $x = c$. Then the graph is always falling before c and always rising after c, since to change direction would require the presence of a second critical point (Figure 3.41). Thus, the relative minimum is also the absolute minimum. These observations suggest that any test for relative extrema becomes a test for absolute extrema in this special case. Here is a statement of the second derivative test for absolute extrema.

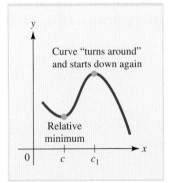

FIGURE 3.41 The relative minimum is not the absolute minimum because of the effect of another critical point.

The Second Derivative Test for Absolute Extrema ■ Suppose that $f(x)$ is continuous on an interval I where $x = c$ is the only critical number and that $f'(c) = 0$. Then,

if $f''(c) > 0$, the absolute minimum of $f(x)$ on I is $f(c)$

if $f''(c) < 0$, the absolute maximum of $f(x)$ on I is $f(c)$

Example 3.4.5 illustrates how the second derivative test for absolute extrema can be used in practice.

(Photo: PhotoDisc/Alamy)

EXAMPLE 3.4.5

A manufacturer estimates that when q thousand bird feeders are produced each month, the total cost will be $C(q) = 0.4q^2 + 3q + 40$ thousand dollars, and all q bird feeders can be sold at a price of $p(q) = 22.2 - 1.2q$ dollars per unit.

a. Determine the level of production that results in maximum profit. What is the maximum profit?

b. At what level of production is the average cost per bird feeder $A(q) = \dfrac{C(q)}{q}$ minimized?

c. At what level of production is the average cost equal to the marginal cost $C'(q)$?

Solution

a. The revenue is

$$R(q) = qp(q) = q(22.2 - 1.2q) = -1.2q^2 + 22.2q$$

thousand dollars, so the profit is

$$\begin{aligned}P(q) = R(q) - C(q) &= -1.2q^2 + 22.2q - (0.4q^2 + 3q + 40)\\ &= -1.6q^2 + 19.2q - 40\end{aligned}$$

thousand dollars. We have

$$P'(q) = -3.2q + 19.2 = 0$$

when

$$-3.2q + 19.2 = 0$$
$$q = \frac{19.2}{3.2} = 6$$

Since $P''(q) = -3.2$, it follows that $P''(6) < 0$, and the second derivative test tells us that the maximum profit occurs when $q = 6$ (thousand) bird feeders are produced. The maximum profit is

$$\begin{aligned}P(6) &= -1.6(6)^2 + 19.2(6) - 40\\ &= 17.6\end{aligned}$$

thousand dollars ($17 600). The graph of the profit function is shown in Figure 3.42a.

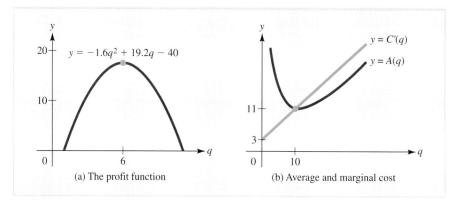

(a) The profit function   (b) Average and marginal cost

FIGURE 3.42   Graphs of profit, average cost, and marginal cost for Example 3.4.5.

b. The average cost is

$$A(q) = \frac{C(q)}{q} = \frac{0.4q^2 + 3q + 40}{q} \quad \frac{\text{thousand dollars}}{\text{thousand units}}$$

$$= 0.4q + 3 + \frac{40}{q} \quad \frac{\text{dollars}}{\text{unit}}$$

for $q > 0$ (the level of production cannot be negative or zero). We find

$$A'(q) = 0.4 - \frac{40}{q^2} = \frac{0.4q^2 - 40}{q^2},$$

which is 0 for $q > 0$ only when $q = 10$. Since

$$A''(q) = \frac{80}{q^3} > 0 \quad \text{when } q > 0,$$

it follows from the second derivative test for absolute extrema that the average cost $A(q)$ is minimized when $q = 10$ (thousand) units. The minimal average cost is

$$A(10) = 0.4(10) + 3 + \frac{40}{10} = 11 \quad \frac{\text{dollars}}{\text{unit}}.$$

c. The marginal cost is $C'(q) = 0.8q + 3$, and it equals the average cost when

$$0.8q + 3 = 0.4q + 3 + \frac{40}{q}$$

$$0.4q = \frac{40}{q}$$

$$0.4q^2 = 40$$

$$q = 10 \text{ (thousand) units}$$

which equals the optimal level of production in part (b). The graphs of the marginal cost $C'(q)$ and average cost $A(q) = \dfrac{C(q)}{q}$ are shown in Figure 3.42b.

Two General Principles of Marginal Analysis

If the revenue derived from the sale of q units is $R(q)$ and the cost of producing those units is $C(q)$, then the profit is $P(q) = R(q) - C(q)$. Since

$$P'(q) = [R(q) - C(q)]' = R'(q) - C'(q),$$

it follows that $P'(q) = 0$ when $R'(q) = C'(q)$. If it is also true that $P''(q) < 0$ or, equivalently, that $R''(q) < C''(q)$, then the profit will be maximized.

Marginal Analysis Criterion for Maximum Profit ■ The profit $P(q) = R(q) - C(q)$ is maximized at a level of production q where marginal revenue equals marginal cost and the rate of change of marginal cost exceeds the rate of change of marginal revenue, that is, where

$$R'(q) = C'(q) \quad \text{and} \quad R''(q) < C''(q)$$

For instance, in Example 3.4.5, the revenue is $R(q) = -1.2q^2 + 22.2q$ and the cost is $C(q) = 0.4q^2 + 3q + 40$, so the marginal revenue is $R'(q) = -2.4q + 22.2$ and the marginal cost is $C'(q) = 0.8q + 3$. Thus, marginal revenue equals marginal cost when

$$R'(q) = C'(q)$$
$$-2.4q + 22.2 = 0.8q + 3$$
$$3.2q = 19.2$$
$$q = 6$$

which is the level of production for maximum profit found in part (a) of Example 3.4.5. Note that $R'' < C''$ is also satisfied since $R'' = -2.4$ and $C'' = 0.8$.

In part (c) of Example 3.4.5, you found that marginal cost equals average cost at the level of production where average cost is minimized. This, too, is no accident. To see why, let $C(q)$ be the cost of producing q units of a commodity. Then the average cost per unit is $A(q) = \dfrac{C(q)}{q}$, and applying the quotient rule gives

$$A'(q) = \frac{qC'(q) - C(q)}{q^2}$$

Thus, $A'(q) = 0$ when the numerator on the right is zero, that is, when

$$qC'(q) = C(q)$$

or, equivalently, when the marginal cost $C'(q) = \dfrac{C(q)}{q}$. But $\dfrac{C(q)}{q} = A(q)$, which is the average cost.

To show that the average cost is actually minimized where average cost equals marginal cost, it is necessary to make a few reasonable assumptions about total cost (see Exercise 56).

Marginal Analysis Criterion for Minimal Average Cost ■ Average cost is minimized at the level of production where average cost equals marginal cost, that is, when $A(q) = C'(q)$.

Here is an informal explanation of the relationship between average and marginal cost that is often given in economics texts. The marginal cost (MC) is approximately the same as the cost of producing one additional unit. If the additional unit costs less to produce than the average cost (AC) of the existing units (if MC < AC), then this less-expensive unit will cause the average cost per unit to decrease. On the other hand, if the additional unit costs more than the average cost of the existing units (if MC > AC), then this more-expensive unit will cause the average cost per unit to increase. However, if the cost of the additional unit is equal to the average cost of the existing units (if MC = AC), then the average cost will neither increase nor decrease, which means (AC)' = 0.

The relationship between average cost and marginal cost can be generalized to apply to any pair of average and marginal quantities. The only possible modification involves the nature of the critical point that occurs when the average quantity equals the marginal quantity. For example, average revenue usually has a relative *maximum* (instead of a minimum) when average revenue equals marginal revenue.

Price Elasticity of Demand

In Section 1.1, we introduced the economic concept of demand as a means for expressing the relationship between the unit price p of a commodity and the number of units q that will be demanded (that is, produced and purchased) by consumers at that price. Previously, we have expressed unit price p as a function of production level q, but for the present discussion it is more convenient to turn things around and write the demand relationship as $q = D(p)$. When we turn things around in this way we are finding the inverse function.

In general, an increase in the unit price of a commodity will result in decreased demand, but the sensitivity or responsiveness of demand to a change in price varies from one product to another. For instance, the demand for products such as soap, flashlight batteries, and salt will not be much affected by a small percentage change in unit price, while a comparable percentage change in the price of airline tickets or home loans can affect demand dramatically.

Sensitivity of demand is commonly measured by the ratio of the percentage rate of change in quantity demanded to the percentage rate of change in price. This is approximately the same as the change in demand produced by a 1% change in unit price. Recall from Section 2.2 that the percentage rate of change of a quantity $Q(x)$ is given by $\dfrac{Q'(x)}{Q(x)}$, which is a fraction that can be expressed as a percentage by multiplying by 100. In particular, if the demand function $q = D(p)$ is differentiable, then

$$\text{Percentage rate of change of demand } q = \frac{\dfrac{dq}{dp}}{q}$$

and

$$\text{Percentage rate of change of price } p = \frac{\dfrac{dp}{dp}}{p} = \frac{1}{p}$$

Thus, sensitivity to change in price is measured by the ratio

$$\frac{\text{Percentage rate of change in } q}{\text{Percentage rate of change in } p} = \frac{\left(\dfrac{dq}{dp}\right)/q}{1/p} = \frac{p}{q}\frac{dp}{dq}$$

which, in economics, is called the *price elasticity of demand*. To summarize:

Price Elasticity of Demand ■ If $q = D(p)$ units of a commodity are demanded by the market at a unit price p, where D is a differentiable function, then the **price elasticity of demand** for the commodity is given by

$$E(p) = \frac{p}{q}\frac{dq}{dp}$$

and has the interpretation

$$E(p) \approx \left[\begin{array}{l}\text{percentage rate of change in demand } q \\ \text{produced by a 1\% rate of change in price } p\end{array}\right].$$

NOTE Since demand q decreases as unit price p increases, we have $\dfrac{dq}{dp} < 0$. Therefore, since $q > 0$ and $p > 0$, it follows that the price elasticity of demand will be negative, that is,

$$E(p) = \frac{p}{q}\frac{dq}{dp} < 0.$$

For instance, to say that a particular commodity has a price elasticity of demand of -0.5 at a certain unit price p means that a 10% rise in price for the commodity will result in a decline of approximately 5% in the number of units demanded (sold). These ideas are illustrated further in Example 3.4.6.

EXAMPLE 3.4.6

Suppose the demand q and price p for a certain commodity are related by the linear equation $q = 240 - 2p$ (for $0 \le p \le 120$).

a. Express the elasticity of demand as a function of p.

b. Calculate the elasticity of demand when the price is $p = 100$. Interpret your answer.

c. Calculate the elasticity of demand when the price is $p = 50$. Interpret your answer.

d. At what price is the elasticity of demand equal to -1? What is the economic significance of this price?

Solution

a. The elasticity of demand is

$$E(p) = \frac{p}{q}\frac{dq}{dp} = \frac{p}{q}(-2) = \frac{-2p}{240 - 2p} = \frac{-p}{120 - p}$$

b. When $p = 100$, the elasticity of demand is

$$E(100) = \frac{-100}{120 - 100} = -5$$

That is, when the price is $p = 100$, a 1% increase in price will produce a decrease in demand of approximately 5%.

c. When $p = 50$, the elasticity of demand is

$$E(50) = \frac{-50}{120 - 50} \approx -0.71$$

That is, when the price is $p = 50$, a 1% increase in price will produce a decrease in demand of approximately 0.71%.

d. The elasticity of demand will be equal to -1 when

$$-1 = \frac{-p}{120 - p} \qquad 120 - p = p \qquad 2p = 120 \qquad p = 60$$

At this price, a 1% increase in price will result in a decrease in demand of approximately the same percentage.

There are three levels of elasticity, depending on whether $|E(p)|$ is greater than, less than, or equal to 1. Here is a description and economic interpretation of each level.

Levels of Elasticity

$|E(p)| > 1$ **Elastic demand.** The percentage decrease in demand is greater than the percentage increase in price that caused it. Thus, demand is relatively sensitive to changes in price.

$|E(p)| < 1$ **Inelastic demand.** The percentage decrease in demand is less than the percentage increase in price that caused it. When this occurs, demand is relatively insensitive to changes in price.

$|E(p)| = 1$ **Demand is of unit elasticity (or unitary).** The percentage changes in price and demand are (approximately) equal.

For instance, in Example 3.4.6b, we found that $E(100) = -5$. Thus, the demand is elastic with respect to price when $p = 100$ since $|E(100)| = 5 > 1$. In part (c) of the same example, we found that $|E(50)| = |-0.71| < 1$, so the demand is inelastic when $p = 50$. Finally, in part (d), we found $|E(60)| = |-1| = 1$, which means that the demand is of unit elasticity when $p = 60$.

The level of elasticity of demand for a commodity gives useful information about the total revenue R obtained from the sale of q units of the commodity at p dollars per unit. Assuming that the demand q is a differentiable function of unit price p, the revenue is $R(p) = pq(p)$ and, by differentiating implicitly with respect to p, we find that

$$\frac{dR}{dp} = p\frac{dq}{dp} + q \quad \text{by the product rule.}$$

To get the elasticity $E(p) = \dfrac{p}{q}\dfrac{dq}{dp}$ into the picture, simply multiply the expression on the right-hand side by $\dfrac{q}{q}$ as follows:

$$\frac{dR}{dp} = \frac{q}{q}\left(p\frac{dq}{dp} + q\right) = q\left(\frac{p}{q}\frac{dq}{dp} + 1\right) = q[E(p) + 1]$$

Suppose demand is elastic so that $|E(p)| > 1$. It follows that $E(p) < -1$ (since $E(p) < 0$) so that $E(p) + 1 < 0$ and

$$\frac{dR}{dp} = q(p)[E(p) + 1] < 0$$

which means that the result of a small increase in price will be to decrease revenue. Similarly, when demand is inelastic, we have $-1 < E(p) < 0$ so that $E(p) + 1 > 0$ and $\dfrac{dR}{dp} > 0$. In this case, a small increase in price results in increased revenue. If the demand is of unit elasticity, then $E(p) = -1$, so $\dfrac{dR}{dp} = 0$, and a small increase in price leaves revenue approximately unchanged (neither increased nor decreased). These observations are summarized as follows:

Levels of Elasticity and the Effect on Revenue

If demand is **elastic** ($|E(p)| > 1$), revenue $R = pq(p)$ decreases as price p increases.

If demand is **inelastic** ($|E(p)| < 1$), revenue R increases as price p increases.

If demand is of **unit elasticity** ($|E(p)| = 1$), revenue is unaffected by a small change in price.

The relationship between revenue and price is shown in Figure 3.43. Note that the revenue curve is rising where demand is inelastic, is falling where demand is elastic, and has a horizontal tangent line where the demand is of unit elasticity.

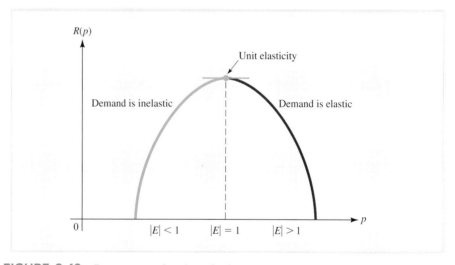

FIGURE 3.43   Revenue as a function of price.

The relationship between elasticity of demand and total revenue is illustrated in Example 3.4.7.

EXAMPLE 3.4.7

The manager of a bookstore determines that when a certain new paperback novel is priced at p dollars per copy, the daily demand will be $q = 300 - p^2$ copies, where $0 \le p \le \sqrt{300}$.

a. Determine where the demand is elastic, inelastic, and of unit elasticity with respect to price.

b. Interpret the results of part (a) in terms of the behaviour of total revenue as a function of price.

Solution

a. The elasticity of demand is

$$E(p) = \frac{p}{q}\frac{dq}{dp} = \frac{p}{300 - p^2}(-2p) = \frac{-2p^2}{300 - p^2}$$

and, since $0 \le p \le \sqrt{300}$,

$$|E(p)| = \frac{2p^2}{300 - p^2}$$

The demand is of unit elasticity when $|E| = 1$, that is, when

$$\frac{2p^2}{300 - p^2} = 1$$
$$2p^2 = 300 - p^2$$
$$3p^2 = 300$$
$$p = \pm 10$$

of which only $p = 10$ is in the relevant interval $0 \le p \le \sqrt{300}$. If $0 \le p < 10$, then

$$|E| = \frac{2p^2}{300 - p^2} < \frac{2(10)^2}{300 - (10)^2} = 1$$

so the demand is inelastic. Likewise, if $10 < p < \sqrt{300}$, then

$$|E| = \frac{2p^2}{300 - p^2} > \frac{2(10)^2}{300 - (10)^2} = 1$$

and the demand is elastic.

b. The total revenue $R = pq$ increases when demand is inelastic, that is, when $0 \le p < 10$. For this range of prices, a specified percentage increase in price results in a smaller percentage decrease in demand, so the bookstore will take in more money for each increase in price up to $10 per copy.

However, for the price range $10 < p \le \sqrt{300}$, the demand is elastic, so the revenue is decreasing. If the book is priced in this range, a specified percentage increase in price results in a larger percentage decrease in demand. Thus, if the bookstore increases the price beyond $10 per copy, it will lose revenue.

This means that the optimal price is $10 per copy, which corresponds to unit elasticity. The graphs of the demand and revenue functions are shown in Figure 3.44.

EXPLORE!

Suppose the demand/price equation in Example 3.4.7 is $q = 300 - ap^2$. Describe how the price is affected when the demand is of unit elasticity for $a = 0, 1, 3$, or 5 by examining the graph of $x(300 - ax^2)$ for these values of a.

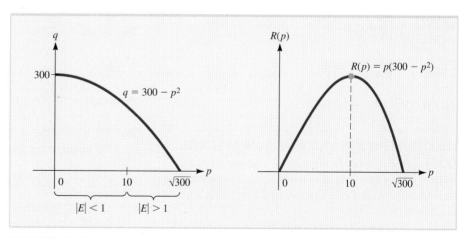

FIGURE 3.44   Demand and revenue curves for Example 3.4.7.

OPTION NOTE: Example 4.3.12 is an elasticity of demand question using an exponential function. Section 4.3 Exercises 39 to 46 are exponential and logarithmic absolute maxima and minima questions.

EXERCISES ■ 3.4

In Exercises 1 through 16, find the absolute maximum and absolute minimum (if any) of the given function on the specified interval.

1. $f(x) = x^2 + 4x + 5; -3 \le x \le 1$

2. $f(x) = x^3 + 3x^2 + 1; -3 \le x \le 2$

3. $f(x) = \frac{1}{3}x^2 - 9x + 2; 0 \le x \le 2$

4. $f(x) = x^5 - 5x^4 + 1; 0 \le x \le 5$

5. $f(t) = 3t^5 - 5t^3; -2 \le t \le 0$

6. $f(x) = 10x^6 + 24x^5 + 15x^4 + 3; -1 \le x \le 1$

7. $f(x) = (x^2 - 4)^5; -3 \le x \le 2$

8. $f(t) = \frac{t^2}{t-1}; -2 \le t \le -\frac{1}{2}$

9. $g(x) = x + \frac{1}{x}; \frac{1}{2} \le x \le 3$

10. $g(x) = \frac{1}{x^2 - 9}; 0 \le x \le 2$

11. $f(u) = u + \frac{1}{u}; u > 0$

12. $f(u) = 2u + \frac{32}{u}; u > 0$

13. $f(x) = \frac{1}{x}; x > 0$

14. $f(x) = \frac{1}{x^2}; x > 0$

15. $f(x) = \frac{1}{x+1}; x \ge 0$

16. $f(x) = \frac{1}{(x+1)^2}; x \ge 0$

MAXIMUM PROFIT AND MINIMUM AVERAGE COST
In Exercises 17 through 22, you are given the price $p(q)$ at which q units of a particular commodity can be sold and the total cost $C(q)$ of producing the q units. In each case:

(a) *Find the revenue function $R(q)$, the profit function $P(q)$, the marginal revenue $R'(q)$, and the marginal cost $C'(q)$. Sketch the graphs of $P(q)$, $R'(q)$, and $C'(q)$ on the same coordinate axes*

and determine the level of production q where $P(q)$ is maximized.

(b) *Find the average cost $A(q) = C(q)/q$ and sketch the graphs of $A(q)$ and the marginal cost $C'(q)$ on the same axes. Determine the level of production q at which $A(q)$ is minimized.*

17. $p(q) = 49 - q; C(q) = \frac{1}{8}q^2 + 4q + 200$

18. $p(q) = 37 - 2q; C(q) = 3q^2 + 5q + 75$

19. $p(q) = 180 - 2q; C(q) = q^3 + 5q + 162$

20. $p(q) = 710 - 1.1q^2;$
    $C(q) = 2q^3 - 23q^2 + 90.7q + 151$

21. $p(q) = 1.0625 - 0.0025q; C(q) = \frac{q^2 + 1}{q + 3}$

22. $p(q) = 81 - 3q; C(q) = \frac{q + 1}{q + 3}$

ELASTICITY OF DEMAND
In Exercises 23 through 28, compute the elasticity of demand for the given demand function $D(p)$ and determine whether the demand is elastic, inelastic, or of unit elasticity at the indicated price p.

23. $D(p) = -1.3p + 10; p = 4$

24. $D(p) = -1.5p + 25; p = 12$

25. $D(p) = 200 - p^2; p = 10$

26. $D(p) = \sqrt{400 - 0.01p^2}; p = 120$

27. $D(p) = \frac{3000}{p} - 100; p = 10$

28. $D(p) = \frac{2000}{p^2} - 100; p = 5$

29. For what value of x in the interval $-1 \le x \le 4$ is the graph of the function

$$f(x) = 2x^2 - \frac{1}{3}x^3$$

steepest? What is the slope of the tangent at this point?

30. At what point does the tangent to the curve $y = 2x^3 - 3x^2 + 6x$ have the smallest slope? What is the slope of the tangent at this point?

In Exercises 31 through 38, solve the practical optimization problem and use one of the techniques from this section to verify that you have actually found the desired absolute extremum.

31. **AVERAGE PROFIT** A manufacturer estimates that when q units of a certain commodity are produced, the profit obtained is $P(q)$ thousand dollars, where

$$P(q) = -2q^2 + 68q - 128.$$

   a. Find the average profit and the marginal profit functions.

   b. At what level of production $\bar{q}$ is average profit equal to marginal profit?

   c. Show that average profit is maximized at the level of production $\bar{q}$ found in part (b).

   d. On the same set of axes, graph the relevant portions of the average and marginal profit functions.

32. **MARGINAL ANALYSIS** A manufacturer estimates that if x units of a particular commodity are produced, the total cost will be $C(x)$ dollars, where

$$C(x) = x^3 - 24x^2 + 350x + 338.$$

   At what level of production will the marginal cost $C'(x)$ be minimized?

33. **GROUP MEMBERSHIP** A national consumers' association determines that x years after its founding in 1995, it will have $P(x)$ members, where

$$P(x) = 100(2x^3 - 45x^2 + 264x).$$

   a. At what time between 1997 and 2010 was the membership largest? Smallest?

   b. What were the largest and smallest membership levels between 1997 and 2010?

34. **BROADCASTING** An all-news radio station has surveyed the listening habits of adults driving in their vehicles between the hours of 5:00 P.M. and midnight. The survey indicates that the percentage of the drivers that are tuned in to the station x hours after 5:00 P.M. is

$$f(x) = \frac{1}{8}(-2x^3 + 27x^2 - 108x + 240).$$

   a. At what time between 5:00 P.M. and midnight is the largest percentage listening to the station? What percentage of drivers are listening at this time?

   b. At what time between 5:00 P.M. and midnight is the lowest percentage listening? What percentage of drivers are listening at this time?

35. **DISPLACEMENT** A skateboarder moves in a straight line so that his position is $s(t) = \dfrac{5}{1 + t^2}$ metres, where t is measured in seconds. At $t = 0$ he drops his ice cream cone. Find the maximum displacement on the interval from 1 second before he dropped his ice cream cone to 2 seconds afterward.

36. **GROWTH OF A SPECIES** The percentage of codling moths* that survive the pupa stage at a given temperature T (degrees Celsius) is modelled by the formula

$$P(T) = -1.42T^2 + 68T - 746$$
$$\text{for } 20 \le T \le 30.$$

   Find the temperatures at which the greatest and smallest percentages of moths survive.

37. **BLOOD CIRCULATION** Poiseuille's law asserts that the speed of blood that is r centimetres from the central axis of an artery of radius R is $V(r) = c(R^2 - r^2)$, where c is a positive constant. Where is the speed of the blood greatest?

38. **POLITICS** A poll indicates that x months after a particular candidate for public office declares her candidacy, she will have the support of $S(x)$ percent of the voters, where

$$S(x) = \frac{1}{29}(-x^3 + 6x^2 + 63x + 1080)$$
$$\text{for } 0 \le x \le 12.$$

   If the election is held in November, when should the politician announce her candidacy? Should she expect to win if she needs at least 50% of the vote?

39. **ELASTICITY OF DEMAND** When a particular commodity is priced at p dollars per unit, consumers demand q units, where p and q are related by the equation $q^2 + 3pq = 22$.

   a. Find the elasticity of demand for this commodity.

   b. For a unit price of \$3, is the demand elastic, inelastic, or of unit elasticity?

---

*For more on codling moths, see Think About It at the end of Chapter 2.

40. **ELASTICITY OF DEMAND** When an electronics store prices a certain brand of DVD player at p hundred dollars, it is found that q DVD players will be sold each month, where $q^2 + 2p^2 = 41$.
    a. Find the elasticity of demand for the DVD players.
    b. For a unit price of $p = 4$ ($400), is the demand elastic, inelastic, or of unit elasticity?

41. **DEMAND FOR ART** An art gallery offers 50 prints by a famous artist. If each print in the limited edition is priced at p dollars, it is expected that $q = 500 - 2p$ prints will be sold.
    a. What limitations are there on the possible range of the price p?
    b. Find the elasticity of demand. Determine the values of p for which the demand is elastic, inelastic, and of unit elasticity.
    c. Interpret the results of part (b) in terms of the behaviour of the total revenue as a function of unit price p.
    d. If you were the owner of the gallery, what price would you charge for each print? Explain your decision.

42. **DEMAND FOR AIRLINE TICKETS** An airline determines that when a one-way ticket between Toronto and Ottawa costs p dollars ($0 \leq p \leq 160$), the daily demand for tickets is $q = 256 - 0.01p^2$.
    a. Find the elasticity of demand. Determine the values of p for which the demand is elastic, inelastic, and of unit elasticity.
    b. Interpret the results of part (a) in terms of the behaviour of the total revenue as a function of unit price p.
    c. What price would you advise the airline to charge for each ticket? Explain your reasoning.

43. **A* PRODUCTION CONTROL** A toy manufacturer produces a basic dump truck and an expensive electronic dump truck with sounds and lights, in units of x hundreds and y hundreds, respectively. Suppose that it is possible to produce the dump trucks in such a way that $y = \dfrac{82 - 10x}{10 - x}$ for $0 \leq x \leq 8$ and that the company receives *twice* as much for selling an electronic dump truck as for selling a basic one. Find the level of production (both x and y) for which the total revenue derived from selling these dump trucks is maximized. We can assume that the company sells every dump truck it produces.

44. **A* VOTING PATTERN** After a presidential election in the United States, the proportion $h(p)$ of seats in the House of Representatives won by the party of the winning presidential candidate may be modelled by the cube rule
$$h(p) = \frac{p^3}{p^3 + (1 - p)^3} \quad \text{for } 0 \leq p \leq 1,$$
where p is the proportion of the popular vote received by the winning presidential candidate.
    a. Find $h'(p)$ and $h''(p)$.
    b. Sketch the graph of $h(p)$.
    c. In 2008, Barack Obama, the Democratic candidate, received 52.9% of the popular vote. What percentage of seats in the House does the cube rule predict should have gone to Democrats? (They actually won 59.4%.)

45. **A* WORKER EFFICIENCY** An efficiency study of the morning shift at a certain manufacturing plant indicates that an average worker who is on the job at 8.00 A.M. will have assembled $f(x) = -x^3 + 6x^2 + 15x$ units x hours later. The study indicates further that after a 15-minute coffee break the worker can assemble $g(x) = -\dfrac{1}{3}x^3 + x^2 + 23x$ units in x hours. Determine the time between 8.00 A.M. and noon at which a 15-minute coffee break should be scheduled so that the worker will assemble the maximum number of units by lunchtime at 12:15 P.M. [*Hint:* If the coffee break begins x hours after 8:00 A.M., $4 - x$ hours will remain after the break.]

46. **NATIONAL CONSUMPTION** Assume that total national consumption is given by a function $C(x)$, where x is the total national income. The derivative $C'(x)$ is called the **marginal propensity to consume.** Then $S = x - C$ represents total national savings, and $S'(x)$ is called the **marginal propensity to save.** Suppose the consumption function is $C(x) = 8 - 0.8x - 0.8\sqrt{x}$. Find the marginal propensity to consume, and determine the value of x that results in the smallest total savings.

47. **PROFIT** The weekly profit function for a certain small shoe company is $P(x) = -\dfrac{1}{10}x^2 + 30x - 500$ dollars, where x is the number of boxes of shoes produced. How many boxes of shoes must the company produce in order to maximize its profit?

48. AERODYNAMICS   In designing airplanes, an important feature is the so-called drag factor, that is, the retarding force exerted on the plane by the air. One model measures drag by a function of the form

$$F(v) = Av^2 + \frac{B}{v^2}$$

where v is the velocity of the plane and A and B are constants. Find the velocity (in terms of A and B) that minimizes $F(v)$. Show that you have found the minimum rather than a maximum.

49. SURVIVAL OF AQUATIC LIFE   It is known that a quantity of water that occupies 1 L at 0°C will occupy

$$V(T) = \left(\frac{-6.8}{10^8}\right)T^3 + \left(\frac{8.5}{10^6}\right)T^2 - \left(\frac{6.4}{10^5}\right)T + 1$$

litres when the temperature is T degrees Celsius, for $0 \le T \le 30$.

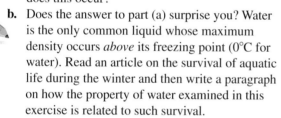

  a. Use a graphing utility to graph $V(T)$ for $0 \le T \le 10$. The density of water is maximized when $V(T)$ is minimized. At what temperature does this occur?

  b. Does the answer to part (a) surprise you? Water is the only common liquid whose maximum density occurs *above* its freezing point (0°C for water). Read an article on the survival of aquatic life during the winter and then write a paragraph on how the property of water examined in this exercise is related to such survival.

50. A★ AMPLITUDE OF OSCILLATION   In physics, it can be shown that a particle forced to oscillate in a resisting medium has amplitude $A(r)$ given by

$$A(r) = \frac{1}{(1 - r^2)^2 + kr^2}$$

where r is the ratio of the forcing frequency to the natural frequency of oscillation and k is a positive constant that measures the damping effect of the resisting medium. Show that $A(r)$ has exactly one positive critical number. Does it correspond to a relative maximum or a relative minimum? Can anything be said about the *absolute* extrema of $A(r)$?

51. A★ ELASTICITY AND REVENUE   Suppose the demand for a certain commodity is given by $q = b - ap$, where a and b are positive constants and $0 \le p \le \dfrac{b}{a}$.

  a. Express elasticity of demand as a function of p.

  b. Show that the demand is of unit elasticity at the midpoint $p = \dfrac{b}{2a}$ of the interval $0 \le p \le \dfrac{b}{a}$.

  c. For what values of p is the demand elastic? Inelastic?

52. INCOME ELASTICITY OF DEMAND   **Income elasticity of demand** is defined to be the percentage change in quantity purchased divided by the percentage change in real income.

  a. Write a formula for income elasticity of demand E in terms of real income I and quantity purchased Q.

  b. In Canada, which would you expect to be greater, the income elasticity of demand for cars or for food? Explain your reasoning.

  c. What do you think is meant by a *negative* income elasticity of demand? Which of the following goods would you expect to have $E < 0$: used clothing, computers, bus tickets, refrigerators, used cars? Explain your reasoning.

53. A★ ELASTICITY   Suppose that the demand equation for a certain commodity is $q = \dfrac{a}{p^m}$, where a and m are positive constants. Show that the elasticity of demand is equal to $-m$ for all values of p. Interpret this result.

54. A★ MARGINAL ANALYSIS   Let $R(x)$ be the revenue obtained from the production and sale of x units of a commodity, and let $C(x)$ be the total cost of producing the x units. Show that the ratio

$$Q(x) = \frac{R(x)}{C(x)}$$ is optimized when the relative rate of change of revenue equals the relative rate of change of cost. Would you expect this optimum to be a maximum or a minimum?

55. A★ BLOOD PRODUCTION   A useful model for the production $p(x)$ of blood cells involves a function of the form

$$p(x) = \frac{Ax}{B + x^m}$$

where x is the number of cells present and A, B, and m are positive constants.★

  a. Find the rate of blood production $R(x) = p'(x)$ and determine where $R(x) = 0$.

★M. C. Mackey and L. Glass, "Oscillations and Chaos in Physiological Control Systems," *Science*, Vol. 197, pp. 287–289.

b. Find the rate at which $R(x)$ is changing with respect to x and determine where $R'(x) = 0$.

c. If $m > 1$, does the nonzero critical number you found in part (b) correspond to a relative maximum or a relative minimum? Explain.

56. A* MARGINAL ANALYSIS Suppose $q > 0$ units of a commodity are produced at a total cost of $C(q)$ dollars and an average cost of $A(q) = \dfrac{C(q)}{q}$.

In this section, we showed that $q = q_c$ satisfies $A'(q_c) = 0$ if and only if $C'(q_c) = A(q_c)$, that is,

when marginal cost equals average cost. The purpose of this exercise is to show that $A(q)$ is *minimized* when $q = q_c$.

a. Generally speaking, the cost of producing a commodity increases at an increasing rate as more and more goods are produced. Using this economic principle, what can be said about the sign of $C''(q)$ as q increases?

b. Show that $A''(q_c) > 0$ if and only if $C''(q_c) > 0$. Then use part (a) to argue that average cost $A(q)$ is minimized when $q = q_c$.

L05

Construct equations for applied optimization problems and solve to optimize one variable.

Additional Applied Optimization

In Section 3.4, you saw a number of applications in which you were given a formula and required to determine either a maximum or a minimum value. In practice, things are often not that simple, and it is necessary first to gather information about a quantity of interest and then to formulate and analyze an appropriate mathematical model.

In this section, you will learn how to combine the techniques of model-building from Section 1.4 with the optimization techniques of Section 3.4. Here is a procedure for dealing with such problems.

Guidelines for Solving Optimization Problems

Step 1. Begin by deciding precisely what you want to maximize or minimize. Once you have done this, assign names to all variables of interest. It may help to pick letters that suggest the nature or role of the variable, such as "*R*" for "revenue."

Step 2. After assigning variables, express the relationships between the variables in terms of equations or inequalities. A figure may help.

Step 3. Express the quantity to be optimized (maximized or minimized) in terms of just one variable (the independent variable). To do this, you may need to use one or more of the available equations from step 2 to eliminate other variables. Also determine any necessary restrictions on the independent variable.

Step 4. If $f(x)$ is the quantity to be optimized, find $f'(x)$ and determine all critical numbers of f. Then find the required maximum or minimum value using the methods of Section 3.4 (the extreme value theorem or the second derivative test for absolute extrema). Remember, you may have to check the value of $f(x)$ at the endpoints of an interval.

Step 5. Interpret your results in a sentence, using appropriate units.

This procedure is illustrated in Examples 3.5.1 through 3.5.3.

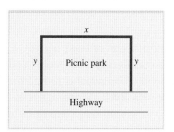

FIGURE 3.45 Rectangular picnic park.

EXAMPLE 3.5.1

The highway department is planning to build a picnic park for motorists along a major highway. The park is to be rectangular with an area of 5000 m² and is to be fenced off on the three sides not adjacent to the highway. What is the least amount of fencing required for this job? How long and wide should the park be for the fencing to be minimized?

Solution

Step 1. Draw the picnic area, as in Figure 3.45. Let x (metres) be the length of the park (the side parallel to the highway) and let y (metres) be the width.

Step 2. Since the park is to have area 5000 m², we must have $xy = 5000$.

Step 3. The length of the fencing is $F = x + 2y$, where x and y are both positive, since otherwise there would be no park. Since

$$xy = 5000 \quad \text{or} \quad y = \frac{5000}{x}$$

we can eliminate y from the formula for F to obtain a formula in terms of x alone:

$$F(x) = x + 2y = x + 2\left(\frac{5000}{x}\right) = x + \frac{10\ 000}{x} \quad \text{for } x > 0.$$

Step 4. The derivative of $F(x)$ is

$$F'(x) = 1 - \frac{10\ 000}{x^2}$$

and we get the critical numbers for $F(x)$ by setting $F'(x) = 0$ and solving for x:

$$F'(x) = 1 - \frac{10\ 000}{x^2} = 0$$

$$\frac{x^2 - 10\ 000}{x^2} = 0 \qquad \text{put the fraction over a common denominator}$$

$$x^2 = 10\ 000 \qquad \text{equate the numerator to 0}$$

$$x = 100 \qquad \text{reject } -100 \text{ since } x > 0$$

Since $x = 100$ is the only critical number in the interval $x > 0$, we can apply the second derivative test for absolute extrema. The second derivative of $F(x)$ is

$$F''(x) = \frac{20\ 000}{x^3}$$

so $F''(100) > 0$ and an absolute minimum of $F(x)$ occurs where $x = 100$. For reference, the graph of the fencing function $F(x)$ is shown in Figure 3.46.

Step 5. We have shown that the minimal amount of fencing is

$$F(100) = 100 \text{ m} + \frac{10\ 000 \text{ m}^2}{100 \text{ m}} = 200 \text{ m},$$

which is achieved when the park is $x = 100$ m long and

$$y = \frac{5000 \text{ m}^2}{100 \text{ m}} = 50 \text{ m}$$

wide.

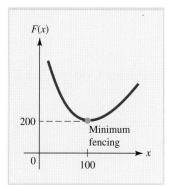

FIGURE 3.46 The graph of $F(x) = x + \dfrac{10\ 000}{x}$ for $x > 0$.

EXAMPLE 3.5.2

A decorative cylindrical container is to be constructed to hold 100 cm^3 of loose tea leaves. The cost of the material used for the top and bottom of the container is 3 cents/cm^2, and the cost of the material used for the curved side is 2 cents/cm^2. Use calculus to find the radius and height of the container that is the least costly to construct.

Solution

Let r denote the radius, h the height, C the cost (in cents), and V the volume. The goal is to minimize total cost, which comes from three sources:

$$\text{Cost} = \text{cost of top} + \text{cost of bottom} + \text{cost of side},$$

where, for each component of the cost,

$$\text{Cost} = (\text{cost/square centimetre})(\text{area}).$$

Hence,

$$\text{Cost of top} = \text{cost of bottom} = 3(\pi r^2)$$

and

$$\text{Cost of side} = 2(2\pi rh) = 4\pi rh,$$

so the total cost is $C = \underbrace{3\pi r^2}_{\text{top}} + \underbrace{3\pi r^2}_{\text{bottom}} + \underbrace{4\pi rh}_{\text{sides}} = 6\pi r^2 + 4\pi rh.$

Before you can apply calculus, you must write the cost in terms of just one variable. To do this, use the fact that the container is to have a volume of 100 cm^3 and solve the equation $V = \pi r^2 h$ for h to get

$$h = \frac{100}{\pi r^2}.$$

Then substitute this expression for h into the formula for C to express the cost in terms of r alone:

$$C(r) = 6\pi r^2 + 4\pi r\left(\frac{100}{\pi r^2}\right) = 6\pi r^2 + \frac{400}{r}$$

The radius r can be any positive number, so the goal is to find the absolute minimum of $C(r)$ for $r > 0$. Differentiating $C(r)$ gives

$$C'(r) = 12\pi r - \frac{400}{r^2}$$

Just-In-Time

A cylinder of radius r and height h has lateral (curved) area $A = 2\pi rh$ and volume $V = \pi r^2 h$.

Since $C'(r)$ exists for all $r > 0$, any critical number $r = R$ must satisfy $C'(R) = 0$; that is,

$$C'(R) = 12\pi R - \frac{400}{R^2} = 0$$

$$12\pi R = \frac{400}{R^2}$$

$$R^3 = \frac{400}{12\pi}$$

$$R = \sqrt[3]{\frac{400}{12\pi}} \approx 2.197$$

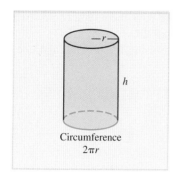

Circumference
$2\pi r$

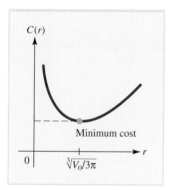

$C(r)$

Minimum cost

0 $\sqrt[3]{V_0/3\pi}$ r

FIGURE 3.47   The cost

function $C(r) = 6\pi r^2 + \dfrac{400}{r}$

for $r > 0$.

If H is the height of the container that corresponds to the radius R, then $V = \pi r^2 H$. Since R must satisfy $V = \pi r^2 H$, then

$$100 \approx \pi(2.197)^2 H$$

$$H \approx \frac{100}{\pi(2.197)^2}$$

$$H \approx 6.59$$

Finally, note that the second derivative of $C(r)$ satisfies $C''(r) = 12\pi + \dfrac{800}{r^3} > 0$

for all $r > 0$.

Therefore, since $r = R$ is the only critical number for $C(r)$ and since $C''(R) > 0$, the second derivative test for absolute extrema assures you that when the total cost of constructing the container is minimized, the height is 6.59 cm and the radius is 2.197 cm. The graph of the cost function $C(r)$ is shown in Figure 3.47.

EXAMPLE 3.5.3

A manufacturer can produce souvenir T-shirts at a cost of $4 each. The shirts have been selling for $10 each, and at this price, tourists have been buying 4000 shirts a month. The manufacturer is planning to raise the price of the shirts and estimates that for each $1 increase in the price, 400 fewer shirts will be sold each month. At what price should the manufacturer sell the shirts to maximize profit?

Solution

Let x be the new price at which the shirts will be sold and let $P(x)$ be the corresponding profit. The goal is to maximize the profit. Begin by stating the formula for profit in words:

$$\text{Profit} = (\text{number of shirts sold})(\text{profit per shirt}).$$

Since 4000 shirts are sold each month when the price is $10 and 400 fewer will be sold each month for each $1 increase in the price, it follows that

$$\text{Number of shirts sold} = 4000 - 400(\text{number of \$1 increases}).$$

The number of $1 increases in the price is the difference $x - 10$ between the new and old selling prices. Hence,

$$\begin{aligned}\text{Number of shirts sold} &= 4000 - 400(x - 10)\\ &= 400[10 - (x - 10)]\\ &= 400(20 - x)\end{aligned}$$

The profit per shirt is simply the difference between the selling price x and the cost $4. That is,

$$\text{Profit per shirt} = x - 4.$$

Putting it all together,

$$P(x) = 400(20 - x)(x - 4).$$

The goal is to find the absolute maximum of the profit function $P(x)$. To determine the relevant interval for this problem, note that since the new price x is to be at least as high as the old price $10 we must have $x \geq 10$. On the other hand, the number of shirts sold is $400(20 - x)$, which will be negative if $x > 20$. If you assume that the manufacturer will not price the shirts so high that no one buys them, you can restrict the optimization problem to the closed interval $10 \le x \le 20$.

To find the critical numbers, compute the derivative using the product and constant multiple rules to get

$$P'(x) = 400[(20 - x)(1) + (x - 4)(-1)]$$
$$= 400(20 - x - x + 4) = 400(24 - 2x)$$

which is zero when

$$24 - 2x = 0 \quad \text{or} \quad x = 12.$$

Comparing the values of the profit function

$$P(10) = 24\ 000 \qquad P(12) = 25\ 600 \quad \text{and} \quad P(20) = 0$$

at the critical number and at the endpoints of the interval, the maximum possible profit is \$25 600, which will be generated if the shirts are sold for \$12 each. For reference, the graph of the profit function is shown in Figure 3.48.

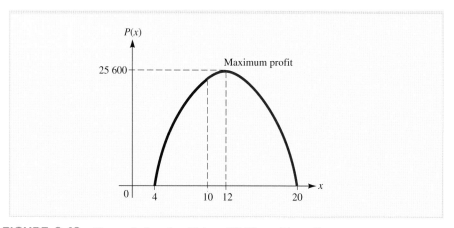

FIGURE 3.48   The profit function $P(x) = 400(20 - x)(x - 4)$.

NOTE **An Alternative Solution to the Profit-Maximization Problem in Example 3.5.3**

Since the number of shirts sold in Example 3.5.3 is described in terms of the number N of \$1 increases in price, you may wish to use N as the independent variable in your solution rather than the new price itself. With this choice,

$$\text{Number of shirts} = 4000 - 400N$$
$$\text{Profit per shirt} = (N + 10) - 4 = N + 6$$

Thus, the total profit is

$$P(N) = (4000 - 400N)(N + 6) = 400(10 - N)(N + 6),$$

and the relevant interval is $0 \le N \le 10$. (Do you see why?) The absolute maximum in this case will occur when $N = 2$ (provide the details), that is, when the old price is increased from \$10 to \$10 + \$2 = \$12. As you would expect, this is the same as the result obtained using price as the independent variable in Example 3.5.3.

Where residential areas are developed near industrial plants, it is important to carefully monitor and control the emission of pollutants. In Example 3.5.4, we examine a modelling problem in which calculus is used to determine the location in a community where pollution is minimized.

EXPLORE!

Refer to Example 3.5.4. Plot the graph of

$$P(x) = \frac{75}{x} + \frac{300}{15 - x}.$$

Move the cursor along the graph to verify the point with minimum pollution.

EXAMPLE 3.5.4

Two industrial plants, A and B, are located 15 km apart and emit 75 ppm (parts per million) and 300 ppm of particulate matter, respectively. Each plant is surrounded by a restricted area of radius 1 km in which no housing is allowed, and the concentration of pollutant arriving at any other point Q from each plant decreases with the reciprocal of the distance between that plant and Q. Where should a house be located on a road joining the two plants to minimize the total pollution arriving from both plants?

Solution

Suppose a house H is located x kilometres from plant A and hence $15 - x$ kilometres from plant B, where x satisfies $1 \le x \le 14$, since there is a 1-km restricted area around each plant (Figure 3.49). Since the concentration of particulate matter arriving at H from each plant decreases with the reciprocal of the distance from the plant to H, the concentration of pollutant from plant A is $75/x$ and from plant B is $300/(15 - x)$. Thus, the total concentration of particulate matter arriving at H is given by the function

$$P(x) = \frac{75}{x} + \frac{300}{15 - x}$$
$$\text{from A} \qquad \text{from B}$$

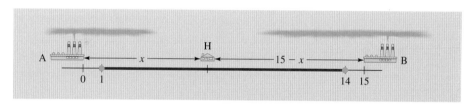

FIGURE 3.49 Pollution at a house located between two industrial plants.

To minimize the total pollution $P(x)$, first find the derivative $P'(x)$ and solve $P'(x) = 0$. Applying the chain rule gives

$$P'(x) = \frac{-75}{x^2} + \frac{-300(-1)}{(15 - x)^2} = \frac{-75}{x^2} + \frac{300}{(15 - x)^2}$$

Just-In-Time

Cross-multiplying means that if

$$\frac{A}{B} = \frac{C}{D}$$

then we can move any of A, B, C, and D along diagonal lines over the equal sign so that

$$AD = CB$$
$$\frac{A}{C} = \frac{B}{D}$$
$$\frac{D}{B} = \frac{C}{A}$$

This is very useful when solving equations

Solving $P'(x) = 0$ gives

$$\frac{-75}{x^2} + \frac{300}{(15 - x)^2} = 0$$

$$\frac{75}{x^2} = \frac{300}{(15 - x)^2}$$

$$75(15 - x)^2 = 300x^2 \qquad \text{cross-multiply}$$

$$x^2 - 30x + 225 = 4x^2 \qquad \text{divide by 75 and expand}$$

$$3x^2 + 30x - 225 = 0 \qquad \text{combine terms}$$

$$3(x - 5)(x + 15) = 0 \qquad \text{factor}$$

$$x = 5, x = -15$$

Just-In-Time

The symbol ≈ means "approximately equal to." Thus, $a \approx b$ means "*a* is approximately equal to *b*." Although a calculator gives many decimal places, it is not exact.

The only critical number in the allowable interval $1 \leq x \leq 14$ is $x = 5$. Evaluating $P(x)$ at $x = 5$ and at the two endpoints of the interval, $x = 1$ and $x = 14$, we get

$$P(1) \approx 96.43 \text{ ppm}$$
$$P(5) = 45 \text{ ppm}$$
$$P(14) \approx 305.36 \text{ ppm}$$

Thus, total pollution is minimized when the house is located 5 km from plant A.

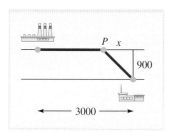

FIGURE 3.50   Relative positions of factory, river, and power plant.

EXAMPLE 3.5.5

A cable is to be run from a power plant on one side of a river that is 900 m wide to a factory on the other side, 3000 m downstream. The cost of running the cable under the water is \$5 per metre, while the cost over land is \$4 per metre. What is the most economical route over which to run the cable?

Solution

To help you visualize the situation, begin by drawing a diagram as shown in Figure 3.50. (Notice that in drawing the diagram in Figure 3.50, we have already assumed that the cable should be run in a *straight line* from the power plant to some point P on the opposite bank. Do you see why this assumption is justified?)

The goal is to minimize the cost of installing the cable. Let C denote this cost and represent C as follows:

$$C = 5(\text{number of metres of cable under water})$$
$$+ 4(\text{number of metres of cable over land})$$

Since you wish to describe the optimal route over which to run the cable, it will be convenient to choose a variable in terms of which you can easily locate the point P. Two reasonable choices for the variable x are illustrated in Figure 3.51.

Just-In-Time

The Pythagorean theorem says that the square of the hypotenuse of a right triangle equals the sum of the squares of the other two sides.

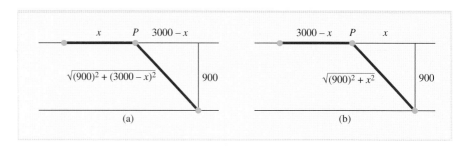

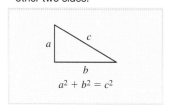

FIGURE 3.51   Two choices for the variable x.

Before plunging into the calculations, take a minute to decide which choice of variables is better. In Figure 3.51a, the distance across the water from the power plant to the point P is (by the Pythagorean theorem) $\sqrt{(900)^2 + (3000 - x)^2}$, and the corresponding total cost function is

$$C(x) = 5\sqrt{(900)^2 + (3000 - x)^2} + 4x.$$

In Figure 3.51b, the distance across the water is $\sqrt{(900)^2 + x^2}$, and the total cost function is

$$C(x) = 5\sqrt{(900)^2 + x^2} + 4(3000 - x).$$

The second function is more attractive since the term $3000 - x$ is merely multiplied by 4, while in the first function it is squared and appears under the square root. Hence, you should choose x as in Figure 3.51b and work with the total cost function

$$C(x) = 5\sqrt{(900)^2 + x^2} + 4(3000 - x).$$

Since the distances x and $3000 - x$ cannot be negative, the relevant interval is $0 \le x \le 3000$, and your goal is to find the absolute minimum of the function $C(x)$ on this closed interval. To find the critical values, compute the derivative,

$$C'(x) = \frac{5}{2}[(900)^2 + x^2]^{-1/2}(2x) - 4 = \frac{5x}{\sqrt{(900)^2 + x^2}} - 4,$$

and set it equal to zero to get

$$\frac{5x}{\sqrt{(900)^2 + x^2}} - 4 = 0 \quad \text{or} \quad \sqrt{(900)^2 + x^2} = \frac{5}{4}x.$$

Square both sides of the equation and solve for x to get

$$(900)^2 + x^2 = \frac{25}{16}x^2 \qquad \text{subtract } \frac{25x^2}{16} \text{ and}$$
$$\qquad\qquad\qquad\qquad\qquad (900)^2 \text{ from each side}$$

$$x^2 - \frac{25}{16}x^2 = -(900)^2$$

$$-\frac{9}{16}x^2 = -(900)^2 \qquad \text{combine terms on the left}$$

$$x^2 = \frac{16}{9}(900)^2 \qquad \text{cross-multiply by } -\frac{16}{9}$$

$$x = \pm\frac{4}{3}(900) = \pm 1200 \qquad \text{take square roots on}$$
$$\qquad\qquad\qquad\qquad\qquad\qquad \text{each side}$$

Since only the positive value $x = 1200$ is in the interval $0 \le x \le 3000$, compute $C(x)$ at this critical value and at the endpoints $x = 0$ and $x = 3000$. Since

$$C(0) = 5\sqrt{(900)^2 + 0} + 4(3000 - 0) = 16\,500$$
$$C(1200) = 5\sqrt{(900)^2 + (1200)^2} + 4(3000 - 1200) = 14\,700$$
$$C(3000) = 5\sqrt{(900)^2 + (3000)^2} + 4(3000 - 3000) = 15\,660$$

it follows that the minimal installation cost is \$14 700, which will occur if the cable reaches the opposite bank 1200 m downstream from the power plant.

In Example 3.5.6, the function to be maximized has practical meaning only when its independent variable is a whole number. However, the optimization procedure leads to a fractional value of this variable, and additional analysis is needed to obtain a meaningful solution.

EXAMPLE 3.5.6

A bus company will charter a bus that holds 50 people to groups of 35 or more. If a group contains exactly 35 people, each person pays $60. In large groups, everybody's fare is reduced by $1 for each person in excess of 35. Determine the size of the group for which the bus company's revenue will be greatest.

Solution

Let R denote the bus company's revenue. Then,

$$R = \text{(number of people in the group)(fare per person)}.$$

You could let x denote the total number of people in the group, but it is slightly more convenient to let x denote the number of people in excess of 35. Then,

$$\text{Number of people in the group} = 35 + x$$

and

$$\text{Fare per person} = 60 - x,$$

so the revenue function is

$$R(x) = (35 + x)(60 - x).$$

Since x represents the number of people in excess of 35 but less than 50, you want to maximize $R(x)$ for a positive integer x in the interval $0 \leq x \leq 15$ (Figure 3.52a). However, to use the methods of calculus, consider the *continuous* function $R(x) = (35 + x)(60 - x)$ defined on the entire interval $0 \leq x \leq 15$ (Figure 3.52b).

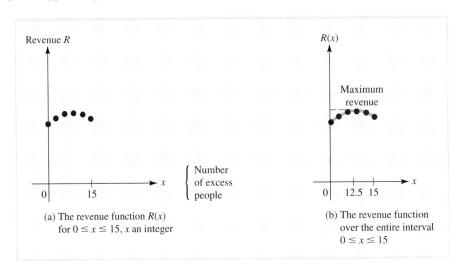

(a) The revenue function $R(x)$ for $0 \leq x \leq 15$, x an integer

(b) The revenue function over the entire interval $0 \leq x \leq 15$

FIGURE 3.52   The revenue function $R(x) = (35 + x)(60 - x)$.

The derivative is

$$R'(x) = (35 + x)(-1) + (60 - x)(1) = 25 - 2x,$$

which is zero when $x = 12.5$. Since

$$R(0) = 2100 \quad R(12.5) = 2256.25 \quad R(15) = 2250$$

it follows that the absolute maximum of $R(x)$ on the interval $0 \leq x \leq 15$ occurs when $x = 12.5$.

But x represents a certain number of people and must be a whole number. Hence, $x = 12.5$ cannot be the solution to this practical optimization problem. To find the optimal *integer* value of x, observe that R is increasing for $0 < x < 12.5$ and decreasing for $x > 12.5$, as shown in this diagram (see also Figure 3.52b).

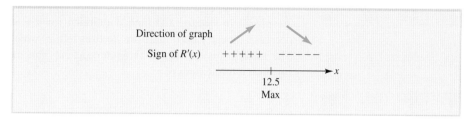

It follows that the optimal integer value of x is either $x = 12$ or $x = 13$. Since

$$R(12) = 2256 \quad \text{and} \quad R(13) = 2256,$$

the bus company's revenue will be greatest when the group contains either 12 or 13 people in excess of 35, that is, for groups of 47 or 48. The revenue in either case will be $2256.

Inventory Control

Inventory control is an important consideration in business. In particular, for each shipment of raw materials, a manufacturer must pay an ordering fee to cover handling and transportation. When the raw materials arrive, they must be stored until needed, and storage costs result. If each shipment of raw materials is large, few shipments will be needed, so ordering costs will be low, while storage costs will be high. On the other hand, if each shipment is small, ordering costs will be high because many shipments will be needed, but storage costs will be low. Example 3.5.7 shows how the methods of calculus can be used to determine the shipment size that minimizes total cost.

EXAMPLE 3.5.7

A bicycle manufacturer buys 6000 tires a year from a distributor. The ordering fee is $20 per shipment, the storage cost is 96 cents per tire per year, and each tire costs $21. Suppose that the tires are used at a constant rate throughout the year and that each shipment arrives just as the preceding shipment is being used up. How many tires should the manufacturer order each time to minimize cost?

Solution

The goal is to minimize the total cost, which can be written as

$$\text{Total cost} = \text{storage cost} + \text{ordering cost} + \text{purchase cost.}$$

Let x denote the number of tires in each shipment and $C(x)$ the corresponding total cost in dollars. Then

$$\text{Ordering cost} = (\text{ordering cost per shipment})(\text{number of shipments}).$$

Since 6000 tires are ordered during the year and each shipment contains x tires, the number of shipments is $\dfrac{6000}{x}$, and so

$$\text{Ordering cost} = 20\left(\frac{6000}{x}\right) = \frac{120\,000}{x}$$

Moreover,

$$\text{Purchase cost} = (\text{total number of tires ordered})(\text{cost per tire})$$
$$= 6000(21) = 126\ 000$$

The storage cost is slightly more complicated. When a shipment arrives, all x tires are placed in storage and then withdrawn for use at a constant rate. The inventory decreases linearly until there are no tires left, at which time the next shipment arrives. The situation is illustrated in Figure 3.53a. This is sometimes called **just-in-time inventory management.**

EXPLORE!

Refer to Example 3.5.7 and the cost function $C(x)$, but vary the ordering fee $q = \$20$. Graph the cost functions for $q = 10, 15, 20,$ and 25. Describe the difference in the graphs for these values of q. Find the minimum cost in each case. Describe how the minimum changes with the changing values of q.

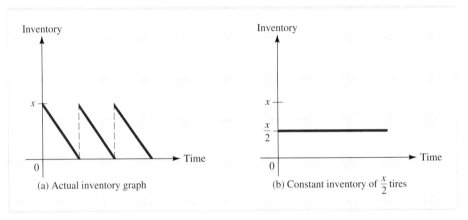

(a) Actual inventory graph   (b) Constant inventory of $\frac{x}{2}$ tires

FIGURE 3.53   Inventory graphs.

The average number of tires in storage during the year is $\frac{x}{2}$, and the total yearly storage cost is the same as if $\frac{x}{2}$ tires were kept in storage for the entire year (Figure 3.53b). This assertion, although reasonable, is not really obvious, and you have every right to be unconvinced. In Chapter 5, you will learn how to prove this fact mathematically using integral calculus. It follows that

$$\text{Storage cost} = (\text{average number of tires stored})(\text{storage cost per tire})$$
$$= \frac{x}{2}(0.96) = 0.48x$$

Putting it all together, the total cost is

$$C(x) = \underbrace{0.48x}_{\substack{\text{storage}\\\text{cost}}} + \underbrace{\frac{120\ 000}{x}}_{\substack{\text{ordering}\\\text{cost}}} + \underbrace{126\ 000}_{\substack{\text{purchase}\\\text{cost}}}$$

and the goal is to find the absolute minimum of $C(x)$ on the interval

$$0 < x \le 6000.$$

The derivative of $C(x)$ is

$$C'(x) = 0.48 - \frac{120\ 000}{x^2}$$

which is zero when

$$x^2 = \frac{120\ 000}{0.48} = 250\ 000 \quad \text{or} \quad x = \pm 500.$$

Since $x = 500$ is the only critical number in the relevant interval $0 < x \le 6000$, you can apply the second derivative test for absolute extrema. The second derivative of the cost function is

$$C''(x) = \frac{240\ 000}{x^3}$$

which is positive when $x > 0$. Hence, the absolute minimum of the total cost $C(x)$ on the interval $0 < x \le 6000$ occurs when $x = 500$, that is, when the manufacturer orders the tires in lots of 500. For reference, the graph of the total cost function is sketched in Figure 3.54.

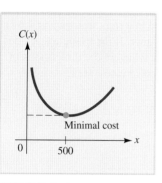

FIGURE 3.54   Total cost
$$C(x) = 0.48x + \frac{120\ 000}{x} + 126\ 000.$$

NOTE   Since the derivative of the (constant) purchase price $126\,000$ in Example 3.5.7 was zero, this component of the total cost had no bearing on the optimization problem. In general, economists distinguish between **fixed costs** (such as the total purchase price) and **variable** costs (such as the storage and ordering costs). To minimize total cost, it is sufficient to minimize the sum of all the variable components of cost. ∎

ALGEBRA WARM-UP

Solve for x.

1. $5x = \dfrac{405}{x^2}$

2. $8x = \dfrac{648}{x}$

3. $0 = 30 - \dfrac{750}{x^2}$

4. $10 = 20 - \dfrac{160}{x^2}$

5. $\dfrac{10}{(2 - x)} - \dfrac{20}{x} = 0$

6. $\dfrac{5}{6 - x} - \dfrac{9}{2x} = 0$

7. $\dfrac{x}{2x + 3} = \dfrac{5}{7}$

8. $2 = \dfrac{10}{x^2 - 3}$

EXERCISES ▪ 3.5

1. What number exceeds its square by the largest amount? [*Hint:* Find the number x that maximizes $f(x) = x - x^2$.]

2. What number is exceeded by its square root by the largest amount?

3. Find two positive numbers whose sum is 50 and whose product is as large as possible.

4. Find two positive numbers x and y with a sum of 30 and such that xy^2 is as large as possible.

5. **RETAIL SALES**   A store has been selling a video game at a price of $40 per unit, and at this price, players have been buying 50 units per month. The owner of the store wishes to raise the price of the game and estimates that for each $1 increase in price, three fewer units will be sold each month. If each unit costs the store $25, at what price should the game be sold to maximize profit?

6. **RETAIL SALES**   A bookstore can obtain a certain gift book from the publisher at a cost of $3 per book. The bookstore has been offering the book at a price of $15 per copy and, at this price, has been selling 200 copies per month. The bookstore is planning to lower its price to stimulate sales and estimates that for each $1 reduction in the price, 20 more books will be sold each month. At what price should the bookstore sell the book to generate the greatest possible profit?

7. **AGRICULTURAL YIELD**   A citrus grower estimates that if 60 orange trees are planted, the average yield per tree will be 400 oranges. The average yield will decrease by 4 oranges per tree for each additional tree planted on the same acreage. How many trees should the grower plant to maximize the total yield?

8. **HARVESTING**   One year a potato farmer can get $200 per tonne for his potatoes on July 1, and after that, the price drops by $1.25 per tonne per day. On July 1, a farmer has 4 tonnes of potatoes in the field and estimates that the crop is increasing at a rate of 0.03 tonnes per day. When should the farmer harvest the potatoes to maximize revenue?

9. **PROFIT**   A specialty ice cream store can make a new rich mocha flavour at a cost of $5 per carton. The store has been selling 2-L cartons of the ice cream for $10 and, at this price, has been selling 25 cartons per month. The store is planning to lower the price to stimulate sales and estimates that for each 25-cent reduction in the price, 5 more cartons will be sold each month. At what price should the 2-L cartons be sold in order to maximize total monthly profit?

10. **PROFIT**   A manufacturer has been selling flashlights at $6 each, and at this price, consumers have been buying 3000 flashlights per month. The manufacturer wishes to raise the price and estimates that for each $1 increase in the price, 1000 fewer flashlights will be sold each month. The manufacturer can produce the flashlights at a cost of $4 per flashlight. At what price should the manufacturer sell the flashlights to generate the greatest possible profit?

11. **FENCING**   A city recreation department plans to build a rectangular playground with an area of 3600 m^2 and surround it by a fence. How can this be done using the least amount of fencing?

12. **FENCING**   There are 320 m of fencing available to enclose a rectangular field. How should this fencing be used so that the enclosed area is as large as possible?

13. A rectangle has sides x and y and perimeter P. Show that when P is constant, the square with $x = y$ has the largest area of all the possible rectangles.

14. A rectangle has area A and sides x and y. Prove that of all rectangles with this area, the square has the smallest perimeter.

15. A rectangle is inscribed in a right triangle, as shown in the accompanying figure. If the triangle has sides of length 5, 12, and 13, what are the dimensions of the inscribed rectangle of greatest area?

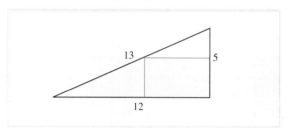

EXERCISE 15

16. A triangle is positioned with its hypotenuse on a diameter of a circle, as shown in the accompanying figure. If the circle has radius 4, what are the dimensions of the triangle of greatest area?

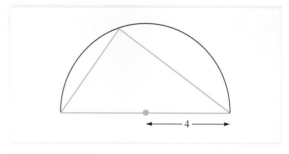

EXERCISE 16

17. **CONSTRUCTION COST**   A carpenter has been asked to build an open box with a square base. The sides of the box will cost $3/\text{m}^2$, and the base will cost $4/\text{m}^2$. What are the dimensions of the box of greatest volume that can be constructed for $48?

18. **CONSTRUCTION COST**   A closed box with a square base is to have a volume of 250 m^3. The material for the top and bottom of the box costs $2/\text{m}^2$, and the material for the sides costs $1/\text{m}^2$. Can the box be constructed for less than $300?

19. **SPY STORY**   It is noon, and the spy is back from space (see Exercise 72 in Section 2.2) and driving a jeep through the sandy desert in the tiny principality of Alta Loma. He is 32 km from the nearest point on a straight paved road. Down the road 16 km is an abandoned power plant where a group of rival spies

are holding captive his superior, code name "N." If the spy doesn't arrive with a ransom by 12:50 P.M., the bad guys have threatened to do N in. The jeep can travel at 48 km/h in the sand and at 80 km/h on the paved road. Can the spy make it in time, or is this the end of N? [*Hint:* The goal is to minimize time, which is distance divided by speed.]

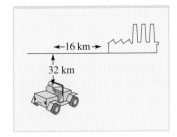

EXERCISE 19

20. **DISTANCE BETWEEN MOVING OBJECTS** A truck is 300 km due east of a car and is travelling west at a constant speed of 30 km/h. Meanwhile, the car is going north at a constant speed of 60 km/h. At what time will the car and truck be closest to each other? [*Hint:* You will simplify the calculation if you minimize the *square* of the distance between the car and truck rather than the distance itself. Can you explain why this simplification is justified?]

21. **INSTALLATION COST** A cable is to be run from a power plant on one side of a river that is 1200 m wide to a factory on the other side, 1500 m downstream. The cost of running the cable under the water is $25/m, while the cost over land is $20/m. What is the most economical route over which to run the cable?

22. **INSTALLATION COST** Find the most economical route in Exercise 21 if the power plant is 2000 m downstream from the factory.

23. **POSTER DESIGN** A printer receives an order to produce a rectangular poster containing 648 cm^2 of print surrounded by margins of 2 cm on each side and 4 cm on the top and bottom. What are the dimensions of the smallest piece of paper that can be used to make the poster? [*Hint:* Choose x and y as the dimensions of the printed part of the page.]

24. **PACKAGING** A cylindrical can is to hold 70π cm^3 of frozen orange juice. The cost per square centimetre of constructing the metal top and bottom is twice the cost per square centimetre of

constructing the cardboard side. What are the dimensions of the least expensive can?

25. **PACKAGING** Find the dimensions of the 350-mL pop can that can be constructed using the least amount of metal. Compare the shape resulting from these dimensions with that of a pop can in your refrigerator. What do you think accounts for the difference?

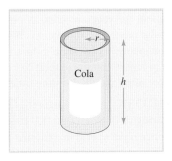

EXERCISE 25

26. **A* PACKAGING** A cylindrical can (with top) is to be constructed using a fixed amount of metal. Use calculus to derive a relationship between the radius and height of the can with the greatest volume.

27. **A* CONSTRUCTION COST** A cylindrical container with no top is to be constructed to hold a fixed volume of liquid. The cost of the material used for the bottom is 3 cents/cm^2, and that for the curved side is 2 cents/cm^2. Use calculus to derive a relationship between the radius and the height of the least expensive container.

28. **PRODUCTION COST** Each machine at a certain factory can produce 50 units per hour. The setup cost is $80 per machine, and the operating cost is $5 per hour. How many machines should be used to produce 8000 units at the least possible cost? (Remember that the answer should be a whole number.)

29. **COST ANALYSIS** It is estimated that the cost of constructing an office building that is n floors high is $C(n) = 2n^2 + 500n + 600$ thousand dollars. How many floors should the building have in order to minimize the average cost per floor? (Remember that your answer should be a whole number.)

30. **INVENTORY** An electronics firm uses 600 cases of components each year. Each case costs $1000. The cost of storing one case for a year is 90 cents, and the ordering fee is $30 per shipment. How many cases should the firm order each time to keep total

cost at a minimum? (Assume that the components are used at a constant rate throughout the year and that each shipment arrives just as the preceding shipment is being used up.)

31. **INVENTORY**   A store expects to sell 800 bottles of perfume this year. The perfume costs $20 per bottle, the ordering fee is $10 per shipment, and the cost of storing the perfume is 40 cents per bottle per year. The perfume is consumed at a constant rate throughout the year, and each shipment arrives just as the preceding shipment is being used up.
    a.   How many bottles should the store order in each shipment to minimize total cost?
    b.   How often should the store order the perfume?

32. **INVENTORY**   A manufacturer of medical monitoring devices uses 36 000 cases of components per year. The ordering cost is $54 per shipment, and the annual cost of storage is $1.20 per case. The components are used at a constant rate throughout the year, and each shipment arrives just as the preceding shipment is being used up. How many cases should be ordered in each shipment in order to minimize total cost?

33. **PRODUCTION COST**   A plastics firm has received an order from the city recreation department to manufacture 8000 special Styrofoam kickboards for its summer swimming program. The firm owns 10 machines, each of which can produce 30 kickboards an hour. The cost of setting up the machines to produce the kickboards is $20 per machine. Once the machines have been set up, the operation is fully automated and can be overseen by a single production supervisor earning $15 per hour.
    a.   How many of the machines should be used to minimize the cost of production?
    b.   How much will the supervisor earn during the production run if the optimal number of machines are used?
    c.   How much will it cost to set up the optimal number of machines?

34. **RECYCLING**   To raise money and raise awareness of recycling, a student group has been collecting used aluminum pop cans. They have collected 120 kg in the last 80 days. The price they will be paid is 70 cents/kg. Aluminum prices are falling at 0.4 cents/kg each day. Their access to the truck is limited so that they can make only one trip to the recycling depot. If they continue to collect at the same rate, what is the best time to deliver the pop cans?

35. **RETAIL SALES**   A retailer has bought several cases of a certain imported wine. As the wine ages, its value initially increases, but eventually the wine will pass its prime and its value will decrease. Suppose that x years from now, the value of a case will be changing at the rate of $53 - 10x$ dollars per year. Suppose, in addition, that storage rates will remain fixed at $3 per case per year. When should the retailer sell the wine to obtain the greatest possible profit?

36. **CONSTRUCTION**   An open box is to be made from a square piece of cardboard, 18 cm by 18 cm, by removing a small square from each corner and folding up the flaps to form the sides. What are the dimensions of the box of greatest volume that can be constructed in this way?

37. **POSTAL REGULATIONS**   According to Canada Post, for a regular parcel (not subject to oversize costs), the girth plus the length of parcel may not exceed 3 m. What is the largest possible volume of a rectangular parcel with two square sides that can be sent as a regular parcel?

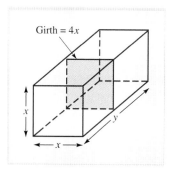

EXERCISE 37

38. **POSTAL REGULATIONS**   Refer to Exercise 37. What is the largest volume of a cylindrical parcel that can be sent as a regular parcel?

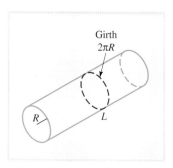

EXERCISE 38

39. MINIMAL COST A manufacturer finds that in producing x units per day (for $0 < x < 100$), three different kinds of cost are involved:
   a. A fixed cost of $1200 per day in wages
   b. A production cost of $1.20 per day for each unit produced
   c. An ordering cost of $\dfrac{100}{x^2}$ dollars per day.

Express the total cost as a function of x and determine the level of production that results in minimal total cost.

40. TRANSPORTATION COST For speeds between 64 km/h and 104 km/h, a truck gets $\dfrac{308}{x}$ kilometres per litre of fuel at a constant speed of x kilometres per hour. Diesel costs $1.01 per litre, and the driver is paid $24 per hour. What is the most economical constant speed between 64 km/h and 104 km/h at which to drive the truck?

41. A* AVIAN BEHAVIOUR Homing pigeons will rarely fly over large bodies of water unless forced to do so, presumably because it requires more energy to maintain altitude in flight in the heavy air over cool water.* Suppose a pigeon is released from a boat B floating on a lake 5 km from a point A on the shore and 13 km from the pigeon's loft L, as shown in the accompanying figure. Assuming the pigeon requires twice as much energy to fly over water as over land, what path should it follow to minimize the total energy expended in flying from the boat to its loft? Assume the shoreline is straight and describe the bird's path as a line from B to a point P on the shore followed by a line from P to L.

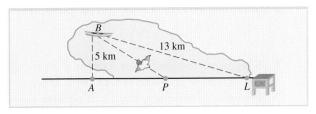

EXERCISE 41

*E. Batschelet, *Introduction to Mathematics for Life Sciences,* 3rd ed., New York: Springer-Verlag, 1979, pp. 276–277.

42. A* CONSTRUCTION The strength of a rectangular beam is proportional to the product of its width and the square of its depth. Find the dimensions of the strongest beam that can be cut from a wooden log of diameter 40 cm.

43. A* CONSTRUCTION The stiffness of a rectangular beam is proportional to the product of its width w and the cube of its depth h. Find the dimensions of the stiffest beam that can be cut from a wooden log of diameter 40 cm. (Note the accompanying figure.)

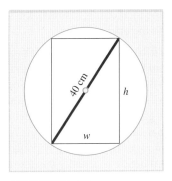

EXERCISE 43

44. A* TRANSPORTATION COST A truck is hired to transport goods from a factory to a warehouse. The driver's wages are figured by the hour and so are inversely proportional to the speed at which the truck is driven. The amount of gasoline used is directly proportional to the speed at which the truck is driven, and the price of gasoline remains constant during the trip. Show that the total cost is smallest at the speed for which the driver's wages are equal to the cost of the gasoline used.

45. A* URBAN PLANNING Two industrial plants, A and B, are located 18 km apart, and respectively emit 80 ppm (parts per million) and 720 ppm of particulate matter each day. Plant A is surrounded by a restricted area of radius 1 km, while the restricted area around plant B has a radius of 2 km. The concentration of particulate matter arriving at any other point Q from each plant decreases with the reciprocal of the distance between that plant and Q. Where should a house be located on a road joining the two plants to minimize the total concentration of particulate matter arriving from both plants?

46. A* URBAN PLANNING   In Exercise 45, suppose the concentration of particulate matter arriving at a point Q from each plant decreases with the reciprocal of the *square* of the distance between that plant and Q. With this alteration, now where should a house be located to minimize the total concentration of particulate matter arriving from both plants?

47. A* OPTIMAL SETUP COST   Suppose that at a certain factory, setup cost is directly proportional to the number N of machines used and operating cost is inversely proportional to N. Show that when the total cost is minimal, the setup cost is equal to the operating cost.

48. A* AVERAGE PRODUCTIVITY   The output Q at a certain factory is a function of the number L of worker-hours of labour that are used. Use calculus to prove that when the average output per worker-hour is greatest, the average output is equal to the marginal output per worker-hour. You may assume without proof that the critical point of the average output function is actually the desired absolute maximum. [*Hint:* The marginal output per worker-hour is the derivative of output Q with respect to labour L.]

49. A* INSTALLATION COST   For the summer, the company installing the cable in Example 3.5.5 has hired Frank Kornercutter as a consultant. Frank, recalling a problem from first-year calculus, asserts that no matter how far downstream the factory is located (beyond 1200 m), it would be most economical to have the cable reach the opposite bank 1200 m downstream from the power plant. The supervisor, amused by Frank's naivety, replies, "Any fool can see that if the factory is farther away, the cable should reach the opposite bank farther downstream. It's just common sense!" Of course, Frank is no common fool, but is he right? Why or why not?

50. A* CONSTRUCTION   As part of a construction project, it is necessary to carry a pipe around a corner as shown in the accompanying figure. What is the length of the longest pipe that will fit horizontally?

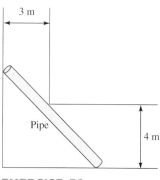

EXERCISE 50

51. A* PRODUCTION COST   A manufacturing firm receives an order for q units of a certain commodity. Each of the firm's machines can produce n units per hour. The setup cost is s dollars per machine, and the operating cost is p dollars per hour.
  a. Derive a formula for the number of machines that should be used to keep total cost as low as possible.
  b. Prove that when the total cost is minimal, the cost of setting up the machines is equal to the cost of operating the machines.

52. A* INVENTORY   The inventory model analyzed in Example 3.5.7 is not the only such model possible. Suppose a company must supply N units per time period at a uniform rate. Assume that the storage cost per unit is D_1 dollars per time period and that the setup (ordering) cost is D_2 dollars. If production is at a uniform rate of m units per time period (with no items in inventory at the end of each period), it can be shown that the total storage cost is

$$C_1 = \frac{D_1 x}{2}\left(1 - \frac{N}{m}\right),$$

where x is the number of items produced in each run.
  a. Show that the total average cost per period is

$$C = \frac{D_1 x}{2}\left(1 - \frac{N}{m}\right) + \frac{D_2 N}{x}.$$

  b. Find an expression for the number of items that should be produced in each run in order to minimize the total average cost per time period.

c. The optimal quantity found in the inventory problem in Example 3.5.7 is sometimes called the **economic order quantity (EOQ)**, while the optimum found in part (b) of this exercise is called the **economic production quantity (EPQ)**. Modern inventory management goes far beyond the simple conditions in the EOQ and EPQ models, but elements of these models are still very important. For instance, the just-in-time inventory management described in Example 3.5.7 fits well with the production philosophy of the Japanese. Read an article on Japanese production methods and write a paragraph on why the Japanese regard using space for the storage of materials as undesirable.

53. **A* EFFECT OF TAXATION ON A MONOPOLY** A **monopolist** is a manufacturer who can manipulate the price of a commodity and usually does so with an eye toward maximizing profit. When the government taxes output, the tax effectively becomes an additional cost item, and the monopolist is forced to decide how much of the tax to absorb and how much to pass on to the consumer.

Suppose a particular monopolist estimates that when x units are produced, the total cost will be

$$C(x) = \frac{7}{8}x^2 + 5x + 100$$ dollars and the market

price of the commodity will be $p(x) = 15 - \frac{3}{8}x$

dollars per unit. Further assume that the government imposes a tax of t dollars on each unit produced.

a. Show that profit is maximized when

$$x = \frac{2}{5}(10 - t).$$

b. Suppose the government assumes that the monopolist will always act so as to maximize total profit. What value of t should be chosen to guarantee maximum total tax revenue?

c. If the government chooses the optimum rate of taxation found in part (b), how much of this tax will be absorbed by the monopolist and how much will be passed on to the consumer?

d. Read an article on taxation and write a paragraph on how it affects consumer spending.

54. **A* ACCOUNT MANAGEMENT** Tom requires $10 000 in spending money each year, which he takes from his savings account by making N equal withdrawals. Each withdrawal incurs a transaction fee of $8, and money in his account earns interest at the simple interest rate of 4%.

a. The total cost C of managing the account is the transaction cost plus the loss of interest due to withdrawn funds. Express C as a function of N. [*Hint:* You may need the fact that $1 + 2 + \cdots + N = N(N + 1)/2$.]

b. How many withdrawals should Tom make each year in order to minimize the total transaction cost $C(N)$?

Concept Summary Chapter 3

Comparison of First and Second Derivative Tests and Absolute Extrema

For all of these, first find $f'(x)$, and then set $f'(x) = 0$ to find critical points.

First Derivative Test

Draw a line marking the critical numbers.
Pick a number on either side of each critical point, and find $f'(x)$ at these points.

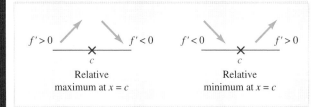

Second Derivative Test

Find $f''(x)$, and then substitute critical numbers into $f''(x)$.
When $f''(c)$ is positive, there is a relative minimum at $x = c$.
When $f''(c)$ is negative, there is a relative maximum at $x = c$.

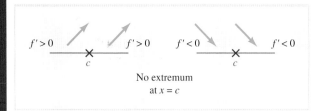

$$f(x) = x^3 - 12x + 4$$
$$f'(x) = 3x^2 - 12$$
$$f'(x) = 0$$
$$3x^2 - 12 = 0$$
$$3(x - 2)(x + 2) = 0$$
$$x = 2,\ x = -2$$
$$f''(x) = 6x$$

$f''(2) = 12$ is positive, so there is a relative minimum at $x = 2$.
$f''(-2) = -12$ is negative, so there is a relative maximum at $x = 2$.

Absolute Extrema

Substitute the critical numbers and the endpoints of the interval into $f(x)$ to find the largest and smallest values. Consider the function below on the interval $[-1, 2]$.

$$f(x) = x^3 - 6x^2$$
$$f'(x) = 3x^2 - 12x$$
$$f'(x) = 0$$
$$3x(x - 4) = 0$$
$$x = 0,\ x = 4 \text{ (not in domain)}$$
$$f(0) = 0,\text{ absolute maximum}$$
$$f(-1) = -7$$
$$f(2) = -16,\text{ absolute minimum}$$

Inflection Point

Set $f''(x) = 0$. Concavity must change on either side of the inflection point.

$$f(x) = (x + 4)^6$$
$$f'(x) = 6(x + 4)^5$$
$$f'(x) = 0 \text{ when } x = -4$$
$$f'(-5) = 6(-1)^5 = -6,\text{ negative}$$
$$f'(0) = 6(4)^5,\text{ positive}$$

Therefore, there is a relative minimum at $x = -4$.

Checkup for Chapter 3

1. Of the two curves shown here, one is the graph of a function $f(x)$ and the other is the graph of its derivative $f'(x)$. Determine which graph is the derivative and give reasons for your decision.

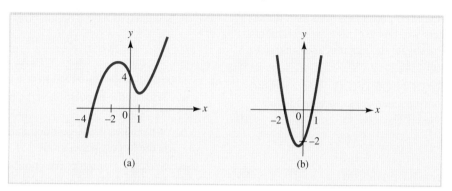

(a)   (b)

2. Find the intervals of increase and the intervals of decrease for each function and determine whether each critical number corresponds to a relative maximum, a relative minimum, or neither.

   a. $f(x) = -x^4 + 4x^3 + 5$

   b. $f(t) = 2t^3 - 9t^2 + 12t + 5$

   c. $g(t) = \dfrac{t}{t^2 + 9}$

   d. $g(x) = \dfrac{4 - x}{x^2 + 9}$

3. Determine where the graph of each function is concave upward and where it is concave downward. Find the x (or t) coordinate of each point of inflection.

   a. $f(x) = 3x^5 - 10x^4 + 2x - 5$

   b. $f(x) = 3x^5 + 20x^4 - 50x^3$

   c. $f(t) = \dfrac{t^2}{t - 1}$

   d. $g(t) = \dfrac{3t^2 + 5}{t^2 + 3}$

4. Determine all vertical and horizontal asymptotes for the graph of each function.

   a. $f(x) = \dfrac{2x - 1}{x + 3}$

   b. $f(x) = \dfrac{x}{x^2 - 1}$

   c. $f(x) = \dfrac{x^2 + x - 1}{2x^2 + x - 3}$

   d. $f(x) = \dfrac{1}{x} - \dfrac{1}{\sqrt{x}}$

5. Sketch the graph of each function. Be sure to show key features such as intercepts, asymptotes, high and low points, points of inflection, cusps, and vertical tangents.

   a. $f(x) = 3x^4 - 4x^3$

   b. $f(x) = x^4 - 3x^3 + 3x^2 + 1$

   c. $f(x) = \dfrac{x^2 + 2x + 1}{x^2}$

   d. $f(x) = \dfrac{1 - 2x}{(x - 1)^2}$

6. Sketch the graph of a function $f(x)$ with all of these properties:

   - $f'(x) > 0$ for $x < 0$ and for $0 < x < 2$
   - $f'(x) < 0$ for $x > 2$
   - $f'(0) = f'(2) = 0$
   - $f''(x) < 0$ for $x < 0$ and for $x > 1$
   - $f''(x) > 0$ for $0 < x < 1$
   - $f(-1) = f(4) = 0, f(0) = 1, f(1) = 2, f(2) = 3$

7. In each of the following cases, find the largest and smallest values of the given function on the specified interval.

   a. $f(x) = x^3 - 3x^2 - 9x + 1, [-2, 4]$

   b. $g(t) = -4t^3 + 9t^2 + 12t - 5, [-1, 4]$

   c. $h(u) = 8\sqrt{u} - u + 3$ on $0 \le u \le 25$

8. **WORKER EFFICIENCY** The mail room of a large company sorts mail into internal, local, provincial, Canada-wide, and international. An employee comes to work at 6 A.M. and t hours later has sorted

approximately $f(t) = -t^3 + 7t^2 + 200t$ pieces of mail. At what time during the period from 6 A.M. to 10 A.M. is the employee performing at peak efficiency?

9. **MAXIMIZING PROFIT**   A manufacturer can produce MP3 players at a cost of $90 each. It is estimated that if the MP3 players are sold for x dollars each, consumers will buy $20(180 - x)$ each month. What unit price should the manufacturer charge to maximize profit?

10. **CONCENTRATION OF DRUG**   The concentration of a drug in a patient's bloodstream t hours after it is injected is given by

$$C(t) = \frac{0.05t}{t^2 + 27}$$

milligrams per cubic centimetre.

a.   Sketch the graph of the concentration function.

b.   At what time is the concentration decreasing most rapidly?

c.   What happens to the concentration in the long run (as $t \to +\infty$)?

11. **BACTERIAL POPULATION**   A bacterial colony is estimated to have a population of

$$P(t) = \frac{15t^2 + 10}{t^3 + 6} \text{ million}$$

t hours after the introduction of a toxin.

a.   What is the population at the time the toxin is introduced?

b.   When does the largest population occur? What is the largest population?

c.   Sketch the graph of the population curve. What happens to the population in the long run (as $t \to +\infty$)?

Review Exercises

In Exercises 1 through 10, determine intervals of increase and decrease and intervals of concavity for the given function. Then sketch the graph of the function. Be sure to show all key features, such as intercepts, asymptotes, high and low points, points of inflection, cusps, and vertical tangents.

1. $f(x) = -2x^3 + 3x^2 + 12x - 5$

2. $f(x) = x^2 - 6x + 1$

3. $f(x) = 3x^3 - 4x^2 - 12x + 17$

4. $f(x) = x^3 - 3x^2 + 2$

5. $f(t) = 3t^5 - 20t^3$

6. $f(x) = \dfrac{x^2 + 3}{x - 1}$

7. $g(t) = \dfrac{t^2}{t + 1}$

8. $G(x) = (2x - 1)^2(x - 3)^3$

9. $F(x) = 2x + \dfrac{8}{x} + 2$

10. $f(x) = \dfrac{1}{x^3} + \dfrac{2}{x^2} + \dfrac{1}{x}$

In Exercises 11 and 12, one of the two curves shown is the graph of a certain function $f(x)$ and the other is the graph of its derivative $f'(x)$. Determine which curve is the graph of the derivative and give reasons for your decision.

11.

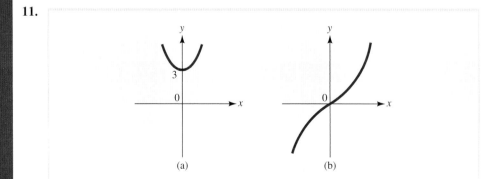

(a)   (b)

12.

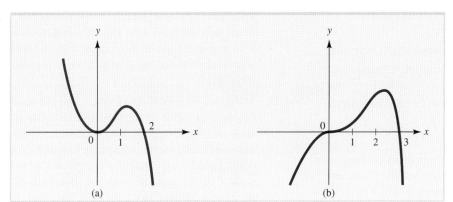

(a)　　　　　　(b)

In Exercises 13 through 16, the derivative $f'(x)$ of a function is given. Use this information to classify each critical number of $f(x)$ as a relative maximum, a relative minimum, or neither.

13. $f'(x) = x^3(2x - 3)^2(x + 1)^5(x - 7)$

14. $f'(x) = \sqrt[3]{x}(3 - x)(x + 1)^2$

15. $f'(x) = \dfrac{x(x - 2)^2}{x^4 + 1}$

16. $f'(x) = \dfrac{x^2 + 2x - 3}{x^2(x^2 + 1)}$

In Exercises 17 through 20, sketch the graph of a function f that has all of the given properties.

17. • $f'(x) > 0$ when $x < 0$ and when $x > 5$

　• $f'(x) < 0$ when $0 < x < 5$

　• $f''(x) > 0$ when $-6 < x < -3$ and when $x > 2$

　• $f''(x) < 0$ when $x < -6$ and when $-3 < x < 2$

18. • $f'(x) > 0$ when $x < -2$ and when $-2 < x < 3$

　• $f'(x) < 0$ when $x > 3$

　• $f'(-2) = 0$ and $f'(3) = 0$

19. • $f'(x) > 0$ when $1 < x < 2$

　• $f'(x) < 0$ when $x < 1$ and when $x > 2$

　• $f''(x) > 0$ for $x < 2$ and for $x > 2$

　• $f'(1) = 0$ and $f'(2)$ is undefined

20. • $f'(x) > 0$ when $x < 1$

　• $f'(x) < 0$ when $x > 1$

　• $f''(x) > 0$ when $x < 1$ and when $x > 1$

　• $f'(1)$ is undefined

In Exercises 21 through 24, find all critical numbers for the given function $f(x)$ and use the second derivative test to determine which (if any) critical points are relative maxima or relative minima.

21. $f(x) = -2x^3 + 3x^2 + 12x - 5$

22. $f(x) = x(2x - 3)^2$

23. $f(x) = \dfrac{x^2}{x + 1}$

24. $f(x) = \dfrac{1}{x} - \dfrac{1}{x + 3}$

In Exercises 25 through 28, find the absolute maximum and the absolute minimum values (if any) of the given function on the specified interval.

25. $f(x) = -2x^3 + 3x^2 + 12x - 5; -3 \le x \le 3$

26. $f(t) = -3t^4 + 8t^3 - 10; 0 \le t \le 3$

27. $g(s) = \dfrac{s^2}{s + 1}; -\dfrac{1}{2} \le s \le 1$

28. $f(x) = 2x + \dfrac{8}{x} + 2; x > 0$

29. The first derivative of a certain function is
$f'(x) = x(x - 1)^2$.

　a. On what intervals is f increasing? Decreasing?

　b. On what intervals is the graph of f concave up? Concave down?

　c. Find the x coordinates of the relative extrema and inflection points of f.

　d. Sketch a possible graph of $f(x)$.

30. The first derivative of a certain function is
$f'(x) = x^2(5 - x)$.

  a. On what intervals is f increasing? Decreasing?

  b. On what intervals is the graph of f concave up? Concave down?

  c. Find the x coordinates of the relative extrema and inflection points of f.

  d. Sketch a possible graph of $f(x)$.

31. **PROFIT**   A manufacturer can produce sunglasses at a cost of $5 each and estimates that if they are sold for x dollars, consumers will buy $100(20 - x)$ pairs of sunglasses per day. At what price should the manufacturer sell the sunglasses to maximize profit?

32. **CONSTRUCTION COST**   A box with a rectangular base is to be constructed of material costing 2 cents/cm^2 for the sides and bottom and 3 cents/cm^2 for the top. If the box is to have volume 1215 cm^3 and the length of its base is to be twice its width, what dimensions of the box will minimize the cost of construction? What is the minimal cost?

33. **A* CONSTRUCTION COST**   A cylindrical container with no top is to be constructed for a fixed amount of money. The cost of the material used for the bottom is 3 cents/cm^2, and the cost of the material used for the curved side is 2 cents/cm^2. Use calculus to derive a simple relationship between the radius and height of the container with the greatest volume.

34. **REAL ESTATE DEVELOPMENT**   A real estate developer estimates that if 60 luxury houses are built in a certain area, the average profit will be $47 500 per house. The average profit will decrease by $500 per house for each additional house built in the area. How many houses should the developer build to maximize the total profit? (Remember, the answer must be an integer.)

35. **OPTIMAL DESIGN**   A farmer wishes to enclose a rectangular pasture with 320 m of fence. What dimensions give the maximum area in each of the following situations?

  a. The fence is on all four sides of the pasture.

  b. The fence is on three sides of the pasture and the fourth side is bounded by a wall.

36. **TRAFFIC CONTROL**   It is estimated that between the hours of noon and 7:00 P.M., the speed of highway traffic flowing past a certain downtown exit is approximately
$$S(t) = t^3 - 9t^2 + 15t + 45$$

kilometres per hour, where t is the number of hours past noon. At what time between noon and 7:00 P.M. is the traffic moving the fastest, and at what time between noon and 7:00 P.M. is it moving the slowest?

37. **MINIMIZING TRAVEL TIME**   You are standing beside your canoe on the bank of a river that is 1 km wide and want to get back to your campsite on the opposite bank, 1 km upstream, before dark. You plan to paddle the canoe in a straight line to some point P on the opposite bank and then walk the remaining distance along the bank on a trail. To what point P should you paddle to reach the campsite in the shortest possible time if you can paddle at 4 km/h and walk at 5 km/h?

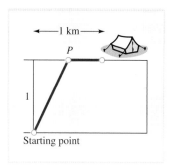

EXERCISE 37

38. **CONSTRUCTION**   You wish to use 300 m of fencing to surround two identical adjacent rectangular plots, as shown in the accompanying figure. How should you do this to make the combined area of the plots as large as possible?

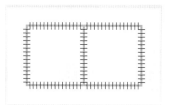

EXERCISE 38

39. **PRODUCTION**   A manufacturing firm has received an order to make 400 000 small souvenir wings for children flying on a certain airline. The firm owns 20 machines, each of which can produce 200 wings per hour. The cost of setting up the machines to produce the wings is $80 per machine, and the total operating cost is $5.76 per hour. How many machines should be used to minimize the cost of producing the 400 000 wings? (Remember, the answer must be an integer.)

40. ELASTICITY OF DEMAND Suppose that the demand equation for a certain commodity is $q = 60 - 0.1p$ (for $0 \le p \le 600$).

a. Express the elasticity of demand as a function of p.

b. Calculate the elasticity of demand when the price is $p = 200$. Interpret your answer.

c. At what price is the elasticity of demand equal to -1?

41. ELASTICITY OF DEMAND Suppose that the demand equation for a certain commodity is $q = 200 - 2p^2$ (for $0 \le p \le 10$).

a. Express the elasticity of demand as a function of p.

b. Calculate the elasticity of demand when the price is $p = 6$. Interpret your answer.

c. At what price is the elasticity of demand equal to -1?

42. ELASTICITY AND REVENUE Suppose that $q = 500 - 2p$ units of a certain commodity are demanded when p dollars per unit are charged, for $0 \le p \le 250$.

a. Determine where the demand is elastic, inelastic, and of unit elasticity with respect to price.

b. Use the results of part (a) to determine the intervals of increase and decrease of the revenue function and the price at which revenue is maximized.

c. Find the total revenue function explicitly and use its first derivative to determine its intervals of increase and decrease and the price at which revenue is maximized.

d. Graph the demand and revenue functions.

43. DEMAND FOR CRUISE TICKETS A cruise line estimates that when each deluxe balcony stateroom on a particular cruise is priced at p thousand dollars, then q tickets for staterooms will be demanded by travellers, where $q = 300 - 0.7p^2$.

a. Find the elasticity of demand for the stateroom tickets.

b. When the price is $8000 ($p = 8$) per stateroom, should the cruise line raise or lower the price in order to increase total revenue?

44. PACKAGING What is the maximum possible volume of a cylindrical can with no top that can be made from an area of 27π cm^2 of metal?

45. ARCHITECTURE A window is in the form of an equilateral triangle surmounted on a rectangle. The rectangle is of clear glass and transmits twice as much light as does the triangle, which is made of stained glass. If the entire window has a perimeter of 8 m, find the dimensions of the window that will admit the most light.

46. CONSTRUCTION COST Oil from an offshore rig located 3 km from the shore is to be pumped to a location on the edge of the shore that is 8 km east of the rig. The cost of constructing a pipe in the ocean from the rig to the shore is 1.5 times the cost of construction on land. How should the pipe be laid to minimize cost?

47. INVENTORY Through its franchised stations, an oil company gives out 16 000 road maps per year. The cost of setting up a press to print the maps is $100 for each production run. In addition, production costs are 6 cents per map and storage costs are 20 cents per map per year. The maps are distributed at a uniform rate throughout the year and are printed in equal batches timed so that each arrives just as the preceding batch has been used up. How many maps should the oil company print in each batch to minimize cost?

48. A* MARINE BIOLOGY When a fish swims upstream at a speed v against a constant current with speed v_w, the energy it expends in travelling to a point upstream is given by a function of the form

$$E(v) = \frac{Cv^k}{v - v_w},$$

where $C > 0$ and $k > 2$ is a number that depends on the species of fish involved.*

a. Show that $E(v)$ has exactly one critical number. Does it correspond to a relative maximum or a relative minimum?

b. Note that the critical number in part (a) depends on k. Let $F(k)$ be the critical number. Sketch the graph of $F(k)$. What can be said about $F(k)$ if k is very large?

---

*E. Batschelet, *Introduction to Mathematics for Life Scientists,* 2nd ed., New York: Springer-Verlag, 1976, p. 280.

CHAPTER SUMMARY

49. A* INVENTORY A manufacturing firm receives raw materials in equal shipments arriving at regular intervals throughout the year. The cost of storing the raw materials is directly proportional to the size of each shipment, while the total yearly ordering cost is inversely proportional to the shipment size. Show that the total cost is lowest when the total storage cost and total ordering cost are equal.

50. A* PHYSICAL CHEMISTRY In physical chemistry, the pressure P of a gas is related to the volume V and temperature T by *van der Waals' equation:*

$$\left(P + \frac{a}{V^2}\right)(V - b) = nRT,$$

where a, b, n, and R are constants. The *critical temperature* T_c of the gas is the highest temperature at which the gaseous and liquid phases can exist as separate phases.

  a. When $T = T_c$, the pressure P is given as a function $P(V)$ of volume alone. Sketch the graph of $P(V)$.

  b. The critical volume V_c is the volume for which $P'(V_c) = 0$ and $P''(V_c) = 0$. Show that $V_c = 3b$.

  c. Find the critical pressure $P_c = P(V_c)$ and then express T_c in terms of a, b, n, and R.

51. A* CRYSTALLOGRAPHY A fundamental problem in crystallography is the determination of the **packing fraction** of a crystal lattice, which is the fraction of space occupied by the atoms in the lattice, assuming that the atoms are hard spheres. When the lattice contains exactly two different kinds of atoms, it can be shown that the packing fraction is given by the formula*

$$f(x) = \frac{K(1 + c^2x^3)}{(1 + x)^3},$$

where $x = \dfrac{r}{R}$ is the ratio of the radii, r and R, of the two kinds of atoms in the lattice, and c and K are positive constants.

  a. The function $f(x)$ has exactly one critical number. Find it and use the second derivative test to classify it as a relative maximum or a relative minimum.

  b. The numbers c and K and the domain of $f(x)$ depend on the cell structure in the lattice. For

ordinary rock salt, $c = 1$, $K = \dfrac{2\pi}{3}$, and the domain is the interval $(\sqrt{2} - 1) \le x \le 1$. Find the largest and smallest values of $f(x)$.

  c. Repeat part (b) for β-cristobalite, for which $c = \sqrt{2}$, $K = \dfrac{\sqrt{3}\pi}{16}$, and the domain is $0 \le x \le 1$.

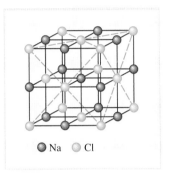

Na   Cl

EXERCISE 51

  d. What can be said about the packing fraction $f(x)$ if r is much larger than R? Answer this question by computing $\lim_{x \to \infty} f(x)$.

 e. Read the article on which this exercise is based and write a paragraph on how packing factors are computed and used in crystallography.

52. A* BOTANY An experimental garden plot contains N annual plants, each of which produces S seeds that are dropped within the same plot. A botanical model measures the number of offspring plants $A(N)$ that survive until the next year by the function

$$A(N) = \frac{NS}{1 + (cN)^p},$$

where c and p are positive constants.

  a. For what value of N is $A(N)$ maximized? What is the maximum value? Express your answer in terms of S, c, and p.

  b. For what value of N is the offspring survival rate $A'(N)$ minimized?

  c. The function $F(N) = \dfrac{A(N)}{N}$ is called the *net reproductive rate*. It measures the number of surviving offspring per plant. Find $F'(N)$ and use it to show that the greater the number of plants, the lower the number of surviving offspring per plant. This is called the principle of *density-dependent mortality*.

---

*John C. Lewis and Peter P. Gillis, "Packing Factors in Diatomic Crystals," *American Journal of Physics,* Vol. 61, No. 5, 1993, pp. 434–438.

53. A* ELIMINATION OF HAZARDOUS WASTE

Certain hazardous waste products have the property that as the concentration of substrate (the substance undergoing change by enzymatic action) increases, a toxic inhibition effect occurs. A mathematical model for this behaviour is the **Haldane equation***

$$R(S) = \frac{cS}{a + S + bS^2},$$

where R is the specific growth rate of the substance (the rate at which cells divide); S is the substrate concentration; and a, b, and c are positive constants.

  a. Sketch the graph of $R(S)$. Does the graph appear to have a highest point? A lowest point? A point of inflection? What happens to the growth rate as S grows larger and larger? Interpret your observations.

  b. Read an article on hazardous waste management and write a paragraph on how mathematical models are used to develop methods for eliminating waste. A good place to start is the reference cited here.

54. A* FIREFIGHTING

If air resistance is neglected, it can be shown that the stream of water emitted by a fire hose will have height

$$x = -4.9(1 + m^2)\left(\frac{x}{v}\right)^2 + mx$$

metres above a point located x metres from the nozzle, where m is the slope of the nozzle and v is

the velocity of the stream of water as it leaves the nozzle, with units of metres per second. Assume that v is constant.

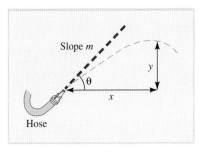

EXERCISE 54

  a. Suppose m is also constant. What is the maximum height reached by the stream of water? How far away from the nozzle does the stream reach (that is, what is x when $y = 0$)?

  b. If m is allowed to vary, find the slope that allows a firefighter to spray water on a fire from the greatest distance.

  c. Suppose the firefighter is $x = x_0$ metres from the base of a building. If m is allowed to vary, what is the highest point on the building that the firefighter can reach with the water from her hose?

---

*Michael D. La Grega, Philip L. Buckingham, and Jeffrey C. Evans, *Hazardous Waste Management*, New York: McGraw-Hill, 1994, p. 578.

THINK ABOUT IT

MODELLING DATA—SUNFLOWER EXPORTS FROM MANITOBA

(Photo: CORBIS/Royalty Free)

Mathematical models can be used to study trends in infectious diseases, in inflation and unemployment, and in enrollment in university, and to study any statistical data that can be approximated by a continuous function. We will look at making mathematical models of some agricultural data, in particular sunflower product exports from Manitoba.*

Year	Exports (millions of dollars)
2002	44.5
2003	45.7
2004	32.0
2005	17.6
2006	28.6
2007	70.8
2008	72.6
2009	61.2
2010	30.0

The goal in modelling is to find a relatively simple function $f(t)$ that provides a close approximation to the data, in this case the value of the exports of sunflower products. One of the simplest approaches to constructing such functions is to use *polynomial regression,* a technique that produces best-fit polynomials of specified degree that are good approximations to observed data. The higher the degree of the polynomial, the better the approximation, but the more complicated the function. Care

*Janet Honey, "Crops in Manitoba 2009–2010," Department of Agribusiness and Agricultural Economics, University of Manitoba.

must be taken to avoid overfitting, where random error is fitted instead of the underlying relationship. Suppose we decide to use a polynomial of degree 3, so that $f(t)$ has the general form $f(t) = at^3 + bt^2 + ct + d$. The coefficients a, b, c, and d of the best-fitting cubic curve we seek are then determined by requiring the sum of squares of the vertical distances between data points in the table and corresponding points on the cubic polynomial $y = f(t)$ to be as small as possible. Calculus methods for carrying out this optimization procedure are developed in Section 7.4. In practice, however, regression polynomials are almost always found using a computer. It is important to realize that there are no underlying reasons why a polynomial should provide a good approximation to the value of the sunflower exports. Nevertheless, we begin with such models because they are easy to construct.

The modelling here was done with the software Maple, with $t = 0$ as 2002 and the vertical scale, $E(t)$, in millions of dollars.

The first graph shows an approximation to a linear relationship. The data show considerable scatter away from the line, although there appears to be an increasing trend in exports. The variations are so large that we can make no conclusions from this line, so the linear model does not represent the data adequately.

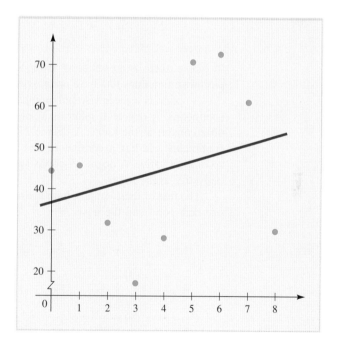

We go to a polynomial fit. This curve follows the data closely, with the implication that there is a sharp downward trend after 2008. The polynomial generated by Maple is $E(t) = -0.0198t^8 + 0.607t^7 - 7.502t^6 + 48.1t^5 - 171.8t^4 + 341.6t^3 - 354.9t^2 + 145.04t + 44.5$.

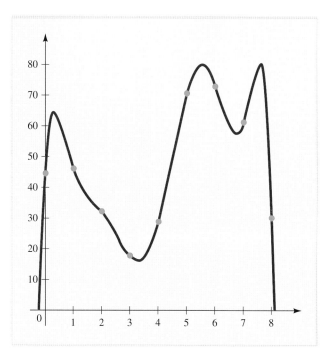

In Maple, there is an option of using splines, which effectively break the curve into piecewise functions that have certain properties where they join, that is, at the "knots."

A spline is a piecewise polynomial such that a certain number of derivatives are continuous. A degree-1 spline is a series of straight lines that is continuous but not differentiable at the knots. A degree-2 spline is a piecewise quadratic function that is continuous and has a continuous first derivative at the knots but not a continuous second derivative. A degree-3, cubic, spline is piecewise cubic with continuous second derivative at the knots.

The first of these graphs, with spline degree 1, shows a linear relationship between one data point and the next.

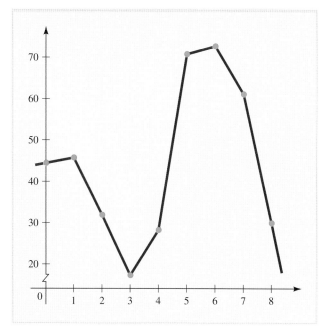

$$E(t) = \begin{cases} 44.5 + 1.2t, & t < 0.5 \\ 59.4 - 13.7t, & t < 2.5 \\ 60.8 - 14.4t, & t < 3.5 \\ -15.4 + 11t, & t < 4.5 \\ 140.2 + 42.2t, & t < 5.5 \\ 61.8 + 1.8t, & t < 6.5 \\ 141 - 11.4t, & t < 7.5 \\ 279.6 - 31.2t & \text{otherwise} \end{cases}$$

The next graph was made with spline degree 2, with a piecewise quadratic function. The function has a continuous first derivative at each knot. The third has a spline of 3, and a continuous second derivative at the knots, and the next has a spline of 4.

$$E(t) = \begin{cases} 44.5 + 5.149t^2, & t < 0.5 \\ 51.2 - 5.497t - 10.6(t-1)^2, & t < 1.5 \\ 66 - 17t - 0.879(t-2)^2, & t < 2.5 \\ 31.9 - 4.98t + 13.1(t-3)^2, & t < 3.5 \\ -99.8 + 32.1t + 23.8(t-4)^2, & t < 4.5 \\ -54.1 + 24.98t - 32.9(t-5)^2, & t < 5.5 \\ 108.6 - 5.997t - 1.04(t-6)^2, & t < 6.5 \\ 253 - 27.4t - 21.3(t-7)^2, & t < 7.5 \\ 30 + 48.7(t-8)^2 & \text{otherwise} \end{cases}$$

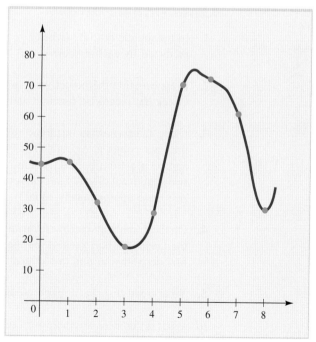

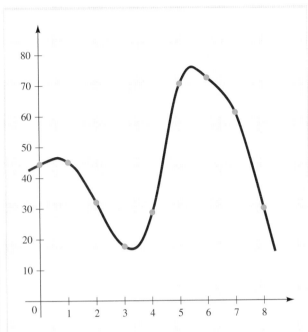

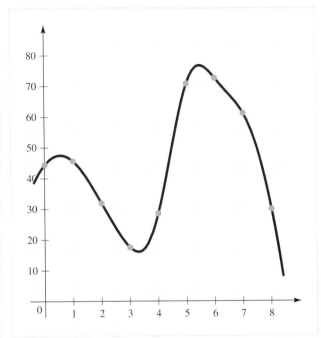

These graphs can be used to analyze the given data and to make predictions. A few examples of this kind of analysis are provided in the accompanying questions.

Questions

1. In the middle of 2007 (when $t = 5.5$), what is the approximate value of exports predicted by each model?

2. It appears that the model with spline degree 1 might be differentiable at $t = 2$. Check the piecewise function and investigate whether this is true.

3. Using the piecewise quadratic function, notice that there are "bends" between the data points at 5.5 and 0.5. Find the derivative at $t = 0.5$ as the function approaches from both directions. Are these derivatives approximately equal, to two significant digits? Find the second derivative at $t = 0.5$, from both directions. Are these approximately equal?

4. On the graph with piecewise quadratic functions, find an approximate time where the tangent is horizontal. Find the part of the function that applies and take the first derivative to find this critical point.

5. What trend does each graph predict for the years following 2010?

6. What differences are there between the cubic piecewise functions and the polynomial function?

 Practise and learn online with Connect.

Sneezing ejects a virus from the nose, potentially spreading it to other people. Exponential functions can be used to describe the number of people infected in an epidemic. In Example 4.4.6, we will see that the number infected does not increase without bound but depends on the number vaccinated and the number susceptible to the disease. (Photo: © StockDisc/PunchStock).

EXPONENTIAL AND LOGARITHMIC FUNCTIONS

LEARNING OBJECTIVES

After completing this chapter, you should be able to

L01 Solve problems, including continuous compounding, using the exponential rules.

L02 Simplify logarithmic and exponential expressions. Solve equations using natural logarithms and exponents.

L03 Differentiate both logarithmic and exponential functions.

L04 Formulate equations for exponential applications and calculate all the unknown constants.

SECTION 4.1

L01

Solve problems, including continuous compounding, using the exponential rules.

Exponential Functions; Continuous Compounding

A population $Q(t)$ is said to grow **exponentially** if whenever it is measured at equally spaced time intervals, the population at the end of any particular interval is a fixed multiple (greater than 1) of the population at the end of the previous interval. For instance, according to the United Nations, in the year 2000, the population of the world was 6.1 billion people and was growing at an annual rate of about 1.4%. If this pattern were to continue, then every year, the population would be 1.014 times the population of the previous year. Thus, if $P(t)$ represented the world population (in billions) t years after the base year 2000, the population would grow as follows:

2000	$P(0) = 6.1$
2001	$P(1) = 6.1(1.014) = 6.185$
2002	$P(2) = 6.185(1.014) = [6.1(1.014)](1.014) = 6.1(1.014)^2 = 6.272$
2003	$P(3) = 6.272(1.014) = [6.1(1.014)^2](1.014) = 6.1(1.014)^3 = 6.360$
$\vdots$	$\vdots$
$2000 + t$	$P(t) = 6.1(1.014)^t$

The graph of $P(t)$ is shown in Figure 4.1a. Notice that according to this model, the world population grows gradually at first but doubles after about 50 years (to 12.22 billion in 2050).

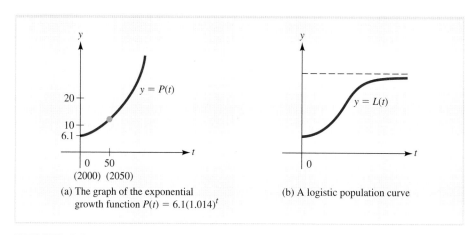

(a) The graph of the exponential growth function $P(t) = 6.1(1.014)^t$

(b) A logistic population curve

FIGURE 4.1   Two models for population growth.

Exponential population models are sometimes referred to as *Malthusian,* after Thomas Malthus (1766–1834), a British economist who predicted that mass starvation would result if a population grew exponentially while the food supply grew at a constant rate (linearly). Fortunately, world population does not continue to grow exponentially as predicted by Malthus's model, and models that take into account various restrictions on the growth rate actually provide more accurate predictions. The population curve that results from one such model, the *logistic* model, is shown in Figure 4.1b. Note how the logistic growth curve rises steeply at first, like an exponential curve, but then eventually turns over and flattens out as environmental factors act to brake the growth rate. We will examine logistic curves in Section 4.4 (Example 4.4.6) and again in Chapter 8 as part of a more detailed study of population models.

A function of the general form $f(x) = b^x$, where b is a positive number, is called an **exponential function.** Such functions can be used to describe exponential and logistic growth and a variety of other important quantities. For instance, exponential functions are used in demography to forecast population size, in finance to calculate the value of investments, in archaeology to date ancient artifacts, in psychology to study learning patterns, and in industry to estimate the reliability of products.

In this section, we will explore the properties of exponential functions and introduce a few basic models in which such functions play a prominent role. We will examine additional applications, such as the logistic model, in subsequent sections.

Working with exponential functions requires the use of exponential notation and the algebraic laws of exponents. Solved examples and practice problems involving this notation can be found in Appendix A1. Here is a brief summary of the notation.

Definition of b^n for Rational Values of n (and $b > 0$) ■ Integer powers: If n is a positive integer, then

$$b^n = b \cdot b \cdot b \ldots b \quad \text{for } n \text{ factors}$$

Fractional powers: If n and m are positive integers, then

$$b^{n/m} = (\sqrt[m]{b})^n = \sqrt[m]{b^n},$$

where $\sqrt[m]{b}$ denotes the positive mth root.

Negative powers: $b^{-n} = \dfrac{1}{b^n}$

Zero power: $b^0 = 1$

For example,

$$3^4 = 3 \cdot 3 \cdot 3 \cdot 3 = 81 \qquad\qquad 3^{-4} = \frac{1}{3^4} = \frac{1}{81}$$

$$4^{1/2} = \sqrt{4} = 2 \qquad\qquad 4^{3/2} = (\sqrt{4})^3 = 2^3 = 8$$

$$4^{-3/2} = \frac{1}{4^{3/2}} = \frac{1}{8} \qquad\qquad 27^{-2/3} = \frac{1}{(\sqrt[3]{27})^2} = \frac{1}{3^2} = \frac{1}{9}$$

We know what is meant by b^r for any rational number r, but if we try to graph $y = b^x$ there will be a hole in the graph for each value of x that is not rational, such as $x = \sqrt{2}$. However, using methods beyond the scope of this book, it can be shown that irrational numbers can be approximated as closely as we like by rational numbers, which in turn implies that there is only one unbroken curve passing through all points (r, b^r) for r rational. In other words, there exists a unique continuous function $f(x)$ that is defined for all real numbers x and is equal to b^r when r is rational. It is this function we define as $f(x) = b^r$.

Exponential Functions ■ If b is a positive number other than 1 ($b > 0$, $b \neq 1$), there is a unique function called the *exponential function* with base b that is defined by

$$f(x) = b^x \qquad \text{for every real number } x.$$

To get a feeling for the appearance of the graph of an exponential function, consider Example 4.1.1.

EXAMPLE 4.1.1

Sketch the graphs of $y = 2^x$ and $y = \left(\dfrac{1}{2}\right)^x$.

Solution

Begin by constructing a table of values for $y = 2^x$ and $y = \left(\dfrac{1}{2}\right)^x$:

x	−15	−10	−1	0	1	3	5	10	15
$y = 2^x$	0.00003	0.001	0.5	1	2	8	32	1024	32 768
$y = \left(\dfrac{1}{2}\right)^x$	32 768	1024	2	1	0.5	0.125	0.313	0.001	0.00003

The pattern of values in this table suggests that the functions $y = 2^x$ and $y = \left(\dfrac{1}{2}\right)^x$ have the following features:

The function $y = 2^x$	The function $y = \left(\dfrac{1}{2}\right)^x$
always increasing	always decreasing
$\displaystyle\lim_{x \to -\infty} 2^x = 0$	$\displaystyle\lim_{x \to -\infty} \left(\dfrac{1}{2}\right)^x = +\infty$
$\displaystyle\lim_{x \to +\infty} 2^x = +\infty$	$\displaystyle\lim_{x \to +\infty} \left(\dfrac{1}{2}\right)^x = 0$

Using this information, we sketch the graphs shown in Figure 4.2. Notice that each graph has $(0, 1)$ as its y intercept, has the x axis as a horizontal asymptote, and appears to be concave upward for all x. The graphs also appear to be reflections of one another in the y axis. You are asked to verify this observation in Exercise 74.

EXPLORE!

Graph $y = b^x$ for $b = 1, 2, 3,$ and 4. Explain what you observe. Conjecture and then check where the graph of $y = 4^x$ will lie relative to that of $y = 2^x$ and $y = 6^x$. Where does $y = e^x$ lie, assuming e is a value between 2 and 3?

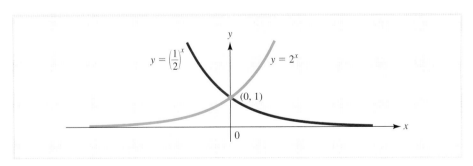

FIGURE 4.2   The graphs of $y = 2^x$ and $y = \left(\dfrac{1}{2}\right)^x$.

Figure 4.3 shows graphs of various members of the family of exponential functions $y = b^x$. Notice that the graph of any function of the form $y = b^x$ resembles that of $y = 2^x$ if $b > 1$ or that of $y = \left(\dfrac{1}{2}\right)^x$ if $0 < b < 1$. In the special case where $b = 1$, the function $y = b^x$ becomes the constant function $y = 1$.

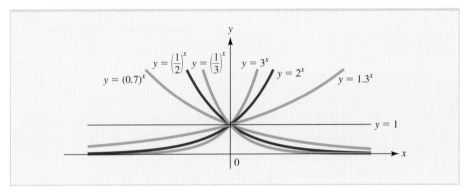

FIGURE 4.3   Graphs of the exponential form $y = b^x$.

Important graphical and analytical properties of exponential functions are summarized in the following box.

Properties of an Exponential Function   ■   The exponential function $f(x) = b^x$ for $b > 0$, $b \neq 1$, has these properties:

1. It is defined, continuous, and positive ($b^x > 0$) for all x.
2. The x axis is a horizontal asymptote of the graph of f.
3. The y intercept of the graph is $(0, 1)$; there is no x intercept.
4. If $b > 1$, $\displaystyle\lim_{x \to -\infty} b^x = 0$ and $\displaystyle\lim_{x \to +\infty} b^x = +\infty$.
    If $0 < b < 1$, $\displaystyle\lim_{x \to -\infty} b^x = +\infty$ and $\displaystyle\lim_{x \to +\infty} b^x = 0$.
5. For all x, the function is increasing (graph rising) if $b > 1$ and decreasing (graph falling) if $0 < b < 1$.

NOTE   Be careful not to confuse the *power* function $p(x) = x^b$ with the *exponential* function $f(x) = b^x$. Remember that in x^b, the variable x is the base and the exponent b is constant, while in b^x, the base b is constant and the variable x is the exponent. The graphs of $y = x^2$ and $y = 2^x$ are shown in Figure 4.4. Notice that after the crossover point $(4, 16)$, the exponential curve $y = 2^x$ rises much more steeply than the power curve $y = x^2$. For instance, when $x = 10$, the y value on the power curve is $y = 10^2 = 100$, while the corresponding y value on the exponential curve is $y = 2^{10} = 1024$. ■

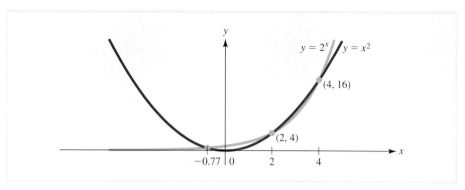

FIGURE 4.4   Comparing the power curve $y = x^2$ with the exponential curve $y = 2^x$.

Exponential functions obey the same algebraic rules as the rules for exponential numbers reviewed in Appendix A1. These rules are summarized in the following box.

Exponential Rules ■ For bases a, b ($a > 0$, $b > 0$) and any real numbers x, y, we have the following:

The **equality rule:** $b^x = b^y$ if and only if $x = y$.

The **product rule:** $b^x b^y = b^{x+y}$

The **quotient rule:** $\dfrac{b^x}{b^y} = b^{x-y}$

The **power rule:** $(b^x)^y = b^{xy}$

The **multiplication rule:** $(ab)^x = a^x b^x$

The **division rule:** $\left(\dfrac{a}{b}\right)^x = \dfrac{a^x}{b^x}$

EXAMPLE 4.1.2

Evaluate each exponential expression.

a. $(3)^2(3)^3$ **c.** $(5^{1/3})(2^{1/3})$ **e.** $\left(\dfrac{4}{7}\right)^3$

b. $(2^3)^2$ **d.** $\dfrac{2^3}{2^5}$

Solution

a. $(3)^2(3)^3 = 3^{2+3} = 3^5 = 243$

b. $(2^3)^2 = 2^{(3)(2)} = 2^6 = 64$

c. $(5^{1/3})(2^{1/3}) = [(5)(2)]^{1/3} = 10^{1/3} = \sqrt[3]{10}$

d. $\dfrac{2^3}{2^5} = 2^{3-5} = 2^{-2} = \dfrac{1}{4}$

e. $\left(\dfrac{4}{7}\right)^3 = \dfrac{4^3}{7^3} = \dfrac{64}{343}$

EXAMPLE 4.1.3

If $f(x) = 5^{x^2+2x}$, find all values of x such that $f(x) = 125$.

Solution

The equation $f(x) = 125 = 5^3$ is satisfied if and only if

$$5^{x^2+2x} = 5^3$$
$$x^2 + 2x = 3 \qquad \text{since } b^x = b^y \text{ only when } x = y$$
$$x^2 + 2x - 3 = 0$$
$$(x - 1)(x + 3) = 0 \qquad \text{factor}$$
$$x = 1, x = -3$$

Thus, $f(x) = 125$ if and only if $x = 1$ or $x = -3$.

The Natural Exponential Base e

In algebra, it is common practice to use the base $b = 10$ for exponential functions or, in some cases, $b = 2$, but in calculus, it turns out to be more convenient to use a number denoted by e and defined by the limit

$$e = \lim_{n \to +\infty} \left(1 + \frac{1}{n}\right)^n.$$

"Hold on!" you may say. "That limit has to be 1, since $1 + \frac{1}{n}$ certainly tends to 1 as n increases without bound, and $1^n = 1$ for any n." Not so. The limit process does not work this way, as you can see from this table:

EXPLORE!

Graph $\left(1 + \frac{1}{x}\right)^x$ with x from 0 to 500. Can you see the horizontal asymptote?

n	10	100	1000	10 000	100 000	1 000 000
$\left(1 + \frac{1}{n}\right)^n$	2.59374	2.70481	2.71692	2.71816	2.71827	2.71828

The number e is one of the most important numbers in all mathematics, and its value has been computed with great precision. To twelve decimal places, its value is

$$e = \lim_{n \to +\infty} \left(1 + \frac{1}{n}\right)^n = 2.718281828459\ldots.$$

The function $y = e^x$ is called the **natural exponential function.** Use the e^x key on a calculator to find exponential functions. For example, to find $e^{1.217}$, press the e^x key and then enter the number 1.217 to obtain $e^{1.217} \approx 3.37704$.

EXAMPLE 4.1.4

Biologists have determined that the number of bacteria in a culture is given by

$$P(t) = 5000e^{0.015t},$$

where t is the number of minutes after observation begins. What is the average rate of change of the bacterial population during the second hour?

Solution

During the second hour (from time $t = 60$ to $t = 120$), the population changes by $P(120) - P(60)$, so the average rate of change during this time period is given by

$$A = \frac{P(120) - P(60)}{120 - 60}$$

$$= \frac{[5000e^{0.015(120)}] - [5000e^{0.015(60)}]}{60}$$

$$\approx \frac{30\ 248 - 12\ 298}{60}$$

$$\approx 299$$

Thus, the population increases at an average rate of roughly 299 bacteria per minute during the second hour.

EXAMPLE 4.1.5

A manufacturer estimates that when x units of a particular commodity are produced, they can all be sold when the market price is p dollars per unit, where p is given by the demand function $p = 200e^{-0.01x}$. How much revenue is obtained when 100 units of the commodity are produced?

Solution

The revenue is given by the product (price/unit)(number of units sold), that is,

$$R(x) = p(x)x = (200e^{-0.01x})x = 200xe^{-0.01x}.$$

Use a calculator to find that the revenue obtained by producing $x = 100$ units is

$$R(100) = 200(100)\ e^{-0.01(100)} \approx 7357.59,$$

or approximately $7357.59.

Continuous Compounding of Interest

The number e is called the natural exponential base, although it may seem anything but natural to you. As an illustration of how this number appears in practical situations, we use it to describe the accounting practice known as continuous compounding of interest.

First, let us review the basic ideas behind compound interest. Suppose a sum of money is invested and the interest is compounded only once. If P is the initial investment (the *principal*) and r is the interest rate (expressed as a decimal), the balance B after the interest is added is

$$B = P + Pr = P(1 + r) \text{ dollars.}$$

That is, to compute the balance at the end of an interest period, you multiply the balance at the beginning of the period by the expression $1 + r$.

At most banks, interest is compounded more than once a year. The interest that is added to the account during one period will itself earn interest during the subsequent periods. If the annual interest rate is r and interest is compounded k times per year, then the year is divided into k equal compounding periods and the interest rate in each period is $\dfrac{r}{k}$. Hence, the balance at the end of the first period is

$$P_1 = \text{principal} + \text{interest}$$
$$= P + P\left(\frac{r}{k}\right) = P\left(1 + \frac{r}{k}\right)$$

At the end of the second period, the balance is

$$P_2 = P_1 + P_1\left(\frac{r}{k}\right) = P_1\left(1 + \frac{r}{k}\right)$$
$$= \left[P\left(1 + \frac{r}{k}\right)\right]\left(1 + \frac{r}{k}\right) = P\left(1 + \frac{r}{k}\right)^2$$

and, in general, the balance at the end of the mth period is

$$P_m = P\left(1 + \frac{r}{k}\right)^m.$$

Since there are k periods in a year, the balance after 1 year is

$$P\left(1 + \frac{r}{k}\right)^k.$$

At the end of t years, interest has been compounded kt times and the balance is given by the function

$$B(t) = P\left(1 + \frac{r}{k}\right)^{kt}.$$

As the frequency with which interest is compounded increases, the corresponding balance $B(t)$ also increases. Hence, a bank that compounds interest frequently may attract more customers than one that offers the same interest rate but compounds interest less often. But what happens to the balance at the end of t years as the compounding frequency increases without bound? More specifically, what will the balance be at the end of t years if interest is compounded not quarterly, not monthly, not daily, but continuously? In mathematical terms, this question is equivalent to asking what happens to the expression $P\left(1 + \dfrac{r}{k}\right)^{kt}$ as k increases without bound. The answer turns out to involve the number e. Here is the argument.

To simplify the calculation, let $n = \dfrac{k}{r}$. Then, $k = nr$ and so

$$P\left(1 + \frac{r}{k}\right)^{kt} = P\left(1 + \frac{1}{n}\right)^{nrt} = P\left[\left(1 + \frac{1}{n}\right)^n\right]^{rt}.$$

Since n increases without bound as k does, and since $\left(1 + \dfrac{1}{n}\right)^n$ approaches e as n increases without bound, it follows that the balance after t years is

$$B(t) = \lim_{k \to +\infty} P\left(1 + \frac{r}{k}\right)^{kt} = P\left[\lim_{n \to +\infty}\left(1 + \frac{1}{n}\right)^n\right]^{rt} = Pe^{rt}.$$

To summarize:

> **Compound Interest Formulas** ■ Suppose P dollars are invested at an annual interest rate r and the accumulated value (called the *future value*) in the account after t years is $B(t)$ dollars. If interest is compounded k times per year, then
>
> $$B(t) = P\left(1 + \frac{r}{k}\right)^{kt},$$
>
> and if interest is compounded continuously, then
>
> $$B(t) = Pe^{rt}.$$

EXPLORE!

Compare the following two formulas for the amount after x years by plotting the two functions on the same graph:

$$A(x) = 1000\left(1 + \frac{0.06}{x}\right)^{10x}$$

and

$$B(x) = 1000e^{0.06(10)}.$$

Both of these are compounded over 10 years at 6%, where x is the number of compounding periods per year. Trace the values of x from 1 to 24. The continuously compounded function will be a straight line. Examine how the amounts $A(x)$ and $B(x)$ approach each other.

EXAMPLE 4.1.6

Suppose $1000 is invested at an annual interest rate of 6%. Compute the balance after 10 years if the interest is compounded

a. quarterly   **b.** monthly   **c.** daily   **d.** continuously

Solution

a. To compute the balance after 10 years if the interest is compounded quarterly, use the formula $B(t) = P\left(1 + \dfrac{r}{k}\right)^{kt}$, with $t = 10$, $P = 1000$, $r = 0.06$, and $k = 4$:

$$B(10) = \$1000\left(1 + \frac{0.06}{4}\right)^{40} \approx \$1814.02.$$

b. This time, take $t = 10$, $P = 1000$, $r = 0.06$, and $k = 12$ to get

$$B(10) = \$1000\left(1 + \frac{0.06}{12}\right)^{120} \approx \$1819.40.$$

c. Take $t = 10$, $P = 1000$, $r = 0.06$, and $k = 365$ to obtain

$$B(10) = \$1000\left(1 + \frac{0.06}{365}\right)^{3650} \approx \$1822.03.$$

d. For continuously compounded interest, use the formula $B(t) = Pe^{rt}$, with $t = 10$, $P = 1000$, and $r = 0.06$:

$$B(10) = 1000e^{0.6} \approx \$1822.12.$$

This value, $1822.12, is an upper bound for the possible balance. No matter how often interest is compounded, $1000 invested at an annual interest rate of 6% cannot grow to more than $1822.12 in 10 years.

Present Value   In many situations, it is useful to know how much money P must be invested at a fixed compound interest rate in order to obtain a desired accumulated (future) value B over a given period of time T. This investment P is called the **present value of the**

Just-In-Time

If $C \neq 0$, the equation

$$A = PC$$

can be solved for P by dividing both sides by C so that $\dfrac{A}{C} = P$.

For convenience, we can change the form of C, remembering that

$$\frac{1}{C} = C^{-1},$$

to obtain $P = AC^{-1}$. In this example, $\dfrac{B}{\left(1 + \dfrac{r}{k}\right)^{kT}} = P$, so

$$B = P\left(1 + \frac{r}{k}\right)^{-kT}.$$

amount B to be received in T years. Present value may be regarded as a measure of the current worth of an investment and is used by economists to compare different investment possibilities.

To derive a formula for present value, we need only solve an appropriate future value formula for P. In particular, if the investment is compounded k times per year at an annual rate r for the term of T years, then

$$B = P\left(1 + \frac{r}{k}\right)^{kT}.$$

and the present value of B dollars in T years is obtained by multiplying both sides of the equation by $\left(1 + \dfrac{r}{k}\right)^{-kT}$ to get

$$P = B\left(1 + \frac{r}{k}\right)^{-kT}.$$

Likewise, if the compounding is continuous, then

$$B = Pe^{rT},$$

and the present value is given by

$$P = Be^{-rT}.$$

To summarize:

Present Value ■ The **present value** of B dollars in T years invested at an annual rate of r compounded k times per year is given by

$$P = B\left(1 + \frac{r}{k}\right)^{-kT}.$$

If interest is compounded continuously at the same rate, the present value in T years is given by

$$P = Be^{-rT}.$$

EXAMPLE 4.1.7

Sara is about to enter university. When she graduates 4 years from now, she wants to take a trip to Europe that she estimates will cost $5000. How much should she invest now at 7% to have enough for the trip if interest is compounded

  a. quarterly?   **b.** continuously?

Solution

The required future value is $B = \$5000$ in $T = 4$ years with $r = 0.07$.

  a. If the compounding is quarterly, then $k = 4$ and the present value is

$$P = \$5000\left(1 + \frac{0.07}{4}\right)^{-4(4)} = \$3788.08.$$

b. For continuous compounding, the present value is

$$P = \$5000e^{-0.07(4)} = \$3778.92.$$

Thus, Sara would have to invest about $9 more if interest were compounded quarterly than if the compounding were continuous.

Effective Interest Rate

Which is better, an investment that earns 10% compounded quarterly, one that earns 9.95% compounded monthly, or one that earns 9.9% compounded continuously? One way to answer this question is to determine the simple annual interest rate that is equivalent to each investment. This is known as the **effective interest rate,** and it can easily be obtained from the compound interest formulas.

Suppose interest is compounded k times per year at the annual rate r. This is called the **nominal** rate of interest. Then the balance at the end of 1 year is

$$A = P(1 + i)^k, \quad \text{where } i = \frac{r}{k}.$$

On the other hand, if x is the effective interest rate, the corresponding balance at the end of 1 year is $A = P(1 + x)$. Equating the two expressions for A, we get

$$P(1 + i)^k = P(1 + x) \quad \text{or} \quad x = (1 + i)^k - 1.$$

For continuous compounding, we have

$$Pe^r = P(1 + x), \quad \text{so} \quad x = e^r - 1.$$

To summarize:

Effective Interest Rate Formulas ■ If interest is compounded at the nominal rate r, the effective interest rate is the simple annual interest rate r_e that yields the same interest after 1 year. If the compounding is k times per year, the effective rate is given by the formula

$$r_e = (1 + i)^k - 1, \quad \text{where } i = \frac{r}{k},$$

while continuous compounding yields

$$r_e = e^r - 1.$$

Example 4.1.8 answers the question raised in the introduction to this subsection.

EXAMPLE 4.1.8

Which is better, an investment that earns 10% compounded quarterly, one that earns 9.95% compounded monthly, or one that earns 9.9% compounded continuously?

Solution

We answer the question by comparing the effective interest rates of the three investments. For the first, the nominal rate is 10% and compounding is quarterly, so we have $r = 0.10$, $k = 4$, and

$$i = \frac{r}{k} = \frac{0.10}{4} = 0.025.$$

Substituting into the formula for effective rate, we get

$$\textbf{First effective rate} = (1 + 0.025)^4 - 1 = 0.10381.$$

For the second investment, the nominal rate is 9.95% and compounding is monthly, so $r = 0.0995$, $k = 12$, and

$$i = \frac{r}{k} = \frac{0.0995}{12} = 0.008292.$$

We find that

$$\textbf{Second effective rate} = (1 + 0.008292)^{12} - 1 = 0.10417.$$

Finally, if compounding is continuous with nominal rate 9.9%, we have $r = 0.099$ and the effective rate is

$$\textbf{Third effective rate} = e^{0.099} - 1 = 0.10407.$$

The effective rates are, respectively, 10.38%, 10.42%, and 10.41%, so the second investment is best.

EXERCISES ■ 4.1

In Exercises 1 and 2, use a calculator to find the indicated powers of e. (Round your answers to three decimal places.)

1. e^2, e^{-2}, $e^{0.05}$, $e^{-0.05}$, e^0, e, $\sqrt{e}$, and $\dfrac{1}{\sqrt{e}}$

2. e^3, e^{-1}, $e^{0.01}$, $e^{-0.1}$, e^2, $e^{-1/2}$, $e^{1/3}$, and $\dfrac{1}{\sqrt[3]{e}}$

3. Sketch the curves $y = 3^x$ and $y = 4^x$ on the same set of axes.

4. Sketch the curves $y = \left(\dfrac{1}{3}\right)^x$ and $y = \left(\dfrac{1}{4}\right)^x$ on the same set of axes.

In Exercises 5 through 12, evaluate the given expressions by simplifying without using a calculator.

5. a. $27^{2/3}$

  b. $\left(\dfrac{1}{9}\right)^{3/2}$

6. a. $(-128)^{3/7}$

  b. $\left(\dfrac{27}{64}\right)^{2/3}\left(\dfrac{64}{25}\right)^{3/2}$

7. a. $8^{2/3} + 16^{3/4}$

  b. $\left(\dfrac{27 + 36}{121}\right)^{3/2}$

8. a. $(2^3 - 3^2)^{11/7}$

  b. $(27^{2/3} + 8^{4/3})^{-3/2}$

9. a. $(3^3)(3^{-2})$ **b.** $(4^{2/3})(2^{2/3})$

10. a. $\dfrac{5^2}{5^3}$

  b. $\left(\dfrac{\pi^2}{\sqrt{\pi}}\right)^{4/3}$

11. a. $(3^2)^{5/2}$

  b. $(e^2 e^{3/2})^{4/3}$

12. a. $\dfrac{(3^{1.2})(3^{2.7})}{3^{4.1}}$

  b. $\left(\dfrac{16}{81}\right)^{1/4}\left(\dfrac{125}{8}\right)^{-2/3}$

In Exercises 13 through 18, use the properties of exponents to simplify the given expressions.

13. a. $(27x^6)^{2/3}$

  b. $(8x^2y^3)^{1/3}$

14. a. $(x^{1/3})^{3/2}$

  b. $(x^{2/3})^{-3/4}$

15. a. $\dfrac{(x + y)^0}{(x^2y^3)^{1/6}}$

  b. $(x^{1.1}y^2)(x^2 + y^3)^0$

16. a. $(-2t^{-3})(3t^{2/3})$
   b. $(t^{-2/3})(t^{3/4})$

17. a. $(t^{5/6})^{-6/5}$
   b. $(t^{-3/2})^{-2/3}$

18. a. $(x^2y^{-3}z)^3$

   b. $\left(\dfrac{x^3y^{-2}}{z^4}\right)^{1/6}$

In Exercises 19 through 28, find all real numbers x that satisfy the given equation.

19. $4^{2x-1} = 16$

20. $3^x2^{2x} = 144$ $= 3^2 2^4$

21. $2^{3-x} = 4^x$

22. $4^x\left(\dfrac{1}{2}\right)^{3x} = 8$

23. $(2.14)^{x-1} = (2.14)^{1-x}$

24. $(3.2)^{2x-3} = (3.2)^{2-x}$

25. $10^{x^2-1} = 10^3$

26. $\left(\dfrac{1}{10}\right)^{1-x^2} = 1000$

27. $\left(\dfrac{1}{8}\right)^{x-1} = 2^{3-2x^2}$

28. $\left(\dfrac{1}{9}\right)^{1-3x^2} = 3^{4x}$

In Exercises 29 through 32, use a graphing utility to sketch the graph of the given exponential function.

29. $y = 3^{1-x}$

30. $y = e^{x+2}$

31. $y = 4 - e^{-x}$

32. $y = 2^{x/2}$

In Exercises 33 and 34, find the values of the constants C and b so that the curve $y = Cb^x$ contains the indicated points.

33. $(2, 12)$ and $(3, 24)$

34. $(2, 3)$ and $(3, 9)$

35. COMPOUND INTEREST   Suppose $1000 is invested at an annual interest rate of 7%. Compute the balance after 10 years if the interest is compounded

   a. annually   **b.** quarterly
   c. monthly   **d.** continuously

36. COMPOUND INTEREST   Suppose $5000 is invested at an annual interest rate of 10%. Compute the balance after 10 years if the interest is compounded

   a. annually
   b. semiannually
   c. daily (using 365 days per year)
   d. continuously

37. PRESENT VALUE   How much money should be invested today at 7% compounded quarterly so that it will be worth $5000 in 5 years?

38. PRESENT VALUE   How much money should be invested today at an annual interest rate of 7% compounded continuously so that 20 years from now it will be worth $20 000?

39. PRESENT VALUE   How much money should be invested now at 7% to obtain $9000 in 5 years if interest is compounded

   a. quarterly?
   b. continuously?

40. PRESENT VALUE   What is the present value of $10 000 over a 5-year period if interest is compounded continuously at an annual rate of 7%? What is the present value of $20 000 under the same conditions?

41. DEMAND   A manufacturer estimates that when x lawnmowers are produced, the market price p (dollars per unit) is given by the demand function

$$p = 300e^{-0.02x}.$$

   a. What market price corresponds to the production of $x = 100$ units?
   b. How much revenue is obtained when 100 lawnmowers are produced?
   c. How much more (or less) revenue is obtained when $x = 100$ lawnmowers are produced than when $x = 50$ are produced?

42. DEMAND   A manufacturer estimates that when x units of a particular commodity are produced, the market price p (dollars per unit) is given by the demand function

$$p = 7 + 50e^{-x/200}.$$

   a. What market price corresponds to the production of $x = 0$ units?
   b. How much revenue is obtained when 200 units of the commodity are produced?
   c. How much more (or less) revenue is obtained when $x = 100$ units are produced than when $x = 50$ are produced?

43. POPULATION GROWTH It is projected that t years from now, the population of a certain country will be $P(t) = 50e^{0.02t}$ million.
a. What is the current population?
b. What will the population be 30 years from now?

44. POPULATION GROWTH It is estimated that t years after 2000, the population of a certain country will be $P(t)$ million people, where
$$P(t) = (2)5^{0.018t}.$$
a. What was the population in 2000?
b. What was the population in 2010?

45. DRUG CONCENTRATION The concentration of drug in a patient's bloodstream t hours after an injection is given by $C(t) = (3)2^{-0.75t}$ milligrams per millilitre.
a. What is the concentration when $t = 0$? After 1 hour?
b. What is the average rate of change of concentration during the second hour?

46. DRUG CONCENTRATION The concentration of drug in a patient's bloodstream t hours after an injection is given by $C(t) = Ae^{-0.87t}$ milligrams per millilitre for constant A. The concentration is 4 mg/mL after 1 hour.
a. What is A?
b. What is the initial concentration ($t = 0$)? The concentration after 2 hours?
c. What is the average rate of change of concentration during the first 2 hours?

47. BACTERIAL GROWTH The size of a bacterial culture grows so that after t minutes, there are $P(t) = (A)2^{0.001t}$ bacteria present, for some constant A. After 10 minutes, there are 10 000 bacteria.
a. What is A?
b. How many bacteria are initially present ($t = 0$)? After 20 minutes? After 1 hour?
c. At what average rate does the bacterial population change over the second hour?

48. ADVERTISING A marketing manager estimates that t days after the end of an advertising campaign for a new video game, the sales will be $S(t)$ units, where
$$S(t) = 4000\, e^{-0.015t}.$$
a. How many video games units are being sold at the time the advertising campaign ends?
b. How many video games will be sold 30 days after the campaign ends? 60 days after the campaign ends?

c. At what average rate do sales change over the first 3 months (90 days) after advertising ends?

49. REAL ESTATE INVESTMENT In 1626, Peter Minuit traded trinkets worth $24 to a tribe of Native Americans for land on Manhattan Island. Assume that in 2010 the same land was worth $38 billion. If the sellers in this transaction had invested their $24 at 7% annual interest compounded continuously during the entire 364-year period, who would have gotten the better end of the deal? By how much?

50. GROWTH OF GDP The gross domestic product (GDP) of a certain country was $500 billion at the beginning of 2000 and increases at a rate of 2.7% per year. (*Hint:* Think of this as a compounding problem.)
a. Express the GDP of this country as a function of the number of years t after 2000.
b. What does this formula predict the GDP of the country would have been at the beginning of 2010?

51. POPULATION GROWTH The size of a bacterial population $P(t)$ grows at a rate of 3.1% per day. If the initial population is 10 000, what is the population after 10 days? (*Hint:* Think of this as a compounding problem.)

52. SUPPLY A manufacturer will supply $S(x) = 300e^{0.03x} - 310$ units of a particular commodity when the price is x dollars per unit.
a. How many units will be supplied when the unit price is $10?
b. How many more units will be supplied when the unit price is $100 than when it is $80?

53. DRUG CONCENTRATION The concentration of a certain drug in an organ t minutes after an injection is given by
$$C(t) = 0.065(1 + e^{-0.025t})$$
grams per cubic centimetre.
a. What is the initial concentration of drug (when $t = 0$)?
b. What is the concentration 20 minutes after an injection? 1 hour after the injection?
c. What is the average rate of change of concentration during the first minute?
d. What happens to the concentration of the drug in the long run (as $t \to \infty$)?
e. Sketch the graph of $C(t)$.

54. DRUG CONCENTRATION   The concentration of a certain drug in an organ t minutes after an injection is given by

$$C(t) = 0.05 - 0.04(1 - e^{-0.03t})$$

grams per cubic centimetre.
a. What is the initial concentration of drug (when $t = 0$)?
b. What is the concentration 10 minutes after an injection? After 1 hour?
c. What is the average rate of change of concentration during the first hour?
d. What happens to the concentration of the drug in the long run (as $t \to \infty$)?
 e. Sketch the graph of $C(t)$.

In Exercises 55 through 58, find the effective interest rate r_e for the given investment.

55. Annual interest rate 6%, compounded quarterly

56. Annual interest rate 8%, compounded daily (use $k = 365$)

57. Nominal annual rate of 5%, compounded continuously

58. Nominal annual rate of 7.3%, compounded continuously

59. RANKING INVESTMENTS   In terms of effective interest rate, order the following nominal rate investments from lowest to highest:
a. 7.9% compounded semiannually
b. 7.8% compounded quarterly
c. 7.7% compounded monthly
d. 7.65% compounded continuously

60. RANKING INVESTMENTS   In terms of effective interest rate, order the following nominal rate investments from lowest to highest:
a. 4.87% compounded quarterly
b. 4.85% compounded monthly
c. 4.81% compounded daily (365 days)
d. 4.79% compounded continuously

61. EFFECT OF INFLATION   Tom buys a rare stamp for $500. If the annual rate of inflation is 4%, how much should he ask for when he sells it in 5 years in order to break even?

62. EFFECT OF INFLATION   Suppose during a 10-year period of rapid inflation, it is estimated that prices inflate at an annual rate of 5% per year. If a bottle of shampoo costs $3 at the beginning of the period, what would you expect to pay for the same item 10 years later?

63. PRODUCT RELIABILITY   A statistical study indicates that the fraction of the toasters manufactured by a certain company that are still in working condition after t years of use is approximately $f(t) = e^{-0.2t}$.
a. What fraction of the toasters can be expected to work for at least 3 years?
b. What fraction of the toasters can be expected to fail before 1 year of use?
c. What fraction of the toasters can be expected to fail during the third year of use?

64. LEARNING   According to the Ebbinghaus model, the fraction $F(t)$ of subject matter you will remember from this course t months after the final exam can be estimated by the formula

$$F(t) = B + (1 - B)e^{-kt}$$

where B is the fraction of the material you will never forget and k is a constant that depends on the quality of your memory. Suppose you are tested and it is found that $B = 0.3$ and $k = 0.2$. What fraction of the material will you remember 1 month after the class ends? What fraction will you remember after 1 year?

65. POPULATION DENSITY   The population density x kilometres from the centre of a certain city is $D(x) = 12e^{-0.07x}$ thousand people per square kilometre.
a. What is the population density at the centre of the city?
b. What is the population density 10 km from the centre of the city?

66. RADIOACTIVE DECAY   The amount of a sample of a radioactive substance remaining after t years is given by a function of the form $Q(t) = Q_0 e^{-0.0001t}$. At the end of 5000 years, 200 g of the substance remain. How many grams were present initially?

67. AQUATIC PLANT LIFE   Plant life exists only in the top 10 m of a lake or sea, primarily because the intensity of sunlight decreases exponentially with depth. Specifically, the Bouguer-Lambert law says that a beam of light that strikes the surface of a body of water with intensity I_0 will have intensity I at a depth of x metres, where $I = I_0 e^{-kx}$ with $k > 0$. The constant k, called the absorption coefficient, depends on the wavelength of the light and the density of the water. Suppose a beam of sunlight is only 10% as intense at a depth of 3 m as at the surface. How intense is the beam at a depth of 1 m? (Express your answer in terms of I_0.)

68. **LINGUISTICS** Glottochronology, a theory used in the past by linguists, stated that the relationship between the proportion of words still used now, $P(t)$, to those used t thousand years previously, P_0, and time t is

$$P(t) = P_0 e^{-0.14t}$$

According to the theory, which is no longer accepted, how many of a set of 500 basic words used in classical Latin in 300 B.C. would you expect still to be in use in modern Italian in the year 2012?

69. **POPULATION GROWTH** It is estimated that t years after 1995, the population of a certain country will be $P(t)$ million people, where

$$P(t) = Ae^{0.03t} - Be^{0.005t}$$

for certain constants A and B. The population was 100 million in 1997 and 200 million in 2010.
   a. Use the given information to find A and B.
   b. What was the population in 1995?
   c. What will the population be in 2015?

Amortization of Debt ■ *If a loan of A dollars is amortized over n years at an annual interest rate r (expressed as a decimal) compounded monthly, the monthly payments are given by*

$$M = \frac{Ai}{1 - (1 + i)^{-12n}}$$

where $i = \dfrac{r}{12}$ *is the monthly interest rate. Use this formula in Exercises 70 through 73.*

---

70. **FINANCE PAYMENTS** Determine the monthly car payment for a new car costing $15 675 if the down payment is $4000 and the car is financed over a 5-year period at an annual rate of 6% compounded monthly.

71. **MORTGAGE PAYMENTS** A home loan is made for $150 000 at 9% annual interest, compounded monthly, for 30 years. What is the monthly mortgage payment on this loan?

72. **MORTGAGE PAYMENTS** Suppose a family figures it can handle monthly mortgage payments of no more than $1200. What is the largest amount of money the family can borrow, assuming the lender is willing to amortize over 30 years at 8% annual interest compounded monthly?

73. **TRUTH IN LENDING** You are selling your car for $6000. A potential buyer says, "I will pay you $1000 now for the car and pay off the rest at 12% interest with monthly payments for 3 years. Let's see . . . 12% of the $5000 is $600 and $5600 divided by 36 months is $155.56, but I'll pay you $160 per month for the trouble of carrying the loan. Is it a deal?"
   a. Explain why the deal is suspicious and compute a fair monthly payment (assuming you still plan to amortize the debt of $5000 over 3 years at 12%).
   b. Read an article on truth in lending and think up some examples of plausible yet shady deals, such as the proposed used-car transaction in this exercise.

74. **A*** Two graphs, $y = f(x)$ and $y = g(x)$, are reflections of one another in the y axis if whenever (a, b) is a point on one of the graphs, then $(-a, b)$ is a point on the other, as indicated in the accompanying figure. Use this criterion to show that the graphs of $y = b^x$ and $y = \left(\dfrac{1}{b}\right)^x$ for $b > 0$, $b \neq 1$, are reflections of one another in the y axis.

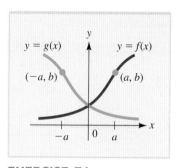

EXERCISE 74

SECTION 4.2

L02

Simplify logarithmic and
exponential expressions.
Solve equations using natural
logarithms and exponents.

Logarithmic Functions

Suppose you invest $1000 at 8% compounded continuously and wish to know how much time must pass for your investment to double in value to $2000. According to the formula derived in Section 4.1, the value of your account after t years will be $1000e^{0.08t}$, so to find the doubling time for your account, you must solve for t in the equation

$$1000e^{0.08t} = 2000$$

or, by dividing both sides by 1000,

$$e^{0.08t} = 2.$$

We will answer the question about doubling time in Example 4.2.10. Solving an exponential equation such as this involves using *logarithms,* which reverse the process of exponentiation. The logarithmic function is the inverse of the exponential function. Logarithms play an important role in a variety of applications, such as measuring the capacity of a transmission channel and measuring earthquake intensity using the famous Richter scale. In this section, we examine the basic properties of logarithmic functions and a few applications. We begin with a definition.

> **Logarithmic Functions**   ■   If x is a positive number, then the **logarithm** of x to the base b ($b > 0$, $b \neq 1$), denoted $\log_b x$, is the number y such that $b^y = x$; that is,
>
> $$y = \log_b x \qquad \text{if and only if} \qquad b^y = x \quad \text{for} \quad x > 0$$

EXAMPLE 4.2.1

Evaluate

a. $\log_{10} 1000$ **b.** $\log_2 32$ **c.** $\log_5\left(\dfrac{1}{125}\right)$

Solution

a. $\log_{10} 1000 = 3$ since $10^3 = 1000$.
b. $\log_2 32 = 5$ since $2^5 = 32$.
c. $\log_5 \dfrac{1}{125} = -3$ since $5^{-3} = \dfrac{1}{125}$.

EXAMPLE 4.2.2

Solve each of the following equations for x:

a. $\log_4 x = \dfrac{1}{2}$ **b.** $\log_{64} 16 = x$ **c.** $\log_x 27 = 3$

Solution

a. By definition, $\log_4 x = \dfrac{1}{2}$ is equivalent to $x = 4^{1/2} = 2$.

b. $\log_{64} 16 = x$ means

$$16 = 64^x$$
$$2^4 = (2^6)^x = 2^{6x}$$
$$4 = 6x \qquad\qquad b^m = b^n \text{ implies } m = n$$
$$x = \frac{2}{3}$$

c. $\log_x 27 = 3$ means

$$x^3 = 27$$
$$x = (27)^{1/3} = 3$$

Logarithms were introduced in the 17th century as a computational device, primarily because they can be used to convert expressions involving products and quotients into much simpler expressions involving sums and differences. Here are the rules for logarithms that facilitate such simplification.

Logarithmic Rules ■ Let b be any logarithmic base ($b > 0$, $b \neq 1$). Then
$$\log_b 1 = 0 \text{ and } \log_b b = 1$$

and if u and v are any positive numbers, then

The **equality rule:** $\log_b u = \log_b v$, if and only if $u = v$

The **product rule:** $\log_b (uv) = \log_b u + \log_b v$

The **power rule:** $\log_b u^r = r \log_b u$, for any real number r

The **quotient rule:** $\log_b \left(\dfrac{u}{v}\right) = \log_b u - \log_b v$

The **inversion rule:** $\log_b b^u = u$, because the logarithmic function is the inverse of the exponential function.

All of these logarithmic rules follow from corresponding exponential rules. For example,

$$\log_b 1 = 0 \qquad \text{because} \qquad b^0 = 1.$$
$$\log_b b = 1 \qquad \text{because} \qquad b^1 = b.$$

To prove the equality rule, let

$$m = \log_b u \quad \text{and} \quad n = \log_b v,$$

so that by definition,

$$b^m = u \quad \text{and} \quad b^n = v.$$

Therefore, if

$$\log_b u = \log_b v,$$

then $m = n$, so

$$b^m = b^n \quad \text{equality rule for exponentials}$$

or, equivalently,

$$u = v,$$

as stated in the equality rule for logarithms. Similarly, to prove the product rule for logarithms, note that

$$\log_b u + \log_b v = m + n$$
$$= \log_b (b^{m+n}) \qquad \text{inversion rule for logarithms}$$
$$= \log_b (b^m b^n) \qquad \text{product rule for exponentials}$$
$$= \log_b (uv) \qquad \text{since } b^m = u \text{ and } b^n = v$$

Proofs of the power rule and the quotient rule are left as exercises (see Exercise 78). Table 4.1 displays the correspondence between basic properties of exponential and logarithmic functions.

TABLE 4.1 Comparison of Exponential and Logarithmic Rules

Exponential Rule	Logarithmic Rule
$b^x b^y = b^{x+y}$	$\log_b (xy) = \log_b x + \log_b y$
$\dfrac{b^x}{b^y} = b^{x-y}$	$\log_b \left(\dfrac{x}{y}\right) = \log_b x - \log_b y$
$b^{xp} = (b^x)^p$	$\log_b x^p = p \log_b x$

EXAMPLE 4.2.3

Use logarithm rules to rewrite each expression in terms of $\log_5 2$ and $\log_5 3$.

a. $\log_5 \left(\dfrac{5}{3}\right)$ **b.** $\log_5 8$ **c.** $\log_5 36$

Solution

a. $\log_5 \left(\dfrac{5}{3}\right) = \log_5 5 - \log_5 3 \qquad$ quotient rule

$\qquad\qquad\quad = 1 - \log_5 3 \qquad\qquad \log_5 5 = 1$

b. $\log_5 8 = \log_5 2^3 = 3 \log_5 2 \qquad$ power rule

c. $\log_5 36 = \log_5 (2^2 3^2)$

$\qquad\qquad\quad = \log_5 2^2 + \log_5 3^2 \qquad$ product rule

$\qquad\qquad\quad = 2 \log_5 2 + 2 \log_5 3 \qquad$ power rule

EXAMPLE 4.2.4

Use logarithmic rules to expand each expression.

a. $\log_3 (x^3 y^{-4})$ **b.** $\log_2 \left(\dfrac{y^5}{x^2}\right)$ **c.** $\log_7 (x^3 \sqrt{1 - y^2})$

Solution

a. $\log_3 (x^3 y^{-4}) = \log_3 x^3 + \log_3 y^{-4}$ product rule

$\qquad\qquad\quad = 3 \log_3 x + (-4)\log_3 y$ power rule

$\qquad\qquad\quad = 3 \log_3 x - 4 \log_3 y$

b. $\log_2 \left(\dfrac{y^5}{x^2}\right) = \log_2 y^5 - \log_2 x^2$ quotient rule

$\qquad\qquad\quad = 5 \log_2 y - 2 \log_2 x$ power rule

c. $\log_7 (x^3 \sqrt{1 - y^2}) = \log_7 [x^3(1 - y^2)^{1/2}]$

$\qquad\qquad\qquad\quad = \log_7 x^3 + \log_7 (1 - y^2)^{1/2}$ product rule

$\qquad\qquad\qquad\quad = 3 \log_7 x + \dfrac{1}{2} \log_7 (1 - y^2)$ power rule

$\qquad\qquad\qquad\quad = 3 \log_7 x + \dfrac{1}{2} \log_7 [(1 - y)(1 + y)]$ factor $1 - y^2$

$\qquad\qquad\qquad\quad = 3 \log_7 x + \dfrac{1}{2} [\log_7 (1 - y) + \log_7 (1 + y)]$ product rule

$\qquad\qquad\qquad\quad = 3 \log_7 x + \dfrac{1}{2} \log_7 (1 - y) + \dfrac{1}{2} \log_7 (1 + y)$

Graphs of Logarithmic Functions

There is an easy way to obtain the graph of the logarithmic function $y = \log_b x$ from the graph of the exponential function $y = b^x$. The idea is that since $y = \log_b x$ is equivalent to $x = b^y$, the graph of $y = \log_b x$ is the same as the graph of $y = b^x$ with the roles of x and y reversed. That is, if (u, v) is a point on the curve $y = \log_b x$, then $v = \log_b u$, or, equivalently, $u = b^v$, which means that (v, u) is on the graph of $y = b^x$. As illustrated in Figure 4.5a, the points (u, v) and (v, u) are mirror images of one another in the line $y = x$ (see Exercise 79). Thus, the graph of $y = \log_b x$ can be obtained by simply *reflecting* the graph of $y = b^x$ in the line $y = x$, as shown in Figure 4.5b for the case where $b > 1$. To summarize:

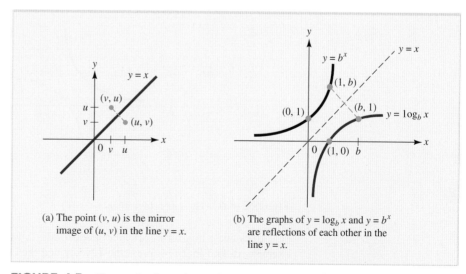

(a) The point (v, u) is the mirror image of (u, v) in the line $y = x$.

(b) The graphs of $y = \log_b x$ and $y = b^x$ are reflections of each other in the line $y = x$.

FIGURE 4.5 The graph of $y = \log_b x$ for $b > 1$ is obtained by reflecting the graph of $y = b^x$ in the line $y = x$.

Relationship between the Graphs of $y = \log_b x$ and $y = b^x$ ■ The graphs of $y = \log_b x$ and $y = b^x$ are mirror images of one another in the line $y = x$. Therefore, the graph of $y = \log_b x$ can be obtained by reflecting the graph of $y = b^x$ in the line $y = x$.

Figure 4.5b reveals important properties of the logarithmic function $f(x) = \log_b x$ for the case where $b > 1$. The following box lists these properties along with similar properties for the case where $0 < b < 1$.

Properties of a Logarithmic Function ■ The logarithmic function $f(x) = \log_b x$ ($b > 0$, $b \neq 1$) has these properties:

1. It is defined and continuous for all $x > 0$.

2. The y axis is a vertical asymptote.

3. The x intercept is $(1, 0)$; there is no y intercept.

4. If $b > 1$, then $\lim\limits_{x \to 0^+} \log_b x = -\infty$ and $\lim\limits_{x \to +\infty} \log_b x = +\infty$.

  If $0 < b < 1$, then $\lim\limits_{x \to 0^+} \log_b x = +\infty$ and $\lim\limits_{x \to +\infty} \log_b x = -\infty$.

5. For all $x > 0$, the function is increasing (graph rising) if $b > 1$ and decreasing (graph falling) if $0 < b < 1$.

The Natural Logarithm

In calculus, the most frequently used logarithmic base is e. In this case, the logarithm $\log_e x$ is called the **natural logarithm** of x and is denoted by $\ln x$ (read as "lon x"); that is, for $x > 0$,

$$y = \ln x \qquad \text{if and only if} \qquad e^y = x$$

The graph of the natural logarithm is shown in Figure 4.6.

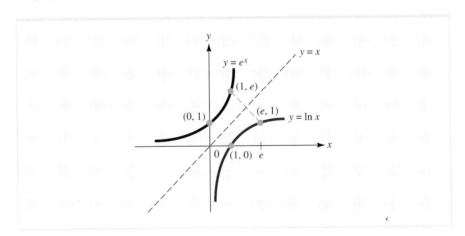

FIGURE 4.6 The graph of $y = \ln x$.

To evaluate $\ln a$ for a particular number $a > 0$, use the **LN** key on a calculator. For example, to find $\ln 2.714$, press the **LN** key and then enter the number 2.714 to get

$$\ln 2.714 = 0.9984 \qquad \text{(to four decimal places).}$$

Here is an example illustrating the computation of natural logarithms.

EXAMPLE 4.2.5

Find

a. $\ln e$　　　　　**b.** $\ln 1$　　　　　**c.** $\ln\sqrt{e}$　　　　　**d.** $\ln 2$

Solution

a. According to the definition, $\ln e$ is the unique number c such that $e = e^c$. Clearly this number is $c = 1$. Hence, $\ln e = 1$.

b. $\ln 1$ is the unique number c such that $1 = e^c$. Since $e^0 = 1$, it follows that $\ln 1 = 0$.

c. $\ln\sqrt{e} = \ln e^{1/2}$ is the unique number c such that $e^{1/2} = e^c$; that is, $c = \dfrac{1}{2}$. Hence, $\ln\sqrt{e} = \dfrac{1}{2}$.

d. $\ln 2$ is the unique number c such that $2 = e^c$. The value of this number is not obvious, and you will have to use a calculator to find that $\ln 2 \approx 0.69315$.

EXAMPLE 4.2.6

a. Find $\ln\sqrt{ab}$ if $\ln a = 3$ and $\ln b = 7$.

b. Show that $\ln\dfrac{1}{x} = -\ln x$.

c. Find x if $2^x = e^3$.

Solution

a. $\ln\sqrt{ab} = \ln(ab)^{1/2} = \dfrac{1}{2}\ln ab = \dfrac{1}{2}(\ln a + \ln b) = \dfrac{1}{2}(3 + 7) = 5$

b. $\ln\dfrac{1}{x} = \ln 1 - \ln x = 0 - \ln x = -\ln x$, or, alternatively,

$\ln\dfrac{1}{x} = \ln x^{-1} = (-1)\ln x = -\ln x$.

c. Take the natural logarithm of each side of the equation $2^x = e^3$ and solve for x to get

$$x \ln 2 = \ln e^3 = 3 \ln e = 3 \qquad \text{since } \ln e = 1$$

Thus,

$$x = \frac{3}{\ln 2} \approx 4.33.$$

Two functions f and g with the property that $f(g(x)) = x$ and $g(f(x)) = x$, whenever both composite functions are defined, are said to be **inverses** of one another. Such an inverse relationship exists between exponential and logarithmic functions with base b. For instance, we have

$$\ln e^x = x \ln e = x(1) = x \qquad \text{for all } x.$$

Similarly, if $y = e^{\ln x}$ for $x > 0$, then by definition, $\ln y = \ln x$, so $y = x$; that is,

$$e^{\ln x} = y = x.$$

This inverse relationship between the natural exponential and logarithmic functions is especially useful. It is summarized in the following box and used in Example 4.2.7.

The Inverse Relationship between e^x and ln x ▪

$$e^{\ln x} = x \text{ for } x > 0 \qquad \text{and} \qquad \ln e^x = x \qquad \text{for all } x.$$

Just-In-Time

In order to solve an equation involving e^u or ln u, where u is a function of x, first isolate that term on one side of the equation, and then solve by taking the inverse of both sides. Take the natural logarithm of both sides if the equation involves e^u. Take everything to the power of e when the equation involves ln u.
But remember to isolate the expression e^u or ln u first.

EXAMPLE 4.2.7

Solve each equation for x.

a. $3 = e^{20x}$ **b.** $2 \ln x = 1$ **c.** $10 - 6 \ln(4x) = 3$

Solution

a. Take the natural logarithm of each side of the equation to get

$$\ln 3 = \ln e^{20x} \quad \text{or} \quad \ln 3 = 20x$$

Solve for x, using a calculator to find $\ln 3$:

$$x = \frac{\ln 3}{20} \approx \frac{1.0986}{20} \approx 0.0549.$$

b. First isolate $\ln x$ on the left side of the equation by dividing both sides by 2:

$$\ln x = \frac{1}{2}$$

Then apply the exponential function to both sides of the equation to get

$$e^{\ln x} = e^{1/2} \quad \text{or} \quad x = e^{1/2} = \sqrt{e} \approx 1.6487.$$

c. $10 - 6 \ln 4x = 3$

$$-6 \ln 4x = 3 - 10$$

$$\ln 4x = \frac{-7}{-6}$$

$$e^{\ln 4x} = e^{7/6}$$

$$4x = e^{7/6}$$

$$x = \frac{e^{7/6}}{4}$$

$$x \approx 0.8028$$

Example 4.2.8 illustrates how to use logarithms to find an exponential function that fits certain specified information.

EXAMPLE 4.2.8

The population density x kilometres from the centre of a city is given by a function of the form $Q(x) = Ae^{-kx}$. Find this function if it is known that the population density at the centre of the city is 15 000 people/km^2 and the density 10 km from the centre is 9000 people/km^2.

Solution

For simplicity, express the density in units of 1000 people/km^2. The fact that $Q(0) = 15$ tells you that $A = 15$. The fact that $Q(10) = 9$ means that

$$9 = 15e^{-10k} \quad \text{or} \quad \frac{3}{5} = e^{-10k}.$$

Taking the logarithm of each side of this equation gives

$$\ln \frac{3}{5} = -10k \quad \text{or} \quad k = -\frac{\ln 3/5}{10} \approx 0.051.$$

Hence, the exponential function for the population density is $Q(x) = 15e^{-0.051x}$.

You have already seen how to use the **LN** or **ln** key to compute natural logarithms, and most calculators have a **LOG** or **log** key for computing logarithms to base 10, but what about logarithms to bases other than e or 10? To be specific, suppose you wish to calculate the logarithmic number $c = \log_b a$. You have

$$c = \log_b a$$
$$b^c = a$$
$$\ln b^c = \ln a$$
$$c \ln b = \ln a$$
$$c = \frac{\ln a}{\ln b}$$

Thus, the logarithm $\log_b a$ can be computed by finding the ratio of two natural logarithms, $\ln a$ and $\ln b$. To summarize:

> **Conversion Formula for Logarithms** ■ If a and b are positive numbers with $b \neq 1$, then
>
> $$\log_b a = \frac{\ln a}{\ln b}$$

EXAMPLE 4.2.9

Find $\log_5 3.172$.

Solution

Use the conversion formula to find

$$\log_5 3.172 = \frac{\ln 3.172}{\ln 5} \approx 0.7172.$$

Compounding Applications

In the paragraph at the beginning of this section, we asked how long it would take for a particular investment to double in value. This question is answered in Example 4.2.10.

EXAMPLE 4.2.10

If $1000 is invested at 8% annual interest, compounded continuously, how long will it take for the investment to double? Would the doubling time change if the principal were something other than $1000?

Solution

With a principal of $1000, the balance after t years is $B(t) = 1000e^{0.08t}$, so the investment doubles when $B(t) = \$2000$, that is, when

$$2000 = 1000e^{0.08t}.$$

Dividing by 1000 and taking the natural logarithm on each side of the equation gives

$$2 = e^{0.08t}$$
$$\ln 2 = 0.08t$$
$$t = \frac{\ln 2}{0.08} \approx 8.66 \text{ years}$$

If the principal had been P_0 dollars instead of $1000, the doubling time would satisfy

$$2P_0 = P_0 e^{0.08t}$$
$$2 = e^{0.08t}$$

which is exactly the same equation we had with $P_0 = \$1000$, so, once again, the doubling time is 8.66 years.

The situation illustrated in Example 4.2.10 applies to any quantity $Q(t) = Q_0 e^{kt}$ with $k > 0$. In particular, since at time $t = 0$ we have $Q(0) = Q_0 e^0 = Q_0$, the quantity doubles when

$$2Q_0 = Q_0 e^{kt}$$
$$2 = e^{kt}$$
$$\ln 2 = kt$$
$$t = \frac{\ln 2}{k}$$

To summarize:

Doubling Time ■ A quantity $Q(t) = Q_0 e^{kt}$ $(k > 0)$ doubles when $t = d$, where

$$d = \frac{\ln 2}{k}.$$

Determining the time it takes for an investment to double is just one of several issues an investor may address when comparing various investment opportunities. Two additional issues are illustrated in Examples 4.2.11 and 4.2.12.

| EXAMPLE 4.2.11 |

How long will it take $5000 to grow to $7000 in an investment earning interest at an annual rate of 6% if the compounding is

a. quarterly?   **b.** continuous?

Solution

a. Use the future value formula $B = P(1 + i)^{kt}$ with $i = \dfrac{r}{k}$. We have $B = 7000$, $P = 5000$, and $i = \dfrac{0.06}{4} = 0.015$, since $r = 0.06$ and there are $k = 4$ compounding periods per year. Substitute to find that

$$7000 = 5000(1.015)^{4t}$$

$$(1.015)^{4t} = \frac{7000}{5000} = 1.4$$

Taking the natural logarithm on each side of this equation gives

$$\ln (1.015)^{4t} = \ln 1.4$$

$$4t \ln 1.015 = \ln 1.4$$

$$4t = \frac{\ln 1.4}{\ln 1.015} \approx 22.6$$

$$t \approx \frac{22.6}{4} = 5.65$$

Thus, it will take roughly 5.65 years for $5000 to grow to $7000 with quarterly compounding.

b. With continuous compounding, use the formula $B = Pe^{rt}$:

$$7000 = 5000e^{0.06t}$$

$$e^{0.06t} = \frac{7000}{5000} = 1.4$$

Taking logarithms gives

$$\ln e^{0.06t} = \ln 1.4$$

$$0.06t = \ln 1.4$$

$$t = \frac{\ln 1.4}{0.06} \approx 5.61$$

So, with continuous compounding, it takes only 5.61 years to reach the investment objective.

| EXAMPLE 4.2.12 |

An investor has $1500 and wishes it to grow to $2000 in 5 years. At what annual rate r compounded continuously must he invest to achieve this goal?

Solution

If the interest rate is r, the future value of $1500 in 5 years is given by $1500\, e^{r(5)}$. In order for this to equal $2000,

$$1.500e^{r(5)} = 2000$$

$$e^{5r} = \frac{2000}{1500} = \frac{4}{3}$$

Taking natural logarithms on both sides of this equation, we get

$$\ln e^{5r} = \ln \frac{4}{3}$$

$$5r = \ln \frac{4}{3}$$

so

$$r = \frac{1}{5}\ln \frac{4}{3} \approx 0.0575.$$

The annual interest rate is approximately 5.75%.

Radioactive Decay and Carbon Dating

It has been experimentally determined that a radioactive sample of initial size Q_0 grams will decay to $Q(t) = Q_0e^{-kt}$ grams in t years. The positive constant k in this formula measures the rate of decay, but this rate is usually given by specifying the amount of time $t = h$ required for half of a given sample to decay. The time h is called the **half-life** of the radioactive substance. Example 4.2.13 shows how half-life is related to k.

EXAMPLE 4.2.13

Show that a radioactive substance that decays according to the formula $Q(t) = Q_0e^{-kt}$ has half-life $h = \dfrac{\ln 2}{k}$.

Solution

The goal is to find the value of h for which $Q(h) = \dfrac{1}{2}Q_0$, that is,

$$\frac{1}{2}Q_0 = Q_0e^{-kh}.$$

Divide by Q_0 and take the natural logarithm of each side to get

$$\ln \frac{1}{2} = -kh.$$

Thus, the half-life is

$$h = \frac{\ln \dfrac{1}{2}}{-k}$$

$$= \frac{-\ln 2}{-k} = \frac{\ln 2}{k} \qquad \text{since } \ln \frac{1}{2} = \ln 2^{-1} = -\ln 2$$

as required.

The carbon dioxide in the air contains the radioactive isotope ^{14}C (carbon-14) as well as the stable isotope ^{12}C (carbon-12). Living plants absorb carbon dioxide from the air, which means that the ratio of ^{14}C to ^{12}C in a living plant (or in an animal that eats plants) is the same as that in the air itself. When a plant or an animal dies, the absorption of carbon dioxide ceases. The ^{12}C already in the plant or animal remains the same as at the time of death, while the ^{14}C decays, so the ratio of ^{14}C to ^{12}C decreases exponentially. It is reasonable to assume that the ratio R_0 of ^{14}C to ^{12}C in the atmosphere is the same today as it was in the past, so that the ratio of ^{14}C to ^{12}C in a sample (e.g., a fossil or an artifact) is given by a function of the form $R(t) = R_0 e^{-kt}$. The half-life of ^{14}C is 5730 years. By comparing $R(t)$ to R_0, archaeologists can estimate the age of the sample. Example 4.2.14 illustrates the dating procedure.

EXAMPLE 4.2.14

An archaeologist has found a fossil in which the ratio of ^{14}C to ^{12}C is $\dfrac{1}{5}$ the ratio found in the atmosphere. Approximately how old is the fossil?

Solution

The age of the fossil is the value of t for which $R(t) = \dfrac{1}{5} R_0$, that is, for which

$$\frac{1}{5}R_0 = R_0 e^{-kt}.$$

Dividing by R_0 and taking logarithms gives

$$\frac{1}{5} = e^{-kt}$$

$$\ln\frac{1}{5} = -kt$$

and

$$t = \frac{-\ln\dfrac{1}{5}}{k} = \frac{\ln 5}{k}.$$

In Example 4.2.13, you found that the half-life h satisfies $h = \dfrac{\ln 2}{k}$, and since ^{14}C has half-life $h = 5730$ years,

$$k = \frac{\ln 2}{h} = \frac{\ln 2}{5730} \approx 0.000121.$$

Therefore, the age of the fossil is

$$t = \frac{\ln 5}{k} = \frac{\ln 5}{0.000121} \approx 13\ 300.$$

That is, the fossil is approximately 13 300 years old.

ALGEBRA WARM-UP

Exponential functions often occur in expressions such as those in Exercises 35 and 36 below. Here are some equations of this form, but without the exponential function, for you to solve.

a. $\dfrac{10}{6 + x} = 3$ **c.** $\dfrac{2x + 1}{3 - 4x} = 11$ **e.** $40 = \dfrac{90\,p^2}{3p^2 - 50}$

b. $\dfrac{5x}{x - 7} = 2$ **d.** $\dfrac{2x}{9 - 3x} = 100$

EXERCISES ■ 4.2

In Exercises 1 and 2, use a calculator to find the indicated natural logarithms.

1. Find $\ln 1$, $\ln 2$, $\ln e$, $\ln 5$, $\ln \dfrac{1}{5}$, and $\ln e^2$. What happens if you try to find $\ln 0$ or $\ln(-2)$? Why?

2. Find $\ln 7$, $\ln \dfrac{1}{3}$, $\ln e^{-3}$, $\ln \dfrac{1}{e^{2.1}}$, and $\ln \sqrt[5]{e}$. What happens if you try to find $\ln(-7)$ or $\ln(-e)$?

In Exercises 3 through 8, evaluate the given expression using properties of the natural logarithm.

3. $\ln e^3$

4. $\ln \sqrt{e}$

5. $e^{\ln 5}$

6. $e^{2 \ln 3}$

7. $e^{3 \ln 2 - 2 \ln 5}$

8. $\ln \dfrac{e^3 \sqrt{e}}{e^{1/3}}$

In Exercises 9 through 12, use logarithmic rules to rewrite each expression in terms of $\log_3 2$ and $\log_3 5$.

9. $\log_3 270$

10. $\log_3 2.5$

11. $\log_3 100$

12. $\log_3\left(\dfrac{64}{125}\right)$

In Exercises 13 through 20, use logarithmic rules to simplify each expression.

13. $\log_2 (x^4 y^3)$

14. $\log_3 (x^5 y^{-2})$

15. $\ln \sqrt[3]{x^2 - x}$

16. $\ln(x^2 \sqrt{4 - x^2})$

17. $\ln\left[\dfrac{x^2(3 - x)^{2/3}}{\sqrt{x^2 + x + 1}}\right]$

18. $\ln\left[\dfrac{1}{x} + \dfrac{1}{x^2}\right]$

19. $\ln(x^3 e^{-x^2})$

20. $\ln\left[\dfrac{\sqrt[4]{x}}{x^3 \sqrt{1 - x^2}}\right]$

In Exercises 21 through 36, solve the given equation for x.

21. $4^x = 53$

22. $\log_2 x = 4$

23. $\log_3 (2x - 1) = 2$

24. $3^{2x-1} = 17$

25. $2 = e^{0.06x}$

26. $\dfrac{1}{2}Q_0 = Q_0 e^{-1.2x}$

27. $3 = 2 + 5e^{-4x}$

28. $-2 \ln x = b$

29. $-\ln x = \dfrac{t}{50} + C$

30. $5 = 3 \ln x - \dfrac{1}{2} \ln x$

31. $\ln x = \dfrac{1}{3}(\ln 16 + 2 \ln 2)$

32. $\ln x = 2 (\ln 3 - \ln 5)$

33. $3^x = e^2$

34. $a^k = e^{kx}$

35. $\dfrac{25e^{0.1x}}{e^{0.1x} + 3} = 10$

36. $\dfrac{5}{1 + 2e^{-x}} = 3$

37. If $\log_2 x = 5$, what is $\ln x$?

38. If $\log_{10} x = -3$, what is $\ln x$?

39. If $\log_5 2x = 7$, what is $\ln x$?

40. If $\log_3 (x - 5) = 2$, what is $\ln x$?

41. Find $\ln \dfrac{1}{\sqrt{ab^3}}$ if $\ln a = 2$ and $\ln b = 3$.

42. Find $\dfrac{1}{a} \ln \left(\dfrac{\sqrt{b}}{c}\right)^a$ if $\ln b = 6$ and $\ln c = -2$.

43. COMPOUND INTEREST How quickly will money double if it is invested at an annual interest rate of 6% compounded continuously?

44. COMPOUND INTEREST How quickly will money double if it is invested at an annual interest rate of 7% compounded continuously?

45. COMPOUND INTEREST Money deposited in a certain bank doubles every 13 years. The bank compounds interest continuously. What annual interest rate does the bank offer?

46. TRIPLING TIME How long will it take for a quantity of money A_0 to triple in value if it is invested at an annual interest rate of r compounded continuously?

47. TRIPLING TIME If an account that earns interest compounded continuously takes 12 years to double in value, how long will it take to triple in value?

48. INVESTMENT The Morenos invest $10 000 in an account that grows to $12 000 in 5 years. What is the annual interest rate r if interest is compounded
a. quarterly?　　　**b.** continuously?

49. COMPOUND INTEREST A certain bank offers an interest rate of 6% per year compounded annually. A competing bank compounds its interest continuously. What (nominal) interest rate should the competing bank offer so that the effective interest rates of the two banks will be equal?

50. CONCENTRATION OF DRUG A drug is injected into a patient's bloodstream and, t seconds later, the concentration of the drug is C grams per cubic centimetre, where
$$C(t) = 0.1(1 + 3e^{-0.03t}).$$

a. What is the drug concentration after 10 seconds?
b. How long does it take for the drug concentration to reach 0.12 g/cm^3?

51. CONCENTRATION OF DRUG The concentration of a drug in a patient's kidneys at time t (seconds) is C grams per cubic centimetre, where
$$C(t) = 0.4(2 - 0.13e^{-0.02t}).$$

a. What is the drug concentration after 20 seconds? After 60 seconds?
b. How long does it take for the drug concentration to reach 0.75 g/cm^3?

52. RADIOACTIVE DECAY The amount of a certain radioactive substance remaining after t years is given by a function of the form $Q(t) = Q_0 e^{-0.003t}$. Find the half-life of the substance.

53. RADIOACTIVE DECAY The half-life of radium is 1690 years. How long will it take for a 50-g sample of radium to be reduced to 5 g?

54. ADVERTISING The editor at a major publishing house estimates that if x thousand complimentary copies are distributed to instructors, the first-year sales of a new textbook will be approximately $f(x) = 20 - 12e^{-0.03x}$ thousand copies. According to this estimate, approximately how many complimentary copies should the editor send out to generate first-year sales of 12 000 copies?

55. GROWTH OF BACTERIA A medical student studying the growth of bacteria in a certain culture has compiled the following data:

Number of minutes	0	20
Number of bacteria	6000	9000

Use these data to find an exponential function of the form $Q(t) = Q_0 e^{kt}$ expressing the number of bacteria in the culture as a function of time. How many bacteria are present after 1 hour?

56. GROSS DOMESTIC PRODUCT An economist has compiled the following data on the gross domestic product (GDP) of a certain country:

Year	2000	2010
GDP (in billions)	100	180

Use these data to predict the GDP in the year 2020 if the GDP is growing
a. linearly, so that GDP $= at + b$
b. exponentially, so that GDP $= Ae^{kt}$

57. WORKER EFFICIENCY An efficiency expert hired by a manufacturing firm has compiled the following data relating workers' output to their experience:

Experience t (months)	0	6
Output Q (units per hour)	300	410

Suppose output Q is related to experience t by a function of the form $Q(t) = 500 - Ae^{-kt}$. Find the function of this form that fits the data. What output is expected from a worker with 1 year's experience?

58. ARCHAEOLOGY An archaeologist has found a fossil in which the ratio of ^{14}C to ^{12}C is $\frac{1}{3}$ the ratio found in the atmosphere. Approximately how old is the fossil?

59. ARCHAEOLOGY Tests of an artifact discovered at the Debert site in Nova Scotia show that 28% of the original ^{14}C is still present. Approximately how old is the artifact?

60. ARCHAEOLOGY The Dead Sea Scrolls were written on parchment in about 100 B.C. What percentage of the original ^{14}C in the parchment remained when the scrolls were discovered in 1947?

61. ART FORGERY A forged painting allegedly painted by Rembrandt in 1640 is found to have 99.7% of its original ^{14}C. When was it actually painted? What percentage of the original ^{14}C would remain if it were legitimate?

62. ARCHAEOLOGY In 1389, Pierre d'Arcis, the bishop of Troyes, wrote a memo to the pope, accusing a colleague of passing off "a certain cloth, cunningly painted," as the burial shroud of Jesus Christ. Despite this early testimony of forgery, the image on the cloth is so compelling that many people regard it as a sacred relic. Known as the Shroud of Turin, the cloth was subjected to carbon dating in 1988. If authentic, the cloth would have been approximately 1960 years old at that time.
  a. If the Shroud were actually 1960 years old, what percentage of the ^{14}C would have remained?
  b. Scientists determined that 92.3% of the Shroud's original ^{14}C remained. Based on this information alone, what was the likely age of the Shroud in 1988?

63. COOLING Instant coffee is made by adding boiling water (100°C) to coffee mix. If the air

temperature is 22°C, Newton's law of cooling says that after t minutes, the temperature of the coffee will be given by a function of the form $f(t) = 22 + Ae^{-kt}$. After cooling for 4 minutes, the coffee is at 80°C. What is the temperature after 6 minutes?

64. DEMOGRAPHICS The world's population grows at a rate of approximately 2% per year. If it is assumed that the population growth is exponential, then the population t years from now will be given by a function of the form $P(t) = P_0e^{0.02t}$, where P_0 is the current population. (This formula is derived in Chapter 8.) Assuming that this model of population growth is correct, how long will it take for the world's population to double?

65. SUPPLY AND DEMAND A manufacturer determines that the supply function for x units of a particular commodity is $S(x) = \ln(x + 2)$ and the corresponding demand function is $D(x) = 10 - \ln(x + 1)$.
  a. Find the demand price $p = D(x)$ when the level of production is $x = 10$ units.
  b. Find the supply price $p = S(x)$ when $x = 100$ units.
  c. Find the level of production and unit price that correspond to market equilibrium (where supply = demand).

66. SUPPLY AND DEMAND A manufacturer determines that the supply function for x units of a particular commodity is $S(x) = e^{0.02x}$ and the corresponding demand function is $D(x) = 3e^{-0.03x}$.
  a. Find the demand price $p = D(x)$ when the level of production is $x = 10$ units.
  b. Find the supply price $p = S(x)$ when $x = 12$ units.
  c. Find the level of production and unit price that correspond to market equilibrium (where supply = demand).

67. SPY STORY Having ransomed his superior in Exercise 19 of Section 3.5, the spy returns home, only to learn that his best friend, Marc, has been murdered in Halifax. The police say that Marc's body was discovered at 1 P.M. on Thursday, stuffed in a freezer where the temperature was $-12°C$. He is also told that the temperature of the body at the time of discovery was 5°C, and he remembers that t hours after death a body has temperature

$$T = T_a + (37 - T_a)(0.97)^t,$$

where T_a is the air temperature adjacent to the body. The spy knows the dark deed was done by either Twoface Tony or Slippery Stan. If Tony was in jail until noon on Wednesday and Stan was seen in Toronto from noon Wednesday until Friday, who killed Marc, and when?

68. SOUND LEVELS   A decibel, named for Alexander Graham Bell, is the smallest increase of the loudness of sound that is detectable by the human ear. In physics, it is shown that when two sounds of intensity I_1 and I_2 (in watts per cubic centimetre, or W/cm^3) occur, the difference in loudness is D decibels, where

$$D = 10 \log_{10}\left(\frac{I_1}{I_2}\right).$$

When sound is rated in relation to the threshold of human hearing ($I_0 = 10^{-12}$), the level of normal conversation is about 60 decibels, while a rock concert may be 50 times as loud (110 decibels).

  a. How much more intense is the rock concert than normal conversation?
  b. The threshold of pain is reached at a sound level roughly 10 times as loud as a rock concert. What is the decibel level of the threshold of pain?

69. SEISMOLOGY   The magnitude formula for the Richter scale is

$$R = \frac{2}{3} \log_{10}\left(\frac{E}{E_0}\right),$$

where E is the energy released by the earthquake (in joules), and $E_0 = 10^{4.4}$ J is the energy released by a small reference earthquake used as a standard of measurement.

  a. The 1906 San Francisco earthquake released approximately 5.96×10^{16} joules (J) of energy. What was its magnitude on the Richter scale?
  b. How much energy was released by the Haitian earthquake in 2010, which measured 7 on the Richter scale?

70. SEISMOLOGY   On the Richter scale, the magnitude R of an earthquake of intensity I is given by

$$R = \frac{\ln I}{\ln 10}.$$

  a. Find the intensity of the 2010 Chilean earthquake, which measured $R = 8.8$ on the Richter scale.

  b. How much more intense was the earthquake in Chile than the devastating Haitian earthquake in the same year, which measured $R = 7$? Because the epicentre of the Haitian earthquake was in a very populated area, it caused a lot more damage than in Chile.

71. RADIOLOGY   Radioactive iodine ^{133}I has a half-life of 20.9 hours. If injected into the bloodstream, the iodine accumulates in the thyroid gland.

  a. After 24 hours, a medical technician scans a patient's thyroid gland to determine whether thyroid function is normal. If the thyroid has absorbed all of the iodine, what percentage of the original amount should be detected?
  b. A patient returns to the medical clinic 25 hours after having received an injection of ^{133}I. The medical technician scans the patient's thyroid gland and detects the presence of 41.3% of the original iodine. How much of the original ^{133}I remains in the rest of the patient's body?

72. AIR PRESSURE   The air pressure $f(s)$ at a height of s metres above sea level is given by

$$f(s) = e^{-0.000125s} \text{ atmospheres.}$$

  a. The atmospheric pressure outside an airplane is 0.25 atmospheres (atm). How high is the plane?
  b. A mountain climber decides she will wear an oxygen mask once she has reached an altitude of 7000 m. What is the atmospheric pressure at this altitude?

73. A* ALLOMETRY   Suppose that for the first 6 years of a moose's life, its shoulder height $H(t)$ and tip-to-tip antler length $A(t)$ increase with time t (years) according to the formulas $H(t) = 125e^{0.08t}$ and $A(t) = 50e^{0.16t}$, where H and A are both measured in centimetres.

  a. On the same graph, plot $H(t)$ and $A(t)$ for the applicable period $0 \leq t \leq 6$.
  b. Express antler length A as a function of height H. [*Hint*: First take logarithms on both sides of the equation $H = 125e^{0.08t}$ to express time t in terms of H and then substitute into the formula for $A(t)$.]
  c. Plot the graph of $A(H)$.

74. INVESTMENT   An investment firm estimates that the value of its portfolio after t years is A million dollars, where

$$A(t) = 300 \ln (t + 3).$$

a. What is the value of the account when $t = 0$?
b. How long does it take for the account to double its initial value?
c. How long does it take before the account is worth a billion dollars?

75. POPULATION GROWTH   A community grows in such a way that t years from now, its population is $P(t)$ thousand, where

$$P(t) = 51 + 100 \ln (t + 3).$$

a. What is the population when $t = 0$?
b. How long does it take for the population to double its initial value?
c. What is the average rate of growth of the population over the first 10 years?

76. ENTOMOLOGY   It is determined that the volume of the yolk of a house fly egg shrinks according to the formula $V(t) = 5e^{-1.3t}$ cubic millimetres, where t is the number of days from the time the egg is produced. The egg hatches after 4 days.

a. What is the volume of the yolk when the egg hatches?
b. Sketch the graph of the volume of the yolk over the time period $0 \le t \le 4$.
c. Find the half-life of the volume of the yolk, that is, the time it takes for the volume of the yolk to shrink to half its original size.

77. A* Show that if y is a power function of x, so that $y = Cx^k$, where C and k are constants, then $\ln y$ is a linear function of $\ln x$. (*Hint:* Take the logarithm on both sides of the equation $y = Cx^k$.)

78. A* In each case, use one of the laws of exponents to prove the indicated law of logarithms.

a. The quotient rule: $\ln \dfrac{u}{v} = \ln u - \ln v$
b. The power rule: $\ln u^r = r \ln u$

79. A* Let a and b be any positive numbers other than 1.
a. Show that $(\log_a b)(\log_b a) = 1$.
b. Show that $\log_a x = \dfrac{\log_b x}{\log_b a}$ for any $x > 0$.

80. A* Sketch the graph of $y = \log_b x$ for $0 < b < 1$ by reflecting the graph of $y = b^x$ in the line $y = x$. Then answer the following questions:
a. Is the graph of $y = \log_b x$ rising or falling for $x > 0$?
b. Is the graph concave upward or concave downward for $x > 0$?
c. What are the intercepts of the graph? Does the graph have any horizontal or vertical asymptotes?

d. What can be said about

$$\lim_{x \to +\infty} \log_b x \quad \text{and} \quad \lim_{x \to 0^+} \log_b x?$$

81. Use a graphing utility to graph $y = 10^x$, $y = x$, and $y = \log_{10} x$ on the same coordinate axes. How are these graphs related?

SECTION 4.3

L03

Differentiate both logarithmic and exponential functions.

Differentiation of Exponential and Logarithmic Functions

In the examples and exercises examined so far in this chapter, we have seen how exponential functions can be used to model a variety of situations, ranging from compound interest to population growth and radioactive decay. In order to discuss rates of change and to determine extreme values in such situations, we need derivative formulas for exponential functions and their logarithmic counterparts. We obtain the formulas in this section and examine a few basic applications. Additional exponential and logarithmic models will be explored in Section 4.4. We begin by showing that the natural exponential function $f(x) = e^x$ has the remarkable property of being its own derivative.

The Derivative of e^x   ■   For every real number x,

$$\frac{d}{dx}[e^x] = e^x.$$

To obtain this formula, let $f(x) = e^x$ and note that

$$f'(x) = \lim_{h \to 0} \frac{f(x + h) - f(x)}{h}$$

$$= \lim_{h \to 0} \frac{e^{x+h} - e^x}{h}$$

$$= \lim_{h \to 0} \frac{e^x e^h - e^x}{h} \qquad \text{since } e^{A+B} = e^A e^B$$

$$= e^x \lim_{h \to 0} \frac{e^h - 1}{h} \qquad \text{factor } e^x \text{ out of the limit}$$

It can be shown that

$$\lim_{h \to 0} \frac{e^h - 1}{h} = 1,$$

as indicated in Table 4.2. Thus, we have

$$f'(x) = e^x \lim_{h \to 0} \frac{e^h - 1}{h}$$

$$= e^x(1)$$

$$= e^x$$

as claimed. This derivative formula is used in Example 4.3.1.

TABLE 4.2

h	$\dfrac{e^h - 1}{h}$
0.01	1.005017
0.001	1.000500
0.0001	1.000050
−0.00001	0.999995
−0.0001	0.999950

EXAMPLE 4.3.1

Differentiate each function.

a. $f(x) = x^2 e^x$

b. $g(x) = \dfrac{x^3}{e^x + 2}$

Solution

a. Using the product rule gives

$$f'(x) = x^2(e^x)' + (x^2)'e^x$$

$$= x^2 e^x + (2x)e^x \qquad \text{power rule and exponential rule}$$

$$= xe^x(x + 2) \qquad \text{factor out } x \text{ and } e^x$$

b. Use the quotient rule to differentiate this function:

$$g'(x) = \frac{(e^x + 2)(x^3)' - x^3(e^x + 2)'}{(e^x + 2)^2}$$

$$= \frac{(e^x + 2)(3x^2) - x^3(e^x + 0)}{(e^x + 2)^2}$$

$$= \frac{x^2(3e^x - xe^x + 6)}{(e^x + 2)^2}$$

The fact that e^x is its own derivative means that at each point $P(c, e^c)$ on the curve $y = e^x$, the slope is equal to e^c, the y coordinate of P (Figure 4.7). This is one of the most important reasons for using e as the base for exponential functions in calculus.

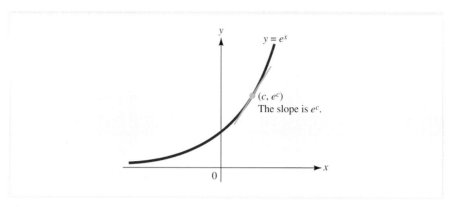

FIGURE 4.7   At each point $P(c, e^c)$ on the graph of $y = e^x$, the slope equals e^c.

By using the chain rule in conjunction with the differentiation formula

$$\frac{d}{dx}[e^x] = e^x,$$

we obtain the following formula for differentiating general exponential functions.

The Chain Rule for e^u   ■   If $u(x)$ is a differentiable function of x, then

$$\frac{d}{dx}[e^{u(x)}] = e^{u(x)}\frac{du}{dx}.$$

EXAMPLE 4.3.2

Differentiate the function $f(x) = e^{x^2+1}$.

Solution
Using the chain rule with $u = x^2 + 1$ gives

$$f'(x) = e^{x^2+1}\left(\frac{d}{dx}[x^2 + 1]\right) = 2xe^{x^2+1}.$$

EXAMPLE 4.3.3

Differentiate the function

$$f(x) = \frac{e^{-3x}}{x^2 + 1}.$$

Solution

Use the chain rule together with the quotient rule to get

$$f'(x) = \frac{(x^2 + 1)(-3e^{-3x}) - (2x)e^{-3x}}{(x^2 + 1)^2}$$

$$= -e^{-3x}\left[\frac{3(x^2 + 1) + 2x}{(x^2 + 1)^2}\right]$$

$$= -e^{-3x}\left[\frac{3x^2 + 2x + 3}{(x^2 + 1)^2}\right]$$

EXAMPLE 4.3.4

Find the largest and the smallest values of the function $f(x) = xe^{2x}$ on the interval $-1 \le x \le 1$.

Solution

By the product rule,

$$f'(x) = x\frac{d}{dx}[e^{2x}] + e^{2x}\frac{d}{dx}[x] = x(2e^{2x}) + e^{2x}(1) = (2x + 1)e^{2x},$$

so $f'(x) = 0$ when

$$(2x + 1)e^{2x} = 0$$
$$2x + 1 = 0 \qquad \text{since } e^{2x} > 0 \text{ for all } x$$
$$x = -\frac{1}{2}$$

Evaluate $f(x)$ at the critical number $x = -\frac{1}{2}$ and at the endpoints of the interval, $x = -1$ and $x = 1$, to find that

$$f(-1) = (-1)e^{-2} \approx -0.135$$
$$f\left(-\frac{1}{2}\right) = \left(-\frac{1}{2}\right)e^{-1} \approx -0.184 \qquad \text{minimum}$$
$$f(1) = (1)e^{2} \approx 7.389 \qquad \text{maximum}$$

Thus, $f(x)$ has its largest value 7.389 at $x = 1$ and its smallest value -0.184 at $x = -\frac{1}{2}$.

Derivatives of Logarithmic Functions

Here is the derivative formula for the natural logarithmic function.

The Derivative of ln x   ■   For all $x > 0$,

$$\frac{d}{dx}[\ln x] = \frac{1}{x}.$$

A proof using the definition of the derivative is outlined in Exercise 88. The formula can also be obtained as follows using implicit differentiation. Consider the equation

$$e^{\ln x} = x$$

Differentiating both sides with respect to x, we find that

$$\frac{d}{dx}[e^{\ln x}] = \frac{d}{dx}[x]$$

$$e^{\ln x}\frac{d}{dx}[\ln x] = 1 \qquad \text{chain rule}$$

$$x\frac{d}{dx}[\ln x] = 1 \qquad \text{since } e^{\ln x} = x$$

so

$$\frac{d}{dx}[\ln x] = \frac{1}{x} \qquad \text{divide both sides by } x$$

as claimed. The derivative formula for the natural logarithmic function is used in Examples 4.3.5 through 4.3.7.

EXAMPLE 4.3.5

Differentiate the function $f(x) = x \ln x$.

Solution

Combine the product rule with the formula for the derivative of $\ln x$ to get

$$f'(x) = x\left(\frac{1}{x}\right) + \ln x = 1 + \ln x.$$

Using the rules for logarithms can simplify the differentiation of complicated expressions. In Example 4.3.6, we use the power rule for logarithms before differentiating.

EXAMPLE 4.3.6

Differentiate $f(x) = \dfrac{\ln \sqrt[3]{x^2}}{x^4}$.

Solution

First, since $\sqrt[3]{x^2} = x^{2/3}$, use the power rule for logarithms to write

$$f(x) = \frac{\ln \sqrt[3]{x^2}}{x^4} = \frac{\ln x^{2/3}}{x^4} = \frac{\frac{2}{3}\ln x}{x^4}.$$

Then, by the quotient rule,

$$f'(x) = \frac{2}{3}\left[\frac{x^4(\ln x)' - (x^4)'\ln x}{(x^4)^2}\right]$$

$$= \frac{2}{3}\left[\frac{x^4\left(\dfrac{1}{x}\right) - 4x^3 \ln x}{x^8}\right]$$

$$= \frac{2}{3}\left[\frac{1 - 4\ln x}{x^5}\right] \qquad \text{cancel common } x^3 \text{ terms}$$

EXAMPLE 4.3.7

Differentiate $g(t) = (t + \ln t)^{3/2}$.

Solution

The function has the form $g(t) = u^{3/2}$, where $u = t + \ln t$, so apply the general power rule:

$$g'(t) = \frac{d}{dt}(u^{3/2}) = \frac{3}{2}u^{1/2}\frac{du}{dt}$$

$$= \frac{3}{2}(t + \ln t)^{1/2}\frac{d}{dt}[t + \ln t]$$

$$= \frac{3}{2}(t + \ln t)^{1/2}\left(1 + \frac{1}{t}\right)$$

If $f(x) = \ln u(x)$, where $u(x)$ is a differentiable function of x, then the chain rule yields the following formula for $f'(x)$.

The Chain Rule for ln u ■ If $u(x)$ is a differentiable function of x, then

$$\frac{d}{dx}[\ln u(x)] = \frac{1}{u(x)}\frac{du}{dx} \qquad \text{for } u(x) > 0$$

or

$$f(x) = \ln(u(x))$$

$$f'(x) = \frac{u'(x)}{u(x)}$$

EXAMPLE 4.3.8

Differentiate the function $f(x) = \ln(2x^3 + 1)$.

Solution

Here, $f(x) = \ln u$, where $u(x) = 2x^3 + 1$. Thus,

$$f'(x) = \frac{u'}{u} = \frac{6x^2}{2x^3 + 1}.$$

EXAMPLE 4.3.9

Find the equation of the tangent line to the graph of $f(x) = x - \ln\sqrt{x}$ at the point where $x = 1$.

Solution

When $x = 1$,

$$y = f(1) = 1 - \ln(\sqrt{1}) = 1 - 0 = 1,$$

so the point of tangency is $(1, 1)$. To find the slope of the tangent line at this point, write

$$f(x) = x - \ln\sqrt{x} = x - \frac{1}{2}\ln x$$

and compute the derivative

$$f'(x) = 1 - \frac{1}{2}\left(\frac{1}{x}\right) = 1 - \frac{1}{2x}.$$

Thus, the tangent line passes through the point $(1, 1)$ with slope

$$f'(x) = 1 - \frac{1}{2(1)} = \frac{1}{2}$$

so it has the equation

$$y - 1 = \frac{1}{2}(x - 1) \qquad \text{point-slope formula}$$

or, equivalently,

$$y = \frac{1}{2}x + \frac{1}{2}$$

Formulas for differentiating exponential and logarithmic functions with bases other than e are similar to those obtained for $y = e^x$ and $y = \ln x$. These formulas are given in the following box.

Derivatives of b^x and $\log_b x$ for Base $b > 0$, $b \neq 1$

$$\frac{d}{dx}[b^x] = (\ln b)b^x \qquad \text{for all } x$$

and

$$\frac{d}{dx}[\log_b x] = \frac{1}{x \ln b} \qquad \text{for all } x > 0.$$

For instance, to obtain the derivative formula for $y = \log_b x$, recall that

$$\log_b x = \frac{\ln x}{\ln b}$$

so

$$\frac{d}{dx}[\log_b x] = \frac{d}{dx}\left[\frac{\ln x}{\ln b}\right] = \frac{1}{\ln b}\frac{d}{dx}[\ln b]$$

$$= \frac{1}{x \ln b}$$

You are asked to obtain the derivative formula for $y = b^x$ in Exercise 89.

EXAMPLE 4.3.10

Differentiate each function.

  a. $f(x) = 5^{2x-3}$ **b.** $g(x) = (x^2 + \log_7 x)^4$

Solution

Using the chain rule:

  a. $f'(x) = [(\ln 5)5^{2x-3}][2x - 3]' = (\ln 5)5^{2x-3}(2)$

  b. $g'(x) = 4(x^2 + \log_7 x)^3[x^2 + \log_7 x]'$

$$= 4(x^2 + \log_7 x)^3\left(2x + \frac{1}{x \ln 7}\right)$$

Applications   Next, we shall examine several applications of calculus involving exponential and logarithmic functions. In Example 4.3.11, we compute the marginal revenue for a commodity with logarithmic demand.

EXAMPLE 4.3.11

A manufacturer of specialty perfumes determines that x bottles of perfume, requiring a special flower, will be sold when the price is $p(x) = 112 - x \ln x^3$ dollars per bottle.

  a. Find the revenue and marginal revenue functions.

  b. Use marginal analysis to estimate the revenue obtained from producing the fifth bottle. What is the actual revenue obtained from producing the fifth bottle?

Solution

  a. The revenue is

$$R(x) = xp(x) = x(112 - x \ln x^3) = 112x - x^2(3 \ln x)$$

dollars, and the marginal revenue is

$$R'(x) = 112 - 3\left[x^2\left(\frac{1}{x}\right) + (2x) \ln x\right] = 112 - 3x - 6x \ln x.$$

  b. The revenue obtained from producing the fifth bottle is estimated by the marginal revenue evaluated at $x = 4$, that is, by

$$R'(4) = 112 - 3(4) - 6(4) \ln 4 \approx 66.73.$$

Thus, marginal analysis suggests that the manufacturer will receive approximately $66.73 in revenue by producing the additional unit. The actual revenue obtained by producing the fifth unit is

$$R(5) - R(4) = [112(5) - 3(5)^2 \ln 5] - [112(4) - 3(4)^2 \ln 4]$$
$$\approx 439.29 - 381.46 = 57.83$$

or $57.83.

In Example 4.3.12, we examine exponential demand and use marginal analysis to determine the price at which the revenue associated with such demand is maximized. Part of the example deals with the concept of elasticity of demand, which was introduced in Section 3.4.

EXAMPLE 4.3.12

A manufacturer determines that $D(p) = 5000e^{-0.02p}$ units of a particular commodity will be demanded (sold) when the price is p dollars per unit.

a. Find the elasticity of demand for this commodity. For what values of p is the demand elastic, inelastic, and of unit elasticity?

b. If the price is increased by 3% from $40, what is the expected effect on demand?

c. Find the revenue $R(p)$ obtained by selling $q = D(p)$ units at p dollars per unit. For what value of p is the revenue maximized?

Solution

a. According to the formula derived in Section 3.4, the elasticity of demand is given by

$$E(p) = \frac{p}{q}\frac{dq}{dp}$$
$$= \left(\frac{p}{5000e^{-0.02p}}\right)[5000e^{-0.02p}(-0.02)]$$
$$= \frac{p[5000(-0.02)e^{-0.02p}]}{5000e^{-0.02p}} = -0.02p$$

Thus,

$$|E(p)| = |-0.02p| = 0.02p,$$

so, the **demand is of unit elasticity** when $|E(p)| = 0.02p = 1$, that is, when $p = 50$; the **demand is elastic** when $|E(p)| = 0.02p > 1$, or $p > 50$; and the **demand is inelastic** when $|E(p)| = 0.02p < 1$, or $p < 50$.

The graph of the demand function, showing the levels of elasticity, is displayed in Figure 4.8a.

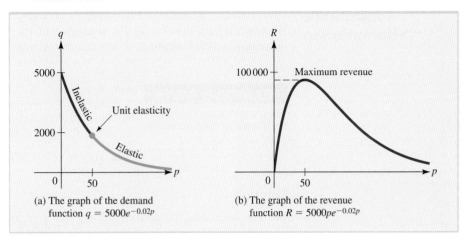

FIGURE 4.8   Demand and revenue curves for the commodity in Example 4.3.12.

b. When $p = 40$, the demand is

$$q(40) = 5000e^{-0.02(40)} \approx 2247 \text{ units}$$

and the elasticity of demand is

$$E(p) = -0.02(40) = -0.8.$$

Thus, an increase of 1% in price from $p = \$40$ will result in a decrease in the quantity demanded by approximately 0.8%. Consequently, an increase of 3% in price, from \$40 to \$41.20, results in a decrease in demand of approximately $2247[3(0.008)] = 54$ units, from 2247 to 2193 units.

c. The revenue function is

$$R(p) = pq = 5000pe^{-0.02p}$$

for $p \geq 0$ (only nonnegative prices have economic meaning), with derivative

$$R'(p) = 5000(-0.02pe^{-0.02p} + e^{-0.02p})$$
$$= 5000(1 - 0.02p)e^{-0.02p}$$

Since $e^{-0.02p}$ is always positive, $R'(p) = 0$ if and only if

$$1 - 0.02p = 0 \quad \text{or} \quad p = \frac{1}{0.02} = 50.$$

To verify that $p = 50$ actually gives the absolute maximum, note that

$$R''(p) = 5000(0.0004p - 0.04)e^{-0.02p},$$

so

$$R''(50) = 5000[0.0004(50) - 0.04]e^{-0.02(50)} \approx -37 < 0.$$

Thus, the second derivative test says that the absolute maximum of $R(p)$ does indeed occur when $p = 50$ (Figure 4.8b).

Logarithmic Differentiation

Differentiating a function that involves products, quotients, or powers can often be simplified by first taking the logarithm of the function. This technique, called **logarithmic differentiation,** is illustrated in Example 4.3.13.

EXAMPLE 4.3.13

Differentiate each function.

a. $f(x) = \dfrac{\sqrt[3]{x+1}}{(1-3x)^4}$ **b.** $g(x) = (x^2 - 3)^4(2x + 5)^2(3x - 1)^7$

Solution

a. You could find the derivative using the quotient rule and the chain rule, but the resulting computation would be somewhat tedious. (Try it!)

A more efficient approach is to take natural logarithms of both sides of the expression for f:

$$\ln f(x) = \ln\left[\frac{\sqrt[3]{x+1}}{(1-3x)^4}\right]$$

$$= \ln\sqrt[3]{x+1} - \ln(1-3x)^4$$

$$= \frac{1}{3}\ln(x+1) - 4\ln(1-3x)$$

Notice that by introducing the logarithm, you eliminate the quotient, the cube root, and the fourth power.

Now use the chain rule for logarithms to differentiate both sides of this equation to get

$$\frac{f'(x)}{f(x)} = \frac{1}{3}\left(\frac{1}{x+1}\right) - 4\left(\frac{-3}{1-3x}\right) = \frac{1}{3}\left(\frac{1}{x+1}\right) + \frac{12}{1-3x},$$

so that

$$f'(x) = f(x)\left[\frac{1}{3}\left(\frac{1}{x+1}\right) + \frac{12}{1-3x}\right]$$

$$= \left[\frac{\sqrt[3]{x+1}}{(1-3x)^4}\right]\left[\frac{1}{3}\left(\frac{1}{x+1}\right) + \frac{12}{1-3x}\right]$$

b.
$$g(x) = (x^2 - 3)^4(2x + 5)^2(3x - 1)^7$$

$$\ln(g(x)) = 4\ln(x^2 - 3) + 2\ln(2x + 5) + 7\ln(3x - 1)$$

$$\frac{g'(x)}{g(x)} = \frac{8x}{x^2 - 3} + \frac{4}{2x + 5} + \frac{21}{3x - 1}$$

$$g'(x) = (x^2 - 3)^4(2x + 5)^2(3x - 1)^7\left(\frac{8x}{x^2 - 3} + \frac{4}{2x + 5} + \frac{21}{3x - 1}\right)$$

This question would have needed the product rule several times. It can be simplified further by multiplying out to get rid of the fractions if desired.

If $Q(x)$ is a differentiable function of x, note that

$$\frac{d}{dx}[\ln Q] = \frac{Q'(x)}{Q(x)}$$

where the ratio on the right is the relative rate of change of $Q(x)$. That is, *the relative rate of change of a quantity $Q(x)$ can be computed by finding the derivative of $\ln Q$.* This special kind of logarithmic differentiation can be used to simplify the computation of various growth rates, as illustrated in Example 4.3.14.

EXAMPLE 4.3.14

A country exports three goods, wheat W, steel S, and oil O. Suppose at a particular time $t = t_0$, the revenue (in billions of dollars) derived from each of these goods is

$$W(t_0) = 4 \qquad S(t_0) = 7 \qquad O(t_0) = 10$$

and that S is growing at 8%, O is growing at 15%, and W is declining at 3%. At what relative rate is total export revenue growing at this time?

Solution

Let $R = W + S + O$. At time $t = t_0$, we know that

$$R(t_0) = W(t_0) + S(t_0) + O(t_0) = 4 + 7 + 10 = 21.$$

The percentage growth rates can be expressed as

$$\frac{W'(t_0)}{W(t_0)} = -0.03 \qquad \frac{S'(t_0)}{S(t_0)} = 0.08 \qquad \frac{O'(t_0)}{O(t_0)} = 0.15$$

so that

$$W'(t_0) = -0.03W(t_0) \qquad S'(t_0) = 0.08S(t_0) \qquad O'(t_0) = 0.15O(t_0)$$

Thus, at $t = t_0$, the relative rate of growth of R is

$$\begin{aligned}
\frac{R'(t_0)}{R(t_0)} &= \frac{d(\ln R)}{dt} = \frac{d}{dt}[\ln(W + S + O)]\bigg|_{t=t_0} \\
&= \frac{[W'(t_0) + S'(t_0) + O'(t_0)]}{[W(t_0) + S(t_0) + O(t_0)]} \\
&= \frac{-0.03W(t_0) + 0.08S(t_0) + 0.15O(t_0)}{W(t_0) + S(t_0) + O(t_0)} \\
&= \frac{-0.03W(t_0) + 0.08S(t_0) + 0.15O(t_0)}{R(t_0)} \\
&= \frac{-0.03W(t_0)}{R(t_0)} + \frac{0.08S(t_0)}{R(t_0)} + \frac{0.15O(t_0)}{R(t_0)} \\
&= \frac{-0.03(4)}{21} + \frac{0.08(7)}{21} + \frac{0.15(10)}{21} \\
&\approx 0.0924
\end{aligned}$$

That is, at time $t = t_0$, the total revenue obtained from the three exported goods is increasing at a rate of 9.24%.

EXERCISES ■ 4.3

In Exercises 1 through 38, differentiate the given function.

1. $f(x) = e^{5x}$

2. $f(x) = 3e^{4x+1}$

3. $f(x) = xe^x$

4. $f(x) = \dfrac{e^x}{x}$

5. $f(x) = 30 + 10e^{-0.05x}$

6. $f(x) = e^{x^2 + 2x - 1}$

7. $f(x) = (x^2 + 3x + 5)e^{6x}$

8. $f(x) = xe^{-x^2}$

9. $f(x) = (1 - 3e^x)^2$

10. $(x) = \sqrt{1 + e^x}$

11. $f(x) = e^{\sqrt{3x}}$

12. $f(x) = e^{1/x}$

13. $f(x) = \ln x^3$

14. $f(x) = \ln 2x$

15. $f(x) = x^2 \ln x$

16. $f(x) = x \ln \sqrt{x}$

17. $f(x) = \sqrt[3]{e^{2x}}$

18. $f(x) = \dfrac{\ln x}{x}$

19. $f(x) = \ln\left(\dfrac{x+1}{x-1}\right)$

20. $f(x) = e^x \ln x$

21. $f(x) = e^{-2x} + x^3$

22. $f(t) = t^2 \ln \sqrt[3]{t}$

23. $g(s) = (e^s + s + 1)(2e^{-s} + s)$

24. $F(x) = \ln(2x^3 - 5x + 1)$

25. $h(t) = \dfrac{e^t + t}{\ln t}$

26. $g(u) = \ln(u^2 - 1)^3$

27. $f(x) = \dfrac{e^x + e^{-x}}{2}$

28. $h(x) = \dfrac{e^{-x}}{x^2}$

29. $f(t) = \sqrt{\ln t + t}$

30. $f(x) = \dfrac{e^x + e^{-x}}{e^x - e^{-x}}$

31. $f(x) = \ln(e^{-x} + x)$

32. $f(s) = e^{s + \ln s}$

33. $g(u) = \ln(u + \sqrt{u^2 + 1})$

34. $L(x) = \ln\left[\dfrac{x^2 + 2x - 3}{x^2 + 2x + 1}\right]$

35. $f(x) = \dfrac{2^x}{x}$

36. $f(x) = x^2 3^{x^2}$

37. $f(x) = x \log_{10} x$

38. $f(x) = \dfrac{\log_2 x}{\sqrt{x}}$

In Exercises 39 through 46, find the largest and smallest values of the given function over the prescribed closed, bounded interval.

39. $f(x) = e^{1-x}$ for $0 \le x \le 1$

40. $F(x) = e^{x^2 - 2x}$ for $0 \le x \le 2$

41. $f(x) = (3x - 1)e^{-x}$ for $0 \le x \le 2$

42. $g(x) = \dfrac{e^x}{2x + 1}$ for $0 \le x \le 1$

43. $g(t) = t^{3/2}e^{-2t}$ for $0 \le t \le 1$

44. $f(x) = e^{-2x} - e^{-4x}$ for $0 \le x \le 1$

45. $f(x) = \dfrac{\ln(x + 1)}{x + 1}$ for $0 \le x \le 2$

46. $h(s) = 2s \ln s - s^2$ for $0.5 \le s \le 2$

In Exercises 47 through 52, find the equation of the tangent line to $y = f(x)$ at the specified point.

47. $f(x) = xe^{-x}$, where $x = 0$

48. $f(x) = (x + 1)e^{-2x}$, where $x = 0$

49. $f(x) = \dfrac{e^{2x}}{x^2}$, where $x = 1$

50. $f(x) = \dfrac{\ln x}{x}$, where $x = 1$

51. $f(x) = x^2 \ln \sqrt{x}$, where $x = 1$

52. $f(x) = x - \ln x$, where $x = e$

In Exercises 53 through 56, find the second derivative of the given function.

53. $f(x) = e^{2x} + 2e^{-x}$

54. $f(x) = \ln 2x + x^2$

55. $f(t) = t^2 \ln t$

56. $g(t) = t^2 e^{-t}$

In Exercises 57 through 64, use logarithmic differentiation to find the derivative $f'(x)$.

57. $f(x) = (2x + 3)^2(x - 5x^2)^{1/2}$

58. $f(x) = x^2 e^{-x}(3x + 5)^3$

59. $f(x) = \dfrac{(x + 2)^5}{\sqrt[6]{3x - 5}}$

60. $f(x) = \sqrt[4]{\dfrac{2x + 1}{1 - 3x}}$

61. $f(x) = (x + 1)^3(6 - x)^2 \sqrt[3]{2x + 1}$

62. $f(x) = \dfrac{e^{-3x}\sqrt{2x - 5}}{(6 - 5x)^4}$

63. $f(x) = 5^{x^2}$

64. $f(x) = \log_2(\sqrt{x})$

MARGINAL ANALYSIS *In Exercises 65 through 68, the demand function $q = D(p)$ for a particular commodity is given in terms of a price p per unit at which all q units can be sold. In each case:*

(a) *Find the elasticity of demand and determine the values of p for which the demand is elastic, inelastic, and of unit elasticity.*

(b) *If the price is increased by 2% from $15, what is the approximate effect on demand?*

(c) *Find the revenue $R(p)$ obtained by selling q units at the unit price p. For what value of p is revenue maximized?*

65. $D(p) = 3000e^{-0.04p}$

66. $D(p) = 10\,000e^{-0.025p}$

67. $D(p) = 5000(p + 11)e^{-0.1p}$

68. $D(p) = \dfrac{10\,000e^{-p/10}}{p + 1}$

MARGINAL ANALYSIS *In Exercises 69 through 72, the cost $C(x)$ of producing x units of a particular commodity is given. In each case:*

(a) *Find the marginal cost $C'(x)$.*

(b) *Determine the level of production x for which the average cost $A(x) = \dfrac{C(x)}{x}$ is minimized.*

69. $C(x) = e^{0.2x}$

70. $C(x) = 100e^{0.01x}$

71. $C(x) = 12\sqrt{x}\, e^{x/10}$

72. $C(x) = x^2 + 10xe^{-x}$

73. DEPRECIATION A certain industrial machine depreciates so that its value after t years becomes $Q(t) = 20\,000e^{-0.4t}$ dollars.

a. At what rate is the value of the machine changing with respect to time after 5 years?

b. At what percentage rate is the value of the machine changing with respect to time after t years? Does this percentage rate depend on t or is it constant?

74. COMPOUND INTEREST Money is deposited in a bank offering interest at an annual rate of 6% compounded continuously. Find the percentage rate of change of the balance with respect to time.

75. POPULATION GROWTH It is projected that t years from now, the population of a certain country will become $P(t) = 50e^{0.02t}$ million.

a. At what rate will the population be changing with respect to time 10 years from now?

b. At what percentage rate will the population be changing with respect to time t years from now? Does this percentage rate depend on t or is it constant?

76. COOLING A cool drink is removed from a refrigerator on a hot summer day and placed in a room whose temperature is 30°C. According to Newton's law of cooling, the temperature of the drink t minutes later is given by a function of the form $f(t) = 30 - Ae^{-kt}$. Show that the rate of change of the temperature of the drink with respect to time is proportional to the difference between the temperature of the room and that of the drink.

77. MARGINAL ANALYSIS The mathematics editor at a major publishing house estimates that if x thousand complimentary copies are distributed to professors, the first-year sales of a certain new textbook will be $f(x) = 20 - 15e^{-0.2x}$ thousand copies. Currently, the editor is planning to distribute 10 000 complimentary copies.

a. Use marginal analysis to estimate the increase in first-year sales that will result if 1000 additional complimentary copies are distributed.

b. Calculate the actual increase in first-year sales that will result from the distribution of the additional 1000 complimentary copies. Is the estimate in part (a) a good one?

78. **POPULATION GROWTH**   It is projected that t years from now, the population of a certain town will be approximately $P(t)$ thousand people, where

$$P(t) = \frac{100}{1 + e^{-0.2t}}.$$

At what rate will the population be changing 10 years from now? At what percentage rate will the population be changing at that time?

79. **LEARNING**   According to the Ebbinghaus model (recall Exercise 64, Section 4.1), the fraction $F(t)$ of subject matter that you will remember from this course t months after the final exam can be estimated by the formula $F(t) = B + (1 - B)e^{-kt}$, where B is the fraction of the material you will never forget and k is a constant that depends on the quality of your memory,
   a. Find $F'(t)$ and explain what this derivative represents.
   b. Show that $F'(t)$ is proportional to $F - B$ and interpret this result. [*Hint:* What does $F - B$ represent in terms of what you remember?]
   c. Sketch the graph of $F(t)$ for the case where $B = 0.3$ and $k = 0.2$.

80. **ALCOHOL ABUSE CONTROL**   Suppose the percentage of alcohol in the blood t hours after consumption is given by

$$C(t) = 0.12te^{-t/2}.$$

   a. At what rate is the blood alcohol level changing at time t?
   b. How much time passes before the blood alcohol level begins to decrease?
   c. Suppose the legal limit for blood alcohol is 0.04%. How much time must pass before the blood alcohol reaches this level? At what rate is the blood alcohol level decreasing when it reaches the legal limit?

81. **CONSUMER EXPENDITURE**   The demand for a certain commodity is $D(p) = 3000e^{-0.01p}$ units per month when the market price is p dollars per unit.
   a. At what rate is the consumer expenditure $E(p) = pD(p)$ changing with respect to price p?
   b. At what price does consumer expenditure stop increasing and begin to decrease?
   c. At what price does the *rate* of consumer expenditure begin to increase? Interpret this result.

82. **A* LEARNING**   In an experiment to test memory learning, a subject is confronted by a series of tasks, and it is found that t minutes after the experiment begins, the number of tasks successfully completed is

$$R(t) = \frac{15(1 - e^{-0.01t})}{1 + 1.5e^{-0.01t}}.$$

   a. For what values of t is the learning function $R(t)$ increasing? For what values is it decreasing?
   b. When is the rate of change of the learning function $R(t)$ increasing? When is it decreasing? Interpret your results.

83. **ENDANGERED SPECIES**   An international agency determines that the number of individuals of an endangered species that remain in the wild t years after a protection policy is instituted may be modelled by

$$N(t) = \frac{600}{1 + 3e^{-0.02t}}.$$

   a. At what rate is the population changing at time t? When is the population increasing? When is it decreasing?
   b. When is the rate of change of the population increasing? When is it decreasing? Interpret your results.
   c. What happens to the population in the long run (as $t \to +\infty$)?

84. **A* ENDANGERED SPECIES**   The agency in Exercise 83 studies a second endangered species but fails to receive funding to develop a policy of protection. The population of the species is modelled by

$$N(t) = \frac{30 + 500e^{-0.3t}}{1 + 5e^{-0.3t}}.$$

   a. At what rate is the population changing at time t? When is the population increasing? When is it decreasing?
   b. When is the rate of change of the population increasing? When is it decreasing? Interpret your results.
   c. What happens to the population in the long run (as $t \to +\infty$)?

85. **PLANT GROWTH**   Two plants grow in such a way that t days after planting, they are $P_1(t)$ and $P_2(t)$ centimetres tall, respectively, where

$$P_1(t) = \frac{21}{1 + 25e^{-0.3t}} \quad \text{and} \quad P_2(t) = \frac{20}{1 + 17e^{-0.6t}}.$$

a. At what rate is the first plant growing at time $t = 10$ days? Is the rate of growth of the second plant increasing or decreasing at this time?

b. At what time do the two plants have the same height? What is this height? Which plant is growing more rapidly when they have the same height?

86. A* PER CAPITA GROWTH The national income $I(t)$ of a particular country is increasing by 2.3% per year, while the population $P(t)$ of the country is decreasing at an annual rate of 1.75%. The per capita income C is defined to be

$$C(t) = \frac{I(t)}{P(t)}.$$

a. Find the derivative of $\ln C(t)$.

b. Use the result of part (a) to determine the percentage rate of growth of per capita income.

87. A* REVENUE GROWTH A country exports electronic components E and textiles T. Suppose at a particular time $t = t_0$, the revenue (in billions of dollars) derived from these goods is

$$E(t_0) = 11 \quad \text{and} \quad T(t_0) = 8$$

and that E is growing at 9%, while T is declining at 2%. At what relative rate is total export revenue $R = E + T$ changing at this time?

88. A* DERIVATIVE FORMULA FOR ln x Prove that the derivative of $f(x) = \ln x$ is $f'(x) = \dfrac{1}{x}$ by completing these steps.

a. Show that the difference quotient of $f(x)$ can be expressed as

$$\frac{f(x + h) - f(x)}{h} = \ln\left(1 + \frac{h}{x}\right)^{1/h}.$$

b. Let $n = \dfrac{x}{h}$ so that $x = nh$. Show that the difference quotient in part (a) can be rewritten as

$$\ln\left[\left(1 + \frac{1}{n}\right)^n\right]^{1/x}$$

c. Show that the limit of the expression in part (b) as $n \to \infty$ is $\ln e^{1/x} = \dfrac{1}{x}$. [*Hint:* What is $\displaystyle\lim_{n\to\infty}\left(1 + \frac{1}{n}\right)^n$?]

d. Complete the proof by finding the limit of the difference quotient in part (a) as $h \to 0$. [*Hint:* How is this related to the limit you found in part c?]

89. A* For base $b > 0$, $b \neq 1$, show that

$$\frac{d}{dx}[b^x] = (\ln b)b^x$$

a. by using the fact that $b^x = e^{x \ln b}$

b. by using logarithmic differentiation

90. A quantity grows so that $Q(t) = Q_0\dfrac{e^{kt}}{t}$. Find the percentage rate of change of Q with respect to t.

SECTION 4.4

L04

Formulate equations for exponential applications and calculate all the unknown constants.

Applications; Exponential Models

Earlier in this chapter we saw how continuous compounding and radioactive decay can be modelled using exponential functions. In this section, we introduce several additional exponential models from a variety of areas, such as business and economics, biology, psychology, demography, and sociology. We begin with two examples illustrating the particular issues that arise when finding intervals where the functions increase and decrease and finding the concavity. The key to graphing a function $f(x)$ involving e^x or $\ln x$ is to use the derivative $f'(x)$ to find intervals of increase and decrease and then use the second derivative $f''(x)$ to determine concavity.

EXAMPLE 4.4.1

Find the intervals where the function $f(x) = x^2 - 8 \ln x$ is increasing and intervals where it is decreasing. Then determine the concavity and whether the graph has asymptotes.

Solution

The function $f(x)$ is defined only for $x > 0$. Its derivative is

$$f'(x) = 2x - \frac{8}{x} = \frac{2x^2 - 8}{x}$$

and $f'(x) = 0$ if and only if $2x^2 = 8$ or $x = 2$ (since $x > 0$). Testing the sign of $f'(x)$ for $0 < x < 2$ and for $x > 2$ gives the intervals of increase and decrease shown in the figure.

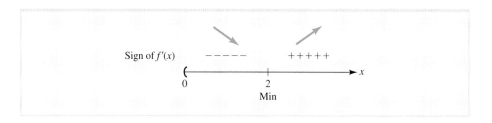

Notice that the arrow pattern indicates that there is a relative minimum at $x = 2$, and since $f(2) = 2^2 - 8 \ln 2 \approx -1.5$, the minimum point is $(2, -1.5)$.

The second derivative

$$f''(x) = 2 + \frac{8}{x^2}$$

satisfies $f''(x) > 0$ for all $x > 0$, so the graph of $f(x)$ is always concave up and there are no inflection points.

Check for asymptotes:

$$\lim_{x \to 0^+} (x^2 - 8 \ln x) = +\infty \quad \text{and} \quad \lim_{x \to +\infty} (x^2 - 8 \ln x) = +\infty.$$

So the y axis ($x = 0$) is a vertical asymptote, but there is no horizontal asymptote. Find the x intercepts by using a graphing utility to solve the equation

$$x^2 - 8 \ln x = 0$$
$$x \approx 1.2 \quad \text{and} \quad x \approx 2.9$$

To summarize, the graph falls from the vertical asymptote to the minimum at $(2, -1.5)$, after which it rises indefinitely, while maintaining a concave-up shape. It passes through the x intercept $(1.2, 0)$ on the way down and through $(2.9, 0)$ as it rises. The graph is shown in Figure 4.9.

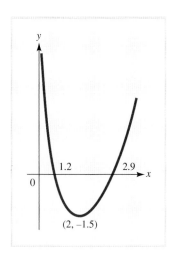

FIGURE 4.9   The graph of $f(x) = x^2 - 8 \ln x$.

Just-In-Time

Remember that $\sqrt{2\pi}$ is just a constant, a number like 6 or 25. Although it looks different, just ignore it and think about the derivatives!

EXAMPLE 4.4.2

Determine where the function

$$f(x) = \frac{1}{\sqrt{2\pi}} e^{-x^2/2}$$

is increasing and where it is decreasing, and where its graph is concave up and where it is concave down. Find the relative extrema and inflection points and draw the graph.

Solution

The first derivative is

$$f'(x) = \frac{-x}{\sqrt{2\pi}} e^{-x^2/2}.$$

Since $e^{-x^2/2}$ is always positive, $f'(x)$ is zero if and only if $x = 0$. Since $f(0) = \dfrac{1}{\sqrt{2\pi}} \approx 0.4$, the only critical point is $(0, 0.4)$. By the product rule, the second derivative is

$$f''(x) = \frac{x^2}{\sqrt{2\pi}} e^{-x^2/2} - \frac{1}{\sqrt{2\pi}} e^{-x^2/2} = \frac{1}{\sqrt{2\pi}} (x^2 - 1) e^{-x^2/2}$$

which is zero if $x = \pm 1$. Since

$$f(1) = \frac{e^{-1/2}}{\sqrt{2\pi}} \approx 0.24 \quad \text{and} \quad f(-1) = \frac{e^{-1/2}}{\sqrt{2\pi}} \approx 0.24,$$

the potential inflection points are $(1, 0.24)$ and $(-1, 0.24)$.

Plot the critical points and then check the signs of the first and second derivatives on each of the intervals defined by the x coordinates of these points:

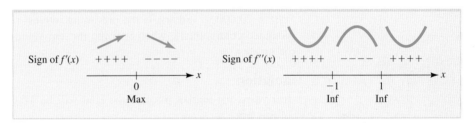

The arrow pattern indicates that there is a relative maximum at $(0, 0.4)$, and since the concavity changes at $x = -1$ (from up to down) and at $x = 1$ (from down to up), both $(-1, 0.24)$ and $(1, 0.24)$ are inflection points.

Complete the graph, as shown in Figure 4.10, by connecting the key points with a curve of appropriate shape on each interval. Notice that the graph has no x intercepts since $e^{-x^2/2}$ is always positive and that the graph approaches the x axis as a horizontal asymptote since $e^{-x^2/2}$ approaches zero as $|x|$ increases without bound.

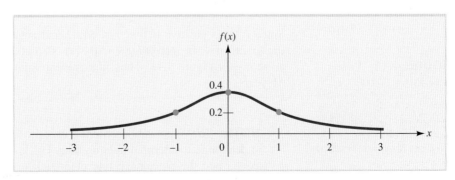

FIGURE 4.10 The standard normal density function: $f(x) = \dfrac{1}{\sqrt{2\pi}} e^{-x^2/2}$.

NOTE   The function $f(x) = \dfrac{1}{\sqrt{2\pi}}e^{-x^2/2}$ from Example 4.4.2 is known as the **standard normal probability density function** and plays a vital role in probability and statistics. The famous bell-shape of the graph is used by physicists and social scientists to describe distributions of IQ scores, measurements on large populations of living organisms, the velocity of a molecule in a gas, and numerous other important phenomena.

Optimal Holding Time

Suppose you own an asset whose value increases with time. The longer you hold the asset, the more it will be worth, but there may come a time when you could do better by selling the asset and reinvesting the proceeds. Economists determine the optimal time for selling by maximizing the present value of the asset in relation to the prevailing rate of interest, compounded continuously. The application of this criterion is illustrated in Example 4.4.3.

EXAMPLE 4.4.3

Suppose you own a parcel of land whose market price t years from now will be $V(t) = 20\,000e^{\sqrt{t}}$ dollars. If the prevailing interest rate remains constant at 7% compounded continuously, when will the present value of the market price of the land be greatest?

Solution

In t years, the market price of the land will be $V(t) = 20\,000e^{\sqrt{t}}$. The present value of this price is

$$P(t) = V(t)e^{-0.07t} = 20\,000e^{\sqrt{t}}e^{-0.07t} = 20\,000e^{\sqrt{t}-0.07t}.$$

The goal is to maximize $P(t)$ for $t \geq 0$. The derivative of P is

$$P'(t) = 20\,000e^{\sqrt{t}-0.07t}\left(\frac{1}{2\sqrt{t}} - 0.07\right).$$

Thus, $P'(t)$ is undefined when $t = 0$ and $P'(t) = 0$ when

$$\frac{1}{2\sqrt{t}} - 0.07 = 0 \qquad \text{or} \qquad t = \left[\frac{1}{2(0.07)}\right]^2 \approx 51.02.$$

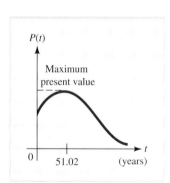

FIGURE 4.11   Present value $P(t) = 20\,000e^{\sqrt{t}-0.07t}$.

Since $P'(t)$ is positive if $0 < t < 51.02$ and negative if $t > 51.02$, it follows that the graph of P is increasing for $0 < t < 51.02$ and decreasing for $t > 51.02$, as shown in Figure 4.11. Thus, the present value is maximized in approximately 51 years.

NOTE   The optimization criterion used in Example 4.4.3 is not the only way to determine the optimal holding time. For instance, you may decide to sell the asset when the percentage rate of growth of the asset's value just equals the prevailing rate of interest (7% in the example). Which criterion seems more reasonable to you? Actually, it does not matter, for the two criteria yield exactly the same result! A proof of this equivalence is outlined in Exercise 56.

Exponential Growth and Decay

A quantity $Q(t)$ for $k > 0$ is said to experience **exponential growth** if $Q(t) = Q_0 e^{kt}$ and **exponential decay** if $Q(t) = Q_0 e^{-kt}$. Many important quantities in business and economics and in the physical, social, and biological sciences can be modelled in terms of exponential growth and decay. The future value of a continuously compounded investment grows exponentially, as does population in the absence of restrictions. We discussed the decay of radioactive substances in Section 4.2. Other examples of exponential decay are the present value of a continuously compounded investment, sales of certain commodities once advertising is discontinued, and the concentration of drug in a patient's bloodstream.

Exponential Growth and Decay ■ A quantity $Q(t)$ *grows exponentially* if $Q(t) = Q_0 e^{kt}$ for $k > 0$ and *decays exponentially* if $Q(t) = Q_0 e^{-kt}$ for $k > 0$.

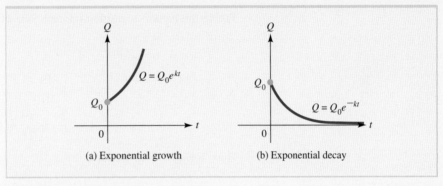

(a) Exponential growth   (b) Exponential decay

FIGURE 4.12 Exponential change.

Typical exponential growth and decay graphs are shown in Figure 4.12. It is customary to display such graphs only for $t \geq 0$. Note that the graph of $Q(t) = Q_0 e^{kt}$ begins at Q_0 on the vertical axis since

$$Q(0) = Q_0 e^{k(0)} = Q_0.$$

Note also that the graph of $Q(t) = Q_0 e^{kt}$ rises sharply since

$$Q'(t) = Q_0 k e^{kt} = kQ(t),$$

which means that $Q(t)$ always increases at a rate proportional to its current value, so the larger the value of Q, the larger the slope. The graph of $Q(t) = Q_0 e^{-kt}$ also begins at Q_0, but falls sharply, approaching the t axis asymptotically. Here is an example of a business model involving exponential decay.

EXAMPLE 4.4.4

A marketing manager determines that sales of a frozen pizza will decline exponentially once the advertising campaign and coupon availability end. It is found that 21 000 pizzas are being sold at the end of the campaign, and 5 weeks later the sales level is 19 000 units.

a. Find an expression for the sales $S(t)$ after time t weeks later.

b. What sales should be expected 8 weeks after the advertising campaign ends?

c. At what rate are sales changing t weeks after the end of the campaign? What is the percentage rate of change?

Solution

We note that the question states "decline exponentially" so we write the form of the equation $S(t) = S_0 e^{-kt}$.

For simplicity, express sales S in terms of thousands of units.

a. We know that $S = 21$ when $t = 0$ and $S = 19$ when $t = 5$. Substituting $t = 0$ and $S = 21$ into the formula $S(t) = S_0 e^{-kt}$ gives

$$21 = S_0 e^{-k(0)},$$

so $S_0 = 21$ and $S(t) = 21 e^{-kt}$ for all t. Substituting $S = 19$ and $t = 5$, we find that $19 = 21 e^{-k(5)}$ or

$$e^{-5k} = \frac{19}{21}.$$

Taking the natural logarithm on each side of this equation, we find that

$$\ln e^{-5k} = \ln \frac{19}{21}$$

$$-5k = \ln \frac{19}{21}$$

$$k = -\frac{1}{5} \ln \frac{19}{21} \approx 0.02$$

Thus, for all $t > 0$, we have

$$S(t) = 21 e^{-0.02t}$$

b. For $t = 8$,

$$S(8) = 21 e^{-0.02(8)} \approx 17.9,$$

so the model predicts sales of about 17 900 pizzas 8 weeks after the advertising campaign ends.

c. The rate of change of sales is given by the derivative

$$S'(t) = 21[e^{-0.02t}(-0.02)] = -0.02(21)e^{-0.02t}$$

and the percentage rate of change PR is

$$PR = \frac{S'(t)}{S(t)} = \frac{-0.02(21)e^{-0.02t}}{21 e^{-0.02t}}$$

$$= -0.02$$

That is, sales are declining at a rate of 2% per week.

Notice that the percentage rate of change obtained in Example 4.4.4b is the same as k, expressed as a percentage. This is no accident, since for any function of the form $Q(t) = Q_0 e^{rt}$, the percentage rate of change is

$$PR = \frac{Q'(t)}{Q(t)} = \frac{Q_0 e^{rt}(r)}{Q_0 e^{rt}} = r.$$

For instance, an investment in an account that earns interest at an annual rate of 5% compounded continuously has a future value of $B = Pe^{0.05t}$. Thus, the percentage rate of change of the future value is 0.05 or 5%, which is exactly what we would expect.

Learning Curves

The graph of a function of the form $Q(t) = B - Ae^{-kt}$, where A, B, and k are positive constants, is sometimes called a **learning curve.** The name arose when psychologists discovered that for $t \geq 0$, functions of this form often realistically model the relationship between the efficiency with which an individual performs a task and the amount of training time or experience the "learner" has had.

To sketch the graph of $Q(t) = B - Ae^{-kt}$ for $t \geq 0$, note that

$$Q'(t) = -Ae^{-kt}(-k) = Ake^{-kt}$$

and

$$Q''(t) = Ake^{-kt}(-k) = -Ak^2e^{-kt}.$$

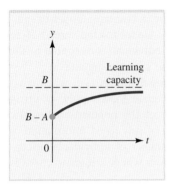

FIGURE 4.13 A learning curve $y = B - Ae^{-kt}$.

Since A and k are positive, it follows that $Q'(t) > 0$ and $Q''(t) < 0$ for all t, so the graph of $Q(t)$ is always rising and is always concave down. Furthermore, the vertical (Q axis) intercept is $Q(0) = B - A$, and $Q = B$ is a horizontal asymptote since

$$\lim_{t \to +\infty} Q(t) = \lim_{t \to +\infty} (B - Ae^{-kt}) = B - 0 = B.$$

A graph with these features is sketched in Figure 4.13. The behaviour of the learning curve as $t \to \infty$ reflects the fact that in the long run, an individual approaches his or her learning capacity, and additional training time will result in only marginal improvement in performance efficiency.

EXAMPLE 4.4.5

The rate at which a server can serve coffee and water at an all-inclusive resort with a large buffet and continual need for those drinks is a function of the server's experience. Suppose the manager estimates that after t weeks on the job, the average server can serve coffee and water continuously to $Q(t) = 40 - 15e^{-0.5t}$ tables.

a. How many tables can a new employee serve?

b. How many tables can a server with 4 weeks' experience serve?

c. Approximately how many tables will the average server ultimately be able to serve?

Solution

a. The number of tables a new employee can serve coffee to is

$$Q(0) = 40 - 15e^0 = 25.$$

b. After 4 weeks, the average server can serve

$$Q(4) = 40 - 15e^{-0.5(4)} = \approx 38 \text{ tables.}$$

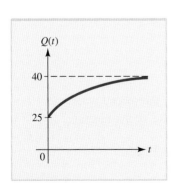

FIGURE 4.14 Worker efficiency $Q(t) = 40 - 15e^{-0.5t}$.

c. As t increases without bound, $Q(t)$ approaches 40. Hence, the average server will ultimately be able to serve approximately 40 tables. The graph of the function $Q(t)$ is sketched in Figure 4.14.

Logistic Curves

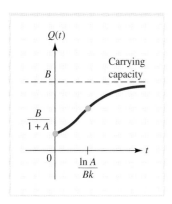

FIGURE 4.15   A logistic curve $Q(t) = \dfrac{B}{1 + Ae^{-Bkt}}$.

The graph of a function of the form $Q(t) = \dfrac{B}{1 + Ae^{-Bkt}}$, where A, B, and k are positive constants, is called a **logistic curve.** A typical logistic curve is shown in Figure 4.15. Notice that it rises steeply like an exponential curve at first, and then turns over and flattens out, approaching a horizontal asymptote in much the same way as a learning curve. The asymptotic line represents a saturation level for the quantity represented by the logistic curve and is called the **carrying capacity** of the quantity. For instance, in population models, the carrying capacity represents the maximum number of individuals the environment can support, while in a logistic model for the spread of an epidemic, the carrying capacity is the total number of individuals susceptible to the disease, say, those who are unvaccinated or, at worst, the entire community.

Logistic curves often provide accurate models of population growth when environmental factors such as restricted living space, inadequate food supply, or urban pollution impose an upper bound on the possible size of the population. Logistic curves are also often used to describe the dissemination of privileged information or rumours in a community, where the restriction is the number of individuals susceptible to receiving such information. Here is an example in which a logistic curve is used to describe the spread of an infectious disease.

EXAMPLE 4.4.6

Public health records indicate that t weeks after the outbreak of a certain form of influenza, approximately $Q(t) = \dfrac{20}{1 + 19e^{-1.2t}}$ thousand people had caught the disease.

a. How many people had the disease when it broke out? How many had it 2 weeks later?

b. What is the rate of change of the number of people infected 2 weeks later? 3 weeks later? Interpret these results.

c. If the trend continues, approximately how many people will eventually contract the disease?

Solution

a. Since $Q(0) = \dfrac{20}{1 + 19} = 1$, it follows that 1000 people initially had the disease. When $t = 2$,

$$Q(2) = \dfrac{20}{1 + 19e^{-1.2(2)}} \approx 7.343,$$

so about 7343 people had contracted the disease by the second week.

b. The rate of infection is found by taking the derivative of the function:

$$Q(t) = \dfrac{20}{1 + 19e^{-1.2t}} = 20(1 + 19e^{-1.2t})^{-1}$$
$$Q'(t) = -20(1 + 19e^{-1.2t})^{-2}(19(-1.2)e^{-1.2t})$$
$$Q'(t) = \dfrac{456e^{-1.2t}}{(1 + 19e^{-1.2t})^2}$$
$$Q'(2) = 5.576$$
$$Q'(3) = 5.399$$

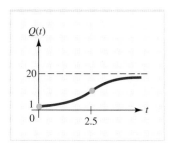

FIGURE 4.16   The spread of an epidemic $Q(t) = \dfrac{20}{1 + 19e^{-1.2t}}$.

The rate of change is an increase of 5576 per week after 2 weeks and 5399 per week after 3 weeks. The rate of change is decreasing.

c. Since $Q(t)$ approaches 20 as t increases without bound, it follows that approximately 20 000 people will eventually contract the disease. For reference, the graph is sketched in Figure 4.16.

EXERCISES ■ 4.4

Each of the curves shown in Exercises 1 through 4 is the graph of one of the six functions listed here. In each case, match the given curve to the correct function.

$$f_1(x) = 2 - e^{-2x} \qquad f_2(x) = x \ln x^5$$

$$f_3(x) = \frac{2}{1 - e^{-x}} \qquad f_4(x) = \frac{2}{1 + e^{-x}}$$

$$f_5(x) = \frac{\ln x^5}{x} \qquad f_6(x) = (x - 1)e^{-2x}$$

1.

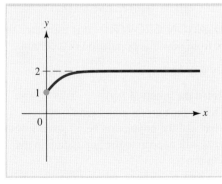

2.

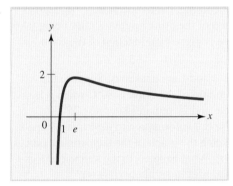

3.

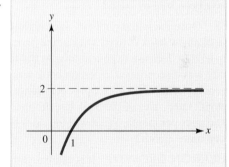

4.

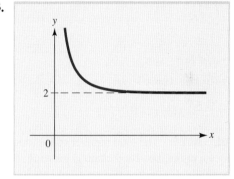

In Exercises 5 through 20, determine where the given function is increasing and where it is decreasing and where its graph is concave upward and where it is concave downward. Sketch the graph of the function. Show as many key features as possible (high and low points, points of inflection, vertical and horizontal asymptotes, intercepts, cusps, vertical tangents).

5. $f(t) = 2 + e^t$

6. $g(x) = 3 + e^{-x}$

7. $g(x) = 2 - 3e^x$

8. $f(t) = 3 - 2e^t$

9. $f(x) = \dfrac{2}{1 + 3e^{-2x}}$

10. $h(t) = \dfrac{2}{1 + 3e^{2t}}$

11. $f(x) = xe^{x}$

12. $f(x) = xe^{-x}$

13. $f(x) = xe^{2-x}$

14. $f(x) = e^{-x^2}$

15. $f(x) = x^2 e^{-x}$

16. $f(x) = e^{x} + e^{-x}$

17. $f(x) = \dfrac{6}{1 + e^{-x}}$

18. $f(x) = x - \ln x$ (for $x > 0$)

19. $f(x) = (\ln x)^2$ (for $x > 0$)

20. $f(x) = \dfrac{\ln x}{x}$ (for $x > 0$)

21. **RETAIL SALES**   The total number of hamburgers sold by a national fast-food chain is growing exponentially. If 4 billion had been sold by 2005 and 12 billion had been sold by 2010, how many will have been sold by 2015?

22. **SALES**   Once the initial publicity surrounding the release of a new book is over, sales of the hardcover edition tend to decrease exponentially. At the time publicity was discontinued, a certain book was experiencing sales of 25 000 copies per month. One month later, sales of the book had dropped to 10 000 copies per month. What will the sales be after one more month?

23. **POPULATION GROWTH**   It is estimated that the population of a certain country grows exponentially. If the population was 60 million in 1997 and 90 million in 2002, what will the population be in 2012?

24. **POPULATION GROWTH**   Based on the estimate that there are 1.4 billion hectares (ha) of arable land on Earth and 0.07 ha is needed as a minimum per person for a vegetarian diet, some demographers believe that Earth can support a population of no more than 40 billion people. The population of Earth was approximately 5.3 billion in 1990 and 6.5 billion in 2005. If the population of Earth were growing exponentially, when would it reach the theoretical limit of 40 billion?

25. **PRODUCT RELIABILITY**   A manufacturer of toys has found that the fraction of its plastic battery-operated toy boats that sink in fewer than t days is approximately $f(t) = 1 - e^{-0.03t}$.
   a. Sketch this reliability function. What happens to the graph as t increases without bound?
   b. What fraction of the boats can be expected to float for at least 10 days?
   c. What fraction of the boats can be expected to sink between the 15th and 20th days?

26. **DEPRECIATION**   When a certain industrial machine becomes t years old, its resale value will be $V(t) = 4800e^{-t/5} + 400$ dollars.
   a. Sketch the graph of $V(t)$. What happens to the value of the machine as t increases without bound?
   b. How much is the machine worth when it is new?
   c. How much will the machine be worth after 10 years?

27. **COOLING**   A hot drink is taken outside on a cold winter day when the air temperature is $-5°C$. According to a principle of physics called Newton's law of cooling, the temperature T (in degrees Celsius) of the drink t minutes after being taken outside is given by a function of the form
$$T(t) = -5 + Ae^{-kt},$$
where A and k are constants. Suppose the temperature of the drink is $80°C$ when it is taken outside, and 20 minutes later it is $25°C$.
   a. Use this information to determine A and k.
   b. Sketch the graph of the temperature function $T(t)$. What happens to the temperature as t increases indefinitely ($t \to \infty$)?
   c. What will the temperature be after 30 minutes?
   d. When will the temperature reach $0°C$?

28. **POPULATION GROWTH**   It is estimated that t years from now, the population of a certain country will be $P(t) = \dfrac{20}{2 + 3e^{-0.06t}}$ million.
   a. Sketch the graph of $P(t)$.
   b. What is the current population?
   c. What will the population be 50 years from now?
   d. What will happen to the population in the long run?

29. **THE SPREAD OF AN EPIDEMIC**   Public health records indicate that t weeks after the outbreak of a certain form of influenza,

approximately $f(t) = \dfrac{2}{1 + 3e^{-0.8t}}$ thousand people

had caught the disease.
a. Sketch the graph of $f(t)$.
b. How many people had the disease initially?
c. How many had caught the disease by the end of 3 weeks?
d. If the trend continues, approximately how many people in all will contract the disease?

30. **RECALL FROM MEMORY**   Psychologists believe that when a person is asked to recall a set of facts, the number of facts recalled after t minutes is given by a function of the form $Q(t) = A(1 - e^{-kt})$, where k is a positive constant and A is the total number of relevant facts in the person's memory.
a. Sketch the graph of $Q(t)$.
b. What happens to the graph as t increases without bound? Explain this behaviour in practical terms.

31. **EFFICIENCY**   The daily output of a worker who has been doing inventory in a very large warehouse for t weeks is given by a function of the form $Q(t) = 40 - Ae^{-kt}$. Initially the worker could inventory 20 shelves a day, and after 1 week the worker can inventory 30 shelves a day. How many shelves will the worker be able to inventory per day after 3 weeks?

32. **ADVERTISING**   When professors select textbooks for their courses, they usually choose from among the books already on their shelves. For this reason, most publishers send complimentary copies of new textbooks to professors teaching related courses. The mathematics editor at a major publishing house estimates that if x thousand complimentary copies are distributed, the first-year sales of a certain new mathematics textbook will be approximately $f(x) = 20 - 15e^{-0.2x}$ thousand copies.
a. Sketch this sales function.
b. How many copies can the editor expect to sell in the first year if no complimentary copies are sent out?
c. How many copies can the editor expect to sell in the first year if 10 000 complimentary copies are sent out?
d. If the editor's estimate is correct, what is the most optimistic projection for the first-year sales of the textbook?

33. **MARGINAL ANALYSIS**   The economics editor at a major publishing house estimates that if x thousand complimentary copies are distributed to professors, the first-year sales of a certain new textbook will be $f(x) = 15 - 20e^{-0.3x}$ thousand copies. Currently, the editor is planning to distribute 9000 complimentary copies.
a. Use marginal analysis to estimate the increase in first-year sales that will result if 1000 additional complimentary copies are distributed.
b. Calculate the actual increase in first-year sales that will result from the distribution of the additional 1000 complimentary copies. Is the estimate in part (a) a good one?

34. **LABOUR MANAGEMENT**   A business manager estimates that when x thousand people are employed at her firm, the profit will be $P(x)$ million dollars, where
$$P(x) = \ln(4x + 1) + 3x - x^2.$$
What level of employment maximizes profit? What is the maximum profit?

35. **CHILDHOOD LEARNING**   A psychologist measures a child's capability to learn and remember by the function
$$L(t) = \frac{\ln(t + 1)}{t + 1}$$
where t is the child's age in years, for $0 \leq t \leq 5$. Answer the following questions about this model.
a. At what age does a child have the greatest learning capability?
b. At what age is a child's learning capability increasing most rapidly?

36. **AEROBIC RATE**   The aerobic rating of a person who is x years old is modelled by the function
$$A(x) = \frac{110(\ln x - 2)}{x} \quad \text{for } x \geq 10.$$
a. At what age is a person's aerobic rating largest?
b. At what age is a person's aerobic rating decreasing most rapidly?

37. **STRUCTURAL DESIGN**   When a chain, a telephone line, or a TV cable is strung between supports, the curve it forms is called a **catenary.** A typical catenary curve is
$$y = 10(e^{0.05x} + e^{-0.05x}).$$
a. Find the critical point, and determine whether it is a maximum or a minimum.

b. If the catenary joined two poles 20 m apart, with x and y measured in metres, how much sag would there be in the middle?

c. Plot the graph of the catenary.

38. **WORLD POPULATION** According to a certain logistic model, the world's population (in billions) t years after 1960 will be approximately

$$P(t) = \frac{40}{1 + 12e^{-0.08t}}.$$

a. If this model is correct, at what rate was the world's population increasing with respect to time in the year 2000? At what *percentage* rate was the population increasing at this time?

b. When will the population be growing most rapidly?

c. Sketch the graph of $P(t)$. What feature occurs on the graph at the time found in part (b)? What happens to $P(t)$ in the long run?

d. Do you think such a population model is reasonable? Why or why not?

39. **MARGINAL ANALYSIS** A manufacturer can produce LCD monitors at a cost of $125 each and estimates that if they are sold for x dollars each, consumers will buy approximately $1000e^{-0.02x}$ each week.

a. Express the profit P as a function of x. Sketch the graph of $P(x)$.

b. At what price should the manufacturer sell the monitors to maximize profit?

40. **THE SPREAD OF AN EPIDEMIC** An epidemic spreads through a community so that t weeks after its outbreak, the number of people who have been infected is given by a function of the form

$$f(t) = \frac{B}{1 + Ce^{-kt}},$$ where B is the number of

residents in the community who are susceptible to the disease. If $\frac{1}{5}$ of the susceptible residents were infected initially and $\frac{1}{2}$ had been infected by the end of the fourth week, what fraction of the susceptible residents will have been infected by the end of the eighth week?

41. **OZONE DEPLETION** It is known that fluorocarbons have the effect of depleting ozone in the upper atmosphere. Suppose it is found that the amount of original ozone Q_0 that remains after t years is given by

$$Q = Q_0 e^{-0.0015t}.$$

a. At what percentage rate is the ozone level decreasing at time t?

b. How many years will it take for 10% of the ozone to be depleted? At what percentage rate is the ozone level decreasing at this time?

42. **BUSINESS TRAINING** A house-painting company organizes a training program in which it is determined that after t weeks, the average painter can paint

$$P(t) = 50(1 - e^{-0.15t})$$

walls per week, while a typical new worker without special training paints

$$W(t) = \sqrt{150t}$$

walls per week.

a. How many walls does the average trainee paint during the third week of the training period?

b. Explain how the function $F(t) = P(t) - W(t)$ can be used to evaluate the effectiveness of the training program. Is the program effective if it lasts just 5 weeks? What if it lasts at least 7 weeks? Explain your reasoning.

43. **OPTIMAL HOLDING TIME** Suppose you own a parcel of land whose value t years from now will be $V(t) = 8000e^{\sqrt{t}}$ dollars. If the prevailing interest rate remains constant at 6% per year compounded continuously, when should you sell the land to maximize its present value?

44. **OPTIMAL HOLDING TIME** Suppose your family owns a rare book whose value t years from now will be $V(t) = 200e^{\sqrt{2t}}$ dollars. If the prevailing interest rate remains constant at 6% per year compounded continuously, when will it be most advantageous for your family to sell the book and invest the proceeds?

45. **A* ACCOUNTING** The **double declining balance** formula in accounting is

$$V(t) = V_0\left(1 - \frac{2}{L}\right)^t,$$

where $V(t)$ is the value after t years of an article that originally cost V_0 dollars and L is a constant, called the "useful life" of the article. A refrigerator costs $875 and has a useful life of 8 years. What is its value after 5 years? What is its annual rate of depreciation?

46. **FISHERY MANAGEMENT** The manager of a fishery determines that t weeks after 3000 fish of a particular species are hatched, the average weight of

an individual fish will be $w(t) = 0.8te^{-0.05t}$ kilograms, for $0 \le t \le 20$. Moreover, the proportion of the fish that will still be alive after t weeks is estimated to be

$$p(t) = \frac{10}{10 + t}.$$

a. The expected yield $E(t)$ from harvesting after t weeks is the total weight of the fish that are still alive. Express $E(t)$ in terms of $w(t)$ and $p(t)$.

b. For what value of t is the expected yield $E(t)$ the largest? What is the maximum expected yield?

c. Sketch the yield curve $y = E(t)$ for $0 \le t \le 20$.

47. FISHERY MANAGEMENT The manager of a fishery determines that t days after 1000 fish of a particular species are released into a pond, the average weight of an individual fish will be $w(t)$ kilograms and the proportion of the fish still alive after t days will be $p(t)$, where

$$w(t) = \frac{10}{1 + 15e^{-0.05t}} \quad \text{and} \quad p(t) = e^{-0.01t}.$$

a. The expected yield $E(t)$ from harvesting after t days is the total weight of the fish that are still alive. Express $E(t)$ in terms of $w(t)$ and $p(t)$.

b. For what value of t is the expected yield $E(t)$ the largest? What is the maximum expected yield?

c. Sketch the yield curve $y = E(t)$.

48. OPTIMAL HOLDING TIME Suppose you own a stamp collection that is currently worth $1200 and whose value increases linearly at a rate of $200 per year. If the prevailing interest rate remains constant at 8% per year compounded continuously, when will it be most advantageous for you to sell the collection and invest the proceeds?

49. THE SPREAD OF A RUMOUR A traffic accident was witnessed by 10% of the residents of a small town, and 25% of the residents had heard about the accident 2 hours later. Suppose the number $N(t)$ of residents who had heard about the accident t hours after it occurred is given by a function of the form

$$N(t) = \frac{B}{1 + Ce^{-kt}},$$

where B is the population of the town and C and k are constants.

a. Use this information to find C and k.

b. How long does it take for half the residents of the town to know about the accident?

50. EFFECT OF A TOXIN A medical researcher determines that t hours from the time a toxin is introduced to a bacterial colony, the population will be

$$P(t) = 10\ 000(7 + 15e^{-0.05t} + te^{-0.05t}).$$

a. What is the population at the time the toxin is introduced?

b. When does the maximum bacterial population occur? What is the maximum population?

c. What eventually happens to the bacterial population as $t \to +\infty$?

51. CORPORATE ORGANIZATION A **Gompertz curve** is the graph of a function of the general form

$$N(t) = CA^{B^t},$$

where A, B, and C are constants. Such curves are used by psychologists and others to describe such things as learning and growth within an organization.

a. Suppose the personnel director of a large corporation conducts a study that indicates that after t years, the corporation will have

$$N(t) = 500(0.03)^{(0.4)^t}$$

employees. How many employees are there originally (at time $t = 0$)? How many are there after 5 years? When will there be 300 employees? How many employees will there be in the long run?

b. Sketch the graph of $N(t)$. Then, on the same graph sketch the graph of the Gompertz function

$$F(t) = 500(0.03)^{-(0.4)^{-t}}.$$

How would you describe the relationship between the two graphs?

52. A* LEARNING THEORY In a learning model proposed by C. L. Hull, the habit strength H of an individual is related to the number r of reinforcements by the equation

$$H(r) = M(1 - e^{-kr}).$$

a. Sketch the graph of $H(r)$. What happens to $H(r)$ as $r \to +\infty$?

b. Show that if the number of reinforcements is doubled from r to $2r$, the habit strength is multiplied by $1 + e^{-kr}$.

53. CONCENTRATION OF DRUG A function of the form $C(t) = Ate^{-kt}$, where A and k are positive constants, is called a **surge function** and is sometimes used to model the concentration of drug in a patient's bloodstream t hours after the drug is administered. Assume $t \ge 0$.

a. Find $C'(t)$ and determine the time interval where the drug concentration is increasing and where it is decreasing. For what value of t is the concentration maximized? What is the maximum concentration? (Your answers will be in terms of A and k.)

b. Find $C''(t)$ and determine time intervals of concavity for the graph of $C(t)$. Find all points of inflection and explain what is happening to the rate of change of drug concentration at the times that correspond to the inflection points.

c. Plot the graph of $C(t) = te^{-kt}$ for $k = 0.2$, $k = 0.5$, $k = 1.0$, and $k = 2.0$. Describe how the shape of the graph changes as k increases.

54. CONCENTRATION OF DRUG The concentration of a certain drug in a patient's bloodstream t hours after being administered orally is assumed to be given by the surge function $C(t) = Ate^{-kt}$, where C is measured in micrograms of drug per millilitre of blood. Monitoring devices indicate that a maximum concentration of 5 μg/mL occurs 20 minutes after the drug is administered.

a. Use this information to find A and k.

b. What is the drug concentration in the patient's blood after 1 hour?

c. At what time after the maximum concentration occurs will the concentration be half the maximum?

d. If you double the time in part (c), will the resulting concentration of drug be $\dfrac{1}{4}$ the maximum? Explain.

55. MARKET RESEARCH A company is trying to use television advertising to expose as many people as possible to a new product in a large metropolitan area with a million possible viewers. A model for the number of people N (in millions) who are aware of the product after t days is found to be

$$N = 2(1 - e^{-0.037t}).$$

Use a graphing utility to graph this function. What happens as $t \to +\infty$?

56. OPTIMAL HOLDING TIME Let $V(t)$ be the value of an asset t years from now and assume that the prevailing annual interest rate remains fixed at r (expressed as a decimal) compounded continuously.

a. Show that the present value of the asset $P(t) = V(t)e^{-rt}$ has a critical number where $V'(t) = V(t)r$. (Using economic arguments, it can be shown that the critical number corresponds to a maximum.)

b. Explain why the present value of $V(t)$ is maximized at a value of t where the percentage rate of change (expressed in decimal form) equals r.

57. THE SPREAD OF AN EPIDEMIC An epidemic spreads throughout a community so that t weeks after its outbreak, the number of residents who have been infected is given by a function of the form

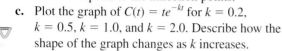

$$f(t) = \frac{A}{1 + Ce^{-kt}} \text{ hundred people, where } A$$

hundred is the total number of susceptible residents.

To individualize this question, make A your age and C your age minus one. For example, if you are 25, then $A = 25$ and $C = 24$. Also, k is $1 +$ the last two digits of your phone number divided by 100. For example, if your phone number is 212-3847, then $k = 1.47$.

a. Plot graphs of $f(t)$, $f'(t)$, and $f''(t)$, either under each other with the same scale, or on the same graph.

b. By hand, find $f''(t)$, factor the numerator, and find the inflection point.

c. On each graph, label the point at which the epidemic is spreading most rapidly.

d. From the graph, show that the epidemic is spreading most rapidly when half of the susceptible residents have been infected.

e. What happens in the long run? Show this on the graph.

58. Use a graphing utility to sketch the graph of $f(x) = x(e^{-x} + e^{-2x})$. Find the largest value of $f(x)$. What happens to $f(x)$ as $x \to +\infty$?

59. A* PROBABILITY DENSITY FUNCTION The general probability density function has the form

$$f(x) = \frac{1}{\sigma\sqrt{2\pi}}e^{-(x-\mu)^2/2\sigma^2},$$

where μ and σ are constants, with $\sigma > 0$.

a. Show that $f(x)$ has an absolute maximum at $x = \mu$ and inflection points at $x = \mu + \sigma$ and $x = \mu - \sigma$.

b. Show that $f(\mu + c) = f(\mu - c)$ for every number c. What does this tell you about the graph of $f(x)$?

60. OPTIMAL HOLDING TIME Suppose you win a parcel of land whose market value t years from now is estimated to be $V(t) = 20\ 000te^{\sqrt{0.4t}}$ dollars. If the prevailing interest rate remains constant at 7% compounded continuously, graphically find the time when it will be most advantageous to sell the land.

Concept Summary Chapter 4

Solving Equations with e^x and $\ln x$

$$9 - 10e^{3x} = 2$$
$$-10e^{3x} = -7$$
$$e^{3x} = \frac{-7}{-10}$$
$$3x = \ln 0.7$$
$$x = \frac{\ln 0.7}{3}$$

$$5 \ln 6x - 1 = 11$$
$$5 \ln 6x = 12$$
$$\ln 6x = \frac{12}{5} = 2.4$$
$$6x = e^{2.4}$$
$$x = \frac{e^{2.4}}{6}$$

Manipulating Logarithms

$$\ln\left(\frac{AB}{C}\right)^n = n\ln\left(\frac{AB}{C}\right)$$
$$= n(\ln AB - \ln C)$$
$$= n(\ln A + \ln B - \ln C)$$

$$\ln e = 1$$
$$\ln 1 = 0$$
$$\ln e^u = u$$
$$e^{\ln u} = u$$
$$\log_b a = \frac{\ln a}{\ln b}$$

Differentiating

$$f(x) = e^{u(x)} \text{ then } f'(x) = u'(x)e^{u(x)}$$
$$g(x) = \ln u(x) \text{ then } g'(x) = \frac{u'(x)}{u(x)}$$
$$h(x) = b^x \text{ then } h'(x) = (\ln b)b^x$$

Business Formulas

k times a year, t years, r annual interest rate

$$B = P\left(1 + \frac{r}{k}\right)^{kt}$$
$$P = B\left(1 + \frac{r}{k}\right)^{-kt}$$
$$r_e = \left(1 + \frac{r}{k}\right)^k - 1$$

Continuous Compounding

$$B = Pe^{rt}, P = Be^{-rt}, r_e = e^r - 1$$

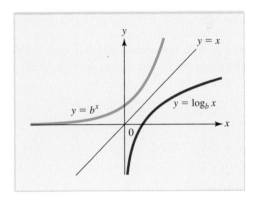

Checkup for Chapter 4

1. Evaluate each expression.

   a. $\dfrac{(3^{-2})(9^2)}{(27)^{2/3}}$

   b. $\sqrt[3]{(25)^{1.5}\left(\dfrac{8}{27}\right)}$

   c. $\log_2 4 + \log_4 16^{-1}$

   d. $\left(\dfrac{8}{27}\right)^{-2/3}\left(\dfrac{16}{81}\right)^{3/2}$

2. Simplify each expression.

   a. $(9x^4y^2)^{3/2}$

   b. $(3x^2y^{4/3})^{-1/2}$

   c. $\left(\dfrac{y}{x}\right)^{3/2}\left(\dfrac{x^{2/3}}{y^{1/6}}\right)^2$

   d. $\left(\dfrac{x^{0.2}y^{-1.2}}{x^{1.5}y^{0.4}}\right)^5$

3. Find all real numbers x that satisfy each equation.

a. $4^{2x-x^2} = \dfrac{1}{64}$

b. $e^{1/x} = 4$

c. $\log_4 x^2 = 2$

d. $\dfrac{25}{1 + 2e^{-0.5t}} = 3$

4. In each case, find the derivative $\dfrac{dy}{dx}$. (In some cases, it may help to use logarithmic differentiation.)

a. $y = \dfrac{e^x}{x^2 - 3x}$

b. $y = \ln(x^3 + 2x^2 - 3x)$

c. $y = x^3 \ln x$

d. $y = \dfrac{e^{-2x}(2x - 1)^3}{1 - x^2}$

5. Find any relative extrema and determine their nature with either the first or second derivative test.

a. $y = x^2 e^{-x}$

b. $y = \dfrac{\ln \sqrt{x}}{x^2}$

c. $y = \ln(\sqrt{x} - x)^2$

d. $y = \dfrac{4}{1 + e^{-x}}$

6. If you invest \$2000 at 5% compounded continuously, how much will your account be worth in 3 years? How long does it take before your account is worth \$3000?

7. PRESENT VALUE Find the present value of \$8000 payable 10 years from now if the annual interest rate is 6.25% and interest is compounded

a. semiannually

b. continuously

8. PRICE ANALYSIS A product is introduced, and t months later, its unit price is $p(t)$ hundred dollars, where

$$p = \dfrac{\ln(t + 1)}{t + 1} + 5.$$

a. For what values of t is the price increasing? When is it decreasing?

b. When is the price decreasing most rapidly?

c. What happens to the price in the long run (as $t \to +\infty$)?

9. MAXIMIZING REVENUE It is determined that q units of a commodity can be sold when the price is p hundred dollars per unit, where

$$q(p) = 1000(p + 2)e^{-p}.$$

a. Verify that the demand function $q(p)$ decreases as p increases for $p \geq 0$.

b. For what price p is revenue $R = pq$ maximized? What is the maximum revenue?

10. CARBON DATING An archaeological artifact is found to have 45% of its original ^{14}C. How old is the artifact? (Use 5730 years as the half-life of ^{14}C.)

11. BACTERIAL GROWTH A toxin is introduced into a bacterial colony, and t hours later, the population is given by

$$N(t) = 10\,000(8 + t)e^{-0.1t}.$$

a. What was the population when the toxin was introduced?

b. When is the population maximized? What is the maximum population?

c. What happens to the population in the long run (as $t \to +\infty$)?

Review Exercises

In Exercises 1 through 4, sketch the graph of the given exponential or logarithmic function without using calculus.

1. $f(x) = 5^x$

2. $f(x) = -2e^{-x}$

3. $f(x) = \ln x^2$

4. $f(x) = \log_3 x$

5. a. Find $f(4)$ if $f(x) = Ae^{-kx}$, $f(0) = 10$, and $f(1) = 25$.

b. Find $f(3)$ if $f(x) = Ae^{kx}$, $f(1) = 3$, and $f(2) = 10$.

c. Find $f(9)$ if $f(x) = 30 + Ae^{-kx}$, $f(0) = 50$, and $f(3) = 40$.

d. Find $f(10)$ if $f(t) = \dfrac{6}{1 + Ae^{-kt}}$, $f(0) = 3$, and $f(5) = 2$.

6. Evaluate each expression without using a calculator.
  a. $\ln e^5$
  b. $e^{\ln 2}$
  c. $e^{3 \ln 4 - \ln 2}$
  d. $\ln 9e^2 + \ln 3e^{-2}$

In Exercises 7 through 14, find all real numbers x that satisfy the given equation.

7. $8 = 2e^{0.04x}$

8. $5 = 1 + 4e^{-6x}$

9. $4 \ln x = 8$

10. $5^x = e^3$

11. $\log_9 (4x - 1) = 2$

12. $\ln (x - 2) + 3 = \ln (x + 1)$

13. $e^{2x} + e^x - 2 = 0$ [*Hint:* Let $u = e^x$.]

14. $e^{2x} + 2e^x - 3 = 0$ [*Hint:* Let $u = e^x$.]

In Exercises 15 through 30, find the derivative $\dfrac{dy}{dx}$. In some of these problems, you may need to use implicit differentiation or logarithmic differentiation.

15. $y = x^2 e^{-x}$

16. $y = 2e^{3x+5}$

17. $y = x \ln x^2$

18. $y = \ln \sqrt{x^2 + 4x + 1}$

19. $y = \log_3 x^2$

20. $y = \dfrac{x}{\ln 2x}$

21. $y = \dfrac{e^{-x} + e^x}{1 + e^{-2x}}$

22. $y = \dfrac{e^{3x}}{e^{3x} + 2}$

23. $y = \ln (e^{-2x} + e^{-x})$

24. $y = (1 + e^{-x})^{4/5}$

25. $y = \dfrac{e^{-x}}{x + \ln x}$

26. $y = \ln \left(\dfrac{e^{3x}}{1 + x} \right)$

27. $ye^{x-x^2} = x + y$

28. $xe^{-y} + ye^{-x} = 3$

29. $y = \dfrac{(x^2 + e^{2x})^3 e^{-2x}}{(1 + x - x^2)^{2/3}}$

30. $y = \dfrac{e^{-2x}(2 - x^3)^{3/2}}{\sqrt{1 + x^2}}$

In Exercises 31 through 38, determine where the given function is increasing and where it is decreasing and where its graph is concave upward and where it is concave downward. Sketch the graph, showing as many key features as possible (high and low points, points of inflection, asymptotes, intercepts, cusps, vertical tangents).

31. $f(x) = e^x - e^{-x}$

32. $f(x) = xe^{-2x}$

33. $f(t) = t + e^{-t}$

34. $f(x) = \dfrac{4}{1 + e^{-x}}$

35. $F(u) = u^2 + 2 \ln (u + 2)$

36. $g(t) = \dfrac{\ln(t + 1)}{t + 1}$

37. $G(x) = \ln(e^{-2x} + e^{-x})$

38. $f(u) = e^{2u} + e^{-u}$

In Exercises 39 through 42, find the largest and smallest values of the given function over the prescribed closed, bounded interval.

39. $f(x) = \ln (4x - x^2)$ for $1 \le x \le 3$

40. $f(x) = \dfrac{e^{-x/2}}{x^2}$ for $-5 \le x \le -1$

41. $h(t) = (e^{-t} + e^t)^5$ for $-1 \le t \le 1$

42. $g(t) = \dfrac{\ln(\sqrt{t})}{t^2}$ for $1 \le t \le 2$

In Exercises 43 through 46, find the equation of the tangent line to the given curve at the specified point.

43. $y = x \ln x^2$, where $x = 1$

44. $y = (x^2 - x)e^{-x}$, where $x = 0$

45. $y = x^3 e^{2-x}$, where $x = 2$

46. $y = (x + \ln x)^3$, where $x = 1$

47. Find $f(9)$ if $f(x) = e^{kx}$ and $f(3) = 2$.

48. Find $f(8)$ if $f(x) = A(2^{kx})$, $f(0) = 20$, and $f(2) = 40$.

49. COMPOUND INTEREST A sum of money is invested at a certain fixed interest rate, and the interest is compounded quarterly. After 15 years, the money has doubled. How will the balance at the end of 30 years compare with the initial investment?

50. COMPOUND INTEREST A bank pays 5% interest compounded quarterly, and a savings institution pays 4.9% interest compounded continuously. Over a 1-year period, which account pays more interest? What about a 5-year period?

51. RADIOACTIVE DECAY A radioactive substance decays exponentially. If 500 g of the substance were present initially and 400 g are present 50 years later, how many grams will be present after 200 years?

52. COMPOUND INTEREST A sum of money is invested at a certain fixed interest rate, and the interest is compounded continuously. After 10 years, the money has doubled. How will the balance at the end of 20 years compare with the initial investment?

53. GROWTH OF BACTERIA The following data were compiled by a researcher during the first 10 minutes of an experiment designed to study the growth of bacteria:

Number of minutes	0	10
Number of bacteria	5000	8000

Assuming that the number of bacteria grows exponentially, how many bacteria will be present after 30 minutes?

54. RADIOACTIVE DECAY The following data were compiled by a researcher during an experiment designed to study the decay of a radioactive substance:

Number of hours	0	5
Grams of substance	1000	700

Assuming that the sample of radioactive substance decays exponentially, how much is left after 20 hours?

55. SALES FROM ADVERTISING It is estimated that if x thousand dollars are spent on advertising a new steak and egg sandwich, approximately $Q(x) = 50 - 40e^{-0.1x}$ thousand will be sold.

a. Sketch the sales curve for $x \geq 0$.

b. How many steak and egg sandwiches will be sold if no money is spent on advertising?

c. How many steak and egg sandwiches will be sold if $8000 is spent on advertising?

d. How much should be spent on advertising to generate sales of 35 000 steak and egg sandwiches?

e. According to this model, what is the most optimistic sales projection?

56. WORKER PRODUCTION An employer determines that the daily output of an employee processing store credit card applications on the job for t weeks is $Q(t) = 120 - Ae^{-kt}$ units. Initially, the employee can process 30 applications per day, and after 8 weeks, the employee can process 80 applications per day. How many applications can the employee process per day after 4 weeks?

57. COMPOUND INTEREST How quickly will $2000 grow to $5000 when invested at an annual interest rate of 8% if interest is compounded

a. quarterly?

b. continuously?

58. COMPOUND INTEREST How much should you invest now at an annual interest rate of 6.25% so that your balance 10 years from now will be $2000 if interest is compounded

a. monthly?

b. continuously?

59. DEBT REPAYMENT You have a debt of $10 000, which is scheduled to be repaid at the end of 10 years. If you want to repay your debt now, how much should your creditor demand if the prevailing interest rate is

a. 7% compounded monthly?

b. 6% compounded continuously?

60. COMPOUND INTEREST A bank compounds interest continuously. What (nominal) interest rate does it offer if $1000 grows to $2054.44 in 12 years?

61. EFFECTIVE RATE OF INTEREST Which investment has the greater effective interest rate: 8.25% per year compounded quarterly or 8.20% per year compounded continuously?

62. **DEPRECIATION** The value of a certain industrial machine decreases exponentially. If the machine was originally worth \$50 000 and, 5 years later, was worth \$20 000, how much will it be worth when it is 10 years old?

63. **POPULATION GROWTH** It is estimated that t years from now the population of a certain country will be P million people, where

$$P(t) = \frac{30}{1 + 2e^{-0.05t}}$$

   a. Sketch the graph of $P(t)$.
   b. What is the current population?
   c. What will the population be in 20 years?
   d. What happens to the population in the long run?

64. **BACTERIAL GROWTH** The number of bacteria in a certain culture grows exponentially. If 5000 bacteria were initially present and 8000 were present 10 minutes later, how long will it take for the number of bacteria to double?

65. **AIR POLLUTION** An environmental study of a certain suburban community suggests that t years from now, the average level of carbon monoxide in the air will be $Q(t) = 4e^{0.03t}$ parts per million.

   a. At what rate will the carbon monoxide level be changing with respect to time 2 years from now?
   b. At what percentage rate will the carbon monoxide level be changing with respect to time t years from now? Does this percentage rate of change depend on t or is it constant?

66. **PROFIT** A manufacturer of cameras estimates that when cameras are sold for x dollars each, consumers will buy $8000e^{-0.02x}$ cameras each week. He also determines that profit is maximized when the selling price x is 1.4 times the cost of producing each unit. What price maximizes weekly profit? How many units are sold each week at this optimal price?

67. **OPTIMAL HOLDING TIME** Suppose you own an asset whose value t years from now will be $V(t) = 2000e^{\sqrt{2t}}$ dollars. If the prevailing interest rate remains constant at 5% per year compounded continuously, when will it be most advantageous to sell the collection and invest the proceeds?

68. **RULE OF 70** Investors are often interested in knowing how long it takes for a particular investment to double. A simple way to make this

determination is the "rule of 70," which says that the doubling time of an investment with an annual interest rate r (expressed as a decimal) compounded continuously is given by $d = \dfrac{70}{r}$.

   a. For interest rate r, use the formula $B = Pe^{rt}$ to find the doubling time for $r = 4, 6, 9, 10,$ and 12. In each case, compare the value with the value obtained from the rule of 70.
   b. Some people prefer a "rule of 72" and others use a "rule of 69." Test these alternative rules as in part (a) and write a paragraph on which rule you would prefer to use.

69. **RADIOACTIVE DECAY** A radioactive substance decays exponentially with half-life λ. Suppose the amount of the substance initially present (when $t = 0$) is Q_0.

   a. Show that the amount of the substance that remains after t years will be $Q(t) = Q_0 e^{-(\ln 2/\lambda)t}$.
   b. Find a number k so that the amount in part (a) can be expressed as $Q(t) = Q_0(0.5)^{kt}$.

70. **ANIMAL DEMOGRAPHY** A naturalist at an animal sanctuary has determined that the function

$$f(x) = \frac{4e^{-(\ln x)^2}}{\sqrt{\pi x}}$$

provides a good measure of the number of animals in the sanctuary that are x years old. Sketch the graph of $f(x)$ for $x > 0$ and find the most "likely" age among the animals, that is, the age for which $f(x)$ is largest.

71. **ARCHAEOLOGY** "Lucy," the famous prehuman whose skeleton was discovered in Ethiopia in 1974, has been found to be approximately 3.8 million years old.

   a. Approximately what percentage of original ^{14}C would you expect to find if you tried to apply carbon dating to Lucy? Why would this be a problem if you were actually trying to "date" Lucy?
   b. In practice, carbon dating works well only for relatively recent samples—those that are no more than approximately 50 000 years old. For older samples, such as Lucy, variations on carbon dating have been developed, such as potassium-argon and rubidium-strontium dating. Research these other methods and write a few sentences describing them.

72. FICK'S LAW Fick's law says that $f(t) = C(1 - e^{-kt})$, where $f(t)$ is the concentration of solute inside a cell at time t, C is the (constant) concentration of solute surrounding the cell, and k is a positive constant. Suppose that for a particular cell, the concentration on the inside of the cell after 2 hours is 0.8% of the concentration outside the cell. What is k?

73. COOLING A child falls into the ocean where the water temperature is $-3°C$. Her body temperature after t minutes in the water is $T(t) = 35e^{-0.32t}$. She will lose consciousness when her body temperature reaches $27°C$. How long do rescuers have to save her? How fast is her body temperature dropping at the time it reaches $27°C$?

74. FORENSIC SCIENCE The temperature T of the body of a murder victim found in a room where the air temperature is $20°C$ is given by

$$T(t) = 20 + 17e^{-0.07t} \text{ degrees Celsius}$$

where t is the number of hours after the victim's death.

   a. Graph the body temperature $T(t)$ for $t \geq 0$. What is the horizontal asymptote of this graph and what does it represent?

   b. What is the temperature of the victim's body after 10 hours? How long does it take for the body's temperature to reach $25°C$?

   c. Abel Baker works at the firm of Dewey, Cheatum, and Howe. He comes to work early one morning and finds the corpse of his boss, Will Cheatum, draped across his desk. He calls the police, and at 8 A.M., they determine that the temperature of the corpse is $33°C$. Since the last item entered on the victim's notepad was, "Fire that idiot Baker," Abel is considered the prime suspect. Actually, Abel is bright enough to have been reading this textbook in his spare time. He glances at the thermostat to confirm that the room temperature is $20°C$. For what time will he need an alibi in order to establish his innocence?

75. CONCENTRATION OF DRUG Suppose that t hours after an antibiotic is administered orally, its concentration in the patient's bloodstream is given by a surge function of the form $C(t) = Ate^{-kt}$, where A and k are positive constants and C is measured in micrograms per millilitre of blood. Blood samples are taken periodically, and it is determined that the maximum concentration of drug occurs 2 hours after it is administered and is $10 \mu g/mL$.

   a. Use this information to determine A and k.

   b. A new dose will be administered when the concentration falls to $1 \mu g/mL$. When does this occur?

76. A* CHEMICAL REACTION RATE The effect of temperature on the reaction rate of a chemical reaction is given by the **Arrhenius equation**

$$k = Ae^{-E_0/RT},$$

where k is the rate constant, T (in degrees Kelvin) is the temperature, and R is the gas constant. The quantities A and E_0 are fixed once the reaction is specified. Let k_1 and k_2 be the reaction rate constants associated with temperatures T_1 and T_2. Find an expression for $\ln\left(\dfrac{k_1}{k_2}\right)$ in terms of E_0, R, T_1, and T_2.

77. POPULATION GROWTH According to a logistic model based on the assumption that Earth can support no more than 40 billion people, the world's population (in billions) t years after 1960 is given by a function of the form $P(t) = \dfrac{40}{1 + Ce^{-kt}}$, where C and k are positive constants. Find the function of this form that is consistent with the fact that the world's population was approximately 3 billion in 1960 and 4 billion in 1975. What does your model predict for the population in the year 2010?

 The actual population in 2010 was 6.8 billion. Can you explain the reasons for any difference between this and your answer?

78. THE SPREAD OF AN EPIDEMIC Public health records indicate that t weeks after the outbreak of a certain form of influenza, approximately

$$Q(t) = \frac{80}{4 + 76e^{-1.2t}}$$

thousand people had caught the disease. At what rate was the disease spreading at the end of the second week?

79. MORTALITY RATES It is sometimes useful for actuaries to be able to project mortality rates within a given population. A formula sometimes used for computing the mortality rate $D(t)$ for women in the age group 25–29 is

$$D(t) = (D_0 - 0.00046)e^{-0.162t} + 0.00046,$$

where t is the number of years after a fixed base year and D_0 is the mortality rate when $t = 0$.

a. Suppose the initial mortality rate of a particular group is 0.008 (8 deaths per 1000 women). What is the mortality rate of this group 10 years later? What is the rate 25 years later?

 b. Sketch the graph of the mortality function $D(t)$ for the group in part (a) for $0 \leq t \leq 25$.

80. **CARBON DATING** A Cro-Magnon cave painting at Lascaux, France, is approximately 15 000 years old. Approximately what ratio of ^{14}C to ^{12}C would you expect to find in a fossil of approximately the same age, found in the same cave as the painting?

81. **GROSS DOMESTIC PRODUCT** The gross domestic product (GDP) of a certain country was $100 billion in 2000 and $165 billion in 2010. Assuming that the GDP is growing exponentially, what will it be in 2020?

82. **RADIOLOGY** The radioactive isotope gallium-67 (^{67}Ga), used in the diagnosis of malignant tumours, has a half-life of 46.5 hours. If we start with 100 mg of the isotope, how many milligrams will be left after 24 hours? When will there be only 25 mg left?

Answer these questions by first using a graphing utility to graph an appropriate exponential function and then moving along the graph curve checking the coordinates.

83. Use a graphing utility to draw the graphs of $y = \sqrt{3^x}$, $y = \sqrt{3^{-x}}$, and $y = 3^{-x}$ on the same set of axes. How do these graphs differ?

84. Use a graphing utility to graph $y = 2^{-x}$, $y = 3^{-x}$, $y = 5^{-x}$, and $y = (0.5)^{-x}$ on the same set of axes. How does a change in base affect the graph of the exponential function?

85. Use a graphing utility to draw the graphs of $y = 3^x$ and $y = 4 - \ln \sqrt{x}$ on the same axes. Then find all points of intersection of the two graphs by expanding the scale near the intersection points, moving the cursor along, and checking the coordinates.

86. Use a graphing utility to draw the graphs of $y = \ln(1 + x^2)$ and $y = \dfrac{1}{x}$ on the same axes. Do these graphs intersect?

THINK ABOUT IT

FORENSIC ACCOUNTING: BENFORD'S LAW

You might guess that the first digit of each number in a collection of numbers has an equal chance of being any of the digits 1 through 9, but it was discovered in 1938 by physicist Frank Benford that the chance that the digit is a 1 is more than 30%! Naturally occurring numbers exhibit a curious pattern in the proportions of the first digit: smaller digits such as 1, 2, and 3 occur much more often than larger digits, as seen in the following table:

First Digit	Proportion
1	30.1
2	17.6
3	12.5
4	9.7
5	7.9
6	6.7
7	5.8
8	5.1
9	4.6

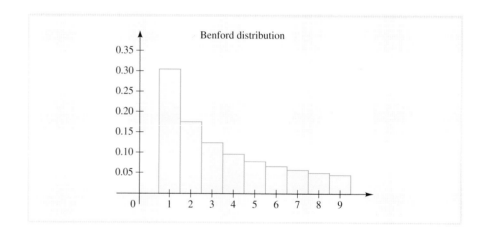

Naturally occurring, in this case, means that the numbers arise without explicit bound and describe similar quantities, such as the populations of cities or the amounts paid out on cheques. This pattern also holds for exponentially growing numbers and some types of randomly sampled (but not randomly generated) numbers, and for this reason it is a powerful tool for determining if these types of data are genuine. The distribution of digits generally conforms closely to the following rule: the proportion of numbers such that the first digit is n is given by

$$\log_{10}(n + 1) - \log_{10} n = \log_{10}\frac{n + 1}{n}.$$

This rule, known as Benford's law, is used to detect fraud in accounting and is one of several techniques in a field called *forensic accounting*. Often people who write fraudulent cheques, such as an embezzler at a corporation, try to make the first digits (or even all the digits) occur *equally often* so as to appear to be random. Benford's law predicts that the first digits of such a set of accounting data should not be evenly proportioned, but rather show a higher occurrence of smaller digits. If an employee is writing a large number of cheques or committing many monetary transfers to a suspicious target, the cheque values can be analyzed to see if they follow Benford's law, indicating possible fraud if they do not.

(Photo: Getty/Royalty Free)

This technique can be applied to various types of accounting data (such as taxes and expenses) and has been used to analyze socioeconomic data such as the gross domestic products of many countries. Benford's law is also used by the U.S. Internal Revenue Service to detect fraud and has been applied to locate errors in data entry and analysis.

Questions

1. Verify that the formula given for the proportions of digits

$$P(n) = \log_{10}\frac{n + 1}{n}$$

produces the values in the given table. Use calculus to show that the proportion is a decreasing function of n.

2. The proportions of first digits depend on the base of the number system used. Computers generally use number systems that are powers of 2. Benford's law for base b is

$$P(n) = \log_b\frac{n + 1}{n}$$

where n ranges from 1 to b. Compute a table like the one given for base $b = 10$ for the bases 4 and 8. Using these computed tables and the given table, do the proportions seem to be evening out or becoming more uneven as the size of the base increases?

Use calculus to justify your assertion by viewing $P(n)$ as a function of b, for particular values of n. For instance, for $n = 1$,

$$f(b) = \log_b 2 = \frac{\ln 2}{\ln b}.$$

Use this new function to determine if the proportion of numbers with leading digit 1 is increasing or decreasing as the size of the base b increases. What happens for the other digits?

THINK ABOUT IT

3. In the course of a potential fraud investigation, it is found that an employee wrote cheques with the following values to a suspicious source: $234, $444, $513, $1120, $2201, $3614, $4311, $5557, $5342, $6710, $8712, and $8998. Compute the proportions corresponding to each of the first digits. Do you think that fraud may be occurring? (In actual investigations, statistical tests are used to determine if the deviation is statistically significant.)

4. Select a collection of numbers arbitrarily from a newspaper or magazine and record the first digit (the first nonzero digit if it is a decimal less than 1). Do the numbers appear to follow Benford's law?

5. The following table is an alphabetical list of the populations of the first 50 settlements in Southern Ontario beginning with the letter A, from Aberfoyle to Ayton, on the official Ontario Road Map. Do they appear to follow Benford's law?

140	40	345	6110	331
119	14 030	8225	81	36
7632	118	444	23 920	239
104	3531	4611	9723	36
38	1228	1025	60	20
50	198	19	620	92
42	158	1020	50	125
989	70	32 745	244	126
63 552	69	198	256	512
340	486	19 303	205	7113

References

T. P. Hill, "The First Digit Phenomenon," *American Scientist*, Vol. 86, 1998, p. 358.
Steven W. Smith, "The Scientist's and Engineer's Guide to Signal Processing," Chapter 34. Available online at http://www.dspguide.com/ch34/1.htm.

 Practise and learn online with Connect.

Computing the area under a curve, such as the area of the region spanned by the scaffolding under a roller coaster track, is an application of integration. In Example 5.3.4, we calculate the area under such a roller coaster. (Photo: © Richard Clune/CORBIS)

INTEGRATION

CHAPTER OUTLINE

1 Antidifferentiation: The Indefinite Integral
2 Integration by Substitution
3 The Definite Integral and the Fundamental Theorem of Calculus
4 Applying Definite Integration: Area between Curves and Average Value
5 Additional Applications to Business and Economics
6 Additional Applications to the Life and Social Sciences
   Chapter Summary
     Concept Summary
     Checkup for Chapter 5
     Review Exercises
   Think About It

LEARNING OBJECTIVES

After completing this chapter, you should be able to

L01 Evaluate indefinite integrals.

L02 Choose part of a function for use in substitution and integration.

L03 Evaluate definite integrals using the substitution method where necessary.

L04 Calculate the area between the curves of two functions. Calculate the average value of a function.

L05 Solve business and economics problems using integration.

L06 Solve life science and social science problems using integration.

SECTION 5.1

Antidifferentiation: The Indefinite Integral

How can a known rate of inflation be used to determine future prices? What is the velocity of an object moving along a straight line with known acceleration? How can knowing the rate at which a population is changing be used to predict future population levels? In all these situations, the derivative (rate of change) of a quantity is known and the quantity itself is required. Here is the terminology we will use in connection with obtaining a function from its derivative.

Antidifferentiation ■ A function $F(x)$ is said to be an *antiderivative* of $f(x)$ if

$$F'(x) = f(x)$$

for every x in the domain of $f(x)$. The process of finding antiderivatives is called *antidifferentiation* or *indefinite integration*.

NOTE Sometimes we write the equation

$$F'(x) = f(x)$$

as

$$\frac{dF}{dx} = f(x).$$

Later in this section, you will learn techniques for finding antiderivatives. Once you have found what you believe to be an antiderivative of a function, you can always check your answer by differentiating. You should get the original function back. Here is an example.

EXAMPLE 5.1.1

Verify that $F(x) = \frac{1}{3}x^3 + 5x + 2$ is an antiderivative of $f(x) = x^2 + 5$.

Solution

$F(x)$ is an antiderivative of $f(x)$ if and only if $F'(x) = f(x)$. Differentiate F to find that

$$F'(x) = \frac{1}{3}(3x^2) + 5$$
$$= x^2 + 5 = f(x)$$

as required.

The General Antiderivative of a Function

A function has more than one antiderivative. For example, one antiderivative of the function $f(x) = 3x^2$ is $F(x) = x^3$, since

$$F'(x) = 3x^2 = f(x),$$

but $x^3 + 12$ and $x^3 - 5$ and $x^3 + \pi$ are also antiderivatives of $F(x)$, since

$$\frac{d}{dx}(x^3 + 12) = 3x^2 \qquad \frac{d}{dx}(x^3 - 5) = 3x^2 \qquad \frac{d}{dx}(x^3 + \pi) = 3x^2$$

In general, if F is one antiderivative of f, then so is any function of the form $G(x) = F(x) + C$ for constant C, since

$$
\begin{aligned}
G'(x) &= [F(x) + C]' \\
&= F'(x) + C' \quad \text{sum rule for derivatives} \\
&= F'(x) + 0 \quad \text{derivative of a constant is 0} \\
&= f(x) \quad \text{F is an antiderivative of f}
\end{aligned}
$$

Conversely, it can be shown that if F and G are both antiderivatives of f, then $G(x) = F(x) + C$ for some constant C (Exercise 64). To summarize:

> **Fundamental Property of Antiderivatives** ■ If $F(x)$ is an antiderivative of the continuous function $f(x)$, then any other antiderivative of $f(x)$ has the form $G(x) = F(x) + C$ for some constant C.

There is a simple geometric interpretation for the fundamental property of antiderivatives. If F and G are both antiderivatives of f, then

$$
G'(x) = F'(x) = f(x).
$$

Just-In-Time

Recall that two lines are parallel if and only if their slopes are equal.

This means that the slope $F'(x)$ of the tangent line to $y = F(x)$ at the point $(x, F(x))$ is the same as the slope $G'(x)$ of the tangent line to $y = G(x)$ at $(x, G(x))$. Since the slopes are equal, it follows that the tangent lines at $(x, F(x))$ and $(x, G(x))$ are parallel, as shown in Figure 5.1a. Since this is true for all x, the entire curve $y = G(x)$ must be parallel to the curve $y = F(x)$, so that

$$
y = G(x) = F(x) + C.
$$

In general, the collection of graphs of all antiderivatives of a given function f is a family of parallel curves that are vertical translations of one another. This is illustrated in Figure 5.1b for the family of antiderivatives of $f(x) = 3x^2$.

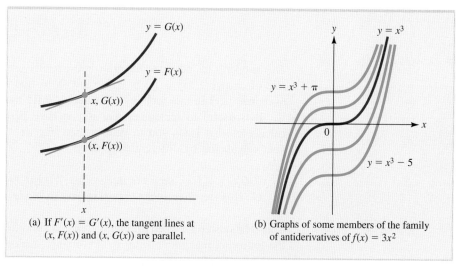

(a) If $F'(x) = G'(x)$, the tangent lines at $(x, F(x))$ and $(x, G(x))$ are parallel.

(b) Graphs of some members of the family of antiderivatives of $f(x) = 3x^2$

FIGURE 5.1 Graphs of antiderivatives of a function f form a family of parallel curves.

The Indefinite Integral

You have just seen that if $F(x)$ is one antiderivative of the continuous function $f(x)$, then all such antiderivatives may be characterized by $F(x) + C$ for constant C. The family of all antiderivatives of $f(x)$ is written

$$\int f(x)\, dx = F(x) + C$$

and is called the **indefinite integral** of $f(x)$. The integral is "indefinite" because it involves a constant C that can take on any value. In Section 5.3, we introduce a **definite integral** that has a specific numerical value and is used to represent a variety of quantities, such as area, average value, and present value of an income flow. The connection between definite and indefinite integrals is made in Section 5.3 through a result so important that it is referred to as the **fundamental theorem of calculus.**

In the context of the indefinite integral $\int f(x)\, dx = F(x) + C$, the **integral symbol** is $\int$, the function $f(x)$ is called the **integrand,** C is the **constant of integration,** and dx is a differential that specifies x as the **variable of integration.** These features are displayed in this diagram for the indefinite integral of $f(x) = 3x^2$:

Just-In-Time

Notice that a differential, dx, is used here. Differentials form an essential part of the integration process and notation. They were first introduced in Section 2.5.

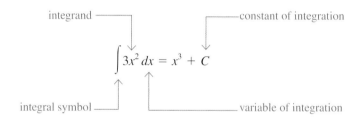

For any differentiable function F, we have

$$\int F'(x)\, dx = F(x) + C$$

since, by definition, $F(x)$ is an antiderivative of $F'(x)$. Equivalently,

$$\int \frac{dF}{dx}\, dx = F(x) + C.$$

This property of indefinite integrals is especially useful in applied problems where a rate of change $F'(x)$ is given and we wish to find $F(x)$. Several such problems are examined later in this section, in Examples 5.1.5 through 5.1.8.

It is useful to remember that if you have performed an indefinite integration calculation that leads you to believe that $\int f(x)\, dx = G(x) + C$, then you can check your calculation by differentiating $G(x)$:

If $G'(x) = f(x)$, then the integration $\int f(x)\, dx = G(x) + C$ is correct, but if $G'(x)$ is anything other than $f(x)$, you've made a mistake.

This relationship between differentiation and antidifferentiation enables us to establish the following integration rules by "reversing" analogous differentiation rules.

Rules for Integrating Common Functions

The **constant rule:** $\int k\,dx = kx + C$ for constant k.

The **power rule:** $\int x^n\,dx = \dfrac{x^{n+1}}{n+1} + C$ for all $n \neq -1$.

The **logarithmic rule:** $\int \dfrac{1}{x}\,dx = \ln|x| + C$ for all $x \neq 0$.

The **exponential rule:** $\int e^{kx}\,dx = \dfrac{1}{k}e^{kx} + C$ for constant $k \neq 0$.

To verify the power rule, it is enough to show that the derivative of $\dfrac{x^{n+1}}{n+1}$ is x^n:

$$\frac{d}{dx}\left[\frac{x^{n+1}}{n+1}\right] = \frac{1}{n+1}[(n+1)x^n] = x^n.$$

For the logarithmic rule, if $x > 0$, then $|x| = x$ and

$$\frac{d}{dx}[\ln|x|] = \frac{d}{dx}[\ln x] = \frac{1}{x}.$$

If $x < 0$, then $-x > 0$ and $\ln|x| = \ln(-x)$, and it follows from the chain rule that

$$\frac{d}{dx}[\ln|x|] = \frac{d}{dx}[\ln(-x)] = \frac{1}{(-x)}(-1) = \frac{1}{x}.$$

Thus, for all $x \neq 0$,

$$\frac{d}{dx}[\ln|x|] = \frac{1}{x},$$

so

$$\int \frac{1}{x}\,dx = \ln|x| + C.$$

You are asked to prove the constant rule and the exponential rule in Exercise 66.

NOTE Notice that the logarithm rule "fills the gap" in the power rule, namely, the case where $n = -1$. You may wish to blend the two rules into this combined form:

$$\int x^n\,dx = \begin{cases} \dfrac{x^{n+1}}{n+1} + C & \text{if } n \neq -1 \\[2mm] \ln|x| + C & \text{if } n = -1 \end{cases}$$

EXAMPLE 5.1.2

Find each integral:

a. $\displaystyle\int 3\,dx$ **b.** $\displaystyle\int x^{17}\,dx$ **c.** $\displaystyle\int \frac{1}{\sqrt{x}}\,dx$ **d.** $\displaystyle\int e^{-3x}\,dx$

Solution

a. Use the constant rule with $k = 3$: $\displaystyle\int 3\,dx = 3x + C$.

b. Use the power rule with $n = 17$: $\displaystyle\int x^{17}\,dx = \frac{1}{18}x^{18} + C$.

c. Use the power rule with $n = -\dfrac{1}{2}$: Since $n + 1 = \dfrac{1}{2}$,

$$\int \frac{dx}{\sqrt{x}} = \int x^{-1/2}\,dx = \frac{1}{1/2}x^{1/2} + C = 2\sqrt{x} + C.$$

d. Use the exponential rule with $k = -3$:

$$\int e^{-3x}\,dx = \frac{1}{-3}e^{-3x} + C = -\frac{1}{3}e^{-3x} + C.$$

Example 5.1.2 illustrates how certain basic functions can be integrated, but what about combinations of functions, such as the polynomial $x^5 + 2x^3 + 7$ or an expression like $5e^{-x} + \sqrt{x}$? Here are algebraic rules that will enable you to handle such expressions in a natural fashion.

Algebraic Rules for Indefinite Integration

The **constant multiple rule:** $\displaystyle\int kf(x)\,dx = k\int f(x)\,dx$ for constant k

The **sum rule:** $\displaystyle\int [f(x) + g(x)]\,dx = \int f(x)\,dx + \int g(x)\,dx$

The **difference rule:** $\displaystyle\int [f(x) - g(x)]\,dx = \int f(x)\,dx - \int g(x)\,dx$

To prove the constant multiple rule, note that if $\dfrac{dF}{dx} = f(x)$, then

$$\frac{d}{dx}[kF(x)] = k\frac{dF}{dx} = kf(x),$$

which means that

$$\int kf(x)\,dx = k\int f(x)\,dx.$$

The sum and difference rules can be established in a similar fashion.

EXAMPLE 5.1.3

Find each integral:

a. $\displaystyle\int (2x^5 + 8x^3 - 3x^2 + 5)\, dx$

b. $\displaystyle\int \left(\frac{x^3 + 2x - 7}{x}\right) dx$

c. $\displaystyle\int (3e^{-5t} + \sqrt{t})\, dt$

Just-In-Time

> Be careful when there are constants in an integral that involves $\dfrac{1}{x}$, resulting in $\ln|x|$ as part of the answer. Keep the constant separate, as in the following examples:
>
> $\displaystyle\int \frac{3}{5x}\, dx = \int \frac{3}{5}\left(\frac{1}{x}\right) dx = \frac{3}{5}\ln|x| + C$
>
> $\displaystyle\int \frac{7}{4x}\, dx = \frac{7}{4}\ln|x| + C$

Solution

a. Use the power rule in conjunction with the sum and difference rules and the constant multiple rule to get

$$\int (2x^5 + 8x^3 - 3x^2 + 5)\, dx = 2\int x^5\, dx + 8\int x^3\, dx - 3\int x^2\, dx + \int 5\, dx$$

$$= 2\left(\frac{x^6}{6}\right) + 8\left(\frac{x^4}{4}\right) - 3\left(\frac{x^3}{3}\right) + 5x + C$$

$$= \frac{1}{3}x^6 + 2x^4 - x^3 + 5x + C$$

b. There is no "quotient rule" for integration, but at least in this case, you can still divide the denominator into the numerator and then integrate using the method in part (a):

$$\int \left(\frac{x^3 + 2x - 7}{x}\right) dx = \int \left(x^2 + 2 - \frac{7}{x}\right) dx$$

$$= \frac{1}{3}x^3 + 2x - 7\ln|x| + C$$

c. $\displaystyle\int (3e^{-5t} + \sqrt{t})\, dt = \int (3e^{-5t} + t^{1/2})\, dt$

$$= 3\left(\frac{1}{-5}e^{-5t}\right) + \frac{1}{3/2}t^{3/2} + C$$

$$= -\frac{3}{5}e^{-5t} + \frac{2}{3}t^{3/2} + C$$

EXAMPLE 5.1.4

Find the function $f(x)$ whose tangent has slope $3x^2 + 1$ for each value of x and whose graph passes through the point $(2, 6)$.

Solution

The slope of the tangent at each point $(x, f(x))$ is the derivative $f'(x)$. Thus,

$$f'(x) = 3x^2 + 1,$$

and so $f(x)$ is the antiderivative

$$f(x) = \int f'(x)\, dx = \int (3x^2 + 1)\, dx = x^3 + x + C.$$

The graph of $y = x^3 + x - 4$.

To find C, use the fact that the graph of f passes through $(2, 6)$. That is, substitute $x = 2$ and $f(2) = 6$ into the equation for $f(x)$ and solve for C to get

$$6 = (2)^3 + 2 + C \quad \text{or} \quad C = -4$$

Thus, the desired function is $f(x) = x^3 + x - 4$. The graph of this function is shown in the accompanying figure.

Applied Initial Value Problems

A **differential equation** is an equation that involves differentials or derivatives. Such equations are of great importance in modelling and occur in a variety of applications. An **initial value problem** is a problem that involves solving a differential equation subject to a specified initial condition. For instance, in Example 5.1.4, we were required to find $y = f(x)$ so that

$$\frac{dy}{dx} = 3x^2 + 1 \qquad \text{subject to the condition } y = 6 \text{ when } x = 2.$$

We solved this initial value problem by finding the antiderivative

$$y = \int (3x^2 + 1)\, dx = x^3 + x + C$$

and then using the initial condition to evaluate C. The same approach is used in Examples 5.1.5 through 5.1.8 to solve a selection of applied initial value problems from business, economics, biology, and physics. Similar initial value problems appear in examples and exercises throughout this chapter.

EXAMPLE 5.1.5

A manufacturer has found that the marginal cost is $3q^2 - 60q + 400$ dollars per unit when q units have been produced. The total cost of producing the first 2 units is \$900. What is the total cost of producing the first 5 units?

Solution

Recall that the marginal cost is the derivative of the total cost function $C(q)$. Thus,

$$\frac{dC}{dq} = 3q^2 - 60q + 400,$$

and so $C(q)$ must be the antiderivative

$$C(q) = \int \frac{dC}{dq}\, dq = \int (3q^2 - 60q + 400)\, dq = q^3 - 30q^2 + 400q + K$$

for some constant K. (The letter K was used for the constant to avoid confusion with the cost function C.)

The value of K is determined by the fact that $C(2) = 900$. In particular,

$$900 = (2)^3 - 30(2)^2 + 400(2) + K \quad \text{or} \quad K = 212.$$

Hence, $C(q) = q^3 - 30q^2 + 400q + 212$, and the cost of producing the first 5 units is

$$C(5) = (5)^3 - 30(5)^2 + 400(5) + 212 = 1587,$$

or \$1587.

EXPLORE!

Graph the function $P(t)$ from Example 5.1.6. Display the population 9 hours from now. When will the population reach 300 000?

EXAMPLE 5.1.6

The population $P(t)$ of a bacterial colony t hours after observation begins is found to be changing at a rate of

$$\frac{dP}{dt} = 200e^{0.1t} + 150e^{-0.03t}.$$

If the population is 200 000 bacteria when observation begins, what will the population be 12 hours later?

Solution

The population $P(t)$ is found by antidifferentiating $\dfrac{dP}{dt}$ as follows:

$$P(t) = \int \frac{dP}{dt}\, dt = \int (200e^{0.1t} + 150e^{-0.03t})\, dt$$

$$= \frac{200e^{0.1t}}{0.1} + \frac{150e^{-0.03t}}{-0.03} + C \qquad \text{exponential and sum rules}$$

$$= 2000e^{0.1t} - 5000e^{-0.03t} + C$$

Since the population is 200 000 when $t = 0$,

$$P(0) = 200\ 000$$
$$= 2000e^0 - 5000e^0 + C$$
$$= -3000 + C$$

so $C = 203\ 000$ and

$$P(t) = 2000e^{0.1t} - 5000e^{-0.03t} + 203\ 000.$$

Thus, after 12 hours, the population is

$$P(12) = 2000e^{0.1(12)} - 5000e^{-0.03(12)} + 203\ 000$$
$$\approx 206\ 152$$

EXAMPLE 5.1.7

A retailer receives a shipment of 10 000 kg of rice that will be used up over a 5-month period at a constant rate of 2000 kg per month. If storage costs are 1 cent per kilogram per month, how much will the retailer pay in storage costs over the next 5 months?

Solution

Let $S(t)$ denote the total storage cost (in dollars) over t months. Since the rice is used up at a constant rate of 2000 kg per month, the number of kilograms of rice in storage after t months is $10\ 000 - 2000t$. Therefore, since storage costs are 1 cent per kilogram per month, the rate of change of the storage cost with respect to time is

$$\frac{dS}{dt} = (\text{monthly cost per kilogram})(\text{number of kilograms}) = 0.01(10\ 000 - 2000t).$$

It follows that $S(t)$ is an antiderivative of

$$0.01(10\ 000 - 2000t) = 100 - 20t.$$

That is,

$$S(t) = \int \frac{dS}{dt}\, dt = \int (100 - 20t)\, dt$$
$$= 100t - 10t^2 + C$$

for some constant C. To determine C, use the fact that at the time the shipment arrives (when $t = 0$) there is no cost, so that

$$0 = 100(0) - 10(0)^2 + C \quad \text{or} \quad C = 0.$$

Hence, $S(t) = 100t - 10t^2$, and the total storage cost over the next 5 months will be

$$S(5) = 100(5) - 10(5)^2 = 250$$

or $250.

Motion along a Line

Recall from Section 2.2 that if an object moving along a straight line is at the position $s(t)$ at time t, then its velocity is given by $v = \dfrac{ds}{dt}$ and its acceleration is given by $a = \dfrac{dv}{dt}$. Turning things around, if the acceleration of the object is given, then its velocity and position can be found by integration. Here is an example.

EXPLORE!

Refer to Example 5.1.8. Graph the position function $s(t) = -3t^2 + 66t$. Locate the stopping time, 4 seconds, and the corresponding position on the graph. Work the problem again for a car travelling at 30 m/s. Enter the new position function and compare the graphs. In this case, what is happening at 4 seconds?

EXAMPLE 5.1.8

A car is travelling along a straight, level road at 24 m/s when the driver is forced to apply the brakes to avoid an accident. If the brakes supply a constant deceleration of 6 m/s^2, how far does the car travel before coming to a complete stop?

Solution

Let $s(t)$ denote the distance travelled by the car in t seconds after the brakes are applied. Since the car decelerates at 6 m/s^2, it follows that $a(t) = -6$; that is,

$$\frac{dv}{dt} = a(t) = -6.$$

Integrating, the velocity at time t is given by

$$v(t) = \int \frac{dv}{dt}\, dt = \int -6\, dt = -6t + C_1.$$

To evaluate C_1, note that $v = 24$ when $t = 0$ so that

$$24 = v(0) = -6(0) + C_1$$

and $C_1 = 24$. Thus, the velocity at time t is $v(t) = -6t + 24$.

Next, to find the distance $s(t)$, begin with the fact that

$$\frac{ds}{dt} = v(t) = -6t + 24$$

and use integration to show that

$$s(t) = \int \frac{ds}{dt}\, dt = \int (-6t + 24)\, dt = -\frac{6}{2}t^2 + 24t + C_2 = -3t^2 + 24t + C_2.$$

Since $s(0) = 0$ (do you see why?), it follows that $C_2 = 0$ and

$$s(t) = -3t^2 + 24t.$$

Finally, to find the stopping distance, note that the car stops when $v(t) = 0$, and this occurs when

$$v(t) = -6t + 24 = 0.$$

Solve this equation to find that the car stops after 4 seconds of deceleration, and in that time it has travelled

$$s(4) = -3(4)^2 + 24(4) = 48$$

or 48 m.

EXERCISES ▪ 5.1

In Exercises 1 through 30, find the indefinite integral. Check your answers by differentiation.

1. $\displaystyle\int -3 \, dx$

2. $\displaystyle\int dx$

3. $\displaystyle\int x^5 \, dx$

4. $\displaystyle\int \sqrt{t} \, dt$

5. $\displaystyle\int \frac{1}{x^2} \, dx$

6. $\displaystyle\int 3e^x \, dx$

7. $\displaystyle\int \frac{2}{\sqrt{t}} \, dt$

8. $\displaystyle\int x^{-0.3} \, dx$

9. $\displaystyle\int u^{-2/5} \, du$

10. $\displaystyle\int \left(\frac{1}{x^2} - \frac{1}{x^3} \right) dx$

11. $\displaystyle\int (3t^2 - \sqrt{5t} + 2) \, dt$

12. $\displaystyle\int (x^{1/3} - 3x^{-2/3} + 6) \, dx$

13. $\displaystyle\int (3\sqrt{y} - 2y^{-3}) \, dy$

14. $\displaystyle\int \left(\frac{1}{2y} - \frac{2}{y^2} + \frac{3}{\sqrt{y}} \right) dy$

15. $\displaystyle\int \left(\frac{e^x}{2} + x\sqrt{x} \right) dx$

16. $\displaystyle\int \left(\sqrt{x^3} - \frac{1}{2\sqrt{x}} + \sqrt{2} \right) dx$

17. $\displaystyle\int u^{1.1} \left(\frac{1}{3u} - 1 \right) du$

18. $\displaystyle\int \left(2e^u + \frac{6}{5u} + \ln 2 \right) du$

19. $\displaystyle\int \left(\frac{x^2 + 2x + 1}{x^2} \right) dx$

20. $\displaystyle\int \frac{x^2 + 3x - 2}{\sqrt{x}} \, dx$

21. $\displaystyle\int (x^3 - 2x^2)\left(\frac{1}{x} - 5 \right) dx$

22. $\displaystyle\int y^3 \left(2y + \frac{1}{y} \right) dy$

23. $\displaystyle\int \sqrt{t}(t^2 - 1) \, dt$

24. $\displaystyle\int x(2x + 1)^2 \, dx$

25. $\displaystyle\int (e^t + 1)^2 \, dt$

26. $\displaystyle\int e^{-0.02t}(e^{-0.13t} + 4) \, dt$

27. $\displaystyle\int \left(\frac{1}{3y} - \frac{5}{\sqrt{y}} + e^{-y/2} \right) dy$

28. $\displaystyle\int \frac{1}{x}(x + 1)^2 \, dx$

29. $\int t^{-1/2}(t^2 - t + 2)\, dt$

30. $\int \ln e^{-x^2}\, dx$

In Exercises 31 through 34, solve the given initial value problem for $y = f(x)$.

31. $\dfrac{dy}{dx} = 3x - 2$,   where $y = 2$ when $x = -1$

32. $\dfrac{dy}{dx} = e^{-x}$,   where $y = 3$ when $x = 0$

33. $\dfrac{dy}{dx} = \dfrac{2}{x} - \dfrac{1}{x^2}$,   where $y = -1$ when $x = 1$

34. $\dfrac{dy}{dx} = \dfrac{x + 1}{\sqrt{x}}$,   where $y = 5$ when $x = 4$

In Exercises 35 through 42, the slope $f'(x)$ at each point (x, y) on a curve $y = f(x)$ is given along with a particular point (a, b) on the curve. Use this information to find $f(x)$.

35. $f'(x) = 4x + 1;\ (1, 2)$

36. $f'(x) = 3 - 2x;\ (0, -1)$

37. $f'(x) = -x(x + 1);\ (-1, 5)$

38. $f'(x) = 3x^2 + 6x - 2;\ (0, 6)$

39. $f'(x) = x^3 - \dfrac{2}{x^2} + 2;\ (1, 3)$

40. $f'(x) = x^{-1/2} + x;\ (1, 2)$

41. $f'(x) = e^{-x} + x^2;\ (0, 4)$

42. $f'(x) = \dfrac{3}{x} - 4;\ (1, 0)$

43. MARGINAL COST   A manufacturer estimates that the marginal cost of producing q units of a certain commodity is $C'(q) = 3q^2 - 24q + 48$ dollars per unit. If the cost of producing 10 units is $5000, what is the cost of producing 30 units?

44. MARGINAL REVENUE   The marginal revenue derived from producing q units of a certain commodity is $R'(q) = 4q - 1.2q^2$ dollars per unit. If the revenue derived from producing 20 units is $30 000, how much revenue should be expected from producing 40 units?

45. MARGINAL PROFIT   A manufacturer estimates the marginal revenue to be $R'(q) = 100q^{-1/2}$ dollars per unit when the level of production is q units. The corresponding marginal cost has been found to be

0.4q dollars per unit. Suppose the manufacturer's profit is $520 when the level of production is 16 units. What is the manufacturer's profit when the level of production is 25 units?

46. SALES   The monthly sales at an import store are currently $10 000 but are expected to be declining at a rate of

$$S'(t) = -10t^{2/5} \text{ dollars per month}$$

t months from now. The store is profitable as long as the sales level is above $8000 per month.
  a. Find a formula for the expected sales in t months.
  b. What sales figure should be expected 2 years from now?
  c. For how many months will the store remain profitable?

47. ADVERTISING   After initiating an advertising campaign in an urban area, a new cell phone company estimates that the number of new subscribers will grow at a rate given by

$$N'(t) = 154t^{2/3} + 37 \text{ subscribers per month,}$$

where t is the number of months after the advertising begins. How many new subscribers should be expected 8 months from now?

48. TREE GROWTH   An arborist finds that a certain type of tree grows so that its height $h(t)$ after t years is changing at a rate of

$$h'(t) = 0.05t^{2/3} + 0.3\sqrt{t}$$

metres per year. If the tree was 0.8 m tall when it was planted, how tall will it be in 27 years?

49. POPULATION GROWTH   It is estimated that t months from now the population of a certain town will be increasing at a rate of $4 + 5t^{2/3}$ people per month. If the current population is 10 000, what will the population be 8 months from now?

50. NET CHANGE IN A BIOMASS   A biomass is growing at a rate of $M'(t) = 0.5e^{0.2t}$ grams per hour. By how much does the mass change during the second hour?

51. LEARNING   Tyler is taking a learning test in which the time he takes to memorize items from a given list is recorded. Let $M(t)$ be the number of items he can memorize in t minutes. His learning rate is found to be

$$M'(t) = 0.4t - 0.005t^2.$$

  a. How many items can Tyler memorize during the first 10 minutes?

b. How many additional items can he memorize during the next 10 minutes (from time $t = 10$ to $t = 20$)?

52. **ENDANGERED SPECIES** A conservationist finds that the population $P(t)$ of a certain endangered species is growing at a rate given by $P'(t) = 0.51e^{-0.03t}$, where t is the number of years after records began to be kept. If the population is $P_0 = 500$ now (at time $t = 0$), what will it be in 10 years?

53. **DEFROSTING** A thick steak is removed from the freezer and left on the counter to defrost. The temperature of the steak was $-4°C$ when it was removed from the freezer. After t hours, the temperature was increasing at a rate of

    $T'(t) = 7e^{-0.35t}$ degrees Celsius per hour.

    a. Find a formula for the temperature of the steak after t hours.
    b. What is the temperature after 2 hours?
    c. Assume the steak is defrosted when its temperature reaches $10°C$. How long does it take for the steak to defrost?

54. **MARGINAL REVENUE** Suppose it has been determined that the marginal revenue associated with the production of x units of a particular commodity is $R'(x) = 240 - 4x$ dollars per unit. What is the revenue function $R(x)$? You may assume that $R(0) = 0$. What price will be paid for each unit when the level of production is $x = 5$ units?

55. **MARGINAL PROFIT** The marginal profit of a certain commodity is $P'(q) = 100 - 2q$ when q units are produced. When 10 units are produced, the profit is $700.

    a. Find the profit function $P(q)$.
    b. What production level q results in maximum profit? What is the maximum profit?

56. **PRODUCTION** At a certain factory, when K thousand dollars are invested in the plant, the production Q is changing at a rate given by

    $$Q'(K) = 200K^{-2/3}$$

    units per thousand dollars invested. When $8000 is invested, the level of production is 5500 units.

    a. Find a formula for the level of production Q to be expected when K thousand dollars are invested.
    b. How many units will be produced when $27\ 000$ is invested?
    c. What capital investment K is required to produce 7000 units?

57. **MARGINAL PROPENSITY TO CONSUME** Suppose the consumption function for a particular country is $c(x)$, where x is national disposable income. Then the **marginal propensity to consume** is $c'(x)$. Suppose x and c are both measured in billions of dollars and

    $$c'(x) = 0.9 + 0.3\sqrt{x}.$$

    If consumption is $10 billion when $x = 0$, find $c(x)$.

58. **MARGINAL ANALYSIS** A manufacturer estimates marginal revenue to be $200q^{-1/2}$ dollars per unit when the level of production is q units. The corresponding marginal cost has been found to be $0.4q$ dollars per unit. If the manufacturer's profit is $2000 when the level of production is 25 units, what is the profit when the level of production is 36 units?

59. **SPY STORY** Our spy, intent on avenging the death of his friend Marc, is driving a sports car toward the lair of the man who killed his friend. To remain as inconspicuous as possible, he is travelling at the legal speed of 28 m/s (100 km/h) when he sees a moose standing in the road, 60 m in front of him. It takes him 0.07 seconds to react to the crisis. Then he hits the brakes, and the car decelerates at a constant rate of 10 m/s^2. Does he stop before hitting the moose?

60. **CANCER THERAPY** A new medical procedure is applied to a cancerous tumour with volume 30 cm^3, and t days later the volume is found to be changing at a rate of

    $V'(t) = 0.15 - 0.09e^{0.006t}$ cubic centimetres per day.

    a. Find a formula for the volume of the tumour after t days.
    b. What is the volume after 60 days? After 120 days?
    c. For the procedure to be successful, it should take no longer than 90 days for the tumour to begin to shrink. Based on this criterion, does the procedure succeed?

61. **LEARNING** Let $f(x)$ represent the total number of items a subject has memorized x minutes after being presented with a long list of items to learn. Psychologists refer to the graph of $y = f(x)$ as a **learning curve** and to $f'(x)$ as the **learning rate.** The time of **peak efficiency** is the time when the learning rate is maximized. Suppose the learning rate is

    $$f'(x) = 0.1(10 + 12x - 0.6x^2) \quad \text{for } 0 \le x \le 25.$$

    a. When does peak efficiency occur? What is the learning rate at peak efficiency?

b. What is $f(x)$? You may assume that $f(0) = 0$.

c. What is the largest number of items memorized by the subject?

62. CORRECTION FACILITY MANAGEMENT
In a certain small country, statistics indicate that x years from now the number of inmates in prisons will be increasing at the rate of $280e^{0.2x}$ per year. Currently, 2000 inmates are housed in prisons. How many inmates should the country expect 10 years from now?

63. FLOW OF BLOOD   One of Poiseuille's laws for the flow of blood in an artery says that if $v(r)$ is the velocity of flow r centimetres from the central axis of the artery, then the velocity decreases at a rate proportional to r. That is,

$$v'(r) = -ar,$$

where a is a positive constant. Find an expression for $v(r)$. Assume that $v(R) = 0$, where R is the radius of the artery.

Artery

EXERCISE 63

64. A* If $H'(x) = 0$ for all real numbers x, what must be true about the graph of $H(x)$? Explain how your observation can be used to show that if $G'(x) = F'(x)$ for all x, then $G(x) = F(x) + C$ for constant C.

65. DISTANCE AND VELOCITY   An object is moving so that its velocity after t minutes is $v(t) = 3 + 2t + 6t^2$ metres per minute. How far does the object travel during the second minute?

66. A* a. Prove the constant rule: $\int k\,dx = kx + C$.

b. Prove the exponential rule: $\int e^{kx}dx = \dfrac{1}{k}e^{kx} + C$.

67. What is $\int b^x\,dx$ for $b > 0$, $b \ne 1$? [*Hint:* Recall that $b^x = e^{x\ln b}$.]

68. It is estimated that x months from now, the population of a certain town will be changing at a rate of $P'(x) = 2 + 1.5\sqrt{x}$ people per month. The current population is 5000.

a. Find a function $P(x)$ that satisfies these conditions. Use a graphing utility to graph this function.

b. Using the graph, determine the population 9 months from now. When will the population be 7590?

c. Suppose the current population were 2000 (instead of 5000). Plot the graph of $P(x)$ with this assumption. Then plot the graph of $P(x)$ assuming current populations of 4000 and 6000. What is the difference between the graphs?

69. A large carpenter ant is moving at a steady speed of 67 cm/min in a straight line. It then senses possible danger and slows down, decelerating at a constant rate of 23 cm/min^2.

a. Find the velocity $v(t)$ of the ant t minutes later. Then find its distance, $s(t)$, from the point where it started to slow down.

b. Use a graphing utility to plot $v(t)$ and $s(t)$ on the same graph.

c. Use the graph to determine when the ant comes to a complete stop and how far it has moved in that time. How fast is the ant moving when it has travelled 45 cm?

SECTION 5.2   Integration by Substitution

L02

Choose part of a function for use in substitution and integration.

The majority of functions that occur in practical situations can be differentiated by applying rules and formulas such as those you learned in Chapter 2. However, integration sometimes appears deceptively simple but may actually require a special technique or clever insight.

For example, we easily find that

$$\int x^7\,dx = \frac{1}{8}x^8 + C$$

by applying the power rule, but suppose we wish to compute

$$\int (3x + 5)^7 \, dx.$$

We could proceed by expanding the integrand $(3x + 5)^7$ and then integrating term by term, but the algebra involved in this approach is daunting. Instead, we make the change of variable

$$u = 3x + 5 \quad \text{so that} \quad du = 3\,dx \quad \text{or} \quad dx = \frac{1}{3}\,du.$$

Then, substituting these quantities into the given integral gives

$$\int (3x + 5)^7 \, dx = \int u^7 \left(\frac{1}{3}\,du \right)$$

$$= \frac{1}{3} \left(\frac{1}{8}\,u^8 \right) + C = \frac{1}{24}\,u^8 + C \quad \text{power rule}$$

$$= \frac{1}{24}(3x + 5)^8 + C \qquad \text{since } u = 3x + 5$$

We can check this computation by differentiating using the chain rule (Section 2.4):

$$\frac{d}{dx}\left[\frac{1}{24}(3x + 5)^8 \right] = \frac{1}{24}[8(3x + 5)^7(3)] = (3x + 5)^7,$$

which verifies that $\frac{1}{24}(3x + 5)^8$ is indeed an antiderivative of $(3x + 5)^7$.

The change of variable procedure we have just demonstrated is called **integration by substitution,** and it amounts to reversing the chain rule for differentiation. To see why, consider an integral that can be written as

$$\int f(x) \, dx = \int g(u(x))u'(x) \, dx.$$

Suppose G is an antiderivative of g, so that $G' = g$. Then, according to the chain rule

$$\frac{d}{dx}[G(u(x))] = G'(u(x))\,u'(x)$$

$$= g(u(x))\,u'(x) \qquad \text{since } G' = g$$

Therefore, we find that

$$\int f(x) \, dx = \int g(u(x))\,u'(x) \, dx$$

$$= \int \left(\frac{d}{dx}[G(u(x))] \right) dx$$

$$= G(u(x)) + C \qquad \text{since } \int G'\,dx = G$$

In other words, once we have an antiderivative for $g(u)$, we also have one for $f(x)$.

A useful device for remembering the substitution procedure is to think of $u = u(x)$ as a change of variable whose differential $du = u'(x)\,dx$ can be manipulated algebraically. Then

$$\int f(x)\ dx = \int g(u(x))\ u'(x)\ dx$$

$$= \int g(u)\ du \qquad \text{substitute } du \text{ for } u'(x)\ dx$$

$$= G(u) + C \qquad \text{where } G \text{ is an antiderivative of } g$$

$$= G(u(x)) + C \qquad \text{substitute } u\,(x) \text{ for } u$$

Here is a step-by-step procedure for integrating by substitution.

Using Substitution to Integrate $\int f(x)\ dx$

Step 1. Choose a substitution $u = u(x)$ that simplifies the integrand $f(x)$.

Step 2. Express the entire integral in terms of u and $du = u'(x)\ dx$. This means that *all* terms involving x and dx must be transformed to terms involving u and du.

Step 3. When step 2 is complete, the given integral should have the form

$$\int f(x)\ dx = \int g(u)\ du.$$

If possible, evaluate this transformed integral by finding an antiderivative $G(u)$ for $g(u)$.

Step 4. Replace u by $u(x)$ in $G(u)$ to obtain an antiderivative $G(u(x))$ for $f(x)$, so that

$$\int f(x)\ dx = G(u(x)) + C.$$

The first step in integrating by substitution is to find a suitable change of variable $u = u(x)$ that simplifies the integrand of the given integral $\int f(x)\ dx$ without adding undesired complexity when dx is replaced by $du = u'(x)\ dx$. Here are a few guidelines for choosing $u(x)$:

1. If possible, try to choose u so that $u'(x)$ is part of the integrand $f(x)$.
2. Consider choosing u as the part of the integrand that makes $f(x)$ difficult to integrate directly, **such as the quantity inside a radical or bracket, the denominator of a fraction, or the exponent of an exponential function.**
3. Don't "oversubstitute." For instance, in our introductory example $\int (3x + 5)^7\ dx$, a common mistake is to use $u = (3x + 5)^7$. This certainly simplifies the integrand, but then $du = 7(3x + 5)^6(3)\ dx$, and you are left with a transformed integral that is more complicated than the original.
4. Persevere. If you try a substitution that does not result in a transformed integral that you can evaluate, try a different substitution.

Examples 5.2.1 through 5.2.6 illustrate how substitutions are chosen and used in various kinds of integrals.

EXAMPLE 5.2.1

Find $\int \sqrt{2x + 7}\, dx$.

Solution

Choose $u = 2x + 7$ and obtain

$$du = 2\, dx \qquad \text{so that} \qquad dx = \frac{1}{2}\, du.$$

Then the integral becomes

$$\int \sqrt{2x + 7}\, dx = \int \sqrt{u}\left(\frac{1}{2}\, du\right)$$

$$= \frac{1}{2}\int u^{1/2}\, du \qquad\qquad \text{since } \sqrt{u} = u^{1/2}$$

$$= \frac{1}{2}\frac{u^{3/2}}{3/2} + C$$

$$= \frac{1}{3}u^{3/2} + C \qquad\qquad \text{power rule}$$

$$= \frac{1}{3}(2x + 7)^{3/2} + C \qquad \text{substitute } 2x + 7 \text{ for } u$$

EXAMPLE 5.2.2

Find $\int 8x(4x^2 - 3)^5\, dx$.

Solution

First, note that the integrand $8x(4x^2 - 3)^5$ is a product in which one of the factors, $8x$, is the derivative of the expression $4x^2 - 3$, which appears in the other factor. This suggests making the substitution

$$u = 4x^2 - 3 \qquad \text{with} \qquad du = 8x\, dx$$

to obtain

$$\int 8x(4x^2 - 3^5)\, dx = \int (4x^2 - 3)^5(8x\, dx)$$

$$= \int u^5 du$$

$$= \frac{1}{6}u^6 + C \qquad\qquad \text{power rule}$$

$$= \frac{1}{6}(4x^2 - 3)^6 + C \qquad \text{substitute } 4x^2 - 3 \text{ for } u$$

EXAMPLE 5.2.3

Find $\int x^3 e^{x^4+2}\, dx$.

Solution

If the integrand of an integral contains an exponential function, it is often useful to substitute for the exponent. In this case, we choose

$$u = x^4 + 2 \quad \text{so that} \quad du = 4x^3\, dx$$

and

$$
\begin{aligned}
\int x^3 e^{x^4+2}\, dx &= \int e^{x^4+2} x^3\, dx \\
&= \int e^u \left(\frac{1}{4}\, du \right) && \text{since } du = 4x^3\, dx \\
&= \frac{1}{4} e^u + C && \text{exponential rule} \\
&= \frac{1}{4} e^{x^4+2} + C && \text{substitute } x^4 + 2 \text{ for } u
\end{aligned}
$$

EXAMPLE 5.2.4

Find $\int \dfrac{x}{x-1}\, dx$.

Solution

Following our guidelines, we substitute for the denominator of the integrand, so that $u = x - 1$ and $du = dx$. Since $u = x - 1$, we also have $x = u + 1$. Thus,

$$
\begin{aligned}
\int \frac{x}{x-1}\, dx &= \int \frac{u+1}{u}\, du \\
&= \int \left[1 + \frac{1}{u} \right] du && \text{divide} \\
&= u + \ln |u| + C && \text{constant and logarithmic rules} \\
&= x - 1 + \ln |x - 1| + C && \text{substitute } x - 1 \text{ for } u
\end{aligned}
$$

EXAMPLE 5.2.5

Find $\int \dfrac{3x + 6}{\sqrt{2x^2 + 8x + 3}}\, dx$.

Solution

This time, our guidelines suggest substituting for the quantity inside the radical in the denominator, that is,

$$u = 2x^2 + 8x + 3 \quad \text{and} \quad du = (4x + 8)\, dx$$

At first glance, it may seem that this substitution fails, since $du = (4x + 8)\,dx$ appears quite different from the term $(3x + 6)\,dx$ in the integral. However, note that

$$(3x + 6)\,dx = 3(x + 2)\,dx = \frac{3}{4}(4)[(x + 2)\,dx]$$

$$= \frac{3}{4}[(4x + 8)\,dx] = \frac{3}{4}\,du$$

Substituting gives

$$\int \frac{3x + 6}{\sqrt{2x^2 + 8x + 3}}\,dx = \int \frac{1}{\sqrt{2x^2 + 8x + 3}}[(3x + 6)\,dx]$$

$$= \int \frac{1}{\sqrt{u}}\left(\frac{3}{4}\,du\right) = \frac{3}{4}\int u^{-1/2}\,du$$

$$= \frac{3}{4}\left(\frac{u^{1/2}}{1/2}\right) + C = \frac{3}{2}\sqrt{u} + C$$

$$= \frac{3}{2}\sqrt{2x^2 + 8x + 3} + C \qquad \text{substitute } u = 2x^2 + 8x + 3$$

EXAMPLE 5.2.6

Find $\displaystyle\int \frac{(\ln x)^2}{x}\,dx$.

Solution

Because

$$\frac{d}{dx}[\ln x] = \frac{1}{x},$$

the integrand

$$\frac{(\ln x)^2}{x} = (\ln x)^2\left(\frac{1}{x}\right)$$

is a product in which the factor $\dfrac{1}{x}$ is the derivative of the expression $\ln x$ that appears in the other factor. This suggests substituting $u = \ln x$ and $du = \dfrac{1}{x}\,dx$ so that

$$\int \frac{(\ln x)^2}{x}\,dx = \int (\ln x)^2\left(\frac{1}{x}\,dx\right)$$

$$= \int u^2\,du = \frac{1}{3}u^3 + C$$

$$= \frac{1}{3}(\ln x)^3 + C \qquad \text{substitute } \ln x \text{ for } u$$

Sometimes an integral looks like it should be evaluated using a substitution but closer examination reveals a more direct approach. Consider Example 5.2.7.

EXAMPLE 5.2.7

Find $\int e^{5x+2}\, dx$.

Solution

You can certainly handle this integral using the substitution

$$u = 5x + 2 \quad \text{and} \quad du = 5\, dx,$$

but it is not really necessary since $e^{5x+2} = e^{5x}e^2$, and e^2 is a constant. Thus,

$$
\begin{aligned}
\int e^{5x+2}\, dx &= \int e^{5x}e^2\, dx \\
&= e^2 \int e^{5x}\, dx && \text{factor constant } e^2 \text{ outside integral} \\
&= e^2 \left[\frac{e^{5x}}{5} \right] + C && \text{exponential rule} \\
&= \frac{1}{5}e^{5x+2} + C && \text{since } e^2 e^{5x} = e^{5x+2}
\end{aligned}
$$

In Example 5.2.7, we used algebra to put the integrand into a form where substitution was not necessary. In Examples 5.2.8 and 5.2.9, we use algebra as a first step, before making a substitution.

EXAMPLE 5.2.8

Find $\int \dfrac{x^2 + 3x + 5}{x + 1}\, dx$.

Solution

There is no easy way to approach this integral as it stands (remember, there is no quotient rule for integration). However, try dividing the denominator into the numerator:

$$
\begin{array}{r}
x + 2 \\
x + 1 \overline{)\, x^2 + 3x + 5} \\
\underline{x^2 + x } \\
2x + 5 \\
\underline{2x + 2} \\
3
\end{array}
$$

that is,

$$\frac{x^2 + 3x + 5}{x + 1} = x + 2 + \frac{3}{x + 1}.$$

Then integrate $x + 2$ directly using the power rule. For the term $\dfrac{3}{x + 1}$, use the substitution $u = x + 1$, which gives $du = dx$:

$$\int \frac{x^2 + 3x + 5}{x + 1}\, dx = \int \left[x + 2 + \frac{3}{x + 1} \right] dx$$

$$= \int x\, dx + \int 2\, dx + \int \frac{3}{u}\, du \qquad u = x + 1,\ du = dx$$

$$= \frac{1}{2}x^2 + 2x + 3 \ln |u| + C$$

$$= \frac{1}{2}x^2 + 2x + 3 \ln |x + 1| + C \qquad \text{substitute } x + 1 \text{ for } u$$

EXAMPLE 5.2.9

Find $\displaystyle\int \frac{1}{1 + e^{-x}}\, dx$.

Solution

You may try to substitute $w = 1 + e^{-x}$. However, this will not work because $dw = -e^{-x}\, dx$ but there is no e^{-x} term in the numerator of the integrand. Instead, note that

$$\frac{1}{1 + e^{-x}} = \frac{1}{1 + \dfrac{1}{e^x}} = \frac{1}{\dfrac{e^x + 1}{e^x}}$$

$$= \frac{e^x}{e^x + 1}$$

Now, substituting $u = e^x + 1$ and $du = e^x dx$ into the given integral results in

$$\int \frac{1}{1 + e^{-x}}\, dx = \int \frac{e^x}{e^x + 1}\, dx = \int \frac{1}{e^x + 1}(e^x\, dx)$$

$$= \int \frac{1}{u}\, du$$

$$= \ln |u| + C$$

$$= \ln |e^x + 1| + C \qquad \text{substitute } e^x + 1 \text{ for } u$$

When Substitution Fails

The method of substitution does not always succeed. In Example 5.2.10, we consider an integral very similar to the one in Example 5.2.3 but just different enough that no substitution will work.

EXAMPLE 5.2.10

Evaluate $\int x^4 e^{x^4+2}\, dx$.

Solution

Just-In-Time

Notice that if $u = x^4 + 2$, then $x^4 = u - 2$, so $x = \pm(u - 2)^{1/4} = \pm\sqrt[4]{u - 2}$.

The natural substitution is $u = x^4 + 2$, as in Example 5.2.3. As before, $du = 4x^3\, dx$, so $x^3\, dx = \dfrac{1}{4}\, du$, but this integrand involves x^4, not x^3. The "extra" factor of x satisfies $x = \sqrt[4]{u - 2}$, so substituting the positive root gives

$$\int x^4 e^{x^4+2}\, dx = \int x e^{x^4+2}(x^3\, dx) = \int \sqrt[4]{u - 2}\,(e^u)\left(\frac{1}{4}\, du\right),$$

which is hardly an improvement on the original integral! Try a few other possible substitutions (say, $u = x^2$ or $u = x^3$) to convince yourself that nothing works.

An Application Involving Substitution

EXAMPLE 5.2.11

The price p (dollars) of each unit of a particular commodity is estimated to be changing at a rate of

$$\frac{dp}{dx} = \frac{-135x}{\sqrt{9 + x^2}}$$

where x (hundred) units is the consumer demand (the number of units purchased at that price). Suppose 400 units ($x = 4$) are demanded when the price is $30 per unit.

a. Find the demand function $p(x)$.

b. At what price will 300 units be demanded? At what price will no units be demanded?

c. How many units are demanded at a price of $20 per unit?

Solution

a. The price per unit $p(x)$ is found by integrating $p'(x)$ with respect to x. To perform this integration, use the substitution

$$u = 9 + x^2, \quad du = 2x\, dx, \quad x\, dx = \frac{1}{2}\, du$$

to get

$$p(x) = \int \frac{-135x}{\sqrt{9 + x^2}}\, dx = \int \frac{-135}{u^{1/2}}\left(\frac{1}{2}\right)\, du$$

$$= \frac{-135}{2}\int u^{-1/2}\, du$$

$$= \frac{-135}{2}\left(\frac{u^{1/2}}{1/2}\right) + C$$

$$= -135\sqrt{9 + x^2} + C \qquad \text{substitute } 9 + x^2 \text{ for } u$$

Since $p = 30$ when $x = 4$,

$$30 = -135\sqrt{9 + 4^2} + C$$

$$C = 30 + 135\sqrt{25} = 705$$

so

$$p(x) = -135\sqrt{9 + x^2} + 705.$$

b. When 300 units are demanded, $x = 3$ and the corresponding price is

$$p(3) = -135\sqrt{9 + 3^2} + 705 = 132.24,$$

or \$132.24 per unit. No units are demanded when $x = 0$, and the corresponding price is

$$p(0) = -135\sqrt{9 + 0} + 705 = 300,$$

or \$300 per unit.

c. To determine the number of units demanded at a unit price of \$20 per unit, solve the equation

$$-135\sqrt{9 + x^2} + 705 = 20.$$

Thus,

$$-135\sqrt{9 + x^2} + 705 = 20$$
$$135\sqrt{9 + x^2} = 685$$
$$\sqrt{9 + x^2} = \frac{685}{135}$$
$$9 + x^2 \approx 25.75 \qquad \text{square both sides}$$
$$x^2 \approx 16.75$$
$$x \approx 4.09$$

That is, roughly 409 units will be demanded when the price is \$20 per unit.

EXERCISES ■ 5.2

In Exercises 1 and 2, fill in the table by specifying the substitution you would choose to find each of the four given integrals.

1.

Integral	Substitution u
a. $\int (3x + 4)^{5/2}\, dx$	
b. $\int \dfrac{4}{3 - x}\, dx$	
c. $\int t e^{2 - t^2}\, dt$	
d. $\int t(2 + t^2)^3\, dt$	

2.

Integral	Substitution u
a. $\int \dfrac{3}{(2x - 5)^4}\, dx$	
b. $\int x^2 e^{-x^3}\, dx$	
c. $\int \dfrac{e^t}{e^t + 1}\, dx$	
d. $\int \dfrac{t + 3}{\sqrt[3]{t^2 + 6t + 5}}\, dt$	

In Exercises 3 through 36, find the indicated integral and check your answer by differentiation.

3. $\displaystyle\int (2x + 6)^5 \, dx$

4. $\displaystyle\int e^{5x+3} \, dx$

5. $\displaystyle\int \sqrt{4x - 1} \, dx$

6. $\displaystyle\int \frac{1}{3x + 5} \, dx$

7. $\displaystyle\int e^{1-x} \, dx$

8. $\displaystyle\int [(x - 1)^5 + 3(x - 1)^2 + 5] \, dx$

9. $\displaystyle\int xe^{x^2} \, dx$

10. $\displaystyle\int 2xe^{x^2-1} \, dx$

11. $\displaystyle\int t(t^2 + 1)^5 \, dt$

12. $\displaystyle\int 3t\sqrt{t^2 + 8} \, dt$

13. $\displaystyle\int x^2(x^3 + 1)^{3/4} \, dx$

14. $\displaystyle\int x^5 e^{1-x^6} \, dx$

15. $\displaystyle\int \frac{2y^4}{y^5 + 1} \, dy$

16. $\displaystyle\int \frac{y^2}{(y^3 + 5)^2} \, dy$

17. $\displaystyle\int (x + 1)(x^2 + 2x + 5)^{12} \, dx$

18. $\displaystyle\int (3x^2 - 1)e^{x^3-x} \, dx$

19. $\displaystyle\int \frac{3x^4 + 12x^3 + 6}{x^5 + 5x^4 + 10x + 12} \, dx$

20. $\displaystyle\int \frac{10x^3 - 5x}{\sqrt{x^4 - x^2 + 6}} \, dx$

21. $\displaystyle\int \frac{3u - 3}{(u^2 - 2u + 6)^2} \, du$

22. $\displaystyle\int \frac{6u - 3}{4u^2 - 4u + 1} \, du$

23. $\displaystyle\int \frac{\ln 5x}{x} \, dx$

24. $\displaystyle\int \frac{1}{x \ln x} \, dx$

25. $\displaystyle\int \frac{1}{x(\ln x)^2} \, dx$

26. $\displaystyle\int \frac{\ln x^2}{x} \, dx$

27. $\displaystyle\int \frac{2x \ln (x^2 + 1)}{x^2 + 1} \, dx$

28. $\displaystyle\int \frac{e^{\sqrt{x}}}{\sqrt{x}} \, dx$

29. $\displaystyle\int \frac{e^x + e^{-x}}{e^x - e^{-x}} \, dx$

30. $\displaystyle\int e^{-x}(1 + e^{2x}) \, dx$

31. $\displaystyle\int \frac{x}{2x + 1} \, dx$

32. $\displaystyle\int \frac{t - 1}{t + 1} \, dt$

33. $\displaystyle\int x\sqrt{2x + 1} \, dx$

34. $\displaystyle\int \frac{x}{\sqrt[3]{4 - 3x}} \, dx$

35. $\displaystyle\int \frac{1}{\sqrt{x} \, (\sqrt{x} + 1)} \, dx$

   [*Hint:* Let $u = \sqrt{x} + 1$.]

36. $\displaystyle\int \frac{1}{x^2}\left(\frac{1}{x} - 1\right)^{2/3} \, dx$

   $\left[\text{\textit{Hint}: Let } u = \dfrac{1}{x} - 1.\right]$

In Exercises 37 through 42, solve the given initial value problem for $y = f(x)$.

37. $\dfrac{dy}{dx} = (3 - 2x)^2$, where $y = 0$ when $x = 0$

38. $\dfrac{dy}{dx} = \sqrt{4x + 5}$, where $y = 3$ when $x = 1$

39. $\dfrac{dy}{dx} = \dfrac{1}{x + 1}$, where $y = 1$ when $x = 0$

40. $\dfrac{dy}{dx} = e^{2-x}$, where $y = 0$ when $x = 2$

41. $\dfrac{dy}{dx} = \dfrac{x + 2}{x^2 + 4x + 5}$, where $y = 3$ when $x = -1$

42. $\dfrac{dy}{dx} = \dfrac{\ln\sqrt{x}}{x}$, where $y = 2$ when $x = 1$

In Exercises 43 through 46, the slope f′(x) at each point (x, y) on a curve y = f(x) is given, along with a particular point (a, b) on the curve. Use this information to find f(x).

43. $f'(x) = (1 - 2x)^{3/2}$; $(0, 0)$

44. $f'(x) = x\sqrt{x^2 + 5}$; $(2, 10)$

45. $f'(x) = xe^{4-x^2}$; $(-2, 1)$

46. $f'(x) = \dfrac{2x}{1 + 3x^2}$; $(0, 5)$

In Exercises 47 through 50, the velocity v(t) = x′(t) at time t of an object moving along the x axis is given, along with the initial position x(0) of the object. In each case, find

    (a) the position x(t) at time t
    (b) the position of the object at time t = 4
    (c) the time when the object is at x = 3

47. $x'(t) = -2(3t + 1)^{1/2}$; $x(0) = 4$

48. $x'(t) = \dfrac{-1}{1 + 0.5t}$; $x(0) = 5$

49. $x'(t) = \dfrac{1}{\sqrt{2t + 1}}$; $x(0) = 0$

50. $x'(t) = \dfrac{-2t}{(1 + t^2)^{3/2}}$; $x(0) = 4$

51. MARGINAL COST At a certain factory, the marginal cost is $3(q - 4)^2$ dollars per unit when the level of production is q units.
    a. Express the total production cost in terms of the overhead (the cost of producing 0 units) and the number of units produced.
    b. What is the cost of producing 14 units if the overhead is $436?

52. DEPRECIATION The resale value of a certain industrial machine decreases at a rate that depends on its age. When the machine is t years old, the rate at which its value is changing is $-960e^{-t/5}$ dollars per year.
    a. Express the value of the machine in terms of its age and initial value.
    b. If the machine was originally worth $5200, how much will it be worth when it is 10 years old?

53. TREE GROWTH A tree has been transplanted and after x years is growing at a rate of
$$1 + \frac{1}{(x + 1)^2}$$
metres per year. After 2 years, it has reached a height of 5 m. How tall was it when it was transplanted?

54. RETAIL PRICES In a certain section of the country, it is estimated that t weeks from now, the price of chicken will be increasing at a rate of $p'(t) = 3\sqrt{t} + 1$ cents per kilogram per week. If chicken currently costs $2.30 per kilogram, what will it cost 8 weeks from now?

55. REVENUE The marginal revenue from the sale of x units of a particular commodity is estimated to be $R'(x) = 50 + 3.5xe^{-0.01x^2}$ dollars per unit, where $R(x)$ is the revenue in dollars.
    a. Find $R(x)$, assuming that $R(0) = 0$.
    b. What revenue should be expected from the sale of 10 units?

56. WATER POLLUTION An oil spill in the ocean is roughly circular in shape, with radius $R(t)$ metres, t minutes after the spill begins. The radius is increasing at a rate of
$$R'(t) = \frac{21}{0.07t + 5} \text{ metres per minute.}$$
    a. Find an expression for the radius $R(t)$, assuming that $R = 0$ when $t = 0$.
    b. What is the area $A = \pi R^2$ of the spill after 1 hour?

57. DRUG CONCENTRATION The concentration $C(t)$ in milligrams per cubic centimetre of a drug in a patient's bloodstream is 0.5 mg/cm^3 immediately after an injection. After t minutes, it is decreasing at a rate of
$$C'(t) = \frac{-0.01e^{0.01t}}{(e^{0.01t} + 1)^2}$$
milligrams per cubic centimetre per minute. A new injection is given when the concentration drops below 0.05 mg/cm^3.
    a. Find an expression for $C(t)$.
    b. What is the concentration after 1 hour? After 3 hours?

    c. Use a graphing utility to plot the function and determine how much time passes before the next injection must be given.

58. LAND VALUE It is estimated that x years from now, the value $V(x)$ of a hectare of farmland will be increasing at a rate of
$$V'(x) = \frac{40x^3}{\sqrt{0.2x^4 + 8000}}$$
dollars per year. The land is currently worth $2500 per hectare.
    a. Find $V(x)$.
    b. How much will the land be worth in 10 years?

59. AIR POLLUTION   One day in Mississauga, the level of ozone $L(t)$ at 7:00 A.M. is 0.25 parts per million (ppm). A 12-hour weather forecast predicts that the ozone level t hours later will be changing at a rate of

$$L'(t) = \frac{0.24 - 0.03t}{\sqrt{36 + 16t - t^2}}$$

parts per million per hour (ppm/h).

a. Express the ozone level $L(t)$ as a function of t. When does the peak ozone level occur? What is the peak level?

b. Use a graphing utility to sketch the graph of $L(t)$. Answer the questions in part (a) using the graph. Then determine at what other time the ozone level will be the same as it is at 11:00 A.M.

60. SUPPLY   The owner of a fast-food chain determines that if x thousand units of a new meal item are supplied, then the marginal price at that level of supply is given by

$$p'(x) = \frac{x}{(x + 3)^2} \text{ dollars per meal,}$$

where $p(x)$ is the price (in dollars) per unit at which all x meal units will be sold. Currently, 5000 units are being supplied at a price of $2.20 per unit.

a. Find the supply (price) function $p(x)$.

b. If 10 000 meal units are supplied to restaurants in the chain, what unit price should be charged so that all of the units will be sold?

61. DEMAND   The manager of a shoe store determines that the price p (dollars) for each pair of a popular brand of running shoes is changing at a rate of

$$p'(x) = \frac{-300x}{(x^2 + 9)^{3/2}}$$

when x (hundred) pairs are demanded by consumers. When the price is $75 per pair, 400 pairs ($x = 4$) are demanded by consumers.

a. Find the demand (price) function $p(x)$.

b. At what price will 500 pairs of running shoes be demanded? At what price will no running shoes be demanded?

c. How many pairs will be demanded at a price of $90 per pair?

62. SUPPLY   The price p (dollars per unit) of a particular commodity is increasing at a rate of

$$p'(x) = \frac{20x}{(7 - x^2)}$$

when x hundred units of the commodity are supplied to the market. The manufacturer supplies 200 units ($x = 2$) when the price is $2 per unit.

a. Find the supply function $p(x)$.

b. What price corresponds to a supply of 500 units?

63. MARGINAL PROFIT   A company determines that the marginal revenue from the production of x units is $R'(x) = 7 - 3x - 4x^2$ hundred dollars per unit, and the corresponding marginal cost is $C'(x) = 5 + 2x$ hundred dollars per unit. By how much does the profit change when the level of production is raised from 5 to 9 units?

64. MARGINAL PROFIT   Repeat Exercise 63 for marginal revenue $R'(x) = \dfrac{11 - x}{\sqrt{14 - x}}$ and for marginal cost $C'(x) = 2 + x + x^2$.

65. A* Find $\displaystyle\int x^{1/3}(x^{2/3} + 1)^{3/2}dx$. [*Hint:* Substitute $u = x^{2/3} + 1$ and use $x^{2/3} = u - 1$.]

66. A* Find $\displaystyle\int x^3(4 - x^2)^{-1/2}dx$. [*Hint:* Substitute $u = 4 - x^2$ and use the fact that $x^2 = 4 - u$.]

67. A* Find $\displaystyle\int \frac{e^{2x}}{1 + e^x} dx$. [*Hint:* Let $u = 1 + e^x$.]

68. A* Find $\displaystyle\int e^{-x}(1 + e^x)^2 dx$. [*Hint:* Is it better to set $u = 1 + e^x$ or $u = e^x$? Or is it better not to use the method of substitution?]

SECTION 5.3

L03

Evaluate definite integrals using the substitution method where necessary.

The Definite Integral and the Fundamental Theorem of Calculus

Suppose a real estate agent wants to evaluate an unimproved parcel of land that is 100 m wide and is bounded by streets on three sides and by a stream on the fourth side. The agent determines that if a coordinate system is set up as shown in Figure 5.2, the stream can be described by the curve $y = x^3 + 1$, where x and y are measured in hundreds of metres. If the area of the parcel is A square metres and the agent estimates that its land is worth \$120/m², then the total value of the parcel is $120A$ dollars. If the parcel were rectangular in shape or triangular or even trapezoidal, its area A could be found by substituting into a well-known formula, but the upper boundary of the parcel is curved, so how can the agent find the area and hence the total value of the parcel?

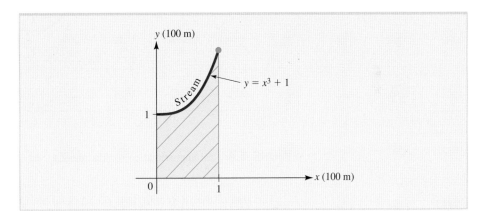

FIGURE 5.2   Determining land value by finding the area under a curve.

Our goal in this section is to show how area under a curve, such as the area A in our real estate example, can be expressed as the limit of a sum of terms called a **definite integral.** We will then introduce a result called the **fundamental theorem of calculus** that allows us to compute *definite* integrals and thus find area and other quantities by using the *indefinite* integration (antidifferentiation) methods of Sections 5.1 and 5.2. In Example 5.3.3, we will illustrate this procedure by expressing the area A in our real estate example as a definite integral and evaluating it using the fundamental theorem of calculus.

Area as the Limit of a Sum

Consider the area of the region under the curve $y = f(x)$ over an interval $a \le x \le b$, where $f(x) \ge 0$ and f is continuous, as illustrated in Figure 5.3. To find this area, we will follow a useful general policy:

When faced with something you don't know how to handle, try to relate it to something you do know how to handle.

In this particular case, we may not know the area under the given curve, but we do know how to find the area of a rectangle. Thus, we proceed by subdividing the region into a number of rectangular regions and then approximating the area A under the curve $y = f(x)$ by adding the areas of the approximating rectangles.

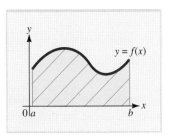

FIGURE 5.3   The region under the curve $y = f(x)$ over the interval $a \le x \le b$.

To be more specific, begin the approximation by dividing the interval $a \leq x \leq b$ into n equal subintervals, each of length $\Delta x = \dfrac{b - a}{n}$, and let x_j denote the left endpoint of the jth subinterval, for $j = 1, 2, \ldots, n$. Then draw n rectangles such that the jth rectangle has the jth subinterval as its base and $f(x_j)$ as its height. The approximation scheme is illustrated in Figure 5.4.

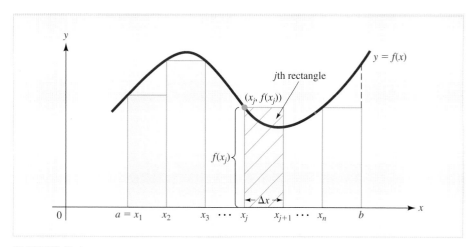

FIGURE 5.4 An approximation of area under a curve by rectangles.

The area of the jth rectangle is $f(x_j)\,\Delta x$ and approximates the area under the curve above the subinterval $x_j \leq x \leq x_{j+1}$. The sum of the areas of all n rectangles is

$$S_n = f(x_1)\Delta x + f(x_2)\Delta x + \cdots + f(x_n)\Delta x$$
$$= [f(x_1) + f(x_2) + \cdots + f(x_n)]\Delta x$$

which approximates the total area A under the curve.

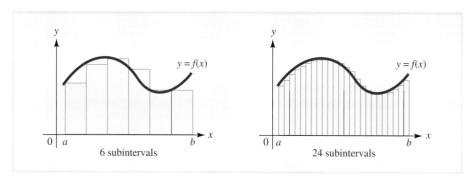

FIGURE 5.5 The approximation improves as the number of subintervals increases.

As the number of subintervals n increases, the approximating sum S_n gets closer and closer to what we intuitively think of as the area under the curve, as illustrated in Figure 5.5. Therefore, it is reasonable to define the actual area A under the curve as the limit of the sums. To summarize:

> **Area under a Curve**   ■   Let $f(x)$ be continuous and satisfy $f(x) \geq 0$ on the interval $a \leq x \leq b$. Then the region under the curve $y = f(x)$ over the interval $a \leq x \leq b$ has area
>
> $$A = \lim_{n \to +\infty} [f(x_1) + f(x_2) + \cdots + f(x_n)] \Delta x,$$
>
> where x_j is the left endpoint of the jth subinterval if the interval $a \leq x \leq b$ is divided into n equal parts, each of length $\Delta x = \dfrac{b - a}{n}$.

> **NOTE**   At this point, you may ask, "Why use the left endpoint of the subintervals rather than, say, the right endpoint or even the midpoint?" The answer is that there is no reason we can't use those other points to compute the height of the approximating rectangles. In fact, the interval $a \leq x \leq b$ can be subdivided arbitrarily and arbitrary points chosen in each subinterval, and the result will still be the same in the limit as the number of rectangles approaches infinity. However, proving this equivalence is difficult, and well beyond the scope of this text.   ■

Here is an example in which area is computed as the limit of a sum and then checked using a geometric formula.

Just-In-Time

A trapezoid is a four-sided polygon with at least two parallel sides. Its area is

$$A = \frac{1}{2}(s_1 + s_2)h,$$

where s_1 and s_2 are the lengths of the two parallel sides and h is the distance between them.

EXAMPLE 5.3.1

Let R be the region under the graph of $f(x) = 2x + 1$ over the interval $1 \leq x \leq 3$, as shown in Figure 5.6a. Compute the area of R as the limit of a sum.

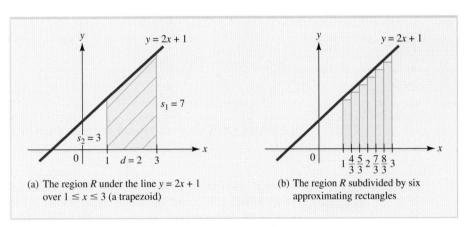

(a) The region R under the line $y = 2x + 1$ over $1 \leq x \leq 3$ (a trapezoid)

(b) The region R subdivided by six approximating rectangles

FIGURE 5.6   Approximating the area under a line with rectangles.

Solution

The region R is shown in Figure 5.6b with six approximating rectangles, each of width $\Delta x = \dfrac{3 - 1}{6} = \dfrac{1}{3}$. The left endpoints in the partition of $1 \leq x \leq 3$ are $x_1 = 1$,

$x_2 = 1 + \dfrac{1}{3} = \dfrac{4}{3}$, and, similarly, $x_3 = \dfrac{5}{3}$, $x_4 = 2$, $x_5 = \dfrac{7}{3}$, and $x_6 = \dfrac{8}{3}$. The corresponding values of $f(x) = 2x + 1$ are given in the following table:

x_j	1	$\frac{4}{3}$	$\frac{5}{3}$	2	$\frac{7}{3}$	$\frac{8}{3}$
$f(x_j) = 2x_j + 1$	3	$\frac{11}{3}$	$\frac{13}{3}$	5	$\frac{17}{3}$	$\frac{19}{3}$

Thus, the area A of the region R is approximated by the sum

$$S = \left(3 + \frac{11}{3} + \frac{13}{3} + 5 + \frac{17}{3} + \frac{19}{3}\right)\left(\frac{1}{3}\right) = \frac{28}{3} \approx 9.333.$$

If you continue to subdivide the region R using more and more rectangles, the corresponding approximating sums S_n approach the actual area A of the region. The sum we have already computed for $n = 6$ is listed in the following table, along with those for $n = 10, 20, 50, 100,$ and 500.

Number of rectangles n	6	10	20	50	100	500
Approximating sum S_n	9.333	9.600	9.800	9.920	9.960	9.992

The numbers on the bottom line of this table seem to be approaching 10 as n gets larger and larger. Thus, it is reasonable to conjecture that the region R has area

$$A = \lim_{n \to +\infty} S_n = 10.$$

Note that the sum in the limit as $n \to \infty$ can be shown precisely, but the calculation is omitted here for simplicity.

Notice in Figure 5.6a that the region R is a trapezoid of width $d = 3 - 1 = 2$ with parallel sides of lengths

$$s_1 = 2(3) + 1 = 7 \quad \text{and} \quad s_2 = 2(1) + 1 = 3.$$

Such a trapezoid has area

$$A = \frac{1}{2}(s_1 + s_2)h = \frac{1}{2}(7 + 3)(2) = 10,$$

the same result we just obtained using the limit of a sum procedure.

The Definite Integral

Area is just one of many quantities that can be expressed as the limit of a sum. To handle all such cases, including those for which $f(x) \geq 0$ is *not* required and left endpoints are not used, we require the terminology and notation introduced in the following definition.

The Definite Integral ▪ Let $f(x)$ be a function that is continuous on the interval $a \leq x \leq b$. Subdivide the interval $a \leq x \leq b$ into n equal parts, each of width $\Delta x = \dfrac{b - a}{n}$, and choose a number x_k from the kth subinterval for $k = 1$, $2, \ldots, n$. Form the sum

$$[f(x_1) + f(x_2) + \cdots + f(x_n)]\Delta x,$$

called a **Riemann sum.**

Then the **definite integral** of f on the interval $a \leq x \leq b$, denoted by $\int_a^b f(x)\, dx$, is the limit of the Riemann sum as $n \to +\infty$; that is,

$$\int_a^b f(x)\, dx = \lim_{n \to +\infty} [f(x_1) + f(x_2) + \cdots + f(x_n)]\Delta x.$$

The function $f(x)$ is called the **integrand,** and the numbers a and b are called the **lower and upper limits of integration,** respectively. The process of finding a definite integral is called **definite integration.**

Notice that the symbol $\int$ is like an elongated 'S' for a continuous sum. (Compare this to the symbol Σ, the Greek uppercase S, for a discrete sum.) Also, the Δx's in the sum have become infinitesimally small increments dx.

Consider some more Riemann sums generated with Maple.

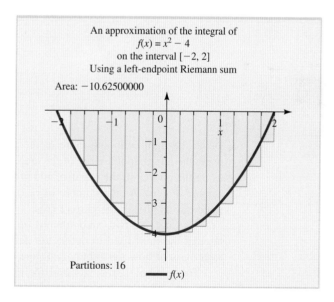

Notice that the graph above is below the x axis on the interval $[-2, 2]$, resulting in each of the $f(x)$ heights of the rectangles being negative. The Riemann sum is then negative.

The next two graphs show the same function, $f(x) = x^2 - 4$, on the interval $[-4, 4]$, one with 100 rectangles, resulting in a Riemann sum of 10.6752, and one with 200 rectangles, resulting in a Riemann sum of 10.6688.

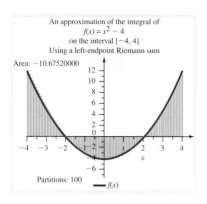

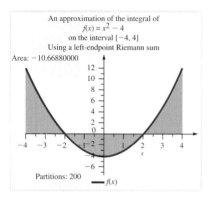

The next three graphs show the effect of using the left side, the right side, and the midpoint of the rectangle when calculating a Riemann sum, using the function $f(x) = x^3$ on the interval $[-4, 4]$ with 16 rectangles.

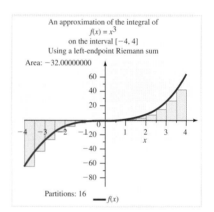

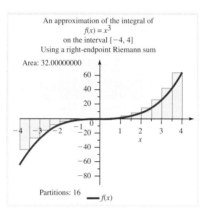

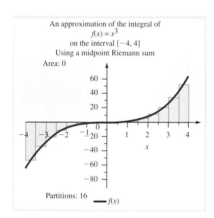

Using the right endpoint, more of the area of the rectangles is above the x axis, and using the left endpoint more of the area of the rectangles is below the x axis. However, using the midpoint, the negative and positive balance each other, resulting in the exact answer of zero. As the number of rectangles increases, the left and right approximations also approach the correct area.

The fact that $f(x)$ is continuous on $a \leq x \leq b$ turns out to be enough to guarantee that the limit used to define the definite integral $\int_a^b f(x)\, dx$ exists and is the same regardless of how the subinterval representatives x_k are chosen.

The symbol $\int_a^b f(x)\, dx$ used for the definite integral is essentially the same as the symbol $\int f(x)\, dx$ for the indefinite integral, even though the definite integral is a specific number while the indefinite integral is a family of functions, the antiderivatives of f. You will soon see that these two apparently very different concepts are intimately related. Here is a compact form for the definition of area using the integral notation.

Area as a Definite Integral ■ If $f(x)$ is continuous and $f(x) \geq 0$ on the interval $a \leq x \leq b$, then the region R under the curve $y = f(x)$ over the interval $a \leq x \leq b$ has area A given by the definite integral $A = \int_a^b f(x)\, dx$.

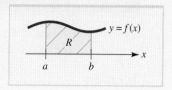

The Fundamental Theorem of Calculus

If computing the limit of a sum were the only way of evaluating a definite integral, the integration process would probably be little more than a mathematical novelty. Fortunately, there is an easier way of performing this computation, thanks to this remarkable result connecting the definite integral to antidifferentiation.

The Fundamental Theorem of Calculus ■ If the function $f(x)$ is continuous on the interval $a \leq x \leq b$, then

$$\int_a^b f(x)\, dx = F(b) - F(a),$$

where $F(x)$ is any antiderivative of $f(x)$ on $a \leq x \leq b$.

A special case of the fundamental theorem of calculus is verified at the end of this section. When applying the fundamental theorem, we use the notation

$$F(x)\Big|_a^b = F(b) - F(a).$$

Thus,

$$\int_a^b f(x)\, dx = F(x)\Big|_a^b = F(b) - F(a).$$

NOTE You may wonder how the fundamental theorem of calculus can promise that if $F(x)$ is *any* antiderivative of $f(x)$, then

$$\int_a^b f(x)\, dx = F(b) - F(a).$$

To see why this is true, suppose $G(x)$ is another such antiderivative. Then $G(x) = F(x) + C$ for some constant C, so $F(x) = G(x) - C$ and

$$\int_a^b f(x)\, dx = F(b) - F(a)$$
$$= [G(b) - C] - [G(a) - C]$$
$$= G(b) - G(a)$$

since the C's cancel. Thus, the result is the same regardless of which antiderivative is used. ■

In Example 5.3.2, we demonstrate the computational value of the fundamental theorem of calculus by using it to compute the same area we estimated as the limit of a sum in Example 5.3.1.

EXAMPLE 5.3.2

Use the fundamental theorem of calculus to find the area of the region under the line $y = 2x + 1$ over the interval $1 \leq x \leq 3$.

Solution

Since $f(x) = 2x + 1$ satisfies $f(x) \geq 0$ on the interval $1 \leq x \leq 3$, the area is given by the definite integral $A = \int_1^3 (2x + 1)\, dx$. Since an antiderivative of $f(x) = 2x + 1$ is $F(x) = x^2 + x$, the fundamental theorem of calculus says that

$$A = \int_1^3 (2x + 1)\, dx = x^2 + x \Big|_1^3$$
$$= [(3)^2 + (3)] - [(1)^2 + (1)]$$
$$= 10$$

as estimated in Example 5.3.1.

EXAMPLE 5.3.3

Find the area of the parcel of land described in the introduction to this section; that is, find the area under the curve $y = x^3 + 1$ over the interval $0 \leq x \leq 1$, where x and y are in hundreds of metres. If the land in the parcel is appraised at \$120/m^2, what is the total value of the parcel?

Solution

The area of the parcel is given by the definite integral

$$A = \int_0^1 (x^3 + 1)\, dx.$$

Since an antiderivative of $f(x) = x^3 + 1$ is $F(x) = \dfrac{1}{4}x^4 + x$, the fundamental theorem of calculus says that

$$A = \int_0^1 (x^3 + 1)\, dx = \frac{1}{4}x^4 + x \Big|_0^1$$
$$= \left[\frac{1}{4}(1)^4 + 1\right] - \left[\frac{1}{4}(0)^4 + 0\right] = \frac{5}{4}$$

Because x and y are measured in hundreds of metres, the total area is

$$\frac{5}{4} \times 100\text{ m} \times 100\text{ m} = 12\,500\text{ m}^2,$$

and since the land in the parcel is worth \$120/m^2, the total value of the parcel is

$$V = (\$120/\text{m}^2)(12\,500\text{ m}^2) = \$1\,500\,000.$$

EXAMPLE 5.3.4

At the beginning of the chapter, we considered the area under a roller coaster. If the roller coaster has the form of a piecewise function

$$f(x) = \begin{cases} -0.15x^2 + 6x & 0 \leq x \leq 20 \\ -0.75x + 75 & 20 < x \leq 100 \end{cases}$$

where units are metres above the ground, represented by the x axis, find the area under the roller coaster.

(Photo: © Richard Clune/CORBIS)

Solution

The area is the sum of the area under the curve and the area under the straight line. The function is continuous (check for yourself), so 20 can be used in the integration for both pieces of the function.

$$\text{Area} = \int_0^{20} (-0.15x^2 + 6x)\,dx + \int_{20}^{100} (-0.75x + 75)\,dx$$

$$= \frac{-0.15}{3}x^3 + \frac{6}{2}x^2 \Big|_0^{20} + \frac{-0.75}{2}x^2 + 75x \Big|_{20}^{100}$$

$$= -0.05(20)^3 + 3(20)^2 - 0 + (-0.375(100)^2 + 75(100) - (-0.375(20)^2 + 75(20)))$$

$$= 3200$$

The area under the roller coaster is 3200 m^2.

Just-In-Time

When the fundamental theorem of calculus,

$$\int_a^b f(x)\,dx = F(b) - F(a),$$

is used to evaluate a definite integral, remember to compute **both** $F(b)$ and $F(a)$, even when $a = 0$.

When $a = 0$ results in a function such as e^0, then the result is not zero.

EXAMPLE 5.3.5

Evaluate the definite integral $\int_0^1 (e^{-x} + \sqrt{x})\,dx$.

Solution

An antiderivative of $f(x) = e^{-x} + \sqrt{x}$ is $F(x) = -e^{-x} + \frac{2}{3}x^{3/2}$, so the definite integral is

$$\int_0^1 (e^{-x} + \sqrt{x})\,dx = \left(-e^{-x} + \frac{2}{3}x^{3/2} \right)\Big|_0^1$$

$$= \left[-e^{-1} + \frac{2}{3}(1)^{3/2} \right] - \left[-e^0 + \frac{2}{3}(0) \right]$$

$$= -e^{-1} + \frac{2}{3} + 1$$

$$= \frac{5}{3} - \frac{1}{e}$$

$$\approx 1.299$$

Our definition of the definite integral was motivated by computing area, which is a nonnegative quantity. However, since the definition does not require $f(x) \geq 0$, it is quite possible for a definite integral to be negative, as illustrated in Example 5.3.6.

When $f(x)$ is negative as in the graph, then $f(x)\,\Delta x$ is negative for each rectangle of the Riemann sum, resulting in a negative definite integral.

EXAMPLE 5.3.6

Evaluate $\displaystyle\int_1^4 \left(\frac{1}{x} - x^2\right) dx$.

Solution

An antiderivative of $f(x) = \dfrac{1}{x} - x^2$ is $F(x) = \ln|x| - \dfrac{1}{3}x^3$, so

$$\int_1^4 \left(\frac{1}{x} - x^2\right) dx = \left(\ln|x| - \frac{1}{3}x^3\right)\Big|_1^4$$

$$= \left[\ln 4 - \frac{1}{3}(4)^3\right] - \left[\ln 1 - \frac{1}{3}(1)^3\right]$$

$$= \ln 4 - \frac{64}{3} - 0 + \frac{1}{3}$$

$$= \ln 4 - \frac{63}{3}$$

$$= \ln 4 - 21 \approx -19.6137$$

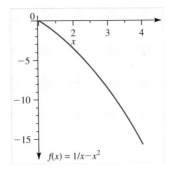

$f(x) = 1/x - x^2$

Integration Rules

The following list of rules can be used to simplify the computation of definite integrals.

Rules for Definite Integrals

Let f and g be any functions that are continuous on $a \leq x \leq b$. Then we have the following:

1. Constant multiple rule: $\displaystyle\int_a^b k f(x)\,dx = k\int_a^b f(x)\,dx$ for constant k.

2. Sum rule: $\displaystyle\int_a^b [f(x) + g(x)]\,dx = \int_a^b f(x)\,dx + \int_a^b g(x)\,dx$

3. Difference rule: $\displaystyle\int_a^b [f(x) - g(x)]\,dx = \int_a^b f(x)\,dx - \int_a^b g(x)\,dx$

4. $\displaystyle\int_a^a f(x)\,dx = 0$

5. $\displaystyle\int_a^b f(x)\,dx = -\int_b^a f(x)\,dx$

6. Subdivision rule: $\displaystyle\int_a^b f(x)\,dx = \int_a^c f(x)\,dx + \int_c^b f(x)\,dx$

Rules 4 and 5 are really special cases of the definition of the definite integral. The first three rules can be proven by using the fundamental theorem of calculus along with an analogous rule for indefinite integrals. For instance, to verify the constant multiple rule, suppose $F(x)$ is an antiderivative of $f(x)$. Then, according to the constant multiple rule for indefinite integrals, $kF(x)$ is an antiderivative of $kf(x)$ and the fundamental theorem of calculus tells us that

$$\int_a^b k f(x)\, dx = kF(x)\Big|_a^b$$

$$= kF(b) - kF(a) = k[F(b) - F(a)]$$

$$= k\int_a^b f(x)\, dx$$

You are asked to verify the sum rule using similar reasoning in Exercise 70.

In the case where $f(x) \geq 0$ on the interval $a \leq x \leq b$, the subdivision rule is a geometric reflection of the fact that the area under the curve $y = f(x)$ over the interval $a \leq x \leq b$ is the sum of the areas under $y = f(x)$ over the subintervals $a \leq x \leq c$ and $c \leq x \leq b$, as illustrated in Figure 5.7. However, it is important to remember that the subdivision rule is true even if $f(x)$ does *not* satisfy $f(x) \geq 0$ on $a \leq x \leq b$.

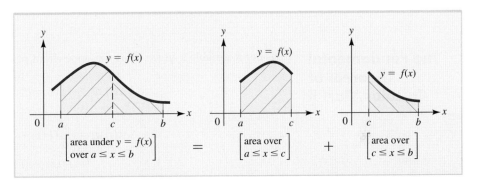

FIGURE 5.7   The subdivision rule for definite integrals (case where $f(x) \geq 0$).

EXAMPLE 5.3.7

Let $f(x)$ and $g(x)$ be functions that are continuous on the interval $-2 \leq x \leq 5$ and that satisfy

$$\int_{-2}^5 f(x)\, dx = 3 \qquad \int_{-2}^5 g(x)\, dx = -4 \qquad \int_3^5 f(x)\, dx = 7$$

Use this information to evaluate each of these definite integrals:

a. $\displaystyle\int_{-2}^5 [2f(x) - 3g(x)]\, dx$

b. $\displaystyle\int_{-2}^3 f(x)\, dx$

Solution

a. Combining the difference rule and constant multiple rule and substituting the given information results in

$$\int_{-2}^{5} [2f(x) - 3g(x)] \, dx = \int_{-2}^{5} 2f(x) \, dx - \int_{-2}^{5} 3g(x) \, dx \quad \text{difference rule}$$

$$= 2\int_{-2}^{5} f(x) \, dx - 3\int_{-2}^{5} g(x) \, dx \quad \begin{array}{l} \text{constant} \\ \text{multiple rule} \end{array}$$

$$= 2(3) - 3(-4) = 18 \quad \begin{array}{l} \text{substitute given} \\ \text{information} \end{array}$$

b. According to the subdivision rule,

$$\int_{-2}^{5} f(x) \, dx = \int_{-2}^{3} f(x) \, dx + \int_{3}^{5} f(x) \, dx$$

Solving this equation for the required integral $\int_{-2}^{3} f(x) \, dx$ and substituting the given information results in

$$\int_{-2}^{3} f(x) \, dx = \int_{-2}^{5} f(x) \, dx - \int_{3}^{5} f(x) \, dx$$

$$= 3 - 7 = -4$$

The Fundamental Theorem of Calculus, Part II

Consider a variable x in $[a, b]$ as the upper limit of integration.

$$\int_{a}^{x} f(t) \, dt = F(x) - F(a)$$

$$\frac{d}{dx} \int_{a}^{x} f(t) \, dt = \frac{d}{dx} (F(x) - F(a))$$

$$= F'(x) - 0$$

$$= f(x)$$

Therefore, the derivative of a definite integral with respect to the upper limit, here x, is the integrand evaluated at this upper limit. The definite integral then becomes a function. We have the following result.

The Fundamental Theorem of Calculus, Part II ■ If $f(t)$ is continuous over $[a, b]$, x is in the interval $[a, b]$, and $g(x) = \int_{a}^{x} f(t) \, dt$, then $g'(x) = f(x)$.

Substituting in a Definite Integral

When using a substitution $u = g(x)$ to evaluate a definite integral $\int_{a}^{b} f(x) \, dx$, you can proceed in either of these two ways:

1. Use the substitution to find an antiderivative $F(x)$ for $f(x)$ and then evaluate the definite integral using the fundamental theorem of calculus.

2. Use the substitution to express the integrand and dx in terms of u and du and to replace the original limits of integration, a and b, with transformed limits $c = g(a)$ and $d = g(b)$. The original integral can then be evaluated by applying the fundamental theorem of calculus to the transformed definite integral.

These procedures are illustrated in Examples 5.3.7 and 5.3.8.

Just-In-Time

Only one member of the family of antiderivatives of $f(x)$ is needed to evaluate $\int_a^b f(x)\, dx$ by the fundamental theorem of calculus. Therefore, the "$+\ C$" may be left out of intermediate integrations.

EXAMPLE 5.3.8

Evaluate $\displaystyle\int_0^1 8x(x^2 + 1)^3\, dx$.

Solution

The integrand is a product in which one factor, $8x$, is a constant multiple of the derivative of the expression $x^2 + 1$ that appears in the other factor. This suggests letting $u = x^2 + 1$. Then $du = 2x\, dx$, and so

$$\int 8x(x^2 + 1)^3\, dx = \int 4u^3\, du = u^4.$$

The limits of integration, 0 and 1, refer to the variable x and not to u. You can, therefore, proceed in one of two ways. Either rewrite the antiderivative in terms of x or find the values of u that correspond to $x = 0$ and $x = 1$.

For the first alternative,

$$\int 8x(x^2 + 1)^3\, dx = u^4 = (x^2 + 1)^4,$$

and so

$$\int_0^1 8x(x^2 + 1)^3\, dx = (x^2 + 1)^4\Big|_0^1 = 16 - 1 = 15.$$

For the second alternative, use the fact that $u = x^2 + 1$ to conclude that $u = 1$ when $x = 0$ and $u = 2$ when $x = 1$. Hence,

$$\int_0^1 8x(x^2 + 1)^3\, dx = \int_1^2 4u^3\, du = u^4\Big|_1^2 = 16 - 1 = 15.$$

EXPLORE!

Graph the function $f(x) = \dfrac{\ln x}{x}$ from Example 5.3.9. Use the interval $x = 0.1$ to 3. Explain in terms of area why the integral of $f(x)$ over $\dfrac{1}{4} \le x \le 2$ is negative.

EXAMPLE 5.3.9

Evaluate $\displaystyle\int_{1/4}^2 \left(\frac{\ln x}{x}\right) dx$.

Solution

Let $u = \ln x$, so $du = \dfrac{1}{x}\, dx$. Then

$$\int \frac{\ln x}{x}\, dx = \int \ln x\left(\frac{1}{x}\, dx\right) = \int u\, du$$

$$= \frac{1}{2}u^2 = \frac{1}{2}(\ln x)^2$$

Thus,

$$\int_{1/4}^{2} \frac{\ln x}{x} \, dx = \left[\frac{1}{2} (\ln x)^2 \right] \Big|_{1/4}^{2} = \frac{1}{2} (\ln 2)^2 - \frac{1}{2} \left(\ln \frac{1}{4} \right)^2$$

$$= -\frac{3}{2} (\ln 2)^2 \approx -0.721$$

Alternatively, use the substitution $u = \ln x$ to transform the limits of integration:

$$\text{When } x = \frac{1}{4}, \text{ then } \qquad u = \ln \frac{1}{4}.$$

$$\text{When } x = 2, \text{ then } \qquad u = \ln 2.$$

Substituting gives

$$\int_{1/4}^{2} \frac{\ln x}{x} \, dx = \int_{\ln 1/4}^{\ln 2} u \, du = \frac{1}{2} u^2 \Big|_{\ln 1/4}^{\ln 2}$$

$$= \frac{1}{2} (\ln 2)^2 - \frac{1}{2} \left(\ln \frac{1}{4} \right)^2 \approx -0.721$$

Net Change

In certain applications, we are given the rate of change $Q'(t)$ of a quantity $Q(t)$ and required to compute the **net change** $Q(b) - Q(a)$ in $Q(t)$ as t varies from $t = a$ to $t = b$. We did this in Section 5.1 by solving initial value problems (recall Examples 5.1.5 through 5.1.8). The total change is the change per unit time, multiplied by a small change in time Δt. In the limit of summing all the parts, we get the following.

> **Net Change** ■ If $Q'(t)$ is continuous on the interval $a \leq t \leq b$, then the **net change** in $Q(t)$ as t varies from $t = a$ to $t = b$ is given by
>
> $$Q(b) - Q(a) = \int_{a}^{b} Q'(t) \, dt.$$

Here are two examples involving net change.

EXAMPLE 5.3.10

At a certain factory, the marginal cost is $3(q - 4)^2$ dollars per unit when the level of production is q units. By how much will the total manufacturing cost increase if the level of production is raised from 6 units to 10 units?

Solution

Let $C(q)$ denote the total cost of producing q units. Then the marginal cost is the derivative $\dfrac{dC}{dq} = 3(q - 4)^2$, and the increase in cost if production is raised from 6 units to 10 units is given by the definite integral

$$C(10) - C(6) = \int_6^{10} \frac{dC}{dq}\, dq$$

$$= \int_6^{10} 3(q - 4)^2\, dq = (q - 4)^3 \Big|_6^{10}$$

$$= (10 - 4)^3 - (6 - 4)^3$$

$$= 208$$

or \$208.

EXAMPLE 5.3.11

A protein with mass m (in grams) disintegrates into amino acids at a rate given by

$$\frac{dm}{dt} = \frac{-30}{(t + 3)^2} \qquad \text{grams per hour}$$

What is the net change in mass of the protein during the first 2 hours?

Solution

The net change is given by the definite integral

$$m(2) - m(0) = \int_0^2 \frac{dm}{dt}\, dt = \int_0^2 \frac{-30}{(t + 3)^2}\, dt.$$

Substitute $u = t + 3$ and $du = dt$, and change the limits of integration accordingly ($t = 0$ becomes $u = 3$ and $t = 2$ becomes $u = 5$) to find

$$m(2) - m(0) = \int_0^2 \frac{-30}{(t + 3)^2}\, dt = \int_3^5 -30u^{-2}\, du$$

$$= -30 \left(\frac{u^{-1}}{-1} \right) \Big|_3^5 = 30 \left[\frac{1}{5} - \frac{1}{3} \right]$$

$$= -4$$

Thus, the mass of the protein has a net decrease of 4 g over the first 2 hours.

Area Justification of the Fundamental Theorem of Calculus

We close this section with a justification of the fundamental theorem of calculus for the case where $f(x) \geq 0$. In this case, the definite integral $\int_a^b f(x)\, dx$ represents the area under the curve $y = f(x)$ over the interval $[a, b]$. For fixed x between a and b, let $A(x)$ denote the area under $y = f(x)$ over the interval $[a, x]$. Then the difference quotient of $A(x)$ is

$$\frac{A(x + h) - A(x)}{h}$$

and the expression $A(x + h) - A(x)$ in the numerator is just the area under the curve $y = f(x)$ between x and $x + h$. If h is small, this area is approximately the same as the area of the rectangle with height $f(x)$ and width h as indicated in Figure 5.8. That is,

$$A(x + h) - A(x) \approx f(x)h$$

or, equivalently,

$$\frac{A(x + h) - A(x)}{h} \approx f(x).$$

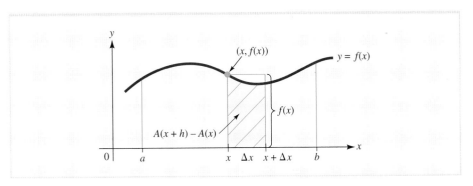

FIGURE 5.8 The area $A(x + h) - A(x)$.

As h approaches 0, the error in the approximation approaches 0, and it follows that

$$\lim_{h \to 0} \frac{A(x + h) - A(x)}{h} = f(x).$$

But by the definition of the derivative,

$$\lim_{h \to 0} \frac{A(x + h) - A(x)}{h} = A'(x),$$

so that

$$A'(x) = f(x).$$

In other words, $A(x)$ is an antiderivative of $f(x)$.

Suppose $F(x)$ is any other antiderivative of $f(x)$. Then, according to the fundamental property of antiderivatives (Section 5.1), we have

$$A(x) = F(x) + C$$

for some constant C and all x in the interval $a \leq x \leq b$. Since $A(x)$ represents the area under $y = f(x)$ between a and x, it is certainly true that $A(a)$, the area between a and a, is 0, so that

$$A(a) = 0 = F(a) + C,$$

and $C = -F(a)$. The area under $y = f(x)$ between $x = a$ and $x = b$ is $A(b)$, which satisfies

$$A(b) = F(b) + C = F(b) - F(a).$$

Finally, since the area under $y = f(x)$ above $a \leq x \leq b$ is also given by the definite integral $\int_a^b f(x)\, dx$, it follows that

$$\int_a^b f(x)\, dx = A(b) = F(b) - F(a),$$

as claimed in the fundamental theorem of calculus.

EXERCISES ▪ 5.3

In Exercises 1 through 30, evaluate the given definite integral using the fundamental theorem of calculus.

1. $\int_{-1}^{2} 5\,dx$

2. $\int_{-2}^{1} \pi\,dx$

3. $\int_{0}^{5} (3x + 2)\,dx$

4. $\int_{1}^{4} (5 - 2t)\,dt$

5. $\int_{-1}^{1} 3t^4\,dt$

6. $\int_{1}^{4} 2\sqrt{u}\,du$

7. $\int_{-1}^{1} (2u^{1/3} - u^{2/3})\,du$

8. $\int_{4}^{9} x^{-3/2}\,dx$

9. $\int_{0}^{1} e^{-x}(4 - e^x)\,dx$

10. $\int_{-1}^{1} \left(\frac{1}{e^x} - \frac{1}{e^{-x}}\right)dx$

11. $\int_{0}^{1} (x^4 + 3x^3 + 1)\,dx$

12. $\int_{-1}^{0} (-3x^5 - 3x^2 + 2x + 5)\,dx$

13. $\int_{2}^{5} (2 + 2t + 3t^2)\,dt$

14. $\int_{1}^{9} \left(\sqrt{t} - \frac{4}{\sqrt{t}}\right)dt$

15. $\int_{1}^{3} \left(1 + \frac{1}{x} + \frac{1}{x^2}\right)dx$

16. $\int_{0}^{\ln 2} (e^t - e^{-t})\,dt$

17. $\int_{-3}^{-1} \frac{t + 1}{t^3}\,dt$

18. $\int_{1}^{6} x^2(x - 1)\,dx$

19. $\int_{1}^{2} (2x - 4)^4\,dx$

20. $\int_{-3}^{0} (2x + 6)^4\,dx$

21. $\int_{0}^{4} \frac{1}{\sqrt{6t + 1}}\,dt$

22. $\int_{1}^{2} \frac{x^2}{(x^3 + 1)^2}\,dx$

23. $\int_{0}^{1} (x^3 + x)\sqrt{x^4 + 2x^2 + 1}\,dx$

24. $\int_{0}^{1} \frac{6t}{t^2 + 1}\,dt$

25. $\int_{2}^{e+1} \frac{x}{x - 1}\,dx$

26. $\int_{1}^{2} (t + 1)(t - 2)^6\,dt$

27. $\int_{1}^{e^2} \frac{(\ln x)^2}{x}\,dx$

28. $\int_{e}^{e^2} \frac{1}{x \ln x}\,dx$

29. $\int_{1/3}^{1/2} \frac{e^{1/x}}{x^2}\,dx$

30. $\int_{1}^{4} \frac{(\sqrt{x} - 1)^{3/2}}{\sqrt{x}}\,dx$

In Exercises 31 through 38, f(x) and g(x) are functions that are continuous on the interval −3 ≤ x ≤ 2 and satisfy

$$\int_{-3}^{2} f(x)\,dx = 5 \quad \int_{-3}^{2} g(x)\,dx = -2 \quad \int_{-3}^{1} f(x)\,dx = 0 \quad \int_{-3}^{1} g(x)\,dx = 4$$

In each case, use this information along with the rules for definite integrals to evaluate the indicated integral.

31. $\int_{-3}^{2} [-2f(x) + 5g(x)]\, dx$

32. $\int_{-3}^{1} [4f(x) - 3g(x)]\, dx$

33. $\int_{4}^{4} g(x)\, dx$

34. $\int_{2}^{-3} f(x)\, dx$

35. $\int_{1}^{2} f(x)\, dx$

36. $\int_{1}^{2} g(x)\, dx$

37. $\int_{1}^{2} [3f(x) + 2g(x)]\, dx$

38. $\int_{-3}^{1} [2f(x) + 3g(x)]\, dx$

In Exercises 39 through 46, find the area of the region R that lies under the given curve $y = f(x)$ over the indicated interval $a \le x \le b$.

39. Under $y = x^4$, over $-1 \le x \le 2$

40. Under $y = \sqrt{x}(x + 1)$, over $0 \le x \le 4$

41. Under $y = (3x + 4)^{1/2}$, over $0 \le x \le 4$

42. Under $y = \dfrac{3}{\sqrt{9 - 2x}}$ over $-8 \le x \le 0$

43. Under $y = e^{2x}$, over $0 \le x \le \ln 3$

44. Under $y = xe^{-x^2}$, over $0 \le x \le 3$

45. Under $y = \dfrac{3}{5 - 2x}$, over $-2 \le x \le 1$

46. Under $y = \dfrac{3}{x}$, over $1 \le x \le e^2$

47. LAND VALUES It is estimated that t years from now the value of a certain parcel of land will be increasing at a rate of $V'(t)$ dollars per year. Find an expression for the amount by which the value of the land will increase during the next 5 years.

48. ADMISSION TO EVENTS The promoters of a county fair estimate that t hours after the gates open at 9:00 A.M., visitors will be entering the fair at a rate of $N'(t)$ people per hour. Find an expression for the number of people who will enter the fair between 11:00 A.M. and 1:00 P.M.

49. STORAGE COST A retailer receives a shipment of 12 000 kg of soybeans that will be used at a constant rate of 300 kg per week. If the cost of storing the soybeans is 0.2 cents per kilogram per week, how much will the retailer have to pay in storage costs over the next 40 weeks?

50. OIL PRODUCTION A certain oil well that yields 400 barrels of crude oil a month will run dry in 2 years. The price of crude oil is currently $95 per barrel and is expected to rise at a constant rate of 30 cents per barrel per month. If the oil is sold as soon as it is extracted from the ground, what will the total future revenue from the well be?

51. AIR POLLUTION An environmental study of a certain community suggests that t years from now the level $L(t)$ of carbon monoxide in the air will be changing at a rate of $L'(t) = 0.1t + 0.1$ parts per million (ppm) per year. By how much will the pollution level change during the next 3 years?

52. WATER POLLUTION It is estimated that t years from now the population of a certain lakeside community will be changing at a rate of $0.6t^2 + 0.2t + 0.5$ thousand people per year. Environmentalists have found that the level of pollution in the lake increases at a rate of approximately 5 units per 1000 people. By how much will the pollution in the lake increase during the next 2 years?

53. NET GROWTH OF POPULATION A study indicates that t months from now the population of a certain town will be growing at a rate of $P'(t) = 5 + 3t^{2/3}$ people per month. By how much will the population of the town increase over the next 8 months?

54. MARGINAL COST The marginal cost of producing a certain commodity is $C'(q) = 6q + 1$ dollars per unit when q units are being produced.
   a. What is the total cost of producing the first 10 units?
   b. What is the cost of producing the *next* 10 units?

55. FARMING It is estimated that t days from now a farmer's crop will be increasing at a rate of $0.3t^2 + 0.6t + 1$ kilograms per day. By how much will the value of the crop increase during the next 5 days if the market price remains fixed at $3/kg?

56. SALES REVENUE   It is estimated that the demand for a manufacturer's product is increasing exponentially at a rate of 2% per year. If the current demand is 5000 units per year and if the price remains fixed at $400 per unit, how much revenue will the manufacturer receive from the sale of the product over the next 2 years?

57. PRODUCTION   Bejax Corporation has set up a production line to manufacture a new type of cell phone. The rate of production of the cell phone is

$$\frac{dP}{dt} = 1500\left(2 - \frac{t}{2t + 5}\right) \text{ units per month.}$$

How many cell phones are produced during the 3rd month?

58. PRODUCTION   The output of a factory is changing at a rate of

$$Q'(t) = 2t^3 - 3t^2 + 10t + 3 \text{ units per hour,}$$

where t is the number of hours after the morning shift begins at 8 A.M. How many units are produced between 10 A.M. and noon?

59. INVESTMENT   An investment portfolio changes value at a rate of

$$V'(t) = 12e^{-0.05t}(e^{0.3t} - 3),$$

where V is in thousands of dollars and t is the number of years after 2006. By how much does the value of the portfolio change between the years
a.   2006 and 2010?
b.   2010 and 2012?

60. ADVERTISING   An advertising agency begins a campaign to promote a new product and determines that t days later, the number of people $N(t)$ who have heard about the product is changing at a rate given by

$$N'(t) = 5t^2 - \frac{0.04t}{t^2 + 3} \text{ people per day.}$$

How many people learn about the product during the first week? During the second week?

61. CONCENTRATION OF DRUG   The concentration of a drug in a patient's bloodstream t hours after an injection is decreasing at a rate of

$$C'(t) = \frac{-0.33t}{\sqrt{0.02t^2 + 10}}$$

milligrams per cubic centimetre per hour.

By how much does the concentration change over the first 4 hours after the injection?

62. ENDANGERED SPECIES   A study conducted by an environmental group in 2010 determined that t years later, the population of a certain endangered bird species will be decreasing at a rate of $P'(t) = -0.75t\sqrt{10 - 0.2t}$ individuals per year. By how much is the population expected to change during the decade 2010–2020?

63. DEPRECIATION   The resale value of a certain industrial machine decreases over a 10-year period at a rate that changes with time. When the machine is x years old, the rate at which its value is changing is $220(x - 10)$ dollars per year. By how much does the machine depreciate during the second year?

64. WATER CONSUMPTION   The water department estimates that water is being consumed by a certain community at a rate of $C'(t) = 10 + 0.3e^{0.03t}$ billion litres per year, where $C(t)$ is the water consumption t years after the year 2010. How much water will be consumed by the community during the decade 2010–2020?

65. CHANGE IN BIOMASS   A protein with mass m (in grams) disintegrates into amino acids at a rate given by

$$\frac{dm}{dt} = \frac{-2}{t + 1}$$

grams per hour. How much more protein is there after 2 hours than after 5 hours?

66. CHANGE IN BIOMASS   Answer the question in Exercise 65 if the rate of disintegration is given by

$$\frac{dm}{dt} = -(0.1t + e^{0.1t}).$$

67. RATE OF LEARNING   In a learning experiment, subjects are given a series of facts to memorize, and it is determined that t minutes after the experiment begins, the average subject is learning at a rate of

$$L'(t) = \frac{4}{\sqrt{t + 1}}$$

facts per minute, where $L(t)$ is the total number of facts memorized by time t. About how many facts does the typical subject learn during the second 5 minutes (between $t = 5$ and $t = 10$)?

68. DISTANCE AND VELOCITY A driver, travelling at a constant speed of 45 km/h, decides to speed up in such a way that her velocity t hours later is $v(t) = 45 + 12t$ kilometres per hour. How far does she travel in the first 2 hours?

69. PROJECTILE MOTION A ball is thrown upward from the top of a building, and t seconds later has velocity $v(t) = -9.8t + 30$ metres per second. What is the difference in the ball's position after 3 seconds?

70. A* Verify the sum rule for definite integrals; that is, if $f(x)$ and $g(x)$ are continuous on the interval $a \le x \le b$, then

$$\int_a^b [f(x) + g(x)]\, dx = \int_a^b f(x)\, dx + \int_a^b g(x)\, dx.$$

71. A* You have seen that the definite integral can be used to compute the area under a curve, but sometimes knowledge of a geometric area can be used to evaluate an integral.

   a. Compute $\int_0^1 \sqrt{1 - x^2}\, dx$. [*Hint:* Note that the integral is part of the area under the circle $x^2 + y^2 = 1$.]

   b. Compute $\int_1^2 \sqrt{2x - x^2}\, dx$. [*Hint:* Describe the graph of $y = \sqrt{2x - x^2}$ and look for a geometric solution as in part (a).]

72. A* Given the function $f(x) = 2\sqrt{x} + \dfrac{1}{x + 1}$, approximate the value of the integral $\int_0^2 f(x)\, dx$ by completing these steps:

   a. Graph the function and find the numbers $x_1, x_2, x_3, x_4,$ and x_5 that subdivide the interval $0 \le x \le 2$ into four equal subintervals. Use these numbers to form four rectangles that approximate the area under the curve $y = f(x)$ over $0 \le x \le 2$.

   b. Estimate the value of the given integral by computing the sum of the areas of the four approximating rectangles in part (a).

   c. Repeat steps (a) and (b) with eight subintervals instead of four.

SECTION 5.4

L04

Calculate the area between the curves of two functions. Calculate the average value of a function.

Applying Definite Integration: Area between Curves and Average Value

We have seen that area can be expressed as a special kind of limit of a sum called a definite integral and then computed by applying the fundamental theorem of calculus. This procedure, called **definite integration,** was introduced through area because area is easy to visualize, but the integration process plays an important role in many applications other than area.

In this section, we extend the ideas introduced in Section 5.3 to find the area between two curves and the average value of a function. As part of our study of area between curves, we will examine an important socioeconomic device called a Lorentz curve, which is used to measure relative wealth within a society.

Applying the Definite Integral

Intuitively, definite integration can be thought of as a process that accumulates an infinite number of small pieces of a quantity to obtain the total quantity. Here is a step-by-step description of how to use this process in applications.

A Procedure for Using Definite Integration in Applications

To use definite integration to accumulate a quantity Q over an interval $a \leq x \leq b$, proceed as follows:

Step 1. Divide the interval $a \leq x \leq b$ into n equal subintervals, each of length $\Delta x = \dfrac{b - a}{n}$. Choose a number x_j from the jth subinterval, for $j = 1$, $2, \ldots, n$.

Step 2. Approximate small parts of the quantity Q by products of the form $f(x_j)\,\Delta x$, where $f(x)$ is an appropriate function that is continuous on $a \leq x \leq b$.

For example:

Area: $Q =$ (heights f)(small pieces of width dx)

Distance: $Q =$ (velocities f)(small increments of time dt)

Wealth: $Q =$ (returns per unit time f)(small increments of time dt)

Step 3. Add the individual approximating products to estimate the total quantity Q by the Riemann sum

$$[f(x_1) + f(x_2) + \cdots + f(x_n)]\,\Delta x.$$

Step 4. Make the approximation in step 3 exact by taking the limit of the Riemann sum as $n \to +\infty$ to express Q as a definite integral, that is,

$$Q = \lim_{n \to +\infty} [f(x_1) + f(x_2) + \cdots + f(x_n)]\,\Delta x = \int_a^b f(x)\,dx.$$

Then use the fundamental theorem of calculus to compute $\displaystyle\int_a^b f(x)\,dx$ and thus to obtain the required quantity Q.

Just-In-Time

Summation notation is reviewed in Appendix A4, which includes examples. Note that there is nothing special about using j for the index in the notation. The most commonly used indices are i, j, and k.

NOTATION: We can use *summation notation* to represent the Riemann sums that occur when quantities are modelled using definite integration. Specifically, to describe the sum

$$a_1 + a_2 + \cdots + a_n,$$

it suffices to specify the general term a_j in the sum and to indicate that n terms of this form are to be added, starting with the term where $j = 1$ and ending with the term where $j = n$. For this purpose, it is customary to use the uppercase Greek letter sigma (Σ) and to write the sum as $\displaystyle\sum_{j=1}^{n} a_j$, that is,

$$\sum_{j=1}^{n} a_j = a_1 + a_2 + \cdots + a_n.$$

In particular, the Riemann sum

$$[f(x_1) + f(x_2) + \cdots + f(x_n)]\,\Delta x$$

can be written in the compact form

$$\sum_{j=1}^{n} f(x_j)\,\Delta x.$$

Thus, the limit statement

$$\lim_{n \to +\infty} [f(x_1) + f(x_2) + \cdots + f(x_n)] \Delta x = \int_a^b f(x) \, dx$$

used to define the definite integral can be expressed as

$$\lim_{n \to +\infty} \sum_{j=1}^n f(x_j) \Delta x = \int_a^b f(x) \, dx.$$

Area between Two Curves

In certain practical applications, you may find it useful to represent a quantity of interest in terms of the area between two curves. First, suppose that f and g are continuous and nonnegative (that is, $f(x) \geq 0$ and $g(x) \geq 0$) and satisfy $f(x) \geq g(x)$ on the interval $a \leq x \leq b$, as shown in Figure 5.9a.

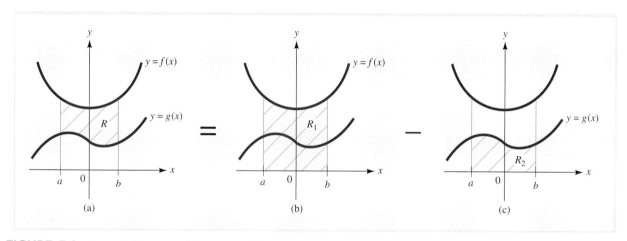

FIGURE 5.9   Area of R = area of R_1 − area of R_2.

Then, to find the area of the region R between the curves $y = f(x)$ and $y = g(x)$ over the interval $a \leq x \leq b$, simply subtract the area under the lower curve $y = g(x)$ (Figure 5.9c) from the area under the upper curve $y = f(x)$ (Figure 5.9b), so that

$$\text{Area of } R = [\text{area under } y = f(x)] - [\text{area under } y = g(x)]$$

$$= \int_a^b f(x) \, dx - \int_a^b g(x) \, dx$$

$$= \int_a^b [f(x) - g(x)] \, dx$$

This formula still applies whenever $f(x) \geq g(x)$ on the interval $a \leq x \leq b$, even when the curves $y = f(x)$ and $y = g(x)$ are not always both above the x axis. We will show that this is true by using the procedure for applying definite integration described on page 403.

Step 1. Subdivide the interval $a \leq x \leq b$ into n equal subintervals, each of width $\Delta x = \dfrac{b - a}{n}$. For $j = 1, 2, \ldots, n$, let x_j be the left endpoint of the jth subinterval.

Step 2. Construct approximating rectangles of width Δx and height $f(x_j) - g(x_j)$. This is possible since $f(x) \geq g(x)$ on $a \leq x \leq b$, which guarantees that the height is nonnegative, that is, $f(x_j) - g(x_j) \geq 0$. For $j = 1, 2, \ldots, n$, the area $[f(x_j) - g(x_j)]\Delta x$ of the jth rectangle you have just constructed is approximately the same as the area between the two curves over the jth subinterval, as shown in Figure 5.10a.

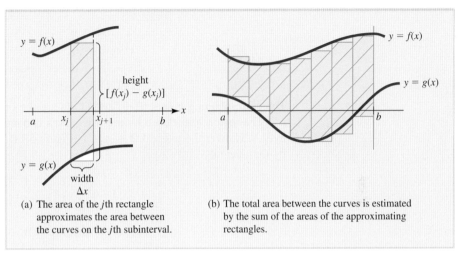

(a) The area of the jth rectangle approximates the area between the curves on the jth subinterval.

(b) The total area between the curves is estimated by the sum of the areas of the approximating rectangles.

FIGURE 5.10   Computing area between curves by definite integration.

Step 3. Add the individual approximating areas $[f(x_j) - g(x_j)]\Delta x$ to estimate the total area A between the two curves over the interval $a \leq x \leq b$ by the Riemann sum

$$A \approx [f(x_1) - g(x_1)]\Delta x + [f(x_2) - g(x_2)]\Delta x + \cdots + [f(x_n) - g(x_n)]\Delta x$$

$$= \sum_{j=1}^{n} [f(x_j) - g(x_j)]\Delta x$$

(see Figure 5.10b).

Step 4. Make the approximation exact by taking the limit of the Riemann sum in step 3 as $n \to +\infty$ to express the total area A between the curves as a definite integral, that is,

$$A = \lim_{n \to +\infty} \sum_{j=1}^{n} [f(x_j) - g(x_j)]\Delta x = \int_{a}^{b} [f(x) - g(x)]\,dx.$$

To summarize:

The Area between Two Curves ■ If $f(x)$ and $g(x)$ are continuous with $f(x) \geq g(x)$ on the interval $a \leq x \leq b$, then the area A between the curves $y = f(x)$ and $y = g(x)$ over the interval is given by

$$A = \int_{a}^{b} [f(x) - g(x)]\,dx.$$

Just-In-Time

Note that $x^2 \geq x^3$ for $0 \leq x \leq 1$.

For example, $\left(\dfrac{1}{3}\right)^2 > \left(\dfrac{1}{3}\right)^3$.

It is best to look at a graph when determining which function is to be subtracted.

EXAMPLE 5.4.1

Find the area of the region R enclosed by the curves $y = x^3$ and $y = x^2$.

Solution

To find the points where the curves intersect, solve the equations simultaneously as follows:

$$
\begin{aligned}
x^3 &= x^2 \\
x^3 - x^2 &= 0 \qquad &\text{subtract } x^2 \text{ from both sides} \\
x^2(x - 1) &= 0 \qquad &\text{factor out } x^2 \\
x &= 0, 1 \qquad &uv = 0 \text{ if and only if } u = 0 \text{ or } v = 0
\end{aligned}
$$

The corresponding points $(0, 0)$ and $(1, 1)$ are the only points of intersection.

The region R enclosed by the two curves is bounded above by $y = x^2$ and below by $y = x^3$, over the interval $0 \leq x \leq 1$ (Figure 5.11). The area of this region is given by the integral

$$
\begin{aligned}
A &= \int_0^1 (x^2 - x^3)\, dx = \left. \frac{1}{3}x^3 - \frac{1}{4}x^4 \right|_0^1 \\
&= \left[\frac{1}{3}(1)^3 - \frac{1}{4}(1)^4 \right] - \left[\frac{1}{3}(0)^3 - \frac{1}{4}(0)^4 \right] = \frac{1}{12}
\end{aligned}
$$

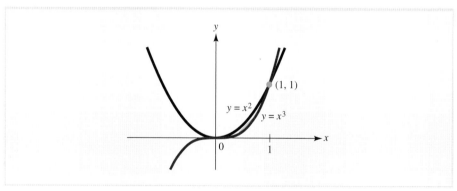

FIGURE 5.11   The region enclosed by the curves $y = x^2$ and $y = x^3$.

In certain applications, you may need to find the area A between the two curves $y = f(x)$ and $y = g(x)$ over an interval $a \leq x \leq b$, where $f(x) \geq g(x)$ for $a \leq x \leq c$ but $g(x) \geq f(x)$ for $c \leq x \leq b$. In this case, we have

$$
A = \underbrace{\int_a^c [f(x) - g(x)]\, dx}_{f(x)\,\geq\, g(x) \text{ on } a\,\leq\, x\,\leq\, c} + \underbrace{\int_c^b [g(x) - f(x)]\, dx}_{g(x)\,\geq\, f(x) \text{ on } c\,\leq\, x\,\leq\, b}
$$

Consider Example 5.4.2.

EXAMPLE 5.4.2

Find the area of the region enclosed by the line $y = 4x$ and the curve $y = x^3 + 3x^2$.

Solution

To find where the line and curve intersect, solve the equations simultaneously as follows:

$$x^3 + 3x^2 = 4x$$
$$x^3 + 3x^2 - 4x = 0 \qquad \text{subtract } 4x \text{ from both sides}$$
$$x(x^2 + 3x - 4) = 0 \qquad \text{factor out } x$$
$$x(x - 1)(x + 4) = 0 \qquad \text{factor } x^2 + 3x - 4$$
$$x = 0, 1, -4 \quad uv = 0 \text{ if and only if } u = 0 \text{ or } v = 0$$

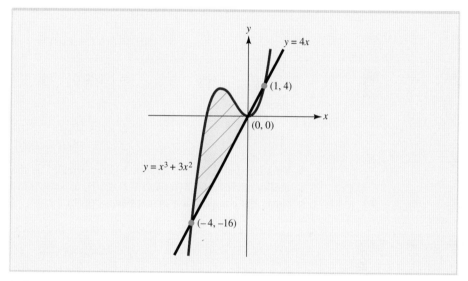

FIGURE 5.12   The region enclosed by the line $y = 4x$ and the curve $y = x^3 + 3x^2$.

The corresponding points of intersection are $(0, 0)$, $(1, 4)$, and $(-4, -16)$. The curve and the line are sketched in Figure 5.12.

Over the interval $-4 \le x \le 0$, the curve is above the line, so $x^3 + 3x^2 \ge 4x$, and the region enclosed by the curve and line has area

$$A_1 = \int_{-4}^{0} [(x^3 + 3x^2) - 4x]\, dx = \frac{1}{4}x^4 + x^3 - 2x^2 \Big|_{-4}^{0}$$

$$= \left[\frac{1}{4}(0)^4 + (0)^3 - 2(0)^2 \right] - \left[\frac{1}{4}(-4)^4 + (-4)^3 - 2(-4)^2 \right] = 32$$

Over the interval $0 \le x \le 1$, the line is above the curve and the enclosed region has area

$$A_2 = \int_{0}^{1} [4x - (x^3 + 3x^2)]\, dx = 2x^2 - \frac{1}{4}x^4 - x^3 \Big|_{0}^{1}$$

$$= \left[2(1)^2 - \frac{1}{4}(1)^4 - (1)^3 \right] - \left[2(0)^2 - \frac{1}{4}(0)^4 - (0)^3 \right]$$

$$= \frac{3}{4}$$

Therefore, the total area enclosed by the line and the curve is given by the sum

$$A = A_1 + A_2 = 32 + \frac{3}{4} = 32.75.$$

Net Excess Profit   The area between curves can sometimes be used as a way of measuring the amount of a quantity that has been accumulated during a particular procedure. For instance, suppose that t years from now, two investment plans will be generating profit $P_1(t)$ and $P_2(t)$, respectively, and that their respective rates of profitability, $P_1'(t)$ and $P_2'(t)$, are expected to satisfy $P_2'(t) \geq P_1'(t)$ for the next N years, that is, over the time interval $0 \leq t \leq N$. Then $E(t) = P_2(t) - P_1(t)$ represents the **excess profit** of plan 2 over plan 1 at time t, and the **net excess profit** $NE = E(N) - E(0)$ over the time interval $0 \leq t \leq N$ is given by the definite integral

$$NE = E(N) - E(0) = \int_0^N E'(t)\,dt$$

$$= \int_0^N [P_2'(t) - P_1'(t)]\,dt$$

since

$$E'(t) = [P_2(t) - P_1(t)]'$$
$$= P_2'(t) - P_1'(t)$$

This integral can be interpreted geometrically as the area between the rate of profitability curves $y = P_1'(t)$ and $y = P_2'(t)$, as shown in Figure 5.13. Example 5.4.3 illustrates the computation of net excess profit.

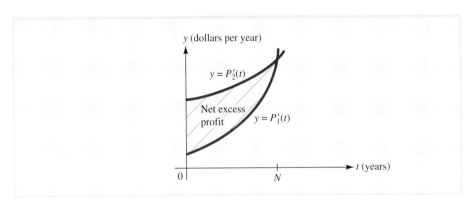

FIGURE 5.13   Net excess profit as the area between rate of profitability curves.

EXAMPLE 5.4.3

Suppose that t years from now, one investment will be generating profit at a rate of $P_1'(t) = 50 + t^2$ hundred dollars per year, while a second investment will be generating profit at a rate of $P_2'(t) = 200 + 5t$ hundred dollars per year.

a. For how many years does the rate of profitability of the second investment exceed that of the first?

b. Compute the net excess profit for the time period determined in part (a). Interpret the net excess profit as an area.

Solution

a. To find the time at which the rate of profitability of the second investment exceeds that of the first, solve the equation $P_2'(t) = P_1'(t)$

$$P_1'(t) = P_2'(t)$$

$$50 + t^2 = 200 + 5t$$

$$t^2 - 5t - 150 = 0 \qquad \text{subtract } 200 + 5t \text{ from both sides}$$

$$(t - 15)(t + 10) = 0 \qquad \text{factor}$$

$$t = 15, -10 \qquad \text{since } uv = 0 \text{ if and only if } u = 0 \text{ or } v = 0$$

$$t = 15 \text{ years} \qquad \text{reject the negative time } t = -10$$

b. The excess profit of plan 2 over plan 1 is $E(t) = P_2(t) - P_1(t)$, and the net excess profit NE over the time period $0 \leq t \leq 15$ determined in part (a) is given by the definite integral

$$\text{NE} = E(15) - E(0) = \int_0^{15} E'(t)\, dt \qquad \text{fundamental theorem of calculus}$$

$$= \int_0^{15} [P_2'(t) - P_1'(t)]\, dt \qquad \text{since } E(t) = P_2(t) - P_1(t)$$

$$= \int_0^{15} [(200 + 5t) - (50 + t^2)]\, dt$$

$$= \int_0^{15} [150 + 5t - t^2]\, dt \qquad \text{combine terms}$$

$$= 150t + 5\left(\frac{1}{2}t^2\right) - \frac{1}{3}t^3 \Big|_0^{15}$$

$$= \left[150(15) + \frac{5}{2}(15)^2 - \frac{1}{3}(15)^3\right] - \left[150(0) + \frac{5}{2}(0)^2 - \frac{1}{3}(0)^3\right]$$

$$= 1687.50$$

hundred dollars. Thus, the net excess profit is $168 750.

The graphs of the rate of profitability functions $P_1'(t)$ and $P_2'(t)$ are shown in Figure 5.14. The net excess profit

$$\text{NE} = \int_0^{15} [P_2'(t) - P_1'(t)]\, dt$$

can be interpreted as the area of the (shaded) region between the rate of profitability curves over the interval $0 \leq t \leq 15$.

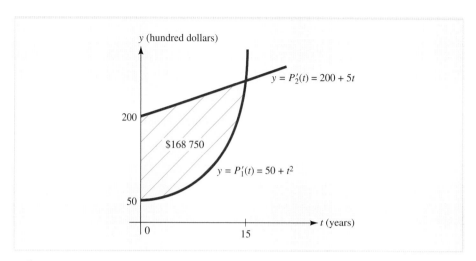

FIGURE 5.14   Net excess profit for one investment plan over another.

Lorentz Curves   Area also plays an important role in the study of **Lorentz curves,** a device used by both economists and sociologists to measure the percentage of a society's wealth that is possessed by a given percentage of its people. To be more specific, the Lorentz curve for a particular society's economy is the graph of the function $L(x)$, which denotes the fraction of total annual national income earned by the lowest-paid $100x$ percent of the wage-earners in the society, for $0 \leq x \leq 1$. For instance, if the lowest-paid 30% of all wage-earners receive 23% of the society's total income, then $L(0.3) = 0.23$.

Note that $L(x)$ is an increasing function on the interval $0 \leq x \leq 1$ and has these properties:

1. $0 \leq L(x) \leq 1$ because $L(x)$ is a percentage
2. $L(0) = 0$ because no wages are earned when no wage-earners are employed
3. $L(1) = 1$ because 100% of wages are earned by 100% of the wage-earners
4. $L(x) \leq x$ because the lowest-paid $100x$ percent of wage-earners cannot receive more than $100x$ percent of total income

A typical Lorentz curve is shown in Figure 5.15a.

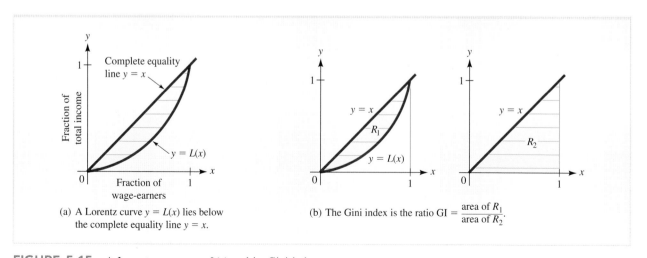

(a) A Lorentz curve $y = L(x)$ lies below the complete equality line $y = x$.

(b) The Gini index is the ratio $\text{GI} = \dfrac{\text{area of } R_1}{\text{area of } R_2}$.

FIGURE 5.15   A Lorentz curve $y = L(x)$ and its Gini index.

The line $y = x$ represents the ideal case corresponding to complete equality in the distribution of income (wage-earners with the lowest $100x$ percent of income receive $100x$ percent of the society's wealth). The closer a particular Lorentz curve is to this line, the more equitable is the distribution of wealth in the corresponding society. We represent the total deviation of the actual distribution of wealth in the society from complete equality by the area of the region R_1 between the Lorentz curve $y = L(x)$ and the line $y = x$. The ratio of this area to the area of the region R_2 under the complete equality line $y = x$ over $0 \leq x \leq 1$ is used as a measure of the inequality in the distribution of wealth in the society. This ratio, called the **Gini index,** denoted GI (also called the **index of income inequality**), may be computed by the formula

$$\text{GI} = \frac{\text{area of } R_1}{\text{area of } R_2} = \frac{\text{area between } y = L(x) \text{ and } y = x}{\text{area under } y = x \text{ over } 0 \leq x \leq 1}$$

$$= \frac{\displaystyle\int_0^1 [x - L(x)]\, dx}{\displaystyle\int_0^1 x\, dx} = \frac{\displaystyle\int_0^1 [x - L(x)]\, dx}{1/2}$$

$$= 2\int_0^1 [x - L(x)]\, dx$$

(see Figure 5.15b). To summarize:

Gini Index   ■   If $y = L(x)$ is the equation of a Lorentz curve, then the inequality in the corresponding distribution of wealth is measured by the *Gini index,* which is given by the formula

$$\text{Gini index} = 2\int_0^1 [x - L(x)]\, dx.$$

The Gini index always lies between 0 and 1. An index of 0 corresponds to total equity in the distribution of income, while an index of 1 corresponds to total inequity (all income belongs to 0% of the population). The smaller the index, the more equitable the distribution of income, and the larger the index, the more the wealth is concentrated in only a few hands. Example 5.4.4 illustrates how Lorentz curves and the Gini index can be used to compare the relative equity of income distribution for two professions.

EXAMPLE 5.4.4

A government agency determines that the Lorentz curves for the distribution of income for dentists and contractors in a certain province are given by the functions

$$L_1(x) = x^{1.7} \quad \text{and} \quad L_2(x) = 0.8x^2 + 0.2x,$$

respectively. For which profession is the distribution of income more fairly distributed?

Solution

The respective Gini indices are

$$G_1 = 2\int_0^1 (x - x^{1.7})\,dx = 2\left(\frac{x^2}{2} - \frac{x^{2.7}}{2.7}\right)\Bigg|_0^1 = 0.2593$$

and

$$G_2 = 2\int_0^1 [x - (0.8x^2 + 0.2x)]\,dx$$

$$= 2\left[\frac{x^2}{2} - 0.8\left(\frac{x^3}{3}\right) - 0.2\left(\frac{x^2}{2}\right)\right]\Bigg|_0^1$$

$$= 2\left[-0.8\left(\frac{x^3}{3}\right) + 0.8\left(\frac{x^2}{2}\right)\right]\Bigg|_0^1$$

$$= 0.2667$$

Since the Gini index for dentists is smaller, it follows that in this province, the incomes of dentists are more evenly distributed than those of contractors.

Using the Gini index, we can see how the distribution of income in Canada compares to that in other countries. Table 5.1 lists the Gini indices for selected industrial and developing nations. Note that with an index of 0.321, the distribution of income in Canada is more equitable than that of seven of the countries. Only Germany and Denmark are more equitable.

TABLE 5.1   Gini Indices For Selected Countries

Country	Gini Index
Germany	0.27
Denmark	0.29
Canada	0.321
France	0.327
United Kingdom	0.34
India	0.368
Japan	0.381
United States	0.45
Brazil	0.567
South Africa	0.65

SOURCE: "Distribution of Family Income—Gini Index," *The World Factbook 2009*. Washington, DC: Central Intelligence Agency, 2009. https://www.cia.gov/library/publications/the-world-factbook/fields/2172.html.

Average Value of a Function

As a second illustration of how definite integration can be used in applications, we will compute the **average value of a function,** which is of interest in a variety of situations. First, let us take a moment to clarify our thinking about what we mean by "average value." A teacher who wants to compute the average score on an examination simply adds all the individual scores and then divides by the number of students taking the exam, but how should one go about finding, say, the average pollution level in a city during the daytime hours? The difficulty is that since time is continuous, there are too many pollution levels to add up in the usual way, so how should we proceed?

Consider the general case in which we wish to find the average value of the function $f(x)$ over an interval $a \le x \le b$ on which f is continuous. We begin by subdividing the interval $a \le x \le b$ into n equal parts, each of length $\Delta x = \dfrac{b - a}{n}$. If x_j is a number taken from the jth subinterval for $j = 1, 2, \ldots, n$, then the average of the corresponding functional values $f(x_1), f(x_2), \ldots, f(x_n)$ is

$$
\begin{aligned}
V_n &= \frac{f(x_1) + f(x_2) + \cdots + f(x_n)}{n} \\[2mm]
&= \frac{b - a}{b - a}\left[\frac{f(x_1) + f(x_2) + \cdots + f(x_n)}{n}\right] && \text{multiply and divide by } (b - a) \\[2mm]
&= \frac{1}{b - a}[f(x_1) + f(x_2) + \cdots + f(x_n)]\left(\frac{b - a}{n}\right) && \text{factor out the expression } \frac{b - a}{n} \\[2mm]
&= \frac{1}{b - a}[f(x_1) + f(x_2) + \cdots + f(x_n)]\,\Delta x && \text{since } \Delta x = \frac{b - a}{n} \\[2mm]
&= \frac{1}{b - a}\sum_{j=1}^{n} f(x_j)\,\Delta x
\end{aligned}
$$

which we recognize as a Riemann sum.

If we refine the partition of the interval $a \le x \le b$ by taking more and more subdivision points, then V_n becomes more and more like what we may intuitively think of as the average value V of $f(x)$ over the entire interval $a \le x \le b$. Thus, it is reasonable to *define* the average value V as the limit of the Riemann sum V_n as $n \to +\infty$, that is, as the definite integral

$$
\begin{aligned}
V &= \lim_{n \to +\infty} V_n = \lim_{n \to +\infty} \frac{1}{b - a}\sum_{j=1}^{n} f(x_j)\,\Delta x \\[2mm]
&= \frac{1}{b - a}\lim_{n \to +\infty}\sum_{j=1}^{n} f(x_j)\,\Delta x \\[2mm]
&= \frac{1}{b - a}\int_a^b f(x)\,dx
\end{aligned}
$$

To summarize:

The Average Value of a Function ■ Let $f(x)$ be a function that is continuous on the interval $a \le x \le b$. Then the *average value* V of $f(x)$ over $a \le x \le b$ is given by the definite integral

$$
V = \frac{1}{b - a}\int_a^b f(x)\,dx.
$$

EXAMPLE 5.4.5

A manufacturer determines that t months after introducing a new product, the company's sales will be $S(t)$ thousand dollars, where

$$S(t) = \frac{750t}{\sqrt{4t^2 + 25}}.$$

What are the average monthly sales of the company over the first 6 months after the introduction of the new product?

Solution

The average monthly sales V (in thousands of dollars) over the time period $0 \leq t \leq 6$ is given by the integral

$$V = \frac{1}{6 - 0} \int_0^6 \frac{750t}{\sqrt{4t^2 + 25}}\, dt.$$

To evaluate this integral, make the substitution

$$u = 4t^2 + 25 \qquad \text{limits of integration:}$$
$$du = 4(2t\, dt) \qquad \text{if } t = 0, \text{ then } u = 4(0)^2 + 25 = 25$$
$$t\, dt = \frac{1}{8}\, du \qquad \text{if } t = 6, \text{ then } u = 4(6)^2 + 25 = 169$$

to obtain

$$V = \frac{1}{6} \int_0^6 \frac{750}{\sqrt{4t^2 + 25}}(t\, dt)$$
$$= \frac{1}{6} \int_{25}^{169} \frac{750}{\sqrt{u}}\left(\frac{1}{8}\, du\right) = \frac{750}{6(8)} \int_{25}^{169} u^{-1/2}\, du$$
$$= \frac{750}{6(8)} \left(\frac{u^{1/2}}{1/2}\right)\Big|_{25}^{169} = \frac{750(2)}{6(8)}(169^{1/2} - 25^{1/2})$$
$$= 250$$

Thus, for the 6-month period immediately after the introduction of the new product, the company's sales average $250\ 000$ per month.

EXAMPLE 5.4.6

A researcher models the temperature T (in degrees Celsius) during the time period from 6 A.M. to 6 P.M. in Calgary by the function

$$T(t) = 3 - \frac{1}{3}(t - 4)^2 \qquad \text{for } 0 \leq t \leq 12,$$

where t is the number of hours after 6 A.M.

a. What is the average temperature in the city during the workday, from 8 A.M. to 5 P.M.?

b. At what time (or times) during the workday is the temperature in the city the same as the average temperature found in part (a)?

Solution

a. Since 8 A.M. and 5 P.M. are, respectively, $t = 2$ hours and $t = 11$ hours after 6 A.M., compute the average of the temperature $T(t)$ for $2 \leq t \leq 11$, which is given by the definite integral

$$T_{ave} = \frac{1}{11 - 2} \int_2^{11} \left[3 - \frac{1}{3}(t - 4)^2 \right] dt$$

$$= \frac{1}{9} \left[3t - \frac{1}{3}\left(\frac{1}{3}(t - 4)^3\right) \right]\Big|_2^{11}$$

$$= \frac{1}{9} \left[3(11) - \frac{1}{9}(11 - 4)^3 \right] - \frac{1}{9} \left[3(2) - \frac{1}{9}(2 - 4)^3 \right]$$

$$= -\frac{4}{3} \approx -1.33$$

Thus, the average temperature during the workday is approximately $-1.33°C$.

b. Find a time $t = t_a$ with $2 \leq t_a \leq 11$ such that $T(t_a) = -\frac{4}{3}$. Solving this equation gives

$$3 - \frac{1}{3}(t_a - 4)^2 = -\frac{4}{3}$$

$$-\frac{1}{3}(t_a - 4)^2 = -\frac{4}{3} - 3 = -\frac{13}{3} \qquad \text{subtract 3 from both sides}$$

$$(t_a - 4)^2 = (-3)\left(-\frac{13}{3}\right) = 13 \qquad \text{multiply both sides by } -3$$

$$t_a - 4 = \pm\sqrt{13} \qquad \text{take square roots on both sides}$$

$$t_a = 4 \pm \sqrt{13}$$

$$\approx 0.39 \quad \text{or} \quad 7.61$$

Since $t = 0.39$ is outside the time interval $2 \leq t_a \leq 11$ (8 A.M. to 5 P.M.), it follows that the temperature in the city is the same as the average temperature only when $t = 7.61$, that is, at approximately 1:37 P.M.

Just-In-Time

Since there are 60 minutes in an hour, 0.61 hour is the same as 0.61(60) $\approx$ 37 minutes. Thus, 7.61 hours after 6 A.M. is 37 minutes past 1 P.M. or 1:37 P.M.

Two Interpretations of Average Value

The average value of a function has several useful interpretations. First, note that if $f(x)$ is continuous on the interval $a \leq x \leq b$ and $F(x)$ is any antiderivative of $f(x)$ over the same interval, then the average value V of $f(x)$ over the interval satisfies

$$V = \frac{1}{b - a} \int_a^b f(x)\, dx$$

$$= \frac{1}{b - a}[F(b) - F(a)] \qquad \text{fundamental theorem of calculus}$$

$$= \frac{F(b) - F(a)}{b - a}$$

We recognize this difference quotient as the average rate of change of $F(x)$ over $a \leq x \leq b$ (see Section 2.1). Thus, we have this interpretation:

Rate Interpretation of Average Value ■ The average value of a function $f(x)$ over an interval $a \leq x \leq b$ where $f(x)$ is continuous is the same as the average rate of change of any antiderivative $F(x)$ of $f(x)$ over the same interval.

For instance, since the total cost $C(x)$ of producing x units of a commodity is an antiderivative of marginal cost $C'(x)$, it follows that the *average rate of change of cost over a range of production* $a \leq x \leq b$ *equals the average value of the marginal cost over the same range.*

The average value of a function $f(x)$ on an interval $a \leq x \leq b$ where $f(x) \geq 0$ can also be interpreted geometrically by rewriting the integral formula for average value

$$V = \frac{1}{b-a} \int_a^b f(x)\, dx$$

in the form

$$(b-a)V = \int_a^b f(x)\, dx.$$

In the case where $f(x) \geq 0$ on the interval $a \leq x \leq b$, the integral on the right can be interpreted as the area under the curve $y = f(x)$ over $a \leq x \leq b$, and the product on the left as the area of a rectangle of height V and width $b - a$ equal to the length of the interval. It can also be said that the average height of $f(x)$ on $[a, b]$ is the area inside the curve of $f(x)$ divided by the width.

Geometric Interpretation of Average Value ■ The average value V of $f(x)$ over an interval $a \leq x \leq b$ where $f(x)$ is continuous and satisfies $f(x) \geq 0$ is equal to the height of a rectangle whose base is the interval and whose area is the same as the area under the curve $y = f(x)$ over $a \leq x \leq b$.

This geometric interpretation is illustrated in Figure 5.16.

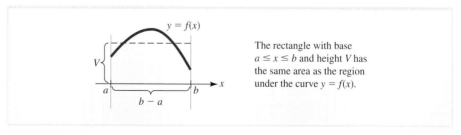

FIGURE 5.16   Geometric interpretation of average value V.

EXERCISES ■ 5.4

In Exercises 1 through 4, find the area of the shaded region.

1.

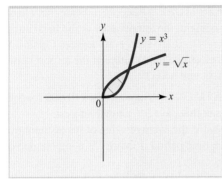

2.

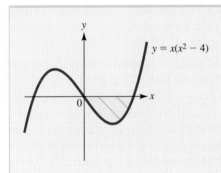

3.

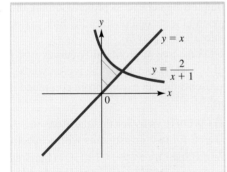

4.

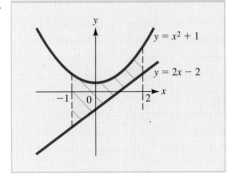

In Exercises 5 through 18, sketch the given region R and then find its area.

5. R is the region bounded by the lines $y = x$, $y = 2x$, and the line $x = 1$.

6. R is the region bounded by the curves $y = x^2$, $y = -x^2$, and the line $x = 1$.

7. R is the region bounded by the x axis and the curve $y = -x^2 + 4x - 3$.

8. R is the region bounded by the curves $y = e^x$, $y = e^{-x}$, and the line $x = \ln 2$.

9. R is the region bounded by the curve $y = x^2 - 2x$ and the x axis. [*Hint:* Note that the region is below the x axis.]

10. R is the region bounded by the curve $y = \dfrac{1}{x^2}$ and the lines $y = x$ and $y = \dfrac{x}{8}$.

11. R is the region bounded by the curves $y = x^2 - 2x$ and $y = -x^2 + 4$.

12. R is the region between the curve $y = x^3$ and the line $y = 9x$, for $x \geq 0$.

13. R is the region between the curves $y = x^3 - 3x^2$ and $y = x^2 + 5x$.

14. R is the triangle bounded by the line $y = 4 - 3x$ and the coordinate axes.

15. R is the triangle with vertices $(-4, 0)$, $(2, 0)$, and $(2, 6)$.

16. R is the rectangle with vertices $(1, 0)$, $(-2, 0)$, $(-2, 5)$, and $(1, 5)$.

17. R is the trapezoid bounded by the lines $y = x + 6$ and $x = 2$ and the coordinate axes.

18. R is the trapezoid bounded by the lines $y = x + 2$, $y = 8 - x$, $x = 2$, and the y axis.

In Exercises 19 through 24, find the average value of the given function f(x) over the specified interval $a \leq x \leq b$.

19. $f(x) = 1 - x^2$ over $-3 \leq x \leq 3$

20. $f(x) = x^2 - 3x + 5$ over $-1 \leq x \leq 2$

21. $f(x) = e^{-x}(4 - e^{2x})$ over $-1 \leq x \leq 1$

22. $f(x) = e^{2x} + e^{-x}$ over $0 \leq x \leq \ln 2$

23. $f(x) = \dfrac{e^x - e^{-x}}{e^x + e^{-x}}$ over $0 \leq x \leq \ln 3$

24. $f(x) = \dfrac{x + 1}{x^2 + 2x + 6}$ over $-1 \le x \le 1$

In Exercises 25 through 28, find the average value V of the given function over the specified interval. In each case, sketch the graph of the function along with the rectangle whose base is the given interval and whose height is the average value V.

25. $f(x) = 2x - x^2$ over $0 \le x \le 2$

26. $f(x) = x$ over $0 \le x \le 4$

27. $h(u) = \dfrac{1}{u}$ over $2 \le u \le 4$

28. $g(t) = e^{-2t}$ over $-1 \le t \le 2$

LORENTZ CURVES   *In Exercises 29 through 34, find the Gini index for the given Lorentz curve.*

29. $L(x) = x^3$

30. $L(x) = x^2$

31. $L(x) = 0.55x^2 + 0.45x$

32. $L(x) = 0.7x^2 + 0.3x$

33. $L(x) = \dfrac{2}{3}x^{3.7} + \dfrac{1}{3}x$

34. $L(x) = \dfrac{e^x - 1}{e - 1}$

35. AVERAGE SUPPLY   A manufacturer supplies $S(p) = 0.5p^2 + 3p + 7$ hundred units of a certain commodity to the market when the price is p dollars per unit. Find the average supply as the price varies from $p = \$2$ to $p = \$5$.

36. EFFICIENCY   After t months on the job, a postal clerk can sort $Q(t) = 700 - 400e^{-0.5t}$ letters per hour. What is the average rate at which the clerk sorts mail during the first 3 months on the job?

37. INVENTORY   An inventory of 60 000 kg of a certain commodity is used at a constant rate and is exhausted after 1 year. What is the average inventory for the year?

38. FOOD PRICES   Records indicate that t months after the beginning of the year, the price of 100 g of lean ground beef in local supermarkets was

$$P(t) = 0.09t^2 - 0.2t + 4$$

dollars per 100 g. What was the average price of ground beef during the first 3 months of the year?

39. BACTERIAL GROWTH   The number of bacteria present in a certain culture after t minutes of an experiment was $Q(t) = 2000e^{0.05t}$. What was the average number of bacteria present during the first 5 minutes of the experiment?

40. TEMPERATURE   Records indicate that t hours past midnight, the temperature at the local airport was $f(t) = -0.3t^2 + 4t + 10$ degrees Celsius. What was the average temperature at the airport between 9:00 A.M. and noon?

41. INVESTMENT   Vijay invests $10 000 for 5 years in a bank that pays 5% annual interest.
   a. What is the average value of his account over this time period if interest is compounded continuously?
   b. How would you find the average value of the account if interest is compounded quarterly? Write a paragraph to explain how you would proceed.

42. INVESTMENT   Suppose that t years from now, one investment plan will be generating profit at a rate of $P_1'(t) = 100 + t^2$ hundred dollars per year, while a second investment will be generating profit at a rate of $P_2'(t) = 220 + 2t$ hundred dollars per year.
   a. For how many years does the rate of profitability of the second investment exceed that of the first?
   b. Compute the net excess profit assuming that you invest in the second plan for the time period determined in part (a).
   c. Sketch the rate of profitability curves $y = P_1'(t)$ and $y = P_2'(t)$ and shade the region whose area represents the net excess profit computed in part (b).

43. INVESTMENT   Answer the questions in Exercise 42 for two investments with respective rates of profitability $P_1'(t) = 130 + t^2$ hundred dollars per year and $P_2'(t) = 306 + 5t$ hundred dollars per year.

44. INVESTMENT   Answer the questions in Exercise 42 for two investments with respective rates for profitability $P_1'(t) = 60e^{0.12t}$ thousand dollars per year and $P_2'(t) = 160e^{0.08t}$ thousand dollars per year.

45. INVESTMENT   Answer the questions in Exercise 42 for two investments with respective rates of profitability $P_1'(t) = 90e^{0.1t}$ thousand dollars per year and $P_2'(t) = 140e^{0.07t}$ thousand dollars per year.

46. EFFICIENCY   After t hours on the job, one factory worker is producing $Q_1'(t) = 60 - 2(t - 1)^2$ units per hour, while a second worker is producing $Q_2'(t) = 50 - 5t$ units per hour.
   a. If both arrive on the job at 8:00 A.M., how many more units will the first worker have produced by noon than the second worker?
   b. Interpret the answer in part (a) as the area between two curves.

47. AVERAGE POPULATION   The population of a certain community t years after the year 2000 is given by
$$P(t) = \frac{e^{0.2t}}{4 + e^{0.2t}} \quad \text{million people.}$$
What was the average population of the community during the decade from 2000 to 2010?

48. AVERAGE COST   The cost of producing x units of a new product is $C(x) = 3x\sqrt{x} + 10$ hundred dollars. What is the average cost of producing the first 81 units?

49. AVERAGE DRUG CONCENTRATION   A patient is injected with a drug, and t hours later, the concentration of the drug remaining in the patient's bloodstream is given by
$$C(t) = \frac{3t}{(t^2 + 36)^{3/2}}$$
milligrams per cubic centimetre. What is the average concentration of drug during the first 8 hours after the injection?

50. AVERAGE PRODUCTION   A company determines that if L worker-hours of labour are employed, then Q units of a particular commodity will be produced, where
$$Q(L) = 500L^{2/3}.$$
   a. What is the average production as labour varies from 1000 to 2000 worker-hours?
   b. What labour level between 1000 and 2000 worker-hours results in the average production found in part (a)?

51. TEMPERATURE   A researcher models the temperature T (in degrees Celsius) during the time period from 6 A.M. to 6 P.M. on a day in Ottawa by the function
$$T(t) = 3 - \frac{1}{3}(t - 5)^2 \quad \text{for } 0 \le t \le 12,$$
where t is the number of hours after 6 A.M.

   a. What is the average temperature in the city during the workday, from 8 A.M. to 5 P.M.?
   b. At what time (or times) during the workday is the temperature in the city the same as the average temperature found in part (a)?

52. ADVERTISING   An advertising firm is hired to promote a new television show for 3 weeks before its debut and 2 weeks afterward. After t weeks of the advertising campaign, it is found that $P(t)$ percent of the viewing public is aware of the show, where
$$P(t) = \frac{59t}{0.7t^2 + 16} + 6.$$
   a. What is the average percentage of the viewing public that is aware of the show during the 5 weeks of the advertising campaign?
   b. At what time during the first 5 weeks of the campaign is the percentage of viewers the same as the average percentage found in part (a)?

53. TRAFFIC MANAGEMENT   For several weeks, the highway department has been recording the speed of traffic flowing past a certain downtown Edmonton coffee shop. The data suggest that between the hours of 1:00 P.M. and 6:00 P.M. on a normal weekday, the speed of traffic past the coffee shop is approximately $v(t) = t^3 - 10.5t^2 + 30t + 20$ kilometres per hour, where t is the number of hours past noon.
   a. Compute the average speed of the traffic between the hours of 1:00 P.M. and 6:00 P.M.
   b. At what time between 1:00 P.M. and 6:00 P.M. is the traffic speed the same as the average speed found in part (a)?

54. AVERAGE AEROBIC RATING   The aerobic rating of a person who is x years old is given by
$$A(x) = \frac{110(\ln x - 2)}{x} \quad \text{for } x \ge 10$$
What is a person's average aerobic rating from age 15 to age 25? From age 60 to age 70?

55. THERMAL EFFECT OF FOOD   Normally, the metabolism of an organism functions at an essentially constant rate, called the *basal metabolic rate* of the organism. However, the metabolic rate may increase or decrease depending on the activity of the organism. In particular, after ingesting nutrients, the organism often experiences a surge in its metabolic rate, which then gradually returns to the basal level.

Michelle has just finished her Thanksgiving dinner, and her metabolic rate has surged from its basal level M_0. She then works off the meal over the next 12 hours. Suppose that t hours after the meal, her metabolic rate is given by

$$f(t) = 50te^{-0.1t^2} \quad 0 \le t \le 12$$

kilojoules per hour (kJ/h).

a. What is Michelle's average metabolic rate over the 12-hour period?

b. Sketch the graph of $f(t) = 50te^{-0.1t^2}$ for $0 \le t \le 12$. Use this graph to find the peak metabolic rate and the time it occurs. [*Note:* You will need to add M_0 to the value found from the graph.]

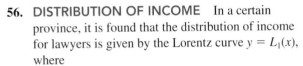

56. DISTRIBUTION OF INCOME   In a certain province, it is found that the distribution of income for lawyers is given by the Lorentz curve $y = L_1(x)$, where

$$L_1(x) = \frac{4}{5}x^2 + \frac{1}{5}x,$$

while that of surgeons is given by $y = L_2(x)$, where

$$L_2(x) = \frac{5}{8}x^4 + \frac{3}{8}x.$$

Compute the Gini index for each Lorentz curve. Which profession has the more equitable income distribution?

57. DISTRIBUTION OF INCOME   Suppose a study indicates that the distribution of income for professional baseball players is given by the Lorentz curve $y = L_1(x)$, where

$$L_1(x) = \frac{2}{3}x^3 + \frac{1}{3}x,$$

while those of professional football players and basketball players are $y = L_2(x)$ and $y = L_3(x)$, respectively, where

$$L_2(x) = \frac{5}{6}x^2 + \frac{1}{6}x$$

and

$$L_3(x) = \frac{3}{5}x^4 + \frac{2}{5}x.$$

Find the Gini index for each professional sport and determine which has the most equitable income distribution. Which has the least equitable distribution?

58. COMPARATIVE GROWTH   The population of a developing country grows exponentially at an unsustainable rate of

$$P_1'(t) = 10e^{0.02t}$$

thousand people per year, where t is the number of years after 2010. A study indicates that if certain socioeconomic changes are instituted in this country, then the population will instead grow at a restricted rate of

$$P_2'(t) = 10 + 0.02t + 0.002t^2$$

thousand people per year. How much smaller will the population of this country be in 2020 if the changes are made than if they are not?

59. COMPARATIVE GROWTH   A second study of the country in Exercise 58 indicates that natural restrictive forces are at work that make the actual rate of growth

$$P_3'(t) = \frac{20e^{0.02t}}{1 + e^{0.02t}},$$

instead of the exponential rate $P_1'(t) = 10e^{0.02t}$. If this rate is correct, how much smaller will the population be in 2020 than if the exponential rate were correct?

60. REAL ESTATE   The territory occupied by a certain community is bounded on one side by a river and on all other sides by mountains, forming the shaded region shown in the accompanying figure. If a coordinate system is set up as indicated, the mountainous boundary is given roughly by the curve $y = 4 - x^2$, where x and y are measured in kilometres. What is the total area occupied by the community?

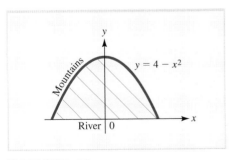

EXERCISE 60

61. REAL ESTATE EVALUATION   A square cabin with a plot of land adjacent to a lake is shown in the accompanying figure. If a coordinate system is set up as indicated, with distances given in metres, the lakefront boundary of the property is part of the curve $y = 10e^{0.04x}$. Assuming that the cabin costs $2000/m² and the land in the plot outside the cabin (the shaded region in the figure) costs $800/m², what is the total cost of this vacation property?

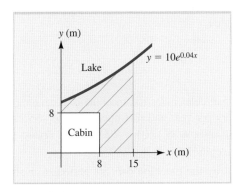

EXERCISE 61

62. A* VOLUME OF BLOOD DURING SYSTOLE
A model* of the cardiovascular system relates the stroke volume $V(t)$ of blood in the aorta at time t during systole (the contraction phase) to the pressure $P(t)$ in the aorta at the same time by the equation

$$V(t) = [C_1 + C_2 P(t)]\left(\frac{3t^2}{T^2} - \frac{2t^3}{T^3}\right),$$

where C_1 and C_2 are positive constants and T is the period of the systolic phase (a fixed time). Assume that aortic pressure $P(t)$ rises at a constant rate from P_0 when $t = 0$ to P_1 when $t = T$.
a. Show that

$$P(t) = \left(\frac{P_1 - P_0}{T}\right)t + P_0.$$

b. Find the average volume of blood in the aorta during the systolic phase $(0 \le t \le T)$. [*Note:* Your answer will be in terms of C_1, C_2, P_0, P_1, and T.]

---

*J. G. Defares, J. J. Osborn, and H. H. Hura, *Acta Physiol. Pharm. Neerl.,* Vol. 12, 1963, pp. 189–265.

63. A* REACTION TO MEDICATION In certain biological models, the human body's reaction to a particular kind of medication is measured by a function of the form

$$F(M) = \frac{1}{3}(kM^2 - M^3) \quad 0 \le M \le k,$$

where k is a positive constant and M is the amount of medication absorbed in the blood. The sensitivity of the body to the medication is measured by the derivative $S = F'(M)$.
a. Show that the body is most sensitive to the medication when $M = \dfrac{k}{3}$.
b. What is the average reaction to the medication for $0 \le M \le \dfrac{k}{3}$?

64. Use a graphing utility to draw the graphs of the curves $y = x^2 e^{-x}$ and $y = \dfrac{1}{x}$ on the same screen. Find the intersection points. With technology, compute the area of the region bounded by the curves.

65. Repeat Exercise 64 for the curves
$$\frac{x^2}{5} - \frac{y^2}{2} = 1 \quad \text{and} \quad y = x^3 - 8.9x^2 + 26.7x - 27.$$

66. A* Show that the average value V of a continuous function $f(x)$ over the interval $a \le x \le b$ may be computed as the slope of the line joining the points $(a, F(a))$ and $(b, F(b))$ on the curve $y = F(x)$, where $F(x)$ is any antiderivative of $f(x)$ over $a \le x \le b$.

67. A* Consider an object moving along a straight line. Explain why the object's average velocity (change in displacement divided by time taken) equals the average of its velocity over this time interval as calculated in Section 5.4.

L05

Solve business and economics problems using integration.

Additional Applications to Business and Economics

In this section, we examine some important applications of definite integration to business and economics, such as future and present value of an income flow. We begin by showing how integration can be used to measure the value of an asset.

Useful Life of a Machine

Suppose that t years after being put into use, a machine has generated total revenue $R(t)$ and that the total cost of operating and servicing the machine up to this time is $C(t)$. Then the total profit generated by the machine up to time t is $P(t) = R(t) - C(t)$. Profit declines when costs accumulate at a higher rate than revenue, that is, when $C'(t) > R'(t)$. Thus, a manager may consider disposing of the machine at the time T when $C'(T) = R'(T)$, and for this reason, the time period $0 \le t \le T$ is called the **useful life** of the machine. The net profit over the useful life of the machine provides the manager with a measure of its value.

EXAMPLE 5.5.1

Suppose that when it is t years old, a particular industrial machine is generating revenue at a rate of $R'(t) = 5000 - 20t^2$ dollars per year and that operating and servicing costs related to the machine are accumulating at a rate of $C'(t) = 2000 + 10t^2$ dollars per year.

a. What is the useful life of this machine?

b. Compute the net profit generated by the machine over its period of useful life.

Solution

a. To find the machine's useful life T, solve

$$C'(t) = R'(t)$$
$$2000 + 10t^2 = 5000 - 20t^2$$
$$30t^2 = 3000$$
$$t^2 = 100$$
$$t = 10$$

b. Since $P'(t)$ is the rate of change of profit per unit time, the earnings are roughly $P'(t)\, dt$ in time dt. The total profits are a continuous sum of profit per unit time multiplied by infinitesimally small time increments, $\int P'(t)\, dt$.

Profit $P(t)$ is given by $P(t) = R(t) - C(t)$, so $P'(t) = R'(t) - C'(t)$, and the net profit generated by the machine over its useful life $0 \le t \le 10$ is

$$\text{NP} = P(10) - P(0) = \int_0^{10} P'(t)\, dt$$
$$= \int_0^{10} [R'(t) - C'(t)]\, dt$$
$$= \int_0^{10} [(5000 - 20t^2) - (2000 + 10t^2)]\, dt$$
$$= \int_0^{10} [3000 - 30t^2]\, dt$$
$$= 3000t - 10t^3 \Big|_0^{10} = 20\,000$$

or \$20 000.

The rate of revenue and rate of cost curves are sketched in Figure 5.17. The net earnings of the machine over its useful life are represented by the area of the (shaded) region between the two rate curves.

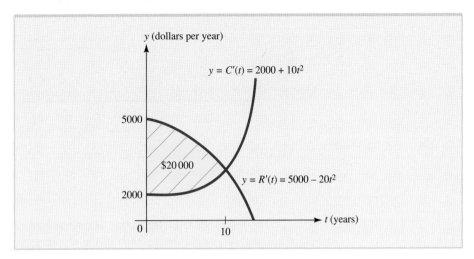

FIGURE 5.17   Net profit over the useful life of a machine.

Future Value and Present Value of an Income Flow

In our next application, we consider a stream of income transferred continuously into an account in which it earns interest over a specified time period (the **term**). Then the **future value of the income stream** is the total amount (money transferred into the account plus interest) that is accumulated in this way during the specified term.

The calculation of the amount of an income stream is illustrated in Example 5.5.2. The strategy is to approximate the continuous income stream by a sequence of discrete deposits called an **annuity.** The amount of the approximating annuity is a certain sum whose limit (a definite integral) is the future value of the income stream.

EXAMPLE 5.5.2

Money is transferred continuously into an account at a constant rate of \$1200 per year. The account earns interest at an annual rate of 8% compounded continuously. How much will be in the account at the end of 2 years?

Solution

Recall from Section 4.1 that P dollars invested at 8% compounded continuously will be worth $Pe^{0.08t}$ dollars t years later.

To approximate the future value of the income stream, divide the 2-year time interval $0 \le t \le 2$ into n equal subintervals of length Δt years and let t_j denote the beginning of the jth subinterval. Then, during the jth subinterval (of length Δt years),

   Money deposited = (dollars per year)(number of years) = $1200\Delta t$.

If all of this money were deposited at the beginning of the subinterval (at time t_j), it would remain in the account for $2 - t_j$ years and therefore would grow to $(1200\Delta t)e^{0.08(2 - t_j)}$ dollars. Thus,

$$\text{Future value of money deposited during } j\text{th subinterval} \approx 1200e^{0.08(2 - t_j)}\Delta t.$$

The situation is illustrated in Figure 5.18.

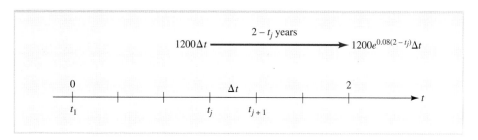

FIGURE 5.18   The (approximate) future value of the money deposited during the *j*th subinterval.

The future value of the entire income stream is the sum of the future values of the money deposited during each of the *n* subintervals. Hence,

$$\text{Future value of income stream} \approx \sum_{j=1}^{n} 1200e^{0.08(2-t_j)}\Delta t.$$

(Note that this is only an approximation because it is based on the assumption that all $1200\,\Delta t_n$ dollars are deposited at time t_j rather than continuously throughout the *j*th subinterval.)

As *n* increases without bound, the length of each subinterval approaches zero and the approximation approaches the true future value of the income stream. Hence,

$$\begin{aligned}
\text{Future value of income stream} &= \lim_{n \to +\infty} \sum_{j=1}^{n} 1200e^{0.08(2-t_j)}\Delta t \\
&= \int_{0}^{2} 1200e^{0.08(2-t)}\,dt = 1200e^{0.16}\int_{0}^{2} e^{-0.08t}\,dt \\
&= -\frac{1200}{0.08}\,e^{0.16}(e^{-0.08t})\Big|_{0}^{2} = -15\,000e^{0.16}(e^{-0.16} - 1) \\
&= -15\,000 + 15\,000e^{0.16} \approx 2602.66
\end{aligned}$$

or \$2602.66.

Generalizing the reasoning illustrated in Example 5.5.2 leads to the following integration formula for the future value of an income stream with rate of flow given by $f(t)$ for a term of T years:

$$\begin{aligned}
\text{FV} &= \int_{0}^{T} f(t)e^{r(T-t)}\,dt \\
&= \int_{0}^{T} f(t)e^{rT}e^{-rt}\,dt \\
&= e^{rT}\int_{0}^{T} f(t)e^{-rt}\,dt \qquad \text{factor constant } e^{rT} \text{ outside integral}
\end{aligned}$$

The first and last forms of the formula for future value are both listed next for future reference.

Future Value of an Income Stream ■ Suppose money is being transferred continuously into an account over a time period $0 \le t \le T$ at a rate given by the function $f(t)$ and that the account earns interest at an annual rate r compounded continuously. Then the future value FV of the income stream over the term T is given by the definite integral

$$\text{FV} = \int_0^T f(t)e^{r(T-t)}\,dt = e^{rT}\int_0^T f(t)e^{-rt}\,dt.$$

In Example 5.5.2, we had $f(t) = 1200$, $r = 0.08$, and $T = 2$, so that

$$\text{FV} = e^{0.08(2)}\int_0^2 1200e^{-0.08t}\,dt.$$

The **present value** of an income stream generated at a continuous rate $f(t)$ over a specified term of T years is the amount of money A that must be deposited now at the prevailing interest rate to generate the same income as the income stream over the same T-year period. Since A dollars invested at an annual interest rate r compounded continuously will be worth Ae^{rT} dollars in T years, we must have

$$Ae^{rT} = e^{rT}\int_0^T f(t)e^{-rt}\,dt$$

$$A = \int_0^T f(t)e^{-rt}\,dt \qquad \text{divide both sides by } e^{rT}$$

This can also be considered as the continuous sum of discounted amounts.
    To summarize:

Present Value of an Income Stream ■ The **present value PV** of an income stream that is deposited continuously at a rate of $f(t)$ into an account that earns interest at an annual rate r compounded continuously for a term of T years is given by

$$\text{PV} = \int_0^T f(t)e^{-rt}\,dt.$$

Example 5.5.3 illustrates how present value can be used in making certain financial decisions.

EXAMPLE 5.5.3

Ali is trying to decide between two investments. The first costs $1000 and is expected to generate a continuous income stream at a rate of $f_1(t) = 3000e^{0.03t}$ dollars per year. The second investment costs $4000 and is estimated to generate income at a constant rate of $f_2(t) = 4000$ dollars per year. If the prevailing annual interest rate remains fixed at 5% compounded continuously over the next 5 years, which investment is better over this time period?

Solution

The net value of each investment over the 5-year time period is the present value of the investment less its initial cost. For each investment, $r = 0.05$ and $T = 5$.

For the first investment:

$$\text{PV} - \text{cost} = \int_0^5 (3000e^{0.03t})e^{-0.05t}\, dt - 1000$$

$$= 3000 \int_0^5 e^{0.03t - 0.05t}\, dt - 1000$$

$$= 3000 \int_0^5 e^{-0.02t}\, dt - 1000$$

$$= 3000 \left(\frac{e^{-0.02t}}{-0.02} \right) \Big|_0^5 - 1000$$

$$= -150\,000[e^{-0.02(5)} - e^0] - 1000$$

$$= 13\,274.39$$

For the second investment:

$$\text{PV} - \text{cost} = \int_0^5 (4000)e^{-0.05t}\, dt - 4000$$

$$= 4000 \left(\frac{e^{-0.05t}}{-0.05} \right) \Big|_0^5 - 4000$$

$$= -80\,000[e^{-0.05(5)} - e^0] - 4000$$

$$= 13\,695.94$$

Thus, the net income generated by the first investment is \$13 274.39, while the second generates net income of \$13 695.94. The second investment is slightly better.

EXERCISES ▪ 5.5

1. **THE AMOUNT OF AN INCOME STREAM** Money is transferred continuously into an account at a constant rate of \$1000 per year. The account earns interest at an annual rate of 10% compounded continuously. How much will be in the account at the end of 10 years?

2. **PROFIT OVER THE USEFUL LIFE OF A MACHINE** Suppose that when it is t years old, a particular industrial machine generates revenue at a rate of $R'(t) = 6025 - 8t^2$ dollars per year and that operating and servicing costs accumulate at a rate of $C'(t) = 4681 + 13t^2$ dollars per year.
   a. How many years pass before the profitability of the machine begins to decline?
   b. Compute the net profit generated by the machine over its useful life.

   c. Sketch the revenue rate curve $y = R'(t)$ and the cost rate curve $y = C'(t)$ and shade the region whose area represents the net profit computed in part (b).

3. **PROFIT OVER THE USEFUL LIFE OF A MACHINE** Answer the question in Exercise 2 for a machine that generates revenue at a rate of $R'(t) = 7250 - 18t^2$ dollars per year and for which costs accumulate at a rate of $C'(t) = 3620 + 12t^2$ dollars per year.

4. **FUNDRAISING** It is estimated that t weeks from now, contributions in response to a fundraising campaign will be coming in at a rate of $R'(t) = 5000e^{-0.2t}$ dollars per week, while campaign expenses are expected to accumulate at a constant rate of \$676 per week.

a. For how many weeks does the rate of revenue exceed the rate of cost?

b. What net earnings will be generated by the campaign during the period of time determined in part (a)?

c. Interpret the net earnings in part (b) as an area between two curves.

5. **FUNDRAISING** Answer the questions in Exercise 4 for a charity campaign in which contributions are made at a rate of $R'(t) = 6537e^{-0.3t}$ dollars per week and expenses accumulate at a constant rate of $593 per week.

6. **CONSTRUCTION DECISION** Magda wants to expand and renovate her import store and is presented with two plans for making the improvements. The first plan will cost her $40 000 and the second will cost only $25 000. However, she expects the improvements resulting from the first plan to provide income at a continuous rate of $10 000 per year, while the income flow from the second plan provides $8000 per year. Which plan will result in more net income over the next 3 years if the prevailing rate of interest is 5% per year compounded continuously?

7. **RETIREMENT ANNUITY** At age 25, Mohi starts making annual deposits of $2500 into an RRSP account that pays interest at an annual rate of 5% compounded continuously. Assuming that his payments are made as a continuous income flow, how much money will be in his account if he retires at age 60? At age 65?

8. **RETIREMENT ANNUITY** When she is 30, Juanita starts making annual deposits of $2000 into a bond fund that pays 8% annual interest compounded continuously. Assuming that her deposits are made as a continuous income flow, how much money will be in her account if she retires at age 55?

9. **PRESENT VALUE OF AN INVESTMENT** An investment will generate income continuously at a constant rate of $1200 per year for 5 years. If the prevailing annual interest rate remains fixed at 5% compounded continuously, what is the present value of the investment?

10. **PRESENT VALUE OF A FRANCHISE** The management of a national chain of fast-food outlets is selling a 10-year franchise in Regina. Past experience in similar locations suggests that t years from now the franchise will be generating profit

at a rate of $f(t) = 10\,000$ dollars per year. If the prevailing annual interest rate remains fixed at 4% compounded continuously, what is the present value of the franchise?

11. **INVESTMENT ANALYSIS** Josh is trying to choose between two investment opportunities. The first will cost $50 000 and is expected to produce income at a continuous rate of $15 000 per year. The second will cost $30 000 and is expected to produce income at a rate of $9000 per year. If the prevailing rate of interest stays constant at 6% per year compounded continuously, which investment is better over the next 5 years?

12. **INVESTMENT ANALYSIS** Deepak spends $4000 for an investment that generates a continuous income stream at a rate of $f_1(t) = 3000$ dollars per year. His friend Melissa makes a separate investment that also generates income continuously, but at a rate of $f_2(t) = 2000e^{0.04t}$ dollars per year. The friends discover that their investments have exactly the same net value over a 4-year period. If the prevailing annual interest rate stays fixed at 5% compounded continuously, how much did Melissa pay for her investment?

13. **DEPLETION OF ENERGY RESOURCES** Oil is being pumped from an oil field t years after its opening at a rate of $P'(t) = 1.3e^{0.04t}$ billion barrels per year. The field has a reserve of 20 billion barrels, and the price of oil holds steady at $112 per barrel.

a. Find $P(t)$, the amount of oil pumped from the field at time t. How much oil is pumped from the field during the first 3 years of operation? The next 3 years?

b. For how many years T does the field operate before it runs dry?

c. If the prevailing annual interest rate stays fixed at 5% compounded continuously, what is the present value of the continuous income stream $V = 112P'(t)$ over the period of operation of the field $0 \le t \le T$?

d. If the owner of the oil field decided to sell on the first day of operation, do you think the present value determined in part (c) would be a fair asking price? Explain your reasoning.

14. **DEPLETION OF ENERGY RESOURCES** Answer the questions in Exercise 13 for another oil field with a pumping rate of $P'(t) = 1.5e^{0.03t}$ and with a reserve of 16 billion barrels. You may assume that the price of oil is still $112 per barrel and that the prevailing annual interest rate is 5%.

15. DEPLETION OF ENERGY RESOURCES
Answer the questions in Exercise 13 for an oil field with a pumping rate of $P'(t) = 1.2e^{0.02t}$ and with a reserve of 12 billion barrels. Assume that the prevailing interest rate is 5% as before but that the price of oil after t years is given by $A(t) = 112e^{0.015t}$.

16. LOTTERY PAYOUT   A $2 million provincial lottery winner is given a $250 000 lump sum now and a continuous income flow at a rate of $200 000 per year for 10 years. If the prevailing rate of interest is 5% per year compounded continuously, is this a good deal for the winner? Explain.

17. LOTTERY PAYOUT   The winner of a provincial lottery is offered a choice of either receiving $10 million now as a lump sum or receiving A dollars a year for the next 6 years as a continuous income stream. If the prevailing annual interest rate is 5% compounded continuously and the two payouts are worth the same, what is A?

18. SPORTS CONTRACTS   A hockey player is the object of a bidding war between two rival teams. The first team offers a $3 million signing bonus and a 5-year contract guaranteeing him $8 million this year and an increase of 3% per year for the remainder of the contract. The second team offers $9 million per year for 5 years with no incentives. If the prevailing annual interest rate stays fixed at 4% compounded continuously, which offer is worth more? [*Hint:* Assume that with both offers, the salary is paid as a continuous income stream.]

19. PRESENT VALUE OF AN INVESTMENT   An investment produces a continuous income stream at a rate of $A(t)$ thousand dollars per year at time t, where
$$A(t) = 10e^{1-0.05t}.$$
The prevailing rate of interest is 5% per year compounded continuously.
   a. What is the future value of the investment over a term of 5 years ($0 \le t \le 5$)?
   b. What is the present value of the investment over the time period $1 \le t \le 3$?

20. A* PROFIT FROM AN INVENTION   A marketing survey indicates that t months after a new type of computerized air purifier is introduced to the market, sales will be generating profit at a rate of $P'(t)$ thousand dollars per month, where
$$P'(t) = \frac{500[1.4 - \ln(0.5t + 1)]}{t + 2}.$$

a. When is the rate of profitability positive and when is it negative? When is the rate increasing and when is it decreasing?
b. At what time $t = t_m$ is monthly profit maximized? Find the net change in profit over the time period $0 \le t \le t_m$.
c. It costs the manufacturer $100 000 to develop the purifier product, so $P(0) = -100$. Use this information along with integration to find $P(t)$.

21. A* TOTAL REVENUE   Consider the following problem: A certain oil well that yields 300 barrels of crude oil a month will run dry in 3 years. It is estimated that t months from now the price of crude oil will be $P(t) = 118 + 0.3\sqrt{t}$ dollars per barrel. If the oil is sold as soon as it is extracted from the ground, what will be the total future revenue from the well? Solve the problem using definite integration. [*Hint:* Divide the 3-year (36-month) time interval $0 \le t \le 36$ into n equal subintervals of length Δt and let t_j denote the beginning of the jth subinterval. Find an expression that estimates the revenue $R(t_j)$ obtained during the jth subinterval. Then express the total revenue as the limit of a sum.]

22. A* INVENTORY STORAGE COSTS   A manufacturer receives N units of a certain raw material that are initially placed in storage and then withdrawn and used at a constant rate until the supply is exhausted 1 year later. Suppose storage costs remain fixed at p dollars per unit per year. Use definite integration to find an expression for the total storage cost the manufacturer will pay during the year. [*Hint:* Let $Q(t)$ denote the number of units in storage after t years and find an expression for $Q(t)$. Then subdivide the interval $0 \le t \le 1$ into n equal subintervals and express the total storage cost as the limit of a sum.]

23. A* FUTURE VALUE OF AN INVESTMENT
A constant income stream of M dollars per year is invested at an annual rate r compounded continuously for a term of T years. Show that the future value of such an investment is
$$FV = \frac{M}{r}(e^{rT} - 1).$$

24. A* PRESENT VALUE OF AN INVESTMENT
A constant income stream of M dollars per year is invested at an annual rate r compounded continuously for a term of T years. Show that the present value of such an investment is
$$PV = \frac{M}{r}(1 - e^{rT}).$$

SECTION 5.6

Solve life science and social science problems using integration.

Additional Applications to the Life and Social Sciences

We have already seen how definite integration can be used to compute quantities of interest in the social and life sciences, such as net change, average value, and the Gini index of a Lorentz curve. In this section, we examine several additional such applications, including survival and renewal within a group and blood flow through an artery. We shall also discuss how volume can be computed using integration and used for purposes such as measuring the size of a lake or a tumour.

Survival and Renewal

In Example 5.6.1, a **survival function** gives the fraction of individuals in a group or population that can be expected to remain in the group for any specified period of time. A **renewal function** giving the rate at which new members arrive is also known, and the goal is to predict the size of the group at some future time. Problems of this type arise in many fields, including sociology, ecology, demography, and even finance, where the "population" is the number of dollars in an investment account and "survival and renewal" refer to features of an investment strategy.

EXAMPLE 5.6.1

A new mental health clinic has just opened. Statistics from similar facilities suggest that the fraction of patients who will still be receiving treatment at the clinic t months after their initial visit is given by the function $f(t) = e^{-t/20}$. The clinic initially accepts 300 people for treatment and plans to accept new patients at a constant rate of $g(t) = 10$ patients per month. Approximately how many people will be receiving treatment at the clinic 15 months from now?

Solution

Since $f(15)$ is the fraction of patients whose treatment continues at least 15 months, it follows that of the current 300 patients, only $300f(15)$ will still be receiving treatment 15 months from now.

To approximate the number of *new* patients who will be receiving treatment 15 months from now, divide the 15-month time interval $0 \leq t \leq 15$ into n equal subintervals of length Δt months and let t_j denote the beginning of the jth subinterval. Since new patients are accepted at a rate of 10 per month, the number of new patients accepted during the jth subinterval is $10\Delta t$. Fifteen months from now, approximately $15 - t_j$ months will have elapsed since these $10\Delta t$ new patients had their initial visits, and so approximately $(10\Delta t) f(15 - t_j)$ of them will still be receiving treatment at that time (Figure 5.19). It follows that the total number of new patients still receiving treatment 15 months from now can be approximated by the sum

$$\sum_{j=1}^{n} 10 f(15 - t_j)\Delta t.$$

Adding this to the number of current patients who will still be receiving treatment in 15 months gives

$$P \approx 300 f(15) + \sum_{j=1}^{n} 10 f(15 - t_j)\Delta t,$$

where P is the total number of patients (current and new) who will be receiving treatment 15 months from now.

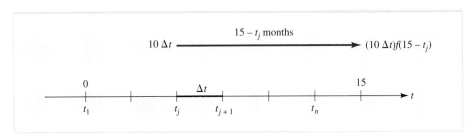

FIGURE 5.19 New members arriving during the jth subinterval.

As n increases without bound, the approximation improves and approaches the true value of P. It follows that

$$P = 300 f(15) + \lim_{n \to +\infty} \sum_{j=1}^{n} 10 f(15 - t_j) \Delta t$$

$$= 300 f(15) + \int_0^{15} 10 f(15 - t) \, dt$$

Since $f(t) = e^{-t/20}$, $f(15) = e^{-3/4}$ and $f(15 - t) = e^{-(15-t)/20} = e^{-3/4} e^{t/20}$. Hence,

$$P = 300 e^{-3/4} + 10 e^{-3/4} \int_0^{15} e^{t/20} \, dt$$

$$= 300 e^{-3/4} + 10 e^{-3/4} \left(\frac{e^{t/20}}{1/20} \right) \Big|_0^{15}$$

$$= 300 e^{-3/4} + 200 e^{-3/4} (e^{3/4} - e^0)$$

$$= 300 e^{-3/4} + 200 (1 - e^{-3/4})$$

$$= 100 e^{-3/4} + 200$$

$$\approx 247.24$$

That is, 15 months from now, the clinic will be treating approximately 247 patients.

In Example 5.6.1, we considered a variable survival function $f(t)$ and a constant renewal rate function $g(t)$. Essentially the same analysis applies when the renewal function also varies with time. Note that time in this example is given in years, but the same basic formula also applies for other units of time, for example, minutes, weeks, or months, as in Example 5.6.1. The result follows.

Survival and Renewal ■ Suppose a population initially has P_0 members and that new members are added at a (renewal) rate of $R(t)$ individuals per year. Further suppose that the fraction of the population that remains for at least t years after arriving is given by the (survival) function $S(t)$. Then, at the end of a term of T years, the population will be

$$P(T) = P_0 S(T) + \int_0^T R(t) S(T - t) \, dt.$$

In Example 5.6.1, each time period is 1 month, the initial population (membership) is $P_0 = 300$, the renewal rate is $R = 10$, the survival function is $f(t) = e^{-t/20}$, and the term is $T = 15$ months. Here is another example of survival/renewal from biology.

EXAMPLE 5.6.2

A mild toxin is introduced to a bacterial colony with a current population of 600 000. Observations indicate that $R(t) = 200e^{0.01t}$ bacteria per hour are born in the colony at time t and that the fraction of the population that survives for t hours after birth is $S(t) = e^{-0.015t}$. What is the population of the colony after 10 hours?

Solution

Substituting $P_0 = 600\ 000$, $R(t) = 200e^{0.01t}$, and $S(t) = e^{-0.015t}$ into the formula for survival and renewal, the population at the end of the term of $T = 10$ hours is

$$P(10) = \underbrace{600\ 000}_{P_0}\underbrace{e^{-0.015(10)}}_{S(10)} + \int_0^{10} \underbrace{200e^{0.01t}}_{R(t)}\ \underbrace{e^{-0.015(10-t)}}_{S(T-t)}\ dt$$

$$\approx 516\ 425 + \int_0^{10} 200e^{0.01t}[e^{-0.015(10)}e^{0.015t}]\,dt \qquad \text{since } e^{a-b} = e^a e^{-b}$$

$$\approx 516\ 425 + 200e^{-0.015(10)}\int_0^{10}[e^{0.01t}e^{0.015t}]\,dt \qquad \begin{array}{l}\text{factor } 200e^{-0.015(10)}\\ \text{outside the integral}\end{array}$$

$$\approx 516\ 425 + 172.14\int_0^{10} e^{0.025t}\,dt \qquad \begin{array}{l}\text{since } e^{a+b} = e^a e^b \text{ and}\\ 200e^{-0.015(10)} \approx 172.14\end{array}$$

$$\approx 516\ 425 + 172.14\left[\frac{e^{0.025t}}{0.025}\right]\Big|_0^{10} \qquad \begin{array}{l}\text{exponential rule for}\\ \text{integration}\end{array}$$

$$\approx 516\ 425 + \frac{172.14}{0.025}[e^{0.025(10)} - e^0]$$

$$\approx 518\ 381$$

Thus, the population of the colony declines from 600 000 to about 518 381 during the first 10 hours after the toxin is introduced.

The Flow of Blood through an Artery

Biologists have found that the speed of blood in an artery is a function of the distance of the blood from the artery's central axis. According to Poiseuille's law, the speed (in centimetres per second) of blood that is r centimetres from the central axis of the artery is $S(r) = k(R^2 - r^2)$, where R is the radius of the artery and k is a constant. In Example 5.6.3, you will see how to use this information to compute the rate at which blood flows through the artery.

EXAMPLE 5.6.3

Find an expression for the rate (in cubic centimetres per second) at which blood flows through an artery of radius R if the speed of the blood r centimetres from the central axis is $V(r) = k(R^2 - r^2)$, where k is a constant.

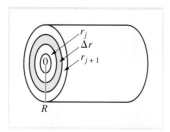

FIGURE 5.20 Subdividing a cross-section of an artery into concentric rings.

Solution

To approximate the volume of blood that flows through a cross-section of the artery each second, divide the interval $0 \leq r \leq R$ into n equal subintervals of width Δr centimetres and let r_j denote the beginning of the jth subinterval. These subintervals determine n concentric rings, as illustrated in Figure 5.20.

If Δr is small, the area of the jth ring is approximately the area of a rectangle whose length is the circumference of the (inner) boundary of the ring and whose width is Δr. That is,

$$\text{Area of } j\text{th ring} \approx 2\pi r_j \Delta r.$$

Multiplying the area of the jth ring (square centimetres) by the speed (centimetres per second) of the blood flowing through this ring gives the volume rate (cubic centimetres per second) at which blood flows through the jth ring. Since the speed of blood flowing through the jth ring is approximately $V(r_j)$ centimetres per second, it follows that

$$\begin{pmatrix} \text{Volume rate of flow} \\ \text{through } j\text{th ring} \end{pmatrix} \approx \begin{pmatrix} \text{area of} \\ j\text{th ring} \end{pmatrix}\begin{pmatrix} \text{speed of blood} \\ \text{through } j\text{th ring} \end{pmatrix}$$
$$\approx (2\pi r_j \Delta r)V(r_j)$$
$$\approx (2\pi r_j \Delta r)[k(R^2 - r_j^2)]$$
$$\approx 2\pi k(R^2 r_j - r_j^3)\Delta r$$

The volume rate of flow of blood through the entire cross-section is the sum of n such terms, one for each of the n concentric rings. That is,

$$\text{Volume rate of flow} \approx \sum_{j=1}^{n} 2\pi k(R^2 r_j - r_j^3)\Delta r.$$

As n increases without bound, this approximation approaches the true value of the rate of flow. In other words,

$$\text{Volume rate of flow} = \lim_{n \to +\infty} \sum_{j=1}^{n} 2\pi k(R^2 r_j - r_j^3)\Delta r$$
$$= \int_0^R 2\pi k(R^2 r - r^3)\, dr$$
$$= 2\pi k\left(\frac{R^2}{2}r^2 - \frac{1}{4}r^4\right)\Bigg|_0^R$$
$$= \frac{\pi k R^4}{2}$$

Thus, the blood is flowing at a rate of $\dfrac{\pi k R^4}{2}$ cubic centimetres per second.

Population Density

The **population density** of an urban area is the number of people $p(r)$ per square kilometre located a distance r kilometres from the city centre. We can determine the total population P of the portion of the city that lies within R kilometres of the city centre by using integration.

Our approach to using population density to determine total population will be similar to the approach used in Example 5.6.3 to determine the flow of blood through an artery. In particular, divide the interval $0 \leq r \leq R$ into n subintervals, each of width $\Delta r = \dfrac{R}{n}$, and let r_k denote the beginning (left endpoint) of the kth subinterval, for $k = 1, 2, \ldots, n$. These subintervals determine n concentric rings, centred on the city centre as shown in Figure 5.21.

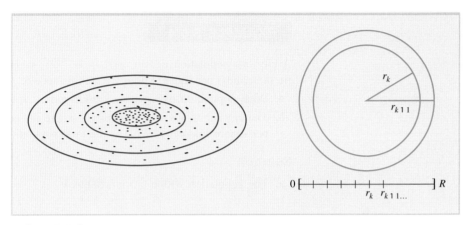

FIGURE 5.21   Subdividing an urban area into concentric rings.

The area of the kth ring is approximately the area of a rectangle whose length is the circumference of the inner boundary of the ring and whose width is Δr. That is,

$$\text{Area of } k\text{th ring} \approx 2\pi r_k \Delta r,$$

and since the population density is $p(r)$ people per square kilometre, it follows that

$$\text{Population within } k\text{th ring} \approx \underbrace{p(r_k)}_{\substack{\text{population} \\ \text{per unit area}}} \cdot \underbrace{[2\pi r_k \Delta r]}_{\substack{\text{area} \\ \text{of ring}}} = 2\pi r_k p(r_k)\Delta r.$$

We can estimate the total area with the bounding radius R by adding up the populations within the approximating rings, that is, by the Riemann sum

$$\begin{bmatrix} \text{Total population} \\ \text{within radius } R \end{bmatrix} = P(R) \approx \sum_{k=1}^{n} 2\pi r_k \, p(r_k)\Delta r.$$

Taking the limit as $n \to \infty$, the estimate approaches the true value of the total population P, and since the limit of a Riemann sum is a definite integral,

$$P(R) = \lim_{n \to \infty} \sum_{k=1}^{n} 2\pi r_k p(r_k)\Delta r = \int_0^R 2\pi r p(r) \, dr.$$

To summarize:

Total Population from Population Density   ■   If a concentration of individuals has population density $p(r)$ individuals per square unit at a distance r from the centre of concentration, then the total population $P(R)$ located within distance R from the centre is given by

$$P(R) = \int_0^R 2\pi r p(r) \, dr.$$

NOTE It is convenient to derive the population density formula by considering the population of a city. However, the formula also applies to more general population concentrations, such as bacterial colonies or even the population of water drops from a sprinkler system.

EXAMPLE 5.6.4

A city has population density $p(r) = 3e^{-0.01r^2}$, where $p(r)$ is the number of people (in thousands) per square kilometre at a distance of r kilometres from the city centre.

a. What population lives within a 5-km radius of the city centre?

b. The city limits are set at a radius R where the population density is 1000 people per square kilometre. What is the total population within the city limits?

Solution

a. The population within a 5-km radius is

$$P(5) = \int_0^5 2\pi r \, (3e^{-0.01r^2}) \, dr = 6\pi \int_0^5 e^{-0.01r^2} r \, dr.$$

Using the substitution $u = -0.01r^2$ gives

$$du = -0.01(2r \, dr) \quad \text{or} \quad r \, dr = \frac{du}{-0.02} = -50 \, du.$$

In addition, the limits of integration are transformed as follows:

When $r = 5$, then $u = -0.01(5)^2 = -0.25$.
When $r = 0$, then $u = -0.01(0)^2 = 0$.

Therefore,

$$P(5) = 6\pi \int_0^5 e^{-0.01r^2} r \, dr$$

$$= 6\pi \int_0^{-0.25} e^u(-50 \, du) \qquad \text{since } r \, dr = -50 \, du$$

$$= -6\pi \int_{-0.25}^0 e^u(-50 \, du) \quad \text{since } \int_b^a f(u) \, du = -\int_a^b f(u) \, du$$

$$= 300\pi(e^u)\Big|_{u=-0.25}^{u=0}$$

$$= 300\pi(e^0 - e^{-0.25})$$

$$\approx 208.5$$

So roughly 208 500 people live within a 5-km radius of the city centre.

b. To find the radius R that corresponds to the city limits, the population density must be 1 (one thousand), so solve

$$3e^{-0.01R^2} = 1$$

$$e^{-0.01R^2} = \frac{1}{3}$$

$$-0.01R^2 = \ln\left(\frac{1}{3}\right) \qquad \text{take logarithms on both sides}$$

$$R^2 = \frac{\ln\left(\frac{1}{3}\right)}{-0.01} = 109.86$$

$$R \approx 10.48$$

Finally, using the substitution $u = -0.01r^2$ from part (a), the population within the city limits is

$$P(10.48) = 6\pi\int_0^{10.48} e^{-0.01r^2}r\,dr \qquad \text{Limits of integration:}$$

$$= 6\pi\int_0^{-1.1} e^u(-50\,du) \qquad \text{when } r = 10.48, \text{ then } u = -0.01(10.48)^2 \approx -1.1$$

$$= -6\pi\int_{-1.1}^0 e^u(-50\,du) \qquad \text{when } r = 0, \text{ then } u = 0$$

$$\approx 300\pi(e^u)\Big|_{u=-1.1}^{u=0}$$

$$\approx 300\pi(e^0 - e^{-1.1})$$

$$\approx 628.75$$

Thus, approximately 628 750 people live within the city limits.

The Volume of a Solid of Revolution

In the next application, the definite integral is used to find the volume of a **solid of revolution** formed by revolving a region R in the xy plane about the x axis.

The technique is to express the volume of the solid as the limit of a sum of the volumes of approximating disks. In particular, suppose that S is the solid formed by rotating the region R under the curve $y = f(x)$ between $x = a$ and $x = b$ about the x axis, as shown in Figure 5.22a. Divide the interval $a \le x \le b$ into n equal subintervals

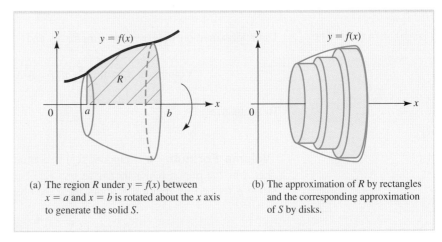

(a) The region R under $y = f(x)$ between $x = a$ and $x = b$ is rotated about the x axis to generate the solid S.

(b) The approximation of R by rectangles and the corresponding approximation of S by disks.

FIGURE 5.22   A solid S formed by rotating the region R about the x axis.

of length Δx. Then approximate the region R by n rectangles and the solid S by the corresponding n cylindrical disks formed by rotating these rectangles about the x axis. The general approximation procedure is illustrated in Figure 5.22b for the case where $n = 3$.

If x_j denotes the beginning (left endpoint) of the jth subinterval, then the jth rectangle has height $f(x_j)$ and width Δx, as shown in Figure 5.23a. The jth approximating disk formed by rotating this rectangle about the x axis is shown in Figure 5.23b.

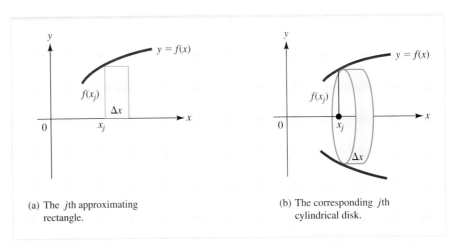

(a) The jth approximating rectangle.

(b) The corresponding jth cylindrical disk.

FIGURE 5.23   The volume of the solid S is approximated by adding volumes of approximating disks.

Since the jth approximating cylindrical disk has radius $r_j = f(x_j)$ and thickness Δx, its volume is

$$\text{Volume of } j\text{th disk} = (\text{area of circular cross-section})(\text{width})$$
$$= \pi r_j^2(\text{width}) = \pi[f(x_j)]^2 \Delta x$$

The total volume of S is approximately the sum of the volumes of the n disks; that is,

$$\text{Volume of } S \approx \sum_{j=1}^{n} \pi[f(x_j)]^2 \Delta x.$$

The approximation improves as n increases and

$$\text{Volume of } S = \lim_{n \to \infty} \sum_{j=1}^{n} \pi[f(x_j)]^2 \Delta x = \pi \int_a^b [f(x)]^2 \, dx.$$

To summarize:

Volume Formula ▪ Suppose $f(x)$ is continuous and $f(x) \geq 0$ on $a \leq x \leq b$. Let R be the region under the curve $y = f(x)$ between $x = a$ and $x = b$. Then the solid S formed by rotating R about the x axis has volume

$$\text{Volume of } S = \pi \int_a^b [f(x)]^2 \, dx.$$

Here are two examples.

EXAMPLE 5.6.5

Find the volume of the solid S formed by revolving the region under the curve $y = x^2 + 1$ from $x = 0$ to $x = 2$ about the x axis.

Solution

The region, the solid of revolution, and the jth disk are shown in Figure 5.24.

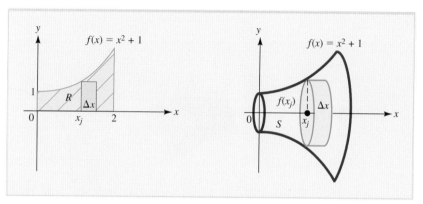

FIGURE 5.24   The solid formed by rotating the region under the curve $y = x^2 + 1$ between $x = 0$ and $x = 2$ about the x axis.

The radius of the jth disk is $f(x_j) = x_j^2 + 1$. Hence,

$$\text{Volume of } j\text{th disk} = \pi[f(x_j)]^2 \Delta x = \pi(x_j^2 + 1)^2 \Delta x$$

and

$$\text{Volume of } S = \lim_{n \to \infty} \sum_{j=1}^{n} \pi(x_j^2 + 1)^2 \Delta x$$

$$= \pi \int_0^2 (x^2 + 1)^2 \, dx$$

$$= \pi \int_0^2 (x^4 + 2x^2 + 1) \, dx$$

$$= \pi \left(\frac{1}{5}x^5 + \frac{2}{3}x^3 + x \right) \Big|_0^2 = \frac{206}{15}\pi \approx 43.14$$

EXAMPLE 5.6.6

A tumour has approximately the same shape as the solid formed by rotating the region under the curve $y = \frac{1}{3}\sqrt{16 - 4x^2}$ about the x axis, where x and y are measured in centimetres. Find the volume of the tumour.

Solution

The curve intersects the x axis where $y = 0$; that is, where

$$\frac{1}{3}\sqrt{16 - 4x^2} = 0$$

$$16 = 4x^2 \qquad \text{since } \sqrt{a - b} = 0 \text{ only if } a = b$$

$$x^2 = 4 \qquad \text{divide both sides by 4}$$

$$x = \pm 2$$

The curve (called an *ellipse*) is shown in Figure 5.25.

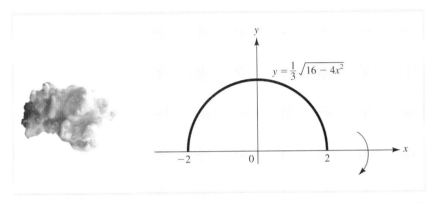

FIGURE 5.25   Tumour with the approximate shape of the solid formed by rotating the curve $y = \dfrac{1}{3}\sqrt{16 - 4x^2}$ about the x axis.

Let $f(x) = \dfrac{1}{3}\sqrt{16 - 4x^2}$. Then the volume of the solid of revolution is given by

$$V = \int_{-2}^{2} \pi [f(x)]^2 \, dx = \int_{-2}^{2} \pi \left[\frac{1}{3}\sqrt{16 - 4x^2}\right]^2 dx$$

$$= \int_{-2}^{2} \frac{\pi}{9}(16 - 4x^2) \, dx$$

$$= \frac{\pi}{9}\left[16x - \frac{4}{3}x^3\right]\Bigg|_{-2}^{2}$$

$$= \frac{\pi}{9}\left[16(2) - \frac{4}{3}(2)^3\right] - \frac{\pi}{9}\left[16(-2) - \frac{4}{3}(-2)^3\right]$$

$$\approx 14.89$$

Thus, the volume of the tumour is approximately 15 cm³.

EXERCISES ■ 5.6

SURVIVAL AND RENEWAL   *In Exercises 1 through 6, an initial population P_0 is given along with a renewal rate $R(t)$ and a survival function $S(t)$. In each case, use the given information to find the population at the end of the indicated term T.*

1. $P_0 = 50\,000$; $R(t) = 40$; $S(t) = e^{-0.1t}$, t in months; term $T = 5$ months

2. $P_0 = 100\,000$; $R(t) = 300$; $S(t) = e^{-0.02t}$, t in days; term $T = 10$ days

3. $P_0 = 500\,000$; $R(t) = 800$; $S(t) = e^{-0.011t}$, t in years; term $T = 3$ years

4. $P_0 = 800\,000$; $R(t) = 500$; $S(t) = e^{-0.005t}$, t in months; term $T = 5$ months

5. $P_0 = 500\,000$; $R(t) = 100e^{0.01t}$; $S(t) = e^{-0.013t}$, t in years; term $T = 8$ years

6. $P_0 = 300\,000$; $R(t) = 150e^{0.012t}$; $S(t) = e^{-0.02t}$, t in months; term $T = 20$ months

VOLUME OF A SOLID OF REVOLUTION   *In Exercises 7 through 14, find the volume of the solid of revolution formed by rotating the region R about the x axis.*

7. R is the region under the line $y = 3x + 1$ from $x = 0$ to $x = 1$.

8. R is the region under the curve $y = \sqrt{x}$ from $x = 1$ to $x = 4$.

9. R is the region under the curve $y = x^2 + 2$ from $x = -1$ to $x = 3$.

10. R is the region under the curve $y = 4 - x^2$ from $x = -2$ to $x = 2$.

11. R is the region under the curve $y = \sqrt{4 - x^2}$ from $x = -2$ to $x = 2$.

12. R is the region under the curve $y = \dfrac{1}{x}$ from $x = 1$ to $x = 10$.

13. R is the region under the curve $y = \dfrac{1}{\sqrt{x}}$ from $x = 1$ to $x = e^2$.

14. R is the region under the curve $y = e^{-0.1x}$ from $x = 0$ to $x = 10$.

15. **NET POPULATION GROWTH**   It is projected that t years from now the population of a certain country will be changing at a rate of $e^{0.02t}$ million per year. If the current population is 50 million, what will the population be 10 years from now?

16. **NET POPULATION GROWTH**   A study indicates that x months from now, the population of a certain town will be increasing at a rate of $10 + 2\sqrt{x}$ people per month. By how much will the population increase over the next 9 months?

17. **GROUP MEMBERSHIP**   A national consumers' association has compiled statistics suggesting that the fraction of its members who are still active t months after joining is given by $f(t) = e^{-0.2t}$. A new local chapter has 200 charter members and expects to attract new members at a rate of 10 per month. How many members can the chapter expect to have at the end of 8 months?

18. **POLITICAL TRENDS**   Samantha Gupta is running for mayor. Polls indicate that the fraction of those who support her t weeks after first learning of her candidacy is given by $f(t) = e^{-0.03t}$. At the time she declared her candidacy, 25 000 people supported her, and new supporters are being added at a constant rate of 100 people per week. Approximately how many people are likely to vote for her if the election is held 20 weeks from the day she entered the race?

19. **SPREAD OF DISEASE**   A new strain of influenza has just been declared an epidemic by health officials. Currently, 5000 people have the disease and 60 more victims are added each day. If the fraction of infected people who still have the disease t days after contracting it is given by $f(t) = e^{-0.02t}$, how many people will have the flu 30 days from now?

20. **NUCLEAR WASTE**   A certain nuclear power plant produces radioactive waste in the form of strontium-90 at a constant rate of 500 kg per year. The waste decays exponentially with a half-life of 28 years. How much of the radioactive waste from the nuclear plant will be present after 140 years? [*Hint:* Think of this as a survival and renewal problem.]

21. **ENERGY CONSUMPTION**   The administration of a small country estimates that the demand for oil is increasing exponentially at a rate of 10% per year. If the demand is currently 30 billion barrels per year,

how much oil will be consumed in this country during the next 10 years?

22. **POPULATION GROWTH**   The administrators of a town estimate that the fraction of people who will still be residing in the town t years after they arrive is given by the function $f(t) = e^{-0.04t}$. If the current population is 20 000 people and new townspeople arrive at a rate of 500 per year, what will the population be 10 years from now?

23. **COMPUTER DATING**   The operators of a new computer dating service estimate that the fraction of people who will retain their membership in the service for at least t months is given by the function $f(t) = e^{-t/10}$. Initially there are 8000 members, and the operators expect to attract 200 new members per month. How many members will the service have 10 months from now?

24. **FLOW OF BLOOD**   Calculate the rate (in cubic centimetres per second) at which blood flows through an artery of radius 0.1 cm if the speed of the blood r centimetres from the central axis is $8 - 800r^2$ centimetres per second.

25. **CHOLESTEROL REGULATION**   Fat travels through the bloodstream attached to protein in a combination called a *lipoprotein*. Low-density lipoprotein (LDL) picks up cholesterol from the liver and delivers it to the cells, dropping off any excess cholesterol on the artery walls. Too much LDL in the bloodstream increases the risk of heart disease and stroke. A patient with a high level of LDL receives a drug that is found to reduce the level at a rate given by
$$L'(t) = -0.3t(49 - t^2)^{0.4}$$
milligrams per decilitre per day, where t is the number of days after the drug is administered, for $0 \le t \le 7$.
   a. By how much does the patient's LDL level change during the first 3 days after the drug is administered?
   b. Suppose the patient's LDL level is 120 mg/dL at the time the drug is administered. Find $L(t)$.
   c. The recommended safe LDL level is 100 mg/dL. How many days does it take for the patient's LDL level to be safe?

26. **POPULATION DENSITY**   The population density r kilometres from the centre of a certain city is $D(r) = 5000(1 + 0.5r^2)^{-1}$ people per square kilometre.

   a. How many people live within 5 km of the city centre?
   b. The city limits are set at a radius L where the population density is 1000 people per square kilometre. What is L and what is the total population within the city limits?

27. **POPULATION DENSITY**   The population density r kilometres from the centre of a certain city is $D(r) = 25\,000e^{-0.05r^2}$ people per square kilometre. How many people live between 1 km and 2 km from the city centre?

28. **A* POISEUILLE'S LAW**   Blood flows through an artery of radius R. At a distance r centimetres from the central axis of the artery, the speed of the blood is given by $V(r) = k(R^2 - r^2)$. Show that the average speed of the blood is one-half the maximum speed.

29. **BACTERIAL GROWTH**   An experiment is conducted with two bacterial colonies, each of which initially has a population of 100 000. In the first colony, a mild toxin is introduced that restricts growth so that only 50 new individuals are added per day and the fraction of individuals that survive at least t days is given by $f(t) = e^{-0.011t}$. The growth of the second colony is restricted indirectly by limiting food supply and space for expansion, and after t days, it is found that this colony contains
$$P(t) = \frac{5000}{1 + 49e^{0.009t}}$$
thousand individuals. Which colony is larger after 50 days? After 100 days? After 300 days?

30. **GROUP MEMBERSHIP**   A group has just been formed with an initial membership of 10 000. Suppose that the fraction of the membership of the group that remain members for at least t years after joining is $S(t) = e^{-0.03t}$, and that at time t, new members are being added at a rate of $R(t) = 10e^{0.017t}$ members per year. How many members will the group have 5 years from now?

31. **GROWTH OF AN ENDANGERED SPECIES**   Environmentalists estimate that the population of a certain endangered species is currently 3000. The population is expected to be growing at a rate of $R(t) = 10e^{0.01t}$ individuals per year t years from now, and the fraction that survive t years is given by $S(t) = e^{-0.07t}$. What will the population of the species be in 10 years?

32. POPULATION TRENDS The population of a small town is currently 85 000. A study commissioned by the mayor's office finds that people are settling in the town at a rate of $R(t) = 1200e^{0.01t}$ per year and that the fraction of the population who continue to live in the town t years after arriving is given by $S(t) = e^{-0.02t}$. How many people will live in the town in 10 years?

33. POPULATION TRENDS Answer the question in Exercise 32 for a constant renewal rate of $R = 1000$ and the survival function

$$S(t) = \frac{1}{t + 1}.$$

34. A* EVALUATING DRUG EFFECTIVENESS A pharmaceutical firm has been granted permission by Health Canada to test the effectiveness of a new drug in combating a virus. The firm administers the drug to a test group of uninfected but susceptible individuals and, using statistical methods, determines that t months after the test begins, people in the group are becoming infected at a rate of $D'(t)$ hundred individuals per month, where

$$D'(t) = 0.2 - 0.04t^{1/4}.$$

Government figures indicate that without the drug, the infection rate would have been $W'(t)$ hundred individuals per month, where

$$W'(t) = \frac{0.8e^{0.13t}}{(1 + e^{0.13t})^2}.$$

If the test is evaluated 1 year after it begins, how many people does the drug protect from infection? What percentage of the people who would have been infected if the drug had not been used were protected from infection by the drug?

35. A* EVALUATING DRUG EFFECTIVENESS Repeat Exercise 34 for another drug for which the infection rate is

$$D'(t) = 0.12 + \frac{0.08}{t + 1}.$$

Assume the government comparison rate stays the same, that is,

$$W'(t) = \frac{0.8e^{0.13t}}{(1 + e^{0.13t})^2}.$$

36. A* LIFE EXPECTANCY In a certain developing country, the life expectancy of a person who is t years old is $L(t)$ years, where

$$L(t) = 41.6(1 + 1.07t)^{0.13}.$$

a. What is the life expectancy of a person in this country at birth? At age 50?

b. What is the average life expectancy of all people in this country between the ages of 10 and 70?

c. Find the age T such that $L(T) = T$. Call T the *life limit* for this country. What can be said about the life expectancy of a person older than T years?

d. Find the average life expectancy L_e over the age interval $0 \leq t \leq T$. Why is it reasonable to call L_e the *expected length of life* for people in this country?

37. A* LIFE EXPECTANCY Answer the questions in Exercise 36 for a country with life expectancy function

$$L(t) = \frac{110e^{0.015t}}{1 + e^{0.015t}}.$$

38. A* ENERGY EXPENDED IN FLIGHT In an investigation by V. A. Tucker and K. Schmidt-Koenig,* it was determined that the energy E expended by a bird in flight varies with the speed v (in kilometres per hour) of the bird. For a particular kind of parakeet, the energy expenditure changes at a rate given by

$$\frac{dE}{dv} = \frac{0.31v^2 - 471.75}{v^2} \quad \text{for } v > 0,$$

where E is given in joules per gram mass per kilometre. Observations indicate that the parakeet tends to fly at the speed v_{min} that minimizes E.

a. What is the most economical speed v_{min}?

b. Suppose that when the parakeet flies at the most economical speed v_{min} its energy expenditure is E_{min}. Use this information to find $E(v)$ for $v > 0$ in terms of E_{min}.

39. MEASURING RESPIRATION A pneumotachograph is a device used by physicians to graph the rate of air flow into and out of the lungs as a patient breathes. The graph in the accompanying figure shows the rate of inspiration (breathing in) for a particular patient. The area under the graph measures the total volume of air inhaled by the patient during the inspiration phase of one breathing cycle. Assume the inspiration rate is given by

$$R(t) = -0.41t^2 + 0.97t \quad \text{litres per second.}$$

---

*E. Batschelet, *Introduction to Mathematics for Life Scientists,* 3rd ed., New York, Springer-Verlag, 1979, p. 299.

a. How long is the inspiration phase?
b. Find the volume of air taken into the patient's lungs during the inspiration phase.
c. What is the average flow rate of air into the lungs during the inspiration phase?

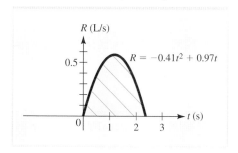

EXERCISE 39

40. **MEASURING RESPIRATION**   Repeat Exercise 39 with the inspiration rate function $R(t) = -1.2t^3 + 5.72t$ litres per second and sketch the graph of $R(t)$.

41. **WATER POLLUTION**   A ruptured pipe in an offshore oil rig produces a circular oil slick that is T metres thick at a distance r metres from the rupture, where

$$T(r) = \frac{3}{2 + r}.$$

At the time the spill is contained, the radius of the slick is 7 m. Do the following to find the volume of oil that has been spilled.

a. Sketch the graph of $T(r)$. Notice that the volume can be obtained by rotating the curve $T(r)$ about the T *axis* (vertical axis) rather than the r axis (horizontal axis).
b. Solve the equation $T = \dfrac{3}{2 + r}$ for r in terms of T. Sketch the graph of $r(T)$, with T on the horizontal axis.
c. Find the required volume by rotating the graph of $r(T)$ found in part (b) about the T axis.

42. **WATER POLLUTION**   Rework Exercise 41 for a situation with spill thickness

$$T(r) = \frac{2}{1 + r^2}$$

(T and r in metres) and radius of containment 9 m.

43. **A* AIR POLLUTION**   Particulate matter emitted from a smokestack is distributed so that r kilometres from the stack, the pollution density is $p(r)$ units per square kilometre, where

$$p(r) = \frac{200}{5 + 2r^2}.$$

a. What is the total amount of pollution within a 3-km radius of the smokestack?
b. Suppose a health agency determines that it is unsafe to live within a radius L of the smokestack where the pollution density is at least four units per square kilometre. What is L, and what is the total amount of pollution in the unsafe zone?

44. **A* VOLUME OF A SPHERE**   Use integration to show that a sphere of radius r has volume

$$V = \frac{4}{3}\pi r^3.$$

[*Hint:* Think of the sphere as the solid formed by rotating the region under the semicircle shown in the accompanying figure about the x axis.]

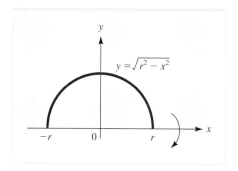

EXERCISE 44

45. **A* VOLUME OF A CONE**   Use integration to show that a right circular cone of height h and top radius r has volume

$$V = \frac{1}{3}\pi r^2 h.$$

[*Hint:* Think of the cone as a solid formed by rotating the triangle shown in the accompanying figure about the x axis.]

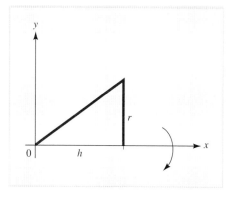

EXERCISE 45

Concept Summary Chapter 5

Rules for Integrating Common Functions

Constant rule: $\int k\,dx = kx + C$ for constant k

Power rule: $\int x^n\,dx = \dfrac{x^{n+1}}{n+1} + C$ for all $n \neq -1$

Logarithmic rule: $\int \dfrac{1}{x}\,dx = \ln|x| + C$ for all $x \neq 0$

Exponential rule: $\int e^{kx}\,dx = \dfrac{1}{k}e^{kx} + C$ for constant $k \neq 0$

Substitution

$$\int g(u)u'(x)\,dx = \int g(u)\,du, \text{ where } du = u'(x)\,dx$$

Area under a Curve

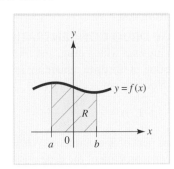

Area of $R = \displaystyle\int_a^b f(x)\,dx$

Fundamental Theorem of Calculus

$$\int_a^b f(x)\,dx = F(b) - F(a), \quad \text{where } F'(x) = f(x)$$

Net Change of $Q(x)$ over the Interval $a \leq x \leq b$

$$Q(b) - Q(a) = \int_a^b Q'(x)\,dx$$

Area between Two Curves

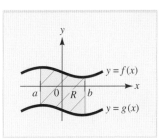

Area of $R = \displaystyle\int_a^b [f(x) - g(x)]\,dx$ (always use the order top function − bottom function)

Gini Index $= 2\displaystyle\int_0^1 [x - L(x)]\,dx$

Average Value of a Function $= \dfrac{1}{b-a}\displaystyle\int_a^b f(x)\,dx$

Future Value (Amount) of an Income Stream

$$\text{FV} = e^{rT}\int_0^T f(t)e^{-rt}\,dt$$

Present Value of an Income Stream $\text{PV} = \displaystyle\int_0^T f(t)e^{-rt}\,dt$

Survival and Renewal

$$P(T) = P_0 S(T) + \int_0^T R(T)S(T - t)\,dt$$

Population from Population Density

$$P(R) = \int_0^R 2\pi r p(r)\,dr$$

Volume of a Solid of Revolution $V = \displaystyle\int_a^b \pi[f(x)^2]\,dx$

Checkup for Chapter 5

1. Find each indefinite integral (antiderivative).

   a. $\displaystyle\int x^3 - \sqrt{3x} + 5e^{-2x}\,dx$

   b. $\displaystyle\int \frac{x^2 - 2x + 4}{x}\,dx$

   c. $\displaystyle\int \sqrt{x}\left(x^2 - \frac{1}{x}\right)dx$

   d. $\displaystyle\int \frac{x}{(3 + 2x^2)^{3/2}}\,dx$

   e. $\displaystyle\int \frac{\ln\sqrt{x}}{x}\,dx$

   f. $\displaystyle\int xe^{1+x^2}\,dx$

2. Evaluate each definite integral.

a. $\displaystyle\int_1^4 x^{3/2} + \frac{2}{x}\, dx$

b. $\displaystyle\int_0^3 e^{3-x}\, dx$

c. $\displaystyle\int_0^1 \frac{x}{x+1}\, dx$

d. $\displaystyle\int_0^3 \frac{(x+3)}{\sqrt{x^2 + 6x + 4}}\, dx$

3. In each case, find the area of the specified region.

a. The region bounded by the curve $y = x + \sqrt{x}$, the x axis, and the lines $x = 1$ and $x = 4$

b. The region bounded by the curve $y = x^2 - 3x$ and the line $y = x + 5$

4. Find the average value of the function $f(x) = \dfrac{x-2}{x}$ over the interval $1 \le x \le 2$.

5. **NET CHANGE IN REVENUE** The marginal revenue of producing q units of a certain commodity is $R'(q) = q(10 - q)$ hundred dollars per unit. How much additional revenue is generated as the level of production is increased from 4 units to 9 units?

6. **BALANCE OF TRADE** The government of a certain country estimates that t years from now, imports will be increasing at a rate $I'(t)$ and exports at a rate $E'(t)$, both in billions of dollars per year,

where

$$I'(t) = 12.5e^{0.2t} \quad \text{and} \quad E'(t) = 1.7t + 3.$$

The trade deficit is $D(t) = I(t) - E(t)$. By how much will the trade deficit for this country change over the next 5 years? Will it increase or decrease during this time period?

7. **AMOUNT OF AN INCOME STREAM** Money is transferred continuously into an account at a constant rate of $5000 per year. The account earns interest at an annual rate of 5% compounded continuously. How much will be in the account at the end of 3 years?

8. **POPULATION GROWTH** Demographers estimate that the fraction of people who will still be residing in a particular town t years after they arrive is given by the function $f(t) = e^{-0.02t}$. If the current population is 50 000 and new townspeople arrive at a rate of 700 per year, what will the population be 20 years from now?

9. **AVERAGE DRUG CONCENTRATION** A patient is injected with a drug, and t hours later, the concentration of the drug remaining in the patient's bloodstream is given by

$$C(t) = \frac{0.3t}{(t^2 + 16)^{1/2}}$$

milligrams per cubic centimetre. What is the average concentration of the drug during the first 3 hours after the injection?

Review Exercises

In Exercises 1 through 20, find the indefinite integral.

1. $\displaystyle\int (x^3 + \sqrt{x} - 9)\, dx$

2. $\displaystyle\int \left(x^{2/3} - \frac{1}{x} + 5 + \sqrt{x} \right) dx$

3. $\displaystyle\int (x^4 - 5e^{-2x})\, dx$

4. $\displaystyle\int \left(2\sqrt[3]{x} + \frac{5}{s} \right) ds$

5. $\displaystyle\int \left(\frac{5x^3 - 3}{x} \right) dx$

6. $\displaystyle\int \left(\frac{3e^{-x} + 2e^{3x}}{e^{2x}} \right) dx$

7. $\displaystyle\int \left(t^5 - 3t^2 + \frac{1}{t^2} \right) dt$

8. $\displaystyle\int (x + 1)(2x^3 + \sqrt{x})\, dx$

9. $\displaystyle\int \sqrt{3x + 1}\, dx$

10. $\displaystyle\int (3x + 1)\sqrt{3x^2 + 2x + 5}\, dx$

11. $\displaystyle\int (x + 2)(x^2 + 4x + 2)^5\, dx$

12. $\displaystyle\int \frac{x + 2}{x^2 + 4x + 2}\, dx$

13. $\displaystyle\int \frac{3x + 6}{(2x^2 + 8x + 3)^2}\, dx$

14. $\displaystyle\int (t - 5)^{12}\, dt$

15. $\displaystyle\int v(v - 5)^{12}\, dv$

16. $\displaystyle\int \frac{\ln 3x}{x}\, dx$

17. $\displaystyle\int 5xe^{-x^2}\, dx$

18. $\displaystyle\int \left(\frac{x}{x-4}\right) dx$

19. $\displaystyle\int \left(\frac{\sqrt{\ln x}}{x}\right) dx$

20. $\displaystyle\int \left(\frac{e^x}{e^x+5}\right) dx$

In Exercises 21 through 30, evaluate the definite integral.

21. $\displaystyle\int_0^1 (5x^4 - 8x^3 + 1)\, dx$

22. $\displaystyle\int_1^4 \sqrt{t} + t^{-3/2})\, dt$

23. $\displaystyle\int_0^1 (e^{2x} + 4\sqrt[3]{x})\, dx$

24. $\displaystyle\int_1^9 \frac{x^2 + \sqrt{x} - 5}{x}\, dx$

25. $\displaystyle\int_{-1}^2 30(5x-2)^2\, dx$

26. $\displaystyle\int_{-1}^1 \frac{(3x+6)}{(x^2+4x+5)^2}\, dx$

27. $\displaystyle\int_0^1 2te^{t^2-1}\, dt$

28. $\displaystyle\int_0^1 e^{-x}(e^{-x}+1)^{1/2}\, dx$

29. $\displaystyle\int_0^{e-1} \left(\frac{x}{x+1}\right) dx$

30. $\displaystyle\int_e^{e^2} \frac{1}{x\,(\ln x)^2}\, dx$

AREA BETWEEN CURVES *In Exercises 31 through 38, sketch the indicated region R and find its area by integration.*

31. *R* is the region under the curve $y = x + 2\sqrt{x}$ over the interval $1 \le x \le 4$.

32. *R* is the region under the curve $y = e^x + e^{-x}$ over the interval $-1 \le x \le 1$.

33. *R* is the region under the curve $y = \dfrac{1}{x} + x^2$ over the interval $0 \le x \le 2$.

34. *R* is the region under the curve $y = \sqrt{9 - 5x^2}$ over the interval $0 \le x \le 1$.

35. *R* is the region bounded by the curve $y = \dfrac{4}{x}$ and the line $x + y = 5$.

36. *R* is the region bounded by the curves $y = \dfrac{8}{x}$ and $y = \sqrt{x}$ and the line $x = 8$.

37. *R* is the region bounded by the curve $y = 2 + x - x^2$ and the *x* axis.

38. *R* is the triangular region with vertices $(0, 0)$, $(2, 4)$, and $(0, 6)$.

AVERAGE VALUE OF A FUNCTION *In Exercises 39 through 42, find the average value of the given function over the indicated interval.*

39. $f(x) = x^3 - 3x + \sqrt{2x}$; over $1 \le x \le 8$

40. $f(t) = t\sqrt[3]{8 - 7t^2}$; over $0 \le t \le 1$

41. $g(v) = ve^{-v^2}$; over $0 \le v \le 2$

42. $h(x) = \dfrac{e^x}{1 + 2e^x}$; over $0 \le x \le 1$

LORENTZ CURVES *In Exercises 43 through 46, sketch the Lorentz curve $y = L(x)$ and find the corresponding Gini index.*

43. $L(x) = x^{3/2}$

44. $L(x) = x^{1.2}$

45. $L(x) = 0.3x^2 + 0.7x$

46. $L(x) = 0.75x^2 + 0.25x$

SURVIVAL AND RENEWAL *In Exercises 47 through 50, an initial population P_0 is given along with a renewal rate $R(t)$ and a survival function $S(t)$. In each case, use the given information to find the population at the end of the indicated term T.*

47. $P_0 = 75\,000$; $R(t) = 60$; $S(t) = e^{-0.09t}$, t in months; term $T = 6$ months

48. $P_0 = 125\,000$; $R(t) = 250$; $S(t) = e^{-0.015t}$, t in years; term $T = 5$ years

49. $P_0 = 100\,000$; $R(t) = 90e^{0.1t}$; $S(t) = e^{-0.2t}$, t in years; term $T = 10$ years

50. $P_0 = 200\,000$; $R(t) = 50\,e^{0.12t}$; $S(t) = e^{-0.017t}$, t in hours; term $T = 20$ hours

VOLUME OF SOLID OF REVOLUTION   *In Exercises 51 through 54, find the volume of the solid of revolution formed by rotating the specified region R about the x axis.*

51. R is the region under the curve $y = x^2 + 1$ from $x = -1$ to $x = 2$.

52. R is the region under the curve $y = e^{-x/10}$ from $x = 0$ to $x = 10$.

53. R is the region under the curve $y = \dfrac{1}{\sqrt{x}}$ from $x = 1$ to $x = 3$.

54. R is the region under the curve $y = \dfrac{x + 1}{\sqrt{x}}$ from $x = 1$ to $x = 4$.

In Exercises 55 through 58, solve the given initial value problem.

55. $\dfrac{dy}{dx} = 2$, where $y = 4$ when $x = -3$

56. $\dfrac{dy}{dx} = x(x - 1)$, where $y = 1$ when $x = 1$

57. $\dfrac{dx}{dt} = e^{-2t}$, where $x = 4$ when $t = 0$

58. $\dfrac{dy}{dt} = \dfrac{t + 1}{t}$, where $y = 3$ when $t = 1$

59. Find the function whose tangent line has slope $x(x^2 + 1)^{-1}$ for each x and whose graph passes through the point $(1, 5)$.

60. Find the function whose tangent line has slope xe^{-2x^2} for each x and whose graph passes through the point $(0, -3)$.

61. NET ASSET VALUE   It is estimated that t days from now a farmer's crop will be increasing at a rate of $0.5t^2 + 4(t + 1)^{-1}$ kilograms per day. By how much will the value of the crop increase during the next 6 days if the market price remains fixed at \$2/kg?

62. DEPRECIATION   The resale value of a certain industrial machine decreases at a rate that changes with time. When the machine is t years old, the rate at which its value is changing is $200(t - 6)$ dollars per year. If the machine was bought new for \$12 000, how much will it be worth 10 years later?

63. TICKET SALES   The promoters of a county fair estimate that t hours after the gates open at 9:00 A.M., visitors will be entering the fair at a rate of $-4(t + 2)^3 + 54(t + 2)^2$ people per hour. How many people will enter the fair between 10:00 A.M. and noon?

64. MARGINAL COST   At a certain factory, the marginal cost is $6(q - 5)^2$ dollars per unit when the level of production is q units. By how much will the total manufacturing cost increase if the level of production is raised from 10 units to 13 units?

65. PUBLIC TRANSPORTATION   It is estimated that x weeks from now, the number of commuters using a new subway line will be increasing at a rate of $18x^2 + 500$ per week. Currently, 8000 commuters use the subway. How many will be using it 5 weeks from now?

66. NET CHANGE IN BIOMASS   A protein with mass m (in grams) disintegrates into amino acids at a rate given by

$$\frac{dm}{dt} = \frac{-15t}{t^2 + 5}.$$

What is the net change in mass of the protein during the first 4 hours?

67. CONSUMPTION OF OIL   It is estimated that t years from the beginning of 2010, the demand for oil in a certain country will be changing at a rate of $D'(t) = (1 + 2t)^{-1}$ billion barrels per year. Will more oil be consumed (demanded) during 2011 or during 2014? How much more?

68. FUTURE VALUE OF AN INCOME STREAM   Money is transferred continuously into an account at a rate of $5000e^{0.015t}$ dollars per year at time t (years). The account earns interest at an annual rate of 5% compounded continuously. How much will be in the account at the end of 3 years?

69. FUTURE VALUE OF AN INCOME STREAM   Money is transferred continuously into an account at a constant rate of \$1200 per year. The account earns interest at an annual rate of 8% compounded continuously. How much will be in the account at the end of 5 years?

70. PRESENT VALUE OF AN INCOME STREAM   What is the present value of an investment that will generate income continuously at a constant rate of \$1000 per year for 10 years if the prevailing annual interest rate remains fixed at 7% compounded continuously?

71. REAL ESTATE INVENTORY In a certain community the fraction of the houses placed on the market that remain unsold for at least t weeks is approximately $f(t) = e^{-0.2t}$. If 200 houses are currently on the market and if additional houses are placed on the market at a rate of 8 per week, approximately how many houses will be on the market 10 weeks from now?

72. AVERAGE REVENUE A bicycle manufacturer expects that x months from now consumers will be buying 5000 bicycles per month at a price of $P(x) = 200 + 3\sqrt{x}$ dollars per bicycle. What is the average revenue the manufacturer can expect from the sale of the bicycles over the next 16 months?

73. NUCLEAR WASTE A nuclear power plant produces radioactive waste at a constant rate of 300 kilograms per year. The waste decays exponentially with a half-life of 35 years. How much of the radioactive waste from the plant will remain after 200 years?

74. GROWTH OF A TREE A tree has been transplanted and after x years is growing at a rate of

$$h'(x) = 0.5 + \frac{1}{(x + 1)^2}$$

metres per year. By how much does the tree grow during the second year?

75. FUTURE REVENUE A certain oil well that yields 900 barrels of crude oil per month will run dry in 3 years. The price of crude oil is currently $92 per barrel and is expected to rise at a constant rate of 80 cents per barrel per month. If the oil is sold as soon as it is extracted from the ground, what will the total future revenue from the well be?

76. SURFACE AREA OF A HUMAN BODY The surface area S of the body of an average person 1.2 m tall who weighs w kilograms changes at a rate of

$$S'(w) = 0.01403w^{-0.4622}$$

square metres per kilogram. The body of a particular child who is 1.2 m tall and weighs 22 kg has surface area 1.1 m². If the child gains 1 kg while remaining the same height, by how much will the surface area of the child's body increase?

77. TEMPERATURE CHANGE At t hours past midnight, the temperature T (degrees Celsius) in

Yorkton, Saskatchewan, is found to be changing at a rate given by

$$T'(t) = -0.02(t - 7)(t - 14)$$

degrees Celsius per hour. By how much does the temperature change between 8 A.M. and 8 P.M.?

78. A* EFFECT OF A TOXIN A toxin is introduced to a bacterial colony, and t hours later, the population $P(t)$ of the colony is changing at a rate of

$$\frac{dP}{dt} = -(\ln 3)3^{4-t}.$$

If 1 million bacteria were in the colony when the toxin was introduced, what is $P(t)$? [*Hint:* Note that $3^x = e^{x \ln 3}$.]

79. MARGINAL ANALYSIS In a certain section of the country, the price of large Grade A eggs is currently $2.50 per dozen. Studies indicate that x weeks from now, the price $p(x)$ will be changing at a rate of $p'(x) = 0.2 + 0.003x^2$ cents per week.
  a. Use integration to find $p(x)$ and then use a graphing utility to sketch the graph of $p(x)$. How much will the eggs cost 10 weeks from now?
  b. Suppose the rate of change of the price were $p'(x) = 0.3 + 0.003x^2$. How does this affect $p(x)$? Check your conjecture by graphing the new price function on the same screen as the original. Now how much will the eggs cost in 10 weeks?

80. INVESTING IN A DOWN MARKET PERIOD Danielle opens a stock account with $5000 at the beginning of January and then deposits $200 per month. Unfortunately, the market is depressed, and she finds that t months after depositing a dollar, only $100f(t)$ cents remain, where $f(t) = e^{-0.01t}$. If this pattern continues, what will her account be worth after 2 years? [*Hint:* Think of this as a survival and renewal problem.]

81. DISTANCE AND VELOCITY After t minutes, an object moving along a line has velocity $v(t) = 1 + 4t + 3t^2$ metres per minute. How far does the object travel during the third minute?

82. AVERAGE POPULATION The population (in thousands) of a certain city t years after January 1, 2000, is given by the function

$$P(t) = \frac{150e^{0.03t}}{1 + e^{0.03t}}.$$

What was the average population of the city during the decade 2000–2010?

83. **DISTRIBUTION OF INCOME**   A study suggests that the distribution of incomes for social workers and physical therapists may be represented by the Lorentz curves $y = L_1(x)$ and $y = L_2(x)$, respectively, where

$L_1(x) = x^{1.6}$ and $L_2(x) = 0.65x^2 + 0.35x$.

For which profession is the distribution of income more equitable?

84. **AVERAGE PRICE**   Records indicate that t months after the beginning of the year, the price of bacon in local supermarkets was $P(t) = 0.06t^2 - 0.2t + 6.2$ dollars per kilogram. What was the average price of bacon during the first 6 months of the year?

85. **CONSERVATION**   A lake has roughly the same shape as the bottom half of the solid formed by rotating the curve $2x^2 + 3y^2 = 6$ about the x axis, for x and y measured in kilometres. Conservationists want the lake to contain 1000 trout/km^3. If the lake currently contains 5000 trout, how many more must be added to meet this requirement?

86. **HORTICULTURE**   A sprinkler system sprays water onto a garden so that $11e^{-r^2/10}$ centimetres of water per hour are delivered at a distance of r metres from the sprinkler. What is the total amount of water laid down by the sprinkler within a 5-m radius during a 20-minute watering period?

87. **A* SPEED AND DISTANCE**   A car is driven so that after t hours its speed is $v(t)$ kilometres per hour.
    a. Write a definite integral that gives the average speed of the car during the first N hours.
    b. Write a definite integral that gives the total distance the car travels during the first N hours.
    c. Discuss the relationship between the integrals in parts (a) and (b).

88. Use a graphing utility to draw the graphs of the curves $y = -x^3 - 2x^2 + 5x - 2$ and $y = x \ln x$ on the same screen. Use technology to find where the curves intersect, and then compute the area of the region bounded by the curves.

89. Repeat Exercise 88 for the curves

$$y = \frac{x - 2}{x + 1} \quad \text{and} \quad y = \sqrt{25 - x^2}.$$

THINK ABOUT IT

JUST-NOTICEABLE DIFFERENCES IN PERCEPTION

Calculus can help us answer questions about human perception, including questions relating to the number of different frequencies of sound or the number of different hues of light people can distinguish. When white light is split by refraction through a prism, we can all see blue, green, yellow, and red, but how do we perceive the many variations between green and blue? Questions 3 and 4 below will answer this, but first we look at human perception of differences in sound frequencies. Our goal is to show how integral calculus can be used to estimate the number of steps a person can distinguish as the frequency of sound increases from the lowest audible frequency of 15 hertz (Hz) to the highest audible frequency of 18 000 Hz. (The unit hertz, abbreviated Hz, represents cycles per second.)

(Photo: Getty Images/Science Photo Library RF)

A mathematical model* for human auditory perception uses the formula $y = 0.767x^{0.439}$, where y hertz is the smallest change in frequency that is detectable at frequency x hertz. Thus, at the low end of the range of human hearing, 15 Hz, the smallest change of frequency a person can detect is $y = 0.767 \times 15^{0.439}$ Hz ≈ 2.5 Hz, while at the upper end of human hearing, near 18 000 Hz, the least noticeable difference is approximately $y = 0.767 \times 18\ 000^{0.439}$ Hz ≈ 57 Hz. If the smallest noticeable change of frequency were the same for all frequencies that people can hear, we could find the number of noticeable steps in human hearing by simply dividing the total frequency range by the size of this smallest noticeable change. Unfortunately, we have

---

*Part of this essay is based on *Applications of Calculus: Selected Topics from the Environmental and Life Sciences,* by Anthony Barcellos, New York: McGraw-Hill, 1994, pp. 21–24.

just seen that the smallest noticeable change of frequency increases as frequency increases, so the simple approach will not work. However, we can estimate the number of distinguishable steps using integration.

To this end, let $y = f(x)$ represent the just-noticeable difference of frequency people can distinguish at frequency x. Next, choose numbers $x_0, x_1, \ldots, x_n$ beginning at $x_0 = 15$ Hz and working up through higher frequencies to $x_n = 18\ 000$ Hz so that for $j = 0, 2, \ldots, n - 1$,

$$x_j + f(x_j) = x_{j+1}.$$

In other words, x_{j+1} is the number we get by adding the just-noticeable difference at x_j to x_j itself. Thus, the jth step has length

$$\Delta x_j = x_{j+1} - x_j = f(x_j).$$

Dividing by $f(x_j)$, we get

$$\frac{\Delta x_j}{f(x_j)} = \frac{x_{j+1} - x_j}{f(x_j)} = 1,$$

and it follows that

$$\sum_{j=0}^{n-1} \frac{\Delta x_j}{f(x_j)} = \sum_{j=0}^{n-1} \frac{x_{j+1} - x_j}{f(x_j)} = \frac{x_1 - x_0}{f(x_0)} + \frac{x_2 - x_1}{f(x_1)} + \cdots + \frac{x_n - x_{n-1}}{f(x_n)}$$

$$= \underbrace{1 + 1 + \cdots + 1}_{n \text{ terms}} = n.$$

The sum on the left side of this equation is a Riemann sum, and since the step sizes $\Delta x_j = x_{j+1} - x_j$ are very small, the sum is approximately equal to a definite integral. Specifically, we have

$$\int_{x_0}^{x_n} \frac{dx}{f(x)} \approx \sum_{j=0}^{n-1} \frac{\Delta x_j}{f(x_j)} = n.$$

Finally, using the modelling formula $f(x) = 0.767x^{0.439}$ along with $x_0 = 15$ and $x_n = 18\ 000$, we find that

$$\int_{x_0}^{x_n} \frac{dx}{f(x)} = \int_{15}^{18\ 000} \frac{dx}{0.767x^{0.439}}$$

$$= \frac{1}{0.767}\left(\frac{x^{0.561}}{0.561}\right)\Bigg|_{15}^{18\ 000}$$

$$= 2.324(18\ 000^{0.561} - 15^{0.561})$$

$$= 556.2$$

Thus, there are approximately 556 just-noticeable steps in the audible range from 15 Hz to 18 000 Hz.

Here are some questions in which you are asked to apply these principles to issues involving both auditory and visual perception.

Questions

1. The 88 keys of a piano range from 15 Hz to 4186 Hz. If the number of keys were based on the number of just-noticeable differences, how many keys would a piano have?

2. An 8-bit grey-scale monitor can display 256 shades of grey. Let x represent the darkness of a shade of grey, where $x = 0$ for white and $x = 1$ for totally black. One model for grey-scale perception uses the formula $y = Ax^{0.3}$, where A is a positive constant and y is the smallest change detectable by the human eye at grey-level x. Experiments show that the human eye is incapable of distinguishing as many as 256 different shades of grey, so the number n of just-noticeable shading differences from $x = 0$ to $x = 1$ must be less than 256. Using the assumption that $n < 256$, find a lower bound for the constant A in the modelling formula $y = Ax^{0.3}$.

3. One model of the ability of human vision to distinguish colours of different hue uses the formula $y = (2.9 \times 10^{-24})x^{8.52}$, where y is the just-noticeable difference for a colour of wavelength x, with both x and y measured in nanometres (nm).
   a. Blue-green light has a wavelength of 580 nm. What is the least noticeable difference at this wavelength?
   b. Red light has a wavelength of 760 nm. What is the least noticeable difference at this wavelength?
   c. How many just-noticeable steps are there in hue from blue-green light to red light?

4. Find a model of the form $y = ax^k$ for just-noticeable differences in hue for the colour spectrum from blue-green light at 580 nm to violet light at 400 nm. Use the fact that the minimum noticeable difference at the wavelength of blue-green light is 1 nm, while at the wavelength of violet light, the minimum noticeable difference is 0.043 nm.

Chapter

6

Nuclear power does not produce carbon emissions, although nuclear waste is produced. The radioactivity in this waste decays exponentially. The long-term effects of accumulating nuclear waste as time increases without bound can be determined by one of the methods in this chapter. See Example 6.2.5, which uses improper integration, to find the effects in the long run. (Photo: Steve Allan/ Brand X Pictures)

ADDITIONAL TOPICS IN INTEGRATION

CHAPTER OUTLINE

LEARNING OBJECTIVES

After completing this chapter, you should be able to

L01 Evaluate integrals using integration by parts. Evaluate integrals using tables of integrals.

L02 Evaluate improper integrals.

L03 Approximate areas and integrals using Simpson's rule and the trapezoidal rule.

Integration by Parts; Integral Tables

Evaluate integrals using integration by parts. Evaluate integrals using tables of integrals.

Integration by parts is a technique of integration based on the product rule for differentiation. In particular, if $u(x)$ and $v(x)$ are both differentiable functions of x, then

$$\frac{d}{dx}[u(x)v(x)] = u(x)\frac{dv}{dx} + v(x)\frac{du}{dx}$$

so that

$$u(x)\frac{dv}{dx} = \frac{d}{dx}[u(x)v(x)] - v(x)\frac{du}{dx}$$

Integrating both sides of this equation with respect to x gives

$$\int \left[u(x)\frac{dv}{dx} \right] dx = \int \frac{d}{dx}[u(x)v(x)]\, dx - \int \left[v(x)\frac{du}{dx} \right] dx$$

$$= u(x)v(x) - \int \left[v(x)\frac{du}{dx} \right] dx$$

since $u(x)v(x)$ is an antiderivative of $\dfrac{d}{dx}[u(x)v(x)]$. Moreover, we can write this integral formula in the more compact form

$$\int u\, dv = uv - \int v\, du,$$

since

$$dv = \frac{dv}{dx}\, dx \quad \text{and} \quad du = \frac{du}{dx}\, dx.$$

The equation $\int u\, dv = uv - \int v\, du$ is called the **integration by parts formula.** The great value of this formula is that if we can find functions u and v so that a given integral $\int f(x)\, dx$ can be expressed in the form $\int f(x)\, dx = \int u\, dv$, then we have

$$\int f(x)\, dx = \int u\, dv = uv - \int v\, du,$$

and the given integral is effectively exchanged for the integral $\int v\, du$. If the integral $\int v\, du$ is easier to compute than $\int u\, dv$, the exchange facilitates finding $\int f(x)\, dx$. Here is an example.

EXAMPLE 6.1.1

Find $\int x^2 \ln x\, dx$.

Solution

Our strategy is to express $\int x^2 \ln x \, dx$ as $\int u \, dv$ by choosing u and v so that $\int v \, du$ is easier to evaluate than $\int u \, dv$. This strategy suggests that we choose

$$u = \ln x \quad \text{and} \quad dv = x^2 \, dx,$$

since

$$du = \frac{1}{x} dx$$

is a simpler expression than $\ln x$, while v can be obtained by the relatively easy integration

$$v = \int x^2 \, dx = \frac{1}{3} x^3.$$

(For simplicity, we leave the "$+ C$" out of the calculation until the final step.) Substituting this choice for u and v into the integration by parts formula gives

$$\int \underbrace{x^2}_{} \underbrace{\ln x \, dx}_{} = \int \underbrace{(\ln x)}_{u} \underbrace{(x^2 dx)}_{dv} = \underbrace{(\ln x)}_{u} \underbrace{\left(\frac{1}{3} x^3\right)}_{v} - \int \underbrace{\left(\frac{1}{3} x^3\right)}_{v} \underbrace{\left(\frac{1}{x} dx\right)}_{du}$$

$$= \frac{1}{3} x^3 \ln x - \frac{1}{3} \int x^2 \, dx = \frac{1}{3} x^3 \ln x - \frac{1}{3}\left(\frac{1}{3} x^3\right) + C$$

$$= \frac{1}{3} x^3 \ln x - \frac{1}{9} x^3 + C$$

Here is a summary of the procedure we have just illustrated.

Integration by Parts

To find an integral $\int f(x) \, dx$ using the integration by parts formula, follow these steps:

Step 1. Choose functions u and v so that $f(x) \, dx = u \, dv$. Try to pick u so that du is simpler than u, and choose a dv that is easy to integrate.

Step 2. Organize the computation of du and v as follows:

$$u \qquad dv$$
$$du \qquad v = \int dv$$

Step 3. Complete the integration by finding $\int v \, du$. Then

$$\int f(x) \, dx = \int u \, dv = uv - \int v \, du.$$

Add "$+ C$" only at the end of the computation.

Choosing a suitable u and dv for integration by parts requires insight and experience. For instance, in Example 6.1.1, things would not have gone so smoothly if we had chosen $u = x^2$ and $dv = \ln x \, dx$. Certainly $du = 2x \, dx$ is simpler than $u = x^2$, but what is $v = \int \ln x \, dx$? In fact, finding this integral is just as hard as finding the original integral $\int x^2 \ln x \, dx$ (see Example 6.1.4). Examples 6.1.2, 6.1.3, and 6.1.4 illustrate several ways of choosing u and dv in integrals that can be handled using integration by parts.

EXAMPLE 6.1.2

Find $\int xe^{2x} \, dx$.

Solution

Although both factors x and e^{2x} are easy to integrate, only x becomes simpler when differentiated. Therefore, we choose $u = x$ and $dv = e^{2x} \, dx$ and find

$$u = x \qquad dv = e^{2x} \, dx$$
$$du = dx \qquad v = \frac{1}{2}e^{2x}$$

Substituting into the integration by parts formula, we obtain

$$\int \underbrace{x}_{u}\underbrace{(e^{2x}\,dx)}_{dv} = \underbrace{x}_{u}\underbrace{\left(\frac{1}{2}e^{2x}\right)}_{v} - \int \underbrace{\left(\frac{1}{2}e^{2x}\right)}_{v}\underbrace{dx}_{du}$$

$$= \frac{1}{2}xe^{2x} - \frac{1}{2}\left(\frac{1}{2}e^{2x}\right) + C$$

$$= \frac{1}{2}\left(x - \frac{1}{2}\right)e^{2x} + C$$

EXAMPLE 6.1.3

Find $\int x\sqrt{x + 5} \, dx$.

Solution

Again, both factors x and $\sqrt{x + 5}$ are easy to differentiate and to integrate, but x is simplified by differentiation, while the derivative of $\sqrt{x + 5}$ is even more complicated than $\sqrt{x + 5}$ itself. This observation suggests choosing

$$u = x \quad \text{and} \quad dv = \sqrt{x + 5} \, dx = (x + 5)^{1/2} \, dx,$$

so that

$$du = dx \quad \text{and} \quad v = \frac{2}{3}(x + 5)^{3/2}.$$

Substituting into the integration by parts formula gives

$$\int \underbrace{x}_{u} \underbrace{(\sqrt{x+5}\,dx)}_{dv} = \underbrace{x}_{u}\underbrace{\left[\frac{2}{3}(x+5)^{3/2}\right]}_{v} - \int \underbrace{\left[\frac{2}{3}(x+5)^{3/2}\right]}_{v}\underbrace{dx}_{du}$$

$$= \frac{2}{3}x(x+5)^{3/2} - \frac{2}{3}\left[\frac{2}{5}(x+5)^{5/2}\right] + C$$

$$= \frac{2}{3}x(x+5)^{3/2} - \frac{4}{15}(x+5)^{5/2} + C$$

NOTE   Some integrals can be evaluated by either substitution or integration by parts. For instance, the integral in Example 6.1.3 can be found by substituting as follows:

Let $u = x + 5$. Then $du = dx$ and $x = u - 5$, and

$$\int x\sqrt{x+5}\,dx = \int (u-5)\sqrt{u}\,du = \int (u^{3/2} - 5u^{1/2})\,du$$

$$= \frac{u^{5/2}}{5/2} - \frac{5u^{3/2}}{3/2} + C$$

$$= \frac{2}{5}(x+5)^{5/2} - \frac{10}{3}(x+5)^{3/2} + C$$

This form of the integral is not the same as that found in Example 6.1.3. To show that the two forms are equivalent, note that both antiderivatives can be reduced to the same expression:

$$\frac{2x}{3}(x+5)^{3/2} - \frac{4}{15}(x+5)^{5/2} = (x+5)^{3/2}\left[\frac{2x}{3} - \frac{4}{15}(x+5)\right]$$

$$= (x+5)^{3/2}\left(\frac{2x}{5} - \frac{4}{3}\right)$$

and

$$\frac{2}{5}(x+5)^{5/2} - \frac{10}{3}(x+5)^{3/2} = (x+5)^{3/2}\left[\frac{2}{5}(x+5) - \frac{10}{3}\right]$$

$$= (x+5)^{3/2}\left(\frac{2x}{5} - \frac{4}{3}\right)$$

This example shows that it is quite possible for you to do everything right and still not get the answer given at the back of the book.

Definite Integration by Parts

The integration by parts formula can be applied to definite integrals by noting that

$$\int_{a}^{b} u\,dv = uv\Big|_{a}^{b} - \int_{a}^{b} v\,du.$$

Definite integration by parts is used in Example 6.1.4 to find an area.

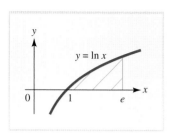

FIGURE 6.1 The region under $y = \ln x$ over $1 \leq x \leq e$.

EXAMPLE 6.1.4

Find the area of the region bounded by the curve $y = \ln x$, the x axis, and the lines $x = 1$ and $x = e$.

Solution

The region is shown in Figure 6.1. Since $\ln x \geq 0$ for $1 \leq x \leq e$, the area is given by the definite integral

$$A = \int_1^e \ln x \, dx.$$

To evaluate this integral using integration by parts, think of $\ln x \, dx$ as $(\ln x)(1 \, dx)$ and use

$$u = \ln x \qquad dv = 1 \, dx$$
$$du = \frac{1}{x} \, dx \qquad v = \int 1 \, dx = x$$

Thus, the required area is

$$
\begin{aligned}
A &= \int_1^e \ln x \, dx = x \ln x \Big|_1^e - \int_1^e x\left(\frac{1}{x} dx\right) \\
&= x \ln x \Big|_1^e - \int_1^e 1 \, dx = x \ln x - x \Big|_1^e \\
&= [e \ln e - e] - [1 \ln 1 - 1] \\
&= [e(1) - e] - [1(0) - 1] \qquad\qquad \ln e = 1 \text{ and } \ln 1 = 0 \\
&= 1
\end{aligned}
$$

As another illustration of the role played by integration by parts in applications, we use it in Example 6.1.5 to compute the future value of a continuous income stream (see Section 5.5).

EXAMPLE 6.1.5

Emily is considering a 5-year investment and estimates that t years from now it will be generating a continuous income stream of $3000 + 50t$ dollars per year. If the prevailing annual interest rate remains fixed at 4% compounded continuously during the entire 5-year term, what should the investment be worth in 5 years?

Solution

We measure the worth of Emily's investment by the future value of the income flow over the 5-year term. Recall (from Section 5.5) that an income stream deposited continuously at a rate of $f(t)$ into an account that earns interest at an annual rate r compounded continuously for a term of T years has future value FV given by the integral

$$\text{FV} = \int_0^T f(t) e^{r(T-t)} \, dt.$$

For this investment, $f(t) = 3000 + 50t$, $r = 0.04$, and $T = 5$, so the future value is given by the integral

$$FV = \int_0^5 (3000 + 50t)e^{0.04(5-t)}\, dt.$$

Integrating by parts with

$$u = 3000 + 50t \qquad dv = e^{0.04(5-t)}\, dt$$
$$du = 50dt \qquad v = \frac{e^{0.04(5-t)}}{-0.04} = -25e^{0.04(5-t)}$$

gives

$$FV = (3000 + 50t)(-25)e^{0.04(5-t)}\Big|_0^5 - \int_0^5 (50)(-25)e^{0.04(5-t)}\, dt$$

$$= (-75\,000 - 1250t)e^{0.04(5-t)}\Big|_0^5 + 1250\left[\frac{e^{0.04(5-t)}}{-0.04}\right]\Big|_0^5$$

$$= (-106\,250 - 1250t)e^{0.04(5-t)}\Big|_0^5 \qquad\qquad \text{combine terms}$$

$$= [-106\,250 - 1250(5)]e^0 - [-106\,250 - 1250(0)]e^{0.04(5)}$$

$$\approx 17\,274.04$$

Thus, in 5 years, Emily's investment will be worth roughly \$17 274.

Repeated Application of Integration by Parts

Sometimes integration by parts leads to a new integral that also must be integrated by parts. This situation is illustrated in Example 6.1.6.

EXAMPLE 6.1.6

Find $\int x^2 e^{2x}\, dx$.

Solution

Since the factor e^{2x} is easy to integrate and x^2 is simplified by differentiation, we choose

$$u = x^2 \quad \text{and} \quad dv = e^{2x}\, dx$$

so that

$$du = 2x\, dx \quad \text{and} \quad v = \int e^{2x}dx = \frac{1}{2}e^{2x}.$$

Integrating by parts, we get

$$\int x^2 e^{2x}\, dx = x^2\left(\frac{1}{2}e^{2x}\right) - \int\left(\frac{1}{2}e^{2x}\right)(2x\, dx)$$

$$= \frac{1}{2}x^2 e^{2x} - \int xe^{2x}\, dx$$

The integral $\int xe^{2x}dx$ that remains can also be obtained using integration by parts. Indeed, in Example 6.1.2, we found that

$$\int xe^{2x}dx = \frac{1}{2}\left(x - \frac{1}{2}\right)e^{2x} + C.$$

Thus,

$$\int x^2e^{2x}dx = \frac{1}{2}x^2e^{2x} - \int xe^{2x}dx$$

$$= \frac{1}{2}x^2e^{2x} - \left[\frac{1}{2}\left(x - \frac{1}{2}\right)e^{2x}\right] + C$$

$$= \frac{1}{2}x^2e^{2x} - \frac{1}{2}xe^{2x} + \frac{1}{4}e^{2x} + C$$

$$= \frac{1}{4}(2x^2 - 2x + 1)e^{2x} + C$$

Using Integral Tables

Most integrals you will encounter in the social, managerial, and life sciences can be evaluated using the basic formulas given in Section 5.1 along with substitution and integration by parts. However, occasionally you may encounter an integral that cannot be handled by these methods. Computer software can be used to solve integrals such as $\int \frac{e^x}{x}\,dx$, but others can be found by using a **table of integrals.**

Table 6.1, a short table of integrals, is listed on pages 462 and 463. Note that the table is divided into sections such as "forms involving $\sqrt{u^2 - a^2}$," and that formulas are given in terms of constants denoted a, b, and n. The use of this table is demonstrated in Examples 6.1.7 through 6.1.11.

EXAMPLE 6.1.7

Find $\int \frac{1}{x(3x - 6)}\,dx$.

Solution
Apply Formula 6 with $a = -6$ and $b = 3$ to obtain

$$\int \frac{1}{x(3x - 6)}\,dx = -\frac{1}{6}\ln\left|\frac{x}{3x - 6}\right| + C.$$

EXAMPLE 6.1.8

Find $\int \frac{1}{6 - 3x^2}\,dx$.

Solution

If the coefficient of x^2 were 1 instead of 3, you could use Formula 16. This suggests that you first rewrite the integrand as

$$\frac{1}{6 - 3x^2} = \frac{1}{3}\left(\frac{1}{2 - x^2}\right)$$

and then apply Formula 16 with $a = \sqrt{2}$:

$$\int \frac{1}{6 - 3x^2}\,dx = \frac{1}{3}\int \frac{1}{2 - x^2}\,dx$$

$$= \frac{1}{3}\left(\frac{1}{2\sqrt{2}}\right)\ln\left|\frac{\sqrt{2} + x}{\sqrt{2} - x}\right| + C$$

EXAMPLE 6.1.9

Find $\displaystyle\int \sqrt{x^2 + 9}\,dx$.

Solution

This one can be matched to Formula 9 with $a = 3$:

$$\int \sqrt{x^2 + 9}\,dx = \frac{x}{2}\sqrt{x^2 + 9} + \frac{9}{2}\ln\left|x + \sqrt{x^2 + 9}\right| + C.$$

EXAMPLE 6.1.10

Find $\displaystyle\int \frac{1}{\sqrt{4x^2 - 9}}\,dx$.

Solution

To put this integral in the form of Formula 20, rewrite the integrand as

$$\frac{1}{\sqrt{4x^2 - 9}} = \frac{1}{\sqrt{4(x^2 - 9/4)}} = \frac{1}{2\sqrt{x^2 - 9/4}}.$$

Then apply the formula with $a^2 = \dfrac{9}{4}$ to get

$$\int \frac{1}{\sqrt{4x^2 - 9}}\,dx = \frac{1}{2}\int \frac{1}{\sqrt{x^2 - 9/4}}\,dx = \frac{1}{2}\ln\left|x + \sqrt{x^2 - 9/4}\right| + C.$$

Our final example involves the use of Formula 26. This is called a **reduction formula** because it allows us to express a given integral in terms of a simpler integral of the same form.

TABLE 6.1 A Short Table of Integrals

Forms Involving $a + bu$

1. $\displaystyle\int \frac{u\,du}{a + bu} = \frac{1}{b^2}[a + bu - a\ln|a + bu|] + C$

2. $\displaystyle\int \frac{u^2\,du}{a + bu} = \frac{1}{2b^3}[(a + bu)^2 - 4a(a + bu) + 2a^2\ln|a + bu|] + C$

3. $\displaystyle\int \frac{u\,du}{(a + bu)^2} = \frac{1}{b^2}\left[\frac{a}{a + bu} + \ln|a + bu|\right] + C$

4. $\displaystyle\int \frac{u\,du}{\sqrt{a + bu}} = \frac{2}{3b^2}(bu - 2a)\sqrt{a + bu} + C$

5. $\displaystyle\int \frac{du}{u\sqrt{a + bu}} = \frac{1}{\sqrt{a}}\ln\left|\frac{\sqrt{a + bu} - \sqrt{a}}{\sqrt{a + bu} + \sqrt{a}}\right| + C, a > 0$

6. $\displaystyle\int \frac{du}{u(a + bu)} = \frac{1}{a}\ln\left|\frac{u}{a + bu}\right| + C$

7. $\displaystyle\int \frac{du}{u^2(a + bu)} = -\frac{1}{a}\left[\frac{1}{u} + \frac{b}{a}\ln\left|\frac{u}{a + bu}\right|\right] + C$

8. $\displaystyle\int \frac{du}{u^2(a + bu)^2} = -\frac{1}{a^2}\left[\frac{a + 2bu}{u(a + bu)} + \frac{2b}{a}\ln\left|\frac{u}{a + bu}\right|\right] + C$

Forms Involving $\sqrt{a^2 + u^2}$

9. $\displaystyle\int \sqrt{a^2 + u^2}\,du = \frac{u}{2}\sqrt{a^2 + u^2} + \frac{a^2}{2}\ln|u + \sqrt{a^2 + u^2}| + C$

10. $\displaystyle\int \frac{du}{\sqrt{a^2 + u^2}} = \ln|u + \sqrt{a^2 + u^2}| + C$

11. $\displaystyle\int \frac{du}{u\sqrt{a^2 + u^2}} = -\frac{1}{a}\ln\left|\frac{\sqrt{a^2 + u^2} + a}{u}\right| + C$

12. $\displaystyle\int \frac{du}{(a^2 + u^2)^{3/2}} = \frac{u}{a^2\sqrt{a^2 + u^2}} + C$

13. $\displaystyle\int u^2\sqrt{a^2 + u^2}\,du = \frac{u}{8}(a^2 + 2u^2)\sqrt{a^2 + u^2} - \frac{a^4}{8}\ln|u + \sqrt{a^2 + u^2}| + C$

Forms Involving $\sqrt{a^2 - u^2}$

14. $\displaystyle \int \frac{du}{u\sqrt{a^2 - u^2}} = -\frac{1}{a} \ln\left|\frac{a + \sqrt{a^2 - u^2}}{u}\right| + C$

15. $\displaystyle \int \frac{du}{u^2\sqrt{a^2 - u^2}} = \frac{\sqrt{a^2 - u^2}}{a^2 u} + C$

16. $\displaystyle \int \frac{du}{a^2 - u^2} = \frac{1}{2a} \ln\left|\frac{a + u}{a - u}\right| + C$

17. $\displaystyle \int \frac{\sqrt{a^2 - u^2}}{u} du = \sqrt{a^2 - u^2} - a \ln\left|\frac{a + \sqrt{a^2 - u^2}}{u}\right| + C$

Forms Involving $\sqrt{u^2 - a^2}$

18. $\displaystyle \int \sqrt{u^2 - a^2}\, du = \frac{u}{2}\sqrt{u^2 - a^2} - \frac{a^2}{2} \ln\left|u + \sqrt{u^2 - a^2}\right| + C$

19. $\displaystyle \int \frac{\sqrt{u^2 - a^2}}{u^2} du = -\frac{\sqrt{u^2 - a^2}}{u} + \ln\left|u + \sqrt{u^2 - a^2}\right| + C$

20. $\displaystyle \int \frac{du}{\sqrt{u^2 - a^2}} = \ln\left|u + \sqrt{u^2 - a^2}\right| + C$

21. $\displaystyle \int \frac{du}{u^2\sqrt{u^2 - a^2}} = \frac{\sqrt{u^2 - a^2}}{a^2 u} + C$

Forms Involving e^{au} and $\ln u$

22. $\displaystyle \int u e^{au} du = \frac{1}{a^2}(au - 1)e^{au} + C$

23. $\displaystyle \int \ln u \, du = u \ln|u| - u + C$

24. $\displaystyle \int \frac{du}{u \ln u} = \ln|\ln u| + C$

25. $\displaystyle \int u^m \ln u \, du = \frac{u^{m+1}}{m + 1}\left(\ln u - \frac{1}{m + 1}\right), \quad m \neq -1$

Reduction Formulas

26. $\displaystyle \int u^n e^{au} \, du = \frac{1}{a}u^n e^{au} - \frac{n}{a}\int u^{n-1}e^{au}du$

27. $\displaystyle \int (\ln u)^n \, du = u(\ln u)^n - n \int (\ln u)^{n-1} du$

28. $\displaystyle \int u^n\sqrt{a + bu}\, du = \frac{2}{b(2n + 3)}[u^n(a + bu)^{3/2} - na\int u^{n-1}\sqrt{a + bu}\, du] \quad \text{for } n \neq -\frac{3}{2}$

EXAMPLE 6.1.11

Find $\int x^3 e^{5x}\, dx$.

Solution

Apply Formula 26 with $n = 3$ and $a = 5$ to get

$$\int x^3 e^{5x}\, dx = \frac{1}{5} x^3 e^{5x} - \frac{3}{5} \int x^2 e^{5x}\, dx.$$

Now apply Formula 26 again, this time with $n = 2$ and $a = 5$, to get

$$\int x^2 e^{5x}\, dx = \frac{1}{5} x^2 e^{5x} - \frac{2}{5} \int x e^{5x}\, dx.$$

Then apply the formula one more time (with $n = 1$, $a = 5$) to evaluate the integral on the right:

$$\int x e^{5x}\, dx = \frac{1}{5} x e^{5x} - \frac{1}{5} \int x^0 e^{5x}\, dx$$

$$= \frac{1}{5} x e^{5x} - \frac{1}{5} \int (1) e^{5x}\, dx$$

$$= \frac{1}{5} x e^{5x} - \frac{1}{5} \left(\frac{1}{5} e^{5x} \right) + C$$

Finally, evaluate the original integral by combining these results:

$$\int x^3 e^{5x}\, dx = \frac{1}{5} x^3 e^{5x} - \frac{3}{5} \left[\frac{1}{5} x^2 e^{5x} - \frac{2}{5} \int x e^{5x}\, dx \right]$$

$$= \frac{1}{5} x^3 e^{5x} - \frac{3}{25} x^2 e^{5x} + \frac{6}{25} \left[\frac{1}{5} x e^{5x} - \frac{1}{5} \left(\frac{1}{5} e^{5x} \right) \right] + C$$

$$= \frac{1}{5} x^3 e^{5x} - \frac{3}{25} x^2 e^{5x} + \frac{6}{125} x e^{5x} - \frac{6}{625} e^{5x} + C$$

$$= \left[\frac{1}{5} x^3 - \frac{3}{25} x^2 + \frac{6}{125} x - \frac{6}{625} \right] e^{5x} + C$$

COMPUTER ALGEBRA SYSTEMS   Many calculators, as well as computer software such as Maple, MathCad, and Mathematica, contain a computer algebra system (CAS) that will compare a given integrand to integrands in a stored table.

EXERCISES ■ 6.1

In Exercises 1 through 26, use integration by parts to find the given integral.

1. $\int x e^{-x}\, dx$

2. $\int x e^{x/2}\, dx$

3. $\int (1 - x) e^x\, dx$

4. $\int (3 - 2x) e^{-x}\, dx$

5. $\int t \ln 2t\, dt$

6. $\displaystyle \int t \ln t^2 dt$

7. $\displaystyle \int v e^{-v/5} dv$

8. $\displaystyle \int w e^{0.1w} dw$

9. $\displaystyle \int x \sqrt{x-6}\, dx$

10. $\displaystyle \int x \sqrt{1-x}\, dx$

11. $\displaystyle \int x(x+1)^8 dx$

12. $\displaystyle \int (x+1)(x+2)^6 dx$

13. $\displaystyle \int \frac{x}{\sqrt{x+2}}\, dx$

14. $\displaystyle \int \frac{x}{\sqrt{2x+1}}\, dx$

15. $\displaystyle \int_{-1}^{4} \frac{x}{\sqrt{x+5}}\, dx$

16. $\displaystyle \int_{0}^{2} \frac{x}{\sqrt{4x+1}}\, dx$

17. $\displaystyle \int_{0}^{1} \frac{x}{e^{2x}}\, dx$

18. $\displaystyle \int_{1}^{e} \frac{\ln x}{x^2}\, dx$

19. $\displaystyle \int_{1}^{e^2} x \ln \sqrt[3]{x}\, dx$

20. $\displaystyle \int_{0}^{1} x(e^{-2x} + e^{-x})\, dx$

21. $\displaystyle \int_{1/2}^{e/2} t \ln 2t\, dt$

22. $\displaystyle \int_{1}^{2} (t-1)e^{1-t} dt$

23. $\displaystyle \int \frac{\ln x}{x^2}\, dx$

24. $\displaystyle \int x(\ln x)^2 dx$

25. $\displaystyle \int x^3 e^{x^2} dx$

[*Hint:* Use $dv = x e^{x^2} dx$.]

26. $\displaystyle \int \frac{x^3}{\sqrt{x^2+1}}\, dx$

$\left[\textit{Hint: Use } dv = \dfrac{x}{\sqrt{x^2+1}}\, dx. \right]$

Use the table of integrals (Table 6.1) to find the integrals in Exercises 27 through 38.

27. $\displaystyle \int \frac{x\, dx}{3 - 5x}$

28. $\displaystyle \int \sqrt{x^2 - 9}\, dx$

29. $\displaystyle \int \frac{\sqrt{4x^2 - 9}}{x^2}\, dx$

30. $\displaystyle \int \frac{dx}{(9 + 2x^2)^{3/2}}$

31. $\displaystyle \int \frac{dx}{x(2 + 3x)}$

32. $\displaystyle \int \frac{t\, dt}{\sqrt{4 - 5t}}$

33. $\displaystyle \int \frac{du}{16 - 3u^2}$

34. $\displaystyle \int w e^{-3w} dw$

35. $\displaystyle \int (\ln x)^3 dx$

36. $\displaystyle \int x^2 \sqrt{2 + 5x}\, dx$

37. $\displaystyle \int \frac{dx}{x^2(5 + 2x)^2}$

38. $\displaystyle \int \frac{\sqrt{9 - x^2}}{x}\, dx$

39. Find the function whose tangent line has slope $(x+1)e^{-x}$ for each value of x and whose graph passes through the point $(1, 5)$.

40. Find the function whose tangent line has slope $x \ln \sqrt{x}$ for each value of $x > 0$ and whose graph passes through the point $(2, -3)$.

41. **FUNDRAISING**   After t weeks, contributions in response to a local fundraising campaign were coming in at a rate of $2000t e^{-0.2t}$ dollars per week. How much money was raised during the first 5 weeks?

42. **MARGINAL COST**   A manufacturer has found that the marginal cost is $(0.1q + 1)e^{0.03q}$ dollars per unit when q units have been produced. The total cost of producing 10 units is $200. What is the total cost of producing the first 20 units?

43. **EFFICIENCY**   After t hours on the job, a factory worker can produce $100te^{-0.5t}$ units per hour. How many units does the worker produce during the first 3 hours?

44. **DISTANCE**   After t seconds, an object is moving with velocity $te^{-t/2}$ metres per second. Express the position of the object as a function of time.

45. **POPULATION GROWTH**   It is projected that t years from now the population of a certain city will be changing at a rate of $t \ln\sqrt{t + 1}$ thousand people per year. If the current population is 2 million, what will the population be 5 years from now?

46. **POPULATION GROWTH**   The population $P(t)$ (thousands) of a bacterial colony t hours after the introduction of a toxin is changing at a rate of $P'(t) = (1 - 0.5t)e^{0.5t}$ thousand bacteria per hour. By how much does the population change during the fourth hour?

Exercises 47 through 59 involve applications developed in Sections 5.4, 5.5, and 5.6 of Chapter 5.

47. **AVERAGE DRUG CONCENTRATION**   The concentration of a drug t hours after injection into a patient's bloodstream is $C(t) = 4\,te^{(2-0.3t)}$ milligrams per millilitre. What is the average concentration of drug in the patient's bloodstream over the first 6 hours after the injection?

48. **AVERAGE DEMAND**   A manufacturer determines that when x hundred units of a particular commodity are produced, the profit generated is $P(x)$ thousand dollars, where

$$P(x) = \frac{500 \ln (x + 1)}{(x + 1)^2}.$$

What is the average profit over the production range $0 \le x \le 10$?

49. **FUTURE VALUE OF AN INVESTMENT**   Money is transferred into an account at a rate of $R(t) = 3000 + 5t$ dollars per year for 10 years, where t is the number of years after 2010. If the account pays 5% interest compounded continuously, how much will be in the account at the end of the 10-year investment period (in 2020)?

50. **FUTURE VALUE OF AN INVESTMENT**   Money is transferred into an account at a rate of $R(t) = 1000te^{-0.3t}$ dollars per year for 5 years. If the account pays 4% interest compounded continuously, how much will accumulate in the account over a 5-year period?

51. **PRESENT VALUE OF AN INVESTMENT**   An investment will generate income continuously at a rate of $R(t) = 20 + 3t$ hundred dollars per year for 5 years. If the prevailing interest rate is 7% compounded continuously, what is the present value of the investment?

52. **INVESTMENT EVALUATION**   The management of a national chain of pizza parlours is selling a 6-year franchise to operate its newest outlet in Winnipeg, Manitoba. Experience in similar locations suggests that t years from now, the franchise will be generating profit continuously at a rate of $R(t) = 300 + 5t$ thousand dollars per year. If the prevailing rate of interest remains fixed during the next 6 years at 6% compounded continuously, what would be a fair price to charge for the franchise? [*Hint:* Use present value to measure what the franchise is worth.]

53. **GROUP MEMBERSHIP**   A book club has compiled statistics suggesting that the fraction of its members who are still active t months after joining is given by the function $S(t) = e^{-0.02t}$. The club currently has 5000 members and expects to attract new members at a rate of $R(t) = 5t$ per month. How many members can the club expect to have at the end of 9 months? [*Hint:* Think of this as a survival/renewal problem.]

54. **LORENTZ CURVE**   Find the Gini index for an income distribution whose Lorentz curve is the graph of the function $L(x) = xe^{x-1}$ for $0 \le x \le 1$.

55. **COMPARING INCOME DISTRIBUTIONS**   In a certain province, the Lorentz curves for the distributions of income for lawyers and engineers are $y = L_1(x)$ and $y = L_2(x)$, respectively, where

$$L_1(x) = 0.6x^2 + 0.4x \qquad \text{and} \qquad L_2(x) = x^2e^{x-1}.$$

Find the Gini index for each curve. Which profession has the more equitable distribution of income?

56. **A*** Use integration by parts to verify reduction Formula 27:

$$\int (\ln u)^n \, du = u(\ln u)^n - n \int (\ln u)^{n-1} du.$$

57. A* Use integration by parts to verify reduction Formula 26:

$$\int u^n e^{au} du = \frac{1}{a} u^n e^{au} - \frac{n}{a} \int u^{n-1} e^{au} du.$$

58. A* Find a reduction formula for

$$\int u^n (\ln u)^m du.$$

CENTRE OF A REGION *Let $f(x)$ be a continuous function with $f(x) \geq 0$ for $a \leq x \leq b$. Let R be the region bounded by the curve $y = f(x)$, the x axis, and the lines $x = a$ and $x = b$. Then the centroid (or centre) of R is the point $(\bar{x}, \bar{y})$, where*

$$\bar{x} = \frac{1}{A} \int_a^b x f(x)\, dx \quad \text{and} \quad \bar{y} = \frac{1}{2A} \int_a^b [f(x)]^2\, dx$$

and A is the area of R. In Exercises 59 and 60, find the centroid of the region shown. You may need to use one or more formulas from Table 6.1.

59. A*

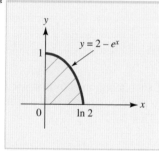

$y = 2 - e^x$

EXERCISE 59

60. A*

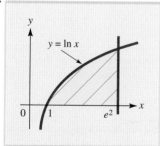

$y = \ln x$

EXERCISE 60

61. A* SHOPPING MALL SECURITY   The shaded region shown in the accompanying figure is a parking lot for a shopping mall. The dimensions are in hundreds of metres. To improve parking security, the mall manager plans to place a surveillance kiosk in the centre of the lot (see the definition preceding Exercises 59 and 60).

   a. Where would you place the kiosk? Describe the location you choose in terms of the coordinate system indicated in the figure. [*Hint:* You may require a formula from Table 6.1.]

   b. Write a paragraph on the security of shopping mall parking lots. In particular, do you think geometric centrality would be the prime consideration in locating a security kiosk or are there other more important issues?

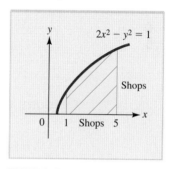

$2x^2 - y^2 = 1$

Shops

Shops   5

EXERCISE 61

Using computer software, evaluate each of the integrals in Exercises 62 through 65. In each case, verify your result by applying an appropriate integration formula from Table 6.1, and graph the function over the given interval.

62. $\displaystyle\int_1^2 x^2 \ln\sqrt{x}\, dx$

63. $\displaystyle\int_2^3 \sqrt{4x^2 - 7}\, dx$

64. $\displaystyle\int_0^1 x^3 \sqrt{4 + 5x}\, dx$

65. $\displaystyle\int_0^1 \frac{\sqrt{x^2 + 2x}}{(x + 1)^2}\, dx$

SECTION 6.2

L02

Evaluate improper integrals.

Improper Integrals

The definition of the definite integral $\displaystyle\int_a^b f(x)\, dx$, given in Section 5.3, requires the interval of integration $a \leq x \leq b$ to be bounded, but in certain applications it is useful to consider integrals over *unbounded* intervals such as $x \geq a$. We will define such **improper integrals** in this section and examine a few properties and applications.

The Improper
Integral $\int_{a}^{+\infty} f(x)\, dx$

We denote the improper integral of $f(x)$ over the unbounded interval $x \geq a$ by $\int_{a}^{+\infty} f(x)\, dx$. If $f(x) \geq 0$ for $x \geq a$, this integral can be interpreted as the area of the region under the curve $y = f(x)$ to the right of $x = a$, as shown in Figure 6.2a. Although this region has infinite extent, its area may be finite or infinite, depending on how quickly $f(x)$ approaches zero as x increases indefinitely.

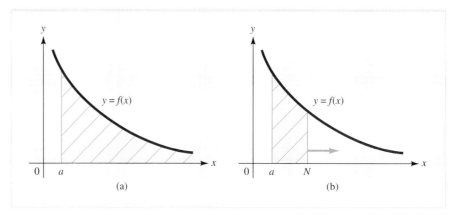

FIGURE 6.2   Area $= \int_{0}^{+\infty} f(x)\, dx = \lim_{N \to +\infty} \int_{0}^{N} f(x)\, dx.$

A reasonable strategy for finding the area of such a region is to first use a definite integral to compute the area from $x = a$ to some finite number $x = N$ and then to let N approach infinity in the resulting expression. That is,

$$\text{Total area} = \lim_{N \to +\infty} (\text{area from } a \text{ to } N) = \lim_{N \to +\infty} \int_{a}^{N} f(x)\, dx.$$

This strategy is illustrated in Figure 6.2b and motivates the following definition of the improper integral.

The Improper Integral   ■   If $f(x)$ is continuous for $x \geq a$, then

$$\int_{a}^{+\infty} f(x)\, dx = \lim_{N \to +\infty} \int_{a}^{N} f(x)\, dx.$$

If the limit exists, the improper integral is said to **converge** to the value of the limit. If the limit does not exist, the improper integral **diverges.**

EXAMPLE 6.2.1

Either evaluate the improper integral

$$\int_{1}^{+\infty} \frac{1}{x^2}\, dx$$

or show that it diverges.

Solution

First compute the integral from 1 to N and then let N approach infinity. Arrange your work as follows:

$$\int_1^{+\infty} \frac{1}{x^2}\,dx = \lim_{N \to +\infty} \int_1^N \frac{1}{x^2}\,dx = \lim_{N \to +\infty} \left(-\frac{1}{x}\Big|_1^N \right) = \lim_{N \to +\infty} \left(-\frac{1}{N} + 1 \right) = 1.$$

EXAMPLE 6.2.2

Either evaluate the improper integral

$$\int_1^{+\infty} \frac{1}{x}\,dx$$

or show that it diverges.

Solution

$$\int_1^{+\infty} \frac{1}{x}\,dx = \lim_{N \to +\infty} \int_1^N \frac{1}{x}\,dx \lim_{N \to +\infty} \left(\ln|x|\,\Big|_1^N \right) = \lim_{N \to +\infty} (\ln N - \ln 1) = \lim_{N \to +\infty} \ln N = +\infty.$$

Since the limit does not exist (as a finite number), the improper integral diverges.

NOTE You have just seen that the improper integral $\int_1^{+\infty} \frac{1}{x^2}\,dx$ converges (Example 6.2.1), while $\int_1^{+\infty} \frac{1}{x}\,dx$ diverges (Example 6.2.2). In geometric terms, this says that the area under the curve $y = \frac{1}{x^2}$ to the right of $x = 1$ is *finite,* while the corresponding area under the curve $y = \frac{1}{x}$ to the right of $x = 1$ is *infinite.* The difference is because as x increases without bound, $\frac{1}{x^2}$ approaches zero more quickly than does $\frac{1}{x}$. These observations are demonstrated in Figure 6.3.

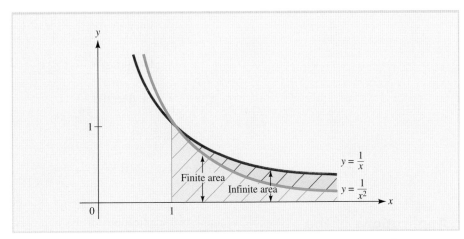

FIGURE 6.3   Comparison of the area under $y = \frac{1}{x}$ with that under $y = \frac{1}{x^2}$.

Evaluating improper integrals arising from practical problems often involves limits of the form

$$\lim_{N \to +\infty} \frac{N^p}{e^{kN}} = \lim_{N \to +\infty} N^p e^{-kN} \quad \text{(for } k > 0\text{)}.$$

In general, an exponential term such as e^{kN} grows much faster than *any* power term N^p, so

$$N^p e^{-kN} = \frac{N^p}{e^{kN}}$$

will become very small in the long run. To summarize:

A Useful Limit for Improper Integrals ■ For any power p and positive number k,

$$\lim_{N \to +\infty} N^p e^{-kN} = 0.$$

EXAMPLE 6.2.3

Either evaluate the improper integral

$$\int_0^{+\infty} xe^{-2x}\, dx$$

or show that it diverges.

Solution

$$\int_0^{+\infty} xe^{-2x}\, dx = \lim_{N \to +\infty} \int_0^N xe^{-2x}\, dx$$

$$= \lim_{N \to +\infty} \left(-\frac{1}{2}xe^{-2x}\Big|_0^N + \frac{1}{2}\int_0^N e^{-2x}\, dx \right) \quad \text{integration by parts}$$

$$= \lim_{N \to +\infty} \left(-\frac{1}{2}xe^{-2x} - \frac{1}{4}e^{-2x} \right)\Big|_0^N$$

$$= \lim_{N \to +\infty} \left(-\frac{1}{2}Ne^{-2N} - \frac{1}{4}e^{-2N} + 0 + \frac{1}{4} \right)$$

$$= 0 + 0 + 0 + \frac{1}{4}$$

$$= \frac{1}{4}$$

since $e^{-2N} \to 0$ and $Ne^{-2N} \to 0$ as $N \to +\infty$.

Applications of the Improper Integral

Next, we examine two applications of the improper integral that generalize applications developed in Chapter 5. In each, the strategy is to express a quantity as a definite integral with a variable upper limit of integration that is then allowed to increase without bound. As you read through these applications, you may find it helpful to refer back to the analogous examples in Chapter 5.

Present Value of a
Perpetual Income Flow

In Example 5.5.3 of Section 5.5, we showed how the present value of an investment that generates income continuously over a finite time period can be computed by a definite integral. If the generation of income continues in perpetuity, then an improper integral is required to compute its present value, as illustrated in Example 6.2.4.

EXAMPLE 6.2.4

A donor wishes to endow a scholarship at a local college with a gift that provides a continuous income stream at a rate of 25 000 + 1200t dollars per year in perpetuity. Assuming the prevailing annual interest rate stays fixed at 5% compounded continuously, what donation is required to finance the endowment?

Solution

The donor's gift should equal the present value of the income stream in perpetuity. Recall from Section 5.5 that an income stream $f(t)$ deposited continuously for a term of T years into an account that earns interest at an annual rate r compounded continuously has a present value given by the integral

$$PV = \int_0^T f(t)e^{-rt}\,dt.$$

For the donor's gift, $f(t) = 25\ 000 + 1200t$ and $r = 0.05$, so for a term of T years, the present value is

$$PV = \int_0^T (25\ 000 + 1200t)e^{-0.05t}\,dt.$$

Integrating by parts, with

$$u = 25\ 000 + 1200t \qquad dv = e^{-0.05t}\,dt$$

$$du = 1200dt \qquad v = \frac{e^{-0.05t}}{-0.05} = -20e^{-0.05t}$$

gives the present value as follows:

$$PV = \int_0^T (25\ 000 + 1200t)e^{-0.05t}\,dt$$

$$= \left[(25\ 000 + 1200t)(-20e^{-0.05t})\right]\Big|_0^T - \int_0^T 1200(-20e^{-0.05t})\,dt$$

$$= \left[(-500\ 000 - 24\ 000t)e^{-0.05t}\right]\Big|_0^T + 24\ 000\left(\frac{e^{-0.05t}}{-0.05}\right)\Big|_0^T$$

$$= \left[(-980\ 000 - 24\ 000t)e^{-0.05t}\right]\Big|_0^T$$

$$= \left[(-980\ 000 - 24\ 000T)e^{-0.05T}\right] - \left[(-980\ 000 - 24\ 000(0))e^{0}\right]$$

$$= (-980\ 000 - 24\ 000T)e^{-0.05T} + 980\ 000$$

To find the present value in perpetuity, take the limit as $T \to +\infty$ by computing the improper integral:

$$\int_0^{+\infty} (25\,000 + 1200t)e^{-0.05t}\,dt = \lim_{T \to +\infty} [(-980\,000 - 24\,000T)e^{-0.05T} + 980\,000]$$

$$= 0 + 980\,000 \qquad \text{since } e^{-0.05T} \to 0 \text{ and}$$
$$= 980\,000 \qquad \qquad Te^{-0.05T} \to 0 \text{ as } T \to +\infty$$

Therefore, the endowment is established with a gift of $980 000.

Nuclear Waste   In Example 5.6.1 of Section 5.6, we examined a survival and renewal problem over a term of finite length in which renewal occurred at a constant rate. In Example 6.2.5, we consider the problem of nuclear waste that is continuously generated but decays exponentially in perpetuity. This is a more general case where the term is no longer finite.

EXAMPLE 6.2.5

It is estimated that t years from now, a certain nuclear power plant will be producing radioactive waste at a rate of $f(t) = 400t$ kilograms per year. The waste decays exponentially at a rate of 2% per year. What will happen to the accumulation of radioactive waste from the plant in the long run?

Solution

To find the amount of radioactive waste present after N years, divide the N-year interval $0 \le t \le N$ into n equal subintervals of length Δt years and let t_j denote the beginning of the jth subinterval (Figure 6.4). Then

Amount of waste produced during jth subinterval $\approx 400t_j \Delta t$.

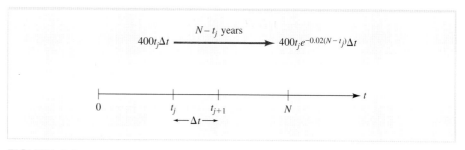

FIGURE 6.4 Radioactive waste generated during the jth subinterval.

Since the waste decays exponentially at a rate of 2% per year, and since there are $N - t_j$ years between times $t = t_j$ and $t = N$, it follows that

$$\begin{array}{l} \text{Amount of waste produced} \\ \text{during } j\text{th subinterval} \\ \text{still present at } t = N \end{array} \approx 400t_j e^{-0.02(N - t_j)} \Delta t$$

Thus,

$$\text{Amount of waste present after } N \text{ years} = \lim_{n \to +\infty} \sum_{j=1}^{n} 400 t_j e^{-0.02(N-t_j)} \Delta t$$

$$= \int_{0}^{N} 400 t e^{-0.02(N-t)} \, dt$$

$$= 400 e^{-0.02N} \int_{0}^{N} t e^{0.02t} \, dt$$

The amount of radioactive waste present in the long run is the limit of this expression as N approaches infinity. That is,

$$\begin{aligned}
\text{Amount of waste present in the long run} &= \lim_{N \to +\infty} 400 e^{-0.02N} \int_{0}^{N} t e^{0.02t} \, dt \\[2mm]
&= \lim_{N \to +\infty} 400 e^{-0.02N} \left(\frac{1}{(0.02)^2} (0.02t - 1) e^{0.02t} \right) \Big|_{0}^{N} \quad \text{using Formula 22 from Table 6.1} \\[2mm]
&= \lim_{N \to +\infty} 400 e^{-0.02N} (50 t e^{0.02t} - 2500 e^{0.02t}) \Big|_{0}^{N} \\[2mm]
&= \lim_{N \to +\infty} 400 e^{-0.02N} (50 N e^{0.02N} - 2500 e^{0.02N} + 2500) \\[2mm]
&= \lim_{N \to +\infty} 400 (50 N - 2500 + 2500 e^{-0.02N})
\end{aligned}$$

That is, in the long run, the accumulation of radioactive waste from the plant will increase without bound.

EXERCISES ■ 6.2

In Exercises 1 through 24, either evaluate the given improper integral or show that it diverges.

1. $\displaystyle\int_{1}^{+\infty} \frac{1}{x^3} \, dx$

2. $\displaystyle\int_{1}^{+\infty} x^{-3/2} \, dx$

3. $\displaystyle\int_{1}^{+\infty} \frac{1}{\sqrt{x}} \, dx$

4. $\displaystyle\int_{1}^{+\infty} x^{-2/3} \, dx$

5. $\displaystyle\int_{3}^{+\infty} \frac{1}{2x - 1} \, dx$

6. $\displaystyle\int_{3}^{+\infty} \frac{1}{\sqrt[3]{2x - 1}} \, dx$

7. $\displaystyle\int_{3}^{+\infty} \frac{1}{(2x - 1)^2} \, dx$

8. $\displaystyle\int_{0}^{+\infty} e^{-x} \, dx$

9. $\displaystyle\int_{0}^{+\infty} 5 e^{-2x} \, dx$

10. $\displaystyle\int_{1}^{+\infty} e^{1-x} \, dx$

11. $\displaystyle\int_{1}^{+\infty} \frac{x^2}{(x^3 + 2)^2} \, dx$

12. $\displaystyle\int_{1}^{+\infty} \frac{x^2}{x^3 + 2} \, dx$

13. $\displaystyle\int_{1}^{+\infty} \frac{x^2}{\sqrt{x^3 + 2}} \, dx$

14. $\displaystyle\int_0^{+\infty} xe^{-x^2}\, dx$

15. $\displaystyle\int_1^{+\infty} \frac{e^{-\sqrt{x}}}{\sqrt{x}}\, dx$

16. $\displaystyle\int_0^{+\infty} xe^{-x}\, dx$

17. $\displaystyle\int_0^{+\infty} 2xe^{-3x}\, dx$

18. $\displaystyle\int_0^{+\infty} xe^{1-x}\, dx$

19. $\displaystyle\int_1^{+\infty} \frac{\ln x}{x}\, dx$

20. $\displaystyle\int_1^{+\infty} \frac{\ln x}{x^2}\, dx$

21. $\displaystyle\int_2^{+\infty} \frac{1}{x \ln x}\, dx$

22. $\displaystyle\int_2^{+\infty} \frac{1}{x\sqrt{\ln x}}\, dx$

23. $\displaystyle\int_0^{+\infty} x^2 e^{-x}\, dx$

24. $\displaystyle\int_1^{+\infty} \frac{e^{1/x}}{x^2}\, dx$

25. **PRESENT VALUE OF AN INVESTMENT**
An investment will generate $2400 per year in perpetuity. If the money is dispensed continuously throughout the year and if the prevailing annual interest rate remains fixed at 4% compounded continuously, what is the present value of the investment?

26. **PRESENT VALUE OF A RENTAL PROPERTY**
It is estimated that t years from now an apartment complex will be generating profit for its owner at a rate of $f(t) = 10\,000 + 500t$ dollars per year. If the profit is generated in perpetuity and the prevailing annual interest rate remains fixed at 5% compounded continuously, what is the present value of the apartment complex?

27. **PRESENT VALUE OF A FRANCHISE**   The management of a national chain of fast-food outlets is selling a permanent franchise in Moose Jaw, Saskatchewan. Past experience suggests that t years from now, the franchise will be generating profit at

a rate of $f(t) = 12\,000 + 900t$ dollars per year. If the prevailing interest rate remains fixed at 5% compounded continuously, what is the present value of the franchise?

28. **NUCLEAR WASTE**   A certain nuclear power plant produces radioactive waste at a rate of 600 kg per year. The waste decays exponentially at a rate of 2% per year. How much radioactive waste from the plant will be present in the long run?

29. **HEALTH CARE**   The fraction of patients who will still be receiving treatment at a certain health clinic t months after their initial visit is $f(t) = e^{-t/20}$. If the clinic accepts new patients at a rate of 10 per month, approximately how many patients will be receiving treatment at the clinic in the long run?

30. **POPULATION GROWTH**   Demographic studies conducted in a certain city indicate that the fraction of the residents that will remain in the city for at least t years is $f(t) = e^{-t/20}$. The current population of the city is 200 000, and it is estimated that new residents will be arriving at a rate of 100 people per year. If this estimate is correct, what will happen to the population of the city in the long run?

31. **MEDICINE**   A hospital patient receives intravenously 5 units of a certain drug per hour. The drug is eliminated exponentially, so that the fraction that remains in the patient's body for t hours is $f(t) = e^{-t/10}$. If the treatment is continued indefinitely, approximately how many units of the drug will be in the patient's body in the long run?

32. **ENDOWMENT**   A wealthy alumnus wishes to endow a mathematics student scholarship for an overseas student to study in Canada. He will give a gift of G thousand dollars to the university for this purpose. The student winning this scholarship is to receive $70 000. If money costs 8% per year compounded continuously, what is the smallest possible value for G?

33. **CAPITALIZED COST OF AN ASSET**   The capitalized cost of an asset is the sum of the original cost of the asset and the present value of maintaining the asset. Suppose a company is considering the purchase of two different machines. Machine 1 costs $10 000 and t years from now will cost $M_1(t) = 1000(1 + 0.06t)$ dollars to maintain. Machine 2 costs only $8000, but its maintenance cost at time t is $M_2(t) = 1100$ dollars.

a. If the cost of money is 9% per year compounded continuously, what is the capitalized cost of each machine? Which one should the company purchase?

b. Research various methods used by economists to make comparisons between competing assets. Write a paragraph comparing these methods.

34. SPY STORY   Still groggy from his encounter with the moose (Exercise 59 of Section 5.1), our spy stumbles into a trap. In the ensuing firefight, the fraction of villains that remain in fighting condition t minutes after first entering the fray is $e^{-t/5}$. Unfortunately for the spy, two new villains arrive every 10 minutes. Assuming the spy has enough bullets to continue fighting with this efficiency, how many villains will be left standing in the long run?

35. A* PRESENT VALUE   An investment will generate income continuously at a constant rate of Q dollars per year in perpetuity. Assuming a fixed annual interest rate r compounded continuously, use an improper integral to show that the present value of the investment is $\dfrac{Q}{r}$ dollars.

36. A* PRESENT VALUE   In t years, an investment will be generating $f(t) = A + Bt$ dollars per year,

where A and B are constants. Suppose the income is generated in perpetuity, with a fixed annual interest rate r compounded continuously. Show that the present value of this investment is $\dfrac{A}{r} + \dfrac{B}{r^2}$ dollars.

37. A* EPIDEMIOLOGY   The proportion P of susceptible people who are infected t weeks after the outbreak of an epidemic is given by the integral

$$\int_0^t C(e^{-ax} - e^{-bx})\, dx,$$

where a and b are parameters that depend on the disease and C is a constant. Assuming that all susceptible people are eventually infected, find C (in terms of a and b).

38. PRODUCT RELIABILITY   The manager of an electronics firm estimates that the proportion of components that last longer than t months is given by the improper integral

$$\int_t^\infty 0.008 e^{-0.008x}\, dx.$$

Which is larger, the proportion of components that last longer than 5 years (60 months) or the proportion that fail in less than 10 years?

SECTION 6.3

L03

Approximate areas and integrals using Simpson's rule and the trapezoidal rule.

Numerical Integration

In this section, you will see some techniques that you can use to approximate definite integrals. Numerical methods such as these are needed when the function to be integrated does not have an elementary antiderivative. For instance, neither $\sqrt{x^3 + 1}$ nor $\dfrac{e^x}{x}$ has an elementary antiderivative.

Approximation by Rectangles

If $f(x)$ is positive on the interval $a \le x \le b$, the definite integral $\displaystyle\int_a^b f(x)\, dx$ is equal to the area under the graph of f between $x = a$ and $x = b$. As you saw in Section 5.3, one way to approximate this area is to use n rectangles and a Riemann sum as shown in Figure 6.5. In particular, divide the interval $a \le x \le b$ into n equal subintervals of width $\Delta x = \dfrac{b - a}{n}$ and let x_j denote the beginning of the jth subinterval.

The base of the jth rectangle is the jth subinterval, and its height is $f(x_j)$. Hence, the area of the jth rectangle is $f(x_j)\Delta x$. The sum of the areas of all n rectangles is an approximation to the area under the curve, so an approximation to the corresponding definite integral is

$$\int_a^b f(x)\, dx \approx f(x_1)\Delta x + \cdots + f(x_n)\Delta x.$$

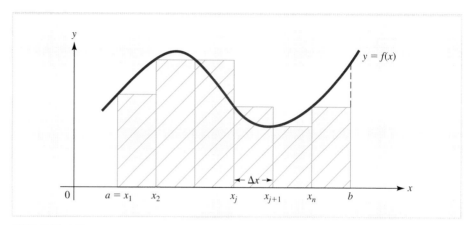

FIGURE 6.5   Approximation by rectangles.

This approximation improves as the number of rectangles increases, and you can estimate the integral to any desired degree of accuracy by making n large enough. However, since fairly large values of n are usually required to achieve reasonable accuracy, approximation by rectangles is rarely used in practice.

Approximation by Trapezoids

The accuracy of the approximation improves significantly if trapezoids are used instead of rectangles. Figure 6.6a shows the area from Figure 6.5 approximated by n trapezoids. Notice how much better the approximation is in this case.

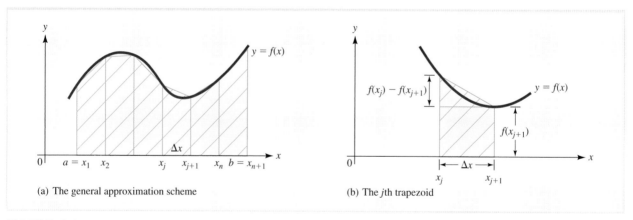

(a) The general approximation scheme

(b) The jth trapezoid

FIGURE 6.6   Approximation by trapezoids.

The jth trapezoid is shown in greater detail in Figure 6.6b. Notice that it consists of a rectangle with a right triangle on top of it. Since

$$\text{Area of rectangle} = f(x_{j+1})\Delta x$$

and

$$\text{Area of triangle} = \frac{1}{2}[f(x_j) - f(x_{j+1})]\Delta x,$$

Just-In-Time

Recall that a trapezoid is a quadrilateral with exactly two parallel sides. If the two parallel sides are h units apart and have lengths a and b, then the trapezoid has area

$$A = \frac{h}{2}(a + b).$$

When the area is divided up into trapezoids, most of the trapezoids share a common side. This results in the factor of 2 for all of the terms except those representing the outer sides of the beginning and ending trapezoids.

it follows that

$$\text{Area of } j\text{th trapezoid} = f(x_{j+1})\Delta x + \frac{1}{2}[f(x_j) - f(x_{j+1})]\Delta x$$

$$= \frac{1}{2}[f(x_j) + f(x_{j+1})]\Delta x$$

The sum of the areas of all n trapezoids is an approximation to the area under the curve and hence an approximation to the corresponding definite integral. Thus,

$$\int_a^b f(x)\,dx \approx \frac{1}{2}[f(x_1) + f(x_2)]\Delta x + \frac{1}{2}[f(x_2) + f(x_3)]\Delta x + \cdots + \frac{1}{2}[f(x_n) + f(x_{n+1})]\Delta x$$

$$= \frac{\Delta x}{2}[f(x_1) + 2f(x_2) + \cdots + 2f(x_n) + f(x_{n+1})]$$

This approximation formula is known as the **trapezoidal rule** and applies even if the function f is not positive.

The Trapezoidal Rule

$$\int_a^b f(x)\,dx \approx \frac{\Delta x}{2}[f(x_1) + 2f(x_2) + \cdots + 2f(x_n) + f(x_{n+1})].$$

The trapezoidal rule is illustrated in Example 6.3.1.

EXAMPLE 6.3.1

Use the trapezoidal rule with $n = 10$ to approximate $\int_1^2 \frac{1}{x}\,dx$.

Solution

Since $\Delta x = \dfrac{2-1}{10} = 0.1$, the interval $1 \le x \le 2$ is divided into 10 subintervals by the points

$$x_1 = 1,\ x_2 = 1.1,\ x_3 = 1.2,\ \ldots,\ x_{10} = 1.9,\ x_{11} = 2,$$

as shown in Figure 6.7.

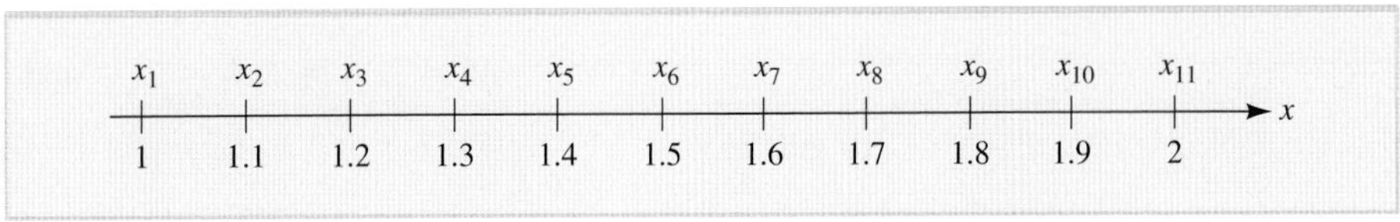

FIGURE 6.7 Division of the interval $1 \le x \le 2$ into 10 subintervals.

Then, by the trapezoidal rule,

$$\int_1^2 \frac{1}{x}\,dx \approx \frac{0.1}{2}\left(\frac{1}{1} + \frac{2}{1.1} + \frac{2}{1.2} + \frac{2}{1.3} + \frac{2}{1.4} + \frac{2}{1.5} + \frac{2}{1.6} + \frac{2}{1.7} + \frac{2}{1.8} + \frac{2}{1.9} + \frac{1}{2}\right)$$

$$\approx 0.693771$$

The definite integral in Example 6.3.1 can be evaluated directly:

$$\int_1^2 \frac{1}{x}\, dx = \ln|x| \Big|_1^2 = \ln 2 \approx 0.693147.$$

Thus, the approximation of this integral by the trapezoidal rule with $n = 10$ is accurate (after round-off) to two decimal places.

Accuracy of the Trapezoidal Rule

The difference between the true value of the integral $\int_a^b f(x)\, dx$ and the approximation generated by the trapezoidal rule when n subintervals are used is denoted by E_n. Here is an estimate for the absolute value of E_n that is proven in more advanced courses.

> **Error Estimate for the Trapezoidal Rule** ■ If M is the maximum value of $|f''(x)|$ on the interval $a \le x \le b$, then
>
> $$|E_n| \le \frac{M(b-a)^3}{12n^2}.$$

The use of this formula is illustrated in Example 6.3.2.

EXAMPLE 6.3.2

Estimate the accuracy of the approximation of $\int_1^2 \frac{1}{x}\, dx$ by the trapezoidal rule with $n = 10$.

Solution

Starting with $f(x) = \frac{1}{x}$, compute the derivatives

$$f'(x) = -\frac{1}{x^2} \quad \text{and} \quad f''(x) = \frac{2}{x^3},$$

and observe that the largest value of $|f''(x)|$ for $1 \le x \le 2$ is $|f''(1)| = 2$.
Apply the error formula with

$$M = 2,\, a = 1,\, b = 2, \quad \text{and} \quad n = 10$$

to get

$$|E_{10}| \le \frac{2(2-1)^3}{12(10)^2} \approx 0.00167.$$

That is, the error in the approximation in Example 6.3.1 is guaranteed to be no greater than 0.00167. (In fact, to five decimal places, the error is 0.00062, as you can see by comparing the approximation obtained in Example 6.3.1 with the decimal representation of ln 2.)

It is important to keep the approximation error small when using the trapezoidal rule to estimate a definite integral. If you wish to achieve a certain degree of accuracy, you can determine how many intervals are required to attain your goal, as illustrated in Example 6.3.3.

EXAMPLE 6.3.3

How many subintervals are required to guarantee that the error will be less than 0.00005 in the approximation of $\int_1^2 \frac{1}{x}\, dx$ using the trapezoidal rule?

Solution

From Example 6.3.2 you know that $M = 2$, $a = 1$, and $b = 2$, so that

$$|E_n| \le \frac{2(2-1)^3}{12n^2} = \frac{1}{6n^2}.$$

The goal is to find the smallest positive integer n for which

$$\frac{1}{6n^2} < 0.00005.$$

Equivalently,

$$n^2 > \frac{1}{6(0.00005)}$$

or

$$n > \sqrt{\frac{1}{6(0.00005)}} \approx 57.74.$$

The smallest such integer is $n = 58$, so 58 subintervals are required to ensure the desired accuracy.

Approximation Using Parabolas: Simpson's Rule

The relatively large number of subintervals required in Example 6.3.3 to ensure accuracy to within 0.00005 suggests that approximation by trapezoids may not be efficient enough for some applications. Another approximation formula, called **Simpson's rule,** is no harder to use than the trapezoidal rule but often requires substantially fewer calculations to achieve a specified degree of accuracy. Like the trapezoidal rule, it is based on the approximation of the area under a curve by columns, but unlike the trapezoidal rule, it uses parabolic arcs rather than line segments at the top of the columns.

To be more specific, the approximation of a definite integral using parabolas is based on the following construction (illustrated in Figure 6.8 for $n = 6$). Divide the interval $a \le x \le b$ into an **even** number of subintervals so that adjacent subintervals can be paired with none left over. Approximate the portion of the graph that lies above the first pair of subintervals by the (unique) parabola that passes through the three points $(x_1, f(x_1))$, $(x_2, f(x_2))$, and $(x_3, f(x_3))$ and use the area under this parabola between x_1 and x_3 to approximate the corresponding area under the curve. Do the same for the remaining pairs of subintervals and use the sum of the resulting areas to approximate the total area under the graph. Here is the approximation formula that results from this construction.

Simpson's Rule ■ For an even integer n,

$$\int_a^b f(x)\, dx \approx \frac{\Delta x}{3}[f(x_1) + 4f(x_2) + 2f(x_3) + 4f(x_4) + \cdots + 2f(x_{n-1}) + 4f(x_n) + f(x_{n+1})].$$

Notice that the first and last function values in the approximating sum in Simpson's rule are multiplied by 1, while the others are multiplied alternately by 4 and 2.

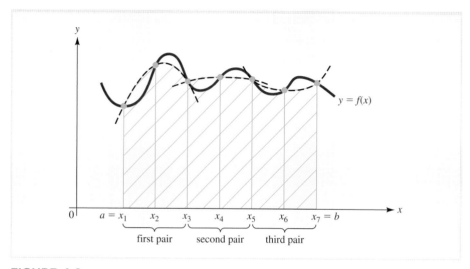

FIGURE 6.8   Approximation using parabolas.

EXAMPLE 6.3.4

Use Simpson's rule with $n = 10$ to approximate $\int_1^2 \frac{1}{x}\, dx$.

Solution

As in Example 6.3.1, $\Delta x = 0.1$, and hence the interval $1 \le x \le 2$ is divided into 10 subintervals by

$$x_1 = 1,\ x_2 = 1.1,\ x_3 = 1.2,\ \ldots,\ x_{10} = 1.9,\ x_{11} = 2.$$

Then, by Simpson's rule,

$$\int_1^2 \frac{1}{x}\, dx \approx \frac{0.1}{3}\left(\frac{1}{1} + \frac{4}{1.1} + \frac{2}{1.2} + \frac{4}{1.3} + \frac{2}{1.4} + \frac{4}{1.5} + \frac{2}{1.6} + \frac{4}{1.7} + \frac{2}{1.8} + \frac{4}{1.9} + \frac{1}{2}\right)$$
$$\approx 0.693150$$

Notice that this is an excellent approximation to the true value to 6 decimal places, namely, $\ln 2 = 0.693147$.

Accuracy of Simpson's Rule   The error estimate for Simpson's rule uses the fourth derivative $f^{(4)}(x)$ in much the same way as the second derivative $f''(x)$ was used in the error estimate for the trapezoidal rule. Here is the estimation formula.

> **Error Estimate for Simpson's Rule** ■ If M is the maximum value of $|f^{(4)}(x)|$ on the interval $a \le x \le b$, then
> $$|E_n| \le \frac{M(b-a)^5}{180n^4}.$$

Here is an application of the formula.

EXAMPLE 6.3.5

Estimate the accuracy of the approximation of $\int_1^2 \frac{1}{x}\, dx$ by Simpson's rule with $n = 10$.

Solution

Starting with $f(x) = \dfrac{1}{x}$, compute the derivatives

$$f'(x) = -\frac{1}{x^2} \quad f''(x) = \frac{2}{x^3} \quad f^{(3)}(x) = -\frac{6}{x^4} \quad f^{(4)}(x) = \frac{24}{x^5}$$

and observe that the largest value of $|f^{(4)}(x)|$ on the interval $1 \le x \le 2$ is $|f^{(4)}(1)| = 24$. Now apply the error formula with $M = 24$, $a = 1$, $b = 2$, and $n = 10$ to get

$$|E_{10}| \le \frac{24(2-1)^5}{180(10)^4} \approx 0.000013.$$

Thus, the error in the approximation in Example 6.3.4 is guaranteed to be no greater than 0.000013.

In Example 6.3.6, the error estimate is used to determine the number of subintervals that are required to ensure a specified degree of accuracy.

EXAMPLE 6.3.6

How many subintervals are required to ensure accuracy to within 0.00005 in the approximation of $\int_1^2 \frac{1}{x}\, dx$ by Simpson's rule?

Solution

From Example 6.3.5 you know that $M = 24$, $a = 1$, and $b = 2$. Hence,

$$|E_n| \le \frac{24(2-1)^5}{180n^4} = \frac{2}{15n^4}.$$

The goal is to find the smallest positive (even) integer for which

$$\frac{2}{15n^4} < 0.00005.$$

Equivalently,

$$n^4 > \frac{2}{15(0.00005)}$$

or

$$n > \left[\frac{2}{15(0.00005)}\right]^{1/4} \approx 7.19.$$

The smallest such (even) integer is $n = 8$, and so eight subintervals are required to ensure the desired accuracy. Compare with the result of Example 6.3.3, where we found that 58 subintervals are required to ensure the same degree of accuracy using the trapezoidal rule.

Interpreting Data with Numerical Integration

Numerical integration can often be used to estimate a quantity $\int_a^b f(x)\, dx$ when all that is known about $f(x)$ is a set of experimentally determined data. Here is an example.

EXAMPLE 6.3.7

Mark needs to know the area of his swimming pool in order to buy a pool cover, but this is difficult because of the pool's irregular shape. Suppose Mark makes the measurements shown in Figure 6.9 at 1-m intervals along the base of the pool (all measurements are in metres). How can he use the trapezoidal rule to estimate the area?

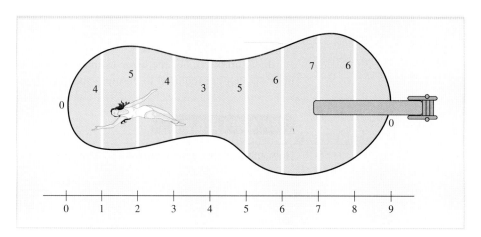

FIGURE 6.9   Measurements across a pool.

Solution

If Mark could find functions $f(x)$ for the top rim of the pool and $g(x)$ for the bottom rim, then the area would be given by the definite integral $A = \int_0^9 [f(x) - g(x)]\, dx$.

The irregular shape makes it impossible or at least impractical to find formulas for f and g, but Mark's measurements tell him that

$$f(0) - g(0) = 0 \qquad f(1) - g(1) = 4 \qquad f(2) - g(2) = 5 \ldots f(9) - g(9) = 0$$

Substituting this information into the trapezoidal rule approximation and using $\Delta x = \dfrac{9 - 0}{9} = 1$, Mark obtains

$$A = \int_0^9 [f(x) - g(x)]\, dx$$

$$= \frac{1}{2}[0 + 2(4) + 2(5) + 2(4) + 2(3) + 2(5) + 2(6) + 2(7) + 2(6) + 0]$$

$$= \frac{1}{2}(80) = 40$$

Thus, Mark estimates the area of the pool to be approximately 40 m^2.

EXAMPLE 6.3.8

The management of a chain of pet supply stores is selling a 10-year franchise. Past records in similar locations suggest that t years from now, the franchise will be generating income at a rate of $f(t)$ thousand dollars per year, where $f(t)$ is as indicated in the following table for a typical decade.

Year t	0	1	2	3	4	5	6	7	8	9	10
Rate of income flow $f(t)$	510	580	610	625	654	670	642	610	590	573	550

If the prevailing rate of interest remains at 5% per year compounded continuously over the 10-year term, what is a fair price for the franchise, based on the given information?

Solution

If the rate of income flow $f(t)$ were a continuous function, a fair price for the franchise might be determined by computing the present value of the income flow over the 10-year term. According to the formula developed in Section 5.5, this present value would be given by the definite integral

$$PV = \int_0^{10} f(t)e^{-0.05t}\, dt$$

since the prevailing rate of interest is 5% ($r = 0.05$). Since we don't have such a continuous function $f(t)$, we will use Simpson's rule with $n = 10$ and $\Delta t = 1$ to *estimate* the present value integral:

$$PV = \int_0^{10} f(t)e^{-0.05t}\, dt$$

$$\approx \frac{\Delta t}{3}[f(0)e^{-0.05(0)} + 4f(1)e^{-0.05(1)} + 2f(2)e^{-0.05(2)} + \cdots + 4f(9)e^{-0.05(9)}$$
$$+ f(10)e{-0.05(10)}]$$

$$\approx \frac{1}{3}[(510)e^{-0.05(0)} + 4(580)e^{-0.05(1)} + 2(610)e^{-0.05(2)} + 4(625)e^{-0.05(3)}$$
$$+ 2(654)e^{-0.05(4)} + 4(670)e^{-0.05(5)} + 2(642)e^{-0.05(6)} + 4(610)e^{-0.05(7)}$$
$$+ 2(590)e^{-0.05(8)} + 4(573)e^{-0.05(9)} + (550)e^{-0.05(10)}]$$

$$\approx \frac{1}{3}(14\ 387) \approx 4796$$

Thus, the present value of the income stream over the 10-year term is approximately $4796 thousand dollars ($4 796 000). The company may use this estimate as a fair asking price for the franchise.

EXERCISES ■ 6.3

In Exercises 1 through 14, approximate the given integral using (a) the trapezoidal rule and (b) Simpson's rule with the specified number of subintervals.

1. $\int_1^2 x^2\, dx;\ n = 4$

2. $\int_4^6 \frac{1}{\sqrt{x}}\, dx;\ n = 10$

3. $\int_0^1 \frac{1}{1 + x^2}\, dx;\ n = 4$

4. $\int_2^3 \frac{1}{x^2 - 1}\, dx;\ n = 4$

5. $\int_{-1}^0 \sqrt{1 + x^2}\, dx;\ n = 4$

6. $\int_0^3 \sqrt{9 - x^2}\, dx;\ n = 6$

7. $\int_0^1 e^{-x^2}\, dx;\ n = 4$

8. $\int_0^2 e^{x^2}\, dx;\ n = 10$

9. $\int_2^4 \frac{dx}{\ln x};\ n = 6$

10. $\int_1^2 \frac{\ln x}{x + 2}\, dx;\ n = 4$

11. $\int_0^1 \sqrt[3]{1 + x^2}\, dx;\ n = 4$

12. $\int_0^1 \frac{dx}{\sqrt{1 + x^3}};\ n = 6$

13. $\int_0^2 e^{-\sqrt{x}}\, dx;\ n = 8$

14. $\int_1^2 \frac{e^x}{x}\, dx;\ n = 4$

In Exercises 15 through 20, approximate the given integral and estimate the error $|E_n|$ using (a) the trapezoidal rule and (b) Simpson's rule with the specified number of subintervals.

15. $\int_1^2 \frac{1}{x^2}\, dx;\ n = 4$

16. $\int_0^2 x^3\, dx;\ n = 8$

17. $\int_1^3 \sqrt{x}\, dx;\ n = 10$

18. $\int_1^2 \ln x\, dx;\ n = 4$

19. $\int_0^1 e^{x^2}\, dx;\ n = 4$

20. $\int_0^{0.6} e^{x^3}\, dx;\ n = 6$

In Exercises 21 through 26, determine how many sub-intervals are required to guarantee accuracy to within 0.00005 in the approximation of the given integral by (a) the trapezoidal rule and (b) Simpson's rule.

21. $\int_1^3 \frac{1}{x}\, dx$

22. $\int_0^4 (x^4 + 2x^2 + 1)\, dx$

23. $\int_1^2 \frac{1}{\sqrt{x}}\, dx$

24. $\int_1^2 \ln(1 + x)\, dx$

25. $\int_{1.2}^{2.4} e^x\, dx$

26. $\int_0^2 e^{x^2}\, dx$

27. A quarter circle of radius 1 has equation

$y = \sqrt{1 - x^2}$ for $0 \le x \le 1$ and has area $\frac{\pi}{4}$. Thus,

$\int_0^1 \sqrt{1 - x^2}\, dx = \frac{\pi}{4}$. Use this formula to estimate π by applying

a. the trapezoidal rule

b. Simpson's rule

In each case, use $n = 8$ subintervals.

28. Use the trapezoidal rule with $n = 8$ to estimate the area bounded by the curve $y = \sqrt{x^3 + 1}$, the x axis, and the lines $x = 0$ and $x = 1$.

29. Use the trapezoidal rule with $n = 10$ to estimate the average value of the function $f(x) = \dfrac{e^{-0.4x}}{x}$ over the interval $1 \le x \le 6$.

30. Use the trapezoidal rule with $n = 6$ to estimate the average value of the function $y = \sqrt{\ln x}$ over the interval $1 \le x \le 4$.

31. Use the trapezoidal rule with $n = 7$ to estimate the volume of the solid generated by rotating the region under the curve $y = \dfrac{x}{1 + x}$ between $x = 0$ and $x = 1$ about the x axis.

32. Use Simpson's rule with $n = 6$ to estimate the volume of the solid generated by rotating the region under the curve $y = \ln x$ between $x = 1$ and $x = 2$ about the x axis.

33. FUTURE VALUE OF AN INVESTMENT   An investment generates income continuously at a rate of $f(t) = \sqrt{t}$ thousand dollars per year at time t (years). If the prevailing rate of interest is 6% per year compounded continuously, use the trapezoidal rule with $n = 5$ to estimate the future value of the investment over a 10-year term. (See Example 5.5.2 in Section 5.5.)

34. PRESENT VALUE OF A FRANCHISE   The management of a national chain of fast-food restaurants is selling a 5-year franchise to operate its newest outlet in Truro, Nova Scotia. Past experience in similar locations suggests that t years from now, the franchise will be generating profit at a rate of $f(t) = 12\,000\sqrt{t}$ dollars per year. Suppose the prevailing annual interest rate remains fixed during the next 5 years at 5% compounded continuously. Use Simpson's rule with $n = 10$ to estimate the present value of the franchise.

35. DISTANCE AND VELOCITY   Leah and Tom are travelling in a car with a broken odometer. In order to determine the distance they travel between 2 P.M. and 3 P.M., Tom (the passenger) takes speedometer readings (in kilometres per hour) every 5 minutes:

Minutes after 2:00 P.M.	0	5	10	15	20	25	30	35	40	45	50	55	60
Speedometer reading	45	48	37	39	55	60	60	55	50	67	58	45	49

Use the trapezoidal rule to estimate the total distance travelled by the pair during the hour in question.

36. MENTAL HEALTH CARE   A mental health clinic has just opened. The clinic initially accepts 300 people for treatment and plans to accept new patients at a rate of 10 per month. Let $f(t)$ denote the fraction of people receiving treatment continuously for at least t days. For the first 60 days, records are kept and these values of $f(t)$ are obtained:

t (days)	0	5	10	15	20	25	30	35	40	45	50	55	60
$f(t)$	1	$\frac{3}{4}$	$\frac{3}{5}$	$\frac{1}{2}$	$\frac{1}{3}$	$\frac{3}{10}$	$\frac{1}{5}$	$\frac{1}{6}$	$\frac{1}{7}$	$\frac{1}{9}$	$\frac{1}{12}$	$\frac{1}{15}$	$\frac{1}{20}$

Use this information together with the trapezoidal rule to estimate the number of patients in the clinic at the end of the 60-day period. [*Hint:* This is a survival/renewal problem. Recall Example 5.6.1 of Section 5.6.]

37. FUTURE VALUE OF AN INVESTMENT　Matt has a small investment providing a variable income stream that is deposited continuously into an account earning interest at an annual rate of 4% compounded continuously. He spot-checks the monthly flow rate of the investment on the first day of every other month for a 1-year period, obtaining the results in this table:

Month	Jan.	Mar.	May	Jul.	Sep.	Nov.	Jan.
Rate of income flow	$437	$357	$615	$510	$415	$550	$593

For instance, income is entering the account at a rate of $615 per month on the first of May, but 2 months later, the rate of income flow is only $510 per month. Use this information together with Simpson's rule to estimate the future value of the income flow during this 1-year period. [*Hint:* Recall Example 5.5.2 of Section 5.5.]

38. SPREAD OF A DISEASE　A new strain of influenza has just been declared an epidemic by health officials. Currently, 3000 people have the disease and new victims are being added at a rate of $R(t) = 50\sqrt{t}$ people per week. Moreover, the fraction of infected people who still have the disease t weeks after contracting it is given by $S(t) = e^{-0.01t}$. Use Simpson's rule with $n = 8$ to estimate the number of people who have the flu 8 weeks from now. (Think of this as a survival/renewal problem, like Example 5.6.2 in Section 5.6.)

39. POLLUTION CONTROL　An industrial plant spills pollutant into a river. The pollutant spreads out as it is carried downstream by the current of the river, and 3 hours later, the spill forms the pattern shown in the accompanying figure. Measurements (in metres) across the spill are made at 5-m intervals, as indicated in the figure. Use this information together with the trapezoidal rule to estimate the area of the spill.

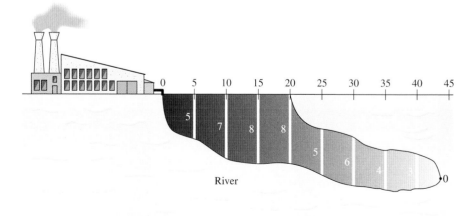

EXERCISE 39

40. **NET PROFIT FROM DATA**   On the first day of each month, the manager of a small company estimates the rate at which profit is expected to increase during that month. The results are listed in the accompanying table for the first 6 months of the year, where $P'(t)$ is the rate of profit growth in thousands of dollars per month expected during the tth month ($t = 1$ for January, $t = 6$ for June). Use this information together with the trapezoidal rule to estimate the total profit earned by the company during this 6-month period (January through June).

t (month)	1	2	3	4	5	6
Rate of profit $P'(t)$	0.65	0.43	0.72	0.81	1.02	0.97

41. **POPULATION DENSITY**   A demographic study determines that the population density of a certain city at a distance of r kilometres from the city centre is $D(r)$ people per square kilometre, where D is as indicated in the following table for $0 \le r \le 10$ at 2-km intervals.

Distance r (km) from city centre	0	2	4	6	8	10
Population density $D(r)$ (people/km^2)	3120	2844	2087	1752	1109	879

Use the trapezoidal rule to estimate the total population of the city that is located within a 10-km radius of the city centre. (See Example 5.6.4 of Section 5.6.)

42. **A* MORTALITY RATE FROM AIDS**   The accompanying table gives the number of reported deaths due to AIDS in the United States, during the tth year after 1995, for the period 1995 to 2006. (Source: *Centers for Disease Control and Prevention, National Center for HIV, STD, and TB Prevention.*)

Year	t	Reported AIDS Deaths	Year	t	Reported AIDS Deaths
1995	0	51 670	2001	6	15 603
1996	1	38 396	2002	7	16 948
1997	2	22 245	2003	8	16 690
1998	3	18 823	2004	9	16 395
1999	4	18 249	2005	10	16 268
2000	5	16 672	2006	11	14 016

However, the table does not tell the complete story because many deaths actually due to AIDS go unreported or are attributed to other causes. Let $D(t)$ be the function that gives the cumulative number of AIDS deaths at time t. Then the data in the table can be thought of as rates of change of $D(t)$ at various times, that is, as mortality rates. For instance, the table tells us that in 1997 ($t = 2$), AIDS deaths were increasing at a rate of 22 245 per year.

a. Assuming that $D(t)$ is differentiable, explain why the total number of AIDS deaths N during the period 1995–2006 is given by the integral

$$N = \int_0^{11} D'(t)\, dt.$$

b. Estimate N by using the data in the table together with the trapezoidal rule to approximate the integral in part (a).

c. Why is the mortality function $D(t)$ probably *not* differentiable? Does this invalidate the estimate of N made in part (b)? Write a paragraph either defending or challenging the procedure used to estimate N in this exercise.

43. **DISTRIBUTION OF INCOME**   A sociologist studying the distribution of income in an industrial society compiles the data displayed in the accompanying table, where $L(x)$ denotes the fraction of the society's total wealth earned by the lowest-paid $100x$ percent of the wage-earners in the society. Use this information together with the trapezoidal rule to estimate the Gini index (GI) for this society, namely,

$$\text{GI} = 2 \int_0^1 [x - L(x)]\, dx.$$

x	0	0.125	0.25	0.375	0.5	0.625	0.75	0.875	1
$L(x)$	0	0.0063	0.0631	0.1418	0.2305	0.3342	0.4713	0.6758	1

Concept Summary Chapter 6

CHAPTER SUMMARY

Integration by Parts $\displaystyle \int u\, dv = uv - \int v\, du$

Improper Integral $\displaystyle \int_a^{+\infty} f(x)\, dx = \lim_{N \to +\infty} \int_a^N f(x)\, dx$

A Useful Limit   For any power p and $k > 0$, $\displaystyle \lim_{N \to +\infty} N^p e^{-kN} = 0$.

Trapezoidal Rule $\displaystyle \int_a^b f(x)\, dx \approx \frac{\Delta x}{2}[f(x_1) + 2f(x_2) + \cdots + 2f(x_n) + f(x_{n+1})]$

Error Estimate $\displaystyle |E_n| \leq \frac{M(b-a)^3}{12n^2}$, where M is the maximum value of $|f''(x)|$ on $[a, b]$.

Simpson's Rule   For n even, $\displaystyle \int_a^b f(x)\, dx \approx \frac{\Delta x}{3}[f(x_1) + 4f(x_2) + 2f(x_3) + \cdots + 2f(x_{n-1}) + 4f(x_n) + f(x_{n+1})]$.

Error Estimate $\displaystyle |E_n| \leq \frac{M(b-a)^5}{180n^4}$, where M is the maximum value of $|f^{(4)}(x)|$ on $[a, b]$.

Checkup for Chapter 6

1. Use integration by parts to find each indefinite or definite integral.

   a. $\displaystyle \int \sqrt{2x}\, \ln x^2\, dx$

   b. $\displaystyle \int_0^1 xe^{0.2x}\, dx$

   c. $\displaystyle \int_{-4}^0 x\sqrt{1-2x}\, dx$

   d. $\displaystyle \int \frac{x-1}{e^x}\, dx$

2. In each case, either evaluate the improper integral or show that it diverges.

   a. $\displaystyle \int_1^{+\infty} \frac{1}{x^{1.1}}\, dx$

   b. $\displaystyle \int_1^{+\infty} xe^{-2x}\, dx$

   c. $\displaystyle \int_1^{+\infty} \frac{x}{(x+1)^2}\, dx$

   d. $\displaystyle \int_0^{+\infty} xe^{-x^2}\, dx$

3. Use the integral table (Table 6.1) to find each integral.

   a. $\int (\ln\sqrt{3x})^2\, dx$

   b. $\int \dfrac{dx}{x\sqrt{4 + x^2}}$

   c. $\int \dfrac{dx}{x^2\sqrt{x^2 - 9}}$

   d. $\int \dfrac{dx}{3x^2 - 4x}$

4. **NUCLEAR WASTE** After t years of operation, a certain nuclear power plant produces radioactive waste at a rate of $R(t)$ kilograms per year, where

$$R(t) = 300e^{0.001t}.$$

The waste decays exponentially at a rate of 3% per year. How much radioactive waste from the plant will be present in the long run?

5. **PRESENT VALUE OF AN ASSET** It is estimated that t years from now an office building will be generating profit for its owner at a rate of $R(t) = 50 + 3t$ thousand dollars per year. If the profit is generated in perpetuity and the prevailing annual interest rate remains fixed at 6%

compounded continuously, what is the present value of the office building?

6. **DRUG CONCENTRATION** A patient in a hospital receives 0.7 mg of a certain drug intravenously every hour. The drug is eliminated exponentially so that the fraction of drug that remains in the patient's body after t hours is $f(t) = e^{-0.2t}$. If the treatment is continued indefinitely, approximately how many units will remain in the patient's bloodstream in the long run (as $t \to +\infty$)?

7. **WATER POLLUTION** Suppose that pollutant is being discharged into a river by an industrial plant so that at time t (hours), the rate of discharge is $R(t) = 800e^{-0.05t}$ units per hour. If the discharge continues indefinitely, what is the total amount of pollutant that will be discharged into the river?

8. Use the trapezoidal rule with $n = 8$ to estimate the value of the integral

$$\int_3^4 \frac{\sqrt{25 - x^2}}{x}\, dx.$$

Then use Table 6.1 to compute the exact value of the integral and compare with your approximation.

Review Exercises

In Exercises 1 through 10, use integration by parts to find the given integral.

1. $\int te^{1-t}\, dt$

2. $\int (5 + 3x)e^{-x/2}\, dx$

3. $\int x\sqrt{2x + 3}\, dx$

4. $\int_{-9}^{-1} \dfrac{y\, dy}{\sqrt{4 - 5y}}$

5. $\int_1^4 \dfrac{\ln\sqrt{s}}{\sqrt{s}}\, ds$

6. $\int (\ln x)^2\, dx$

7. $\int_{-2}^{1} (2x + 1)(x + 3)^{3/2}\, dx$

8. $\int \dfrac{w^3}{\sqrt{1 + w^2}}\, dw$

9. $\int x^3\sqrt{3x^2 + 2}\, dx$

10. $\int_0^1 \dfrac{x + 2}{e^{3x}}\, dx$

In Exercises 11 through 16, use Table 6.1 to find the given integral.

11. $\int \dfrac{5\, dx}{8 - 2x^2}$

12. $\int \dfrac{2\, dt}{\sqrt{9t^2 + 16}}$

13. $\int w^2 e^{-w/3}\, dw$

CHAPTER SUMMARY

14. $\int \dfrac{4\,dx}{x(9 + 5x)}$

15. $\int (\ln 2x^3)\,dx$

16. $\int \dfrac{dx}{x\sqrt{4 - x^2}}$

In Exercises 17 through 26, either evaluate the given improper integral or show that it diverges.

17. $\displaystyle\int_0^{+\infty} \dfrac{1}{\sqrt[3]{1 + 2x}}\,dx$

18. $\displaystyle\int_0^{+\infty} (1 + 2x)^{-3/2}\,dx$

19. $\displaystyle\int_0^{+\infty} \dfrac{3t}{t^2 + 1}\,dt$

20. $\displaystyle\int_0^{+\infty} 3e^{-5x}\,dx$

21. $\displaystyle\int_0^{+\infty} xe^{-2x}\,dx$

22. $\displaystyle\int_0^{+\infty} 2x^2 e^{-x^3}\,dx$

23. $\displaystyle\int_0^{+\infty} x^2 e^{-2x}\,dx$

24. $\displaystyle\int_2^{+\infty} \dfrac{1}{t(\ln t)^2}\,dt$

25. $\displaystyle\int_1^{+\infty} \dfrac{\ln x}{\sqrt{x}}\,dx$

26. $\displaystyle\int_0^{+\infty} \dfrac{x - 1}{x + 2}\,dx$

27. PRESENT VALUE OF AN INVESTMENT   It is estimated that t years from now, a certain investment will be generating income at a rate of $f(t) = 8000 + 400t$ dollars per year. If the income is generated in perpetuity and the prevailing annual interest rate remains fixed at 5% compounded continuously, find the present value of the investment.

28. PRODUCTION   After t hours on the job, a factory worker can produce $100te^{-0.5t}$ units per hour. How many units does a worker who arrives on the job at 8:00 A.M. produce between 10:00 A.M. and noon?

29. DEMOGRAPHICS (SURVIVAL/RENEWAL)   Demographic studies indicate that the fraction of the residents that will remain in a certain city for at least t years is $f(t) = e^{-t/20}$. The current population of the city is 100 000, and it is estimated that t years from now, new people will be arriving at a rate of $100t$ people per year. If this estimate is correct, what will happen to the population of the city in the long run?

30. SUBSCRIPTION GROWTH   The publishers of a national magazine have found that the fraction of subscribers who remain subscribers for at least t years is $f(t) = e^{-t/10}$. Currently, the magazine has 20 000 subscribers and it is estimated that new subscriptions will be sold at a rate of 1000 per year. Approximately how many subscribers will the magazine have in the long run?

31. NUCLEAR WASTE   After t years of operation, a certain nuclear power plant produces radioactive waste at a rate of $R(t)$ kilograms per year, where

$$R(t) = 300 - 200e^{-0.03t}.$$

The waste decays exponentially at a rate of 2% per year. How much radioactive waste from the plant will be present in the long run?

32. PSYCHOLOGICAL TESTING   In a psychological experiment, it is found that the proportion of participants who require more than t minutes to finish a particular task is given by

$$\int_t^{+\infty} 0.07e^{-0.07u}\,du.$$

a. Find the proportion of participants who require more than 5 minutes to finish the task.

b. What proportion requires between 10 minutes and 15 minutes to finish?

NUMERICAL INTEGRATION   *In Exercises 33 through 36, approximate the given integral and estimate the error with the specified number of subintervals using*

   (a) the trapezoidal rule
   (b) Simpson's rule

33. $\displaystyle\int_1^3 \dfrac{1}{x}\,dx;\ n = 10$

34. $\displaystyle\int_0^2 e^{x^2}\,dx;\ n = 8$

35. $\int_1^2 \dfrac{e^x}{x} \, dx; \; n = 10$

36. $\int_1^2 x e^{1/x} \, dx; \; n = 8$

NUMERICAL INTEGRATION   *In Exercises 37 and 38, determine how many subintervals are required to guarantee accuracy to within 0.00005 of the actual value of the given integral using*
  (a) the trapezoidal rule
  (b) Simpson's rule

37. $\int_1^3 \sqrt{x} \, dx$

38. $\int_{0.5}^1 e^{-1.1x} \, dx$

39. TOTAL COST FROM MARGINAL COST   A manufacturer determines that the marginal cost of producing q units of a particular commodity is $C'(q) = \sqrt{q}e^{0.01q}$ dollars per unit.
  a. Express the total cost of producing the first 8 units as a definite integral.
  b. Estimate the value of the total cost integral in part (a) using the trapezoidal rule with $n = 8$ subintervals.

40. REVENUE FROM DEMAND DATA   An economist studying the demand for a particular commodity gathers the data in the accompanying table, which lists the number of units q (in thousands) of the commodity that will be demanded (sold) at a price of p dollars per unit.

q (1000 units)	0	4	8	12	16	20	24
p (\$/unit)		49.12	42.90	31.32	19.83	13.87	10.58	7.25

Use this information together with Simpson's rule to estimate the total revenue

$$R = \int_0^{24} qp(q) \, dq \text{ thousand dollars}$$

obtained as the level of production is increased from 0 to 24 000 units ($q = 0$ to $q = 24$).

 Using computer software, evaluate the integrals in Exercises 41 and 42. Graph to view the area found. In each case, verify your result by applying an appropriate integration formula from Table 6.1.

41. $\int_{-1}^1 \dfrac{2}{9 - x^2} \, dx$

42. $\int_0^1 x^2 \sqrt{9 + 4x^2} \, dx$

43. Use technology to compute

$$I(N) = \int_0^N \dfrac{1}{\sqrt{\pi}} e^{-x^2} \, dx$$

for $N = 1$, 10, and 50. Based on your results, do you think the improper integral

$$\int_0^{+\infty} \dfrac{1}{\sqrt{\pi}} e^{-x^2} \, dx$$

converges? If so, to what value?

44. Use technology to compute

$$I(N) = \int_0^N \dfrac{\ln(x + 1)}{x} \, dx$$

for $N = 10$, 100, 1000, and 10 000. Based on your results, do you think the improper integral

$$\int_1^{+\infty} \dfrac{\ln(x + 1)}{x} \, dx$$

converges? If so, to what value?

THINK ABOUT IT

MEASURING CARDIAC OUTPUT

The **cardiac output** F of a person's heart is the rate at which the heart pumps blood over a specified period of time. Measuring cardiac output helps physicians treat heart attack patients by diagnosing the extent of heart damage. Specifically, for a healthy person at rest, cardiac output is between 4 L and 5 L per minute and is relatively steady, but the cardiac output of someone with serious cardiovascular disease may be as low as 2 L to 3 L per minute. Because cardiac output is so important in cardiology, medical researchers have devised a variety of ways to measure it. We will examine two methods for using integration to measure cardiac output, one of which involves improper integration. (Our coverage is adapted from *The UMAP Journal* article and the UMAP Module listed in the references at the end of this essay.)

(Photo: Getty Images/MedicalRF.com)

A traditional method for measuring cardiac output is a technique known as **dye dilution.** In this method, dye is injected into a main vein near the heart. Blood carrying this dye circulates through the right ventricle, to the lungs, through the left auricle and left ventricle, and then into the arterial system, returning to the left auricle and left ventricle. The concentration of dye passing through an artery or vein is measured at specific time intervals, and from these concentrations, cardiac output is computed. The dye that is typically used in dye dilution is indocyanine green. Very little of this dye is recirculated back to the heart after it passes through the body because it is metabolized by the liver.

Now suppose that X_0 units of dye is inserted into a vein near the heart where it mixes with a patient's blood. Let $C(t)$ denote the concentration of dye at time t at the beginning of the pulmonary artery. Note that $C(0) = 0$ and that $C(t)$ rises relatively quickly to a maximum, and then decreases to zero after a relatively short time period, say, T seconds. The rate at which dye leaves the heart at time t equals $F \cdot C(t)$.

To derive a formula for F, we begin by splitting the interval between $t = 0$ and $t = T$ into n subintervals of equal length, using the points $0 = t_0, t_1 = \dfrac{T}{n}$, $t_2 = \dfrac{2T}{n}, \ldots, t_n = T$. Note that $C_j(t)$, the concentration of dye during the jth subinterval, is nearly constant, so that the amount of dye that leaves the heart during the jth subinterval is approximately

$$F \cdot C_j(t)\Delta t.$$

To find the total amount of dye leaving the heart, we sum the terms for all n subintervals and obtain

$$X_0 = \sum_{j=0}^{n-1} FC_j(t)\Delta t.$$

Taking the limit as n approaches infinity, we derive the estimate

$$X_0 = F\int_0^T C(t)\, dt,$$

which means that

$$F = \frac{X_0}{\displaystyle\int_0^T C(t)\, dt}.$$

To see how this formula for F is used, suppose that $X_0 = 5$ mg of indocyanine green dye is injected into a major vein near the heart of a patient who has been exercising and that the concentration of dye $C(t)$ (in milligrams per litre) measured each 4 seconds is as shown in Table 6.1.

TABLE 6.2 Dye Dilution Data for Measuring Cardiac Output

Time t (seconds)	Concentration $C(t)$ (mg/L)
0	0
4	1.2
8	4.5
12	3.1
16	1.4
20	0.5
24	0

Applying Simpson's rule to estimate the integral of the concentration $C(t)$ over the time interval $0 \le t \le 24$, we find that

$$\int_0^{24} C(t)\, dt \approx \frac{1}{3}[0 + 4(1.2) + 2(4.5) + 4(3.1) + 2(1.4) + 4(0.5) + 0](4) \approx 41.33$$

so that

$$F = \frac{X_0}{\displaystyle\int_0^{24} C(t)\, dt} \approx \frac{5}{41.33} \approx 0.121.$$

Thus, the cardiac output F is approximately 0.121 L/s or, equivalently,

$$(60 \text{ s/min})(0.121 \text{ L/s}) \approx 7.26 \text{ L/min},$$

which is normal for someone who has been exercising.

Because of the need to sample blood repeatedly, measuring cardiac output using blood flow can be difficult. Dye dilution may also produce inaccurate results because dyes can be unstable and a considerable amount of a dye may recirculate through the body or dye may dissipate slowly. Consequently, other methods of determining cardiac output may be preferable. In the 1970s, a method for measuring cardiac output known as **thermodilution** was introduced. To use this method, a catheter with a balloon tip is inserted into the arm vein of a patient. This catheter is guided to the superior vena cava. Then a thermistor, a device that measures temperatures, is guided through the right side of the heart and positioned several centimetres inside the pulmonary artery. Once the thermistor is in place, a cold solution is injected, typically 10 mL of a 5% solution of dextrose in water at 0°C. As this cold solution mixes with blood in the right side of the heart, the blood temperature decreases. The temperature of the blood mixed with this cold solution is detected by the thermistor. The thermistor is connected to a computer that records the temperatures and displays the graph of these temperatures on a monitor. Figure 6.10 shows the typical temperature variation of the blood at the beginning of the pulmonary artery t seconds after the injection of 10 mL of dextrose solution at 0°C.

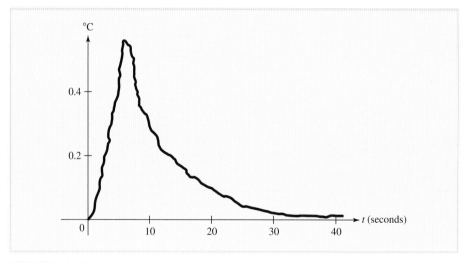

FIGURE 6.10 Variation in blood temperature T_{var} measured in the pulmonary artery t seconds after injection of cold solution into the right atrium of the heart.

Next, we show how to determine cardiac output using the temperature data collected during thermodilution. Let T_i denote the temperature of the injectant, and let T_b denote the body temperature, that is, the temperature of the blood before the cold solution is injected. Then Q_{in}, the cooling effect of the injectant, may be measured by the product of the volume V of the injectant and the temperature difference $T_b - T_i$ between the body and the injectant, that is,

$$Q_{\text{in}} = V(T_b - T_i).$$

Over a short time interval Δt, the cooling effect from the chilled blood passing the thermistor is given by the product of the temperature variation T_{var} that we recorded and the total amount of blood $F\Delta t$ that flows through the pulmonary artery during this interval. Thus, the *total* cooling effect Q_{out} (for all time $t \geq 0$) may be measured by the improper integral

$$Q_{out} = \int_0^\infty F\, T_{var}\, dt = F \int_0^\infty T_{var}\, dt.$$

The key fact for computing cardiac output is that $Q_{out} = rQ_{in}$, where r is a correction factor related to how much cooling effect is lost as the chilled solution passes into the body. (The value of r for a particular way the injectant enters the bloodstream can be calculated through data collection.) Inserting the formulas we have found for Q_{in} and Q_{out}, we obtain

$$rV(T_b - T_i) = F \int_0^\infty T_{var}\, dt.$$

Solving for the cardiac output rate F, we find that

$$F = \frac{rV(T_b - T_i)}{\displaystyle\int_0^\infty T_{var}\, dt}.$$

To illustrate how this method can be used, we will model the temperature variation by the continuous function $T_{var} = 0.1t^2 e^{-0.31t}$, whose graph is shown in Figure 6.11.

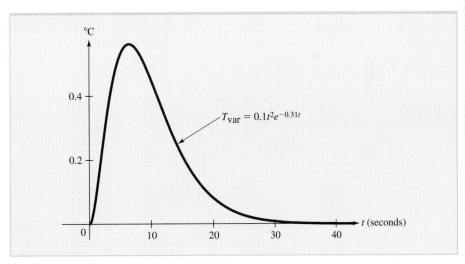

FIGURE 6.11   Approximation of temperature variation using $T_{var} = 0.1t^2 e^{-0.31t}$.

Using this approximation, suppose that $T_b = 37°C$, $T_i = 0°C$, $V = 10$, and $r = 0.9$ (so that 10% of the cooling effect of the injectant is lost as it passes into the bloodstream). Substituting into the equation for F, we get

$$F = \frac{(0.9)(10)(37 - 0)}{\displaystyle\int_0^\infty 0.1t^2 e^{-0.31t}\, dt}.$$

Computing this improper integral using integration by parts (the details of this computation are left to the reader) shows that

$$F = \frac{333}{0.1[2/(0.31)^3]} \approx 49.6 \text{ mL/s} = 2.98 \text{ L/min.}$$

Note that in our computations the numerator in the formula for F has units of millilitres times degrees Celsius and the denominator has units of degrees Celsius times seconds, so that F has units of millilitres per second. To convert this to units of litres per minute we multiply by $\frac{60}{1000} = 0.06$. A cardiac output rate of 2.98 L/min indicates that the patient has a serious cardiovascular problem.

Questions

1. Cardiac output is measured by injecting 10 mg of dye into a vein near the heart of a patient, and the dye concentration after t seconds is C milligrams per litre, where C is modelled by the function $C(t) = 4te^{-0.15t}$ for $0 \le t \le 30$. Use integration by parts to find the cardiac output F (in litres per minute). Based on your result, do you think this patient is a healthy person at rest, a healthy person exercising, or a person with a serious cardiovascular problem?

2. Use Simpson's rule to estimate a patient's cardiac output if 7 mg of indocyanine green dye is injected into a major vein near the heart and the concentration of dye measured every 3 seconds is as shown in the following table.

Time t (seconds)	Concentration $C(t)$ (mg/L)
0	0
3	1
6	3.5
9	5.8
12	4.4
15	3.2
18	2.1
21	1.2
24	0.5

Based on your result, do you think this patient is a healthy person at rest, a healthy person exercising, or a person with a serious cardiovascular problem?

3. Suppose that when the thermodilution method is applied to a patient, we find that $T_{\text{var}} = 0.2t^2e^{-0.43t}$ can be used to model the temperature variation. Suppose that the patient's body temperature is 36°C, that 12 mL of injectant at temperature 5°C is inserted into a major artery near the patient's heart, and that the correction factor is $r = 0.9$. Estimate the cardiac output F for this patient.

4. The dye-dilution approach to measuring cardiac output can be modified to apply to other flow problems. Use this approach to find a formula for P, the total daily discharge of a pollutant into a river flowing at a constant rate of F cubic metres

per hour if it is known that at a fixed point downstream from the industrial plant where the discharge takes place, the concentration of pollutant at time t (hours) is given by the function $C(t)$.

5. A battery plant begins discharging cadmium into a river at midnight on a certain day and the concentration C of cadmium at a point 1000 m downriver is subsequently measured at 3-hour intervals over the next 24 hours, as indicated in the accompanying table. If the flow rate of the river is $F = 2.3 \times 10^5$ m^3/h, use the formula you found in Question 4 to estimate the total amount of cadmium discharged from the plant during this 24-hour period. You may assume that the concentration is 0 when the cadmium is first discharged, that is, $C(0) = 0$.

Time	Concentration (mg/m^3)
3 A.M.	8.5
6 A.M.	12.0
9 A.M.	14.5
Noon	13.0
3 P.M.	12.5
6 P.M.	10.0
9 P.M.	7.5
Midnight	6.0

References

M. R. Cullen, *Linear Models in Biology,* Chichester: Ellis Horwood Ltd., 1985.

Arthur Segal, "Flow System Integrals," *The UMAP Journal,* Vol. III, No. 1, 1982.

Brindell Horelick and Sinan Koont, "Measuring Cardiac Output," UMAP Module 71, Birkhauser-Boston Inc., 1979.

William Simon, *Mathematical Techniques for Biology and Medicine,* Mineola, NY: Dover Publications, 1986, pp. 194–210 (Chapter IX).

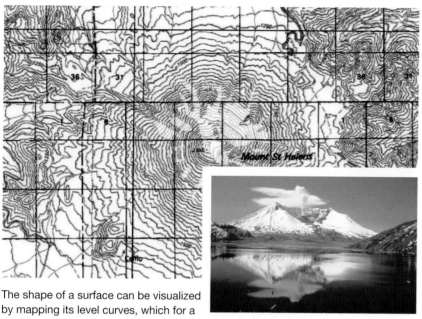

The shape of a surface can be visualized by mapping its level curves, which for a mountain are curves of constant elevation. The level curves are shown on the topographical map corresponding to Mount St. Helens in the state of Washington. Where the lines are close together, the terrain is very steep. For details on level curves, see Figure 7.6 and Example 7.1.8. (Photos: Maps a la carte, Inc. and U.S. Geological Survey)

Chapter 7

CALCULUS OF SEVERAL VARIABLES

CHAPTER OUTLINE

1 Functions of Several Variables
2 Partial Derivatives
3 Optimizing Functions of Two Variables
4 The Method of Least Squares
5 Constrained Optimization: The Method of Lagrange Multipliers
6 Double Integrals
    Chapter Summary
        Concept Summary
        Checkup for Chapter 7
        Review Exercises
    Think About It

LEARNING OBJECTIVES

After completing this chapter, you should be able to

LO1 Evaluate functions of several variables at a given point. Sketch level curves and interpret representations of computer-generated surfaces.

LO2 Compute partial derivatives.

LO3 Find the relative extrema of functions of several variables using the second partials test.

LO4 Fit a set of data to a line using the least squares method.

LO5 Optimize when there is a constraint on the variables using the method of Lagrange multipliers.

LO6 Evaluate double integrals.

L01

Evaluate functions of several variables at a given point. Sketch level curves and interpret representations of computer-generated surfaces.

Functions of Several Variables

In business, if a manufacturer determines that x units of a particular commodity can be sold domestically for $90 per unit, and y units can be sold to foreign markets for $110 per unit, then the total revenue obtained from all sales is given by

$$R = 90x + 110y.$$

In psychology, a person's intelligence quotient (IQ) is measured by the ratio

$$IQ = \frac{100m}{a}$$

where a and m are the person's actual age and mental age, respectively. A carpenter constructing a storage box x metres long, y metres wide, and z metres deep knows that the box will have volume V and surface area S, where

$$V = xyz \quad \text{and} \quad S = 2xy + 2xz + 2yz.$$

These are typical of practical situations in which a quantity of interest depends on the values of two or more variables. Other examples include the volume of water in a community's reservoir, which may depend on the amount of rainfall and the population of the community, and the output of a factory, which may depend on the amount of capital invested in the plant, the size of the labour force, and the cost of raw materials.

In this chapter, we will extend the methods of calculus to include functions of two or more independent variables. Most of our work will be with functions of two variables, which can be represented geometrically by surfaces in three-dimensional space instead of curves in the plane. We begin with a definition and some terminology.

Function of Two Variables ■ A function f of the two independent variables x and y is a rule that assigns to each ordered pair (x, y) in a given set D (the **domain** of f) exactly one real number, denoted by $f(x, y)$.

NOTE **Domain Convention:** Unless otherwise stated, we assume that the domain of f is the set of all (x, y) for which the expression $f(x, y)$ is defined.

As in the case of a function of one variable, a function of two variables $f(x, y)$ can be thought of as a machine in which there is a unique output $f(x, y)$ for each input (x, y), as illustrated in Figure 7.1. The domain of f is the set of all possible

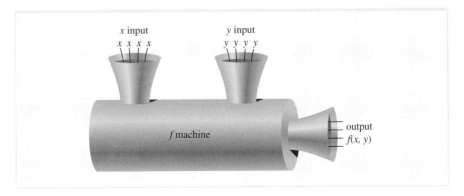

FIGURE 7.1   A function of two variables as a machine.

inputs, and the set of all possible corresponding outputs is the **range** of f. Functions of three independent variables $f(x, y, z)$, four independent variables $f(x, y, z, t)$, and so on can be defined in a similar fashion.

EXAMPLE 7.1.1

Suppose $f(x, y) = \dfrac{3x^2 + 5y}{x - y}$.

a. Find the domain of f.

b. Compute $f(1, -2)$.

Solution

a. Since division by any real number except zero is possible, the expression $f(x, y)$ can be evaluated for all ordered pairs (x, y) with $x - y \neq 0$ or $x \neq y$. Geometrically, this domain is the set of all points in the xy plane except for those on the line $y = x$.

b. $f(1, -2) = \dfrac{3(1)^2 + 5(-2)}{1 - (-2)} = \dfrac{3 - 10}{1 + 2} = -\dfrac{7}{3}$.

EXAMPLE 7.1.2

Suppose $f(x, y) = xe^y + \ln x$.

a. Find the domain of f.

b. Compute $f(e^2, \ln 2)$.

Solution

a. Since xe^y is defined for all real numbers x and y and since $\ln x$ is defined only for $x > 0$, the domain of f consists of all ordered pairs (x, y) of real numbers for which $x > 0$.

b. $f(e^2, \ln 2) = e^2 e^{\ln 2} + \ln e^2 = 2e^2 + 2 = 2(e^2 + 1) \approx 16.78$

EXAMPLE 7.1.3

Given the function of three variables $f(x, y, z) = xy + xz + yz$, evaluate $f(-1, 2, 5)$.

Solution

Substituting $x = -1$, $y = 2$, $z = 5$ into the formula for $f(x, y, z)$ gives

$$f(-1, 2, 5) = (-1)(2) + (-1)(5) + (2)(5) = 3.$$

Applications Here are four applications involving functions of several variables in business, economics, finance, and life science.

EXAMPLE 7.1.4

A sports store in Winnipeg carries two kinds of tennis racquets, the Frank Dancevic and the Aleksandra Wosniak autograph brands. The consumer demand for each brand depends not only on its own price, but also on the price of the competing brand. Sales figures indicate that if the Dancevic brand sells for x dollars per racquet and the Wosniak brand for y dollars per racquet, the demand for Dancevic racquets will be $D_1 = 300 - 20x + 30y$ racquets per year and the demand for Wosniak racquets will be $D_2 = 200 + 40x - 10y$ racquets per year. Express the store's total annual revenue from the sale of these racquets as a function of the prices x and y.

Solution

Let R denote the total monthly revenue. Then

$$R = \text{(number of Dancevic racquets sold)(price per Dancevic racquet)}$$
$$+ \text{(number of Wosniak racquets sold)(price per Wosniak racquet)}$$

Hence,

$$R(x, y) = (300 - 20x + 30y)(x) + (200 + 40x - 10y)(y)$$
$$= 300x + 200y + 70xy - 20x^2 - 10y^2$$

Output Q at a factory is often regarded as a function of the amount K of capital investment and the size L of the labour force. Output functions of the form

$$Q(K, L) = AK^\alpha L^\beta,$$

where A, α, and β are positive constants with $\alpha + \beta = 1$, have proven to be especially useful in economic analysis and are known as **Cobb-Douglas production functions.** Here is an example involving such a function.

EXAMPLE 7.1.5

Suppose that at a certain factory, output is given by the Cobb-Douglas production function $Q(K, L) = 60K^{1/3}L^{2/3}$ units, where K is the capital investment measured in units of \$1000 and L is the size of the labour force measured in worker-hours.

a. Compute the output if the capital investment is \$512 000 and 1000 worker-hours of labour are used.

b. Show that the output in part (a) will double if both the capital investment and the size of the labour force are doubled.

Solution

a. Evaluate $Q(K, L)$ with $K = 512$ (thousand) and $L = 1000$ to get

$$Q(512, 1000) = 60(512)^{1/3}(1000)^{2/3}$$
$$= 60(8)(100) = 48\ 000 \text{ units}$$

b. Evaluate $Q(K, L)$ with $K = 2(512)$ and $L = 2(1000)$ as follows to get

$$
\begin{aligned}
Q[2(512),\, 2(1000)] &= 60[2(512)]^{1/3}[2(1000)]^{2/3} \\
&= 60(2)^{1/3}(512)^{1/3}(2)^{2/3}(1000)^{2/3} \\
&= (2)^{1/3}(2)^{2/3}(60)(512)^{1/3}(1000)^{2/3} \\
&= 2[(60)(512)^{1/3}(1000)^{2/3}] \\
&= 96\,000
\end{aligned}
$$

that is, twice the output when $K = 512$ and $L = 1000$.

EXAMPLE 7.1.6

Recall (from Section 4.1) that the present value of B dollars in t years invested at an annual rate of r compounded k times per year is given by

$$
P(B, r, k, t) = B\left(1 + \frac{r}{k}\right)^{-kt}.
$$

Find the present value of \$10 000 in 5 years invested at 6% per year compounded quarterly.

Solution

We have $B = 10\,000$, $r = 0.06$ (6% per year), $k = 4$ (compounded 4 times per year), and $t = 5$, so the present value is

$$
P(10\,000,\, 0.06,\, 4,\, 5) = 10\,000\left(1 + \frac{0.06}{4}\right)^{-4(5)}
$$

$$
\approx 7424.7
$$

or approximately \$7425.

EXAMPLE 7.1.7

A population that grows exponentially satisfies

$$
P(A, k, t) = Ae^{kt},
$$

where P is the population at time t, A is the initial population (when $t = 0$), and k is the relative (per capita) growth rate. The population of a certain country is currently 5 million people and is growing at a rate of 3% per year. What will the population be in 7 years?

Solution

Let P be measured in millions of people. Substituting $A = 5$, $k = 0.03$ (3% annual growth), and $t = 7$ into the population function gives

$$
P(5,\, 0.03,\, 7) = 5e^{0.03(7)} \approx 6.16839.
$$

Therefore, in 7 years, the population will be approximately 6 168 400 people.

Graphs of Functions of Two Variables

The **graph** of a function of two variables $f(x, y)$ is the set of all triples (x, y, z) such that (x, y) is in the domain of f and $z = f(x, y)$. To picture such graphs, we need to construct a **three-dimensional** coordinate system. The first step in this construction is to add a third coordinate axis (the z axis) perpendicular to the familiar xy coordinate plane, as shown in Figure 7.2. Note that the xy plane is taken to be horizontal, and the positive z axis is up.

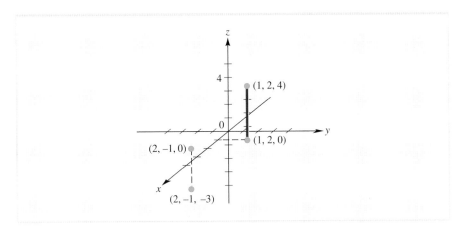

FIGURE 7.2   A three-dimensional coordinate system.

You can describe the location of a point in three-dimensional space by specifying three coordinates. For example, the point that is 4 units above the xy plane and lies directly above the point with xy coordinates $(x, y) = (1, 2)$ is represented by the ordered triple $(x, y, z) = (1, 2, 4)$. Similarly, the ordered triple $(2, -1, -3)$ represents the point that is 3 units directly below the point $(2, -1)$ in the xy plane. These points are shown in Figure 7.2.

To graph a function $f(x, y)$ of the two independent variables x and y, it is customary to introduce the letter z to stand for the dependent variable and to write $z = f(x, y)$ (Figure 7.3). The ordered pairs (x, y) in the domain of f are thought of as points in the xy plane, and the function f assigns a height z to each such point (depth if z is negative). Thus, if $f(1, 2) = 4$, you would express this fact geometrically by plotting the point $(1, 2, 4)$ in a three-dimensional coordinate space. The function may assign different heights to different points in its domain, and, in general, its graph will be a surface in three-dimensional space.

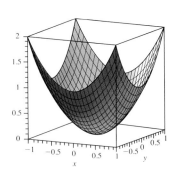

FIGURE 7.3   The graph of $z = f(x, y)$.

Four such surfaces are shown in Figure 7.4 on page 505. The surface in Figure 7.4a is a **cone,** Figure 7.4b shows a **paraboloid,** Figure 7.4c shows an **ellipsoid,** and Figure 7.4d shows what is commonly called a **saddle surface.** Surfaces such as these play an important role in examples and exercises in this chapter.

Level Curves

It is usually not easy to sketch the graph of a function of two variables. One way to visualize a surface is shown in Figure 7.5 on page 505. Notice that when the plane $z = C$ intersects the surface $z = f(x, y)$, the result is a curve in space. The corresponding set of points (x, y) in the xy plane that satisfy $f(x, y) = C$ is called the **level curve** of f at C, and an entire family of level curves is generated as C varies over a set of numbers. By sketching members of this family in the xy plane, you can obtain a useful representation of the surface $z = f(x, y)$.

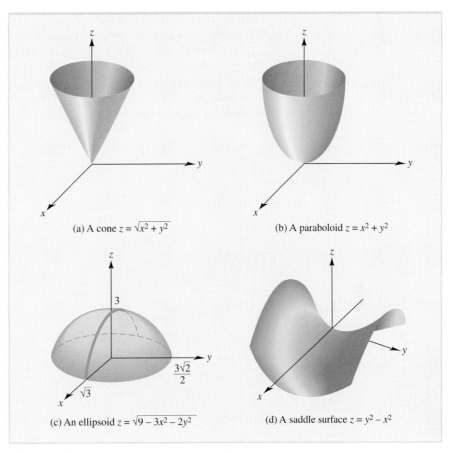

(a) A cone $z = \sqrt{x^2 + y^2}$　　　　　　　(b) A paraboloid $z = x^2 + y^2$

(c) An ellipsoid $z = \sqrt{9 - 3x^2 - 2y^2}$　　　(d) A saddle surface $z = y^2 - x^2$

FIGURE 7.4　Several surfaces in three-dimensional space.

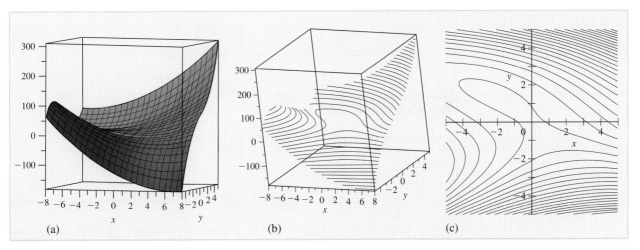

FIGURE 7.5　The function $z = f(x, y)$, the 3D contour plot, and the level curves of the surface of this function.

For instance, imagine that the surface $z = f(x, y)$ is a mountain whose elevation at the point (x, y) is given by $f(x, y)$, as shown in Figure 7.6a. The level curve $f(x, y) = C$ lies directly below a path on the mountain where the elevation is always C. To graph the mountain, indicate the paths of constant elevation by sketching the family of level curves in the plane and pinning a flag to each curve to show the elevation to which it corresponds (Figure 7.6b). This flat figure is called a **topographical map** of the surface $z = f(x, y)$.

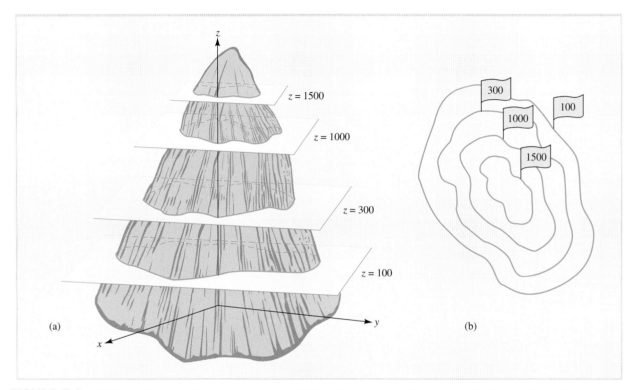

(a)

(b)

FIGURE 7.6   (a) The surface $z = f(x, y)$ as a mountain. (b) Level curves provide a topographical map of $z = f(x, y)$.

EXAMPLE 7.1.8

Discuss the level curves of the function $f(x, y) = x^2 + y^2$.

Solution

The level curve $f(x, y) = C$ has equation $x^2 + y^2 = C$. If $C = 0$, this is the point $(0, 0)$, and if $C > 0$, it is a circle of radius $\sqrt{C}$. If $C < 0$, there are no points that satisfy $x^2 + y^2 = C$.

The graph of the surface $z = x^2 + y^2$ is shown in Figure 7.7. The level curves correspond to cross-sections perpendicular to the z axis. It can be shown that cross-sections perpendicular to the x axis and the y axis are parabolas. (Try to see why this is true.) For this reason, the surface is shaped like a bowl. It is called a **circular paraboloid** or a **paraboloid of revolution**.

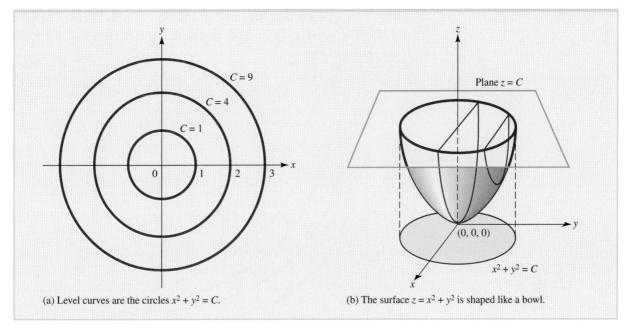

(a) Level curves are the circles $x^2 + y^2 = C$.   (b) The surface $z = x^2 + y^2$ is shaped like a bowl.

FIGURE 7.7   Level curves help to visualize the shape of a surface.

Level Curves in Economics: Isoquants and Indifference Curves

Level curves appear in many different applications. For instance, in economics, if the output $Q(x, y)$ of a production process is determined by two inputs x and y (say, hours of labour and capital investment), then the level curve $Q(x, y) = C$ is called the **curve of constant product** C or, more briefly, an **isoquant** ("iso" means "equal").

Another application of level curves in economics involves the concept of indifference curves. A consumer who is considering the purchase of a number of units of each of two commodities is associated with a **utility function** $U(x, y)$, which measures the total satisfaction (or **utility**) the consumer derives from having x units of the first commodity and y units of the second. A level curve $U(x, y) = C$ of the utility function is called an **indifference curve** and gives all of the combinations of x and y that lead to the same level of consumer satisfaction. These terms are illustrated in Example 7.1.9.

EXAMPLE 7.1.9

Suppose the utility derived by a consumer from x units of one commodity and y units of a second commodity is given by the utility function $U(x, y) = x^{3/2}y$. If the consumer currently owns $x = 16$ units of the first commodity and $y = 20$ units of the second, find the consumer's current level of utility and sketch the corresponding indifference curve.

Solution

The current level of utility is

$$U(16, 20) = (16)^{3/2}(20) = 1280,$$

and the corresponding indifference curve is

$$x^{3/2}y = 1280$$

or $y = 1280x^{-3/2}$. This curve consists of all points (x, y) where the level of utility $U(x, y)$ is 1280. The curve $x^{3/2}y = 1280$ and several other curves of the family $x^{3/2}y = C$ are shown in Figure 7.8.

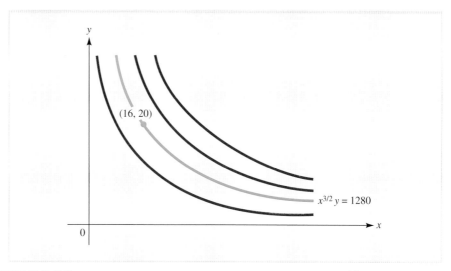

FIGURE 7.8 Indifference curves for the utility function $U(x, y) = x^{3/2}y$.

Computer Graphics

In practical work, computer software is used to graph functions of two variables. Hence, we will spend no more time developing graphing procedures for such functions.

Such software often allows you to choose different scales along each coordinate axis and may also enable you to rotate the surface to visualize it from different viewpoints. These features permit you to obtain a detailed picture of the graph. A variety of computer-generated graphs are displayed in Figure 7.9. These programs can also give a contour plot, showing the level curves of the function.

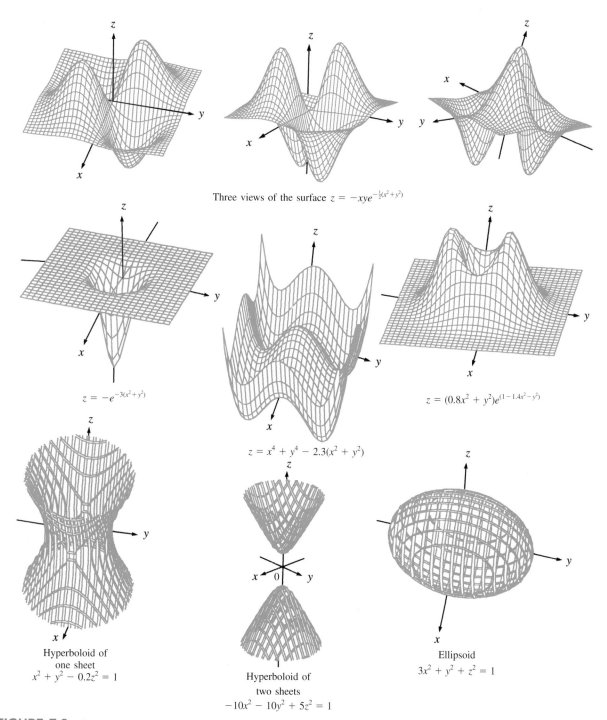

Three views of the surface $z = -xye^{-\frac{1}{2}(x^2 + y^2)}$

$z = -e^{-3(x^2 + y^2)}$

$z = (0.8x^2 + y^2)e^{(1 - 1.4x^2 - y^2)}$

$z = x^4 + y^4 - 2.3(x^2 + y^2)$

Hyperboloid of
one sheet
$x^2 + y^2 - 0.2z^2 = 1$

Hyperboloid of
two sheets
$-10x^2 - 10y^2 + 5z^2 = 1$

Ellipsoid
$3x^2 + y^2 + z^2 = 1$

FIGURE 7.9   Some computer-generated surfaces.

EXERCISES ■ 7.1

In Exercises 1 through 16, compute the indicated functional values.

1. $f(x, y) = 5x + 3y; f(-1, 2), f(3, 0)$

2. $f(x, y) = x^2 + x - 4y; f(1, 3), f(2, -1)$

3. $g(x, y) = x(y - x^3); g(1, 1), g(-1, 4)$

4. $g(x, y) = xy - x(y + 1); g(1, 0), g(-2, 3)$

5. $f(x, y) = (x - 1)^2 + 2xy^3; f(2, -1), f(1, 2)$

6. $f(x, y) = \dfrac{3x + 2y}{2x + 3y}; f(1, 2), f(-4, 6)$

7. $g(x, y) = \sqrt{y^2 - x^2}; g(4, 5), g(-1, 2)$

8. $g(u, v) = 10u^{1/2}v^{2/3}; g(16, 27), g(4, -1331)$

9. $f(r, s) = \dfrac{s}{\ln r}; f(e^2, 3), f(\ln 9, e^3)$

10. $f(x, y) = xye^{xy}; f(1, \ln 2), f(\ln 3, \ln 4)$

11. $g(x, y) = \dfrac{y}{x} + \dfrac{x}{y}; g(1, 2), g(2, -3)$

12. $f(s, t) = \dfrac{e^{st}}{2 - e^{st}}; f(1, 0), f(\ln 2, 2)$

13. $f(x, y, z) = xyz; f(1, 2, 3), f(3, 2, 1)$

14. $g(x, y, z) = (x + y)e^{yz}; g(1, 0, -1), g(1, 1, 2)$

15. $f(r, s, t) = \dfrac{\ln (r + t)}{r + s + t}; f(1, 1, 1), f(0, e^2, 3e^2)$

16. $f(x, y, z) = xye^z + xze^y + yze^x;$
$f(1, 1, 1), f(\ln 2, \ln 3, \ln 4)$

In Exercises 17 through 22, describe the domain of the given function.

17. $f(x, y) = \dfrac{5x + 2y}{4x + 3y}$

18. $f(x, y) = \sqrt{9 - x^2 - y^2}$

19. $f(x, y) = \sqrt{x^2 - y}$

20. $f(x, y) = \dfrac{x}{\ln (x + y)}$

21. $f(x, y) = \ln (x + y - 4)$

22. $f(x, y) = \dfrac{e^{xy}}{\sqrt{x - 2y}}$

In Exercises 23 through 30, sketch the indicated level curve $f(x, y) = C$ for each choice of constant C.

23. $f(x, y) = x + 2y; C = 1, C = 2, C = -3$

24. $f(x, y) = x^2 + y; C = 0, C = 4, C = 9$

25. $f(x, y) = x^2 - 4x - y; C = -4, C = 5$

26. $f(x, y) = \dfrac{x}{y}; C = -2, \ C = 2$

27. $f(x, y) = xy; C = 1, C = -1, C = 2, C = -2$

28. $f(x, y) = ye^x; C = 0, C = 1$

29. $f(x, y) = xe^y; C = 1, C = e$

30. $f(x, y) = \ln (x^2 + y^2); C = 4, C = \ln 4$

31. PRODUCTION   Using x skilled workers and y unskilled workers, a manufacturer can produce $Q(x, y) = 10x^2y$ units per day. Currently there are 20 skilled workers and 40 unskilled workers on the job.

  a. How many units are currently being produced each day?

  b. By how much will the daily production level change if 1 more skilled worker is added to the current workforce?

  c. By how much will the daily production level change if 1 more unskilled worker is added to the current workforce?

  d. By how much will the daily production level change if 1 more skilled worker *and* 1 more unskilled worker are added to the current workforce?

32. PRODUCTION COST   A manufacturer can produce graphing calculators at a cost of $40 each and business calculators for $20 each.

  a. Express the manufacturer's total monthly production cost as a function of the number of graphing calculators and the number of business calculators produced.

  b. Compute the total monthly cost if 500 graphing and 800 business calculators are produced.

  c. The manufacturer wants to increase the output of graphing calculators by 50 a month from the level in part (b). What corresponding change should be made in the monthly output of business calculators so the total monthly cost will not change?

33. **RETAIL SALES** A paint store carries two brands of latex paint. Sales figures indicate that if the first brand is sold for x_1 dollars per 4-L can and the second for x_2 dollars per 4-L can, the demand for the first brand will be $D_1(x_1, x_2) = 200 - 10x_1 + 20x_2$ cans per month and the demand for the second brand will be $D_2(x_1, x_2) = 100 + 5x_1 - 10x_2$ cans per month.
    a. Express the paint store's total monthly revenue from the sale of the paint as a function of the prices x_1 and x_2.
    b. Compute the revenue in part (a) if the first brand is sold for $21 per can and the second for $16 per can.

34. **PRODUCTION** The output at a certain factory is $Q(K, L) = 120K^{2/3}L^{1/3}$ units, where K is the capital investment measured in units of $1000 and L the size of the labour force measured in worker-hours.
    a. Compute the output if the capital investment is $125 000 and the size of the labour force is 1331 worker-hours.
    b. What will happen to the output in part (a) if both the level of capital investment and the size of the labour force are cut in half?

35. **PRODUCTION** The Easy-Gro agricultural company estimates that when $100x$ worker-hours of labour are employed on y hectares of land, the number of kilograms of wheat produced is $f(x, y) = Ax^ay^b$, where A, a, and b are positive constants. Suppose the company decides to double the production factors x and y. Determine how this decision affects the production of wheat in each of these cases:
    a. $a + b > 1$
    b. $a + b < 1$
    c. $a + b = 1$

36. **PRODUCTION** Suppose that when x machines and y worker-hours are used each day, a certain factory will produce $Q(x, y) = 10xy$ cell phones. Describe the relationship between the inputs x and y that results in an output of 1000 phones each day. (Note that you are finding a level curve of Q.)

37. **RETAIL SALES** A manufacturer with exclusive rights to a sophisticated new industrial machine is planning to sell a limited number of the machines to both foreign and domestic firms. The price the manufacturer can expect to receive for the machines will depend on the number of machines made available. It is estimated that if the manufacturer supplies x machines to the domestic market and y machines to the foreign market, the machines will sell for $60 - \dfrac{x}{5} + \dfrac{y}{20}$ thousand dollars each domestically and $50 - \dfrac{y}{10} + \dfrac{x}{20}$ thousand dollars each abroad. Express the revenue R as a function of x and y.

38. **RETAIL SALES** A manufacturer is planning to sell a new product at a price of A dollars per unit and estimates that if x thousand dollars are spent on development and y thousand dollars on promotion, consumers will buy approximately $\dfrac{320y}{y + 2} + \dfrac{160x}{x + 4}$ units of the product. If manufacturing costs are $50 per unit, express profit in terms of x, y, and A. [*Hint:* Profit = revenue − total cost of manufacture, development, and promotion.]

39. **A* SURFACE AREA OF THE HUMAN BODY** Pediatricians and medical researchers sometimes use the following empirical formula* relating the surface area S (m^2) of a person to the person's weight W (kg) and height H (cm):
    $$S(W, H) = 0.0072W^{0.425}H^{0.725}.$$

    a. Find $S(15.83, 87.11)$. Sketch the level curve of $S(W, H)$ that passes through (15.83, 87.11). Sketch several additional level curves of $S(W, H)$. What do these level curves represent?
    b. If Matt weighs 18.37 kg and has surface area 0.648 m^2, approximately how tall would you expect him to be?
    c. Suppose at some time in her life, Jenny weighs six times as much and is twice as tall as she was at birth. What is the corresponding percentage change in the surface area of her body?

40. **PSYCHOLOGY** A person's intelligence quotient (IQ) is measured by the function
    $$I(m, a) = \frac{100m}{a}$$

---

*B. Reading and B. Freeman, "Simple Formula for the Surface Area of the Body and a Simple Model for Anthropometry," *Clinical Anatomy* 18:126–130 (2005). Available online at http://www.med.unsw.edu.au/somsweb.nsf/resources/POM0501/$file/POMMar_05.pdf.

where a is the person's actual age and m is his or her mental age.

a. Find $I(12, 11)$ and $I(16, 17)$.

b. Sketch the graphs of several level curves of $I(m, a)$. How would you describe these curves?

41. CONSTANT PRODUCTION CURVES Using x skilled and y unskilled workers, a manufacturer can produce $Q(x, y) = 3x + 2y$ units per day. Currently the workforce consists of 10 skilled workers and 20 unskilled workers.

a. Compute the current daily output.

b. Find an equation relating the levels of skilled and unskilled labour if the daily output is to remain at its current level.

c. On a two-dimensional coordinate system, draw the isoquant (constant production curve) that corresponds to the current level of output.

d. What change should be made in the level of unskilled labour y to offset an increase in skilled labour x of two workers so that the output will remain at its current level?

42. A* INDIFFERENCE CURVES Suppose the utility derived by a consumer from x units of one commodity and y units of a second commodity is given by the utility function $U(x, y) = 2x^3y^2$. The consumer currently owns $x = 5$ units of the first commodity and $y = 4$ units of the second. Find the consumer's current level of utility and sketch the corresponding indifference curve.

43. A* INDIFFERENCE CURVES The utility derived by a consumer from x units of one commodity and y units of a second is given by the utility function $U(x, y) = (x + 1)(y + 2)$. The consumer currently owns $x = 25$ units of the first commodity and $y = 8$ units of the second. Find the current level of utility and sketch the corresponding indifference curve.

44. A* AIR POLLUTION Dumping and other material-handling operations near a landfill may result in contaminated particles being emitted into the surrounding air. To estimate such particulate emission, the following empirical formula* can be used:

$$E(V, M) = k(0.00163)\left(\frac{8V}{25}\right)^{1.3}\left(\frac{M}{2}\right)^{-1.4}$$

where E is the emission factor (kilograms of particles released into the air per tonne of soil moved), V is the mean speed of the wind (kilometres per hour), M is the moisture content of the material (given as a percentage), and k is a constant that depends on the size of the particles.

a. For a small particle (diameter 5 mm), it turns out that $k = 0.2$. Find $E(10, 13)$.

b. The emission factor E can be multiplied by the number of tonnes of material handled to obtain a measure of total emissions. Suppose 19 tonnes of the material in part (a) is handled. How many tonnes of a second kind of material with $k = 0.48$ (diameter 15 mm) and moisture content 27% must be handled to achieve the same level of total emissions if the wind velocity stays the same?

c. Sketch several level curves of $E(V, M)$, assuming the size of the particle stays fixed. What is represented by these curves?

45. A* FLOW OF BLOOD One of Poiseuille's laws** says that the speed of blood V (cm/s) flowing at a distance r (cm) from the axis of a blood vessel of radius R (cm) and length L (cm) is given by

$$V(P, L, R, r) = \frac{9.3 \times 10^5\, P}{L}(R^2 - r^2),$$

where P (newtons per square centimetre (N/cm²)) is the pressure in the vessel. Suppose a particular vessel has radius 0.0075 cm and is 1.675 cm long.

a. How fast is the blood flowing at a distance of 0.004 cm from the axis of this vessel if the pressure in the vessel is 0.03875 N/cm²?

b. Since R and L are fixed for this vessel, V is a function of P and r alone. Sketch several level curves of $V(P, r)$. Explain what they represent.

46. A* CONSTANT RETURNS TO SCALE Suppose output Q is given by the Cobb-Douglas production function $Q(K, L) = AK^\alpha L^{1-\alpha}$, where A and α are positive constants and $0 < \alpha < 1$. Show that if K and L are both multiplied by the same positive number m, then the output Q will also be multiplied by m; that is, show that $Q(mK, mL) = mQ(K, L)$. A production function with this property is said to have *constant returns to scale*.

---

*M. D. LaGrega, P. L. Buckingham, and J. C. Evans, *Hazardous Waste Management,* New York: McGraw-Hill, 1994, p. 140. Note that units have been converted, resulting in different constants.

**E. Batschelet, *Introduction to Mathematics for Life Scientists,* 2nd ed., New York: Springer-Verlag, 1979, pp. 102–103.

47. A* CHEMISTRY   Van der Waal's equation of state says that 1 mole of a confined gas satisfies the equation

$$T(P, V) = 0.0122\left(P + \frac{a}{V^2}\right)(V - b) - 273.15,$$

where T (°C) is the temperature of the gas, V (cm^3) is its volume, P (atmospheres (atm)) is the pressure of the gas on the walls of its container, and a and b are constants that depend on the nature of the gas.

a. Sketch the graphs of several level curves of T. These curves are called **curves of constant temperature** or **isotherms.**

b. If the confined gas is chlorine, experiments show that $a = 6.49 \times 10^6$ and $b = 56.2$. Find $T(1.13, 31.275 \times 10^3)$, that is, the temperature that corresponds to 31 275 cm^3 of chlorine under 1.13 atm of pressure.

48. ICE AGE PATTERNS OF ICE AND TEMPERATURE   The level curves in the land areas of the accompanying figure indicate ice elevations above sea level (in metres) during the last major ice age (approximately 18 000 years ago). The level curves in the sea areas indicate sea surface temperature. For instance, the ice was 1000 m thick above Toronto, and the sea temperature near the British Columbia coast was about 12°C. Where on Earth was the ice pack the thickest? What ice-bound land area was adjacent to the warmest sea?

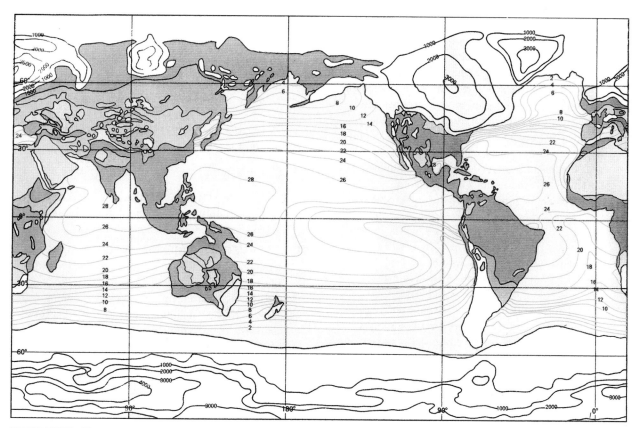

EXERCISE 48

SOURCE: *The Cambridge Encyclopedia of Earth Sciences*, New York: Crown/Cambridge Press, 1981, p. 302.

49. A* DAILY ENERGY EXPENDITURE Suppose a person of age A years has weight w in kilograms and height h in centimetres. Then the *Harris-Benedict equations* say that the daily basal energy expenditure in kilocalories (kCal) is

$$B_m(w, h, A) = 66.47 + 13.75w + 5.00h - 6.77A$$

for a male and

$$B_f(w, h, A) = 655.10 + 9.60w + 1.85h - 4.68A$$

for a female.

a. Find the basal energy expenditure of a man who weighs 90 kg, is 190 cm tall, and is 22 years old.
b. Find the basal energy expenditure of a woman who weighs 61 kg, is 170 cm tall, and is 27 years old.
c. A man maintains a weight of 85 kg and a height of 193 cm throughout his adult life. At what age will his daily basal energy expenditure be 2018 kCal?
d. A woman maintains a weight of 67 kg and a height of 173 cm throughout her adult life. At what age will her daily basal energy expenditure be 1504 kCal?

50. A* INVENTORY Suppose a company requires N units per year of a certain commodity. Suppose further that it costs D dollars to order a shipment of the commodity and that the storage cost per unit is S dollars per year. The units are used (or sold) at a constant rate throughout the year, and each shipment is used up just as a new shipment arrives.

a. Show that the total cost $C(x)$ of maintaining inventory when x units are ordered in each shipment is minimized when $x = Q$, where

$$Q(N, D, S) = \sqrt{\frac{2DN}{S}}.$$

(See Example 3.5.7 in Section 3.5.)
b. The optimum order size Q found in part (a) is called the *economic order quantity* (EOQ). What is the EOQ when 9720 units per year are ordered with an ordering cost of $35 per shipment and a storage cost of 84 cents per unit per year?

51. AMORTIZATION OF DEBT Suppose a loan of A dollars is amortized over n years at an annual interest rate r compounded monthly. Let $i = \dfrac{r}{12}$ be the equivalent monthly rate of interest. Then the monthly payments will be M dollars, where

$$M(A, n, i) = \frac{Ai}{1 - (1 + i)^{-12n}}.$$

(See Exercises 70 through 73 in Section 4.1.)

a. Allison has a home mortgage of $250 000 at a fixed rate of 5.2% per year for 15 years. What are her monthly payments? How much total interest does she pay for this loan?
b. Jorge also has a mortgage of $250 000 but at a fixed rate of 5.6% per year for 30 years. What are his monthly payments? How much total interest does he pay?

52. WIND POWER A machine based on wind energy generally converts the kinetic energy of moving air to mechanical energy by means of a device such as a rotating shaft, as in a windmill. Suppose a wind of velocity v is travelling through a wind-collecting machine with cross-sectional area A. Then, in physics,* it is shown that the total power generated by the wind is given by a formula of the form

$$P(v, A) = abAv^3,$$

where $b = 1.2$ kg/m^3 is the density of the air and a is a positive constant.

a. If a wind machine is perfectly efficient, then $a = \dfrac{1}{2}$ in the formula for $P(v, A)$. How much power would be produced by an ideal windmill with blade radius 15 m if the wind speed is 22 m/s?
b. No wind machine is perfectly efficient. In fact, it has been shown that the best we can hope for is about 59% of ideal efficiency, and a good empirical formula for power is obtained by taking $a = \dfrac{8}{27}$.

Compute $P(v, A)$ using this value of a if the blade radius of the windmill in part (a) is doubled and the wind velocity is halved.

53. A* REVERSE OSMOSIS In manufacturing semiconductors, it is necessary to use water with an extremely low mineral content. Thus, to separate water from contaminants, it is common practice to use a membrane process called **reverse osmosis**. A key to the effectiveness of such a process is the **osmotic pressure**, which may be determined by the **van't Hoff equation****

$$P(N, C, T) = 0.075NC(273.15 + T),$$

where P is the osmotic pressure (atmospheres (atm)), N is the number of ions in each molecule of solute, C is the concentration of solute (grams times moles per litre (g·mol/L)), and T is the temperature

---

*Raymond A. Serway, *Physics*, 3rd ed., Philadelphia: Saunders, 1992, pp. 408–410.
**M. D. LaGrega, P. L. Buckingham, and J. C. Evans, *Hazardous Waste Management*, New York: McGraw-Hill, 1994, pp. 530–543.

of the solute (°C). Find the osmotic pressure for a sodium chloride brine solution with concentration 0.53 g·mol/L at a temperature of 23°C. (You will need to know that each molecule of sodium chloride contains two ions: NaCl = Na$^+$ + Cl$^-$.)

54. **A*** Output in a certain factory is given by the Cobb-Douglas production function $Q(K, L) = 57K^{1/4}L^{3/4}$, where K is the capital in units of \$1000 and L is the size of the labour force, measured in worker-hours.

    a. Use a calculator to obtain $Q(K, L)$ for the values of K and L in this table:

K (\$1000)	277	311	493	554	718
L	743	823	1221	1486	3197
Q(K, L)					

    b. Note that the output $Q(277, 743)$ is doubled when K is doubled from 277 to 554 and L is doubled from 743 to 1486. In a similar manner, verify that output is tripled when K and L are both tripled, and that output is halved when K and L are both halved. Does anything interesting happen if K is doubled and L is halved? Verify your response with a calculator.

55. **A* CES PRODUCTION** A **constant elasticity of substitution** (CES) production function is one with the general form

    $$Q(K, L) = A[aK^{-\beta} + (1 - \alpha)L^{-\beta}]^{-1/\beta},$$

    where K is capital expenditure; L is the level of labour; and A, α, β are constants that satisfy $A > 0, 0 < \alpha < 1$, and $\beta > -1$. Show that such a function has *constant returns to scale*, that is,

    $$Q(sK, sL) = sQ(K, L)$$

    for any constant multiplier s. (Compare with Exercise 46.)

Partial Derivatives

In many problems involving functions of two variables, the goal is to find the rate of change of the function with respect to one of its variables when the other is held constant. That is, the goal is to differentiate the function with respect to the particular variable in question while keeping the other variable fixed. This process is known as **partial differentiation**, and the resulting derivative is said to be a **partial derivative** of the function.

For example, suppose a manufacturer finds that

$$Q(x, y) = 5x^2 + 7xy$$

units of a certain commodity will be produced when x skilled workers and y unskilled workers are employed. Then if the number of unskilled workers remains fixed, the production rate with respect to the number of skilled workers is found by differentiating $Q(x, y)$ with respect to x while holding y constant. We call this the **partial derivative of Q with respect to x** and denote it by $Q_x(x, y)$; thus,

$$Q_x(x, y) = 5(2x) + 7(1)y = 10x + 7y.$$

Similarly, if the number of skilled workers remains fixed, the production rate with respect to the number of unskilled workers is given by the **partial derivative of Q with respect to y,** which is obtained by differentiating $Q(x, y)$ with respect to y, holding x constant, that is, by

$$Q_y(x, y) = (0) + 7x(1) = 7x.$$

Here is a general definition of partial derivatives and some alternative notation.

Partial Derivatives ■ Suppose $z = f(x, y)$. The partial derivative of f with respect to x is denoted by

$$\frac{\partial z}{\partial x} \quad \text{or} \quad f_x(x, y)$$

and is the function obtained by differentiating f with respect to x, treating y as a constant. The partial derivative of f with respect to y is denoted by

$$\frac{\partial z}{\partial y} \quad \text{or} \quad f_y(x, y)$$

and is the function obtained by differentiating f with respect to y, treating x as a constant.

> **NOTE** Recall from Chapter 2 that the derivative of a function of one variable $f(x)$ is defined by the limit of a difference quotient, namely,
>
> $$f'(x) = \lim_{h \to 0} \frac{f(x + h) - f(x)}{h}.$$
>
> With this definition in mind, the partial derivative $f_x(x, y)$ is given by
>
> $$f_x(x, y) = \lim_{h \to 0} \frac{f(x + h, y) - f(x, y)}{h}$$
>
> and the partial derivative $f_y(x, y)$ by
>
> $$f_y(x, y) = \lim_{h \to 0} \frac{f(x, y + h) - f(x, y)}{h}.$$

Computation of Partial Derivatives

No new rules are needed for the computation of partial derivatives. To compute f_x, simply differentiate f with respect to the single variable x, pretending that y is a constant. To compute f_y, differentiate f with respect to y, pretending that x is a constant. Here are some examples.

EXAMPLE 7.2.1

Find the partial derivatives f_x and f_y if $f(x, y) = x^2 + 2xy^2 + \dfrac{2y}{3x}$.

Solution

To simplify the computation, begin by rewriting the function as

$$f(x, y) = x^2 + 2xy^2 + \frac{2}{3}yx^{-1}.$$

To compute f_x, think of f as a function of x and differentiate the sum term by term, treating y as a constant, to get

$$f_x(x, y) = 2x + 2(1)y^2 + \frac{2}{3}y(-x^{-2}) = 2x + 2y^2 - \frac{2y}{3x^2}.$$

To compute f_y, think of f as a function of y and differentiate term by term, treating x as a constant, to get

$$f_y(x, y) = 0 + 2x(2y) + \frac{2}{3}(1)x^{-1} = 4xy + \frac{2}{3x}.$$

EXAMPLE 7.2.2

Find the partial derivatives $\dfrac{\partial z}{\partial x}$ and $\dfrac{\partial z}{\partial y}$ if $z = (x^2 + xy + y)^5$.

Solution

Holding y fixed and using the chain rule to differentiate z with respect to x results in

$$\frac{\partial z}{\partial x} = 5(x^2 + xy + y)^4 \frac{\partial}{\partial x}(x^2 + xy + y)$$
$$= 5(x^2 + xy + y)^4(2x + y)$$

Holding x fixed and using the chain rule to differentiate z with respect to y gives

$$\frac{\partial z}{\partial y} = 5(x^2 + xy + y)^4 \frac{\partial}{\partial y}(x^2 + xy + y)$$
$$= 5(x^2 + xy + y)^4(x + 1)$$

EXAMPLE 7.2.3

Find the partial derivatives f_x and f_y if $f(x, y) = xe^{-2xy}$.

Solution

From the product rule,

$$f_x(x, y) = x(-2ye^{-2xy}) + e^{-2xy} = (-2xy + 1)e^{-2xy},$$

and from the constant multiple rule,

$$f_y(x, y) = x(-2xe^{-2xy}) = -2x^2e^{-2xy}.$$

Geometric Interpretation of Partial Derivatives

Recall from Section 7.1 that functions of two variables can be represented graphically as surfaces drawn on three-dimensional coordinate systems. In particular, if $z = f(x, y)$, an ordered pair (x, y) in the domain of f can be identified with a point in the xy plane, and the corresponding function value $z = f(x, y)$ can be thought of as assigning a height to this point. The graph of f is the surface consisting of all points (x, y, z) in three-dimensional space whose height z is equal to $f(x, y)$.

The partial derivatives of a function of two variables can be interpreted geometrically as follows. For each fixed number y_0, the points (x, y_0, z) form a vertical plane whose equation is $y = y_0$. If $z = f(x, y)$ and if y is kept fixed at $y = y_0$, then the corresponding points $(x, y_0, f(x, y_0))$ form a curve in three-dimensional space that is the intersection of the surface $z = f(x, y)$ with the plane $y = y_0$. At each point on this curve, the partial derivative $\dfrac{\partial z}{\partial x}$ is simply the slope of the line in the plane $y = y_0$ that is tangent to the curve at the point in question. That is, $\dfrac{\partial z}{\partial x}$ is the slope of the tangent line in the x direction. The situation is illustrated in Figure 7.10a.

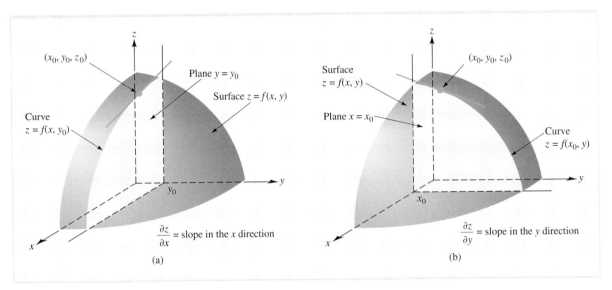

FIGURE 7.10   Geometric interpretation of partial derivatives.

Similarly, if x is kept fixed at $x = x_0$, the corresponding points $(x_0, y, f(x_0, y))$ form a curve that is the intersection of the surface $z = f(x, y)$ with the vertical plane $x = x_0$. At each point on this curve, the partial derivative $\dfrac{\partial z}{\partial y}$ is the slope of the tangent line in the plane $x = x_0$. That is, $\dfrac{\partial z}{\partial y}$ is the slope of the tangent line in the y direction. The situation is illustrated in Figure 7.10b.

Marginal Analysis

In economics, the term **marginal analysis** refers to the practice of using a derivative to estimate the change in the value of a function resulting from a 1-unit increase in one of its variables. In Section 2.5, you saw some examples of marginal analysis involving ordinary derivatives of functions of one variable. Here is an example of how partial derivatives can be used in a similar fashion.

EXAMPLE 7.2.4

It is estimated that the weekly output of a certain plant is given by the function $Q(x, y) = 1200x + 500y + x^2y - x^3 - y^2$ units, where x is the number of skilled workers and y is the number of unskilled workers employed at the plant. Currently the workforce consists of 30 skilled workers and 60 unskilled workers. Use marginal analysis to estimate the change in the weekly output that will result from the addition of 1 more skilled worker if the number of unskilled workers is not changed.

Solution

The partial derivative

$$Q_x(x, y) = 1200 + 2xy - 3x^2$$

is the rate of change of output with respect to the number of skilled workers. For any values of x and y, this is an approximation of the number of additional units that will be produced each week if the number of skilled workers is increased from x to $x + 1$ while the number of unskilled workers is kept fixed at y. In particular, if the workforce is increased from 30 skilled and 60 unskilled workers to 31 skilled and 60 unskilled workers, the resulting change in output is approximately

$$Q_x(30, 60) = 1200 + 2(30)(60) - 3(30)^2 = 2100 \text{ units.}$$

For practice, compute the exact change $Q(31, 60) - Q(30, 60)$. Is the approximation a good one?

If $Q(K, L)$ is the output of a production process involving the expenditure of K units of capital and L units of labour, then the partial derivative $Q_K(K, L)$ is called the **marginal productivity of capital** and measures the rate at which output Q changes with respect to capital expenditure when the labour force is held constant. Similarly, the partial derivative $Q_L(K, L)$ is called the **marginal productivity of labour** and measures the rate of change of output with respect to the labour level when capital expenditure is held constant. Example 7.2.5 illustrates one way these partial derivatives can be used in economic analysis.

EXAMPLE 7.2.5

A manufacturer estimates that the monthly output at a certain factory is given by the Cobb-Douglas function

$$Q(K, L) = 50K^{0.4}L^{0.6},$$

where K is the capital expenditure in units of \$1000 and L is the size of the labour force, measured in worker-hours.

a. Find the marginal productivity of capital Q_K and the marginal productivity of labour Q_L when the capital expenditure is \$750 000 and the level of labour is 991 worker-hours.

b. Should the manufacturer consider adding capital or increasing the labour level in order to increase output?

Solution

a. We have

$$Q_K(K, L) = 50(0.4K^{-0.6})L^{0.6} = 20K^{-0.6}L^{0.6}$$

and

$$Q_L(K, L) = 50K^{0.4}(0.6L^{-0.4}) = 30K^{0.4}L^{-0.4},$$

so with $K = 750$ (\$750 000) and $L = 991$,

$$Q_K(750, 991) = 20(750)^{-0.6}(991)^{0.6} \approx 23.64$$

and

$$Q_L(750, 991) = 30(750)^{0.4}(991)^{-0.4} \approx 26.84.$$

b. From part (a), an increase in 1 unit of capital (that is, \$1000) results in an increase in output of 23.64 units, which is less than the 26.84-unit increase in output that results from a unit increase in the labour level. Therefore, the manufacturer should increase the labour level by 1 worker-hour (from 991 worker-hours to 992) to increase output as quickly as possible from the current level.

Substitute and Complementary Commodities

Two commodities are said to be **substitute commodities** if an increase in the demand for either results in a decrease in demand for the other. Substitute commodities are competitive, like butter and margarine.

On the other hand, two commodities are said to be **complementary commodities** if a decrease in the demand of either results in a decrease in the demand of the other. An example is provided by cameras and printer paper for photos. If consumers buy fewer cameras, they will likely buy fewer packets of photo printing paper.

We can use partial derivatives to obtain criteria for determining whether two commodities are substitute or complementary. Suppose $D_1(p_1, p_2)$ units of the first commodity and $D_2(p_1, p_2)$ of the second are demanded when the unit prices of the commodities are p_1 and p_2, respectively. It is reasonable to expect demand to decrease with increasing price, so

$$\frac{\partial D_1}{\partial p_1} < 0 \quad \text{and} \quad \frac{\partial D_2}{\partial p_2} < 0.$$

For substitute commodities, the demand for each commodity increases with respect to the price of the other, so

$$\frac{\partial D_1}{\partial p_2} > 0 \quad \text{and} \quad \frac{\partial D_2}{\partial p_1} > 0.$$

However, for complementary commodities, the demand for each decreases with respect to the price of the other, and

$$\frac{\partial D_1}{\partial p_2} < 0 \quad \text{and} \quad \frac{\partial D_2}{\partial p_1} < 0.$$

Example 7.2.6 illustrates how these criteria can be used to determine whether a given pair of commodities are complementary, substitute, or neither.

EXAMPLE 7.2.6

Suppose the demand function for flour in a certain community is given by

$$D_1(p_1, p_2) = 500 + \frac{10}{p_1 + 2} - 5p_2$$

while the corresponding demand for bread is given by

$$D_2(p_1, p_2) = 400 - 2p_1 + \frac{7}{p_2 + 3}$$

where p_1 is the dollar price of a kilogram of flour and p_2 is the price of a loaf of bread. Determine whether flour and bread are substitute commodities, complementary commodities, or neither.

Solution

We find that

$$\frac{\partial D_1}{\partial p_2} = -5 < 0 \quad \text{and} \quad \frac{\partial D_2}{\partial p_1} = -2 < 0.$$

Since both partial derivatives are negative for all p_1 and p_2, it follows that flour and bread are complementary commodities.

Second-Order Partial Derivatives

Partial derivatives can themselves be differentiated. The resulting functions are called **second-order partial derivatives.** Here is a summary of the definition and notation for the four possible second-order partial derivatives of a function of two variables.

Second-Order Partial Derivatives ■ If $z = f(x, y)$, the partial derivative of f_x with respect to x is

$$f_{xx} = (f_x)_x \quad \text{or} \quad \frac{\partial^2 z}{\partial x^2} = \frac{\partial}{\partial x}\left(\frac{\partial z}{\partial x}\right).$$

The partial derivative of f_x with respect to y is

$$f_{xy} = (f_x)_y \quad \text{or} \quad \frac{\partial^2 z}{\partial y \partial x} = \frac{\partial}{\partial y}\left(\frac{\partial z}{\partial x}\right).$$

The partial derivative of f_y with respect to x is

$$f_{yx} = (f_y)_x \quad \text{or} \quad \frac{\partial^2 z}{\partial x \partial y} = \frac{\partial}{\partial x}\left(\frac{\partial z}{\partial y}\right).$$

The partial derivative of f_y with respect to y is

$$f_{yy} = (f_y)_y \quad \text{or} \quad \frac{\partial^2 z}{\partial y^2} = \frac{\partial}{\partial y}\left(\frac{\partial z}{\partial y}\right).$$

The computation of second-order partial derivatives is illustrated in Example 7.2.7.

EXAMPLE 7.2.7

Compute the four second-order partial derivatives of the function

$$f(x, y) = xy^3 + 5xy^2 + 2x + 1$$

Solution

Since

$$f_x = y^3 + 5y^2 + 2,$$

it follows that

$$f_{xx} = 0 \quad \text{and} \quad f_{xy} = 3y^2 + 10y.$$

Since

$$f_y = 3xy^2 + 10xy,$$

we have

$$f_{yy} = 6xy + 10x \quad \text{and} \quad f_{yx} = 3y^2 + 10y.$$

NOTE   The two partial derivatives f_{xy} and f_{yx} are sometimes called the **mixed second-order partial derivatives** of f. Notice that the mixed partial derivatives in Example 7.2.7 are equal. This is not an accident. It turns out that for virtually all functions $f(x, y)$ that you will encounter in practical work, the mixed partials will be equal, that is,

$$f_{xy} = f_{yx}.$$

This means you will get the same answer if you first differentiate $f(x, y)$ with respect to x and then differentiate the resulting function with respect to y as you would if you performed the differentiation in the reverse order. However, always calculate both of these, as a check on your calculations.

Example 7.2.8 illustrates how a second-order partial derivative can convey useful information in a practical situation.

EXAMPLE 7.2.8

Suppose the output Q at a factory depends on the amount K of capital invested in the plant and equipment and also on the size L of the labour force, measured in worker-hours. Give an economic interpretation of the sign of the second-order partial derivative $\dfrac{\partial^2 Q}{\partial L^2}$.

Solution

If $\dfrac{\partial^2 Q}{\partial L^2} < 0$, the marginal product of labour $\dfrac{\partial Q}{\partial L}$ decreases as L increases. This implies that for a fixed level of capital investment, the effect on output of one additional worker-hour of labour is greater when the workforce is small than when the workforce is large.

Similarly, if $\dfrac{\partial^2 Q}{\partial L^2} > 0$, it follows that for a fixed level of capital investment, the effect on output of one additional worker-hour of labour is greater when the workforce is large than when it is small.

Typically, for a factory operating with an adequate workforce, the derivative $\dfrac{\partial^2 Q}{\partial L^2}$ will be negative. Can you give an economic explanation for this fact?

The Chain Rule for Partial Derivatives

In many practical situations, a particular quantity is given as a function of two or more variables, each of which can be thought of as a function of yet another variable, and the goal is to find the rate of change of the quantity with respect to this other variable. For example, the demand for a certain commodity may depend on the price of the commodity itself and on the price of a competing commodity, both of which are increasing with time, and the goal may be to find the rate of change of the demand with respect to time. You can solve problems of this type by using the following generalization of the chain rule

$$\frac{dz}{dt} = \frac{dz}{dx}\frac{dx}{dt}$$

obtained in Section 2.4.

> **Chain Rule for Partial Derivatives** ■ Suppose z is a function of x and y, each of which is a function of t. Then z can be regarded as a function of t and
>
> $$\frac{dz}{dt} = \frac{\partial z}{\partial x}\frac{dx}{dt} + \frac{\partial z}{\partial y}\frac{dy}{dt}.$$

Observe that the expression for $\dfrac{dz}{dt}$ is the sum of two terms, each of which can be interpreted using the chain rule for a function of one variable. In particular,

$$\frac{\partial z}{\partial x}\frac{dx}{dt} = \text{rate of change of } z \text{ with respect to } t \text{ for fixed } y.$$

and

$$\frac{\partial z}{\partial y}\frac{dy}{dt} = \text{rate of change of } z \text{ with respect to } t \text{ for fixed } x.$$

The chain rule for partial derivatives says that the total rate of change of z with respect to t is the sum of these two partial rates of change. Here is a practical example illustrating the use of the chain rule for partial derivatives.

EXAMPLE 7.2.9

A health food store carries two kinds of vitamin water, brand A and brand B. Sales figures indicate that if brand A is sold for x dollars per bottle and brand B for y dollars per bottle, the demand for brand A will be

$$Q(x, y) = 300 - 20x^2 + 30y \quad \text{bottles per month.}$$

It is estimated that t months from now the price of brand A will be

$$x = 2 + 0.05t \quad \text{dollars per bottle}$$

and the price of brand B will be

$$y = 2 + 0.1\sqrt{t} \quad \text{dollars per bottle.}$$

At what rate will the demand for brand A be changing with respect to time 4 months from now?

Solution

The goal is to find $\dfrac{dQ}{dt}$ when $t = 4$. Using the chain rule gives

$$\frac{dQ}{dt} = \frac{\partial Q}{\partial x}\frac{dx}{dt} + \frac{\partial Q}{\partial y}\frac{dy}{dt} = -40x(0.05) + 30(0.05t^{-1/2}).$$

When $t = 4$,

$$x = 2 + 0.05(4) = 2.2,$$

and hence

$$\frac{dQ}{dt} = -40(2.2)(0.05) + 30(0.05)(4^{-1/2}) = -3.65.$$

That is, 4 months from now the monthly demand for brand A will be decreasing at a rate of 3.65 bottles per month.

In Section 2.5, you learned how to use increments to approximate the change in a function resulting from a small change in its independent variable. In particular, if y is a function of x, then

$$\Delta y \approx \frac{dy}{dx}\Delta x,$$

where Δx is a small change in the variable x and Δy is the corresponding change in y. Here is the analogous approximation formula for functions of two variables, based on the chain rule for partial derivatives.

Incremental Approximation Formula for Functions of Two Variables

■ Suppose z is a function of x and y. If Δx denotes a small change in x and Δy a small change in y, the corresponding change in z is

$$\Delta z \approx \frac{\partial z}{\partial x}\Delta x + \frac{\partial z}{\partial y}\Delta y.$$

The marginal analysis approximations made earlier in Examples 7.2.4 and 7.2.5 involved unit increments. However, the incremental approximation formula allows a more flexible range of marginal analysis computations, as illustrated in Example 7.2.10.

EXAMPLE 7.2.10

At a certain factory, the daily output is $Q = 60K^{1/2}L^{1/3}$ units, where K denotes the capital investment measured in units of \$1000 and L the size of the labour force measured in worker-hours. The current capital investment is \$900 000, and 1000 worker-hours of labour are used each day. Estimate the change in output that will result if capital investment is increased by \$1000 and labour is increased by 2 worker-hours.

Solution

Apply the approximation formula with $K = 900$, $L = 1000$, $\Delta K = 1$, and $\Delta L = 2$ to get

$$\Delta Q \approx \frac{\partial Q}{\partial K} \Delta K + \frac{\partial Q}{\partial L} \Delta L$$

$$= 30K^{-1/2}L^{1/3}\Delta K + 20K^{1/2}L^{-2/3}\Delta L$$

$$= 30\left(\frac{1}{30}\right)(10)(1) + 20(30)\left(\frac{1}{100}\right)(2)$$

$$= 22$$

That is, output will increase by approximately 22 units.

EXERCISES ▪ 7.2

In Exercises 1 through 20, compute all first-order partial derivatives of the given function.

1. $f(x, y) = 7x - 3y + 4$

2. $f(x, y) = x - xy + 3$

3. $f(x, y) = 4x^3 - 3x^2y + 5x$

4. $f(x, y) = 2x(y - 3x) - 4y$

5. $f(x, y) = 2xy^5 + 3x^2y + x^2$

6. $z = 5x^2y + 2xy^3 + 3y^2$

7. $z = (3x + 2y)^5$

8. $f(x, y) = (x + xy + y)^3$

9. $f(s, t) = \dfrac{3t}{2s}$

10. $z = \dfrac{t^2}{s^3}$

11. $z = xe^{xy}$

12. $f(x, y) = xye^x$

13. $f(x, y) = \dfrac{e^{2-x}}{y^2}$

14. $f(x, y) = xe^{x+2y}$

15. $f(x, y) = \dfrac{2x + 3y}{y - x}$

16. $z = \dfrac{xy^2}{x^2y^3 + 1}$

17. $z = u \ln v$

18. $f(u, v) = u \ln uv$

19. $f(x, y) = \dfrac{\ln(x + 2y)}{y^2}$

20. $z = \ln\left(\dfrac{x}{y} + \dfrac{y}{x}\right)$

In Exercises 21 through 28, evaluate the partial derivatives $f_x(x, y)$ and $f_y(x, y)$ at the given point $P_0(x_0, y_0)$.

21. $f(x, y) = x^2 + 3y$ at $P_0(1, -1)$

22. $f(x, y) = x^3y - 2(x + y)$ at $P_0(1, 0)$

23. $f(x, y) = \dfrac{y}{2x + y}$ at $P_0(0, -1)$

24. $f(x, y) = x + \dfrac{x}{y - 3x}$ at $P_0(1, 1)$

25. $f(x, y) = 3x^2 - 7xy + 5y^3 - 3(x + y) - 1$ at $P_0(-2, 1)$

26. $f(x, y) = (x - 2y)^2 + (y - 3x)^2 + 5$ at $P_0(0, -1)$

27. $f(x, y) = xe^{-2y} + ye^{-x} + xy^2$ at $P_0(0, 0)$

28. $f(x, y) = xy \ln\left(\dfrac{y}{x}\right) + \ln(2x - 3y)^2$ at $P_0(1, 1)$

In Exercises 29 through 34, find the second partials (including the mixed partials).

29. $f(x, y) = 5x^4y^3 + 2xy$

30. $f(x, y) = \dfrac{x + 1}{y - 1}$

31. $f(x, y) = e^{x^2y}$

32. $f(u, v) = \ln(u^2 + v^2)$

33. $f(s, t) = \sqrt{s^2 + t^2}$

34. $f(x, y) = x^2 y e^x$

SUBSTITUTE AND COMPLEMENTARY COMMODITIES

In Exercises 35 through 40, the demand functions for a pair of commodities are given. Use partial derivatives to determine whether the commodities are substitute, complementary, or neither.

35. $D_1 = 500 - 6p_1 + 5p_2$;
$D_2 = 200 + 2p_1 - 5p_2$

36. $D_1 = 1000 - 0.02p_1^2 - 0.05p_2^2$;
$D_2 = 800 - 0.001p_1^2 - p_1 p_2$

37. $D_1 = 3000 + \dfrac{400}{p_1 + 3} + 50p_2$;

$D_2 = 2000 - 100p_1 + \dfrac{500}{p_2 + 4}$

38. $D_1 = 2000 + \dfrac{100}{p_1 + 2} - 25p_2$;

$D_2 = 1500 - \dfrac{p_2}{p_1 + 7}$

39. $D_1 = \dfrac{7p_2}{1 + p_1^2}$; $D_2 = \dfrac{p_1}{1 + p_2^2}$

40. $D_1 = 200p_1^{-1/2} p_2^{-1/2}$; $D_2 = 300p_1^{-1/2} p_2^{-3/2}$

LAPLACE'S EQUATION

*The function $z = f(x, y)$ is said to satisfy **Laplace's equation** if $z_{xx} + z_{yy} = 0$. Functions that satisfy such an equation play an important role in a variety of applications in the physical sciences, especially in the theory of electricity and magnetism. In Exercises 41 through 44, determine whether the given function satisfies Laplace's equation.*

41. $z = x^2 - y^2$

42. $z = xy$

43. $z = xe^y - ye^x$

44. $z = [(x - 1)^2 + (y + 3)^2]^{-1/2}$

45. MARGINAL ANALYSIS At a certain factory, the daily output is $Q(K, L) = 60K^{1/2}L^{1/3}$ units, where K denotes the capital investment measured in units of $1000 and L the size of the labour force measured in worker-hours. Suppose that the current capital investment is $900 000 and that 1000 worker-hours of labour are used each day. Use marginal analysis to estimate the effect of an additional capital investment of $1000 on the daily output if the size of the labour force is not changed.

46. MARGINAL PRODUCTIVITY A manufacturer estimates that the annual output at a certain factory is given by

$$Q(K, L) = 30K^{0.3}L^{0.7}$$

units, where K is the capital expenditure in units of $1000 and L is the size of the labour force in worker-hours.

  a. Find the marginal productivity of capital Q_K and the marginal productivity of labour Q_L when the capital expenditure is $630 000 and the labour level is 830 worker-hours.

  b. Should the manufacturer consider adding a unit of capital or a unit of labour in order to increase output more rapidly?

47. NATIONAL PRODUCTIVITY The annual productivity of a certain country is

$$Q(K, L) = 150[0.4K^{-1/2} + 0.6L^{-1/2}]^{-2}$$

units, where K is capital expenditure in millions of dollars and L measures the labour force in thousands of worker-hours.

  a. Find the marginal productivity of capital Q_K and the marginal productivity of labour Q_L.

  b. Currently, capital expenditure is $5.041 billion ($K = 5041$), and 4 900 000 worker-hours ($L = 4900$) are being employed. Find the marginal productivities Q_K and Q_L at these levels.

  c. Should the government of the country encourage capital investment or additional labour employment to increase productivity as rapidly as possible?

48. MARGINAL ANALYSIS A grocer's daily profit from the sale of two brands of cat food is

$$P(x, y) = (x - 30)(70 - 5x + 4y) + (y - 40)(80 + 6x - 7y)$$

cents, where x is the price per can of the first brand and y is the price per can of the second. Currently the first brand sells for 50 cents per can and the second for 52 cents per can. Use marginal analysis to estimate the change in the daily profit that will result if the grocer raises the price of the second brand by one cent per can but keeps the price of the first brand unchanged.

49. FLOW OF BLOOD The smaller the resistance to flow in a blood vessel, the less energy is expended by the pumping heart. One of Poiseuille's laws*

*E. Batschelet, *Introduction to Mathematics for Life Scientists,* 2nd ed., New York: Springer-Verlag, 1979, p. 279.

says that the resistance to the flow of blood in a blood vessel satisfies

$$F(L, r) = \frac{kL}{r^4}$$

where L is the length of the vessel, r is its radius, and k is a constant that depends on the viscosity of blood.

a. Find F, $\dfrac{\partial F}{\partial L}$, and $\dfrac{\partial F}{\partial r}$ in the case where $L = 3.17$ cm and $r = 0.085$ cm. Leave your answer in terms of k.

b. Suppose the vessel in part (a) is constricted and lengthened so that its new radius is 20% smaller than before and its new length is 20% greater. How do these changes affect the flow $F(L, r)$? How do they affect the values of $\dfrac{\partial F}{\partial L}$ and $\dfrac{\partial F}{\partial r}$?

50. **CONSUMER DEMAND** The monthly demand for a certain brand of toaster is given by the function $f(x, y)$, where x is the amount of money (measured in units of $1000) spent on advertising and y is the selling price (in dollars) of the toasters. Give economic interpretations of the partial derivatives f_x and f_y. Under normal economic conditions, what will be the sign of each of these derivatives?

51. **CONSUMER DEMAND** A bicycle dealer has found that if commuter bicycles are sold for x dollars each and the price of gasoline is y cents per litre, approximately $F(x, y)$ bicycles will be sold each month, where

$$F(x, y) = 200 - 24\sqrt{x} + 4(0.4y + 3)^{3/2}.$$

Currently, the bicycles sell for $324 each and gasoline sells for $1.10 per litre. Use marginal analysis to estimate the change in the demand for bicycles that results when the price of bicycles is kept fixed but the price of gasoline decreases by one cent per litre.

52. **SURFACE AREA OF THE HUMAN BODY** Recall from Exercise 39 of Section 7.1 that the surface area of a person's body may be measured by the empirical formula

$$S(W, H) = 0.0072W^{0.425}H^{0.725}$$

where W (kg) and H (cm) are the person's weight and height, respectively. Currently, a child weighs 34 kg and is 120 cm tall.

a. Compute the partial derivatives $S_W(34, 120)$ and $S_H(34, 120)$, and interpret each as a rate of change.

b. Estimate the change in surface area that results if the child's height stays constant but her weight increases by 1 kg.

53. **PACKAGING** A pop can is a cylinder H centimetres tall with radius R centimetres. Its volume is given by the formula $V = \pi R^2 H$. A particular can is 12 cm tall with radius 3 cm. Use calculus to estimate the change in volume that results if the radius is increased by 1 cm while the height remains constant at 12 cm.

54. **PACKAGING** For the pop can in Exercise 53, the surface area is given by $S = 2\pi R^2 + 2\pi RH$. Use calculus to estimate the change in surface area that results if

a. The radius is increased from 3 cm to 4 cm while the height stays at 12 cm.

b. The height is decreased from 12 cm to 11 cm while the radius stays at 3 cm.

55. **A* CONSUMER DEMAND** Two competing brands of power lawn mowers are sold in the same town. The price of the first brand is x dollars per mower, and the price of the second brand is y dollars per mower. The local demand for the first brand of mower is given by the function $D(x, y)$.

a. How would you expect the demand for the first brand of mower to be affected by an increase in x? By an increase in y?

b. Translate your answers in part (a) into conditions on the signs of the partial derivatives of D.

c. If $D(x, y) = a + bx + cy$, what can you say about the signs of the coefficients b and c if your conclusions in parts (a) and (b) are to hold?

56. **CHEMISTRY** The **ideal gas law** says that for n moles of an ideal gas, $PV = nRT$, where P is the pressure exerted by the gas, V is the volume of the gas, T is the temperature of the gas, and R is a constant (the **gas constant**). Compute the product

$$\frac{\partial V}{\partial T} \frac{\partial T}{\partial P} \frac{\partial P}{\partial V}.$$

57. **A* CARDIOLOGY** To estimate the amount of blood that flows through a patient's lung, cardiologists use the empirical formula

$$P(x, y, u, v) = \frac{100xy}{xy + uv}$$

where P is a percentage of the total blood flow, x is the carbon dioxide output of the lung, y is the arteriovenous carbon dioxide difference in the lung, u is the carbon dioxide output of the lung, and v is the arteriovenous carbon dioxide difference in the other lung.

It is known that blood flows into the lungs to pick up oxygen and dump carbon dioxide, so the arteriovenous carbon dioxide difference measures the extent to which this exchange is accomplished. (The actual measurement is accomplished by a device called a **cardiac shunt**.) The carbon dioxide is then exhaled from the lungs so that oxygen-bearing air can be inhaled.

Compute the partial derivatives P_x, P_y, P_u, and P_v, and give a physiological interpretation of each derivative.

58. **A* ELECTRIC CIRCUIT**   In an electric circuit with two resistors of resistance R_1 and R_2 connected in parallel, the total resistance R is given by the formula

$$\frac{1}{R} = \frac{1}{R_1} + \frac{1}{R_2}.$$

Show that

$$R_1 \frac{\partial R}{\partial R_1} + R_2 \frac{\partial R}{\partial R_2} = R.$$

59. **BLOOD CIRCULATION**   The flow of blood from an artery into a small capillary is given by the formula

$$F(x, y, z) = \frac{c\pi x^2}{4} \sqrt{y - z}$$

cubic centimetres per second, where c is a positive constant, x is the diameter of the capillary, y is the pressure in the artery, and z is the pressure in the capillary. What function gives the rate of change of blood flow with respect to capillary pressure, assuming fixed arterial pressure and capillary diameter? Is this rate increasing or decreasing?

60. **MARGINAL PRODUCTIVITY**   Suppose the output Q of a factory depends on the amount K of capital investment measured in units of $1000 and on the size L of the labour force measured in worker-hours. Give an economic interpretation of the second-order partial derivative $\frac{\partial^2 Q}{\partial K^2}$.

61. **MARGINAL PRODUCTIVITY**   At a certain factory, the output is $Q = 120K^{1/2}L^{1/3}$ units, where K denotes the capital investment measured in units of $1000 and L the size of the labour force measured in worker-hours.

   a. Determine the sign of the second-order partial derivative $\frac{\partial^2 Q}{\partial L^2}$ and give an economic interpretation.

   b. Determine the sign of the second-order partial derivative $\frac{\partial^2 Q}{\partial K^2}$ and give an economic interpretation.

62. **LAW OF DIMINISHING RETURNS**   Suppose the daily output Q of a factory depends on the amount K of capital investment and on the size L of the labour force. A **law of diminishing returns** states that in certain circumstances, there is a value L_0 such that the marginal product of labour will be increasing for $L < L_0$ and decreasing for $L > L_0$.

   a. Translate this law of diminishing returns into statements about the sign of a certain second order partial derivative.

   b. Read about the principle of diminishing returns in an economics text. Then write a paragraph discussing the economic factors that might account for this phenomenon.

63. It is estimated that the weekly output at a certain plant is given by

$$Q(x, y) = 1175x + 483y + 3.1x^2y - 1.2x^3 - 2.7y^2$$

units, where x is the number of skilled workers and y is the number of unskilled workers employed at the plant. Currently the workforce consists of 37 skilled and 71 unskilled workers.

Use the partial derivative $Q_y(x, y)$ to estimate the change in output that results when the number of unskilled workers is increased from 71 to 72 while the number of skilled workers stays at 37. Compare with the actual change $Q(37, 72) - Q(37, 71)$.

64. Repeat Exercise 63 with the output function

$$Q(x, y) = 1731x + 925y + x^2y - 2.7x^2 - 1.3y^{3/2}$$

and initial employment levels of $x = 43$ and $y = 85$.

In Exercises 65 through 70, use the chain rule to find $\dfrac{dz}{dt}$.

Express your answer in terms of x, y, and t.

65. $z = 2x + 3y; x = t^2, y = 5t$

66. $z = x^2y; x = 3t + 1, y = t^2 - 1$

67. $z = \dfrac{3x}{y}; x = t, y = t^2$

68. $z = x^{1/2}y^{1/3}; x = 2t, y = 2t^2$

69. $z = xy; x = e^{2t}, y = e^{-3t}$

70. $z = \dfrac{x + y}{x - y}; x = t^3 + 1, y = 1 - t^2$

Each of Exercises 71 through 83 involves either the chain rule for partial derivatives or the incremental approximation formula for functions of two variables.

71. ALLOCATION OF LABOUR Using x hours of skilled labour and y hours of unskilled labour, a manufacturer can produce $Q(x, y) = 10xy^{1/2}$ units. Currently 30 hours of skilled labour and 36 hours of unskilled labour are being used. Suppose the manufacturer reduces the skilled labour level by 3 hours and increases the unskilled labour level by 5 hours. Use calculus to determine the approximate effect of these changes on production.

72. DEMAND FOR HYBRID CARS A car dealer determines that if gasoline–electric hybrid automobiles are sold for x dollars each and the price of gasoline is y cents per litre, then approximately H hybrid cars will be sold each year, where

$$H(x, y) = 3500 - 19x^{1/2} + 6(0.4y + 16)^{3/2}.$$

She estimates that t years from now, the hybrid cars will be selling for

$$x(t) = 35\ 050 + 350t$$

dollars each and that gasoline will cost

$$y(t) = 80 + 2.5(3t)^{1/2}$$

cents per litre. At what rate will the annual demand for hybrid cars be changing with respect to time 3 years from now? Will it be increasing or decreasing?

73. CONSUMER DEMAND The demand for a certain product is

$$Q(x, y) = 200 - 10x^2 + 20xy$$

units per month, where x is the price of the product

and y is the price of a competing product. It is estimated that t months from now, the price of the product will be

$$x(t) = 10 + 0.5t$$

dollars per unit while the price of the competing product will be

$$y(t) = 12.8 + 0.2t^2$$

dollars per unit.

a. At what rate will the demand for the product be changing with respect to time 4 months from now?

b. At what percentage rate $\dfrac{100Q'(t)}{Q(t)}$ will the demand for the product be changing with respect to time 4 months from now?

74. ALLOCATION OF RESOURCES At a certain factory, when the capital expenditure is K thousand dollars and L worker-hours of labour are employed, the daily output will be $Q = 120K^{1/2}L^{1/3}$ units. Currently capital expenditure is $400\ 000 ($K = 400$) and is increasing at a rate of $9000 per day, while 1000 worker-hours are being employed and labour is being decreased at a rate of 4 worker-hours per day. At what rate is production currently changing? Is it increasing or decreasing?

75. ALLOCATION OF LABOUR The output at a certain plant is

$$Q(x, y) = 0.08x^2 + 0.12xy + 0.03y^2$$

units per day, where x is the number of hours of skilled labour used and y is the number of hours of unskilled labour used. Currently, 80 hours of skilled labour and 200 hours of unskilled labour are used each day. Use calculus to estimate the change in output that will result if an additional $\dfrac{1}{2}$ hour of skilled labour is used each day, along with an additional 2 hours of unskilled labour.

76. PUBLISHING SALES An editor estimates that if x thousand dollars are spent on development and y thousand dollars are spent on promotion, approximately $Q(x, y) = 20x^{3/2}y$ copies of a new book will be sold. Current plans call for the expenditure of $36\ 000 on development and $25\ 000 on promotion. Use calculus to estimate how sales will be affected if the amount spent on development is increased by $500 and the amount on promotion is decreased by $1000.

77. RETAIL SALES   A grocer's daily profit from the sale of two brands of flavoured iced tea is

$$P(x, y) = (x - 40)(55 - 4x + 5y)$$
$$+ (y - 45)(70 + 5x - 7y)$$

cents, where x is the price per bottle of the first brand and y is the price per bottle of the second, both in cents. Currently the first brand sells for 70 cents per bottle and the second sells for 73 cents per bottle.

a. Find the marginal profit functions P_x and P_y.
b. Evaluate P_x and P_y for the current values of x and y.
c. Use calculus to estimate the change in daily profit that will result if the grocer decides to raise the price of the first brand by 1 cent and the price of the second by 2 cents.
d. Estimate the change in profit if the price of the first brand is increased by 2 cents and the price of the second is decreased by 1 cent.

78. LANDSCAPING   A rectangular garden that is 30 m long and 40 m wide is bordered by a concrete path that is 0.8 m wide. Use calculus to estimate the area of the concrete path.

79. PACKAGING   A pop can is H centimetres tall and has a radius of R centimetres. The cost of material in the can is 0.0005 cents per square centimetre and the pop itself costs 0.001 cents per cubic centimetre.

a. Find a function $C(R, H)$ for the cost of the materials and contents of a can of pop. (You will need the formulas for volume and surface area given in Exercises 53 and 54.)
b. The cans are currently 12 cm tall and have a radius of 3 cm. Use calculus to estimate the effect on cost of increasing the radius by 0.3 cm and decreasing the height by 0.2 cm.

80. SLOPE OF A LEVEL CURVE   Suppose $y = h(x)$ is a differentiable function of x and that $f(x, y) = C$ for some constant C. Use the chain rule (with x taking the role of t) to show that

$$\frac{\partial f}{\partial x} + \frac{\partial f}{\partial y}\frac{\partial y}{\partial x} = 0.$$

Conclude that the slope at each point (x, y) on the level curve $f(x, y) = C$ is given by

$$\frac{dy}{dx} = -\frac{f_x}{f_y}.$$

81. Use the formula obtained in Exercise 80 to find the slope of the level curve

$$x^2 + xy + y^3 = 1$$

at the point $(-1, 1)$. What is the equation of the tangent line to the level curve at this point?

82. Use the formula obtained in Exercise 80 to find the slope of the level curve

$$x^2y + 2y^3 - 2e^{-x} = 14$$

at the point $(0, 2)$. What is the equation of the tangent line to the level curve at this point?

83. A* INVESTMENT SATISFACTION   Suppose a particular investor derives $U(x, y)$ units of satisfaction from owning x stock units and y bond units, where

$$U(x, y) = (2x + 3)(y + 5).$$

The investor currently owns $x = 27$ stock units and $y = 12$ bond units.

a. Find the marginal utilities U_x and U_y.
b. Evaluate U_x and U_y for the current values of x and y.
c. Use calculus to estimate how the investor's satisfaction changes if she adds 3 stock units and removes 2 bond units from her portfolio.
d. Estimate how many bond units the investor could substitute for 1 stock unit without affecting her total satisfaction with her portfolio.

SECTION 7.3   # Optimizing Functions of Two Variables

L03

Find the relative extrema of functions of several variables using the second partials test.

Suppose a manufacturer produces two DVD player models, the deluxe and the standard, and that the total cost of producing x units of the deluxe and y units of the standard is given by the function $C(x, y)$. How would you find the level of production $x = a$ and $y = b$ that results in minimal cost? Or perhaps the output of a certain production process is given by $Q(K, L)$, where K and L measure capital and labour expenditure, respectively. What levels of expenditure K_0 and L_0 result in maximum output?

In Section 3.4, you learned how to use the derivative $f'(x)$ to find the largest and smallest values of a function of a single variable, $f(x)$, and the goal of this section is to extend those methods to functions of two variables, $f(x, y)$. We begin with a definition.

> **Relative Extrema** ■ The function $f(x, y)$ is said to have a **relative maximum** at the point $P(a, b)$ in the domain of f if $f(a, b) \geq f(x, y)$ for all points (x, y) in a circular disk centred at P. Similarly, if $f(c, d) \leq f(x, y)$ for all points (x, y) in a circular disk centred at Q, then $f(x, y)$ has a **relative minimum** at $Q(c, d)$.

In geometric terms, there is a relative maximum of $f(x, y)$ at $P(a, b)$ if the surface $z = f(x, y)$ has a peak at the point $(a, b, f(a, b))$, that is, if $(a, b, f(a, b))$ is at least as high as any nearby point on the surface. Similarly, a relative minimum of $f(x, y)$ occurs at $Q(c, d)$ if the point $(c, d, f(c, d))$ is at the bottom of a valley, so $(c, d, f(c, d))$ is at least as low as any nearby point on the surface. For example, in Figure 7.11, the function $f(x, y)$ has a relative maximum at $P(a, b)$ and a relative minimum at $Q(c, d)$.

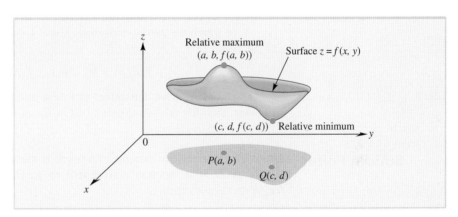

FIGURE 7.11 Relative extrema of the function $f(x, y)$.

Critical Points

The points (a, b) in the domain of $f(x, y)$ for which both $f_x(a, b) = 0$ and $f_y(a, b) = 0$ are said to be **critical points** of f. Like the critical numbers for functions of one variable, these critical points play an important role in the study of relative maxima and minima.

To see the connection between critical points and relative extrema, suppose $f(x, y)$ has a relative maximum at (a, b). Then the curve formed by intersecting the surface $z = f(x, y)$ with the vertical plane $y = b$ has a relative maximum and hence a horizontal tangent line when $x = a$ (Figure 7.12a). Since the partial derivative $f_x(a, b)$ is the slope of this tangent line, it follows that $f_x(a, b) = 0$. Similarly, the curve formed by intersecting the surface $z = f(x, y)$ with the plane $x = a$ has a relative maximum when $y = b$ (Figure 7.12b), and so $f_y(a, b) = 0$. This shows that a point at which a function of two variables has a relative maximum must be a critical point. A similar argument shows that a point at which a function of two variables has a relative minimum must also be a critical point.

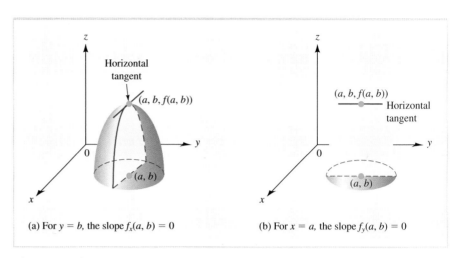

(a) For $y = b$, the slope $f_x(a, b) = 0$ (b) For $x = a$, the slope $f_y(a, b) = 0$

FIGURE 7.12   The partial derivatives are zero at a relative extremum.

Here is a more precise statement of the situation.

Critical Points and Relative Extrema   ■   A point (a, b) in the domain of $f(x, y)$ for which the partial derivatives f_x and f_y both exist is called a *critical point* of f if both

$$f_x(a, b) = 0 \quad \text{and} \quad f_y(a, b) = 0.$$

If the first-order partial derivatives of f exist at all points in some region R in the xy plane, then the relative extrema of f in R can occur only at critical points.

Saddle Points   Although all the relative extrema of a function must occur at critical points, not every critical point of a function corresponds to a relative extremum. For example, if $f(x, y) = y^2 - x^2$, then

$$f_x(x, y) = -2x \quad \text{and} \quad f_y(x, y) = 2y,$$

so $f_x(0, 0) = f_y(0, 0) = 0$. Thus, the origin $(0, 0)$ is a critical point for $f(x, y)$, and the surface $z = y^2 - x^2$ has horizontal tangents at the origin along both the x axis and the y axis. However, in the xz plane (where $y = 0$) the surface has the equation $z = -x^2$, which is a downward-opening parabola, while in the yz plane (where $x = 0$), we have the upward-opening parabola $z = y^2$. This means that at the origin, the surface $z = y^2 - x^2$ has a *relative maximum* in the x direction and a *relative minimum* in the y direction.

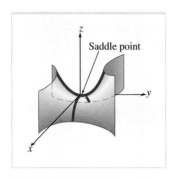

FIGURE 7.13   The saddle surface $z = y^2 - x^2$.

Instead of having a peak or a valley above the critical point $(0, 0)$, the surface $z = y^2 - x^2$ is shaped like a saddle, as shown in Figure 7.13, and for this reason the surface is called a **saddle surface.** For a critical point to correspond to a relative extremum, the same extreme behaviour (maximum or minimum) must occur in *all directions.* Any critical point (like the origin in this example) where there is a relative maximum in one direction and a relative minimum in another direction is called a **saddle point.** Using a computer program such as Maple to plot this function, and rotating the saddle point about the z axis, the maximum and minimum can be seen easily. This saddle point shape is also the shape of a popular brand of potato chip.

The Second Partials Test

Here is a procedure involving second-order partial derivatives that you can use to decide whether a given critical point is a relative maximum, a relative minimum, or a saddle point. This procedure is the two-variable version of the second derivative test for functions of a single variable that you saw in Section 3.2.

The Second Partials Test

Let $f(x, y)$ be a function of x and y whose partial derivatives f_x, f_y, f_{xx}, f_{yy}, and f_{xy} all exist, and let $D(x, y)$ be the function

$$D(x, y) = f_{xx}(x, y)f_{yy}(x, y) - [f_{xy}(x, y)]^2.$$

Step 1. Find all critical points of $f(x, y)$, that is, all points (a, b) so that

$$f_x(a, b) = 0 \quad \text{and} \quad f_y(a, b) = 0.$$

Step 2. For each critical point (a, b) found in step 1, evaluate $D(a, b)$.

Step 3. If $D(a, b) < 0$, there is a **saddle point** at (a, b).

Step 4. If $D(a, b) > 0$, compute $f_{xx}(a, b)$:

If $f_{xx}(a, b) > 0$, there is a **relative minimum** at (a, b).
If $f_{xx}(a, b) < 0$, there is a **relative maximum** at (a, b).

If $D(a, b) = 0$, the test is inconclusive and f may have either a relative extremum or a saddle point at (a, b).

Notice that there is a saddle point at the critical point (a, b) only when the quantity D in the second partials test is negative. If D is positive, there is either a relative maximum or a relative minimum *in all directions*. To decide which, you can restrict your attention to any one direction (say, the x direction) and use the sign of the second partial derivative f_{xx} in exactly the same way as the single-variable second derivative was used in the second derivative test given in Chapter 3, namely,

$$\text{a relative minimum if } f_{xx}(a, b) > 0$$
$$\text{a relative maximum if } f_{xx}(a, b) < 0$$

You may find the following tabular summary a convenient way of remembering the conclusions of the second partials test:

Sign of D	Sign of f_{xx}	Behaviour at (a, b)
+	+	Relative minimum
+	−	Relative maximum
−		Saddle point

The proof of the second partials test involves ideas beyond the scope of this text and is omitted. Examples 7.3.1 through 7.3.3 illustrate how the test can be used.

EXAMPLE 7.3.1

Find all critical points for the function $f(x, y) = x^2 + y^2$ and classify each as a relative maximum, a relative minimum, or a saddle point.

Solution

Since

$$f_x = 2x \quad \text{and} \quad f_y = 2y,$$

the only critical point of f is $(0, 0)$. To test this point, use the second-order partial derivatives

$$f_{xx} = 2, f_{yy} = 2, \quad \text{and} \quad f_{xy} = 0$$

to get

$$D(x, y) = f_{xx}f_{yy} - (f_{xy})^2 = (2)(2) - 0^2 = 4.$$

That is, $D(x, y) = 4$ for *all* points (x, y) and, in particular,

$$D(0, 0) = 4 > 0.$$

Hence, f has a relative extremum at $(0, 0)$. Moreover, since

$$f_{xx}(0, 0) = 2 > 0,$$

it follows that the relative extremum at $(0, 0)$ is a relative minimum. For reference, the graph of f is sketched in Figure 7.14.

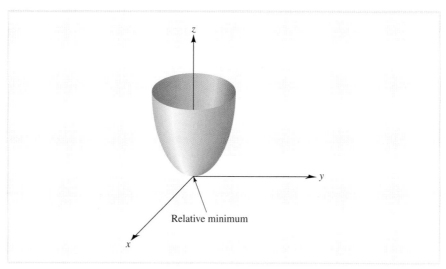

FIGURE 7.14   The surface $z = x^2 + y^2$ with a relative minimum at $(0, 0)$.

EXAMPLE 7.3.2

Find all critical points for the function $f(x, y) = 12x - x^3 - 4y^2$ and classify each as a relative maximum, a relative minimum, or a saddle point.

Solution

Since

$$f_x = 12 - 3x^2 \quad \text{and} \quad f_y = -8y,$$

find the critical points by solving simultaneously the two equations

$$12 - 3x^2 = 0$$
$$-8y = 0$$

From the second equation, $y = 0$, and from the first,

$$3x^2 = 12$$
$$x = 2 \quad \text{or} \quad -2$$

Thus, there are two critical points, $(2, 0)$ and $(-2, 0)$.

To determine the nature of these points, first compute

$$f_{xx} = -6x, \qquad f_{yy} = -8, \qquad \text{and} \qquad f_{xy} = 0,$$

and then form the function

$$D = f_{xx}f_{yy} - (f_{xy})^2 = (-6x)(-8) - 0 = 48x.$$

Applying the second partials test to the two critical points results in

$$D(2, 0) = 48(2) = 96 > 0 \quad \text{and} \quad f_{xx}(2, 0) = -6(2) = -12 < 0$$

and

$$D(-2, 0) = 48(-2) = -96 < 0$$

so a relative maximum occurs at $(2, 0)$ and a saddle point occurs at $(-2, 0)$. These results are summarized in the following table.

Critical point (a, b)	Sign of $D(a, b)$	Sign of $f_{xx}(a, b)$	Behaviour at (a, b)
$(2, 0)$	$+$	$-$	Relative maximum
$(-2, 0)$	$-$		Saddle point

EXAMPLE 7.3.3

Find all critical points for the function $f(x, y) = e^{2xy}$ (shown in the margin) and classify each as a relative maximum, a relative minimum, or a saddle point.

Solution

Since

$$f_x = 2ye^{2xy} \quad \text{and} \quad f_y = 2xe^{2xy},$$

there is only one critical point of f, at $x = 0$, $y = 0$, or $(0, 0)$.

The second-order partial derivatives f_{xx} and f_{yy} of $f(x, y)$ are

$$f_{xx} = 4y^2e^{2xy} \quad \text{and} \quad f_{yy} = 4x^2e^{2xy}.$$

But to find f_{xy}, we need to use the product rule.

$$f_{xy} = 4xye^{2xy} + 2e^{2xy}.$$

Thus,

$$D(x, y) = f_{xx}f_{yy} - (f_{xy})^2 = (4y^2e^{2xy})(4x^2e^{2xy}) - (4xye^{2xy} + 2e^{2xy})^2.$$

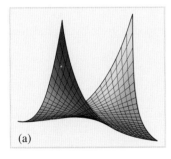

(a)

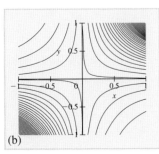

(b)

$f(x, y) = e^{2xy}$

Substituting $(0, 0)$, all of the terms are zero except the last one:

$$D(0, 0) = 0 - (0 + 2e^0)^2 = -4 < 0.$$

So f has a saddle point at $(0, 0)$.

The figure in the margin on the previous page shows a plot of the surface and a contour plot of this function.

Practical Optimization Problems

In Example 7.3.4, you will use the theory of relative extrema to solve an optimization problem from economics. Actually, you will be trying to find the *absolute* maximum of a certain function, which turns out to coincide with the *relative* maximum of the function. This is typical of two-variable optimization problems in the social and life sciences, and *in this text, you can assume that a relative extremum you find as the solution to any practical optimization problem is actually the absolute extremum.*

EXAMPLE 7.3.4

A grocery store carries two brands of cat food, a local brand that it obtains at a cost of 30 cents per can and a well-known national brand it obtains at a cost of 40 cents per can. The grocer estimates that if the local brand is sold for x cents per can and the national brand is sold for y cents per can, then approximately $70 - 5x + 4y$ cans of the local brand and $80 + 6x - 7y$ cans of the national brand will be sold each day. How should the grocer price each brand to maximize total daily profit from the sale of cat food?

Solution

Since

$$\begin{pmatrix} \text{Total} \\ \text{profit} \end{pmatrix} = \begin{pmatrix} \text{profit from the sale} \\ \text{of the local brand} \end{pmatrix} + \begin{pmatrix} \text{profit from the sale} \\ \text{of the national brand} \end{pmatrix},$$

it follows that the total daily profit from the sale of the cat food is given by the function

$$f(x, y) = \underbrace{(70 - 5x + 4y)}_{\substack{\text{items sold} \\ \text{local brand}}} \cdot \underbrace{(x - 30)}_{\text{profit per item}} + \underbrace{(80 + 6x - 7y)}_{\substack{\text{items sold} \\ \text{national brand}}} \cdot \underbrace{(y - 40)}_{\text{profit per item}}$$

$$= 70x - 2100 - 5x^2 + 150x + 4xy - 120y + 80y - 3200 + 6xy - 240x - 7y^2 + 280y$$

$$= -5x^2 + 10xy - 20x - 7y^2 + 240y - 5300$$

Compute the first-order partial derivatives,

$$f_x = -10x + 10y - 20 \qquad \text{and} \qquad f_y = 10x - 14y + 240,$$

and set them equal to zero to get

$$-10x + 10y - 20 = 0 \qquad \text{and} \qquad 10x - 14y + 240 = 0$$

or

$$-x + y = 2 \qquad \text{and} \qquad 5x - 7y = -120$$

Then solve these equations simultaneously to get

$$x = 53 \qquad \text{and} \qquad y = 55.$$

It follows that (53, 55) is the only critical point of f.

Next apply the second partials test. Since

$$f_{xx} = -10, \quad f_{yy} = -14, \quad \text{and} \quad f_{xy} = 10,$$

then

$$D(x, y) = f_{xx}f_{yy} - (f_{xy})^2 = (-10)(-14) - (10)^2 = 40$$

Because

$$D(53, 55) = 40 > 0 \quad \text{and} \quad f_{xx}(53, 55) = -10 < 0,$$

it follows that f has a (relative) maximum when $x = 53$ and $y = 55$. That is, the grocer can maximize profit by selling the local brand of cat food for 53 cents per can and the national brand for 55 cents per can.

EXAMPLE 7.3.5

A business manager plots a grid on a map of the region his company serves and determines that the company's three most important customers are located at points $A(1, 5)$, $B(0, 0)$, and $C(8, 0)$, where units are in kilometres. At what point $W(x, y)$ should a warehouse be located in order to minimize the sum of the squares of the distances from W to A, B, and C (Figure 7.15)?

Solution

The sum of the squares of the distances from W to A, B, and C is given by the function

$$S(x, y) \;=\; \underbrace{(x-1)^2 + (y-5)^2}_{\substack{\text{square of distance} \\ \text{from } W \text{ to } A}} \;+\; \underbrace{x^2 + y^2}_{\substack{\text{square of distance} \\ \text{from } W \text{ to } B}} \;+\; \underbrace{(x-8)^2 + y^2}_{\substack{\text{square of distance} \\ \text{from } W \text{ to } C}},$$

where the left grouping is labelled "sum of squares of distances."

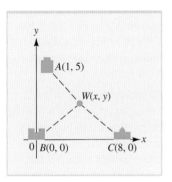

FIGURE 7.15 Locations of businesses A, B, and C and warehouse W.

To minimize $S(x, y)$, begin by computing the partial derivatives

$$S_x = 2(x - 1) + 2x + 2(x - 8) = 6x - 18$$
$$S_y = 2(y - 5) + 2y + 2y = 6y - 10$$

Then $S_x = 0$ and $S_y = 0$ when

$$6x - 18 = 0$$
$$6y - 10 = 0$$

or $x = 3$ and $y = \dfrac{5}{3}$. Since $S_{xx} = 6$, $S_{xy} = 0$, and $S_{yy} = 6$, then

$$D = S_{xx}S_{yy} - S_{xy}^2 = (6)(6) - 0^2 = 36 > 0$$

and

$$S_{xx}\left(3, \frac{5}{3}\right) = 6 > 0.$$

Thus, the sum of squares is minimized at the map point $W\left(3, \dfrac{5}{3}\right)$.

EXERCISES ■ 7.3

In Exercises 1 through 22, find the critical points of the given functions and classify each as a relative maximum, a relative minimum, or a saddle point. (Note: The algebra in Exercises 19 through 22 is challenging.)

1. $f(x, y) = 5 - x^2 - y^2$

2. $f(x, y) = 2x^2 - 3y^2$

3. $f(x, y) = xy$

4. $f(x, y) = x^2 + 2y^2 - xy + 14y$

5. $f(x, y) = \dfrac{16}{x} + \dfrac{6}{y} + x^2 - 3y^2$

6. $f(x, y) = xy + \dfrac{8}{x} + \dfrac{8}{y}$

7. $f(x, y) = 2x^3 + y^3 + 3x^2 - 3y - 12x - 4$

8. $f(x, y) = (x - 1)^2 + y^3 - 3y^2 - 9y + 5$

9. $f(x, y) = x^3 + y^2 - 6xy + 9x + 5y + 2$

10. $f(x, y) = -x^4 - 32x + y^3 - 12y + 7$

11. $f(x, y) = xy^2 - 6x^2 - 3y^2$

12. $f(x, y) = x^2 - 6xy - 2y^3$

13. $f(x, y) = (x^2 + 2y^2)e^{1 - x^2 - y^2}$

14. $f(x, y) = e^{-(x^2 + y^2 - 6y)}$

15. $f(x, y) = x^3 - 4xy + y^3$

16. $f(x, y) = (x - 4) \ln (xy)$

17. $f(x, y) = 4xy - 2x^4 - y^2 + 4x - 2y$

18. $f(x, y) = 2x^4 + x^2 + 2xy + 3x + y^2 + 2y + 5$

19. A* $f(x, y) = \dfrac{1}{x^2 + y^2 + 3x - 2y + 1}$

20. A* $f(x, y) = xye^{-(16x^2 + 9y^2)/288}$

21. A* $f(x, y) = x \ln \left(\dfrac{y^2}{x} \right) + 3x - xy^2$

22. A* $f(x, y) = \dfrac{x}{x^2 + y^2 + 4}$

23. **RETAIL SALES** A discount tourist souvenir T-shirt shop carries two competing shirts, one with a picture of a peaceful sunset over Niagara Falls, and the other with a mermaid climbing out of Niagara Falls. The owner of the store can obtain both types at a cost of $2 per shirt and estimates that if sunset shirts are sold for x dollars each and mermaid shirts for y dollars each, consumers will buy $40 - 50x + 40y$ sunset shirts and $20 + 60x - 70y$ mermaid shirts each day. The profit is then $P(x, y) = (x - 2)(40 - 50x + 40y) + (y - 2)(20 + 60x - 70y)$. How should the owner price the shirts in order to generate the largest possible profit?

24. **PRICING** A telephone company is planning to introduce two new types of executive communications systems that it hopes to sell to its largest commercial customers. It is estimated that if the first type of system is priced at x hundred dollars per system and the second type at y hundred dollars per system, approximately $40 - 8x + 5y$ consumers will buy the first type and $50 + 9x - 7y$ will buy the second type. If the cost of manufacturing the first type is $1000 per system and the cost of manufacturing the second type is $3000 per system, the profit is $P(x, y) = (x - 10)(40 - 8x + 5y) + (y - 30)(50 + 9x - 7y)$. How should the telephone company price the systems to generate the largest possible profit?

25. **CONSTRUCTION** Suppose you wish to construct a rectangular shipping crate with a volume of 32 m³. Three different materials will be used in the construction. The material for the sides costs $10/m², the material for the bottom costs $30/m², and the material for the top costs $50/m². What are the dimensions of the least expensive such box?

26. **CONSTRUCTION** A farmer wishes to fence off a rectangular pasture along the bank of a river. The area of the pasture is to be 6400 m², and no fencing is needed along the river bank. Find the dimensions of the pasture that will require the least amount of fencing.

27. **RETAIL SALES** A company produces x units of commodity A and y units of commodity B. All of the units can be sold for $p = 100 - x$ dollars per unit of A and $q = 100 - y$ dollars per unit of B. The profit is $P(x, y) = x(100 - x) + y(100 - y) - (x^2 - xy + y^2)$. What should x and y be to maximize profit?

28. **RETAIL SALES** Repeat Exercise 27 for the case where $p = 20 - 5x$ and $q = 4 - 2y$.

29. **RESPONSE TO STIMULI** Consider an experiment in which a subject performs a task while being exposed to two different stimuli (for example, sound and light). For low levels of the stimuli, the subject's performance might actually improve, but as the stimuli increase, they eventually

become a distraction and the performance begins to deteriorate. Suppose in a certain experiment in which x units of stimulus A and y units of stimulus B are applied, the performance of a subject is measured by the function

$$f(x, y) = C + xye^{1-x^2-y^2}$$

where C is a positive constant. How many units of each stimulus result in maximum performance?

30. **SOCIAL CHOICES** The social desirability of an enterprise often involves making a choice between the commercial advantage of the enterprise and the social or ecological loss that may result. For instance, the lumber industry provides paper products to society and income to many workers and entrepreneurs, but the gain may be offset by the destruction of habitat for spotted owls and other endangered species. Suppose the social desirability of a particular enterprise is measured by the function

$$S(x, y) = (16 - 6x)x - (y^2 - 4xy + 40),$$

where x measures commercial advantage (profit and jobs) and y measures ecological disadvantage (species displacement, as a percentage) with $x \geq 0$ and $y \geq 0$. The enterprise is deemed desirable if $S \geq 0$ and undesirable if $S < 0$.

   a. What values of x and y will maximize social desirability? Interpret your result. Is it possible for this enterprise to be desirable?

   b. The function given in part (a) is artificial, but the ideas are not. Research the topic of ethics in industry and write a paragraph on how you feel these choices should be made.

31. **A* PARTICLE PHYSICS** A particle of mass m in a rectangular box with dimensions x, y, and z has ground state energy

$$E(x, y, z) = \frac{k^2}{8m}\left(\frac{1}{x^2} + \frac{1}{y^2} + \frac{1}{z^2}\right),$$

where k is a physical constant. If the volume of the box satisfies $xyz = V_0$ for constant V_0, find the values of x, y, and z that minimize the ground state energy.

32. **A* ALLOCATION OF FUNDS** A manufacturer is planning to sell a new product at a price of $210 per unit and estimates that if x thousand dollars are spent on development and y thousand dollars are spent on promotion, consumers will buy approximately

$$\frac{640y}{y + 3} + \frac{216x}{x + 5}$$

units of the product. If manufacturing costs for this product are $135 per unit, how much should the manufacturer spend on development and

how much on promotion to generate the largest possible profit from the sale of this product? [*Hint:* Profit = (number of units)(price per unit − cost per unit) − total amount spent on development and promotion.]

33. **PROFIT UNDER MONOPOLY** A manufacturer with exclusive rights to a sophisticated new industrial machine is planning to sell a limited number of the machines to both foreign and domestic firms. The price the manufacturer can expect to receive for the machines will depend on the number of machines made available. (For example, if only a few of the machines are placed on the market, competitive bidding among prospective purchasers will tend to drive the price up.) It is estimated that if the manufacturer supplies x machines to the domestic market and y machines to the foreign market, the machines will sell for $60 - \dfrac{x}{5} + \dfrac{y}{20}$ thousand dollars each domestically and for $50 - \dfrac{y}{10} + \dfrac{x}{20}$ thousand dollars each abroad. If the manufacturer can produce the machines at a cost of $10 000 each, how many should be supplied to each market to generate the largest possible profit?

34. **PROFIT UNDER MONOPOLY** A manufacturer with exclusive rights to a new industrial machine is planning to sell a limited number of them and estimates that if x machines are supplied to the domestic market and y to the foreign market, the machines will sell for $150 - \dfrac{x}{6}$ thousand dollars each domestically and for $100 - \dfrac{y}{20}$ thousand dollars each abroad.

   a. How many machines should the manufacturer supply to the domestic market to generate the largest possible profit at home?

   b. How many machines should the manufacturer supply to the foreign market to generate the largest possible profit abroad?

   c. How many machines should the manufacturer supply to each market to generate the largest possible *total* profit?

   d. Is the relationship between the answers in parts (a), (b), and (c) accidental? Explain. Does a similar relationship hold in Exercise 33? What accounts for the difference between these two problems in this respect?

35. CITY PLANNING Four small towns in a rural area wish to pool their resources to build a television station. If the towns are located at the points $(-5, 0)$, $(1, 7)$, $(9, 0)$, and $(0, -8)$ on a rectangular map grid, where units are in kilometres, at what point $S(a, b)$ should the station be located to minimize the sum of the squares of the distances from the towns?

36. MAINTENANCE In relation to a rectangular map grid, four oil rigs are located at the points $(-300, 0)$, $(-100, 500)$, $(0, 0)$, and $(400, 300)$, where units are in metres. Where should a maintenance shed $M(a, b)$ be located to minimize the sum of the squares of the distances from the rigs?

37. GENETICS Alternative forms of a gene are called *alleles*. Three alleles, designated A, B, and O, determine the four human blood types A, B, O, and AB. Suppose that p, q, and r are the proportions of A, B, and O in a particular population, so that

$p + q + r = 1$. Then according to the Hardy-Weinberg law in genetics, the proportion of individuals in the population who carry two different alleles is given by $P = 2pq + 2pr + 2rq$. What is the largest value of P?

38. LEARNING In a learning experiment, a subject is first given x minutes to examine a list of facts. The fact sheet is then taken away and the subject is allowed y minutes to prepare mentally for an exam based on the fact sheet. Suppose it is found that the score achieved by a particular subject is related to x and y by the formula

$$S(x, y) = -x^2 + xy + 10x - y^2 + y + 15.$$

a. What score does the subject achieve if he takes the test cold (with no study or contemplation)?

b. How much time should the subject spend in study and contemplation to maximize his score? What is the maximum score?

39. A* Tom, Dan, and Maria are participating in a cross-country relay race. Tom will trudge as fast as he can through thick woods to the edge of a river; then Dan will take over and row to the opposite shore. Finally, Maria will take the baton and walk-run along the very pot-holed and muddy river road to the finish line. The course is shown in the accompanying figure. Teams must start at point S and finish at point F, but they may position one member anywhere along the shore of the river and another anywhere along the river road.

a. Suppose Tom can trudge at 2 km/h, Dan can row at 4 km/h, and Maria can walk-run at 6 km/h. Where should Dan and Maria wait to receive the baton in order for the team to finish the course as quickly as possible?

b. The main competition for Tom, Dan, and Maria is the team of Anna, Steph, and Wes. If Anna can trudge at 1.7 km/h, Steph can row at 3.5 km/h, and Wes can walk-run at 6.3 km/h, which team should win? By how much?

c. Does this exercise remind you of the spy story in Exercise 19 of Section 3.5? Create your own spy story problem based on the mathematical ideas in this exercise.

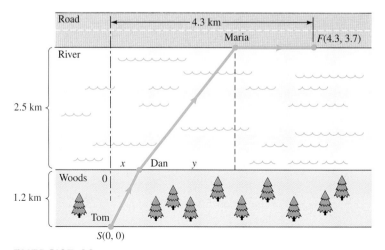

EXERCISE 39

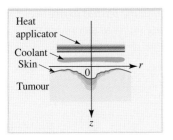

Heat
applicator
Coolant
Skin
Tumour

EXERCISE 40

40. A* CANCER THERAPY   Certain malignant tumours that do not respond to conventional methods of treatment such as surgery or chemotherapy may be treated by **hyperthermia**, which involves applying extreme heat to tumours using microwave transmissions.* One particular kind of microwave applicator used in this therapy produces an absorbed energy density that falls off exponentially. Specifically, the temperature at each point located r units from the central axis of a tumour and z units inside the tumour is given by a formula of the form

$$T(r, z) = Ae^{-pr^2}(e^{-qz} - e^{-sz}),$$

where A, p, q, and s are positive constants that depend on properties of both blood and the heating appliance. At what depth inside the tumour does the maximum temperature occur? Express your answer in terms of A, p, q, and r.

41. A* LIVEABLE SPACE   Define the *liveable space* of a building to be the volume of space in the building where a person 2 m tall can walk (upright). An A-frame cabin is y metres long and has equilateral triangular ends x metres on a side, as shown in the accompanying figure. If the surface area of the cabin (roof and two ends) is to be 55 m^2, what dimensions x and y will maximize the liveable space?

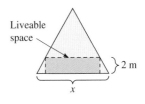

Liveable
space

2 m

EXERCISE 41

42. A* BUTTERFLY WING PATTERNS   The beautiful patterns on the wings of butterflies have long been a subject of curiosity and scientific study. Mathematical models used to study these patterns often focus on determining the level of morphogen (a chemical that effects change). In a model dealing with eyespot patterns,** a quantity of morphogen is released from an eyespot and the morphogen concentration t days later is given by

$$S(r, t) = \frac{1}{\sqrt{4\pi t}}e^{-(\gamma kt + r^2/4t)}, \quad t > 0,$$

where r measures the radius of the region on the wing affected by the morphogen and k and γ are positive constants.

a. Find t_m so that $\dfrac{\partial s}{\partial t} = 0$. Show that the function $S_m(t)$ formed from $S(r, t)$ by fixing r has a relative maximum at t_m. Is this the same as saying that the function of two variables $S(r, t)$ has a relative maximum?

b. Let $M(r)$ denote the maximum found in part (a); that is, $M(r) = S(r, t_m)$. Find an expression for M in terms of $z = (1 + 4\gamma kr^2)^{1/2}$.

*The ideas in this exercise are based on the article by Leah Edelstein-Keshet, "Heat Therapy for Tumours," *UMAP Modules 1991: Tools for Teaching,* Lexington, MA: Consortium for Mathematics and Its Applications, Inc., 1992, pp. 73–101.
**J. D. Murray, *Mathematical Biology,* 2nd ed., New York: Springer-Verlag, 1993, pp. 461–468.

c. It turns out that $M(z)$ is what is really needed to analyze the eyespot wing pattern. Write a paragraph on how biology and mathematics are blended in the study of butterfly wing patterns.

43. A* Let $f(x, y) = x^2 + y^2 - 4xy$. Show that f does *not* have a relative minimum at its critical point $(0, 0)$, even though it does have a relative minimum at $(0, 0)$ in both the x and y directions. [*Hint:* Consider the direction defined by the line $y = x$. That is, substitute x for y in the formula for f and analyze the resulting function of x.]

In Exercises 44 through 47, find the partial derivatives f_x and f_y and then use a graphing utility to determine the critical points of each function.

44. $f(x, y) = (x^2 + 3y - 5)e^{-x^2 - 2y^2}$

45. $f(x, y) = \dfrac{x^2 + xy + 7y^2}{x \ln y}$

46. $f(x, y) = 6x^2 + 12xy + y^4 + x - 16y - 3$

47. $f(x, y) = 2x^4 + y^4 - x^2(11y - 18)$

48. LEVEL CURVES Sometimes you can classify the critical points of a function by inspecting its level curves. In each case shown in the accompanying figure, determine the nature of the critical point of f at $(0, 0)$.

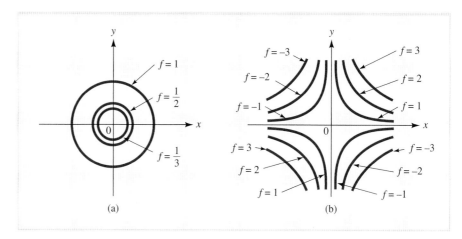

(a)

(b)

EXERCISE 48

L04

Fit a set of data to a line using the least squares method.

The Method of Least Squares

Throughout this text, you have seen applied functions, many of which are derived from published research, and you may have wondered how researchers come up with such functions. A common procedure for associating a function with an observed physical phenomenon is to gather data, plot the data points on a graph, and then find a function whose graph best fits the data in some mathematically meaningful way. We will now develop such a procedure, called the **method of least squares** or **regression analysis,** which was first mentioned in Example 1.3.7 of Section 1.3.

The Least-Squares Procedure

Suppose you wish to find a function $y = f(x)$ that fits a particular data set reasonably well. The first step is to decide what type of function to try. Sometimes this can be done by a theoretical analysis of the observed phenomenon and sometimes by inspecting the plotted data. Two sets of data are plotted in Figure 7.16. These are called **scatter diagrams.** In Figure 7.16a, the points lie roughly along a straight line, suggesting that a linear function $y = mx + b$ be used. However, in Figure 7.16b, the points appear to follow an exponential curve, and a function of the form $y = Ae^{-kx}$ would be more appropriate.

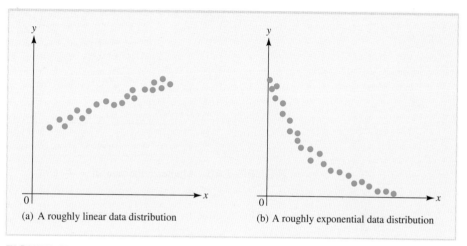

(a) A roughly linear data distribution   (b) A roughly exponential data distribution

FIGURE 7.16   Two scatter diagrams.

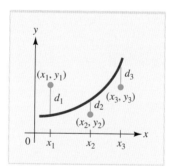

FIGURE 7.17   Sum of the squares of the vertical distances $d_1^2 + d_2^2 + d_3^2$.

Once the type of function has been chosen, the next step is to determine the particular function of this type whose graph is closest to the given set of points. A convenient way to measure how close a curve is to a set of points is to compute the sum of the squares of the vertical distances from the points to the curve. In Figure 7.17, for example, this is the sum $d_1^2 + d_2^2 + d_3^2$. The closer the curve is to the points, the smaller this sum will be, and the curve for which this sum is smallest is said to best fit the data according to the **least-squares criterion.**

The use of the least-squares criterion to fit a linear function to a set of points is illustrated in Example 7.4.1. The computation involves the technique from Section 7.3 for minimizing a function of two variables.

EXAMPLE 7.4.1

Use the least-squares criterion to find the equation of the line that is closest to the three points (1, 1), (2, 3), and (4, 3).

Solution

As indicated in Figure 7.18, the sum of the squares of the vertical distances from the three given points to the line $y = mx + b$ is

$$d_1^2 + d_2^2 + d_3^2 = (m + b - 1)^2 + (2m + b - 3)^2 + (4m + b - 3)^2.$$

This sum depends on the coefficients m and b that define the line, and so the sum can be thought of as a function $S(m, b)$ of the two variables m and b. The goal, therefore, is to find the values of m and b that minimize the function

$$S(m, b) = (m + b - 1)^2 + (2m + b - 3)^2 + (4m + b - 3)^2.$$

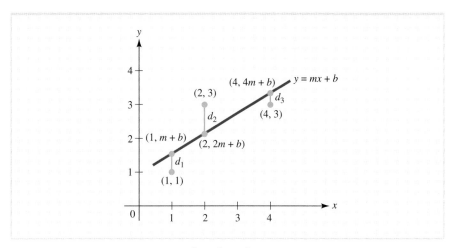

FIGURE 7.18 Minimize the sum $d_1^2 + d_2^2 + d_3^2$.

This is done by setting the partial derivatives $\dfrac{\partial S}{\partial m}$ and $\dfrac{\partial S}{\partial b}$ equal to zero to get

$$\frac{\partial S}{\partial m} = 2(m + b - 1) + 4(2m + b - 3) + 8(4m + b - 3)$$
$$= 42m + 14b - 38 = 0$$

and

$$\frac{\partial S}{\partial b} = 2(m + b - 1) + 2(2m + b - 3) + 2(4m + b - 3)$$
$$= 14m + 6b - 14 = 0$$

Solving the resulting equations

$$42m + 14b = 38$$
$$14m + 6b = 14$$

simultaneously for m and b gives

$$m = \frac{4}{7} \quad \text{and} \quad b = 1.$$

It can be shown that the critical point $(m, \ b) = \left(\dfrac{4}{7}, \ 1\right)$ does indeed minimize the function $S(m, b)$, and so it follows that

$$y = \frac{4}{7}x + 1$$

is the equation of the line that is closest to the three given points.

The Least-Squares Line

The line that is closest to a set of points according to the least-squares criterion is called the **least-squares line** for the points. (The term **regression line** is also used, especially in statistical work.) The procedure used in Example 7.4.1 can be generalized to give formulas for the slope m and the y intercept b of the least-squares line for an arbitrary set of n points (x_1, y_1), (x_2, y_2), ..., (x_n, y_n). The formulas involve sums of the x and y values. All the sums run from $j = 1$ to $j = n$, and to simplify the notation, the indices are omitted. For example, Σx is used instead of $\displaystyle\sum_{j=1}^{n} x_j$.

The Least-Squares Line ■ The equation of the least-squares line for the n points (x_1, y_1), (x_2, y_2), ..., (x_n, y_n) is $y = mx + b$, where

$$m = \frac{n\Sigma xy - \Sigma x \Sigma y}{n\Sigma x^2 - (\Sigma x)^2} \quad \text{and} \quad b = \frac{\Sigma x^2 \Sigma y - \Sigma x \Sigma xy}{n\Sigma x^2 - (\Sigma x)^2}.$$

EXAMPLE 7.4.2

Use the formulas to find the least-squares line for the points $(1, 1)$, $(2, 3)$, and $(4, 3)$ from Example 7.4.1.

Solution

Arrange your calculations as follows:

x	y	xy	x^2
1	1	1	1
2	3	6	4
4	3	12	16
$\Sigma x = 7$	$\Sigma y = 7$	$\Sigma xy = 19$	$\Sigma x^2 = 21$

Then use the formulas with $n = 3$ to get

$$m = \frac{3(19) - 7(7)}{3(21) - (7)^2} = \frac{4}{7} \quad \text{and} \quad b = \frac{21(7) - 7(19)}{3(21) - (7)^2} = 1,$$

from which it follows that the equation of the least-squares line is

$$y = \frac{4}{7}x + 1.$$

Least-Squares Prediction

The least-squares line (or curve) that best fits the data collected in the past can be used to make rough predictions about the future. This is illustrated in Example 7.4.3.

EXAMPLE 7.4.3

A university using a grade point system, with 0 to 4 corresponding to grades F, D, C, B, and A, has compiled these data relating students' high school and first-year university grade point averages (GPAs):

High school GPA	2.0	2.5	3.0	3.0	3.5	3.5	4.0	4.0
First-year GPA	1.5	2.0	2.5	3.5	2.5	3.0	3.0	3.5

Find the equation of the least-squares line for these data and use it to predict the first-year GPA of a student whose high school GPA is 3.7.

Solution

Let x denote the high school GPA and y the first-year GPA and arrange the calculations as follows:

x	y	xy	x²
2.0	1.5	3.0	4.0
2.5	2.0	5.0	6.25
3.0	2.5	7.5	9.0
3.0	3.5	10.5	9.0
3.5	2.5	8.75	12.25
3.5	3.0	10.5	12.25
4.0	3.0	12.0	16.0
4.0	3.5	14.0	16.0
$\Sigma x = 25.5$	$\Sigma y = 21.5$	$\Sigma xy = 71.25$	$\Sigma x^2 = 84.75$

Use the least-squares formula with $n = 8$ to get

$$m = \frac{8(71.25) - 25.5(21.5)}{8(84.75) - (25.5)^2} \approx 0.78$$

and

$$b = \frac{84.75(21.5) - 25.5(71.25)}{8(84.75) - (25.5)^2} \approx 0.19.$$

The equation of the least-squares line is therefore

$$y = 0.78x + 0.19.$$

To predict the first-year GPA y of a student whose high school GPA x is 3.7, substitute $x = 3.7$ into the equation of the least-squares line. This gives

$$y = 0.78(3.7) + 0.19 \approx 3.08,$$

which suggests that the student's first-year GPA might be about 3.1.

The original data are plotted in Figure 7.19, together with the least-squares line $y = 0.78x + 0.19$. Actually, in practice, it is a good idea to plot the data before proceeding with the calculations. By looking at the graph you will usually be able to tell whether approximation by a straight line is appropriate or whether a better fit might be possible with a curve of some sort. It is also important to note that although a line can be made mathematically with any data, this line may not have any validity to predict. The data may be too scattered or periodic. Statistical methods can be used to give several measures of how well the data fit the line. When software finds the least-squares line, these statistical measures are given with the resulting equation of the line.

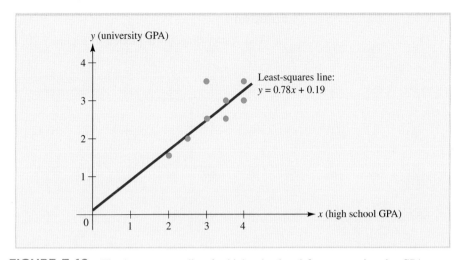

FIGURE 7.19 The least-squares line for high school and first-year university GPAs.

Nonlinear Curve-Fitting

In Examples 7.4.1, 7.4.2, and 7.4.3, the least-squares criterion was used to fit a linear function to a set of data. With appropriate modifications, the procedure can also be used to fit nonlinear functions to data. One kind of modified curve-fitting procedure is illustrated in Example 7.4.4.

EXAMPLE 7.4.4

A manufacturer gathers these data relating the level of production x (hundred units) of a particular commodity to the demand price p (dollars per unit) at which all x units will be sold:

k	Production x (hundred units)	Demand Price p (dollars per unit)
1	6	743
2	10	539
3	17	308
4	22	207
5	28	128
6	35	73

  a. Plot a scatter diagram for the data on a graph with production level on the x axis and demand price on the y axis.

  b. Notice that the scatter diagram in part (a) suggests that the demand function is exponential. Modify the least-squares procedure to find a curve of the form $p = Ae^{mx}$ that best fits the data in the table.

  c. Use the exponential demand function you found in part (b) to predict the revenue the manufacturer should expect if 4000 ($x = 40$) units are produced.

Solution

a. The scatter diagram is plotted in Figure 7.20.

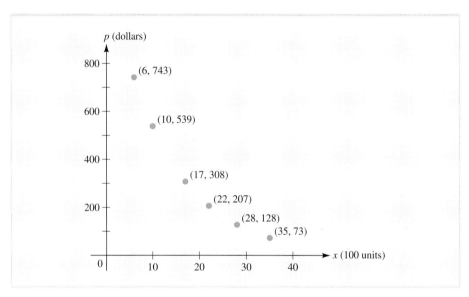

FIGURE 7.20   A scatter diagram for the demand data in Example 7.4.4.

Just-In-Time

Recall that
$\ln(ab) = \ln a + \ln b$.

b. Taking logarithms on both sides of the equation $p = Ae^{mx}$ gives

$$\ln p = \ln(Ae^{mx})$$
$$= \ln A + \ln(e^{mx}) \qquad \text{product rule for logarithms}$$
$$= \ln A + mx \qquad \ln e^u = u$$

or, equivalently, $y = mx + b$, where $y = \ln p$ and $b = \ln A$. Thus, to find the curve of the form $p = Ae^{mx}$ that best fits the given data points (x_k, p_k) for $k = 1, \ldots, 6$, we first find the least-squares line $y = mx + b$ for the data points $(x_k, \ln p_k)$. Arrange the calculations as follows:

k	x_k	p_k	$y_k = \ln p_k$	$x_k y_k$	x_k^2
1	6	743	6.61	39.66	36
2	10	539	6.29	62.90	100
3	17	308	5.73	97.41	289
4	22	207	5.33	117.26	484
5	28	128	4.85	135.80	784
6	35	73	4.29	150.15	1225
	$\Sigma x = \overline{118}$		$\Sigma y = \overline{33.10}$	$\Sigma xy = \overline{603.18}$	$\Sigma x^2 = \overline{2918}$

Use the least-squares formula with $n = 6$ to get

$$m = \frac{6(603.18) - (118)(33.10)}{6(2918) - (118)^2} = -0.08$$

and

$$b = \frac{2918(33.10) - (118)(603.18)}{6(2918) - (118)^2} = 7.09,$$

which means the least-squares line has the equation

$$y = -0.08x + 7.09.$$

Finally, returning to the exponential curve $p = Ae^{mx}$, recall that $\ln A = b$, so that

$$\ln A = b = 7.09$$
$$A = e^{7.09} \approx 1200$$

Thus, the exponential function that best fits the given demand data is

$$p = Ae^{mx} = 1200e^{-0.08x}.$$

c. Use the exponential demand function $p = 1200e^{-0.08x}$ found in part (b) to find that when $x = 40$ (hundred) units are produced, they can all be sold at a unit price of

$$p = 1200e^{-0.08(40)} = 48.91$$

or \$48.91. Therefore, when 4000 ($x = 40$) units are produced, we would expect the revenue generated to be approximately

$$R = xp(x) = (4000 \text{ units})(\$48.91 \text{ per unit})$$
$$= \$195\ 640$$

The procedure illustrated in Example 7.4.4 is sometimes called **log-linear regression.** A curve of the form $y = Ax^k$ that best fits given data can also be found by log-linear regression. The specific procedure is outlined in Exercise 30, where it is used to verify an allometric formula first mentioned in the Think About It essay at the end of Chapter 1.

The least-squares procedure can also be used to fit other nonlinear functions to data. For instance, to find the quadratic function $y = Ax^2 + Bx + C$ whose graph (a parabola) best fits a particular set of data, you would proceed as in Example 7.4.1, minimizing the sum of squares of vertical distances from the given points to the graph. Such computations are algebraically complicated and usually require the use of a computer or graphing calculator.

EXERCISES ■ 7.4

In Exercises 1 through 4, plot the given points and use the method of Example 7.4.1 to find the corresponding least-squares line.

1. $(0, 1), (2, 3), (4, 2)$

2. $(1, 1), (2, 2), (6, 0)$

3. $(1, 2), (2, 4), (4, 4), (5, 2)$

4. $(1, 5), (2, 4), (3, 2), (6, 0)$

In Exercises 5 through 12, plot the given points and use the formula to find the corresponding least-squares line.

5. $(1, 2), (2, 2), (2, 3), (5, 5)$

6. $(-4, -1), (-3, 0), (-1, 0), (0, 1), (1, 2)$

7. $(-2, 5), (0, 4), (2, 3), (4, 2), (6, 1)$

8. $(-6, 2), (-3, 1), (0, 0), (0, -3), (1, -1), (3, -2)$

9. $(0, 1)$, $(1, 1.6)$, $(2.2, 3)$, $(3.1, 3.9)$, $(4, 5)$

10. $(3, 5.72)$, $(4, 5.31)$, $(6.2, 5.12)$, $(7.52, 5.32)$, $(8.03, 5.67)$

11. $(-2.1, 3.5)$, $(-1.3, 2.7)$, $(1.5, 1.3)$, $(2.7, -1.5)$

12. $(-1.73, -4.33)$, $(0.03, -2.19)$, $(0.93, 0.15)$, $(3.82, 1.61)$

In Exercises 13 through 16, modify the least-squares procedure as illustrated in Example 7.4.4 to find the curve of the form $y = Ae^{mx}$ that best fits the given data.

13. $(1, 15.6)$, $(3, 17)$, $(5, 18.3)$, $(7, 20)$, $(10, 22.4)$

14. $(5, 9.3)$, $(10, 10.8)$, $(15, 12.5)$, $(20, 14.6)$, $(25, 17)$

15. $(2, 13.4)$, $(4, 9)$, $(6, 6)$, $(8, 4)$, $(10, 2.7)$

16. $(5, 33.5)$, $(10, 22.5)$, $(15, 15)$, $(20, 10)$, $(25, 6.8)$, $(30, 4.5)$

17. **UNIVERSITY ADMISSIONS**   Over the past 4 years, a university admissions officer has compiled the following data (measured in units of 1000) relating the number of university catalogues requested by high school students to the number of completed applications received:

Catalogues requested	4.5	3.5	4.0	5.0
Applications received	1.0	0.8	1.0	1.5

a. Plot these data on a graph.
b. Find the equation of the least-squares line.
c. Use the least-squares line to predict the number of completed applications if 4800 catalogues are requested.

18. **SALES**   A company's annual sales (in units of $1 billion) for its first 5 years of operation are shown in the following table:

Year	1	2	3	4	5
Sales	0.9	1.5	1.9	2.4	3.0

a. Plot these data on a graph.
b. Find the equation of the least-squares line.
c. Use the least-squares line to predict the company's sixth-year sales.

19. **DEMAND AND REVENUE**   A manufacturer gathers the data listed in the accompanying table relating the level of production x (hundred units) of a particular commodity to the demand price p (dollars per unit) at which all the units will be sold:

Production x (hundreds of units)	5	10	15	20	25	30	35
Demand price p (dollars per unit)	44	38	32	25	18	12	6

a. Plot these data on a graph.
b. Find the equation of the least-squares line for the data.
c. Use the linear demand equation you found in part (b) to predict the revenue the manufacturer should expect if 4000 units ($x = 40$) are produced.

20. **DRUG ABUSE**   For each of 5 different years, the accompanying table gives the percentage of high school students in the United States who had used cocaine at least once in their lives up to that year:

Year	1991	1993	1995	1997	1999	2001	2003	2005
Percentage who had used cocaine at least once	6.0	4.9	7.0	8.2	9.5	9.4	8.7	7.6

SOURCE: The White House Office of National Drug Control Policy, "2002 National Drug Control Strategy," (http://www.whitehousedrugpolicy.gov).

   a. Plot these data on a graph, with the number of years after 1991 on the x axis and the percentage of cocaine users on the y axis.
   b. Find the equation of the least-squares line for the data.
   c. Use the least-squares line to predict the percentage of high school students who used cocaine at least once by the year 2009.

21. **VOTER TURNOUT**   On election day, the polls in a certain province open at 8:00 A.M. Every 2 hours after that, an election official determines what percentage of the registered voters have already cast their ballots. The data through 6:00 P.M. are shown here:

Time	10:00	12:00	2:00	4:00	6:00
Percentage turnout	12	19	24	30	37

   a. Plot these data on a graph.
   b. Find the equation of the least-squares line. (Let x denote the number of hours after 8:00 A.M.)
   c. Use the least-squares line to predict what percentage of the registered voters will have cast their ballots by the time the polls close at 8:00 P.M.

22. **POPULATION PREDICTION**   The accompanying table gives the Canadian census figures (in millions) for the period 1981 to 2006:

Year	1981	1986	1991	1996	2001	2006
Population	24.3	25.3	27.3	28.8	30	31.6

   SOURCE: Statistics Canada.
   a. Find the least-squares line $y = mt + b$ for these data, where y is the Canadian population t years after 1981.
   b. Use the least-squares line found in part (a) to predict the Canadian population for the year 2020.

23. **POPULATION PREDICTION**   Use technology to fit the data of Exercise 22 to a quadratic and a cubic model.
   a. What population is predicted in the year 2020 in Canada with each of the models?
   b. Write a sentence about which model you think is most appropriate.

24. **PUBLIC HEALTH**   In a study of five industrial areas, a researcher obtained these data relating the average number of units of a certain pollutant in the air and the incidence (per 100 000 people) of a certain disease:

Units of pollutant	3.4	4.6	5.2	8.0	10.7
Incidence of disease	48	52	58	76	96

   a. Plot these data on a graph.
   b. Find the equation of the least-squares line.
   c. Use the least-squares line to estimate the incidence of the disease in an area with an average pollution level of 7.3 units.

25. **INVESTMENT ANALYSIS**   Jennifer has several different kinds of investments, whose total value $V(t)$ (in thousands of dollars) at the beginning of the tth year after she began investing is given in the following table, for $1 \le t \le 10$:

Year t	1	2	3	4	5	6	7	8	9	10
Value $V(t)$ of all investments	57	60	62	65	62	65	70	75	79	85

a. Modify the least-squares procedure, as illustrated in Example 7.4.4, to find a function of the form $V(t) = Ae^{rt}$ whose graph best fits these data. Roughly at what annual rate, compounded continuously, is her account growing?

b. Use the function you found in part (a) to predict the total value of her investments at the beginning of the 20th year after she began investing.

c. Jennifer estimates that she will need $300 000 for retirement. Use the function from part (a) to determine how long it will take her to attain this goal.

d. Jennifer's friend, Frank, looks over her investment analysis and snorts, "What a waste of time! You can find the A and r in your function $V(t) = Ae^{rt}$ by just using $V(1) = 57$ and $V(10) = 85$ and a little algebra." Find A and r using Frank's method and comment on the relative merits of the two approaches.

26. **DISPOSABLE INCOME AND CONSUMPTION**   The accompanying table gives the personal consumption expenditure and the corresponding disposable income (in billions of dollars) for the United States in the period 1996–2001:

Year	1996	1997	1998	1999	2000	2001
Disposable income	5677.7	5968.2	6355.6	6627.4	7120.0	7393.2
Personal consumption	5237.5	5529.3	5856.0	6246.5	6683.7	6967.0

SOURCE: U.S. Department of Commerce, Bureau of Economic Analysis, "Personal Consumption Expenditures by Major Type of Product" (http://www.dea.doc.gov).

a. Plot these data on a graph, with disposable income on the x axis and consumption expenditure on the y axis.

b. Find the equation of the least-squares line for the data.

c. Use the least-squares line to predict the consumption that would correspond to $8000 billion ($8 trillion) of disposable income.

d. Write a paragraph on the relationship between disposable income and consumption.

27. **GROSS DOMESTIC PRODUCT PER CAPITA**   The following table lists the gross domestic product (GDP) figures for China (dollars) for the period 2002–2007:

Year	2002	2003	2004	2005	2006	2007
GDP/person	4600	4400	5000	5600	6800	7700

SOURCE: http://www.indexmundi.com.

a. Find the least-squares line $y = mt + b$ for these data, where y is the GDP of China t years after 2002.

b. Use the least-squares line found in part (a) to predict the GDP of China for the year 2008.

28. **BACTERIAL GROWTH**   A biologist studying a bacterial colony measures its population each hour and records these data:

Time t (hours)	1	2	3	4	5	6	7	8
Population $P(t)$ (thousands)	280	286	292	297	304	310	316	323

a. Plot these data on a graph. Does the scatter diagram suggest that the population growth is linear or does it suggest that it is exponential?

b. If you think the scatter diagram in part (a) suggests linear growth, find the population function of the form $P(t) = mt + b$ that best fits the data. However, if you think the scatter diagram suggests exponential growth, modify the least-squares procedure, as illustrated in Example 7.4.4, to obtain a best-fitting population function of the form $P(t) = Ae^{kt}$.

  c. Use the population function you obtained in part (b) to predict how long it will take for the population to reach 400 000. How long will it take for the population to double from 280 000?

29. SPREAD OF AIDS The number of reported cases of AIDS in the United States by year of reporting at 4-year intervals since 1980 is given in the following table:

Year	1980	1984	1988	1992	1996	2000	2004
Reported cases of AIDS	99	6360	36 064	79 477	61 109	42 156	37 726

SOURCE: World Health Organization and the United Nations (http://www.unaids.org).

  a. Plot these data on a graph with time t (years after 1980) on the x axis.

  b. Find the equation of the least-squares line for the given data.

  c. How many cases of AIDS does the least-squares line in part (b) predict would be reported in 2008?

  d. Do you think the least-squares line fits the given data well? If not, write a paragraph explaining which (if any) of the following four curve types would fit the data better:

    1. (quadratic) $y = At^2 + Bt + C$
    2. (cubic) $y = At^3 + Bt^2 + Ct + D$
    3. (exponential) $y = Ae^{kt}$
    4. (power-exponential) $y = Ate^{kt}$

30. A* ALLOMETRY The determination of relationships between measurements of various parts of a particular organism is a topic of interest in the branch of biology called *allometry*.* (Recall the Think About It essay at the end of Chapter 1.) Suppose a biologist observes that the shoulder height h and antler size w of an elk, both in centimetres, are related as indicated in the following table:

Shoulder Height h (cm)	Antler Size w (cm)
87.9	52.4
95.3	60.3
106.7	73.1
115.4	83.7
127.2	98.0
135.8	110.2

  a. For each data point (h, w) in the table, plot the point $(\ln h, \ln w)$ on a graph. Note that the scatter diagram suggests that $y = \ln w$ and $x = \ln h$ are linearly related.

  b. Find the least-squares line $y = mx + b$ for the data $(\ln h, \ln w)$ you obtained in part (a).

  c. Find numbers a and c so that $w = ah^c$. [*Hint:* Substitute $y = \ln w$ and $x = \ln h$ from part (a) into the least-squares equation found in part (b).]

31. A* ALLOMETRY The accompanying table relates the weight C of the large claw of a fiddler crab to the weight W of the rest of the crab's body, both measured in milligrams.

Weight W (mg) of the body	57.6	109.2	199.7	300.2	355.2	420.1	535.7	743.3
Weight C (mg) of the claw	5.3	13.7	38.3	78.1	104.5	135.0	195.6	319.2

---

*Roger V. Jean, "Differential Growth, Huxley's Allometric Formula, and Sigmoid Growth," *UMAP Modules 1983: Tools for Teaching,* Lexington, MA: Consortium for Mathematics and Its Applications, Inc., 1984.

a. For each data point (W, C) in the table, plot the point $(\ln W, \ln C)$ on a graph. Note that the scatter diagram suggests that $y = \ln C$ and $x = \ln W$ are linearly related.

b. Find the least-squares line $y = mx + b$ for the data $(\ln W, \ln C)$ you obtained in part (a).

c. Find positive numbers a and k so that $C = aW^k$. [*Hint:* Substitute $y = \ln C$ and $x = \ln W$ from part (a) into the least-squares equation found in part (b).]

SECTION 7.5

L05

Optimize when there is a constraint on the variables using the method of Lagrange multipliers.

Constrained Optimization: The Method of Lagrange Multipliers

In many applied problems, a function of two variables is to be optimized subject to a restriction or **constraint** on the variables. For example, an editor, constrained to stay within a fixed budget of $60 000, may wish to decide how to divide this money between development and promotion in order to maximize the future sales of a new book. If x denotes the amount of money allocated to development, y the amount allocated to promotion, and $f(x, y)$ the corresponding number of books that will be sold, the editor would like to maximize the sales function $f(x, y)$ subject to the budgetary constraint that $x + y = 60\ 000$.

For a geometric interpretation of the process of optimizing a function of two variables subject to a constraint, think of the function itself as a surface in three-dimensional space and of the constraint (which is an equation involving x and y) as a curve in the xy plane. When you find the maximum or minimum of the function subject to the given constraint, you are restricting your attention to the portion of the surface that lies directly above the constraint curve. The highest point on this portion of the surface is the constrained maximum, and the lowest point is the constrained minimum. The situation is illustrated in Figure 7.21.

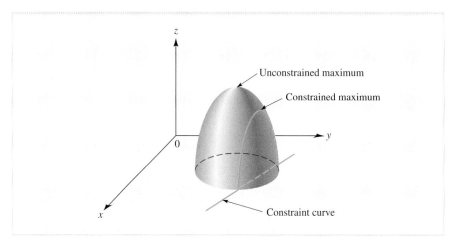

FIGURE 7.21   Constrained and unconstrained extrema.

You have already seen some constrained optimization problems in Chapter 3. (For instance, recall Example 3.5.1 of Section 3.5.) The technique you used in Chapter 3 to solve such a problem involved reducing it to a problem of a single variable by solving the constraint equation for one of the variables and then substituting the resulting expression into the function to be optimized. The success of this technique depended on solving the constraint equation for one of the variables, which is often difficult or even impossible to do in practice. In this section, you will see a more versatile technique called the **method of Lagrange multipliers,** in which the introduction of a *third* variable (the multiplier) enables you to solve constrained optimization problems without first solving the constraint equation for one of the variables.

More specifically, the method of Lagrange multipliers uses the fact that any relative extremum of the function $f(x, y)$ subject to the constraint $g(x, y) = k$ must occur at a critical point (a, b) of the function

$$F(x, y) = f(x, y) - \lambda[g(x, y) - k],$$

where λ is a new variable (the **Lagrange multiplier**). To find the critical points of F, compute its partial derivatives

$$F_x = f_x - \lambda g_x, \qquad F_y = f_y - \lambda g_y, \qquad F_\lambda = -(g - k),$$

and solve the equations $F_x = 0$, $F_y = 0$, and $F_\lambda = 0$ simultaneously, as follows:

$$\begin{aligned}
F_x = f_x - \lambda g_x = 0 \qquad &\text{or} \qquad f_x = \lambda g_x \\
F_y = f_y - \lambda g_y = 0 \qquad &\text{or} \qquad f_y = \lambda g_y \\
F_\lambda = -(g - k) = 0 \qquad &\text{or} \qquad g = k
\end{aligned}$$

Finally, evaluate $f(a, b)$ at each critical point (a, b) of F.

> **NOTE**   The method of Lagrange multipliers tells you only that any constrained extrema must occur at critical points of the function $F(x, y)$. The method cannot be used to show that constrained extrema exist or to determine whether any particular critical point (a, b) corresponds to a constrained maximum, a constrained minimum, or neither. However, *for the functions considered in this text, you can assume that if f has a constrained maximum (minimum) value, it will be given by the largest (smallest) of the critical values f(a, b).*

Here is a summary of the Lagrange multiplier procedure for finding the largest and smallest values of a function of two variables subject to a constraint.

The Method of Lagrange Multipliers

Step 1. (*Formulation*) Find the largest (or smallest) value of $f(x, y)$ subject to the constraint $g(x, y) = k$, assuming that this extreme value exists.

Step 2. Compute the partial derivatives f_x, f_y, g_x, and g_y, and find all numbers $x = a$, $y = b$, and λ that satisfy the system of equations

$$\begin{aligned}
f_x(a, b) &= \lambda g_x(a, b) \\
f_y(a, b) &= \lambda g_y(a, b) \\
g(a, b) &= k
\end{aligned}$$

These are the *Lagrange equations.*

Step 3. Evaluate f at each point (a, b) that satisfies the system of equations in step 2.

Step 4. (*Interpretation*) If $f(x, y)$ has a largest (smallest) value subject to the constraint $g(x, y) = k$, it will be the largest (smallest) of the values found in step 3.

A geometric justification of the multiplier method is given at the end of this section. In Example 7.5.1, the method is used to solve the problem from Example 3.5.1 in Section 3.5.

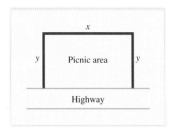

FIGURE 7.22   Rectangular picnic area.

EXAMPLE 7.5.1

The highway department is planning to build a picnic area for motorists along a major highway. It is to be rectangular with an area of 5000 m^2 and is to be fenced off on the three sides not adjacent to the highway. What is the least amount of fencing that will be needed to complete the job?

Solution

Label the sides of the picnic area as indicated in Figure 7.22 and let f denote the amount of fencing required. Then

$$f(x, y) = x + 2y.$$

The goal is to minimize f given the requirement that the area must be 5000 m^2, that is, subject to the constraint

$$g(x, y) = xy = 5000.$$

Find the partial derivatives

$$f_x = 1, \quad f_y = 2, \quad g_x = y, \quad \text{and} \quad g_y = x,$$

and obtain the three Lagrange equations

$$1 = \lambda y, \quad 2 = \lambda x, \quad \text{and} \quad xy = 5000.$$

From the first and second equations,

$$\lambda = \frac{1}{y} \quad \text{and} \quad \lambda = \frac{2}{x}$$

(since $y \neq 0$ and $x \neq 0$), which implies that

$$\frac{1}{y} = \frac{2}{x} \quad \text{or} \quad x = 2y.$$

Now substitute $x = 2y$ into the third Lagrange equation to get

$$2y^2 = 5000 \quad \text{or} \quad y = \pm 50,$$

and use the positive solution (it's a length) $y = 50$ in the equation $x = 2y$ to get $x = 100$. Thus, $x = 100$ and $y = 50$ are the values that minimize the function $f(x, y) = x + 2y$ subject to the constraint $xy = 5000$. The optimal picnic area is 100 m wide (along the highway), extends 50 m back from the road, and requires 100 m + 50 m + 50 m = 200 m of fencing.

EXAMPLE 7.5.2

Find the maximum and minimum values of the function $f(x, y) = xy$ subject to the constraint $x^2 + y^2 = 8$.

Solution

Let $g(x, y) = x^2 + y^2$ and use the partial derivatives

$$f_x = y, \quad f_y = x, \quad g_x = 2x, \quad \text{and} \quad g_y = 2y$$

to get the three Lagrange equations

$$y = 2\lambda x, \quad x = 2\lambda y, \quad \text{and} \quad x^2 + y^2 = 8.$$

Neither x nor y can be zero if all three of these equations are to hold (do you see why?), and so you can rewrite the first two equations as

$$2\lambda = \frac{y}{x} \quad \text{and} \quad 2\lambda = \frac{x}{y},$$

which implies that $\quad \dfrac{y}{x} = \dfrac{x}{y} \quad$ or $\quad x^2 = y^2.$

Now substitute $x^2 = y^2$ into the third equation to get

$$2x^2 = 8 \quad \text{or} \quad x = \pm 2.$$

If $x = 2$, it follows from the equation $x^2 = y^2$ that $y = 2$ or $y = -2$. Similarly, if $x = -2$, it follows that $y = 2$ or $y = -2$. Hence, the four points at which the constrained extrema can occur are $(2, 2)$, $(2, -2)$, $(-2, 2)$, and $(-2, -2)$. Since

$$f(2, 2) = f(-2, -2) = 4 \quad \text{and} \quad f(2, -2) = f(-2, 2) = -4,$$

it follows that when $x^2 + y^2 = 8$, the maximum value of $f(x, y)$ is 4, which occurs at the points $(2, 2)$ and $(-2, -2)$, and the minimum value is -4, which occurs at $(2, -2)$ and $(-2, 2)$.

For practice, check these answers by solving the optimization problem using the methods of Chapter 3.

> **NOTE**   In Examples 7.5.1 and 7.5.2, the first two Lagrange equations were used to eliminate the new variable λ, and then the resulting expression relating x and y was substituted into the constraint equation. For most constrained optimization problems you encounter, this particular sequence of steps will often lead quickly to the desired solution. ∎

Maximization of Utility

A *utility function* $U(x, y)$ measures the total satisfaction or *utility* a consumer receives from having x units of one particular commodity and y units of another. Example 7.5.3 illustrates how the method of Lagrange multipliers can be used to determine how many units of each commodity the consumer should purchase to maximize utility while staying within a fixed budget.

EXAMPLE 7.5.3

A consumer has \$600 to spend on two commodities, the first of which costs \$20 per unit and the second \$30 per unit. Suppose that the utility derived by the consumer from x units of the first commodity and y units of the second commodity is given by the **Cobb-Douglas utility function** $U(x, y) = 10x^{0.6}y^{0.4}$. How many units of each commodity should the consumer buy to maximize utility?

Solution

The total cost of buying x units of the first commodity at \$20 per unit and y units of the second commodity at \$30 per unit is $20x + 30y$. Since the consumer has only \$600 to spend, the goal is to maximize utility $U(x, y)$ subject to the budgetary constraint $20x + 30y = 600$.

The three Lagrange equations are

$$6x^{-0.4}y^{0.4} = 20\lambda, \quad 4x^{0.6}y^{-0.6} = 30\lambda, \quad \text{and} \quad 20x + 30y = 600.$$

From the first two equations,

$$\frac{6x^{-0.4}y^{0.4}}{20} = \frac{4x^{0.6}y^{-0.6}}{30} = \lambda$$

$$9x^{-0.4}y^{0.4} = 4x^{0.6}y^{-0.6}$$

$$9y = 4x \quad \text{or} \quad y = \frac{4}{9}x$$

Substituting $y = \dfrac{4x}{9}$ into the third Lagrange equation gives

$$20x + 30\left(\frac{4}{9}x\right) = 600$$

$$\left(\frac{100}{3}\right)x = 600$$

so that

$$x = 18 \quad \text{and} \quad y = \frac{4}{9}(18) = 8.$$

That is, to maximize utility, the consumer should buy 18 units of the first commodity and 8 units of the second.

Recall from Section 7.1 that the level curves of a utility function are known as *indifference curves*. A graph showing the relationship between the optimal indifference curve $U(x, y) = C$, where $C = U(18, 8)$, and the budgetary constraint $20x + 30y = 600$, is sketched in Figure 7.23.

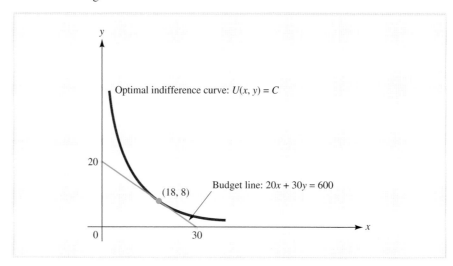

FIGURE 7.23   Budgetary constraint and optimal indifference curve.

Allocation of Resources   An important class of problems in business and economics involves determining an optimal allocation of resources subject to a constraint on those resources. Here is an example in which production is maximized subject to a cost constraint.

EXAMPLE 7.5.4

A manufacturer has $600 000 to spend on the production of a certain product and determines that if x units of capital and y units of labour are allocated to production, then P units will be produced, where P is given by the Cobb-Douglas production function

$$P(x, y) = 120x^{4/5}y^{1/5}.$$

Suppose each unit of labour costs $3000 and each unit of capital costs $5000. How many units of labour and capital should be allocated in order to maximize production?

Solution

The cost of capital is $3000x$ and the cost of labour is $5000y$, so the total cost of resources is $g(x, y) = 3000x + 5000y$. The goal is to maximize the production function $P(x, y) = 120x^{4/5}y^{1/5}$ subject to the cost constraint $g(x, y) = 600 000$. The corresponding Lagrange equations are

$$120\left(\frac{4}{5}\right)x^{-1/5}y^{1/5} = 3000\lambda, \qquad 120\left(\frac{1}{5}\right)x^{4/5}y^{-4/5} = 5000\lambda,$$

and

$$3000x + 5000y = 600\ 000,$$

or, equivalently,

$$96x^{-1/5}y^{1/5} = 3000\lambda, \qquad 24x^{4/5}y^{-4/5} = 5000\lambda, \quad \text{and} \quad 3x + 5y = 600.$$

Solving for λ in the first two equations results in

$$\lambda = 0.032x^{-1/5}y^{1/5} = 0.0048x^{4/5}y^{-4/5}.$$

Multiply both sides of this equation by $x^{1/5}y^{4/5}$ to obtain

$$[0.032x^{-1/5}y^{1/5}]x^{1/5}y^{4/5} = [0.0048x^{4/5}y^{-4/5}]x^{1/5}y^{4/5}$$
$$0.032y = 0.0048x$$

so

$$y = 0.15x.$$

Substituting into the cost constraint equation $3x + 5y = 600$ gives

$$3x + 5(0.15x) = 600.$$

Thus,

$$x = 160$$

and

$$y = 0.15x = 0.15(160) = 24.$$

That is, to maximize production, the manufacturer should allocate 160 units to capital and 24 units to labour. If this is done,

$$p(160, 24) = 120(160)^{4/5}(24)^{1/5} \approx 13\ 138 \text{ units}$$

will be produced.

A graph showing the relationship between the cost constraint and the level curve for optimal production is shown in Figure 7.24.

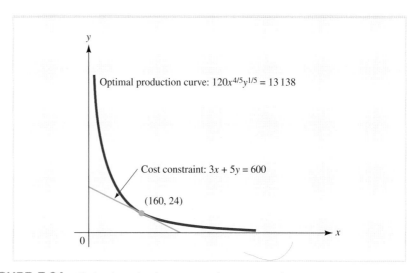

FIGURE 7.24 Optimal production curve and cost constraint.

The Significance of the Lagrange Multiplier λ

Usually, solving a constrained optimization problem by the method of Lagrange multipliers does not require actually finding a numerical value for the Lagrange multiplier λ. In some problems, however, you may want to compute λ, which has this useful interpretation.

The Lagrange Multiplier as a Rate ■ Suppose M is the maximum (or minimum) value of $f(x, y)$, subject to the constraint $g(x, y) = k$. The Lagrange multiplier λ is the rate of change of M with respect to k. That is,

$$\lambda = \frac{dM}{dk}.$$

Hence,

$$\lambda \approx \text{change in } M \text{ resulting from a 1-unit increase in } k.$$

EXAMPLE 7.5.5

Suppose the manufacturer is given an extra $1000 to spend on capital and labour for the production of the commodity in Example 7.5.4, that is, a total of $601 000. Estimate the effect on the maximum production level.

Solution

In Example 7.5.4, we found the maximum value M of the production function $P(x, y) = 120x^{4/5}y^{1/5}$ subject to the cost constraint $3000x + 5000y = 600\ 000$ by solving the three Lagrange equations

$$96x^{-1/5}y^{1/5} = 3000\lambda, \quad 24x^{4/5}y^{-4/5} = 5000\lambda, \quad \text{and} \quad 3x + 5y = 600$$

to obtain $x = 160$ and $y = 24$ and the maximum production level

$$p(160, 24) \approx 13\ 138 \text{ units.}$$

The multiplier λ can be found by substituting these values of x and y into either the first or the second Lagrange equation. Using the first equation,

$$\lambda = 0.032x^{-1/5}y^{1/5} = 0.032(160)^{-1/5}(24)^{1/5} \approx 0.0219.$$

This means that the maximum production with the new cost constraint increases by approximately 0.0219 units for each \$1 increase in the constraint. Since the constraint increases by \$1000, the maximum production increases by approximately

$$(0.0219)(1000) = 21.9 \text{ units,}$$

that is, to

$$13\ 138 + 21.9 = 13\ 159.9 \text{ units.}$$

To check, if we repeat Example 7.5.4 with the revised constraint

$$3000x + 5000y = 601\ 000,$$

it can be shown that the maximum occurs when $x = 160.27$ and $y = 24.04$ (verify this result), so the maximum productivity is

$$P(160.27, 24.04) = 120(160.27)^{4/5}(24.04)^{1/5} \approx 13\ 159.82.$$

This is essentially the estimate obtained using the Lagrange multiplier.

> **NOTE**   A problem like Example 7.5.4, where production is maximized subject to a cost constraint, is called a **fixed budget problem** (see Exercise 45). In the context of such a problem, the Lagrange multiplier λ is called the **marginal productivity of money.** Similarly, the multiplier in a utility problem like that in Example 7.5.3 is called the **marginal utility of money** (see Exercise 41). ∎

Lagrange Multipliers for Functions of Three Variables

The method of Lagrange multipliers can be extended to constrained optimization problems involving functions of more than two variables and more than one constraint. For instance, to optimize $f(x, y, z)$ subject to the constraint $g(x, y, z) = k$, solve

$$f_x = \lambda g_x, \quad f_y = \lambda g_y, \quad f_z = \lambda g_z, \quad \text{and} \quad g = k.$$

Here is an example of a problem involving this kind of constrained optimization.

EXAMPLE 7.5.6

A jewellery box is to be constructed of material that costs $1/cm^2 for the bottom, $2/cm^2 for the sides, and $5/cm^2 for the top. If the total volume is to be 96 cm^3, what dimensions will minimize the total cost of construction?

Solution

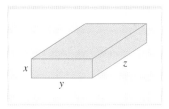

Let the box be x centimetres deep, y centimetres long, and z centimetres wide, where x, y, and z are all positive, as indicated in the accompanying figure. Then the volume of the box is $V = xyz$ and the total cost of construction is given by

$$C = \underbrace{1yz}_{\text{bottom}} + \underbrace{2(2xy + 2xz)}_{\text{sides}} + \underbrace{5yz}_{\text{top}} = 6yz + 4xy + 4xz.$$

Now, minimize $C = 6yz + 4xy + 4xz$ subject to $V = xyz = 96$. The Lagrange equations are

$$C_x = \lambda V_x \quad \text{or} \quad 4y + 4z = \lambda(yz)$$
$$C_y = \lambda V_y \quad \text{or} \quad 6z + 4x = \lambda(xz)$$
$$C_z = \lambda V_z \quad \text{or} \quad 6y + 4x = \lambda(xy)$$

and $xyz = 96$. Solving each of the first three equations for λ results in

$$\frac{4y + 4z}{yz} = \frac{6z + 4x}{xz} = \frac{6y + 4x}{xy} = \lambda.$$

Multiply each expression by xyz to obtain

$$4xy + 4xz = 6yz + 4yx$$
$$4xy + 4xz = 6yz + 4xz$$
$$6yz + 4yx = 6yz + 4xz$$

which can be further simplified by cancelling common terms on both sides of each equation to get

$$4xz = 6yz$$
$$4xy = 6yz$$
$$4yx = 4xz$$

Dividing z from both sides of the first equation, y from the second, and x from the third (which we can do because x, y, and z are all positive, and thus nonzero) results in

$$4x = 6y \quad \text{and} \quad 4x = 6z \quad \text{and} \quad 4y = 4z,$$

so that $y = \dfrac{2}{3}x$ and $z = \dfrac{2}{3}x$. Substitute these values into the constraint equation $xyz = 96$ to find that

$$x\left(\frac{2}{3}x\right)\left(\frac{2}{3}x\right) = 96$$

$$\frac{4}{9}x^3 = 96$$

$$x^3 = 216 \quad \text{or} \quad x = 6$$

and then

$$y = z = \frac{2}{3}(6) = 4.$$

Thus, the minimal cost occurs when the jewellery box is 6 cm deep with a square base, 4 cm on a side.

Why the Method of Lagrange Multipliers Works

Although a rigorous explanation of why the method of Lagrange multipliers works involves advanced ideas beyond the scope of this text, there is a rather simple geometric argument that you should find convincing. This argument depends on the fact that for the level curve $F(x, y) = C$, the slope at each point (x, y) is given by

$$\frac{dy}{dx} = -\frac{F_x}{F_y},$$

provided $F_y \neq 0$. We obtained this formula using the chain rule for partial derivatives in Exercise 80 in Section 7.2. Exercises 54 and 55 illustrate the use of this formula.

Now, consider the following constrained optimization problem:

Maximize $f(x, y)$ subject to $g(x, y) = k$.

Geometrically, this means you must find the highest level curve of f that intersects the constraint curve $g(x, y) = k$. As Figure 7.25 suggests, the critical intersection will occur at a point where the constraint curve is tangent to a level curve, that is, where the slope of the constraint curve $g(x, y) = k$ is equal to the slope of a level curve $f(x, y) = C$.

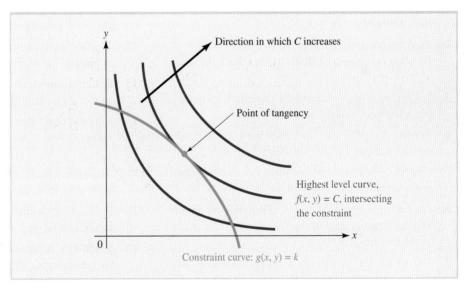

FIGURE 7.25   Increasing level curves and the constraint curve.

According to the formula stated at the beginning of this discussion,

slope of constraint curve = slope of level curve

$$-\frac{g_x}{g_y} = -\frac{f_x}{f_y},$$

or, equivalently,

$$\frac{f_x}{g_x} = \frac{f_y}{g_y}.$$

Letting λ denote this common ratio gives

$$\frac{f_x}{g_x} = \lambda \quad \text{and} \quad \frac{f_y}{g_y} = \lambda,$$

from which the first two Lagrange equations result:

$$f_x = \lambda g_x \quad \text{and} \quad f_y = \lambda g_y.$$

The third Lagrange equation,

$$g(x, y) = k,$$

is simply a statement of the fact that the point of tangency actually lies on the constraint curve.

EXERCISES ■ 7.5

In Exercises 1 through 16, use the method of Lagrange multipliers to find the indicated extremum. You may assume the extremum exists.

1. Find the maximum value of $f(x, y) = xy$ subject to the constraint $x + y = 1$.

2. Find the maximum and minimum values of the function $f(x, y) = xy$ subject to the constraint $x^2 + y^2 = 1$.

3. Let $f(x, y) = x^2 + y^2$. Find the minimum value of $f(x, y)$ subject to the constraint $xy = 1$.

4. Let $f(x, y) = x^2 + 2y^2 - xy$. Find the minimum value of $f(x, y)$ subject to the constraint $2x + y = 22$.

5. Find the minimum value of $f(x, y) = x^2 - y^2$ subject to the constraint $x^2 + y^2 = 4$.

6. Let $f(x, y) = 8x^2 - 24xy + y^2$. Find the maximum and minimum values of the function $f(x, y)$ subject to the constraint $8x^2 + y^2 = 1$.

7. Let $f(x, y) = x^2 - y^2 - 2y$. Find the maximum and minimum values of the function $f(x, y)$ subject to the constraint $x^2 + y^2 = 1$.

8. Find the maximum value of $f(x, y) = xy^2$ subject to the constraint $x + y^2 = 1$.

9. Let $f(x, y) = 2x^2 + 4y^2 - 3xy - 2x - 23y + 3$. Find the minimum value of the function $f(x, y)$ subject to the constraint $x + y = 15$.

10. Let $f(x, y) = 2x^2 + y^2 + 2xy + 4x + 2y + 7$. Find the minimum value of the function $f(x, y)$ subject to the constraint $4x^2 + 4xy = 1$.

11. Find the maximum and minimum values of $f(x, y) = e^{xy}$ subject to $x^2 + y^2 = 4$.

12. Find the maximum value of $f(x, y) = \ln (xy^2)$ subject to $2x^2 + 3y^2 = 8$ for $x > 0$ and $y > 0$.

13. Find the maximum value of $f(x, y, z) = xyz$ subject to $x + 2y + 3z = 24$.

14. Find the maximum and minimum values of $f(x, y, z) = x + 3y - z$ subject to $z = 2x^2 + y^2$.

15. Let $f(x, y, z) = x + 2y + 3z$. Find the maximum and minimum values of $f(x, y, z)$ subject to the constraint $x^2 + y^2 + z^2 = 16$.

16. Find the minimum value of $f(x, y, z) = x^2 + y^2 + z^2$ subject to $4x^2 + 2y^2 + z^2 = 4$.

17. **PROFIT**   A manufacturer of television sets makes two models, the Deluxe and the Standard. The manager estimates that when x hundred Deluxe sets and y hundred Standard sets are produced each year, the annual profit will be $P(x, y)$ thousand dollars, where

$$P(x, y) = -0.3x^2 - 0.5xy - 0.4y^2$$
$$+ 85x + 125y - 2500.$$

The company can produce exactly 30 000 sets each year. How many Deluxe and how many Standard sets should be produced each year in order to maximize annual profit?

18. **PROFIT** A manufacturer supplies refrigerators to stores in Calgary and Edmonton. The manager estimates that if x refrigerators are delivered to Calgary and y refrigerators to Edmonton each month, the monthly profit will be $P(x, y)$ hundred dollars, where

$$P(x, y) = -0.02x^2 - 0.03xy - 0.04y^2 \\ + 15x + 40y - 3000.$$

Each month, the company can produce exactly 700 refrigerators. How many refrigerators should be supplied to Calgary and how many to Edmonton to maximize monthly profit?

19. **SALES** An editor has been allotted $60 000 to spend on the development and promotion of a new book. It is estimated that if x thousand dollars are spent on development and y thousand dollars on promotion, approximately

$$S(x, y) = 20x^{3/2}y$$

copies of the book will be sold.
   a. How much money should the editor allocate to development and how much to promotion in order to maximize sales?
   b. Suppose the editor is allotted an extra $1000 for development and promotion. Use the Lagrange multiplier λ to estimate the change in the maximum sales level.

20. **SALES** A manager has been allotted $8000 to spend on the development and promotion of a new product. It is estimated that if x thousand dollars are spent on development and y thousand dollars on promotion, approximately

$$f(x, y) = 50x^{1/2}y^{3/2}$$

units of the product will be sold.
   a. How much money should the manager allocate to development and how much to promotion in order to maximize sales?
   b. Suppose the manager is allotted an extra $1000 for development and promotion. Use the Lagrange multiplier λ to estimate the change in the maximum sales level.

21. **CONSTRUCTION** A farmer wishes to fence off a rectangular pasture along the bank of a river. The area of the pasture is to be 3200 m^2, and no fencing is needed along the river bank. Find the dimensions of the pasture that will require the least amount of fencing.

22. **CONSTRUCTION** There are 320 m of fencing available to enclose a rectangular field. How should the fencing be used so that the enclosed area is as large as possible?

23. **POSTAL PACKAGING** According to postal regulations in a certain country, the girth plus length of parcels may not exceed 108 cm to avoid falling into another price category. What is the largest possible volume of a rectangular parcel with two square sides that can be sent in this way? (Refer to the accompanying figure.)

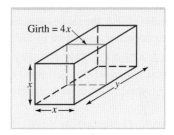

EXERCISE 23

24. **POSTAL PACKAGING** According to the postal regulation given in Exercise 23, what is the largest volume of a cylindrical can that can be sent in this way? (A cylinder of radius R and length H has volume $\pi R^2 H$.)

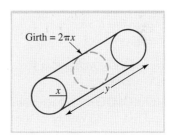

EXERCISE 24

25. **PACKAGING** Find the dimensions of the 450-mL pop can that can be constructed using the least amount of metal. (Recall that the volume of a cylinder of radius r and height h is $\pi r^2 h$, that a circle of radius r has area πr^2 and circumference $2\pi r$, and that 1 mL = 1 cm^3.)

26. **PACKAGING** A cylindrical can is to hold 40π cm^3 of frozen orange juice. The cost per square centimetre of constructing the metal top and bottom is twice the cost per square centimetre of constructing the cardboard side. What are the dimensions of the least expensive can? (See the measurement information in Exercise 25.)

27. ALLOCATION OF FUNDS   When x thousand dollars are spent on labour and y thousand on equipment, the output of a certain factory is Q units, where

$$Q(x, y) = 60x^{1/3}y^{2/3}.$$

Suppose $120\,000$ is available for labour and equipment.

  a. How should the money be allocated between labour and equipment to generate the largest possible output?

  b. Use the Lagrange multiplier λ to estimate the change in the maximum output of the factory that will result if the money available for labour and equipment is increased to $121\,000$.

28. SURFACE AREA OF THE HUMAN BODY   Recall from Exercise 39 of Section 7.1 that an empirical formula for the surface area of a person's body is

$$S(W, H) = 0.0072W^{0.425}H^{0.725},$$

where W (kg) is the person's weight and H (cm) is his or her height. Suppose that for a short period of time, Maria's weight adjusts as she grows taller so that $W + H = 160$. With this constraint, what height and weight will maximize the surface area of Maria's body?

In Exercises 29 and 30, you will need to know that a closed cylinder of radius R and length L has volume $V = \pi R^2 L$ and surface area $S = 2\pi RL + 2\pi R^2$. The volume of a hemisphere of radius R is $V = \frac{2}{3}\pi R^3$ and its surface area is $S = 2\pi R^2$.

29. A* MICROBIOLOGY   A bacterium is shaped like a cylindrical rod. If the volume of the bacterium is fixed, what relationship between the radius R and length H of the bacterium will result in minimum surface area?

 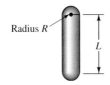

Radius R

L

R

H

EXERCISE 29   **EXERCISE 30**

30. A* MICROBIOLOGY   A bacterium is shaped like a cylindrical rod with two hemispherical caps on the ends. If the volume of the bacterium is fixed, what must be true about its radius R and length L to achieve minimum surface area?

31. A* OPTICS   The thin lens formula in optics says that the focal length L of a thin lens is related to the object distance d_o and image distance d_i by the equation

$$\frac{1}{d_o} + \frac{1}{d_i} = \frac{1}{L}.$$

If L remains constant while d_o and d_i are allowed to vary, what is the maximum distance $s = d_o + d_i$ between the object and the image?

32. CONSTRUCTION   A jewellery box is constructed by partitioning a box with a square base as shown in the accompanying figure. If the box is designed to have volume 800 cm^3, what dimensions should it have to minimize its total surface area (top, bottom, sides, and interior partitions)? Notice that we have said nothing about where the partitions are located. Does it matter?

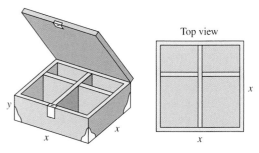

Top view

x

y

x x x

x x

EXERCISE 32

33. CONSTRUCTION   Suppose the jewellery box in Exercise 32 is designed so that the material in the top costs twice as much as the material in the bottom and sides and three times as much as the material in the interior partitions. What dimensions minimize the total cost of constructing the box?

34. SPY STORY   Having disposed of the villains in Chapter 6, the spy goes looking for his enemy. He enters a room and the door slams behind him. Immediately, he begins to feel warm and, too late, he realizes he is trapped inside the dreaded broiler room. Searching desperately for a way to survive, he notices that the room is shaped like the circle $x^2 + y^2 = 60$ and that he is standing at the centre $(0, 0)$. He presses a button on his special heat-detecting smartphone and sees that the temperature at each point (x, y) in the room is given by

$$T(x, y) = x^2 + y^2 + 3xy + 5x + 15y + 130.$$

From an informant's report, he knows that somewhere in the walls of this room there is a trap door leading outside the castle, and he reasons that it must be located at the coolest point. Where is it? Just how cool will the spy be when he gets there?

35. A* PARTICLE PHYSICS   A particle of mass m in a rectangular box with dimensions x, y, and z has ground state energy

$$E(x, y, z) = \frac{k^2}{8m}\left(\frac{1}{x^2} + \frac{1}{y^2} + \frac{1}{z^2}\right),$$

where k is a physical constant. In Section 7.3, Exercise 31, you were asked to minimize the ground state energy subject to the fixed volume constraint $V_0 = xyz$ using substitution. Solve the same constrained optimization problem using the method of Lagrange multipliers.

36. CONSTRUCTION   A rectangular building is to be constructed of material that costs $31/m^2$ for the roof, $27/m^2$ for the sides and the back, and $55/m^2$ for the facing and glass used in constructing the front. If the building is to have a volume of $16\,000$ m^3, what dimensions will minimize the total cost of construction?

37. CONSTRUCTION   A storage shed is to be constructed of material that costs $15/m^2$ for the roof, $12/m^2$ for the two sides and back, and $20/m^2$ for the front. What are the dimensions of the largest shed (in volume) that can be constructed for $8000?

38. ALLOCATION OF FUNDS   A manufacturer is planning to sell a new product at a price of $150 per unit and estimates that if x thousand dollars are spent on development and y thousand dollars are spent on promotion, approximately $\dfrac{320y}{y + 2} + \dfrac{160x}{x + 4}$ units of the product will be sold. The cost of manufacturing the product is $50 per unit. If the manufacturer has a total of $8000 to spend on development and promotion, how should this money be allocated to generate the largest possible profit? [*Hint:* Profit = (number of units)(price per unit − cost per unit) − amount spent on development and production.]

39. MARGINAL ANALYSIS   Suppose the manufacturer in Exercise 38 decides to spend $8100 instead of $8000 on the development and promotion of the new product. Use the Lagrange multiplier λ to estimate how this change will affect the maximum possible profit.

40. A* ALLOCATION OF UNRESTRICTED FUNDS
   a. If unlimited funds are available, how much should the manufacturer in Exercise 38 spend on development and how much on promotion in order to generate the largest possible profit? [*Hint:* Use the methods of Section 7.3.]
   b. Suppose the allocation problem in part (a) is solved by the method of Lagrange multipliers. What is the value of λ that corresponds to the optimal budget? Interpret your answer in terms of $\dfrac{dM}{dk}$.
   c. Your answer to part (b) should suggest another method for solving the problem in part (a). Solve the problem using this new method.

41. UTILITY   A consumer has $280 to spend on two commodities, the first of which costs $2 per unit and the second $5 per unit. Suppose that the utility derived by the consumer from x units of the first commodity and y units of the second is given by $U(x, y) = 100x^{0.25}y^{0.75}$.
   a. How many units of each commodity should the consumer buy to maximize utility?
   b. Compute the Lagrange multiplier λ and interpret it in economic terms. (In the context of maximizing utility, λ is called the **marginal utility of money.**)

42. A* UTILITY   A consumer has k dollars to spend on two commodities, the first of which costs a dollars per unit and the second b dollars per unit. Suppose that the utility derived by the consumer from x units of the first commodity and y units of the second commodity is given by the Cobb-Douglas utility function $U(x, y) = x^\alpha y^\beta$, where $0 < \alpha < 1$ and $\alpha + \beta = 1$. Show that utility is maximized when $x = \dfrac{k\alpha}{a}$ and $y = \dfrac{k\beta}{b}$.

43.   In Exercise 42, how does the maximum utility change if k is increased by 1 dollar?

In Exercises 44 through 46, let $Q(x, y)$ be a production function, where x and y represent units of labour and capital, respectively. If unit costs of labour and capital are given by p and q, respectively, then $px + qy$ represents the total cost of production.

44. A* MINIMUM COST   Use Lagrange multipliers to show that subject to a fixed production level c, the total cost is minimized when

$$\frac{Q_x}{p} = \frac{Q_y}{q} \quad \text{and} \quad Q(x, y) = c,$$

provided Q_x and Q_y are not both 0 and $p \neq 0$ and $q \neq 0$. (This is often referred to as the **minimum-cost problem**, and its solution is called the **least-cost combination of inputs**.)

45. **A* FIXED BUDGET** Show that the inputs x and y that maximize the production level $Q(x, y)$ subject to a fixed cost k satisfy

$$\frac{Q_x}{p} = \frac{Q_y}{q} \quad \text{with} \quad px + qy = k.$$

(Assume that neither p nor q is 0.) This is called a **fixed-budget problem.**

46. **A* MINIMUM COST** Show that with the fixed production level $Ax^\alpha y^\beta = k$, where k is a constant and α and β are positive with $\alpha + \beta = 1$, the joint cost function $C(x, y) = px + qy$ is minimized when

$$x = \frac{k}{A}\left(\frac{\alpha q}{\beta p}\right)^\beta \quad \text{and} \quad y = \frac{k}{A}\left(\frac{\beta p}{\alpha q}\right)^\alpha.$$

CES PRODUCTION *A constant elasticity of substitution (CES) production function is one with the general form*

$$Q(K, L) = A[\alpha K^{-\beta} + (1 - \alpha)L^{-\beta}]^{-1/\beta},$$

where K is capital expenditure; L is the level of labour; and A, α, and β are constants that satisfy $A > 0$, $0 < \alpha < 1$, and $\beta > -1$. Exercises 47 through 49 are concerned with such production functions.

47. **A*** Use the method of Lagrange multipliers to maximize the CES production function

$$Q = 55[0.6K^{-1/4} + 0.4L^{-1/4}]^{-4}$$

subject to the constraint

$$2K + 5L = 150.$$

48. **A*** Use the method of Lagrange multipliers to maximize the CES production function

$$Q = 50[0.3K^{-1/5} + 0.7L^{-1/5}]^{-5}$$

subject to the constraint

$$5K + 2L = 140.$$

49. **A*** Suppose you wish to maximize the CES production function

$$Q(K, L) = A[\alpha K^{-\beta} + (1 - \alpha)L^{-\beta}]^{-1/\beta}$$

subject to the linear constraint $c_1 K + c_2 L = B$. Show that the values of K and L at the maximum must satisfy

$$\left(\frac{K}{L}\right)^{\beta + 1} = \frac{c_2}{c_1}\left(\frac{\alpha}{1 - \alpha}\right).$$

50. **A* SATELLITE CONSTRUCTION** A space probe has the shape of the surface

$$4x^2 + y^2 + 4z^2 = 16,$$

where x, y, and z are in metres. When it re-enters Earth's atmosphere, the probe begins to heat up so that the temperature at each point $P(x, y, z)$ on the probe's surface is given by

$$T(x, y, z) = 8x^2 + 4yz - 16x + 600,$$

where T is in degrees Celsius. Use the method of Lagrange multipliers to find the hottest and coolest points on the probe's surface. What are the extreme temperatures?

51. **A*** Use Lagrange multipliers to find the possible maximum or minimum points on the part of the surface $z = x - y$ for which $y = x^5 + x - 2$. Then use a graphing utility to sketch the curve $y = x^5 + x - 2$ and the level curves to the surface $f(x, y) = x - y$ and show that the points you have just found do not represent relative maxima or minima. What do you conclude from this observation?

52. **A* HAZARDOUS WASTE MANAGEMENT** A study conducted at a waste disposal site reveals soil contamination over a region that may be described roughly as the interior of the ellipse

$$\frac{x^2}{4} + \frac{y^2}{9} = 1,$$

where x and y are in kilometres. The manager of the site plans to build a circular enclosure to contain all polluted territory.

a. If the office at the site is at the point $S(1, 1)$, what is the radius of the smallest circle centred at S that contains the entire contaminated region? [*Hint:* The function

$$f(x, y) = (x - 1)^2 + (y - 1)^2$$

measures the square of the distance from $S(1, 1)$ to the point $P(x, y)$. The required radius can be found by maximizing $f(x, y)$ subject to a certain constraint.] Use software to solve.

b. Read an article on waste management, and write a paragraph on how management decisions are made regarding landfills and other disposal sites.

53. **A* MARGINAL ANALYSIS** Let $P(K, L)$ be a production function, where K and L represent the capital and labour required for a certain manufacturing procedure. Suppose we wish to maximize $P(K, L)$ subject to the cost constraint $C(K, L) = A$ for constant A. Use the method of

Lagrange multipliers to show that optimal production is attained when

$$\frac{\dfrac{\partial P}{\partial K}}{\dfrac{\partial C}{\partial K}} = \frac{\dfrac{\partial P}{\partial L}}{\dfrac{\partial C}{\partial L}}$$

that is, when the ratio of marginal production from capital to the marginal cost of capital equals the ratio of marginal production of labour to the marginal cost of labour.

54. **A*** Let $F(x, y) = x^2 + 2xy - y^2$.
    a. If $F(x, y) = k$ for constant k, use the method of implicit differentiation developed in Chapter 2 to find $\dfrac{dy}{dx}$.
    b. Find the partial derivatives F_x and F_y and verify that
    $$\frac{dy}{dx} = -\frac{F_x}{F_y}.$$

55. **A*** Repeat Exercise 54 for the function
    $$F(x, y) = xe^{xy^2} + \frac{y}{x} + x \ln (x + y).$$

In Exercises 56 through 59, use the method of Lagrange multipliers to find the indicated maximum or minimum. You will need to use a computer program to solve functions.

56. Maximize $f(x, y) = e^{x+y} - x \ln\left(\dfrac{y}{x}\right)$
    subject to $x + y = 4$.

57. Minimize $f(x, y) = \ln (x + 2y)$ subject to $xy + y = 5$.

58. Minimize $f(x, y) = \dfrac{1}{x^2} + \dfrac{3}{xy} + \dfrac{1}{y^2}$
    subject to $x + 2y = 7$.

59. Maximize $f(x, y) = xe^{x^2-y}$ subject to $x^2 + 2y^2 = 1$.

SECTION 7.6   Double Integrals

L06

Evaluate double integrals.

In Chapters 5 and 6, you integrated a function of one variable $f(x)$ by reversing the process of differentiation, and a similar procedure can be used to integrate a function of two variables $f(x, y)$. However, since two variables are involved, we shall integrate $f(x, y)$ by holding one variable fixed and integrating with respect to the other. (Recall when doing partial derivatives that we kept one variable fixed and differentiated with respect to the other.)

For instance, to evaluate the partial integral $\displaystyle\int_1^2 xy^2\, dx$ you would integrate with respect to x, using the fundamental theorem of calculus, with y held constant:

$$\int_1^2 xy^2\, dx = \frac{1}{2}x^2y^2 \Big|_{x=1}^{x=2}$$

$$= \left[\frac{1}{2}(2)^2y^2\right] - \left[\frac{1}{2}(1)^2y^2\right]$$

$$= \frac{3}{2}y^2$$

Similarly, to evaluate $\displaystyle\int_{-1}^1 xy^2\, dy,$ integrate with respect to y, holding x constant:

$$\int_{-1}^1 xy^2\, dy = x\left(\frac{1}{3}y^3\right)\Big|_{y=-1}^{y=1}$$

$$= \left[x\left(\frac{1}{3}(1)^3\right)\right] - \left[x\left(\frac{1}{3}(-1)^3\right)\right]$$

$$= \frac{2}{3}x$$

In general, partially integrating a function $f(x, y)$ with respect to x results in a function of y alone, which can then be integrated as a function of a single variable, thus producing what we call an **iterated integral** $\int\left[\int f(x, y)\, dx\right] dy$. Similarly, the iterated integral $\int\left[\int f(x, y)\, dy\right] dx$ is obtained by first integrating with respect to y, holding x constant, and then with respect to x. Returning to our example,

$$\int_{-1}^{1}\left(\int_{1}^{2} xy^2\, dx\right) dy = \int_{-1}^{1} \frac{3}{2}y^2\, dy = \frac{1}{2}y^3\bigg|_{y=-1}^{y=1} = 1$$

and

$$\int_{1}^{2}\left(\int_{-1}^{1} xy^2\, dy\right) dx = \int_{1}^{2} \frac{2}{3}x\, dx = \frac{1}{3}x^2\bigg|_{x=1}^{x=2} = 1.$$

In our example, the two iterated integrals turned out to have the same value, and you can assume this will be true for all iterated integrals considered in this text. The double integral of $f(x, y)$ over a rectangular region in the xy plane has the following definition in terms of iterated integrals.

The Double Integral over a Rectangular Region ■ The **double integral** $\iint_R f(x, y)\, dA$ over the rectangular region $R: a \le x \le b,\ c \le y \le d$ is given by the common value of the two iterated integrals

$$\int_{a}^{b}\left[\int_{c}^{d} f(x, y)\, dy\right] dx \quad \text{and} \quad \int_{c}^{d}\left[\int_{a}^{b} f(x, y)\, dx\right] dy,$$

that is,

$$\iint_R f(x, y)\, dA = \int_{a}^{b}\left[\int_{c}^{d} f(x, y)\, dy\right] dx = \int_{c}^{d}\left[\int_{a}^{b} f(x, y)\, dx\right] dy$$

$R: a \le x \le b, c \le y \le d$

Example 7.6.1 illustrates the computation of this kind of double integral.

EXAMPLE 7.6.1

Evaluate the double integral

$$\iint_R xe^{-y}\, dA,$$

where R is the rectangular region $-2 \le x \le 1,\ 0 \le y \le 5$, using

a. x integration first   **b.** y integration first

Solution

a. With x integration first:

$$\iint_R xe^{-y}\, dA = \int_0^5 \left(\int_{-2}^1 xe^{-y}\, dx \right) dy$$

$$= \int_0^5 \frac{1}{2} x^2 e^{-y} \bigg|_{x=-2}^{x=1} dy$$

$$= \int_0^5 \frac{1}{2} e^{-y} [(1)^2 - (-2)^2]\, dy = \int_0^5 -\frac{3}{2} e^{-y}\, dy$$

$$= -\frac{3}{2}(-e^{-y}) \bigg|_{y=0}^{y=5} = \frac{3}{2}(e^{-5} - e^0) = \frac{3}{2}(e^{-5} - 1)$$

b. Integrating with respect to y first results in

$$\iint_R xe^{-y}\, dA = \int_{-2}^1 \left(\int_0^5 xe^{-y}\, dy \right) dx$$

$$= \int_{-2}^1 x(-e^{-y}) \bigg|_{y=0}^{y=5} dx = \int_{-2}^1 [-x(e^{-5} - e^0)]\, dx$$

$$= -(e^{-5} - 1)\left[\left(\frac{1}{2} x^2 \right) \bigg|_{x=-2}^{x=1} \right]$$

$$= -\frac{1}{2}(e^{-5} - 1)[(1)^2 - (-2)^2] = \frac{3}{2}(e^{-5} - 1)$$

In Example 7.6.1, the order of integration made no difference. Not only do the computations yield the same result, but the integrations are essentially of the same level of difficulty. However, sometimes the order does matter, as illustrated in Example 7.6.2.

EXAMPLE 7.6.2

Evaluate the double integral

$$\iint_R xe^{xy}\, dA,$$

where R is the rectangular region $0 \le x \le 2,\ 0 \le y \le 1$.

Solution

To evaluate the integral in the order

$$\int_0^1 \left(\int_0^2 xe^{xy}\, dx \right) dy,$$

it is necessary to use integration by parts for the inner integration:

$$u = x \qquad dv = e^{xy}\, dx$$

$$du = dx \qquad v = \frac{1}{y} e^{xy}$$

$$\int_0^2 xe^{xy}\, dx = \frac{x}{y}e^{xy}\Big|_{x=0}^{x=2} - \int_0^2 \frac{1}{y}e^{xy}\, dx$$

$$= \left(\frac{x}{y} - \frac{1}{y^2}\right)e^{xy}\Big|_{x=0}^{x=2} = \left(\frac{2}{y} - \frac{1}{y^2}\right)e^{2y} - \left(\frac{-1}{y^2}\right)$$

Then the outer integration becomes

$$\int_0^1 \left[\left(\frac{2}{y} - \frac{1}{y^2}\right)e^{2y} + \frac{1}{y^2}\right] dy.$$

Now what? Any ideas?

On the other hand, using y integration first, both computations are easy:

$$\int_0^2 \left(\int_0^1 xe^{xy}\, dy\right) dx = \int_0^2 \frac{xe^{xy}}{x}\Big|_{y=0}^{y=1}\, dx$$

$$= \int_0^2 (e^x - 1)\, dx = (e^x - x)\Big|_{x=0}^{x=2}$$

$$= (e^2 - 2) - (e^0 - 0) = e^3 - 3$$

Double Integrals over Nonrectangular Regions

In each of the preceding examples, the region of integration is a rectangle, but double integrals can also be defined over nonrectangular regions. Before presenting this definition, however, we will introduce an efficient procedure for describing certain such regions in terms of inequalities.

Vertical Cross-Sections

The region R shown in Figure 7.26 is bounded below by the curve $y = g_1(x)$, above by the curve $y = g_2(x)$, and on the sides by the vertical lines $x = a$ and $x = b$. This region can be described by the inequalities

$$R: a \leq x \leq b, g_1(x) \leq y \leq g_2(x).$$

The first inequality specifies the interval in which x must lie, while the second indicates the lower and upper bounds of the vertical cross-section of R for each x in this interval. In words:

R is the region such that for each x between a and b,
y varies from $g_1(x)$ to $g_2(x)$.

This method for describing a region is illustrated in Example 7.6.3.

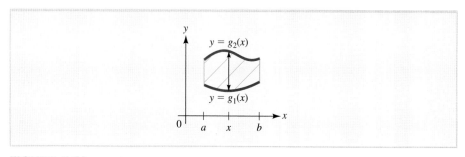

FIGURE 7.26   Vertical cross-sections. The region $R: a \leq x \leq b, g_1(x) \leq y \leq g_2(x)$.

EXAMPLE 7.6.3

Let R be the region bounded by the curve $y = x^2$ and the line $y = 2x$. Use inequalities to describe R in terms of its vertical cross-sections.

Solution

Begin with a sketch of the curve and line as shown in Figure 7.27. Identify the region R, and, for reference, draw a vertical cross-section. Solve the equations $y = x^2$ and $y = 2x$ simultaneously to find the points of intersection, $(0, 0)$ and $(2, 4)$. Observe that in the region R, the variable x takes on all values from $x = 0$ to $x = 2$ and that for each such value of x, the vertical cross-section is bounded below by $y = x^2$ and above by $y = 2x$. Hence, R can be described by the inequalities

$$0 \le x \le 2 \quad \text{and} \quad x^2 \le y \le 2x.$$

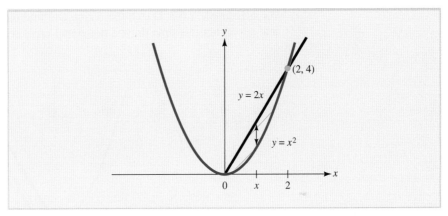

FIGURE 7.27   The region R between $y = x^2$ and $y = 2x$ described by vertical cross-sections as $R: 0 \le x \le 2$, $x^2 \le y \le 2x$.

Horizontal Cross-Sections   The region R in Figure 7.28 is bounded on the left by the curve $x = h_1(y)$, on the right by $x = h_2(y)$, below by the horizontal line $y = c$, and above by $y = d$. This region can be described by the pair of inequalities

$$R: c \le y \le d, \, h_1(y) \le x \le h_2(y).$$

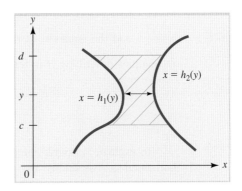

FIGURE 7.28   Horizontal cross-sections. The region $R: c \le y \le d$, $h_1(y) \le x \le h_2(y)$.

The first inequality specifies the interval in which y must lie, and the second indicates the left-hand (trailing) and right-hand (leading) bounds of a horizontal cross-section. In words:

R is the region such that for each y between c and d,
x varies from $h_1(y)$ to $h_2(y)$.

This method of description is illustrated in Example 7.6.4 for the same region described using vertical cross-sections in Example 7.6.3.

EXAMPLE 7.6.4

Describe the region R bounded by the curve $y = x^2$ and the line $y = 2x$ in terms of inequalities using horizontal cross-sections.

Solution

As in Example 7.6.3, sketch the region and find the points of intersection of the line and curve, but this time draw a horizontal cross-section (Figure 7.29).

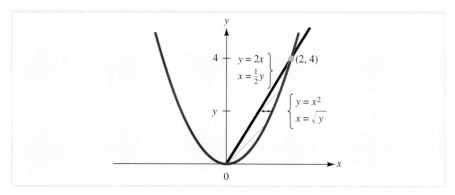

FIGURE 7.29   The region R between $y = x^2$ and $y = 2x$ described by horizontal cross-sections as $R: 0 \le y \le 4, \frac{1}{2}y \le x \le \sqrt{y}$.

In the region R, the variable y takes on all values from $y = 0$ to $y = 4$. For each such value of y, the horizontal cross-section extends from the line $y = 2x$ on the left to the curve $y = x^2$ on the right. Since the equation of the line can be rewritten as $x = \frac{1}{2}y$ and the equation of the curve as $x = \sqrt{y}$ (since $x \ge 0$ in this region), the inequalities describing R in terms of its horizontal cross-sections are

$$0 \le y \le 4 \quad \text{and} \quad \frac{1}{2}y \le x \le \sqrt{y}.$$

To evaluate a double integral over a region R described using either vertical or horizontal cross-sections, use an iterated integral whose limits of integration come from the inequalities describing the region. Here is a more precise description of how the limits of integration are determined.

Limits of Integration for Double Integrals ▪ If R can be described by the inequalities

$$a \le x \le b \quad \text{and} \quad g_1(x) \le y \le g_2(x),$$

then

$$\iint\limits_{R} f(x, y)\, dA = \int_a^b \left[\int_{g_1(x)}^{g_2(x)} f(x, y)\, dy \right] dx.$$

If R can be described by the inequalities

$$c \le y \le d \quad \text{and} \quad h_1(y) \le x \le h_2(y),$$

then

$$\iint\limits_{R} f(x, y)\, dA = \int_c^d \left[\int_{h_1(y)}^{h_2(y)} f(x, y)\, dx \right] dy.$$

EXAMPLE 7.6.5

Let I be the double integral

$$I = \int_0^1 \int_0^y y^2 e^{xy}\, dx\, dy.$$

a. Sketch the region of integration and rewrite the integral with the order of integration reversed.

b. Evaluate I using either order of integration.

Solution

a. Comparing I with the general form for the order $dx\, dy$, we see that the region of integration is

$$R: \underbrace{0 \le y \le 1,}_{\substack{\text{outer limits} \\ \text{of integration}}} \underbrace{0 \le x \le y.}_{\substack{\text{inner limits} \\ \text{of integration}}}$$

Thus, if y is a number in the interval $0 \le y \le 1$, the horizontal cross-section of R at y extends from $x = 0$ on the left to $x = y$ on the right. The region is the triangle shown in Figure 7.30a. As shown in Figure 7.30b, the same region R can be described by taking vertical cross-sections at each number x in the interval $0 \le x \le 1$ that are bounded below by $y = x$ and above by $y = 1$. Expressed in terms of inequalities, this means that

$$R: 0 \le x \le 1, x \le y \le 1,$$

so the integral can also be written as

$$I = \int_0^1 \int_x^1 y^2 e^{xy}\, dy\, dx.$$

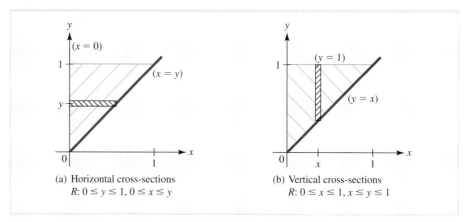

FIGURE 7.30   The region of integration for $I = \int_0^1 \int_0^y y^2 e^{xy} \, dx \, dy$.

b. Here is the evaluation of I using the given order of integration. Try to compute I by reversing the order of integration. What happens?

$$\int_0^1 \int_0^y y^2 e^{xy} \, dx \, dy = \int_0^1 \left(y e^{xy} \Big|_{x=0}^{x=y} \right) dy \qquad \text{since } \int e^{xy} \, dx = \frac{1}{y} e^{xy}$$

$$= \int_0^1 (y e^{y^2} - y) \, dy$$

$$= \left(\frac{1}{2} e^{y^2} - \frac{1}{2} y^2 \right) \Big|_0^1$$

$$= \left(\frac{1}{2} e - \frac{1}{2} \right) - \left(\frac{1}{2} - 0 \right) = \frac{1}{2} e - 1$$

Applications

Next, we will examine a few applications of double integrals, all of which are generalizations of familiar applications of definite integrals of functions of one variable. Specifically, we will see how double integration can be used to compute area, volume, and average value.

The Area of a Region in the Plane

The area of a region R in the xy plane can be computed as the double integral over R of the constant function $f(x, y) = 1$.

Area Formula ■ The area of a region R in the xy plane is given by the formula

$$\text{Area of } R = \iint_R 1 \, dA.$$

To get a feel for why the area formula holds, consider the elementary region R shown in Figure 7.31, which is bounded above by the curve $y = g_2(x)$ and below by the curve $y = g_1(x)$, and which extends from $x = a$ to $x = b$. According to the double integral formula for area,

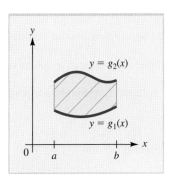

FIGURE 7.31

Area of $R = \displaystyle\iint_R 1 \, dA$.

$$\text{Area of } R = \iint_R 1 \, dA$$

$$= \int_a^b \int_{g_1(x)}^{g_2(x)} 1 \, dy \, dx$$

$$= \int_a^b \left[y \Big|_{y=g_1(x)}^{y=g_2(x)} \right] dx$$

$$= \int_a^b [g_2(x) - g_1(x)] \, dx$$

which is precisely the formula for the area between two curves that you saw in Section 5.4. Here is an example illustrating the use of the area formula.

EXAMPLE 7.6.6

Find the area of the region R bounded by the curves $y = x^3$ and $y = x^2$.

Solution

The region is shown in Figure 7.32. Using the area formula results in

$$\text{Area of } R = \iint_R 1 \, dA = \int_0^1 \int_{x^3}^{x^2} 1 \, dy \, dx$$

$$= \int_0^1 \left(y \Big|_{y=x^3}^{y=x^2} \right) dx$$

$$= \int_0^1 (x^2 - x^3) \, dx$$

$$= \left[\frac{1}{3}x^3 - \frac{1}{4}x^4 \right] \Big|_0^1$$

$$= \frac{1}{12}$$

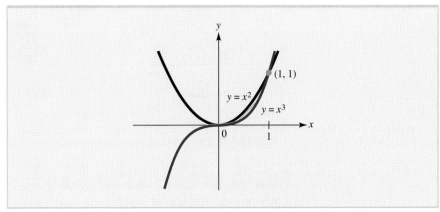

FIGURE 7.32 The region bounded by $y = x^2$ and $y = x^3$.

Volume as a
Double Integral

Recall from Section 5.3 that the region under the curve $y = f(x)$ over an interval $a \leq x \leq b$, where $f(x)$ is continuous and $f(x) \geq 0$, has area given by the definite integral $A = \int_a^b f(x) \, dx$. An analogous argument for a continuous, nonnegative function of two variables $f(x, y)$ yields the following formula for volume as a double integral.

Volume as a Double Integral ■ If $f(x, y)$ is continuous and $f(x, y) \geq 0$ on the region R, then the solid region under the surface $z = f(x, y)$ over R has volume given by

$$V = \int \int_R f(x, y) \, dA.$$

EXAMPLE 7.6.7

A biomass covers the triangular bottom of a container with vertices $(0, 0)$, $(6, 0)$, and $(3, 3)$, to a depth $h(x, y) = \dfrac{x}{y + 2}$ at each point (x, y) in the region, where all dimensions are in centimetres. What is the total volume of the biomass?

Solution

The volume is given by the double integral $V = \int \int_R h(x, y) \, dA$, where R is the triangular region shown in Figure 7.33.

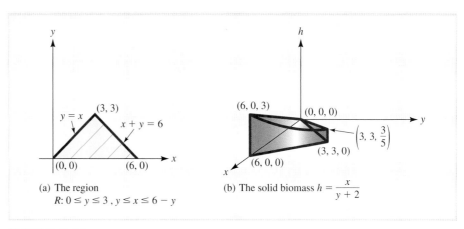

(a) The region
$R: 0 \leq y \leq 3, y \leq x \leq 6 - y$

(b) The solid biomass $h = \dfrac{x}{y + 2}$

FIGURE 7.33 The volume of a biomass.

Note that this region is bounded by the x axis ($y = 0$) and the lines $x = y$ and $x + y = 6$, so it can be described as

$$R: 0 \leq y \leq 3, y \leq x \leq 6 - y.$$

Therefore, the volume of the biomass is given by

$$V = \int_0^3 \int_y^{6-y} \frac{x}{y+2} \, dx \, dy$$

$$= \int_0^3 \frac{1}{y+2} \left(\frac{x^2}{2} \right) \Big|_y^{6-y} \, dy$$

$$= \int_0^3 \frac{1}{2(y+2)} [(6-y)^2 - y^2] \, dy = \int_0^3 \frac{1}{2(y+2)} (36 - 12y) \, dy$$

$$= \int_0^3 \frac{1}{(y+2)} (18 - 6y) \, dy$$

$$= \int_0^3 \left[-6 + \left(\frac{30}{y+2} \right) \right] dy \qquad \text{divide } y+2 \text{ into } -6y + 18$$

$$= [-6y + 30 \ln|y+2|] \Big|_0^3$$

$$= [-6(3) + 30 \ln 5] - [-6(0) + 30 \ln 2]$$

$$\approx 9.489$$

The volume of the biomass is approximately 9.5 cm^3.

Average Value of a Function f(x, y)

In Section 5.4, you saw that the average value of a function $f(x)$ over an interval $a \le x \le b$ is given by the integral formula

$$AV = \frac{1}{b-a} \int_a^b f(x) \, dx.$$

That is, to find the average value of a function of one variable over an interval, integrate the function over the interval and divide by the length of the interval. The two-variable procedure is similar. In particular, to find the average value of a function of two variables $f(x, y)$ over a region R, integrate the function over R and divide by the area of R.

> **Average Value Formula** ■ The average value of the function $f(x, y)$ over the region R is given by the formula
>
> $$AV = \frac{1}{\text{area of } R} \iint_R f(x, y) \, dA.$$

EXAMPLE 7.6.8

In a certain factory, output is given by the Cobb-Douglas production function

$$Q(K, L) = 50K^{3/5}L^{2/5},$$

where K is the capital investment in units of $1000 and L is the size of the labour force measured in worker-hours. Suppose that monthly capital investment varies between $10 000 and $12 000, while monthly use of labour varies between 2800 and 3200 worker-hours. Find the average monthly output for the factory.

Solution

It is reasonable to estimate the average monthly output by the average value of $Q(K, L)$ over the rectangular region R: $10 \leq K \leq 12$, $2800 \leq L \leq 3200$. The region has area

$$A = \text{area of } R = (12 - 10) \times (3200 - 2800)$$
$$= 800$$

so the average output is

$$AV = \frac{1}{800} \iint_R 50K^{3/5}L^{2/5}\, dA$$

$$= \frac{1}{800} \int_{2800}^{3200} \left(\int_{10}^{12} 50K^{3/5}L^{2/5}\, dK \right) dL$$

$$= \frac{1}{800} \int_{2800}^{3200} 50L^{2/5}\left(\frac{5}{8}K^{8/5} \right)\bigg|_{K=10}^{K=12} dL$$

$$= \frac{1}{800}(50)\left(\frac{5}{8}\right)\int_{2800}^{3200} L^{2/5}(12^{8/5} - 10^{8/5})\, dL$$

$$= \frac{1}{800}(50)\left(\frac{5}{8}\right)(12^{8/5} - 10^{8/5})\left(\frac{5}{7}L^{7/5}\right)\bigg|_{L=2800}^{L=3200}$$

$$= \frac{1}{800}(50)\left(\frac{5}{8}\right)\left(\frac{5}{7}\right)(12^{8/5} - 10^{8/5})(3200^{7/5} - 2800^{7/5})$$

$$\approx 5181.23$$

Thus, the average monthly output is approximately 5181 units.

EXERCISES ▪ 7.6

Evaluate the double integrals in Exercises 1 through 18.

1. $\int_0^1 \int_1^2 x^2 y \, dx \, dy$

2. $\int_1^2 \int_0^1 x^2 y \, dy \, dx$

3. $\int_0^{\ln 2} \int_{-1}^0 2xe^y \, dx \, dy$

4. $\int_2^3 \int_{-1}^1 (x + 2y) \, dy \, dx$

5. $\int_1^3 \int_0^1 \frac{2xy}{x^2 + 1} \, dx \, dy$

6. $\int_0^1 \int_0^1 x^2 e^{xy} \, dy \, dx$

7. $\int_0^4 \int_{-1}^1 x^2 y \, dy \, dx$

8. $\int_0^1 \int_1^5 y\sqrt{1 - y^2} \, dx \, dy$

9. $\int_2^3 \int_1^2 \frac{x + y}{xy} \, dy \, dx$

10. $\int_1^2 \int_2^3 \left(\frac{y}{x} + \frac{x}{y} \right) dy \, dx$

11. $\int_0^4 \int_0^{\sqrt{x}} x^2 y \, dx \, dy$

12. $\int_0^1 \int_1^5 xy\sqrt{1 - y^2} \, dx \, dy$

13. $\int_0^1 \int_{y-1}^{1-y} (2x + y)\, dx\, dy$

14. $\int_0^1 \int_{x^2}^{x} 2xy\, dy\, dx$

15. $\int_0^1 \int_0^4 \sqrt{xy}\, dy\, dx$

16. $\int_0^1 \int_x^{2x} e^{y-x}\, dy\, dx$

17. $\int_1^e \int_0^{\ln x} xy\, dy\, dx$

18. $\int_1^3 \int_{y^2/4}^{\sqrt{10-y^2}} xy\, dx\, dy$

In Exercises 19 through 24, use inequalities to describe R in terms of its vertical and horizontal cross-sections.

19. R is the region bounded by $y = x^2$ and $y = 3x$.

20. R is the region bounded by $y = \sqrt{x}$ and $y = x^2$.

21. R is the rectangle with vertices $(-1, 1)$, $(2, 1)$, $(2, 2)$, and $(-1, 2)$.

22. R is the triangle with vertices $(1, 0)$, $(1, 1)$, and $(2, 0)$.

23. R is the region bounded by $y = \ln x$, $y = 0$, and $x = e$.

24. R is the region bounded by $y = e^x$, $y = 2$, and $x = 0$.

In Exercises 25 through 36, evaluate the given double integral for the specified region R.

25. $\iint_R 3xy^2\, dA$, where R is the rectangle bounded by the lines $x = -1$, $x = 2$, $y = -2$, and $y = 0$

26. $\iint_R (x + 2y)\, dA$, where R is the triangle with vertices $(0, 0)$, $(1, 0)$, and $(0, 2)$

27. $\iint_R xe^y\, dA$, where R is the triangle with vertices $(0, 0)$, $(1, 0)$, and $(1, 1)$

28. $\iint_R 48\, xy\, dA$, where R is the region bounded by $y = x^3$ and $y = \sqrt{x}$

29. $\iint_R (2y - x)\, dA$, where R is the region bounded by $y = x^2$ and $y = 2x$

30. $\iint_R 12x\, dA$, where R is the region bounded by $y = x^2$ and $y = 6 - x$

31. $\iint_R (2x + 1)\, dA$, where R is the triangle with vertices $(-1, 0)$, $(1, 0)$, and $(0, 1)$

32. $\iint_R 2x\, dA$, where R is the region bounded by $y = \dfrac{1}{x^2}$, $y = x$, and $x = 2$

33. $\iint_R \dfrac{1}{y^2 + 1}\, dA$, where R is the triangle bounded by the lines $y = \dfrac{1}{2}x$, $y = -x$, and $y = 2$

34. $\iint_R e^{y^3}\, dA$, where R is the region bounded by $y = \sqrt{x}$, $y = 1$, and $x = 0$

35. $\iint_R 12x^2 e^{y^2}\, dA$, where R is the region in the first quadrant bounded by $y = x^3$ and $y = x$

36. $\iint_R y\, dA$, where R is the region bounded by $y = \ln x$, $y = 0$, and $x = e$

In Exercises 37 through 44, sketch the region of integration for the given integral and set up an equivalent integral with the order of integration reversed.

37. $\int_0^2 \int_0^{4-x^2} f(x, y)\, dy\, dx$

38. $\int_0^1 \int_0^{2y} f(x, y)\, dx\, dy$

39. $\int_0^1 \int_{x^3}^{\sqrt{x}} f(x, y)\, dy\, dx$

40. $\int_0^4 \int_{y/2}^{\sqrt{y}} f(x, y)\, dx\, dy$

41. $\int_1^{e^2} \int_{\ln x}^{2} f(x, y)\, dy\, dx$

42. $\int_0^{\ln 3} \int_{e^x}^{3} f(x, y)\, dy\, dx$

43. $\int_{-1}^1 \int_{x^2+1}^{2} f(x, y)\, dy\, dx$

44. $\int_{-1}^1 \int_{-\sqrt{y+1}}^{\sqrt{y+1}} f(x, y)\, dx\, dy$

In Exercises 45 through 54, use a double integral to find the area of R.

45. *R* is the triangle with vertices $(-4, 0)$, $(2, 0)$, and $(2, 6)$.

46. *R* is the triangle with vertices $(0, -1)$, $(-2, 1)$, and $(2, 1)$.

47. *R* is the region bounded by $y = \dfrac{1}{2}x^2$ and $y = 2x$.

48. *R* is the region bounded by $y = \sqrt{x}$ and $y = x^2$.

49. *R* is the region bounded by $y = x^2 - 4x + 3$ and the *x* axis.

50. *R* is the region bounded by $y = x^2 + 6x + 5$ and the *x* axis.

51. *R* is the region bounded by $y = \ln x$, $y = 0$, and $x = e$.

52. *R* is the region bounded by $y = x$, $y = \ln x$, $y = 0$, and $y = 1$.

53. *R* is the region in the first quadrant bounded by $y = 4 - x^2$, $y = 3x$, and $y = 0$.

54. *R* is the region bounded by $y = \dfrac{16}{x}$, $y = x$, and $x = 8$.

In Exercises 55 through 64, find the volume of the solid under the surface $z = f(x, y)$ and over the given region R.

55. $f(x, y) = 6 - 2x - 2y$; $R: 0 \le x \le 1, 0 \le y \le 2$

56. $f(x, y) = 9 - x^2 - y^2$; $R: -1 \le x \le 1, -2 \le y \le 2$

57. $f(x, y) = \dfrac{1}{xy}$; $R: 1 \le x \le 2, 1 \le y \le 3$

58. $f(x, y) = e^{x+y}$; $R: 0 \le x \le 1, 0 \le y \le \ln 2$

59. $f(x, y) = xe^{-y}$; $R: 0 \le x \le 1, 0 \le y \le 2$

60. $f(x, y) = (1 - x)(4 - y)$; $R: 0 \le x \le 1, 0 \le y \le 4$

61. $f(x, y) = 2x + y$; *R* is bounded by $y = x$, $y = 2 - x$, and $y = 0$.

62. $f(x, y) = e^{y^2}$; *R* is bounded by $x = 2y$, $x = 0$, and $y = 1$.

63. $f(x, y) = x + 1$; *R* is bounded by $y = 8 - x^2$ and $y = x^2$.

64. $f(x, y) = 4xe^y$; *R* is bounded by $y = 2x$, $y = 2$, and $x = 0$.

In Exercises 65 through 72, find the average value of the function $f(x, y)$ over the given region R.

65. $f(x, y) = xy(x - 2y)$; $R: -2 \le x \le 3, -1 \le y \le 2$

66. $f(x, y) = \dfrac{y}{x} + \dfrac{x}{y}$; $R: 1 \le x \le 4, 1 \le y \le 3$

67. $f(x, y) = xye^{x^2y}$; $R: 0 \le x \le 1, 0 \le y \le 2$

68. $f(x, y) = \dfrac{\ln x}{xy}$; $R: 1 \le x \le 2, 2 \le y \le 3$

69. $f(x, y) = 6xy$; *R* is the triangle with vertices $(0, 0)$, $(0, 1)$, $(3, 1)$.

70. $f(x, y) = e^{x^2}$; *R* is the triangle with vertices $(0, 0)$, $(1, 0)$, $(1, 1)$.

71. $f(x, y) = x$; *R* is the region bounded by $y = 4 - x^2$ and $y = 0$.

72. $f(x, y) = e^x y^{-1/2}$; *R* is the region bounded by $x = \sqrt{y}$, $y = 0$, and $x = 1$.

In Exercises 73 through 76, evaluate the double integral over the specified region R. Choose the order of integration carefully.

73. $\displaystyle\iint_R \dfrac{\ln (xy)}{y} \, dA$; $R: 1 \le x \le 3, 2 \le y \le 5$

74. $\displaystyle\iint_R ye^{xy} \, dA$; $R: -1 \le x \le 1, 1 \le y \le 2$

75. $\displaystyle\iint_R x^3 e^{x^2y} \, dA$; $R: 0 \le x \le 1, 0 \le y \le 1$

76. $\displaystyle\iint_R e^{x^3} \, dA$; $R: \sqrt{y} \le x \le 1, 0 \le y \le 1$

77. PRODUCTION   At a certain factory, output *Q* is related to inputs *x* and *y* by the expression
$$Q(x, y) = 2x^3 + 3x^2y + y^3.$$
If $0 \le x \le 5$ and $0 \le y \le 7$, what is the average output of the factory?

78. PRODUCTION   A bicycle dealer has found that if commuter bicycles are sold for *x* dollars each and the price of a city transit token is *y* cents, then approximately
$$Q(x, y) = 200 - 24\sqrt{x} + 4(0.1y + 3)^{3/2}$$
bicycles will be sold in that time period. If the price of a bicycle varies between \$289 and \$324 during this time, and the price of a city transit token varies between \$2.96 and \$3.05, approximately how many bicycles will be sold in this time period on average?

79. AVERAGE PROFIT   A manufacturer estimates that when *x* units of a particular commodity are sold domestically and y units are sold to foreign markets, the profit is given by

$$P(x, y) = (x - 30)(70 + 5x - 4y)$$
$$+ (y - 40)(80 - 6x + 7y)$$

hundred dollars. If monthly domestic sales vary between 100 units and 125 units and foreign sales between 70 units and 89 units, what is the average monthly profit?

80. AVERAGE RESPONSE TO STIMULI   In a psychological experiment, x units of stimulus A and y units of stimulus B are applied to a subject, and the subject's performance on a certain task is then measured by the function

$$P(x, y) = 10 + xye^{1 - x^2 - y^2}.$$

Suppose x varies between 0 and 1 while y varies between 0 and 3. What is the subject's average response to the stimuli?

81. AVERAGE ELEVATION   A map of a small regional park is a rectangular grid, bounded by the lines $x = 0$, $x = 4$, $y = 0$, and $y = 3$, where units are in kilometres. It is found that the elevation above sea level at each point (x, y) in the park is given by

$$E(x, y) = 90(2x + y^2)$$

metres. Find the average elevation in the park.

82. PROPERTY VALUE   A community is laid out as a rectangular grid in relation to two main streets that intersect at the city centre. Each point in the community has coordinates (x, y) in this grid, for $-10 \leq x \leq 10$, $-8 \leq y \leq 8$, with x and y measured in kilometres. Suppose the value of the land located at the point (x, y) is V thousand dollars, where

$$V(x, y) = (250 + 17x)e^{-0.01x - 0.05y}.$$

Estimate the value of the block of land occupying the rectangular region $1 \leq x \leq 3$, $0 \leq y \leq 2$.

83. PROPERTY VALUE   Repeat Exercise 82 for

$$V(x, y) = (300 + x + y)e^{-0.01x}$$

and the region $-1 \leq x \leq 1$, $-1 \leq y \leq 1$.

84. PROPERTY VALUE   Repeat Exercise 82 for

$$V(x, y) = 400xe^{-y}$$

and the region R: $0 \leq y \leq x$, $0 \leq x \leq 1$.

POPULATION FROM POPULATION DENSITY

Suppose a rectangular coordinate grid is superimposed on a map and that the population density at the point (x, y) is $f(x, y)$ people per square kilometre. Then the

total population P inside the region R is given by the double integral

$$P = \iint_R f(x, y) \, dA$$

Use this formula in Exercises 85 and 86.

85. POPULATION   The population density is $f(x, y) = 2500e^{-0.01x - 0.02y}$ people per square kilometre at each point (x, y) within the triangular region R with vertices $(-5, -2)$, $(0, 3)$, and $(5, -2)$.
  a. Find the total population in the region R.
  b. Find the average population density of R.
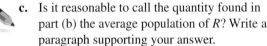
  c. Is it reasonable to call the quantity found in part (b) the average population of R? Write a paragraph supporting your answer.

86. POPULATION   The population density is $f(x, y) = 1000y^2e^{-0.01x}$ people per square kilometre at each point (x, y) within the region R bounded by the parabola $x = y^2$ and the vertical line $x = 4$.
  a. Find the total population in the region R.
  b. Find the average population density of R.

87. A* AVERAGE SURFACE AREA OF THE HUMAN BODY   Recall from Exercise 39, Section 7.1, that the surface area of a person's body may be estimated by the empirical formula

$$S(W, H) = 0.0072W^{0.425}H^{0.725},$$

where W is the person's weight in kilograms, H is the person's height in centimetres, and the surface area S is measured in square metres.
  a. Find the average value of the function $S(W, H)$ over the region

  $$R: 3.2 \leq W \leq 80, 0 \leq H \leq 180.$$

  b. A child weighs 3.2 kg and is 38 cm tall at birth. As an adult, this person has a stable weight of 80 kg and a height of 180 cm. Can the average value in part (a) be interpreted as the average lifetime surface area of this person's body? Explain.

88. ARCHITECTURAL DESIGN   A building is to have a curved roof above a rectangular base. In relation to a rectangular grid, the base is the rectangular region $-30 \leq x \leq 30$, $-20 \leq y \leq 20$, where x and y are measured in metres. The height of the roof above each point (x, y) in the base is given by

$$h(x, y) = 12 - 0.003x^2 - 0.005y^2.$$

  a. Find the volume of the building.
  b. Find the average height of the roof.

89. A* CONSTRUCTION A storage bin is to be constructed in the shape of the solid bounded above by the surface

$$z = 20 - x^2 - y^2,$$

below by the xy plane, and on the sides by the plane $y = 0$ and the parabolic cylinder $y = 4 - x^2$, where x, y, and z are in metres. Find the volume of the bin.

90. A* CONSTRUCTION A fancy jewellery box has the shape of the solid bounded above by the plane

$$3x + 4y + 2z = 12,$$

below by the xy plane, and on the sides by the planes $x = 0$ and $y = 0$, where x, y, and z are in centimetres. Find the volume of the box.

91. A* EXPOSURE TO DISEASE The likelihood that a person with a contagious disease will infect others in a social situation may be assumed to be a function $f(s)$ of the distance s between individuals. Suppose contagious individuals are uniformly distributed throughout a rectangular region R in the xy plane. Then the likelihood of infection for someone at the origin $(0, 0)$ is proportional to the exposure index E, given by the double integral

$$E = \iint_R f(s)\, dA,$$

where $s = \sqrt{x^2 + y^2}$ is the distance between $(0, 0)$ and (x, y). Find E for the case where

$$f(s) = 1 - \frac{s^2}{9}$$

and R is the square region defined by

$$R: -2 \le x \le 2, \; -2 \le y \le 2.$$

In Exercises 92 through 94, use double integration to find the required quantity. In some cases, you may need to use software to solve the equations.

92. A* Find the area of the region bounded above by the curve (ellipse) $4x^2 + 3y^2 = 7$ and below by the parabola $y = x^2$.

93. A* Find the volume of the solid bounded above by the graph of $f(x, y) = x^2 e^{-xy}$ and below by the rectangular region $R: 0 \le x \le 2, 0 \le y \le 3$.

94. A* Find the average value of $f(x, y) = xy \ln\left(\dfrac{y}{x}\right)$ over the rectangular region bounded by the lines $x = 1$, $x = 2$, $y = 1$, and $y = 3$.

Concept Summary Chapter 7

Partial Derivatives of $z = f(x, y)$

$$f_x = \frac{\partial z}{\partial x}, \; f_y = \frac{\partial z}{\partial y}$$

Second-Order Partial Derivatives

$$f_{xx} = \frac{\partial^2 z}{\partial x^2}, \quad f_{yy} = \frac{\partial^2 z}{\partial y^2}, \quad f_{yx} = \frac{\partial^2 z}{\partial x \partial y}, \quad f_{xy} = \frac{\partial^2 z}{\partial y \partial x}$$

Chain Rule for Partial Derivatives

$$\frac{\partial z}{\partial t} = \frac{\partial z}{\partial x}\frac{dx}{dt} + \frac{\partial z}{\partial y}\frac{dy}{dt}$$

Second Partials Test
Critical point: $f_x = 0, f_y = 0$
At critical point (a, b):
$D(a, b) = f_{xx} f_{yy} - (f_{xy})^2$
If $D < 0$, f has a saddle point at (a, b).
If $D > 0$ and $f_{xx} < 0$, f has a relative maximum at (a, b).
If $D > 0$ and $f_{xx} > 0$, f has a relative minimum at (a, b).
If $D = 0$, the test is inconclusive.

Least Squares Line $y = mx + b$

$$m = \frac{n\Sigma xy - \Sigma x \Sigma y}{n\Sigma x^2 - (\Sigma x)^2} \quad \text{and} \quad b = \frac{\Sigma x^2 \Sigma y - \Sigma x \Sigma xy}{n\Sigma x^2 - (\Sigma x)^2}$$

Lagrange Multipliers
To find extreme values of $f(x, y)$ subject to $g(x, y) = k$, solve the equations

$$f_x = \lambda g_x, \qquad f_y = \lambda g_y, \qquad g = k.$$

The Lagrange multiplier is $\lambda = \dfrac{dM}{dk}$, where M is the optimal value of $f(x, y)$ subject to $g(x, y) = k$.

Double Integrals
Over the region $R: a \le x \le b, g_1(x) \le y \le g_2(x)$:

$$\iint_R f(x, y)\, dA = \int_a^b \left[\int_{g_1(x)}^{g_2(x)} f(x, y)\, dy \right] dx.$$

Over the region R: $c \leq y \leq d$, $h_1(y) \leq x \leq h_2(y)$:

$$\iint_R f(x, y)\, dA = \int_c^d \left[\int_{h_1(y)}^{h_2(y)} f(x, y)\, dx \right] dy.$$

Area, Volume, Average

$$\text{Area of } R = \iint_R 1\, dA$$

$$\text{Volume} = \iint_R f(x, y)\, dA, \text{ where } f(x, y) \geq 0.$$

$$\text{Average} = \frac{1}{\text{area of } R} \iint_R f(x, y)\, dA$$

Checkup for Chapter 7

1. In each case, first describe the domain of the given function and then find the partial derivatives f_x, f_y, f_{xx}, and f_{yx}.

   a. $f(x, y) = x^3 + 2xy^2 - 3y^4$

   b. $f(x, y) = \dfrac{2x + y}{x - y}$

   c. $f(x, y) = e^{2x-y} + \ln(y^2 - 2x)$

2. Describe the level curves of each function.

   a. $f(x, y) = x^2 + y^2$

   b. $f(x, y) = x + y^2$

3. In each case, find all critical points of the given function $f(x, y)$ and use the second partials test to classify each as a relative maximum, a relative minimum, or a saddle point.

   a. $f(x, y) = 4x^3 + y^3 - 6x^2 - 6y^2 + 5$

   b. $f(x, y) = x^2 - 4xy + 3y^2 + 2x - 4y$

   c. $f(x, y) = xy - \dfrac{1}{y} - \dfrac{1}{x}$

4. Use the method of Lagrange multipliers to find each constrained extremum.

   a. The smallest value of $f(x, y) = x^2 + y^2$ subject to $x + 2y = 4$

   b. The largest and the smallest values of the function $f(x, y) = xy^2$ subject to $2x^2 + y^2 = 6$

5. Evaluate each double integral.

   a. $\displaystyle\int_{-1}^3 \int_0^2 x^3 y\, dx\, dy$

   b. $\displaystyle\int_0^2 \int_{-1}^1 x^2 e^{xy}\, dx\, dy$

   c. $\displaystyle\int_1^2 \int_1^y \frac{y}{x}\, dx\, dy$

   d. $\displaystyle\int_0^2 \int_0^{2-x} xe^{-y}\, dy\, dx$

6. **MARGINAL PRODUCTIVITY** A company will produce $Q(K, L) = 120K^{3/4}L^{1/4}$ hundred units of a particular commodity when the capital expenditure is K thousand dollars and the size of the workforce is L worker-hours. Find the marginal productivity of capital Q_K and the marginal productivity of labour Q_L when the capital expenditure is \$1 296 000 and the labour level is 20 736 worker-hours.

7. **UTILITY** Everett has just received \$500 as a birthday gift and has decided to spend it on DVDs and video games. He has determined that the utility (satisfaction) derived from the purchase of x DVDs and y video games is

   $$U(x, y) = \ln(x^2 \sqrt{y}).$$

   If each DVD costs \$20 and each video game costs \$50, how many DVDs and video games should he purchase in order to maximize utility?

8. **MEDICINE** A certain disease can be treated by administering at least 70 units of drug C, but that level of medication sometimes results in serious side effects. Looking for a safer approach, a physician decides instead to use drugs A and B, which result in no side effects as long as their combined dosage is less than 60 units. Moreover, she determines that when x units of drug A and y units of drug B are administered to a patient, the effect is equivalent to administering E units of drug C, where

   $$E = 0.05(xy - 2x^2 - y^2 + 95x + 20y).$$

   What dosages of drugs A and B will maximize the equivalent level E of drug C? If the physician administers the optimal dosages of drugs A and B, will the combined effect be enough to help the patient without risking side effects?

9. **AVERAGE TEMPERATURE** A flat metal plate lying in the xy plane is heated so that the temperature at the point (x, y) is T degrees Celsius, where

   $$T(x, y) = 10ye^{-xy}.$$

   Find the average temperature over the rectangular portion of the plate for which $0 \leq x \leq 2$ and $0 \leq y \leq 1$.

10. LEAST-SQUARES APPROXIMATION OF PROFIT DATA A company's annual profit (in millions of dollars) for the first 5 years of operation is shown in the following table:

Year	1	2	3	4	5
Profit (millions of dollars)	1.03	1.52	2.03	2.41	2.84

a. Plot these data on a graph.

b. Find the equation of the least-squares line through the data.

c. Use the least-squares line to predict the company's sixth-year profit.

Review Exercises

In Exercises 1 through 10, find the partial derivatives f_x and f_y.

1. $f(x, y) = 2x^3y + 3xy^2 + \dfrac{y}{x}$

2. $f(x, y) = (xy^2 + 1)^5$

3. $f(x, y) = \sqrt{x}(x - y^2)$

4. $f(x, y) = xe^{-y} + ye^{-x}$

5. $f(x, y) = \sqrt{\dfrac{x}{y}} + \sqrt{\dfrac{y}{x}}$

6. $f(x, y) = x \ln (x^2 - y) + y \ln (y - 2x)$

7. $f(x, y) = \dfrac{x^3 - xy}{x + y}$

8. $f(x, y) = xye^{xy}$

9. $f(x, y) = \dfrac{x^2 - y^2}{2x + y}$

10. $f(x, y) = \ln \left(\dfrac{xy}{x + 3y} \right)$

For each function in Exercises 11 through 14, compute the second-order partial derivatives $f_{xx}, f_{yy}, f_{xy},$ and f_{yx}.

11. $f(x, y) = e^{x^2 + y^2}$

12. $f(x, y) = x^2 + y^3 - 2xy^2$

13. $f(x, y) = x \ln y$

14. $f(x, y) = (5x^2 - y)^3$

15. For each function, sketch the indicated level curves.

   a. $f(x, y) = x^2 - y; f = 2, f = -2$

   b. $f(x, y) = 6x + 2y; f = 0, f = 1, f = 2$

16. For each function, find the slope of the indicated level curve at the specified value of x.

   a. $f(x, y) = x^2 - y^3; f = 2; x = 1$

   b. $f(x, y) = xe^y; f = 2; x = 2$

In Exercises 17 through 24, find all critical points of the given function and use the second partials test to classify each as a relative maximum, a relative minimum, or a saddle point.

17. $f(x, y) = (x + y)(2x + y - 6)$

18. $f(x, y) = (x + y + 3)^2 - (x + 2y - 5)^2$

19. $f(x, y) = x^3 + y^3 + 3x^2 - 3y^2$

20. $f(x, y) = x^3 + y^3 + 3x^2 - 18y^2 + 81y + 5$

21. $f(x, y) = x^2 + y^3 + 6xy - 7x - 6y$

22. $f(x, y) = 3x^2y + 2xy^2 - 10xy - 8y^2$

23. $f(x, y) = xe^{2x^2 + 5xy + 2y^2}$

24. $f(x, y) = 8xy - x^4 - y^4$

In Exercises 25 through 28, use the method of Lagrange multipliers to find the maximum and minimum values of the given function $f(x, y)$ subject to the indicated constraint.

25. $f(x, y) = x^2 + 2y^2 + 2x + 3; x^2 + y^2 = 4$

26. $f(x, y) = 4x + y; \dfrac{1}{x} + \dfrac{1}{y} = 1$

27. $f(x, y) = x + 2y; 4x^2 + y^2 = 68$

28. $f(x, y) = x^2 + y^3; x^2 + 3y = 4$

29. At a certain factory, the daily output is approximately $40K^{1/3}L^{1/2}$ units, where K denotes the capital investment measured in units of $1000 and L denotes the size of the labour force measured in worker-hours. Suppose that the current capital investment is $125 000 and that 900 worker-hours of labour are used each day. Use marginal analysis to estimate the effect that an additional capital investment of $1000 will have on the daily output if the size of the labour force is not changed.

30. In economics, the marginal product of labour is the rate at which output Q changes with respect to labour L for a fixed level of capital investment K. An economic law states that under certain circumstances, the marginal product of labour increases as the level of capital investment increases. Translate this law into a mathematical statement involving a second-order partial derivative.

31. MARGINAL ANALYSIS Using x skilled workers and y unskilled workers, a manufacturer can produce $Q(x, y) = 60x^{1/3}y^{2/3}$ units per day. Currently the manufacturer employs 10 skilled workers and 40 unskilled workers and is planning to hire 1 additional skilled worker. Use calculus to estimate the corresponding change that the manufacturer should make in the level of unskilled labour so that the total output will remain the same.

32. Use the method of Lagrange multipliers to prove that of all isosceles triangles with a given perimeter, the equilateral triangle has the largest area.

33. Use the method of Lagrange multipliers to prove that of all rectangles with a given perimeter, the square has the largest area.

34. ALLOCATION OF FUNDS A manufacturer is planning to sell a new product at a price of $350 per unit and estimates that if x thousand dollars are spent on development and y thousand dollars are spent on promotion, consumers will buy approximately

$$\frac{250y}{y + 2} + \frac{100x}{x + 5}$$

units of the product. If manufacturing costs for the product are $150 per unit, how much should the manufacturer spend on development and how much on promotion to generate the largest possible profit? Assume unlimited funds are available.

35. ALLOCATION OF FUNDS Suppose the manufacturer in Exercise 34 has only $11 000 to spend on the development and promotion of the new product. How should this money be allocated to generate the largest possible profit?

36. ALLOCATION OF FUNDS Suppose the manufacturer in Exercise 35 decides to spend $12 000 instead of $11 000 on the development and promotion of the new product. Use the Lagrange multiplier λ to estimate how this change will affect the maximum possible profit.

37. Let $f(x, y) = \dfrac{12}{x} + \dfrac{18}{y} + xy$, where $x > 0$, $y > 0$.

How do you know that f must have a minimum in the region $x > 0$, $y > 0$? Find the minimum.

In Exercises 38 through 45, evaluate the double integral. You may need to exchange the order of integration.

38. $\displaystyle\int_0^1 \int_{-2}^0 (2x + 3y)\, dy\, dx$

39. $\displaystyle\int_0^1 \int_0^2 e^{-x-y}\, dy\, dx$

40. $\displaystyle\int_0^1 \int_0^2 x\sqrt{1 - y}\, dx\, dy$

41. $\displaystyle\int_0^1 \int_{-1}^1 xe^{2y}\, dy\, dx$

42. $\displaystyle\int_0^2 \int_{-1}^1 \frac{6xy^2}{x^2 + 1}\, dy\, dx$

43. $\displaystyle\int_1^e \int_1^e \ln(xy)\, dy\, dx$

44. $\displaystyle\int_0^1 \int_0^{1-x} x(y - 1)^2\, dy\, dx$

45. $\displaystyle\int_1^2 \int_0^x e^{y/x}\, dy\, dx$

In Exercises 46 and 47, evaluate the given double integral for the specified region R.

46. $\displaystyle\iint_R 6x^2y\, dA$, where R is the rectangle with vertices $(-1, 0)$, $(2, 0)$, $(2, 3)$, and $(-1, 3)$

47. $\displaystyle\iint_R (x + 2y)\, dA$, where R is the rectangular region bounded by $x = 0$, $x = 1$, $y = -2$, and $y = 2$

48. Find the volume under the surface $z = 2xy$ and above the rectangle with vertices $(0, 0)$, $(2, 0)$, $(0, 3)$, and $(2, 3)$.

49. Find the volume under the surface $z = xe^{-y}$ and above the rectangle bounded by the lines $x = 1$, $x = 2$, $y = 2$, and $y = 3$.

50. Find the average value of $f(x, y) = xy^2$ over the rectangular region with vertices $(-1, 3)$, $(-1, 5)$, $(2, 3)$, and $(2, 5)$.

51. Find three positive numbers x, y, and z so that $x + y + z = 20$ and the product $P = xyz$ is a maximum. [*Hint:* Use the fact that $z = 20 - x - y$ to express P as a function of only two variables.]

52. Find three positive numbers x, y, and z so that $2x + 3y + z = 60$ and the sum $S = x^2 + y^2 + z^2$ is minimized. [See the hint to Exercise 51.]

53. Find the shortest distance from the origin to the surface $y^2 - z^2 = 10$. [*Hint:* Express the distance $\sqrt{x^2 + y^2 + z^2}$ from the origin to a point (x, y, z) on the surface in terms of the two variables x and y, and minimize the *square* of the resulting distance function.]

54. Plot the points $(1, 1)$, $(1, 2)$, $(3, 2)$, and $(4, 3)$ and use partial derivatives to find the corresponding least-squares line.

55. SALES The marketing manager for a certain company has compiled the following data relating monthly advertising expenditure and monthly sales (both measured in units of $1000):

Advertising	3	4	7	9	10
Sales	78	86	138	145	156

a. Plot these data on a graph.

b. Find the least-squares line.

c. Use the least-squares line to predict monthly sales if the monthly advertising expenditure is $5000.

56. UTILITY Suppose the utility derived by a consumer from x units of one commodity and y units of a second commodity is given by the utility function $U(x, y) = x^3 y^2$. The consumer currently owns $x = 5$ units of the first commodity and $y = 4$ units of the second. Use calculus to estimate how many units of the second commodity the consumer could substitute for 1 unit of the first commodity without affecting total utility.

57. CONSUMER DEMAND A paint company makes two brands of latex paint. Sales figures indicate that if the first brand is sold for x dollars per litre and the second for y dollars per litre, the demand for the first brand will be Q litres per month, where

$$Q(x, y) = 200 + 10x^2 - 20y.$$

It is estimated that t months from now the price of the first brand will be $x(t) = 18 + 0.02t$ dollars

per litre and the price of the second will be $y(t) = 21 + 0.4\sqrt{t}$ dollars per litre. At what rate will the demand for the first brand of paint be changing with respect to time 9 months from now?

58. COOLING AN ANIMAL'S BODY The difference between an animal's surface temperature and that of the surrounding air causes a transfer of energy by convection. The coefficient of convection h is given by

$$h = \frac{kV^{1/3}}{D^{2/3}},$$

where V (cm/s) is wind velocity, D (cm) is the diameter of the animal's body, and k is a constant.

a. Find the partial derivatives h_V and h_D. Interpret these derivatives as rates.

b. Compute the ratio $\dfrac{h_V}{h_D}$.

59. CONSUMER DEMAND Suppose that when apples sell for x cents per kilogram and bakers earn y dollars per hour, the price of apple pies at a certain supermarket chain is

$$p(x, y) = \frac{1}{4} x^{1/3} y^{1/2}$$

dollars per pie. Suppose also that t months from now, the price of apples will be

$$x = 129 - \sqrt{8t}$$

cents per kilogram and bakers' wages will be

$$y = 15.60 + 0.2t$$

dollars per hour. If the supermarket chain can sell

$$Q = \frac{4184}{p}$$ pies per week when the price is p dollars

per pie, at what rate will the weekly demand Q for pies be changing with respect to time 2 months from now?

60. Arnold, the heat-seeking mussel, is the world's smartest mollusk. Arnold likes to stay warm, and by using the crustacean coordinate system he learned from a passing crab, he has determined that at each nearby point (x, y) on the ocean floor, the temperature (°C) is

$$T(x, y) = 2x^2 - xy + y^2 - 2y + 1.$$

Arnold's world consists of a rectangular portion of ocean bed with vertices $(-1, -1)$, $(-1, 1)$, $(1, -1)$, and $(1, 1)$, and since it is very hard for him to move, he plans to stay where he is as long as the average temperature of this region is at least 5°C. Does Arnold move or stay put?

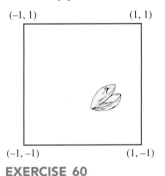

(−1, 1)   (1, 1)

(−1, −1)   (1, −1)

EXERCISE 60

61. AIR POLLUTION At a certain factory, the amount of air pollution generated each day is measured by the function $Q(E, T) = 125E^{2/3}T^{1/2}$, where E is the number of employees and T (°C) is the average temperature during the workday. Currently, there are $E = 151$ employees and the average temperature is $T = 10$°C. If the average daily temperature is falling at a rate of 0.21°C per day and the number of employees is increasing at a rate of 2 per month, use calculus to estimate the corresponding effect on the rate of pollution. Express your answer in units per day. You may assume that there are 22 workdays per month.

62. POPULATION A demographer sets up a grid to describe location within a suburb of a major metropolitan area. In relation to this grid, the population density at each point (x, y) is given by
$$f(x, y) = 1 + 3y^2$$
hundred people per square kilometre, where x and y are in kilometres. A subdivision occupies the region R bounded by the curve $y^2 = 4 - x$ and the y axis $(x = 0)$. What is the total population within the project region R?

63. POLLUTION Two sources of air pollution affect the health of a certain community. Health officials have determined that at a point located r kilometres from source A and s kilometres from source B, there will be
$$N(r, s) = 40e^{-r/2}e^{-s/3}$$
units of pollution. A subdivision lies in a region R for which
$$2 \le r \le 3 \quad \text{and} \quad 1 \le s \le 2.$$
What is the total pollution within the region R?

64. A* NUCLEAR WASTE DISPOSAL Nuclear waste is often disposed of by sealing it into

containers that are then dumped into the ocean. It is important to dump the containers into water shallow enough to ensure that they do not break when they hit bottom. Suppose as the container falls through the water, it experiences a drag force that is proportional to the container's velocity. Then it can be shown that the depth s (in metres) of a container of weight W newtons (N) at time t seconds is given by the formula
$$s(W, t) = \left(\frac{W - B}{k}\right)t + \frac{W(W - B)}{k^2 g}[e^{-(kgt/W)} - 1],$$
where B is a (constant) buoyancy force, k is the drag constant, and $g = 9.8$ m/s^2 is the constant acceleration due to gravity.

a. Find $\frac{\partial s}{\partial W}$ and $\frac{\partial s}{\partial t}$. Interpret these derivatives as rates. Do you think it is possible for either partial derivative to ever be zero?

b. For a fixed weight, the speed of the container is $\frac{\partial s}{\partial t}$. Suppose the container will break when its speed on impact with the ocean floor exceeds 10 m/s. If $B = 1983$ N and $k = 0.597$ kg/s, what is the maximum depth for safely dumping a container of weight $W = 2417$ N?

c. Research the topic of nuclear waste disposal and write a paragraph on whether you think it is best done on land or at sea.

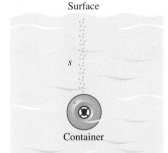

EXERCISE 64

65. A* PRODUCTION For the production function given by $Q = x^a y^b$, where $a > 0$ and $b > 0$, show that
$$x\frac{\partial Q}{\partial x} + y\frac{\partial Q}{\partial y} = (a + b)Q.$$
In particular, if $b = 1 - a$ with $0 < a < 1$, then
$$x\frac{\partial Q}{\partial x} + y\frac{\partial Q}{\partial y} = Q.$$

THINK ABOUT IT

MODELLING POPULATION DIFFUSION

(Photo: © NPS photo by Bryan Harry)

In 1905, five muskrats were accidentally released near Prague in the current Czech Republic. Subsequent to 1905, the range of the muskrat population expanded and the front (the outer limit of the population) moved as indicated in Figure 7.34. In the figure, the closed curves labelled with dates are equipopulation contours, that is, curves of constant, minimally detectable populations of muskrats. For instance, the curve labelled 1920 indicates that the muskrat population had expanded from Prague to the gates of Vienna in the 15 years after their release. A population dispersion such as this can be studied using mathematical models based on *partial differential equations,* that is, equations involving functions of two or more variables and their partial derivatives. We will examine such a model and then return to our illustration to see how well the model can be used to describe the dispersion of the muskrats.

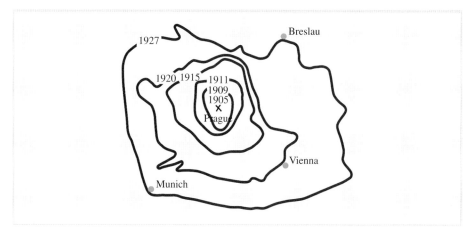

FIGURE 7.34   Equipopulation curves for a muskrat population in Europe.

SOURCE: Leah Edelstein-Keshet, *Mathematical Models in Biology,* Boston: McGraw-Hill, 1988, p. 439.

The model we will discuss is based on the **diffusion equation,** an extremely versatile partial differential equation with important applications in the physical and life sciences as well as economics. *Diffusion* is the name used for the process by which particles spread out as they continually collide and randomly change direction after being inserted at a source. Suppose the particles can move in only one spatial direction (say, along a thin rod or tube). Then $C(x, t)$, the concentration of particles at time t located x units from the source (the point of insertion), satisfies the one-dimensional diffusion equation

$$\frac{\partial C}{\partial t} = \alpha \frac{\partial^2 C}{\partial x^2},$$

where α is a positive constant called the *diffusion coefficient*. Similarly, the two-dimensional diffusion equation

$$\frac{\partial C}{\partial t} = \alpha\left(\frac{\partial^2 C}{\partial x^2} + \frac{\partial^2 C}{\partial y^2}\right)$$

is used to model the dispersion of particles moving randomly in a plane, where $C(x, y, t)$ is the concentration of particles at the point (x, y) at time t.

Mathematical biologists have adapted the diffusion equation to model the spread of living organisms, including both plants and animals. We will examine such a model, due to J. G. Skellam. First, suppose that at a particular time $(t = 0)$, an organism is introduced at a point (called a source) where it had previously not been present. Skellam assumed that the population of the organism disperses from the source in two ways:

a. By growing exponentially at a continuous reproduction rate r

b. By moving randomly in an xy coordinate plane, with the source at the origin

Based on these assumptions, he then modelled the dispersion of the population by the modified two-dimensional diffusion equation

$$(1) \quad \frac{\partial N}{\partial t} = D\underbrace{\left(\frac{\partial^2 N}{\partial x^2} + \frac{\partial^2 N}{\partial y^2}\right)}_{\substack{\text{expansion} \\ \text{by random} \\ \text{movement}}} + \underbrace{rN}_{\substack{\text{growth by} \\ \text{exponential} \\ \text{reproduction}}}$$

where $N(x, y, t)$ is the population density at the point (x, y) at time t and D is a positive constant, called the *dispersion coefficient,* that is analogous to the diffusion coefficient.

It can be shown that one solution of Skellam's equation is

$$(2) \quad N(x, y, t) = \frac{M}{4\pi Dt}e^{rt-(x^2+y^2)/(4Dt)}$$

where M is the number of individuals initially introduced at the source (see Question 5). The *asymptotic rate of population expansion V* is the distance between locations with equal population densities in successive years, and Skellam's model can be used to show that

$$(3) \quad V = \sqrt{4rD}$$

(see Question 4). Likewise, the *intrinsic rate of growth r* can be estimated using data on the growth of existing populations, and the dispersion coefficient D can be estimated using the formula

$$(4) \quad D \approx \frac{2A^2(t)}{\pi t}$$

where $A(t)$ is the average distance organisms have travelled at time t.

Skellam's model has been used to study the spread of a variety of organisms, including oak trees, cereal leaf beetles, and cabbage butterflies. As an illustration of how the model can be applied, we return to the population of Central European muskrats introduced in the opening paragraph and Figure 7.34. Population studies indicate that r, the intrinsic rate of muskrat population increase, was no greater than 1.1 per year, and that D, the dispersion coefficient, was no greater than 230 km^2/year. Consequently, the solution to Skellam's model stated in Equation (2) predicts that the distribution of muskrats, under the best circumstances for the species, is given by

$$(5) \quad N(x, y, t) = \frac{5}{4\pi(230)t} e^{1.1t - (x^2 + y^2)/(920t)}$$

where (x, y) is the point x kilometres east and y kilometres north from the release point near Prague and t is the time in years (after 1905). Equation (3) predicts that the maximum rate of population expansion is

$$V = \sqrt{4rD} = \sqrt{4(1.1)(230)} \approx 31.8 \text{ km/year},$$

which is a little greater than the observed rate of 25.4 km/year.

The derivation of the diffusion equation may be found in many differential equations texts. It is important to emphasize that Skellam's modified diffusion equation given in Equation (1) has solutions other than Equation (2). In general, solving partial differential equations is very difficult, and often the best that can be done is to focus on finding solutions with certain specified forms. Such solutions can then be used to analyze practical situations, as we did with the muskrat problem.

Questions

1. Verify that $C(x, t) = \dfrac{M}{2\sqrt{\pi Dt}} e^{-(x^2/4Dt)}$ satisfies the diffusion equation

$$\frac{\partial C}{\partial t} = \alpha \frac{\partial^2 C}{\partial x^2}.$$

   Do this by calculating the partial derivatives and inserting them into the equation.

2. What relationship must hold between the coefficients a and b for $C(x, t) = e^{ax + bt}$ to be a solution of the diffusion equation

$$\frac{\partial C}{\partial t} = \alpha \frac{\partial^2 C}{\partial x^2}?$$

3. Suppose that a population of organisms spreads out along a one-dimensional line according to the partial differential equation

$$\frac{\partial N}{\partial t} = D \frac{\partial^2 N}{\partial x^2} + rN.$$

   Show that the function $N(x, t) = \dfrac{M}{2\sqrt{\pi Dt}} e^{rt - (x^2/4Dt)}$ is a solution to this partial differential equation, where M is the initial population of organisms located at the point $x = 0$ when $t = 0$.

4. Show that on the contours of equal population density (that is, the curves of the form $N(x, t) = A$, where A is a constant) the ratio $\dfrac{x}{t}$ is

$$\frac{x}{t} = \pm \left[4rD - \frac{2D}{t} \ln t - \frac{4D}{t} \ln \left(\sqrt{2\pi D} \, \frac{A}{M} \right) \right]^{1/2}.$$

Using this formula, it can be shown that the ratio $\dfrac{x}{t}$ can be approximated by $\dfrac{x}{t} \approx \pm 2\sqrt{rD}$, which is a formula for the rate of population expansion.

5. Verify that $N(x, y, t) = \dfrac{M}{4\pi Dt} e^{rt - (x^2 + y^2)/(4Dt)}$ is a solution of the partial differential equation

$$\frac{\partial N}{\partial t} = D\left(\frac{\partial^2 N}{\partial x^2} + \frac{\partial^2 N}{\partial y^2}\right) + rN.$$

6. Use Equation (5), which we obtained using Skellam's model, to find the population density for muskrats in 1925 at the location 50 km north and 50 km west of the release point near Prague.

7. Use Skellam's model to construct a function that estimates the population density of the small cabbage white butterfly if the largest diffusion coefficient observed is 129 km^2/year and the largest intrinsic rate of increase observed is 31.5/year. What is the predicted maximum rate of population expansion? How does this compare with the largest observed rate of population expansion of 170 km/year?

8. In this question, you are asked to explore an alternative approach to analyzing the muskrat problem using Skellam's model. Recall that the intrinsic rate of growth of the muskrat population was $r = 1.1$ and that the maximum rate of dispersion was observed to be $V = 25.4$.
   a. Use these values for r and V in Equation (3) to estimate the dispersion coefficient D.
   b. By substituting $r = 1.1$ into Equation (2) along with the value for D you obtained in part (a), find the population density in 1925 for the muskrats at the location 50 km north and 50 km west of the release point (source) near Prague. Compare your answer with the answer to Question 6.
   c. Use Equation (4) to estimate the average distance A of the muskrat population in 1925 from its source near Prague.

References

D. A. Andow, P. M. Kareiva, Simon A. Levin, and Akira Okubo, "Spread of Invading Organisms," *Landscape Ecology,* Vol. 4, nos. 2/3, 1990, pp. 177–188.
Leah Edelstein-Keshet, *Mathematical Models in Biology,* Boston: McGraw-Hill, 1988.
J. G. Skellam, "The Formulation and Interpretation of Mathematical Models of Diffusionary Processes in Population Biology," in *The Mathematical Theory of the Dynamics of Biological Populations,* edited by M. S. Bartlett and R. W. Hiorns, New York: Academic Press, 1973, pp. 63–85.
J. G. Skellam, "Random Dispersal in Theoretical Populations," *Biometrika,* Vol. 28, 1951, pp. 196–218.

 Practise and learn online with Connect.

The rate of removal of a pollutant in a lake may depend on the amount of pollutant already in the lake, the rate at which more pollutant is added to the lake, or the rate at which cleaner water runs into the lake. The concentration at any time can be determined using differential equations.
See Exercises 38 and 39 of Section 8.3 for details.
(Photos: David Frazier/CORBIS and © Digital Vision)

DIFFERENTIAL EQUATIONS

CHAPTER OUTLINE

LEARNING OBJECTIVES

After completing this chapter, you should be able to

L01 Evaluate general and particular solutions to separable differential equations. Formulate differential equations corresponding to given applications.

L02 Solve first-order linear differential equations.

L03 Solve application problems using differential equations.

L04 Approximate solutions to differential equations with slope fields and Euler's method.

L05 Formulate and solve difference equations.

L01

Evaluate general and particular solutions to separable differential equations. Formulate differential equations corresponding to given applications.

Introduction to Differential Equations

In Section 1.4, we introduced mathematical modelling as a dynamic process involving three stages:

1. A real-world problem is given a mathematical formulation, called a *mathematical model.* This often involves making simplifying assumptions about the problem to make it more accessible.
2. The mathematical model is analyzed or solved by using tools from such areas as algebra, calculus, or statistics, among others, or by numerical methods involving graphing software.
3. The solution of the mathematical model is interpreted in terms of the original real-world problem. This often leads to adjustments in the assumptions of the model.

The process may then be repeated, each time with a more refined model, until a satisfactory understanding of the real-world problem is attained.

We have used calculus to analyze mathematical models throughout this text, and many of these models have involved rates of change. Sometimes the mathematical formulation of a problem involves an equation in which a quantity and the rate of change of that quantity are related by an equation. Since rates of change are expressed in terms of derivatives or differentials, such an equation is appropriately called a **differential equation.** For example,

$$\frac{dy}{dx} = 3x^2 + 5, \qquad \frac{dP}{dt} = kP, \qquad \text{and} \qquad \left(\frac{dy}{dx}\right)^2 + 3\frac{dy}{dx} + 2y = e^x$$

are all differential equations. Differential equations are among the most useful tools for modelling continuous phenomena, including important situations that occur in business and economics as well as the social and life sciences. In this section, we introduce techniques for solving basic differential equations and examine a few practical applications.

A complete characterization of all possible solutions of a given differential equation is called the **general solution** of the equation. A differential equation coupled with a side condition is referred to as an **initial value problem,** and a solution that satisfies both the differential equation and the side condition is called a **particular solution** of the initial value problem. In Examples 8.1.1 through 8.1.3, we illustrate this terminology while examining differential equations of the form

$$\frac{dy}{dx} = g(x),$$

which can be solved by finding the indefinite integral of the function $g(x)$.

EXAMPLE 8.1.1

Find the general solution of the differential equation

$$\frac{dy}{dx} = x^2 + 3x$$

and the particular solution that satisfies $y = 2$ when $x = 1$.

Solution

Integrating results in

$$y = \int \left(\frac{dy}{dx}\right) dx = \int (x^2 + 3x)\, dx$$

$$= \frac{1}{3}x^3 + \frac{3}{2}x^2 + C$$

This is the general solution since all solutions can be expressed in this form. For the particular solution, substitute $x = 1$ and $y = 2$ into the general solution:

$$2 = \frac{1}{3}(1)^3 + \frac{3}{2}(1)^2 + C$$

$$C = 2 - \frac{1}{3} - \frac{3}{2} = \frac{1}{6}$$

Thus, the required particular solution is $y = \frac{1}{3}x^3 + \frac{3}{2}x^2 + \frac{1}{6}$.

EXAMPLE 8.1.2

The resale value of a certain industrial machine decreases over a 10-year period at a rate that depends on the age of the machine. When the machine is x years old, the rate at which its value is changing is $220(x - 10)$ dollars per year. Express the value of the machine as a function of its age and initial value. If the machine was originally worth \$12 000, how much will it be worth when it is 10 years old?

Solution

Let $V(x)$ denote the value of the machine when it is x years old. The derivative $\dfrac{dV}{dx}$ is equal to the rate $220(x - 10)$ at which the value of the machine is changing. Hence, this problem can be modelled using the differential equation

$$\frac{dV}{dx} = 220(x - 10) = 220x - 2200.$$

To find V, solve this differential equation by integration:

$$V(x) = \int (220x - 2200)\, dx = 110x^2 - 2200x + C.$$

Notice that C is equal to $V(0)$, the initial value of the machine. A more descriptive symbol for this constant is V_0. Using this notation, write the general solution as

$$V(x) = 110x^2 - 2200x + V_0.$$

If $V_0 = 12\,000$, the corresponding particular solution is

$$V(x) = 110x^2 - 2200x + 12\,000.$$

Thus, when the machine is $x = 10$ years old, its value is

$$V(10) = 110(10)^2 - 2200(10) + 12\,000 = 1000$$

or \$1000.

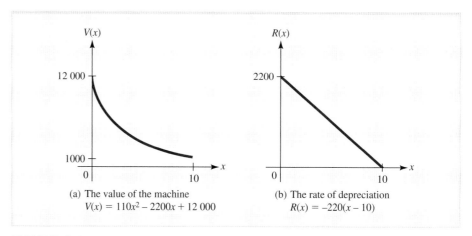

FIGURE 8.1   The value of the machine and its rate of depreciation.

The negative of the rate of change of resale value of the machine,

$$R = \frac{dV}{dx} = -220(x - 10),$$

is called the **rate of depreciation.** Graphs of the resale value V of the machine and the rate of depreciation R are shown in Figure 8.1.

EXAMPLE 8.1.3

An oil well that has just been opened is expected to yield 300 barrels of crude oil per month and, at that rate, is expected to run dry in 3 years. It is estimated that t months from now, the price of crude oil will be $p(t) = 28 + 0.3\sqrt{t}$ dollars per barrel. If the oil is sold as soon as it is extracted from the ground, what is the total revenue generated by the well during its operation?

Solution

Let $R(t)$ be the revenue generated during the first t months after the well is opened, so that $R(0) = 0$. To construct the relevant differential equation, use the rate relationship

$$\begin{pmatrix} \text{Rate of change of revenue with} \\ \text{respect to time (dollars/month)} \end{pmatrix} = \begin{pmatrix} \text{dollars per} \\ \text{barrel} \end{pmatrix} \begin{pmatrix} \text{barrels per} \\ \text{month} \end{pmatrix}$$

It follows that we can model this problem using the differential equation

$$\underbrace{\frac{dR}{dt}}_{\text{\$/month}} = \underbrace{(28 + 0.3\sqrt{t})}_{\text{\$/barrel}} \underbrace{(300)}_{\text{barrels/month}}$$

and it follows that

$$\frac{dR}{dt} = 300(28 + 0.3\sqrt{t})$$

$$= 8400 + 90t^{1/2}$$

Integrating gives the general solution of this differential equation as

$$R(t) = \int (8400 + 90t^{1/2})\, dt = 8400t + 90\left(\frac{t^{3/2}}{3/2}\right) + C$$

$$= 8400t + 60t^{3/2} + C$$

Since $R(0) = 0$,

$$R(0) = 0 = 8400(0) + 60(0)^{3/2} + C,$$

so $C = 0$, and the appropriate particular solution is

$$R(t) = 8400t + 60t^{3/2}.$$

Therefore, since the well will run dry in 36 months, the total revenue generated by the well during its operation is

$$R(36) = 8400(36) + 60(36)^{3/2}$$

$$= 315\ 360$$

or \$315 360.

Separable Equations

Many useful differential equations can be formally rewritten so that all the terms containing the independent variable appear on one side of the equation and all the terms containing the dependent variable appear on the other. Differential equations with this special property are said to be **separable** and can be solved by the following procedure involving two integrations.

> **Separable Differential Equations** ■ A differential equation that can be written in the form
>
> $$\frac{dy}{dx} = \frac{h(x)}{g(y)}$$
>
> or
>
> $$g(y)\, dy = h(x)\, dx$$
>
> is said to be **separable.** The general solution of such an equation can be obtained by separating the variables and integrating both sides as follows:
>
> $$\int g(y)\, dy = \int h(x)\, dx.$$

EXAMPLE 8.1.4

Find the general solution of the differential equation $\dfrac{dy}{dx} = \dfrac{2x}{y^2}$.

Solution

To separate the variables, pretend that the derivative $\dfrac{dy}{dx}$ is actually a quotient and write

$$y^2 dy = 2x\, dx.$$

Now integrate both sides of this equation to get

$$\int y^2 dy = \int 2x \, dx$$

$$\frac{1}{3}y^3 + C_1 = x^2 + C_2$$

where C_1 and C_2 are arbitrary constants. Solving for y results in

$$\frac{1}{3}y^3 = x^2 + (C_2 - C_1) = x^2 + C_3 \qquad C_3 = C_2 - C_1$$

$$y^3 = 3x^2 + 3C_3 = 3x^2 + C \qquad C = 3C_3$$

$$y = (3x^2 + C)^{1/3}$$

Note that we changed the constant of integration to the simplest form on each line. The constant $3C_3$ is unknown, so it is easier to refer to this constant as C. In Example 8.1.5 and those that follow, only one C is used when integrating both sides of the equation. We do not need to write C_1 and C_2; instead we write C on the right side to represent $C_2 - C_1$.

EXAMPLE 8.1.5

Amanda starts to ride her bicycle along a straight road, for a distance x, where $x > 0$, so that at t seconds, her velocity is given by the differential equation

$$\frac{dx}{dt} = \frac{te^{0.1t^2}}{x}$$

metres per second. If Amanda is at $x = 2$ m when $t = 0$ s, where is she when $t = 3$ s?

Solution

Separating the variables in the given differential equation and integrating gives

$$\int x \, dx = \int te^{0.1t^2} \, dt.$$

Using substitution with $u = 0.1t^2$ and $du = 0.2t \, dt$ results in

$$\frac{x^2}{2} = \frac{e^{0.1t^2}}{0.2} + C_1$$

$$x^2 = \frac{e^{0.1t^2}}{0.1} + 2C_1$$

$$x^2 = 10e^{0.1t^2} + C \qquad \text{substitute } C \text{ for } 2C_1$$

$$x = \sqrt{10e^{0.1t^2} + C} \qquad \text{take the positive root because } x > 0$$

Since $x = 2$ when $t = 0$,

$$2 = \sqrt{10 + C}$$

$$4 = 10 + C$$

$$C = -6$$

so that

$$x = \sqrt{10e^{0.1t^2} - 6}.$$

In particular, when $t = 3$,

$$x = \sqrt{10e^{0.1(9)} - 6}$$
$$x = 4.312$$

which means that Amanda has moved 4.31 m after 3 seconds.

Exponential Growth and Decay

In Section 4.1, we obtained the formulas $Q = Q_0 e^{kt}$ for exponential growth and $Q = Q_0 e^{-kt}$ for exponential decay, and then in Section 4.3 we showed that the rate of change of a quantity undergoing either kind of exponential change (growth or decay) is proportional to its size; that is, $\dfrac{dQ}{dt} = mQ$ for some constant m. Example 8.1.6 shows that the converse of this result is also true.

EXAMPLE 8.1.6

Show that a quantity Q that changes at a rate proportional to its size satisfies $Q(t) = Q_0 e^{mt}$, where $Q_0 = Q(0)$.

Solution
Since Q changes at a rate proportional to its size,

$$\frac{dQ}{dt} = mQ$$

for constant m. Separate the variables and integrate to get

$$\int \frac{1}{Q}\, dQ = \int m\, dt$$
$$\ln Q = mt + C_1$$

Then take exponentials on both sides:

$$Q(t) = e^{\ln Q} = e^{mt + C_1} = e^{C_1} e^{mt}$$
$$= Ce^{mt} \qquad \text{where } C = e^{C_1}$$

Since $Q(0) = Q_0$, it follows that

$$Q_0 = Q(0) = Ce^0 = C \qquad \text{since } e^0 = 1.$$

Therefore, $Q(t) = Q_0 e^{mt}$, as required.

Learning Models

In Chapter 4, we referred to the graphs of functions of the form $Q(t) = B - Ae^{-kt}$ as *learning curves* because functions of this form often describe the relationship between the efficiency with which an individual performs a given task and the amount of training or experience the learner has had. In general, any quantity that grows at a rate proportional to the difference between its size and a fixed upper limit is represented by a function of this form. Here is an example involving such a learning model.

EXAMPLE 8.1.7

The rate at which people hear about a new increase in postal rates is proportional to the number of people in the country who have not yet heard about it. Express the number of people who have already heard about the increase as a function of time.

Solution

Let $Q(t)$ denote the number of people who have already heard about the rate increase at time t and let B be the total population of the country. Then

$$\text{Rate at which people hear about the increase} = \frac{dQ}{dt}$$

and

$$\text{Number of people who have not yet heard about the increase} = B - Q.$$

Because the rate at which people hear about the increase is proportional to the number of people who have not yet heard about the increase, it follows that

$$\frac{dQ}{dt} = k(B - Q),$$

where k is the constant of proportionality. Notice that the constant k must be positive because $\frac{dQ}{dt} > 0$ (since Q is an increasing function of t) and $B - Q > 0$ (since $B > Q$).

Separate the variables by writing

$$\frac{1}{B - Q} dQ = k \, dt$$

and integrate to get

$$\int \frac{1}{B - Q} dQ = \int k \, dt$$
$$-\ln|B - Q| = kt + C$$

(Be sure you understand where the minus sign came from.) This time you can drop the absolute value sign immediately since $B - Q > 0$. Hence,

$$-\ln(B - Q) = kt + C$$
$$\ln(B - Q) = -kt - C$$
$$B - Q = e^{-kt-C} = e^{-kt}e^{-C}$$
$$Q = B - e^{-C}e^{-kt}$$

Denote the constant e^{-C} by A and use functional notation to obtain

$$Q(t) = B - Ae^{-kt}$$

which is precisely the general equation of a learning curve. For reference, the graph of Q is sketched in Figure 8.2.

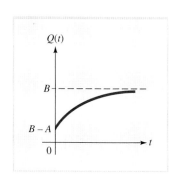

FIGURE 8.2   A learning curve: $Q(t) = B - Ae^{-kt}$.

Logistic Growth   The relative rate of growth of a quantity $Q(t)$ is given by the ratio $\dfrac{Q'(t)}{Q(t)}$. In exponential growth and decay, where $Q'(t) = kQ(t)$, the relative rate of growth is constant; namely,

$$\frac{Q'(t)}{Q(t)} = k.$$

A population with no restrictions on environment or resources can often be modelled as having exponential growth. However, when such restrictions exist, it is reasonable to assume there is a largest population B supportable by the environment (called the **carrying capacity**) and that at any time t, the relative rate of population growth is proportional to the *potential* population $B - Q(t)$, that is,

$$\frac{Q'(t)}{Q(t)} = k(B - Q(t)),$$

where k is a positive constant. This rate relationship leads to the differential equation

$$\frac{dQ}{dt} = kQ(B - Q).$$

This is called the **logistic equation,** and the corresponding model of restricted population growth is known as the **logistic model.**

The logistic differential equation is separable, but before solving it analytically, let us see what we can deduce using qualitative methods. First, note that the logistic equation can be written as $\dfrac{dQ}{dt} = R(Q)$, where the population rate $R(Q) = kQ(B - Q)$ is a quadratic function in Q. Since $k > 0$, the graph of $R(Q)$ is a downward-opening parabola (Figure 8.3), and since $R(0) = R(B) = 0$, the graph intersects the Q axis at 0 and at B. The high point (vertex) of the graph occurs at $Q = \dfrac{B}{2}$.

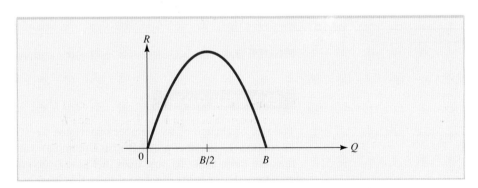

FIGURE 8.3   The graph of the population rate function $R(Q) = kQ(B - Q)$.

If the initial population $Q(0)$ is 0, then $Q(t) = 0$ for all $t > 0$, and, likewise, if the initial population is $Q(0) = B$, then the population stays at the level of the carrying capacity for all $t > 0$. The two constant solutions $Q = 0$ and $Q = B$, called *equilibrium solutions* of the logistic equation, are shown in Figure 8.4a.

If the initial population $Q(0)$ satisfies $0 < Q(0) < B$, then

$$R(Q) = kQ(B - Q) > 0,$$

since k, Q, and $B - Q$ are all positive. Since the rate of growth is positive, the population $Q(t)$ itself is increasing. As $Q(t)$ approaches B, the population growth rate $\dfrac{dQ}{dt}$ approaches 0, which suggests that the graph of the population function $Q(t)$ flattens out, approaching the carrying capacity asymptotically. If the initial population satisfies $Q(0) > B$, then

$$R(Q) = kQ(B - Q) < 0,$$

since k and Q are positive and $B - Q$ is negative. In this case, the population *decreases* from its initial value, once again approaching the carrying capacity asymptotically as Q approaches B. Some nonconstant solutions of the logistic equation are shown in Figure 8.4b.

In addition to its role in modelling restricted population growth, the logistic equation can also be used to describe economic behaviour as well as interactions within a social group, such as the spread of rumours (see Exercise 40). In Example 8.1.8, we use a logistic model to describe the spread of an epidemic.

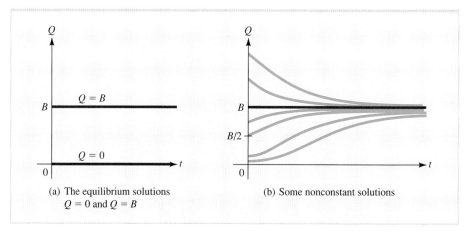

(a) The equilibrium solutions
$Q = 0$ and $Q = B$

(b) Some nonconstant solutions

FIGURE 8.4 Typical solutions of the logistic equation $\dfrac{dQ}{dt} = kQ(B - Q)$.

EXAMPLE 8.1.8

The rate of change of the number of infested trees in a forest that has a pine beetle infestation (as in Alberta and British Columbia) can be considered as jointly proportional to the number of pine trees that have been infested and the number of healthy pine trees. Express the number of trees that have been infested as a function of time.

Solution

Let $Q(t)$ denote the number of susceptible trees that have been infested at time t, and let B denote the total number of susceptible trees. Then

$$\text{Rate susceptible trees are being infested at time } t = \frac{dQ}{dt}$$

and

$$\text{Number of susceptible trees not yet infested} = B - Q.$$

Just-In-Time

To say that the quantity z is jointly proportional to the quantities x and y means that there is a constant k such that $z = kxy$.

"Jointly proportional" means "proportional to the product," so the differential equation that describes the spread of the pine beetle is

$$\frac{dQ}{dt} = kQ(B - Q),$$

where $k > 0$ is the constant of proportionality. (Do you see why k must be positive?) This is a separable differential equation with solution

$$\int \frac{dQ}{Q(B - Q)} = \int k\, dt,$$

where the integration on the left can be performed by a method called "partial fractions" as follows. Let

$$\frac{1}{Q(B - Q)} = \frac{M}{Q} + \frac{N}{B - Q}.$$

Multiply both sides by $Q(B - Q)$ to obtain

$$1 = M(B - Q) + NQ$$
$$1 = MB - MQ + NQ$$
$$1 = MB + (N - M)Q$$
$$1 + 0Q = MB + (N - M)Q$$

Compare both sides of the equation. Because B is a constant, $1 = MB$, or $M = 1/B$. Because Q is a variable, $0 = N - M$, or $N = M = 1/B$. This results in

$$\frac{1}{Q(B - Q)} = \frac{1}{BQ} + \frac{1}{B(B - Q)}.$$

(Verify that this is true by putting the fractions on the right over a common denominator and simplifying.) Substitute these fractions back into the integral and solve.

$$\int \frac{dQ}{Q(B - Q)} = \int k\, dt$$

$$\int \frac{1}{BQ} + \frac{1}{B(B - Q)}\, dQ = \int k\, dt$$

$$\frac{1}{B}\ln|Q| - \frac{1}{B}\ln|B - Q| = kt + C$$

$$\frac{1}{B}\ln\left|\frac{Q}{B - Q}\right| = kt + C$$

Since $Q > 0$ and $B > Q$, remove the absolute value bars from the solution and write

$$\frac{1}{B}\ln\left(\frac{Q}{B - Q}\right) = kt + C$$

$$\ln\left(\frac{Q}{B - Q}\right) = Bkt + BC$$

$$\frac{Q}{B - Q} = e^{Bkt + BC} = e^{Bkt}e^{BC}$$

$$= A_1 e^{Bkt} \quad \text{where } A_1 = e^{BC}$$

Multiply both sides of the last equation by $B - Q$ and solve for Q to get

$$Q = (B - Q)A_1 \, e^{Bkt} = (BA_1 - QA_1)e^{Bkt}$$

$$Qe^{-Bkt} = BA_1 - QA_1 \qquad \text{multiply both sides by } e^{-Bkt}$$

$$Q(A_1 + e^{-Bkt}) = BA_1 \qquad \text{add } QA_1 \text{ to both sides}$$

$$Q = \frac{BA_1}{A_1 + e^{-Bkt}}$$

$$= \frac{B}{1 + \dfrac{1}{A_1}e^{-Bkt}} \qquad \text{divide all terms on the right by } A_1$$

Finally, setting $A = \dfrac{1}{A_1}$, Q has the logistic form

$$Q(t) = \frac{B}{1 + Ae^{-Bkt}}.$$

The graph of $Q(t)$, the number of infested trees, is shown in Figure 8.5. Note that this graph has the characteristic S shape of a logistic curve with carrying capacity B. At the beginning of the epidemic, only $Q(0) = \dfrac{B}{1 + A}$ trees are infested. The number of infested trees increases rapidly at first and is spreading most rapidly at the inflection point on the graph. It is not difficult to show that this occurs when half the susceptible population is infested (see Exercise 50). The infection rate then begins to decrease as the total number of infested trees asymptotically approaches the level B where all susceptible trees would be infested.

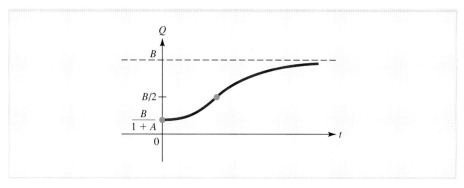

FIGURE 8.5   The logistic curve $y = \dfrac{B}{1 + Ae^{-Bkt}}$ showing the number of trees $Q(t)$ infested with the pine beetle at time t.

Why the Method of Separation of Variables Works

Consider the separable differential equation

$$\frac{dy}{dx} = \frac{h(x)}{g(y)}$$

or, equivalently,

$$g(y)\frac{dy}{dx} - h(x) = 0.$$

The left-hand side of this equation can be rewritten in terms of the antiderivatives of g and h. In particular, if G is an antiderivative of g and H an antiderivative of h, it follows from the chain rule that

$$\frac{d}{dx}[G(y) - H(x)] = G'(y)\frac{dy}{dx} - H'(x) = g(y)\frac{dy}{dx} - h(x).$$

Hence, the differential equation $g(y)\dfrac{dy}{dx} - h(x) = 0$ says that

$$\frac{d}{dx}[G(y) - H(x)] = 0.$$

But constants are the only functions whose derivatives are identically zero, and so

$$G(y) - H(x) = C$$

for some constant C. That is,

$$G(y) = H(x) + C$$

or, equivalently,

$$\int g(y)\, dy = \int h(x)\, dx + C,$$

and the proof is complete.

EXERCISES ■ 8.1

In Exercises 1 through 20, find the general solution of the given differential equation. You will need integration by parts in Exercises 17 through 20.

1. $\dfrac{dy}{dx} = 3x^2 + 5x - 6$

2. $\dfrac{dp}{dt} = \sqrt{t} + e^{-t}$

3. $\dfrac{dy}{dx} = 3y$

4. $\dfrac{dy}{dx} = y^2$

5. $\dfrac{dy}{dx} = e^y$

6. $\dfrac{dy}{dx} = e^{x+y}$

7. $\dfrac{dy}{dx} = \dfrac{x}{y}$

8. $\dfrac{dy}{dx} = \dfrac{y}{x}$

9. $\dfrac{dy}{dx} = \sqrt{xy}$

10. $\dfrac{dy}{dx} = \dfrac{y^2 + 4}{xy}$

11. $\dfrac{dy}{dx} = \dfrac{y}{x - 1}$

12. $\dfrac{dy}{dx} = e^y\sqrt{x + 1}$

13. $\dfrac{dy}{dx} = \dfrac{y + 3}{(2x - 5)^6}$

14. $\dfrac{dy}{dx} = e^{-2y}(x - 2)^9$

15. $\dfrac{dx}{dt} = \dfrac{xt}{2t + 1}$

16. $\dfrac{dy}{dt} = \dfrac{te^y}{2t - 1}$

17. $\dfrac{dy}{dx} = xe^{x-y}$

18. $\dfrac{dw}{ds} = \dfrac{se^{2w}}{w}$

19. $\dfrac{dy}{dt} = y \ln \sqrt{t}$

20. $\dfrac{dx}{dt} = \dfrac{\ln t}{\ln x}$

In Exercises 21 through 28, find the particular solution of the differential equation satisfying the indicated condition.

21. $\dfrac{dy}{dx} = e^{5x};\ y = 1$ when $x = 0$

22. $\dfrac{dy}{dx} = 5x^4 - 3x^2 - 2;\ y = 4$ when $x = 1$

23. $\dfrac{dy}{dx} = \dfrac{x}{y^2};\ y = 3$ when $x = 2$

24. $\dfrac{dy}{dx} = 4x^3 y^2;\ y = 2$ when $x = 1$

25. $\dfrac{dy}{dx} = y^2 \sqrt{4 - x};\ y = 2$ when $x = 4$

26. $\dfrac{dy}{dx} = xe^{y - x^2};\ y = 0$ when $x = 1$

27. $\dfrac{dy}{dt} = \dfrac{y + 1}{t(y - 1)};\ y = 2$ when $t = 1$

$\left[\textit{Hint:}\ \dfrac{y - 1}{y + 1} = 1 - \dfrac{2}{y + 1} \right]$

28. $\dfrac{dx}{dt} = xt\sqrt{t + 1};\ x = 1$ when $t = 0$

In Exercises 29 through 40, write a differential equation describing the given situation. Define all variables you introduce. (Do not try to solve the differential equation at this time.)

29. INVESTMENT GROWTH   An investment grows at a rate equal to 7% of its size.

30. MARGINAL COST   A manufacturer's marginal cost is $60 per unit.

31. GROWTH OF BACTERIA   The number of bacteria in a culture grows at a rate that is proportional to the number present.

32. CONCENTRATION OF DRUGS   The rate at which the concentration of a drug in the blood stream decreases is proportional to the concentration.

33. POPULATION GROWTH   The population of a certain town increases at a constant rate of 500 people per year.

34. RADIOACTIVE DECAY   A sample of radium decays at a rate that is proportional to its size.

35. TEMPERATURE CHANGE   The rate at which the temperature of an object changes is proportional to the difference between its own temperature and the temperature of the surrounding medium.

36. DISSOLUTION OF SUGAR   After being placed in a container of water, sugar dissolves at a rate proportional to the amount of undissolved sugar remaining in the container.

37. RECALL FROM MEMORY   When a person is asked to recall a set of facts, the rate at which the facts are recalled is proportional to the number of relevant facts in the person's memory that have not yet been recalled.

38. MARKET SHARE   The rate at which a new product is replacing an old, obsolete product is jointly proportional to the market share of the new product and that of the old product. [*Hint:* Express market share as a percentage.]

39. CORRUPTION IN GOVERNMENT   The rate at which people are implicated in a government scandal is jointly proportional to the number of people already implicated and the number of people involved who have not yet been implicated.

40. THE SPREAD OF A RUMOUR   The rate at which a rumour spreads through a community is jointly proportional to the number of people in the community who have heard the rumour and the number who have not.

41. Verify that the function $y = Ce^{kx}$ is a solution of the differential equation $\dfrac{dy}{dx} = ky$.

42. Verify that the function $Q = B - Ce^{-kt}$ is a solution of the differential equation $\dfrac{dQ}{dt} = k(B - Q)$.

43. Verify that $y = C_1 e^x + C_2 xe^x$ is a solution of the differential equation $\dfrac{d^2 y}{dx^2} - 2\dfrac{dy}{dx} + y = 0$.

44. Verify that the function $y = \dfrac{1}{20}x^4 - \dfrac{C_1}{x} + C_2$ is a solution of the differential equation $x\dfrac{d^2 y}{dx^2} + 2\dfrac{dy}{dx} = x^3$.

45. OIL PRODUCTION A certain oil well that yields 400 barrels of crude oil per month will run dry in 2 years. The price of crude oil is currently $130 per barrel and is expected to rise at a constant rate of 4 cents per barrel per month. If the oil is sold as soon as it is extracted from the ground, what will the total future revenue from the well be?

46. AGRICULTURAL PRODUCTION The Mitscherlich model for agricultural production specifies that the size $Q(t)$ of a crop changes at a rate proportional to the difference between the maximum possible crop size B and Q; that is,

$$\frac{dQ}{dt} = k(B - Q).$$

a. Solve this equation for $Q(t)$. Express your answer in terms of k and the initial crop size $Q_0 = Q(0)$.

b. A particular crop has a maximum size of 200 kg per hectare. At the start of the growing season ($t = 0$), the crop size is 50 kg, and 1 month later, it is 60 kg. How large is the crop 3 months later ($t = 3$)?

 c. Note that this model is similar to the learning model discussed in Example 8.1.7. Is this just a coincidence or is there some meaningful analogy linking the two situations? Explain.

47. DILUTION A tank holds 200 L of brine containing 2 g of salt per litre. Clear water flows into the tank at a rate of 5 L per minute, and the mixture, kept uniform by stirring, runs out at the same rate.

a. If $S(t)$ is the amount of salt in solution at time t, then the amount of salt in a typical litre of solution is given by

$$\frac{\text{Amount of salt}}{\text{Amount of fluid}} = \frac{S(t)}{200}.$$

At what rate is salt flowing out of the tank at time t?

b. Write a differential equation for the rate of change of $S(t)$ using the fact that

$$\frac{dS}{dt} = \begin{bmatrix} \text{rate at which} \\ \text{salt enters tank} \end{bmatrix} - \begin{bmatrix} \text{rate at which} \\ \text{salt leaves tank} \end{bmatrix}.$$

c. Solve the differential equation in part (b) to obtain $S(t)$. [*Hint:* What is $S(0)$?]

48. THE SPREAD OF AN EPIDEMIC The rate at which an epidemic spreads through a community with 2000 susceptible residents is jointly proportional to the number of residents who have been infected and the number of susceptible residents who have not. Express the number of residents who have been infected as a function of time (in weeks), if 500 residents had the disease initially and 855 residents had been infected by the end of the first week.

49. A* THE SPREAD OF AN EPIDEMIC The rate at which an epidemic spreads through a community is jointly proportional to the number of residents who have been infected and the number of susceptible residents who have not. Show that the epidemic is spreading most rapidly when one-half of the susceptible residents have been infected. [*Hint:* You do not have to solve a differential equation to do this. Just start with a formula for the *rate* at which the epidemic is spreading and use calculus to maximize this rate.]

50. A* LOGISTIC CURVES Show that if a quantity Q satisfies the differential equation $\dfrac{dQ}{dt} = kQ(B - Q)$, where k and B are positive constants, then the rate of change $\dfrac{dQ}{dt}$ is greatest when $Q(t) = \dfrac{B}{2}$. What does this result tell you about the inflection point of a logistic curve? Explain. [*Hint:* See the hint for Exercise 49.]

51. A* WORKER EFFICIENCY For $0 \le p \le 1$, let $p(t)$ be the likelihood that an assembly line worker will make a mistake t hours into the worker's 8-hour shift. A particular worker, Tom, never makes a mistake at the beginning of his shift and is only 5% likely to make a mistake at the end. Set up and solve a differential equation assuming that at each time t, the likelihood of Tom's making an error increases at a rate proportional to the likelihood $1 - p(t)$ that he has not already made a mistake.

52. A* WORKER EFFICIENCY Sue, a co-worker of Tom's in Exercise 51, has a 1% likelihood of making an error at the beginning of her shift but only a 3% likelihood of making an error at the end. Set up and solve a differential equation assuming that at each time t, the likelihood $p(t)$ of Sue's making an error increases at a rate jointly proportional to $p(t)$ and the likelihood $1 - p(t)$ that she has not already made a mistake.

53. A* CORRUPTION IN GOVERNMENT The number of people implicated in a certain government scandal increases at a rate jointly proportional to the number of people already implicated and the number involved who have not yet been implicated. Suppose that 7 people were implicated when a Capital City newspaper first made the scandal public, that 9 more were implicated over the next 3 months, and that another 12 were implicated during the following 3 months. Approximately how many people are involved in the scandal? [*Warning:* This problem will test your algebraic ingenuity!]

54. **A* ALLOMETRY**   The different members or organs of an individual often grow at different rates, and an important part of **allometry** involves the study of relationships among these growth rates (recall the Think About It essay at the end of Chapter 1). Suppose $x(t)$ is the size (length, volume, or weight) at time t of one organ or member of an individual organism and $y(t)$ is the size of another organ or member of the same individual. Then the **law of allometry** states that the relative growth rates of x and y are proportional; that is,

$$\frac{y'(t)}{y(t)} = k\frac{x'(t)}{x(t)} \quad \text{for some constant } k > 0.$$

First show that the allometry law can be written as

$$\frac{dy}{dx} = k\frac{y}{x}$$

and then solve this equation for y in terms of x.

55. **A* RESPONSE TO STIMULUS**   The Weber-Fechner law in experimental psychology specifies that the rate of change of a response R with respect to the level of stimulus S is inversely proportional to the stimulus; that is,

$$\frac{dR}{dS} = \frac{k}{S}$$

Let S_0 be the threshold stimulus, that is, the highest level of stimulus for which there is no response, so that $R = 0$ when $S = S_0$.
a. Solve this differential equation for $R(S)$. Express your answer in terms of k and S_0.
b. Sketch the graph of the response function $R(S)$ found in part (a).

56. **A* FICK'S LAW**   When a cell is placed in a liquid containing a solute, the solute passes through the cell wall by diffusion. As a result, the concentration of the solute inside the cell changes, increasing if the concentration of the solute outside the cell is greater than the concentration inside and decreasing

if the opposite is true. In biology, Fick's law asserts that the concentration of the solute inside the cell changes at a rate that is jointly proportional to the area of the cell wall and the difference between the concentrations of the solute inside and outside the cell. Assuming that the concentration of the solute outside the cell is constant and greater than the concentration inside, derive a formula for the concentration of the solute inside the cell.

57. **A* INVENTORY DYNAMICS**   The producer of a certain commodity determines that to protect profits, the price p should decrease at a rate equal to half the inventory (surplus) $S - D$, where S and D are, respectively, the supply and demand for the commodity; that is,

$$\frac{dp}{dt} = -\frac{1}{2}(S - D).$$

Suppose that supply and demand vary with price so that

$$S(p) = 80 + 3p \quad \text{and} \quad D(p) = 120 - 2p$$

and so that the price is \$5 per unit when $t = 0$.
a. Determine the price $p(t)$ at any time t.
b. Find the equilibrium price, that is, the price p_e at which supply equals demand.
c. Show that the price $p(t)$ you determined in part (a) tends toward the equilibrium price in the long run (as $t \to \infty$).
[*Note:* This price-adjustment inventory model will be examined in more detail in Example 8.3.2 in Section 8.3.]

58. **A* INVENTORY DYNAMICS**   Suppose a competitor of the producer in Exercise 57 decides that the price of his commodity should decrease at a rate equal to a third of the inventory $S - D$; that is,

$$\frac{dp}{dt} = -\frac{1}{3}(S - D).$$

If supply and demand for this commodity are given by

$$S(p) = 30 + 5p \quad \text{and} \quad D(p) = 86 - 3p$$

and the price is \$9 when $t = 0$, determine the price $p(t)$ at any time t. What happens to the price $p(t)$ in the long run (as $t \to \infty$)?

59. **A* ELIMINATION OF HAZARDOUS WASTE**   To study the degradation of certain hazardous wastes with a high toxic content, biological researchers sometimes use the Haldane equation

$$\frac{dS}{dt} = \frac{aS}{b + cS + S^2}$$

where a, b, and c are positive constants and $S(t)$ is the concentration of substrate (the substance acted on by bacteria in the waste material).* Find the general solution of the Haldane equation. Express your answer in implicit form (as an equation involving S and t).

60. **A* DEMOGRAPHY**   The differential equation

$$\frac{dQ}{dt} = Q(a - b \ln Q),$$

where a and b are constants, is called the *Gompertz equation*, and a solution of the equation is called a *Gompertz function*. Such functions are used to describe restricted growth in populations as well as matters such as learning and growth within an organization.

---

*Michael D. LaGrega, Philip L. Buckingham, and Jeffery C. Evans, *Hazardous Waste Management,* New York: McGraw-Hill, 1994, p. 578.

a. Use the Gompertz equation to show that a Gompertz function is growing most rapidly when $\ln Q = \dfrac{a - b}{b}$. [*Hint:* What derivative measures how fast the growth rate $\dfrac{dQ}{dt}$ is changing?]
b. Solve the Gompertz equation. [*Hint:* Substitute $u = \ln Q$.]
c. Compute $\lim_{t \to \infty} Q(t)$.
d. Sketch the graph of a typical Gompertz function.

61. **A* PARETO'S LAW**   *Pareto's law* in economics says that the rate of change (decrease) of the number of people P in a stable economy who have an income of at least x dollars is directly proportional to the number of such people and inversely proportional to their income. Express this law as a differential equation and solve for P in terms of x.

Solve first-order linear differential equations.

First-Order Linear Differential Equations

A **first-order linear** differential equation is one with the general form

$$\frac{dy}{dx} + p(x)y = q(x).$$

For instance,

$$\frac{dy}{dx} + \frac{y}{x} = e^{-x} \quad \text{for } x > 0$$

is a first-order linear differential equation with $p(x) = \dfrac{1}{x}$ and $q(x) = e^{-x}$. In its current form, this equation cannot be solved by simple integration or by separation of variables. However, if you multiply both sides of the equation by x, it becomes

$$x\frac{dy}{dx} + y = xe^{-x}.$$

Notice that the left side of this equation is the derivative (with respect to x) of the product xy since

$$\frac{d}{dx}(xy) = x\frac{dy}{dx} + y\frac{dx}{dx} = x\frac{dy}{dx} + y.$$

Thus, the given differential equation now reads

$$\frac{d}{dx}(xy) = xe^{-x},$$

which you can solve by integrating both sides with respect to x:

$$\int \frac{d}{dx}(xy)\, dx = \int xe^{-x}\, dx$$

$$xy = \int xe^{-x}\, dx = -(x + 1)e^{-x} + C \qquad \text{using integration by parts}$$

Solving for y, the given first-order linear differential equation has the general solution

$$y = \frac{1}{x}[e^{-x}(-x - 1) + C].$$

In this example, multiplying by the function x converted the given first-order linear differential equation into one that could be solved by simple integration. Such a function is called an **integrating factor** of the differential equation. In more advanced texts, it is shown that every first-order linear differential equation $\frac{dy}{dx} + p(x)y = q(x)$ has an integrating factor equal to $I(x) = e^{\int p(x)\, dx}$. For instance, in the introductory example, where $p(x) = \frac{1}{x}$, the integrating factor is

$$I(x) = e^{\int (1/x)\, dx} = e^{\ln x} = x \qquad \text{since } x > 0.$$

To summarize:

Solution of a First-Order Linear Differential Equation ■ The first-order linear differential equation $\frac{dy}{dx} + p(x)y = q(x)$ has the general solution

$$y = \frac{1}{I(x)}\left[\int I(x)q(x)\, dx + C\right],$$

where C is an arbitrary constant and $I(x)$ is the integrating factor

$$I(x) = e^{\int p(x)\, dx}.$$

EXAMPLE 8.2.1

Find the general solution of the differential equation

$$\frac{dy}{dx} + 2y = 2x.$$

Solution
This is a first-order linear differential equation with $p(x) = 2$ and $q(x) = 2x$. The integrating factor is

$$I(x) = e^{\int 2\, dx} = e^{2x},$$

and the general solution of the first-order linear equation is

$$y = \frac{1}{e^{2x}}\left[\int 2x\, e^{2x}\, dx + C\right].$$

The integral on the right can be found using integration by parts:

$$\int 2x\, e^{2x}\, dx = xe^{2x} - \frac{1}{2}e^{2x}.$$

(Verify the details.) Thus, the general solution is

$$y = \frac{1}{e^{2x}}\left[xe^{2x} - \frac{1}{2}e^{2x} + C\right] = x - \frac{1}{2} + Ce^{-2x}.$$

EXAMPLE 8.2.2

Find the general solution of the differential equation

$$\frac{dy}{dx} = \frac{2x + y}{x}$$

and the particular solution that satisfies $y = -2$ when $x = 1$.

Solution

The given differential equation can be rewritten as

$$\frac{dy}{dx} = \frac{2x + y}{x} = 2 + \frac{y}{x} \qquad \text{divide } 2x + y \text{ by } x$$

$$\frac{dy}{dx} - \frac{y}{x} = 2 \qquad \text{subtract } \frac{y}{x} \text{ from both sides}$$

We recognize this as a first-order linear differential equation with $p(x) = -\frac{1}{x}$ and $q(x) = 2$. The integrating factor is

$$I(x) = e^{\int(-1/x)dx} = e^{-\ln x}$$

$$= e^{\ln x^{-1}} = x^{-1} = \frac{1}{x}$$

and the general solution of the equation is

$$y = \frac{1}{1/x}\left[\int \frac{1}{x}(2)\,dx + C\right] = x\left[\int \frac{2}{x}\,dx + C\right]$$

$$= x[2 \ln x + C]$$

To find the particular solution, substitute $x = 1$ and $y = -2$ into the general solution to get

$$-2 = (1)[2 \ln (1) + C] = 2(0) + C = C.$$

Thus, the particular solution is

$$y = x(2 \ln x - 2) = 2x(\ln x - 1).$$

In Section 4.1, we used a limit to compute the amount of money in an account in which interest is compounded continuously. Example 8.2.3 illustrates how such computations can be performed using a first-order linear differential equation.

EXAMPLE 8.2.3

David deposits $20 000 into an account in which interest accumulates at a rate of 5% per year, compounded continuously. He plans to withdraw $3000 per year.

a. Set up and solve a differential equation to determine the value $Q(t)$ of his account t years after the initial deposit.

b. How long does it take for his account to be exhausted?

Solution

a. If no withdrawals were made, the value of the account would change at a percentage rate equal to the annual interest rate; that is,

$$\frac{Q'(t)}{Q(t)} = 0.05,$$

or, equivalently, $Q'(t) = 0.05Q(t)$. This is the rate at which interest is added to the account, and by subtracting the annual withdrawal rate of $3000, we obtain the net rate of change of the account, that is,

$$\underbrace{\frac{dQ}{dt}}_{\substack{\text{net rate} \\ \text{of change}}} = \underbrace{0.05Q}_{\substack{\text{rate interest} \\ \text{is added}}} - \underbrace{3000}_{\substack{\text{rate money} \\ \text{is withdrawn}}}$$

Rewriting this equation as

$$\frac{dQ}{dt} - 0.05Q = -3000$$

gives a first-order linear differential equation with $p(t) = -0.05$ and $q(t) = -3000$, which we wish to solve subject to the initial condition that $Q(0) = 20\ 000$. The integrating factor for this equation is

$$I(t) = e^{\int -0.05\,dt} = e^{-0.05t}$$

so the general solution is

$$Q(t) = \frac{1}{e^{-0.05t}}\left[\int e^{-0.05t}(-3000)\,dt + C\right]$$

$$= e^{0.05t}\left[-3000\left(\frac{e^{-0.05t}}{-0.05}\right) + C\right]$$

$$= 60\ 000 + Ce^{0.05t}$$

Since $Q(0) = 20\ 000$, then

$$Q(0) = 20\ 000 = 60\ 000 + Ce^0 = 60\ 000 + C,$$

so that

$$C = -40\ 000$$

and

$$Q(t) = 60\ 000 - 40\ 000e^{0.05t}.$$

b. The account becomes exhausted when $Q(t) = 0$. Solve the equation

$$0 = 60\ 000 - 40\ 000e^{0.05t}$$

to get

$$40\ 000e^{0.05t} = 60\ 000 \qquad \text{add } 40\ 000e^{0.05t} \text{ to both sides}$$

$$e^{0.05t} = \frac{60\ 000}{40\ 000} = 1.5 \qquad \text{divide both sides by } 40\ 000$$

$$0.05t = \ln 1.5 \qquad \text{take logarithms on both sides}$$

$$t = \frac{\ln 1.5}{0.05} \qquad \text{divide both sides by 0.05}$$

$$\approx 8.11$$

Thus, the account is exhausted in approximately 8 years.
(Note that this problem could also be solved by separation of variables.)

Dilution Models   First-order linear differential equations can be used to model situations in which a quantity is diluted. Such models appear in areas such as finance, ecology, medicine, and chemistry. The general procedure for modelling dilution is illustrated in Example 8.2.4.

EXAMPLE 8.2.4

A 70-L tank initially contains 20 g of salt dissolved in 50 L of water. Suppose that 3 L of brine containing 2 g of dissolved salt per litre runs into the tank every minute and that the mixture (kept uniform by stirring) runs out of the tank at a rate of 2 L per minute.

a. Set up and solve a differential equation to find an expression for the amount of salt in the tank after t minutes.

b. How much salt will be in the tank at the instant it begins to overflow?

Solution

a. Let $S(t)$ be the amount of salt in the tank at time t. Since 3 L of brine flows into the tank each minute and each litre contains 2 g of salt, then (3 L)(2 g/L) = 6 g of salt flows into the tank each minute. To find the number of grams of salt flowing out of the tank each minute, note that at time t, there are $S(t)$ grams of salt and $50 + (3 - 2)t = 50 + t$ litres of solution in the tank (because there is a net increase of 3 L − 2 L = 1 L of solution every minute). Thus, the concentration of salt in solution at time t is $\dfrac{S(t)}{50 + t}$ grams/litre, and salt is leaving the tank at a rate of

$$\left[\frac{S(t)}{50 + t} \text{ grams/litre} \right][2 \text{ litres/minute}] = \frac{2S(t)}{50 + t} \text{ grams/minute.}$$

It follows that the net rate of change $\dfrac{dS}{dt}$ of salt in the tank is given by

$$\underbrace{\frac{dS}{dt}}_{\substack{\text{net rate of} \\ \text{change}}} = \underbrace{6}_{\text{inflow}} - \underbrace{\frac{2S}{50 + t}}_{\text{outflow}}$$

(see Figure 8.6). This equation can be written as

$$\frac{dS}{dt} + \frac{2S}{50 + t} = 6,$$

which is a first-order linear differential equation with $p(t) = \dfrac{2}{50 + t}$ and $q(t) = 6$.

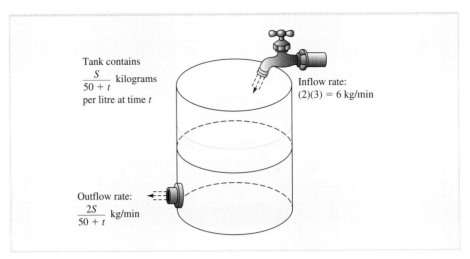

Tank contains
$\dfrac{S}{50 + t}$ kilograms
per litre at time t

Inflow rate:
$(2)(3) = 6$ kg/min

Outflow rate:
$\dfrac{2S}{50 + t}$ kg/min

FIGURE 8.6 The net rate of change of salt in solution equals the inflow rate minus the outflow rate.

The integrating factor is

$$
\begin{aligned}
I(t) &= e^{\int p(t)\,dt} = e^{\int [2/(50 + t)]\,dt} \\
&= e^{2\ln|50 + t|} = e^{\ln|50 + t|^2} \\
&= (50 + t)^2
\end{aligned}
$$

Hence, the general solution is

$$
\begin{aligned}
S(t) &= \frac{1}{(50 + t)^2}\left[\int (6)(50 + t)^2\,dt + C\right] \\
&= \frac{1}{(50 + t)^2}[2(50 + t)^3 + C] \\
&= 2(50 + t) + \frac{C}{(50 + t)^2}
\end{aligned}
$$

To find C, note that 20 g of salt is initially in the tank. Thus,

$$
20 = S(0) = 2(50 + 0) + \frac{C}{(50 + 0)^2}
$$

so that $C = -80(50)^2$. It follows that the amount of salt at time t is

$$
S(t) = 2(50 + t) - \frac{80(50)^2}{(50 + t)^2}.
$$

b. After t minutes, the tank contains $50 + t$ litres of fluid, and it begins to overflow when the amount of fluid reaches its capacity, 70 L. Solving the equation $50 + t = 70$, we see that this occurs when $t = 20$ minutes. By substituting $t = 20$ into the formula obtained in part (a), we find that

$$
\begin{aligned}
S(20) &= 2(50 + 20) - \frac{80(50)^2}{(50 + 20)^2} \\
&\approx 99.18
\end{aligned}
$$

Thus, the tank contains approximately 99 g of salt when it begins to overflow.

First-order linear differential equations can be used to model a variety of situations in the life sciences. Example 8.2.5 illustrates how such a model can be used to assist a physician in making a decision regarding the medication administered to a patient.

EXAMPLE 8.2.5

A drug is infused into a patient's bloodstream at a constant rate r (the *infusion rate*) and is eliminated from the bloodstream at a rate proportional to the amount of drug present at time t. Initially (when $t = 0$), the patient's blood contains no drug.

a. Set up and solve a differential equation for the amount of drug $A(t)$ present in the patient's bloodstream at time t.

b. Show that $A(t)$ approaches a constant value as $t \to \infty$. This is the *steady-state* level of medication.

c. Suppose an attending physician sets the infusion rate for a particular patient at 2 mg per hour and that a sample is taken indicating that 1.69 mg of the drug is in the patient's blood after 1 hour. How much medication will be present after 6 hours? What is the steady-state level of medication?

d. Suppose the physician in part (c) wants the steady-state level of medication in the patient's bloodstream to be 8 mg. How should the infusion rate be adjusted to achieve this goal?

Solution

a. The amount of drug $A(t)$ in the bloodstream at time t changes at a rate of

$$\frac{dA}{dt} = \underbrace{r}_{\text{infusion}} - \underbrace{kA}_{\text{elimination}}$$

where k is a constant (the *elimination coefficient*). Rewriting this equation as

$$\frac{dA}{dt} + kA = r$$

gives a first-order linear differential equation with $p(t) = k$ and $q(t) = r$. The integrating factor is

$$I(t) = e^{\int k\, dt} = e^{kt},$$

so the general solution is

$$A(t) = \frac{1}{e^{kt}}\left[\int re^{kt}\, dt + C\right]$$

$$= e^{-kt}\left[\frac{re^{kt}}{k} + C\right]$$

$$= \frac{r}{k} + Ce^{-kt}$$

Since there is no drug initially present in the blood, $A(0) = 0$ so that

$$A(0) = 0 = \frac{r}{k} + Ce^{0}.$$

Thus, $C = -\dfrac{r}{k}$, and

$$A(t) = \frac{r}{k} - \frac{r}{k}e^{-kt} = \frac{r}{k}(1 - e^{-kt}).$$

b. Since

$$\lim_{t\to\infty} A(t) = \lim_{t\to\infty} \frac{r}{k}(1 - e^{-kt}) = \frac{r}{k}(1 - 0) = \frac{r}{k},$$

the steady-state level of medication is $\dfrac{r}{k}$.

c. In this case, $r = 2$ and $A(1) = 1.69$. Substituting this information into the formula derived for $A(t)$ in part (a) results in

$$A(1) = 1.69 = \frac{2}{k}[1 - e^{-k(1)}].$$

Using a computer, solve this equation to obtain $k \approx 0.35$. Thus, after 6 hours, the amount of drug in the patient's bloodstream is

$$A(6) = \frac{2}{0.35}[1 - e^{-0.35(6)}] \approx 5.01 \text{ mg}.$$

The steady-state level of medication is

$$\frac{r}{k} = \frac{2}{0.35} \approx 5.71 \text{ mg}.$$

d. Assume that the elimination coefficient k depends on the patient's metabolism and not on the infusion rate r. (Does this assumption seem reasonable?) Therefore, with k fixed at 0.35, a steady-state level of 8 mg is achieved by adjusting the infusion rate r to satisfy

$$\frac{r}{0.35} = 8,$$

so that

$$r = 2.8 \text{ mg/h}.$$

EXERCISES ■ 8.2

In Exercises 1 through 14, find the general solution of the given first-order linear differential equation.

1. $\dfrac{dy}{dx} + \dfrac{3y}{x} = x$

2. $\dfrac{dy}{dx} + \dfrac{2y}{x} = \sqrt{x} + 1$

3. $\dfrac{dy}{dx} + \dfrac{y}{2x} = \sqrt{x}e^x$

4. $x^4\dfrac{dy}{dx} + 2x^3y = 5$

5. $x^2\dfrac{dy}{dx} + xy = 2$

6. $x\dfrac{dy}{dx} + 2y = xe^{x^3}$

7. $\dfrac{dy}{dx} + \left(\dfrac{2x + 1}{x}\right)y = e^{-2x}$

8. $\dfrac{dy}{dx} + \dfrac{xy}{x^2 + 1} = x$

9. $\dfrac{dy}{dx} = \dfrac{1 + xy}{1 + x}$

10. $\dfrac{dy}{dx} = \dfrac{x + y}{2x + 1}$

11. $\dfrac{dx}{dt} + \dfrac{x}{1 + t} = t$

12. $\dfrac{dy}{dt} - \dfrac{y}{t} = te^{-t}$

13. $\dfrac{dy}{dt} + \dfrac{ty}{t + 1} = t + 1$

14. $\dfrac{dx}{dt} + \dfrac{x}{1 + 2t} = 5$

In Exercises 15 through 20, find the particular solution of the given differential equation that satisfies the given condition.

15. $x\dfrac{dy}{dx} - 2y = 2x^3;\ y = 2$ when $x = 1$

16. $\dfrac{dy}{dx} + y = x^2;\ y = 2$ when $x = 0$

17. $\dfrac{dy}{dx} + xy = x + e^{-x^2/2};\ y = -1$ when $x = 0$

18. $\dfrac{dy}{dx} + y = x;\ y = 4$ when $x = 0$

19. $\dfrac{dy}{dx} + \dfrac{y}{x} = \dfrac{1}{x^2};\ y = -2$ when $x = 1$

20. $\dfrac{dy}{dx} - \dfrac{y}{x} = \ln x;\ y = 3$ when $x = 1$

Differential equations can often be solved in several different ways. In Exercises 21 through 24, solve the given differential equation by regarding it as (a) separable and (b) first-order linear.

21. $\dfrac{dy}{dx} + 3y = 5$

22. $\dfrac{dy}{dx} + \dfrac{y}{x} = 0$

23. $\dfrac{dy}{dx} + \dfrac{y}{x + 1} = \dfrac{2}{x + 1}$

24. $\dfrac{dy}{dx} = 1 + x + y + xy$ (*Hint:* Factor.)

25. GEOMETRY Find a function $f(x)$ whose graph passes through the point $(-1, 2)$ and has the property that at each point (x, y) on the graph, the slope of the tangent line equals the sum $x + y$ of the coordinates.

26. GEOMETRY Find a function $f(x)$ whose graph passes through the point $(1, 1)$ and has the property that at each point (x, y) on the graph, the slope of the tangent line equals $x^2 + 2y$.

27. REAL ESTATE The price of a certain house is currently $200 000. Suppose it is estimated that after t months, the price $p(t)$ will be increasing at a rate of $0.01p(t) + 1000t$ dollars per month. How much will the house cost 9 months from now?

28. RETAIL PRICES The price of a certain commodity is currently $3 per unit. It is estimated that t weeks from now, the price $p(t)$ will be increasing at a rate of $0.02p(t) + e^{0.1t}$ cents per week. How much will the commodity cost in 10 weeks?

29. INVESTMENT PLAN An investor makes regular deposits totalling D dollars each year into an account that earns interest at an annual rate r compounded continuously.

a. Explain why the account grows at a rate of

$$\dfrac{dV}{dt} = rV + D,$$

where $V(t)$ is the value of the account t years after the initial deposit. Solve this differential equation to express $V(t)$ in terms of r and D.

b. Amanda wants to retire in 20 years. To build up a retirement fund, she makes regular annual deposits of $8000. If the prevailing interest rate stays constant at 4% compounded continuously, how much will she have in her account at the end of the 20-year period?

c. Ray estimates he will need $800 000 to retire. If the prevailing rate of interest is 5% compounded continuously, how large should his regular annual deposits be so that he can retire in 30 years?

30. SAVINGS   Chris has a starting salary of $47 000 per year and figures that with salary increases and bonuses, her compensation will increase at an average annual rate of 9%. She regularly deposits 5% of her salary in a savings account that earns interest at an annual rate of 8% compounded continuously. How much will be in her account in 20 years? (See Exercise 29.)

31. SAVINGS   Mike is saving to take an $8000 tour of Europe in 4 years. His parents want to help by making a lump sum deposit of A dollars into a savings account that earns 4% interest per year compounded continuously. Mike plans to add to this amount by making frequent deposits into the account totalling $800 per year. What must A be in order for him to meet his goal? (See Exercise 29.)

32. DILUTION   A tank contains 10 g of salt dissolved in 30 L of water. Suppose that 2 L of brine containing 1 g of dissolved salt per litre runs into the tank every minute and that the mixture, kept uniform by stirring, runs out at a rate of 1 L per minute.

   a. Set up and solve an initial value problem for the amount of salt $S(t)$ in the tank at time t.

   b. Suppose the tank has a capacity of 50 L. How much salt is in the tank when it is full?

33. DILUTION   A tank contains 5 g of salt dissolved in 40 L of water. Pure water runs into the tank at a rate of 1 L per minute, and the mixture, kept uniform by stirring, runs out at a rate of 3 L per minute.

   a. Set up and solve an initial value problem for the amount of salt $S(t)$ in the tank at time t.

   b. How long does it take for the amount of salt in solution to reach 4 g?

   c. How much salt is in the tank just as it drains empty?

34. AIR PURIFICATION   A 2400-m^3 warehouse contains an activated charcoal air filter through which air passes at a rate of 400 m^3 per minute. The ozone in the air is absorbed by the charcoal as the air flows through the filter, and the purified air is recirculated in the warehouse. Assuming that the remaining ozone is evenly distributed throughout the room at all times, determine how long it takes the filter to remove 50% of the ozone from the room.

35. HAZARDOUS EMISSIONS   Noxious fumes from a garage enter an adjacent 2000-m^3 workroom through a vent at a rate of 0.2 m^3 per minute, and the

mixed air (fresh air and fumes) escapes through a crack under the door at the same rate.

   a. Set up and solve a differential equation for the amount of noxious fumes $g(t)$ in the room t minutes after they begin to enter the room (so $g(0) = 0$).

   b. The air in the room becomes hazardous when it contains 1% noxious fumes. How long does it take for this to occur?

36. SAVINGS   Ella deposits $10 000 into an account in which interest accumulates at a rate of 4% per year, compounded continuously. She plans to withdraw $2000 per year.

   a. Set up and solve a differential equation to determine the value $Q(t)$ of her account t years after the initial deposit.

   b. How long does it take for her account to be exhausted?

37. SAVINGS   Alonzo plans to make an initial deposit into an account that earns 5% interest compounded continuously and then to make $5000 withdrawals each year. How much must he deposit for his account to be exhausted in 10 years?

38. SAVINGS   Janine has a starting salary of $40 000 and figures that with salary increases and bonuses her compensation will increase at an average annual rate of 5%. She regularly deposits 10% of her salary into a savings account that earns interest at a rate of 4% per year compounded continuously and withdraws $3000 per year for expenses.

   a. Set up and solve a differential equation for the amount of money $A(t)$ in her account after t years.

   b. How much will Janine have in her account when she retires in 40 years?

39. MEDICATION   A drug injected into a patient's bloodstream at a constant infusion rate of 3 mg per hour is eliminated from the blood at a rate proportional to the amount of the drug currently present. Initially (when $t = 0$), the patient's blood contains no drug.

   a. Set up and solve a differential equation for the amount of drug $A(t)$ present in the patient's bloodstream at time t.

   b. Suppose that a sample is taken indicating that 2.3 mg of the drug is in the patient's blood after 1 hour. How much medication will be present after 8 hours? How much medication is present as $t \to \infty$ (the steady-state level)?

c. Suppose the patient's physician wants the steady-state level of medication in the patient's bloodstream to be 9 mg. How should the infusion rate be adjusted to achieve this goal?

40. MEDICATION   A drug injected into a patient's bloodstream at a constant infusion rate of 2.2 mg per hour is eliminated from the blood at a rate proportional to the amount of the drug currently present. Initially (when $t = 0$), the patient's blood contains 0.5 mg of the drug.

a. Set up and solve a differential equation for the amount of drug $A(t)$ present in the patient's bloodstream at time t.

b. Suppose that a sample is taken indicating that 2 mg of the drug is in the patient's blood after 1 hour. How much medication will be present after 6 hours? How much medication is present as $t \to \infty$ (the steady-state level)?

c. Suppose the patient's physician wants the steady-state level of medication in the patient's bloodstream to be 8 mg. How should the infusion rate be adjusted to achieve this goal?

41. EQUILIBRIUM PRICE ADJUSTMENT   The supply S and demand D of a particular commodity are given by

$$S(p) = 34 + 3p + 2p'$$

and

$$D(p) = 25 - 2p + 5p'$$

where $p(t)$ is the price of the commodity as a function of time t (weeks). Suppose the price is $8 per unit at time $t = 0$ and adjusts continuously so that supply always equals demand, that is, $S(p) = D(p)$.

The pricing of the commodity is *stable* if $p(t)$ approaches a finite value as $t \to \infty$ and *unstable* (or *inflated*) if $p(t) \to \infty$.

a. What is $p(t)$?

b. Is the pricing of this commodity stable or unstable?

42. EQUILIBRIUM PRICE ADJUSTMENT   The supply S and demand D of a particular commodity are given by

$$S(p) = 6 + 5p + 3p'$$

and

$$D(p) = 30 - p + p'$$

where $p(t)$ is the price of the commodity as a function of time t (weeks). Suppose the price is $6 per unit at time $t = 0$ and adjusts continuously so that supply always equals demand, that is, $S(p) = D(p)$.

The pricing of the commodity is *stable* if $p(t)$ approaches a finite value as $t \to \infty$ and *unstable* (or *inflated*) if $p(t) \to \infty$.

a. What is $p(t)$?

b. Is the pricing of this commodity stable or unstable?

SECTION 8.3

Solve application problems using differential equations.

Additional Applications of Differential Equations

In the first two sections of this chapter, you have seen applications of differential equations to finance, learning, population growth, the spread of an epidemic, and the amount of medication in a patient's bloodstream. In all these models, a specified rate of change is interpreted as a derivative that is then related to other features of the model by a differential equation. In this section, we examine several additional models involving separable and first-order linear differential equations. We begin with an applied model involving public health.

A Public Health Dilution Model

EXAMPLE 8.3.1

The residents of a certain community have voted to discontinue the fluoridation of their water supply. The local reservoir currently holds 200 million litres of fluoridated water that contains 160 kg of fluoride. The fluoridated water is flowing out of the reservoir at a rate of 4 million litres per day and is being replaced at the same rate by unfluoridated water. At all times, the remaining fluoride is evenly distributed in the reservoir. Express the amount of fluoride in the reservoir as a function of time.

Solution

Let $Q(t)$ be the amount of fluoride (in kilograms) in the reservoir t days after the fluoridation ends. To model the flow of fluoride, start with the rate relationship

$$\begin{bmatrix} \text{Net rate of change of fluoride} \\ \text{with respect to time} \end{bmatrix} = \begin{bmatrix} \text{daily rate of fluoride} \\ \text{flowing in} \end{bmatrix} - \begin{bmatrix} \text{daily rate of fluoride} \\ \text{flowing out} \end{bmatrix}$$

The net rate of change of the fluoride with respect to time is $\dfrac{dQ}{dt}$, and since the fluoridation has been terminated, the rate of fluoride flowing in is 0. Since the volume of fluoridated water in the reservoir stays fixed at 200 million litres and the fluoride is always evenly distributed in the reservoir, the concentration of fluoride in the reservoir at time t is given by the ratio

$$\frac{Q(t) \text{ kilograms of fluoride}}{200 \text{ million litres of fluoridated water}}.$$

Therefore, since 4 million litres of fluoridated water are being removed each day, the daily outflow rate of fluoride is given by the product

$$\begin{matrix} \text{Daily rate of fluoride} \\ \text{flowing out} \end{matrix} = \left(\frac{Q(t) \text{ kilograms}}{200 \text{ million litres}} \right) \left(\frac{4 \text{ million litres}}{\text{day}} \right) = \frac{4Q}{200} \text{ kilograms per day}.$$

The flow-rate relationships are summarized in the following diagram:

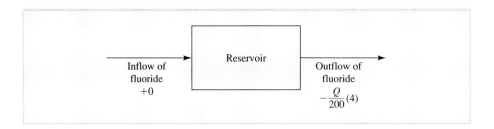

Since the rate of change of the amount of fluoride in the reservoir equals the inflow rate minus the outflow rate, it follows that

$$\frac{dQ}{dt} = \underbrace{0}_{\text{inflow}} - \underbrace{\frac{4Q}{200}}_{\text{outflow}} = -\frac{Q}{50}.$$

Solve this differential equation by separation of variables to get

$$\int \frac{1}{Q}\,dQ = -\int \frac{1}{50}\,dt$$

$$\ln Q = -\frac{t}{50} + C$$

$$Q = e^{C-t/50} = e^C e^{-t/50}$$

$$= Q_0 e^{-t/50} \qquad \text{where } Q_0 = e^C$$

Initially, the reservoir contained 160 kg of fluoride, so that

$$160 = Q(0) = Q_0 e^0 = Q_0;$$

thus,

$$Q(t) = 160e^{-t/50},$$

and the amount of fluoride decreases exponentially, as illustrated in Figure 8.7.

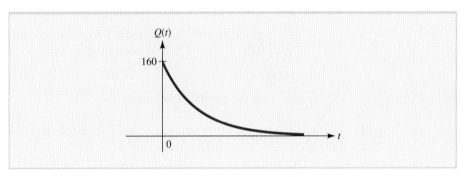

FIGURE 8.7   The amount of fluoride $Q(t) = 160e^{-t/50}$.

A Price Adjustment Model

Let $S(p)$ denote the number of units of a particular commodity supplied to the market at a price of p dollars per unit, and let $D(p)$ denote the corresponding number of units demanded by the market at the same price. In static circumstances, market equilibrium occurs at the price where demand equals supply (recall the discussion in Section 1.4). However, certain economic models consider a more dynamic kind of economy in which price, supply, and demand are assumed to vary with time. One of these, the *Evans price adjustment model*, assumes that the rate of change of price with respect to time t is proportional to the shortage $D - S$, so that

$$\frac{dp}{dt} = k(D - S),$$

where k is a positive constant. Here is an example involving this model.

EXAMPLE 8.3.2

A commodity is introduced with an initial price of $5 per unit, and t months later the price is $p(t)$ dollars per unit. A study indicates that at time t, the demand for the commodity will be $D(t) = 3 + 10e^{-0.01t}$ thousand units and that $S(t) = 2 + p$ thousand units will be supplied. Suppose that at each time t, the price is increasing at a rate equal to 2% of the shortage $D(t) - S(t)$.

  a. Set up and solve a differential equation for the unit price $p(t)$.

  b. What is the unit price of the commodity after 6 months?

  c. At what time is the unit price the largest? What are the maximum unit price and the corresponding supply and demand?

  d. What happens to the price in the long run (as $t \to \infty$)?

Solution

a. According to the given information, the unit price $p(t)$ satisfies

$$\frac{dp}{dt} = 0.02[D(t) - S(t)] = 0.02[(3 + 10e^{-0.01t}) - (2 + p)]$$

$$= 0.02 + 0.2e^{-0.01t} - 0.02p$$

or, equivalently,

$$\frac{dp}{dt} + 0.02p = 0.02 + 0.2e^{-0.01t}$$

which is a first-order linear differential equation with $P(t) = 0.02$ and $Q(t) = 0.02 + 0.2e^{-0.01t}$. The integrating factor is

$$I(t) = e^{\int 0.02\, dt} = e^{0.02t}$$

so the general solution is

$$p(t) = \frac{1}{e^{0.02t}}\left[\int e^{0.02t}(0.02 + 0.2e^{-0.01t})\, dt + C\right]$$

$$= e^{-0.02t}\left[\int 0.02e^{0.02t} + 0.2e^{0.01t})\, dt + C\right]$$

$$= e^{-0.02t}\left[\frac{0.02e^{0.02t}}{0.02} + \frac{0.2e^{0.01t}}{0.01} + C\right]$$

$$= 1 + 20e^{-0.01t} + Ce^{-0.02t}$$

Since the initial price is $p(0) = 5$, we have

$$p(0) = 5 = 1 + 20e^0 + Ce^0 = 1 + 20 + C,$$

so that $C = 5 - 21 = -16$ and

$$p(t) = 1 + 20e^{-0.01t} - 16e^{-0.02t}.$$

b. After 6 months ($t = 6$), the price is

$$p(6) = 1 + 20e^{-0.01(6)} - 16e^{-0.02(6)} \approx 5.6446,$$

that is, approximately \$5.64 per unit.

c. To maximize the price, compute the derivative

$$p'(t) = 20(-0.01e^{-0.01t}) - 16(-0.02e^{-0.02t})$$

$$= -0.2e^{-0.01t} + 0.32e^{-0.02t}$$

and then note that $p'(t) = 0$ when

$$p'(t) = -0.2e^{-0.01t} + 0.32e^{-0.02t} = 0 \qquad \text{add } 0.2e^{-0.01t} \text{ to}$$

$$0.32e^{-0.02t} = 0.2e^{-0.01t} \qquad \text{both sides}$$

$$\frac{e^{-0.01t}}{e^{-0.02t}} = \frac{0.32}{0.2}$$

$$e^{-0.01t - (-0.02t)} = 1.6$$

$$e^{0.01t} = 1.6$$

$$0.01t = \ln 1.6 \qquad \text{take logarithms}$$

$$t \approx 47 \qquad \text{on both sides}$$

This critical number corresponds to a maximum (verify that $p''(47) < 0$), so the largest price occurs after roughly 47 months. Since

$$p(47) = 1 + 20e^{-0.01(47)} - 16e^{-0.02(47)} \approx 7.25,$$

the largest unit price is approximately \$7.25. Since

$$D(47) = 3 + 10e^{-0.01(47)} \approx 9.25$$

and

$$S(47) = 2 + p(47) = 2 + 7.25 \approx 9.25,$$

it follows that approximately 9250 units will be both demanded and supplied at the maximum price of \$7.25 per unit.

d. Since

$$\lim_{t \to \infty} p(t) = \lim_{t \to \infty} (1 + 20e^{-0.01t} - 16e^{-0.02t})$$
$$= 1 + 20(0) - 16(0) = 1$$

the price tends toward \$1 per unit in the long run. The graph of the unit price function $p(t)$ is shown in Figure 8.8.

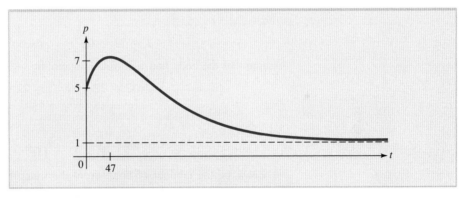

FIGURE 8.8   The graph of the unit price function in Example 8.3.2.

Newton's Law of Cooling: A Forensics Model

Newton's law of cooling says that the rate of change of the temperature $T(t)$ of an object is proportional to the difference between $T(t)$ and the temperature T_m of the surrounding medium, that is,

$$\frac{dT}{dt} = k(T - T_m).$$

The law of cooling can be used to model a variety of situations involving temperature change. For instance, when a dead body is discovered relatively quickly (within 2 days of the time of death) and the temperature of the air surrounding the body is essentially constant, the time of death can often be determined using the law of cooling, as illustrated in Example 8.3.3.

EXAMPLE 8.3.3

A body is discovered at noon on Friday in a room where the air temperature is 22°C. The temperature of the body at the time of discovery is 25°C, and 1 hour later it is 24.5°C. Use this information to determine the time of death.

Solution

Since the temperature of the medium surrounding the body is $T_m = 22$, Newton's law of cooling says that

$$\frac{dT}{dt} = k(T - 22),$$

where t is the number of hours since the time of death. Separate the variables and integrate to obtain

$$\int \frac{dT}{T - 22} = \int k \, dt$$

$$\ln|T - 22| = kt + C_1$$

$$\ln(T - 22) = kt + C_1 \qquad \text{since } T > 22$$

Then take exponentials on both sides:

$$T - 22 = e^{kt + C_1} = e^{kt} e^{C_1},$$

so that

$$T = 22 + Ce^{kt} \qquad \text{where } C = e^{C_1}.$$

Assume that the body had a normal temperature of 37°C at the time of death. Then

$$37 = 22 + Ce^{k(0)} = 22 + C$$

$$C = 37 - 22 = 15$$

and

$$T = 22 + 15e^{kt}.$$

Let t_0 be the number of hours from the time of death to the time of discovery. Then the given information says that $T(t_0) = 25$ and $T(t_0 + 1) = 24.5$. Since

$$T(t_0) = 25 = 22 + 15e^{kt_0},$$

then

$$15e^{kt_0} = 25 - 22 = 3.$$

Also,

$$T(t_0 + 1) = 24.5 = 22 + 15 \, e^{k(t_0 + 1)}$$

$$= 22 + 15e^{kt_0} e^k$$

so that

$$(15e^{kt_0})e^k = 2.5$$

$$(3)e^k = 2.5 \qquad\qquad \text{since } 15e^{kt_0} = 3$$

$$e^k = \frac{2.5}{3} = 0.8333$$

$$k = \ln(0.8333) \approx -0.1823 \qquad \text{take logarithms on both sides}$$

Since $15e^{kt_0} = 3$, it follows that

$$e^{kt_0} = \frac{3}{15} = 0.2$$

$$e^{-0.1823t_0} \approx 0.2 \qquad \text{since } k \approx -0.1823$$

$$-0.1823t_0 \approx \ln 0.2 \approx -1.609$$

$$t_0 \approx \frac{-1.609}{-0.1823} \approx 8.826 \qquad \text{take logarithms on both sides}$$

Thus, the body had been dead for approximately 8.83 hours (8 hours and 50 minutes) at the time of discovery at noon on Friday. This places the time of death at 3:10 A.M. on Friday morning.

Orthogonal Trajectories: Chemotaxis

A curve that intersects every curve in a given family of curves at right angles is said to be an **orthogonal trajectory** for that family (Figure 8.9). Orthogonal trajectories appear in a variety of applications. For instance, in aerodynamics, the orthogonal trajectories of streamlines (curves of airflow direction) are *equipotential* lines, and in thermodynamics, the orthogonal trajectories of heat flow lines are the *isothermal* lines (curves of constant temperature).

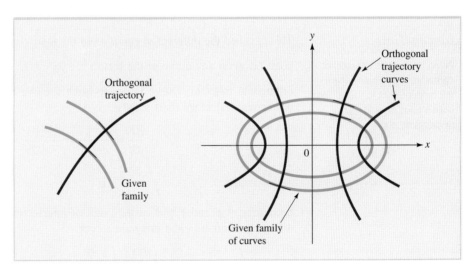

FIGURE 8.9   Orthogonal trajectories of a family of curves.

The general procedure for finding the orthogonal trajectories of a given family of curves involves solving a differential equation. This procedure is illustrated in Example 8.3.4 as we address an applied problem from biology.

EXAMPLE 8.3.4

In biology, an organism moving on a flat surface in a nutrient bed ideally follows a path along which the level of nutrient concentration increases. The organism receives cues to its progress by continually crossing curves of constant nutrient concentration, that is, curves along which the concentration stays the same. This type of movement, called *chemotaxis,* is performed most efficiently by following a path that is orthogonal to the curves of constant nutrient concentration. In other words, in its quest for greater and greater nutrient levels, the organism will proceed along an orthogonal trajectory of the family of curves of constant nutrient concentration.

Suppose a particular nutrient bed is distributed in the xy plane so that the nutrient concentration is always the same along any curve of the form $x^2 + 2y^2 = C$. If a chemotactic organism is introduced to the nutrient bed at the point $(1, 2)$, what path does it follow?

Solution

It is necessary to find an orthogonal trajectory of the family of curves $x^2 + 2y^2 = C$ that passes through the point $(1, 2)$. Use uppercase letters X and Y when representing the orthogonal trajectory curves to distinguish those curves from curves in the given family $x^2 + 2y^2 = C$.

Differentiate implicitly with respect to x in the equation $x^2 + 2y^2 = C$ to find that

$$2x + 2(2y)\frac{dy}{dx} = 0,$$

so that

$$\frac{dy}{dx} = \frac{-2x}{4y} = \frac{-x}{2y}.$$

Just-In-Time

When two curves intersect at right angles at a point where the tangent lines to the curves have slopes m_1 and m_2, respectively, then

$$m_2 = \frac{-1}{m_1}.$$

This is called the *differential equation of the family $x^2 + 2y^2 = C$*. It says that at each point (x, y) on any curve of the form $x^2 + 2y^2 = C$, the slope is $\frac{dy}{dx} = \frac{-x}{2y}$. Since an orthogonal trajectory curve intersects such a curve at right angles, the slope of the orthogonal trajectory must satisfy

$$\frac{dY}{dX} = \frac{-1}{dy/dx} = \frac{-1}{-x/2y} = \frac{2y}{x}$$

$$= \frac{2Y}{X}$$

because $x = X$ and $y = Y$ at the point of intersection. Separate the variables in this differential equation and integrate to get

$$\int \frac{dY}{Y} = \int \frac{2 \, dX}{X}$$

$$\ln Y = 2 \ln X + K_1 = \ln X^2 + K_1$$

for some constant K_1. Taking exponentials on each side of this equation results in

$$Y = KX^2 \qquad \text{where } K = e^{K_1}.$$

This is the formula for the family of orthogonal trajectories of the given family of curves $x^2 + 2y^2 = C$. To find the particular member of the orthogonal trajectory family that passes through the point $(1, 2)$ where the organism begins its search for food, substitute $X = 1$ and $Y = 2$ into the formula $Y = KX^2$ and note that

$$2 = K(1)^2,$$

so that

$$K = 2.$$

Thus, the organism travels along the orthogonal trajectory curve $Y = 2X^2$. Members of the given family of curves (ellipses) are shown in Figure 8.10 along with the orthogonal trajectory curve $Y = 2X^2$ (a parabola).

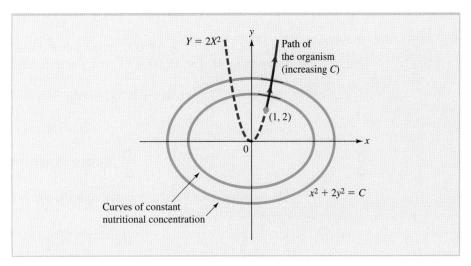

FIGURE 8.10   Motion of an organism in a nutrient bed.

A Financial Model: Linked Differential Equations

Certain applications involve variables whose rates of change are linked by a system of differential equations. In Example 8.3.5, we examine a model in which the flow of money between two components of a financial portfolio is determined by such a system.

EXAMPLE 8.3.5

Jessica currently has $15 000 invested in a money market fund earning 3% per year compounded continuously. Market conditions are improving and she decides to continuously transfer 20% of her account into a stock fund earning 8% per year. How much will be in each account 4 years from now?

Solution

Let $M(t)$ be the value of the money market account after t years and let $S(t)$ be the corresponding value of the stock account. According to the given information, the money market account changes at a rate of

$$\frac{dM}{dt} = \underbrace{0.03M}_{\text{growth rate}} - \underbrace{0.2M}_{\text{transfer rate}} = -0.17M,$$

while the rate of change of the stock account is

$$\frac{dS}{dt} = \underbrace{0.08S}_{\text{growth rate}} + \underbrace{0.2M}_{\text{transfer from } M}$$

The differential equation $\frac{dM}{dt} = -0.17M$ is an exponential growth equation. In Example 8.1.6 of Section 8.1, we showed that such an equation has solution

$$M(t) = M_0 e^{-0.17t}$$

where $M_0 = M(0) = 15\ 000$ is the initial value of the money market account. Substituting $M(t) = 15\ 000e^{-0.17t}$ into the differential equation for $S(t)$ results in

$$\frac{dS}{dt} = 0.08S + 0.20[15\ 000e^{-0.17t}] = 0.08S + 3000e^{-0.17t}$$

or, equivalently,

$$\frac{dS}{dt} - 0.08S = 3000e^{-0.17t}.$$

This is a first-order linear differential equation with $p(t) = -0.08$ and $q(t) = 3000e^{-0.17t}$. The integrating factor is

$$I(t) = e^{\int -0.08dt} = e^{-0.08t}$$

and the general solution is

$$S(t) = \frac{1}{e^{-0.08t}}\left[\int e^{-0.08t}(3000e^{-0.17t})\ dt + C\right]$$

$$= e^{0.08t}\left[\int 3000e^{-0.25t}dt + C\right]$$

$$= e^{0.08t}\left[\frac{3000e^{-0.25t}}{-0.25} + C\right]$$

$$= -12\ 000e^{-0.17t} + Ce^{0.08t}$$

Since the stock account initially contains no money, $S(0) = 0$ and

$$S(0) = 0 = -12\ 000e^0 + Ce^0 = -12\ 000 + C,$$

so that $C = 12\ 000$ and

$$S(t) = 12\ 000(-e^{-0.17t} + e^{0.08t}).$$

Therefore, when $t = 4$,

$$M(4) = 15\ 000e^{-0.17(4)} \approx 7599.25$$

and

$$S(4) = 12\ 000(-e^{-0.17(4)} + e^{0.08(4)})$$
$$\approx 10\ 446.13$$

That is, after 4 years, the money market account contains approximately \$7599, while the stock account contains about \$10 446.

Interaction of Species: The Predator-Prey Model

An important area of interest in ecology involves the study of interactions among various species occupying the same environment. We shall examine a classic case, called the **predator-prey model,** in which one species (the predator) feeds on a second species (the prey) that in turn has an adequate supply of food and space. To simplify the discussion of the model, we shall call the predators "foxes" and the prey "rabbits" and denote their respective populations at time t by $F(t)$ and $R(t)$.

If there were no foxes, then since rabbits have access to ample food and space, we would expect their population to grow exponentially, that is,

$$\frac{dR}{dt} = aR, \quad \text{where } a > 0.$$

On the other hand, if there were no rabbits, the foxes would have no food supply and their population would be expected to decline exponentially, so that

$$\frac{dF}{dt} = -bF, \quad \text{where } b > 0.$$

If both foxes and rabbits are present, then encounters between them result in a decline in the rabbit population (rabbit deaths) and an increase in the fox population (more food). We assume such encounters are proportional to the product FR, since an increase in either population should increase the frequency of encounters. Combining these assumptions, we model the growth rate of rabbits by

$$\frac{dR}{dt} \;=\; \underbrace{aR}_{\substack{\text{natural} \\ \text{growth} \\ \text{rate}}} \;-\; \underbrace{cFR}_{\substack{\text{accelerated} \\ \text{death rate due} \\ \text{to encounters}}}$$

and the growth rate of foxes by

$$\frac{dF}{dt} \;=\; \underbrace{-bF}_{\substack{\text{starvation} \\ \text{rate}}} \;+\; \underbrace{dFR}_{\substack{\text{accelerated} \\ \text{growth rate due} \\ \text{to encounters}}}$$

where c and d are positive constants called *interaction coefficients.*

Solving this linked pair of differential equations would allow us to determine the two populations at each time t. Unfortunately, however, it is usually impossible to find explicit formulas for $F(t)$ and $R(t)$. Instead, we analyze the populations indirectly using the differential equations themselves. Here is an example.

EXAMPLE 8.3.6

An ecologist studying a forested region models the dynamics of the fox and rabbit populations within the region by the predator-prey equations

$$\frac{dR}{dt} = 0.045R - 0.0015FR$$

$$\frac{dF}{dt} = -0.12F + 0.0003FR$$

a. An *equilibrium solution* is the constant populations $R(t) = R_e$ and $F(t) = F_e$ that satisfy the modelling equations. Find and interpret all equilibrium solutions of the given model.

b. Find an expression for $\dfrac{dR}{dF}$ and separate variables to obtain an implicit solution for the predator-prey model.

c. Suppose the ecologist determines that there are 300 rabbits and 18 foxes at time $t = 0$. Use graphing software along with the implicit solution found in part (b) to sketch the solution curve that passes through the point (300, 18). Then use your curve to describe how the two populations change for increasing time t.

Solution

a. An equilibrium solution (R_e, F_e) must satisfy

$$0.045R_e - 0.0015F_eR_e = 0$$
$$-0.12F_e + 0.0003F_eR_e = 0$$

One solution is $R_e = 0$ and $F_e = 0$, that is, no rabbits or foxes. Clearly, if this situation ever occurs, it will never change. If $R_e \neq 0$ and $F_e \neq 0$, then R_e can be cancelled from the first equation and F_e from the second to obtain

$$0.045 - 0.0015F_e = 0$$
$$-0.12 + 0.0003R_e = 0$$

so that

$$F_e = \frac{0.045}{0.0015} = 30$$

$$R_e = \frac{0.12}{0.0003} = 400$$

If the rabbit population is ever 400 at the same time as the fox population is 30, then these populations will never change because the rates of change $R'(t)$ and $F'(t)$ are both zero.

b. According to the chain rule,

$$\frac{dF}{dt} = \frac{dF}{dR}\frac{dR}{dt}$$

so that

$$\frac{dF}{dR} = \frac{dF/dt}{dR/dt} = \frac{-0.12F + 0.0003FR}{0.045R - 0.0015FR}$$

$$= \frac{0.0003F(-400 + R)}{0.0015R(30 - F)} = \frac{F(-400 + R)}{5R(30 - F)}$$

Separating the variables and integrating gives

$$\int \frac{5(30 - F)}{F}\,dF = \int \frac{-400 + R}{R}\,dR \qquad \begin{array}{l}\text{cross-multiply}\\\text{and integrate}\end{array}$$

$$5\int \left[\frac{30}{F} - 1\right]dF = \int \left[\frac{-400}{R} + 1\right]dR$$

$$5(30 \ln F - F) = -400 \ln R + R + C$$

This equation provides an implicit general solution of the given predator-prey system of equations.

c. Substituting the initial conditions $R = 300$ when $F = 18$ into the equation in part (b) results in

$$5[30 \ln (18) - 18] = -400 \ln (300) + 300 + C$$
$$C = 400 \ln (300) + 150 \ln (18) - 300 - 90$$
$$\approx 2325$$

so the required solution curve has the equation

$$5(30 \ln F - F) = -400 \ln R + R + 2325$$

or, equivalently,

$$150 \ln F + 400 \ln R - 5F - R = 2325.$$

A computer sketch of this solution curve is shown in Figure 8.11. Note that the solution is an oval-shaped closed curve that contains the equilibrium point $(400, 30)$ in its interior.

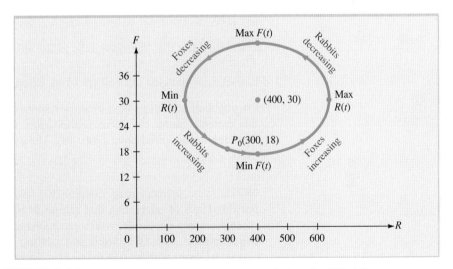

FIGURE 8.11   The predator-prey solution curve through the point $(300, 18)$:
$$150 \text{ LN } F + 400 \text{ LN } R - 5F - R = 2325.$$

Next, we shall interpret the behaviour of the fox and rabbit populations in terms of the solution curve shown in Figure 8.11. To this end, we need to know whether the curve is traversed clockwise or counterclockwise with increasing time. To make this determination, we substitute the initial values $R = 300$ and $F = 18$ into the given predator-prey equations and find that

$$\frac{dR}{dt} = 0.045(300) - 0.0015(18)(300) = 5.4$$

and

$$\frac{dF}{dt} = -0.12(18) + 0.0003(18)(300) = -0.54.$$

Since $\dfrac{dR}{dt} > 0$ and $\dfrac{dF}{dt} < 0$ at the initial point (300, 18) on the solution curve, it follows that as t increases, the rabbit population *increases* while the fox population *decreases*. That is, we initially move to the right (R increasing) and down (F decreasing) along the solution curve, so the curve is *traversed counterclockwise*, as indicated by the arrows in Figure 8.11.

Since the fox population is initially decreasing, 18 foxes are not enough to keep the rabbit population in check. Moving from the initial point (300, 18) to the right along the solution curve (rabbits increasing), the curve keeps falling (fox population is decreasing) until the rabbit population reaches 400 (where $\dfrac{dR}{dt} = 0$). At this time, there are so many rabbits that hunting becomes almost a sure thing, and the fox population begins to increase. For a while, the curve rises to the right (both populations are increasing), but eventually, there are so many foxes that the rabbits have trouble foraging for food without being killed. This corresponds to the extreme right end of the curve ($F = 30$, where $\dfrac{dR}{dt} = 0$). The curve then moves to the left and up (rabbits are decreasing while foxes are increasing) until it reaches its highest point. At this time, the rabbit population has fallen to the level where hunting ceases to be automatic ($R = 400$, where $\dfrac{dF}{dt} = 0$). After that, the curve falls to the left (both populations decreasing) until it reaches its extreme left end ($F = 30$, where $\dfrac{dR}{dt} = 0$), which corresponds to a fox population small enough to allow rabbits to again forage freely. Finally, from this extreme point, the curve moves again to the right and down, eventually returning to the initial point (300, 18). The cycle then repeats indefinitely.

NOTE The predator-prey situation we have analyzed is only one of several possible kinds of interaction that can occur between species. For instance, some species actually cooperate, with interactions acting to increase both growth rates. Other species, such as squirrels and rabbits, compete for food within the same ecological space without one feeding on the other. Sometimes such species coexist within their environment, and sometimes one species dominates while the other becomes extinct.

EXERCISES ■ 8.3

ORTHOGONAL TRAJECTORIES *In Exercises 1 through 4, find an equation for the orthogonal trajectories of the given family of curves. In each case, sketch a few members of both the given family and the family of orthogonal trajectories.*

1. $xy^2 = C$

2. $x^2 + y^2 = C$

3. $y = Cx^2$

4. $y = Ce^{-x}$

PRICE ADJUSTMENTS FOR SUPPLY/DEMAND *In Exercises 5 through 8, the supply S(t) and demand D(t) functions for a commodity are given in terms of the unit price p(t) at time t. Assume that price changes at a rate proportional to the shortage D(t) − S(t), with the indicated constant of proportionality k and initial price p_0. In each exercise, do the following:*

(a) Set up and solve a differential equation for p(t).

(b) Find the unit price of the commodity when t = 4.
(c) Determine what happens to the price as t → ∞.

5. $S(t) = 2 + 3p(t)$; $D(t) = 10 - p(t)$; $k = 0.02$; $p_0 = 1$

6. $S(t) = 1 + 4p(t)$; $D(t) = 15 - 3p(t)$; $k = 0.015$; $p_0 = 3$

7. $S(t) = 2 + p(t)$; $D(t) = 3 + 7e^{-1}$; $k = 0.02$; $p_0 = 4$

8. $S(t) = 1 + p(t)$; $D(t) = 2 + 8e^{-t/2}$; $k = 0.03$; $p_0 = 2$

9. **SAVINGS**   Marcia currently has \$30 000 invested in a money market fund earning 2% per year compounded continuously. Market conditions are improving, and she decides to continuously transfer 25% of her account into a stock fund earning 10% per year. How much will be in each account 5 years from now?

10. **SAVINGS**   Tyler currently has \$40 000 invested in a savings account earning 3% per year compounded continuously. Market conditions are improving, and he decides to continuously transfer 15% of his account into a stock fund earning 12% per year. He also withdraws \$5000 from the savings account each year for expenses.
    a. Set up and solve a linked pair of differential equations to determine the amount of money in each account t years from now.
    b. How long does it take for the savings account to be exhausted? How much money is in the stock account at this time?

11. **DISSOLUTION OF SUGAR**   When placed in a container of water, sugar dissolves at a rate proportional to the amount $Q(t)$ of undissolved sugar remaining in the container at that time (see Exercise 36, Section 8.1). If half the sugar has dissolved after 3 minutes, how long does it take for $\dfrac{3}{4}$ of the sugar to dissolve?

12. **NEWTON'S LAW OF COOLING**   When a cold drink is taken from the refrigerator, its temperature is 5°C. It is taken outside, where the air temperature is 22°C, and 20 minutes later, its temperature is 10°C.
    a. What will its temperature be 30 minutes after being removed from the refrigerator?
    b. How long will it take before the temperature of the drink is 15°C?

13. **NEWTON'S LAW OF COOLING**   At the Grey Cup in Edmonton one year, an insulated cup of hot chocolate is taken outside, where the temperature is −5°C. After 10 minutes, its temperature is 70°C, and 10 minutes after that, its temperature is 50°C. What was the original temperature of the drink?

14. **NEWTON'S LAW OF COOLING**   At noon, a turkey is removed from the oven where its temperature was 80°C and is placed in a room where it takes 1 hour for it to cool to 52°C. If the turkey is ready to be served at 2:00 P.M. when its temperature reaches 38°C, what is the air temperature in the room?

15. **FORENSICS**   A dead body is discovered at 3:00 P.M. on Monday in a storage room where the air temperature is 10°C. The temperature of the body at the time of discovery is 27°C, and 20 minutes later, it is 26°C. Assume normal body temperature is 37°C. What was the time of death?

16. **ATMOSPHERIC PRESSURE**   Assuming constant temperature, atmospheric pressure P decreases at a rate proportional to P with respect to altitude h above sea level. Suppose the pressure is 101.3 kPa at sea level, and at an altitude of 1600 m it is 83.1 kPa.
    a. Set up and solve a differential equation to express P as a function of h.
    b. What is the pressure at the top of Mount Everest, at an altitude of 8848 m?
    c. A healthy human adult becomes distressed when the atmospheric pressure decreases to half its value at sea level. At what altitude does this occur?

17. **AMOUNT OF DRUG IN AN ORGAN**   A drug with concentration 0.16 g/cm³ is delivered into an organ at a rate of 6 cm³ per second and is dispersed at the same rate. Suppose the organ has volume 200 cm³ and initially contains none of the drug. Set up and solve a differential equation for the amount of drug $A(t)$ in the organ at time t.

18. **AMOUNT OF DRUG IN AN ORGAN**   A drug with concentration 0.08 g/cm³ is delivered into an organ at a rate of 12 cm³ per second and is dispersed at the same rate. Suppose the organ has a volume of 600 cm³ and initially contains none of the drug.
    a. Set up and solve a differential equation for the amount of drug $A(t)$ in the organ at time t.
    b. What is the concentration of the drug in the organ after 30 seconds? After 1 minute?
    c. How long does it take for the concentration of drug in the organ to reach 0.06 g/cm³?

19. **A* AMOUNT OF DRUG IN AN ORGAN**   Suppose that a drug with concentration c_0 grams per cubic centimetre enters an organ of volume V cubic centimetres at a rate of A cubic centimetres per second and is dispersed at the same rate. If the

organ initially contains none of the drug and it can safely accommodate a concentration no greater than L grams per cubic centimetre, where $L < c_0$, what is the maximum length of time for which the drug can be allowed to enter the organ? Your answer will be an expression involving V, c_0, L, and A.

20. **ABSORPTION OF LIGHT** The intensity of light $I(d)$ at a depth d below the surface of a body of water changes at a rate proportional to I. If the intensity at a depth of $d = 0.5$ m is half the surface intensity I_0, at what depth is the intensity 5% of I_0?

21. **RESTAURANT MAINTENANCE** As part of its decor, a seafood restaurant features a 3000-L aquarium filled with exotic fish. To keep the fish healthy and active, a solution containing $0.2te^{-t/40}$ grams of soluble nutrient is pumped into the tank and circulated at a rate of 75 L per minute.
    a. Set up and solve a differential equation for the amount of nutrient $N(t)$ in solution at time t.
    b. What is the maximum concentration of nutrient in the tank and when does it occur?

22. **DIFFUSION ACROSS A MEMBRANE** A chemical in a solution diffuses from a compartment with concentration $C_1(t)$ across a membrane to a second compartment where the concentration is $C_2(t)$. Experiments suggest that the rate of change of $C_2(t)$ with respect to time is proportional to the difference $C_1 - C_2$ in concentrations. Set up and solve a differential equation for $C_2(t)$ for the case where $C_2(0) = 0$, $C_1(t) = 100e^{-t}$, and $k = 1.5$ is the constant of proportionality.

EMIGRATION FROM A POPULATION *Suppose a population $P(t)$ grows exponentially at a natural rate r and that $E(t)$ individuals are emigrating from the population at time t so that*

$$\frac{dP}{dt} = rP - E.$$

Solve this equation to determine the population at time t for each of the cases in Exercises 23 and 24.

23. $r = 0.03$, $E(t) = 10t$, and $P(0) = 100\,000$.

24. $r = 0.015$, $E(t) = 200e^{-t}$, and $P(0) = 250\,000$.

25. **IMMIGRATION TO A POPULATION** Suppose a population $P(t)$ grows exponentially at a rate r and that $I(t)$ individuals are immigrating into the population at time t. Set up and solve a differential equation for $P(t)$ in the case where t is in years, $r = 0.02$, $I(t) = 100e^{-t}$, and $P(0) = 300\,000$.

26. **MANAGEMENT OF A FISHERY** Suppose a population of fish in a fishery grows exponentially with natural growth rate k and that harvesting (fishing) is allowed at a constant, continuous rate h.
    a. Set up and solve a differential equation for the population $P(t)$ at time t. [Your answer will involve k, h, and the initial population $P_0 = P(0)$.]
    b. Find $P(t)$ for the case where t is in years, $k = 0.2$, the initial population is $P(0) = 2000$, and the harvesting rate is $h = 300$ fish per year.
    c. Suppose the fishery in part (b) decides on a policy of harvesting just enough fish so the population remains constant at 2000. What harvesting rate h will implement this policy? Use technology to help with this answer.

27. **SPY STORY** The spy's brush with death in the broiler room in Chapter 7 convinces him that he needs more information about the villain's operations before resuming his quest for vengeance. He enters a village near the villain's chateau, disguised as a travel blog writer. On the day of his arrival, his true identity is known only to his operatives. However, that night one of the operatives tells his girlfriend. Soon, word concerning his identity begins to spread among the 60 citizens of the village at a rate jointly proportional to the number of people $N(t)$ who know who he is t days after his arrival and the $60 - N$ people in the village who still do not know at that time. The spy thinks that the villain will learn of his identity as soon as at least 20 villagers know. If he needs a week to gather the required information, does he complete his mission?

28. **AMOUNT OF DRUG IN AN ORGAN** Answer the question in Exercise 17 for the case where the initial concentration of drug in the organ is d_0 grams per cubic centimetre rather than 0. Your answer will now be an expression involving V, c_0, d_0, L, and A.

29. **AMOUNT OF DRUG IN AN ORGAN** A certain drug is injected into an organ of a patient at a rate of 5 units per hour and is dispersed at a rate of 4 units per hour. This causes the organ to expand at a rate of 2 cm^3 per hour. Initially, the organ contains 50 units of the drug and has volume 800 cm^3.
    a. Set up and solve a differential equation to find the concentration $C(t)$ of drug in the organ after t hours.
    b. The organ can safely accommodate a concentration of no more than 0.13 units/cm^3. What is the maximum length of time the drug can be allowed to enter the organ?

30. CURRENCY RENEWAL A nation has \$5 billion in currency. Each day, about \$18 million comes into the banks and the same amount is paid out. Suppose the country decides that whenever a dollar bill comes into a bank, it is destroyed and replaced by a new style of currency. How long will it take for 90% of the currency in circulation to be the new style? [*Hint:* Think of this situation as if it were a dilution problem.]

31. CHEMOTAXIS Suppose a particular nutrient bed is distributed in the xy plane so that the nutrient concentration is always the same along any curve of the form $3x^2 + y^2 = C$. If a chemotactic organism is introduced to the nutrient bed at the point $(1, 1)$, what path does it follow?

32. CHEMOTAXIS Suppose a particular nutrient bed is distributed in the xy plane so that the nutrient concentration is always the same along any curve of the form $x^2 - y^2 = C$. If a chemotactic organism is introduced to the nutrient bed at the point $(3, 1)$, what path does it follow?

33. A* SPREAD OF AN EPIDEMIC Let $N(t)$ be the number of people from a susceptible population N_s who are infected by a certain disease t days after the onset of an epidemic. Researchers model the epidemic using the differential equation

$$\frac{dN}{dt} = k(N_s - N)(N - m),$$

where $k > 0$ and m are constants, with $0 < m < N_s$. Call m the *epidemic threshold* of the disease.

Suppose that $k = 0.04$, that $N_s = 250\,000$ people are susceptible to the disease, and that the epidemic threshold m is 1% of the susceptible population.

a. Solve the differential equation for this epidemic in terms of $N_0 = N(0)$, the number of people who are initially infected. (*Hint:* Integration formula 6 in Table 6.1 may help.)

b. Show that if $N_0 = 5000$, then $N(t)$ approaches N_s in the long run. That is, the entire susceptible population is eventually infected.

c. Show that if $N_0 = 1000$, then there is a finite time T such that $N(T) = 0$. That is, the epidemic eventually dies out.

d. In general, what relationship between N_0 and m determines whether an epidemic modelled by a differential equation of the given form dies out in finite time? What effect (if any) does changing the value of k have on this determination?

34. A* PREDATOR-PREY Suppose a large wildlife preserve contains W thousand wolves (predators) and E thousand elk (prey). An ecologist finds that there are currently 20 000 elk and 5000 wolves and models the growth rates of these two populations with respect to time t by the following pair of differential equations:

$$\frac{dE}{dt} = 96E - 8EW$$

$$\frac{dW}{dt} = -123W + 3EW$$

a. Find equilibrium populations for this model, that is, populations E_e and W_e that satisfy

$$\frac{dE}{dt} = \frac{dW}{dt} = 0.$$

b. Note that

$$\frac{dE}{dW} = \frac{96E - 8EW}{-123W + 3EW}.$$

Separate the variables in this equation, and solve to obtain an implicit solution for the system.

35. A* COMPETITION BETWEEN SPECIES A certain ecological territory contains S thousand squirrels and R thousand rabbits. Currently, there are 4000 of each species and the growth rates of the populations with respect to time t are given by the following pair of differential equations:

$$\frac{dR}{dt} = 63R - 3RS$$

$$\frac{dS}{dt} = 26S - RS$$

a. Find equilibrium populations for this model, that is, populations R_e and S_e that satisfy

$$\frac{dR}{dt} = \frac{dS}{dt} = 0.$$

b. Note that

$$\frac{dR}{dS} = \frac{63R - 3RS}{26S - RS}.$$

Separate the variables in this equation and solve to obtain an implicit solution for the system.

36. A* DOMAR DEBT MODEL Let D and I denote the national debt and national income of a certain country, and assume that both are functions of time t. One of several **Domar debt models** assumes that the time rates of change of D and I are both proportional to I, so that

$$\frac{dD}{dt} = aI \quad \text{and} \quad \frac{dI}{dt} = bI.$$

Suppose that $I(0) = I_0$ and $D(0) = D_0$.

a. Solve both of these differential equations and express $D(t)$ and $I(t)$ in terms of a, b, I_0, and D_0.

b. The economist who first studied this model, Evsey Domar, was interested in the ratio of national debt to national income. Determine what happens to this ratio in the long run by computing the limit

$$\lim_{t \to \infty} \frac{D(t)}{I(t)}.$$

37. A* TIME ADJUSTMENT OF SUPPLY AND DEMAND   The supply $S(t)$ and demand $D(t)$ of a certain commodity vary with time t (months) so that

$$\frac{dD}{dt} = -kD \quad \text{and} \quad \frac{dS}{dt} = 2kS$$

for some constant $k > 0$. It is known that $D(0) = 50$ units and $S(0) = 5$ units and that equilibrium occurs when $t = 10$ months, that is, when $D(10) = S(10)$.

a. Use this information to find k.

b. Find $D(t)$ and $S(t)$.

c. How many units are supplied and demanded at equilibrium?

38. A* WATER POLLUTION   A lake holds 4 billion cubic metres of water, and initially, its pollutant content is 0.19%. A river whose water contains 0.04% pollutant flows into the lake at a rate of 250 million cubic metres per day and then flows out of the lake at the same rate. Assuming that the water in the lake and the two rivers is always well mixed, how long does it take for the pollutant content in the lake to be reduced to 0.1%?

39. A* WATER POLLUTION   Two lakes are connected by a river as shown in the accompanying figure. The lakes both contain pure water until 2000 kg of pollutant is dumped into the upper lake. Suppose the upper lake contains 700 000 L of water, the lower lake contains 400 000 L, and the river flows at a rate of 1500 L per hour. Assume that the pollutant disperses rapidly enough so that the mixture of pollutant and water is well mixed at all times.

a. Set up and solve a differential equation for the amount of pollutant $P_1(t)$ in the upper lake at time t.

b. If $P_2(t)$ is the amount of pollutant in the lower lake at time t, then there are constants a and b such that

$$\frac{dP_2}{dt} = aP_1 - bP_2.$$

Use the given information to determine a and b. Then solve the equation for $P_2(t)$.

c. What is the maximum amount of pollutant contained in the lower lake? When does this maximum amount occur?

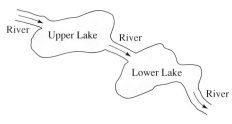

EXERCISE 39

40. A* CRIMINAL INVESTIGATION   At midnight in Toronto, a surveillance camera at an intersection records a man driving erratically enough to suggest that he may be under the influence of alcohol. Two hours later, the police locate the driver and determine that the alcohol concentration in his blood is 0.06%. An hour later at police headquarters, a follow-up test detects a blood-alcohol level of 0.05%. Evidence indicates that the suspect consumed no alcohol after midnight. If the legal blood-alcohol level is 0.08%, can the suspect be charged with driving under the influence of alcohol? (Assume that the percentage P of alcohol in the bloodstream satisfies the decay equation $P' = -kP$.)

41. A* RUNAWAY GROWTH   For certain prolific species, the population $P(t)$ grows at a rate proportional to P^{1+c} for $c > 0$; that is,

$$\frac{dP}{dt} = kP^{1+c}.$$

a. Solve this differential equation in terms of k, c, and the initial population $P_0 = P(0)$.

b. Show that there is a finite time t_a such that
$$\lim_{t \to t_a} P(t) = \infty.$$

c. Suppose $c = 0.02$ for a certain population of rodents. Find $P(t)$ if the population begins with a mated pair and two months later has doubled (that is, $P(0) = 2$ and $P(2) = 4$).

d. How long does it take for the population in part (c) to become essentially infinite?

SECTION 8.4

L04

Approximate solutions to differential equations with slope fields and Euler's method.

Approximate Solutions of Differential Equations

Certain differential equations, including many that arise in practical applications, cannot be solved by any known method. However, it is often possible to approximate solutions to such equations. In this section, we will discuss several procedures for obtaining approximate solutions of differential equations.

Slope Fields

Suppose we wish to solve the initial value problem

$$\frac{dy}{dx} = f(x, y), \quad y(x_0) = y_0.$$

We may not be able to find the exact solution, but we can obtain a rough sketch by using the fact that since $y' = f(x, y)$ at each point (x, y) on a solution curve $y = y(x)$, the slope of the tangent line to the curve at (x, y) is given by $f(x, y)$. Here is a description of the procedure we will follow:

Slope Field Procedure for Sketching a Solution Curve of $y' = f(x, y)$ That Contains the Point (x_0, y_0)

Step 1. At each point (x, y) on a grid in the xy plane, draw a short line segment with slope equal to $f(x, y)$. The collection of all such points and segments is called a **slope field** for the differential equation $y' = f(x, y)$.

Step 2. Starting with the point (x_0, y_0), use the slopes indicated by the slope field to sketch the required solution, as demonstrated in the following figure:

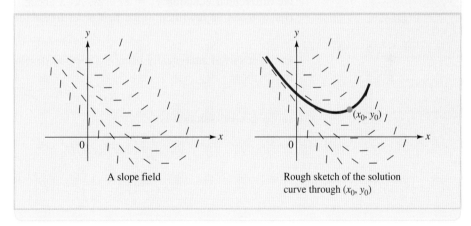

A slope field   Rough sketch of the solution curve through (x_0, y_0)

The slope field procedure for visualizing a solution of a differential equation is illustrated in Example 8.4.1. Slope fields can be produced by computer software such as Maple.

EXAMPLE 8.4.1

Use a slope field to sketch several solutions of the differential equation

$$\frac{dy}{dx} = x - 2y.$$

In particular, sketch the solution that passes through the point $(0, 1)$.

Solution

Note that if (x, y) is a point on the line $x - 2y = m$, then any solution curve $y = y(x)$ of the differential equation $y' = x - 2y$ that passes through (x, y) will have slope $y' = m$. Thus, construct the slope field for this differential equation by drawing the family of parallel lines $x - 2y = m$ and placing a small line segment of slope m at selected points along each such line. This is done in Figure 8.12a.

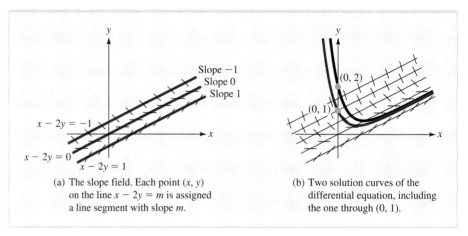

(a) The slope field. Each point (x, y) on the line $x - 2y = m$ is assigned a line segment with slope m.

(b) Two solution curves of the differential equation, including the one through $(0, 1)$.

FIGURE 8.12   Using a slope field to construct solutions for $y' = x - 2y$.

In Figure 8.12b, the slope field in Figure 8.12a is used to draw two solutions to the differential equation $y' = x - 2y$. As a check, notice that the given differential equation can be rewritten in the first-order linear form

$$y' + 2y = x$$

and then solved analytically using the method developed in Section 8.2. The integrating factor is

$$I(x) = e^{\int 2\,dx} = e^{2x},$$

so the general solution is

$$y = \frac{1}{e^{2x}}\left[\int x\,e^{2x}dx + C\right]$$

$$= e^{-2x}\left[\left(\frac{x}{2} - \frac{1}{4}\right)e^{2x} + C\right] \qquad \begin{array}{l}\text{integration by parts:}\\ u = x \qquad dv = e^{2x}\,dx\end{array}$$

$$= \frac{x}{2} - \frac{1}{4} + Ce^{-2x}$$

For the particular solution with $y = 1$ when $x = 0$,

$$1 = 0 - \frac{1}{4} + Ce^0 = -\frac{1}{4} + C,$$

so that $C = \dfrac{5}{4}$ and

$$y = \frac{x}{2} - \frac{1}{4} + \frac{5}{4}e^{-2x}.$$

This is the curve sketched in red in Figure 8.12b. A computer plot of the same problem is shown in the margin.

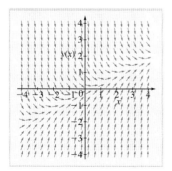

Maple field plot Example 8.4.1.

The slope field procedure was not really necessary in Example 8.4.1 because it was fairly easy to solve the differential equation using the methods of Section 8.2. However, in Example 8.4.2, we use the procedure to obtain approximate solutions to a differential equation that cannot be solved directly.

EXAMPLE 8.4.2

Use a slope field to sketch solutions of the differential equation

$$\frac{dy}{dx} = \sqrt{x + y}.$$

Solution

If (x, y) is a point on the line $x + y = m$ for $m \geq 0$, then any solution curve $y = y(x)$ of the differential equation $y' = \sqrt{x + y}$ that passes through (x, y) will have slope $y' = \sqrt{m}$. Construct the slope field for this equation by drawing the family of parallel lines $x + y = m$ and assigning a small line segment of slope $\sqrt{m}$ at selected points along each such line. Figure 8.13a shows the slope field. Note that since $\sqrt{x + y}$ is not defined for $x + y < 0$, the entire slope field lies on or above the line $x + y = 0$. Three solution curves are sketched in Figure 8.13b. Notice that each solution curve is flat (has slope 0) on the line $x + y = 0$ and then rises rapidly to the right.

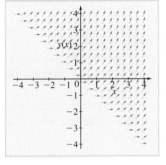

Maple field plot Example 8.4.2.

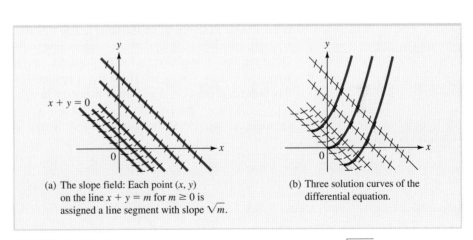

(a) The slope field: Each point (x, y) on the line $x + y = m$ for $m \geq 0$ is assigned a line segment with slope $\sqrt{m}$.

(b) Three solution curves of the differential equation.

FIGURE 8.13   A slope field construction of solutions for $y' = \sqrt{x + y}$.

Euler's Method

The slope field procedure we have just examined enables us to obtain a rough sketch of the solution to an initial value problem, and our next goal is to show how such a solution can be generated numerically. The great advantage of a numerical approximation procedure is that it can be programmed for solution by a computer. Indeed, using computers and numerical approximation schemes, it is now possible to analyze accurate models of complex situations such as weather patterns, ecosystems, and economic interactions among nations.

The particular numerical approximation procedure we will discuss is called Euler's method, after its originator, the great Swiss mathematician Leonhard Euler (1707–1783). Although Euler's method is much less accurate than the numerical procedures used in practical applications, it deals with the same issues as the more sophisticated procedures and its relatively simple structure makes it easier to demonstrate.

To illustrate the ideas behind Euler's method, suppose we wish to approximate the solution to the initial value problem

$$\frac{dy}{dx} = f(x, y) \qquad \text{for } y(x_0) = y_0.$$

The key to Euler's approach is that once the solution value $y(c)$ is known for some number c, then we can compute $f(c, y(c))$, which equals $y'(c)$, the slope of the tangent line to the graph of the solution at $(c, y(c))$. Using this fact, we can approximate values of the solution $y = y(x)$ at successive, evenly spaced points.

To begin the approximation process, choose the *step size h*. This is the distance along the x axis between successive points on the approximate solution curve. That is, we will approximate values of the solution at the points

$$x_0, \, x_1 = x_0 + h, \, x_2 = x_0 + 2h, \ldots, x_n = x_0 + nh.$$

Starting at (x_0, y_0), compute $f(x_0, y_0)$ and draw the line through (x_0, y_0) with slope $f(x_0, y_0)$. We know that this line is tangent to the solution curve $y = y(x)$ at (x_0, y_0), and we proceed along this line to the point (x_1, y_1), where

$$y_1 = y_0 + f(x_0, y_0)(x_1 - x_0) = y_0 + f(x_0, y_0)h$$

is the y value of the point on the tangent line that corresponds to x_1. Next, we compute $f(x_1, y_1)$, which is the slope of the tangent line to the solution curve through (x_1, y_1). Draw the line through (x_1, y_1) with slope $f(x_1, y_1)$ and proceed to the point (x_2, y_2), where $y_2 = y_1 + f(x_1, y_1)h$. These two steps are shown in Figure 8.14a. Continue in this fashion, at the $(k + 1)$th step drawing the line through the point (x_k, y_k) with slope $f(x_k, y_k)$ to locate the point (x_{k+1}, y_{k+1}) where $y_{k+1} = y_k + f(x_k, y_k)h$. The complete approximate solution is the polygonal path shown in Figure 8.14b.

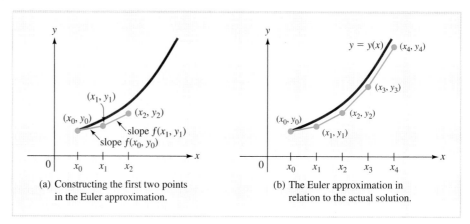

(a) Constructing the first two points
in the Euler approximation.

(b) The Euler approximation in
relation to the actual solution.

FIGURE 8.14   Euler's method for approximating the solution to the initial value problem $y'(x) = f(x, y)$ with $y(x_0) = y_0$.

Here is a step-by-step summary of how to use Euler's method to obtain an approximate solution of an initial value problem. The method is illustrated in Example 8.4.3.

> **Euler's Method for Approximating the Solution to the Initial Value Problem $y'(x) = f(x, y)$ with $y(x_0) = y_0$**
>
> **Step 1.** Choose the step size h.
>
> **Step 2.** For $k = 1, 2, \ldots, n$, compute the successive points (x_k, y_k), where $x_k = x_{k-1} + h$ and $y_k = y_{k-1} + f(x_{k-1}, y_{k-1})h$.
>
> **Step 3.** Construct the Euler approximation by plotting the points (x_k, y_k) for $k = 0, 1, 2, \ldots, n$ in the xy plane and connecting successive pairs of points with line segments.

EXAMPLE 8.4.3

Use Euler's method with step size $h = 0.1$ to approximate the solution of the initial value problem

$$\frac{dy}{dx} = x - y \quad \text{with } y(0) = 1.$$

Compare the approximation with the actual solution.

Solution

The given differential equation can be written in the first-order linear form

$$\frac{dy}{dx} + y = x.$$

Using the methods of Section 8.2, it can be shown that the general solution is $y = x - 1 + Ce^{-x}$, and the particular solution with $y(0) = 1$ is

$$g(x) = x - 1 + 2e^{-x}.$$

Compare this solution with the approximate solution obtained by applying Euler's method with $h = 0.1, f(x, y) = x - y, x_0 = 0, y_0 = 1$, and 10 steps. The first step yields

$$y_1 = y_0 + f(x_0, y_0)h = 1 + (0 - 1)(0.1) = 0.9,$$

and comparing this with the actual value of the solution $y = g(x)$ at $x_1 = 0.1$ results in $y_1 - g(x_1) = 0.9 - 0.90967 = -0.00967$. The rest of the computations are shown in the following table:

k	x_k	y_k	Slope $f(x_k, y_k)$	y_{k+1}	Actual Solution $g(x_k)$	Difference $y_k - g(x_k)$
0	0	1	-1	0.9	1	0
1	0.1	0.9	-0.8	0.820	0.910	-0.010
2	0.2	0.82	-0.62	0.758	0.837	-0.017
3	0.3	0.758	-0.458	0.712	0.782	-0.024
4	0.4	0.712	-0.312	0.681	0.741	-0.029
5	0.5	0.681	-0.181	0.663	0.713	-0.032
6	0.6	0.663	-0.063	0.657	0.698	-0.035
7	0.7	0.657	0.043	0.661	0.693	-0.036
8	0.8	0.661	0.139	0.675	0.699	-0.038
9	0.9	0.675	0.225	0.698	0.713	-0.038
10	1.0	0.698			0.736	-0.038

In Figure 8.15, the Euler approximation (polygonal path) is shown along with the actual solution curve for comparison.

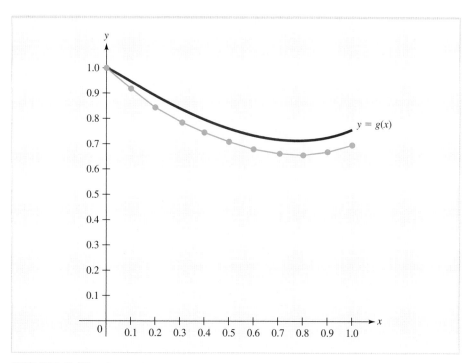

FIGURE 8.15 Comparing the solution curve for the initial value problem $y' = x - y$ with $y(0) = 1$ to the Euler approximation.

Applied models often involve initial value problems whose exact solution is either difficult or impossible to obtain. In Example 8.4.4, we approximate a solution to a population model using Euler's method.

EXAMPLE 8.4.4

After t months of observation, the population $P(t)$ of an animal species isolated on an island is found to be changing at a rate of

$$\frac{dP}{dt} = 0.03\sqrt{P} - 0.005t.$$

Suppose the population was 1000 when observations began ($t = 0$). Use Euler's method to estimate the population after 1 year.

Solution

The slope function is $f(t, P) = 0.03\sqrt{P} - 0.005t$, with $P = 1000$ when $t = 0$. Since time is measured in months, we wish to estimate $P(12)$. Assume a step size $h = 2$. Then the population P_{k+1} at the $(k + 1)$th step is related to the population at the kth step by the formula

$$P_{k+1} = P_k + f(t_k, P_k)h$$
$$= P_k + [0.03\sqrt{P_k} - 0.005t_k](2)$$

where $t_k = 0 + 2k = 2k$. The following table summarizes the computations for the approximation.

k	t_k	P_k	Slope $f(t_k, P_k)$	P_{k+1}
0	0	1000.00	0.9487	1001.90
1	2	1001.90	0.9396	1003.78
2	4	1003.78	0.9305	1005.64
3	6	1005.64	0.9214	1007.48
4	8	1007.48	0.9122	1009.30
5	10	1009.30	0.9031	1011.11
6	12	1011.11		

Therefore, based on this Euler approximation, we would estimate the population at the end of 1 year to be approximately 1011 animals.

EXERCISES ■ 8.4

In Exercises 1 through 12, plot the slope field for the given differential equation and use it to sketch the graph of the solution that passes through the specified point.

1. $y' = x + y$; $(1, 1)$

2. $y' = x - y$; $(1, 1)$

3. $y' = y^2$; $(0, 1)$

4. $y' = \sqrt{x}$; $(4, 0)$

5. $y' = y(2 - y)$; $(0, -1)$

6. $y' = \dfrac{x}{y + 1}$; $(0, 0)$

7. $y' = x^2 + y^2$; $(1, -1)$

8. $y' = x - y^2$; $(1, 1)$

9. $y' = \dfrac{2y - 3x}{x + y}$; $(0, 1)$

10. $y' = 2 - x - y$; $(0, 0)$

11. $y' = e^{xy}$; $(0, 0)$

12. $y' = e^{x/y}$; $(1, 1)$

In Exercises 13 through 20, use Euler's method with the indicated step size h and number of increments n to obtain an approximate solution of the given initial value problem.

13. $y' = 2x + y$, $y(0) = 1$, $h = 0.2$, $n = 5$

14. $y' = x - 3y$, $y(0) = 0$, $h = 0.4$, $n = 5$

15. $y' = \dfrac{y}{x} + x$, $y(1) = 0$, $h = 0.2$, $n = 5$

16. $y' = xy + y$, $y(0) = 1$, $h = 0.5$, $n = 6$

17. $y' = \sqrt{x} - y$, $y(1) = 1$, $h = 0.5$, $n = 8$

18. $y' = \dfrac{e^x}{y}$, $y(0) = -1$, $h = 0.5$, $n = 8$

19. $y' = \dfrac{x^2 - y^2}{xy}$, $y(1) = 1$, $h = 0.4$, $n = 5$

20. $y' = \dfrac{x}{\sqrt{x^2 + y^2}}$, $y(0) = 1$, $h = 0.2$, $n = 5$

In Exercises 21 through 26, use Euler's method with the indicated step size h to estimate the required solution of the given initial value problem.

21. $y' = x + 2y$, $y(0) = 1$, $h = 0.2$; estimate $y(1)$

22. $y' = \dfrac{x}{y}$, $y(1) = 1$, $h = 0.2$; estimate $y(2)$

23. $y' = \dfrac{x - y}{2x + y}$, $y(0) = 1$, $h = 0.2$; estimate $y(1)$

24. $y' = x\sqrt{1 + y^2}$, $y(0) = 0$, $h = 0.5$; estimate $y(4)$

25. $y' = x^2 + 2y^2$, $y(0) = 0$, $h = 0.4$; estimate $y(2)$

26. $y' = \ln(x + y)$, $y(0) = 1$, $h = 0.5$; estimate $y(4)$

27. The length $L(t)$ of the curve $y = \ln x$ over the interval $1 \le x \le t$ satisfies

$$\frac{dL}{dt} = \frac{\sqrt{1 + t^2}}{t}, \qquad L(1) = 0.$$

Estimate the length $L(4)$ using Euler's method with step size $h = 0.5$.

28. **SUPPLY** Suppose the number of units $S(t)$ of a particular commodity that will be supplied by producers at time t increases at a rate of

$$\frac{dS}{dt} = 0.3Se^{-0.1t}.$$

If $S(0) = 3$, use Euler's method with $h = 0.5$ to estimate $S(4)$.

29. **PRODUCTION** The output Q at a certain factory changes at a rate of

$$\frac{dQ}{dK} = 0.2Q + K^{1/2},$$

where K is the capital expenditure (in units of 1000). Suppose when the capital expenditure is 2000, there will be $Q(2) = 500$ units produced. Use Euler's method with $h = 1$ to estimate the number of units produced when the capital expenditure is 5000.

30. **GROWTH OF A SPECIES** The fox population $F(t)$ on a large island satisfies the modified logistic initial value problem

$$\frac{dF}{dt} = 0.01(80 - F)(F - 76), \qquad F(0) = 70.$$

Use Euler's method with $h = 1$ to estimate $F(10)$, the fox population after 10 years.

31. **GROWTH OF A SPECIES** The length $L(t)$ of a species of fish at age t is modelled by the von Bertalanffy initial value problem

$$\frac{dL}{dt} = 1.03(5.3 - L), \qquad L(0) = 0.2,$$

where L is measured in centimetres and t in months.
a. Use Euler's method with $h = 0.4$ to estimate the length of the fish at time $t = 2$.
b. Solve the initial value problem and evaluate $L(2)$. Compare this exact value with the estimate you obtained in part (a).

32. **POPULATION EXTINCTION** When the population density of a sexually reproducing species falls below a certain threshold, individuals have difficulty finding a suitable mate, which has the effect of depressing the population still further, eventually leading to extinction. If the population at time t is $P(t)$, this effect can be modelled by the modified logistic equation

$$\frac{dP}{dt} = kP(P - A)\left(1 - \frac{P}{K}\right),$$

where k is the natural growth rate of the species, K is the carrying capacity, and A is the threshold population. Assume that $k > 0$ and $0 < A < K$.

For a particular species, suppose P is measured in thousands and t in years, and that $k = 0.2$, $K = 20$, and $A = 2$. Assume further that the initial population is $P(0) = 1$. Use Euler's method with $h = 0.25$ to estimate $P(1)$, the population after 1 year.

SECTION 8.5

L05

Formulate and solve difference equations.

Difference Equations; The Cobweb Model

We have already seen how differential equations can be used to model a variety of dynamic situations involving continuous behaviour. However, it is often more natural or convenient to describe certain situations in terms of behaviour measured at a collection of discrete points instead of continuously. For example, the population of Canada is measured by a census taken every 5 years, a biologist interested in knowing the population of a bacterial colony might take measurements every hour, or an economist charting the supply and demand of a commodity might gather the pertinent data once a week.

A **difference equation** is an equation involving the differences between values of a variable at discrete times. For instance, if f is a function and $y_n = f(n)$ for $n = 0, 1, 2, \ldots$, then

$$y_n - y_{n-1} = 3$$

is a difference equation, as are

$$y_n - y_{n-1} = y_{n-2}$$ the Fibonacci equation

and

$$y_n - y_{n-1} = ky_{n-1}\left(1 - \frac{1}{K}y_{n-1}\right)$$ the logistic equation

Such equations play the same role in the study of discrete dynamical situations as differential equations play in modelling continuous phenomena. Our goal in this section is to examine a few properties and applications of difference equations. We begin with some terminology.

A **solution** of a given difference equation is a sequence $\{y_n\}$ whose terms satisfy the equation, and a **general solution** is a formula that characterizes all possible solutions. An **initial value problem** consists of a difference equation together with a collection of initial conditions that must be satisfied, and a solution of such a problem is called a **particular solution.**

For instance, suppose a bacterial colony initially containing 100 000 individuals is observed every hour. If it is found that the population P_{k+1} at time $k + 1$ is always 1000 more than 3% of the population P_k at time k, we can express this information in terms of the initial value problem

$$P_{k+1} = 0.03P_k + 1000 \qquad \text{with } P_0 = 100\ 000.$$

Examples 8.5.1 and 8.5.2 further illustrate this notation and terminology.

EXAMPLE 8.5.1

Write out the first five terms of the sequence $y_1, y_2, \ldots$ that satisfies the initial value problem

$$y_n = -3y_{n-1} + 1 \qquad \text{with } y_0 = 2.$$

Solution
The initial term is $y_0 = 2$, and each successive term is generated by substituting into the difference equation. Therefore,

$$
\begin{aligned}
y_0 &= 2 \\
y_1 &= -3y_0 + 1 = -3(2) + 1 = -5 \\
y_2 &= -3y_1 + 1 = -3(-5) + 1 = 16 \\
y_3 &= -3y_2 + 1 = -3(16) + 1 = -47 \\
y_4 &= -3y_3 + 1 = -3(-47) + 1 = 142
\end{aligned}
$$

Thus, the first five terms of the sequence are 2, -5, 16, -47, and 142.

Difference equations play an important role in mathematical modelling. For instance, such equations can be used to relate the values of a particular quantity at different discrete times or periods, such as generations. This is illustrated in Example 8.5.2 for an application involving genetics.

EXAMPLE 8.5.2

Population genetics is concerned with how the distribution of genotypes in a population changes from generation to generation. Suppose random mating occurs in a population, and it is assumed that individuals who inherit a particular recessive gene from both parents will themselves have no offspring. Under these conditions, the frequency f_n of the gene recurring in the nth generation may be modelled by the difference equation

$$f_{n+1} = \frac{f_n}{1 + f_n}.$$

Verify that

$$f_n = \frac{f_0}{1 + nf_0}$$

satisfies this difference equation, where f_0 is the initial frequency.

Solution

To verify that $f_n = \dfrac{f_0}{1 + nf_0}$ satisfies the given difference equation, note that

$$f_{n+1} = \frac{f_0}{1 + (n+1)f_0}$$

while

$$\frac{f_n}{1 + f_n} = \frac{\dfrac{f_0}{1 + nf_0}}{1 + \left(\dfrac{f_0}{1 + nf_0}\right)} = \frac{\dfrac{f_0}{1 + nf_0}}{\dfrac{1 + nf_0 + f_0}{1 + nf_0}}$$

$$= \frac{f_0}{1 + f_0 + nf_0} = \frac{f_0}{1 + (n+1)f_0}$$

Thus,

$$f_{n+1} = \frac{f_0}{1 + (n+1)f_0} = \frac{f_n}{1 + f_n}$$

as required.

Solution of a Difference Equation

We can always find solutions of a given initial value problem by starting with the initial condition and using the difference equation to generate one value after another as we did in Example 8.5.1. However, what we really want is a formula for y_n in terms of n, which will enable us to find any particular term y_k in the solution sequence by simply substituting k for n in the formula without having to first find $y_1, y_2, \ldots, y_{k-1}$. For instance, the result in Example 8.5.2 tells us that the 20th term in the solution sequence for the difference equation

$$f_{n+1} = \frac{f_n}{1 + f_n}.$$

can be found by substituting $n = 20$ into the formula

$$f_n = \frac{f_0}{1 + nf_0}.$$

That is,

$$f_{20} = \frac{f_0}{1 + 20f_0}.$$

A **first-order linear** difference equation is one of the general form

$$y_n = ay_{n-1} + b,$$

where a and b are constants. Our next goal is to show how the general term y_n of the solution sequence of such an equation can be expressed in terms of the initial term y_0. We will use an iterative procedure in which y_n is expressed in terms of y_{n-1}, then y_{n-1} in terms of y_{n-2}, and so on until we have a formula for y_n in terms of y_0. Here are the specific steps:

$$y_0$$
$$y_1 = ay_0 + b$$
$$y_2 = ay_1 + b = a(ay_0 + b) + b = a^2y_0 + ab + b$$
$$y_3 = ay_2 + b = a(a^2y_0 + ab + b) + b = a^3y_0 + a^2b + ab + b$$
$$y_4 = ay_3 + b = a^4y_0 + a^3b + a^2b + ab + b$$

and, in general,

$$y_n = ay_{n-1} + b = a^ny_0 + a^{n-1}b + a^{n-2}b + \cdots + a^2b + ab + b$$
$$= a^ny_0 + (a^{n-1} + a^{n-2} + \cdots + a + 1)b$$

For $a = 1$, all powers of a also equal 1, and we have $y_n = y_0 + nb$. If $a \neq 1$, the expression $a^{n-1} + a^{n-2} + \cdots + a + 1$ can be summed by the **geometric series** formula

$$a^{n-1} + a^{n-2} + \cdots + a + 1 = \frac{1 - a^n}{1 - a}$$

(see Exercise 48 or the general discussion of geometric series in Section 9.1 of Chapter 9). In this case, our formula for y_n can be written as

$$y_n = a^ny_0 + \left(\frac{1 - a^n}{1 - a}\right)b.$$

To summarize:

Solving a First-Order Linear Difference Equation with an Initial Condition

For constants a and b, consider the initial value problem

$$y_n = ay_{n-1} + b \qquad \text{with initial value } y_0.$$

If $a = 1$, the general solution is

$$y_n = y_0 + nb.$$

If $a \neq 1$, the general solution is

$$y_n = a^ny_0 + \left(\frac{1 - a^n}{1 - a}\right)b.$$

EXAMPLE 8.5.3

Solve the first-order linear initial value problem

$$y_n = -\frac{1}{2}y_{n-1} + 5 \qquad \text{with initial value } y_0 = 3.$$

Solution

Comparing the given difference equation with the first-order linear form $y_n = ay_{n-1} + b$ gives $a = -\frac{1}{2}$ and $b = 5$. Since $a \neq 1$, the solution is

$$
\begin{aligned}
y_n &= a^n y_0 + \left(\frac{1 - a^n}{1 - a}\right)b \\
&= \left(-\frac{1}{2}\right)^n (3) + \left[\frac{1 - (-1/2)^n}{1 - (-1/2)}\right](5) \\
&= \left(-\frac{1}{2}\right)^n (3) + \frac{10}{3}\left[1 - \left(-\frac{1}{2}\right)^n\right] \\
&= \frac{10}{3} + \left(3 - \frac{10}{3}\right)\left(-\frac{1}{2}\right)^n \\
&= \frac{10}{3} - \frac{1}{3}\left(-\frac{1}{2}\right)^n
\end{aligned}
$$

For example, the fifth term in the solution sequence is

$$y_5 = \frac{10}{3} - \frac{1}{3}\left(-\frac{1}{2}\right)^5 = \frac{107}{32}.$$

EXAMPLE 8.5.4

There are 12 000 people in a northern mining town. Each year, 3% of the people in the town die and 720 babies are born.

  a. What will the population be in 5 years?
  b. What will the population be in the long run?

Solution

  a. Let P_n be the population after n years. Since $0.03P_{n-1}$ people die during the nth year, it follows that $0.97P_{n-1}$ people survive, and to that number is added the 720 births so that

$$\underbrace{P_n}_{\substack{\text{current} \\ \text{population}}} = \underbrace{0.97P_{n-1}}_{\substack{\text{surviving} \\ \text{population}}} + \underbrace{720}_{\text{births}}$$

Substituting $a = 0.97$, $b = 720$, and $P_0 = 12\ 000$ into the formula for the general solution of the equation $y_n = ay_{n-1} + b$ we get

$$P_n = (0.97)^n(12\ 000) + \frac{[1 - (0.97)^n](720)}{1 - 0.97}.$$

Just-In-Time

Recall that if $-1 < a < 1$, we are effectively multiplying a fraction by itself, resulting in a smaller and smaller number; then $\lim\limits_{n\to\infty} a^n = 0$.

Thus, when $n = 5$, the population is

$$P_5 = (0.97)^5(12\ 000) + \frac{[1 - (0.97)^5](720)}{1 - 0.97}$$

$$= 13\ 695$$

b. To find the long-run population, compute the limit

$$\lim_{n\to\infty} P_n = \lim_{n\to\infty}\left\{(0.97)^n(12\ 000) + \frac{[1 - (0.97)^n](720)}{1 - 0.97}\right\}$$

$$= 0 + \frac{[1 - 0](720)}{0.03} = 24\ 000$$

Thus, over a long period of time, the population tends toward 24 000 people.

Applications of Difference Equations

Difference equations are used in models involving business and economics, psychology, sociology, biology, information theory, and a variety of other areas. Two such applications are examined in Examples 8.5.5 and 8.5.6.

EXAMPLE 8.5.5   *(Amortization of Debt)*

When money is borrowed at an interest rate r, the total debt increases in the same way as a bank account paying the same rate of interest. **Amortization** is a procedure for repaying the debt, including interest, by making a sequence of payments. Usually the payments are of equal size and are made at equal time intervals (annually, quarterly, or monthly).

Suppose a debt of D dollars at an annual interest rate of r is amortized over a period of n years by making annual payments of A dollars. Set up and solve a difference equation to show that

$$A = \frac{rD}{1 - (1 + r)^{-n}}.$$

Solution

Let y_n be the total debt after n years. Then $y_n - y_{n-1}$ is the change in debt during the nth year, which equals the interest ry_{n-1} charged during the nth year less the annual payment A. That is,

$$\underbrace{y_n - y_{n-1}}_{\substack{\text{change}\\\text{in debt}}} = \underbrace{ry_{n-1}}_{\substack{\text{interest}\\\text{charged}}} - \underbrace{A}_{\substack{\text{annual}\\\text{payment}}}$$

Combining the y_{n-1} terms and using the fact that $y_0 = D$, the initial debt results in the initial value problem

$$y_n = (1 + r)y_{n-1} - A \qquad \text{with } y_0 = D.$$

This difference equation has the first-order linear form, with $a = 1 + r$ and $b = -A$, so the general solution is

$$y_n = a^n y_0 + \left(\frac{1 - a^n}{1 - a}\right)b$$

$$= (1 + r)^n D + \left[\frac{1 - (1 + r)^n}{1 - (1 + r)}\right](-A)$$

$$= (1 + r)^n D + \left(\frac{-A}{-r}\right)[1 - (1 + r)^n]$$

$$= (1 + r)^n D + \frac{A}{r}[1 - (1 + r)^n]$$

Since the debt is to be paid off in n years, find A so that $y_n = 0$, that is,

$$y_n = (1 + r)^n D + \frac{A}{r}[1 - (1 + r)^n] = 0.$$

Subtract $\frac{A}{r}[1 - (1 + r)^n]$ from both sides of this equation to get

$$(1 + r)^n D = -\frac{A}{r}[1 - (1 + r)^n] = \frac{A}{r}[(1 + r)^n - 1],$$

so that

$$A = rD\left[\frac{(1 + r)^n}{(1 + r)^n - 1}\right].$$

Finally, multiplying both numerator and denominator by $(1 + r)^{-n}$ gives the amortization formula

$$A = \frac{rD}{1 - (1 + r)^{-n}}.$$

EXAMPLE 8.5.6   *(Biology-Fishery Management)*

A biologist is employed by the Ministry of Natural Resources to help regulate commercial fishing of trout in a lake. The biologist estimates that the lake initially contains 150 000 trout and that the trout population is growing at a rate of 15% per year. Currently, commercial fishing harvests trout at a rate of 30 000 fish per year.

a. Set up and solve a difference equation for the fish population after n years.

b. How long would it take before the lake contained no trout, if fishing continued in this way?

c. Suppose the fishing policy changes so that the lake will still contain trout 20 years from now. What is the maximum harvesting rate that will allow this to occur?

d. What is the maximum harvesting rate that can be allowed if the lake is to always contain some trout?

Solution

a. Let p_n denote the trout population at the end of the nth year. Then the change in population $p_n - p_{n-1}$ during the nth year satisfies

$$\underbrace{p_n - p_{n-1}}_{\substack{\text{change in} \\ \text{population}}} = \underbrace{0.15 p_{n-1}}_{\substack{\text{natural} \\ \text{growth}}} - \underbrace{30\ 000}_{\substack{\text{harvested} \\ \text{fish}}} \qquad \text{where } p_0 = 150\ 000.$$

Combining the p_{n-1} terms on the right results in a difference equation with the first-order linear form

$$p_n = (1 + 0.15)p_{n-1} - 30\ 000 = 1.15 p_{n-1} - 30\ 000,$$

where $a = 1.15$ and $b = -30\ 000$. The solution of this equation is

$$p_n = (1.15)^n (150\ 000) + \left[\frac{1 - (1.15)^n}{1 - 1.15} \right] (-30\ 000)$$

$$= (1.15)^n (150\ 000) + \left(\frac{-30\ 000}{-0.15} \right) [1 - (1.15)^n]$$

$$= (1.15)^n (150\ 000) + 200\ 000 [1 - (1.15)^n]$$

$$= (1.15)^n [150\ 000 - 200\ 000] + 200\ 000$$

$$= -50\ 000 (1.15)^n + 200\ 000$$

b. The lake contains no trout when $p_n = 0$, that is, when

$$-50\ 000 (1.15)^n + 200\ 000 = 0$$

$$(1.15)^n = \frac{200\ 000}{50\ 000} = 4$$

Taking logarithms on both sides of this equation gives

$$\ln (1.15)^n = \ln 4,$$

so that

$$n \ln 1.15 = \ln 4$$

$$n = \frac{\ln 4}{\ln 1.15} = 9.92$$

Therefore, no fish will remain in the lake after roughly 10 years.

c. Suppose the harvesting rate is h. Then the difference equation becomes $p_n = 1.15 p_{n-1} - h$, and its solution is

$$p_n = (1.15)^n (150\ 000) + \left[\frac{1 - (1.15)^n}{1 - 1.15} \right] (-h)$$

$$= (1.15)^n (150\ 000) + \left(\frac{-h}{-0.15} \right) [1 - (1.15)^n]$$

$$= (1.15)^n \left(150\ 000 - \frac{h}{0.15} \right) + \frac{h}{0.15}$$

Find h so that the lake will be fished out when $n = 20$. Set $p_{20} = 0$ and solve for h:

$$p_{20} = (1.15)^{20}\left(150\ 000 - \frac{h}{0.15}\right) + \frac{h}{0.15} = 0$$

$$\frac{h}{0.15}[1 - (1.15)^{20}] = -(1.15)^{20}(150\ 000)$$

$$h = \frac{-0.15(1.15)^{20}(150\ 000)}{1 - (1.15)^{20}} = 23\ 964$$

Therefore, approximately 24 000 trout may be harvested each year.

d. If the lake is always to contain some trout, the year n in which all the trout disappear must satisfy $n \to \infty$. According to the computation in part (c), the fish population will be 0 in year n if

$$p_n = (1.15)^n\left(150\ 000 - \frac{h}{0.15}\right) + \frac{h}{0.15} = 0.$$

Solve this equation for h:

$$
\begin{aligned}
h &= \frac{-0.15(1.15)^n(150\ 000)}{1 - (1.15)^n} \\
&= \frac{0.15(1.15)^n(150\ 000)}{(1.15)^n - 1} \qquad \text{multiply numerator and} \\
&\qquad\qquad\qquad\qquad\qquad\quad \text{denominator by } -1 \\
&= \frac{0.15(150\ 000)}{1 - (1.15)^{-n}} \qquad \text{multiply numerator and} \\
&\qquad\qquad\qquad\qquad\qquad\quad \text{denominator by } (1.15)^{-n}
\end{aligned}
$$

To find the fishing rate for the case where the lake always contains some trout, compute h as $n \to \infty$, that is,

$$
\begin{aligned}
h &= \lim_{n \to \infty} \frac{0.15(150\ 000)}{1 - (1.15)^{-n}} = \frac{0.15(150\ 000)}{1 - 0} \qquad \text{since } \lim_{n \to \infty}(1.15)^{-n} = 0 \\
&= 22\ 500
\end{aligned}
$$

Therefore, if the fishing rate is set at no more than 22 500 trout per year, the lake will always contain some trout.

The Cobweb Model

In Chapter 1, we introduced the economic concepts of supply and demand and defined market equilibrium as the situation that occurs when supply equals demand. In Section 8.3, we discussed a price adjustment model in which supply and demand are regarded as functions of price, which changes continuously with time as market conditions adjust. However, equilibrium does not occur instantaneously by mutual agreement between producers and consumers. Instead, it is reached by a sequence of discrete adjustments, and our next goal is to describe a model involving difference equations for dealing with discrete supply-demand adjustments in a dynamic market.

We will consider a situation involving a single commodity, which for purposes of illustration we assume to be a crop. Suppose the price is currently high, inspiring farmers to plant a large crop. However, by the time the crop is harvested and brought to market, the demand may be less than the farmers were expecting. If so, the price drops, prompting farmers to plant a smaller crop the next year. This time, when the crop is harvested, supply may be smaller than demand. The price will then rise, farmers will be encouraged to grow larger crops again, supply will exceed demand, the price will fall, and so on. The pattern continues, with supply in each period determined by the price of the *previous* period.

To describe this supply-demand adjustment mathematically, let p_n be the price of the crop during the nth period, and let S_n and D_n be the corresponding supply and demand. Assume for simplicity that the supply and demand functions are linear, that is,

$$D = a - bp \qquad \text{and} \qquad S = c + dp$$

for constants a, b, c, and d. During the nth period, the demand satisfies

$$D_n = a - bp_n,$$

but since supply in the nth period is determined by the price p_{n-1} of the previous period,

$$S_n = c + bp_{n-1}.$$

During each period, the price is determined by equating supply and demand, so

$$\underbrace{c + dp_{n-1}}_{\text{supply } S_n} = \underbrace{a - bp_n}_{\text{demand } D_n}$$

This is a difference equation that can be rewritten in the first-order linear form as follows:

$$c + dp_{n-1} = a - bp_n$$
$$bp_n = a - (c + dp_{n-1})$$
$$p_n = \left(-\frac{d}{b}\right)p_{n-1} + \left(\frac{a-c}{b}\right)$$

Substitute into the formula for the general solution of a first-order linear difference equation to find that

$$p_n = \left(-\frac{d}{b}\right)^n p_0 + \left[\frac{1 - (-d/b)^n}{1 - (-d/b)}\right]\left(\frac{a-c}{b}\right)$$

$$= \left(-\frac{d}{b}\right)^n p_0 + \left[\frac{1 - (-d/b)^n}{\dfrac{b+d}{b}}\right]\left(\frac{a-c}{b}\right)$$

$$= \left(-\frac{d}{b}\right)^n p_0 + [1 - (-d/b)^n]\left(\frac{b}{b+d}\right)\left(\frac{a-c}{b}\right)$$

$$= \left(-\frac{d}{b}\right)^n p_0 + \left(\frac{a-c}{b+d}\right)\left[1 - \left(-\frac{d}{b}\right)^n\right]$$

$$= \left(-\frac{d}{b}\right)^n\left[p_0 - \left(\frac{a-c}{b+d}\right)\right] + \left(\frac{a-c}{b+d}\right)$$

where p_0 is the initial price of the commodity (the crop in our example).

The model we have just described is called the **cobweb model**. To see why, look at Figure 8.16. The supply and demand lines, $y = c + dp$ and $y = a - bp$, are shown in two different cases. In both, the point of market equilibrium occurs where the lines intersect, that is, at the price p_e where

$$a - bp_e = c + dp_e$$
$$p_e = \frac{a-c}{b+d}$$

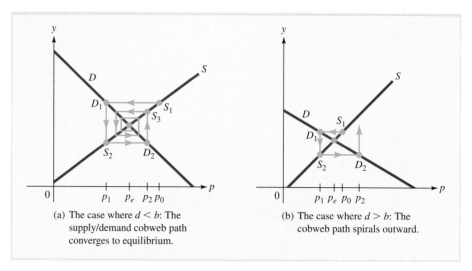

FIGURE 8.16   The progression of supply and demand in the cobweb model.

In Figure 8.16a, we plot the path of supply and demand over several time periods in the case where $d < b$. Notice that the initial price p_0 determines the supply S_1 in the first period, which in turn determines the demand D_1 in that period and the price p_1 such that $S_1 = D_1$. Then p_1 determines the supply S_2 in the second period, and so on. The path traced out in this fashion resembles a cobweb converging toward the equilibrium point. Figure 8.16b shows the case where $d > b$. This time, the cobweb spirals outward, diverging away from the equilibrium point.

Analytically, notice that if $d < b$, then $\left(-\dfrac{d}{b}\right)^n$ tends to 0 as $n \to \infty$. Therefore, no matter what the initial price p_0 may be, the price p_n will approach the equilibrium price $p_e = \dfrac{a-c}{b+d}$. However, if $d > b$, the term $\left(-\dfrac{d}{b}\right)^n$ will grow larger and larger in absolute value, while if $d = b$, then $\left(-\dfrac{d}{b}\right)^n = (-1)^n$ oscillates between 1 and -1.

In both these latter cases, the price sequence $p_0, p_1, p_2, \ldots$ will not converge (approach a finite number as $n \to \infty$).

Since d is the slope of the supply line, it measures the rate of increase of supply, just as b measures the rate of decrease of demand. Thus, the result we have obtained can be interpreted as saying that the price sequence $p_0, p_1, p_2, \ldots$ converges to the equilibrium p_e if and only if the rate of decrease in demand exceeds the rate of increase in supply. In the language of economics, *market stability occurs in the long run only when suppliers are less sensitive to price changes than consumers.*

Difference Equations Versus Differential Equations

When modelling with difference equations, the difference $y_n - y_{n-1}$ between consecutive terms is used in much the same way as the derivative $\dfrac{dy}{dx}$ is used when modelling with a differential equation. Continuing the analogy, the ratio

$$\frac{y_n - y_{n-1}}{y_{n-1}}$$

is used to represent a relative or percentage rate of change in a discrete model.

For instance, suppose a species of fish is observed to reproduce at a rate of 15% per year. Then the fish population F_n after n years may be modelled by the difference equation

$$\frac{F_n - F_{n-1}}{F_{n-1}} = 0.15$$

or, equivalently, by

$$F_n - F_{n-1} = 0.15F_{n-1}.$$

On the other hand, if we wish to model the fish population as a continuous function $F(t)$ of time t, we could represent its growth by the analogous differential equation

$$\frac{dF}{dt} = 0.15F.$$

Results obtained with difference equations may be quite different from those obtained using the analogous differential equations. For instance, in Example 8.5.5, we used the difference equation

$$y_n - y_{n-1} = ry_{n-1} - A$$

to model the total debt y_n accumulated after n years when interest is compounded continuously at an annual rate r and A dollars are spent annually to reduce debt. An analogous model in which the debt $Q(t)$ is regarded as a continuous function of time t would involve the differential equation

$$\frac{dQ}{dt} = rQ - A.$$

Using the methods discussed in Section 8.2 to solve this equation subject to the initial condition that $Q(0) = D$ (the initial debt), we find that

$$Q(t) = \left(D - \frac{A}{r}\right)e^{rt} + \frac{A}{r}$$

(for practice, you should verify this result). If the debt is to be amortized over a term of n years, then we want $Q(n) = 0$, so that

$$0 = Q(n) = \left(D - \frac{A}{r}\right)e^{rt} + \frac{A}{r}.$$

Solving for A in this equation gives

$$\left(D - \frac{A}{r}\right)e^{rt} = -\frac{A}{r} \qquad \text{subtract } \frac{A}{r} \text{ from both sides}$$

$$D - \frac{A}{r} = -\frac{A}{r}e^{-rt} \qquad \text{multiply both sides by } e^{-rt}$$

$$D = (1 - e^{-rt})\frac{A}{r} \qquad \text{add } \frac{A}{r} \text{ to both sides and combine}$$

$$A = \frac{rD}{1 - e^{-rt}} \qquad \text{multiply both sides by } \frac{r}{1 - e^{-rt}}$$

Notice the similarities and differences between this continuous amortization formula and the discrete formula found in Example 8.5.5. For instance, if the initial debt is $D = \$10\,000$ and the annual interest rate is 8% ($r = 0.08$), then the discrete model tells us the debt will be amortized over a term of $n = 10$ years by making annual payments of

$$A = \frac{0.08(10\,000)}{1 - (1 + 0.08)^{-10}} = 1490.29,$$

or $1490.29, while the continuous model yields annual payments of

$$A = \frac{0.08(10\,000)}{1 - e^{-0.08(10)}} = 1452.77,$$

or $1452.77. The difference is that in the discrete case, the payment is made once a year, as a lump sum, while in the continuous case, the annual payment is assumed to be spread out continuously over the entire year.

EXERCISES ■ 8.5

In Exercises 1 through 8, write the first five terms of the given initial value problem.

1. $y_n = 3y_{n-1}; y_0 = 1$

2. $y_n = y_{n-1} + 2; y_0 = 0$

3. $y_n = y_{n-1}^2; y_0 = 1$

4. $y_n = y_{n-1}^3 + (1 - y_{n-1})y_{n-1}; y_0 = -0.1$

5. $y_{n+1} = y_n + y_{n-1}; y_0 = 1, y_1 = 1$

6. $y_{n+1} - y_n^2 - 2y_{n-1}; y_0 = -1, y_1 = 0$

7. $y_{n+1} = \dfrac{y_n}{y_n + 2}; y_0 = 1$

8. $y_{n+1} = \dfrac{y_n}{y_{n-1} + 1}; y_0 = 0, y_1 = 1$

In Exercises 9 and 10, find constants A and B so that the given expression y_n satisfies the specified difference equation.

9. $ny_n + (n - 1)y_{n-1} = 2n - 3; y_n = A + \dfrac{B}{n}$

10. $y_{n+1} = y_n + n; y_n = An^2 + Bn$

In Exercises 11 through 17, solve the given first-order linear initial value problem.

11. $y_n = -y_{n-1}; y_0 = 1$

12. $y_n = 2y_{n-1} + 3; y_0 = 0$

13. $y_n = \dfrac{1}{2}y_{n-1} + 1; y_0 = 0$

14. $y_n = \dfrac{-1}{2}y_{n-1} - \dfrac{1}{4}; y_0 = 10$

15. $y_n - y_{n-1} = \dfrac{1}{8}y_{n-1} + \dfrac{1}{4}; y_0 = 0$

16. $y_n - y_{n-1} = -\dfrac{1}{2}y_{n-1} + \dfrac{1}{3}; y_0 = 1$

17. $y_n - y_{n-1} = \dfrac{1}{10}y_{n-1}; y_0 = 10$

18. **SAVINGS ACCOUNT**   Veejay deposits $1000 in an account that pays 5% interest, compounded annually. Set up and solve an initial value problem for the amount of money A_n in the account after n years.

19. **SAVINGS ACCOUNT**   Like Veejay in Exercise 18, Taylor invests $1000 in an account that pays 5% interest, compounded annually. However, she also makes additional deposits of $50 at the end of each year. Set up and solve an initial value problem for the amount of money A_n in the account after n years.

20. **STOCK MARKET SPECULATION**   Maxime invests in a speculative stock whose price p_n at the end of each month satisfies the initial value problem

$$p_n = -0.1p_{n-1} + 300, \quad \text{where } p_0 = 100.$$

How much will his investment be worth after n months?

21. **ANIMAL POPULATION WITH HUNTING**   The fox population in a certain region is initially 50 000 and is found to grow at a rate of 8% per year. Hunters are allowed to harvest h foxes per year.

  a. If $h = 2000$, what is the size of the fox population after n years?

  b. What should h be for the fox population to remain the same from year to year?

22. SPREAD OF A DISEASE A disease spreads so that the number of people exposed during month n is proportional to the number of susceptible people in the population who have not yet acquired the disease. That is, if S_n is the number of sick people in a susceptible population P, then

$$S_{n+1} - S_n = k(P - S_n).$$

Suppose 10 000 people in a community are susceptible to a new strain of influenza. If 500 people initially have the disease and 700 are newly infected during the first month, how many will have the disease n months after the outbreak? What happens to S_n in the long run (as $n \to \infty$)?

23. SPREAD OF A DISEASE WITH A CURE In Exercise 22, suppose 20% of the people who have the disease at the end of the $(n - 1)$th month are cured during the nth month.

  a. Rewrite the difference equation in Exercise 22 to take into account this new feature.

  b. Now how many people will have the disease n months after the outbreak? What happens to S_n as $n \to \infty$ in this case?

24. RECYCLING Allison has a library of 1200 paperback books. Each year, she gives away 15% of the books she has and buys 135 new books. Set up and solve a difference equation for the number of books B_n in her library at the end of the nth year. How many books will she have at the end of the 10th year? How many will she have in the long run?

25. BUSINESS MANAGEMENT A car rental company currently owns 300 cars. Each year, the company plans to dispose of 25% of its car fleet and to add 50 new cars.

  a. Set up and solve a difference equation for the number of cars C_n in the fleet at the end of the nth year.

  b. How many cars will be in the fleet after 10 years? How large will the fleet be in the long run?

26. AMORTIZATION OF DEBT In Example 8.5.5, we obtained an amortization formula for the case where interest is compounded annually and annual payments are made. In practice, monthly compounding and monthly payments are more common.

Set up and solve a difference equation to show that a debt of D dollars due in t years at an annual interest rate of r compounded monthly will be amortized by making equal monthly payments of M dollars, where

$$M = \frac{iD}{1 - (1 + i)^{-12t}} \quad \text{with } i = \frac{r}{12}.$$

27. AMORTIZATION OF DEBT Generalize the result in Exercise 26 to obtain a formula for amortizing a debt of D dollars due in t years at an annual interest rate of r compounded k times per year by making equal payments of M dollars, also k times per year.

In Exercises 28 through 31, use the formula in Exercise 27 to answer the given question.

28. MORTGAGE PAYMENTS A homeowner puts $35 000 down on a $350 000 home and takes out a 30-year mortgage for the balance. If the interest rate is 6% per year, compounded monthly, what are the monthly mortgage payments?

29. INSTALMENT PAYMENTS Martin pays $3000 down on a $25 000 new car and will pay off the balance by making monthly instalment payments over the next 5 years. What are his monthly payments if the interest rate is 8% per year, compounded monthly?

30. INSTALMENT PAYMENTS Kylie borrows $5000 for 24 months at 9% interest, compounded monthly. If the loan is to be repaid in monthly instalments of $200 with a final balloon payment of A dollars at the end of the 24-month term, what is A?

31. SINKING FUND Evan has a payroll savings plan at work in which a certain fixed amount A is taken from his monthly paycheque and placed into an account earning 5% annual interest compounded monthly. What should A be in order for Evan to save $7000 for a trip to China in 3 years?

32. A* POPULATION Suppose the number of females in the nth generation of the population of a certain species is given by P_n. Further assume that on average, each female member of the $(n - 1)$th generation gives birth to b female offspring and that in each generation, d females can be expected to die, where b and d are positive constants.

a. Set up and solve a difference equation to find P_n in terms of b, d, and the initial population of females $P_0 = P(0)$.

b. Suppose $P_0 = 100$, $b = 0.2$, and $d = 15$. What will the female population of the 10th generation be? What happens to the female population in the long run (as $n \to \infty$)?

33. **A* NATIONAL INCOME** In the classic Harrod-Hicks model* of an expanding national economy, it is assumed that if I_n denotes the national income at the end of the nth year, then the change in national income during that year is given by the difference equation

$$I_n - I_{n-1} = cI_{n-1} - h,$$

where c and h are positive constants (c is the *marginal propensity to consume*). Solve this difference equation to find I_n in terms of c, h, and the initial national income $I_0 = I(0)$.

34. **A* GENETICS** Certain genetic diseases can be modelled by a difference equation of the form

$$f_{n+1} = \frac{f_n}{1 + f_n}$$

where f_n is the relative frequency of a recessive, disease-producing gene in the nth generation, assuming that individuals who carry the recessive gene do not reproduce (recall Example 8.5.2). This difference equation is not first-order linear, but you can solve it by proceeding as follows:

a. Show that if $g_n = \dfrac{1}{f_n}$, then g_n satisfies the first-order linear difference equation $g_n = Ag_{n-1} + B$ for certain constants A and B.

b. Solve the difference equation for g_n in part (a), and then use this result to find a formula for f_n.

35. **A* CURRENCY RENEWAL** A nation has 5 billion dollar bills in currency. Each day, about $18 million comes into the nation's banks and the same amount is paid out. Suppose the country decides that whenever a dollar bill comes into a bank, it is destroyed and replaced by a new style of currency.

a. Set up and solve a difference equation for the amount of new currency C_n in circulation at the beginning of the nth day after the currency renewal begins. You may assume that $C_0 = 0$.

b. How long will it take before 90% of the currency in circulation is new?

c. Compare the results you have just obtained for your discrete currency model with the analogous continuous model analyzed in Exercise 30 in Section 8.3. In particular, do the two models yield the same time period for 90% renewal of the currency?

d. Write a paragraph on which of the two models (discrete or continuous) you think is more appropriate for describing the renewal process.

COBWEB MODEL *In Exercises 36 through 39, p is the price of a particular commodity and S(p) and D(p) are the corresponding supply and demand functions. Assume these quantities are recorded at a sequence of regular time periods, and let p_n, S_n, and D_n be the price, supply, and demand, respectively, during the nth period. In each case, do the following:*

(a) Find the equilibrium price p_e.

(b) Set up and solve a difference equation to find a formula for p_n.

(c) Find p_n, S_n, and D_n for $n = 1$, 2, and 3, and then sketch the corresponding cobweb graph for the first three time periods.

36. $S_n = 1.5p_{n-1}$; $D_n = -2p_n + 3$; $p_0 = 1.5$

37. $S_n = 3p_{n-1} - 2$; $D_n = -p_n + 1.3$; $p_0 = 1$

38. $S_n = 1.5p_{n-1} + 3$; $D_n = -p_n + 4$; $p_0 = 3$

39. $S_n = p_{n-1} + 1$; $D_n = -2p_n + 7$; $p_0 = 3$

40. **INSTALMENT LOANS** A loan of P dollars is being paid off with monthly payments of M dollars each to a bank that charges interest at an annual rate of r compounded monthly. Set up and solve a difference equation for the balance owed B_n after n payments have been made.

41. **LOTTERY WINNINGS** Robyn wins a million-dollar lottery that is to be paid in equal annual amounts over a period of years.

a. Set up and solve a difference equation to find the present value P_n of the payout if the prevailing annual interest rate is 5% and the payments are spread over n years.

b. Robyn does the math in part (a) and sues the lottery for false advertising, claiming that she was promised a million dollars but will receive less. The court agrees with her and orders the lottery to make an extra payment of A dollars each year in order to bring the present value up to the promised million-dollar level. What must A be if the payout is spread over 20 years?

42. FUTURE VALUE OF AN ANNUITY An annuity consists of equal deposits of D dollars made annually into an account that pays interest at an annual rate r compounded annually.

a. Set up and solve a difference equation to obtain a formula for the accumulated value A_n of the annuity after n deposits.

b. Tom deposits $1000 annually into an account that pays 5% interest compounded annually. How much is his account worth after 20 years?

43. DRUG ELIMINATION A patient receives an intravenous injection of a certain drug, which is then eliminated from his body at a rate of 6% per hour.

a. Set up and solve a difference equation for the amount D_n of the drug remaining in the patient's body after n hours.

b. If the patient receives 150 mg of the drug at 6 A.M., how much remains in his body at 6 P.M.?

44. UNIVERSITY FUNDING A department at a large university was initially funded with a budget of $10 million. Each year, the department requests $300 000 more than it received the previous year, but the final budget approved by the board of trustees is always 5% less than the amount requested.

a. Set up and solve a difference equation for the budget B_n of the department n years after its initial funding.

b. How large will the department budget be 20 years after its initial funding? How large will it be in the long run?

45. VALUE OF AN INHERITANCE Charlotte receives an inheritance in which she is paid $25 000 each year on her birthday until she retires. Suppose she deposits the money into a retirement annuity that pays annual interest at an effective rate of 7%. (Recall the definition of effective rate given in Section 4.1.)

a. Set up and solve a difference equation for the accumulated value A_n in her account n years after the payments begin. [*Note:* $A_0 = 25\,000$]

b. She decides she will retire on the first birthday when her retirement account is worth a million dollars. If the inheritance payments begin on her 21st birthday, when does she retire?

c. What is the equivalent cash value (that is, the present value) of Charlotte's inheritance?

46. A* LEARNING Consider an experiment in which a subject is repeatedly presented with a stimulus, which may be thought of as a test. Each time the test is presented, the subject is required to respond, and if the response is positive, a reward is given. For example, a rat may be presented with a maze (the stimulus). If it responds by choosing the correct path, it is rewarded with food, while an incorrect path leads to no food or a small electric shock.*

Suppose in this experiment that p_n (for $0 \leq p_n \leq 1$) is the likelihood that the subject succeeds on the test on the nth trial. Assume that the change in performance $p_n - p_{n-1}$ between the $(n-1)$th and nth trials satisfies the difference equation

$$p_n - p_{n-1} = A(1 - p_{n-1}) - Bp_{n-1},$$

where A and B are parameters that satisfy $0 < A < 1$ and $0 < B < 1$. The parameter A can be thought of as a measure of the value of a reward to the subject, while $-B$ measures the negative impact of a punishment.

a. Solve the difference equation for p_n in terms of A, B, and the likelihood p_0 that the subject enjoys immediate success.

b. How likely is the subject to succeed on the test in the long run (as the number of trials increases indefinitely)?

c. Suppose the likelihood $p(t)$ varies continuously with time t. Set up and solve an analogous differential equation to find $p(t)$. Compare this solution with the solution to the discrete problem you found in part (a). In particular, do both models yield the same eventual likelihood of success?

d. Write a paragraph on which model (discrete or continuous) you think is more appropriate for this experiment.

47. A* LOGISTIC GROWTH An alternative form of the logistic differential equation discussed in Section 8.1 is

$$\frac{dP}{dt} = kP\left(1 - \frac{P}{K}\right).$$

---

*The learning model in Exercise 46 is adapted from a classic model developed by R. R. Bush and F. Mosteller, as described in *Introduction to Difference Equations*, by Samuel Goldberg, New York: Wiley, 1958, pp. 103–107.

Such equations are frequently used in continuous models for populations $P(t)$ that grow rapidly at first but then slow down and eventually approach the carrying capacity K as $t \to \infty$. The analogous difference equation

$$P_n - P_{n-1} = kP_{n-1}\left(1 - \frac{1}{K}P_{n-1}\right)$$

is called the *discrete logistic equation* and is a useful tool in discrete modelling, especially in biology.

a. Suppose a particular population is modelled by a discrete logistic equation. The population is said to reach an *equilibrium* level L if $P_m = P_{m+1} = L$ for some m. Show that a population at an equilibrium level stays there; that is, $P_n = L$ for all $n > m$.

b. The population of a certain bacterial colony is recorded at discrete time intervals. Let P_n be the population (in thousands) at the nth observation, and suppose that $P_0 = 50$ (that is, 50 000), $P_1 = 55$, and $P_2 = 60$. If the population is

modelled by a discrete logistic equation, use this information to find K and k and the equilibrium level L.

c. Set up and solve an analogous logistic differential equation for the bacterial population in part (b), regarded as a continuous function $P(t)$ of time t, with $P(0) = 50$, $P(1) = 55$, and $P(2) = 60$. Compare with the results obtained for the discrete model in part (b).

48. A* a. Show that for $a \neq 1$,

$$S_n = a^{n-1} + a^{n-2} + \cdots + a + 1 = \frac{1 - a^n}{1 - a}.$$

[*Hint:* What is $S_n - aS_{n-1}$?]

b. Show that

$$1 + a + a^2 + a^3 + \cdots = \begin{cases} \dfrac{1}{1-a} & \text{if } |a| < 1 \\ \infty & \text{if } a < 1 \end{cases}$$

What happens when $a = -1$?

Concept Summary Chapter 8

Separable Differential Equation

$$\frac{dy}{dx} = \frac{h(x)}{g(y)}$$

Solution: $\displaystyle\int g(y)\, dy = \int h(x)\, dx$

Exponential Growth and Decay

$$\frac{dQ}{dt} = mQ \quad \text{with } Q(0) = Q_0$$

Solution: $Q(t) = Q_0(t)e^{mt}$

Learning Models

$$\frac{dQ}{dt} = k(B - Q)$$

Solution: $Q(t) = B - Ae^{-kt}$

Logistic Growth Models

$$\frac{dQ}{dt} = kQ(B - Q)$$

Solution: $Q(t) = \dfrac{B}{1 + Ae^{-Bkt}}$

where B is the carrying capacity.

First-Order Linear Differential Equation

$$\frac{dy}{dx} + p(x)y = q(x)$$

Integrating factor: $I(x) = e^{\int p(x)\, dx}$

Solution: $y = \dfrac{1}{I(x)}\left[\displaystyle\int I(x)q(x)\, dx + C\right]$

Euler's Method

$$y_k = y_{k-1} + f(x_{k-1}, y_{k-1})h, \text{ where } \frac{dy}{dx} = f(x,y)$$

First-Order Linear Difference Equation

$$y_n = ay_{n-1} + b \quad \text{with initial value } y_0$$

Solution: $y_n = a^n y_0 + \left(\dfrac{1 - a^n}{1 - a}\right)b$

Checkup for Chapter 8

1. Find the general solution for each differential equation.

   a. $\dfrac{dy}{dx} = \dfrac{xy}{x^2 + 1}$

   b. $\dfrac{dy}{dx} = \dfrac{2y}{x} + x^3$

   c. $\dfrac{dy}{dx} = xe^{y - x}$

2. In each case, find the particular solution of the given differential equation that satisfies the indicated condition.

   a. $\dfrac{dy}{dx} = \dfrac{-2}{x^2 y}$, where $y = 1$ when $x = -1$

   b. $\dfrac{dy}{dx} + \dfrac{y}{x} = 5$, where $y = 0$ when $x = 1$

   c. $\dfrac{dy}{dx} + \dfrac{xy}{x^2 + 1} = x$, where $y = 0$ when $x = 0$

3. Solve each difference equation subject to the indicated initial condition.

   a. $y_n = -\dfrac{1}{2} y_{n-1};\ y_0 = -1$

   b. $y_n - y_{n-1} = \dfrac{1}{3} y_{n-1};\ y_0 = 2$

   c. $y_n - y_{n-1} = \dfrac{1}{4} y_{n-1} - \dfrac{1}{2};\ y_0 = 0$

4. Plot a slope field for each differential equation and use the field to sketch the graph of the solution that passes through the specified point.

   a. $y' = 2x - y;\ (0, 0)$
   b. $y' = x^2 - y^2;\ (1, 1)$

   c. $y' = \dfrac{y}{x + 1};\ (0, 0)$

In Exercises 5 and 6, write a differential equation describing the given situation. (Do not try to solve the equation.)

5. **WATER POLLUTION** The amount $P(t)$ of pollution in a stream t days after an industrial spill decreases with respect to time at a rate equal to 11% of the amount present at that time.

6. **PRICE ADJUSTMENT** The price $p(t)$ of a commodity changes with respect to time at a rate that is proportional to the surplus $S(p) - D(p)$, where S and D are the supply and demand, respectively.

7. **MEDICATION** A child is injected with A_0 milligrams of an anaesthetic drug, and t hours later, the amount $A(t)$ of drug still present in the child's bloodstream is decreasing with respect to time at a rate proportional to $A(t)$. It takes 4 hours to eliminate half the original dose of the drug, and the anaesthetic becomes ineffective when the blood contains less than 1800 mg. What dose A_0 should be administered in order to guarantee that the child will remain fully anaesthetized during a 2-hour operation?

8. **GROWTH OF AN INVESTMENT** An investment manager determines that the value $V(t)$ of a certain investment is growing with respect to time at a rate proportional to V^2. The initial value of the investment was $10 000, and it took 10 years to double. How long will it take before the investment is worth $50 000?

9. **DILUTION** A 500-L tank initially contains 200 L of brine containing 75 g of dissolved salt. Brine containing 2 g of salt per litre flows into the tank at a rate of 4 L per minute, and the well stirred mixture flows out at a rate of 1 L per minute. Set up and solve a differential equation for the amount of salt $A(t)$ in the tank at time t. How much salt is in the tank when it is full?

10. **COBWEB MODEL** The supply S_n and demand D_n for a certain commodity during the nth time period for $n = 1, 2, \ldots$ are given by

    $$S_n = 20 + p_{n-1} \quad \text{and} \quad D_n = 50 - 2p_n,$$

    where p_n is the price of the commodity during the nth period. The initial price of the commodity is $p_0 = 20$.

    a. Find the equilibrium price p_e.

    b. Set up and solve a difference equation to find a formula for p_n.

    c. Sketch the cobweb graph for the first three time periods.

Review Exercises

In Exercises 1 through 12, find the general solution of the given differential equation.

1. $\dfrac{dy}{dx} = x^3 - 3x^2 + 5$

2. $\dfrac{dy}{dx} = 0.02y$

3. $\dfrac{dy}{dx} = k(80 - y)$

4. $\dfrac{dy}{dx} = y(1 - y)$

5. $\dfrac{dy}{dx} = e^{x+y}$ [*Hint:* $e^{x+y} = e^x e^y$]

6. $\dfrac{dy}{dx} = ye^{-2x}$

7. $\dfrac{dy}{dx} + \dfrac{4y}{x} = e^{-x}$

8. $\dfrac{dy}{dx} + \dfrac{y}{x} = \dfrac{2}{x + 1}$

9. $\dfrac{dy}{dx} + \dfrac{2y}{x + 1} = \dfrac{4}{x + 2}$

10. $x^3\dfrac{dy}{dx} + xy = 5$

11. $\dfrac{dy}{dx} = \dfrac{\ln x}{xy}$

12. $\dfrac{dy}{dx} = e^{2x-y}$ $\left[\textit{Hint: } e^{2x-y} = \dfrac{e^{2x}}{e^y} \right]$

In Exercises 13 through 24, find the particular solution of the given differential equation that satisfies the specified condition.

13. $\dfrac{dy}{dx} = 5x^4 - 3x^2 - 2$; $y = 4$ when $x = 1$

14. $\dfrac{dy}{dx} = 3 - y$; $y = 2$ when $x = 0$

15. $\dfrac{dy}{dx} = 0.06y$; $y = 100$ when $x = 0$

16. $\dfrac{dy}{dx} - \dfrac{y}{x} = \dfrac{2}{x}$; $y = -1$ when $x = 1$

17. $\dfrac{dy}{dx} - \dfrac{5y}{x} = x^2$; $y = 4$ when $x = -1$

18. $\dfrac{dy}{dx} + \dfrac{y}{x + 1} = x$; $y = 0$ when $x = 3$

19. $\dfrac{dy}{dx} - xy = e^{x^2/2}$; $y = 4$ when $x = 0$

20. $\dfrac{dy}{dx} + 3x^2y = x^2$; $y = 2$ when $x = 0$

21. $\dfrac{dy}{dx} + y \ln \sqrt{x}$; $y = -4$ when $x = 1$

22. $\dfrac{dy}{dx} = \dfrac{\ln x}{y}$; $y = 100$ when $x = 1$

23. $\dfrac{dy}{dx} = \dfrac{xy}{\sqrt{1 - x^2}}$; $y = 2$ when $x = 0$

24. $\dfrac{d^2y}{dx^2} = 2$; $y = 5$ and $\dfrac{dy}{dx} = 3$ when $x = 0$

In Exercises 25 through 30, find the solution of the given first-order linear difference equation that satisfies the specified initial condition.

25. $y_n = y_{n-1} + 2$; $y_0 = 1$

26. $y_n = -8y_{n-1}$; $y_0 = 1$

27. $y_n - y_{n-1} = 3$; $y_0 = 1$

28. $y_n - y_{n-1} = -1$; $y_0 = -2$

29. $y_n - 5y_{n-1} = 2$; $y_0 = 2$

30. $y_n - 2y_{n-1} = 8$; $y_0 = -1$

In Exercises 31 through 36, plot a slope field for the given differential equation and use the field to sketch the graph of the particular solution that passes through the specified point.

31. $y' = 3x - y$; $(1, 0)$

32. $y' = 3 - x + y$; $(0, 0)$

33. $y' = x + y^2$; $(1, 1)$

34. $y' = \dfrac{x - y}{x + y}$; $(1, 1)$

35. $y' = \dfrac{y}{x} + \dfrac{x}{y}$; $(1, 0)$

36. $y' = \dfrac{xy}{x^2 + y^2}$; $(0, 1)$

In Exercises 37 through 40, use Euler's method with the indicated step size h and number of increments n to obtain an approximate solution of the given initial value problem.

37. $y' = 3x - y$, $y(0) = 0$, $h = 0.1$, $n = 5$

38. $y' = x + xy$, $y(0) = 1$, $h = 0.1$, $n = 4$

39. $y' = \dfrac{y}{x} + y$, $y(1) = 1$, $h = 0.2$, $n = 4$

40. $y' = \sqrt{x} - \sqrt{y}$, $y(0) = 0$, $h = 0.2$, $n = 5$

In Exercises 41 through 44, use Euler's method with the indicated step size h to estimate the required solution of the given initial value problem.

41. $y' = x + 4y$, $y(0) = 1$; estimate $y(1)$ using $h = 0.2$

42. $y' = \dfrac{y}{x}$, $y(1) = 0$; estimate $y(2)$ using $h = 0.2$

43. $y' = e^{xy}$, $y(0) = 0$; estimate $y(2)$ using $h = 0.4$

44. $y' = \ln(xy)$, $y(1) = 1$; estimate $y(3)$ using $h = 0.4$

45. RECALL FROM MEMORY Some psychologists believe that when a person is asked to recall a set of facts, the rate at which the facts are recalled is proportional to the number of relevant facts in the subject's memory that have not yet been recalled. Suppose a subject is presented with a list of A facts, and that at time $t = 0$, the subject knows none of these facts. Set up and solve a differential equation to find the number of facts $N(t)$ that have been recalled as a function at time t.

46. DECAY OF A BIOMASS A researcher determines that a certain protein with mass $m(t)$ grams at time t hours disintegrates into amino acids at a rate jointly proportional to $m(t)$ and t. Experiments indicate that half of any given sample will disintegrate in 12 hours.
  a. Set up and solve a differential equation for $m(t)$.
  b. What fraction of a given sample remains after 9 hours?

47. DILUTION A tank currently holds 200 L of brine that contains 3 g of salt per litre. Clear water flows into the tank at a rate of 4 L per minute, while the mixture, kept uniform by stirring, runs out of the tank at the same rate. How much salt is in the tank at the end of 100 minutes?

48. DILUTION A 200-L tank currently holds 30 g of salt dissolved in 50 L of water. Brine containing 2 g of salt per litre runs into the tank at a rate of 2 L per minute, and the mixture, kept uniform by stirring, flows out at a rate of 1 L per minute.
  a. Set up and solve a differential equation for the amount of salt $S(t)$ in solution at time t.
  b. How many litres of solution are in the tank when 40 g of salt is in solution?
  c. How much salt is in the tank when it overflows?

49. POPULATION GROWTH Demographers estimate that environmental factors place an upper bound of 10 million people on the population of a certain country. At each time t, the population $P(t)$ is growing at a rate that is jointly proportional to the current population P and the difference between the upper bound and P. Suppose the population of this country was 4 million in 2006 and 7.4 million in 2010. Set up and solve a differential equation to find $P(t)$. What will the population be in 2016?

50. RETIREMENT INCOME A retiree deposits S dollars into an account that earns interest at an annual rate r compounded continuously and annually withdraws W dollars.
  a. Explain why the account changes at a rate of

  $$\frac{dV}{dt} = rV - W,$$

  where $V(t)$ is the value of the account t years after the account is opened. Solve this differential equation to express $V(t)$ in terms of r, W, and S.
  b. Andrew and Amy deposit $500\,000 into an account that pays 5% interest compounded continuously. If they withdraw $50\,000 annually, what is their account worth at the end of 10 years?
  c. What annual amount W can Andrew and Amy withdraw each year if their goal is to keep their account unchanged at $500\,000?
  d. If Andrew and Amy decide to withdraw $80\,000 annually, how long will it take to exhaust their account? That is, when will $V(t)$ become 0?

51. CONCENTRATION OF BLOOD GLUCOSE Glucose is infused into the bloodstream of a patient at a constant rate R, and at each time t, it is converted and excreted at a rate proportional to the present concentration of glucose $C(t)$.
  a. Set up and solve a differential equation for $C(t)$. Express your answer in terms of R, the constant of proportionality k (the elimination constant), and the initial concentration of glucose $C_0 = C(0)$.

b. Show that in the long run (as $t \rightarrow \infty$) the glucose concentration approaches a level that depends only on r and k and not on the initial concentration C_0.

c. Find an expression for the time T required for half the initial concentration C_0 to be eliminated. (Notice that this half-life depends on C_0 unless $C_0 = 0$.)

52. REDUCTION OF AIR POLLUTANT The air in a conference centre room with volume 3000 m^3 initially contains 0.21% carbon dioxide. Suppose the room receives fresh air containing 0.04% carbon dioxide at a rate of 800 m^3 per minute. All of the air in the room is circulated by a fan, and the mixture leaves at the same rate (800 m^3 per minute).

a. Set up and solve a differential equation for the percentage of carbon dioxide in the room after t minutes.

b. How long does it take before the room contains only 0.1% carbon dioxide?

c. What percentage of carbon dioxide is left in the room in the long run (as $t \rightarrow \infty$)?

53. PERSONNEL MANAGEMENT A certain company began operations with 100 employees, and in subsequent years, the number of employees E_n during the nth year always exceeded the number E_{n-1} during the previous year by 5%. Express this relationship as a first-order linear difference equation and solve the equation to find E_n.

54. MEDICATION A drug administered intravenously to a patient is eliminated from the patient's bloodstream at a rate of 14% per hour. Set up and solve a difference equation for the amount D_n of a dose of D_0 milligrams of drug that remains after n hours. How much of the dose remains after 10 hours?

In Exercises 55 and 56, you will need the formula

$$1 + a + a^2 + a^3 + \cdots = \frac{1}{1 - a} \quad \text{for } |a| < 1$$

derived in Exercise 48 of Section 8.5.

55. A* QUALITY CONTROL Suppose that in a particular industrial assembly line, the likelihood p_n that a defective unit receives its imperfection during the nth stage of production is 30% of the likelihood p_{n-1} of the imperfection being introduced during the $(n - 1)$th stage.

a. Set up and solve a difference equation for p_n in terms of the likelihood p_0 that the raw material of the unit is already defective.

b. Why is it true that $p_0 + p_1 + p_2 + \cdots = 1$? Use this fact to find a formula for p_n. How likely is it that the imperfection occurred during the third stage?

56. A* POLITICAL INFLUENCE Suppose the likelihood p_n that a particular politician influences n colleagues is 60% of the likelihood p_{n-1} that the same politician influences only $n - 1$ colleagues.

a. Set up and solve a difference equation for p_n in terms of the likelihood p_0 that the politician influences no colleagues.

b. Why is it true that $p_0 + p_1 + p_2 + \cdots = 1$? Use this fact to find a formula for p_n.

c. How likely is it that the politician influences as many as five colleagues?

57. A* POPULATION GROWTH A bacterial colony begins with population $P_0 > 0$ and grows so that the population P_n after n hours is related to the population P_{n-1} after $n - 1$ hours by the difference equation

$$P_n = \frac{bP_{n-1}}{a + P_{n-1}}$$

where a and b are positive constants.

a. Show that $P_n < \left(\dfrac{b}{a}\right)P_{n-1}$ and hence that

$$P_n < \left(\frac{b}{a}\right)^n P_0.$$

b. If $a > b$, show that the population eventually becomes extinct, that is, $\lim\limits_{n \to \infty} P_n = 0$.

c. If $a < b$, what happens to the population in the long run? You may assume that $\lim\limits_{n \to \infty} P_n$ exists and that $\lim\limits_{n \to \infty} P_n = \lim\limits_{n \to \infty} P_{n-1}$.

58. A* Solve the difference equation

$$P_n = \frac{bP_{n-1}}{a + P_{n-1}}$$

in Exercise 57 by following these steps:

a. Show that substituting $Q_n = \dfrac{1}{P_n}$ converts the given difference equation into the linear equation

$$Q_n = \frac{a}{b}Q_{n-1} + \frac{1}{b}.$$

b. Solve the linear equation in Q_n obtained in part (a) and then use your solution to solve the given difference equation in terms of a, b, and P_0.

59. A* THE FIBONACCI EQUATION   The 13th-century mathematician Fibonacci (1170–1250), also known as Leonardo of Pisa, posed this problem in his book *Liber Abaci:*

Consider an island where rabbits live forever. Every month each pair of adult rabbits produces a pair of offspring, which themselves become reproductive adults 2 months later. If the rabbit colony begins with one newborn pair, how many pairs will there be after n months?

a. Let R_n be the total number of rabbit pairs (adult, newborn, and adolescent) in the colony after n months. Explain why $R_1 = 1$, $R_2 = 1$, $R_3 = 2$, $R_4 = 3$, and, in general,

$$R_n = R_{n-1} + R_{n-2}.$$

This difference equation is called the *Fibonacci equation.*

b. Find all numbers λ such that $R_n = \lambda^n$ satisfies the Fibonacci equation.

c. If λ_1 and λ_2 are the numbers you found in part (b), it can be shown that the general solution of the Fibonacci equation has the form

$$R_n = A\lambda_1^n + B\lambda_2^n$$

for constants A and B. Find A and B so that the initial conditions $R_1 = 1$ and $R_2 = 1$ are satisfied.

d. The numbers 1, 1, 2, 3, 5, 8, 13, . . . generated by the Fibonacci equation are called *Fibonacci numbers.* Such numbers occur in a remarkable variety of applications, ranging from population genetics to stock market analysis. For instance, in botany, the numbers appear in the study of phyllotaxis, the arrangement of leaves on the stems of plants. Write a paragraph on applications of the Fibonacci numbers, using the Internet as a research source.

60. A* A child, standing at the origin of a coordinate plane, holds a 5-m length of rope that is attached to a sled. She walks along the x axis keeping the rope taut, as shown in the figure (units are in metres).

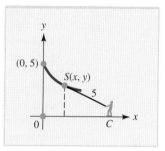

EXERCISE 60

a. If the sled begins at $(0, 5)$ and is at $S(x, y)$ when the child is at C, explain why

$$\frac{dy}{dx} = \frac{-y}{\sqrt{25 - y^2}}.$$

b. Find the equation of the path followed by the sled. The path is called a **tractrix**. [*Hint:* Use an integral formula from Table 6.1.]

c. Use a graphing utility to draw the path. Then, by moving the cursor, find the point on the path that is exactly 3 m from the y axis.

THINK ABOUT IT

MODELLING EPIDEMICS

An outbreak of a disease affecting a large number of people is called an *epidemic* (from the Greek roots *epi* "upon" and *demos* "the people"). Mathematics plays an important role in the study of epidemics, a discipline known as *epidemiology*. Mathematical models have been developed for epidemics of influenza, such as H1N1 in 2009–2010; SARS; bubonic plague; HIV; AIDS; smallpox; gonorrhoea; and various other diseases. By making some simplifying assumptions, we can construct a model based on differential equations, called the **S-I-R model,** that is useful for making predictions about the course of an epidemic. Our model will be fairly straightforward, with enough simplifying assumptions to make the mathematics accessible. Later, we will examine the assumptions, with a view toward improving the model by making it more realistic.

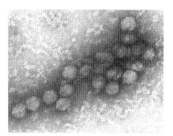

(Photo: CDC/Cynthia Goldsmith)

To build our model, we consider an epidemic affecting a population of N people, each of whom falls into exactly one of the following groups at time t:

Susceptibles, $S(t)$, are people who have not yet become ill, but who could become sick later. When a susceptible person gets sick, it is assumed to happen with no time lag.

Infectives, $I(t)$, are people who have the disease and can infect others. Infectives are not quarantined and continue a normal pattern of interacting with other people.

Removeds, $R(t)$, are people who can no longer infect others. In particular, a recovered person cannot catch the disease again or infect others.

We assume that no people enter or leave the population (including births and deaths), and that at time $t = 0$, there are some susceptibles and infectives but no removeds; that is,

$$N = S(t) + I(t) + R(t) \qquad \text{for all } t \geq 0,$$

where

$$S(0) > 0, \quad I(0) > 0, \quad \text{and } R(0) = 0,$$

so that $S(0) + I(0) = N$.

The key to modelling an epidemic or any other dynamic situation lies in the assumptions made about rates of change. In our model, we assume that the rate at which susceptible people are being infected at any time is proportional to the number of contacts between susceptibles and infectives, that is, to the product SI. Thus, the susceptible population $S(t)$ is *decreasing* (becoming infected) at a rate given by

$$(1) \quad \frac{dS}{dt} = -aSI.$$

We also assume that the rate at which people are being removed from the infected population is proportional to the number of people who are infective, so that

$$(2) \quad \frac{dR}{dt} = bI$$

for some number $b > 0$. Finally, we assume that the rate at which the number of infectives changes is the rate at which susceptibles become infective minus the rate at which infective people are removed from the population of infectives. This means that

$$(3) \quad \frac{dI}{dt} = aSI - bI = aI\left(S - \frac{b}{a}\right) = aI(S - c),$$

with $c = \frac{b}{a}$. (In Question 1, you are asked to obtain this last rate by differentiating both sides of the equation $I(t) = N - S(t) - R(t)$ with respect to t and substituting the rates in Equations (1) and (2).)

The S-I-R model consists of the three simultaneous differential equations we have constructed, namely,

$$\frac{dS}{dt} = -aSI, \quad \frac{dR}{dt} = bI, \quad \text{and} \quad \frac{dI}{dt} = aI(S - c),$$

where $S(t) + I(t) + R(t) = N$ for all t, with $R(0) = 0$, $S(0) > 0$, and $I(0) > 0$. Although we cannot explicitly solve this system to obtain simple formulas for the functions $S(t)$, $I(t)$, and $R(t)$, we can use these differential equations to help us understand the progress of epidemics. We illustrate this by applying our S-I-R model to a historic case, an epidemic of bubonic plague in Eyam, a village near Sheffield, England, during 1665–1666. This analysis is possible because the village quarantined itself during the epidemic and kept sufficiently detailed records. We will use these records to estimate the constants a and b in Equations (1) and (2).

First, note that the constant b in Equation (2) can be interpreted as the rate at which people are removed from the population of infectives. The infective period of the bubonic plague in Eyam was 11 days or 0.367 months. Therefore, if we measure time t in months and assume, for simplicity, that infected people are removed from this population at a constant rate, we can estimate b by the ratio

$$b = \frac{1}{0.367} \approx 2.72.$$

Next, records kept during the epidemic show that on July 19, 1666, there were 201 susceptibles and 22 infectives, while on August 19, 1666, there were 121 susceptibles and 21 infectives. Thus, the number of susceptibles changed by $121 - 201 = -80$ during this month-long period, and this change can be used to estimate $\dfrac{dS}{dt}$ on the day the period began. In other words, on July 19, 1666,

$$S = 201, \quad I = 22, \quad \text{and} \quad \frac{dS}{dt} \approx -80.$$

Substituting into Equation (1) and solving for the constant a, we find that

$$a = \frac{-dS/dt}{SI} = \frac{-(-80)}{(201)(22)} = 0.018.$$

It turns out that the resulting model, with the estimates we have just obtained for the values of a and b, does a good job of estimating the actual data. Figure 8.17 displays the values predicted by the model and the actual data. (This information is adapted from Raggett (1982) and from the discussion in Brauer and Castillo-Chavez (2001).)

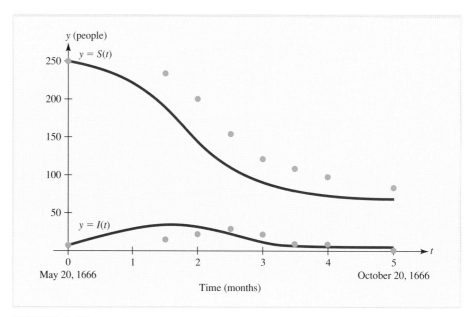

FIGURE 8.17 Comparison of Eyam plague data with values predicted by the S-I-R model.

The S-I-R model can be used to make some general observations about the course of epidemics. For example, unless the number of infectives increases initially, the epidemic will never get started (since the number of infectives does not grow beyond the initial number). In the language of calculus, an epidemic begins only if $\dfrac{dI}{dt} > 0$ at $t = 0$. Examining Equation (3) of the S-I-R model, we see that this happens only if $S(0) > c$. For this reason, c is called the *threshold number of susceptibles*. When the initial number of susceptibles exceeds c, the epidemic spreads; otherwise, it dies out. For example, the records of the 1666 bubonic plague in Eyam indicate that there were initially $S(0) = 254$ susceptibles. Using our estimates of $a \approx 0.018$ and $b \approx 2.72$, we estimate the threshold number for this model to be

$$c = \frac{b}{a} = \frac{2.72}{0.018} \approx 151$$

Since $S(0) = 254 > 151$, the model predicts that the epidemic should spread, which is exactly what happened.

You have probably noticed that in constructing our S-I-R model and applying it to the Eyam epidemic, we have made several fairly strong assumptions, not the least of which is that $S(t) + I(t) + R(t)$ is a constant. Is it realistic to assume that the size of a population subjected to an epidemic would never change? Is it any more realistic to assume that people become infective immediately with no time lag? Certainly not, since diseases usually have incubation periods, but inserting a time lag to account for the incubation period would make the resulting model too complicated to analyze without more sophisticated mathematical tools.

This illustrates the central issue facing anyone who uses mathematical modelling: *A completely realistic model is often difficult if not impossible to analyze, while making assumptions to simplify the analysis can lead to results that are not entirely realistic.* The S-I-R model is indeed based on several simplifying assumptions that are somewhat unrealistic, but we have also seen that the model can be used effectively to explore the dynamics of an epidemic. How do you think the model could be made more realistic without sacrificing too much of its simplicity?

Questions

1. Use the equation $N = S(t) + I(t) + R(t)$ and the rate relations in Equations (1) and (2) of the S-I-R model to verify the rate relation in Equation (3).

2. **a.** Use the chain rule to show that in the S-I-R model, the rate of change of the number of infectives with respect to the number of susceptibles is the ratio of the rate of change of infectives divided by the rate of change of susceptibles, that is,

   $$\frac{dI}{dS} = \frac{dI/dt}{dS/dt}.$$

   b. Find a formula for $\dfrac{dI}{dS}$ using Equations (1) and (3) of the model. Use this to find $\dfrac{dI}{dS}$ for the 1666 Eyam bubonic plague epidemic.

   c. Use your answer in part (b) to find the number of susceptibles when the number of infecteds is a maximum.

3. An influenza epidemic broke out at a British boys' boarding school in 1978. At the start of the epidemic, there were 762 susceptible boys and 1 infective boy, and 1 day later, 2 more boys became ill. Assume that anyone sick with influenza is also infective.

   a. Use the data supplied to estimate the constant a in Equation (1) of the S-I-R model for this epidemic, with t representing time in days.

   b. Suppose that all infected boys are removed from the population the day after they become ill. Use this information to estimate the constant b in Equation (2) of the S-I-R model for this epidemic.

THINK ABOUT IT

   c. Use the values of a and b you found in parts (a) and (b) to find the threshold number of susceptibles needed for this epidemic to get started.

   d. Use the results of Question 2 to find the number of susceptibles when the number of students infected with influenza was the maximum.

4. Modify the S-I-R model for an epidemic to take into account vaccinations of people at a constant rate d, assuming that someone vaccinated is immediately no longer susceptible. Find a formula $\dfrac{dI}{dS}$ for the resulting model.

5. An S-I model can be used to model an epidemic where everyone who gets the disease remains infective. (An S-I model is really the special case of the S-I-R model in which $b = 0$.) Answer these questions about such an epidemic.

   a. Show that $\dfrac{dS}{dt} = -aS(N - S)$, where N is the population size (which we assume is constant).

   b. Verify that $S(t) = \dfrac{N(N - 1)}{(N - 1) + e^{aNt}}$ is a solution of this differential equation and conclude that $I(t) = \dfrac{Ne^{aNt}}{(N - 1) + e^{aNt}}$.

   c. Show that $\lim\limits_{t \to +\infty} S(t) = 0$ and $\lim\limits_{t \to +\infty} I(t) = N$. Draw a graph showing the susceptible and infective curves, $y = S(t)$ and $y = I(t)$, assuming that at time $t = 0$ there are I_0 infected people, where $0 < I_0 < N$.

References

W. O. Kermack and A. G. McKendrick, "A Contribution to the Mathematical Theory of Epidemics," *Proc. Royal Soc. London,* Vol. 115, 1927, pp. 700–721.

G. F. Raggett, "Modeling the Eyam Plague," *IMA Journal,* Vol. 18, 1982, pp. 221–226.

Fred Brauer and Carlos Castillo-Chavez, *Mathematical Models in Population Biology and Epidemiology,* New York: Springer-Verlag, 2001.

Leah Edelstein-Keshet, *Mathematical Models in Biology,* Birkhauser Mathematics Series, Boston: McGraw-Hill, 1988, pp. 243–256.

 Practise and learn online with Connect.

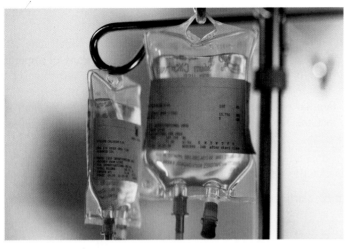

The accumulation of medication in a patient's body can be determined using an infinite series. See Example 9.1.11 for more details.
(Photo: Randy Allbritton/Getty Images)

INFINITE SERIES AND TAYLOR SERIES APPROXIMATIONS

LEARNING OBJECTIVES

After completing this chapter, you should be able to

LO1 Investigate the convergence or divergence of an infinite series. Find the sum of a converging infinite series.

LO2 Determine whether a series converges by choosing from a number of tests for convergence.

LO3 Find power series and Taylor series for given functions. Use Taylor polynomials to approximate quantities and integrals.

SECTION 9.1

L01

Investigate the convergence or divergence of an infinite series. Find the sum of a converging infinite series.

Infinite Series; Geometric Series

The sum of infinitely many numbers may be finite. This statement, which may seem paradoxical at first, plays a central role in mathematics and has a variety of important applications. The purpose of this chapter is to explore its meaning and some of its consequences.

You are already familiar with the phenomenon of a finite-valued infinite sum. You know, for example, that the repeating decimal $0.333\ldots$ stands for the infinite sum $\dfrac{3}{10} + \dfrac{3}{100} + \dfrac{3}{1000} + \cdots$ and that its value is the finite number $\dfrac{1}{3}$. What may not be familiar, however, is exactly what it means to say that this infinite sum adds up to $\dfrac{1}{3}$. The situation will be clarified in this introductory section.

Convergence and Divergence of Infinite Series

An expression of the form

$$a_1 + a_2 + \cdots + a_n + \cdots$$

is called an **infinite series.** It is customary to use summation notation to write the series compactly as follows:

$$a_1 + a_2 + \cdots + a_n + \cdots = \sum_{n=1}^{\infty} a_n.$$

The use of this notation is illustrated in Example 9.1.1.

EXAMPLE 9.1.1

a. Write some representative terms of the series $\displaystyle\sum_{n=1}^{\infty} \dfrac{1}{2^n}$.

b. Use summation notation to write the series $1 - \dfrac{1}{4} + \dfrac{1}{9} - \dfrac{1}{16} + \cdots$ in compact form.

Solution

a. $\displaystyle\sum_{n=1}^{\infty} \dfrac{1}{2^n} = \dfrac{1}{2} + \dfrac{1}{4} + \dfrac{1}{8} + \cdots + \dfrac{1}{2^n} + \cdots$

b. The nth term of this series is

$$\frac{(-1)^{n+1}}{n^2},$$

where the factor $(-1)^{n+1}$ generates the alternating signs, starting with a plus sign when $n = 1$. Thus,

$$1 - \frac{1}{4} + \frac{1}{9} - \frac{1}{16} + \cdots = \sum_{n=1}^{\infty} \frac{(-1)^{n+1}}{n^2}.$$

Roughly speaking, an infinite series

$$a_1 + a_2 + a_3 + \cdots = \sum_{k=1}^{\infty} a_k$$

is said to *converge* if it adds up to a finite number and to *diverge* if it does not.

A more precise statement of the criterion for convergence involves the sum

$$S_n = a_1 + a_2 + \cdots + a_n = \sum_{k=1}^{n} a_k$$

of the first n terms of the infinite series. This finite sum is called the ***n*th partial sum** of the infinite series, and its behaviour as n approaches infinity determines the convergence or divergence of the series. Here is a summary of the criterion.

Convergence or Divergence of an Infinite Series ■ An infinite series $\sum_{k=1}^{\infty} a_k$ with nth partial sum

$$S_n = a_1 + a_2 + \cdots + a_n = \sum_{k=1}^{n} a_k$$

is said to **converge with sum S** if S is a finite number such that

$$\lim_{n\to\infty} S_n = S,$$

and in this case, we write

$$\sum_{k=1}^{\infty} a_k = S.$$

The series is said to **diverge** if $\lim_{n\to\infty} S_n$ does not exist as a finite number.

According to this criterion, to determine the convergence or divergence of an infinite series, start by finding an expression for the sum of the first n terms of the series and then take the limit of this finite sum as n approaches infinity. This procedure is similar to the one used in Section 6.2 to determine the convergence or divergence of an improper integral. Here is an example.

EXAMPLE 9.1.2

Investigate the possible convergence of each series.

a. $\displaystyle\sum_{k=1}^{\infty} (-1)^{k+1} = 1 - 1 + 1 - 1 + \cdots$

b. $\displaystyle\sum_{k=1}^{\infty} \frac{3}{10^k} = \frac{3}{10} + \frac{3}{10^2} + \frac{3}{10^3} + \cdots$

Solution

a. The first three partial sums of the series $\displaystyle\sum_{k=1}^{\infty} (-1)^{k+1}$ are

$$S_1 = 1$$
$$S_2 = 1 - 1 = 0$$
$$S_3 = 1 - 1 + 1 = 1$$

This pattern continues, and in general, the even partial sums are 0 and the odd partial sums are 1. Therefore, $\lim_{n\to\infty} S_n$ does not exist, so the series diverges.

b. Notice that each term of the series

$$\sum_{k=1}^{\infty} \frac{3}{10^k} = \frac{3}{10} + \frac{3}{10^2} + \frac{3}{10^3} + \cdots$$

is $\frac{1}{10}$ times the preceding term. This leads to the following trick for finding a compact formula for S_n.

Begin with the nth partial sum in the form

$$S_n = \frac{3}{10} + \frac{3}{10^2} + \frac{3}{10^3} + \cdots + \frac{3}{10^n}$$

and multiply both sides of this equation by $\frac{1}{10}$ to get

$$\frac{1}{10}S_n = \frac{3}{10^2} + \frac{3}{10^3} + \cdots + \frac{3}{10^n} + \frac{3}{10^{n+1}}$$

Now subtract the expression for $\frac{1}{10}S_n$ from the expression for S_n. Most of the terms cancel, and you are left with only the first term of S_n and the last term of $\frac{1}{10}S_n$; that is,

$$S_n - \frac{1}{10}S_n = \frac{9}{10}S_n = \frac{3}{10} - \frac{3}{10^{n+1}} = \frac{3}{10}\left(1 - \frac{1}{10^n}\right)$$

or

$$S_n = \frac{1}{3}\left(1 - \frac{1}{10^n}\right).$$

Then

$$\lim_{n \to \infty} S_n = \lim_{n \to \infty} \frac{1}{3}\left(1 - \frac{1}{10^n}\right) = \frac{1}{3},$$

and so

$$\sum_{n=1}^{\infty} \frac{3}{10^n} = \frac{1}{3}.$$

EXAMPLE 9.1.3

Investigate the possible convergence of the series

$$\sum_{k=1}^{\infty} \frac{1}{k(k+1)} = \frac{1}{(1)(2)} + \frac{1}{(2)(3)} + \frac{1}{(3)(4)} + \cdots.$$

Solution

First, note that

$$\frac{1}{k(k+1)} = \frac{1}{k} - \frac{1}{k+1}.$$

Therefore, the nth partial sum of the given series can be expressed as

$$S_n = \sum_{n=1}^{n} \frac{1}{k(k+1)} = \sum_{n=1}^{n} \left(\frac{1}{k} - \frac{1}{k+1} \right)$$

$$= \left(\frac{1}{1} - \frac{1}{2} \right) + \left(\frac{1}{2} - \frac{1}{3} \right) + \left(\frac{1}{3} - \frac{1}{4} \right) + \cdots + \left(\frac{1}{n} - \frac{1}{n+1} \right)$$

$$= 1 + \left(-\frac{1}{2} + \frac{1}{2} \right) + \left(-\frac{1}{3} + \frac{1}{3} \right) + \cdots + \left(\frac{-1}{n} + \frac{1}{n} \right) - \frac{1}{n+1}$$

$$= 1 - \frac{1}{n+1}$$

Taking the limit of the partial sums as $n \to \infty$ gives

$$\lim_{n \to \infty} S_n = \lim_{n \to \infty} \left(1 - \frac{1}{n+1} \right) = 1.$$

Therefore, the given series converges with sum $S = 1$.

> **NOTE**   A series like the one in Example 9.1.3 in which the terms of the partial sums cancel in pairs is called a *telescoping series*.   ∎

Convergent infinite series can be added together and multiplied by constants in much the same way as ordinary sums of numbers. Here is a summary of these properties.

> **Sum and Multiple Rules for Convergent Infinite Series** ■ If the series $\sum_{n=1}^{\infty} a_n$ and $\sum_{n=1}^{\infty} b_n$ converge, then the series $\sum_{n=1}^{\infty} (a_n + b_n)$ and $\sum_{n=1}^{\infty} c a_n$ for constant c also converge and
>
> $$\sum_{n=1}^{\infty} (a_n + b_n) = \sum_{n=1}^{\infty} a_n + \sum_{n=1}^{\infty} b_n,$$
>
> $$\sum_{n=1}^{\infty} c a_n = c \sum_{n=1}^{\infty} a_n.$$

EXAMPLE 9.1.4

Investigate the possible convergence of the series

$$\sum_{k=1}^{\infty} \left[\frac{6}{10^k} - \frac{5}{k(k+1)} \right]$$

Solution

In Examples 9.1.2 and 9.1.3, we found that the series $\sum_{k=1}^{\infty} \frac{3}{10^k}$ and $\sum_{k=1}^{\infty} \frac{1}{k(k+1)}$ both converge and that

$$\sum_{k=1}^{\infty} \frac{3}{10^k} = \frac{1}{3} \quad \text{and} \quad \sum_{k=1}^{\infty} \frac{1}{k(k+1)} = 1.$$

Therefore, by applying the sum and multiple rules for series, we find that the given series converges and that

$$\sum_{k=1}^{\infty}\left[\frac{6}{10^k} - \frac{5}{k(k+1)}\right] = \sum_{k=1}^{\infty} 2\left(\frac{3}{10^k}\right) + \sum_{k=1}^{\infty}(-5)\frac{1}{k(k+1)} \qquad \text{sum rule}$$

$$= 2\sum_{k=1}^{\infty} \frac{3}{10^k} + (-5)\sum_{k=1}^{\infty} \frac{1}{k(k+1)} \qquad \text{multiple rule}$$

$$= 2\left(\frac{1}{3}\right) + (-5)(1)$$

$$= -\frac{13}{3}$$

Geometric Series

Notice that the series featured in Example 9.1.2b has the property that the ratio of successive terms is a constant $\left(\text{namely, } \frac{1}{10}\right)$. Any series with this property is called a **geometric series** and must have the general form

$$\sum_{k=0}^{\infty} ar^k = a + ar + ar^2 + \cdots,$$

where a is a nonzero constant and r is the ratio of successive terms. For example,

$$\frac{2}{9} - \frac{2}{27} + \frac{2}{81} - \frac{2}{243} + \cdots = \frac{2}{9}\left(1 - \frac{1}{3} + \frac{1}{3^2} - \frac{1}{3^3} + \cdots\right)$$

$$= \sum_{k=0}^{\infty} \frac{2}{9}\left(-\frac{1}{3}\right)^k$$

is a geometric series with $a = \frac{2}{9}$ and ratio $r = -\frac{1}{3}$.

A geometric series $\sum_{k=0}^{\infty} ar^k$ converges if $|r| < 1$ and diverges if $|r| \geq 1$. To show this, we proceed as in Example 9.1.2b. Begin with the nth partial sum

$$S_n = a + ar + ar^2 + \cdots + ar^{n-1} \qquad (r \neq 0).$$

(Note that the power of r in the nth term is only $n - 1$ since the first term in the sum is $a = ar^0$.) Now multiply both sides of this equation by r to get

$$rS_n = ar + ar^2 + \cdots + ar^{n-1} + ar^n,$$

and subtract the expression for rS_n from the expression for S_n to get

$$S_n - rS_n = a - ar^n \qquad \text{all other terms cancel}$$

$$(1 - r)S_n = a(1 - r^n)$$

$$S_n = \frac{a(1 - r^n)}{1 - r}.$$

If $|r| < 1$, we have $\lim_{n \to \infty} r^n = 0$, and it follows that

$$\lim_{n \to \infty} S_n = \lim_{n \to \infty} \frac{a(1 - r^n)}{1 - r} = \frac{a}{1 - r}.$$

In other words, the given geometric series converges with sum $\dfrac{a}{1-r}$. If $|r| > 1$, then $\lim\limits_{n\to\infty} |r^n| = \infty$, and $\lim\limits_{n\to\infty} S_n$ cannot exist. The series also diverges when $|r| = 1$ (see Exercise 55). To summarize:

Convergence of a Geometric Series ■ The geometric series $\sum\limits_{n=0}^{\infty} ar^n$ with *common ratio* r and $a \neq 0$ converges if $|r| < 1$ with sum

$$S = \sum_{n=0}^{\infty} ar^n = \frac{a}{1-r}.$$

Otherwise (that is, if $|r| \geq 1$), the series diverges.

EXPLORE!

Given the geometric series with $a = 2$ and $r = 0.85$, evaluate S_{15} and S_{50} to five-decimal-place accuracy. Compare S_{15} and S_{50} to S_∞. What is the error in each case?

For example, the following two series both diverge since $|r| > 1$:

$$2 + 2(1.5) + 2(1.5)^2 + \cdots \qquad (r = 1.5)$$

and

$$\frac{1}{2} - \frac{1}{2}\left(\frac{9}{7}\right) + \frac{1}{2}\left(\frac{9}{7}\right)^2 - \frac{1}{2}\left(\frac{9}{7}\right)^3 + \cdots \qquad \left(r = -\frac{9}{7}\right).$$

Examples 9.1.5, 9.1.6, and 9.1.7 illustrate the case where $|r| < 1$.

EXAMPLE 9.1.5

Find the sum of the series $\sum\limits_{n=0}^{\infty} \left(-\dfrac{2}{3}\right)^n$.

Solution

This is a geometric series with $r = -\dfrac{2}{3}$. Since $|r| < 1$, it converges with sum

$$\sum_{n=0}^{\infty} \left(-\frac{2}{3}\right)^n = \frac{1}{1 - \left(-\dfrac{2}{3}\right)} = \frac{1}{\left(\dfrac{5}{3}\right)} = \frac{3}{5}.$$

EXAMPLE 9.1.6

Find the sum of the series $\sum\limits_{n=0}^{\infty} \dfrac{3}{2^n}$.

Solution

$$\sum_{n=0}^{\infty} \frac{3}{2^n} = \sum_{n=0}^{\infty} 3\left(\frac{1}{2}\right)^n = 3\left(\frac{1}{1 - \frac{1}{2}}\right) = 3\left(\frac{1}{\frac{1}{2}}\right) = 6.$$

EXAMPLE 9.1.7

Find the sum of the series $\displaystyle\sum_{n=1}^{\infty} \frac{(-2)^n}{3^{n+1}}$.

Solution

$$\sum_{n=1}^{\infty} \frac{(-2)^n}{3^{n+1}} = \frac{-2}{3^2} + \frac{(-2)^2}{3^3} + \frac{(-2)^3}{3^4} + \cdots$$

$$= \frac{-2}{3^2}\left[1 + \left(-\frac{2}{3}\right) + \left(-\frac{2}{3}\right)^2 + \cdots\right]$$

$$= \left(-\frac{2}{9}\right)\sum_{n=0}^{\infty}\left(-\frac{2}{3}\right)^n$$

$$= -\frac{2}{9}\left(\frac{1}{1 - (-\frac{2}{3})}\right) = -\frac{2}{9}\left(\frac{3}{5}\right) = -\frac{2}{15}$$

Applications of Geometric Series

Geometric series arise in many branches of mathematics and in a variety of applications. Four such applications are considered in Examples 9.1.8 through 9.1.11.

Repeating Decimals

It turns out that any number N whose decimal form contains a pattern that repeats infinitely is rational; that is, $N = \dfrac{p}{q}$ for integers p and q. Example 9.1.8 illustrates a general procedure for using a geometric series to find p and q.

EXAMPLE 9.1.8

Express the repeating decimal $0.232323\ldots$ as a fraction.

Solution

Write the decimal as a geometric series as follows:

$$0.232323\ldots = \frac{23}{100} + \frac{23}{10\ 000} + \frac{23}{1\ 000\ 000} + \cdots$$

$$= \frac{23}{100}\left[1 + \frac{1}{100} + \left(\frac{1}{100}\right)^2 + \cdots\right]$$

$$= \frac{23}{100}\sum_{n=0}^{\infty}\left(\frac{1}{100}\right)^n$$

$$= \frac{23}{100}\left(\frac{1}{1 - \frac{1}{100}}\right) = \frac{23}{100}\left(\frac{100}{99}\right) = \frac{23}{99}$$

The Multiplier Effect in Economics

If a sum of money A is injected into the economy, say by a tax rebate, the total effect on the economy is to induce a level of spending that is much larger than A. The reason for this is that the portion of S that is spent by one individual becomes income for one or more others, who, in turn, spend some of it again, thus creating income for yet other individuals to spend. Economists refer to this kind of multistage spending as the **multiplier effect.** If the fraction of income that is spent at each stage of this process remains constant, then the total amount of spending is the sum of a geometric series, as illustrated in Example 9.1.9.

EXAMPLE 9.1.9

Suppose that nationwide, approximately 90% of all income is spent and 10% is saved. What is the total amount of spending generated by a $40 billion tax rebate if savings habits do not change?

Solution

The government begins by spending $40 billion, and then the original recipients of the rebate spend 0.9(40) billion dollars. This becomes new income, of which

$$0.9[0.9(40)] = 40(0.9)^2$$

billion dollars is spent. This, in turn, generates additional income of

$$0.9[40(0.9)^2] = 40(0.9)^3,$$

and so on. If this process continues indefinitely, the total amount spent is

$$T = 40 + 40(0.9) + 40(0.9)^2 + 40(0.9)^3 + \cdots = \sum_{n=0}^{\infty} 40(0.9)^n,$$

which we recognize as a geometric series with $a = 40$ and $r = 0.9$. Since $|r| < 1$, the geometric series converges with sum

$$T = \frac{40}{1 - 0.9} = 400.$$

Thus, the total amount of spending generated is $400 billion.

In general, suppose A dollars are originally injected into the economy and that each time a person receives a dollar, that person will spend a fixed fraction c of the dollar. For example, $c = \dfrac{9}{10}$ in Example 9.1.9. Then c is called the **marginal propensity to consume,** and by proceeding as in Example 9.1.9, it can be shown that the total amount spent will be

$$T = \frac{A}{1 - c} = \frac{A}{s} = mA,$$

where $s = 1 - c$ is the **marginal propensity to save** and $m = \dfrac{1}{s}$ is the **multiplier** for the spending process.

Present Value of a
Perpetual Annuity

EXAMPLE 9.1.10

Find the amount of money you should invest today at an annual interest rate of 10% compounded continuously so that, starting next year, you can make annual withdrawals of $400 in perpetuity.

Solution

The amount you should invest today to generate the desired sequence of withdrawals is the sum of the present values of the individual withdrawals. Compute the present value of each withdrawal using the formula $P = Be^{-rt}$ from Section 4.1, with $B = 400$, $r = 0.1$, and t the time (in years) at which the withdrawal is made. Thus,

$$P_1 = \text{present value of first withdrawal} = 400e^{-0.1(1)} = 400e^{-0.1}$$
$$P_2 = \text{present value of second withdrawal} = 400e^{-0.1(2)} = 400e^{-0.2}$$
$$\vdots$$
$$P_n = \text{present value of } n\text{th withdrawal} = 400e^{-0.1(n)}$$

and so on. The situation is illustrated in Figure 9.1.

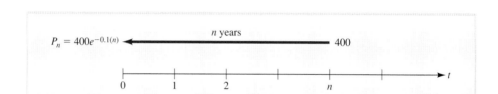

FIGURE 9.1   The present value of the nth withdrawal.

The amount that you should invest today is the sum of these infinitely many present values. That is,

$$\genfrac{}{}{0pt}{}{\text{Amount to}}{\text{be invested}} = \sum_{n=1}^{\infty} P_n = \sum_{n=1}^{\infty} 400e^{-0.1(n)}$$
$$= 400[e^{-0.1} + e^{-0.2} + e^{-0.3} + \cdots]$$
$$= 400e^{-0.1}[1 + e^{-0.1} + (e^{-0.1})^2 + \cdots]$$

This is a geometric series with $a = 400e^{-0.1}$ and $r = e^{-0.1} = 0.9048$. Since $|r| < 1$, the series converges with sum

$$\sum_{n=1}^{\infty} 400e^{-0.1(n)} = \frac{400e^{-0.1}}{1 - e^{-0.1}} \approx 3803.33.$$

Thus, you should invest \$3803.33 in order to achieve the stated objective.

Accumulation of Medication in the Body   In Example 6.2.5 of Section 6.2, we considered a nuclear power plant generating radioactive waste that was decaying exponentially. The problem was to determine the long-run accumulation of waste from the plant. Because the waste was being generated *continuously,* the model used to analyze this situation involved an (improper) integral. The situation in Example 9.1.11 is similar. A patient receives medication that is eliminated from the body exponentially, and the goal is to determine the long-run accumulation of medication in the patient's body. In this case, however, the medication is being administered in *discrete* doses, and the model involves an infinite series rather than an integral.

EXAMPLE 9.1.11

A patient is given an injection of 10 units of a certain drug every 24 hours. The drug is eliminated exponentially so that the fraction that remains in the patient's body after t days is $f(t) = e^{-t/5}$. If the treatment is continued indefinitely, approximately how many units of the drug will eventually be in the patient's body just prior to an injection?

Solution

Of the original dose of 10 units, only $10e^{-1/5}$ units are left in the patient's body after 1 day (just prior to the second injection). That is,

$$\text{Amount in body after 1 day} = S_1 = 10e^{-1/5}.$$

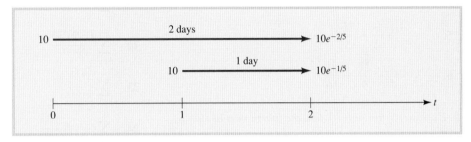

FIGURE 9.2 Amount of medication in the body after 2 days.

The medication in the patient's body after 2 days consists of what remains from the first *two* doses. Of the original dose, only $10e^{-2/5}$ units are left (since 2 days have elapsed), and of the second dose, $10e^{-1/5}$ units remain (Figure 9.2). Hence,

$$\text{Amount in body after 2 days} = S_2 = 10e^{-1/5} + 10e^{-2/5}.$$

Similarly,

$$\begin{array}{c}\text{Amount in body} \\ \text{after } n \text{ days}\end{array} = S_n = 10e^{-1/5} + 10e^{-2/5} + \cdots + 10e^{-n/5}.$$

The amount S of medication in the patient's body in the long run is the limit of S_n as n approaches infinity. That is,

$$S = \lim_{n \to \infty} S_n = \sum_{n=1}^{\infty} 10e^{-n/5}$$

$$= \sum_{n=1}^{\infty} 10(e^{-1/5})^n = 10e^{-1/5} \sum_{n=1}^{\infty} (e^{-1/5})^{n-1} = 10e^{-1/5} \sum_{n=0}^{\infty} (e^{-1/5})^n$$

$$= 10e^{-1/5} \left(\frac{1}{1 - e^{-1/5}} \right) \approx 45.17 \text{ units} \qquad \text{since } e^{-1/5} < 1$$

EXERCISES ■ 9.1

In Exercises 1 through 6, use summation notation to write the given series in compact form.

1. $\dfrac{1}{3} + \dfrac{1}{9} + \dfrac{1}{27} + \dfrac{1}{81} + \cdots$

2. $1 + \dfrac{1}{8} + \dfrac{1}{27} + \dfrac{1}{64} + \cdots$

3. $\dfrac{1}{2} + \dfrac{2}{3} + \dfrac{3}{4} + \dfrac{4}{5} + \cdots$

4. $\dfrac{1}{3} + \dfrac{2}{5} + \dfrac{3}{7} + \dfrac{4}{9} + \cdots$

5. $\dfrac{1}{2} - \dfrac{4}{3} + \dfrac{9}{4} - \dfrac{16}{5} + \cdots$

6. $-3 + \dfrac{9}{4} - \dfrac{27}{9} + \dfrac{81}{16} - \cdots$

In Exercises 7 through 10, find the fourth partial sum S_4 of the given series.

7. $\displaystyle\sum_{n=1}^{\infty} \dfrac{1}{2^n}$

8. $\displaystyle\sum_{n=1}^{\infty} \dfrac{n}{n+1}$

9. $\displaystyle\sum_{n=1}^{\infty} \dfrac{(-1)^n}{n}$

10. $\displaystyle\sum_{n=1}^{\infty} \dfrac{(-1)^{n+1}}{n(n+1)}$

In Exercises 11 through 24, determine whether the given geometric series converges, and if so, find its sum.

11. $\displaystyle\sum_{n=0}^{\infty} \left(\dfrac{4}{5}\right)^n$

12. $\displaystyle\sum_{n=0}^{\infty} \left(-\dfrac{4}{5}\right)^n$

13. $\displaystyle\sum_{n=0}^{\infty} \dfrac{2}{3^n}$

14. $\displaystyle\sum_{n=0}^{\infty} \dfrac{2}{(-3)^n}$

15. $\displaystyle\sum_{n=1}^{\infty} \left(\dfrac{3}{2}\right)^n$

16. $\displaystyle\sum_{n=1}^{\infty} \dfrac{3}{2^n}$

17. $\displaystyle\sum_{n=2}^{\infty} \dfrac{3}{(-4)^n}$

18. $\displaystyle\sum_{n=2}^{\infty} \left(-\dfrac{4}{3}\right)^n$

19. $\displaystyle\sum_{n=1}^{\infty} 5(0.9)^n$

20. $\displaystyle\sum_{n=1}^{\infty} e^{-0.2n}$

21. $\displaystyle\sum_{n=1}^{\infty} \dfrac{3^n}{4^{n+2}}$

22. $\displaystyle\sum_{n=2}^{\infty} \dfrac{(-2)^{n-1}}{3^{n+1}}$

23. $\displaystyle\sum_{n=0}^{\infty} \dfrac{4^{n+1}}{5^{n+1}}$

24. $\displaystyle\sum_{n=2}^{\infty} (-1)^n \dfrac{2^{n+1}}{3^{n-3}}$

In each of Exercises 25 through 28, evaluate the specified series given that $\displaystyle\sum_{k=1}^{\infty} a_k = 5$ and $\displaystyle\sum_{k=1}^{\infty} b_k = -3$.

25. $\displaystyle\sum_{k=1}^{\infty} 2a_k$

26. $\displaystyle\sum_{k=1}^{\infty} (3a_k - 5b_k)$

27. $\displaystyle\sum_{k=1}^{\infty} \left(\dfrac{1}{2^k} - b_k\right)$

28. $\displaystyle\sum_{k=1}^{\infty} \left[a_k - \dfrac{(-1)^k}{3^{k+1}}\right]$

In each of Exercises 29 through 32, find the sum of the given telescoping series.

29. $\displaystyle\sum_{k=0}^{\infty} \dfrac{1}{(k+1)(k+2)}$

30. $\displaystyle\sum_{k=0}^{\infty} \dfrac{1}{(2k+1)(2k+3)}$

31. $\displaystyle\sum_{k=1}^{\infty} \dfrac{1}{(2k-1)(2k+1)}$

32. $\displaystyle\sum_{k=1}^{\infty} \dfrac{\sqrt{k+1} - \sqrt{k}}{\sqrt{k(k+1)}}$

In Exercises 33 through 36, express the given decimal as a fraction.

33. 0.3333 . . .

34. 0.5555 . . .

35. 0.252525 . . .

36. 1.405405405 . . .

37. THE MULTIPLIER EFFECT Suppose that nationwide, approximately 92% of all income is spent and 8% is saved. What is the total amount of spending generated by a $50 billion tax rebate if savings habits do not change?

38. THE MULTIPLIER EFFECT Suppose that nationwide, approximately nine times as much income is spent as is saved and that economic planners wish to generate total spending of $270 billion by instituting a tax cut of N dollars. What is N?

39. PRESENT VALUE An investment guarantees annual payments of $1000 in perpetuity, with the payments beginning immediately. Find the present value of this investment if the prevailing annual interest rate remains fixed at 4% compounded continuously. [*Hint:* The present value of the investment is the sum of the present values of the individual payments.]

40. PRESENT VALUE How much should you invest today at an annual interest rate of 5% compounded continuously so that, starting next year, you can make annual withdrawals of $2000 in perpetuity?

41. PRESENT VALUE An investment guarantees annual payments of A dollars each, to be paid at the end of each year in perpetuity into an account earning a fixed annual interest rate r, compounded annually. Show that the present value of the investment is $\dfrac{A}{r}$.

42. PRESENT VALUE Suppose in Exercise 41, payments of A dollars each are made k times per year (at the end of each period) into an account that earns interest at an annual rate r, compounded k times per year. Show that the present value of the investment is $\dfrac{kA}{r}$.

43. ACCUMULATION OF MEDICATION A patient is given an injection of 20 units of a certain drug every 24 hours. The drug is eliminated exponentially so that the fraction that remains in the patient's body after t days is $f(t) = e^{-t/2}$. If the treatment is continued indefinitely, approximately how many units of the drug will eventually be in the patient's body just prior to an injection?

44. A* ACCUMULATION OF MEDICATION A patient is given an injection of A units of a certain drug every h hours. The drug is eliminated exponentially so that the fraction that remains in the patient's body t hours after being administered is $f(t) = e^{-rt}$, where r is a positive constant.

   a. If this treatment is continued indefinitely, find an expression for the total number of units T of the drug that will accumulate in the patient's body.

   b. Suppose a physician determines that the accumulated amount of drug T found in part (a) is 20% too much for the safety of the patient. Assuming that the elimination constant r is a fixed characteristic of the patient's body, how should h be adjusted to guarantee that the amount of accumulated drug does not become unsafe?

45. PUBLIC HEALTH A health problem, especially in developing nations, is that the food available to people often contains toxins or poisons. Suppose that a person consumes a dose d of a certain toxin every day and that the proportion of accumulated toxin excreted by the person's body each day is q. (For example, if $q = 0.7$, then 70% of the accumulated toxin is excreted each day.) Find an expression for the total amount of toxin that will eventually accumulate in the person's body.

46. A* DISTANCE TRAVELLED BY A BOUNCING BALL A ball has the property that each time it falls from a height h onto a level, rigid surface, it will rebound to a height rh, where r ($0 < r < 1$) is a number called the *coefficient of restitution*.

   a. A super ball with $r = 0.8$ is dropped from a height of 5 m. Show that if the ball continues to bounce indefinitely, it will travel a total distance of 45 m.

   b. Find a formula for the total distance travelled by a ball with coefficient of restitution r that is dropped from a height H.

   c. A certain ball that bounces indefinitely after being dropped from a height H travels a total distance of $2H$. What is r?

47. A* TIME OF TRAVEL BY A BOUNCING BALL Refer to Exercise 46. Assuming no air resistance, an object dropped from a height will fall $s = 4.9t^2$ metres in t seconds.

a. A golf ball with coefficient of restitution $r = 0.36$ is dropped from a height of 4.9 m. Show that if the ball continues to bounce indefinitely, its total time of travel will be 4 seconds.

b. Find a formula for the total time of travel of a ball with coefficient of restitution r that is dropped from a height of H metres.

48. GROUP MEMBERSHIP   Each January 1, a large school board adds six new members to its board of trustees. If the fraction of trustees who remain active for at least t years is $f(t) = 5e^{-0.02t}$, approximately how many active trustees will the school have on December 31 in the long run?

49. DEPLETION OF RESERVES   A certain rare gas used in industrial processes had known reserves of 3×10^{11} m³ in 2000. In 2001, 1.7×10^9 m³ of the gas was consumed with an annual increase of 7.4% each year thereafter. When will all known reserves of the gas be consumed?

50. A* ZENO'S PARADOX   The Greek philosopher Zeno of Elea (495–435 B.C.E.) presented a number of paradoxes that perplexed many mathematicians of his day. Perhaps the most famous is the racecourse paradox, which may be stated as follows:

As a runner runs a racecourse, he must first run $\frac{1}{2}$ its length, then $\frac{1}{2}$ the remaining distance, then $\frac{1}{2}$ the distance that remains after that, and so on, indefinitely. Since there are an infinite number of such half segments and the runner requires finite time to complete each one, he can never finish the race.

Suppose the runner runs at a constant pace and takes T minutes to run the first half of the course. Set up an infinite series that gives the total time required to finish the course and verify that the total time is $2T$, as intuition would suggest.

51. SPY STORY   The spy is plucked from the village (in Section 8.3) by his enemies. "Don't be afraid," he is told icily. "Surely you do not think we would kill you without first having a little . . . chat?" The spy is tied to a chair and given an 800-mg injection of truth serum. Every 4 hours, he is interrogated, and when he doesn't talk, he is injected with an additional 400 mg of the drug. Fortunately, the spy has built up resistance to this particular drug, and his body eliminates 30% of the drug present during each 4-hour period between injections. Still, even the spy has limits, and he figures he will break when 1000 mg of the drug has accumulated in his body. How much time does he have to plan and carry out an escape before he spills the beans and thus becomes expendable?

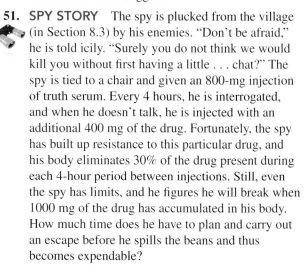

52. Evaluate $\displaystyle\sum_{n=1}^{5} \frac{n^3 - 3n}{2n + 1}$ to three-decimal-place accuracy using a calculator.

53. Repeat Exercise 52 with the sum $\displaystyle\sum_{n=1}^{5} 300e^{-0.2n}$.

54. Find the sum S of the geometric series $\displaystyle\sum_{n=1}^{\infty} 2(0.85)^n$. Then calculate the partial sums S_5 and S_{10} with five-decimal-place accuracy and compare the values of these partial sums to the sum S of the series.

55. A* Show that the geometric series $\displaystyle\sum_{k=0}^{\infty} ar^k$ diverges when $a \neq 0$ and

a. $r = 1$. [*Hint:* Note that the nth partial sum is $S_n = na$.]

b. $r = -1$. [*Hint:* Note that the nth partial sum is $S_n = 0$ if n is odd or $S_n = a$ if n is even.]

Tests for Convergence

L02

Determine whether a series converges by choosing from a number of tests for convergence.

In Section 9.1, we introduced the notions of convergence and divergence of infinite series and found that a geometric series $\displaystyle\sum_{n=b}^{\infty} ar^n$ converges if $|r| < 1$ and diverges otherwise. This is a fairly easy criterion to apply, but what about series that are not geometric? In this section, we shall develop several convergence tests that will enable us to determine the convergence or divergence of a variety of series. We begin by examining a special series that will be used in subsequent examples and applications.

The Harmonic Series

The series

$$\sum_{k=1}^{\infty} \frac{1}{k} = 1 + \frac{1}{2} + \frac{1}{3} + \frac{1}{4} + \frac{1}{5} + \cdots$$

is called the **harmonic series** because it occurs in physics in the study of sound. In Example 9.2.1, we determine whether this series converges by examining its partial sums.

EXAMPLE 9.2.1

Show that the harmonic series $\sum_{k=1}^{\infty} \frac{1}{k}$ diverges.

Solution

Let $S_n = \sum_{k=1}^{n} \frac{1}{k}$ be the nth partial sum of the harmonic series, and note that

$$S_1 = 1$$

$$S_2 = 1 + \frac{1}{2}$$

$$S_4 = 1 + \frac{1}{2} + \left(\frac{1}{3} + \frac{1}{4}\right)$$

$$> 1 + \frac{1}{2} + \left(\frac{1}{4} + \frac{1}{4}\right) = 1 + \frac{1}{2} + \frac{1}{2} = 1 + \frac{2}{2}$$

$$S_8 = 1 + \frac{1}{2} + \left(\frac{1}{3} + \frac{1}{4}\right) + \left(\frac{1}{5} + \frac{1}{6} + \frac{1}{7} + \frac{1}{8}\right)$$

$$> 1 + \frac{1}{2} + \left(\frac{1}{4} + \frac{1}{4}\right) + \left(\frac{1}{8} + \frac{1}{8} + \frac{1}{8} + \frac{1}{8}\right) = 1 + \frac{1}{2} + \frac{1}{2} + \frac{1}{2} = 1 + \frac{3}{2}$$

and in general,

$$S_{2^n} > 1 + \frac{n}{2}.$$

This means that the partial sums S_n can be made as large as desired by taking n sufficiently large. Therefore, $\lim_{n\to\infty} S_n$ does not exist as a finite number, and the harmonic series must diverge.

We have just seen that the partial sums of the harmonic series grow without bound as n becomes large. However, the size of the partial sums grows very slowly. For instance, computations show that $S_{100} \approx 5.19$ and $S_{1000} \approx 7.49$, and that to get a partial sum exceeding 20 requires more than 178 million terms. Example 9.2.2 features an application in which estimating the size of a partial sum of the harmonic series plays a crucial role.

EXAMPLE 9.2.2

Suppose records are kept over a 100-year period of the annual snowfall in Banff, Alberta. For how many years should we expect record-breaking snowfall in the sense that the snowfall in that winter exceeds the snowfall in all previous years during the period?

Solution

In modelling this situation, assume that the snowfall in any given year of the 100-year period is not affected by the snowfall in previous years and that the climate does not change in a systematic way. Also assume that the snowfall is measured quite accurately so that no ties occur. The analysis will involve some basic (intuitive) notions from probability.

Consider the 100 years as a sequence of events in which the snowfall levels are recorded. The snowfall in the first year is a record, since records begin at that time. In the second year, the snowfall is equally likely to be more than the snowfall in the first year or less than the snowfall in the first year. It follows that there is a probability of $\frac{1}{2}$ that the snowfall in the second year sets a record by exceeding the snowfall in the first year. Consequently, in the first two years of the 100-year period, we expect to have $1 + \frac{1}{2}$ records set. For the third year, the probability is $\frac{1}{3}$ that the snowfall that year exceeds the snowfall in both of the first two years, so the expected number of record snowfalls in three years is $1 + \frac{1}{2} + \frac{1}{3}$. Continuing in this fashion for n years, the expected number of years with record snowfalls is

$$1 + \frac{1}{2} + \frac{1}{3} + \cdots + \frac{1}{n},$$

which is the nth partial sum of the harmonic series. In particular, the expected number of record snowfalls over the entire 100-year period is

$$1 + \frac{1}{2} + \frac{1}{3} + \cdots + \frac{1}{100} \approx 5.19$$

On average, about 5 record snowfall years would be expected during the 100-year period.

If you tested this result by checking a specific 100-year period, you might find less than 5 record years or more than 5, perhaps many more. The correct interpretation of the result is that if you look at a large collection (the larger the better) of 100-year periods, then on average, there would be about 5 years of record snowfall per 100-year period.

The Divergence Test

Next, we shall develop several general tests for determining whether a series $\sum_{k=1}^{\infty} a_k$ converges or diverges. The easiest such test to apply can only be used to establish divergence and is appropriately called the *divergence test*.

The Divergence Test ■ If $\lim_{n \to \infty} a_n$ does not exist or if $\lim_{n \to \infty} a_n \neq 0$, then $\sum_{k=1}^{\infty} a_k$ diverges.

We shall establish the divergence test by showing that if the series $\sum_{k=1}^{\infty} a_k$ converges, the terms a_n must tend to 0 as n tends to infinity, that is, $\lim_{n \to \infty} a_n = 0$. (Do you see why this is equivalent to the statement of the divergence test?) To see why this is true, let $S_n = \sum\limits_{k=1}^{n} a_k$ be the nth partial sum of the series, and note that a_n is the difference between S_n and S_{n-1}, that is,

$$a_n = (a_1 + a_2 + \cdots + a_n) - (a_1 + a_2 + \cdots + a_{n-1}) = S_n - S_{n-1}.$$

If $\sum\limits_{k=1}^{\infty} a_k$ converges with sum S, then

$$\lim_{n\to\infty} S_n = \lim_{n\to\infty} S_{n-1} = S,$$

and it follows that

$$\lim_{n\to\infty} a_n = \lim_{n\to\infty}(S_n - S_{n-1}) = \lim_{n\to\infty} S_n - \lim_{n\to\infty} S_{n-1}$$
$$= S - S = 0$$

as claimed.

Example 9.2.3 shows how the divergence test can be used to show that a given series diverges.

EXAMPLE 9.2.3

Show that each of the following series diverges:

a. $\sum\limits_{k=1}^{\infty} \sqrt{k}$ **b.** $\sum\limits_{k=1}^{\infty} \dfrac{k}{k+1}$

Solution

a. Since $\lim\limits_{k\to\infty} \sqrt{k} = \infty$, it follows that $\lim\limits_{k\to\infty} \sqrt{k}$ is not 0, so $\sum\limits_{k=1}^{\infty} \sqrt{k}$ diverges by the divergence test.

b. The series $\sum\limits_{k=1}^{\infty} \dfrac{k}{k+1}$ diverges since

$$\lim_{k\to\infty} \frac{k}{k+1} = \lim_{k\to\infty} \frac{1}{1+1/k} = \frac{1}{1+0} = 1 \neq 0.$$

WARNING   *The divergence test cannot be used to show that a series* $\sum\limits_{n=1}^{\infty} a_n$ *converges.* Specifically, you cannot conclude that $\sum\limits_{n=1}^{\infty} a_n$ converges just because $\lim\limits_{n\to\infty} a_n = 0$.

For instance, for the harmonic series $\sum\limits_{k=1}^{\infty} \dfrac{1}{k}$, we have

$$\lim_{k\to\infty} \frac{1}{k} = 0,$$

yet the harmonic series diverges. ▪

The Integral Test Our second test for convergence, the *integral test,* relates the convergence of the infinite series $\sum_{n=1}^{\infty} a_n$ to the convergence of the improper integral $\int_{1}^{\infty} f(x)\, dx$, where $f(n) = a_n$. Here is a precise statement of the test.

> **The Integral Test** ■ Let $\sum_{n=1}^{\infty} a_n$ be a series with $a_n > 0$ for all n. If $f(x)$ is a function such that $f(n) = a_n$ for all n, and if f is continuous, positive ($f(x) > 0$), and decreasing for $x \geq 1$, then $\sum_{n=1}^{\infty} a_n$ and $\int_{1}^{\infty} f(x)\, dx$ either both converge or both diverge. In particular,
>
> $$\sum_{n=1}^{\infty} a_n \text{ converges if } \int_{1}^{\infty} f(x)\, dx \text{ converges}$$
>
> and
>
> $$\sum_{n=1}^{\infty} a_n \text{ diverges if } \int_{1}^{\infty} f(x)\, dx \text{ diverges.}$$

We will provide a geometric justification of the integral test after illustrating its use in Example 9.2.4.

EXAMPLE 9.2.4

Determine whether each series converges or diverges.

a. $\sum_{n=3}^{\infty} \dfrac{\ln n}{n}$ **b.** $\sum_{n=1}^{\infty} ne^{-n}$

Solution

a. Let $f(x) = \dfrac{\ln x}{x}$, so that $f(n) = \dfrac{\ln n}{n}$. Note that $f(x)$ is continuous and positive for $x \geq 2$, and it is decreasing since its derivative satisfies

$$f'(x) = \frac{x(1/x) - (1)\ln x}{x^2} = \frac{1 - \ln x}{x^2} < 0 \qquad \text{for } x > e > 2.$$

Therefore, the integral test applies, and since

$$\int \frac{\ln x}{x}\, dx = \frac{1}{2}(\ln x)^2 + C \qquad \begin{array}{l} \text{substitution} \\ \text{let } u = \ln x \\ du = \dfrac{1}{x}\, dx \end{array}$$

it follows that

$$\int_{2}^{\infty} \frac{\ln x}{x}\, dx = \lim_{t \to \infty} \int_{2}^{t} \frac{\ln x}{x}\, dx = \lim_{t \to \infty} \left[\frac{1}{2}(\ln t)^2 - \frac{1}{2}(\ln 2)^2 \right] = \infty.$$

Since the improper integral diverges, it follows that the series $\sum_{n=2}^{\infty} \dfrac{\ln n}{n}$ diverges as well.

Just-In-Time

Recall that $\ln x > 1$ when

$$x > e.$$

Also, think of the shape of the graphs of $y = e^x$ and $y = e^{-x}$; these functions are positive for all x.

b. Let $f(x) = xe^{-x}$, so that $f(n) = ne^{-n}$. Then $f(x)$ is continuous and positive for $x \geq 1$ and it is decreasing since its derivative satisfies

$$f'(x) = x(-e^{-x}) + (1)e^{-x} = (1 - x)e^{-x} < 0 \qquad \text{for } x > 1.$$

Therefore, the integral test applies, and since

$$\int xe^{-x}\, dx = (-x - 1)e^{-x} + C \qquad \begin{array}{l}\text{integration by parts with}\\ u = x \qquad dv = e^{-x}\, dx\end{array}$$

it follows that

$$\int_1^\infty xe^{-x}\, dx = \lim_{t \to \infty} \int_1^t xe^{-x}\, dx = \lim_{t \to \infty} [(-t - 1)e^{-t} - (-1 - 1)e^{-1}] = 2e^{-1}.$$

Since the improper integral converges, the series $\displaystyle\sum_{n=1}^\infty ne^{-n}$ also converges.

Geometric Justification of the Integral Test

To see why the integral test must be true, first note that since the test requires $a_k > 0$ for all k, the series $\displaystyle\sum_{k=1}^\infty a_k$ will converge unless it becomes infinitely large; in particular, the series cannot diverge by oscillation as with the geometric series

$$\sum_{k=1}^\infty (-1)^k = 1 - 1 + 1 - 1 + \cdots.$$

Similarly, since $f(x) > 0$ and $f(x)$ is decreasing for $x > 1$, the improper integral $\displaystyle\int_1^\infty f(x)\, dx$ will converge unless it becomes infinite. We shall show that if the improper integral is finite, then so is the series $\displaystyle\sum_{k=1}^\infty a_k$, and that if $\displaystyle\int_1^\infty f(x)\, dx$ is infinite, then $\displaystyle\sum_{k=1}^\infty a_k$ must also be infinite.

Refer to the two parts of Figure 9.3. Both parts show the graph of $f(x)$ together with a number of approximating rectangles built on the intervals $[1, 2], [2, 3], \ldots$. In Figure 9.3a, the rectangles are *inscribed* under the curve $y = f(x)$ and the height

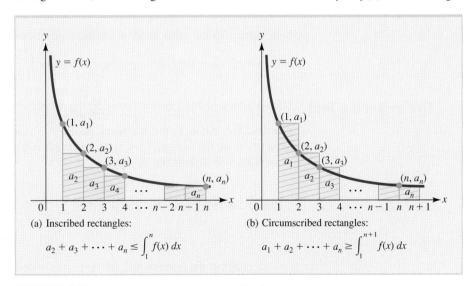

(a) Inscribed rectangles:

$$a_2 + a_3 + \cdots + a_n \leq \int_1^n f(x)\, dx$$

(b) Circumscribed rectangles:

$$a_1 + a_2 + \cdots + a_n \geq \int_1^{n+1} f(x)\, dx$$

FIGURE 9.3   A geometric interpretation of the integral test.

of the approximating rectangle built on the interval $[k - 1, k]$ is $f(k) = a_k$, the height of the curve $y = f(x)$ above the *right* endpoint of the interval. Since this approximating rectangle has height a_k and width 1, its area is $(a_k)(1) = a_k$, and the total area of the first n rectangles is $a_2 + a_3 + \cdots + a_n$. Since the rectangles are inscribed under the curve $y = f(x)$, this area is no greater than the area under the curve above the interval $[1, n]$, so we have

$$a_2 + a_3 + \cdots + a_n \le \int_1^n f(x)\, dx,$$

which means that the nth partial sum of the series $\sum_{k=1}^{\infty} a_k$ satisfies

$$S_n = a_1 + a_2 + a_3 + \cdots + a_n \le a_1 + \int_1^n f(x)\, dx.$$

The rectangles in Figure 9.3b are *circumscribed* over the curve $y = f(x)$ on the interval $[1, n + 1]$. This time, the height of the rectangle based on the interval $[k - 1, k]$ is $f(k - 1) = a_{k-1}$, the height of the curve above the *left* endpoint of the subinterval. The area of this approximating rectangle is $(a_{k-1})(1) = a_{k-1}$, and since the circumscribing rectangles cover an area no less than the area under the curve on $[1, n + 1]$ we have

$$S_n = a_1 + a_2 + \cdots + a_n \ge \int_1^{n+1} f(x)\, dx.$$

Combining these results, we see that the nth partial sum S_n is trapped between two integrals:

$$\int_1^{n+1} f(x)\, dx \le S_n \le a_1 + \int_1^n f(x)\, dx.$$

If the improper integral converges, then the partial sums S_n are bounded from above by the (finite) quantity $a_1 + \int_1^{\infty} f(x)\, dx$, which means that the series $\sum_{k=1}^{\infty} a_k$ converges to a finite value. On the other hand, if the improper integral diverges, the partial sums cannot converge since they are larger than a quantity $\int_1^{n+1} f(x)\, dx$ that is tending to infinity.

The *p*-Series Test

Infinite series of the general form $\sum_{n=1}^{\infty} \dfrac{1}{n^p}$, where p is a positive constant, are called ***p*-series.** For instance, the harmonic series is a p-series with $p = 1$. An important application of the integral test is to establish this simple rule for determining the convergence or divergence of p-series.

> **The *p*-Series Test** ■ The p-series $\sum_{n=1}^{\infty} \dfrac{1}{n^p}$ converges if $p > 1$ and diverges if $p \le 1$.

To establish the p-series test, let $f(x) = \dfrac{1}{x^p} = x^{-p}$, so that $f(n) = \dfrac{1}{n^p}$. Then $f(x)$ is continuous and positive for $x \ge 1$ and it is also decreasing since its derivative satisfies

$$f'(x) = -p\,x^{-p-1} = \frac{-p}{x^{p+1}} < 0.$$

Applying the integral test, we first note that if $p = 1$, then

$$\int_1^\infty \frac{1}{x}\,dx = \lim_{t\to\infty}\int_1^t \frac{1}{x}\,dx = \lim_{t\to\infty}(\ln t - \ln 1) = \infty,$$

which confirms that the harmonic series $\sum_{n=1}^\infty \frac{1}{n}$ diverges (recall Example 9.2.1). For the case where $p \ne 1$, we have

$$\int_1^\infty \frac{1}{x^p}\,dx = \lim_{t\to\infty}\int_1^t x^{-p}\,dx = \lim_{t\to\infty}\left[\frac{t^{-p+1}}{(-p+1)} - \frac{1}{(-p+1)}\right]$$

$$= \begin{cases} \infty & \text{if } p < 1 \\ \dfrac{1}{p-1} & \text{if } p > 1 \end{cases}$$

Therefore, the improper integral and hence the p-series both converge for $p > 1$ and both diverge for $p \le 1$.

EXAMPLE 9.2.5

Determine whether each series converges or diverges:

a. $\sum_{n=1}^\infty \frac{1}{\sqrt{n}}$ **b.** $\sum_{n=1}^\infty \frac{1}{n\sqrt{n^3}}$ **c.** $\sum_{n=1}^\infty \frac{1}{n^{0.999}}$

Solution

a. $\sum_{n=1}^\infty \frac{1}{\sqrt{n}} = \sum_{n=1}^\infty \frac{1}{n^{1/2}}$ is a p-series with $p = \frac{1}{2} < 1$, so the series diverges.

b. $\sum_{n=1}^\infty \frac{1}{n\sqrt{n^3}} = \sum_{n=1}^\infty \frac{1}{n^{5/2}}$ is a p-series with $p = \frac{5}{2} > 1$. Thus, the series converges.

c. $\sum_{n=1}^\infty \frac{1}{n^{0.999}}$ is a p-series with $p = 0.999 < 1$, so the series diverges.

The Direct Comparison Test

Our next convergence test compares the behaviour of a given series to that of another series whose convergence properties are known.

The Direct Comparison Test ■ Consider two series $\sum_{n=1}^\infty a_n$ and $\sum_{n=1}^\infty b_n$ where $0 < b_n \le a_n$ for all n.

If $\sum_{n=1}^\infty a_n$ converges, so does $\sum_{n=1}^\infty b_n$.

If $\sum_{n=1}^\infty b_n$ diverges, so does $\sum_{n=1}^\infty a_n$.

To see why the direct comparison test works, note that a series of positive terms diverges only if it becomes infinitely large. If $b_n \leq a_n$ for all n and $\sum\limits_{n=1}^{\infty} a_n$ converges (exists as a finite number), then $\sum\limits_{n=1}^{\infty} b_n \leq \sum\limits_{n=1}^{\infty} a_n$. It follows that the smaller series $\sum\limits_{n=1}^{\infty} b_n$ must also be finite, so it converges.

Similarly, if $\sum\limits_{n=1}^{\infty} b_n$ diverges, then $\sum\limits_{n=1}^{\infty} b_n$ is infinitely large. Since $\sum\limits_{n=1}^{\infty} b_n \leq \sum\limits_{n=1}^{\infty} a_n$, the series $\sum\limits_{n=1}^{\infty} a_n$ must also be infinitely large and hence diverges.

EXAMPLE 9.2.6

Determine whether each series converges or diverges:

a. $\sum\limits_{n=1}^{\infty} \dfrac{1}{n2^n}$ **b.** $\sum\limits_{n=1}^{\infty} \dfrac{n2^{2n}}{3^n}$

Solution

a. $\sum\limits_{n=1}^{\infty} \dfrac{1}{2^n}$ is a convergent geometric series $\left(r = \dfrac{1}{2} < 1 \right)$, and since $\dfrac{1}{n2^n} \leq \dfrac{1}{2^n}$ for $n \geq 1$, it follows from the direct comparison test that the given series also converges.

b. $\sum\limits_{n=1}^{\infty} \dfrac{2^{2n}}{3^n} = \sum\limits_{n=1}^{\infty} \left(\dfrac{4}{3} \right)^n$ is a divergent geometric series $\left(r = \dfrac{4}{3} > 1 \right)$, and since

$\dfrac{n2^{2n}}{3^n} \geq \dfrac{2^{2n}}{3^n}$ for all $n \geq 1$, the direct comparison test tells us that the given series also diverges.

The Ratio Test   Applying the comparison test can be frustrating since it may require several steps of trial and error to find a suitable comparison. An alternative, the *ratio test,* enables a kind of internal comparison of terms of the given series and has the added advantage of being applicable to series that contain negative terms. Here is a statement of this test.

> **The Ratio Test** ■ For the series $\sum\limits_{n=1}^{\infty} a_n$, let $L = \lim\limits_{n \to \infty} \left| \dfrac{a_{n+1}}{a_n} \right|$. Then:
> **a.** If $L < 1$, the series converges.
> **b.** If $L > 1$, the series diverges.
> **c.** If $L = 1$, the test is inconclusive; that is, when this occurs, the series may converge or it may diverge.

The ratio test is especially useful when applied to a series $\sum\limits_{n=1}^{\infty} a_n$ in which the terms a_n involve powers or factorials. This is illustrated in Example 9.2.7.

EXAMPLE 9.2.7

Determine whether each series converges or diverges:

a. $\displaystyle\sum_{n=1}^{\infty} \frac{5^n}{n^2}$ **b.** $\displaystyle\sum_{n=1}^{\infty} \frac{(-2)^n}{n!}$ **c.** $\displaystyle\sum_{n=1}^{\infty} \frac{n}{n+1}$

Solution

a. To apply the ratio test, compute the limit

$$L = \lim_{n\to\infty} \left| \frac{\dfrac{5^{n+1}}{(n+1)^2}}{\dfrac{5^n}{n^2}} \right|$$

$$= \lim_{n\to\infty} \frac{5^{n+1}n^2}{5^n(n+1)^2}$$

$$= \lim_{n\to\infty} 5 \left(\frac{n}{n+1} \right)^2$$

$$= \lim_{n\to\infty} 5 \left(\frac{1}{1+1/n} \right)^2 = 5\left(\frac{1}{1+0} \right)^2 = 5$$

Since $L = 5 > 1$, the series diverges.

b. Applying the ratio test, we find that

$$L = \lim_{n\to\infty} \left| \frac{\dfrac{(-2)^{n+1}}{(n+1)!}}{\dfrac{(-2)^n}{n!}} \right|$$

$$= \lim_{n\to\infty} \left| \frac{(-2)^{n+1}n!}{(-2)^n(n+1)!} \right|$$

$$= \lim_{n\to\infty} \left| \frac{-2}{n+1} \right| \qquad \text{since } \frac{n!}{(n+1)!} = \frac{1(2)\cdots(n)}{1(2)\cdots(n)(n+1)} = \frac{1}{n+1}$$

$$= 0$$

Since $L = 0 < 1$, the series converges.

c. We have already shown that this series diverges in Example 9.2.3b. However, if we try to confirm this result using the ratio test, we get

$$L = \lim_{n\to\infty} \left| \frac{\dfrac{n+1}{(n+1)+1}}{\dfrac{n}{n+1}} \right|$$

$$= \lim_{n\to\infty} \left| \frac{(n+1)(n+1)}{n(n+2)} \right| = \lim_{n\to\infty} \left| \frac{n^2 + 2n + 1}{n^2 + 2n} \right|$$

$$= \lim_{n\to\infty} \left| \frac{1 + 2/n + 1/n^2}{1 + 2/n} \right| = \frac{1 + 0 + 0}{1 + 0}$$

$$= 1$$

so the ratio test is inconclusive.

Knowing which convergence test to apply to a given series takes experience, and to gain that experience, it helps to work as many practice exercises as possible. In the following exercise set, a group of exercises focuses on each test we have discussed. In addition, in a larger collection of exercises, the test is not specified.

EXERCISES ■ 9.2

In Exercises 1 through 4, use the divergence test to show that the given series diverges.

1. $\displaystyle\sum_{k=1}^{\infty} \frac{2k}{k+5}$

2. $\displaystyle\sum_{k=1}^{\infty} \frac{k+3}{\sqrt{k}+1}$

3. $\displaystyle\sum_{k=1}^{\infty} [1 + (-1)^k]$

4. $\displaystyle\sum_{k=1}^{\infty} \frac{e^{-k}+1}{e^{-k}+2}$

In Exercises 5 through 8, use the integral test to determine whether the given series converges or diverges.

5. $\displaystyle\sum_{k=1}^{\infty} \frac{2}{3k+1}$

6. $\displaystyle\sum_{k=1}^{\infty} ke^{-k^2}$

7. $\displaystyle\sum_{k=1}^{\infty} \frac{k}{3+k^2}$

8. $\displaystyle\sum_{k=1}^{\infty} \frac{1}{\sqrt[3]{5k-1}}$

In Exercises 9 through 12, use the p-series test to determine whether the given series converges or diverges.

9. $\displaystyle\sum_{k=1}^{\infty} \frac{1}{k^{5/6}}$

10. $\displaystyle\sum_{k=1}^{\infty} \frac{1}{k\sqrt[3]{k^2}}$

11. $\displaystyle\sum_{k=1}^{\infty} \frac{1}{k^2\sqrt{k}}$

12. $\displaystyle\sum_{k=1}^{\infty} \frac{1}{k^{1.001}}$

In Exercises 13 through 16, use the comparison test to determine whether the given series converges or diverges.

13. $\displaystyle\sum_{k=1}^{\infty} \frac{3}{2^k+1}$

14. $\displaystyle\sum_{k=1}^{\infty} \frac{3^k+1}{2^{k-1}}$

15. $\displaystyle\sum_{k=2}^{\infty} \frac{1}{\sqrt{k}-1}$

16. $\displaystyle\sum_{k=2}^{\infty} \frac{1}{k\sqrt{k}+1}$

In Exercises 17 through 20, use the ratio test to determine whether the given series converges or diverges.

17. $\displaystyle\sum_{k=1}^{\infty} \frac{2^k}{k^3}$

18. $\displaystyle\sum_{k=1}^{\infty} k\left(\frac{1}{3}\right)^k$

19. $\displaystyle\sum_{k=1}^{\infty} \frac{(-3)^k}{k2^{2k}}$

20. $\displaystyle\sum_{k=1}^{\infty} \frac{2^k}{k!}$

In Exercises 21 through 36, determine whether the given series converges or diverges. You may use any test.

21. $\displaystyle\sum_{k=3}^{\infty} \frac{k}{\ln k}$

22. $\displaystyle\sum_{k=1}^{\infty} \frac{2^k+3^k}{4^k+1}$

23. $\displaystyle\sum_{k=1}^{\infty} \left(\frac{\pi}{2}\right)^k$

24. $\displaystyle\sum_{k=1}^{\infty} \frac{k^3+1}{k^3+2}$

25. $\displaystyle\sum_{k=1}^{\infty} \frac{1}{(3k)^{3/2}}$

26. $\displaystyle\sum_{k=1}^{\infty} \frac{1}{(3k+1)^2}$

27. $\displaystyle\sum_{k=3}^{\infty} \frac{1}{k\sqrt{\ln k}}$

28. $\displaystyle\sum_{k=3}^{\infty} \frac{1}{k(\ln k)^3}$

29. $\displaystyle\sum_{k=1}^{\infty} \frac{e^{1/k}}{k^2}$

30. $\displaystyle\sum_{k=1}^{\infty} \frac{k!}{k^3}$

31. $\displaystyle\sum_{k=2}^{\infty} \frac{\ln k}{e^k}$

32. $\displaystyle\sum_{k=1}^{\infty} \frac{3 + (-1)^k}{2^k}$

33. $\displaystyle\sum_{k=1}^{\infty} \frac{1}{k^{\sqrt{2}}}$

34. $\displaystyle\sum_{k=3}^{\infty} \frac{1}{k^{3/4} - 2}$

35. $\displaystyle\sum_{k=1}^{\infty} e^{1/k}$

36. $\displaystyle\sum_{k=1}^{\infty} \frac{e^k}{k!}$

In Exercises 37 and 38, show that the given series converges by applying the comparison test in two different ways. That is, find two different convergent series with terms greater than the terms of the given series.

37. $\displaystyle\sum_{k=1}^{\infty} \frac{1}{k^2 2^k}$

38. $\displaystyle\sum_{k=1}^{\infty} \frac{1}{k^2 + 3^k}$

In Exercises 39 and 40, use the integral test to determine the values of p for which the given series converges.

39. $\displaystyle\sum_{k=1}^{\infty} \frac{1}{(1 + 2k)^p}$

40. $\displaystyle\sum_{k=3}^{\infty} \frac{1}{k(\ln k)^p}$

41. For what values of the positive number c does the series $\displaystyle\sum_{k=1}^{\infty} \frac{c^k}{k^c}$ converge?

42. Show that the series

$$1 + \frac{1}{3} + \frac{1}{5} + \cdots = \sum_{n=0}^{\infty} \frac{1}{2n + 1}$$

diverges.

43. Show that the series

$$1 + \frac{1}{4} + \frac{1}{7} + \cdots = \sum_{n=0}^{\infty} \frac{1}{3n + 1}$$

diverges.

44. A* TRAFFIC ENGINEERING   Suppose there is no passing on a single-lane road. You observe that a slow car is followed by a group of cars that want to drive faster but are blocked from doing so. Explain why if n cars are driving in the same direction on this road, you would expect to see

$$1 + \frac{1}{2} + \frac{1}{3} + \cdots + \frac{1}{n}$$

bunches of cars that are more widely spaced. About how many bunches of cars will you observe if 20 cars are driving in the same direction? What if 100 cars are driving in the same direction? [*Hint:* Look for record low speeds.]

45. A* MATERIAL TESTING   Suppose you want to estimate the minimum breaking strain of 100 wooden beams of approximately the same dimensions using a machine that applies a gradually increasing force to the centre of a beam supported at its ends. By increasing the force until a beam breaks, you can find the breaking strain of the beam. Show that you can determine the minimum breaking strain of all 100 beams, so that on average, you will expect to break 5.19 beams in testing. [*Hint:* Find the breaking strain of the first beam. Next, test the second beam, increasing the force up to the breaking strain of the first beam. Then test the third beam, increasing the force up to the breaking strain of the first or second beam, whichever is smaller, and so on.]

46. A* DESERT WARFARE   At a base in the desert, an unlimited number of Jeeps are available for action. The gas tanks of the Jeeps are full, but no other gasoline is available. Show that it is possible to have one of the Jeeps travel from the base to a destination at an arbitrary distance by transferring gas from other Jeeps in such a way that all other Jeeps return to base without running out of gas in the desert. [*Hint:* The key idea is to send out a sequence of Jeeps, transferring gas in such a way that the first Jeep gets to its destination, leaving all other Jeeps enough gas to return to base. To develop this strategy, note that if you send out one Jeep, it can travel only as far as it can go on one tank of gas, but you can get the first Jeep farther out by sending a second Jeep. Have both Jeeps stop when they have used $\frac{1}{3}$ of their gas. Then transfer $\frac{1}{3}$ of the second Jeep's tank of gas to the first Jeep, thus leaving the second Jeep enough gas to get back to base while providing the first Jeep with enough gas to travel $1 + \frac{1}{3}$ times its usual range, that is, the distance it can go on one tank of gas. Continue by considering how to use three Jeeps, four Jeeps, and so on. You will need the result of Exercise 42.]

47. **A* ZENO'S PARADOX REVISITED** This is a continuation of Exercise 50, Section 9.1. A supporter of Zeno, called Summo (dates unknown), is not convinced that a racer who keeps running will always finish a racecourse of finite length. "Not so fast!" he challenges. "Suppose the runner takes T minutes for the first half of the course and $\frac{T}{2}$ minutes for the next quarter, but then he tires and it takes $\frac{T}{3}$ minutes for the next eighth, $\frac{T}{4}$ for the next sixteenth, and so on. Now how long does it take him to finish the race?"

*The **limit comparison test** says that if $\sum_{k=1}^{\infty} a_k$ and $\sum_{k=1}^{\infty} b_k$ are positive-term series ($a_k > 0$ and $b_k > 0$ for all k) and the limit $L = \lim_{k\to\infty} \frac{a_k}{b_k}$ is a positive, finite real number ($0 < L < \infty$), then $\sum_{k=1}^{\infty} a_k$ and $\sum_{k=1}^{\infty} b_k$ either both converge or both diverge. Use this test in Exercises 48 through 51 to determine whether the given series converges or diverges.*

48. $\sum_{k=1}^{\infty} \frac{2}{k^3 + 5k + 71}$

49. $\sum_{k=1}^{\infty} \frac{k + 2}{\sqrt[3]{k^5 + k^2}}$

50. $\sum_{k=2}^{\infty} \frac{3}{(2k + 5)(\ln k)^2}$

51. $\sum_{k=1}^{\infty} \frac{1 + 2^k}{3 + 5^k}$

It can be shown that $\ln n < n$ and $e^n > n$ for all $n \geq 1$. Use these facts together with the direct comparison test to determine whether the series in Exercises 52 and 53 converge or diverge.

52. **A*** $\sum_{n=1}^{\infty} \frac{e^n}{n^2}$

53. **A*** $\sum_{n=3}^{\infty} \frac{1}{(\ln n)^2}$

SECTION 9.3

[L03]

Find power series and Taylor series for given functions. Use Taylor polynomials to approximate quantities and integrals.

Functions as Power Series; Taylor Series

A series of the form

$$a_0 + a_1 x + a_2 x^2 + a_3 x^3 + \cdots = \sum_{n=1}^{\infty} a_n x^n$$

is called a **power series** in the variable x. In this section, we will show how functions can be represented by power series. Such representations are useful in a variety of ways, mainly because power series can be thought of as generalized polynomials and polynomials are easily evaluated and manipulated. For instance, the functional values on calculators and computers are obtained by approximating functions by polynomials. A similar procedure is used to evaluate definite integrals and to solve certain differential equations.

Convergence of a Power Series

A power series is a function of x, and as with any function, it is important to know the domain of this function. In other words, what is the set of all x for which a given power series converges? Here is the answer to this question.

Convergence of a Power Series ■ For a power series

$$a_0 + a_1x + a_2x^2 + a_3x^3 + \cdots = \sum_{n=1}^{\infty} a_n x^n,$$

exactly one of the following must be true:
  a. The power series converges only for $x = 0$.
  b. The series converges for all x.
  c. There is a number R such that the series converges for all $|x| < R$ and diverges for $|x| > R$. It may converge or diverge at the endpoints $x = -R$ and $x = R$.

The interval $|x| < R$ or, equivalently, $-R < x < R$ is called the **interval of absolute convergence** of the power series, and R is called the **radius of convergence.** If the power series converges only for $x = 0$, then $R = 0$, and $R = \infty$ in the case where the power series converges for all x. In practice, the radius of convergence of a power series is usually determined by applying the ratio test, as illustrated in Example 9.3.1. Convergence or divergence at the endpoints $x = -R$ and $x = R$ of the interval of absolute convergence must be checked by other tests, some of which are outside the scope of this text.

EXAMPLE 9.3.1

For each power series, find the radius of convergence and the interval of absolute convergence.

  a. $\displaystyle\sum_{k=0}^{\infty} k! x^k$ **b.** $\displaystyle\sum_{k=0}^{\infty} \frac{x^k}{k!}$ **c.** $\displaystyle\sum_{k=0}^{\infty} \frac{x^{2k}}{4^k}$

Solution

  a. Applying the ratio test shows that the limit of the absolute value of the ratio of consecutive terms $k!x^k$ and $(k + 1)!x^{k+1}$ is

$$L = \lim_{k \to \infty} \left| \frac{(k + 1)!x^{k+1}}{k!x^k} \right| = \lim_{k \to \infty} |(k + 1)x|$$

$$= \infty \qquad \text{unless } x = 0$$

Therefore, the power series converges only for $x = 0$. The radius of convergence is $R = 0$, and the interval of absolute convergence is the single point $x = 0$.

  b. For this series,

$$L = \lim_{k \to \infty} \left| \frac{\dfrac{x^{k+1}}{(k + 1)!}}{\dfrac{x^k}{k!}} \right| = \lim_{k \to \infty} \left| \frac{k!x^{k+1}}{(k + 1)!x^k} \right|$$

$$= \lim_{k \to \infty} \left| \frac{x}{k + 1} \right| = 0 \qquad \text{for all } x$$

Since $L = 0 < 1$ for all x, it follows from the ratio test that the power series converges for all x. In this case, the radius of convergence is $R = \infty$, and the interval of absolute convergence is the entire x axis.

c. For this power series, we get

$$L = \lim_{k \to \infty} \left| \frac{\dfrac{x^{2(k+1)}}{4^{k+1}}}{\dfrac{x^{2k}}{4^k}} \right| = \lim_{k \to \infty} \left| \frac{4^k x^{2k+2}}{4^{k+1} x^{2k}} \right|$$

$$= \frac{x^2}{4} \qquad \text{since } \frac{x^{2k+2}}{x^{2k}} = x^{2k+2-2k} = x^2$$

According to the ratio test, the power series converges if $L < 1$ and diverges if $L > 1$. In other words, the series converges if $\dfrac{x^2}{4} < 1$ or, equivalently, if $x^2 < 4$.

Thus, the interval of absolute convergence is $-2 < x < 2$, and the radius of convergence is $R = 2$.

Power Series Representation of Functions

Next, we will see how to use geometric series and functions related to geometric series as power series. Recall that if x is any number such that $|x| < 1$, we have

$$1 + x + x^2 + x^3 + \cdots = \frac{1}{1-x},$$

which may be interpreted as saying that the function $f(x) = \dfrac{1}{1-x}$ can be represented by the power series $1 + x + x^2 + \cdots = \displaystyle\sum_{k=0}^{\infty} x^k$. with interval of convergence $-1 < x < 1$. As illustrated in Example 9.3.2, the geometric series formula can be modified by substitution to find power series representations for other functions as well.

EXAMPLE 9.3.2

In each case, find a power series for the given function and determine its interval of absolute convergence.

a. $f(x) = \dfrac{x}{1 + 3x^2}$

b. $g(x) = \dfrac{1}{2 + 3x}$

Solution

a. Start with the geometric power series

$$\frac{1}{1-x} = \sum_{n=0}^{\infty} x^n \qquad \text{for } |x| < 1$$

and replace x by $-3x^2$ to get

$$\frac{1}{1 + 3x^2} = \sum_{n=0}^{\infty} (-3x^2)^n = \sum_{n=0}^{\infty} (-3)^n x^{2n}.$$

The new interval of absolute convergence is $|-3x^2| < 1$, which can be rewritten as $|x^2| < \dfrac{1}{3}$ or, equivalently, as $-\dfrac{1}{\sqrt{3}} < x < \dfrac{1}{\sqrt{3}}$. Now multiply by x (which does not affect the interval of convergence) to get

$$\frac{x}{1 + 3x^2} = \sum_{n=0}^{\infty} (-3)^n x^{2n+1} \qquad \text{for } |x| < \frac{1}{\sqrt{3}}.$$

b. The form of the geometric series $\dfrac{1}{1-x}$ requires the constant term in the denominator to be 1, not 2. Therefore, begin by factoring 2 from the denominator of the given functional expression to obtain

$$\frac{1}{2+3x} = \frac{1}{2\left(1 + \dfrac{3x}{2}\right)} = \frac{1}{2}\left[\frac{1}{1 - \left(-\dfrac{3x}{2}\right)}\right]$$

$$= \frac{1}{2}\left[1 + \left(\frac{-3x}{2}\right) + \left(\frac{-3x}{2}\right)^2 + \cdots\right] = \frac{1}{2}\sum_{n=0}^{\infty}\left(\frac{-3x}{2}\right)^n$$

$$= \frac{1}{2}\sum_{n=0}^{\infty}(-1)^n \frac{3^n}{2^n} x^n$$

$$= \sum_{n=0}^{\infty} \frac{(-1)^n 3^n}{2^{n+1}} x^n$$

The series converges when

$$\left|\frac{-3x}{2}\right| < 1$$

$$|x| < \frac{2}{3}$$

so the interval of absolute convergence is $-\dfrac{2}{3} < x < \dfrac{2}{3}$.

Polynomials are particularly easy to differentiate and integrate because the operations involve term-by-term application of the power rule, that is,

$$\frac{d}{dx}[a_0 + a_1 x + a_2 x^2 + \cdots + a_n x^n] = a_1 + a_2(2x) + a_3(3x^2) + \cdots + a_n(nx^{n-1})$$

and

$$\int [a_0 + a_1 x + a_2 x^2 + \cdots + a_n x^n]\, dx$$

$$= a_0 x + a_1\left(\frac{x^2}{2}\right) + a_2\left(\frac{x^3}{3}\right) + \cdots + a_n\left(\frac{x^{n+1}}{n+1}\right) + C.$$

Here is a statement of this principle as it applies to power series.

Term-by-Term Differentiation and Integration of Power Series ■

Suppose the power series $a_0 + a_1x + a_2x^2 + \cdots = \sum_{n=0}^{\infty} a_n x^n$ converges for $|x| < R$, and let f be the function defined by

$$f(x) = \sum_{n=0}^{\infty} a_n x^n \qquad \text{for } -R < x < R.$$

Then $f(x)$ is differentiable for $-R < x < R$, and its derivative is given by

$$f'(x) = a_1 + a_2(2x) + a_3(3x^2) + \cdots = \sum_{n=1}^{\infty} na_n x^{n-1}.$$

The function $f(x)$ is also integrable for $-R < x < R$, and its integral is given by

$$\int f(x)\, dx = C + a_0x + a_1\left(\frac{x^2}{2}\right) + a_2\left(\frac{x^3}{3}\right) + \cdots = \sum_{n=0}^{\infty} a_n\left(\frac{x^{n+1}}{n+1}\right) + C.$$

The proof of this result is omitted. Note that the constant of integration C is placed in front of the expanded form of the power series in order to distinguish it from the actual terms of the series.

We can use term-by-term differentiation and integration when we need to generate power series representations of functions in much the same way that we used substitution in Example 9.3.2. This procedure is illustrated in Examples 9.3.3 and 9.3.4.

EXAMPLE 9.3.3

Find a power series for the function

$$g(x) = \frac{1}{(1-x)^2}.$$

Solution

Note that if $f(x) = \dfrac{1}{(1-x)} = (1-x)^{-1}$, then

$$f'(x) = -1(1-x)^{-2}(-1) = \frac{1}{(1-x)^2} = g(x).$$

Therefore, since $f(x)$ is represented by the geometric series

$$f(x) = \frac{1}{1-x} = 1 + x + x^2 + \cdots = \sum_{n=0}^{\infty} x^n,$$

its derivative $f'(x) = g(x)$ can be found by differentiating the geometric series term by term:

$$g(x) = f'(x) = 1 + 2x + 3x^2 + \cdots = \sum_{n=1}^{\infty} nx^{n-1}.$$

Since the geometric series converges for $|x| < 1$, it follows that the power series obtained for $g(x)$ by term-by-term differentiation also converges for $|x| < 1$.

> ### EXAMPLE 9.3.4

Find a power series for the logarithmic function $f(x) = \ln(1 + x)$.

Solution

First, note that

$$\ln(1 + x) = \int \frac{1}{1 + x} \, dx,$$

which suggests that we can find a power series for $\ln(1 + x)$ by finding one for $\dfrac{1}{1 + x}$ and integrating term by term. Substituting $-x$ for x in the (geometric) power series for $\dfrac{1}{1 - x}$ gives

$$\frac{1}{1 + x} = \frac{1}{1 - (-x)} = 1 + (-x) + (-x)^2 + (-x)^3 + \cdots$$

$$= 1 - x + x^2 - x^3 + \cdots = \sum_{n=0}^{\infty} (-1)^n x^n$$

and integrating term by term results in

$$\ln(1 + x) = \int \frac{1}{1 + x} \, dx = \int [1 - x + x^2 - x^3 + \cdots] \, dx$$

$$= C + x - \frac{x^2}{2} + \frac{x^3}{3} - \frac{x^4}{4} + \cdots$$

$$= \sum_{n=1}^{\infty} \frac{(-1)^{n+1} x^n}{n} + C$$

To determine the value of C, put $x = 0$ into this equation and get

$$\ln(1 + 0) = \sum_{n=1}^{\infty} \frac{(-1)^{n+1}(0)^n}{n} + C = 0 + C$$

$$0 = C \qquad \text{since } \ln 1 = 0$$

Thus,

$$\ln(1 + x) = \sum_{n=1}^{\infty} \frac{(-1)^{n+1} x^n}{n}.$$

Since the modified geometric series for $\dfrac{1}{1 + x}$ converges for $|-x| < 1$, it follows that the power series just obtained for $(1 + x)$ converges on the interval $-1 < x < 1$.

Taylor Series

Imagine that a function $f(x)$ is given and that you would like to find the corresponding coefficients a_n such that the power series $\sum_{n=0}^{\infty} a_n x^n$ converges to $f(x)$ on some interval. To discover what these coefficients might be, suppose that

$$f(x) = \sum_{n=0}^{\infty} a_n x^n = a_0 + a_1 x + a_2 x^2 + a_3 x^3 + \cdots + a_n x^n + \cdots.$$

Just-In-Time

Remember that $0! = 1$. Then the term $a_0 = \dfrac{f(0)}{0!}$ reduces to $f(0)$.

If $x = 0$, only the first term in the sum is nonzero and so

$$a_0 = f(0).$$

Now differentiate the series term by term. It can be shown that if the original power series converges to $f(x)$ on the interval $-R < x < R$, then the differentiated series converges to $f'(x)$ on this interval. Hence,

$$f'(x) = a_1 + 2a_2x + 3a_3x^2 + \cdots + na_nx^{n-1} + \cdots,$$

and, if $x = 0$, it follows that

$$a_1 = f'(0).$$

Differentiate again to get

$$f''(x) = 2a_2 + 3(2a_3x) + \cdots + n(n-1)a_nx^{n-2} + \cdots,$$

and let $x = 0$ to conclude that

$$f''(0) = 2a_2 \qquad \text{or} \qquad a_2 = \frac{f''(0)}{2}.$$

Similarly, the third derivative of f is

$$f^{(3)}(x) = 3(2a_3) + \cdots + n(n-1)(n-2)a_nx^{n-3} + \cdots,$$

and

$$f^{(3)}(0) = 3(2a_3) \qquad \text{or} \qquad a_3 = \frac{f^{(3)}(0)}{3!},$$

and so on. In general,

$$f^{(n)}(0) = n!a_n \qquad \text{or} \qquad a_n = \frac{f^{(n)}(0)}{n!},$$

where $f^{(n)}(0)$ denotes the nth derivative of f evaluated at $x = 0$, and $f^{(0)}(0) = f(0)$.

The preceding argument shows that *if* there is any power series that converges to $f(x)$, it must be the one whose coefficients are obtained from the derivatives of f by the formula

$$a_n = \frac{f^{(n)}(0)}{n!}.$$

This series is known as the **Taylor series of f** (about $x = 0$), and the corresponding coefficients a_n are called the **Taylor coefficients of f.**

Taylor Series ■ The Taylor series of $f(x)$ about $x = 0$ is the power series $\displaystyle\sum_{n=0}^{\infty} a_nx^n$, where

$$a_n = \frac{f^{(n)}(0)}{n!}.$$

When we developed the Taylor series representation of a function, we noted that if there is a power series that converges to a given function $f(x)$, then it must be the

Taylor series. For instance, since the geometric series $\displaystyle\sum_{n=0}^{\infty} x^n$ represents $g(x) = \dfrac{1}{1-x}$ for $|x| < 1$, we would expect the Taylor series for $g(x)$ to be $\displaystyle\sum_{n=0}^{\infty} x^n$. This fact is confirmed in Example 9.3.5.

EXAMPLE 9.3.5

Show that the Taylor series about $x = 0$ for the function $f(x) = \dfrac{1}{1-x}$ is

$$\frac{1}{1-x} = \sum_{n=0}^{\infty} x^n \qquad \text{for } |x| < 1.$$

Solution

Compute the Taylor coefficients as follows:

$$f(x) = \frac{1}{1-x} \qquad f(0) = 1 \qquad a_0 = \frac{f(0)}{0!} = \frac{1}{0!} = 1$$

$$f'(x) = \frac{1}{(1-x)^2} \qquad f'(0) = 1 \qquad a_1 = \frac{f'(0)}{1!} = \frac{1}{1!} = 1$$

$$f''(x) = \frac{2(1)}{(1-x)^3} \qquad f''(0) = 2! \qquad a_2 = \frac{f''(0)}{2!} = \frac{2!}{2!} = 1$$

$$f^{(3)}(x) = \frac{3(2)(1)}{(1-x)^4} \qquad f^{(3)}(0) = 3! \qquad a_3 = \frac{f^{(3)}(0)}{3!} = \frac{3!}{3!} = 1$$

$$\vdots \qquad\qquad \vdots \qquad\qquad \vdots$$

$$f^{(n)}(x) = \frac{n!}{(1-x)^{n+1}} \qquad f^{(n)}(0) = n! \qquad a_n = \frac{f^{(n)}(0)}{n!} = \frac{n!}{n!} = 1$$

The corresponding Taylor series is $\displaystyle\sum_{n=0}^{\infty} x^n$, which, as you already know, converges to $f(x) = \dfrac{1}{1-x}$ for $|x| < 1$. Thus,

$$\frac{1}{1-x} = \sum_{n=0}^{\infty} x^n = 1 + x + x^2 + \cdots \qquad \text{for } |x| < 1.$$

In Examples 9.3.2 through 9.3.4, we showed how to modify geometric series to obtain power series representations for several rational functions and a logarithmic function. Sometimes, however, finding the Taylor series for a given function is the most direct way of obtaining a power series representation of the function. For instance, the power series representation for the exponential function $E(x) = e^x$ cannot be found by modifying or manipulating a geometric series, but as is shown in Example 9.3.6, the Taylor series for $E(x)$ can be found quite easily.

EXAMPLE 9.3.6

Show that the Taylor series about $x = 0$ for the function $f(x) = e^x$ is

$$e^x = \sum_{n=0}^{\infty} \frac{x^n}{n!}.$$

Show also that this power series converges for all x.

Solution

Compute the Taylor coefficients as follows:

$$f(x) = e^x \qquad f(0) = 1 \qquad a_0 = \frac{f(0)}{0!} = \frac{1}{0!}$$

$$f'(x) = e^x \qquad f'(0) = 1 \qquad a_1 = \frac{f'(0)}{1!} = \frac{1}{1!}$$

$$f''(x) = e^x \qquad f''(0) = 1 \qquad a_2 = \frac{f''(0)}{2!} = \frac{1}{2!}$$

$$f^{(3)}(x) = e^x \qquad f^{(3)}(0) = 1 \qquad a_3 = \frac{f^{(3)}(0)}{3!} = \frac{1}{3!}$$

$$\vdots \qquad\qquad \vdots \qquad\qquad \vdots$$

$$f^{(n)}(x) = e^x \qquad f^{(n)}(0) = 1 \qquad a_n = \frac{f^{(n)}(0)}{n!} = \frac{1}{n!}$$

The corresponding Taylor series is

$$\sum_{n=0}^{\infty} a_n x^n = \sum_{n=0}^{\infty} \frac{x^n}{n!}.$$

To determine the convergence set for this power series, use the ratio test. Thus,

$$L = \lim_{n \to \infty} \left| \frac{\dfrac{x^{n+1}}{(n+1)!}}{\dfrac{x^n}{n!}} \right| = \lim_{n \to \infty} \left| \frac{n!\, x^{n+1}}{(n+1)!\, x^n} \right|$$

$$= \lim_{n \to \infty} \frac{|x|}{n+1} = 0 \qquad \text{for all } x$$

Since $L = 0 < 1$, it follows that the Taylor series for e^x converges for all x. In addition, it can be shown that the Taylor series converges to e^x for all x; that is,

$$e^x = \sum_{n=0}^{\infty} \frac{x^n}{n!} = 1 + x + \frac{x^2}{2!} + \frac{x^3}{3!} + \cdots$$

for all x.

Although the Taylor series of a function is the only power series that can possibly converge to the function, it need not actually do so. In fact, there are some functions whose Taylor series converge, but not to the value of the function. Fortunately, such functions rarely occur in practice, and most functions you are likely to encounter are equal to their Taylor series wherever the series converge.

Taylor Series
about x = a

Sometimes it is inappropriate (or impossible) to represent a given function by a power series of the form $\sum_{n=0}^{\infty} a_n x^n$, and you will have to use a series of the form $\sum_{n=0}^{\infty} a_n(x - a)^n$ instead, where a is a constant. This is the case, for example, for the function $f(x) = \ln x$, which is undefined for $x \leq 0$ and so cannot possibly be the sum of a series of the form $\sum_{n=0}^{\infty} a_n x^n$ on a symmetric interval about $x = 0$.

Using an argument similar to the one on page 703, you can show that if $f(x) = \sum_{n=0}^{\infty} a_n(x - a)^n$, the coefficients of the series are related to the derivatives of f by the formula

$$a_n = \frac{f^{(n)}(a)}{n!}.$$

The power series in $(x - a)$ with these coefficients is called the **Taylor series of $f(x)$ about $x = a$.** It can be shown that series of this type converge on symmetric intervals of the form $a - R < x < a + R$, including possibly one or both of the endpoints $x = a - R$ and $x = a + R$. Notice that the Taylor series about $x = 0$ is simply a special case of the more general Taylor series about $x = a$.

> **Taylor Series about x = a** ■ The Taylor series of $f(x)$ about $x = a$ is the power series
>
> $$\sum_{n=0}^{\infty} a_n(x - a)^n,$$
>
> where
> $$a_n = \frac{f^{(n)}(a)}{n!}.$$

In Example 9.3.4, we found a power series representation for $\ln(1 + x)$, but there is no Taylor series about $x = 0$ for $\ln x$ since $\ln x$ and its derivatives are not defined at $x = 0$. However, we can find a Taylor series for $\ln x$ about $x = a$ for $a > 0$, as illustrated in Example 9.3.7 for the case where $a = 1$.

EXAMPLE 9.3.7

Show that the Taylor series for $\ln x$ about $x = 1$ is

$$\sum_{n=1}^{\infty} \frac{(-1)^{n+1}(x - 1)^n}{n} = (x - 1) - \frac{(x - 1)^2}{2} + \frac{(x - 1)^3}{3} - \frac{(x - 1)^4}{4} + \cdots.$$

For what values of x does this series converge?

Solution

Compute the Taylor coefficients as follows:

$$f(x) = \ln x \qquad\qquad f(1) = 0 \qquad\qquad a_0 = \frac{f(1)}{0!} = 0$$

$$f'(x) = \frac{1}{x} \qquad\qquad f'(1) = 1 \qquad\qquad a_1 = \frac{f'(1)}{1!} = 1$$

$$f''(x) = \frac{-1}{x^2} \qquad\qquad f''(1) = -1 \qquad\qquad a_2 = \frac{f''(1)}{2!} = -\frac{1}{2}$$

$$f^{(3)}(x) = \frac{2!}{x^3} \qquad\qquad f^{(3)}(1) = 2! \qquad\qquad a_3 = \frac{f^{(3)}(1)}{3!} = \frac{1}{3}$$

$$f^{(4)}(x) = \frac{-3!}{x^4} \qquad\qquad f^{(4)}(1) = -3! \qquad\qquad a_3 = \frac{f^{(4)}(1)}{4!} = -\frac{1}{4}$$

$$\vdots \qquad\qquad\qquad \vdots \qquad\qquad\qquad \vdots$$

$$f^{(n)}(x) = (-1)^{(n+1)}\frac{(n-1)!}{x^n} \quad f^{(n)}(1) = (-1)^{n+1}(n-1)! \quad a_n = \frac{(-1)^{n+1}}{n}$$

The corresponding Taylor series is

$$\sum_{n=1}^{\infty} \frac{(-1)^{n+1}}{n}(x-1)^n.$$

Using the ratio test to determine the convergence set for this power series, the limit of the ratio of consecutive terms is

$$L = \lim_{n\to\infty} \left| \frac{\dfrac{(-1)^{n+2}(x-1)^{n+1}}{n+1}}{\dfrac{(-1)^{n+1}(x-1)^n}{n}} \right| = \lim_{n\to\infty} \left| \frac{n(x-1)^{n+1}}{(n+1)(x-1)^n} \right|$$

$$= \lim_{n\to\infty} \left| \frac{n(x-1)}{n+1} \right| = \lim_{n\to\infty} \left| \frac{x-1}{1+\dfrac{1}{n}} \right|$$

$$= |x-1|$$

so the series converges for

$$L = |x-1| < 1$$

or, equivalently, for

$$-1 < x - 1 < 1$$
$$0 < x < 2$$

and diverges for $|x-1| > 1$. It can be shown that the series also converges at the endpoint $x = 2$ and, indeed, that

$$\ln x = \sum_{n=1}^{\infty} \frac{(-1)^{n+1}}{n}(x-1)^n$$

$$= (x-1) - \frac{1}{2}(x-1)^2 + \frac{1}{3}(x-1)^3 - \cdots \qquad \text{for } 0 < x \leq 2$$

Approximation by Taylor Polynomials

The partial sums of a Taylor series of a function are polynomials that can be used to approximate the function. In general, the more terms there are in the partial sum, the better the approximation will be. Let $P_n(x)$ denote the $(n + 1)$th partial sum, which is a polynomial of degree (at most) n. In particular,

$$P_n(x) = f(a) + f'(a)(x - a) + \frac{f''(a)}{2!}(x - a)^2 + \cdots + \frac{f^{(n)}(a)}{n!}(x - a)^n.$$

The polynomial $P_n(x)$ is known as the **nth Taylor polynomial** of $f(x)$ about $x = a$. The approximation of $f(x)$ by its Taylor polynomials $P_n(x)$ is most accurate near $x = a$ and for large values of n. Here is a geometric argument that should give you some additional insight into the situation.

Observe first that at $x = a$, f and P_n are equal, as are, respectively, their first n derivatives. For example, with $n = 2$,

$$P_2(x) = f(a) + f'(a)(x - a) + \frac{f''(a)}{2!}(x - a)^2 \quad \text{and} \quad P_2(a) = f(a)$$

$$P_2'(x) = f'(a) + f''(a)(x - a) \quad\quad\quad\quad \text{and} \quad P_2'(a) = f'(a)$$

$$P_2''(x) = f''(a) \quad\quad\quad\quad\quad\quad\quad\quad\quad \text{and} \quad P_2''(a) = f''(a)$$

The fact that $P_n(a) = f(a)$ implies that the graphs of P_n and f intersect at $x = a$. The fact that $P_n'(a) = f'(a)$ implies further that the graphs have the same slope at $x = a$. The fact that $P_n''(a) = f''(a)$ imposes the further restriction that the graphs have the same concavity at $x = a$. In general, as n increases, the number of matching derivatives increases, and the graph of P_n approximates more closely that of f near $x = a$. The situation is illustrated in Figure 9.4, which shows the graphs of $f(x) = e^x$ and its first three Taylor polynomials.

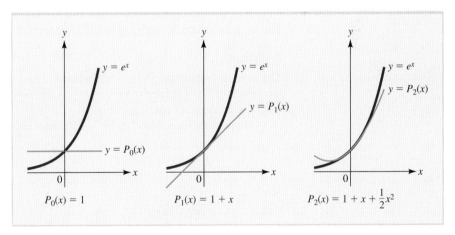

FIGURE 9.4 The graphs of $f(x) = e^x$ and its first three Taylor polynomials.

The use of Taylor polynomials to approximate functions is illustrated in Examples 9.3.8 and 9.3.9.

EXAMPLE 9.3.8

Use a Taylor polynomial of degree 6 to approximate e.

Solution

From Example 9.3.6,

$$e^x = \sum_{n=0}^{\infty} \frac{x}{n!} \text{ for all } x.$$

Hence,

$$P_6(x) = 1 + x + \frac{x^2}{2!} + \frac{x^3}{3!} + \frac{x^4}{4!} + \frac{x^5}{5!} + \frac{x^6}{6!},$$

and

$$e = e^1 \approx P_6(1) = 1 + 1 + \frac{1}{2!} + \frac{1}{3!} + \frac{1}{4!} + \frac{1}{5!} + \frac{1}{6!}$$

$$= 1 + 1 + \frac{1}{2} + \frac{1}{6} + \frac{1}{24} + \frac{1}{120} + \frac{1}{720} \approx 2.71806$$

By analysis of the error that results from this sort of approximation (based on techniques beyond the scope of this text), it can be shown that the actual value of e, rounded off to five decimal places, is 2.71828.

EXAMPLE 9.3.9

Use an appropriate Taylor polynomial of degree 3 to approximate $\sqrt{4.1}$.

Solution

The goal is to estimate $f(x) = \sqrt{x}$ when $x = 4.1$. Since 4.1 is close to 4 and since the values of f and its derivatives at $x = 4$ are easy to compute, we will use a Taylor polynomial about $x = 4$.

Compute the necessary Taylor coefficients about $x = 4$ as follows:

$$f(x) = \sqrt{x} \qquad\qquad f(4) = 2 \qquad\qquad a_0 = \frac{2}{0!} = 2$$

$$f'(x) = \frac{1}{2}x^{-1/2} \qquad f'(4) = \frac{1}{4} \qquad a_1 = \frac{\frac{1}{4}}{1!} = \frac{1}{4}$$

$$f''(x) = -\frac{1}{4}x^{-3/2} \qquad f''(4) = -\frac{1}{32} \qquad a_2 = \frac{-\frac{1}{32}}{2!} = -\frac{1}{64}$$

$$f^{(3)}(x) = \frac{3}{8}x^{-5/2} \qquad f^{(3)}(4) = \frac{3}{256} \qquad a_3 = \frac{\frac{3}{256}}{3!} = \frac{1}{512}$$

The corresponding Taylor polynomial of degree 3 is

$$P_3(x) = 2 + \frac{1}{4}(x-4) - \frac{1}{64}(x-4)^2 + \frac{1}{512}(x-4)^3$$

and so

$$\sqrt{4.1} \approx P_3(4.1) = 2 + \frac{1}{4}(0.1) - \frac{1}{64}(0.1)^2 + \frac{1}{512}(0.1)^3 \approx 2.02485.$$

Rounded off to seven decimal places, the true value of $\sqrt{4.1}$ is 2.0248457.

In Example 9.3.10, a Taylor polynomial is used to estimate the definite integral of a function that has no elementary antiderivative. In Chapter 10, a form of this integral will be used to compute probabilities involving the normal distribution.

EXAMPLE 9.3.10

Use a Taylor polynomial of degree 8 to approximate $\displaystyle\int_0^1 e^{-x^2}dx$.

Solution

The easiest way to get the Taylor series of e^{-x^2} is to start with the series

$$e^x = \sum_{n=0}^{\infty} \frac{x^n}{n!}$$

for e^x and replace x by $-x^2$ to get

$$e^{-x^2} = \sum_{n=0}^{\infty} \frac{(-1)^n x^{2n}}{n!},$$

writing $(-x^2)^n$ as $(-1)^n(x^2)^n$. Thus,

$$e^{-x^2} \approx 1 - x^2 + \frac{1}{2}x^4 - \frac{1}{6}x^6 + \frac{1}{24}x^8,$$

and so

$$\int_0^1 e^{-x^2}dx \approx \int_0^1 \left(1 - x^2 + \frac{1}{2}x^4 - \frac{1}{6}x^6 + \frac{1}{24}x^8\right)dx$$

$$= \left(x - \frac{1}{3}x^3 + \frac{1}{10}x^5 - \frac{1}{42}x^7 + \frac{1}{216}x^9\right)\Big|_0^1$$

$$= 1 - \frac{1}{3} + \frac{1}{10} - \frac{1}{42} + \frac{1}{216} \approx 0.7475$$

This value is correct to only two decimal places. The true value of the integral to seven places is 0.7468241.

EXERCISES ■ 9.3

In Exercises 1 through 8, determine the radius of convergence and the interval of absolute convergence for the given power series.

1. $\displaystyle\sum_{k=0}^{\infty} 5^k x^k$

2. $\displaystyle\sum_{k=1}^{\infty} \sqrt{k} x^k$

3. $\displaystyle\sum_{k=1}^{\infty} \frac{x^k}{\sqrt{k}}$

4. $\displaystyle\sum_{k=1}^{\infty} \frac{3^k x^k}{k}$

5. $\displaystyle\sum_{k=1}^{\infty} \frac{2^{2k} x^k}{k^2}$

6. $\displaystyle\sum_{k=1}^{\infty} \frac{x^k}{(2k)!}$

7. $\displaystyle\sum_{k=1}^{\infty} \frac{5^k x^k}{k!}$

8. $\displaystyle\sum_{k=0}^{\infty} \frac{k! x^k}{3^k}$

In Exercises 9 through 14, find a power series for the given function and determine its interval of absolute convergence.

9. $f(x) = \dfrac{x}{1 + x}$

10. $f(x) = \dfrac{1}{2 - x}$

11. $f(x) = \dfrac{x^2}{1 - x^2}$

12. $f(x) = \dfrac{x}{3 + 2x}$

13. $f(x) = \ln(2 + x)$

14. $f(x) = \ln(1 + 2x)$

In Exercises 15 through 23, find the Taylor series for the given function at the indicated point $x = a$.

15. $f(x) = e^{3x}; a = 0$

16. $f(x) = e^{-2x}; a = 0$

17. $f(x) = \dfrac{1}{2}(e^x + e^{-x}); a = 0$

18. $f(x) = (1 + x)e^x; a = 0$

19. $f(x) = e^{-3x}; a = -1$

20. $f(x) = \dfrac{1}{x}; a = 1$

21. $f(x) = \ln(2x); a = \dfrac{1}{2}$

22. $f(x) = \dfrac{1}{2 - x}; a = 1$

23. $f(x) = \dfrac{x}{1 + x}; a = 2$

24. Find a power series for $f(x) = \dfrac{1}{1 - x^3}$ by using term-by-term differentiation and the result of Example 9.3.3.

25. Find the Taylor series for $f(x) = 3x^2 e^{x^3}$ about $x = 0$ from the Taylor series for e^x obtained in Example 9.3.6 in two different ways:
   a. Substitute x^3 for x and then multiply the resulting series by $3x^2$.
   b. Substitute x^3 for x and then differentiate term by term.

26. Find the Taylor series about $x = 0$ for the indefinite integral

$$\int \frac{1}{1 + x^2} \, dx.$$

[*Hint:* Start with the geometric series for $\dfrac{1}{1 - x}$, and then substitute $-x^2$ for x and integrate term by term.]

27. Find the Taylor series about $x = 0$ for the indefinite integral

$$\int x^2 e^{-x^2} \, dx.$$

[*Hint:* Start by substituting $-x^2$ for x in the Taylor series for e^x obtained in Example 9.3.6. Then multiply the resulting series by x^2 and integrate term by term.]

In Exercises 28 through 35, use a Taylor polynomial of specified degree n to approximate the indicated quantity.

28. $\sqrt{3.8}; n = 3$

29. $\sqrt{1.2}; n = 3$

30. $\ln 1.1$; $n = 5$

31. $\ln 0.7$; $n = 5$

32. $e^{0.3}$; $n = 4$

33. $\dfrac{1}{\sqrt{e}}$; $n = 4$

34. $\ln 2$; $n = 5$ [*Hint*: $\ln 2 = -\ln 0.5$]

35. $\ln \sqrt[3]{4}$; $n = 5$ [*Hint*: See the hint to Exercise 34.]

In Exercises 36 through 39, use a Taylor polynomial of specified degree n together with term-by-term integration to estimate the indicated definite integral.

36. $\displaystyle\int_0^{1/2} e^{-x^2}\, dx$; $n = 6$

37. $\displaystyle\int_{-0.2}^{0.1} e^{-x^2}\, dx$; $n = 4$

38. $\displaystyle\int_0^{0.1} \dfrac{1}{1 + x^2}\, dx$; $n = 4$

39. $\displaystyle\int_{-1/2}^{0.1} \dfrac{1}{1 - x^3}\, dx$; $n = 9$

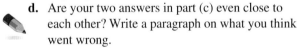

In Exercises 40 through 43, calculate the first four Taylor polynomials $P_0(x)$, $P_1(x)$, $P_2(x)$, and $P_3(x)$ for the given function about the specified point $x = a$. Then use technology to draw the graphs of $f(x)$ and the four approximating polynomials on the same set of axes.

40. $f(x) = e^{2x}$; $a = 0$

41. $f(x) = 1 - e^{-x}$; $a = 0$

42. $f(x) = \ln x$; $a = 1$

43. $f(x) = \dfrac{\ln x}{x}$; $a = 1$

44. **A* MARGINAL ANALYSIS**   The marginal cost of producing x units of a particular commodity is $C'(x)$ thousand dollars per unit, where

$$C'(x) = \ln (1 + 0.01x^2).$$

The net cost of producing the first eight units of the commodity is given by the integral

$$C(8) - C(0) = \int_0^8 C'(x)\, dx = \int_0^8 \ln (1 + 0.01x^2)\, dx.$$

a.   Find the Taylor polynomial of degree 6 for $\ln (1 + 0.01x^2)$ at $x = 0$. [*Hint*: Example 9.3.4 may help.]

b.   Integrate the Taylor polynomial from part (a) to estimate the net change $C(8) - C(0)$. [*Note*: The actual net cost is about 1.452, that is, \$1452.]

45. **MARGINAL ANALYSIS**   The marginal profit derived from producing x units of a particular commodity is $P'(x)$ thousand dollars per unit, where

$$P'(x) = 10x^2 e^{-x^2}.$$

Find the Taylor polynomial of degree 7 for $P'(x)$ at $x = 0$ and integrate to obtain an estimate of the net profit $P(1) - P(0)$ obtained from producing the first unit.

POPULATION DENSITY   *Recall from Section 5.6 that if the population density x kilometres from the centre of an urban area is D(x), then the total population P living within a radius of M kilometres of the city centre is given by the integral*

$$P(M) = \int_0^M [2\pi x D(x)]\, dx.$$

Use this formula in Exercises 46 and 47.

46.   Suppose the population density is

$$D(x) = \dfrac{5000}{1 + 0.5x^2}.$$

a.   Find the Taylor polynomial of degree 6 for $2\pi x D(x)$ at $x = 0$ and integrate to obtain an estimate of the total number of people $P(1)$ living within 1 km of the city centre.

b.   Evaluate the integral for $P(1)$ directly using the substitution $u = 1 + 0.5x^2$. Compare this exact value of $P(1)$ with your estimate in part (a).

c.   Repeat parts (a) and (b) for $P(10)$.

d.   Are your two answers in part (c) even close to each other? Write a paragraph on what you think went wrong.

47.   Suppose the population density is

$$D(x) = \dfrac{5000}{1 + 0.5x^3}.$$

Use a Taylor polynomial of degree 8 to estimate $P(1)$, the population within 1 km of the city centre.

48.   **A* GENETICS**   The Jukes-Cantor model in genetics is used to study mutations from an original DNA sequence. The Jukes-Cantor distance between DNA sequences S_0 and S_1 is the quantity

$$d = -\dfrac{3}{4} \ln \left(1 - \dfrac{4}{3} p \right),$$

where p is the fraction of sites that disagree between the two sequences. It provides a measure for the total number of substitutions per site that occur as S_0 evolves into S_1.

a. Approximate d by a Taylor polynomial of degree 3 at $x = 0$.

b. A researcher finds that out of 43 sites in a particular DNA sequence S_0, 17 have undergone a substitution when the sequence evolves to S_1. Use the Taylor polynomial you obtained in part (a) to estimate the Jukes-Cantor distance for this mutation.

c. Read an article on how the Jukes-Cantor distance and other *phylogenetic distances* are used to measure sequence similarity. Then write a paragraph describing the role played by such ideas in mathematical models of evolution.

49. A* LOGISTIC GROWTH The population of a city grows logistically in such a way that after t years, the population is P million, where

$$P(t) = \frac{1}{1 + e^{-1.2t}}.$$

Thus, the average population over the next 10 years is given by the integral

$$P_{\text{ave}} = \frac{1}{10} \int_0^{10} P(t)\, dt.$$

a. Estimate the value of P_{ave} by integrating the Taylor polynomial of degree 2 for $P(t)$ expanded at $t = c$ for $c = 0$, $c = 5$, and $c = 10$.

b. The actual average population is
$$P_{\text{ave}} = 0.942 \ (942\ 000 \text{ people}).$$
Which choice of c in part (a) resulted in the best approximation?

Concept Summary Chapter 9

Infinite Series

$$\sum_{k=1}^{\infty} a_k = a_1 + a_2 + a_3 + \cdots$$

The nth Partial Sum

$$S_n = \sum_{k=1}^{n} a_k = a_1 + a_2 + \cdots + a_n$$

Convergence and Divergence of a Series

$$\sum_{k=1}^{\infty} a_k$$ converges with sum S if $\lim_{n \to \infty} S_n = S$ and diverges otherwise.

Geometric Series With Common Ratio r ($a \neq 0$)

$$\sum_{n=0}^{\infty} ar^n$$ converges to $\dfrac{a}{1 - r}$ if $|r| < 1$ and diverges if $|r| \geq 1$.

The Harmonic Series $\displaystyle\sum_{k=1}^{\infty} \frac{1}{k}$

The Divergence Test

$$\sum_{k=1}^{\infty} a_k$$ diverges if $\lim_{k \to \infty} a_k \neq 0$.

The Integral Test

If f is decreasing, $f(x) > 0$, and $f(n) = a_n$, then $\displaystyle\sum_{n=1}^{\infty} a_n$ and $\displaystyle\int_1^{\infty} f(x)\, dx$ either both converge or both diverge.

The p-Series Test

$$\sum_{n=1}^{\infty} \frac{1}{n^p}$$ converges if $p > 1$ and diverges if $0 \leq p \leq 1$.

The Direct Comparison Test

Suppose $0 \leq b_n \leq a_n$ for all n. Then

if $\displaystyle\sum_{n=1}^{\infty} a_n$ converges, so does $\displaystyle\sum_{n=1}^{\infty} b_n$, and

if $\displaystyle\sum_{n=1}^{\infty} b_n$ diverges, so does $\displaystyle\sum_{n=1}^{\infty} a_n$.

The Ratio Test

For the series $\displaystyle\sum_{n=1}^{\infty} a_n$, let $L = \lim_{n \to \infty} \left| \dfrac{a_n + 1}{a_n} \right|$.
The series converges if $L < 1$ and diverges if $L > 1$. The test is inconclusive if $L = 1$.

Power Series

A power series $\displaystyle\sum_{n=0}^{\infty} a_n x^n$ can do exactly one of the following:

Converge only for $x = 0$
Converge for all x
Converge for $|x| < R$ and diverge for $|x| > R$

Taylor Series of $f(x)$ About $x = 0$

$$\sum_{n=0}^{\infty} a_n x^n \text{ where } a_n = \frac{f^{(n)}(0)}{n!}$$

Taylor Series of $f(x)$ About $x = a$

$$\sum_{n=0}^{\infty} a_n(x - a)^n \text{ where } a_n = \frac{f^{(n)}(a)}{n!}$$

Important Taylor Series

$$\frac{1}{1 - x} = \sum_{n=0}^{\infty} x^n \text{ for } |x| < 1 \qquad e^x = \sum_{n=0}^{\infty} \frac{x^n}{n!} \text{ for all } x$$

$$\ln x = \sum_{n=1}^{\infty} \frac{(-1)^{n+1}}{n} (x - 1)^n \quad \text{for } 0 < x \le 2$$

Checkup for Chapter 9

1. Determine whether each geometric series converges or diverges. If the series converges, find its sum.

   a. $\displaystyle\sum_{n=0}^{\infty} \frac{(-3)^n}{5^{n+1}}$ **b.** $\displaystyle\sum_{n=1}^{\infty} \frac{2^{2n}}{3^{n-1}}$

2. In each case, show that the given series converges.

   a. $\displaystyle\sum_{n=1}^{\infty} \frac{1}{n^{3/2}}$ **b.** $\displaystyle\sum_{n=1}^{\infty} \frac{3^n}{n!}$

3. In each case, show that the given series diverges.

   a. $\displaystyle\sum_{n=1}^{\infty} \frac{n(n^2 + 1)}{100n^3 + 9}$ **b.** $\displaystyle\sum_{n=2}^{\infty} \frac{\ln n}{\sqrt{n}}$

4. Determine whether each series converges or diverges.

   a. $\displaystyle\sum_{n=1}^{\infty} \frac{1}{\sqrt{n}}$ **b.** $\displaystyle\sum_{n=1}^{\infty} \sqrt{n}\, e^{-n}$

   c. $\displaystyle\sum_{n=1}^{\infty} \frac{n^2}{2^n}$ **d.** $\displaystyle\sum_{n=0}^{\infty} \frac{1}{n^2 + 1}$

5. For each power series, find the interval of absolute convergence.

   a. $\displaystyle\sum_{n=1}^{\infty} \frac{x^n}{n + 1}$ **b.** $\displaystyle\sum_{n=1}^{\infty} \frac{(2x)^n}{n!}$

6. Find a power series for each function.

   a. $f(x) = \dfrac{x}{1 + x^2}$ **b.** $g(x) = \dfrac{5}{2 + 3x}$

7. Find the Taylor series about $x = 0$ for each function.

   a. $f(x) = e^x + e^{-x}$ **b.** $g(x) = \ln(x^2 + 1)$

8. ACCUMULATION OF MEDICATION A patient is given an injection of 25 units of a certain drug every 24 hours. The drug is eliminated exponentially so that the fraction that remains after t days is $f(t) = e^{-t/3}$. If the treatment is continued indefinitely, approximately how many units will eventually be in the patient's body just prior to an injection?

9. PRESENT VALUE How much should you invest today at an annual rate of 4% compounded continuously so that starting next year, you can make annual withdrawals of $5000 in perpetuity?

10. Estimate the value of the integral

$$\int_0^1 \frac{1}{1 + x^2}\, dx$$

by finding a power series for $f(x) = \dfrac{1}{1 + x^2}$ and integrating term by term.

Review Exercises

In Exercises 1 through 8, determine whether the given geometric series converges or diverges. If the series converges, find its sum.

1. $\displaystyle\sum_{n=0}^{\infty} \frac{(-2)^n}{5^{n+1}}$ **2.** $\displaystyle\sum_{n=1}^{\infty} \frac{4^{n-1}}{3^{2n}}$

3. $\displaystyle\sum_{n=0}^{\infty} \frac{13}{(-5)^n}$ **4.** $\displaystyle\sum_{n=0}^{\infty} \left(-\frac{3}{2}\right)^n$

5. $\displaystyle\sum_{n=0}^{\infty} e^{-0.5n}$ **6.** $\displaystyle\sum_{n=1}^{\infty} \frac{2^{n+1}}{3^{n-1}}$

7. $\displaystyle\sum_{n=0}^{\infty} \left[\left(\frac{2}{3}\right)^n + \left(\frac{3}{2}\right)^n\right]$ **8.** $\displaystyle\sum_{n=0}^{\infty} \left(\frac{3}{2}\right)\left(\frac{2}{3}\right)^n$

In Exercises 9 through 17, determine whether the given series converges or diverges.

9. $\displaystyle\sum_{n=0}^{\infty} \frac{1}{2n + 1}$ **10.** $\displaystyle\sum_{n=1}^{\infty} \left(1 + \frac{1}{n}\right)^2$

11. $\displaystyle\sum_{n=1}^{\infty} \ln\left(2 + \frac{1}{n}\right)$ **12.** $\displaystyle\sum_{n=1}^{\infty} \frac{1}{n^2\sqrt{n}}$

13. $\displaystyle\sum_{n=2}^{\infty} \frac{\ln \sqrt{n}}{\sqrt{n}}$

14. $\displaystyle\sum_{n=1}^{\infty} \frac{10^n}{n!}$

15. $\displaystyle\sum_{n=1}^{\infty} \frac{n^3}{3^n}$

16. $\displaystyle\sum_{n=1}^{\infty} \frac{(-2)^n}{n^2}$

17. $\displaystyle\sum_{n=1}^{\infty} \frac{(-3)^{2n}}{n!}$

In Exercises 18 through 21, find the interval of absolute convergence for the given power series.

18. $\displaystyle\sum_{n=0}^{\infty} (3x)^n$

19. $\displaystyle\sum_{n=1}^{\infty} \frac{(2x)^{2n}}{3^{n-1}}$

20. $\displaystyle\sum_{n=1}^{\infty} \frac{(-2x)^n}{n!}$

21. $\displaystyle\sum_{n=1}^{\infty} e^n x^{n-1}$

In Exercises 22 through 25, find the Taylor series at x = 0 for the given function, either by using the definition or by manipulating a known series.

22. $f(x) = \dfrac{1}{(1 + 2x)^2}$

23. $f(x) = e^x - e^{-x}$

24. $f(x) = x^2 e^{-2x}$

25. $f(x) = \ln\left(\dfrac{x + 1}{2x + 1}\right)$

REPEATING DECIMALS   *In Exercises 26 and 27, express the repeating decimal as a fraction $\dfrac{p}{q}$.*

26. $0.1223535\ldots$

27. $51.34747\ldots$

28. BOUNCING BALL   A ball is dropped from a height of 6 m and bounces indefinitely, repeatedly rebounding to 80% of its previous height. How far does the ball travel?

29. BOUNCING BALL   A ball is dropped from a height of H metres and bounces indefinitely, repeatedly rebounding to 75% of its previous height. If it travels a total distance of 70 m, what is H?

30. SAVINGS   How much should you invest today at an annual interest rate of 5% compounded continuously so that starting next year, you can make annual withdrawals of $4000, in perpetuity?

31. PRESENT VALUE   An investment guarantees annual payments of $5000 in perpetuity, with the payments beginning immediately. Find the present value of this investment if the prevailing annual interest rate remains fixed at 5% compounded continuously.

32. THE MULTIPLIER EFFECT   Suppose that nationwide, approximately 91% of all income is spent and 9% is saved. What is the total amount of spending generated by a $60 billion tax rebate if saving habits do not change?

33. ACCUMULATION OF MEDICATION   A patient is given an injection of 25 units of a certain drug every 24 hours. The drug is eliminated exponentially so that the fraction that remains in the patient's body after t days is $f(t) = e^{-kt}$ for some constant k. If 70 units of the drug eventually accumulate in the patient's body just prior to an injection, what is k?

34. DEMOGRAPHICS   A developing country currently has 2500 trained scientists. The government estimates that each year, 6% of the current number of scientists retire, die, or emigrate, while 278 new scientists graduate from college. If these trends continue, how many scientists will there be in 20 years? How many will there be in the long run?

35. LINGUISTICS   Linguists and psychologists who are interested in the evolution of language have noticed an interesting pattern in the frequency of so-called rare words in certain literary works. According to one classic model,* if a book contains a total of T different such words, then approximately $\dfrac{T}{(1)(2)}$ words appear exactly once, $\dfrac{T}{(2)(3)}$ words appear exactly twice, and in general $\dfrac{T}{k(k + 1)}$ words appear exactly k times.

a. Assuming the word frequency pattern in the model is accurate, why should you expect the series

$$\sum_{k=1}^{\infty} \frac{T}{k(k + 1)}$$

to converge? What would you expect its sum to be?

b. Verify your conjecture in part (a) by actually summing the series. [*Hint:* See Example 9.1.3 in Section 9.1.]

   c. Read an article on information and learning theory and write a paragraph on mathematical methods in this subject.

*G. K. Zipf, *Human Behavior and the Principle of Least Effort,* Cambridge, MA: Addison-Wesley, 1949. Another good source is C. E. Shannon and W. Weaver, *The Mathematical Theory of Communication,* Urbana, IL: University of Illinois Press, 1998.

36. GENETICS In a classic genetic model,[†] the average lifespan of a harmful gene is related to the infinite series

$$1 + 2r + 3r^2 + 4r^3 + \cdots = \sum_{k=1}^{\infty} kr^{k-1}$$

for $0 < r < 1$. Show that this series converges and find its sum in terms of r. [*Hint:* If S_n is the nth partial sum of the series, what is $S_n - rS_n$?]

In Exercises 37 through 40, find the Taylor series for the given function at the specified value of $x = a$.

37. $f(x) = \dfrac{1}{(1 - x)^2}; a = 2$

38. $f(x) = \ln(2 + x); a = -1$

39. $f(x) = x \ln x; a = 1$

40. $f(x) = \dfrac{1 - x}{1 + x}; a = 0$

41. Use a Taylor polynomial of degree 3 to approximate $\sqrt{0.9}$.

42. Use a Taylor polynomial of degree 10 to approximate

$$\int_0^{1/2} \frac{x}{1 + x^3} \, dx.$$

43. BEE STORY The story goes that the famous mathematician John von Neumann (1903–1957) was once challenged to solve a version of the following problem:

> Two trains, each travelling at 30 m/s, approach each other on a straight track. When they are 1000 m apart, a super-charged bee begins flying from one train to the other and back again at a rate of 60 m/s, and continues to do so until the trains crash. How far does the bee fly before it is crushed by the crashing trains?

Von Neumann pondered the question only briefly before giving the correct answer. The poser of the problem chuckled appreciatively and said, "You saw the trick. I should have known better than to try to fool you, Professor." Von Neumann looked puzzled. "What trick?" he replied, "I summed the series."

a. Sum a series as von Neumann did to find the distance travelled by the ill-fated bee.

b. Unlike von Neumann, do you see an easy way to solve this problem? (Incidentally, a form of this question is sometimes used by Microsoft to test new employees.)

c. Try this kinder, gentler version of the same problem. Suppose the conductors of the two trains see each other and hit the brakes when they are 180 m apart. If both trains decelerate at a rate of 5 m/s², how far does the super-charged bee fly before the trains come together?

44. Use a calculator to compute the sum

$$S(n) = 1 - \frac{1}{2!} + \frac{1}{4!} - \frac{1}{6!} + \cdots + \frac{(-1)^n}{(2N)!}$$

for $N = 2, 7, 10, 15$, and 30. Based on your results, do you think the series

$$\sum_{k=0}^{\infty} \frac{(-1)^k}{(2k)!}$$

converges or diverges? If it converges, what do you think its sum will be?

45. Calculate the first four Taylor polynomials $P_0(x)$, $P_1(x)$, $P_2(x)$, and $P_3(x)$ at $x = 0$ for the function

$$f(x) = \frac{1}{\sqrt{1 - x^2}}.$$

a. Use technology to graph these four polynomials on the same set of axes.

b. Use $P_3(x)$ to estimate the value of the definite integral

$$\int_0^{1/2} \frac{dx}{\sqrt{1 - x^2}}.$$

46. A* LABOUR MIGRATION Economists refer to the process of moving from one job to another as *labour migration*. Such movement is usually undertaken as a means for social or economic improvement, but it also involves costs, such as the loss of seniority in the old job and the psychological cost of disrupting relationships. Consider the function[*]

$$V = \sum_{n=1}^{N} \frac{E_2(n) - E_1(n)}{(1 + i)^n} - \sum_{n=1}^{N} \frac{C_m(n)}{(1 + i)^n} - C_p,$$

where $E_2(n)$ and $E_1(n)$ denote the earnings from the new and old jobs, respectively, in year n after the move is to be made; i is the prevailing annual

[†]C. C. Li, *Human Genetics: Principles and Methods,* New York: McGraw-Hill, 1961.

[*]C. R. McConnell and S. L. Brue, *Contemporary Labor Economics,* New York: McGraw-Hill, 1992, pp. 440–444.

interest rate (in decimal form); N is the number of years the person is expected to be on the new job; and C_m and C_p are the expected monetary and net psychological costs of the move (psychological gain minus psychological loss).

a. What does V represent? Why is the job move desirable if $V > 0$ and undesirable if $V < 0$?

b. For simplicity, assume that $E_2 - E_1$ and C_m are constant for all n and that the person expects to stay on the new job "forever" once he or she moves (that is, $N \to \infty$). Find a formula for V and use it to obtain a criterion for whether or not the job move should be made. [*Hint*: Your criterion should be an inequality involving $E_1, E_2, C_m, C_p,$ and i.]

c. Read an article on job mobility and labour migration and write a paragraph on mathematical methods for modelling such issues.

47. A* GEOLOGICAL DATING At the time we studied carbon dating in Chapter 4, we noted that radiocarbon methods are used mainly for dating specimens that are not too old.* In order to date rocks or artifacts that are older than 40 000 years, it is necessary to use other methods. Whether ^{14}C or a radioactive isotope of some other element is used for dating, it can be shown that

$$\frac{S(t) - S(0)}{R(t)} + 1 = e^{(\ln 2)t/\lambda}$$

---

*Paul J. Campbell, "*How Old Is the Earth?*" UMAP Modules 1992: Tools for Teaching, Lexington, MA: Consortium for Mathematics and Its Applications, Inc., 1993, pp. 105–137.

where $R(t)$ is the number of atoms of radioactive isotope at time t, $S(t)$ is the number of atoms of the stable product of radioactive decay, $S(0)$ is the number of atoms of stable product initially present (at $t = 0$), and λ is the half-life of the radioactive isotope (the time it takes for half a sample to decay).

a. Approximate the time t in this formula using the Taylor polynomial of degree 2 for e^x at $x = 0$.

b. Suppose a piece of mica is analyzed, and it is found that 5% of the atoms in the rock are radioactive rubidium-87 and 0.04% are strontium-87. If all the strontium-87 was produced by decay of the rubidium-87 in the rock, how old is the rock? Use the approximation obtained in part (a). You will need to know that the half-life of rubidium-87 is 48.6×10^9 years.

c. Read an article on geological dating methods and write a paragraph on how these methods differ from radioactive dating and from each other.

48. A* GEOLOGICAL DATING Reconsider the geological dating equation

$$\frac{S(t) - S(0)}{R(t)} + 1 = e^{(\ln 2)t/\lambda}$$

from Exercise 47.

a. Solve the equation for t. Then use the Taylor polynomial of degree 2 for $\ln(1 + x)$ at $x = 0$ to estimate the value of t that satisfies the equation.

b. What adjustment would be necessary if your approximation were used to measure a value of t as large as or larger than the half-life of the radioactive substance used in the dating?

THINK ABOUT IT

HOW MANY CHILDREN WILL A COUPLE HAVE?

When planning a family, one consideration for some couples is that they prefer at least one boy and at least one girl. Call this the *gender requirement*. On average, how many children must be born to a couple before the gender requirement is met?

(Photo: PhotoDisc Collection/Getty Images)

We will answer this question in terms of the proportion of male births among all births in a particular population. If this proportion is m, where $0 < m < 1$, then the proportion of female births will be $f = 1 - m$. For example, in Canada, birth statistics show that the proportions are 51% male births versus 49% female, so that $m = 0.51$ and $f = 0.49$. (This ratio actually does vary by country from 0.53 in Armenia to 0.5 in Grenada and can be affected by male exposure to environmental toxins as well as other factors.) In the discussion that follows, we assume that the likelihood of a male child being born to a couple is m regardless of the gender of their previous children.

We first determine the possibilities when a couple continue to have children until the gender requirement is met. Suppose their first child is a boy. The couple then continue to have children until they have a girl. Consequently, the couple can have a boy followed by a girl, or two boys followed by a girl, or three boys followed by a girl, and so on. We denote these possibilities by *BG*, *BBG*, *BBBG*, …, where *B* denotes a boy child and *G* denotes a girl.

Our next step is to determine the proportion of each of these possibilities among all couples who continue to have children until the gender requirement is met. The proportion of such families who have a boy followed by a girl (*BG*) is

$$mf = m(1 - m),$$

while the proportion who have two boys followed by a girl (*BBG*) is

$$mmf = m^2(1 - m),$$

and, in general, the proportion of families that have $j - 1$ boys before having their first girl is

$$\underbrace{m\, m \cdots m}_{j-1\ \text{terms}} f = m^{j-1}(1 - m).$$

Similarly, the proportion of such families who have $j - 1$ girls before having their first boy is

$$\underbrace{f f \cdots f\, m}_{j-1\ \text{terms}} = f^{j-1}m = (1 - m)^{j-1}m.$$

Putting things together, we see that the proportion of couples that have $j - 1$ children of the same gender before having a child of the opposite gender is given by the sum

$$p(m, j) = m^{j-1}(1 - m) + (1 - m)^{j-1}m.$$

If we multiply the proportion $p(m, j)$ we have just found by the number of children j in the family at the time the gender requirement is met, we get the weighted proportion

$$p(m, j) = j[m^{j-1}(1 - m) + (1 - m)^{j-1}m].$$

Then the average size $A(m)$ of a family that satisfies the gender requirement is found by adding such weighted proportions for all possible family sizes, that is,

$$A(m) = \sum_{j=2}^{\infty} j[m^{j-1}(1 - m) + (1 - m)^{j-1}m],$$

where the sum begins at $j = 2$ since there must be at least two children in any family that satisfies the gender requirement. To see why such a sum represents an average, consider the simpler problem in which a class of students is asked to rate their instructor on a point scale from 1 (terrible) to 4 (excellent). Suppose the instructor receives 1 point from 10% of the students, 2 points from 40%, 3 points from 20%, and 4 points from 30%. Then the instructor's average score A is given by the sum of weighted proportions

$$A = 1(0.1) + 2(0.4) + 3(0.2) + 4(0.3) = 2.7,$$

so the instructor is rated as slightly above average by the students.

Now we shall determine the average size $A(m)$ of a family that satisfies the gender requirement by finding the sum of the infinite series $A(m)$:

$$A(m) = \sum_{j=2}^{\infty} j[m^{j-1}(1 - m) + (1 - m)^{j-1}m]$$

$$= \sum_{j=2}^{\infty} j[m^{j-1}(1 - m)] + \sum_{j=2}^{\infty} j[(1 - m)^{j-1}m]$$

$$= (1 - m) \sum_{j=2}^{\infty} jm^{j-1} + m \sum_{j=2}^{\infty} j(1 - m)^{j-1}$$

factor $1 - m$ from the left series and m from the right

THINK ABOUT IT

Notice that we have expressed the given series for $A(m)$ as the sum of two subseries, each of which has the form $\sum_{j=2}^{\infty} jx^{j-1}$, where $x = m$ for the first subseries and $x = 1 - m$ for the second. We can sum these series using the formula

$$\frac{1}{(1 - x)^2} = \sum_{j=1}^{\infty} jx^{j-1}$$

derived in Example 9.3.3. Applying this formula, we find that

$$\frac{1}{(1 - m)^2} = \sum_{j=1}^{\infty} jm^{j-1}$$

and

$$\frac{1}{m^2} = \frac{1}{[1 - (1 - m)]^2} = \sum_{j=1}^{\infty} j(1 - m)^{j-1}.$$

Because both these sums are missing their first term (that is, each sum begins with $j = 2$), we subtract off the $j = 1$ term in each sum to compute its value. We find that

$$\sum_{j=2}^{\infty} jm^{j-1} = \left(\sum_{j=1}^{\infty} jm^{j-1} \right) - 1(m^{1-1})$$

$$= \frac{1}{(1 - m)^2} - 1 = \frac{2m - m^2}{(1 - m)^2} = \frac{m(2 - m)}{(1 - m)^2}$$

and

$$\sum_{j=2}^{\infty} j(1 - m)^{j-1} = \sum_{j=1}^{\infty} j(1 - m)^{j-1} - 1(1 - m)^{1-1}$$

$$= \frac{1}{m^2} - 1 = \frac{1 - m^2}{m^2} = \frac{(1 + m)(1 - m)}{m^2}$$

It follows that the average number of children a couple will have if they decide to have children until they have at least one boy and one girl is

$$A(m) = (1 - m)m\frac{2 - m}{(1 - m)^2} + m(1 - m)\frac{1 + m}{m^2}$$

$$= \frac{m(2 - m)}{1 - m} + \frac{1 - m^2}{m} = \frac{1 - m + m^2}{m(1 - m)}$$

So what is the average number of children of such families when boys and girls are born in the same proportion, that is, when $m = \frac{1}{2}$? We find that

$$A\left(\frac{1}{2}\right) = \frac{1 - \frac{1}{2} + \left(\frac{1}{2}\right)^2}{\left(\frac{1}{2}\right)^2} = \frac{\frac{3}{4}}{\frac{1}{4}} = 3.$$

This essay was adapted from a discussion in Paul J. Nahin's fascinating little book, *Duelling Idiots and Other Probability Puzzlers,* Princeton, NJ: Princeton

THINK ABOUT IT

University Press, 2000. In our discussion, we have explicitly avoided using ideas and terminology from probability theory, but you may find it instructive to revisit this essay after encountering probability in Chapter 10.

Questions

1. Find the average number of children that will be born to a couple before the gender requirement is met if $m = 0.51$, the observed proportion of male births in Canada; that is, find $A(0.51)$.

2. Use calculus to show that the average size $A(m)$ of a family that satisfies the gender requirement is minimized when $m = f = \dfrac{1}{2}$.

3. What happens to the average size $A(m)$ of a family that satisfies the gender requirement as the proportion of male births m approaches 0? What happens as m approaches 1?

4. A couple decides to have children until they have a girl. On average, how many children will they have before this requirement is satisfied? Express your answer in terms of a formula involving m, the proportion of male births.

 Practise and learn online with Connect.

Probability can be used to measure waiting times, such as waiting for a red light to change or waiting for a doctor's appointment to start. See Examples 10.2.3 and 10.3.1 to find out more about waiting for red lights to change. (Photo: © Ingram Publishing/SuperStock)

PROBABILITY AND CALCULUS

CHAPTER OUTLINE

LEARNING OBJECTIVES

After completing this chapter, you should be able to

L01 Distinguish between discrete and continuous random variables. Describe the sample space for a given experiment. Find probabilities of discrete random variables.

L02 Determine whether a given function is a probability density function. Find probabilities using probability density functions.

L03 Evaluate the expected value and variance of a continuous random variable.

L04 Find probabilities using the Poisson distribution. Find probabilities using the normal distribution.

SECTION 10.1

Distinguish between discrete and continuous random variables. Describe the sample space for a given experiment. Find probabilities of discrete random variables.

Introduction to Probability; Discrete Random Variables

Uncertainty is a daily feature of every aspect of life. We cannot predict exactly what will happen when we flip a coin, when we make an investment, or when an insurance company underwrites a policy. The price we pay for a laptop may be specified exactly, but the length of time before the disk drive crashes, or we drop it, is uncertain, a random occurrence. **Probability theory** is the area of mathematics that deals with random behaviour. The primary goal of this chapter is to explore the use of calculus in probability, but first we need to introduce some basic concepts and terminology.

Outcomes, Events, and Sample Spaces

Mathematical modelling with probability involves the study of **random experiments,** that is, actions with various possible outcomes that can be observed and recorded. Each repetition of a random experiment is called a **trial.** Here are a few examples of random experiments.

1. Rolling a die and recording the number on the top face.
2. Sampling a randomly chosen production run and recording whether or not the item selected is defective.
3. Choosing a child at random from a particular school and recording the number of siblings that child has.

In describing an experiment, we use the following terminology.

Outcome, Event, and Sample Space ■ For a particular experiment, an **outcome** is a result of the experiment, and the **sample space** S is the collection of all possible outcomes. An **event** is a subset E of the sample space S, and the individual members of S are called **simple events** or **sample points.** The sample space itself called the **certain event,** and the empty set Ø, the subset that contains no outcomes, is an **impossible event.**

EXAMPLE 10.1.1

Suppose a box contains three balls, coloured red, black, and yellow. You conduct an experiment by selecting one of the balls from the box without looking. List the events and the sample space for this experiment.

Solution

Denote the balls by R, B, and Y, for red, black, and yellow, respectively. The sample space is the set of all outcomes $S = \{R, B, Y\}$. Events can be thought of as subsets of S. For instance, the simple events are the subsets $\{R\}$, $\{B\}$, and $\{Y\}$, and the event "the ball picked is red or black" is the subset $\{R, B\}$. The event "the ball picked is not red, not black, and not yellow" is the empty set $\{Ø\}$. Listing the set of all events in this fashion, we obtain

$$\{Ø\}, \{R\}, \{B\}, \{Y\}, \{R, B\}, \{R, Y\}, \{B, Y\}, \{R, B, Y\}.$$

Random Variables

A **random variable** X is a function that assigns a numerical value to each sample point in the sample space of a random experiment. A **discrete random variable** takes on values from a finite set of numbers or an infinite sequence of numbers, such as the positive integers. We shall study the properties of discrete random variables in this section, and in subsequent sections we shall examine **continuous random variables,** which take on a continuum of values, often an interval of real numbers. Example 10.1.2 illustrates how to distinguish between different types of random variables.

EXAMPLE 10.1.2

In each of the following cases, determine whether the given random variable X is discrete or continuous. If the random variable is discrete, state its value at each point of the sample space.

  a. A manufacturing process involves daily production runs of N units. Let X be the number of defective units in a randomly selected production run.

  b. Let X be the age of a cell chosen at random from a particular culture.

  c. Let X denote the time the hard drive of a randomly selected laptop works until it crashes.

  d. Let X count the number of times a randomly chosen coin must be tossed until heads comes up.

Solution

  a. The random variable X is discrete and finite because its possible values are 0, 1, 2, . . . , N.

  b. Since the age of a cell can be any positive real number, this random variable is continuous.

  c. The laptop hard drive may be defective or may crash at any time, so the random variable X can take on any nonnegative real number. Therefore, X must be continuous.

  d. When a coin is tossed repeatedly until it comes up heads, the possible outcomes consist of an immediate head or a string of tails terminating in a head. In other words, the sample space is {H, TH, TTH, TTTH, TTTTH, . . .}, and since X measures the number of times the coin is tossed before heads appear, $X(\text{H}) = 1$, $X(\text{TH}) = 2$, $X(\text{TTH}) = 3$, ... Since the set of values of X is the set of all positive integers, it follows that X is a discrete random variable, but it is not finite.

Probability Functions

The probability of an outcome is a numerical measure of the likelihood of that outcome occurring. Suppose a particular random experiment has outcomes $s_1, s_2, \ldots, s_n$, and that to each outcome s_k we assign a real number $P(s_k)$ in such a way that the following two conditions are satisfied:

$$0 \le P(s_k) \le 1,$$
$$P(s_1) + P(s_2) + \cdots + P(s_n) = 1.$$

Then $P(s_k)$ is called the **probability** of outcome s_k. The first required condition says that the likelihood of any outcome must be somewhere between impossible ($P = 0$) and certain ($P = 1$). The second requirement says that whenever the random experiment is performed, it is certain that exactly one outcome will occur.

More generally, for any event E, consisting of sample points $t_1, t_2, \ldots, t_k$, we have

$$P(E) = P(t_1) + P(t_2) + \cdots + P(t_k).$$

It follows that the **complement** of E, the event $\overline{E}$ consisting of all points in the sample space that are not in E, has probability

$$P(\overline{E}) = 1 - P(E),$$

since every outcome is either in E or in $\overline{E}$ and the sum of the probabilities of all outcomes is 1.

EXAMPLE 10.1.3

A fair die is thrown and the number on the top face is observed. The possible outcomes are 1, 2, 3, 4, 5, and 6, and we assign a probability of $\dfrac{1}{6}$ to each. Find the probability of the following events:

a. The event E_1 that the die shows either a 1 or a 2.
b. The event E_2 that the die shows 3, 4, 5, or 6.

Solution

a. The probability that the die shows a 1 or a 2 is

$$P(E_1) = \frac{1}{6} + \frac{1}{6} = \frac{1}{3}.$$

b. The event that the die shows 3, 4, 5, or 6 is the complement of the event that a 1 or a 2 shows. Thus, $E_2 = \overline{E_1}$ and

$$P(E_2) = P(\overline{E_1}) = 1 - P(E_1) = 1 - \frac{1}{3} = \frac{2}{3}.$$

We shall write $(X = c)$ to denote the event that contains all points of the sample space of an experiment for which the random variable X has the value c. For instance, the sample space for the random experiment of tossing a coin twice is {HH, HT, TH, TT}. If X is the random variable that counts the number of heads that occur, then the event $(X = 2)$ contains the single sample point HH because this is the only way to get two heads. Likewise, the event $(X = 1)$ contains both HT and TH, while $(X = 0)$ contains only TT.

Once probabilities have been assigned to the points in the sample space S of a random experiment, we can examine how the probability varies in relation to a particular discrete random variable X defined on S. This is done by studying the *probability function* for X, which can be defined as follows.

Probability Function ■ Suppose a probability assignment has been made for the sample space S of a particular random experiment, and let X be a discrete random variable defined on S. Then the **probability function** for X is a function p such that

$$p(x) = P(X = x)$$

for each value x assumed by X.

NOTE   A probability function p is also known as a *probability distribution*. Notice that since p is a probability, it must satisfy

$$0 \leq p(x) \leq 1,$$

and if X is a finite discrete random variable, then

$$p(x_1) + p(x_2) + \cdots + p(x_n) = 1,$$

where $x_1, x_2, \ldots, x_n$ are the values assumed by X.

EXAMPLE 10.1.4

Consider the random experiment of tossing a fair coin three times, and let X be the random variable that counts the number of heads that appear. Describe the probability function for X.

Solution

For a fair coin that is tossed three times, each of the eight outcomes

$$\text{HHH, HHT, HTT, HTH, THH, THT, TTH, TTT}$$

occurs with probability $\dfrac{1}{8}$. Find the associated probability function $p(x)$ by adding the probabilities of all the sample points for which X has the value x:

$$p(0) = P(X = 0) = P(\text{TTT}) = \frac{1}{8}$$

$$p(1) = P(X = 1) = P(\text{HTT}) + P(\text{THT}) + P(\text{TTH}) = \frac{1}{8} + \frac{1}{8} + \frac{1}{8} = \frac{3}{8}$$

$$p(2) = P(X = 2) = P(\text{HHT}) + P(\text{HTH}) + P(\text{THH}) = \frac{1}{8} + \frac{1}{8} + \frac{1}{8} = \frac{3}{8}$$

$$p(3) = P(X = 3) = P(\text{HHH}) = \frac{1}{8}$$

Note that

$$p(0) + p(1) + p(2) + p(3) = \frac{1}{8} + \frac{3}{8} + \frac{3}{8} + \frac{1}{8} = 1,$$

as required.

NOTE   There is no such thing as a natural probability assignment. For instance, consider the random experiment of flipping a coin, where the outcomes are heads (H) and tails (T). It may be reasonable to assign $P(\text{H}) = P(\text{T}) = \dfrac{1}{2}$ if the coin is fair. However, when we carry out an experiment of flipping a particular coin a large number of times, we may observe different relative frequencies for heads and tails. For example, if heads comes up 55% of the time in such an experiment, we may assign $P(\text{H}) = 0.55$ and $P(\text{T}) = 0.45$.

Histograms

One way to gain insight into the probability function for a discrete random variable X is to construct a graph showing the values of X and the probability assignments that correspond to those values. A **histogram** is a graphical representation obtained by first placing the values of the random variable X on a number line and then constructing above each such value n a vertical rectangle with width 1 and height equal to the probability $P(X = n)$. Note that

$$\text{Area of the } n\text{th rectangle} = 1 \cdot P(X = n) = P(X = n),$$

so the total area bounded by all the rectangles in the histogram is

$$\text{Total area} = \sum P(X = n) = 1,$$

where the sum is over all positive integers n for which $P(X = n) > 0$. The histogram for the random variable in Example 10.1.4 is shown in Figure 10.1.

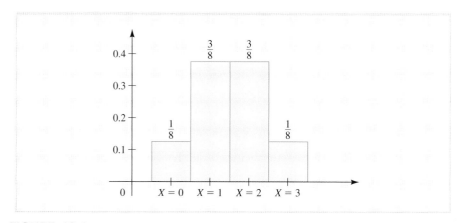

FIGURE 10.1   The histogram for the random variable in Example 10.1.4.

EXAMPLE 10.1.5

A study of the effect of a flu epidemic on the students living on a particular university residence floor shows that during the 30 days of November, there were

    4 days when no students had the flu
    3 days when exactly 1 student had the flu
    1 day when exactly 2 students had the flu
    12 days when exactly 6 students had the flu
    8 days when exactly 7 students had the flu
    2 days when exactly 8 students had the flu

Let X be the random variable that represents the number of students on the residence floor who had the flu on a randomly chosen day in November. Describe the event $(X = x)$ and compute the probabilities $P(X = x)$ for $x = 0, 1, 2, \ldots, 8$. Then draw the histogram for X.

Solution

For each value of x, $(X = x)$ is the event that on a randomly selected day in November, exactly x students on the residence floor had the flu, and $P(X = x)$ is the probability assignment for that event. We find that

$$P(X = 0) = \frac{4}{30} \qquad \text{for 4 days of the 30, } x = 0 \text{ students had the flu}$$

$$P(X = 1) = \frac{3}{30} \qquad \text{for 3 days of the 30, } x = 1 \text{ student had the flu}$$

$$P(X = 2) = \frac{1}{30} \qquad \text{for 1 day of the 30, } x = 2 \text{ students had the flu}$$

$$P(X = 3) = P(X = 4) \qquad \text{there are no days when } x = 3, x = 4, \text{ or } x = 5$$
$$= P(X = 5) = 0 \qquad \text{students had the flu}$$

$$P(X = 6) = \frac{12}{30} \qquad \text{for 12 days of the 30, } x = 6 \text{ students had the flu}$$

$$P(X = 7) = \frac{8}{30} \qquad \text{for 8 days of the 30, } x = 7 \text{ students had the flu}$$

$$P(X = 8) = \frac{2}{30} \qquad \text{for 2 days of the 30, } x = 8 \text{ students had the flu}$$

The histogram for this random variable is shown in Figure 10.2.

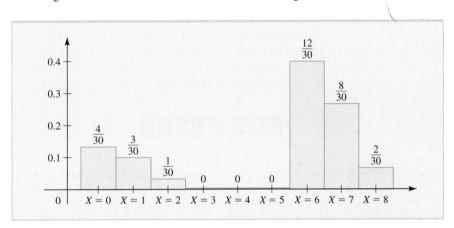

FIGURE 10.2   The histogram for the flu epidemic random variable in Example 10.1.5.

Expected Value   For the flu epidemic in Example 10.1.5, how many students would you expect to be ill on a randomly chosen day in November? From the calculation in the example, we know that 0 students were ill $\frac{4}{30}$ of the time during the month, 1 student was ill $\frac{3}{30}$ of the time, 2 students were ill $\frac{1}{30}$ of the time, and so on. Therefore, we compute the expected number of ill students on our randomly selected day by evaluating the weighted sum

$$0 \cdot P(X = 0) + 1 \cdot P(X = 1) + 2 \cdot P(X = 2) + 3 \cdot P(X = 3) + 4 \cdot P(X = 4)$$
$$+ 5 \cdot P(X = 5) + 6 \cdot P(X = 6) + 7 \cdot P(X = 7) + 8 \cdot P(X = 8)$$

$$= 0\left(\frac{4}{30}\right) + 1\left(\frac{3}{30}\right) + 2\left(\frac{1}{30}\right) + 3(0) + 4(0) + 5(0) + 6\left(\frac{12}{30}\right) + 7\left(\frac{8}{30}\right) + 8\left(\frac{2}{30}\right)$$

$$= \frac{149}{30} \approx 5$$

Thus, if we repeatedly pick days in November at random, on average we would expect to find about 5 students ill.

What we have computed in this illustration is the *expected value* of the random variable X that represents the number of students who were ill on a randomly selected day in April. Here is a definition of expected value.

Expected Value of a Finite Discrete Random Variable ■ Let X be a finite discrete random variable with values $x_1, x_2, \ldots, x_n$, and let p be the probability function of X, so that $p(x_k) = P(X = x_k)$ for $k = 1, 2, \ldots, n$. Then the expected value of X is given by the weighted sum

$$E(X) = x_1 p(x_1) + x_2 p(x_2) + \cdots + x_n p(x_n) = \sum_{k=1}^{n} x_k p(x_k).$$

The *arithmetic average* or *mean* of a set of values is found by adding the values and dividing by the number of values in the sum. Intuitively, the mean may be thought of as the centre of the values in the set. In more advanced treatments of probability theory, it is shown that when an experiment is carried out repeatedly, the arithmetic average or mean of the observed values of a random variable X will grow closer and closer to the expected value $E(X)$. In this sense, the expected value is a measure of *central tendency* for the random variable. Expected value is computed and interpreted in Examples 10.1.6 and 10.1.7.

EXAMPLE 10.1.6

Continuing Example 10.1.4, find the expected value of the random variable X that represents the number of heads that come up when a fair coin is flipped three times.

Solution
Using the formula for $E(X)$ and the probabilities found in Example 10.1.2 gives

$$E(X) = 0 \cdot P(X = 0) + 1 \cdot P(X = 1) + 2 \cdot P(X = 2) + 3 \cdot P(X = 3)$$
$$= 0\left(\frac{1}{8}\right) + 1\left(\frac{3}{8}\right) + 2\left(\frac{3}{8}\right) + 3\left(\frac{1}{8}\right)$$
$$= \frac{3}{2}$$

When the experiment of tossing a coin three times is repeatedly carried out, the average number of heads that come up will grow closer and closer to $\frac{3}{2}$.

EXAMPLE 10.1.7

The grand prize of a lottery is $1000, and there are also three second prizes each worth $500 and five third prizes each worth $100. Suppose that 10 000 lottery tickets are sold, one of which is to you. How much should you be willing to pay for your ticket?

Solution

A fair price to pay would be the expected value $E(X)$ of the random variable X that represents the payoff of a randomly selected ticket. This is because if the lottery were held a large number of times and you paid $E(X)$ dollars for each ticket, then your average winnings or losses would be approximately 0.

Computing the payoff probabilities results in

$$p(1000) = P(X = 1000) = \frac{1}{10\ 000} \qquad \text{since only 1 ticket of the 10 000 wins \$1000}$$

$$p(500) = P(X = 500) = \frac{3}{10\ 000} \qquad \text{since 3 of the 10 000 tickets win \$500}$$

$$p(100) = P(X = 100) = \frac{5}{10\ 000} \qquad \text{since 5 of the 10 000 tickets win \$100}$$

$$p(0) = P(X = 0) = \frac{9991}{10\ 000} \qquad \text{since the other 9991 tickets win nothing}$$

Using this information, the expected value of X is

$$E(X) = 0p(0) + 100p(100) + 500p(500) + 1000p(1000)$$

$$= 0\left(\frac{9991}{10\ 000}\right) + 100\left(\frac{5}{10\ 000}\right) + 500\left(\frac{3}{10\ 000}\right) + 1000\left(\frac{1}{10\ 000}\right)$$

$$= \frac{3}{10}$$

It follows that as more and more of the 10 000 available tickets are purchased, the average amount to be won gets closer and closer to \$0.30. Therefore, a fair price for your ticket is 30 cents.

Variance and Standard Deviation

After we have found the expected value of a random variable X, we can learn more about X by examining how its values are spread around the expected value. We now introduce the notion of the *variance* of a random variable, which is small when the values of the random variable are close to its expected value and large when these values are spread widely around the expected value. In this sense, it is a measure of *dispersion,* as is the square root of the variance, which is called the *standard deviation.*

> **Variance and Standard Deviation of a Finite Discrete Random Variable** ■ Let X be a finite discrete random variable with values $x_1, x_2, \ldots, x_n$ and expected value $E(X) = \mu$, and let p be the probability function for X, so that $p(x_k) = P(X = x_k)$ for $k = 1, 2, \ldots, n$. Then the **variance** of X, denoted by $\mathrm{Var}(X)$, is the weighted sum
>
> $$\mathrm{Var}(X) = (x_1 - \mu)^2 \cdot p(x_1) + (x_2 - \mu)^2 \cdot p(x_2) + \cdots + (x_n - \mu)^2 \cdot p(x_n)$$
>
> $$= \sum_{k=1}^{n} (x_k - \mu)^2 \cdot p(x_k)$$
>
> The square root of the variance is called the **standard deviation** of X and is denoted by $\sigma(X)$; that is,
>
> $$\sigma(X) = \sqrt{\mathrm{Var}(X)}.$$

Note that the variance $\text{Var}(X)$ of a discrete random variable X with values $x_1, x_2, \ldots, x_n$ and probability assignments $p(x_1), \ldots, p(x_n)$ is the weighted sum of squares $(x_k - \mu)^2 p(x_k)$ of the differences between each x_k and the mean $\mu = E(X)$. In this sense, the variance measures the deviation of X from its mean μ. You may wonder why this deviation is not measured by simply adding the weighted differences $(x_k - \mu)p(x_k)$ without squaring, but this sum, $\sum\limits_{k=1}^{n}(x_k - \mu)p(x_k)$, will always be 0 (see Exercise 48).

EXAMPLE 10.1.8

Find the variance and standard deviation of the random variable X that represents the number of heads that come up when a coin is flipped three times.

Solution

In Example 10.1.6, we showed that the expected value of X satisfies $E(X) = \dfrac{3}{2}$. Therefore, the variance of X is given by

$$\text{Var}(X) = \left(0 - \frac{3}{2}\right)^2 \cdot \frac{1}{8} + \left(1 - \frac{3}{2}\right)^2 \cdot \frac{3}{8} + \left(2 - \frac{3}{2}\right)^2 \cdot \frac{3}{8} + \left(3 - \frac{3}{2}\right)^2 \cdot \frac{1}{8} = \frac{3}{4}.$$

Find the standard deviation by taking the square root of the variance, so that

$$\sigma(X) = \sqrt{\text{Var}(X)} = \frac{\sqrt{3}}{2}.$$

An *infinite discrete random variable* X is one whose values $x_1, x_2, \ldots$ form an infinite sequence. If the probability assignment for such a random variable is $p(x_k) = P(X = x_k)$, then the expected value of X is given by the infinite series

$$E(X) = x_1 p(x_1) + x_2 p(x_2) + \cdots = \sum_{k=1}^{\infty} x_k p(x_k).$$

If we set $\mu = E(X)$, then the variance of such a random variable X is given by the infinite series

$$\text{Var}(X) = (x_1 - \mu)^2 \cdot p(x_1) + (x_2 - \mu)^2 \cdot p(x_2) + \cdots = \sum_{k=1}^{\infty} (x_k - \mu)^2 \cdot p(x_k),$$

and, as before, the standard deviation of X is given by $\sigma(X) = \sqrt{\text{Var}(X)}$. These formulas will be used in the following discussion of geometric random variables.

Geometric Random Variables

Certain random experiments that occur in modelling can be characterized as having two possible outcomes, traditionally called **success** and **failure,** and are carried out repeatedly until the first success is achieved. For example, a coin may be tossed repeatedly until tails first appears or newly manufactured computer chips may be examined until a defective chip is found. In such situations, the random variable X that counts the number of repetitions required to produce the first success is known as a **geometric random variable.**

A geometric random variable can take on any positive integer value, so it is discrete but not finite. For any positive integer n, the probability that $X = n$ is the probability that the first $n - 1$ trials of the experiment result in failure and the nth trial is a success:

$$\underbrace{FF \cdots F}_{\substack{n - 1 \\ \text{failures}}} \cdot \underbrace{S}_{\substack{\text{first} \\ \text{success}}}$$

If, on each trial, the probability of success is p, then the probability of failure must be $1 - p$, so the probability of $n - 1$ failures followed by 1 success is the product

$$\underbrace{[(1 - p)(1 - p) \cdots (1 - p)]}_{n - 1 \text{ failures}} \cdot \underbrace{p}_{\text{success}} = (1 - p)^{n-1}p$$

That is,

$$P(X = n) = (1 - p)^{n-1}p.$$

If $p = 1$, success occurs immediately, while if $p = 0$, success is impossible. Otherwise, where $0 < p < 1$, we expect success to eventually occur if the experiment is repeated indefinitely, so $P(X \geq 1)$ ought to be 1. To verify that $P(X \geq 1) = 1$, we compute this probability by adding all the probabilities $P(X = n)$ for $n = 1, 2, \ldots$. This leads to the following infinite geometric series, which we sum using the geometric series formula of Section 9.1:

$$
\begin{aligned}
P(X \geq 1) &= \sum_{n=1}^{\infty} P(X = n) \\
&= \sum_{n=1}^{\infty} (1 - p)^{n-1}p \\
&= p + (1 - p)p + (1 - p)^2 p + \cdots \\
&= p[1 + (1 - p) + (1 - p)^2 + \cdots] \\
&= p \sum_{n=0}^{\infty} (1 - p)^n \\
&= \frac{p}{1 - (1 - p)} = \frac{p}{p} = 1
\end{aligned}
$$

Note that in the last step, we used the geometric series formula $\sum_{n=0}^{\infty} ar^n = \dfrac{a}{1 - r}$ with $a = p$ and $r = 1 - p$, which is valid since $0 < p < 1$ implies that $0 < 1 - p < 1$.

The random variable X such that $P(X = n) = (1 - p)^{n-1}p$ is called a **geometric random variable with parameter p.** The histogram for the first seven values of a geometric random variable with parameter $p = 0.5$ is shown in Figure 10.3. Using techniques for working with power series developed in Chapter 9 (see Exercise 47), it can be shown that a geometric random variable X with parameter p has expected value

$$E(X) = \frac{1}{p}.$$

Similar, but slightly more complicated computations can be used to show that X has variance

$$\text{Var}(X) = \frac{1 - p}{p^2}.$$

Applications of geometric random variables are examined in Examples 10.1.9 and 10.1.10.

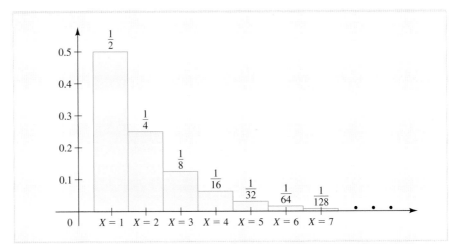

FIGURE 10.3   The histogram for the geometric random variable with parameter $p = 0.5$.

EXAMPLE 10.1.9

In an industrial process, items are sampled repeatedly until a defective item is found. The company claims that any item of its product is 100 times more likely to be effective than defective. Assuming the company's claim is accurate, find the following:

a. The probability that a randomly chosen item is defective.

b. The probability that it takes exactly 100 samplings before a defective item is found.

c. The expected number of samplings before a defective item is found.

Solution

a. If p is the probability of a defective item, then $1 - p$ is the probability of an effective item, and the company claims that

$$1 - p = 100p,$$

so that

$$p = \frac{1}{101} \approx 0.0099.$$

b. Let X be the geometric random variable that counts the number of items that must be sampled before a defective item is found. The probability that it takes exactly 100 samplings before a defective item is found is

$$P(X = 100) = (1 - p)^{100-1}p \approx (1 - 0.0099)^{99}(0.0099) \approx 0.0037.$$

c. From the formula $E(X) = \dfrac{1}{p}$ for the expected value of a geometric random variable

of parameter p, since $p = \dfrac{1}{101}$, the number of expected samplings is

$$E(p) = \frac{1}{\dfrac{1}{101}} = 101.$$

That is, if the random sampling is carried out repeatedly, on average we would expect to find the first defective item on the 101st sampling.

EXAMPLE 10.1.10

Suppose that a fair coin is tossed repeatedly until it comes up heads. Find each of the following quantities:

a. The probability that the coin must be tossed exactly three times until heads comes up.

b. The probability that the coin comes up heads within the first three tosses.

c. The probability that the coin must be tossed more than three times to come up heads.

d. The expected number of times the coin must be tossed before heads appears.

Solution

Let X be the random variable that represents the number of times the coin must be tossed until it comes up heads. This is a geometric random variable with parameter $p = \dfrac{1}{2}$, so

$$P(X = n) = \left(1 - \frac{1}{2}\right)^{n-1}\left(\frac{1}{2}\right) = \left(\frac{1}{2}\right)^n.$$

a. The required probability is $P(X = 3) = \left(\dfrac{1}{2}\right)^3 = \dfrac{1}{8}$.

b. The coin comes up heads within the first three tosses if heads appears on toss 1, toss 2, or toss 3. This is the event $X \leq 3$, so

$$P(X \leq 3) = P(X = 1) + P(X = 2) + P(X = 3)$$
$$= \frac{1}{2} + \left(\frac{1}{2}\right)^2 + \left(\frac{1}{2}\right)^3 = \frac{1}{2} + \frac{1}{4} + \frac{1}{8}$$
$$= \frac{7}{8}$$

c. The event $X > 3$ that the coin must be tossed more than three times to come up heads is the complement of the event $X \leq 3$ that it comes up heads within the first three tosses, so $P(X > 3) = 1 - P(X \leq 3)$. Since $P(X \leq 3) = \dfrac{7}{8}$ from part (b), it follows that

$$P(X > 3) = 1 - P(X \leq 3) = 1 - \frac{7}{8} = \frac{1}{8}.$$

d. From the formula for the expected value of a geometric random variable with parameter $p = \dfrac{1}{2}$, $E(X) = \dfrac{1}{p} = \dfrac{1}{1/2} = 2$. Consequently, if this random experiment is carried out repeatedly, on average the coin should have to be tossed only twice before heads appears.

EXAMPLE 10.1.11

In a certain two-person board game played with 1 die, each player needs to roll a 6 to start play. You and your opponent take turns rolling the die. Find the probability that you will get the first 6 if you roll first.

Solution

In this context, *success* is getting a 6 and *failure* is getting any of the other five possible numbers. Hence, the probability of success is $p = \dfrac{1}{6}$ and the probability of failure is $1 - p = \dfrac{5}{6}$. Let X be the random variable that counts the number of the roll on which the first 6 comes up. The first 6 will occur on one of your rolls if X is odd and on one of your opponent's rolls if X is even. Hence, the probability that you get the first 6 is

$$P(X \text{ is odd}) = P(X = 1) + P(X = 3) + P(X = 5) + \cdots$$

$$= \frac{1}{6} + \left(\frac{5}{6}\right)^2\left(\frac{1}{6}\right) + \left(\frac{5}{6}\right)^4\left(\frac{1}{6}\right) + \cdots$$

$$= \frac{1}{6} + \left(\frac{25}{36}\right)\left(\frac{1}{6}\right) + \left(\frac{25}{36}\right)^2\left(\frac{1}{6}\right) + \cdots$$

$$= \frac{1}{6}\sum_{n=0}^{\infty}\left(\frac{25}{36}\right)^n$$

$$= \frac{1}{6}\left(\frac{1}{1 - \dfrac{25}{36}}\right) = \frac{1}{6}\left(\frac{1}{\dfrac{11}{36}}\right) = \frac{1}{6}\left(\frac{36}{11}\right) = \frac{6}{11} \approx 0.5455$$

EXERCISES ■ 10.1

In Exercises 1 through 10, determine whether the given random variable X is discrete or continuous.

1. X counts the number of tails that come up when a coin is tossed 10 times.

2. X measures the time a randomly selected flashlight battery works before failing.

3. X measures the weight of a randomly selected member of your class.

4. X measures the income in dollars of a randomly selected person in your community.

5. X measures the living area of a randomly selected house in your neighbourhood.

6. X measures the length of time a randomly selected traveller has to wait until the next airport shuttle arrives.

7. X counts the number of players on a hockey team who are injured in a randomly selected game.

8. X counts the number of birds in a randomly selected flock.

9. X counts the number of passengers on a randomly selected airplane.

10. X measures the speed of a randomly selected car passing a police speed trap.

In Exercises 11 through 14, a die is rolled twice. In each case, describe the sample space for the indicated random experiment.

11. Observe how many 6s are thrown.

12. Observe how many 1s and 4s are thrown.

13. Observe how many even numbers are thrown.

14. Observe the sum of the numbers on the two throws.

15. A box contains 4 tickets numbered 1 through 4. A ticket is selected at random from the box, a die is rolled, and the numbers on the ticket and die are recorded.
   a. List the sample space for this experiment.
   b. List the outcomes that make up the event that the sum of the numbers on the ticket and die is even.
   c. List the outcomes that make up the event that the number on the ticket is the same as the number on the die.

16. A number between 15 and 49, inclusive, is selected at random. List the outcomes that make up each event.
   a. The two digits of the selected number are the same.
   b. The sum of the digits of the selected number is 8.

17. Consider the random experiment of tossing a fair coin four times, and let X denote the random variable that counts the number of times tails appears.
   a. List all possible outcomes of this experiment, and the value that X assigns to each outcome.
   b. List the outcomes that make up the event $(X = 3)$.
   c. List the outcomes that make up the event $(X \leq 4)$.
   d. Find each of these probabilities:
      $P(X = 0)$, $P(X \leq 3)$, and $P(X > 3)$

18. Consider the random experiment of rolling a fair die three times, and let X denote the random variable that counts the number of times a 6 appears.
   a. List all possible outcomes of this experiment and the value that X assigns to each outcome.
   b. List the outcomes that make up the event $(X = 2)$.
   c. List the outcomes that make up the event $(X < 3)$.
   d. Find each of these probabilities:
      $P(X = 0)$, $P(X \leq 2)$, and $P(X > 2)$

In Exercises 19 through 22, the outcomes and corresponding probability assignments for a discrete random variable X are given. In each case, find the expected value E(X), variance Var(X), and standard deviation σ(X).

19.

Outcomes for X	0	1	2	4
Probability	$\frac{1}{8}$	$\frac{1}{8}$	$\frac{1}{4}$	$\frac{1}{2}$

20.

Outcomes for X	1	2	3	5
Probability	$\frac{1}{4}$	$\frac{1}{3}$	$\frac{1}{6}$	$\frac{1}{4}$

21.

Outcomes for X	0	2	4	6
Probability	$\frac{1}{7}$	$\frac{3}{7}$	$\frac{2}{7}$	$\frac{1}{7}$

22.

Outcomes for X	0	1	2	$\cdots$	k	$\cdots$
Probability	$\frac{1}{2}$	$\frac{1}{4}$	$\frac{1}{8}$	$\cdots$	$\frac{1}{2^{k+1}}$	$\cdots$

23. Draw the probability histogram for the random variable X in Exercise 17. Then find the expected value, variance, and standard deviation for X.

24. Draw the probability histogram for the random variable X in Exercise 18. Then find the expected value, variance, and standard deviation for X.

25. **LOTTERY** Suppose that the grand prize in a lottery is \$1 000 000. There are also 5 second prizes of \$100 000 and 50 third prizes of \$10 000. If 100 000 tickets are sold, what is a fair price to pay for a ticket to this lottery?

26. **CASINO GAMES** A roulette wheel has 18 red numbers, 18 black numbers, and 2 green numbers. Each time the wheel is spun, Gerry bets that the ball will land on a red number. When the ball lands on a red number, he wins \$1 from the casino for every \$1 bet. When the ball lands on a black or green number, he loses his bet. Find his expected winnings from a \$1 bet on red.

27. **CASINO GAMES** Gerry, the roulette player in Exercise 26, finds another wheel with the same number of red and black numbers but with only 1 green number. Now what are his expected winnings from a \$1 bet on red?

28. **RESTAURANT BUSINESS** The random variable X that represents the number of daily orders for lobster at a particular restaurant has the following probability assignments:
      $P(X = 0) = 0.05$, $P(X = 1) = 0.15$,
      $P(X = 2) = 0.20$, $P(X = 3) = 0.25$,

$$P(X = 4) = 0.12, P(X = 5) = 0.13,$$
$$P(X = 6) = 0.05, P(X = 7) = 0.02,$$
$$P(X = 8) = 0.03$$

a. Describe the event $(X = x)$.

b. Draw the histogram for this random variable.

c. Find $E(X)$, the expected number of daily lobster orders at this restaurant.

d. Find the variance and the standard deviation of the number of daily lobster orders at the restaurant.

29. **HIGHWAY SAFETY** The highway department kept track of the number of accidents on a 10-km stretch of a major highway every day during the month of September. They found

   6 days with no accidents

   8 days with one accident

   7 days with two accidents

   2 days with three accidents

   3 days with four accidents

   0 days with five accidents

   2 days with six accidents

   0 days with seven accidents

   1 day with eight accidents

   1 day with nine accidents

   Let X be the random variable that represents the number of accidents on a randomly chosen day in September.

   a. Construct a histogram for X.

   b. Find the expected value $E(X)$ and interpret this number.

   c. Find the variance and standard deviation of X.

30. Find the variance and standard deviation of the random variable X from Example 10.1.5 that represents the number of students on a particular residence floor who are ill with the flu in November.

31. **AIRLINE TRAVEL** Let X represent the number of passengers bumped from a randomly chosen flight on a particular commercial airline. Suppose the probability $P(X = N)$ that N people ($N = 0$ to 6) are bumped is given in the table.

N	0	1	2	3	4	5	6
$P(X = N)$	0.961	0.015	0.009	0.005	0.004	0.004	0.002

   Based on this information, how many people on average would you expect to be bumped from a randomly chosen flight?

32. **BOARD GAMES** There are 100 tiles in the board game Scrabble; 98 of these tiles have letters of the English alphabet printed on them and 2 are blank. Each tile also has a value, depending on the letter printed on it. Blank tiles have zero value. The following table shows the number of tiles having each value.

Value	Number of Tiles
0	2
1	68
2	7
3	8
4	10
5	1
6	0
7	0
8	2
9	0
10	2

   a. Let X be the random variable that represents the value of a randomly selected Scrabble tile from all 100 tiles. Draw the histogram of X.

   b. Find $E(X)$, the expected value of a randomly selected Scrabble tile.

FAIR BETS AND ODDS *A gaming bet is said to be* ***fair*** *if, on average, a player's expected winnings are 0 (no loss or gain). The bet has m to 1* ***odds*** *if for every $1 wagered, a winning player gains m dollars (in addition to the return of the $1 bet).* ***Fair odds*** *on a bet are the odds that make the bet fair. This terminology is used in Exercises 33 through 35.*

33. For a certain game, the odds are w to 1, and the probability of winning a particular bet is p.

   a. Let X be the random variable representing a player's winnings when betting $1. Express the expected value $E(X)$ in terms of w.

   b. Show that the fair odds for this bet are $\dfrac{1 - p}{p}$ to 1.

34. Use the formula found in Exercise 33 to find fair odds for the game of roulette with two green numbers as described in Exercise 26.

35. Use the formula found in Exercise 33 to find fair odds for the game of roulette with one green number as described in Exercise 27.

36. Let X be a geometric random variable with parameter $p = 0.4$. Compute $P(X = n)$ for $n = 1$ to 5 and then draw the histogram for the first five values of X.

37. Let X be a geometric random variable with parameter $p = 0.8$. Compute $P(X = n)$ for $n = 1$ to 8 and then draw the histogram for the first eight values of X.

38. **BIASED COINS**   A biased coin that comes up heads with probability 0.55 is tossed repeatedly until it comes up heads. Let X be the random variable that measures the number of times the coin must be tossed until it comes up heads. Find

   a. The probability that the coin must be tossed exactly 3 times until heads comes up.

   b. The probability that the coin comes up heads within the first 3 tosses.

   c. The probability that the coin must be tossed more than 3 times to come up heads.

   d. The expected number of times the coin must be tossed until it comes up heads.

   e. The variance of the number of times the coin must be tossed until it comes up heads.

39. **RELIABILITY**   Suppose that 7% of the computer chips manufactured by a certain company are defective. A quality control inspector tests each chip before it goes out of the factory. Using testing equipment, this inspector can always detect a defective chip.

   a. Find the probability that the inspector passes exactly 4 chips as good before detecting a defective chip.

   b. Find the probability that the inspector passes at most 4 chips as good before detecting a defective chip.

   c. Find the probability that the inspector passes at least 4 chips as good before detecting a defective chip.

   d. Find the expected number of chips the inspector examines before detecting a defective chip.

40. Show that for a geometric random variable X, the most likely number of trial runs is 1. That is, $P(X = n)$ takes on its largest value when $n = 1$, as n ranges over all positive integers.

41. **RELIABILITY**   Suppose that a particular printer in a computer lab has a 99.95% chance of successfully printing out a randomly selected page in a document and a 0.05% chance of jamming.

   a. Find the probability that the printer will jam on the 10th page of a 10-page document after printing out the first 9 pages without jamming.

   b. On what page would you expect the printer to jam when a large document is printed out?

   c. Determine the probability that the printer completes a document of 25 pages without jamming.

42. **JURY SELECTION**   Potential jurors from a large jury pool are examined one at a time by the defence and prosecution. Suppose that a person selected at random from the jury pool has probability $p = 0.25$ of being disqualified.

   a. Find the probability that the third potential juror examined is the first to be disqualified.

   b. Find the probability that at least one of the first three jurors is disqualified.

   c. Find the probability that a jury of 12 people can be seated without any potential jurors being disqualified.

43. **POPULATION STUDIES**   Suppose 4% of the students at a large university have red hair. You interview a number of students, one at a time.

   a. Find the probability that the fifth student interviewed is the first to have red hair.

   b. Find the probability that at least one student with red hair is among the first five students interviewed.

   c. Find the probability that there are no red-haired students among the first 12 interviewees.

   d. On average, how many students would you expect to interview before finding one with red hair?

44. **GAMES OF CHANCE**   You and a friend take turns rolling a die until one of you wins by rolling a 3 or a 4. If your friend rolls first, find the probability that you will win.

45. **NUMISMATICS**   Between 1953 and 1965, Canadian coins had a picture of a young Queen Elizabeth with her hair tied back. Today, very few of these older coins are still in circulation. If only 2% of the coins now in circulation were minted in this period, find the probability that you will have to examine at least 100 coins to find one with the young Queen Elizabeth. (Incidentally, there is an easy way to do this exercise that does not involve infinite series. For practice, do it both ways.)

46. **QUALITY CONTROL**   Three inspectors take turns checking electronic components as they come off an assembly line. If 10% of all components produced on the assembly line are defective, find the probability that the inspector who checks the first (and the fourth, seventh, and so on) component will be the one who finds the first defective component.

47. **A*** In Section 9.3, it is shown that

$$1 + 2x + 3x^2 + \cdots = \frac{1}{(1 - x)^2} \text{ for } -1 < x < 1.$$

Use this fact to show that a geometric random variable X with parameter p has expected value $E(X) = \dfrac{1}{p}$.

48. **A*** Let X be a discrete random variable with values $x_1, x_2, \ldots, x_n$ and probability assignments $p(x_1) = p_1, \ldots, p(x_n) = p_n$. If $\mu = E(X)$ is the expected value of X, show that

$$(x_1 - \mu)p_1 + (x_2 - \mu)p_2 + \cdots + (x_n - \mu)p_n = 0.$$

*A random variable X follows a **zeta distribution** if there are positive integers a and b with b > 1 such that the event (X = n) has probability*

$$P(X = n) = \frac{a}{n^b}$$

for all positive integers n. For instance, the frequency of occurrence of words in English ranked according to their use approximately follows a zeta distribution. This definition is used in Exercises 49 and 50.

49. **A*** Let X be a random variable with a zeta distribution. Use the requirement that

$$\sum_{n=1}^{\infty} P(X = n) = 1$$

to show that $a = \dfrac{1}{\zeta(b)}$, where

$$\zeta(b) = \sum_{n=1}^{\infty} \frac{1}{n^b}.$$

50. **A* THE INTERNET**   A website is said to have *rank n* if it is the *n*th most popular site. Consider the random experiment in which someone currently surfing the web selects a website and let X be the random variable that measures the rank of the selected website. Recent studies suggest that X follows a zeta distribution with parameter $b = 2$.
    a. It can be shown that

    $$\zeta(2) = \sum_{n=1}^{\infty} \frac{1}{n^2} = \frac{\pi^2}{6}.$$

    Use this fact and the result of Exercise 49 to find a formula for the probability $P(X = n)$ that a randomly chosen visit to the Internet is to the *n*th most popular site.
    b. Find the probability that a random visit to a website is to the fifth most popular site.
    c. Find the probability that a random visit to a website is to one of the three most popular sites.
    d. Find the probability that a random visit to a website is to a site not ranked among the four most popular.
    e. What expression represents the expected value of X? Use a calculator to estimate $E(X)$, and interpret your result.

SECTION 10.2

L02

Determine whether a given function is a probability density function. Find probabilities using probability density functions.

Continuous Random Variables

In Section 10.1, we studied discrete random variables as a means for introducing basic probabilistic concepts and procedures. Beginning with this section, our focus will be on continuous random variables and the use of calculus in the study of probability.

Recall that a continuous random variable is one whose values are taken from a continuum of numbers, such as an interval. An important difference between discrete and continuous probability distributions is the manner in which probability is assigned.

When a fair coin is tossed, the probability of heads is $\dfrac{1}{2}$, but what probability would you assign to the event that *exactly* 4.31 cm of rain falls in Regina during a particular day? Such a question is unreasonable because the continuous random variable that measures the rainfall can assume any value between, say, 0 cm (no rain) and 30 cm (flash flood), so its probability distribution cannot possibly be given as a list. It makes more sense to ask for the probability that the observed rainfall lies between 4 and 5 cm or that it is less than 6 cm.

To obtain the probability distribution for a continuous random variable X, we can partition the domain of X into subintervals and construct on each a bar representing the probability that X assumes a value in that subinterval. This creates a histogram for the data (Figure 10.4a), and if we take smaller and smaller subintervals, the top boundary of this histogram suggests a curve, the graph of a function $f(x)$ called the **probability density function** for X. We then define the probability that X lies between any two numbers a and b in its range as the area under the graph of the probability density function $f(x)$ over the interval $a \leq x \leq b$. That is,

$$P(a \leq X \leq b) = \{\text{area under the graph of } f(x) \text{ over } a \leq x \leq b\},$$

as indicated in Figure 10.4b.

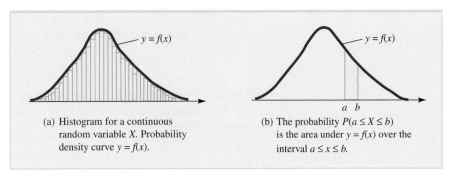

(a) Histogram for a continuous random variable X. Probability density curve $y = f(x)$.

(b) The probability $P(a \leq X \leq b)$ is the area under $y = f(x)$ over the interval $a \leq x \leq b$.

FIGURE 10.4   Determination of the probability density function for a continuous random variable X.

Finding a formula for the most suitable probability density function for a particular continuous random variable is a matter for a course in probability and statistics. However, once this determination has been made, the area that yields probability is computed by integration, specifically,

$$P(a \leq X \leq b) = \int_a^b f(x)\, dx.$$

For example, Figure 10.5 shows the graph of a possible probability density function $f(x)$ for the random variable X that measures the lifespan of an incandescent light bulb selected randomly from a manufacturer's stock. Its shape reflects the fact that most of this type of bulb burn out relatively quickly. The probability $P(0 \leq X \leq 40)$ that a bulb will fail within the first 40 hours is given by the (shaded) area under the curve between $x = 0$ and $x = 40$ and is computed by the definite integral of $f(x)$ over the interval $a \leq x \leq b$:

$$P(0 \leq X \leq 40) = \int_0^{40} f(x)\, dx.$$

The smaller area on the right in the figure represents the probability that the selected bulb fails between the 80th and 120th hours of use and is given by the integral

$$P(80 \leq X \leq 120) = \int_{80}^{120} f(x)\, dx.$$

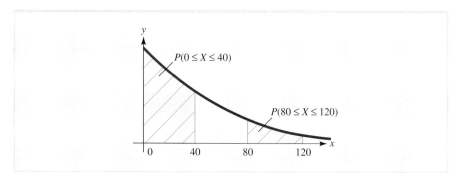

FIGURE 10.5 A possible probability density function for the lifespan of a light bulb.

Here is the definition of the probability density function in terms of the integrals that you will use to compute the areas representing probability.

Continuous Probability Density Function ■ A probability density function for the continuous random variable X is a function f that satisfies the following three conditions:

1. $f(x) \geq 0$ for all x.

2. $\displaystyle\int_{-\infty}^{+\infty} f(x)\,dx = 1$

3. The probability that X lies in the interval $a \leq X \leq b$ is given by the integral

$$P(a \leq X \leq b) = \int_a^b f(x)\,dx.$$

The values of a and b in this formula need not be finite, and if either is infinite, the corresponding probability is given by an improper integral. For example, the probability that $X \geq a$ is

$$P(X \geq a) = P(a \leq X < +\infty) = \int_a^{+\infty} f(x)\,dx.$$

The second condition in the definition of a probability density function expresses the fact that since X must have some value, the event $-\infty < X < +\infty$ is certain to occur. Thus, we have $P(-\infty < X < +\infty) = 1$, which means that

$$\int_{-\infty}^{+\infty} f(x)\,dx = 1,$$

where the improper integral on the left is defined as

$$\int_{-\infty}^{+\infty} f(x)\,dx = \lim_{N \to +\infty} \int_{-N}^{0} f(x)\,dx + \lim_{N \to +\infty} \int_{0}^{N} f(x)\,dx$$

and converges if and only if both limits exist.

EXAMPLE 10.2.1

Determine whether the following function is a probability density function:

$$f(x) = \begin{cases} \dfrac{3}{2}x(x-1) & \text{for } 0 \le x \le 2 \\ 0 & \text{otherwise} \end{cases}$$

Solution

We find that

$$\int_{-\infty}^{+\infty} f(x)\, dx = \int_0^2 \frac{3}{2}x(x-1)\, dx = \frac{3}{2}\left[\frac{1}{3}x^3 - \frac{1}{2}x^2\right]_0^2$$

$$= \frac{3}{2}\left[\frac{1}{3}(2)^3 - \frac{1}{2}(2)^2\right] - \frac{3}{2}[0 - 0] = 1$$

However, the condition $f(x) \ge 0$ for all x is not satisfied ($f(x) < 0$ for $0 < x < 1$), so $f(x)$ is *not* a probability density function.

EXAMPLE 10.2.2

Either find a number k such that the function $f(x)$ is a probability density function or explain why no such number exists:

$$f(x) = \begin{cases} k(x^3 + x + 1) & \text{for } 1 \le x \le 3 \\ 0 & \text{otherwise} \end{cases}$$

Solution

First, note that if $k \ge 0$, then $f(x)$ is positive on the interval $1 \le x \le 3$, which means $f(x) \ge 0$ for all x. Then:

$$\int_{-\infty}^{+\infty} f(x)\, dx = \int_1^3 k(x^3 + x + 1)\, dx = k\left[\frac{1}{4}x^4 + \frac{1}{2}x^2 + x\right]_1^3$$

$$= k\left\{\left[\frac{1}{4}(3)^4 + \frac{1}{2}(3)^2 + 3\right] - \left[\frac{1}{4}(1)^4 + \frac{1}{2}(1)^2 + 1\right]\right\} = 26k$$

Thus, $f(x)$ is a probability density function if $26k = 1$, that is, if $k = \dfrac{1}{26}$.

There are many probability density functions. Next, we consider uniform and exponential density functions, two general forms that prove to be especially useful in modelling applications.

Uniform Density Functions

A **uniform probability density function** (Figure 10.6) is constant over a bounded interval $a \le x \le b$ and zero outside that interval. A random variable that has a uniform density function is said to be **uniformly distributed.** Roughly speaking, for a uniformly distributed random variable, all values in some bounded interval are equally likely. More precisely, a continuous random variable is uniformly distributed if the

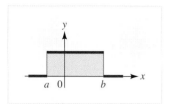

FIGURE 10.6   A uniform density function.

probability that its value will be in a particular subinterval of the bounded interval is equal to the probability that it will be in any other subinterval that has the same length. An example of a uniformly distributed random variable is the waiting time of a motorist at a traffic light that remains red for, say, 40 seconds at a time. This random variable has a uniform distribution because all waiting times between 0 and 40 seconds are equally likely.

If k is the constant value of a uniform density function $f(x)$ on the interval $a \leq x \leq b$, its value may be determined using the requirement that the total area under the graph of f must be equal to 1. In particular,

$$1 = \int_{-\infty}^{+\infty} f(x)\, dx = \int_{a}^{b} f(x)\, dx \qquad \text{since } f(x) = 0 \text{ outside the interval } a \leq x \leq b$$

$$= \int_{a}^{b} k\, dx = kx\Big|_{a}^{b} = k(b - a)$$

and so

$$k = \frac{1}{b - a}.$$

This observation leads to the following formula for a uniform density function.

Uniform Density Function

$$f(x) = \begin{cases} \dfrac{1}{b - a} & \text{if } a \leq x \leq b \\ 0 & \text{otherwise} \end{cases}$$

Here is a typical application involving a uniform density function.

EXAMPLE 10.2.3

A certain traffic light remains red for 40 seconds at a time. You arrive (at random) at the light and find it red. Use an appropriate uniform density function to find the probability that you will have to wait at least 15 seconds for the light to turn green.

Solution

Let X denote the random variable that measures the time (in seconds) that you must wait. Since all waiting times between 0 and 40 are equally likely, X is uniformly distributed over the interval $0 \leq x \leq 40$. The corresponding uniform density function is

$$f(x) = \begin{cases} \dfrac{1}{40} & \text{if } 0 \leq x \leq 40 \\ 0 & \text{otherwise} \end{cases}$$

and the desired probability is

$$P(15 \leq X \leq 40) = \int_{15}^{40} \frac{1}{40}\, dx = \frac{x}{40}\Big|_{15}^{40} = \frac{40 - 15}{40} = \frac{5}{8}.$$

Exponential Density Function

An **exponential probability density function** is a function $f(x)$ that is zero for $x < 0$ and decreases exponentially for $x \geq 0$. That is,

$$f(x) = \begin{cases} Ae^{-\lambda x} & \text{for } x \geq 0 \\ 0 & \text{for } x < 0 \end{cases}$$

where A and λ are positive constants.

The value of A is determined by the requirement that the total area under the graph of f must be equal to 1. Thus,

$$1 = \int_{-\infty}^{+\infty} f(x)\, dx = \int_{0}^{+\infty} Ae^{-\lambda x}\, dx = \lim_{N \to +\infty} \int_{0}^{N} Ae^{-\lambda x}\, dx$$

$$= \lim_{N \to +\infty} \left(-\frac{A}{\lambda} e^{-\lambda x} \Big|_{0}^{N} \right) = \lim_{N \to +\infty} \left(-\frac{A}{\lambda} e^{-\lambda N} + \frac{A}{\lambda} \right) = \frac{A}{\lambda}$$

and so $A = \lambda$.

This calculation leads to the following general formula for an exponential density function. The corresponding graph is shown in Figure 10.7.

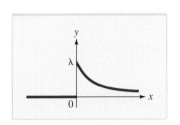

FIGURE 10.7   An exponential density function.

> ### Exponential Density Function
>
> $$f(x) = \begin{cases} \lambda e^{-\lambda x} & \text{if } x \geq 0 \\ 0 & \text{if } x < 0 \end{cases}$$

A random variable that has an exponential density function is said to be **exponentially distributed**. As you can see from the graph in Figure 10.7, the value of an exponentially distributed random variable is much more likely to be small than large. Such random variables include the lifespan of electronic components, the duration of telephone calls, and the interval between the arrivals of successive planes at an airport. Here is an example.

EXAMPLE 10.2.4

Let X be a random variable that measures the duration of telephone calls in a certain city and assume that X has an exponential distribution with density function

$$f(t) = \begin{cases} 0.5e^{-0.5t} & \text{if } t \geq 0 \\ 0 & \text{if } t < 0 \end{cases}$$

where t denotes the duration (in minutes) of a randomly selected call.

a. Find the probability that a randomly selected call will last between 2 minutes and 3 minutes.

b. Find the probability that a randomly selected call will last at least 2 minutes.

Solution

a. $P(2 \leq X \leq 3) = \displaystyle\int_{2}^{3} 0.5e^{-0.5t}\, dt = -e^{-0.5t} \Big|_{2}^{3}$

$$= -e^{-1.5} + e^{-1} \approx 0.1447$$

b. There are two ways to compute this probability. The first method is to evaluate an improper integral:

$$P(X \geq 2) = P(2 \leq X < +\infty) = \int_2^{+\infty} 0.5e^{-0.5t} \, dt$$

$$= \lim_{N \to +\infty} \int_2^N 0.5e^{-0.5t} \, dt = \lim_{N \to +\infty} \left(-e^{-0.5t} \Big|_2^N \right)$$

$$= \lim_{N \to +\infty} (-e^{-0.5N} + e^{-1}) = e^{-1} \approx 0.3679$$

The second method is to compute 1 minus the probability that X is less than 2:

$$P(X \geq 2) = 1 - P(X < 2)$$

$$= 1 - \int_0^2 0.5e^{-0.5t} \, dt = 1 - \left(-e^{-0.5t} \Big|_0^2 \right)$$

$$= 1 - (-e^{-1} + 1) = e^{-1} \approx 0.3679$$

Joint Probability Density Functions

As we have seen, if $f(x)$ is the probability density function for the continuous random variable X, then the probability that X is between a and b is given by the formula

$$P(a \leq X \leq b) = \int_a^b f(x) \, dx,$$

which can be interpreted geometrically as the area under the curve $y = f(x)$ over the interval $a \leq x \leq b$ (Figure 10.8a).

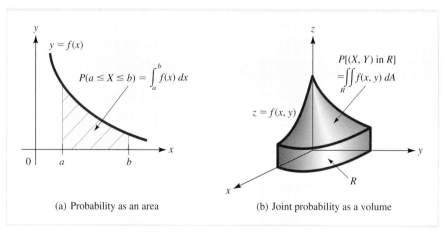

(a) Probability as an area

(b) Joint probability as a volume

FIGURE 10.8   Geometric interpretation of probability.

Just-In-Time

Computations and applications involving double integrals may be found in Section 7.6.

In situations involving two random variables X and Y, we compute probabilities by evaluating double integrals of a two-variable density function. In particular, we integrate a **joint probability density function** $f(x, y)$, which has these properties:

1. $f(x, y) \geq 0$ for all points (x, y) in the xy plane.

2. $\displaystyle\int_{-\infty}^{+\infty} \int_{-\infty}^{+\infty} f(x, y)\, dy\, dx = 1$

3. The probability that the ordered pair (X, Y) lies in a region R is given by

$$P[(X, Y) \text{ in } R] = \int_R \int f(x, y)\, dA.$$

Note that by property (3), the probability that X is between a and b and Y is between c and d is given by

$$P(a \leq X \leq b \text{ and } c \leq Y \leq d) = \int_c^d \int_a^b f(x, y)\, dx\, dy$$

$$= \int_a^b \int_c^d f(x, y)\, dy\, dx$$

In geometric terms, the probability $P[(X, Y) \text{ in } R]$ is the volume under the probability density surface $z = f(x, y)$ above the region R (Figure 10.8b), and condition (2) says that the total volume under the surface is 1.

The techniques for constructing joint probability density functions from experimental data are discussed in most texts on probability and statistics. Joint probability density functions are used to study product reliability, life expectancy depending on different factors, and a variety of other applications. Examples 10.2.5 and 10.2.6 illustrate how integration may be used to compute probability when the joint density function is known.

EXAMPLE 10.2.5

Smoke detectors manufactured by a certain firm contain two independent circuits, one manufactured at the firm's Winnipeg plant and the other at the firm's plant in Oshawa, Ontario. Reliability studies suggest that if X measures the lifespan (in years) of a randomly selected circuit from the Winnipeg plant and Y the lifespan (in years) of a randomly selected circuit from the Oshawa plant, then the joint probability density function for X and Y is

$$f(x, y) = \begin{cases} 0.4e^{-x}e^{-0.4y} & \text{if } x \geq 0 \text{ and } y \geq 0 \\ 0 & \text{otherwise} \end{cases}$$

If the smoke detector will operate as long as either of its circuits is operating, find the probability that a randomly selected smoke detector will fail within 1 year.

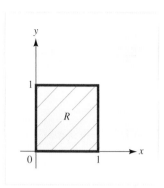

FIGURE 10.9   Square consisting of all points (x, y) for which $0 \le x \le 1$ and $0 \le y \le 1$.

Solution

Since the smoke detector will operate as long as either of its circuits is operating, it will fail within 1 year if and only if *both* of its circuits fail within 1 year. Thus, we want the probability that both $0 \le X \le 1$ and $0 \le Y \le 1$. The points (x, y) for which both of these inequalities hold lie inside and on the boundary of the square R shown in Figure 10.9. The corresponding probability is the double integral of the density function f over this region R. That is,

$$P(0 \le X \le 1 \text{ and } 0 \le Y \le 1) = \int_0^1 \int_0^1 0.4e^{-x}e^{-0.4y}\, dx\, dy$$

$$= \int_0^1 0.4e^{-0.4y}\left(\frac{e^{-x}}{-1}\right)\Bigg|_{x=0}^{x=1}\, dy$$

$$= 0.4(-e^{-1} + 1)\int_0^1 e^{-0.4y}\, dy$$

$$= 0.4(1 - e^{-1})\left(\frac{e^{-0.4y}}{-0.4}\right)\Bigg|_{y=0}^{y=1}$$

$$= (1 - e^{-1})(-e^{-0.4} + 1)$$

$$\approx 0.2084$$

Thus, it is approximately 21% likely that a randomly selected smoke detector will fail within 1 year.

EXAMPLE 10.2.6

Suppose the random variable X denotes the time (in minutes) that a person sits in the waiting room of a certain dentist and Y the time (in minutes) that a person stands in line at a certain grocery store. You have an appointment with the dentist, after which you are planning to go to the grocery store. If $f(x, y)$ is the joint probability density function for X and Y, write a double integral that gives the probability that your *total* waiting time will be no more than 20 minutes.

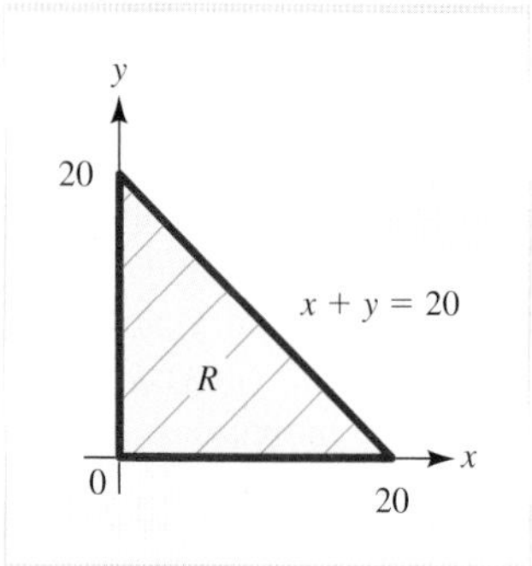

Solution

The goal is to find the probability that $X + Y \le 20$. The points (x, y) for which $x + y \le 20$ lie on or below the line $x + y = 20$. Moreover, since X and Y stand for nonnegative quantities, only those points in the first quadrant are meaningful in this particular context. The problem, then, is to find the probability that the point (X, Y) lies in R, where R is the region in the first quadrant bounded by the line $x + y = 20$ and the coordinate axes (Figure 10.10). This probability is given by the double integral of the density function f over the region R. That is,

$$P[(X, Y) \text{ in } R] = \int_R \int f(x, y)\, dA = \int_0^{20} \int_0^{20-x} f(x, y)\, dy\, dx$$

FIGURE 10.10   Triangle consisting of all points (x, y) for which $x + y \le 20$.

EXERCISES ■ 10.2

In Exercises 1 through 14, determine whether the given function is a probability density function.

1. $f(x) = \begin{cases} \dfrac{1}{10} & \text{for } 30 \leq x \leq 40 \\ 0 & \text{otherwise} \end{cases}$

2. $f(x) = \begin{cases} \dfrac{1}{e} & \text{for } 0 \leq x \leq 1 \\ 0 & \text{otherwise} \end{cases}$

3. $f(x) = \begin{cases} \dfrac{1}{4}(7 - 2x) & \text{for } 1 \leq x \leq 5 \\ 0 & \text{otherwise} \end{cases}$

4. $f(x) = \begin{cases} x - \dfrac{1}{2} & \text{for } 1 \leq x \leq 2 \\ 0 & \text{otherwise} \end{cases}$

5. $f(x) = \begin{cases} \dfrac{10}{(x + 10)^2} & \text{for } x \geq 0 \\ 0 & \text{for } x < 0 \end{cases}$

6. $f(x) = \begin{cases} \dfrac{1}{2}e^{-2x} & \text{for } x \geq 0 \\ 0 & \text{for } x < 0 \end{cases}$

7. $f(x) = \begin{cases} xe^{-x} & \text{for } x \geq 0 \\ 0 & \text{for } x < 0 \end{cases}$

8. $f(x) = \begin{cases} \dfrac{1}{9}\sqrt{x} & \text{for } 0 \leq x \leq 9 \\ 0 & \text{otherwise} \end{cases}$

9. $f(x) = \begin{cases} \dfrac{3}{2}x^2 + 2x & \text{for } -1 \leq x \leq 1 \\ 0 & \text{otherwise} \end{cases}$

10. $f(x) = \begin{cases} \dfrac{1}{2}x^2 + \dfrac{5}{3}x & \text{for } 0 \leq x \leq 1 \\ 0 & \text{otherwise} \end{cases}$

11. $f(x, y) = \begin{cases} 2e^{-(6y + x/3)} & \text{for } x \geq 0, y \geq 0 \\ 0 & \text{otherwise} \end{cases}$

12. $f(x, y) = \begin{cases} 4x^2 & \text{for all } (x, y) \text{ with } 0 \leq y \leq 1 \text{ and } y \leq x \leq 1 \\ 0 & \text{otherwise} \end{cases}$

13. $f(x, y) = \begin{cases} 3x^2 + 6xy & \text{for all } (x, y) \text{ with } 0 \leq x \leq 1 \text{ and } x \leq y \leq 1 \\ 0 & \text{otherwise} \end{cases}$

14. $f(x, y) = \begin{cases} 2e^{-x - y/2} & \text{for all } (x, y) \text{ with } x \geq y \geq 0 \\ 0 & \text{otherwise} \end{cases}$

In Exercises 15 through 24, either find a number k such that the given function is a probability density function or explain why no such number exists.

15. $f(x) = \begin{cases} k - 3x & \text{for } 0 \leq x \leq 1 \\ 0 & \text{otherwise} \end{cases}$

16. $f(x) = \begin{cases} 4 + 3kx & \text{for } 0 \leq x \leq 2 \\ 0 & \text{otherwise} \end{cases}$

17. $f(x) = \begin{cases} (kx - 1)(x - 2) & \text{for } 0 \leq x \leq 2 \\ 0 & \text{otherwise} \end{cases}$

18. $f(x) = \begin{cases} kx(3 - x) & \text{for } 0 \leq x \leq 3 \\ 0 & \text{otherwise} \end{cases}$

19. $f(x) = \begin{cases} x - kx^2 & \text{for } 0 \leq x \leq 1 \\ 0 & \text{otherwise} \end{cases}$

20. $f(x) = \begin{cases} x^3 + kx & \text{for } 0 \leq x \leq 2 \\ 0 & \text{otherwise} \end{cases}$

21. $f(x) = \begin{cases} xe^{-kx} & \text{for } x \geq 0 \\ 0 & \text{otherwise} \end{cases}$

22. $f(x) = \begin{cases} k(x + 2)^{-1} & \text{for } -1 \leq x \leq 2 \\ 0 & \text{otherwise} \end{cases}$

23. $f(x, y) = \begin{cases} ke^{-(x + 3y)} & \text{for } x \geq 0, y \geq 0 \\ 0 & \text{otherwise} \end{cases}$

24. $f(x, y) = \begin{cases} 4e^{-(kx + y)} & \text{for } x \geq 0, y \geq 0 \\ 0 & \text{otherwise} \end{cases}$

In Exercises 25 through 32, f(x) is a probability density function for a particular random variable X. Integrate to find the indicated probabilities.

25. $f(x) = \begin{cases} \dfrac{1}{3} & \text{if } 2 \le x \le 5 \\ 0 & \text{otherwise} \end{cases}$

    a. $P(2 \le X \le 5)$
    b. $P(3 \le X \le 4)$
    c. $P(X \ge 4)$

26. $f(x) = \begin{cases} \dfrac{x}{2} & \text{if } 0 \le x \le 2 \\ 0 & \text{otherwise} \end{cases}$

    a. $P(0 \le X \le 2)$
    b. $P(1 \le X \le 2)$
    c. $P(X \le 1)$

27. $f(x) = \begin{cases} \dfrac{1}{8}(4 - x) & \text{if } 0 \le x \le 4 \\ 0 & \text{otherwise} \end{cases}$

    a. $P(0 \le X \le 4)$
    b. $P(2 \le X \le 3)$
    c. $P(X \ge 1)$

28. $f(x) = \begin{cases} \dfrac{3}{32}(4x - x^2) & \text{if } 0 \le x \le 4 \\ 0 & \text{otherwise} \end{cases}$

    a. $P(0 \le X \le 4)$
    b. $P(1 \le X \le 2)$
    c. $P(X \le 1)$

29. $f(x) = \begin{cases} \dfrac{3}{x^4} & \text{if } x \ge 1 \\ 0 & \text{if } x < 1 \end{cases}$

    a. $P(1 \le X < +\infty)$
    b. $P(X \le 2)$
    c. $P(X \ge 2)$

30. $f(x) = \begin{cases} \dfrac{1}{10}e^{-x/10} & \text{if } x \ge 0 \\ 0 & \text{if } x < 0 \end{cases}$

    a. $P(0 \le X < +\infty)$
    b. $P(X \le 2)$
    c. $P(X \ge 5)$

31. $f(x) = \begin{cases} 2xe^{-x^2} & \text{if } x \ge 0 \\ 0 & \text{if } x < 0 \end{cases}$

    a. $P(X \ge 0)$
    b. $P(1 \le X \le 2)$
    c. $P(X \le 2)$

32. $f(x) = \begin{cases} \dfrac{1}{4}xe^{-x/2} & \text{if } x \ge 0 \\ 0 & \text{if } x < 0 \end{cases}$

    a. $P(0 \le X < +\infty)$
    b. $P(2 \le X \le 4)$
    c. $P(X \ge 6)$

33. USEFUL LIFE OF A MACHINE The useful life X of a particular kind of machine is a random variable with density function

$$f(x) = \begin{cases} \dfrac{3}{28} + \dfrac{3}{x^2} & \text{if } 3 \le x \le 7 \\ 0 & \text{otherwise} \end{cases}$$

where x is the number of years a randomly selected machine stays in use.

    a. Find the probability that a randomly selected machine will be useful for more than 4 years.
    b. Find the probability that a randomly selected machine will be useful for less than 5 years.
    c. Find the probability that a randomly selected machine will be useful for between 4 years and 6 years.

34. TRAFFIC FLOW A certain traffic light remains red for 45 seconds at a time. You arrive (at random) at the light and find it red. Use an appropriate uniform density function to find

    a. The probability that the light will turn green within 15 seconds.
    b. The probability that the light will turn green between 5 seconds and 10 seconds after you arrive.

35. COMMUTING During the morning rush hour, commuter trains run every 20 minutes from the station near your home into the city centre. You arrive (at random) at the station and find no train at the platform. Assuming that the trains are running on schedule, use an appropriate uniform density function to find

    a. The probability that you will have to wait at least 8 minutes for a train.
    b. The probability that you will have to wait between 2 minutes and 5 minutes for a train.

36. EXPERIMENTAL PSYCHOLOGY Suppose the length of time that it takes a laboratory rat to traverse a certain maze is measured by a random variable X that is exponentially distributed with probability density function

$$f(x) = \begin{cases} \dfrac{1}{3}e^{-x/3} & \text{if } x \geq 0 \\ 0 & \text{if } x < 0 \end{cases}$$

where x is the number of minutes a randomly selected rat spends in the maze.

a. Find the probability that a randomly selected rat will take more than 3 minutes to traverse the maze.

b. Find the probability that a randomly selected rat will take between 2 minutes and 5 minutes to traverse the maze.

37. PRODUCT RELIABILITY The lifespan of the light bulbs manufactured by a certain company is measured by a random variable X with probability density function

$$f(x) = \begin{cases} 0.01e^{-0.01x} & \text{if } x \geq 0 \\ 0 & \text{if } x < 0 \end{cases}$$

where x denotes the lifespan (in hours) of a randomly selected bulb.

a. What is the probability that the lifespan of a randomly selected bulb is between 50 hours and 60 hours?

b. What is the probability that the lifespan of a randomly selected bulb is no greater than 60 hours?

c. What is the probability that the lifespan of a randomly selected bulb is greater than 60 hours?

38. PRODUCT RELIABILITY The useful life of a particular type of printer is measured by a random variable X with probability density function

$$f(x) = \begin{cases} 0.02e^{-0.02x} & \text{if } x \geq 0 \\ 0 & \text{if } x < 0 \end{cases}$$

where x denotes the number of months a randomly selected printer has been in use.

a. What is the probability that a randomly selected printer will last between 10 months and 15 months?

b. What is the probability that a randomly selected printer will last less than 8 months?

c. What is the probability that a randomly selected printer will last longer than 1 year?

39. CUSTOMER SERVICE Suppose the time X a customer must spend waiting in a drive-through line at a certain coffee shop before work is a random variable that is exponentially distributed with probability density function

$$f(x) = \begin{cases} \dfrac{1}{4}e^{-x/4} & \text{if } x \geq 0 \\ 0 & \text{if } x < 0 \end{cases}$$

where x is the number of minutes a randomly selected customer spends waiting in line.

a. Find the probability that a customer will have to wait at least 8 minutes.

b. Find the probability that a customer will have to wait between 1 minute and 5 minutes.

40. BIOLOGY Let X be a random variable that measures the age of a randomly selected cell in a particular population. Suppose X is distributed exponentially with a probability density function of the form

$$f(x) = \begin{cases} ke^{-kx} & \text{for } x \geq 0 \\ 0 & \text{otherwise} \end{cases}$$

where x is the age of a randomly selected cell (in days) and k is a positive constant. Experiments indicate that it is twice as likely for a cell to be less than 3 days old as it is for it to be more than 3 days old.

a. Use this information to determine k.

b. Find the probability that a randomly selected cell is at least 5 days old.

41. EXPERIMENTAL PSYCHOLOGY Suppose the length of time that it takes a laboratory rat to traverse a certain maze is measured by a random variable X with probability density function

$$f(x) = \begin{cases} \dfrac{1}{16}xe^{-x/4} & \text{if } x \geq 0 \\ 0 & \text{if } x < 0 \end{cases}$$

where x is the number of minutes a randomly selected rat spends in the maze.

a. Find the probability that a randomly selected rat will take no more than 5 minutes to traverse the maze.

b. Find the probability that a randomly selected rat will take at least 10 minutes to traverse the maze.

42. AIRPLANE ARRIVALS The time interval between the arrivals of successive planes at a certain airport is measured by a random variable X with probability density function

$$f(x) = \begin{cases} 0.2e^{-0.2x} & \text{for } x \geq 0 \\ 0 & \text{for } x < 0 \end{cases}$$

where x is the time (in minutes) between the arrivals of a randomly selected pair of successive planes.

a. What is the probability that two successive planes selected at random will arrive within 5 minutes of one another?

b. What is the probability that two successive planes selected at random will arrive more than 6 minutes apart?

43. **SPY STORY**   The spy manages to escape from the chat room (Exercise 51, Section 9.1) and retrieve his equipment before he is surrounded. He fights back but eventually runs out of bullets and is forced to use his last weapon, a special stun grenade. Pulling the pin from the grenade, he remembers that he received it exactly 1 year ago from his superior, for Valentine's Day. As the grenade arcs through the air toward the enemy, he also recalls that this particular kind of grenade has a 1-year warranty and that its lifespan X is a random variable exponentially distributed with probability density function

$$f(t) = \begin{cases} 0.08e^{-0.08t} & \text{for } t \geq 0 \\ 0 & \text{for } t < 0 \end{cases}$$

where t is the number of months since the grenade left the munitions factory. Assuming the grenade was new when the spy received it and that it had been chosen randomly from the factory stock, what is the probability that the spy will expire along with the warranty?

44. **MEDICINE**   Let X be the random variable that represents the number of years a patient with a particular type of cancer survives after receiving radiation therapy. Suppose X has an exponential distribution with parameter k. Research indicates that a patient with this type of cancer has an 80% chance of surviving for at least 5 years after radiation therapy begins.

a. Use this information to determine k.

b. What is the probability of a cancer patient surviving at least 10 years after beginning radiation therapy?

45. **TRANSPORTATION**   Suppose that the GO Transit bus from Niagara Falls to Toronto leaves every hour in summer. Let X be the random variable that represents the time a person arriving at the bus terminal at a random time must wait until the next bus leaves. Assume X has a uniform distribution.

a. Find the probability density function for X.

b. Find the probability that a person has to wait at least 45 minutes for the next bus to Toronto.

c. Find the probability that a person has to wait between 5 minutes and 15 minutes for the next bus.

46. **ENTERTAINMENT**   A 2-hour movie runs continuously at a local theatre. You leave for the theatre without first checking the show times. Use an appropriate uniform density function to find the probability that you will arrive at the theatre within 10 minutes of the start of the film (before or after).

47. **TROPICAL ECOLOGY**   Let X be the random variable that represents the time (in hours) between successive visits by hummingbirds to feed on the flowers of a particular tropical plant. Suppose X is distributed exponentially with parameter $k = 0.5$. Find the probability that the time between the arrivals of successive hummingbirds is greater than 2 hours.

48. Suppose the joint probability density function for the nonnegative random variables X and Y is

$$f(x, y) = \begin{cases} 2e^{-2x}e^{-y} & \text{if } x \geq 0 \text{ and } y \geq 0 \\ 0 & \text{otherwise} \end{cases}$$

a. Find the probability that $0 \leq X \leq 1$ and $1 \leq Y \leq 2$.

b. Find the probability that $2X + Y \leq 1$.

49. Suppose the joint probability density function for the nonnegative random variables X and Y is

$$f(x, y) = \begin{cases} \dfrac{1}{6}e^{-x/2}e^{-y/3} & \text{if } x \geq 0 \text{ and } y \geq 0 \\ 0 & \text{otherwise} \end{cases}$$

a. Find the probability that $1 \leq X \leq 2$ and $0 \leq Y \leq 2$.

b. Find the probability that $X + Y \leq 1$.

50. Suppose the joint probability density function for the random variables X and Y is

$$f(x, y) = \begin{cases} xe^{-x-y} & \text{if } x \geq 0 \text{ and } y \geq 0 \\ 0 & \text{otherwise} \end{cases}$$

Find the probability that (X, Y) lies in the rectangular region R with vertices $(0, 0)$, $(1, 0)$, $(0, 2)$, and $(1, 2)$.

51. **HEALTH CARE**   Suppose the random variables X and Y measure the length of time (in days) that a patient stays in the hospital after abdominal and orthopaedic surgery, respectively. On Monday, the patient in bed 107A undergoes an emergency

appendectomy while her roommate in bed 107B undergoes orthopaedic surgery for the repair of torn knee cartilage. Suppose the joint probability density function for X and Y is

$$f(x, y) = \begin{cases} \dfrac{1}{12}e^{-x/4}e^{-y/3} & \text{if } x \geq 0 \text{ and } y \geq 0 \\ 0 & \text{otherwise} \end{cases}$$

a. What is the probability that both patients will be discharged from the hospital within 3 days?
b. What is the probability that at least one of the patients is still in the hospital after 3 days?

52. **WARRANTY PROTECTION**   A certain appliance consisting of two independent electronic components will be usable as long as at least one of its components is still operating. The appliance carries a warranty from the manufacturer guaranteeing replacement if the appliance becomes unusable within 1 year of the date of purchase. Let the random variable X measure the lifespan (in years) of the first component and Y measure the lifespan (also in years) of the second component, and suppose that the joint probability density function for X and Y is

$$f(x, y) = \begin{cases} \dfrac{1}{4}e^{-x/2}e^{-y/2} & \text{if } x \geq 0 \text{ and } y \geq 0 \\ 0 & \text{otherwise} \end{cases}$$

You purchase one of these appliances, selected at random from the manufacturer's stock. Find the probability that the warranty will expire before your appliance becomes unusable. [*Hint:* You want the probability that the appliance *does not* fail during the first year. How is this related to the probability that the appliance *does* fail during this period?]

53. **TIME MANAGEMENT**   Suppose the random variable X measures the time (in minutes) that a person spends in the waiting room of a doctor's office and Y measures the time required for a complete physical examination (also in minutes). You arrive at the doctor's office for a physical 50 minutes before you are due for a meeting. If the joint probability distribution for X and Y is

$$f(x, y) = \begin{cases} \dfrac{1}{500}e^{-x/10}e^{-y/50} & \text{if } x \geq 0 \text{ and } y \geq 0 \\ 0 & \text{otherwise} \end{cases}$$

find the probability that you will be late leaving for your meeting. [*Hint:* The probability you want is 1 minus the probability of being on time.]

54. **INSECT POPULATIONS**   Entomologists are studying two insect species that interact within the same ecosystem. Let X be the random variable that measures the population of species A and let Y measure the population of species B. Research indicates that the joint probability density function for X and Y is

$$f(x, y) = \begin{cases} xe^{-x-y} & \text{if } x \geq 0 \text{ and } y \geq 0 \\ 0 & \text{otherwise} \end{cases}$$

What is the probability that species A outnumbers species B?

*The **median** of a random variable X is the real number m such that $P(X \leq m) = \dfrac{1}{2}$. This term is used in Exercises 55 through 58.*

55. Find the median time you have to wait for the light to turn green in Example 10.2.3.

56. Find the median length of a telephone call in Example 10.2.4.

57. Find the median of a uniformly distributed random variable over the interval $A \leq x \leq B$.

58. Find the median of a random variable that has an exponential distribution with parameter λ.

59. **A*** The *reliability function* $r(x)$ of an electronic component is the probability that the component continues to work for more than x days. Suppose that the working life of a component is measured by a random variable X with a probability density function $f(x)$.
a. Find a formula for $r(x)$ in terms of $f(x)$.
b. Suppose the working life (days) of a memory chip has an exponential distribution with a parameter $\lambda = 20$. Find $r(10)$, the probability that the memory chip continues to work for more than 10 days.

A continuous random variable X has a Pareto distribution if its probability density function has the form

$$f(x) = \begin{cases} \dfrac{a\lambda^a}{x^{a+1}} & \text{if } x \geq \lambda \\ 0 & \text{otherwise} \end{cases}$$

where a and λ are parameters (positive real numbers). The Pareto distribution is often used to model the distribution of incomes in a society. Exercises 60 and 61 involve the Pareto distribution.

60. **A*** Verify that for all positive numbers a and λ, the region under the graph of the Pareto probability

density function $f(x)$ has area 1, as required of all probability density functions.

61. **A*** Suppose X is a random variable that has a Pareto distribution with parameters $a = 1$ and $\lambda = 1.2$.
   a. Sketch the graph of the probability density function $f(x)$ for X.
   b. Find the probability $P(10 \leq X \leq 20)$.

   c. Read an article on the Pareto distribution and write a paragraph on how it can be used.

Expected Value and Variance of Continuous Random Variables

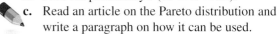

Evaluate the expected value and variance of a continuous random variable.

Suppose it has been determined by medical researchers that when a new drug is used with patients suffering from a certain kind of cancer, the number of years X that a patient survives after receiving the drug is a continuous random variable with a particular probability density function $s(x)$. On average, how long can a patient expect to live after receiving the drug? Reliability studies of a certain product indicate that the lifespan X of the product is a continuous random variable with probability density function $L(x)$. What is the average lifespan of a randomly selected unit? To what extent are the experimental outcomes of the reliability testing spread out in relation to the average lifespan? In Section 10.1, we answered questions of this nature for discrete random variables by examining the expected value and variance of such variables, and in this section, we shall use integration to extend these concepts to random variables that are continuous.

Expected Value of a Continuous Random Variable

An important characteristic of a random variable is its "average" value. Familiar averages of random variables include the average highway gas consumption for a particular car model, the average waiting time for security checks at an airport, and the average lifespan of workers in a hazardous profession. The average value of a random variable X is called its **expected value** or **mean** and is denoted by the symbol $E(X)$. For continuous random variables, you can calculate the expected value from the probability density function using the following formula.

Expected Value ■ If X is a continuous random variable with probability density function f, the expected value (or mean) of x is

$$E(X) = \int_{-\infty}^{\infty} xf(x)\, dx.$$

To see why this integral gives the average value of a continuous random variable, first consider the simpler case in which X is a discrete random variable that takes on the values $x_1, x_2, \ldots, x_n$. If each of these values occurs with equal frequency, the probability of each is $\dfrac{1}{n}$ and the average value of X is

$$\frac{x_1 + x_2 + \cdots + x_n}{n} = x_1\left(\frac{1}{n}\right) + x_2\left(\frac{1}{n}\right) + \cdots + x_n\left(\frac{1}{n}\right).$$

More generally, if the values $x_1, x_2, \ldots, x_n$ occur with probabilities $p_1, p_2, \ldots, p_n$, respectively, the average value of X is the weighted sum

$$x_1 p_1 + x_2 p_2 + \cdots + x_n p_n = \sum_{j=1}^{n} x_j p_j$$

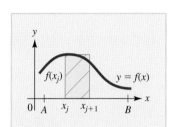

FIGURE 10.11 The jth subinterval with an approximating rectangle.

Now consider the continuous case. For simplicity, restrict your attention to a bounded interval $A \le x \le B$. Imagine that this interval is divided into n subintervals of width Δx, and let x_j denote the beginning of the jth subinterval. Then

$$p_j = \frac{\text{probability that } x \text{ is}}{\text{in the } j\text{th subinterval}} = \frac{\text{area under the graph}}{\text{of } f \text{ from } x_j \text{ to } x_{j+1}} \approx f(x_j)\,\Delta x,$$

where $f(x_j)\Delta x$ is the area of an approximating rectangle (Figure 10.11).

To approximate the average value of X over the interval $A \le x \le B$, treat X as if it were a discrete random variable that takes on the values $x_1, x_2, \ldots, x_n$ with probabilities $p_1, p_2, \ldots, p_n$, respectively. Then

$$\genfrac{}{}{0pt}{}{\text{Average value of } X}{\text{on } A \le x \le B} \approx \sum_{j=1}^{n} x_j p_j \approx \sum_{j=1}^{n} x_j f(x_j)\,\Delta x.$$

The actual average value of X on the interval $A \le x \le B$ is the limit of this approximating sum as n increases without bound. That is,

$$\genfrac{}{}{0pt}{}{\text{Average value of } X}{\text{on } A \le x \le B} = \lim_{n \to \infty} \sum_{j=1}^{n} x_j f(x_j)\,\Delta x = \int_{A}^{B} x f(x)\, dx.$$

An extension of this argument shows that

$$\genfrac{}{}{0pt}{}{\text{Average value of } X}{\text{on } -\infty < x < \infty} = \int_{-\infty}^{\infty} x f(x)\, dx,$$

which is the formula for the expected value of a continuous random variable.

Calculation of the Expected Value

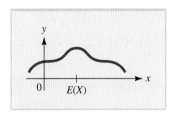

FIGURE 10.12 The expected value of a symmetric random variable.

To get a better feel for the expected value of a continuous random variable X, think of the probability density function for X as describing the distribution of mass on a beam lying on the x axis. Then the expected value of X is the point at which the beam will balance. If the graph of the density function is symmetric, the expected value is the point of symmetry, as illustrated in Figure 10.12.

The use of the integral formula to compute the expected value of a continuous random variable is illustrated in Examples 10.3.1 and 10.3.2.

EXAMPLE 10.3.1

Continuing Example 10.2.3, a certain traffic light remains red for 40 seconds at a time. You arrive (at random) at the light and find it red. Let X denote the random variable that measures the time (in seconds) that you must wait for the light to change and assume that X is distributed uniformly with the probability density function

$$f(x) = \begin{cases} \dfrac{1}{40} & \text{if } 0 \le x \le 40 \\ 0 & \text{otherwise} \end{cases}$$

How long should you expect to wait for the light to turn green?

Solution

The expected value of X is

$$E(X) = \int_{-\infty}^{\infty} x f(x)\, dx = \int_0^{40} \frac{x}{40}\, dx = \frac{x^2}{80}\Big|_0^{40} = \frac{1600}{80} = 20.$$

Therefore, the average waiting time at the red light is 20 seconds, a conclusion that should come as no surprise since the random variable is uniformly distributed between 0 and 40.

EXAMPLE 10.3.2

Continuing Example 10.2.4, let X be the random variable that measures the duration of phone calls in a certain city and assume that X is distributed exponentially with the probability density function

$$f(x) = \begin{cases} 0.5e^{-0.5x} & \text{if } x \ge 0 \\ 0 & \text{if } x < 0 \end{cases}$$

How long would you expect a randomly selected phone call to last?

Solution

The expected value of X is

$$\begin{aligned}
E(X) &= \int_{-\infty}^{\infty} x f(x)\, dx = \int_0^{\infty} 0.5x e^{-0.5x}\, dx \\
&= \lim_{N \to \infty} \int_0^N 0.5x e^{-0.5x}\, dx \\
&= \lim_{N \to \infty} \left(-x e^{-0.5x}\Big|_0^N + \int_0^N e^{-0.5x}\, dx \right) \\
&= \lim_{N \to \infty} \left(-x e^{-0.5x} - 2e^{-0.5x} \right)\Big|_0^N \\
&= \lim_{N \to \infty} \left(-N e^{-0.5N} - 2e^{-0.5N} + 2 \right) \\
&= 2
\end{aligned}$$

Thus, the expected (average) duration of telephone calls in the city is 2 minutes.

Variance of a Continuous Random Variable

The expected value or mean of a random variable tells us the centre of its distribution. Another concept that is useful in describing the distribution of a random variable is the **variance,** which tells us how spread out the distribution is. That is, the variance measures the tendency of the values of a random variable to cluster about their mean. Here is the definition.

> **Variance** ■ If X is a continuous random variable with probability density function f, the variance of X is
>
> $$\text{Var}\,(X) = \int_{-\infty}^{\infty} [x - E(X)]^2 f(x)\,dx.$$

The definition of the variance is not as mysterious as it may seem at first glance. Notice that the formula for the variance is the same as that for the expected value, except that x has been replaced by the expression $[x - E(X)]^2$, which represents the square of the deviation of x from the mean $E(X)$. The variance, therefore, is simply the expected value or average of the squared deviations of the values of X from the mean. If the values of X tend to cluster about the mean as in Figure 10.13a, most of the deviations from the mean will be small and the variance, which is the average of the squares of these deviations, will also be small. On the other hand, if the values of the random variable are widely scattered as in Figure 10.13b, there will be many large deviations from the mean and the variance will be large.

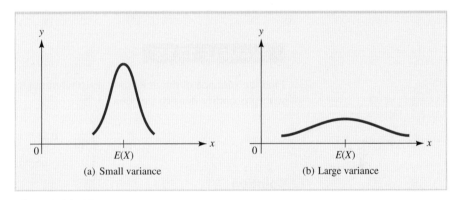

(a) Small variance   (b) Large variance

FIGURE 10.13   The variance of a continuous random variable as a measure of the spread of a distribution.

Calculation of the Variance

As we have just seen, the formula

$$\text{Var}(X) = \int_{-\infty}^{\infty} [x - E(X)]^2 f(x)\,dx$$

defining the variance is fairly easily interpreted as a measure of the spread of the distribution of X. However, for all but the simplest density functions, this formula is cumbersome to use for the actual calculation of the variance because the term $x - E(X)$ must be squared and multiplied by $f(x)$ before the integration can be performed. Here is an equivalent formula for the variance, which is easier to use for computational purposes.

> **Equivalent Formula for Variance**
>
> $$\text{Var}(X) = \int_{-\infty}^{\infty} x^2 f(x)\,dx - [E(X)]^2$$

The derivation of this formula is straightforward. It involves expanding the integrand in the definition of the variance and rearranging the resulting terms until the new formula emerges. As you read the steps, keep in mind that the expected value $E(X)$ is a constant and can be brought outside integrals by the constant multiple rule.

$$\text{Var}(X) = \int_{-\infty}^{\infty} [x - E(X)]^2 f(x)\, dx \qquad \text{definition of variance}$$

$$= \int_{-\infty}^{\infty} \{x^2 - 2xE(X) + [E(X)]^2\} f(x)\, dx \qquad \text{integrand expanded}$$

$$= \int_{-\infty}^{\infty} x^2 f(x)\, dx - 2E(X) \int_{-\infty}^{\infty} x f(x)\, dx + [E(X)]^2 \int_{-\infty}^{\infty} f(x)\, dx$$

$$\text{sum rule and constant multiple rule}$$

$$= \left[\int_{-\infty}^{\infty} x^2 f(x)\, dx \right] - 2E(X) \cdot E(X) + [E(X)]^2 (1)$$

$$\text{definition of } E(X) \text{ and the fact that } \int_{-\infty}^{\infty} f(x)\, dx = 1$$

$$= \int_{-\infty}^{\infty} x^2 f(x)\, dx - [E(X)]^2$$

This formula is used to compute the variance in Examples 10.3.3 and 10.3.4.

EXAMPLE 10.3.3

Find the variance of the uniformly distributed random variable X from Example 10.3.1 with probability density function

$$f(x) = \begin{cases} \dfrac{1}{40} & \text{if } 0 \le x \le 40 \\ 0 & \text{otherwise} \end{cases}$$

Solution

The first step is to compute $E(X)$. This was done in Example 10.3.1, where we found that $E(X) = 20$. Using this value in the variance formula, we get

$$\text{Var}(X) = \int_{-\infty}^{\infty} x^2 f(x)\, dx - [E(X)]^2$$

$$= \int_0^{40} \frac{x^2}{40}\, dx - 400 = \frac{x^3}{120} \Big|_0^{40} - 400$$

$$= \frac{64\,000}{120} - 400 = \frac{16\,000}{120} = \frac{400}{3}$$

EXAMPLE 10.3.4

Find the variance of the exponentially distributed random variable X from Example 10.3.2 with probability density function

$$f(x) = \begin{cases} 0.5e^{-0.5x} & \text{if } x \ge 0 \\ 0 & \text{if } x < 0 \end{cases}$$

Solution

Using the value $E(X) = 2$ obtained in Example 10.3.2 and integrating by parts twice results in

$$
\begin{aligned}
\mathrm{Var}(X) &= \int_{-\infty}^{\infty} x^2 f(x)\, dx - [E(X)]^2 \\
&= \int_{0}^{\infty} 0.5 x^2 e^{-0.5x}\, dx - 4 \\
&= \lim_{N \to \infty} \int_{0}^{N} 0.5 x^2 e^{-0.5x}\, dx - 4 \\
&= \lim_{N \to \infty} \left(-x^2 e^{-0.5x} \Big|_{0}^{N} + 2 \int_{0}^{N} x e^{-0.5x}\, dx \right) - 4 \\
&= \lim_{N \to \infty} \left[\left(-x^2 e^{-0.5x} - 4x e^{-0.5x} \right) \Big|_{0}^{N} + 4 \int_{0}^{N} e^{-0.5x}\, dx \right] - 4 \\
&= \lim_{N \to \infty} (-x^2 - 4x - 8) e^{-0.5x} \Big|_{0}^{N} - 4 \\
&= \lim_{N \to \infty} \left[(-N^2 - 4N - 8) e^{-0.5N} + 8 \right] - 4 \\
&= 4
\end{aligned}
$$

Expected Value for a Joint Probability Distribution

If X and Y are continuous random variables with joint probability density function $f(x, y)$, then the expected values of X and Y are given by the integrals

$$
E(X) = \int_{-\infty}^{+\infty} \int_{-\infty}^{+\infty} x f(x, y)\, dA \quad \text{and} \quad E(Y) = \int_{-\infty}^{+\infty} \int_{-\infty}^{+\infty} y f(x, y)\, dA.
$$

In Example 10.3.5, we compute the expected value of one of the random variables in Example 10.2.5.

EXAMPLE 10.3.5

Smoke detectors manufactured by a certain firm contain two independent circuits, one manufactured at the firm's Winnipeg plant and the other at the firm's plant in Oshawa. Reliability studies suggest that if X measures the lifespan (in years) of a randomly selected circuit from the Winnipeg plant and Y measures the lifespan (in years) of a randomly selected circuit from the Oshawa plant, then the joint probability density function for X and Y is

$$
f(x, y) = \begin{cases} 0.4 e^{-x} e^{-0.4y} & \text{if } x \geq 0 \text{ and } y \geq 0 \\ 0 & \text{otherwise} \end{cases}
$$

How long would you expect a randomly selected circuit from the Oshawa plant to last?

Solution

You wish to compute the expected value of the random variable Y, that is,

$$E(Y) = \int_{-\infty}^{+\infty} \int_{-\infty}^{+\infty} y(0.4e^{-x}e^{-0.4y})\, dx\, dy = \int_{0}^{+\infty} 0.4ye^{-0.4y} \left(\int_{0}^{+\infty} e^{-x}\, dx \right) dy$$

$$= \int_{0}^{+\infty} 0.4ye^{-0.4y} \left[\lim_{N \to +\infty} \left(\frac{e^{-x}}{-1} \right) \Big|_{x=0}^{x=N} \right] dy$$

$$= \int_{0}^{+\infty} 0.4ye^{-0.4y}(1)\, dy \qquad \text{since } \lim_{N \to +\infty} [-(e^{-N} - e^0)] = 1$$

$$= \lim_{N \to +\infty} \left(-ye^{-0.4y} + \frac{e^{-0.4y}}{-0.4} \right) \Big|_{y=0}^{y=N} \qquad \begin{array}{l} \text{integration by parts:} \\ u = 0.4y \quad dv = e^{-0.4y}\, dy \end{array}$$

$$= \lim_{N \to +\infty} [(-Ne^{-0.4N} - 2.5e^{-0.4N}) - (0 - 2.5)]$$

$$= 2.5 \qquad \text{since } \lim_{N \to +\infty} (-N - 2.5)e^{-0.4N} = 0$$

Thus, you would expect a randomly selected circuit from the Oshawa plant to last 2.5 years.

EXERCISES ■ 10.3

In Exercises 1 through 10, the probability density function for a continuous random variable X is given. In each case, find the expected value E(X) and the variance Var(X) of X.

1. $f(x) = \begin{cases} \dfrac{1}{3} & \text{if } 2 \le x \le 5 \\ 0 & \text{otherwise} \end{cases}$

2. $f(x) = \begin{cases} \dfrac{x}{2} & \text{if } 0 \le x \le 2 \\ 0 & \text{otherwise} \end{cases}$

3. $f(x) = \begin{cases} \dfrac{1}{8}(4 - x) & \text{if } 0 \le x \le 4 \\ 0 & \text{otherwise} \end{cases}$

4. $f(x) = \begin{cases} \dfrac{3}{32}(4x - x^2) & \text{if } 0 \le x \le 4 \\ 0 & \text{otherwise} \end{cases}$

5. $f(x) = \begin{cases} \dfrac{3}{x^4} & \text{if } 1 \le x \le \infty \\ 0 & \text{if } x < 1 \end{cases}$

6. $f(x) = \begin{cases} \dfrac{1}{10}e^{-x/10} & \text{if } x \ge 0 \\ 0 & x < 0 \end{cases}$

7. $f(x) = \begin{cases} 4e^{-4x} & \text{if } x \ge 0 \\ 0 & \text{otherwise} \end{cases}$

8. $f(x) = \begin{cases} x & \text{if } 0 \le x < 1 \\ 2 - x & \text{if } 1 \le x \le 2 \\ 0 & \text{otherwise} \end{cases}$

9. $f(x) = \begin{cases} 20(x^3 - x^4) & \text{if } 0 \le x \le 1 \\ 0 & \text{otherwise} \end{cases}$

10. $f(x) = \begin{cases} xe^{-x} & \text{if } x \ge 0 \\ 0 & \text{otherwise} \end{cases}$

In Exercises 11 through 14, the joint probability density function f(x, y) for two continuous random variables X and Y is given. In each case, find the expected values E(X) and E(Y).

11. $f(x, y) = \begin{cases} \dfrac{1}{2}e^{-x/2}e^{-y} & \text{if } x \ge 0 \text{ and } y \ge 0 \\ 0 & \text{otherwise} \end{cases}$

12. $f(x, y) = \begin{cases} \dfrac{1}{3}e^{-2x}e^{-y/6} & \text{if } x \ge 0 \text{ and } y \ge 0 \\ 0 & \text{otherwise} \end{cases}$

13. $f(x, y) = \begin{cases} ye^{-x-y} & \text{if } x \geq 0 \text{ and } y \geq 0 \\ 0 & \text{otherwise} \end{cases}$

14. $f(x, y) = \begin{cases} \dfrac{24}{5}x(1 - y) & \text{if } 0 \leq y \leq x \text{ and } 0 \leq x \leq 1 \\ 0 & \text{otherwise} \end{cases}$

15. USEFUL LIFE OF A MACHINE   The useful life X of a particular kind of machine is a random variable with density function

$$f(x) = \begin{cases} \dfrac{3}{28} + \dfrac{3}{x^2} & \text{if } 3 \leq x \leq 7 \\ 0 & \text{otherwise} \end{cases}$$

where x is the number of years a randomly selected machine stays in use. What is the expected useful life of the machine?

16. TRAFFIC FLOW   A certain traffic light remains red for 45 seconds at a time. You arrive (at random) at the light and find it red. Use an appropriate uniform density function to find the expected waiting time for cars arriving on red at the traffic light.

17. COMMUTING   During the morning rush hour, commuter trains run every 20 minutes from the station near your home into the city centre. You arrive (at random) at the station and find no train at the platform. Assuming that the trains are running on schedule, use an appropriate uniform density function to find the expected wait for rush-hour commuters arriving at the station when no train is at the platform.

18. AIRPLANE ARRIVALS   The time interval between the arrivals of successive planes at a certain airport is measured by a random variable X with probability density function

$$f(x) = \begin{cases} 0.2e^{-0.2x} & \text{for } x \geq 0 \\ 0 & \text{for } x < 0 \end{cases}$$

where x is the time (in minutes) between the arrivals of a randomly selected pair of successive planes. If you arrive at the airport just in time to see a plane landing, how long would you expect to wait for the next arriving flight?

19. PRODUCT RELIABILITY   The useful life of a particular type of printer is measured by a random variable X with probability density function

$$f(x) = \begin{cases} 0.05e^{-0.05x} & \text{if } x \geq 0 \\ 0 & \text{if } x < 0 \end{cases}$$

where x denotes the number of months a randomly selected printer has been in use. What is the expected life of a randomly selected printer?

20. CUSTOMER SERVICE   Suppose the time X a customer must spend waiting in line at a certain bank is a random variable that is exponentially distributed with density function

$$f(x) = \begin{cases} \dfrac{1}{4}e^{-x/4} & \text{if } x \geq 0 \\ 0 & \text{if } x < 0 \end{cases}$$

where x is the number of minutes a randomly selected customer spends waiting in line. Find the expected waiting time for customers at the bank.

21. MEDICAL RESEARCH   A group of patients with a potentially fatal disease has been treated with an experimental drug. Assume the survival time X for a patient receiving the drug is a random variable exponentially distributed with probability density function

$$f(x) = \begin{cases} \lambda e^{-\lambda x} & \text{if } x \geq 0 \\ 0 & \text{if } x < 0 \end{cases}$$

where x is the number of years a patient survives after first receiving the drug.
a. Research indicates that the expected survival time for a patient receiving the drug is 5 years. Based on this information, what is λ?
b. Using the value of λ determined in part (a), what is the probability that a randomly selected patient survives for less than 2 years?
c. What is the probability that a randomly selected patient survives for more than 7 years?

22. A* EXPERIMENTAL PSYCHOLOGY   Suppose the length of time that it takes a laboratory rat to traverse a certain maze is measured by a random variable X that is distributed with a probability density function of the form

$$f(x) = \begin{cases} axe^{-bx} & \text{if } x \geq 0 \\ 0 & \text{otherwise} \end{cases}$$

where x is the number of minutes a randomly selected rat spends in the maze and a and b are positive numbers. Experiments indicate that on average, a rat will take 6 minutes to traverse the maze.
a. Use the given information to find a and b. [*Hint:* Why must it be true that $\dfrac{a}{b^2} = 1$ and $\dfrac{2a}{b^3} = 6$?]
b. Is it more likely that the rat spends less than 5 minutes in the maze or more than 7 minutes?

23. A* VIROLOGY   A research institute for public health has identified a new deadly virus. To study the virus, researchers consider a model in which it is assumed that the lifespan of an individual chosen from a sample population of the virus is a random variable X distributed with density function of the form

$$f(x) = \begin{cases} Ae^{-bx} & \text{for } x \geq 0 \\ 0 & \text{otherwise} \end{cases}$$

where x is time (in weeks) and A and b are positive constants. Experiments suggest that after 1 week, 90% of the population is still alive.

a. Use this information to find A and b.

b. What percentage of the sample population survives for 10 weeks?

c. How long does it take for 90% of the sample population to die off?

d. What is the expected lifespan of a virus selected randomly from the population?

e. What is the variance of X? Interpret this result.

24. **LEARNING** Let X be the random variable that measures the amount of time a randomly selected calculus student in your class spends studying the subject each day. Suppose X is distributed with probability density function

$$f(x) = \begin{cases} \dfrac{5}{324}\sqrt{x}(9 - x) & \text{for } 0 \leq x \leq 9 \\ 0 & \text{otherwise} \end{cases}$$

where x is the number of hours the student studies each day.

a. Sketch the graph of the density function $f(x)$.

b. What is the probability that a randomly selected calculus student studies for at least 2 hours per day?

c. On average, how long should an instructor expect the students to study calculus each day?

25. **TROPICAL ECOLOGY** Let X be the random variable that represents the time (in hours) between successive visits by hummingbirds to feed on the flowers of a particular tropical plant. Suppose X is distributed exponentially with parameter $\lambda = 0.5$. If a hummingbird has just left a flower you are observing, how long would you expect to wait for the next hummingbird to arrive?

26. What is the variance of the random variable X in Exercise 25?

27. **A*** Show that a uniformly distributed random variable X with probability density function

$$f(x) = \begin{cases} \dfrac{1}{b - a} & \text{if } a \leq x \leq b \\ 0 & \text{otherwise} \end{cases}$$

has expected value

$$E(X) = \frac{a + b}{2}.$$

28. **A*** Show that the uniformly distributed random variable in Exercise 27 has variance

$$\text{Var}(X) = \frac{(b - a)^2}{12}.$$

29. **A*** Show that an exponentially distributed random variable X with probability density function

$$f(x) = \begin{cases} \lambda e^{-\lambda x} & \text{if } x \geq 0 \\ 0 & \text{otherwise} \end{cases}$$

has expected value $E(X) = \dfrac{1}{\lambda}$.

30. **A*** Show that the exponentially distributed random variable in Exercise 29 has variance $\text{Var}(X) = \dfrac{1}{\lambda^2}$.

31. **A* THE INTERNET** Suppose that during evening hours, the time between successive hits on a popular web page can be represented by an exponentially distributed random variable X whose expected value is $E(X) = 0.5$ seconds.

a. Use the result of Exercise 29 to find the probability density function $f(x)$ for X.

b. Find the probability that the time between successive hits on the web page is more than 1 second.

c. Find the probability that the time between successive hits on the web page is between $\dfrac{1}{4}$ second and $\dfrac{1}{2}$ second.

32. **A* TELECOMMUNICATIONS** Suppose that during business hours, the time between successive wireless calls at a network switch can be represented by an exponentially distributed random variable X with expected value $E(X) = 0.01$ second.

a. Use the result of Exercise 29 to find the probability density function for X.

b. Find the probability that the time between the arrivals of successive wireless calls at the switch is more than 0.05 seconds.

c. Find the probability that the time between the arrivals of successive wireless calls at the switch is between 0.05 and 0.08 seconds.

33. **A* HEALTH CARE** Suppose the random variables X and Y measure the length of time (in days) that a patient stays in the hospital after abdominal and orthopaedic surgery, respectively.

On Monday, the patient in bed 107A undergoes an emergency appendectomy while her roommate in bed 107B undergoes surgery for the repair of torn knee cartilage. Suppose the joint probability density function for X and Y is

$$f(x, y) = \begin{cases} \dfrac{1}{12}e^{-x/4}e^{-y/3} & \text{if } x \geq 0 \text{ and } y \geq 0 \\ 0 & \text{otherwise} \end{cases}$$

How long should each patient expect to stay in the hospital?

34. A* TIME MANAGEMENT   Let the random variable X measure the time (in minutes) that a person spends in the waiting room of a doctor's office and let Y measure the time required for a complete physical examination (also in minutes). Suppose the joint probability density function for X and Y is

$$f(x, y) = \begin{cases} \dfrac{1}{500}e^{-x/10}e^{-y/50} & \text{if } x \geq 0 \text{ and } y \geq 0 \\ 0 & \text{otherwise} \end{cases}$$

How much time would you expect to spend in the waiting room? Is this more or less than the time you would expect to spend during your examination?

35. A* COST ANALYSIS   A product consists of two components, A and B, that are manufactured separately. To estimate the cost of labour for the process, the manufacturer assigns a random variable X to measure the number of worker-hours required during an 8-hour shift to produce each unit of component A and another random variable Y to measure the number of worker-hours required to produce each unit of component B. Analysis suggests assigning to X and Y the joint probability density function

$$f(x, y) = \begin{cases} \dfrac{1}{5120}(x^2 + xy + 2y^2) & \text{for } 0 \leq x \leq 8, \ 0 \leq y \leq 8 \\ 0 & \text{otherwise} \end{cases}$$

a.   What is the probability that the total number of worker-hours required to produce both components is less than 8?

b.   How many worker-hours should the manufacturer expect to be expended on producing a unit of component A? A unit of component B?

c.   Suppose the workers assigned to producing component A receive $18 per hour, while those producing component B receive $20 per hour. What is the expected cost of producing one complete (combined) unit of the product?

36. A* Recall from the Section 10.2 Exercises that a continuous random variable X has a Pareto distribution if its probability density function has the form

$$f(x) = \begin{cases} \dfrac{a\lambda^a}{x^{a+1}} & \text{if } x \geq \lambda \\ 0 & \text{otherwise} \end{cases}$$

where a and λ are positive real numbers.

a.   Find the expected value $E(X)$ for $a > 1$.

b.   Find the variance $\text{Var}(X)$ for $a > 2$.

SECTION 10.4

L04

Find probabilities using the Poisson distribution. Find probabilities using the normal distribution.

Normal and Poisson Probability Distributions

In this section, we introduce two important families of probability distributions, one continuous and the other discrete. We begin by discussing normal distributions, which are continuous and are arguably the best known and most widely used of all probability distributions. Then we shall examine Poisson distributions, which are discrete and somewhat narrower in scope than normal distributions but are nonetheless extremely useful for modelling certain phenomena. We shall consider both computational issues and practical applications involving these two distribution classes.

Normal Distributions

A **normal density function** is one of the form

$$f(x) = \frac{1}{\sigma\sqrt{2\pi}}e^{-(x-\mu)^2/2\sigma^2},$$

where μ and σ are real numbers, with $\sigma > 0$. A random variable X with a density function of this form is called a **normal random variable,** and its values are said to be **normally distributed.** The graph of a normal density function is called a **normal curve** or, less formally, a **bell curve** because of its distinctive shape (Figure 10.14).

The normal distribution was originally introduced in 1733 by the French mathematician Abraham de Moivre, who used it to study the results of certain games of

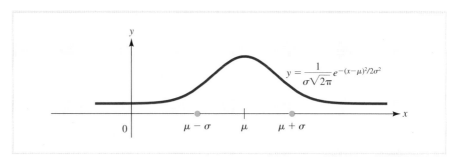

FIGURE 10.14   A bell curve, the graph of a normal density function.

chance. Other applications soon followed. For instance, the Belgian scientist Adolph Quetelet showed that height and chest measurements taken of French and Scottish soldiers were approximately normally distributed.

Indeed, a primary reason for the importance of the normal distribution is that many random variables of practical interest behave as if their distributions are either normal or essentially normal. Examples of quantities closely approximated by normal distributions include time until first failure of product, the height of people of a particular age and gender, the daily caloric intake of a grizzly bear, the number of shark attacks in a given time period, the lifespan of an organism, and the concentration of heavy metals in blood. Moreover, the normal distribution arises frequently in psychological and educational studies. Examples include variables such as the job satisfaction rating of a company's employees, the reading ability of children of a particular age, the number of crimes of a certain type committed annually, and the number of words a person can memorize during a given time period. Finally, other important distributions can be approximated in terms of normal distributions. In fact, a key result in probability theory, called the *central limit theorem,* shows that the sample mean based on a random sampling of a large number of independent observations drawn from a given distribution (e.g., exponential, geometric) is approximately normally distributed.

Our next goal is to examine some of the features and properties of the normal density function. First, recall that any probability density function $f(x)$ must satisfy $\int_{-\infty}^{+\infty} f(x)\, dx = 1$. Using special substitution methods discussed in most engineering calculus texts, it can be shown that

$$\int_{-\infty}^{+\infty} e^{-(x-\mu)^2/2\sigma^2}\, dx = \sigma\sqrt{2\pi},$$

which means that

$$\frac{1}{\sigma\sqrt{2\pi}} \int_{-\infty}^{+\infty} e^{-(x-\mu)^2/2\sigma^2}\, dx = 1,$$

as required.

We have already commented on the bell shape of the graph of the normal density function. However, the bell can be flattened or peaked, depending on the parameters μ and σ. The graph of any normal density function is symmetric about the vertical line $x = \mu$ and has a high point where $x = \mu$ and inflection points where $x = \mu + \sigma$ and $x = \mu - \sigma$ (see Exercise 60). Three different normal curves are shown in Figure 10.15a. The normal density curve with $\mu = 0$ and $\sigma = 1$, called the *standard normal curve,* is shown in Figure 10.15b. Recall that this curve was analyzed using derivative methods in Chapter 4.

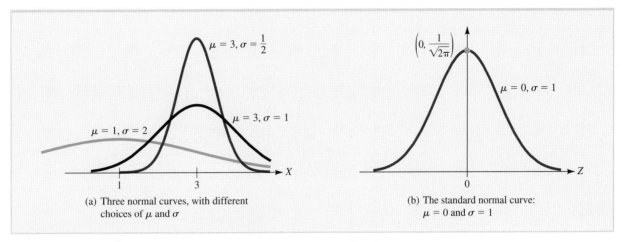

(a) Three normal curves, with different
choices of μ and σ

(b) The standard normal curve:
$\mu = 0$ and $\sigma = 1$

FIGURE 10.15   Graphs of several normal density functions (normal curves).

The parameters μ and σ that appear in the normal density function are closely related to the expected value, variance, and standard deviation of the associated normal random variable, as summarized in the next box.

> **Formulas for the Expected Value, Variance, and Standard Deviation of a Normal Random Variable** ▪ For a normal random variable X with density function
>
> $$f(x) = \frac{1}{\sigma\sqrt{2\pi}}e^{-(x-\mu)^2/2\sigma^2},$$
>
> we have
>
> Expected value: $\quad E(X) = \mu$
> Variance: $\quad \mathrm{Var}(X) = \sigma^2$
> Standard deviation: $\quad \sqrt{\mathrm{Var}(X)} = \sigma$

Most of the area under a normal curve lies relatively close to the mean μ. In fact, it can be shown that roughly 68.3% of normally distributed data lie within one standard deviation of the mean μ and that almost all the data (99.7%) lie within three standard deviations of μ. These observations are part of the **empirical rule** summarized in the following box and displayed in Figure 10.16. The analysis in Example 10.4.1 is typical of how the empirical rule may be used to quickly draw important conclusions about normally distributed data.

> **Empirical Rule for Area Under a Normal Curve**
>
> For any normal curve:
>
> **1.** Roughly 68.3% of the area under the curve lies between $\mu - \sigma$ and $\mu + \sigma$, that is, within one standard deviation of the mean.
> **2.** Roughly 95.4% of the area under the curve lies between $\mu - 2\sigma$ and $\mu + 2\sigma$, that is, within two standard deviations of the mean.
> **3.** Roughly 99.7% of the area under the curve lies between $\mu - 3\sigma$ and $\mu + 3\sigma$, that is, within three standard deviations of the mean.

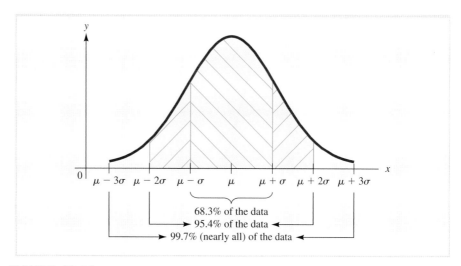

FIGURE 10.16 The empirical rule for normally distributed data.

EXAMPLE 10.4.1

The heights of women between the ages of 18 and 24 in Canada are normally distributed with a mean of 166.4 cm and a standard deviation of 6.4 cm. Describe the distribution of these heights using the empirical rule.

Solution

Using the empirical rule gives the following results:

68.3% of these women have heights between 166.4 cm − 6.4 cm = 160 cm and 166.4 cm + 6.4 cm = 172.8 cm.

95.4% of these women have heights between 166.4 cm − 2(6.4 cm) = 153.6 cm and 166.4 cm + 2(6.4 cm) = 179.2 cm.

99.7% of all these women have heights between 166.4 cm − 3(6.4 cm) = 147.2 cm and 166.4 cm + 3(6.4 cm) = 185.6 cm.

Computations Involving Standard Normal Distributions

There are many normal density distributions, but, fortunately, it is not necessary to study them individually. Indeed, we shall find that any normal distribution can be standardized by transforming its density function into standard normal form

$$S(x) = \frac{1}{\sqrt{2\pi}} e^{-x^2/2},$$

with mean $\mu = 0$ and standard deviation $\sigma = 1$. The graph of the standard normal density function is shown in Figure 10.15b. Notice that it is symmetric about the y axis, with its peak at $\left(0, \frac{1}{\sqrt{2\pi}}\right)$. We shall begin our study of normal distributions by focusing on the standard normal case before turning to more general distributions.

It is common practice to denote the standard normal random variable by Z. To compute the probability $P(a \leq Z \leq b)$ that the standard normal random variable Z lies between a and b, we need to evaluate the integral

$$P(a \leq Z \leq b) = \int_a^b S(x)\, dx = \int_a^b \frac{1}{\sqrt{2\pi}} e^{-x^2/2}\, dx.$$

The bad news is that there is no simple antiderivative for the standard normal density function $S(x)$, so we cannot compute the required probability directly using the fundamental theorem of calculus. The good news is that because computing such probabilities is so important, calculators have been programmed to carry out the calculations numerically, and tables of the probabilities, such as Table 10.1, have been compiled. Notice that Table 10.1 gives probabilities of the form $P(Z \le b)$ or, equivalently, the area under the standard normal density curve to the left of the vertical line $x = b$. These probabilities can then be combined algebraically to obtain probabilities of the form $P(a \le Z \le b)$, as illustrated in Example 10.4.2.

TABLE 10.1 Areas Under the Standard Normal Curve to the Left of Positive Z Values: $P(Z \le b)$

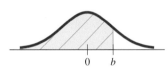

b	0.00	0.01	0.02	0.03	0.04	0.05	0.06	0.07	0.08	0.09
0.0	0.5000	0.5040	0.5080	0.5120	0.5160	0.5199	0.5239	0.5279	0.5319	0.5359
0.1	0.5398	0.5438	0.5478	0.5517	0.5557	0.5596	0.5636	0.5675	0.5714	0.5753
0.2	0.5793	0.5832	0.5871	0.5910	0.5948	0.5987	0.6026	0.6064	0.6103	0.6141
0.3	0.6179	0.6217	0.6255	0.6293	0.6331	0.6368	0.6406	0.6443	0.6480	0.6517
0.4	0.6554	0.6591	0.6628	0.6664	0.6700	0.6736	0.6772	0.6808	0.6844	0.6879
0.5	0.6915	0.6950	0.6985	0.7019	0.7054	0.7088	0.7123	0.7157	0.7190	0.7224
0.6	0.7257	0.7291	0.7324	0.7357	0.7389	0.7422	0.7454	0.7486	0.7517	0.7549
0.7	0.7580	0.7611	0.7642	0.7673	0.7704	0.7734	0.7764	0.7794	0.7823	0.7852
0.8	0.7881	0.7910	0.7939	0.7967	0.7995	0.8023	0.8051	0.8078	0.8106	0.8133
0.9	0.8159	0.8186	0.8212	0.8238	0.8264	0.8289	0.8315	0.8340	0.8365	0.8389
1.0	0.8413	0.8438	0.8461	0.8485	0.8508	0.8531	0.8554	0.8577	0.8599	0.8621
1.1	0.8643	0.8665	0.8686	0.8708	0.8729	0.8749	0.8770	0.8790	0.8810	0.8830
1.2	0.8849	0.8869	0.8888	0.8907	0.8925	0.8944	0.8962	0.8980	0.8997	0.9015
1.3	0.9032	0.9049	0.9066	0.9082	0.9099	0.9115	0.9131	0.9147	0.9162	0.9177
1.4	0.9192	0.9207	0.9222	0.9236	0.9251	0.9265	0.9279	0.9292	0.9306	0.9319
1.5	0.9332	0.9345	0.9357	0.9370	0.9382	0.9394	0.9406	0.9418	0.9429	0.9441
1.6	0.9452	0.9463	0.9474	0.9484	0.9495	0.9505	0.9515	0.9525	0.9535	0.9545
1.7	0.9554	0.9564	0.9573	0.9582	0.9591	0.9599	0.9608	0.9616	0.9625	0.9633
1.8	0.9641	0.9649	0.9656	0.9664	0.9671	0.9678	0.9686	0.9693	0.9699	0.9706
1.9	0.9713	0.9719	0.9726	0.9732	0.9738	0.9744	0.9750	0.9756	0.9761	0.9767
2.0	0.9772	0.9778	0.9783	0.9788	0.9793	0.9798	0.9803	0.9808	0.9812	0.9817
2.1	0.9821	0.9826	0.9830	0.9834	0.9838	0.9842	0.9846	0.9850	0.9854	0.9857
2.2	0.9861	0.9864	0.9868	0.9871	0.9875	0.9878	0.9881	0.9884	0.9887	0.9890
2.3	0.9893	0.9896	0.9898	0.9901	0.9904	0.9906	0.9909	0.9911	0.9913	0.9916
2.4	0.9918	0.9920	0.9922	0.9925	0.9927	0.9929	0.9931	0.9932	0.9934	0.9936
2.5	0.9938	0.9940	0.9941	0.9943	0.9945	0.9946	0.9948	0.9949	0.9951	0.9952
2.6	0.9953	0.9955	0.9956	0.9957	0.9959	0.9960	0.9961	0.9962	0.9963	0.9964
2.7	0.9965	0.9966	0.9967	0.9968	0.9969	0.9970	0.9971	0.9972	0.9973	0.9974
2.8	0.9974	0.9975	0.9976	0.9977	0.9977	0.9978	0.9979	0.9979	0.9980	0.9981
2.9	0.9981	0.9982	0.9982	0.9983	0.9984	0.9984	0.9985	0.9985	0.9986	0.9986
3.0	0.9987	0.9987	0.9987	0.9988	0.9988	0.9989	0.9989	0.9989	0.9990	0.9990
3.1	0.9990	0.9991	0.9991	0.9991	0.9992	0.9992	0.9992	0.9992	0.9993	0.9993
3.2	0.9993	0.9993	0.9994	0.9994	0.9994	0.9994	0.9994	0.9995	0.9995	0.9995
3.3	0.9995	0.9995	0.9995	0.9996	0.9996	0.9996	0.9996	0.9996	0.9996	0.9997
3.4	0.9997	0.9997	0.9997	0.9997	0.9997	0.9997	0.9997	0.9997	0.9997	0.9998

EXAMPLE 10.4.2

Suppose that the random variable Z has the standard normal distribution. Find each probability.

a. $P(Z \leq 1)$ **b.** $P(Z \geq 0.03)$ **c.** $P(Z \leq -1.87)$

d. $P(-1.87 \leq Z \leq 1)$ **e.** $P(Z \leq 4.7)$

Solution

a. $P(Z \leq 1)$ is the area under the normal curve to the left of $b = 1$ (Figure 10.17a). Look up $b = 1$ in the table to find that

$$P(Z \leq 1) = 0.8413.$$

b. $P(Z \geq 0.03)$ is the area under the curve to the right of 0.03 (Figure 10.17b). Since the table only gives areas to the *left* of values of Z, this area cannot be obtained directly from the table. Instead, use the fact that

$$P(Z \geq 0.03) = 1 - P(Z \leq 0.03).$$

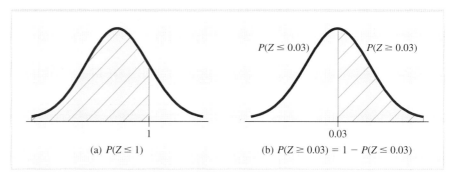

FIGURE 10.17   Probability as area under the standard normal density curve.

That is, the area to the right of 0.03 is 1 minus the area to the left of 0.03. The table gives

$$P(Z \leq 0.03) = 0.5120,$$

and so

$$P(Z \geq 0.03) = 1 - 0.5120 = 0.4880.$$

c. $P(Z \leq -1.87)$ is the area under the curve to the left of -1.87 (Figure 10.18a). Since the curve is symmetric, this is the same as the area to the right of 1.87 (Figure 10.18b). That is,

$$P(Z \leq -1.87) = P(Z \geq 1.87).$$

Proceed as in part (b) to get

$$P(Z \geq 1.87) = 1 - P(Z \leq 1.87) = 1 - 0.9693 = 0.0307.$$

Hence, $P(Z \leq -1.87) = 0.0307.$

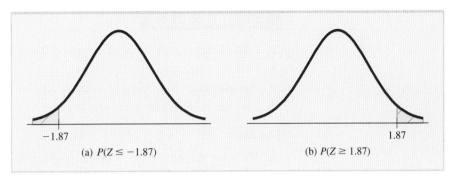

(a) $P(Z \leq -1.87)$ · (b) $P(Z \geq 1.87)$

FIGURE 10.18 $P(Z \leq -1.87) = P(Z \geq 1.87) = 1 - P(Z \leq 1.87)$.

d. $P(-1.87 \leq Z \leq 1)$ is the area under the curve between -1.87 and 1 (Figure 10.19c). This can be computed by subtracting the area to the left of -1.87 (Figure 10.19b) from the area to the left of 1 (Figure 10.19a). Use the results of parts (a) and (c) to obtain

$$P(-1.87 \leq Z \leq 1) = P(Z \leq 1) - P(Z \leq -1.87)$$
$$= 0.8413 - 0.0307 = 0.8106$$

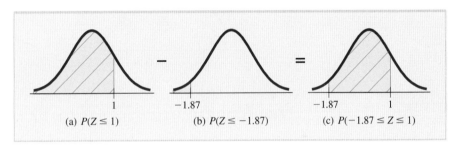

(a) $P(Z \leq 1)$ · (b) $P(Z \leq -1.87)$ · (c) $P(-1.87 \leq Z \leq 1)$

FIGURE 10.19 $P(-1.87 \leq Z \leq 1) = P(Z \leq 1) - P(Z \leq -1.87)$.

e. $P(Z \leq 4.7)$ is the area under the curve to the left of 4.7. However, the table only covers the values $-3.49 \leq Z \leq 3.49$. In fact, outside the interval $-3 \leq Z \leq 3$, the curve is so close to the x axis that there is virtually no area under it (Figure 10.20). Since almost all of the area under the curve is to the left of 4.7, $P(Z \leq 4.7) \approx 1$.

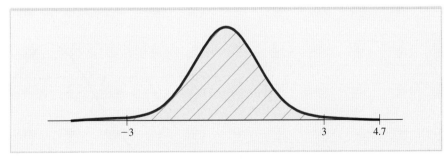

FIGURE 10.20   There is virtually no area to the right of 4.7, so $P(Z \leq 4.7) \approx 1$.

Example 10.4.3 illustrates how to use the table to find Z values corresponding to given probabilities.

EXAMPLE 10.4.3

a. Find the value b so that $P(Z \le b) = 0.9$.
b. Find the value b so that $P(Z \ge b) = 0.01$.
c. Find the value b so that $P(Z \le b) = 0.1$.
d. Find the positive number b so that $P(-b \le Z \le b) = 0.95$.

Solution

a. We want the value of b with the property that the area to the left of b is 0.9 (Figure 10.21a). In the area columns of the normal table, there is no entry equal to 0.9, so we choose the entry closest to it, 0.8997, which corresponds to a b value of 1.28. That is, rounded off to 2 decimal places, $P(Z \le 1.28) = 0.9$, and so the required value is (approximately) $b = 1.28$.

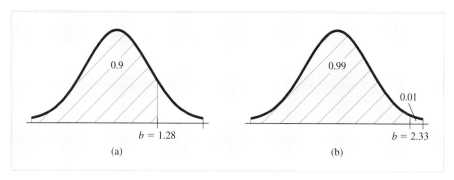

FIGURE 10.21 (a) $P(Z \le b) = 0.9$ and (b) $P(Z \ge b) = 0.01$.

b. We want the value of b with the property that the area to the right of it is 0.01, or, equivalently, the area to the left of it is 0.99 (Figure 10.21b). Searching the table, we find that 0.9901 is the area closest to 0.99, and so we take the corresponding value, $b = 2.33$.

c. We want the value of b so that the area to the left of b is 0.1 (Figure 10.22). Since the area under the curve to the left of 0 is 0.5, it follows that the desired value of b is less than 0 and does not appear in the table. Proceeding indirectly,

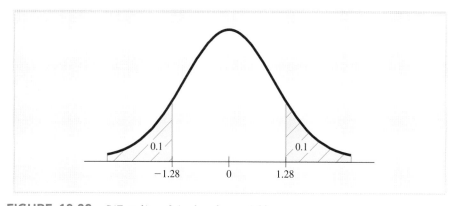

FIGURE 10.22 $P(Z \le b) = 0.1$ when $b = -1.28$.

we first find the value of b with the property that the area to the right is 0.1. Equivalently, this is also the value of b for which the area to the left is 0.9, and from part (a), we know that this value is 1.28. By symmetry, the value of b we seek is the negative of this; that is, the value of b such that $P(Z \le b) = 0.1$ is $b = -1.28$.

d. We want the value of b so that the area to the left of b is

$$0.95 + \frac{1}{2}(1 - 0.95) = 0.95 + 0.025 = 0.975$$

(Figure 10.23). From Table 10.1, we find that $b = 1.96$.

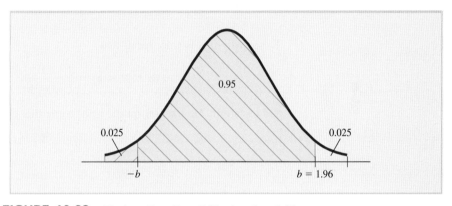

FIGURE 10.23 $P(-b \le Z \le b) = 0.95$ when $b = 1.96$.

Other Normal Distributions

You will not find separate tables of $P(X \le b)$ for nonstandard normal variables. Instead, we compute $P(a \le X \le b)$ using a table for standard normal variables together with the transformations displayed in this box.

> **Transforming a Nonstandard Normal Distribution to Standard Form**
>
> If the random variable X has a normal distribution with mean μ and standard deviation σ, then the transformed variable $Z = \dfrac{X - \mu}{\sigma}$ has a standard normal distribution. Furthermore, we have
>
> $$P(a \le X \le b) = P\left(\frac{a - \mu}{\sigma} \le Z \le \frac{b - \mu}{\sigma}\right).$$

Geometrically, this result says that the area under the normal curve with mean μ and standard deviation σ between a and b is the same as the area under the standard normal curve between $z_1 = \dfrac{a - \mu}{\sigma}$ and $z_2 = \dfrac{b - \mu}{\sigma}$. This is illustrated in Figure 10.24.

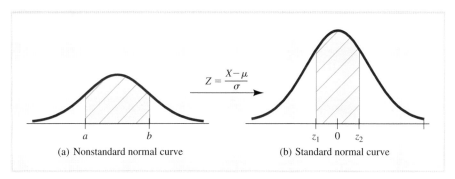

FIGURE 10.24   The area under the nonstandard normal curve between a and b is the same as the area under the standard normal curve between z_1 and z_2.

To see why this is true, recall that

$$P(a \le X \le b) = \int_a^b \frac{1}{\sigma\sqrt{2\pi}} e^{-(x-\mu)^2/2\sigma^2} \, dx,$$

and transform the integral on the right using this change of variable:

limits of integration

$$z = \frac{x - \mu}{\sigma} \qquad \text{when } x = a, \text{ then } z = \frac{a - \mu}{\sigma}$$

$$dz = \frac{1}{\sigma} \, dx \qquad \text{when } x = b, \text{ then } z = \frac{b - \mu}{\sigma}$$

Substituting gives

$$\int_a^b \frac{1}{\sigma\sqrt{2\pi}} e^{-[(x-\mu)^2/\sigma^2]/2} \, dx = \int_{(a-\mu)/\sigma}^{(b-\mu)/\sigma} \frac{1}{\sqrt{2\pi}} e^{-z^2/2} \, dz$$

$$= P\left(\frac{a - \mu}{\sigma} \le Z \le \frac{b - \mu}{\sigma}\right)$$

The use of this transformation is illustrated in Example 10.4.4.

EXAMPLE 10.4.4

Suppose X has a normal distribution with $\mu = 20$ and $\sigma = 4$. Find each probability:

  a. $P(X \le 26)$ **b.** $P(X \ge 18)$ **c.** $P(15 \le X \le 21)$

Solution

In each case, apply the formula $z = \dfrac{x - \mu}{\sigma}$ and then use the values in Table 10.1.

  a. $P(X \le 26) = P\left[Z \le \dfrac{26 - 20}{4}\right] = P(Z \le 1.5) = 0.9332$

  b. $P(x \ge 18) = P\left[Z \ge \dfrac{18 - 20}{4}\right] = P(Z \ge -0.5) = 0.6915$

c. $P(15 \leq X \leq 21) = P\left[\dfrac{15 - 20}{4} \leq Z \leq \dfrac{21 - 20}{4}\right]$

$= P(-1.25 \leq Z \leq 0.25)$

$= 0.5987 - 0.1056 = 0.4931$

Applications of the Normal Distribution

Next, we will examine several practical situations in which the normal distribution can be used. The first example deals with random errors made by measuring devices.

EXAMPLE 10.4.5

On a particular scale, measurement errors are made that are normally distributed with $\mu = 0$ g and $\sigma = 0.1$ g. If you weigh an object on this scale, what is the probability that the measurement error is no greater than 0.15 g?

Solution

Let X denote the measurement error made when an object is weighed on the scale. Our goal is to find $P(-0.15 \leq X \leq 0.15)$. Since $\mu = 0$ and $\sigma = 0.1$, we transform to Z using the formula $Z = \dfrac{X - 0}{0.1}$ and find that

$$P(-0.15 \leq X \leq 0.15) = P\left(\frac{-0.15 - 0}{0.1} \leq Z \leq \frac{0.15 - 0}{0.1}\right)$$

$$= P(-1.5 \leq Z \leq 1.5) = 0.8664$$

That is, there is an 87% chance that the error in measurement is no greater than 0.15 g.

EXAMPLE 10.4.6

The number of loaves of bread that can be sold during a day by a certain supermarket is normally distributed with $\mu = 1000$ loaves and $\sigma = 100$ loaves. If the market stocks 1200 loaves on a given day, what is the probability that the loaves will be sold out before the day is over?

Solution

If X denotes the number of loaves that can be sold during a day, then X is normally distributed with $\mu = 1000$ and $\sigma = 100$. We wish to find $P(X \geq 1200)$. Transforming to Z and using the table gives

$$P(X \geq 1200) = P\left(Z \geq \frac{1200 - 1000}{100}\right) = P(Z \geq 2) = 0.0228.$$

Thus, there is about a 2% chance that the market will run out of loaves before the day is over.

> **EXAMPLE 10.4.7**

The lengths of trout in a certain lake are normally distributed with a mean of 20 cm and a standard deviation of 5 cm. If the Ministry of Natural Resources would like anglers to keep only the largest 20% of the trout, what should the minimum size for keepers be?

Solution

Let X denote the length of a randomly selected trout and c the minimum size for a keeper. Since only the largest 20% are to be keepers, c must satisfy the equation $P(X \geq c) = 0.2$. Using $\mu = 20$ and $\sigma = 5$, transform to Z and obtain

$$P\left(Z \geq \frac{c - 20}{5}\right) = 0.2 \quad \text{or} \quad P\left(Z \leq \frac{c - 20}{5}\right) = 0.8.$$

From Table 10.1,

$$P(Z \leq 0.84) = 0.8.$$

Hence,

$$\frac{c - 20}{5} = 0.84 \quad \text{or} \quad c = 24.2.$$

Thus, the Ministry of Natural Resources should set 24.2 cm as the minimum size for keepers.

Poisson Distributions

Next, we examine the Poisson distribution, which is a discrete distribution named for the 19th-century French mathematician Siméon Denis Poisson. Although Poisson introduced this distribution in theoretical work done in 1837, its first practical application was made in 1898 by Ladislaus Josephowitsch Bortkiewicz, who used it to study the accidental deaths of Prussian soldiers from kicks by horses, a relatively rare event. Since then, Poisson distributions have become an important tool for modelling other rare events, such as the number of people struck by lightning during a year or the number of substitutions that have occurred in a particular amino acid sequence over a million-year period. More generally, Poisson distributions are used for modelling counts in time or space, such as the number of customers that arrive at a store during an hour, or the number of typographical errors on a page in a particular book, or the density of trees in areas of a fixed size throughout a large forest, or the number of bacterial colonies in a Petri dish.

A discrete random variable X is said to be **Poisson distributed** with parameter λ ($\lambda > 0$) if X describes an experiment whose possible outcomes are the nonnegative integers $n = 0, 1, 2, \ldots$ and for each nonnegative integer n, the event $(X = n)$ is assigned the probability

$$P(X = n) = \frac{\lambda^n}{n!} e^{-\lambda}.$$

To verify that this satisfies the requirements of a probability assignment, we must show that $P(X = n) \geq 0$ for $n = 0, 1, 2, \ldots$ and that $\sum_{n=0}^{\infty} P(X = n) = 1$. The condition

$P(X = n) \geq 0$ for all n is satisfied since $\lambda > 0$ and $e^{-\lambda} > 0$. To show that the probability assignments of the distribution sum to 1, we require the Taylor series expansion $e^x = \sum_{n=0}^{\infty} \dfrac{x^n}{n!}$ developed in Section 9.3. Note that

$$\sum_{n=0}^{\infty} P(X = n) = P(X = 0) + P(X = 1) + P(X = 2) + \cdots$$

$$= \frac{\lambda^0}{0!}e^{-\lambda} + \frac{\lambda^1}{1!}e^{-\lambda} + \frac{\lambda^2}{2!}e^{-\lambda} + \cdots$$

$$= e^{-\lambda}\left(1 + \lambda + \frac{\lambda^2}{2!} + \cdots\right) = e^{-\lambda}\sum_{n=0}^{\infty}\frac{\lambda^n}{n!}$$

$$= e^{-\lambda}e^{\lambda} \quad \text{since } \sum_{n=0}^{\infty}\frac{\lambda^n}{n!} \text{ is the Taylor series for } e^{\lambda}$$

$$= 1$$

The use of the Poisson distribution is illustrated in Examples 10.4.8 and 10.4.9.

EXAMPLE 10.4.8

A botanist researching the growth patterns of a particular kind of orchid models the number of orchid plants per square metre as a random variable X with a Poisson distribution. If the distribution has parameter $\lambda = 0.2$, find

a. The probability that there are no orchid plants located within a randomly selected square metre.

b. The probability that there is no more than one orchid plant within a randomly selected square metre.

Solution

a. The probability that there are no orchid plants is

$$P(X = 0) = \frac{(0.2)^0}{0!}e^{-0.2} = e^{-0.2} \qquad \text{since } (0.2)^0 = 1 \text{ and } 0! = 1$$

$$\approx 0.82$$

b. The event that there is no more than one orchid plant is $(X \leq 1)$ and the probability of this event is

$$P(X \leq 1) = P(X = 0) + P(X = 1) = \frac{(0.2)^0}{0!}e^{-0.2} + \frac{(0.2)^1}{1!}e^{-0.2}$$

$$= (1 + 0.2)e^{-0.2}$$

$$\approx 0.98$$

Thus, a randomly selected square metre is not likely to contain many orchids.

EXAMPLE 10.4.9

The number of accidents in a particular industry that occur monthly can be modelled as a random variable X with a Poisson distribution if the industrial accidents tend to occur independently of one another at an approximately constant rate. Suppose studies indicate that there is only a 10% likelihood of no accidents occurring during a randomly chosen month.

a. Determine the parameter λ for the distribution.

b. Find the probability that at least three accidents occur during a randomly chosen month.

Solution

a. Since there is a 10% likelihood that no accidents will occur,

$$P(X = 0) = \frac{\lambda^0}{0!}e^{-\lambda} = 0.10$$

$$e^{-\lambda} = 0.10 \qquad \text{since } \lambda^0 = 1 \text{ and } 0! = 1$$

Taking logarithms on both sides of this equation gives

$$\ln(e^{-\lambda}) = \ln(0.10)$$
$$-\lambda = -2.3$$

so the parameter is $\lambda = 2.3$.

b. The event that at least three industrial accidents occur is $(X \geq 3)$, and the probability of this event satisfies

$$P(X \geq 3) = 1 - P(X < 3),$$

where $(X < 3)$ is the complementary event that less than three accidents occur. Thus,

$$P(X < 3) = P(X = 0) + P(X = 1) + P(X = 2)$$
$$= \frac{(2.3)^0}{0!}e^{-2.3} + \frac{(2.3)^1}{1!}e^{-2.3} + \frac{(2.3)^2}{2!}e^{-2.3}$$
$$\approx 0.6$$

so that

$$P(X \geq 3) \approx 1 - 0.6 = 0.4.$$

Thus, there is approximately a 40% probability that at least three accidents will occur.

If the random variable X is Poisson distributed with parameter λ, then the expected value of X is $E(X) = \lambda$. To show that this is true, note that

$$E(X) = \sum_{n=0}^{\infty} nP(X = n) \qquad \text{definition of expected value}$$

$$= \sum_{n=0}^{\infty} n\left(\frac{\lambda^n e^{-\lambda}}{n!}\right) \qquad \text{definition of the Poisson distribution}$$

$$= 0 \cdot \left(\frac{\lambda^0 e^{-\lambda}}{0!}\right) + 1 \cdot \left(\frac{\lambda^1 e^{-\lambda}}{1!}\right) + 2 \cdot \left(\frac{\lambda^2 e^{-\lambda}}{2!}\right) + \cdots + n \cdot \left(\frac{\lambda^n e^{-\lambda}}{n!}\right) + \cdots$$

$$= \lambda e^{-\lambda} + \frac{\lambda^2}{1}e^{-\lambda} + \frac{\lambda^3}{2!}e^{-\lambda} + \cdots \qquad \text{since } \frac{n}{n!} = \frac{1}{(n-1)!}$$

$$= \lambda e^{-\lambda}\left[1 + \lambda + \frac{\lambda^2}{2!} + \cdots\right] \qquad \text{factor } \lambda e^{-\lambda} \text{ from each term}$$

$$= \lambda e^{-\lambda} \sum_{n=0}^{\infty} \frac{\lambda^n}{n!}$$

$$= \lambda e^{-\lambda} e^{\lambda} \qquad \text{since } \sum_{n=0}^{\infty} \frac{\lambda^n}{n!} \text{ is the Taylor series for } e^{\lambda}$$

$$= \lambda \qquad \text{since } e^{-\lambda} e^{\lambda} = 1$$

We leave it as an exercise (Exercise 58) for you to show that the variance of X also equals λ. To summarize:

Expected Value, Variance, and Standard Deviation of a Random Variable with a Poisson Distribution ■ If the random variable X has a Poisson distribution with parameter λ, then

$$P(X = n) = \frac{\lambda^n}{n!} e^{-\lambda}$$

and

Expected value: $\qquad E(X) = \lambda$

Variance: $\qquad \text{Var}(X) = \lambda$

Standard deviation: $\sigma(X) = \sqrt{\text{Var}(X)} = \sqrt{\lambda}$

EXAMPLE 10.4.10

Suppose that the number of calls you receive from telemarketers during the supper hour on a randomly selected day follows a Poisson distribution, and that on average, you expect to get three such calls.

a. What is the parameter λ of the distribution?

b. What is the probability that you will get exactly three calls?

c. What is the probability that you will receive more than three calls?

Solution

a. Let X be the random variable that counts the number of calls you receive during a randomly selected supper hour. Interpret the statement that "on average, you expect to get three calls" as meaning that the expected value of X is 3. Thus, $E(X) = \lambda = 3$.

b. The probability that you will get exactly three calls is

$$P(X = 3) = \frac{3^3 e^{-3}}{3!} \approx 0.22.$$

c. The event that you receive more than three calls is $(X > 3)$, and the probability of this event satisfies

$$P(X > 3) = 1 - P(X \leq 3),$$

where $(X \leq 3)$ is the complementary event that no more than three calls are received. Thus,

$$P(X \leq 3) = P(X = 0) + P(X = 1) + P(X = 2) + P(X = 3)$$

$$= \frac{3^0 e^{-3}}{0!} + \frac{3^1 e^{-3}}{1!} + \frac{3^2 e^{-3}}{2!} + \frac{3^3 e^{-3}}{3!}$$

$$\approx 0.647$$

so that

$$P(X > 3) = 1 - P(X \le 3) \approx 1 - 0.647$$
$$= 0.353$$

Thus, it is approximately 35% likely that you will receive more than three calls from telemarketers during a randomly selected dinner hour.

EXERCISES ■ 10.4

In Exercises 1 through 4, the density function f(x) of a normal random variable X is given. In each case, find the expected value E(X), the variance Var(X), and the standard deviation σ.

1. $f(x) = \dfrac{1}{2\sqrt{2\pi}}e^{-x^2/8}$

2. $f(x) = \dfrac{1}{7\sqrt{2\pi}}e^{-(x-3)^2/98}$

3. $f(x) = \dfrac{1}{\sqrt{6\pi}}e^{-(x+1)^2/6}$

4. $f(x) = \dfrac{1}{\sqrt{2\pi}}e^{-(x+6)^2/2}$

In Exercises 5 through 10, find the indicated probability assuming that the random variable Z has a standard normal distribution.

5. $P(Z \le 1.24)$

6. $P(Z \le -1.20)$

7. $P(Z \le 4.26)$

8. $P(Z \ge 0.19)$

9. $P(-1 \le Z \le 1)$

10. $P(-1.20 \le Z \le 4.26)$

In Exercises 11 through 14, find the appropriate value of b for the given probability, assuming that the random variable Z has a standard normal distribution.

11. $P(Z \le b) = 0.8413$

12. $P(Z \ge b) = 0.2266$

13. $P(Z \le b) = 0.8643$

14. $P(-b \le Z \le b) = 0.9544$

15. Suppose X has a normal distribution with $\mu = 60$ and $\sigma = 4$. Find each probability.
   a. $P(X \ge 68)$ **b.** $P(X \ge 60)$
   c. $P(56 \le X \le 64)$ **d.** $P(X \le 40)$

16. Suppose X has a normal distribution with $\mu = 35$ and $\sigma = 3$. Find each probability.
   a. $P(X \ge 38)$ **b.** $P(X \ge 35)$
   c. $P(32 \le X \le 38)$ **d.** $P(X \le 20)$

17. Suppose X has a normal distribution with $\mu = 4$ and $\sigma = 0.5$. For each probability, find the appropriate value of x.
   a. $P(X \ge x) = 0.0228$
   b. $P(X \le x) = 0.1587$

18. Suppose X has a normal distribution with $\mu = 5$ and $\sigma = 0.6$. For each probability, find the appropriate value of x.
   a. $P(5 - x \le X \le 5 + x) = 0.6826$
   b. $P(X \le x) = 0.6$

19. A random variable X has a normal distribution with $\mu = 15$ and $\sigma = 4$. Using the empirical rule, find
   a. $P(11 \le X \le 19)$
   b. $P(7 \le X \le 23)$

20. A random variable X has a normal distribution with $\mu = 18$ and $\sigma = 3$. Using the empirical rule, find
   a. $P(15 \le X \le 21)$
   b. $P(9 \le X \le 27)$

21. Suppose that X is a Poisson random variable with parameter $\lambda = 3$. Give the exact formula and the value to three decimal places for each probability.
   a. $P(X = 1)$ **b.** $P(X = 3)$
   c. $P(X = 4)$ **d.** $P(X = 8)$

22. Suppose that X is a Poisson random variable with parameter $\lambda = \dfrac{3}{2}$. Give the exact formula and the value to three decimal places for each probability.
  a. $P(X = 0)$ **b.** $P(X = 1)$
  c. $P(X = 3)$ **d.** $P(X = 5)$

23. Draw the histogram for the Poisson random variable with $\lambda = 2$. Show 5 rectangles.

24. Draw the histogram for the Poisson random variable with $\lambda = \dfrac{1}{2}$. Show 5 rectangles.

25. Decide whether each statement is true or false. Explain your answer.
  a. If X has a normal distribution, then
$$P(X \geq 2.1) = P(X \leq 2.1).$$
  b. If Z has the standard normal distribution, then
$$P(Z \geq 2.1) = P(Z \leq 2.1).$$

26. Decide whether each statement is true or false. Explain your answer.
  a. If X has a normal distribution with $\mu = 8$, then
$$P(X \geq 8.21) = P(X \leq 7.79).$$
  b. If X has a normal distribution, then the probability is about 0.87 that its value will be within 1.5 standard deviations of the mean.

27. VETERINARY MEDICINE The weights of adult basset hounds are normally distributed with $\mu = 23$ kg and standard deviation $\sigma = 2$ kg. Use the empirical rule to describe this distribution of weights.

28. ACADEMIC TESTING The scores on a certain test are normally distributed with $\mu = 78$ and standard deviation $\sigma = 7$. Use the empirical rule to describe the distribution of these scores.

29. EMPLOYEE COMPENSATION The incomes of industrial workers in a certain region are normally distributed with a mean of $42 500 and a standard deviation of $3000. Find the probability that a randomly selected worker has an income between $41 000 and $44 000.

30. PRODUCT MANAGEMENT A meat company produces hot dogs with lengths that are normally distributed with a mean length of 15 cm and standard deviation of 0.4 cm. Find the fraction of hot dogs that are less than 14 cm long.

31. QUALITY CONTROL A certain company produces glass jars. On the average, the jars hold 1 L, but there is some variation among them. Assume the volumes are normally distributed with standard deviation $\sigma = 0.01$ L.
  a. Find the probability that a randomly chosen jar will hold between 0.98 L and 1.02 L.
  b. Find the fraction of jars that hold less than 0.97 L.

32. QUALITY CONTROL The breaking strength of a certain type of rope is normally distributed. If the rope has an average breaking strength of 500 kg, with standard deviation 40 kg, find the probability that a randomly selected piece of rope will break under a strain of less than 480 kg.

33. FARMING A certain carrot farm is famous for growing baby salad carrots of almost identical length. If the length is normally distributed with mean 5 cm, what must the standard deviation be if 99% of the carrots are between 4.9 cm and 5.1 cm long?

34. AGRICULTURE Suppose that the weight of a cherry in a good year from a particular orchard is normally distributed with a weight of 10 g and a standard deviation of 1 g.
  a. Find the probability that a cherry from this orchard weighs at least 9 g.
  b. Find the probability that a cherry from this orchard weighs between 8 g and 12 g.
  c. Find the weight so that 99% of all cherries from this orchard weigh more than this weight.

35. ACADEMIC TESTING Suppose all students in grade 4 in a certain school are taught to read by the same teaching method and that at the end of the year they are tested for reading speed. Suppose reading speed is normally distributed and that average reading speed is 150 words per minute with a standard deviation of 25 words per minute. Find the percentage of students who read more than 180 words per minute.

36. ACADEMIC TESTING The scores on a psychology exam are normally distributed with a mean of 75 and a standard deviation of 10.
  a. If a score of at least 60 is needed to pass, find the fraction of students who pass the exam.
  b. If the instructor wishes to pass 70% of those taking the test, what should the lowest passing score be?
  c. How high must someone score to be in the top 10%?

37. SOCIAL PATTERNS A certain club is having a meeting. A total of 1000 members will attend, but only 200 seats are available. It is decided that the oldest members will get seats. Suppose the age of members is normally distributed. If the average age of members is 40 with a standard deviation of 5, find the age of the youngest person who will get a seat.

38. VETERINARY MEDICINE The weights of healthy 10-week-old domesticated kittens follow a normal distribution with a mean weight of 690 g and standard deviation of 150 g. Find the symmetric intervals around the mean of 690 g that include the weights of 68.3%, 95.4%, and 99.7% of 10-week-old domesticated kittens, respectively.

39. VETERINARY MEDICINE Use the information in Exercise 38 to find
  a. The percentage of healthy 10-week-old domesticated kittens that weigh more than 900 g.
  b. The weight that is exceeded by exactly 90% of 10-week-old domesticated kittens.

40. PEDIATRIC MEDICINE The cholesterol level in children follows a normal distribution with a mean level of 4.5 millimoles/L (mmol/L) and a standard deviation of 0.9 mmol/L. Find the symmetric intervals around the mean of 4.5 mmol/L that respectively include the cholesterol levels of 68.3%, 95.4%, and 99.7% of all children.

41. PEDIATRIC MEDICINE Using the information in Exercise 40, find
  a. The percentage of children with a cholesterol level above 5.8 mmol/L.
  b. The cholesterol level exceeded by 98% of children.

42. WASTE DISPOSAL Consider an environmentally progressive neighbourhood where everyone puts their kitchen and yard waste into their green bin for city composting. The weight of compostable waste follows a normal distribution with an average of 2.5 kg and a standard deviation of 0.4 kg.
  a. Find the percentage of households that put more than 3 kg in the green bin.
  b. Find the percentage of households that put less than 1.5 kg in the green bin.
  c. Find the percentage of households that put between 3 kg and 3.5 kg in the green bin.
  d. Find the weight of waste in the green bin from households that produce the highest 5% of compostable waste.
  e. Find the weight of waste in the green bin from households that produce the lowest 10% of compostable waste.

43. MEDICINE The gestation period of humans approximately follows a normal distribution with a mean of 266 days and a standard deviation of 16 days. Find the probability of each of the following.
  a. A gestation period that is greater than 290 days.
  b. A gestation period that is less than 260 days.
  c. A gestation period that is between 260 days and 280 days.

44. MEDICINE Assume that the human gestation period follows a normal distribution with a mean of 266 days and a standard deviation of 16 days. Find the value of A in each case.
  a. The probability is 0.33 that a gestation period is less than A days.
  b. The probability is 0.74 that a gestation period exceeds A days.
  c. The probability is 0.5 that a gestation period is between $266 - A$ days and $266 + A$ days.

45. MANUFACTURING Suppose that the weight that a processing plant puts in each can of peas is normally distributed with a mean of 280 g and a standard deviation of 1.4 g. The quality control department of the factory has specified that a can must contain at least 277 g so that consumers do not feel cheated, but no more than 284 g to make sure the can is not overfilled. What is the probability that a randomly selected can from the production line passes these specifications?

46. MANUFACTURING When a certain type of engine is built, a metal rod is inserted into a 6.2-cm round hole. If the diameter of the rod is greater than 6.2 cm, the rod cannot be used. Find the percentage of rods that cannot be used in the assembly of the engine if the diameter of these rods follows a normal distribution with a mean of 6.05 cm and a standard deviation of 0.1 cm.

47. WEATHER FORECASTING In spring, in a Saskatchewan town, the temperature can vary considerably. Suppose that during April, the minimum daily temperature has a normal distribution with a mean of $-1.6°C$ and a standard deviation of 8°C.
  a. Find the probability that during a day in April, the minimum daily temperature lies between 0°C and 10°C. What is the probability that the minimum daily temperature lies between $-5°C$ and 5°C?

b. What is the probability that during a day in April, the minimum daily temperature is greater than 5°C?

c. Find the probability that during a day in April, the minimum daily temperature is less than 5°C.

48. POLITICAL POLLS   There is a tight race for mayor. A poll from a survey of potential voters states that Nate Newton leads Patsy Poisson by 51.2% to 48.8%, where the error in the poll results is normally distributed with a mean of 0 percentage points and a standard deviation of 3 percentage points.

   a. Find the probability that Poisson is actually leading. [*Hint:* How large must the error in Newton's share be for the lead to switch? Assume there are no undecided votes.]

   b. Find the probability that Newton's lead is actually at least 3 percentage points instead of the indicated 2.4.

   c. Research polling procedures and write a paragraph on whether you believe political polls provide useful, reliable information.

49. QUALITY CONTROL   A manufacturer needs washers that are between 4.8 mm and 5.2 mm thick. Any other thickness will render a washer unusable. Machine shop A sells washers for $1.00 per thousand, and the thicknesses of its washers are modelled as a normal random variable with $\mu = 5$ mm and $\sigma = 0.1$ mm. Machine shop B sells washers for $0.90 per thousand, and the thicknesses of its washers are also modelled as a normal random variable, but with $\mu = 5$ mm and $\sigma = 0.11$ mm. Using the price per usable washer as the criterion, which shop offers the manufacturer the better deal?

50. QUALITY CONTROL   Suppose, in Exercise 49, that machine shop C wants to compete for the washer contract. Its quality control model is based on the assumption that the thicknesses of its washers are normally distributed with mean $\mu = 4.95$ mm and standard deviation $\sigma = 0.09$ mm. What is the maximum price p it can charge per thousand washers so that its price per usable washer will be better than the deals offered by the competition, machine shops A and B?

51. DRUG TESTING   The recovery time of patients with a certain rash is normally distributed with a mean of 11 days and a standard deviation of 2 days.

   a. Find the probability that a patient takes between 10 days and 18 days to recover.

b. The On-Line-Only-Natural Drugs website claims that its treatment of this rash accelerates recovery. A patient takes the On-Line-Only-Natural medication and recovers in 9 days. Are you impressed? [*Hint:* What is the probability that the patient would have recovered in 9 days or fewer without the medication?]

c. Another patient takes the On-Line-Only-Natural medication and recovers in 6 days. Now are you impressed?

d. Actual drug testing involves conducting carefully monitored experiments using large numbers of subjects, with the goal of achieving consistent results that are statistically significant. Research the policies for drug certification required by Health Canada. Do you think that Health Canada is justified in being so strict?

52. SHARK ATTACKS   Suppose that the number of shark attacks in coastal waters in the Caribbean follows a Poisson distribution with a mean of 0.3 shark attacks per day.

   a. What is the parameter λ of the distribution?

   b. Find the probability that on a randomly selected day, there will be no shark attacks.

   c. Find the probability that on a randomly selected day, there will be at least 2 shark attacks.

53. BACTERIOLOGY   Suppose that the number of *Escherichia coli* bacteria in a water sample drawn from a swimming beach on Lake Ontario follows a Poisson distribution with a mean of 6.1 bacteria per cubic millimetre of water.

   a. What is the parameter λ of the distribution?

   b. Find the probability that a 1-mm^3 random sample of water contains no *E. coli* bacteria.

   c. Find the probability that a random sample of 1 mm^3 of water contains exactly 3 *E. coli* bacteria.

   d. Find the probability that a random sample of 1 mm^3 of water contains fewer than 6 *E. coli* bacteria.

54. DISEASE CONTROL   Assume that the number of people who die from hepatitis A each year follows a Poisson distribution and that an average of 3 people per 100 000 die from hepatitis A each year in North America.

   a. Find the probability that in a city of 100 000, no people die during a randomly selected year from hepatitis A.

b. Find the probability that in a city of 100 000, no more than 4 people die from hepatitis A during a randomly selected year.
c. Find the probability that in a city of 100 000, at least 5 people die from hepatitis A during a randomly selected year.

55. DISEASE CONTROL   Using the information from Exercise 54, in a randomly selected year, how much more likely is it for at least 1 person to die from hepatitis A in a city of 1 000 000 people than in a city of 100 000 people?

56. NUTRITION   Suppose a bakery makes a batch of 2400 chocolate chip cookies using a total of 12 000 chocolate chips. The random variable X that represents the number of chocolate chips in a cookie can be modelled using a Poisson distribution with parameter

$$\lambda = \frac{12\,000}{2400} = 5.$$

a. Find the probability that a cookie has exactly 5 chocolate chips.
b. Find the probability that a cookie has no chocolate chips.
c. Find the probability that a cookie has 3 or fewer chocolate chips.
d. Find the probability that a cookie has 7 or more chocolate chips.

57. WEBSITE HITS   Suppose that the number of hits on a popular website in a 1-minute period follows a Poisson distribution. It is estimated that there is a 5% likelihood of no hits occurring during the period.
a. Find the parameter λ for the distribution.
b. Find the probability that during a randomly chosen 1-minute period, there are exactly 5 hits.
c. Find the probability that the website receives no more than 5 hits during a 1-minute period.
d. Find the probability that the website receives at least 1 hit during a 1-minute period.
e. Determine the most likely number of hits the website receives during a 1-minute interval.

58. A* Show that a Poisson random variable X with parameter λ has variance $\text{Var}(X) = \lambda$.

59. A* Verify that normally distributed data are 95.4% likely to be within 2 standard deviations of the mean and 99.7% likely to be within 3 standard deviations of the mean.

60. A* Show that the probability density function

$$f(x) = \frac{1}{\sigma\sqrt{2\pi}}e^{-(x-\mu)^2/2\sigma^2}$$

has an absolute maximum at $x = \mu$ and inflection points at $x = \mu + \sigma$ and $x = \mu - \sigma$.

Concept Summary Chapter 10

CHAPTER SUMMARY

$0 \le P(s_k) \le 1$
$P(s_1) + P(s_2) + \cdots + P(s_n) = 1$

Features of a Discrete Random Variable X
Event $(X = x_k)$
Probability: $p(x_k) = P(X = x_k)$
Expected value: $\mu = E(X) = \sum x_k p(x_k)$
Variance: $\text{Var}(X) = \sum (x_k - \mu)^2 \cdot p(x_k)$
Standard deviation: $\sigma(X) = \sqrt{\text{Var}(X)}$

Geometric Random Variable X
Probability (p is probability of success):
$P(X = n) = (1 - p)^{n-1}p$

Expected value: $E(X) = \dfrac{1}{p}$

Variance: $\text{Var}(X) = \dfrac{1 - p}{p^2}$

Probability Density Function $f(x)$ for the Continuous Random Variable X
$f(x) \ge 0$ for all x

$$\int_{-\infty}^{\infty} f(x)\,dx = 1$$

$$P(a \le X \le b) = \int_{a}^{b} f(x)\,dx$$

Uniform Density Function

$$f(x) = \begin{cases} \dfrac{1}{(b - a)} & \text{if } a \le X \le b \\ 0 & \text{otherwise} \end{cases}$$

Exponential Density Function

$$f(x) = \begin{cases} \lambda e^{-\lambda x} & \text{if } X \ge 0 \\ 0 & \text{if } X < 0 \end{cases}$$

Joint Probability Density Function $f(x, y)$, for Random Variables X and Y

$f(x, y) \geq 0$ for all (x, y)

$$\int_{-\infty}^{\infty}\int_{-\infty}^{\infty} f(x, y)\, dA = 1$$

$$P[X, Y] \text{ in } R = \int\int_R f(x, y)\, dA$$

Expected Values for a Joint Probability Density Function $f(x, y)$

$$E(X) = \int_{-\infty}^{\infty}\int_{-\infty}^{\infty} xf(x, y)\, dA$$

$$E(Y) = \int_{-\infty}^{\infty}\int_{-\infty}^{\infty} yf(x, y)\, dA$$

Features of a Continuous Random Variable X

Expected value

$$E(x) = \int_{-\infty}^{\infty} xf(x)\, dx$$

Variance

$$\text{Var}(X) = \int_{-\infty}^{+\infty} [x - E(X)]^2 f(x)\, dx$$

Features of a Normal Random Variable X

Normal probability density function:

$$f(x) = \frac{1}{\sigma\sqrt{2\pi}} e^{-(x-\mu)^2/2\sigma^2}$$

Expected value: $E(X) = \mu$

Standard deviation: $\sigma(X) = \sigma$

Standard normal distribution, $\mu = 0$, $\sigma = 1$:

$$Z = \frac{X - \mu}{\sigma}$$

Features of a Poisson Distributed Random Variable X with Parameter λ

Probability: $P(X = n) = \dfrac{\lambda^n}{n!} e^{-\lambda}$

Expected value: $E(X) = \lambda$

Variance: $\text{Var}(X) = \lambda$

Standard deviation $\sigma(X) = \sqrt{\lambda}$

Checkup for Chapter 10

1. The outcomes and corresponding probability assignments for a discrete random variable X are listed in the table. Draw the histogram for X. Then find the expected value $E(X)$, the variance $\text{Var}(X)$, and the standard deviation $\sigma(X)$.

Outcomes for X	0	1	2	3	4
Probability	$\dfrac{1}{6}$	$\dfrac{1}{6}$	$\dfrac{1}{5}$	$\dfrac{2}{5}$	$\dfrac{1}{15}$

2. In each case, determine whether the given function is a probability density function.

   a. $f(x) = \begin{cases} \dfrac{4}{3}\sqrt[3]{x} & \text{if } 0 \leq x \leq 1 \\ 0 & \text{otherwise} \end{cases}$

   b. $f(x) = \begin{cases} 2e^{-x/2} & \text{if } x \geq 0 \\ 0 & \text{if } x < 0 \end{cases}$

3. The probability density function for a particular random variable X is

   $$f(x) = \begin{cases} \dfrac{1}{3}e^{-x/3} & \text{if } x \geq 0 \\ 0 & \text{if } x < 0 \end{cases}$$

Find each probability.

   a. $P(2 \leq X \leq 3)$ b. $P(X \geq 3)$

4. The probability density function for a particular random variable X is

   $$f(x) = \begin{cases} \dfrac{1}{36}(6x - x^2) & \text{if } 0 \leq x \leq 6 \\ 0 & \text{otherwise} \end{cases}$$

   Find the expected value $E(X)$ and variance $\text{Var}(X)$ for X.

5. Suppose the random variable Z has a standard normal distribution. Find each probability.

   a. $P(Z \geq 0.85)$ b. $P(1.30 \leq Z \leq 3.25)$

6. A random variable X has a normal distribution with mean $\mu = 12$ and standard deviation $\sigma = 2$. Use the empirical rule to find

   a. $P(10 \leq X \leq 14)$ b. $P(18 \leq X \leq 16)$

7. **MAINTENANCE** The failure of a certain key component will shut down the operation of a system until a new component is installed. If installation time is distributed uniformly from 1 hour to 4 hours, what is the probability that the system will be inoperable for at least 2 hours?

8. **QUALITY CONTROL**   Suppose that 5% of the items produced by a certain company are defective. A quality control inspector tests each item before it leaves the factory. Find the probability that the inspector passes at least 5 items before detecting a defective item.

9. **MEDICINE**   Let X be the random variable that represents the number of years a person in a test group with a certain kind of virulent cancer lives after receiving an experimental drug. Suppose X has an exponential distribution with parameter $\lambda = 0.5$.
   a. What is the probability of a randomly selected cancer victim in this group surviving at least 3 years after beginning therapy with the drug?
   b. Find the expected value of X. Describe what $E(X)$ represents.

10. **ACADEMIC TESTING**   To qualify for a police college, a candidate must score in the top 20% on a qualifying test. Assuming that the test scores are normally distributed and the average score on the test is 150 points with a standard deviation of 24 points, find the minimum qualifying score.

11. **MEDICINE**   In southern Ontario, the tick that spreads Lyme disease is spreading northward from

Lake Erie. Assume that the number of people who get bitten by a tick carrying Lyme disease each year follows a Poisson distribution and that on average, 5 people per 1 million in a certain area are infected. Find the probability that at least 5 people in 1 million in such an area will be infected and will need to get antibiotics to stop the disease from affecting them in later years.

12. **WARRANTY PROTECTION**   A major appliance contains two components that are vital for its operation in the sense that if either fails, the appliance is rendered useless. Let the random variable X measure the useful life (in years) of the first component and let Y measure the useful life of the second component (also in years). Suppose the joint probability density function for X and Y is

$$f(x, y) = \begin{cases} 0.1e^{-x/2}e^{-y/5} & \text{if } x \geq 0 \text{ and } y \geq 0 \\ 0 & \text{otherwise} \end{cases}$$

   a. Find the probability that the appliance fails within the first 5 years.
   b. Which component of a randomly selected appliance would you expect to last longer? How much longer?

Review Exercises

In Exercises 1 and 2, the outcomes and corresponding probability assignments for a discrete random variable X are listed. Draw the histogram for X. Then find the expected value E(X), the variance Var(X), and the standard deviation σ(X).

1.

Outcomes for X	1	2	3	4	5
Probability	$\frac{1}{9}$	$\frac{2}{9}$	$\frac{1}{3}$	$\frac{1}{9}$	$\frac{2}{9}$

2.

Outcomes for X	0	2	4	6	8
Probability	$\frac{1}{8}$	$\frac{3}{8}$	$\frac{1}{4}$	$\frac{1}{8}$	$\frac{1}{8}$

In each of Exercises 3 through 6, determine whether the given random variable X is discrete or continuous. If it is discrete, specify whether it is finite.

3. X counts the number of eggs laid by a randomly selected fruit fly.

4. X measures the annual distance flown by a randomly selected airplane from a particular airline.

5. X measures the annual salary of a randomly selected player on a particular major league baseball team.

6. X counts the number of books in the library of a randomly selected professor at your school.

In Exercises 7 through 10, f(x) is a probability density function for a particular continuous random variable X. In each case, find the indicated probabilities.

7. $f(x) = \begin{cases} \dfrac{1}{5} & \text{if } 3 \leq x \leq 8 \\ 0 & \text{otherwise} \end{cases}$

   $P(2 \leq X \leq 7)$ and $P(X \geq 5)$

8. $f(x) = \begin{cases} \dfrac{1}{x^2} & \text{if } x \geq 1 \\ 0 & \text{if } x < 1 \end{cases}$

   $P(1 \leq X \leq 3)$ and $P(X \geq 2)$

9. $f(x) = \begin{cases} 0.75x(2x - x^2) & \text{if } 0 \le x \le 2 \\ 0 & \text{otherwise} \end{cases}$

$P(X \le 1)$ and $P(1 \le X \le 3)$

10. $f(x) = \begin{cases} 0.25xe^{-x/2} & \text{if } x \ge 0 \\ 0 & \text{otherwise} \end{cases}$

$P(X \ge 2)$ and $P(0 \le X \le 3)$

11. Find the expected value $E(X)$ and variance $\text{Var}(X)$ for the random variable X in Exercise 9.

12. Find the expected value $E(X)$ and variance $\text{Var}(X)$ for the random variable X in Exercise 10.

13. Find a number c so that $f(x)$ is a probability density function:

$$f(x) = \begin{cases} cxe^{-x/4} & \text{if } x \ge 0 \\ 0 & \text{otherwise} \end{cases}$$

14. Find a number c so that $f(x)$ is a probability density function:

$$f(x) = \begin{cases} \dfrac{c}{x^4} & \text{if } x \ge 1 \\ 0 & \text{otherwise} \end{cases}$$

15. If the random variable X is normally distributed with mean $\mu = 7$ and standard deviation $\sigma = 2$, what is $P(X \ge 9)$?

16. Find b if $P(Z \ge b) = 0.73$, where Z is a random variable with a standard normal distribution ($\mu = 0$, $\sigma = 1$).

17. PORTFOLIO MANAGEMENT   Anika estimates that the probability is 0.6 that a certain stock in her portfolio will go up 25% in price. She also figures that the probability is 0.25 that it will go up only 10%, and that the probability is 0.15 that it will go down 10%. What percentage return should she expect from this stock?

18. MILITARY TRAINING   As part of his training, a paratrooper is required to jump into a circular area marked out on the ground. A bull's eye that occupies 10% of the circle is marked 10, and three other circles, numbered 5, 3, and 1, respectively occupy 20%, 30%, and 40% of the total area of the target. Assuming the trainee lands at a random position inside the target, what score should he expect to receive?

19. LOTTERY   The probability of winning $100 in a particular lottery is 0.08, the probability of winning $20 is 0.12, the probability of winning $5 is 0.2, and the probability of winning nothing is 0.6. What is a fair price to pay for a lottery ticket?

20. WHEEL OF FORTUNE   A wheel of fortune is divided into 20 circular sectors of equal area. The wheel is spun and a payoff is made according to the colour of the region on which the indicator lands. One region is gold and pays $50 when hit, five regions are blue and pay $25, and four other regions are red and pay $10. The remaining 10 regions are black and pay nothing. Would you be willing to pay $10 to play this game?

21. WEIGHT LOSS   There are 200 people in a weight loss program. The weight of the participants is a random variable X that is normally distributed with mean $\mu = 80$ kg and standard deviation $\sigma = 7$ kg.
a.   How many participants weigh more than 90 kg?
b.   How many participants weigh less than 70 kg?
c.   How many participants weigh exactly 80 kg?

22. PRODUCT RELIABILITY   Electronic components made by a certain process have a time to failure that is measured by a normal random variable. If the mean time to failure is 15 000 hours with a standard deviation of 800 hours, what is the probability that a randomly selected component will last no more than 10 000 hours?

23. QUALITY CONTROL   A toy manufacturer makes hollow rubber balls. The thickness of the outer shell of such a ball is normally distributed with mean 0.03 mm and standard deviation 0.0015 mm. What is the probability that the outer shell of a randomly selected ball will be less than 0.025 mm thick?

24. TIME MANAGEMENT   A bakery turns out a fresh batch of chocolate chip cookies every 45 minutes. You arrive (at random) at the bakery, hoping to buy a fresh cookie. Use an appropriate uniform density function to find the probability that you arrive within 5 minutes before or after the time that the cookies come out of the oven.

25. DEMOGRAPHICS   A study recently commissioned by the mayor of a large city indicates that the number of years a current resident will continue to live in the city may be modelled as an exponential random variable with probability density function

$$f(t) = \begin{cases} 0.4e^{-0.4t} & \text{for } t \ge 0 \\ 0 & \text{otherwise} \end{cases}$$

a.   Find the probability that a randomly selected resident will move within 10 years.
b.   Find the probability that a randomly selected resident will remain in the city for more than 20 years.

c. How long should a randomly selected resident be expected to remain in town?

26. **TRAFFIC CONTROL**   Suppose the time (in minutes) between the arrivals of successive cars at a border crossing from Manitoba into North Dakota is measured by the random variable X with probability density function

$$f(t) = \begin{cases} 0.5e^{-0.5t} & \text{if } t \geq 0 \\ 0 & \text{otherwise} \end{cases}$$

a. Find the probability that a randomly selected pair of successive cars will arrive at the border crossing at least 6 minutes apart.

b. Find the average time between the arrivals of successive cars at the border crossing.

27. **TRAFFIC MANAGEMENT**   The distance (in metres) between successive cars on a freeway is modelled by the random variable X with probability density function

$$f(x) = \begin{cases} 0.25xe^{-x/2} & \text{if } x \geq 0 \\ 0 & \text{otherwise} \end{cases}$$

a. Find the probability that a randomly selected pair of cars will be less than 10 m apart.

b. What is the average distance between successive cars on the freeway?

28. **TRAFFIC MANAGEMENT**   Suppose the random variable X in Exercise 27 is normally distributed with mean $\mu = 12$ m and standard deviation $\sigma = 4$ m. Now what is the probability that a randomly selected pair of cars will be less than 10 m apart?

29. **INSURANCE POLICY**   An insurance company charges $10 000 for a policy insuring against a certain kind of accident and pays $100 000 if the accident occurs. Suppose it is estimated that the probability of the accident occurring is $p = 0.02$. Let X be the random variable that measures the insurance company's profit on each policy it sells.

a. What is the probability distribution for X?

b. What is the company's expected profit per policy sold?

c. What should the company charge per policy in order to double its expected profit per policy?

30. **PERSONAL HEALTH**   James decides to go on a diet for 6 months, with a goal of losing between 10 kg and 15 kg. Based on his body configuration and metabolism, his doctor determines that the amount of weight he will lose can be modelled by

a continuous random variable X with probability density function $f(x)$ of the form

$$f(x) = \begin{cases} k(x - 10)^2 & \text{for } 10 \leq x \leq 15 \\ 0 & \text{otherwise} \end{cases}$$

If the doctor's model is valid, how much weight should James expect to lose? [*Hint:* First determine the value of the constant k.]

31. **FISHERY MANAGEMENT**   The manager of a fishery determines that the age X (in weeks) at which a certain species of fish dies follows an exponential distribution with probability density function

$$f(t) = \begin{cases} \lambda e^{-\lambda t} & \text{for } t \geq 0 \\ 0 & \text{otherwise} \end{cases}$$

Observations indicate that it is twice as likely for a randomly selected fish to die during the first 10-week period as during the next 10 weeks (from week 10 to week 20).

a. What is λ?

b. What is the probability that a randomly chosen fish will die within the first 5 weeks?

c. What is the expected lifespan of a randomly chosen fish?

32. **METALLURGY**   The proportion of impurities by weight in samples of copper ore taken from a particular mine is measured by a random variable X with probability density function

$$f(x) = \begin{cases} 21x^2(1 - \sqrt{x}) & \text{for } 0 \leq x \leq 1 \\ 0 & \text{otherwise} \end{cases}$$

a. What is the probability that the proportion of impurities in a randomly selected sample will be less than 5%?

b. What is the probability that the proportion of impurities will be greater than 50%?

c. What proportion of impurities would you expect to find in a randomly selected sample?

33. **BEVERAGES**   Suppose that the volume of pop in a bottle produced at a particular plant is normally distributed with a mean of 350 mL and a standard deviation of 1.5 mL.

a. Find the probability that a bottle filled at this plant contains at least 344 mL.

b. Find the volume of pop so that 95% of all bottles filled at this plant contain less than this amount.

34. **QUALITY CONTROL**   An automobile manufacturer claims that its new cars have an

average gas consumption of 12 L/100 km in city driving. Assume the manufacturer's claim is correct and that gas consumption is normally distributed, with standard deviation 0.8 L/100 km.

a. Find the probability that a randomly selected car will use less than 10 L/100 km.

b. If you test 2 cars, what is the probability that both will use less than 10 L/100 km?

35. ACADEMIC TESTING The results of a calculus exam are normally distributed with a mean of 72.3 and a standard deviation of 16.4. Find the probability that a randomly chosen student's score is between 50 and 75. If there are 82 students in the class, about how many have scores between 50 and 75?

36. MEDICINE Suppose that the number of children who die each year from leukemia follows a Poisson distribution and that, on average, 7.3 children per 100 000 die from leukemia. For a city with 100 000 children, find the probability of each of the following events.

a. Exactly 7 children in the city die from leukemia each year.

b. Fewer than 2 children in the city die from leukemia each year.

c. More than 5 children in the city die from leukemia each year.

37. ECOLOGY The pH level of a liquid measures its acidity and is an important issue in studying the effects of acid rain. Suppose that a test is conducted under controlled conditions that allow the change in pH in a particular lake resulting from acid rain to be recorded. Let X be a random variable that measures the pH of a sample of water taken from the lake, and assume that X has the probability density function

$$f(x) = \begin{cases} 0.75(x - 4)(6 - x) & \text{for } 4 \le x \le 6 \\ 0 & \text{otherwise} \end{cases}$$

a. Find the probability that the pH of a randomly selected sample will be at least 5.

b. Find the expected pH of a randomly selected sample.

38. SPORTS MEDICINE Suppose that the number of injuries a team suffers during a typical football game follows a Poisson distribution with an average of 2.5 injuries.

a. Find the probability that during a randomly chosen game, the team suffers exactly 2 injuries.

b. Find the probability that during a randomly chosen game, the team suffers no injuries.

c. Find the probability that during a randomly chosen game, the team suffers at least 1 injury.

39. PACKAGE DELIVERY Suppose that the number of overnight packages a business receives during a business day follows a Poisson distribution and that, on average, the company receives 4 overnight packages per day.

a. Find the probability that the company receives exactly 4 overnight packages on a randomly selected business day.

b. Find the probability that the company does not receive any overnight packages on a randomly selected business day.

c. Find the probability that the company receives fewer than 4 overnight packages on a randomly selected business day.

40. JOURNALISM Suppose that the number of typographical errors on a page of a local newspaper follows a Poisson distribution with an average of 2.5 errors per page.

a. Find the probability that a randomly selected page is free of typographical errors.

b. Find the probability that a randomly selected page has at least 1 typographical error.

c. Find the probability that a randomly selected page has at least 3 typographical errors.

d. Find the probability that a randomly selected page has fewer than 3 typographical errors.

41. HIGHWAY ACCIDENTS A report models the number of automobile accidents on a particular highway as a random variable with a Poisson distribution. Suppose it is found that on average, there is an accident every 10 hours.

a. Find the probability that there are no accidents on this highway during a randomly selected 24-hour period.

b. Find the probability that there is at least 1 accident on this highway during a randomly selected 12-hour period.

c. Find the probability that there are no accidents on this highway during a randomly selected hour.

42. PUBLIC HEALTH As part of a campaign to combat a new strain of influenza, public health authorities are planning to inoculate 1 million people. It is estimated that the probability of an individual having a bad reaction to the vaccine is 0.0005. Suppose the number of people inoculated who have bad reactions to the vaccine is modelled by a random variable with a Poisson distribution.

a. What is λ for the distribution?
b. What is the probability that of the 1 million people inoculated, exactly 5 will have a bad reaction?
c. What is the probability that of the 1 million people inoculated, more than 10 will have a bad reaction?

43. **A* LABOUR EFFICIENCY**   A company wishes to examine the efficiency of two members of its senior staff, Raoul and Lina, who work independently of one another. Let X and Y be random variables that measure the proportion of the work week that Raoul and Lina, respectively, actually spend performing their duties. Assume that the joint probability density function for X and Y is

$$f(x, y) = \begin{cases} 0.4(2x + 3y) & \text{if } 0 \le x \le 1 \text{ and } 0 \le y \le 1 \\ 0 & \text{otherwise} \end{cases}$$

a. Verify that $f(x, y)$ satisfies the requirements for a joint probability density function.
b. Find the probability that Raoul spends less than half his time working while Lina spends more than half her time working.
c. Find the probability that Raoul and Lina each spend at least 80% of the work week performing their assigned tasks.
d. Find the probability that Raoul and Lina combined work less than a full work week. [*Hint:* This is the event that $X + Y < 1$.]

44. **A* LABOUR EFFICIENCY**   For the situation described in Exercise 43, find the expected values of X and Y. The company expects its senior staff members to spend at least 75% of their time performing their duties. Is either Raoul or Lina in danger of a reprimand?

45. **A* TIME MANAGEMENT**   Suppose the random variable X measures the time (in minutes) that a person stands in line for a latte at a popular gourmet coffee shop in the morning and Y measures the duration (in minutes) of the transaction of making the coffee and paying for it. Assume that the joint probability density function for X and Y is

$$f(x, y) = \begin{cases} 0.125e^{-x/4}e^{-y/2} & \text{if } x \ge 0 \text{ and } y \ge 0 \\ 0 & \text{otherwise} \end{cases}$$

a. What is the probability that neither activity takes more than 5 minutes?

b. What is the probability that you will be able to start drinking your latte (both activities complete) within 8 minutes?

46. **A* INSURANCE SALES**   Let X be a random variable that measures the time (in minutes) that a person spends with an agent choosing a life insurance policy and let Y measure the time (in minutes) the agent spends doing paperwork once the client has selected a policy. Suppose the joint probability density function for X and Y is

$$f(x, y) = \begin{cases} \dfrac{1}{300}e^{-x/30}e^{-y/10} & \text{for } x \ge 0 \text{ and } y \ge 0 \\ 0 & \text{otherwise} \end{cases}$$

a. Find the probability that choosing the policy takes more than 20 minutes.
b. Find the probability that the entire transaction (policy selection and paperwork) will take more than half an hour.
c. How much more time would you expect to spend selecting the policy than completing the paperwork?

47. **A* TIME MANAGEMENT**   A shuttle bus to downtown arrives at an airport at a randomly selected time X within a 1-hour period, and a tourist independently arrives at the same stop also at a randomly selected time Y within the same hour. The tourist has the patience to wait for the shuttle bus for up to 20 minutes before calling a taxi. The joint probability density function for X and Y is

$$f(x, y) = \begin{cases} 1 & \text{if } 0 \le x \le 1 \text{ and } 0 \le y \le 1 \\ 0 & \text{otherwise} \end{cases}$$

a. What is the probability that the shuttle bus takes longer than 20 minutes to arrive?
b. What is the probability that the tourist arrives after the shuttle bus?
c. What is the probability that the tourist connects with the shuttle bus? [*Hint:* The event of this occurring has the form $Y + a \le X \le Y + b$ for suitable numbers a and b.]

48. **A* HAZARDOUS WASTE DISPOSAL**
Probabilistic methods can also be used to study the effectiveness of hazardous waste disposal procedures. Carcinogenic risk associated with polluted soil near a factory may be measured by a quantity called the *slope factor*. Suppose the slope factor at a particular disposal site is modelled by a random variable X with probability density function

$$f(x) = \begin{cases} \dfrac{5.36}{(x + 5.36)^2} & \text{if } x \geq 0 \\ 0 & \text{otherwise} \end{cases}$$

so that

$$P(a \leq X \leq b) = \int_a^b f(x)\, dx$$

is the probability of developing cancer from soil in which the slope factor is a number between a and b.

a. Compute the probability of a person developing cancer from a randomly collected sample of soil whose slope factor is less than 2.

b. Compute the probability of a person developing cancer from a soil sample whose slope factor is greater than 10.

c. What is the expected slope value of a randomly collected sample of soil at this disposal site?

d. Actual research models involving industrial carcinogens often use more complicated probability density functions than the one considered in this exercise. Read an article on the use of statistical methods in determining carcinogenic risk, and write a paragraph summarizing these methods.

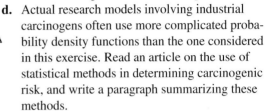

THINK ABOUT IT

RELIABILITY

In Canada in 2007–2008, there were 38 922 knee replacements. Of these, 3328 were for patients between 45 and 54 years old. However, artificial knees wear out. Will these patients outlast their artificial knees, or can they still be active into their senior years? How can we determine the probability that an artificial knee will last more than 20 years? We can answer this question if we have a probability distribution for the time an artificial knee works before failing. But what type of distribution is appropriate for the lifespan of an artificial knee? In this essay, we introduce probability models that can be used to study the reliability, or lifespan, of systems. The study of reliability is important because all human-made systems, ranging from simple components to complex systems, ultimately fail. The same holds for biological systems, as living organisms have finite lifespans. Many practical questions involve determining the reliability of a system, either human-made or biological.

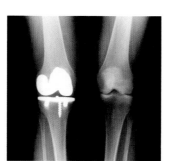 

(Photos: Jim Wehtje/Getty Images and © BananaStock/Alamy)

Using what we have already learned about probability, we can begin studying reliability. Given a particular system, let the random variable X denote the time it operates properly before failing, assuming that it was put into service at time $t = 0$. We can answer many practical questions, such as finding the average time a system operates until it fails and the probability a system operates for at least a specified length of time, by studying properties of the random variable X. In reliability studies, we make use of several important functions related to X. First, we let $f(t)$ denote the probability density function of X. Then the probability that the system fails before time t is

$$F(t) = \int_0^t f(x)\, dx.$$

We call $F(t)$ the **cumulative distribution function** of X. The probability density function $f(t)$ is the derivative of the cumulative distribution $F(t)$ (this follows from one of the two parts of the fundamental theorem of calculus). That is,

$$\frac{dF(t)}{dt} = f(t).$$

To study questions involving reliability, we introduce the function $R(t)$, which is aptly called the **reliability function.** This is the probability that the system survives past time t, that is,

$$R(t) = P(X > t).$$

Note that $R(t) = 1 - P(X \leq t) = 1 - F(t)$ and that if we know any of the three functions $f(t)$, $F(t)$, and $R(t)$, we can easily find the other two functions.

Given the reliability function associated with the random variable X, we would like to find the rate at which failures are occurring at a particular time t. This should be the limit as s approaches zero of the probability that the system fails during the s units of time following time t, divided by s. That is, the failure rate at time t is

$$\lim_{s \to 0} \frac{\text{probability of failure in the time interval from time } t \text{ to time } t + s}{s}.$$

To find this failure rate, first observe that the probability that a system operating at time t continues to operate during the s units of time immediately following time t is the proportion of systems working at time t that are still working at time $t + s$. This proportion is

$$\frac{P(t < X \leq t + s)}{P(t < X)} = \frac{F(t + s) - F(t)}{R(t)}.$$

Consequently, the failure rate at time t is

$$\lim_{s \to 0} \frac{1}{s} \frac{F(t + s) - F(t)}{R(t)} = \frac{1}{R(t)} \lim_{s \to 0} \frac{F(t + s) - F(t)}{s} = \frac{F'(t)}{R(t)} = \frac{f(t)}{R(t)}.$$

This leads us to define the **failure rate function** $r(t)$, which represents the failure rate at time t of the system being studied, by

$$r(t) = \frac{f(t)}{R(t)}.$$

It follows that once we know one of the three functions $f(t)$, $F(t)$, and $R(t)$, we also can find $r(t)$. Furthermore, given the failure rate function $r(t)$, we can find the reliability function using the equation

$$(1) \qquad R(t) = e^{-\int_0^t r(x)\,dx}.$$

To see why Equation (1) is true, first note that because $R(t) = 1 - F(t)$, we have $R'(t) = -F'(t)$. Moreover, because $f(t) = F'(t)$, it follows that

$$r(t) = \frac{f(t)}{R(t)} = \frac{F'(t)}{R(t)} = \frac{-R'(t)}{R(t)}.$$

THINK ABOUT IT

This implies that

$$-\int_0^t r(x)\,dx = \int_0^t\left[\frac{-R'(x)}{R(x)}\right]dx = \ln R(t) - \ln R(0) = \ln R(t) - \ln 1 = \ln R(t),$$

where we have used the fact that $R(0) = P(X > 0) = 1$. We can now conclude that $e^{-\int_0^t r(x)\,dx} = e^{\ln R(t)} = R(t)$.

When modelling the reliability of a system, certain assumptions are made about its failure rate. The simplest model of the reliability of a system assumes that the system has a constant failure rate. We leave it as an exercise (Question 1) to use Equation (1) to show that if $r(t) = \lambda$ where $\lambda > 0$, then $f(t)$ has an exponential distribution. Assuming a constant failure rate produces a simple mathematical model using the exponential distribution. However, the failure rates of most, if not all, human-made products and biological systems are not constant. In practice, many systems have failure rates that increase with time. For example, artificial knees wear out at a more rapid rate as components wear out. Older cars suffer a higher failure rate than new cars as their engines, transmissions, and other systems age. Many electronic components that are put into service after an initial burn-in period exhibit low failure rates for a while but then degrade over time. Even computer software experiences increasing failure rates. Living organisms also have a higher failure (or mortality) rate as they get older. It is more likely that a 100-year-old person will die in the next year than that a 50-year-old person will die in this period. Also note that for some systems, failure rates decrease with time, such as some mechanical components whose failures are mostly caused by manufacturing defects. Such components fail at a faster rate initially, but once the components with flaws have failed, this rate decreases.

How can we model increasing failure rates, as well as failure rate decreases, with time? Finding adequate models challenged mathematicians and engineers for many years. After extensive study, it was found that an extremely wide class of applications can be accurately modelled using a failure rate function of the form

$$r(t) = \alpha\beta t^{\beta-1},$$

where α and β are positive constants. This function looks quite complicated, but by carefully choosing values of α and β, engineers and scientists have had success using failure rate functions of this form in many applications. Note that the function $r(t)$ is an increasing function when $\beta > 1$; values of β greater than 1 are used in models where the failure rate increases with time. The failure $r(t)$ is a decreasing function when $0 < \beta < 1$; values of β less than 1 are used in models where the failure rate decreases with time. When $\beta = 1$, the failure rate is constant. Using this function for $r(t)$ implies (see Question 4) that the reliability function $R(t)$ is given by

$$R(t) = e^{-\alpha t^\beta}$$

and that the cumulative distribution function is given by

$$F(t) = 1 - e^{-\alpha t^\beta}$$

for $t > 0$. By differentiating $F(t)$, we can find the probability density function. That is,

$$f(t) = F'(t) = \alpha\beta t^{\beta-1}e^{-\alpha t^\beta}.$$

A random variable X with this probability density function is said to follow a **Weibull distribution** with parameters α and β, where $\alpha > 0$ and $\beta > 0$. This distribution is widely used in reliability studies and in many other models. It is named for its inventor, Waloddi Weibull, who introduced the distribution in 1939. Weibull was a Swedish scientist, inventor, and statistician. He also consulted extensively on projects for Saab and for the U.S. Air Force. We display graphs of $f(t)$ for three different Weibull distributions in Figure 10.25. In all three of these distributions, the parameter α equals 0.5, while the parameter β takes on the values 1, 2, and 3, respectively.

We can now return to the question of modelling the lifespan of an artificial knee. Suppose that it has been shown that the lifespan of an artificial knee after it is implanted can be modelled using a Weibull distribution with parameters $\alpha = 0.0000939$ (measured in years) and $\beta = 3$. This means that $R(t) = e^{-0.0000939 t^3}$. We can now answer the question posed at the beginning of this essay using the formula we have found for the reliability function for a Weibull distribution. We find that the probability that the artificial knee lasts more than 20 years is $R(20) = e^{-0.0000939 \cdot 20^3} \approx 0.472$.

Weibull models have been used to solve a variety of problems from many different disciplines. For example, the Weibull distribution has been used to determine the expected costs of pumping sand to replenish beaches damaged by hurricanes, to determine how effective new cancer treatments are, to set premiums for insurance policies, to find the best mix of tree species that should be planted for paper production, to set up maintenance procedures for replacing street lights in a city, to estimate the risk of an offshore platform collapsing if it is designed to withstand waves up to a certain height, to determine the viability of windmills, to determine the best blasting strategy for ores, to manage inventories of spare parts, and to study conflicts in highway lane merging. They have also been used to study the yield strength and the fatigue life of steel, the tensile strength of optical fibres, the fineness of coal, the size of droplets in sprays, the pitting corrosion in pipes, the time an organism takes to die after exposure

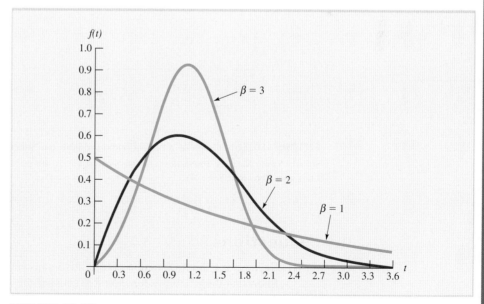

FIGURE 10.25   Three Weibull probability density function graphs, with parameters $\alpha = 0.5$ in each case and $\beta = 1, 2, 3$.

SOURCE: Steven Nahmias, *Production and Operations Analysis*, 4th ed., © 2001. The McGraw-Hill Companies, Inc., New York, NY.

to carcinogens, fracturing in concrete, the pace of technology change, the fracture strength of glass, the size of icebergs, wave heights, flood and earthquake frequency, temperature fluctuations, precipitation amounts, the size of raindrops, inventory lead times, and wind speed distributions, among other things.

For many complex systems, the failure rate is neither an increasing nor a decreasing function. Instead, the failure rate resembles a bathtub, as shown in Figure 10.26. When the product is young, manufacturing defects cause a high initial failure rate in the infant mortality phase. After bad components have failed, the failure rate remains relatively constant until the system reaches a phase where it begins to wear out and the failure rate increases. To build models where the failure rate resembles a bathtub, we need more complicated functions than are introduced here. To learn more about such models, the reader is invited to consult books that focus on reliability theory.

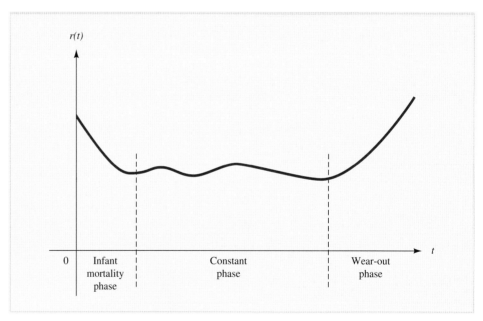

FIGURE 10.26   Graph of a typical bathtub failure rate function for a complex system.
Source: Steven Nahmias, *Production and Operations Analysis*, 4th ed., © 2001. The McGraw-Hill Companies, Inc., New York, NY.

Questions

1. Suppose the failure function is a positive constant; that is, $r(t) = \lambda$ for $\lambda > 0$. Show that the reliability function $R(t)$ satisfies the separable differential equation $R'(t) = -\lambda R(t)$. Solve this differential equation to find $R(t)$. What kind of probability density function $f(t)$ corresponds to $R(t)$ in this case?

2. Show that when the parameter β in a Weibull distribution satisfies $\beta = 1$, then the distribution is exponential.

3. Suppose the random variable X that represents the time to failure of a system (the length of time it works until it fails) has an exponential distribution.

   a. Show that the probability that a system working at time t will fail in the next s units of time does not depend on t. That is, the probability of the system failing in a particular time span does not depend on how long it has been working.

   b. The property in part (a) is described by saying that *an exponential random variable has no memory*. Show that a geometric random variable also has no memory.

4. Verify that the equation given for the reliability function $R(t)$ for a Weibull distribution with parameters α and β is correct.

5. Suppose that the reliability function for a computer model is $R(t) = e^{-0.008t^{1.98}}$. Find the failure rate function for these computers.

6. Suppose that the failure rate function for a copier model is $r(t) = 3.11t^{1.5}$, with time measured in months. Find the probability that one of these copiers continues to operate for at least 6 months.

7. Find the probability that the artificial knee discussed in the text lasts at least 30 years.

8. Let $\Gamma(\alpha) = \int_0^\infty x^{\alpha-1} e^{-x}\, dx$, where $\alpha > 0$. Show that if the random variable X has a Weibull distribution with parameters α and β, then

$$E(X) = \alpha^{-1/\beta}\, \Gamma\!\left(1 + \frac{1}{\beta}\right)$$

and

$$\mathrm{Var}(X) = \alpha^{-2/\beta} \Gamma\!\left(1 + \frac{2}{\beta}\right) - [E(X)]^2.$$

References

Steven Nahmias, *Production and Operations Analysis,* 4th ed., New York: McGraw-Hill, 2001.

Wallace R. Blischke and D. N. Prabhakar Murthy, *Reliability: Modeling, Prediction, and Optimization,* New York: John Wiley and Sons, 2000.

connect™   Practise and learn online with Connect.

11

Periodic behaviour occurs in humans. The periodic motion of a heartbeat takes only a second to complete. Body temperature rises and falls in a 24-hour cycle. Example 11.1.6 deals with a periodic function of blood pressure. Find out more about how trigonometry can help us understand and predict outcomes in this chapter. (Photo: Brand X/PunchStock)

TRIGONOMETRIC FUNCTIONS

CHAPTER OUTLINE

LEARNING OBJECTIVES

After completing this chapter, you should be able to

LO1 Evaluate the three basic trigonometric functions and their reciprocals when given angles in radians. Solve problems using the addition formula and the double angle formula.

LO2 Differentiate trigonometric functions. Integrate trigonometric functions.

LO3 Formulate an application problem in terms of trigonometric functions and differentiate to optimize one variable.

SECTION 11.1

L01

Evaluate the three basic trigonometric functions and their reciprocals when given angles in radians. Solve problems using the addition formula and the double angle formula.

The Trigonometric Functions

In this chapter, you will be introduced to some functions that are widely used in the natural sciences to study periodic or rhythmic phenomena such as oscillations, the periodic motion of planets, and the respiratory cycle and heartbeat of animals. These functions are also related to the measurement of angles and hence play an important role in such fields as architecture, navigation, and surveying.

Angle   An **angle** is formed when one line segment in the plane is rotated into another about their common endpoint. The resulting angle is said to be a **positive angle** if the rotation is in a counterclockwise direction and a **negative angle** if the rotation is in a clockwise direction. The situation is illustrated in Figure 11.1.

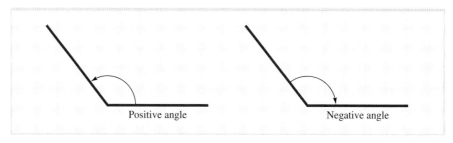

Positive angle   Negative angle

FIGURE 11.1   Positive and negative angles.

Measurement of Angles   You are probably already familiar with the use of degrees to measure angles. A **degree** is the amount by which a line segment must be rotated so that its free endpoint traces out $\dfrac{1}{360}$ of one complete rotation. Thus, for example, a complete counterclockwise rotation generating an entire circle contains 360°, one-half of a complete counterclockwise rotation contains 180°, and one-sixth of a complete counterclockwise rotation contains 60°. Some important angles are shown in Figure 11.2.

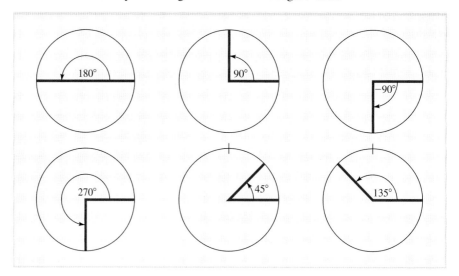

FIGURE 11.2   Angles measured in degrees.

Radian Measure   Although measurement of angles in degrees is convenient for many geometric applications, another unit of angle measurement called the **radian** leads to simpler rules

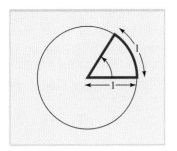

FIGURE 11.3   An angle of 1 radian.

for differentiating and integrating trigonometric functions. One radian is defined to be the amount by which a line segment of length 1 must be rotated so that its free end-point traces out a circular arc of length 1. The situation is illustrated in Figure 11.3.

The total circumference of a circle of radius 1 is 2π. Hence, 2π radians is equal to 360°, π radians is equal to 180°, and, in general, the relationship between radians and degrees is given by the following useful conversion formula.

Conversion Formula

$$\frac{degrees}{180} = \frac{radians}{\pi}$$

The use of this formula is illustrated in Example 11.1.1.

EXAMPLE 11.1.1

a. Convert 45° to radians.   **b.** Convert $\frac{\pi}{6}$ radians to degrees.

Solution

a. From the proportion $\frac{45}{180} = \frac{radians}{\pi}$, it follows that 45° is equivalent to $\frac{\pi}{4}$ radians.

b. From the proportion $\frac{degrees}{180} = \frac{\pi/6}{\pi}$, it follows that $\frac{\pi}{6}$ radians is equivalent to $\frac{180°}{6} = 30°$.

For reference, six of the most important angles are shown in Figure 11.4 along with their measurements in degrees and in radians.

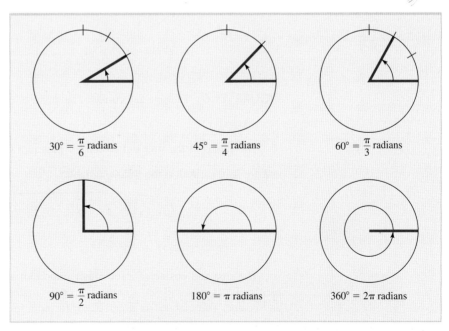

$30° = \frac{\pi}{6}$ radians $45° = \frac{\pi}{4}$ radians $60° = \frac{\pi}{3}$ radians

$90° = \frac{\pi}{2}$ radians $180° = \pi$ radians $360° = 2\pi$ radians

FIGURE 11.4   Six important angles in degrees and radians.

Sine and Cosine

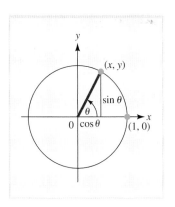

FIGURE 11.5   The sine and cosine of an angle.

Suppose the line segment joining the points $(0, 0)$ and $(1, 0)$ on the x axis is rotated through an angle of θ radians so that the free endpoint of the segment moves from $(1, 0)$ to a point (x, y) as in Figure 11.5. The x and y coordinates of the point (x, y) are known, respectively, as the **cosine** and **sine** of the angle θ. The symbol cos θ is used to denote the cosine of θ, and sin θ is used to denote its sine.

The Sine and Cosine ■ For any angle θ,

$$\cos\theta = x \quad \text{and} \quad \sin\theta = y,$$

where (x, y) is the point to which $(1, 0)$ is carried by a rotation of θ radians about the origin.

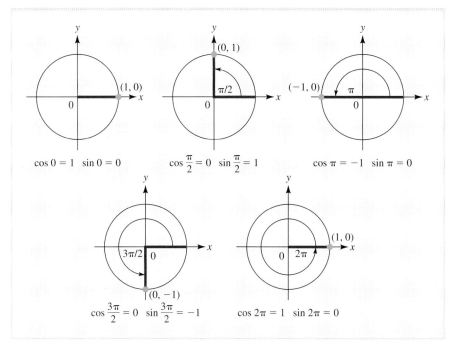

FIGURE 11.6   The sine and cosine of multiples of $\dfrac{\pi}{2}$.

The values of the sine and cosine for multiples of $\dfrac{\pi}{2}$ can be read from Figure 11.6 and are summarized in the following table:

θ	0	$\dfrac{\pi}{2}$	π	$\dfrac{3\pi}{2}$	2π
$\cos\theta$	1	0	-1	0	1
$\sin\theta$	0	1	0	-1	0

The sine and cosine of a few other important angles are also easy to obtain geometrically, as you will see shortly. For other angles, use a calculator to find the sine and cosine. You can set up a calculator to input either degrees or radians. Computer software is usually set up to use radians.

Elementary Properties of Sine and Cosine

Since there are 2π radians in a complete rotation, it follows that

$$\sin(\theta + 2\pi) = \sin\theta \qquad \text{and} \qquad \cos(\theta + 2\pi) = \cos\theta.$$

That is, the sine and cosine functions are **periodic** with period 2π. The situation is illustrated in Figure 11.7.

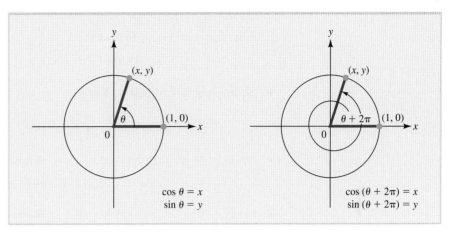

FIGURE 11.7　The periodicity of $\sin\theta$ and $\cos\theta$.

Since negative angles correspond to clockwise rotations, it follows that

$$\sin(-\theta) = -\sin\theta \qquad \text{and} \qquad \cos(-\theta) = \cos\theta.$$

This is illustrated in Figure 11.8.

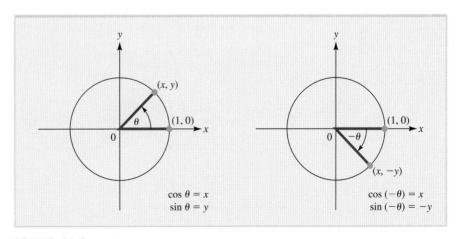

FIGURE 11.8　The sine and cosine of negative angles.

Properties of Sine and Cosine

$$\sin(\theta + 2\pi) = \sin\theta \qquad \text{and} \qquad \cos(\theta + 2\pi) = \cos\theta$$
$$\sin(-\theta) = -\sin\theta \qquad \text{and} \qquad \cos(-\theta) = \cos\theta$$

The use of these properties is illustrated in Example 11.1.2.

EXAMPLE 11.1.2

Evaluate each expression.

a. $\cos(-\pi)$ **b.** $\sin\left(-\dfrac{\pi}{2}\right)$ **c.** $\cos 3\pi$ **d.** $\sin\dfrac{5\pi}{2}$

Solution

a. Since $\cos \pi = -1$, it follows that

$$\cos(-\pi) = \cos \pi = -1.$$

b. Since $\sin\dfrac{\pi}{2} = 1$, it follows that

$$\sin\left(-\frac{\pi}{2}\right) = -\sin\frac{\pi}{2} = -1.$$

c. Since $3\pi = \pi + 2\pi$ and $\cos \pi = -1$, it follows that

$$\cos 3\pi = \cos(\pi + 2\pi) = \cos \pi = -1.$$

d. Since $\dfrac{5\pi}{2} = \dfrac{\pi}{2} + 2\pi$ and $\sin\dfrac{\pi}{2} = 1$, it follows that

$$\sin\frac{5\pi}{2} = \sin\left(\frac{\pi}{2} + 2\pi\right) = \sin\frac{\pi}{2} = 1.$$

The Graphs of $\sin\theta$ and $\cos\theta$

It is obvious from the definitions of sine and cosine that as θ goes from 0 to 2π, the function $\sin \theta$ oscillates between 1 and -1, starting with $\sin 0 = 0$, and the function $\cos\theta$ oscillates between 1 and -1, starting with $\cos 0 = 1$. This observation, together with the elementary properties previously derived, suggests that the graphs of the functions $\sin\theta$ and $\cos\theta$ resemble the curves in Figures 11.9 and 11.10, respectively.

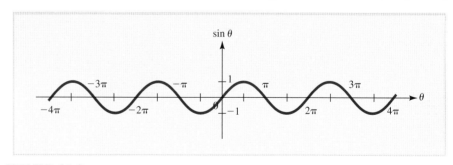

FIGURE 11.9 The graph of $\sin\theta$.

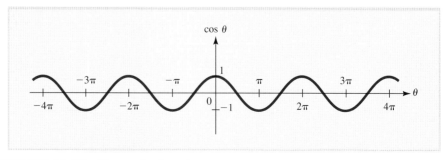

FIGURE 11.10 The graph of $\cos\theta$.

Other Trigonometric Functions

Four other useful trigonometric functions can be defined in terms of the sine and cosine as follows.

> **Tangent, Cotangent, Secant, and Cosecant** ■ For any angle θ,
>
> $$\tan\theta = \frac{\sin\theta}{\cos\theta} \qquad \cot\theta = \frac{1}{\tan\theta} = \frac{\cos\theta}{\sin\theta}$$
>
> $$\sec\theta = \frac{1}{\cos\theta} \qquad \csc\theta = \frac{1}{\sin\theta}$$
>
> provided that the denominators are not zero.

EXAMPLE 11.1.3

Evaluate each expression.

   a. $\tan\pi$ **b.** $\cot\dfrac{\pi}{2}$ **c.** $\sec(-\pi)$ **d.** $\csc\left(-\dfrac{5\pi}{2}\right)$

Solution

   a. Since $\sin\pi = 0$ and $\cos\pi = -1$, it follows that

$$\tan\pi = \frac{\sin\pi}{\cos\pi} = \frac{0}{-1} = 0.$$

   b. Since $\sin\dfrac{\pi}{2} = 1$ and $\cos\dfrac{\pi}{2} = 0$, it follows that

$$\cot\frac{\pi}{2} = \frac{\cos\dfrac{\pi}{2}}{\sin\dfrac{\pi}{2}} = \frac{0}{1} = 0.$$

   c. Since $\cos\pi = -1$, it follows that

$$\sec(-\pi) = \frac{1}{\cos(-\pi)} = \frac{1}{\cos\pi} = -1.$$

   d. Since $\dfrac{5\pi}{2} = \dfrac{\pi}{2} + 2\pi$ and $\sin\dfrac{\pi}{2} = 1$, it follows that

$$\csc\left(-\frac{5\pi}{2}\right) = \frac{1}{\sin\left(-\dfrac{5\pi}{2}\right)} = \frac{1}{-\sin\dfrac{5\pi}{2}} = \frac{1}{-\sin\dfrac{\pi}{2}} = -1.$$

EXPLORE!

Graph on the same scale, one under the other, $f(x) = \sin x$, $f(x) = \cos x$, and $f(x) = \tan x$.

Right Triangles

If you have taken a course in trigonometry, you may remember the following definitions of sine, cosine, and tangent involving the lengths of the sides of a right triangle such as the one in Figure 11.11.

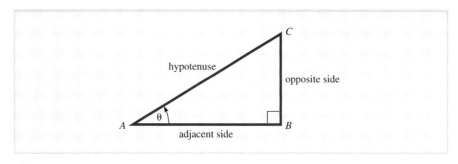

FIGURE 11.11   Triangle used to define trigonometric functions.

The Trigonometry of Right Triangles

$$\sin \theta = \frac{\text{opposite side}}{\text{hypotenuse}} \qquad \cos \theta = \frac{\text{adjacent side}}{\text{hypotenuse}} \qquad \tan \theta = \frac{\text{opposite side}}{\text{adjacent side}}$$

The definitions of trigonometric functions that you have seen in this section involving the coordinates of points on a circle of radius 1 are equivalent to the definitions from trigonometry. To see this, superimpose an xy coordinate system over the triangle ABC as shown in Figure 11.12 and draw the circle of radius 1 that is centred at the origin.

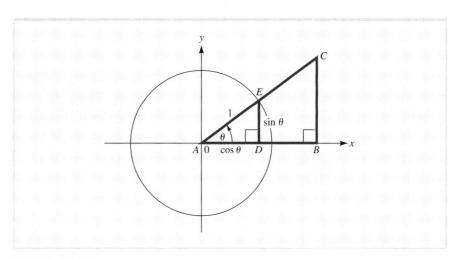

FIGURE 11.12   Similar triangles.

Since ABC and ADE are similar triangles, it follows that

$$\frac{AD}{AE} = \frac{AB}{AC}.$$

But the length AD is the x coordinate of point E on the circle, and so, by the definition of the cosine, $AD = \cos \theta$. Moreover,

$$AE = 1, \qquad AB = \text{adjacent side}, \qquad \text{and} \qquad AC = \text{hypotenuse}.$$

Hence,

$$\frac{\cos\theta}{1} = \frac{\text{adjacent side}}{\text{hypotenuse}} \quad \text{or} \quad \cos\theta = \frac{\text{adjacent side}}{\text{hypotenuse}}.$$

For practice, convince yourself that similar arguments lead to the formulas for sine and tangent.

Calculations with Right Triangles

Many calculations involving trigonometric functions can be performed easily and quickly with the aid of appropriate right triangles. For example, from the well-known 30-60-90 triangle in Figure 11.13, you see immediately that

$$\sin 30° = \frac{1}{2} \quad \text{and} \quad \cos 30° = \frac{\sqrt{3}}{2},$$

or, using radian measure,

$$\sin\frac{\pi}{6} = \frac{1}{2} \quad \text{and} \quad \cos\frac{\pi}{6} = \frac{\sqrt{3}}{2}.$$

Here is an example that further illustrates the use of right triangles.

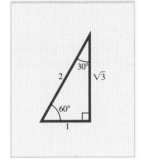

FIGURE 11.13
A 30-60-90 triangle.

EXAMPLE 11.1.4

Find $\tan\theta$ if $\sec\theta = \frac{3}{2}$.

Solution

Since

$$\sec\theta = \frac{1}{\cos\theta} = \frac{\text{hypotenuse}}{\text{adjacent}},$$

begin by drawing a right triangle (Figure 11.14) in which the hypotenuse has length 3 and the side adjacent to the angle θ has length 2.

By the Pythagorean theorem,

$$\text{Length of } BC = \sqrt{3^2 - 2^2} = \sqrt{5},$$

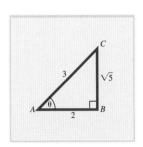

FIGURE 11.14 Right triangle for Example 11.1.4.

and so

$$\tan\theta = \frac{\text{opposite}}{\text{adjacent}} = \frac{\sqrt{5}}{2}.$$

Using appropriate right triangles and the definitions and elementary properties of the sine and cosine, you should now be able to find the sine and cosine of the frequently used angles listed in the following table.

A Table of Frequently Used Values of Sine and Cosine								
θ	0	$\dfrac{\pi}{6}$	$\dfrac{\pi}{4}$	$\dfrac{\pi}{3}$	$\dfrac{\pi}{2}$	$\dfrac{2\pi}{3}$	$\dfrac{3\pi}{4}$	$\dfrac{5\pi}{6}$	π
$\sin\theta$	0	$\dfrac{1}{2}$	$\dfrac{\sqrt{2}}{2}$	$\dfrac{\sqrt{3}}{2}$	1	$\dfrac{\sqrt{3}}{2}$	$\dfrac{\sqrt{2}}{2}$	$\dfrac{1}{2}$	0
$\cos\theta$	1	$\dfrac{\sqrt{3}}{2}$	$\dfrac{\sqrt{2}}{2}$	$\dfrac{1}{2}$	0	$-\dfrac{1}{2}$	$-\dfrac{\sqrt{2}}{2}$	$-\dfrac{\sqrt{3}}{2}$	-1

Trigonometric Identities

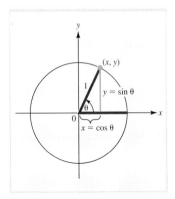

FIGURE 11.15 The identity $\sin^2\theta + \cos^2\theta = 1$.

An **identity** is an equation that holds for all values of its variable or variables. Identities involving trigonometric functions are called **trigonometric identities** and can be used to simplify trigonometric expressions. You have already seen some elementary trigonometric identities in this section. For example, the formulas

$$\cos(-\theta) = \cos\theta \qquad \text{and} \qquad \sin(-\theta) = -\sin\theta$$

are identities because they hold for all values of θ. Most trigonometry texts list dozens of trigonometric identities. Fortunately, only a few of them will be needed in this book.

One of the most important and well-known trigonometric identities is a simple consequence of the Pythagorean theorem. It states that

$$\sin^2\theta + \cos^2\theta = 1,$$

where $\sin^2\theta$ stands for $(\sin\theta)^2$ and $\cos^2\theta$ stands for $(\cos\theta)^2$. To see why this identity holds, look at Figure 11.15, which shows a point (x, y) on a circle of radius 1. By definition,

$$y = \sin\theta \qquad \text{and} \qquad x = \cos\theta,$$

and, by the Pythagorean theorem,

$$y^2 + x^2 = 1,$$

from which the identity follows immediately. Note that if we divide each term in the Pythagorean identity by $\cos^2\theta$, we obtain the identity

$$\frac{\sin^2\theta}{\cos^2\theta} + \frac{\cos^2\theta}{\cos^2\theta} = \frac{1}{\cos^2\theta}$$

or, equivalently,

$$\tan^2\theta + 1 = \sec^2\theta.$$

Here is a summary of these two forms of the Pythagorean identity.

Just-In-Time

Note that any point (x, y) on the circle with radius 1 satisfies $y^2 + x^2 = 1$.

Two Forms of the Pythagorean Identity ■ For any angle θ, we have

$$\sin^2\theta + \cos^2\theta = 1.$$

In addition, for any angle θ for which $\cos\theta \neq 0$, we have

$$\tan^2\theta + 1 = \sec^2\theta.$$

The following formulas for the sine and cosine of the sum of two angles are particularly useful. Proofs of these identities can be found in most trigonometry texts and will be omitted here.

The Addition Formulas for Sine and Cosine ■ For any angles A and B,

$$\cos(A + B) = \cos A \cos B - \sin A \sin B$$

and

$$\sin(A + B) = \sin A \cos B + \cos A \sin B.$$

If the angles A and B in the addition formulas are equal, the identities can be rewritten as follows.

> **The Double-Angle Formulas for Sine and Cosine** ■ For any angle θ,
> $$\sin 2\theta = 2 \sin \theta \cos \theta.$$
> $$\cos 2\theta = \cos^2\theta - \sin^2\theta$$

Some of these trigonometric identities will be used in Section 11.2 to derive the formulas for the derivatives of the sine and cosine functions. They can also be used in the solution of certain trigonometric equations, as illustrated in Example 11.1.5.

EXPLORE!

Refer to Example 11.1.5. Plot the graph of $f(x) = \sin^2 x - \cos^2 x + \sin x$ on $[0, 2\pi]$. Find the x intercepts of the graph of the function. Compare your results with those determined in the example.

EXAMPLE 11.1.5

Find all the values of θ in the interval $0 \le \theta \le 2\pi$ that satisfy the equation $\sin^2\theta - \cos^2\theta + \sin \theta = 0$.

Solution

Use the Pythagorean identity
$$\sin^2\theta + \cos^2\theta = 1 \qquad \text{or} \qquad \cos^2\theta = 1 - \sin^2\theta$$
to rewrite the original equation without any cosine terms as
$$\sin^2\theta - (1 - \sin^2\theta) + \sin \theta = 0$$
or
$$2 \sin^2\theta + \sin \theta - 1 = 0,$$
and factor to get
$$(2 \sin \theta - 1)(\sin \theta + 1) = 0.$$
The only solutions of
$$2 \sin \theta - 1 = 0 \qquad \text{or} \qquad \sin \theta = \frac{1}{2}$$
in the interval $0 \le \theta \le 2\pi$ are
$$\theta = \frac{\pi}{6} \qquad \text{and} \qquad \theta = \frac{5\pi}{6},$$
and the only solution of
$$\sin \theta + 1 = 0 \qquad \text{or} \qquad \sin \theta = -1$$
in the interval is
$$\theta = \frac{3\pi}{2}.$$

Hence, the solutions of the original equation in the specified interval are $\theta = \frac{\pi}{6}$, $\theta = \frac{5\pi}{6}$, and $\theta = \frac{3\pi}{2}$.

Application   Natural phenomena often involve cyclical patterns that can be described by trigonometric functions. In Example 11.1.6, we examine a medical application. Additional modelling applications involving differentiation and integration of trigonometric functions will be examined in Sections 11.2 and 11.3.

EXAMPLE 11.1.6

A device monitoring the vital signs of a hospital patient provides data that suggest modelling the patient's blood pressure by the trigonometric function

$$P(t) = 108 + 27 \cos 8t,$$

where time t is measured in seconds and P is in millimetres of mercury.

a. What is the maximum (systolic) pressure P_s of the patient? What is the minimum (diastolic) pressure P_d?

b. A heartbeat is the interval between successive systolic peaks. What is the patient's pulse rate (heartbeats per minute)?

Solution

a. When $\cos 8t = 1$, the blood pressure function has its maximum value of

$$P_s = 108 + 27(1) = 135.$$

When $\cos 8t = -1$, the blood pressure function has its minimum value of

$$P_d = 108 + 27(-1) = 81.$$

b. Systolic pressure occurs whenever $\cos 8t = 1$. Since the cosine function repeats itself every 2π radians, it follows that each heartbeat takes t_h seconds, where

$$8t_h = 2\pi$$

$$t_h = \frac{\pi}{4}$$

Therefore, the number of heartbeats that occur each minute is given by the product

$$\left(\frac{1 \text{ heartbeat}}{\pi/4 \text{ seconds}}\right)\left(\frac{60 \text{ seconds}}{1 \text{ minute}}\right) = \frac{(60)(4)}{\pi} \text{ heartbeats per minute}$$

$$\approx 76.4 \text{ heartbeats per minute}$$

We conclude that the pulse rate is approximately 76 heartbeats per minute.

EXERCISES ■ 11.1

In Exercises 1 through 4, specify the number of degrees in each indicated angle.

1.   **2.**

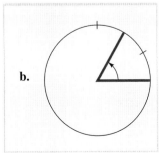

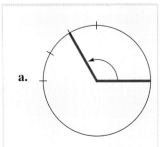

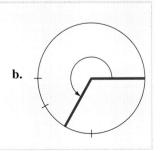

3.

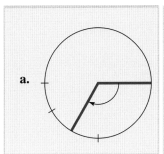 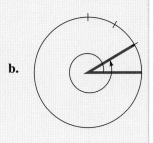

4.

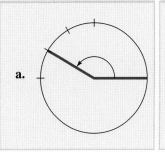 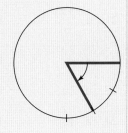

In Exercises 5 through 8, specify the number of radians in each indicated angle.

5.

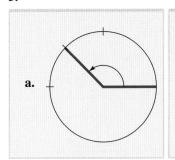

6.

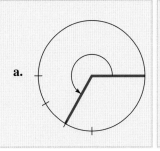

7.

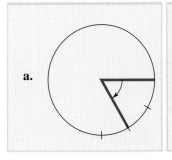

8.

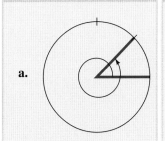

In Exercises 9 through 12, show each angle graphically.

  9. **a.** $60°$ **b.** $120°$

10. **a.** $-45°$ **b.** $135°$

11. **a.** $45°$ **b.** $-150°$

12. **a.** $240°$ **b.** $405°$

In Exercises 13 through 16, show graphically the angles with the indicated radian measure.

13. **a.** $\dfrac{3\pi}{4}$ **b.** $-\dfrac{2\pi}{3}$

14. **a.** $-\dfrac{\pi}{6}$ **b.** $\dfrac{5\pi}{6}$

15. **a.** $-\dfrac{5\pi}{3}$ **b.** $\dfrac{3\pi}{2}$

16. **a.** $-\dfrac{5\pi}{4}$ **b.** $\dfrac{7\pi}{6}$

In Exercises 17 through 20, convert each angle from degrees to radians.

17. **a.** $15°$ **b.** $-240°$

18. **a.** $30°$ **b.** $270°$

19. **a.** $135°$ **b.** $540°$

20. **a.** $-150°$ **b.** $1°$

In Exercises 21 through 24, convert each angle from radians to degrees.

21. a. $\dfrac{5\pi}{6}$ **b.** $-\dfrac{\pi}{12}$

22. a. $\dfrac{2\pi}{3}$ **b.** $-\dfrac{3\pi}{4}$

23. a. 3π **b.** 1

24. a. 3 **b.** $\dfrac{13\pi}{12}$

In Exercises 25 through 44, evaluate the given expression without using a calculator.

25. $\cos \dfrac{7\pi}{2}$

26. $\sin 5\pi$

27. $\sin\left(-\dfrac{7\pi}{2}\right)$

28. $\cos\left(-\dfrac{5\pi}{2}\right)$

29. $\cot \dfrac{5\pi}{2}$

30. $\sec 3\pi$

31. $\csc\left(-\dfrac{7\pi}{2}\right)$

32. $\tan(-\pi)$

33. $\cos\left(-\dfrac{2\pi}{3}\right)$

34. $\sin\left(-\dfrac{\pi}{6}\right)$

35. $\sin\left(-\dfrac{7\pi}{6}\right)$

36. $\cos \dfrac{7\pi}{3}$

37. $\cot \dfrac{\pi}{3}$

38. $\sec \dfrac{\pi}{3}$

39. $\tan\left(-\dfrac{\pi}{4}\right)$

40. $\cot\left(-\dfrac{5\pi}{3}\right)$

41. $\csc \dfrac{2\pi}{3}$

42. $\tan \dfrac{\pi}{6}$

43. $\sec \dfrac{5\pi}{4}$

44. $\sin\left(-\dfrac{3\pi}{4}\right)$

45. Use a right triangle to find $\cos \dfrac{\pi}{3}$ and $\sin \dfrac{\pi}{3}$.

46. Use a right triangle to find $\cos \dfrac{\pi}{4}$ and $\sin \dfrac{\pi}{4}$.

47. Complete the following table:

θ	$\dfrac{7\pi}{6}$	$\dfrac{5\pi}{4}$	$\dfrac{4\pi}{3}$	$\dfrac{3\pi}{2}$	$\dfrac{5\pi}{3}$	$\dfrac{7\pi}{4}$	$\dfrac{11\pi}{6}$	2π
$\sin\theta$								
$\cos\theta$								

In Exercises 48 through 53, find the given angle for $0 \le \theta \le \dfrac{\pi}{2}$.

48. Find $\tan\theta$ if $\sec\theta = \dfrac{5}{4}$.

49. Find $\tan\theta$ if $\csc\theta = \dfrac{5}{3}$.

50. Find $\cos\theta$ if $\tan\theta = \dfrac{3}{4}$.

51. Find $\tan\theta$ if $\cos\theta = \dfrac{3}{5}$.

52. Find $\sin\theta$ if $\tan\theta = \dfrac{2}{3}$.

53. Find $\sec\theta$ if $\cot\theta = \dfrac{4}{3}$.

In Exercises 54 through 61, find all of the values of θ in the specified interval that satisfy the given equation.

54. $\cos\theta - \sin 2\theta = 0;\ 0 \le \theta \le 2\pi$

55. $3\sin^2\theta + \cos 2\theta = 2;\ 0 \le \theta \le 2\pi$

56. $\sin 2\theta = \sqrt{3}\cos\theta;\ 0 \le \theta \le \pi$

57. $\cos 2\theta = \cos\theta;\ 0 \le \theta \le \pi$

58. $2\cos^2\theta = \sin 2\theta;\ 0 \le \theta \le \pi$

59. $3\cos^2\theta - \sin^2\theta = 2; 0 \le \theta \le \pi$

60. $2\cos^2\theta + \sin\theta = 1; 0 \le \theta \le \pi$

61. $\cos^2\theta - \sin^2\theta + \cos\theta = 0; 0 \le \theta \le \pi$

62. REFRACTION OF LIGHT BY WATER   A man is swimming at a depth d under water, but because light is refracted by the water, his *apparent* depth s is less than d. In physics,* it is shown that if the man is viewed from an angle of incidence θ, then

$$s = \frac{3d\cos\theta}{\sqrt{7 + 9\cos^2\theta}}.$$

a. If $d = 3$ m and $\theta = 31°$, what is the apparent depth of the man?

b. If the actual depth is $d = 1.9$ m, what angle of incidence yields an apparent depth of $s = 1.12$ m?

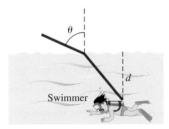

EXERCISE 62

63. A* BIORHYTHMS   Jake believes that his emotional well-being varies periodically with time t so that the pattern is repeated every 28 days. That is, assuming his emotional state was neutral at birth, his emotional level t days later will be

$$E(t) = E_0 \sin\left(\frac{2\pi t}{28}\right),$$

where E_0 is his peak emotional level.

a. What will Jake's emotional level be on his 21st birthday? (Ignore leap years.)

b. Jake also believes that his state of physical well-being follows a similar periodic pattern, only this time the pattern repeats itself every 23 days. If P_0 is Jake's peak level of physical well-being, find a formula for his physical state at time t and determine the level on his 21st birthday.

c. How many days after Jake's 21st birthday should he expect his emotional and physical cycles to *both* be at peak levels?

64. PROFIT   Suppose a company determines that the profit P (in hundreds of dollars) obtained from sales during week t (for $1 \le t \le 52$) of a particular year may be modelled by the trigonometric function

$$P(t) = 75 + 55\cos\left[\frac{\pi(t - 15)}{26}\right].$$

a. What is the highest weekly profit? When does it occur?

b. What is the lowest weekly profit? When does it occur?

65. MEDICINE   A device monitoring a patient's blood pressure (measured in millimetres of mercury) indicates that the *systolic* (maximum) pressure is 130, while the *diastolic* (minimum) pressure is 82. The same patient has a pulse rate of 79 heartbeats per minute. Suppose that t seconds after a systolic reading, the blood pressure is $P(t)$, where P is modelled by a function of the form

$$P(t) = A + B\cos kt,$$

where A, B, and k are positive constants.

a. Use the given information to find A, B, and k.

b. What is the patient's blood pressure after 1 minute?

66. Starting with the addition formulas for sine and cosine, derive the addition formula

$$\tan(A + B) = \frac{\tan A + \tan B}{1 - \tan A \tan B}$$

for tangent.

67. Use the addition formula for sine to derive the identity

$$\sin\left(\frac{\pi}{2} - \theta\right) = \cos\theta.$$

68. Use the addition formula for cosine to derive the identity

$$\cos\left(\frac{\pi}{2} - \theta\right) = \sin\theta.$$

69. Give geometric arguments involving the coordinates of points on a circle of radius 1 to show why the identities in Exercises 67 and 68 are valid.

70. Starting with the identity $\sin^2\theta + \cos^2\theta = 1$, derive the identity $1 + \tan^2\theta = \sec^2\theta$.

---

*R. A. Serway, *Physics,* 3rd ed., Philadelphia: Saunders, 1992, p. 1007.

L02

Differentiate trigonometric functions. Integrate trigonometric functions.

Differentiation and Integration of Trigonometric Functions

The trigonometric functions are relatively easy to differentiate and integrate, and in this section we shall obtain formulas for derivatives and integrals involving such functions. We shall illustrate the use of these formulas by finding rates of change, area, and average value. Then, in Section 11.3, we shall examine a variety of additional applications.

Derivative Formulas for Sine and Cosine

Here are the formulas for the derivatives of the sine and cosine. Since these formulas are often used in conjunction with the chain rule, we also give the generalized version of each formula. The proof of the sine formula will be given at the end of this section, after we have had a chance to see how the formulas are used in practice.

Derivatives for the Sine and Cosine ■ If t is measured in radians, then

$$\frac{d}{dt}[\sin t] = \cos t \quad \text{and} \quad \frac{d}{dt}[\cos t] = -\sin t,$$

and, according to the chain rule, if $u = u(t)$ is a **differentiable function,** then

$$\frac{d}{dt}[\sin u] = \cos u \frac{du}{dt} \quad \text{and} \quad \frac{d}{dt}[\cos u] = -\sin u \frac{du}{dt}.$$

Here are some examples that illustrate the use of these formulas.

EXAMPLE 11.2.1

Differentiate the function $f(t) = \sin(3t + 1)$.

Solution
Using the chain rule for the sine function with $u(t) = 3t + 1$ gives

$$f'(t) = 3\cos(3t + 1).$$

EXAMPLE 11.2.2

Just-In-Time

The nth power of cos x is written as $\cos^n x$; that is, $(\cos x)^n = \cos^n x$. Similarly,
$(\sin x)^n = \sin^n x$
and
$(\tan x)^n = \tan^n x$.

Differentiate the function $f(t) = \cos^2 t$.

Solution
Since
$$\cos^2 t = (\cos t)^2,$$

use the chain rule for powers and the formula for the derivative of the cosine to get

$$f'(t) = 2\cos t(-\sin t) = -2\cos t \sin t.$$

EXAMPLE 11.2.3

An object moves along a straight line so that at time t its position is given by $x = \sin(\pi - 3t) + \cos \pi t$. Where is the object and what are its velocity and acceleration at time $t = 0$?

Solution

The position of the object at time $t = 0$ is given by

$$x(0) = \sin[\pi - 3(0)] + \cos[\pi(0)] = 0 + 1 = 1.$$

The velocity of the object is

$$v(t) = x'(t) = \cos(\pi - 3t)(-3) + [-\sin(\pi t)(\pi)]$$
$$= -3\cos(\pi - 3t) - \pi\sin\pi t$$

and its acceleration is

$$a(t) = v'(t) = -3[-\sin(\pi - 3t)(-3)] - \pi\cos(\pi t)(\pi)$$
$$= -9\sin(\pi - 3t) - \pi^2\cos\pi t$$

Thus, the velocity and acceleration at time $t = 0$ are

$$v(0) = -3\cos[\pi - 3(0)] - \pi\sin[\pi(0)] = -3(-1) - \pi(0) = 3$$

and

$$a(0) = -9\sin[\pi - 3(0)] - \pi^2\cos[\pi(0)] = -9(0) - \pi^2(1) = -\pi^2.$$

The Derivatives of Other Trigonometric Functions

The differentiation formulas for sine and cosine can be used to obtain formulas for the derivatives of other trigonometric functions. For instance, the differentiation formula for tangent may be obtained by using the quotient rule as follows:

$$\frac{d}{dt}[\tan t] = \frac{d}{dt}\left[\frac{\sin t}{\cos t}\right]$$

$$= \frac{\cos t(\cos t) - \sin t(-\sin t)}{\cos^2 t} \qquad \text{quotient rule}$$

$$= \frac{\cos^2 t + \sin^2 t}{\cos^2 t} = \frac{1}{\cos^2 t} \qquad \text{since } \sin^2 t + \cos^2 t = 1$$

$$= \sec^2 t \qquad \text{since } \sec t = \frac{1}{\cos t}$$

Similarly, it can be shown that

$$\frac{d}{dt}[\sec t] = \sec t \tan t$$

(see Exercise 62).

Derivatives for the Tangent and Secant ■ If t is measured in radians, then

$$\frac{d}{dt}[\tan t] = \sec^2 t \qquad \text{and} \qquad \frac{d}{dt}[\sec t] = \sec t \tan t,$$

and if $u = u(t)$ is a differentiable function of t, then

$$\frac{d}{dt}[\tan u] = \sec^2 u \frac{du}{dt} \qquad \text{and} \qquad \frac{d}{dt}[\sec u] = \sec u \tan u \frac{du}{dt}.$$

EXAMPLE 11.2.4

Differentiate the function $f(t) = \tan t^2 + \sec(1 - 3t)$.

Solution

Using the chain rule results in

$$f'(t) = \frac{d}{dt}[\tan t^2] + \frac{d}{dt}[\sec(1 - 3t)]$$

$$= (\sec^2 t^2)\frac{d}{dt}[t^2] + [\sec(1 - 3t)\tan(1 - 3t)]\frac{d}{dt}[1 - 3t]$$

$$= (\sec^2 t^2)(2t) + [\sec(1 - 3t)\tan(1 - 3t)](-3)$$

$$= 2t\sec^2 t^2 - 3\sec(1 - 3t)\tan(1 - 3t)$$

EXAMPLE 11.2.5

An observer watches a plane approach at a speed of 400 km/h and at an altitude of 4 km. At what rate is the angle of elevation of the observer's line of sight changing with respect to time when the horizontal distance between the plane and the observer is 3 km?

Solution

Let x denote the horizontal distance between the plane and the observer, let t denote time (in hours), and draw a diagram representing the situation as in Figure 11.16.

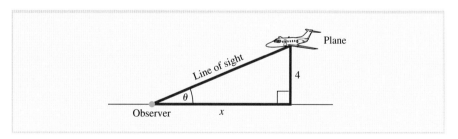

FIGURE 11.16   Observation of an approaching plane.

We know that $\frac{dx}{dt} = -400$ (the minus sign indicates that the distance x is decreasing),

and the goal is to find $\frac{d\theta}{dt}$ when $x = 3$. From the right triangle in Figure 11.16,

$$\tan\theta = \frac{4}{x}.$$

Differentiate both sides of this equation with respect to t (remembering to use the chain rule) to get

$$\sec^2\theta\frac{d\theta}{dt} = -\frac{4}{x^2}\frac{dx}{dt}$$

or

$$\frac{d\theta}{dt} = -\frac{4}{x^2} \cos^2\theta \frac{dx}{dt}.$$

The values of x and $\dfrac{dx}{dt}$ are given. Determine the value of $\cos^2\theta$ when $x = 3$ from the right triangle in Figure 11.17. In particular, when $x = 3$,

$$\cos^2\theta = \left(\frac{3}{5}\right)^2 = \frac{9}{25}.$$

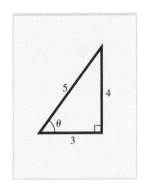

FIGURE 11.17   Triangle for computing $\cos\theta$ when $x = 3$.

Substituting $x = 3$, $\dfrac{dx}{dt} = -400$, and $\cos^2\theta = \dfrac{9}{25}$ into the formula for $\dfrac{d\theta}{dt}$ gives

$$\frac{d\theta}{dt} = -\frac{4}{9}\left(\frac{9}{25}\right)(-400) = 64$$

radians per hour.

Integrals Involving Trigonometric Functions

Since the indefinite integral of a function $f(t)$ is determined by an antiderivative of $f(t)$, each of the differentiation formulas for trigonometric functions we have just obtained corresponds to an integration formula. For instance, since $(\sin u)' = \cos u$, we have

$$\int \cos u \; du = \sin u + C.$$

Integration Formulas Involving Trigonometric Functions

$$\int \cos u \; du = \sin u + C$$

$$\int \sin u \; du = -\cos u + C$$

$$\int \sec^2 u \; du = \tan u + C$$

$$\int \sec u \tan u \; du = \sec u + C$$

EXAMPLE 11.2.6

Find each integral.

a. $\displaystyle\int \sec t(\sec t + \tan t) \; dt$ **b.** $\displaystyle\int x \sin x^2 \; dx$

Solution

a. $\displaystyle\int \sec t(\sec t + \tan t) \; dt = \int \sec^2 t \, dt + \int \sec t \tan t \, dt$

$= \tan t + \sec t + C$

b. Use the substitution $u = x^2$, $du = 2x\,dx$. Then

$$\int x\,\sin x^2\,dx = \int \sin x^2\,(x\,dx)$$

$$= \int \sin u\left(\frac{1}{2}\,du\right) = \frac{1}{2}(-\cos u) + C$$

$$= -\frac{1}{2}\cos x^2 + C$$

EXAMPLE 11.2.7

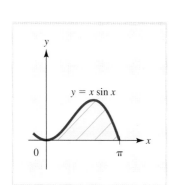

FIGURE 11.18   Area under $y = x\sin x$ over $0 \le x \le \pi$.

Find the area of the region under the curve $y = x\sin x$ between the lines $x = 0$ and $x = \pi$ (Figure 11.18).

Solution

The area is given by the definite integral $A = \displaystyle\int_0^\pi x\,\sin x\,dx$. To evaluate this integral, use integration by parts, as follows (recall Section 6.1):

$$u = x \qquad dv = \sin x\,dx$$
$$du = dx \qquad v = -\cos x$$

so that

$$\int x\,\sin x\,dx = x(-\cos x) - \int (-\cos x)\,dx$$

$$= -x\cos x + \sin x + C$$

Therefore, the area is

$$A = \int_0^\pi x\,\sin x\,dx = (-x\cos x + \sin x)\Big|_0^\pi$$

$$= (-\pi\cos\pi + \sin\pi) - [-(0)\cos 0 + \sin 0]$$

$$= [-\pi(-1) + 0] - (0) = \pi$$

square units.

Just-In-Time

Recall that $\sin 0 = 0$ and $\cos 0 = 1$, and that $\sin\dfrac{\pi}{2} = 1$ and $\cos\dfrac{\pi}{2} = 0$.

EXAMPLE 11.2.8

Find the volume of the solid S formed by revolving the region under the curve $y = \sqrt{\cos x}$ from $x = 0$ to $x = \dfrac{\pi}{2}$ about the x axis.

Solution

According to the formula obtained in Section 5.6, the volume is given by

$$V = \int_0^{\pi/2} \pi y^2\,dx = \int_0^{\pi/2} \pi(\sqrt{\cos x})^2\,dx$$

$$= \int_0^{\pi/2} \pi\cos x\,dx$$

$$= \pi \sin x \Big|_0^{\pi/2}$$

$$= \pi \left(\sin \frac{\pi}{2} - \sin 0 \right) = \pi(1 - 0)$$

$$= \pi$$

Modelling Periodic Phenomena

A *periodic phenomenon* is one that recurs repeatedly in cycles. The motion of a pendulum, the rise and fall of tides, and the circadian rhythms of human metabolism are examples of periodic phenomena.

Periodic phenomena can often be modelled by trigonometric functions of one of these general forms:

$$s(t) = a + b \sin\left[\frac{2\pi}{p}(t - d) \right] \quad \text{or} \quad c(t) = a + b \cos\left[\frac{2\pi}{p}(t - d) \right].$$

For such a function,

p is the *period* and $\dfrac{1}{p}$ is the *frequency.*

b is the *amplitude.*
d is the *phase shift.*
a is the *vertical shift.*

The period p is the length of a cycle in the sense that $f(t + p) = f(t)$ for all t (see Exercise 65), which means that the graph of $f(t)$ over the interval $[0, p]$ is the same as the graph over $[kp, (k + 1)p]$ for $k = 1, 2, \ldots$. The frequency $\dfrac{1}{p}$ provides a measure of how often cycles occur. The amplitude b measures the vertical spread in the graph of $f(t)$ in the sense that $|2b|$ is the difference between the largest and smallest values of $f(t)$. The phase shift d and vertical shift a determine the displacement of the graph of $f(t)$ from the position of a standard sine or cosine curve.

For instance, the function

$$f(t) = 3 + 0.5 \sin \pi(t - 1)$$

has period $p = 2$, frequency $\dfrac{1}{2}$, amplitude 0.5, phase shift 1, and vertical shift 3. The graph of this function is shown in Figure 11.19.

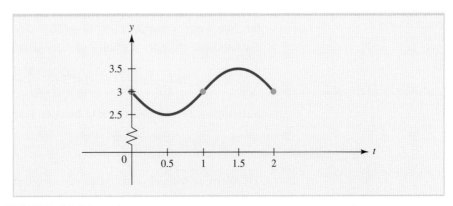

FIGURE 11.19   The graph of the periodic function $f(t) = 3 + 0.5 \sin \pi(t - 1)$.

Here is an example that illustrates how periodic functions can be used to study sales of seasonal products.

EXAMPLE 11.2.9

Suppose the number of bathing suits sold in a large department store during week t of a particular year is modelled by the function

$$B(t) = 40 + 36 \cos\left[\frac{\pi}{52}(t - 24)\right] \qquad \text{for } 1 \le t \le 52.$$

a. During what week are sales the highest?

b. At what rate are sales changing at the end of the year (when $t = 52$)?

c. Past experience indicates that the peak period for sales is from week 20 to week 33. Find the average weekly sales during this period.

Solution

The graph of the sales function is shown in Figure 11.20.

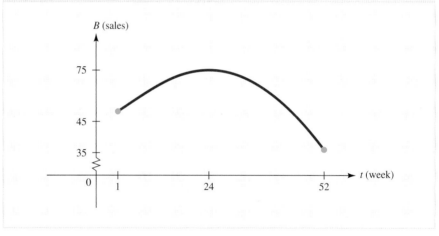

FIGURE 11.20   The graph of the sales function $B(t) = 40 + 36 \cos\left[\frac{\pi}{52}(t - 24)\right]$.

a. Sales are highest when $\cos\left[\frac{\pi}{52}(t - 24)\right] = 1$, and for the time interval $1 \le t \le 52$, this occurs only when $t = 24$. Thus, the highest sales occur during the 24th week, when $B(24) = 40 + 36 = 76$ suits are sold.

b. To determine the rate of change of sales with respect to time, differentiate $B(t)$:

$$B'(t) = 36\left\{-\sin\left[\frac{\pi}{52}(t - 24)\right]\left(\frac{\pi}{52}\right)\right\} = -\frac{9\pi}{13}\sin\left[\frac{\pi}{52}(t - 24)\right].$$

Substituting $t = 52$ into this rate function results in

$$B'(52) = -\frac{9\pi}{13}\sin\left[\frac{\pi}{52}(52 - 24)\right] = -\frac{9\pi}{13}\sin\left(\frac{28\pi}{52}\right) \approx -2.16.$$

Therefore, at the end of the year, sales are decreasing at a rate of approximately 2 suits per week.

c. The average sales figure over the peak period $20 \leq t \leq 33$ is given by the integral

$$AV = \frac{1}{33 - 20} \int_{20}^{33} 40 + 36 \cos\left[\frac{\pi}{52}(t - 24)\right] dt$$

$$= \frac{1}{13}\left\{40t + 36\left(\frac{52}{\pi}\right)\sin\left[\frac{\pi}{52}(t - 24)\right]\right\}\Big|_{20}^{33}$$

$$= \frac{1}{13}\left\{40(33) + 36\left(\frac{52}{\pi}\right)\sin\left[\frac{\pi}{52}(33 - 24)\right]\right\}$$

$$- \frac{1}{13}\left\{40(20) + 36\left(\frac{52}{\pi}\right)\sin\left[\frac{\pi}{52}(20 - 24)\right]\right\}$$

$$\approx \frac{1}{13}(1628.27 - 657.40) \approx 74.68$$

Thus, sales during the peak period average roughly 75 suits per week.

Periodic functions also play an important role in differential equations. In Example 11.2.10, we solve a differential equation to answer a question about a biological population.

EXAMPLE 11.2.10

The size of an animal population often varies with the seasons. Suppose $P(t)$ is the population of a herd of large mammals at time t (months) and that

$$\frac{dP}{dt} = 12 \sin(0.1 \pi t).$$

If the population is initially 3000, what is it 1 year later?

Solution
Separate the variables and integrate to get

$$\int dP = \int 12 \sin(0.1\pi\ t)\ dt$$

$$P(t) = 12\left(\frac{10}{\pi}\right)(-\cos(0.1\ \pi t)) + C \approx -38.2 \cos(0.1\ \pi t) + C$$

Since $P(0) = 3000$,

$$P(0) = 3000 \approx -38.2 \cos 0 + C = -38.2 + C$$
$$C \approx 3000 + 38.2 = 3038.2$$

and the population at time t is

$$P(t) \approx -38.2 \cos(0.1\ \pi t) + 3038.2.$$

Thus, after 1 year ($t = 12$ months),

$$P(12) \approx -38.2 \cos[0.1\ \pi(12)] + 3038.2$$
$$\approx 3069.1$$

so the population increases to about 3069 during the year.

The Proof That

$$\frac{d}{dt}\sin t = \cos t$$

Since the function $f(t) = \sin t$ is unrelated to any of the functions whose derivatives were obtained in previous chapters, it will be necessary to return to the definition of the derivative (from Section 2.1) to find the derivative in this case. In terms of t, the definition states that

$$f'(t) = \lim_{h \to 0} \frac{f(t + h) - f(t)}{h}.$$

In performing the required calculations when $f(t) = \sin t$, we need to use the addition formula

$$\sin(A + B) = \sin A \cos B + \cos A \sin B$$

from Section 11.1 as well as the following two facts about the behaviour of $\sin t$ and $\cos t$ as t approaches zero.

Two Important Limits Involving Sine and Cosine ■ If t is measured in radians, then

$$\lim_{t \to 0} \frac{\sin t}{t} = 1$$

and $$\lim_{t \to 0} \frac{\cos t - 1}{t} = 0.$$

Now we are ready to establish the derivative formula for $\sin t$:

$$\frac{d}{dt}[\sin t] = \lim_{h \to 0} \frac{\sin(t + h) - \sin t}{h}$$

$$= \lim_{h \to 0} \frac{(\sin t \cos h + \cos t \sin h) - \sin t}{h} \qquad \text{addition formula for sine}$$

$$= \lim_{h \to 0} \frac{\sin t(\cos h - 1) + \cos t \sin h}{h} \qquad \text{limits affect only } h, \text{ not } t$$

$$= (\sin t) \lim_{h \to 0} \frac{\cos h - 1}{h} + (\cos t) \lim_{h \to 0} \frac{\sin h}{h}$$

$$= (\sin t)(0) + (\cos t)(1) = \cos t$$

EXPLORE!

Graph $f(x) = \dfrac{\sin x}{x}$. Investigate $\lim\limits_{x \to 0} \dfrac{\sin x}{x}$.

That is,

$$\frac{d}{dt}[\sin t] = \cos t.$$

The derivative formula for cosine can be derived in a similar manner (see Exercise 61).

EXERCISES ■ 11.2

In Exercises 1 through 20, differentiate the given function.

1. $f(t) = \sin 3t$

2. $f(t) = \cos 2t$

3. $f(t) = \sin(1 - 2t)$

4. $f(t) = \sin t^2$

5. $f(t) = \cos(t^3 + 1)$

6. $f(t) = \sin^2 t$

7. $f(t) = \cos^2\left(\dfrac{\pi}{2} - t\right)$

8. $f(t) = \sin(2t + 1)^2$

9. $f(x) = \cos(1 + 3x)^2$

10. $f(x) = e^{-x}\sin x$

11. $f(u) = e^{-u/2}\cos 2\pi u$

12. $f(u) = \dfrac{\cos u}{1 - \cos u}$

13. $f(t) = \dfrac{\sin t}{1 + \sin t}$

14. $f(t) = \tan(5t + 2)$

15. $f(t) = \tan(1 - t^3)$

16. $f(t) = \tan^2 t$

17. $f(t) = \sec\left(\dfrac{\pi}{2} - 2\pi t\right)$

18. $f(t) = \sec(\pi - 4t)^2$

19. $f(t) = \ln(\sin^2 t)$

20. $f(t) = \ln(\tan^2 t)$

In Exercises 21 through 30, find the indicated integral.

21. $\displaystyle\int \sin\dfrac{t}{2}\, dt$

22. $\displaystyle\int (\sin t + \cos 2t)\, dt$

23. $\displaystyle\int x\cos(1 - 3x^2)\, dx$

24. $\displaystyle\int (\tan^2 x + \sec^2 x)\, dx$
[*Hint:* Use the identity $\tan^2 x + 1 = \sec^2 x$.]

25. $\displaystyle\int x\cos 2x\, dx$

26. $\displaystyle\int \dfrac{\sin x}{1 + \cos x}\, dx$

27. $\displaystyle\int \sin x\cos x\, dx$

28. $\displaystyle\int \sqrt{\cos x}\,\sin x\, dx$

29. $\displaystyle\int x\sin x\, dx$

30. $\displaystyle\int \cos^4 x\,\sin x\, dx$

In Exercises 31 through 34, determine the period p, the amplitude b, the phase shift d, and the vertical shift a of the given trigonometric function f(t). Then sketch the graph of f(t).

31. $f(t) = 2 + 1.5\sin\left[\dfrac{\pi}{2}(t - 1)\right]$

32. $f(t) = -1 + 2\cos\left[\dfrac{\pi}{3}(t - 3)\right]$

33. $f(t) = 3 - \cos\left[2\left(t - \dfrac{\pi}{2}\right)\right]$

34. $f(t) = 5 - 3\sin\left[\dfrac{\pi}{4}(t + 1)\right]$

In Exercises 35 through 38, find the area between the given curves over the specified interval.

35. $y = \sin\dfrac{\pi x}{2}$ and $y = \cos \pi x$; $0 \le x \le \dfrac{1}{3}$

36. $y = \sin x$ and $y = \cos(\pi - x)$; $0 \le x \le \dfrac{\pi}{2}$

37. $y = x^2 - x$ and $y = \sin \pi x$; $0 \le x \le 1$

38. $y = x^2 - 1$ and $y = \cos\dfrac{\pi x}{2}$; $0 \le x \le 1$

In Exercises 39 through 42, find the volume of the solid generated when the region under the given curve y = f(x) over the specified region a ≤ x ≤ b is rotated about the x axis.

39. $y = \sqrt{\sin x}$; $0 \le x \le \pi$

40. $y = \sin x\sqrt{\cos x}$; $0 \le x \le \dfrac{\pi}{2}$

41. $y = \sec x$; $-\dfrac{\pi}{3} \le x \le \dfrac{\pi}{3}$

42. $y = \tan x$; $0 \le x \le \dfrac{\pi}{4}$

In Exercises 43 through 46, solve the given differential equation subject to the specified initial condition.

43. $\dfrac{dy}{dt} = \dfrac{\sin 2t}{y + 1}$; $y(0) = 1$

44. $\dfrac{dx}{dt} = x^2\cos t$; $x(\pi) = 1$

45. $\dfrac{dy}{dx} = \dfrac{\sin x}{\cos y}$; $y(\pi) = 0$

46. $\dfrac{dy}{dx} = \dfrac{\sin x - y}{x}$; $y\left(\dfrac{\pi}{2}\right) = 0$

47. **WEATHER**   The average temperature (in degrees Celsius) t hours after midnight in a certain city is given by the formula

$$T(t) = 15 + 12\sin\left(\dfrac{\pi t}{10} - \dfrac{5}{6}\right).$$

At what rate is the temperature changing at noon? Is it getting hotter or cooler at that time?

48. **SALES**   Suppose that the number of pairs of skis sold by a sporting goods store during week t after the beginning of a particular year is modelled by the function

$$S(t) = 125 - 120\cos\left(\dfrac{\pi(t - 18)}{26}\right)$$

for $1 \le t \le 52$.
   a. During what week are the maximum number of skis sold? What is the maximum number?
   b. At what rate are sales changing at the end of the year (when $t = 52$)?

49. **LANDSCAPING**   A large garden is bounded by the positive x and y axes and a cobbled path that follows the curve

$$y = 50\sin(0.3x + 1) + 10\cos(0.6x + 2),$$

where x and y are measured in metres.
   a. Use technology to sketch the graph of the path. What are the endpoints of the path (the points where it intersects the x and y axes)?
   b. Find the area of the garden.

50. **POLLUTION**   A study indicates that on a typical spring day, the level of air pollution in a certain large city is

$$P(t) = 1.5 + 18t - 0.75t^2 + 0.92\sin(2t + 1)$$

units, where t ($0 \le t \le 24$) is the number of hours after midnight.
   a. Find the rate of change of the pollution level at each of the following times: $t = 0$, $t = 9$, $t = 12$, and $t = 24$.
   b. If dawn is at 5 A.M. and dusk is at 6 P.M., what is the average pollution level from dawn to dusk?
   c. Use technology to graph $P(t)$ for $0 \le t \le 24$. When does the maximum level of pollution occur? The minimum level?

51. **BIOLOGY**   Suppose that a biomass $B(t)$ changes with respect to time t (hours) at a rate of

$$\dfrac{dB}{dt} = \sqrt{t} + \sin\dfrac{\pi t}{\sqrt{4}}$$

over a 24-hour period, $0 \le t \le 24$. If the biomass is B_0 at time $t = 0$, how large is it at time $t = 24$?

52. **MEDICINE**   Suppose human metabolism excretes phosphate periodically at a rate of

$$\dfrac{dP}{dt} = -\dfrac{1}{3} + \dfrac{1}{6}\cos\left[\dfrac{\pi}{12}(t - 6)\right]$$

grams per hour at time t (hours) over a 24-hour period ($0 \le t \le 24$), where $P(t)$ is the amount of phosphate in the body at time t.
   a. When does the maximum level of phosphate occur? The minimum level?
   b. If the body of a particular patient contains 380 g of phosphate at time $t = 0$, how much phosphate does it contain 24 hours later?

53. **MALE METABOLISM**   Suppose the body temperature (in degrees Celsius) of a particular man t hours after midnight (for $0 \le t \le 24$) is given by the function

$$T(t) = 37 + 0.2\cos\left[\dfrac{2\pi(t - 15)}{24}\right].$$

   a. What are the maximum and minimum body temperatures and the times when they occur?
   b. At what rate is the man's body temperature changing at noon?
   c. When is the man's body temperature increasing most rapidly?

54. **FEMALE METABOLISM**   The body temperature of a woman is affected by two different cycles, a daily cycle and a monthly cycle. Suppose the body temperature $T(x)$ in degrees Celsius of a woman x days after the start of a monthly cycle is modelled by the formula

$$T(x) = 37 + 0.3\cos\left[\dfrac{2\pi(x - 15)}{28}\right] + 0.4\cos[2\pi(x - 0.6)]$$

for $0 \le x \le 28$.
   a. At what rate is the body temperature changing when $t = 15$? When $t = 28$?
   b. What is the woman's average body temperature over the 28-day period?

55. **SALES**   The sales of a certain product satisfy

$$S(t) = 27.5 + 11.2\sin\dfrac{\pi t}{5} + 2.3\cos\dfrac{\pi t}{3},$$

where t is the time in months measured from January, and S is measured in thousands of units.

a. Find the rate of change of sales at midyear ($t = 6$). Are sales increasing or decreasing at that time?

b. Find the average sales over the entire year ($0 \le t \le 12$).

56. AREA AND VOLUME   Let R be the region bounded by the curves $y = \sin t$ and $y = \cos t$ for $0 \le x \le \dfrac{\pi}{4}$.

a. Find the area of R.

b. Find the volume of the solid formed by revolving R about the x axis. [*Hint:* This is easier than it may seem.]

57. RELATED RATES   A man 2 m tall is watching a streetlight 6 m high while walking toward it at a speed of $\dfrac{5}{3}$ m/s. At what rate is the angle of elevation of the man's line of sight changing with respect to time when he is 3 m from the base of the light?

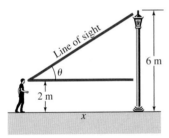

EXERCISE 57

58. RELATED RATES   An attendant is standing at the end of a pier 3 m above a rowboat and is pulling a rope attached to the boat at the rate of 1 m/minute. At what rate is the angle that the rope makes with the surface of the water changing with respect to time when the boat is 4 m from the pier?

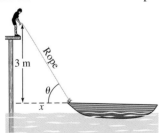

EXERCISE 58

59. RELATED RATES   An observer watches a plane approach at a speed of 500 km/h and at an altitude of 3 km. At what rate is the angle of elevation of the observer's line of sight changing with respect to time t when the horizontal distance between the plane and the observer is 4 km?

60. Find the equation of the tangent line to the graph of $y = x + \sin x$ at the point where $x = 1.5$ (radians). Use technology to graph the function and the tangent.

61. Use the definition of the derivative to show that

$$\frac{d}{dt}[\cos t] = -\sin t.$$

62. A* Use the derivative formulas for $\sin t$ and $\cos t$ to show that

a. $\dfrac{d}{dt}[\sec t] = \sec t \tan t$

b. $\dfrac{d}{dt}[\cot t] = -\csc^2 t$

c. $\dfrac{d}{dt}[\csc t] = -\csc t \cot t$

63. Use technology to sketch the graph of the function

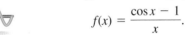

$$f(x) = \frac{\cos x - 1}{x}.$$

Then investigate the limit

$$\lim_{x \to 0} \frac{\cos x - 1}{x}.$$

64. Repeat Exercise 63 for the function

$$f(x) = \frac{\tan x}{x}.$$

65. A* Show that $f(t + p) = f(t)$ for all t, where

a. $f(t) = a + b \sin\left[\dfrac{2\pi}{p}(t - d)\right]$

b. $f(t) = a + b \cos\left[\dfrac{2\pi}{p}(t - d)\right]$

[*Hint:* Use the identities
$\sin(A + B) = \sin A \cos B + \sin B \cos A$ and
$\cos(A + B) = \cos A \cos B - \sin A \sin B$.]

Formulate an application problem in terms of trigonometric functions and differentiate to optimize one variable.

Additional Applications Involving Trigonometric Functions

In this section, we shall examine a variety of applications of calculus involving periodic functions. We begin with several examples in which optimization is the crucial issue.

Periodic Optimization

Using calculus to handle an optimization issue in a model involving periodic behaviour follows the same basic procedure developed in Section 3.5. In particular:

1. Construct a function representing the quantity to be optimized in terms of a convenient variable.
2. Identify an interval on which the function has a practical interpretation.
3. Differentiate the function and set the derivative equal to zero to find the critical numbers.
4. Compare the values of the function at the critical numbers in the interval and at the endpoints (or use the second derivative test) to verify that the desired absolute extremum has been found.

EXAMPLE 11.3.1

A landscape architect is designing a garden that is to be triangular in shape with two equal sides, each 4 m in length. What should the angle be between the two equal sides to make the area of the garden as large as possible?

Solution

The triangle is shown in Figure 11.21. In general, the area of a triangle is given by the formula

$$\text{Area} = \frac{1}{2}(\text{base})(\text{height}).$$

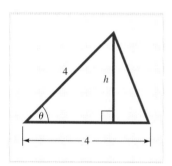

FIGURE 11.21 Triangle for Example 11.3.1.

In this case, the base is $b = 4$, and since $\sin\theta = \dfrac{h}{4}$, the height is $h = 4\sin\theta$. Hence, the area of the triangle is given by the function

$$A(\theta) = \frac{1}{2}(4)(4\sin\theta) = 8\sin\theta.$$

Since only values of θ between 0 and π radians are meaningful in the context of this problem, the goal is to find the absolute maximum of the function $A(\theta)$ on the interval $0 \le \theta \le \pi$.

The derivative of $A(\theta)$ is

$$A'(\theta) = 8\cos\theta,$$

which is zero on the interval $0 \le \theta \le \pi$ only when $\theta = \dfrac{\pi}{2}$. Comparing

$$A(0) = 8\sin 0 = 0$$
$$A\left(\frac{\pi}{2}\right) = 8\sin\frac{\pi}{2} = 8(1) = 8$$
$$A(\pi) = 8\sin\pi = 0$$

we conclude that the area is maximized when the angle θ measures $\dfrac{\pi}{2}$ radians (or 90°), that is, when the triangle is a right triangle.

EXPLORE!

Graph $f(x) = 8\sin x$, and move the cursor along the graph to find the absolute maximum of this function over this interval. Compare to the solution found in Example 11.3.1.

EXAMPLE 11.3.2

A lamp with adjustable height hangs directly above the centre of a circular kitchen table that is 3 m in diameter. The illumination at the edge of the table is directly proportional to the cosine of the angle θ and inversely proportional to the square of the distance d, where θ and d are as shown in Figure 11.22. How close to the table should the lamp be lowered to maximize the illumination at the edge of the table?

Solution

Let L denote the illumination at the edge of the table. Then

$$L = \frac{k \cos \theta}{d^2},$$

where k is a (positive) constant of proportionality. Moreover, from the right triangle in Figure 11.22,

$$\sin \theta = \frac{1.5}{d} \quad \text{or} \quad d = \frac{1.5}{\sin \theta}.$$

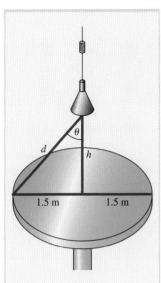

FIGURE 11.22

Hanging lamp and table for Example 11.3.2.

Hence,

$$L(\theta) = \frac{k \cos \theta}{(1.5/\sin \theta)^2} = \frac{k}{2.25} \cos \theta \sin^2 \theta.$$

Only values of θ between 0 and $\dfrac{\pi}{2}$ radians are meaningful in the context of this problem. Hence, the goal is to find the absolute maximum of the function $L(\theta)$ on the interval $0 \le \theta \le \dfrac{\pi}{2}$.

Using the product rule and chain rule to differentiate $L(\theta)$ gives

$$L'(\theta) = \frac{k}{2.25}[\cos \theta (2 \sin \theta \cos \theta) + \sin^2 \theta(-\sin \theta)]$$

$$= \frac{k}{2.25}(2 \cos^2 \theta \sin \theta - \sin^3 \theta)$$

$$= \frac{k}{2.25} \sin \theta (2 \cos^2 \theta - \sin^2 \theta)$$

which is zero when

$$\sin \theta = 0 \quad \text{or} \quad 2 \cos^2 \theta - \sin^2 \theta = 0.$$

The only value of θ in the interval $0 \le \theta \le \dfrac{\pi}{2}$ for which $\sin \theta = 0$ is the endpoint $\theta = 0$. To solve the equation

$$2 \cos^2 \theta - \sin^2 \theta = 0,$$

rewrite it as

$$2 \cos^2 \theta = \sin^2 \theta,$$

divide both sides by $\cos^2\theta$

$$2 = \frac{\sin^2\theta}{\cos^2\theta} = \left(\frac{\sin\theta}{\cos\theta}\right)^2,$$

and take the square root of each side to get

$$\pm\sqrt{2} = \frac{\sin\theta}{\cos\theta} = \tan\theta.$$

Since both $\sin\theta$ and $\cos\theta$ are nonnegative on the interval $0 \le \theta \le \frac{\pi}{2}$, discard the negative square root and conclude that the critical point occurs when

$$\tan\theta = \sqrt{2}.$$

Although it would not be hard to use a calculator to find (or estimate) the angle θ for which $\tan\theta = \sqrt{2}$, it is not necessary to do so. Instead, look at the right triangle in Figure 11.23, in which

$$\tan\theta = \frac{\text{opposite side}}{\text{adjacent side}} = \frac{\sqrt{2}}{1} = \sqrt{2},$$

and observe that

$$\sin\theta = \frac{\text{opposite side}}{\text{hypotenuse}} = \frac{\sqrt{2}}{\sqrt{3}} \qquad \text{and} \qquad \cos\theta = \frac{\text{adjacent side}}{\text{hypotenuse}} = \frac{1}{\sqrt{3}}.$$

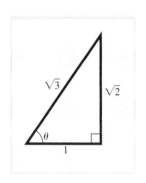

FIGURE 11.23   Right triangle with $\tan\theta = \sqrt{2}$.

Substitute these values into the equation for $L(\theta)$ to conclude that, at the critical point,

$$L(\theta) = \frac{k}{2.25}\cos\theta\,\sin^2\theta = \frac{k}{2.25}\left(\frac{1}{\sqrt{3}}\right)\left(\frac{\sqrt{2}}{\sqrt{3}}\right)^2 = \frac{2k}{6.75\sqrt{3}}.$$

Compare this value with the values of $L(\theta)$ at the endpoints $\theta = 0$ and $\theta = \frac{\pi}{2}$, which are

$$L(\theta) = \frac{k}{2.25}(\cos 0)(\sin^2 0) = 0 \qquad \text{since } \sin 0 = 0$$

and

$$L\left(\frac{\pi}{2}\right) = \frac{k}{2.25}\left(\cos\frac{\pi}{2}\right)\left(\sin^2\frac{\pi}{2}\right) = 0 \qquad \text{since } \cos\frac{\pi}{2} = 0.$$

The largest of these values is $\dfrac{2k}{6.75\sqrt{3}}$, which occurs when $\tan\theta = \sqrt{2}$, so this is the maximum of L.

Finally, to find the height h that maximizes the illumination L, observe from Figure 11.22 that

$$\tan\theta = \frac{1.5}{h}.$$

Since $\tan\theta = \sqrt{2}$ when L is maximized, it follows that

$$h = \frac{1.5}{\tan\theta} = \frac{1.5}{\sqrt{2}} \approx 1.06.$$

So a height of 1.06 m maximizes L.

In Example 11.3.3, calculus is used to establish a general principle about minimizing travel time.

EXAMPLE 11.3.3

Two off-shore oil wells A and B are, respectively, a kilometres and b kilometres out to sea. A motorboat travelling at a constant speed s is to shuttle workers from well A to a depot on the shore and then on to well B. To make this operation cost-efficient, the location of the depot is chosen to minimize the total travel time. Show that at this optimal location P, the angle α between the motorboat's path of arrival and the shoreline will be equal to the angle β between the shoreline and the path of departure.

Solution

Even though the goal is to prove that two angles are equal, it turns out that the easiest way to solve this problem is to let the variable x represent a convenient distance and to introduce trigonometry only at the end.

Begin with a sketch as in Figure 11.24 and define the variable x and the constant distance d as indicated. Then, by the Pythagorean theorem,

$$\text{Distance from first well to } P = \sqrt{a^2 + x^2}$$

and

$$\text{Distance from second well to } P = \sqrt{b^2 + (d - x)^2}.$$

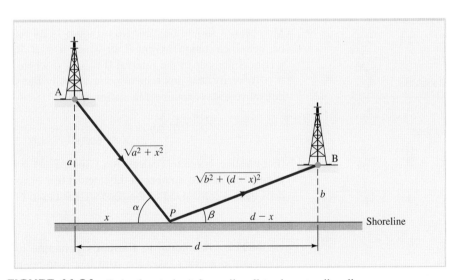

FIGURE 11.24   Path of motorboat from oil well to shore to oil well.

Since (for constant speeds)

$$\text{Time} = \frac{\text{distance}}{\text{speed}},$$

Just-In-Time

Recall that

$$\sqrt{x} = x^{1/2},$$

so that

$$\frac{d}{dx}[\sqrt{x}] = \frac{d}{dx}[x^{1/2}]$$

$$= \frac{1}{2}x^{-1/2}$$

the total travel time is given by the function

$$T(x) = \frac{\sqrt{a^2 + x^2}}{s} + \frac{\sqrt{b^2 + (d - x)^2}}{s},$$

where $0 \le x \le d$. The derivative of $T(x)$ is

$$T'(x) = \frac{x}{s\sqrt{a^2 + x^2}} - \frac{d - x}{s\sqrt{b^2 + (d - x)^2}},$$

and setting $T'(x) = 0$ and multiplying by s results in

$$\frac{x}{\sqrt{a^2 + x^2}} = \frac{d - x}{\sqrt{b^2 + (d - x)^2}}.$$

Now look again at the right triangles in Figure 11.24 and observe that

$$\frac{x}{\sqrt{a^2 + x^2}} = \cos\alpha \quad \text{and} \quad \frac{d - x}{\sqrt{b^2 + (d - x)^2}} = \cos\beta.$$

Hence, $T'(x) = 0$ implies that

$$\cos\alpha = \cos\beta.$$

In the context of this problem, both α and β are between 0 radians and $\dfrac{\pi}{2}$ radians, and for angles in this range, equality of the cosines implies equality of the angles themselves. (See, for example, the graph of the cosine function in Figure 11.10.) Hence, $T'(x) = 0$ implies that

$$\alpha = \beta.$$

Moreover, it should be clear from Figure 11.24 that no matter what a, b, and d may be, there is a (unique) point P for which $\alpha = \beta$. It follows that the function $T(x)$ has a critical number in the interval $0 \le x \le d$ where $\alpha = \beta$.

To verify that this critical point is really the absolute minimum, use the second derivative test. A routine calculation gives

Just-In-Time

Recall that

$$\frac{1}{\sqrt{x}} = x^{-1/2},$$

so that

$$\frac{d}{dx}\left[\frac{1}{\sqrt{x}}\right] = \frac{d}{dx}[x^{-1/2}]$$

$$= \frac{1}{2}x^{-3/2}$$

$$T''(x) = \frac{a^2}{s(a^2 + x^2)^{3/2}} + \frac{b^2}{s[b^2 + (d - x)^2]^{3/2}}.$$

(For practice, check this calculation.) Since $T''(x) > 0$ for all values of x and since there is only one critical number in the interval $0 \le x \le d$, it follows that this critical point does indeed correspond to the absolute minimum of the total time T on this interval.

NOTE According to **Fermat's principle** in optics, light travelling from one point to another takes the path that requires the least amount of time. Suppose, as illustrated in Figure 11.25, that a ray of light is transmitted from a source at a point A, strikes a reflecting surface (such as a mirror) at P, and is subsequently received by an observer at point B. Since by Fermat's principle, the path from A to P to B minimizes time, it follows from the calculation in Example 11.3.3 that the angles α and β are equal. This, in turn, implies the **law of reflection,** which states that *the angle of incidence must equal the angle of reflection* ($\theta_1 = \theta_2$ in Figure 11.25).

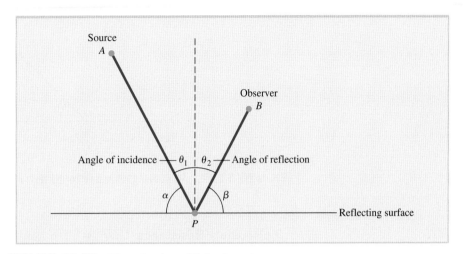

FIGURE 11.25   The reflection of light: $\theta_1 = \theta_2$.

The human vascular system operates so that the circulation of blood from the heart, through the organs, and back to the heart is accomplished with as little expenditure of energy as possible. Thus, it is reasonable to expect that when an artery branches, the angle between the parent and daughter arteries should minimize the total resistance to the flow of blood. In Example 11.3.4, we use calculus to find this optimal branching angle.

EXAMPLE 11.3.4

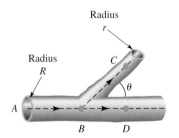

FIGURE 11.26   Vascular branching.

Figure 11.26 shows a small artery of radius r branching at an angle θ $\left(0 < \theta < \dfrac{\pi}{2} \right)$ from a larger artery of radius R $(R > r)$.* Blood flows in the direction of the arrows from point A to the branch at B and then to points C and D. Show that the resistance of the blood flow is minimized when θ satisfies

$$\cos \theta = \frac{r^4}{R^4}.$$

Solution

Figure 11.27 is a geometric version of Figure 11.26. For simplicity, assume that C and D are located so that CD is perpendicular to the main line through A and B, with C located h units above D. Find the value of the branching angle θ that minimizes the total resistance to the flow of blood as it moves from A to B and then to point C.

Before we can solve this problem, we must have a way of measuring the resistance to blood flow, and this is provided by the following result, which was discovered empirically by the French physiologist and physician Jean Louis Poiseuille (1799–1869).

---

* This application is adapted from an example in the text by E. Batschelet, *Introduction to Mathematics for Life Scientists,* 2nd ed., New York: Springer-Verlag, 1975, pp. 278–280. As the original source of the material, the author cites the text by R. Rosen, *Optimality Principles in Biology*, London, England: Butterworths, 1967.

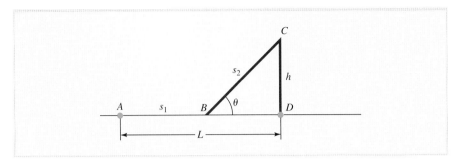

FIGURE 11.27   Minimizing resistance to the flow of blood.

Poiseuille's Resistance Law ■ The resistance to the flow of blood in an artery is directly proportional to the artery's length and inversely proportional to the fourth power of its radius.

According to Poiseuille's law, the resistance to the flow of blood from point A to point B in Figure 11.27 is

$$F_1 = \frac{ks_1}{R^4},$$

and the resistance from point B to point C is

$$F_2 = \frac{ks_2}{r^4},$$

Just-In-Time

Recall that

$$\frac{1}{\sin\theta} = \csc\theta$$

and

$$\frac{1}{\tan\theta} = \cot\theta,$$

and note that $\csc\theta$ is never 0.

where k is a constant that depends on the viscosity of blood, and s_1 and s_2 are the lengths of the portions of artery from A to B and from B to C, respectively. Thus, the total resistance to the flow of blood up to and through the branch is given by the sum

$$F = F_1 + F_2 = \frac{ks_1}{R^4} + \frac{ks_2}{r^4} = k\left(\frac{s_1}{R^4} + \frac{s_2}{r^4}\right).$$

Next, write F as a function of the branching angle θ. To do this, in Figure 11.27, note that

$$\sin\theta = \frac{h}{s_2} \quad\text{and}\quad \tan\theta = \frac{h}{L - s_1},$$

where L is the length of the main artery from A to D. Solve these equations for s_1 and s_2 to get

$$s_2 = \frac{h}{\sin\theta} = h\csc\theta \quad\text{and}\quad s_1 = L - \frac{h}{\tan\theta} = L - h\cot\theta,$$

so that F can be expressed as

$$F = k\left(\frac{s_1}{R^4} + \frac{s_2}{r^4}\right) = k\left(\frac{L - h\cot\theta}{R^4} + \frac{h\csc\theta}{r^4}\right).$$

In Exercise 62 of Section 11.2, you showed that

$$\frac{d}{dx}[\cot u] = -\csc^2 u \frac{du}{dx} \qquad \text{and} \qquad \frac{d}{dx}[\csc u] = -\cot u \csc u \frac{du}{dx}.$$

Use these formulas to differentiate $F(\theta)$:

$$F'(\theta) = k\left[\frac{-h(-\csc^2\theta)}{R^4} + \frac{h(-\csc\theta\,\cot\theta)}{r^4}\right]$$

$$= kh\,\csc\theta\left(\frac{\csc\theta}{R^4} - \frac{\cot\theta}{r^4}\right)$$

Solve $F'(\theta) = 0$ to obtain all critical numbers for F:

$$\csc\theta = 0 \qquad \text{or} \qquad \frac{\csc\theta}{R^4} - \frac{\cot\theta}{r^4} = 0.$$

Because $\csc\theta$ is never 0, it follows that $F'(\theta) = 0$ only when

$$\frac{\csc\theta}{R^4} - \frac{\cot\theta}{r^4} = 0$$

$$\frac{\csc\theta}{R^4} = \frac{\cot\theta}{r^4}$$

$$\frac{1}{R^4\sin\theta} = \frac{\cos\theta}{r^4\sin\theta}$$

$$\frac{r^4}{R^4} = \cos\theta$$

Finally, if θ_0 is the acute angle that satisfies this equation, it can be shown that the resistance F has an absolute minimum when $\theta = \theta_0$ (see Exercise 35). Thus, θ_0 is the optimal angle of vascular branching.

Periodic Flow of Income

The rate of flow of income into an interest-earning account can fluctuate periodically and thus can often be modelled by trigonometric functions. In this context, positive values of the flow rate function correspond to deposits into the account, while negative values correspond to withdrawals from the account.

In Section 5.5, we showed that if income is transferred continuously at a rate $f(t)$ into an account earning interest at an annual rate r compounded continuously, then the accumulated (future) value of the account after a term of T years is given by the integral

$$FV = \int_0^T f(t)e^{r(T-t)}\,dt,$$

and the present value of the same income flow is given by

$$PV = \int_0^T f(t)e^{-rt}\,dt.$$

Recall that present value PV represents the amount that must be deposited now at the prevailing rate for the same term of T years to generate the accumulated value FV of the income flow. Present value is used to assess the value of a periodic income flow in Example 11.3.5. For this example and other similar problems, you may need the following two integration formulas, which may be obtained by two applications of integration by parts.

Integration Formulas Involving $e^{au} \sin bu$ and $e^{au} \cos bu$ ∎

$$\int e^{au} \sin bu \, du = \frac{e^{au}}{a^2 + b^2}(a \sin bu - b \cos bu) + C$$

$$\int e^{au} \cos bu \, du = \frac{e^{au}}{a^2 + b^2}(a \cos bu + b \sin bu) + C$$

EXAMPLE 11.3.5

Jasmine is considering an investment that is projected to be generating income continuously at a rate of $f(t)$ dollars per year t years after it is initiated, where

$$f(t) = 2000 \cos \frac{t}{4}.$$

She plans to hold the investment for a term of 4 years and expects the prevailing interest rate to remain fixed at 5% compounded continuously during that period. What is a fair price for her to pay for the investment?

Solution

A fair price for the investment would be the present value

$$PV = \int_0^4 2000 \cos \frac{t}{4} e^{-0.05t} \, dt$$

$$= 2000 \int_0^4 e^{-0.05t} \cos \frac{t}{4} \, dt$$

Applying the integral formula for $\int e^{au} \cos bu \, du$, with $u = t$, $a = -0.05$, and $b = \frac{1}{4}$, gives

$$\int_0^4 e^{-0.05t} \cos \frac{t}{4} \, dt = \frac{e^{-0.05t}}{(-0.05)^2 + \left(\frac{1}{4}\right)^2}\left[-0.05 \cos \frac{t}{4} + \frac{1}{4} \sin \frac{t}{4}\right]\Bigg|_0^4$$

$$= \frac{e^{-0.2}}{0.065}\left[-0.05 \cos 1 + \frac{1}{4} \sin 1\right] - \frac{e^0}{0.065}\left[-0.05 \cos 0 - \frac{1}{4} \sin 0\right]$$

$$\approx 3.0787$$

Therefore, the present value of the investment is

$$PV = 2000 \int_0^4 e^{-0.05t} \cos \frac{t}{4} \, dt$$

$$= 2000(3.0787)$$

$$\approx 6157$$

so a fair price is roughly $6157.

EXERCISES ■ 11.3

In each of Exercises 1 through 4, find the future value FV and the present value PV of a continuous income flow f(t) into an account earning interest at the specified annual rate r over the specified term T.

1. $f(t) = 2000 \cos \dfrac{t}{5}$; 4%; $T = 5$

2. $f(t) = 1000 \sin \dfrac{\pi t}{3}$; 5%; $T = 6$

3. $f(t) = 3000 \sin \pi t$; 3%; $T = 3$

4. $f(t) = 10\,000\,(\sin t + \cos t)$; 6%; $T = 5$

*An object moving back and forth along a straight line is said to undergo **vibrational motion** if its displacement from equilibrium at time t is given by an equation of the form*

$$x(t) = e^{-kt}(A \cos \omega t + B \sin \omega t),$$

where A, B, k, and ω are constants. In Exercises 5 through 10, answer the following questions about the given function x(t) with this form for the indicated time interval $a \le x \le b$.

   (a) Sketch the graph of x(t). You may wish to use technology.

   (b) Find the values of t for which the largest and smallest values of x(t) occur for $a \le t \le b$.

5. $x(t) = \sin 2t + \cos 2t$ for $0 \le t \le \dfrac{\pi}{2}$

6. $x(t) = 2 \sin 2t + 3 \cos 2t$ for $0 \le t \le \dfrac{\pi}{2}$

7. $x(t) = 5 \sin 3t + 2 \cos t$ for $0 \le t \le \pi$

8. $x(t) = 2 \sin t - 3 \cos 2t$ for $0 \le t \le \pi$

9. $x(t) = e^{-t}(\sin t + \cos t)$ for $\dfrac{\pi}{2} \le t \le \dfrac{3\pi}{2}$

10. $x(t) = e^{-2t}(3 \sin 2t + 2 \cos 3t)$ for $0 \le t \le \pi$

11. Find the largest value of
$$f(t) = \cot t - \sqrt{2} \csc t \quad \text{for } 0 < t < \pi.$$
Does $f(t)$ have a smallest value on this interval?

12. Find the smallest value of
$$f(t) = 27 \sec t - 8 \csc t \quad \text{for } 0 < t < \dfrac{\pi}{2}.$$
Does $f(t)$ have a largest value on this interval?

13. HUMAN RESPIRATION A typical adult male inhales and exhales approximately every

3.5 seconds. Let $B(t)$ be the volume (in litres) of the air in the lungs of an adult male subject t seconds after exhaling. Suppose $B(t)$ can be modelled by the function

$$B(t) = 0.5 - 0.4 \cos \dfrac{\pi t}{1.75}.$$

  a. Find the maximum volume of air in the lungs of the subject and the times when his lungs contain the maximum volume of air.

  b. Answer the questions in part (a) for the minimum volume of air in the subject's lungs.

  c. Sketch the graph of $B(t)$.

14. ENERGY CONSUMPTION Let $H(t)$ be the number of litres of heating oil used to heat a particular house in Kingston, Ontario, during month t after February 2011. Suppose that $H(t)$ is modelled by the function

$$H(t) = 800 + 660 \cos \dfrac{\pi t}{6}.$$

  a. During which month is oil consumption the highest? What is this maximum amount?

  b. During which month is oil consumption the lowest? What is this minimum amount?

  c. What total amount of heating oil will be consumed in 2012?

  d. What is the average monthly consumption of heating oil in 2012?

15. SEASONAL POPULATIONS The death rate and birth rate of many animal and plant species fluctuate periodically with the seasons. The population $P(t)$ of such a species at time t changes at a rate that may be modelled by a differential equation of the form

$$\dfrac{dP}{dt} = (b + a \cos 2\pi t)P,$$

where a and b are constants. Solve this equation to show that

$$P(t) = P_0 e^{bt + \frac{a}{2\pi} \sin 2\pi t},$$

where $P_0 = P(0)$ is the initial population.

16. SEASONAL POPULATIONS A herd of large mammals has a population $P(t)$ that fluctuates periodically with the seasons and satisfies a differential equation of the form given in Exercise 15, where t is measured in years. Initially, there are

$P(0) = 3000$ animals in the herd, and it is observed that the populations after 3 months and 6 months, respectively, are

$$P\left(\frac{1}{4}\right) = 2800 \quad \text{and} \quad P\left(\frac{1}{2}\right) = 3200.$$

a. Use this information to determine $P(t)$. That is, find the parameters a and b in the formula obtained in Exercise 15.

b. What will the population be after 9 months $\left(t = \frac{3}{4}\right)$? After 1 year?

c. When during the first year ($0 \le t \le 1$) is the population of the herd the greatest? What is the maximum population?

d. Answer the questions in part (c) for the minimum population of the herd.

17. A* AVIAN BEHAVIOUR Experiments indicate that homing pigeons will avoid flying over large areas of water whenever possible, perhaps because it takes more effort to fly through the heavy, cool air over the water than the relatively light, warm air over land.* Suppose a pigeon is released from a boat at point P on a lake and flies to its loft at point L on the lakeshore, as shown in the accompanying figure. Assume that E_w units of energy per kilometre are required to fly over water and E_L units of energy per kilometre are required over land. If the boat is s_1 kilometres from the nearest point A on the shore and A is s_2 kilometres from the loft L along the shore, what heading θ should the pigeon take from the boat in order to minimize the total energy expended in getting to L? (Express your answer in terms of $\sin \theta$.)

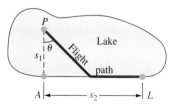

EXERCISE 17

18. LANDSCAPE DESIGN A landscape architect wishes to construct a garden in the shape of an isosceles triangle. The two equal sides are each to

*This exercise is adapted from an example from the text *Introduction to Mathematics for Life Scientists*, 2nd ed., by Edward Batschelet, New York: Springer-Verlag, 1979, pp. 276–278.

have length 20 m. How long should the third side be in order for the area of the garden to be as large as possible? What is the maximum area of a garden that is designed with these specifications?

19. ARCHITECTURE An architect is asked to construct a large conference room in the shape of an isosceles trapezoid, as shown in the accompanying figure. The two equal (nonparallel) walls of the room and the shorter of the two parallel walls are each to be 20 m long. How long should the fourth wall be to maximize the area of the room? What is the maximum area of a room designed with these specifications? [*Hint*: Express the area in terms of the angle θ shown in the figure.]

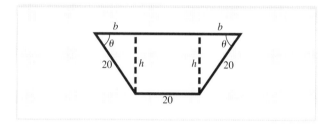

EXERCISE 19

20. WEATHER Suppose Environment Canada determines that the average temperature (°C) of a certain city is given by

$$F(t) = 25 \sin\left[\frac{2\pi}{365}(t - 97)\right] + 3,$$

 where t is the number of days after January 1.

a. Sketch the graph of the temperature function $F(t)$. You may wish to use technology.

b. What are the highest and lowest temperatures of the year in this city? When do these extreme temperatures occur?

21. SEASONAL POPULATIONS The population $P(t)$ of a species fluctuates periodically with the seasons so that

$$\frac{dP}{dt} = 0.03P + 0.004 \cos 2\pi t,$$

where t is time, in years. If the population is initially $P(0) = 10\,000$, what is it 10 years later?

22. A* OPTIMAL ANGLE OF OBSERVATION At an art gallery a very large painting that is 8 m high is hung on the wall of an art museum, 2 m above the eye level of an observer, as shown in the accompanying figure (figure not to scale).

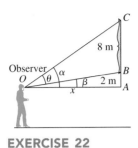

EXERCISE 22

Let θ be the angle subtended by the painting at the observer's eye when he stands x metres from the wall. It is reasonable to assume that the observer gets the best view of the painting when θ is as large as possible.

a. Let β be the angle $\angle AOB$ and let α be $\angle AOC$. Use the trigonometric identity

$$\tan(\alpha - \beta) = \frac{\tan \alpha - \tan \beta}{1 + \tan \alpha \tan \beta}$$

to express $\tan \theta$ in terms of x.

b. How far back from the wall should the observer stand in order to maximize the angle θ? [*Hint:* Use implicit differentiation with the equation for $\tan \theta$ you found in part (a).]

23. **MEDICINE**   A *pneumotachograph* is a device used to measure the rate of air flow in and out of the lungs. Let $V(t)$ be the volume of air in the lungs (in litres) at time t (seconds). Suppose the rate of air flow is modelled by

$$V'(t) = 0.87 \sin 0.65t.$$

a. Inhalation occurs when $V'(t) > 0$ and exhalation when $V'(t) < 0$. The *respiratory period* is the amount of time L required to inhale and then exhale exactly once. What is L?

b. What volume of air is inhaled during the time interval $0 \le t \le L$? Is this the same as the volume exhaled during the same time period?

c. Answer the questions in parts (a) and (b) if the rate of air flow is modelled by

$$V'(t) = 0.87t \sin 0.65t.$$

[*Hint:* Use integration by parts.]

24. **A* MEDICINE**   Repeat Exercise 23 for the rate function

$$V'(t) = 0.31e^{-0.1t} \sin 0.81t.$$

25. **A* CONSTRUCTION**   You have a piece of metal that is 20 cm long and 6 cm wide, which you are going to bend to form a trough as indicated in the accompanying figure. At what angle should the sides

meet so that the volume of the trough is as large as possible? [*Hint:* The volume is the length of the trough times the area of its triangular cross-section.]

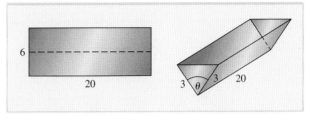

EXERCISE 25

26. **A* CONSTRUCTION**   In a building under construction, a hallway $3\sqrt{2}$ m wide turns at a corner into a long room that is 6 m wide, as shown in the accompanying figure. What is the length of the longest pipe that workers can carry horizontally around this corner? Ignore the thickness of the pipe. [*Hint:* Why is this the same as minimizing the horizontal clearance

$$C(\theta) = \frac{6}{\sin \theta} + \frac{3\sqrt{2}}{\cos \theta}$$

on an appropriate interval $0 \le \theta \le b$?]

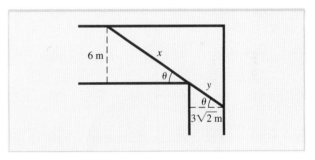

EXERCISE 26

27. **A* REFRACTION OF LIGHT**   Fermat's principle in physics states that light takes the path requiring the least amount of time. The accompanying figure shows a ray of light that is emitted from a source at point A under water and, subsequently, received by an observer at point B above the surface of the water. If v_1 is the speed of light in water and v_2 is the speed of light in air, use Fermat's principle to show that

$$\frac{\sin \theta_1}{\sin \theta_2} = \frac{v_1}{v_2},$$

where θ_1 is the **angle of incidence** and θ_2 is the **angle of refraction**. [*Hint:* Use the solution to Example 11.3.3 as a guide. You may assume without proof that the total time is minimized when its derivative is equal to zero.]

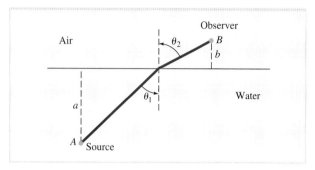

EXERCISE 27

28. **A* ILLUMINATION**   Rework Example 11.3.2, assuming the illumination is directly proportional to the sine of the angle ϕ and is inversely proportional to the square of the distance d, where ϕ is the angle at which the ray of light meets the table, as shown in the accompanying figure. Explain why your answer is the same as before.

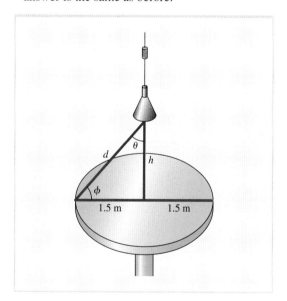

EXERCISE 28

MOTION OF A PROJECTILE   *Suppose a projectile is fired from the top of a cliff s_0 metres high at an angle of θ radians with an initial (muzzle) speed of v_0 metres per second. In physics, it is shown that t seconds later, the projectile will be at the point (x, y), where x and y are in metres and*

$$x = (v_0 \cos \theta)t$$
$$y = -4.9t^2 + (v_0 \sin \theta)t + s_0$$

These formulas are used in Exercises 29 through 31.

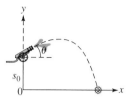

29. **WATER-SKIING**   A water-skier is being pulled by a motorboat toward a jump ramp. The angle of the ramp is 12° and its end is 1 m above the water, as shown in the accompanying figure. If the skier is travelling at 15 m/s as she leaves the ramp, how far from the ramp does she land?

EXERCISE 29

30. **SPY STORY**   Much to his relief, the spy's grenade detonates (Section 10.2), disabling most of the villains. However, one of them is not stunned and lobs a grenade in the spy's direction. The spy estimates that his opponent releases the grenade at an angle of 45° from a height of 2 m with an initial speed of 17 m/s. Once the grenade lands, the explosion will damage everything within a 3-m radius of its point of impact. How far away from his enemy must the spy be when the grenade lands if he wants to escape?

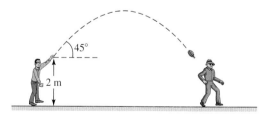

EXERCISE 30

31. **HEIGHT AND RANGE OF A PROJECTILE**
A projectile is fired at ground level ($s_0 = 0$) at an angle θ with initial speed v_0 metres per second.
   a. Find a formula for the maximum height reached by the projectile during its flight. (Your answer will involve θ and v_0.)

b. The **range** of the projectile is the horizontal distance measured from its firing position to the point of impact. Find a formula for the range (in terms of θ and v_0). For what value of θ is the range the largest?

32. SATELLITE ORBIT   A telecommunications satellite is located

$$r(t) = \frac{7730}{1 + 0.15 \cos(0.06t)}$$

kilometres above the centre of Earth t minutes after achieving orbit, as indicated in the accompanying figure.

a. Sketch the graph of $r(t)$.

b. What are the lowest and highest points on the satellite's orbit? (These are called the *perigee* and *apogee* positions, respectively.)

EXERCISE 32

33. ALZHEIMER'S DISEASE   Research* indicates that the body temperature $T(t)$ (in degrees Celsius) of patients with Alzheimer's disease fluctuates periodically over a 24-hour period according to the formula

$$T(t) = 37.29 + 0.46 \cos\left[\frac{\pi(t - 16.37)}{12}\right],$$

where t $(0 \le t \le 24)$ is the number of hours past midnight.

a. Find the derivative $T'(t)$.

b. At what time (or times) during the 24-hour period does the maximum body temperature occur? What is the maximum temperature?

c. Answer the questions in part (b) for the minimum body temperature during the 24-hour period.

d. Suppose that patients in a certain control group are awake from 7 A.M. to 10 P.M. What is the average body temperature of such a patient over this wakeful period?

* L. Volicer et al., "Sundowning and Circadian Rhythms in Alzheimer's Disease," *American Journal of Psychiatry,* Vol. 158, No. 5, May 2001, pp. 704–711.

34. SURFACE AREA OF A BEE'S CELL   A beehive cell is a regular hexagonal prism that is open at one end and has a trihedral angle θ at the other, as shown in the accompanying figure.

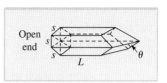

EXERCISE 34

Using trigonometry** (quite a lot!), it can be shown that the cell has surface area

$$S(\theta) = 6sL + 1.5s^2(-\cot\theta + \sqrt{3} \csc\theta)$$

for $0 < \theta < \dfrac{\pi}{2}$, where s is the length of each side in the open hexagonal base, and L is the height of the prism.

a. Assuming that s and L are fixed (constant), what angle minimizes the surface area $S(\theta)$? Express the minimum area in terms of s and L.

b. Read the article cited with this problem and write a paragraph on whether or not you think bees actually do conserve wax by constructing their cells with the "bee angle" found in part (a).

35. BLOOD FLOW   Let $F(\theta)$ be the resistance to blood flow function obtained in Example 11.3.4, namely,

$$F(\theta) = k\left[\frac{L - h \cot\theta}{R^4} + \frac{h \csc\theta}{r^4}\right].$$

We found the derivative $F'(\theta)$ in Example 11.3.4 and showed that $F(\theta)$ has its only critical number at θ_0, where $\cos\theta_0 = \dfrac{r^4}{R^4}$. Show that $F'(\theta) > 0$ if $0 < \theta < \theta_0$ and that $F'(\theta) > 0$ if $\theta > \theta_0$. Why does this mean that $F(\theta)$ has an absolute minimum at θ_0?

36. POLARIZED LIGHT   A polarized light wave travels so that its vertical displacement y at time t is a function of both t and its horizontal displacement x according to the formula

$$y(x, t) = 0.27 \sin\left(10\pi t - 3\pi x + \frac{\pi}{4}\right).$$

a. Find $\dfrac{\partial y}{\partial x}$ and $\dfrac{\partial y}{\partial t}$.

b. For what points (x, t) is $y(x, t)$ maximized? For what points is $y(x, t)$ minimized?

**E. Batschelet, *Introduction to Mathematics for Life Scientists,* 2nd ed., New York: Springer-Verlag, 1979, pp. 274–276.

37. POPULATION GROWTH   The number $P(t)$ of deer ticks per hectare in an Ontario provincial park can be modelled by the function

$$P(t) = 1250e^{0.3t}\left\{1.47 + 1.38\cos\left[\frac{2\pi}{365}(365t - 145)\right]\right\},$$

where t is the number of years after the base year 2008.

a. Find the number of deer ticks per hectare in the park in 2008, 2013, and 2018.

 b. With technology, draw a graph of the function over the 10-year period.

c. Write a few sentences about the graph. What is the significance of the number 145 in the argument of cosine? Is the model realistic? Can you suggest any flaws in the model?

38. Two sides of a triangle have lengths 3 and 4 and are separated by an angle θ, as shown in the accompanying figure.

a. Show that the area of the triangle is given by $A(\theta) = 6\sin\theta$ and use graphing software to graph $A(\theta)$ for $0 \le \theta \le \pi$.

b. Find the angle for which $A(\theta)$ is maximized for $0 \le \theta \le \pi$.

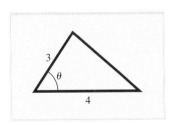

EXERCISE 38

39. ENERGY CONSUMPTION   The amount of household energy consumption for space heating is sometimes measured by *heating degree days* (hdd), which are defined as the difference between the average temperature that day and 18°C.

Suppose the number of hdd in Saint John, New Brunswick, on day t of a particular year is modelled by the function

$$H(t) = 25 + 22\cos\left[\frac{2\pi}{365}(t - 35)\right]$$

for $0 \le t \le 365$, where $t = 0$ corresponds to the beginning of the day on January 1.

a. On what day of the year does this model predict the maximum number of hdd in Saint John will be experienced? What is this maximum number?

b. Answer the questions in part (a) for the minimum number of hdd.

c. At what rate is the energy consumption function $H(t)$ changing on January 1? Is $H(t)$ increasing or decreasing at this time? Repeat for April 1 ($t = 91$) and June 1 ($t = 152$).

d. Sketch the graph of $H(t)$ over the entire year.

e. What is the average hdd over the first 90 days of the year (January through March)? This provides a measure of the average daily energy consumption of a household in Saint John during the first quarter of the year.

40. A* Let I denote the integral $I = \int e^{au}\sin bu\, du$.

a. Use integration by parts to show that

$$I = \frac{1}{b}\left(-e^{au}\cos bu + a\int e^{au}\cos bu\, du\right).$$

b. Use integration by parts a second time to show that

$$I = \frac{1}{b}\left(-e^{au}\cos bu\right) + \frac{a}{b}\left(\frac{1}{b} - e^{au}\sin bu - \frac{a}{b}I\right).$$

c. Solve the equation in part (b) for I to obtain

$$I = \int e^{au}\sin bu\, du = \frac{e^{au}}{a^2 + b^2}(a\sin bu - b\cos bu).$$

d. Substitute the integration formula obtained in part (c) into the equation in part (a) and solve to obtain a formula for $\int e^{au}\cos bu\, du$.

Concept Summary Chapter 11

$$\frac{degrees}{180} = \frac{radians}{\pi}$$

A Table of Frequently Used Values of Sine and Cosine

θ	0	$\frac{\pi}{6}$	$\frac{\pi}{4}$	$\frac{\pi}{3}$	$\frac{\pi}{2}$	$\frac{2\pi}{3}$	$\frac{3\pi}{4}$	$\frac{5\pi}{6}$	π
$\sin\theta$	0	$\frac{1}{2}$	$\frac{\sqrt{2}}{2}$	$\frac{\sqrt{3}}{2}$	1	$\frac{\sqrt{3}}{2}$	$\frac{\sqrt{2}}{2}$	$\frac{1}{2}$	0
$\cos\theta$	1	$\frac{\sqrt{3}}{2}$	$\frac{\sqrt{2}}{2}$	$\frac{1}{2}$	0	$\frac{-1}{2}$	$\frac{-\sqrt{2}}{2}$	$\frac{-\sqrt{3}}{2}$	-1

$$\sin(\theta + 2\pi) = \sin\theta \qquad \cos(\theta + 2\pi) = \cos\theta$$
$$\sin(-\theta) = -\sin\theta \qquad \cos(-\theta) = \cos\theta$$

$$\tan\theta = \frac{\sin\theta}{\cos\theta} \qquad \cot\theta = \frac{\cos\theta}{\sin\theta}$$

$$\sec\theta = \frac{1}{\cos\theta} \qquad \csc\theta = \frac{1}{\sin\theta}$$

$$\sin\theta = \frac{opposite}{hypotenuse}$$

$$\cos\theta = \frac{adjacent}{hypotenuse}$$

$$\tan\theta = \frac{opposite}{adjacent}$$

$$\sin^2\theta + \cos^2\theta = 1$$

$$\sin(A + B) = \sin A \cos B + \sin B \cos A$$
$$\cos(A + B) = \cos A \cos B - \sin A \sin B$$

$$\sin 2A = 2\sin A \cos A$$

$$\cos 2A = \cos^2 A - \sin^2 A$$
$$= 2\cos^2 A - 1$$
$$= 1 - 2\sin^2 A$$

$$\frac{d}{dt}[\sin u(t)] = \cos u \frac{du}{dt}$$

$$\frac{d}{dt}[\cos u(t)] = -\sin u \frac{du}{dt}$$

$$\frac{d}{dt}[\tan u(t)] = \sec^2 u \frac{du}{dt}$$

$$\frac{d}{dt}[\sec u(t)] = \sec u \tan u \frac{du}{dt}$$

$$\int \cos u \, du = \sin u + C$$

$$\int \sin u \, du = -\cos u + C$$

$$\int \sec^2 u \, du = \tan u + C$$

$$\int \sec u \tan u \, du = \sec u + C$$

$$s(t) = a + b\sin\left[\frac{2\pi}{p}(t - d)\right]$$

$$c(t) = a + b\cos\left[\frac{2\pi}{p}(t - d)\right]$$

p is period, $\frac{1}{p}$ is frequency

b is amplitude
d is phase shift
a is vertical shift

Checkup for Chapter 11

1. Evaluate each expression without using tables or a calculator.

   a. $\sin\frac{5\pi}{6}$ **b.** $\cos\left(-\frac{4\pi}{3}\right)$

   c. $\tan\frac{3\pi}{4}$ **d.** $\sec\left(-\frac{5\pi}{6}\right)$

2. Find $\cos\theta$ for $0 \le \theta \le \frac{\pi}{2}$ given that $\tan\theta = \frac{2}{3}$.

3. Differentiate each function.

   a. $f(x) = \sin 2x$ **b.** $f(x) = x\cos x$

   c. $f(x) = \frac{\tan x}{x}$ **d.** $f(x) = x\sin x^2$

CHAPTER SUMMARY

4. Evaluate each integral.

   a. $\int \sin 2x \, dx$

   b. $\int x \cos x^2 \, dx$

   c. $\int_0^{\pi/2} \sec^2 2t \, dt$

   d. $\int_0^{\pi/6} \sin^2 t \cos t \, dt$

5. Find all values of θ such that $\sin 2\theta = \cos \theta$ for $0 \le \theta \le \pi$.

6. Find the area of the region bounded above by $y = \cos x$, below by $y = \sin x$, and on the left by the y axis.

7. Starting with the identity $\sin^2 x + \cos^2 x = 1$, derive the identity $1 + \cot^2 x = \csc^2 x$.

8. **SALES**   The monthly sales of a certain product satisfy

   $$S(t) = 19.3 + 9.4 \sin\frac{\pi t}{6} + 3.4 \cos\frac{\pi t}{3},$$

   where t $(0 \le t \le 12)$ is the time in months measured from the beginning of January and S is measured in thousands of units.

   a. At what rate are sales changing at midyear $(t = 6)$? Are sales increasing or decreasing at this time?

   b. What are the highest and lowest sales levels for the year? In what months do these extreme sales levels occur? [*Hint:* You may need the double-angle identity $\sin 2A = 2 \sin A \cos A$.]

9. **MARGINAL COST**   Suppose $C(x)$ is the cost (in hundreds of dollars) of producing x units of a particular commodity, and that the marginal cost is given by

   $$C'(x) = 3x + 2.5 \sin 2\pi x$$

   for $0 \le x \le 10$.

   a. What is the net cost of producing the first 5 units?

   b. At what rate is the cost changing when 5 units are produced? By how much is the marginal cost changing at this level of production?

10. **BLOOD PRESSURE**   The blood pressure of a particular patient at time t (seconds) is modelled by the function

    $$B(t) = 105 + 31 \cos kt,$$

    where k is a positive constant and B is measured in millimetres of mercury. Suppose the patient has a pulse rate of 80 heartbeats per minute.

    a. Find k.

    b. Using the value of k you found in part (a), sketch the graph of $B(t)$.

    c. Determine the systolic (highest) and diastolic (lowest) pressures for the patient.

    d. What is the patient's blood pressure when it is increasing most rapidly?

Review Exercises

1. Specify the radian measurement and degree measurement for each angle.

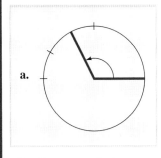
a.

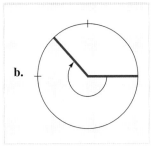
b.

2. Convert each degree measurement to radians.
   a. $50°$ b. $120°$ c. $-15°$

3. Convert each radian measurement to degrees.
   a. 0.25 radians   b. 1 radian   c. -1.5 radians

4. Evaluate each expression without using Table III or a calculator.

   a. $\sin\left(-\dfrac{5\pi}{3}\right)$

   b. $\cos\dfrac{15\pi}{4}$

   c. $\sec\dfrac{7\pi}{3}$

   d. $\cot\dfrac{2\pi}{3}$

5. Find $\tan \theta$ if $\sin \theta = \dfrac{4}{5}$ and $0 \le \theta \le \dfrac{\pi}{2}$.

6. Find $\csc \theta$ if $\cot \theta = \dfrac{\sqrt{5}}{2}$ and $0 \le \theta \le \dfrac{\pi}{2}$.

In Exercises 7 through 10, find all values of θ in the specified interval that satisfy the given equation.

7. $2 \cos \theta + \sin 2\theta = 0; \, 0 \le \theta \le 2\pi$

8. $3 \sin^2 \theta - \cos^2 \theta = 2; \; 0 \le \theta \le \pi$

9. $2 \sin^2 \theta = \cos 2\theta; \; 0 \le \theta \le \pi$

10. $5 \sin \theta - 2 \cos^2 \theta = 1; \; 0 \le \theta \le 2\pi$

11. Starting with the identities
$$\cos 2\theta = \cos^2 \theta - \sin^2 \theta \quad \text{and} \quad \sin^2 \theta + \cos^2 \theta = 1,$$
derive the identities
$$\cos^2 \theta = \frac{1}{2}(1 + \cos 2\theta) \quad \text{and} \quad \sin^2 \theta = \frac{1}{2}(1 - \cos 2\theta).$$

12. Starting with the double-angle formulas for the sine and cosine, derive a double-angle formula for the tangent.

13. a. Starting with the addition formulas for the sine and cosine, derive the identities
$$\cos\left(\frac{\pi}{2} + \theta\right) = -\sin \theta \quad \text{and} \quad \sin\left(\frac{\pi}{2} + \theta\right) = \cos \theta.$$

    b. Give geometric arguments to justify the identities in part (a).

In Exercises 14 through 20, differentiate the given function.

14. $f(x) = \sin(3x + 1)\cos x$

15. $f(x) = \cos^2 x$

16. $f(x) = \tan(3x^2 + 1)$

17. $f(x) = \tan^2(3x^2 + 1)$

18. $f(x) = \dfrac{\sin x}{1 - \cos x}$

19. $f(x) = \ln(\cos^2 x)$

20. $f(x) = e^{-2x} \cos 3x$

In Exercises 21 through 26, find the indicated integral.

21. $\displaystyle\int \sin 2t \, dt$

22. $\displaystyle\int \cos(1 - 20t) \, dt$

23. $\displaystyle\int \sin x \cos x \, dx$

24. $\displaystyle\int x \sin x \, dx$

25. $\displaystyle\int \frac{\sec^2 t}{\tan t} \, dt$

26. $\displaystyle\int \tan^2 t \, dt$ [*Hint:* Use $1 + \tan^2 t = \sec^2 t$.]

27. Angle θ has a terminal side in quadrant IV, as shown in the accompanying figure. Find the reference

angle A in degrees. Then find the smallest positive angle θ so that $\csc \theta = \csc A$.

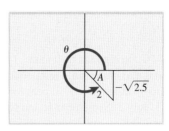

EXERCISE 27

28. In each case, use technology to draw the graphs of the given pair of functions $f(x)$ and $g(x)$ on the same set of axes. Describe the relationship between the graphs of $f(x)$ and $g(x)$.

    a. $f(x) = \sin x$ and $g(x) = 2 \sin x$

    b. $f(x) = \cos x$ and $g(x) = 2 \cos 2x$

    c. $f(x) = \sin x$ and $g(x) = \sin\left(x + \dfrac{\pi}{2}\right)$

    d. $f(x) = \cos x$ and $g(x) = 2 + \cos x$

29. Use technology to solve the equation

$$2 \tan 3x - 5.87 = 2 \sin 2x \quad \text{for } 0 \le x \le \frac{\pi}{2}$$
to three decimal places.

30. Find the area of the region bounded by the curves $y = \sin 2x$ and $y = \cos x$ over the interval $\dfrac{\pi}{6} \le x \le \dfrac{\pi}{2}$.

31. Let R be the region bounded by the x axis, the curve $y = \cos x + \sin x$, and the lines $x = -\dfrac{\pi}{2}$ and $x = \dfrac{\pi}{6}$. Find the volume of the solid generated by rotating R about the x axis.

32. a. Find the period p, the amplitude b, the phase shift d, and the vertical shift a of the function
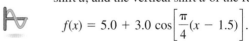
$$f(x) = 5.0 + 3.0 \cos\left[\frac{\pi}{4}(x - 1.5)\right].$$

    b. Sketch the graph of the function $f(x)$ in part (a).

33. a. Find the period p, the amplitude b, the phase shift d, and the vertical shift a of the function
$$f(x) = 33 + 27 \cos\left[\frac{2\pi}{25}(x - 11)\right].$$
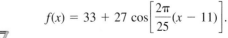

    b. Sketch the graph of the function $f(x)$ in part (a).

34. WEATHER The maximum daily temperature $T(x)$ in degrees Celsius in Winnipeg on day x of the year can be modelled as

$$T(x) = 13 + 33 \cos\left[\frac{2\pi}{365}(x - 271)\right],$$

where $x = 0$ corresponds to January 1.

a. Find the maximum daily temperature in Winnipeg on January 1. Repeat for March 1, May 1, July 1, September 1, and November 1.

b. Find the highest and lowest maximum daily temperatures in Winnipeg during the year.

c. Draw the graph of the maximum daily temperature function $T(x)$.

35. CONSTRUCTION COST A cable is run in a straight line from a power plant on one side of a river that is 900 m wide to a point P on the other side and then along the river bank to a factory, 3000 m downstream from the power plant, as shown in the accompanying figure. The cost of running the cable under the water is $5/m, while the cost over land is $4/m. If θ is the (smaller) angle between the segment of cable under the river and the opposite bank, show that $\cos\theta = \frac{4}{5}$ (the ratio of the per-metre costs) for the route that minimizes the total installation cost. (You may assume without proof that the absolute minimum occurs when the derivative of the cost function is zero.)

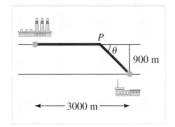

EXERCISE 35

36. SEASONAL POPULATIONS We can model the population $P(t)$ of an animal or plant species that fluctuates seasonally using a function of the form $P(t) = P_0 e^{a \sin 2\pi t}$, where t is the time in years and $P_0 = P(0)$ is the initial population.

The population function $P(t)$ satisfies a differential equation of the form

$$\frac{dP}{dt} = R(t)P(t),$$

where $R(t)$ is the relative growth rate function for the species (birth rate − death rate). What is $R(t)$?

37. SEASONAL POPULATIONS Suppose the population of mosquitoes in a swamp can be modelled by $P(t) = 500e^{7 \sin 2\pi t}$, where time t is in years and $t = 0$ corresponds to May 1.

a. Find the number of mosquitoes in the swamp at time $t = 0.125$. Repeat for times $t = 0.25$, $t = 0.325$, and $t = 0.5$.

b. Find the rate of change of the population of mosquitoes in the swamp at time $t = 0.125$. Repeat for times $t = 0.25$, $t = 0.325$, and $t = 0.5$.

c. Draw the graph of $P(t)$.

38. POLLUTION The ozone levels in parts per million (ppm) in a city can be modelled by the function $F(t) = 0.01t^3 + 0.05t^2 + 1.1t + 56 + 22 \sin 2\pi t$, where t is the time in years after 2000.

a. Find the levels of ozone on July 1, 2000. Repeat for January 1, 2010, and March 1, 2015.

b. Find the rate of change of the level of ozone on the three dates in part (a).

c. Graph $F(t)$ for the time period from 2000 to 2020 ($0 \le t \le 20$).

d. Describe the behaviour of $F(t)$ as t increases from 0 to 20. Interpret the roles of the polynomial part of $F(t)$ and the periodic part.

39. POPULATION GROWTH Suppose that caribou enter the Arctic National Wildlife Reserve at the rate modelled by the derivative

$$C'(t) = 275 + 275 \cos\left[\frac{\pi}{6}(t - 3)\right],$$

where $C(t)$ is the number of caribou in the reserve at time t (months).

a. By how much does the caribou population in the reserve grow during the first year ($0 \le t \le 12$)? What about during the second year ($12 \le t \le 24$)?

b. During what month of the year is the caribou population the greatest? The least?

40. SALES The monthly sales of chicken wings in a large city can be estimated using the function

$$B(t) = 4\,128\,500 + F_1(t) + F_2(t),$$

where

$$F_1(t) = -841\,000 \cos\frac{\pi t}{2} - 111\,500 \sin\frac{\pi t}{2}$$

$$F_2(t) = 234\,500 \cos\frac{\pi t}{3} - 88\,000 \sin\frac{\pi t}{3}$$

and t is measured in months, with $t = 1$ corresponding to January.

a. Use the formula for $B(t)$ to estimate the number of chicken wings sold in this city during June.

b. Find the average monthly sales of chicken wings during the first year ($0 \le t \le 12$).

41. **DAYLIGHT**   The number of hours of daylight in New York City for day t of the year can be modelled using the function

$$D(t) = 12.2 + 3.09 \cos\left[\frac{2\pi}{365}(t - 185)\right],$$

where $t = 0$ corresponds to January 1.

a. How many hours of daylight are there on January 1? On March 15 ($t = 74$)? On June 21 ($t = 172$)?

b. On which day of the year is the number of daylight hours the greatest? When does the least number of daylight hours occur?

c. What is the average number of daylight hours per day over the entire year ($0 \le t \le 365$)?

42. **A* RELATIVE RATE OF CHANGE**   A revolving searchlight in a lighthouse 2 km off shore is following a jogger running along the shore, as shown in the accompanying figure. When the jogger is 1 km from the point A on the shore that is closest to the lighthouse, the searchlight is turning at a rate of 0.25 revolutions per hour. How fast is the jogger running at this moment?

$$\left[\textit{Hint: } \text{Since 0.25 revolutions per hour is } \frac{\pi}{2} \text{ radians}\right.$$
$$\text{per hour, the problem is to find } \frac{dx}{dt} \text{ when } x = 1 \text{ and}$$
$$\left. \frac{d\theta}{dt} = \frac{\pi}{2}\right].$$

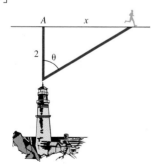

EXERCISE 42

43. **A* RELATIVE RATE OF CHANGE**   On New Year's Eve, a holiday reveller is watching the descent of a lighted ball from the top a tall building

that is 60 m away. The ball is falling at a rate of 2 m/s. At what rate is the angle of elevation of the reveller's line of sight changing with respect to time when the ball is 80 m from the ground?

44. **A* CONSTRUCTION**   A trough 9 m long is to have a cross-section consisting of an isosceles trapezoid in which the base and two sides are all 4 m long, as shown in the accompanying figure. At what angle should the sides of the trapezoid meet the horizontal top to maximize the capacity of the trough?

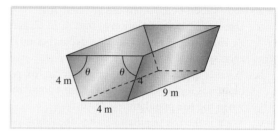

EXERCISE 44

45. **A* CONSTRUCTION**   A piece of paper measuring 8.5 cm by 11 cm is folded so that the lower right-hand corner reaches the left edge, forming an angle θ at the crease, as shown in the accompanying figure.

a. Express the length L of the crease as a function of θ.

b. Find the crease angle θ for which L is minimized. What is the minimum length?

c. What is the smallest area of the triangle $\triangle ABC$ shown in the figure?

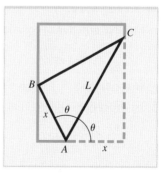

EXERCISE 45

46. **A* PREDATOR-PREY**   In Section 8.3, we examined a competitive species model involving foxes (predators) and rabbits (prey). We found that the populations of the two species behave periodically in relation to one another, as indicated in Figure 8.11. In particular, suppose the fox population $F(t)$ and the rabbit population $R(t)$ at time t (years) are modelled by

$$F(t) = A + B \sin \frac{\pi t}{24}$$

$$R(t) = C + D \cos \frac{\pi t}{24}$$

where A, B, C, and D are constants. Further suppose that the greatest and least fox populations are 54 and 16, respectively, and that the corresponding greatest and least rabbit populations are 581 and 261.

a. Use the given information to find A, B, C, and D. How long is the population cycle (period) for each species?

b. What is the rabbit population when the fox population is highest? What is the rabbit population when the fox population is lowest?

c. What is the fox population when the rabbit population is highest? What is the fox population when the rabbit population is lowest?

d. Sketch the graphs of the two population functions on the same set of axes.

47. A* PREDATOR-PREY   Answer the questions in Exercise 46 for the case where

$$F(t) = A + B \sin \frac{\pi t}{12}$$

$$R(t) = C + D \cos \frac{\pi t}{12}$$

and the highest and lowest fox populations are 70 and 22, while the highest and lowest rabbit populations are 635 and 227.

48. Solve the separable differential equation

$$\frac{dy}{dx} = \sin x \sec y$$

subject to the condition $y = 1$ when $x = 0$.

49. Solve the first-order linear differential equation

$$\frac{dy}{dx} + \frac{y}{x} = \sin x$$

subject to the condition $y = 1$ when $x = \pi$. (You may need to use integration by parts.)

50. A* Use technology to draw the curves $y = \sin x$ and $y = e^{x-2}$ for $x \geq 0$ on the same set of axes. Find all points of intersection of the two curves. Let R be the region enclosed by the two curves.

a. Find the area of the region R.

b. Find the volume of the solid formed by revolving the region R around the x axis. $\left[\textit{Hint:} \text{ It may help to recall the identity } \sin^2 x = \dfrac{1 - \cos 2x}{2}. \right]$

c. Check the integration in part (b) by using technology.

51. The definite integral

$$\int_0^{\pi/2} \sin x^2 \, dx$$

is associated with the diffraction of light, but the integrand $f(x) = \sin x^2$ has no simple antiderivative.

a. Use Simpson's rule (Section 6.3) with $n = 10$ to estimate the value of the given definite integral.

b. Check the result in part (a) with technology.

52. Repeat the steps in Exercise 51 for the definite integral

$$\int_0^{\pi/2} \cos x^2 \, dx.$$

53. Evaluate $\int \tan x \, dx$. [*Hint:* Use the substitution $u = \cos x$.]

54. a. Find $\dfrac{dy}{dx}$ where $y = \ln(\sec x + \tan x)$.

b. Find $\int \sec x \, dx$.

c. The **Mercator projection*** in cartography is a procedure for drawing the map of Earth on a flat surface in which the paths of constant compass direction are all straight lines. It turns out that drawing such a map requires the values of the integral

$$\int_0^x \sec \theta \, d\theta$$

for all $0 \leq x \leq \dfrac{\pi}{2}$. Read about this and write a paragraph on the role played by this integral in the development of the Mercator projection.

* Philip M. Tuchinsky, "Mercator's World Map and Calculus," *UMAP Modules 1977: Tools for Teaching,* Lexington, MA: Consortium for Mathematics and Its Applications, Inc., 1978.

Exercises 55 through 60 deal with topics developed in Chapters 7 and 9.

55. **A*** What is the largest possible value of the product
$$f(A, B, C) = \sin A \, \sin B \, \sin C$$
given that A, B, and C are the angles in a triangle? [*Hint:* It may help to note that $A + B + C = \pi$.]

56. **A*** The integral of the function $f(x) = \dfrac{\sin x}{x}$ appears in certain engineering applications, but $f(x)$ has no simple antiderivative.

   a. Find the Taylor series about $x = 0$ for $\sin x$.

   b. Use the result of part (a) to show that
$$\frac{\sin x}{x} = 1 - \frac{x^2}{3!} + \frac{x^4}{5!} - \frac{x^6}{7!} + \cdots = \sum_{n=0}^{\infty} \frac{(-1)^n x^{2n}}{(2n+1)!}.$$
Then integrate this power series term by term to express the definite integral
$$\int_0^1 \frac{\sin x}{x} \, dx$$
as a numerical infinite series.

   c. Estimate the value of the definite integral
$$\int_0^1 \frac{\sin x}{x} \, dx$$
by computing the sum of the first four terms of the numerical series in part (b).

57. **THE DIFFUSION EQUATION** The following equation involving partial derivatives of the function $u(x, t)$ is called the *diffusion equation*:
$$u_t = c^2 u_{xx}.$$
The diffusion equation is used in modelling a large variety of physical phenomena. For instance, in biology it is used to model the mechanism for butterfly wing patterns, the effects of genetic drift, and macrophage response to bacteria in the lungs, while in physics, it is used to study the motion of molecules and heat conduction.

   a. Show that the function $u = e^{-c^2 k^2 t} \sin kx$ satisfies the diffusion equation.

   b. Read an article on the diffusion equation and write a paragraph on one of its applications.

58. **ECOLOGY** Ground temperature models are important in ecology, where they are used to study phenomena such as frost penetration. Suppose ground temperature T at time t (months) and depth x (centimetres) is modelled by a function of the form
$$T(x, t) = A + Be^{-kt} \sin (at - kx),$$
where $a = \dfrac{\pi}{6}$ and A, B, and k are positive constants.

   a. Find the partial derivatives T_x and T_t.

   b. The partial derivative T_x measures the rate at which the ground temperature drops with increasing depth for fixed time. Give a similar interpretation for the partial derivative T_t.

   c. Show that $T(x, t)$ satisfies the diffusion equation $T_t = c^2 T_{xx}$, where c is a constant involving B and k.

59. **A*** Use the method of Lagrange multipliers to find the maximum value of
$$f(x, y) = \cos x + \cos y$$
subject to the constraint condition
$$y - x = \frac{\pi}{4}.$$

60. **A*** Find the largest and the lowest values of the function
$$f(x, y) = 2 \sin x + 5 \cos y$$
over the rectangle R with vertices $(0, 0)$, $(2, 0)$, $(2, 5)$, and $(0, 5)$.

THINK ABOUT IT

WHAT ANGLE IS BEST FOR VIEWING A RAINBOW?

Every visiting tourist takes a picture of the rainbow over Niagara Falls, but do any of them think about what viewing angle will make the rainbow appear most vivid?

(Photo: © Brand X Pictures/PunchStock)

Rainbows are formed when sunlight travelling through the air is both reflected and refracted by raindrops. Figure 11.28a shows a raindrop, which, for simplicity, is assumed to be spherical in shape. An incoming beam of sunlight strikes the raindrop at point A with angle of incidence α. Some of the light is reflected and some is refracted through angle β. The refracted beam then continues to point B on the other side of the raindrop, where part of it is reflected through angle β back through the raindrop to point C, while part of what remains is refracted through angle α back into the air.

At each interface, the light beam is deflected (from a straight-line path), which causes its intensity to be reduced. In particular, at A, the incoming beam is deflected through a clockwise rotation of $\alpha - \beta$ radians, while at B, the deflection is $\pi - 2\beta$ radians and at C, it is again $\alpha - \beta$ radians. Thus, the total deflection is D radians, where

$$D = (\alpha - \beta) + \pi - 2\beta + \alpha - \beta = \pi + 2\alpha - 4\beta.$$

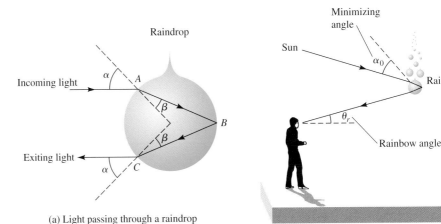

(a) Light passing through a raindrop

(b) A rainbow is brightest when viewed at the rainbow angle.

FIGURE 11.28 Analysis of light passing through a raindrop.

The intensity of the light exiting from the raindrop is greatest for the angle of incidence α_0 that minimizes the total deflection D. As indicated in Figure 11.28b, this minimizing angle α_0 is directly related to the angle of observation θ_r for which the rainbow is brightest. The angle θ_r is appropriately called the **rainbow angle.** You will determine the value of the rainbow angle by using calculus to answer Questions 1 through 3.

This essay was adapted from the article "Somewhere within the Rainbow" by Steven Janke, *UMAP Modules 1992: Tools for Teaching,* Lexington, MA: Consortium for Mathematics and Its Applications, Inc., 1993. The article explores a variety of issues regarding the observation of a rainbow.

Questions

1. The **law of refraction** in optics says that the angle of incidence α is related to the angle of refraction β by the equation $\sin \alpha = k \sin \beta$, where k is a constant that depends on the refracting medium (recall Exercise 27 of Section 11.3). Use this law to express the light deflection D in terms of α, and then differentiate to show that

$$\frac{dD}{d\alpha} = 2 - \frac{4 \cos \alpha}{k \cos \beta}.$$

2. Show that $D'(\alpha_0) = 0$, where α_0 satisfies $\cos \alpha_0 = \sqrt{\dfrac{k^2 - 1}{3}}$. Use the second derivative test to show that $D(\alpha)$ is minimized at $\alpha = \alpha_0$.

3. For the water in our raindrop, it can be shown that $k \approx 1.33$. Compute the angle α_0 that minimizes light deflection for the raindrop. Then use α_0 to find the rainbow angle $\theta_r = \pi - D(\alpha_0)$.

4. There are several important questions we have not explored. For instance, what if the raindrops are not spherical? What is the optimum viewing angle if the rainbow has a secondary arc (a double rainbow)? Research this and write a paragraph on one of these topics.

References

Steven Janke, "Somewhere Within the Rainbow," *UMAP Modules 1992: Tools for Teaching,* Lexington, MA: Consortium for Mathematics and Its Applications, Inc., 1993.

Mc Graw Hill **connect**™   Practise and learn online with Connect.

(Photo: Getty Images/Digital Vision)

ALGEBRA REVIEW, L'HÔPITAL'S RULE, AND SUMMATION NOTATION

APPENDIX OUTLINE

A.1 A Brief Review of Algebra

A.2 Factoring Polynomials and Solving Systems of Equations

A.3 Evaluating Limits with L'Hôpital's Rule

A.4 Summation Notation

Appendix Summary

Concept Summary

Review Exercises

Think About It

LEARNING OBJECTIVES

After completing this appendix, you should be able to

LO1 Find the intervals corresponding to a given inequality that includes an absolute value. Simplify expressions that include exponents and roots. Rationalize a denominator.

LO2 Factor polynomials. Add and subtract rational expressions and compound fractions. Solve equations by factoring, completing the square, and using the quadratic formula. Solve systems of two equations.

LO3 Evaluate limits using L'Hôpital's rule.

LO4 Evaluate sums given the summation notation.

A Brief Review of Algebra

Find the intervals corresponding to a given inequality that includes an absolute value. Simplify expressions that include exponents and roots. Rationalize a denominator.

Many techniques from elementary algebra are needed in calculus. This appendix contains a review of such topics, and we begin by examining numbering systems.

The Real Numbers

An **integer** is a whole number that can be positive, negative, or zero. For example, 1, 2, 875, -15, -83, and 0 are integers, while $\frac{2}{3}$, 8.71, and $\sqrt{2}$ are not.

A **rational number** is a number that can be expressed as the quotient $\frac{a}{b}$ of two integers, where $b \neq 0$. For example, $\frac{2}{3}$, $\frac{8}{5}$, and $-\frac{4}{7}$ are rational numbers, as are

$$-6\frac{1}{2} = \frac{-13}{2} \qquad \text{and} \qquad 0.25 = \frac{25}{100} = \frac{1}{4}.$$

Every integer is a rational number because it can be expressed as itself divided by 1. When expressed in decimal form, rational numbers are either terminating or infinitely repeating decimals. For example,

$$\frac{5}{8} = 0.625, \qquad \frac{1}{3} = 0.33\ldots, \qquad \text{and} \qquad \frac{13}{11} = 1.181818\ldots$$

are rational numbers.

A number that cannot be expressed as the quotient of two integers is called an **irrational number.** For example,

$$\sqrt{2} \approx 1.41421356 \qquad \text{and} \qquad \pi \approx 3.14159265$$

are irrational numbers.

The rational numbers and irrational numbers form the **real numbers** and can be visualized geometrically as points on a line, called the **real number line.** To construct such a representation, choose a point on a line as the location of the number 0. This is called the **origin.** Select a point to represent the number 1. This determines the scale of the number line, and each number is located an appropriate distance (multiple of 1) from the origin. If the line is horizontal, the positive numbers are located to the right of the origin and the negative numbers to the left, as indicated in Figure A.1. The **coordinate** of a particular point on the line is the number associated with it.

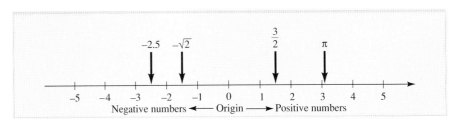

FIGURE A.1   The real number line.

Inequalities

If a and b are real numbers and a is to the right of b on the number line, we say that ***a* is greater than *b*** and write ***a* > *b*.** If a is to the left of b, we say that ***a* is less than *b*** and write ***a* < *b*** (Figure A.2). For example,

$$5 > 2, \qquad -12 < 0, \qquad \text{and} \qquad -8.2 < -2.4.$$

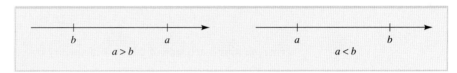

FIGURE A.2 Inequalities.

Moreover,

$$\frac{6}{7} < \frac{7}{8},$$

as you can see by noting that

$$\frac{6}{7} = \frac{48}{56} \qquad \text{and} \qquad \frac{7}{8} = \frac{49}{56}.$$

A few basic properties of inequalities are presented in the following box. Note especially property 3, which states that the direction of an inequality is preserved if both sides are multiplied by a positive number but is *reversed* if the multiplier is negative.

Properties of Inequalities

1. **Transitive property:** If $a > b$ and $b > c$, then $a > c$.
2. **Additive property:** If $a > b$ and $c \geq d$, then $a + c > b + d$.
3. **Multiplicative property:** If $a > b$ and $c > 0$, then $ac > bc$, but if $a > b$ and $c < 0$, then $ac < bc$.

For example, since $7 > 3$, we have $7 - 9 > 3 - 9$ or $-2 > -6$. Since $5 > 2$ and $3 > 0$, it follows that $5(3) > 2(3)$, or $15 > 6$. Since $5 > 2$ and $-2 < 0$, we have $5(-2) < 2(-2)$, or $-10 < -4$.

The symbol $\geq$ stands for **greater than or equal to,** and the symbol $\leq$ stands for **less than or equal to.** Thus, for example,

$$-3 \geq -4, \qquad -3 \geq -3, \qquad -4 \leq -3, \qquad \text{and} \qquad -4 \leq -4.$$

A real number is said to satisfy a particular inequality involving a variable if the inequality is satisfied when the number is substituted for the variable. The inequality is said to be **solved** when all numbers that satisfy it have been found. The set of all solutions is called the **solution set** of the inequality.

EXAMPLE A.1.1

Solve the two-sided inequality $-5 < 2x - 3 \leq 1$.

Solution

Add 3 to both sides of the inequality (property 2) to obtain

$$-2 < 2x \leq 4.$$

Then multiply each side of this new inequality by $\frac{1}{2}$:

$$-1 < x \leq 2.$$

Thus the solution set comprises all real numbers between -1 and 2, including 2 (but not including -1).

Intervals

A set of real numbers that can be represented on the number line by a line segment is called an **interval.** Inequalities can be used to describe intervals. For example, the interval $a \leq x < b$ consists of all real numbers x that are between a and b, including a but excluding b. This interval is shown in Figure A.3. The numbers a and b are known as the **endpoints** of the interval. The square bracket at a indicates that a is included in the interval, while the rounded bracket at b indicates that b is excluded.

Intervals may be finite or infinite in extent and may or may not contain either endpoint. The possibilities (including customary notation and terminology) are illustrated in Figure A.4.

FIGURE A.3   The interval $a \leq x < b$.

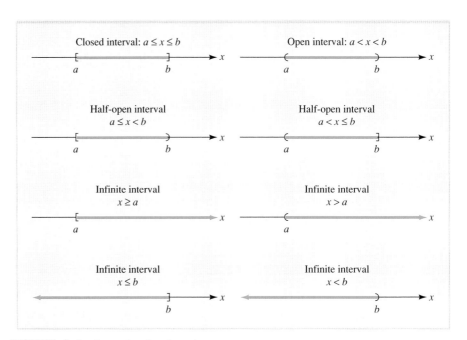

FIGURE A.4   Intervals of real numbers.

EXAMPLE A.1.2

Use inequalities to describe these intervals.

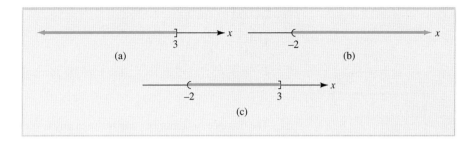

Solution

a. $x \le 3$　　　**b.** $x > -2$　　　**c.** $-2 < x \le 3$

EXAMPLE A.1.3

Represent each interval as a line segment on a number line.

a. $x < -1$　　　　　**b.** $-1 \le x \le 2$　　　**c.** $x > 2$

Solution

a.

b.

c.

Absolute Value

The **absolute value** of a real number x, denoted by $|x|$ is the distance from x to 0 on a number line. Since distance is always nonnegative, it follows that $|x| \ge 0$. For example,

$$|4| = 4 \qquad |-4| = 4 \qquad |0| = 0 \qquad |5 - 9| = 4 \qquad |\sqrt{3}-3| = 3 - \sqrt{3}$$

Here is a general formula for absolute value.

> **Absolute Value** ■ For any real number x, the absolute value of x is
>
> $$|x| = \begin{cases} x & \text{if } x \ge 0 \\ -x & \text{if } x < 0 \end{cases}$$

Notice that $|-a| = |a|$ for any real number a. This is one of several useful properties of absolute value listed in the following box.

Properties of Absolute Value

Let a and b be real numbers. Then

1. $|-a| = |a|$
2. $|ab| = |a||b|$
3. $\left|\dfrac{a}{b}\right| = \dfrac{|a|}{|b|}$ if $b \neq 0$
4. $|a + b| \leq |a| + |b|$ (the triangle inequality)

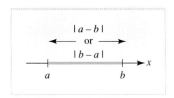

FIGURE A.5   The distance between a and b is equal to $|a - b|$.

The distance on a number line between any two numbers a and b is the absolute value of their difference taken in either order ($a - b$ or $b - a$) as illustrated in Figure A.5. For instance, the distance between $a = -2$ and $b = 3$ is $|-2 - 3| = |3 - (-2)| = 5$ (Figure A.6).

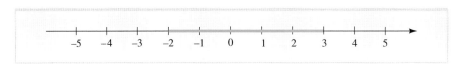

FIGURE A.6   The distance between -2 and 3.

The solution set of an inequality of the form $|x| \leq c$ for $c > 0$ is the interval $-c \leq x \leq c$, that is, $[-c, c]$. This property is used in Example A.1.4.

EXAMPLE A.1.4

Find the interval consisting of all real numbers x such that $|x - 1| \leq 3$.

Solution

In geometric terms, the numbers x for which $|x - 1| \leq 3$ are those whose distance from 1 is less than or equal to 3. As illustrated in Figure A.7, these are the numbers that satisfy $-2 \leq x \leq 4$.

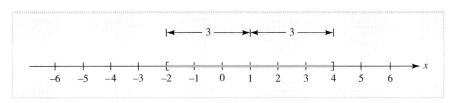

FIGURE A.7   The interval on which $|x - 1| \leq 3$ is $-2 \leq x \leq 4$.

To find this interval algebraically, without relying on geometry, rewrite the inequality $|x - 1| \leq 3$ as

$$-3 \leq x - 1 \leq 3,$$

and add 1 to each part to get

$$-3 + 1 \leq x - 1 + 1 \leq 3 + 1$$

or

$$-2 \leq x \leq 4.$$

Exponents and Roots

If a is a real number and n is a positive integer, the expression

$$a^n = \underbrace{a \cdot a \cdots a}_{n \text{ terms}}$$

indicates that a is to be multiplied by itself n times. The number a is called the **base** of the **exponential expression** a^n, and n is called the **exponent.** If $a \neq 0$, we define

$$a^{-n} = \frac{1}{a^n} \quad \text{and} \quad a^0 = 1.$$

Note that 0^0 is not defined.

If m is a positive integer, then $a^{1/m}$ denotes the number whose mth power is a. This is called the **mth root** of a and is also denoted by $\sqrt[m]{a}$; that is,

$$a^{1/m} = \sqrt[m]{a}.$$

The mth root of a negative number is not defined when m is even. For example, $\sqrt[4]{-5}$ is not defined since there is no real number whose 4th power is -5.

By convention, if m is even, $a^{1/m}$ is taken to be positive even when there is a negative number whose mth power is a. For example, 2^4 and $(-2)^4$ both equal 16, but the 4th root of 16 is defined to be 2. Thus,

$$\sqrt[4]{16} = 16^{1/4} = 2,$$

not ± 2.

Finally, we write $a^{n/m}$ to denote the nth power of the mth root of the real number a, which is the same as the mth root of the nth power. That is,

$$a^{n/m} = (a^{1/m})^n = (a^n)^{1/m}.$$

For example,

$$8^{-2/3} = (8^{-2})^{1/3} = \left(\frac{1}{8^2}\right)^{1/3} = \left(\frac{1}{64}\right)^{1/3} = \frac{1}{4} \quad \text{since } \left(\frac{1}{4}\right)^3 = 64,$$

or, equivalently,

$$8^{-2/3} = (8^{1/3})^{-2} = 2^{-2} = \frac{1}{2^2} = \frac{1}{4}.$$

Here is a summary of the exponential notation.

Exponential Notation ■ Let a be a real number and m and n be positive integers. Then we have the following:

Integer powers: $a^n = \underbrace{a \cdot a \cdots a}_{n \text{ terms}}$ and $a^0 = 1$

Negative integer powers: $a^{-n} = \dfrac{1}{a^n}$

Reciprocal integer powers (roots): $a^{1/m} = \sqrt[m]{a}$

Fractional exponents: $a^{n/m} = (a^{1/m})^n = (a^n)^{1/m}$

EXAMPLE A.1.5

Evaluate each expression without using a calculator.

a. $9^{1/2}$ **b.** $27^{2/3}$ **c.** $8^{-1/3}$ **d.** $\left(\dfrac{1}{100}\right)^{-3/2}$ **e.** 5^0

Solution

a. $9^{1/2} = \sqrt{9} = 3$

b. $27^{2/3} = (\sqrt[3]{27})^2 = 3^2 = 9$

$\qquad = \sqrt[3]{27^2} = \sqrt[3]{729} = 9$

c. $8^{-1/3} = \dfrac{1}{8^{1/3}} = \dfrac{1}{\sqrt[3]{8}} = \dfrac{1}{2}$

d. $\left(\dfrac{1}{100}\right)^{-3/2} = 100^{3/2} = (\sqrt{100})^3 = 10^3 = 1000$

e. $5^0 = 1$

Exponents obey these useful laws.

Laws of Exponents ■ For real numbers a, b and integers m, n, the following laws are valid whenever the quantities are defined.

Identity law: If $a^m = a^n$, then $m = n$.

Product law: $a^m \cdot a^n = a^{m+n}$

Quotient law: $\dfrac{a^m}{a^n} = a^{m-n}$ if $a \neq 0$.

Power laws: $(a^m)^n = a^{mn}$ and $(ab)^n = a^n \cdot b^n$.

The laws of exponents are illustrated in the following four examples.

EXAMPLE A.1.6

Evaluate each expression (without using a calculator).

    a. $(2^{-2})^3$ **b.** $\dfrac{3^3}{3^{1/3}(3^{2/3})}$ **c.** $2^{7/4}(8^{-1/4})$

Solution

    a. $(2^{-2})^3 = 2^{-6} = \dfrac{1}{2^6} = \dfrac{1}{64}$

    b. $\dfrac{3^3}{3^{1/3}(3^{2/3})} = \dfrac{3^3}{3^{1/3+2/3}} = \dfrac{3^3}{3^1} = 3^2 = 9$

    c. $2^{7/4}(8^{-1/4}) = 2^{7/4}(2^3)^{-1/4} = 2^{7/4}(2^{-3/4}) = 2^{7/4-3/4} = 2^1 = 2$

EXAMPLE A.1.7

Solve each equation for n.

    a. $\dfrac{a^5}{a^2} = a^n$ **b.** $(a^n)^5 = a^{20}$

Solution

    a. Since $\dfrac{a^5}{a^2} = a^{5-2} = a^3$, it follows that $n = 3$.

    b. Since $(a^n)^5 = a^{5n}$, it follows that $5n = 20$ or $n = 4$.

EXAMPLE A.1.8

Simplify each expression, and express each in terms of positive exponents.

    a. $(x^3)^{-2}$ **b.** $(x^{-5})^{-2}$ **c.** $(x^{-2}y^{-3})^{-4}$

    d. $\left(\dfrac{x^{-3}}{y^4}\right)^{-2}$ **e.** $\dfrac{4x^{-3}y^2}{2x^2y^{-5}}$

Solution

    a. $(x^3)^{-2} = x^{3(-2)} = x^{-6} = \dfrac{1}{x^6}$

    b. $(x^{-5})^{-2} = x^{(-5)(-2)} = x^{10}$

    c. $(x^{-2}y^{-3})^{-4} = x^{(-2)(-4)}y^{(-3)(-4)} = x^8y^{12}$

    d. $\left(\dfrac{x^{-3}}{y^4}\right)^{-2} = (x^{-3}y^{-4})^{-2} = x^{(-3)(-2)}y^{(-4)(-2)} = x^6y^8$

    e. $\dfrac{4x^{-3}y^2}{2x^2y^{-5}} = \dfrac{4}{2}x^{-3-2}y^{2-(-5)} = 2x^{-5}y^7 = \dfrac{2y^7}{x^5}$

EXAMPLE A.1.9

Simplify each root expression.

a. $3\sqrt{64} + 5\sqrt{72} - 9\sqrt{50}$

b. $\sqrt{a^{-5}b^{-8}c^{10}}$, $a > 0$, $b \neq 0$

c. $\sqrt{\dfrac{36x^3}{y^3}} \sqrt{\dfrac{y^8}{25x}}$, $x > 0$, $y > 0$

Solution

a. $3\sqrt{64} + 5\sqrt{72} - 9\sqrt{50} = 3\sqrt{8^2} + 5\sqrt{6^2 \cdot 2} - 9\sqrt{5^2 \cdot 2}$

$$= 3(8) + 5(6)\sqrt{2} - 9(5)\sqrt{2} = 24 - 15\sqrt{2}$$

b. $\sqrt{a^{-5}b^{-8}c^{10}} = \sqrt{\dfrac{c^{10}}{a^5 b^8}} = \dfrac{c^5}{b^4\sqrt{a^4}\sqrt{a}} = \dfrac{c^5}{b^4 a^2 \sqrt{a}}$

c. $\sqrt{\dfrac{36x^3}{y^3}} \sqrt{\dfrac{y^8}{25x}} = \sqrt{\dfrac{36}{25}} \sqrt{\dfrac{x^3 y^8}{xy^3}} = \dfrac{6}{5}\sqrt{x^2 y^5} = \dfrac{6}{5}\sqrt{x^2(y^4 \cdot y)}$

$$= \dfrac{6}{5}\sqrt{x^2}\sqrt{y^4}\sqrt{y} = \dfrac{6}{5}xy^2\sqrt{y}$$

Rationalizing

Sometimes it is necessary, or at least desirable, to write a fraction so that either the numerator or the denominator contains no roots. The algebraic procedure for achieving this is called **rationalizing.** Here is an example in which a root is removed from the denominator.

EXAMPLE A.1.10

Rationalize the denominator in the expression $\dfrac{5}{3\sqrt{x}}$.

Solution
Multiply both the numerator and the denominator of the given expression by $\sqrt{x}$:

$$\frac{5}{3\sqrt{x}} = \frac{5(\sqrt{x})}{3\sqrt{x}(\sqrt{x})} = \frac{5(\sqrt{x})}{3(\sqrt{x})^2}$$

$$= \frac{5\sqrt{x}}{3x}$$

The algebraic identity

$$(x + y)(x - y) = x^2 - y^2$$

can be used to rationalize fractions when the numerator or denominator contains a factor of the form $a + \sqrt{b}$. The key lies in noting that the root can be removed from $a + \sqrt{b}$ by multiplying by the complementary expression $a - \sqrt{b}$ since

$$(a + \sqrt{b})(a - \sqrt{b}) = a^2 - (\sqrt{b})^2 = a^2 - b.$$

An expression of the form $\sqrt{a} + b$ can be rationalized in a similar fashion by using its complement $\sqrt{a} - b$. This procedure is illustrated in Examples A.1.11 and A.1.12.

EXAMPLE A.1.11

Rationalize the numerator in the expression $\dfrac{4 - \sqrt{3}}{7}$.

Solution
Multiply both the numerator and the denominator by $4 + \sqrt{3}$ and obtain

$$\frac{4 - \sqrt{3}}{7} = \frac{(4 - \sqrt{3})(4 + \sqrt{3})}{7} \cdot \frac{1}{(4 + \sqrt{3})} = \frac{4^2 - (\sqrt{3})^2}{7(4 + \sqrt{3})} = \frac{16 - 3}{7(4 + \sqrt{3})} = \frac{13}{7(4 + \sqrt{3})}$$

EXAMPLE A.1.12

Rationalize the denominator in the expression $\dfrac{1}{\sqrt{2} + \sqrt{x}}$.

Solution
Multiplying and dividing the fraction by $\sqrt{2} - \sqrt{x}$ gives

$$\frac{1}{\sqrt{2} + \sqrt{x}} = \frac{1}{(\sqrt{2} + \sqrt{x})} \frac{(\sqrt{2} - \sqrt{x})}{(\sqrt{2} - \sqrt{x})} = \frac{\sqrt{2} - \sqrt{x}}{(\sqrt{2})^2 - (\sqrt{x})^2} = \frac{\sqrt{2} - \sqrt{x}}{2 - x}.$$

EXERCISES ■ A.1

INTERVALS   *In Exercises 1 through 4, use inequalities to describe the indicated interval.*

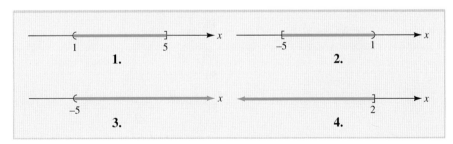

In Exercises 5 through 8, represent the given interval as a line segment on a number line.

5. $x \geq 2$

6. $-6 \leq x < 4$

7. $-2 < x \leq 0$

8. $x > 3$

DISTANCE   *In Exercises 9 through 12, find the distance on the number line between the real numbers in each pair.*

9. 0 and -4

10. 2 and 5

11. -2 and 3

12. -3 and -1

ABSOLUTE VALUE AND INTERVALS *In Exercises 13 through 18, find the interval or intervals consisting of all real numbers x that satisfy the given inequality.*

13. $|x| \leq 3$

14. $|x - 2| \leq 5$

15. $|x + 4| \leq 2$

16. $|1 - x| < 3$

17. $|x + 2| \geq 5$

18. $|x - 1| > 3$

EXPONENTIAL NOTATION *In Exercises 19 through 26, evaluate the given expression without using a calculator.*

19. 5^3

20. 2^{-3}

21. $16^{1/2}$

22. $36^{-1/2}$

23. $8^{2/3}$

24. $27^{-4/3}$

25. $\left(\dfrac{1}{4}\right)^{1/2}$

26. $\left(\dfrac{1}{4}\right)^{-3/2}$

In Exercises 27 through 34, evaluate the given expression without using a calculator.

27. $\dfrac{2^5(2^2)}{2^8}$

28. $\dfrac{3^4(3^3)}{(3^2)^3}$

29. $\dfrac{2^{4/3}(2^{5/3})}{2^5}$

30. $\dfrac{5^{-3}(5^2)}{(5^{-2})^3}$

31. $\dfrac{2(16^{3/4})}{2^3}$

32. $\dfrac{\sqrt{27}(\sqrt{3})^3}{9}$

33. $[\sqrt{8}(2^{5/2})]^{-1/2}$

34. $[\sqrt{27}(3^{5/2})]^{1/2}$

In Exercises 35 through 42, solve the given equation for n. (Assume $a > 0$ and $a \neq 1$.)

35. $a^3 a^7 = a^n$

36. $\dfrac{a^5}{a^2} = a^n$

37. $a^4 a^{-3} = a^n$

38. $a^2 a^n = \dfrac{1}{a}$

39. $(a^3)^n = a^{12}$

40. $(a^n)^5 = \dfrac{1}{a^{10}}$

41. $a^{3/5} a^{-n} = \dfrac{1}{a^2}$

42. $(a^n)^3 = \dfrac{1}{\sqrt{a}}$

In Exercises 43 through 76, simplify the given expression as much as possible. Assume a, b, and c are positive real numbers.

43. $(a^2 b^2 c^5)(a^2 b^6 c^3)$

44. $(a^5 b^2 c)^3$

45. $\left(\dfrac{a^2 c^3}{b}\right)^4$

46. $\left(\dfrac{a^{-2} b}{c^{-3}}\right)^2$

47. $\left(\dfrac{a^2 b^3 c^{-3}}{a^{-3} b^4 c^4}\right)^2$

48. $\left(\dfrac{a^{-3} b^{-2} c^{-4}}{a^4 b^3 c^5}\right)^{-3}$

49. $[(a^3 b^2)^{-2} c^2]^{-3}$

50. $[a^3 (b^3 c^{-1})^{-3}]^{-2}$

51. $\dfrac{a^{-2} b^{-3} + a^{-3} b + b c^{-1}}{ab^2 c^3}$

52. $\left(\dfrac{3a^{-3}}{c^2}\right)^{-1}\left(\dfrac{2c^{-2}}{a^3}\right)^2$

53. $\dfrac{a^{-3} + b^{-1}}{(ab)^{-2}}$

54. $(a^{-1} + b^{-2})^2$

55. $\sqrt[7]{128} + \sqrt[3]{-64}$

56. $\sqrt{18} + \sqrt[3]{-162(27)}$

57. $\sqrt[3]{6^5 5^8 3^6}$

58. $\sqrt[3]{(-2)^{15}(-3)^{18}}$

59. $2\sqrt{32} + 5\sqrt{72}$

60. $3\sqrt{96} + \sqrt{294}$

61. $3\sqrt{24} - 2\sqrt{54} + \sqrt{486}$

62. $3\sqrt[3]{15} - \sqrt[3]{120} + 5\sqrt[3]{405}$

63. $\sqrt[5]{a^{15}b^{20}c^{35}}$

64. $\dfrac{\sqrt[3]{-64a^9b^{-6}}}{\sqrt{a^2b^4}}$

65. $\sqrt{\dfrac{25a^2}{b}}\sqrt{\dfrac{b^3}{49a^4}}$

66. $\sqrt[3]{\dfrac{a^6b^9}{64c^{15}}}$

67. $\sqrt[3]{\dfrac{a^5}{b^7c^9}}$

68. $\sqrt[5]{\dfrac{a^8b^{-16}}{c^7}}$

69. $(a^4b^2c^{12})^{-1/2}$

70. $\dfrac{a^2b}{(a^6b^4)^{-1/4}}$

71. $(a^{1/6}b^{-1/3}c^{1/4})^{12}$

72. $\dfrac{(a^{25}b^{35})^{-3/5}}{(a^{16}b^{12})^{-3/4}}$

73. $(a^{1/2} + b^{1/4})(a^{1/2} - b^{1/4})$

74. $(a^{2/3} + b^{2/3})(a^{2/3} - b^{2/3})$

75. $\sqrt[3]{\dfrac{a^{17}b^9}{c^{11}}}$

76. $\sqrt[5]{(a^{24}b^{-8}c^{11})^4}$

In Exercises 77 through 84, rationalize the numerator or denominator in each expression.

77. $\dfrac{\sqrt{3} - \sqrt{2}}{5}$

78. $\dfrac{\sqrt{7} + 3}{2}$

79. $\dfrac{7}{3 - \sqrt{3}}$

80. $\dfrac{5}{\sqrt{5} + \sqrt{2}}$

81. $\dfrac{\sqrt{5} + 2}{3}$

82. $\dfrac{\sqrt{5} - \sqrt{11}}{4}$

83. $\dfrac{5}{\sqrt{5} + 1}$

84. $\dfrac{3}{2 - \sqrt{7}}$

85. Show that
$$\sqrt{x + h} - \sqrt{x} = \frac{h}{\sqrt{x + h} + \sqrt{x}},$$
where x and h are positive numbers.

86. Simplify the expression
$$\frac{1}{\sqrt{x + h}} - \frac{1}{\sqrt{x}},$$
where x and h are positive constants.

87. A* ECOLOGY   The atmosphere above each square centimetre of Earth's surface weighs 1 kg.

   a. Assuming Earth is a sphere of radius $R = 6440$ km, use the formula $S = 4\pi R^2$ to calculate the surface area of Earth and then find the total mass of the atmosphere.

   b. Oxygen occupies approximately 22% of the total mass of the atmosphere, and it is estimated that plant life produces approximately 0.9×10^{13} kg of oxygen per year. If none of this oxygen were used up by plants or animals (or combustion), how long would it take to build up the total mass of oxygen in the atmosphere (part (a))?*

88. Show that $(\sqrt[n]{x})^m = \sqrt[n]{x^m}$ in the case where m is a negative integer.

---

*Adapted from a problem in E. Batschelet, *Introduction to Mathematics for Life Scientists,* 2nd ed., New York: Springer-Verlag, 1979, p. 31.

SECTION A.2

L02

Factor polynomials. Add and subtract rational expressions and compound fractions. Solve equations by factoring, completing the square, and using the quadratic formula. Solve systems of two equations.

Factoring Polynomials and Solving Systems of Equations

A **polynomial** is an expression of the form

$$a_0 + a_1 x + a_2 x^2 + \cdots + a_n x^n,$$

where n is a nonnegative integer and $a_n, a_{n-1}, \ldots, a_0$ are real numbers known as the **coefficients** of the polynomial. Polynomials appear in a variety of mathematical contexts, and the first goal of this section is to examine some important algebraic properties of polynomials.

If $a_n \neq 0$, n is said to be the **degree** of the polynomial. A nonzero constant is said to be a **polynomial of degree 0.** (Technically, the number 0 is also a polynomial, but it has no degree.) For example, $3x^5 - 7x + 12$ is a polynomial of degree 5, with terms $3x^5$, $-7x$, and 12. **Similar terms** in two polynomials in the variable x are terms with the same degree. Thus, in the fifth-degree polynomial $3x^5 - 5x^2 + 3$ and the third-degree polynomial $-2x^3 + 2x^2 + 7x - 9$, the terms $-5x^2$ and $2x^2$ are similar terms. Polynomials can be multiplied by constants and added and subtracted by combining similar terms, as illustrated in Example A.2.1.

EXAMPLE A.2.1

Let $p(x) = 3x^2 - 5x + 7$ and $q(x) = -4x^2 + 9$. Find the polynomials $2p(x)$ and $p(x) + q(x)$.

Solution

$$2p(x) = 2(3)x^2 - 2(5)x + 2(7) = 6x^2 - 10x + 14$$

$$\begin{aligned} p(x) + q(x) &= [3 + (-4)]x^2 + [-5 + 0]x + [7 + 9] \\ &= -x^2 - 5x + 16 \end{aligned}$$

A convenient way to remember how to multiply two first-degree polynomials $p(x) = ax + b$ and $q(x) = cx + d$ is the FOIL method:

$$
\begin{array}{cccc}
\text{F} & \text{O} & \text{I} & \text{L} \\
\text{First} & \text{Outer} & \text{Inner} & \text{Last} \\
\text{product} & \text{product} & \text{product} & \text{product}
\end{array}
$$

$$(ax + b)(cx + d) = \overbrace{(ac)x^2} + \overbrace{(ad)x} + \overbrace{(bc)x} + \overbrace{(bd)}$$

Here is an example.

EXAMPLE A.2.2

Find $(3x + 5)(-2x + 7)$.

Solution

Applying the FOIL method results in

	F	O	I	L
	First	Outer	Inner	Last
	product	product	product	product

$$(3x + 5)(-2x + 7) = \overbrace{(3)(-2)x^2} + \overbrace{(3)(7)x} + \overbrace{(5)(-2)x} + \overbrace{(5)(7)}$$
$$= -6x^2 + 11x + 35$$

To multiply two polynomials that are not both of degree one, use the distributive laws of real numbers, namely,

$$a(b + c) = ab + ac \qquad \text{and} \qquad (a + b)c = ac + bc.$$

Here is an example of this procedure.

EXAMPLE A.2.3

Find $(-x^2 + 3x + 5)(x^2 + 2x - 4)$.

Solution

To find the required product, multiply each term of $(-x^2 + 3x + 5)$ by each term of $(x^2 + 2x - 4)$, and then combine similar terms:

$$(-x^2 + 3x + 5)(x^2 + 2x - 4)$$
$$= -x^2(x^2 + 2x - 4) + 3x(x^2 + 2x - 4) + 5(x^2 + 2x - 4)$$
$$= (-x^4 - 2x^3 + 4x^2) + (3x^3 + 6x^2 - 12x) + (5x^2 + 10x - 20)$$
$$= -x^4 + (-2 + 3)x^3 + (4 + 6 + 5)x^2 + (-12 + 10)x - 20$$
$$= -x^4 + x^3 + 15x^2 - 2x - 20$$

Factoring Polynomials with Integer Coefficients

Many of the polynomials that arise in practice have integer coefficients (or are closely related to polynomials that do). Techniques for factoring polynomials with integer coefficients are illustrated in the following examples. In each, the goal is to rewrite the given polynomial as a product of polynomials of lower degree that also have integer coefficients.

EXAMPLE A.2.4

Factor the polynomial $x^2 - 2x - 3$ using integer coefficients.

Solution

The goal is to write the polynomial as a product of the form

$$x^2 - 2x - 3 = (x + a)(x + b),$$

where a and b are integers. The distributive law implies that

$$(x + a)(x + b) = x^2 + (a + b)x + ab.$$

Hence, we must find integers a and b such that

$$x^2 - 2x - 3 = x^2 + (a + b)x + ab,$$

or, equivalently, such that

$$a + b = -2 \quad \text{and} \quad ab = -3.$$

From the list

$$1, -3 \quad \text{and} \quad -1, 3$$

of pairs of integers whose product is -3, choose $a = -3$ and $b = 1$ as the only pair whose sum is -2. It follows that

$$x^2 - 2x - 3 = (x - 3)(x + 1),$$

which you should check by multiplying out the right-hand side.

When factoring it is very important to always check by multiplying out the factors. An incorrect sign in factoring can lead the rest of a long solution in a very wrong direction.

EXAMPLE A.2.5

Factor the polynomial $12x^2 - 11x - 15$ using integer coefficients.

Solution
Write the polynomial as a product of the form

$$12x^2 - 11x - 15 = (ax + b)(cx + d).$$

Expanding the product on the right by the FOIL method gives

$$12x^2 - 11x - 15 = (ac)x^2 + (bc + ad)x + bd.$$

The goal is to find integers a, b, c, and d such that

$$ac = 12, \quad bc + ad = -11, \quad \text{and} \quad bd = -15.$$

Since ac is to be positive, there is no harm in assuming that a and c are both positive. (What happens if both are negative?) The factors of the coefficients 12 and -15 are as follows:

\ 12		\ -15	
a	c	b	d
12	1	15	-1
6	2	5	-3
4	3	3	-5
3	4	1	-15
2	6		
1	12		

Try each pair on the left with each pair on the right, with a goal of finding a combination that produces the middle term $bc + ad = -11$. By trial and error, $a = 4$ and $c = 3$

matched with $b = 3$ and $d = -5$ gives the correct middle term, resulting in the following factorization:

$$12x^2 - 11x - 15 = (4x + 3)(3x - 5).$$

Certain polynomial types occur so often that it is useful to have the following formulas for factoring them:

> ### Factorization Formulas
> **Square of sum:** $A^2 + 2AB + B^2 = (A + B)^2$
> **Square of difference:** $A^2 - 2AB + B^2 = (A - B)^2$
> **Difference of squares:** $A^2 - B^2 = (A - B)(A + B)$
> **Difference of cubes:** $A^3 - B^3 = (A - B)(A^2 + AB + B^2)$
> **Sum of cubes:** $A^3 + B^3 = (A + B)(A^2 - AB + B^2)$

EXAMPLE A.2.6

Factor the polynomial $x^3 - 8$ using integer coefficients.

Solution
Since $8 = 2^3$, use the difference of cubes formula, with $A = x$ and $B = 2$, to obtain the factorization

$$x^3 - 8 = x^3 - 2^3 = (x - 2)(x^2 + 2x + 4).$$

Sometimes a polynomial can be factored by grouping terms strategically, as in part (a) of the next example.

EXAMPLE A.2.7

Factor the following polynomials:
a. $p(x) = 4(x - 2)^3 + 3(x - 2)^2$
b. $q(x) = 9x^2 - 49$

Solution
a. Factor out the common term $(x - 2)^2$:

$$\begin{aligned} 4(x - 2)^3 + 3(x - 2)^2 &= (x - 2)^2[4(x - 2) + 3] \\ &= (x - 2)^2(4x - 5) \end{aligned}$$

b. The polynomial $q(x) = 9x^2 - 49$ can be written as a difference of squares $A^2 - B^2$, with $A = 3x$ and $B = 7$. Thus,

$$9x^2 - 49 = (3x)^2 - 7^2 = (3x - 7)(3x + 7).$$

Rational Expressions

The quotient of two polynomials is called a **rational expression.** For instance,

$$\frac{1}{x}, \qquad \frac{4}{2x^2 + 3}, \qquad \frac{-2x^3 + 7x - 1}{5x^2 + 3x + 9}, \qquad \text{and} \qquad \frac{x^3 + x - 6}{2}$$

are all rational expressions. One of our goals in working with rational expressions is to reduce such an expression to *lowest terms,* that is, to eliminate all common factors from the numerator and denominator. The following properties of fractions will be useful in this process.

Properties of Fractions

1. **Sum rule:** $\dfrac{a}{b} + \dfrac{c}{d} = \dfrac{ad + bc}{bd}$

2. **Product rule:** $\left(\dfrac{a}{b}\right)\left(\dfrac{c}{d}\right) = \dfrac{ac}{bd}$

3. **Quotient rule:** $\dfrac{\dfrac{a}{b}}{\dfrac{c}{d}} = \dfrac{a}{b} \cdot \dfrac{d}{c} = \dfrac{ad}{bc}$

EXAMPLE A.2.8

Write each of the following as a rational expression in lowest terms:

a. $\dfrac{-2}{x^2 - 1} + \dfrac{x}{x - 1}$ **b.** $\left(\dfrac{x^3 - 7x^2 + 10x}{x^2 + 6x + 9}\right)\left(\dfrac{x + 3}{x - 5}\right)$

Solution

a. $\dfrac{-2}{x^2 - 1} + \dfrac{x}{x - 1} = \dfrac{-2}{x^2 - 1} + \dfrac{x}{x - 1}\left(\dfrac{x + 1}{x + 1}\right)$

$\qquad = \dfrac{-2}{x^2 - 1} + \dfrac{x^2 + x}{x^2 - 1} = \dfrac{x^2 + x - 2}{x^2 - 1}$

$\qquad = \dfrac{(x + 2)(x - 1)}{(x + 1)(x - 1)} = \dfrac{x + 2}{x + 1} \quad \text{for } x \neq 1, -1$

b. $\left(\dfrac{x^3 - 7x^2 + 10x}{x^2 + 6x + 9}\right)\left(\dfrac{x + 3}{x - 5}\right)$

$\qquad = \dfrac{x(x^2 - 7x + 10)(x + 3)}{(x + 3)^2(x - 5)}$

$\qquad = \dfrac{x(x - 2)(x - 5)(x + 3)}{(x + 3)(x - 5)(x + 3)} = \dfrac{x^2 - 2x}{x + 3} \quad \text{for } x \neq 5, -3$

A rational expression with fractions in both the numerator and the denominator is known as a **compound fraction.** It is often useful to represent a compound fraction as the quotient of two polynomials. This procedure is illustrated in Example A.2.9.

EXAMPLE A.2.9

Simplify the compound fraction

$$\frac{1 + \dfrac{3}{x} - \dfrac{4}{x^2}}{1 + \dfrac{4}{x} - \dfrac{5}{x^2}}.$$

Solution

Write both the numerator and the denominator as rational expressions and then simplify:

$$\frac{1 + \dfrac{3}{x} - \dfrac{4}{x^2}}{1 + \dfrac{4}{x} - \dfrac{5}{x^2}} = \frac{\dfrac{x^2 + 3x - 4}{x^2}}{\dfrac{x^2 + 4x - 5}{x^2}}$$

$$= \frac{(x^2 + 3x - 4)x^2}{(x^2 + 4x - 5)x^2} \quad \text{since} \quad \frac{\dfrac{a}{b}}{\dfrac{c}{d}} = \frac{ad}{bc}$$

$$= \frac{(x + 4)(x - 1)x^2}{(x + 5)(x - 1)x^2}$$

$$= \frac{x + 4}{x + 5} \quad \text{for } x \neq 0, 1, -5$$

Solving Equations by Factoring

The **solutions** of an equation are the values of the variable that make the equation true. For example, $x = 2$ is a solution of the equation

$$x^3 - 6x^2 + 12x - 8 = 0$$

because substitution of 2 for x gives

$$2^3 - 6(2^2) + 12(2) - 8 = 8 - 24 + 24 - 8 = 0.$$

In Examples A.2.10 and A.2.11, you will see how factoring can be used to solve certain equations. The technique is based on the fact that if the product of two (or more) terms is equal to zero, then at least one of the terms must be equal to zero. For example, if $ab = 0$, then either $a = 0$ or $b = 0$ (or both).

EXAMPLE A.2.10

Solve the equation $x^2 - 3x = 10$.

Solution

First subtract 10 from both sides to get

$$x^2 - 3x - 10 = 0,$$

and then factor the resulting polynomial on the left-hand side to get

$$(x - 5)(x + 2) = 0.$$

Since the product $(x - 5)(x + 2)$ can be zero only if one (or both) of its factors is zero, it follows that the solutions are $x = 5$ (which makes the first factor zero) and $x = -2$ (which makes the second factor zero).

EXAMPLE A.2.11

Solve the equation $1 - \dfrac{1}{x} - \dfrac{2}{x^2} = 0$.

Solution
Put the fractions on the left-hand side over the common denominator x^2 and add to get

$$\frac{x^2}{x^2} - \frac{x}{x^2} - \frac{2}{x^2} = 0$$

or

$$\frac{x^2 - x - 2}{x^2} = 0.$$

Now factor the polynomial in the numerator to get

$$\frac{(x + 1)(x - 2)}{x^2} = 0$$

A quotient is zero only if its numerator is zero and its denominator is *not* zero, so it follows that $x = -1$ and $x = 2$ are the required solutions.

Completing the Square

An equation of the form

$$ax^2 + bx + c = 0 \quad \text{for } a \neq 0$$

is called a **quadratic equation.** A quadratic equation can have at most two solutions. As you have seen, one way to find the solutions is to factor the equation. Another is by the algebraic procedure called **completing the square,** in which the equation is rewritten in the form

$$(x + r)^2 = s$$

for real numbers r and s. Here are the steps in the procedure.

Step 1. Divide both sides of the given equation

$$ax^2 + bx + c = 0$$

by a (remember, $a \neq 0$) to obtain

$$x^2 + \left(\frac{b}{a}\right)x + \left(\frac{c}{a}\right) = 0.$$

Then subtract $\dfrac{c}{a}$ from both sides:

$$x^2 + \left(\frac{b}{a}\right)x = -\frac{c}{a}.$$

Step 2. Add the square of $\dfrac{1}{2}\left(\dfrac{b}{a}\right)$ to both sides:

$$x^2 + \left(\frac{b}{a}\right)x + \left(\frac{b}{2a}\right)^2 = -\frac{c}{a} + \left(\frac{b}{2a}\right)^2$$

Step 3. Notice that the left side of the equation can be factored to obtain $\left(x + \dfrac{b}{2a}\right)^2$.

Thus, the equation can be written as

$$\left(x + \frac{b}{2a}\right)^2 = -\frac{c}{a} + \left(\frac{b}{2a}\right)^2$$

EXAMPLE A.2.12

Solve the quadratic equation $x^2 + 5x + 4 = 0$ by completing the square.

Solution

$$x^2 + 5x + 4 = 0$$
$$x^2 + 5x = -4 \qquad \text{subtract 4 from both sides}$$
$$x^2 + 5x + \left(\frac{5}{2}\right)^2 = -4 + \left(\frac{5}{2}\right)^2 \qquad \text{add the square of } \frac{1}{2}(5) \text{ to both sides}$$
$$\left(x + \frac{5}{2}\right)^2 = \frac{9}{4} \qquad \text{since } x^2 + 5x + \left(\frac{5}{2}\right)^2 = \left(x + \frac{5}{2}\right)^2$$

So

$$x + \frac{5}{2} = \sqrt{\frac{9}{4}} = \frac{3}{2} \qquad \text{and} \qquad x + \frac{5}{2} = -\sqrt{\frac{9}{4}} = -\frac{3}{2},$$

and the solutions are

$$x = \frac{3}{2} - \frac{5}{2} = -1 \qquad \text{and} \qquad x = -\frac{3}{2} - \frac{5}{2} = -4.$$

EXAMPLE A.2.13

Solve the quadratic equation $3x^2 + 5x + 7 = 0$ by completing the square.

Solution

$$3x^2 + 5x + 7 = 0$$
$$x^2 + \left(\frac{5}{3}\right)x + \left(\frac{7}{3}\right) = 0 \qquad \text{divide each term by 3}$$
$$x^2 + \left(\frac{5}{3}\right)x = -\frac{7}{3} \qquad \text{subtract } \frac{7}{3} \text{ from each side}$$
$$x^2 + \left(\frac{5}{3}\right)x + \left(\frac{5}{6}\right)^2 = -\frac{7}{3} + \left(\frac{5}{6}\right)^2 \qquad \text{add the square of } \frac{1}{2}\left(\frac{5}{3}\right) \text{ to both sides}$$
$$\left(x + \frac{5}{6}\right)^2 = -\frac{59}{36}$$

Since it is impossible for the square $\left(x + \dfrac{5}{6}\right)^2$ to equal the negative number $-\dfrac{59}{36}$, the given quadratic equation has no (real) solutions.

The Quadratic Formula

By completing the square in the general quadratic equation

$$ax^2 + bx + c = 0 \quad \text{(for } a \neq 0\text{)},$$

we can obtain a general form for the solutions of the equation called the **quadratic formula.** Usually this is the best way of solving a quadratic equation. It is quicker than completing the square or using trial and error when the coefficient of x^2 is different from 1.

> **The Quadratic Formula** ■ The solutions of the quadratic equation
>
> $$ax^2 + bx + c = 0 \qquad \text{(for } a \neq 0\text{)}$$
>
> are given by the formula
>
> $$x = \frac{-b \pm \sqrt{b^2 - 4ac}}{2a}.$$

The term $b^2 - 4ac$ in the quadratic formula is called the **discriminant** of the quadratic equation. If the discriminant is positive, the equation has two solutions, one coming from the formula with the sign $\pm$ replaced by $+$ and the other with $\pm$ replaced by $-$. If the discriminant is zero, the equation has only one solution since the formula reduces to $x = \dfrac{-b}{2a}$. If the discriminant is negative, the equation has no real solutions since negative numbers do not have real square roots.

The use of the quadratic formula is illustrated in Examples A.2.14 through A.2.16.

EXAMPLE A.2.14

Solve the equation $x^2 + 3x + 1 = 0$.

Solution
This is a quadratic equation with $a = 1$, $b = 3$, and $c = 1$. Use the quadratic formula to get

$$x = \frac{-3 + \sqrt{5}}{2} \approx -0.38 \qquad \text{and} \qquad x = \frac{-3 - \sqrt{5}}{2} \approx -2.62.$$

EXAMPLE A.2.15

Solve the equation $x^2 + 18x + 81 = 0$.

Solution
This is a quadratic equation with $a = 1$, $b = 18$, and $c = 81$. Use the quadratic formula to find that the discriminant is zero and that the formula for x gives

$$x = \frac{-18 \pm \sqrt{0}}{2} = -\frac{18}{2} = -9.$$

Note that the left side of the equation $x^2 + 18x + 81 = 0$ is the square of the sum $(x + 9)^2$, so it can also be solved by factoring.

EXAMPLE A.2.16

Solve the equation $x^2 + x + 1 = 0$.

Solution
This is a quadratic equation with $a = 1$, $b = 1$, and $c = 1$. Use the quadratic formula to get

$$x = \frac{-1 \pm \sqrt{-3}}{2}.$$

Since there is no real square root of -3, it follows that the equation has no real solution.

Systems of Equations

A collection of equations that are to be solved simultaneously is called a **system of equations.** Some of the calculus problems in Chapter 7 involve the solution of systems of two (or more) equations in two (or more) unknowns. For example, you may wish to find the real numbers x and y that satisfy the system

$$2x + 3y = 5$$
$$x + 2y = 4$$

The procedure for solving a system of two equations in two unknowns is to (temporarily) eliminate one of the variables, thereby reducing the problem to a single equation in one variable, which you then solve for its variable. Once you have found the value of one of the variables, you can substitute it into either of the original equations and solve to get the value of the other variable.

The most common techniques for the elimination of variables are illustrated in the following two examples.

EXAMPLE A.2.17

Solve the system

$$4x + 3y = 13$$
$$3x + 2y = 7$$

Solution
To eliminate y, multiply both sides of the first equation by 2 and both sides of the second equation by -3 so that the system becomes

$$8x + 6y = 26$$
$$-9x - 6y = -21$$

Then add the equations to get

$$-x + 0 = 5 \qquad \text{or} \qquad x = -5.$$

To find y, substitute $x = -5$ into either of the original equations. For the second equation,

$$3(-5) + 2y = 7$$
$$2y = 22$$
$$y = 11$$

That is, the solution of the system is $x = -5$ and $y = 11$.

To check this answer, substitute $x = -5$ and $y = 11$ into both original equations. From the first equation,

$$4(-5) + 3(11) = -20 + 33 = 13,$$

and from the second equation,

$$3(-5) + 2(11) = -15 + 22 = 7,$$

as required.

EXAMPLE A.2.18

Solve the system

$$2y^2 - x^2 = 14$$
$$x - y = 1$$

Solution

Solve the second equation for x to get

$$x = y + 1,$$

and substitute this into the first equation to eliminate x. This gives

$$2y^2 - (y + 1)^2 = 14$$
$$2y^2 - (y^2 + 2y + 1) = 14$$
$$2y^2 - y^2 - 2y - 1 = 14$$
$$y^2 - 2y - 15 = 0$$

or

$$(y + 3)(y - 5) = 0,$$

from which it follows that

$$y = -3 \quad \text{or} \quad y = 5.$$

If $y = -3$, the second equation gives

$$x - (-3) = 1 \quad \text{or} \quad x = -2,$$

and if $y = 5$, the second equation gives

$$x - 5 = 1 \quad \text{or} \quad x = 6.$$

Hence, the system has two solutions,

$$x = 6, y = 5 \quad \text{and} \quad x = -2, y = -3.$$

To check these answers, substitute each pair x, y into the first equation. If $x = 6$ and $y = 5$,

$$2(5^2) - 6^2 = 50 - 36 = 14,$$

and if $x = -2$ and $y = -3$,

$$2(-3)^2 - (-2)^2 = 18 - 4 = 14,$$

as required.

EXERCISES ■ A.2

In Exercises 1 through 10, find the indicated product.

1. $3x(x - 9)$

2. $-2x^2(3 - 4x)$

3. $(x - 7)(x + 2)$

4. $(x + 1)(x + 5)$

5. $(3x - 7)(4 - 2x)$

6. $(-x - 3)(5 - 3x)$

7. $(x - 1)(x^2 + 2x - 3)$

8. $(3x^2 - 5x + 4)(x + 2)$

9. $(x^3 - 3x + 4)(x^2 - 3x + 2)$

10. $(2x + 3 + x^2 - 5)(x^2 - x - 3)$

In Exercises 11 through 28, simplify the given rational expression.

11. $\dfrac{x + 3}{x - 3} + \dfrac{x}{x + 3}$

12. $\dfrac{4}{x^2 + 5x + 6} + \dfrac{x - 2}{x + 3}$

13. $\dfrac{-5x - 6}{x^2 + 2x - 3} + \dfrac{x + 2}{x - 1}$

14. $\dfrac{x - 6}{x^2 + 3x - 10} - \dfrac{x + 3}{x - 5}$

15. $\dfrac{x - 2}{2x^2 - 7x - 15} - \dfrac{1}{2x + 3}$

16. $\left(\dfrac{x^3 - 8}{x}\right)\left(\dfrac{x^2 - 3x}{x - 2}\right)$

17. $\dfrac{4}{x + 2} - \dfrac{3}{x - 1} - \dfrac{2x}{x^2 + x - 2}$

18. $\dfrac{-2}{x - 4} + \dfrac{1}{x + 4} + \dfrac{1 - 2x}{x^2 - 16}$

19. $\dfrac{4}{x + 3} - \dfrac{2}{x + 4} - \dfrac{2x + 3}{x^2 + 7x + 12}$

20. $\dfrac{7}{x - 1} + \dfrac{5}{2x + 3} - \dfrac{x + 2}{2x^2 + x - 3}$

21. $\dfrac{\dfrac{1}{x} - \dfrac{1}{3}}{\dfrac{1}{x} + \dfrac{1}{3}}$

22. $\dfrac{\dfrac{1}{x}}{1 + \dfrac{1}{x}}$

23. $\dfrac{\dfrac{x - 3}{x + 3} - \dfrac{x + 3}{x - 3}}{\dfrac{x}{x - 3} - \dfrac{x}{x + 3}}$

24. $\dfrac{\dfrac{3x^2 + 5x - 8}{x^3 - 1}}{\dfrac{3x + 8}{x^2 + x + 1}}$

25. $1 - \dfrac{1}{1 + \dfrac{x}{2x - 1}}$

26. $3 + \dfrac{5}{1 - \dfrac{x - 1}{x + 1}}$

27. $\dfrac{\dfrac{1}{x} - 2 + \dfrac{x}{x + 1}}{\dfrac{3x - 1}{x^2 + x}}$

28. $\dfrac{\dfrac{x}{x^2 - 9} - \dfrac{1}{x + 3}}{\dfrac{3}{x - 3}}$

FACTORING POLYNOMIALS WITH INTEGER COEFFICIENTS   *In Exercises 29 through 58, factor the given polynomial using integer coefficients.*

29. $x^2 + x - 2$

30. $x^2 + 3x - 10$

31. $x^2 - 7x + 12$

32. $x^2 + 8x + 12$

33. $x^2 - 2x + 1$

34. $x^2 + 6x + 9$

35. $16x^2 - 25$

36. $3x^2 - x - 14$

37. $x^3 - 1$

38. $x^3 - 27$

39. $x^7 - x^5$

40. $x^3 + 2x^2 + x$

41. $2x^3 - 8x^2 - 10x$

42. $x^4 + 5x^3 - 14x^2$

43. $x^2 + x - 12$

44. $x^2 - 9x + 14$

45. $2x^2 - x - 15$

46. $3x^2 - 22x + 35$

47. $x^2 - 7x - 18$

48. $x^2 + 8x + 15$

49. $28x^2 + 2x - 6$

50. $12x^2 - x - 20$

51. $x^3 + 2x^2 - 15x$

52. $25x^3 - 16x$

53. $x^3 + 27$

54. $25x^2 - 81$

55. $x^5 + x^2$

56. $x^4 - 9x^2$

57. $3(x + 2)^3 - 5(x + 2)^2$

58. $5(x - 1)^4 + 3(x - 1)^2$

SOLUTION OF EQUATIONS BY FACTORING

In Exercises 59 through 74, solve the given equation by factoring.

59. $x^2 - 2x - 8 = 0$

60. $x^2 - 4x + 3 = 0$

61. $x^2 + 10x + 25 = 0$

62. $x^2 + 8x + 16 = 0$

63. $x^2 - 16 = 0$

64. $x^2 - 25 = 0$

65. $2x^2 + 3x + 1 = 0$

66. $x^2 - 2x + 1 = 0$

67. $4x^2 + 12x + 9 = 0$

68. $6x^2 + 7x - 3 = 0$

69. $1 + \dfrac{4}{x} - \dfrac{5}{x^2} = 0$

70. $\dfrac{9}{x^2} - \dfrac{6}{x} + 1 = 0$

71. $2 + \dfrac{2}{x} - \dfrac{4}{x^2} = 0$

72. $\dfrac{3}{x^2} - \dfrac{5}{x} - 2 = 0$

73. $\dfrac{x}{x - 2} - \dfrac{4}{x + 3} - \dfrac{10}{x^2 + x - 6} = 0$

74. $\dfrac{x}{x + 1} + \dfrac{3}{2x + 3} - \dfrac{11x + 10}{2x^2 + 5x + 3} = 0$

SOLUTION OF EQUATIONS BY COMPLETING THE SQUARE

In Exercises 75 through 82, solve the given quadratic equation by completing the square.

75. $x^2 + 2x - 3 = 0$

76. $2x^2 + 11x + 15 = 0$

77. $15x^2 - 14x + 3 = 0$

78. $21x^2 + 11x - 2 = 0$

79. $x^2 + 5x + 11 = 0$

80. $4x^2 + 3x + 1 = 0$

81. $6x^2 + 17x - 4 = 0$

82. $7x^2 + 12x - 5 = 0$

QUADRATIC FORMULA

In Exercises 83 through 88, use the quadratic formula to solve the given equation.

83. $2x^2 + 3x + 1 = 0$

84. $-x^2 + 3x - 1 = 0$

85. $x^2 - 2x + 3 = 0$

86. $x^2 - 2x + 1 = 0$

87. $4x^2 + 12x + 9 = 0$

88. $x^2 + 12 = 0$

SYSTEMS OF EQUATIONS

In Exercises 89 through 94, solve the given system of equations.

89. $\begin{aligned} x + 5y &= 13 \\ 3x - 10y &= -11 \end{aligned}$

90. $\begin{aligned} 2x - 3y &= 4 \\ 3x - 5y &= 2 \end{aligned}$

91. $\begin{aligned} 5x - 4y &= 12 \\ 2x - 3y &= 2 \end{aligned}$

92. $\begin{aligned} 3x^2 - 9y &= 0 \\ 3y^2 - 9x &= 0 \end{aligned}$

93. $\begin{aligned} 2y^2 - x^2 &= 1 \\ x - 2y &= 3 \end{aligned}$

94. $\begin{aligned} 2x^2 - y^2 &= -7 \\ 2x + y &= 1 \end{aligned}$

SECTION A.3

L03

Evaluate limits using L'Hôpital's rule.

L'Hôpital's Rule:

$\dfrac{0}{0}$ and $\dfrac{\infty}{\infty}$ **Forms**

Evaluating Limits with L'Hôpital's Rule

In curve sketching and other applications of calculus, it is often necessary to compute a limit of the form

$$\lim_{x \to c} \frac{f(x)}{g(x)},$$

where c is either a finite number or ∞. If $\lim_{x \to c} g(x) \neq 0$, then the quotient rule for limits may be used, but if both $f(x)$ and $g(x)$ approach 0 as x approaches c, practically anything can happen. For example,

$$\lim_{x \to \infty} \frac{\dfrac{1}{x^3} - \dfrac{1}{x^2}}{\dfrac{1}{x}}, \qquad \lim_{x \to 0} \frac{2x^3 + 3x^2}{x^5 + x^4}, \qquad \text{and} \qquad \lim_{x \to 1} \frac{x - 1}{x^3 - 1}$$

all have this property, but the limit on the left is 0, the one in the centre is ∞, and the one on the right is $\dfrac{1}{3}$.

Limits such as these are called $\dfrac{0}{0}$ **indeterminate forms.** Similarly, limits of quotients in which both the numerator and the denominator increase or decrease without bound as $x \to c$ are called $\dfrac{\infty}{\infty}$ **indeterminate forms.**

A powerful technique, known as **L'Hôpital's rule,** can be used to analyze indeterminate forms. The rule says, in effect, that if an attempt to find the limit of a quotient leads to either a $\dfrac{0}{0}$ or an $\dfrac{\infty}{\infty}$ indeterminate form, then take derivatives of the numerator and the denominator and try again. Here is a more symbolic statement of the procedure.

L'Hôpital's Rule

If $\lim_{x \to c} f(x) = 0$ and $\lim_{x \to c} g(x) = 0$, then

$$\lim_{x \to c} \frac{f(x)}{g(x)} = \lim_{x \to c} \frac{f'(x)}{g'(x)}.$$

If $\lim_{x \to c} f(x) = \infty$ and $\lim_{x \to c} g(x) = \infty$, then

$$\lim_{x \to c} \frac{f(x)}{g(x)} = \lim_{x \to c} \frac{f'(x)}{g'(x)}.$$

The use of L'Hôpital's rule is illustrated in Examples A.3.1 through A.3.4. As you read through these examples, pay particular attention to the following two points:

1. L'Hôpital's rule involves differentiating the numerator and the denominator *separately.* A common mistake is to differentiate the entire quotient using the quotient rule.

2. L'Hôpital's rule applies only to quotients whose limits are indeterminate forms $\frac{0}{0}$ or $\frac{\infty}{\infty}$. Limits of the form $\frac{0}{\infty}$ or $\frac{\infty}{0}$ are *not* indeterminate (the first is 0, and the second is ∞).

EXAMPLE A.3.1

Use L'Hôpital's rule to compute the limit

$$\lim_{x \to \infty} \frac{x}{(x + 1)^2}.$$

Solution

This is a $\frac{\infty}{\infty}$ indeterminate form, so L'Hôpital's rule applies, and

$$\lim_{x \to \infty} \frac{x}{(x + 1)^2} = \lim_{x \to \infty} \frac{(x)'}{[(x + 1)^2]'} = \lim_{x \to \infty} \frac{1}{2(x + 1)} = 0.$$

EXAMPLE A.3.2

Use L'Hôpital's rule to compute the limit

$$\lim_{x \to 1} \frac{x^5 - 3x^4 + 5x - 3}{4x^5 + 2x^3 - 5x^2 - 1}.$$

Solution

Substituting $x = 1$ into the numerator and denominator reveals that this is a $\frac{0}{0}$ indeterminate form. It is possible to evaluate this limit by the factor method developed in Chapter 1, but notice how much easier it is to use L'Hôpital's rule:

$$\lim_{x \to 1} \frac{x^5 - 3x^4 + 5x - 3}{4x^5 + 2x^3 - 5x^2 - 1} = \lim_{x \to 1} \frac{(x^5 - 3x^4 + 5x - 3)'}{(4x^5 + 2x^3 - 5x^2 - 1)'}$$

$$= \lim_{x \to 1} \frac{5x^4 - 12x^3 + 5}{20x^4 + 6x^2 - 10x} = -\frac{2}{16} = -\frac{1}{8}$$

EXAMPLE A.3.3

Evaluate $\lim_{x \to 2} \dfrac{2x + 5}{x^2 + 3x - 10}$.

Solution
Blindly applying L'Hôpital's rule results in

$$\lim_{x \to 2} \frac{2x + 5}{x^2 + 3x - 10} = \lim_{x \to 2} \frac{2}{2x + 3} = \frac{2}{7}.$$

However, using a calculator to evaluate the given quotient at a number very close to 2 (say, at 2.0001) results in a number that is much larger than $\frac{2}{7}$. Why? The answer from L'Hôpital's rule was wrong because the given limit is not indeterminate. In fact, it is only necessary to simply substitute $x = 2$ to get

$$\lim_{x \to 2} \frac{2x + 5}{x^2 + 3x - 10} = \frac{9}{0} = \infty.$$

EXAMPLE A.3.4

Find $\lim\limits_{x \to \infty} \dfrac{3 - e^x}{x^2}$.

Solution

The limit is indeterminate of the form $\dfrac{\infty}{\infty}$. Applying L'Hôpital's rule gives

$$\lim_{x \to \infty} \frac{3 - e^x}{x^2} = \lim_{x \to \infty} \frac{-e^x}{2x}.$$

Since this new limit is also of the form $\dfrac{\infty}{\infty}$, apply L'Hôpital's rule again to get

$$\lim_{x \to \infty} \frac{-e^x}{2x} = \lim_{x \to \infty} \frac{-e^x}{2} = -\infty,$$

so that

$$\lim_{x \to \infty} \frac{3 - e^x}{x^2} = -\infty.$$

Although L'Hôpital's rule only applies to $\dfrac{0}{0}$ and $\dfrac{\infty}{\infty}$ indeterminate forms, other kinds of indeterminate forms can often be computed by combining L'Hôpital's rule with a little algebra. This procedure is illustrated in Examples A.3.5 and A.3.6.

EXAMPLE A.3.5

Find $\lim\limits_{x \to \infty} e^{-x} \ln x$.

Solution

This limit is of the indeterminate form $0 \cdot \infty$ and can be rewritten as

$$\lim_{x \to \infty} \frac{e^{-x}}{\dfrac{1}{\ln x}} \qquad \left(\text{of the form } \frac{0}{0} \right)$$

or as

$$\lim_{x \to \infty} \frac{\ln x}{e^x} \qquad \left(\text{of the form } \frac{\infty}{\infty} \right).$$

Applying L'Hôpital's rule to the simpler second quotient gives

$$\lim_{x \to \infty} e^{-x} \ln x = \lim_{x \to \infty} \frac{\ln x}{e^x} = \lim_{x \to \infty} \frac{\dfrac{1}{x}}{e^x} = 0.$$

In the last calculation, the limit was reduced to the form $\dfrac{0}{\infty}$, which is equal to 0.

As a final illustration of this technique, here is the limit that was used in Section 4.1 to define the number e.

EXAMPLE A.3.6

Find $\lim\limits_{x \to \infty} \left(1 + \dfrac{1}{x} \right)^x$.

Solution

This limit is of the indeterminate form 1^∞. To simplify the problem, let

$$y = \left(1 + \frac{1}{x} \right)^x.$$

Then

$$\ln y = x \ln \left(1 + \frac{1}{x} \right)$$

$$\lim_{x \to \infty} \ln y = \lim_{x \to \infty} x \ln \left(1 + \frac{1}{x} \right) \qquad (\infty \cdot 0)$$

$$\lim_{x \to \infty} \ln y = \lim_{x \to \infty} \frac{\ln \left(1 + \dfrac{1}{x} \right)}{\dfrac{1}{x}} \qquad \left(\frac{0}{0} \right)$$

$$= \lim_{x \to \infty} \frac{\left[\ln \left(1 + \dfrac{1}{x} \right) \right]'}{\left[\dfrac{1}{x} \right]'} = \lim_{x \to \infty} \frac{\left(\dfrac{-\dfrac{1}{x^2}}{1 + \dfrac{1}{x}} \right)}{-\dfrac{1}{x^2}} \qquad \text{L'Hôpital's rule}$$

$$= \lim_{x \to \infty} \frac{1}{1 + \dfrac{1}{x}} \qquad \text{algebraic simplification}$$

$$= 1$$

Since $\ln y \to 1$, it follows that $y \to e^1 = e$. That is,

$$\lim_{x \to \infty} \left(1 + \frac{1}{x} \right)^x = e.$$

EXERCISES ■ A.3

In Exercises 1 through 16, use L'Hôpital's rule to evaluate the given limit if the limit is an indeterminate form.

1. $\displaystyle\lim_{x\to\infty} \frac{x^3 - 3x^2}{3x^4 + 2x}$

2. $\displaystyle\lim_{x\to 0} \frac{x^2(x - 1)}{3x^3 + 2x - 5}$

3. $\displaystyle\lim_{x\to\infty} \frac{x^2 - 2x + 3}{2x^2 + 5x + 1}$

4. $\displaystyle\lim_{x\to\infty} \frac{x^2 + x - 5}{1 - 2x - x^3}$

5. $\displaystyle\lim_{x\to\infty} \frac{\dfrac{1}{x} - \dfrac{2}{x^2}}{\dfrac{1}{x^3} + \dfrac{2}{x^2} - \dfrac{3}{x}}$

6. $\displaystyle\lim_{x\to\infty} \frac{x^2 + 2x - 15}{x^3 - 19x + 3}$

7. $\displaystyle\lim_{x\to -1} \frac{x^3 + 3x^2 + 3x + 1}{2x^3 + 3x^2 - 1}$

    [*Hint:* Use L'Hôpital's rule twice.]

8. $\displaystyle\lim_{x\to 1/2} \frac{-8x^3 + 2x^2 + 3x - 1}{(2x - 1)^3}$

9. $\displaystyle\lim_{x\to\infty} \frac{e^{-x}}{1 + e^{-2x}}$

10. $\displaystyle\lim_{x\to\infty} x^2 e^{-x}$

11. $\displaystyle\lim_{t\to 0} \frac{\sqrt{t}}{e^t}$

12. $\displaystyle\lim_{t\to\infty} \frac{\ln\sqrt{t}}{t}$

13. $\displaystyle\lim_{x\to\infty} \frac{(\ln x)^2}{x}$

14. $\displaystyle\lim_{x\to\infty} x^{1/x}$

15. $\displaystyle\lim_{x\to\infty} (1 + 2x)^{1/x}$

16. $\displaystyle\lim_{x\to\infty} \left(1 + \frac{1}{x}\right)^{x^2}$

SECTION A.4

L04

Evaluate sums given the summation notation.

Summation Notation

Sums of the form $a_1 + a_2 + \cdots + a_n$ appear so often in mathematics that a special notation has been developed to handle them. To describe such a sum, it suffices to characterize the general term a_j and to indicate that n terms of this form are to be added, starting with the term where $j = 1$ and ending with the term where $j = n$. It is customary to use the Greek uppercase letter Σ (sigma) to denote summation and to express the sum compactly as follows.

Summation Notation ■ The sum of the numbers $a_1, \ldots, a_n$ is given by

$$a_1 + a_2 + \cdots + a_n = \sum_{j=1}^{n} a_j.$$

The use of summation notation is illustrated in the next two examples.

EXAMPLE A.4.1

Use summation notation to represent each sum.

  a. $1 + 4 + 9 + 16 + 25 + 36 + 49 + 64$

  b. $(1 - x_1)^2 \Delta x + (1 - x_2)^2 \Delta x + \cdots + (1 - x_{15})^2 \Delta x$

Solution

a. This is a sum of 8 terms of the form j^2, starting with $j = 1$ and ending with $j = 8$. Hence,

$$1 + 4 + 9 + 16 + 25 + 36 + 49 + 64 = \sum_{j=1}^{8} j^2.$$

b. The jth term of this sum is $(1 - x_j)^2 \Delta x$. Hence,

$$(1 - x_1)^2 \Delta x + (1 - x_2)^2 \Delta x + \cdots + (1 - x_{15})^2 \Delta x = \sum_{j=1}^{15} (1 - x_j)^2 \Delta x.$$

EXAMPLE A.4.2

Evaluate each sum.

a. $\displaystyle\sum_{j=1}^{4} (j^2 + 1)$

b. $\displaystyle\sum_{j=1}^{3} (-2)^j$

Solution

a. $\displaystyle\sum_{j=1}^{4} (j^2 + 1) = (1^2 + 1) + (2^2 + 1) + (3^2 + 1) + (4^2 + 1)$

$$= 2 + 5 + 10 + 17 = 34$$

b. $\displaystyle\sum_{j=1}^{3} (-2)^j = (-2)^1 + (-2)^2 + (-2)^3 = -2 + 4 - 8 = -6$

EXERCISES ■ A.4

In Exercises 1 through 4, evaluate the given sum.

1. $\displaystyle\sum_{j=1}^{4} (3j + 1)$

2. $\displaystyle\sum_{j=1}^{5} j^2$

3. $\displaystyle\sum_{j=1}^{10} (-1)^j$

4. $\displaystyle\sum_{j=1}^{5} 2^j$

In Exercises 5 through 10, use summation notation to represent the given sum.

5. $1 + \dfrac{1}{2} + \dfrac{1}{3} + \dfrac{1}{4} + \dfrac{1}{5} + \dfrac{1}{6}$

6. $3 + 6 + 9 + 12 + 15 + 18 + 21 + 24 + 27 + 30$

7. $2x_1 + 2x_2 + 2x_3 + 2x_4 + 2x_5 + 2x_6$

8. $1 - 1 + 1 - 1 + 1 - 1$

9. $1 - 2 + 3 - 4 + 5 - 6 + 7 - 8$

10. $x - x^2 + x^3 - x^4 + x^5$

Concept Summary Appendix

$$|x| = \begin{cases} x & \text{if } x \geq 0 \\ -x & \text{if } x < 0 \end{cases}$$

$$a^n = \underbrace{a \cdot a \cdot \cdots a}_{n \text{ terms}}$$

$$a^{-n} = \frac{1}{a^n} \quad \text{(negative power)}$$

$$a^{n/m} = (\sqrt[m]{a})^n = \sqrt[m]{a^n} \quad \text{(fractional powers)}$$

$$a^r a^s = a^{r+s}$$

$$\frac{a^r}{a^s} = a^{r-s}$$

$$(a^r)^s = a^{rs}$$

$(x + a)(x + b) = x^2 + (a + b)x + ab$

To factor, find two numbers a and b that add to make the coefficient of x and multiply to make the integer ab.

Difference of Two Squares $a^2 - b^2 = (a + b)(a - b)$

Difference of Cubes $a^3 - b^3 = (a - b)(a^2 + ab + b^2)$

Sum of Cubes $a^3 + b^3 = (a + b)(a^2 - ab + b^2)$

$$\frac{a}{b} + \frac{c}{d} = \frac{ad + bc}{bd}$$

$$\left(\frac{a}{b}\right)\left(\frac{c}{d}\right) = \frac{ac}{bd}$$

$$\frac{\dfrac{a}{b}}{\dfrac{c}{d}} = \frac{ad}{bc}$$

The Quadratic Formula

The solutions of $ax^2 + bx + c = 0$ for $a \neq 0$ are

$$x = \frac{-b \pm \sqrt{b^2 - 4ac}}{2a}.$$

The discriminant is $b^2 - 4ac$.

L'Hôpital's Rule

$\left(\dfrac{0}{0} \text{ form}\right)$ If $\lim\limits_{x \to c} f(x) = 0$ and $\lim\limits_{x \to c} g(x) = 0$, then

$$\lim_{x \to c} \frac{f(x)}{g(x)} = \lim_{x \to c} \frac{f'(x)}{g'(x)}.$$

$\left(\dfrac{\infty}{\infty} \text{ form}\right)$ If $\lim\limits_{x \to c} f(x) = \infty$ and $\lim\limits_{x \to c} g(x) = \infty$, then

$$\lim_{x \to c} \frac{f(x)}{g(x)} = \lim_{x \to c} \frac{f'(x)}{g'(x)}.$$

$$\sum_{j=1}^{n} a_j = a_1 + a_2 + \cdots + a_n$$

Review Exercises

In Exercises 1 and 2, use inequalities to describe the given interval.

1. **2.**

In Exercises 3 through 6, represent the given interval as a line segment on a number line.

3. $-3 \leq x < 2$

4. $-1 < x < 5$

5. $x \geq 1$

6. $2 \leq x < 7$

In Exercises 7 and 8, find the distance on the number line between the real numbers in each pair.

7. 0 and 3

8. -5 and -2

In Exercises 9 and 10, find the interval or intervals consisting of all real numbers x that satisfy the given inequality.

9. $|x - 3| \leq 1$

10. $|2x + 1| > 3$

In Exercises 11 through 20, evaluate the given expression without using a calculator.

11. 3^5

12. 4^{-2}

13. $8^{2/3}$

14. $49^{-3/2}$

15. $\dfrac{4(32)^{3/4}}{(\sqrt{2})^3}$

16. $\left(\dfrac{1}{9}\right)^{-5/2}$

17. $16^{3/2} + 27^{2/3}$

18. $\dfrac{2^{3/2}(4^{5/2})}{8^{2/3}}$

19. $\dfrac{\sqrt[3]{54}\,\sqrt[6]{2}}{\sqrt{8}}$

20. $\dfrac{\sqrt[3]{81}\,(6^{2/3})}{2^{4/3}}$

In Exercises 21 through 24, solve the given equation for n (assume a > 0, a ≠ 1).

21. $a^{2/3}a^{1/2} = a^{3n}$

22. $\dfrac{a^3}{(\sqrt{a})^5} = a^{2n}$

23. $a^2 a^{-5} = (a^n)^3$

24. $a^{2n}a^3 = a^{-7}$

In Exercises 25 through 28, evaluate the given sum.

25. $\displaystyle\sum_{k=1}^{3}(2k + 3)$

26. $\displaystyle\sum_{k=1}^{4}(k + 1)^2$

27. $\displaystyle\sum_{k=1}^{5}(2k^2 - k)$

28. $\displaystyle\sum_{k=1}^{4}\left(\dfrac{k - 1}{k + 3}\right)^2$

In Exercises 29 and 30, express the given sum in summation notation.

29. $1 + \dfrac{1}{2} - \dfrac{1}{3} + \dfrac{1}{4} - \dfrac{1}{5} + \dfrac{1}{6} - \dfrac{1}{7}$

30. $3 + 12 + 27 + 48 + 75$

In Exercises 31 through 34, factor the given expression.

31. $x^4 - 9x^2$

32. $x^3 + 3(x - 12)$

33. $x^{16} - (2x)^4$

34. $2(x - 3)^2(x + 1) - 5(x - 3)^3(2x)$

In Exercises 35 and 36, simplify the given quotient as much as possible.

35. $\dfrac{x^2(x - 1)^3 - 2x(x - 1)^2}{x^2 - x - 2}$

36. $\dfrac{x(x + 2)^4 - x^3(x + 2)^2}{x^2 + 3x + 2}$

In Exercises 37 through 42, factor the given polynomial using integral coefficients.

37. $x^2 + 2x - 15$

38. $2x^2 + 5x - 3$

39. $4x^2 + 12x + 9$

40. $12x^2 + 5x - 3$

41. $x^3 + 3x^2 - x - 3$

42. $x^4 - 5x^2 + 4$

In Exercises 43 through 48, solve the given equation by factoring.

43. $x^2 + 3x - 4 = 0$

44. $2x^2 - 3x - 2 = 0$

45. $x^2 + 14x + 49 = 0$

46. $x^2 - 64 = 0$

47. $1 - \dfrac{1}{x} - \dfrac{2}{x^2} = 0$

48. $4 + \dfrac{9}{x^2} = \dfrac{12}{x}$

In Exercises 49 through 54, use the quadratic formula to find all real numbers x that satisfy the given equation.

49. $14x^2 - x - 3 = 0$

50. $24x^2 + x - 10 = 0$

51. $x^2 - 3x + 5 = 0$

52. $7x^2 + 3x - 2 = 0$

53. $3x^2 + 5x - 2 = 0$

54. $2x^2 + 12x + 11 = 0$

In Exercises 55 through 58, solve the given system of equations.

55. $3x + 5y = -1$
$2x + 7y = 3$

56. $2x + y = 7$
$-x + 4y = 1$

57. $3x^2 - y^2 = -1$
$2x + y = 4$

58. $5x^2 - 2y^2 = 2$
$5x - 2y = 4$

In Exercises 59 through 64, use L'Hôpital's rule to evaluate the given limit if the limit is an indeterminate form.

59. $\displaystyle\lim_{x \to 1} \frac{x^2 - 1}{2x^3 + x - 3}$

60. $\displaystyle\lim_{x \to -2} \frac{x^3 + 8}{3x^3 - 7x + 10}$

61. $\displaystyle\lim_{x \to \infty} \frac{e^{-2x}}{3 + 2e^{-2x}}$

62. $\displaystyle\lim_{x \to \infty} \sqrt{x}\, e^{-x}$

63. $\displaystyle\lim_{x \to \infty} x(e^{1/x} - 1)$

64. $\displaystyle\lim_{x \to \infty} \left(1 - \frac{3}{x}\right)^{2x}$

THINK ABOUT IT

(Photo: CDC/Cade Martin)

LIVING SPACE

Problem

For comfortable modern living, it is estimated that each person needs roughly 60 m^2 for housing, 40 m^2 for his or her job, 50 m^2 for public buildings and recreation facilities, 90 m^2 for transportation (e.g., highways), and 4000 m^2 for the production of food.

Questions

1. Switzerland has approximately 11 000 km^2 of liveable space (arable and habitable land). How many people can comfortably live in Switzerland? Look up the actual population of Switzerland. (The *CIA World Fact Book*, online, has current statistics for all countries.) Based on the figures given here, is Switzerland overcrowded or is there still room for comfortable growth?

2. What is the population of India? How much liveable space would there have to be in India to accommodate its current population without overcrowding? Now look up the total area of India. Even if all of India is liveable space, is there enough room for the population to live comfortably?

3. You probably knew that India is overcrowded. Pick another country where the answer is not so obvious (Bolivia? Zimbabwe? San Marino?). Describe the "comfort of living" in that country.

Source: Adapted from a problem in E. Batschelet, *Introduction to Mathematics for Life Scientists,* 2nd ed., New York: Springer-Verlag, 1979, p. 31.

 Practise and learn online with Connect.

ANSWERS TO ODD-NUMBERED EXERCISES, ALGEBRA WARM-UPS, REVIEW EXERCISES, AND CHECKUP EXERCISES

CHAPTER 1   Section 1

Algebra Warm-Up

1. $(x - 6)(x - 1)$

2. $(x - 6)(x - 2)$

3. $3(x^2 + 2x - 16)$

4. $(x - 10)(x + 10)$

5. $(6 - 2x)(6 + 2x)$

Exercises

1. $f(0) = 5; f(-1) = 2; f(2) = 11$

3. $f(0) = -2; f(-2) = 0; f(1) = 6$

5. $g(-1) = -2; g(1) = 2; g(2) = \dfrac{5}{2}$

7. $h(2) = 2\sqrt{3}; h(0) = 2; h(-4) = 2\sqrt{3}$

9. $f(1) = 1; f(5) = \dfrac{1}{27}; f(13) = \dfrac{1}{125}$

11. $f(1) = 0; f(2) = 2; f(3) = 2$

13. $h(3) = 10; h(1) = 2; h(0) = 4; h(-3) = 10$

15. Yes

17. No, $f(t)$ is not defined for $t > 1$

19. All real numbers x except $x = -2$

21. All real numbers x for which $x \geq -3$

23. All real numbers t for which $-3 < t < 3$

25. $f(g(x)) = 3x^2 + 14x + 10$

27. $f(g(x)) = x^3 + 2x^2 + 4x + 2$

29. $f(g(x)) = \dfrac{1}{(x - 1)^2}$

31. $f(g(x)) = |x|$

33. $\dfrac{f(x + h) - f(x)}{h} = -5$

35. $\dfrac{f(x + h) - f(x)}{h} = 4 - 2x - h$

37. $\dfrac{f(x + h) - f(x)}{h} = \dfrac{1}{(x + 1)(x + h + 1)}$

39. $f(g(x)) = \sqrt{1 - 3x}; g(f(x)) = 1 - 3\sqrt{x};$
$f(g(x)) = g(f(x))$ if $x = 0$

41. $f(g(x)) = x; g(f(x)) = x; f(g(x)) = g(f(x))$ for all real numbers except $x = 1$ and $x = 2$

43. $f(x - 2) = 2x^2 - 11x + 15$

45. $f(x - 1) = x^5 - 3x^2 + 6x - 3$

47. $f(x^2 + 3x - 1) = \sqrt{x^2 + 3x - 1}$

49. $f(x + 1) = \dfrac{x}{x + 1}$

Note: **In Exercises 51 to 55, answers may vary.**

51. $h(x) = x - 1; g(u) = u^2 + 2u + 3$

53. $h(x) = x^2 + 1; g(u) = \dfrac{1}{u}$

55. $h(x) = 2 - x; g(u) = \sqrt[3]{u} + \dfrac{4}{u}$

57. **a.** $R(x) = -0.02x^2 + 29x;$
$P(x) = -1.45x^2 + 10.7x - 15.6$
b. $P(x) > 0$ if $2 < x < 5.38$

59. **a.** $R(x) = -0.5x^2 + 39x;$
$P(x) = -2x^2 + 29.8x - 67$
b. $P(x) > 0$ if $2.76 < x < 12.14$

61. **a.** $C(10) = 12$
b. $C(10) - C(9) = 1.09$

63. **a.** All real numbers x except $x = 300$
b. All real numbers x for which $0 \leq x \leq 100$
c. $W(50) = 120$ hours
d. $W(100) = 300$ hours
e. $W(x) = 150$ hours implies that $x = 60\%$

65. **a.** All real numbers x except $x = 200$
b. All real numbers x for which $0 \leq x \leq 100$
c. $C(50) = \$50$ million
d. $C(100) - C(50) = \$100$ million
e. $C(x) = 37.5$ million implies that $x = 40\%$

ANSWERS

67. a. $P(9) = \dfrac{97}{5}$; 19 400 people

    b. $P(9) - P(8) = \dfrac{1}{15}$; 67 people

    c. $P(t)$ approaches 20 (20 000 people)

69. a. $S(0) = 25.344$ cm/s

    b. $S(6 \times 10^{-3}) = 19.008$ cm/s

71. a. $s(12) = 2.59$

    b. $s_2 = \sqrt[3]{2}\,s_1$

    c. Approximately 693 050 km^2

73. a. $Q(p(t)) = \dfrac{4.374}{(0.04t^2 + 0.2t + 12)^2}$

    b. $Q(p(10)) = 13.5$ kilograms/week

    c. $t = 0$

75. a. $c(p(t)) = 4.2 + 0.08t^2$

    b. $c(p(2)) = 4.52$ parts per million

    c. 5 years

77. All real numbers x except $x = 1$ and $x = -1.5$

79. $f(g(2.3)) = 6.31$

81. a. $B = 10t$

       $A = 4t + 10$

       $C = 6t + 9$

    b. Corey will finish first.

    c. The intersection point indicates the time at which each person is finished.

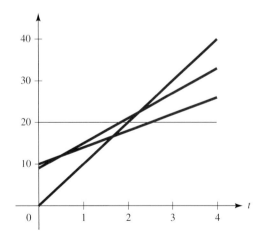

CHAPTER 1   Section 2

1.

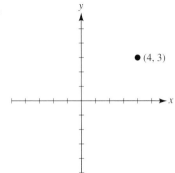

3.

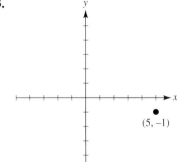

5.

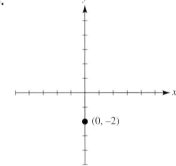

7. $D = 2\sqrt{5}$

9. $D = 2\sqrt{10}$

11. a. Power function

     b. Polynomial

     c. Polynomial

     d. Rational function

13. $f(x) = x$

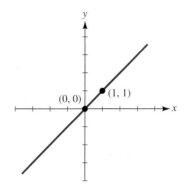

15. $f(x) = \sqrt{x}$

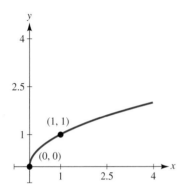

17. $f(x) = 2x - 1$

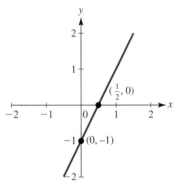

19. $f(x) = x(2x + 5)$

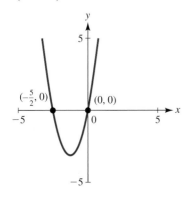

21. $f(x) = -x^2 - 2x + 15$

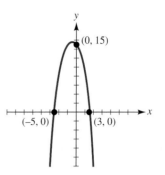

23. $f(x) = x^3$

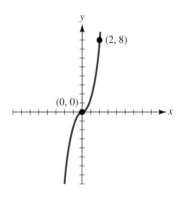

25. $f(x) = \begin{cases} x - 1 & \text{if } x \leq 0 \\ x + 1 & \text{if } x > 0 \end{cases}$

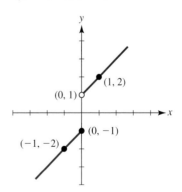

27. $f(x) = \begin{cases} x^2 + x - 3 & \text{if } x < 1 \\ 1 - 2x & \text{if } x \geq 1 \end{cases}$

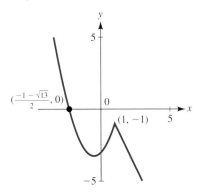

33. $3y - 2x = 5$ and $y + 3x = 9$; $(2, 3)$

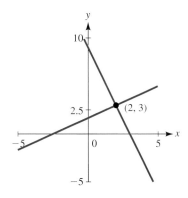

29. $y = 3x + 5$ and $y = -x + 3$; $\left(-\dfrac{1}{2}, \dfrac{7}{2}\right)$

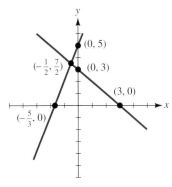

31. $y = x^2$ and $y = 3x - 2$; $(2, 4)$ and $(1, 1)$

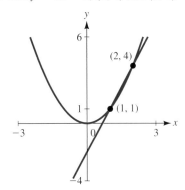

35. **a.** $(0, -1)$
 b. $(1, 0)$
 c. $f(x) = 3$ when $x = 4$
 d. $f(x) = -3$ when $x = -2$

37. **a.** $(0, 2)$
 b. $(-1, 0)$, $(3.5, 0)$
 c. $f(x) = 3$ when $x = 2$
 d. $f(x) = -3$ when $x = 4$

39. $P(p) = (p - 40)(120 - p)$; optimal price is \$80 per backpack.

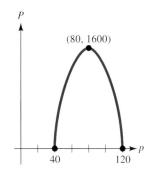

41. $P(x) = (27 - x)(5x - 75)$; optimal price is \$21 per game; 30 sets will be sold each week.

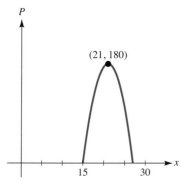

43. a. $E(p) = -200p^2 + 12\ 000p$

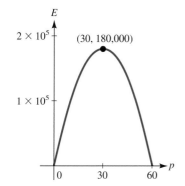

b. The p intercepts represent prices at which consumers spend no money on the commodity.

c. \$30 per unit

45. a. $H(t) = -4.9t^2 + 49t$

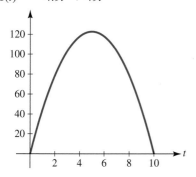

b. After 10 seconds

c. 122.5 m

47. a. $P(x) = -0.07x^2 + 35x - 574.77$

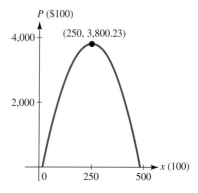

b. 25 000 units; \$25.50

49. $D(v) = 0.0554v^2 + 0.789v$

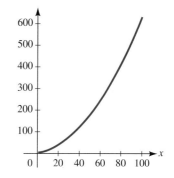

51. a. $R(p) = -0.05p^2 + 210p$

b.

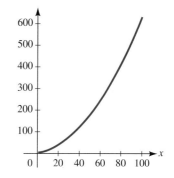

c. \$2100 per month; \$220 500

ANSWERS

53. a.

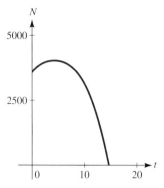

b. 3967 thousand tons

c. 4.27 years after 1990, or in March 1994.

d. No, the formula predicts negative emissions after December 2004.

55. Function

57. Not a function

59. Answers will vary.

61. a. The graph of $y = x^2 + 3$ is the graph of $y = x^2$ shifted upward 3 units.

b.

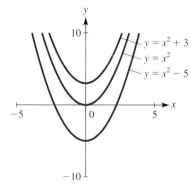

c. The graph of $g(x)$ is the graph of $f(x)$ shifted $|c|$ units upward if $c > 0$ or downward if $c < 0$.

63. a. The graph of $y = (x - 2)^2$ is the graph of $y = x^2$ shifted to the right 2 units.

b.

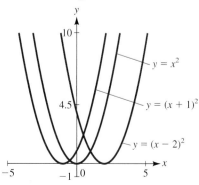

c. The graph of $g(x)$ is the graph of $f(x)$ shifted $|c|$ units to the right if $c > 0$ and to the left if $c < 0$.

65. a.

Days of Training	Mowers per Day
2	6
3	7.23
5	8.15
10	8.69
50	8.96

b. The number of mowers per day approaches 9.

c.

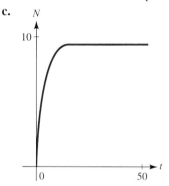

67.

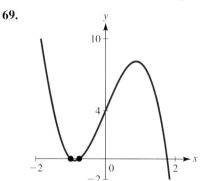

$f(x)$ is defined for $x \neq \dfrac{-1 \pm \sqrt{17}}{8}$.

69.

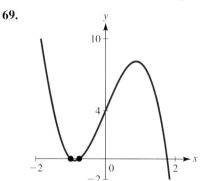

x- intercepts: $x = -1, -0.76, 1.76$.

71. **a.** $(x - 2)^2 + (y + 3)^2 = 16$
    b. Centre: $(2, -3)$; radius; $2\sqrt{6}$
    c. There are no points (x, y) that satisfy the equation.

CHAPTER 1   Section 3

1. $m = -\dfrac{7}{2}$

3. $m = -1$

5. m is undefined.

7. $m = 0$

9. Slope: 2; intercepts: $(0, 0)$; $y = 2x$

11. Slope: $-\dfrac{5}{3}$; intercepts: $(0, 5)$, $(3, 0)$;

    $y = -\dfrac{5}{3}x + 5$

13. Slope: undefined; intercepts: $(3, 0)$

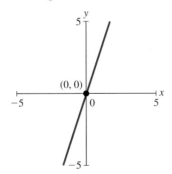

15. Slope: 3; intercepts: $(0, 0)$

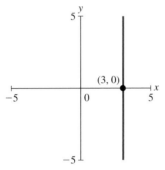

17. Slope: $-\dfrac{3}{2}$, intercepts: $(2, 0)$, $(0, 3)$.

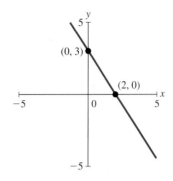

19. Slope: $-\dfrac{5}{2}$, intercepts: $(2, 0)$, $(0, 5)$.

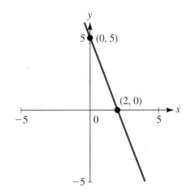

21. $y = x - 2$

23. $y = -\dfrac{1}{2}x + \dfrac{1}{2}$

25. $y = 5$

27. $y = -x + 1$

29. $y = -\dfrac{45}{52}x + \dfrac{43}{52}$

31. $y = 5$

33. $y = -2x + 9$

35. $y = x + 2$

37. $y = C(x) = 60x + 5000$

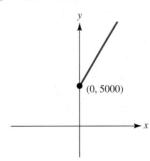

39. a. $y = V(t) = 19t + 125.18$

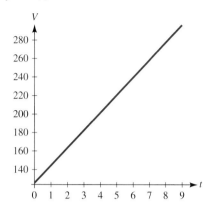

b. $410.18 billion

c. There was a recession in 2009, affecting spending.

41. a. $y = f(t) = 16t + 22$

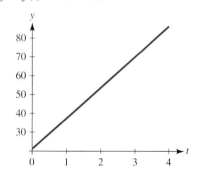

b. 54

c. 22

43. $f(t) = -350t + 3500$

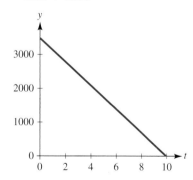

45. a. $y = f(t) = -17t + 1004$

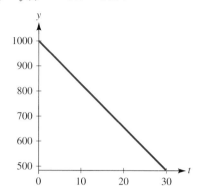

b. $f(8) = 868$ million litres

47.

a.

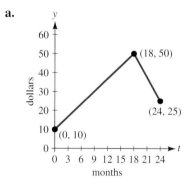

b.

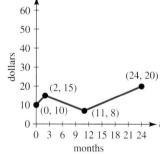

c.

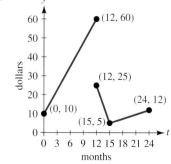

49. a. 95.5 cm

   b. 15.4 years old

   c. 50 cm; yes

   d. 180 cm; yes

51. a. $F = \dfrac{9}{5}C + 32$

   b. 59°F

   c. 20°C

   d. $-40°$ Celsius $= -40°$ Fahrenheit

53. a. $v(1930) = \$800$; $v(1990) = \$51\ 200$; $v(2020) = \$409\ 600$

   b. No, it is not linear.

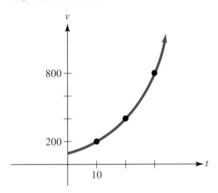

55. a. $t = 0$ in 2000, $y = -3t + 575$

   b. 530

   c. 2016

57. a. $B = \left(\dfrac{S - V}{N}\right)t + V$

   b. $30\ 800

59. The two lines are not parallel; they do not have the same slope.

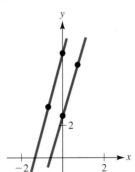

61. a.

Hours rented	2	5	10	t
Total cost	$70.00	$85.00	$110.00	$60 + 5t$

   b. $y = 60 + 5t,\ t \geq 0$

c.

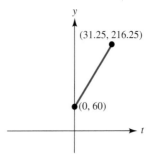

d. Solve $60 + 5t = 216.25$; $t = 31.25$ hours (31 hours, 15 minutes)

63.

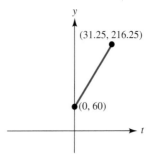

a. The graph does not appear to be linear although there is a definite downward trend.
The main difference between this graph and the graph of Example 1.3.7 is the fact that the slope of this graph is negative, with greenhouse gas per person decreasing with time, where the slope of Example 1.3.7 was positive with total greenhouse gases increasing.

b. Answers will vary.

65. Answers will vary.

CHAPTER 1   Section 4

1. a. $R(x) = 1000x(-6x + 100)$

   b. $150\ 000

3. a. $P(x) = 3x - 17\ 000$

   b. A profit of $43 000; a loss of $2000

5. $S = x + \dfrac{318}{x}$

7. $A = 2w(500 - w)$

9. $A = x(160 - x)$; 80 m by 80 m

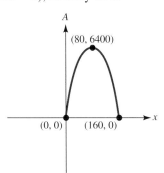

11. $V = x\left(1000 - \dfrac{x^2}{2}\right)$

13. $V = \pi r(60 - r^2)$

15. $C = 0.06\pi\left(2r^2 + \dfrac{70}{r}\right)$

17. $R = kP$; R = rate of population growth; P = size of population

19.

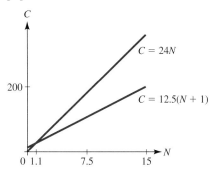

21. **a.** 95.2 mg
b. Since $0.0072(2W)^{0.425}(2H)^{0.725}$
$$= 0.0072(2)^{0.425}(W)^{0.425}(2)^{0.725}(H)^{0.725}$$
$$= (2)^{0.425}(2)^{0.725}0.0072(W)^{0.425}(H)^{0.725}$$
$$\approx 2.22[0.0072(W)^{0.425}(H)^{0.725}]$$

the larger child has 2.22 times the surface area of the smaller. If S is multiplied by 2.22 in the formula

$C = \dfrac{SA}{1.7}$, then C grows by the same factor.

23. $R(x) = \begin{cases} 2400 & \text{if } 1 \le x \le 40 \\ x\left(80 - \dfrac{1}{2}x\right) & \text{if } 40 < x < 80 \\ 40x & \text{if } x \ge 80 \end{cases}$

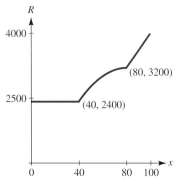

25. **a.** $f(x) = \begin{cases} 0.15x & x \le 40\,726 \\ 0.22x - 2851 & 40\,726 < x \le 81\,452 \\ 0.26x - 6108 & 81\,452 < x \le 126\,264 \\ 0.29x - 9897 & x > 126\,264 \end{cases}$

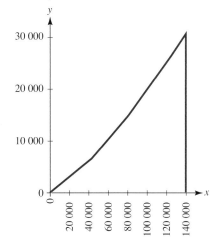

b. The slopes are 0.15, 0.22, 0.26, and 0.29. These represent the increase in amount of income tax paid per dollar increase in earnings.

27. **a.** $V(x) = 20x(160 - x)$

    b. Height: 20 m; sides: 80 m by 80 m

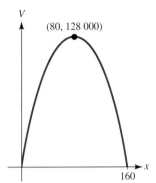

    c. $9 880 000

29. $A(x) = 8x + \dfrac{100}{x} + 57$

31. $V(x) = 4x - \dfrac{x^3}{3}$

33. $P(p) = (53\,000 - 1000p)(p - 29)$

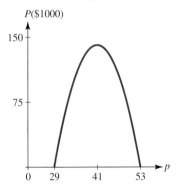

   Optimal price is $41.

35. $C(x) = 20x + \dfrac{5120}{x}$

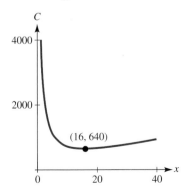

The number of machines is 16.

37. $R(x) = (200 - 1.25x)(4 + 0.03x)$, where $x =$ days from July 1

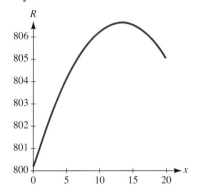

   Farmer should harvest about July 14th.

39. **a.** $x_e = 25;\ p_e = 225$

    b.

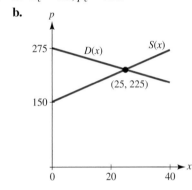

    c. $0 < x < 25;\ x > 25$

41. **a.** $x_e = 9;\ p_e = 25.43$

    b.

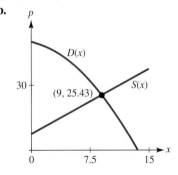

    c. $0 < x < 9;\ x > 9$

43. **a.** $x_e = 10$; $p_e = 35$

b.

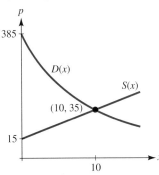

c. $S(0) = 15$; No units will be produced until the price is at least $15.

45. 3 hours 15 min after the first plane leaves.

47. If $N \leq 6000$ or $N \geq 126\,000$, choose publisher A, otherwise publisher B.

49. $I = k\pi r^2$

51. **a.** $M(t) = 0.283t + 47$; $F(t) = 0.314t + 50$

b. Male age is slightly lower than predicted, and female age is slightly higher than predicted.

53. **a.** $a > 0$; $b > 0$; $c < 0$; $d > 0$

b. $q_e = \dfrac{d - b}{a - c}$; $p_e = \dfrac{ad - bc}{a - c}$

c. As a increases, q_e decreases; as d increases, q_e increases.

CHAPTER 1   Section 5

1. $\lim\limits_{x \to a} f(x) = b$

3. $\lim\limits_{x \to a} f(x) = b$

5. Limit does not exist.

7. 4

9. 7

11. 16

13. $\dfrac{4}{7}$

15. Limit does not exist.

17. 2

19. 7

21. $\dfrac{5}{3}$

23. 5

25. $\dfrac{1}{4}$

27. $+\infty$; $-\infty$

29. $-\infty$; $-\infty$

31. $\dfrac{1}{2}, \dfrac{1}{2}$

33. 0; 0

35. $+\infty$; $-\infty$

37. 1; -1

39.

x	1.9	1.99	1.999	2	2.001	2.01	2.1
$f(x)$	1.71	1.9701	1.997001		2.003001	2.0301	2.31

$\lim\limits_{x \to 2} f(x) = 2$

41.

x	0.99	0.999	1	1.001	1.01	1.1
$f(x)$	-197.0299	-1997.002999		2003.003001	203.0301	23.31

$\lim\limits_{x \to 1} f(x)$ does not exist.

43. $\lim\limits_{x \to c} [2f(x) - 3g(x)] = 2(5) - 3(-2) = 16$

45. $\lim\limits_{x \to c} \sqrt{f(x) + g(x)} = \sqrt{5 + (-2)} = \sqrt{3}$

47. $\lim\limits_{x \to c} \dfrac{f(x)}{g(x)} = -\dfrac{5}{2}$

49. $\lim\limits_{x \to \infty} \dfrac{2f(x) + g(x)}{x + f(x)} = \lim\limits_{x \to \infty} \dfrac{\dfrac{2f(x)}{x} + \dfrac{g(x)}{x}}{1 + \dfrac{f(x)}{x}}$

$= \dfrac{0 + 0}{1 + 0} = 0$

51. 1.8 cm

53. **a.** $P(t) = \dfrac{\sqrt{9t^2 + 0.5t + 179}}{0.2t + 1.500}$ thousand dollars per person

b. $\lim\limits_{t \to \infty} P(t) = \$15\,000$ per person

55. $\lim\limits_{s \to +\infty} l(S) = a$. No matter how large the bite size, the required vigilance limits the amount of food intake.

57. $7.50; As the number of units produced increases, the contribution of fixed costs to the average cost decreases to 0.

59. a. Species I: 10 000; species II: 16 000

   b. As t increases, $P(t)$ approaches 0; as t increases to approach 4, $Q(t)$ approaches ∞.

   c.

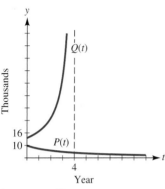

   d. Answers will vary.

61. a. $C(0) = 0.413$ mg/mL

   b. $C(5) - C(4) \approx -0.013$
   The concentration decreases by about 0.013 mg/mL.

   c. $\lim_{t \to \infty} C(t) = 0.013$
   The residual concentration is 0.013 mg/mL.

63. $\lim_{x \to 0} f(x)$ does not exist

65. a. 0

   b. $\dfrac{a_n}{b_m}$

   c. $+\infty$ if a_n and b_m have the same sign, $-\infty$ if a_n and b_m have different signs

CHAPTER 1   Section 6

1. -2; 1; does not exist

3. 2; 2; exists and equals 2

5. 39

7. 0

9. $\dfrac{5}{4}$

11. 0

13. $\dfrac{1}{4}$

15. 15; 0

17. Yes

19. Yes

21. No

23. No

25. No

27. Yes

29. $f(x)$ is continuous for all real numbers x.

31. $f(x)$ is continuous for all real numbers x except $x = 2$.

33. $f(x)$ is continuous for all $x \neq -1$.

35. $f(x)$ is continuous for all $x \neq -3, 6$.

37. $f(x)$ is continuous for all $x \neq 0, 1$.

39. $f(x)$ is continuous for all real numbers x.

41. $f(x)$ is continuous for all $x \neq 0$.

43. a. $WC = -18.7$

   b. Not continuous

45. $P(x)$ is discontinuous at $x = 250$ g, at $x = 500$ g

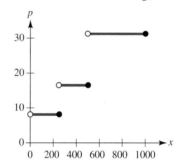

47. The graph is discontinuous at $t = 10$ and $t = 25$. Jan stops at a gas station and purchases some amount of gas at these times.

49. a. 99.76%

   b.

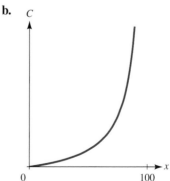

   c. $\lim_{x \to 100^-} C(x) = \infty$; it is not possible to remove all the pollution.

51. **a.** $C(0) \approx 0.333$; $C(100) \approx 7.179$

b. $C(x)$ is not continuous on the interval $0 \le x \le 100$ because $C(80)$ is not defined.

53. $A = 6$

55. $f(x)$ is continuous on the open interval $0 < x < 1$. $f(x)$ is continuous at every point of the closed interval $0 \le x \le 1$ except $x = 0$.

57. Let $f(x) = \sqrt[3]{x - 8} + 9x^{2/3} - 29$. Then $f(x)$ is continuous on $0 \le x \le 8$, $f(0) = -31 < 0$ and $f(8) = 7 > 0$. By the IVT, $f(x) = 0$ for some $0 < x < 8$.

59. **a.** $\lim\limits_{x \to 2} f(x)$ exists at $x = 2$, but $f(x)$ is not continuous there.

b. $\lim\limits_{x \to -2} f(x)$ does not exist, so $f(x)$ is not continuous at $x = -2$.

61. During each hour, the minute hand moves continuously from being behind the hour hand to being ahead. Therefore, at some time, they must coincide.

CHAPTER 1   Checkup

1. All real numbers x such that $-2 < x < 2$

2. $g(h(x)) = \dfrac{2x + 1}{4x + 5}, x \ne -\dfrac{1}{2}$

3. **a.** $y = -\dfrac{1}{2}x + \dfrac{3}{2}$

b. $y = 2x - 3$

4. **a.**

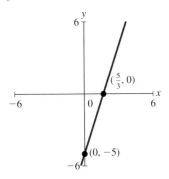

b.

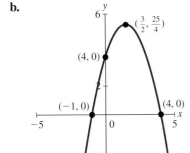

5. **a.** 2

b. 4

c. 1

d. $-\infty$

6. $f(x)$ is not continuous at $x = 1$.

7. **a.** $p(t) = 0.02t + 1.06$ where t is measured from June 1.

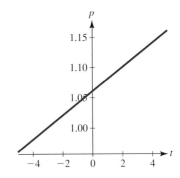

b. $\$0.96$/litre

c. $\$1.16$/ litre

8. $D(t) = 30\sqrt{5r^2 - 20t + 100}$

9. **a.** $A = 3, B = -1$; $\$52$

b.

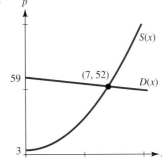

c. $-\$26$; $\$54$

10. **a.** $t = 9$

b. The function $g(t) = f(t) - 10$ is continuous and $g(1) = -2$ and $g(7) = 6$, so by the IVT $g(t) = 0$ or $f(t) = 10$ sometime between $t = 1$ and $t = 7$.

11. $M = 423 - 7.17t$, in 2011

CHAPTER 1   Review Exercises

1. **a.** All real numbers x

b. All real numbers x except $x = 1$ and $x = -2$

c. All real numbers x for which $|x| \ge 3$

3. a. $g(h(x)) = x^2 - 4x + 4$

b. $g(h(x)) = \dfrac{1}{2x + 5}$

5. a. $f(3 - x) = -x^2 + 7x - 8$

b. $f(x^2 - 3) = x^2 - 4$

c. $f(x + 1) - f(x) = \dfrac{-1}{x(x - 1)}$

7. *Note:* Answers will vary.

a. $g(u) = u^5;\ h(x) = x^2 + 3x + 4$

b. $g(u) = (u - 1)^2 + \dfrac{5}{2u^3};\ h(x) = 3x + 2$

9. $f(x) = x^2 + 2x - 8$

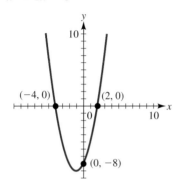

11. a. $m = 3, b = 2$

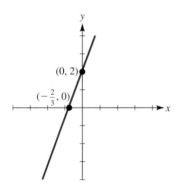

b. $m = \dfrac{5}{4}, b = -5$

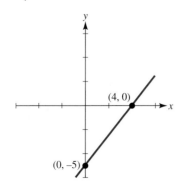

13. a. $y = 5x - 4$

b. $y = -2x + 5$

c. $2x + y = 14$

15. a. $(3, -4)$

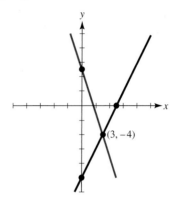

b. No intersection

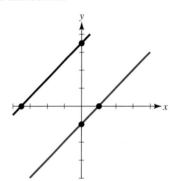

17. $c = -4$

19. $\dfrac{3}{2}$

21. -12

23. Limit does not exist.

25. 0

27. $-\infty$

29. 0

31. 0

33. Not continuous for $x = -3$

35. Continuous for all real numbers x

37. a. $P(5) = \$45$

b. $P(5) - P(4) = -1$
(a \$1 drop)

c. In 9 months

d. The price approaches \$40.

39. a.

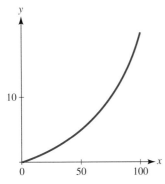

b. 5 weeks

c. 20 weeks

41. $V = \dfrac{4\pi}{3}\left(\sqrt{\dfrac{S}{4\pi}}\right)^3 = \dfrac{S^{3/2}}{6\sqrt{\pi}}$; V is multiplied by $2^{3/2}$.

43. For x machines $C(x) = 80x + \dfrac{11.520}{x}$

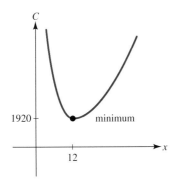

Minimum cost when $x = 12$

45. $P(p) = 2(360 - p)(p - 150)$; optimal price $p = \$255$

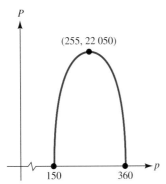

47. Choose Proposition A if $V < 30\,000$, otherwise choose Proposition B.

49. a. 150 units

b. \$1500 profit

c. 180 units

51. $y = k(N - x)$, where y is the recall rate, x is the number of facts that have been recalled, and N is the total number of facts.

53. $C = 1.8x\left(350 - x - \dfrac{\pi}{2}x\right) + 1.5\pi x^2$

55. $C(x) = 1500 + 2x$, for $0 \le x \le 5000$

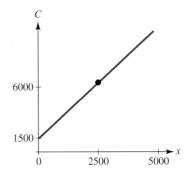

$C(x)$ is continuous for $0 \le x \le 5000$.

57. $A = \dfrac{B}{(6400)^3}$

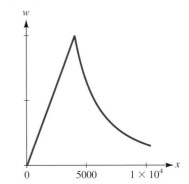

59. The limit exists and is 0.

61.

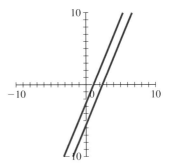

No, because $\dfrac{21}{9} \ne \dfrac{654}{279}$.

63. The function is discontinuous at $x = 1$.

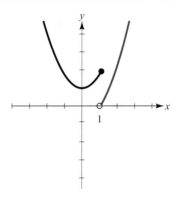

CHAPTER 2   Section 1

1. $f'(x) = 0$; $m = 0$

3. $f'(x) = 5$; $m = 5$

5. $f'(x) = 4x - 3$; $m = -3$

7. $f'(x) = 3x^2$; $m = 12$

9. $g'(t) = -\dfrac{2}{t^2}$; $m = -8$

11. $H'(u) = -\dfrac{1}{2u\sqrt{u}}$; $m = -\dfrac{1}{16}$

13. $f'(x) = 0$; $y = 2$

15. $f'(x) = -2$; $y = -2x + 7$

17. $f'(x) = 2x$; $y = 2x - 1$

19. $f'(x) = \dfrac{2}{x^2}$; $y = 2x + 4$

21. $f'(x) = \dfrac{1}{\sqrt{x}}$; $y = \dfrac{1}{2}x + 2$

23. $f'(x) = -\dfrac{3}{x^4}$; $y = -3x + 4$

25. $\dfrac{dy}{dx} = 0$

27. $\dfrac{dy}{dx} = 3$

29. $\dfrac{dy}{dx} = 3$

31. $\dfrac{dy}{dx} = 2$

33. **a.** $m_{sec} = -3.9$

    b. $m_{tan} = -4$

35. **a.** $m_{sec} = 3.31$

    b. $m_{tan} = 3$

37. **a.** $rate_{ave} = -\dfrac{13}{16}$

    b. $rate_{ins} = -1$

39. **a.** $rate_{ave} = 4$

    b. $rate_{ins} = 8$

41. **a.** The average rate of temperature change between t_0 and $t_0 + h$ hours after midnight. The instantaneous rate of temperature change t_0 hours after midnight.

    b. The average rate of change in blood alcohol level between t_0 and $t_0 + h$ hours after consumption. The instantaneous rate of change in blood alcohol level t_0 hours after consumption.

    c. The average rate of change of the 30-year fixed mortgage rate between t_0 and $t_0 + h$ years after 2005. The instantaneous rate of change of the 30-year fixed mortgage rate t_0 years after 2005.

43. **a.** $V'(30) \approx \dfrac{65 - 50}{50 - 30} = \dfrac{3}{4}$; decreases to 0

    b. The growth rate starts to decrease at this time.

45. Approx. $-0.009°C$/metre; approx. $0°C$/metre

47. **a.** $P'(x) = 4000(17 - 2x)$

    b. $P'(x) = 0$ when $x = \dfrac{17}{2}$ or 850 units. At this level of production, profits are neither increasing nor decreasing.

49. **a.** \$5.94 per unit

    b. \$5.90 per unit; increasing

51. **a.** Answers will vary; approximately -10%

    b. Answers will vary; approximately 2%

53. **a.** $H'(t) = 4.4 - 9.8t$; $H(t)$ is decreasing at the rate of -5.4 m/sec when t is 1 sec.

    b. $H'(t) = 0$ when $t = 0.449$ sec; this is the highest point of the jump.

    c. Lands when $t = 0.898$ sec; -4.4 m/sec; decreasing

55. **a.** 0.0211 mm per mm of mercury

    b. 0.022 mm per mm of mercury; increasing

    c. 72.22 mm of mercury. At this pressure the aortic diameter is neither increasing nor decreasing.

57. **a.** The graph of $y = x^2 - 3$ is the same as the graph of $y = x^2$ shifted down by 3 units. Thus, both curves have the same slope for each x, and their derivatives are the same, both equal to $y' = 2x$.

    b. $y' = 2x$

59. a. $\dfrac{dy}{dx} = 2x; \dfrac{dy}{dx} = 3x^2$

b. $\dfrac{dy}{dx} = 4x^3; \dfrac{dy}{dx} = 27x^{26}$

61. For $x > 0$, we have $f(x) = x$ and

$$f'(x) = \lim_{h \to 0} \frac{(x + h) - (-x)}{h} = 1$$

and for $x < 0$, we have $f(x) = -x$ so that

$$f'(x) = \lim_{h \to 0} \frac{-(x + h) - (-x)}{h} = -1$$

However, the derivative for $x = 0$ would be

$$f'(0) = \lim_{h \to 0} \frac{|0 + h| - 0}{h} = \lim_{h \to 0} \frac{|h|}{h}$$

which does not exist since the two one-sided limits at $x = 0$ are not the same (limit from the left is -1 and from the right is $+1$).

63. $f(x)$ is not continuous at $x = 1$, so it cannot be differentiable there.

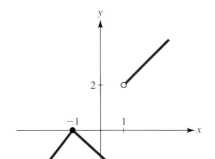

CHAPTER 2   Section 2

Algebra Warm-Up

1. $2x^{-3}$

2. $\dfrac{2}{3}x^{-1/2}$

3. $\dfrac{2}{5}x^{1/2}$

4. $2x^{-1/3}$

5. $7x^{-5} - 10x^{1/2} + 6x^4$

Exercises

1. $\dfrac{dy}{dx} = 0$

3. $\dfrac{dy}{dx} = 5$

5. $\dfrac{dy}{dx} = -4x^{-5}$

7. $\dfrac{dy}{dx} = 3.7x^{2.7}$

9. $\dfrac{dy}{dr} = 2\pi r$

11. $\dfrac{dy}{dx} = \dfrac{\sqrt{2}}{2\sqrt{x}}$

13. $\dfrac{dy}{dt} = \dfrac{-9}{2\sqrt{t^3}}$

15. $\dfrac{dy}{dx} = 2x + 2$

17. $f'(x) = 9x^8 - 40x^7 + 1$

19. $f'(x) = -0.06x^2 + 0.3$

21. $\dfrac{dy}{dt} = -\dfrac{1}{r^2} - \dfrac{2}{t^3} + \dfrac{1}{2\sqrt{t^3}}$

23. $f'(x) = \dfrac{3}{2}\sqrt{x} - \dfrac{3}{2\sqrt{x^5}}$

25. $\dfrac{dy}{dx} = -\dfrac{x}{8} - \dfrac{2}{x^2} - \dfrac{3}{2}x^{1/2} - \dfrac{2}{3x^3} + \dfrac{1}{3}$

27. $\dfrac{dy}{dx} = 2x + \dfrac{4}{x^2}$

29. $y = 10x + 2$

31. $y = -\dfrac{1}{16}x + 2$

33. $y = x + 3$

35. $y = -4x - 1$

37. $y = 3x - 3$

39. $y = -3x + \dfrac{22}{3}$

41. $f'(-1) = -5$

43. $f'(1) = -\dfrac{3}{2}$

45. $f'(1) = \dfrac{1}{2}$

47. The population decreases at 4000 people/month after 1 month.

49. The plant grows at a rate of 3.5 cm/week after 1 week.

51. **a.** $10 800 per year
   b. 17.53%

53. **a.** $f'(x) = -6$
   b. The change in the average SAT score is the same each year. The average SAT score decreases each year.

55. **a.** $T'(0) = \$40$ per year
   b. $480

57. $P'(x) = 2 + 6x^{1/2}$
   a. $P'(9) = 20$ persons per month
   b. 0.39%

59. Approximately 2435 people per day.

61. **a.** 0.2 parts per million per year
   b. 0.15 parts per million
   c. 0.4 parts per million

63. **a.** $f'(t) = 2000$ \$/year
   b. % rate of change $= \dfrac{100(2000)}{45\,000 + 2000t} = \dfrac{200}{45 + 2t}\%$

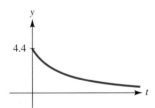

65. **a.** $C(x) = 4.2\left(\dfrac{1920}{x} + \dfrac{5x}{8}\right) + \dfrac{12\,000}{x}$
   b. $C'(64) = -2.27$ cents/km/hour; decreasing

67. **a.** $v(t) = 6t + 2;\ a(t) = 6$
   b. Not stationary

69. **a.** $v(t) = 4t^3 - 12t^2 + 8;\ a(t) = 12t^2 - 24t$
   b. $t = 1,\ 1 + \sqrt{3}$

71. **a.** 9.8 m/s
   b. 39.2 m
   c. -9.8 m/s
   d. 29.4 m/s

73. $a = -1, b = 5, c = 0$

75. $(f + g)'(x)$

$$= \lim_{h \to 0} \frac{(f + g)(x + h) - (f + g)(x)}{h}$$

$$= \lim_{h \to 0} \frac{[f(x + h) + g(x + h)] - [f(x) + g(x)]}{h}$$

$$= \lim_{h \to 0} \frac{[f(x + h) - f(x)] + [g(x + h) - g(x)]}{h}$$

$$= \lim_{h \to 0} \left[\frac{f(x + h) - f(x)}{h} + \frac{g(x + h) - g(x)}{h}\right]$$

$$= \lim_{h \to 0} \frac{f(x + h) - f(x)}{h} + \lim_{h \to 0} \frac{g(x + h) - g(x)}{h}$$

$$= f'(x) + g'(x)$$

77. **a.** 2005 to 2008, approximately
   b. partway through 2003, at beginning 2004, and 2008
   c. slope approximately 3/4%/year; in 2101, 90% will be recycled (answers will vary)
   d. answers will vary

CHAPTER 2   Section 3

1. $f'(x) = 12x - 1$

3. $\dfrac{dy}{du} = -300u - 20$

5. $f'(x) = \dfrac{1}{3}\left(6x^5 - 12x^3 + 4x + 1 + \dfrac{1}{x^2}\right)$

7. $\dfrac{dy}{dx} = \dfrac{-3}{(x - 2)^2}$

9. $f'(t) = \dfrac{-(t^2 + 2)}{(t^2 - 2)^2}$

11. $\dfrac{dy}{dx} = \dfrac{-3}{(x + 5)^2}$

13. $f'(x) = \dfrac{11x^2 - 10x - 7}{(2x^2 + 5x - 1)^2}$

15. $f'(x) = \dfrac{2(x^2 + 2x + 4)}{(x + 1)^2}$

17. $f'(x) = 10(2 + 5x)$

19. $g'(t) = \dfrac{4\sqrt{t^5} + 20\sqrt{t^3} - 2t + 5}{2\sqrt{t}(2t + 5)^2}$

21. $y = 16x - 4$

23. $y = 3x + 2$

25. $y = -\dfrac{11}{2}x + \dfrac{19}{2}$

27. $(1, -4), (-1, 0)$

29. $(0, 1), \left(-2, -\dfrac{1}{3}\right)$

31. $(5, 0), (3, 108), (0, 0)$

33. -18

35. 4

37. $y = \dfrac{2}{5}x + \dfrac{3}{5}$

39. $y = \dfrac{1}{31}(-x - 371)$

41. **a.–d.** $y' = \dfrac{9 - 4x}{x^4}$

43. $f''(x) = 8x^3 - 24x + 18$

45. $\dfrac{d^2y}{dx^2} = \dfrac{4}{3x^3} + \dfrac{\sqrt{2}}{4x^{3/2}} - \dfrac{1}{8x^{5/2}}$

47. $\dfrac{d^2y}{dx^2} = 36x^2 + 30x + 12$

49. **a.** $S'(2) = 378.07$
  b. Sales approach a limit of \$6 666 666.67.

51. **a.** $P'(5) = \dfrac{1900}{441} \approx 4.3\%$ increase per week.
  b. $P(t)$ approaches 100% in the long run; the rate of change approaches 0.

53. **a.** $P'(16) \approx -0.631$; decreasing
  b. Increasing for $0 < x < 10$; decreasing for $x > 10$

55. **a.** $R(t) = -3t^2 + 16t + 15$
  b. 10 units per hour per hour

57. **a.** $v(t) = 15t^4 - 15t^2$; $a(t) = 60t^3 - 30t$
  b. $a(t) = 0$ when $t = 0, \dfrac{\sqrt{2}}{2}$

59. **a.** $v(t) = -3t^2 + 14t + 1$
  $a(t) = -6t + 14$
  b. $a(t) = 0$ when $t = \dfrac{7}{3}$

61. **a.** 9.8 metres per minute
  b. 9.83 metres

63. **a.** $a(t) = -9.8$ m/s^2
  b. $a(t)$ is constant.
  c. The object is accelerating downward.

65. $\dfrac{3}{8x^{5/2}} + \dfrac{3}{x^4}$

67. **a.** $\dfrac{d}{dx}\left(\dfrac{fg}{h}\right) = \dfrac{hfg' + hf'g - fgh'}{h^2}$
  b. $\dfrac{dy}{dx} = \dfrac{12x^3 + 51x^2 + 70x - 33}{(3x + 5)^2}$

71.

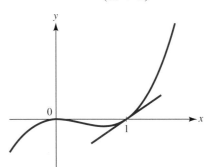

$f'(x) = 0$ where $x = 0$ and $x = \dfrac{2}{3}$.

73.

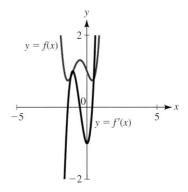

The x intercepts of $f'(x)$ occur at $x = -0.5$,

$x = -\dfrac{1}{2} + \dfrac{\sqrt{3}}{2} \approx 0.366$, and

$x = -\dfrac{1}{2} - \dfrac{\sqrt{3}}{2} \approx -1.366$. The x intercepts are

those points where $f'(x) = 0$, which are the points where the tangent to the graph of $f(x)$ is horizontal, that is, the maxima and minima of $f(x)$.

CHAPTER 2   Section 4

Algebra Warm-Up

  1. $x^2(2x - 1)^4(16x - 1)$
  2. $2x(6x^2(3x + 1)^3 + (2x - 1)^4)$
  3. $5(2 - 5x)^3(3x - 1)^4(10 - 27x)$
  4. $-4(7 - 4x)^2(3 - 2x)^3(24x^2 - 78x + 65)$

Exercises

1. $\dfrac{dy}{dx} = 6(3x - 2)$

3. $\dfrac{dy}{dx} = \dfrac{x + 1}{\sqrt{x^2 + 2x - 3}}$

5. $\dfrac{dy}{dx} = \dfrac{-4x}{(x^2 + 1)^3}$

7. $\dfrac{dy}{dx} = \dfrac{-2x}{(x^2 - 1)^2}$

9. $\dfrac{dy}{dx} = 1 + \dfrac{1}{\sqrt{x}}$

11. $\dfrac{dy}{dx} = -\dfrac{2 + x}{x^3}$

13. 20

15. -160

17. $\dfrac{2}{3}$

19. -16

21. $f'(x) = 2.8(2x + 3)^{0.4}$

23. $f'(x) = 8(2x + 1)^3$

25. $f'(x) = 8x^2(x^5 - 4x^3 - 7)^7(5x^2 - 12)$

27. $f'(t) = \dfrac{-2(5t - 3)}{(5t^2 - 6t + 2)^2}$

29. $g'(x) = \dfrac{-4x}{(4x^2 + 1)^{3/2}}$

31. $f'(x) = \dfrac{24x}{(1 - x^2)^5}$

33. $h'(s) = \dfrac{15(1 + \sqrt{3s})^4}{2\sqrt{3s}}$

35. $f'(x) = (x + 2)^2(2x - 1)^4(16x + 17)$

37. $G'(x) = -\dfrac{5}{2}(3x + 1)^{-1/2}(2x - 1)^{-3/2}$

39. $f'(x) = \dfrac{(x + 1)^4(9 - x)}{(1 - x)^5}$

41. $f'(y) = \dfrac{5 - 6y}{(1 - 4y)^{3/2}}$

43. $y = \dfrac{3}{4}x + 2$

45. $y = -48x - 32$

47. $y = -12x + 13$

49. $y = \dfrac{2}{3}x - \dfrac{1}{3}$

51. $x = 0; x = -1; x = -\dfrac{1}{2}$

53. $x = -\dfrac{2}{3}$

55. $x = 2$

57. $f'(x) = 6(3x + 5)$

59. $f''(x) = 180(3x + 1)^3$

61. $h''(t) = 80(t^2 + 5)^6(3t^2 + 1)$

63. $f''(x) = (1 + x^2)^{-3/2}$

65. **a.** $2295 per year
   b. 10.4% per year

67. **a.** -12 kilograms per dollar
   b. $(-12)(0.5) = -6$ kilograms per week; decreasing

69. **a.** 0.4625 ppm per thousand people
   b. 0.308 ppm per year; increasing

71. **a.** $L'(w) = 0.65w^{1.6}$; $L'(60) \approx 455$ mm/kg
   b. A 100-day-old tiger weighs $w(100) = 24$ kg and is $L(24) \approx 969$ mm long. By the chain rule,
   $$L'(A) = L'(w)w'(A) = (0.65w^{1.6})(0.21)$$
   so that when $A = 100$, $w = 24$, we have
   $$L'(100) = (0.65)(0.21)(24)^{1.6} \approx 22.1$$
   That is, the tiger's length is increasing at the rate of about 22.1 mm per day.

73. **a.** Decreasing at about 0.2254% per day
   b. Increasing
   c. Evntually the oxygen proportion returns to its typical level.

75. **a.** $64 000; 8000 units
   b. 6501 units per month; increasing

77. **a.** $V'(T) = 0.41(-0.02T + 0.4)$
   b. $m'(V) = \dfrac{0.39}{(1 + 0.09V)^2}$
   c. $V(10) = 2.6732$ cm³; 0.02078 g/°C

79. $\dfrac{dT}{dL} = \dfrac{a(3L - 2b)}{2\sqrt{L - b}}$ As L approaches b, the denominator of the rate becomes smaller and the rate of learning new words becomes very steep.

81. **a.** $v(t) = \dfrac{3}{2}(1 - 2t)(3 + t - t^2)^{1/2}$
   $$a(t) = \dfrac{24t^2 - 24t - 33}{4(3 + t - t^2)^{1/2}}$$
   b. $t = \dfrac{1}{2}; s\left(\dfrac{1}{2}\right) = \dfrac{13\sqrt{13}}{8} \approx 5.86$
   $$a\left(\dfrac{1}{2}\right) = \dfrac{-3\sqrt{13}}{2} \approx -5.41$$

ANSWERS

c. $a(t) = 0$ when $t = \dfrac{2 + \sqrt{26}}{4} \approx 1.775$

$s(1.775) \approx 2.07; v(1.775) \approx -4.875$

d.

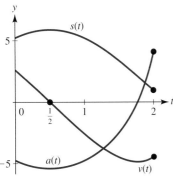

e. The object is slowing down for $0 \le t < 0.5$ and $1.775 < t \le 2$.

83. $\dfrac{dy}{dx} = \dfrac{d}{dx}[h(x)h(x)] = h(x)\dfrac{dh(x)}{dx} + \dfrac{dh(x)}{dx}h(x)$

$= 2h(x)\dfrac{dh(x)}{dx}$

85. One horizontal tangent, $y = 2.687$, where $x = 0$.

CHAPTER 2   Section 5

1. a. $C'(x) = \dfrac{2}{5}x + 4; R'(x) = 9 - \dfrac{x}{2}$

b. $C'(3) = \$5.20$

c. $C(4) - C(3) = \$5.40$

d. $R'(3) = \$7.50$

e. $R(4) - R(3) = \$7.25$

3. a. $C'(x) = \dfrac{2}{3}x + 2; R'(x) = -3x^2 - 8x + 80$

b. $C'(3) = 4$

c. $C(4) - C(3) \approx 4.33$

d. $R'(3) = 29$

e. $R(4) - R(3) = 15$

5. a. $C'(x) = \dfrac{x}{2}; R'(x) = \dfrac{2x^2 + 4x + 3}{(1 + x)^2}$

b. $C'(3) = \$1.50$

c. $C(4) - C(3) = \$1.75$

d. $R'(3) = \dfrac{33}{16} \approx \$2.06$

e. $R(4) - R(3) = \$2.05$

7. 2.1

9. $\dfrac{100f'(4)}{f(4)}(0.3) \approx 0.20; 20\%$

11. a. $C'(3) = \$499.70$

b. $C(4) - C(3) = \$500.20$

13. a. $\$241$

b. $\$244$

15. 200

17. Revenue will decrease by approximately $\$150.80$.

19. Daily output will increase by approximately 8 units.

21. 5.12%

23. Daily output will increase by approximately 825 units.

25. 0.2 units

27. $\dfrac{3c}{|c - b|}$

29. 46.67%

31. 0.0385 ft

CHAPTER 2   Section 6

Algebra Warm-Up

1. $y' = \dfrac{3x - 1 - 2xy}{x^2 + 4}$

2. $y' = \dfrac{7y - 3x^2y^3 - 2}{3x^3y^2 - 7x}$

3. $y' = \dfrac{2xy - 3y^2 - 9}{5 - x^2 - 6xy}$

4. $\dfrac{dy}{dx} = \dfrac{2xy}{1 - x^2}$

5. $\dfrac{dy}{dx} = \dfrac{2y - 15x^2y - 12}{5x^3 + 2x}$

Exercises

1. $\dfrac{dy}{dx} = -\dfrac{2}{3}$

3. a. $\dfrac{dy}{dx} = \dfrac{3x^2}{2y} = \dfrac{3x^2}{2\sqrt{x^3 - 5}}$

b. $\dfrac{dy}{dx} = \dfrac{3x^2}{2\sqrt{x^3 - 5}}$

5. a. $\dfrac{dy}{dx} = -\dfrac{y}{x} = \dfrac{-(4/x)}{x}$

b. $\dfrac{dy}{dx} = -\dfrac{4}{x^2}$

7. a. $\dfrac{dy}{dx} = \dfrac{-y}{x + 2} = \dfrac{-[3/(x + 2)]}{x + 2}$

b. $\dfrac{dy}{dx} = \dfrac{-3}{(x + 2)^2}$

9. $\dfrac{dy}{dx} = -\dfrac{x}{y}$

11. $\dfrac{dy}{dx} = \dfrac{y - 3x^2}{3y^2 - x}$

13. $\dfrac{dy}{dx} = \dfrac{3 - 2y^2}{2y(1 + 2x)}$

15. $\dfrac{dy}{dx} = -\dfrac{\sqrt{y}}{\sqrt{x}}$

17. $\dfrac{dy}{dx} = \dfrac{y - 1}{1 - x}$

19. $\dfrac{dy}{dx} = \dfrac{1}{3(2x + y)^2} - 2$

21. $\dfrac{dy}{dx} = \dfrac{y - 5x(x^2 + 3y^2)^4}{15y(x^2 + 3y^2)^4 - x}$

23. $y = \dfrac{1}{3}x + \dfrac{4}{3}$

25. $y = -\dfrac{1}{2}x + 2$

27. $y = \dfrac{5}{8}x - \dfrac{9}{4}$

29. $y = \dfrac{13}{12}x + \dfrac{11}{12}$

31. **a.** None
    b. $(9, 0)$

33. **a.** None
    b. $(0, 0)$ and $(64, 2)$

35. **a.** $(1, -2), (-1, 2)$
    b. $(-2, 1), (2, -1)$

37. $\dfrac{d^2y}{dx^2} = \dfrac{-3y^2 - x^2}{9y^3} = \dfrac{-5}{9y^3}$

39. $\Delta y \approx -1.704$

41. $\dfrac{dx}{dt} = 1.74$ or 174 units per month

43. $\dfrac{dx}{xt} = 0.15419$ or 15.419 units per month

45. $\dfrac{dR}{dt} = 20$ mm/min

47. 0.476 cm^3 per month

49. **a.** 14.04
    b. -9.87

51. $\dfrac{dK}{dt} = -\dfrac{2}{25}$ ($1000) $= -$400 per week

53. 1.5 m/s

55. $\Delta y \approx 0.566$ units

57. Since $v = \dfrac{KR^2}{L} = $ constant at the centre of the vessel $(r = 0)$, we have

$$0 = v'(t) = K\left[\dfrac{2R}{L}R' - \dfrac{R^2}{L^2}L'\right]$$

Thus,

$$\dfrac{2R}{L}R' = \dfrac{R^2}{L^2}L'$$

$$\dfrac{L'}{L} = 2\dfrac{R'}{R}$$

so the relative rate of change of L is twice that of R.

59. $\dfrac{x^2}{a^2} + \dfrac{y^2}{b^2} = 1; \dfrac{2x}{a^2} + \dfrac{2yy'}{b^2} = 0; 2b^2x + 2a^2yy' = 0,$

$y' = -\dfrac{2b^2x}{2a^2y} = -\dfrac{b^2x}{a^2y}.$ At $P(x_0, y_0), m = -\dfrac{b^2x_0}{a^2y_0}$

so the equation of the tangent line is

$$y - y_0 = -\dfrac{b^2x_0}{a^2y_0}(x - x_0)$$

$$a^2yy_0 - a^2y_0^2 = -b^2xx_0 + b^2x_0^2$$

$$b^2xx_0 + a^2yy_0 = b^2x_0^2 + a^2y_0^2$$

$$\dfrac{x_0x}{a^2} + \dfrac{y_0y}{b^2} = \dfrac{x_0^2}{a^2} + \dfrac{y_0^2}{b^2} = 1$$

since $P(x_0, y_0)$ lies on the curve and thus satisfies the equation of the curve.

61. Let $y = x^{r/s}$, then $y^s = x^r$ and $sy^{s-1}\dfrac{dy}{dx} = rx^{r-1},$

$\dfrac{dy}{dx} = \dfrac{rx^{r-1}}{sy^{s-1}},$ But $y^{s-1} = \dfrac{y^s}{x^{r/s}} = \dfrac{x^r}{x^{r/s}},$ so

$$\dfrac{dy}{dx} = \dfrac{r}{s} \cdot x^{r-1} \cdot \dfrac{x^{r/s}}{x^r}$$

$$= \dfrac{r}{s} \cdot x^{r-1+r/s-r}$$

$$= \dfrac{r}{s} \cdot x^{r/s-1}$$

63. Horizontal tangent lines are $y = 1.24$ and $y = -1.24$.

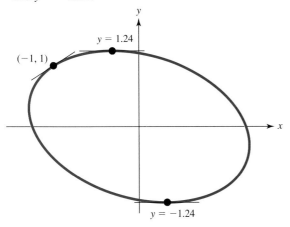

65. Horizontal tangent lines are $y = 1.23$ and $y = -1.23$.

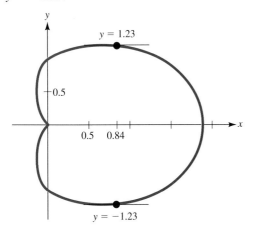

Algebra Insight

1. -7
2. $8(x + 4)^2(x + 1)$
3. $(x + 1)(3 - x)$
4. $\dfrac{-10y^3(1 - 2xy^3)^4 - 1}{4 + 30xy^2(1 - 2xy^3)^4}$
5. $-2(t + 4)(t - 3)$

CHAPTER 2   Checkup

1. **a.** $\dfrac{dy}{dx} = 12x^3 - \dfrac{2}{\sqrt{x}} - \dfrac{10}{x^3}$

   b. $\dfrac{dy}{dx} = -15x^4 + 39x^2 - 2x - 4$

c. $\dfrac{dy}{dx} = \dfrac{-10x^2 + 10x + 1}{(1 - 2x)^2}$

d. $\dfrac{dy}{dx} = (9x - 6)(3 - 4x + 3x^2)^{1/2}$

2. $f''(t) = 24t + 8$
3. $y = -4x$
4. $\dfrac{3}{8}$
5. **a.** 58 dollars per year
   b. 2.98%
6. **a.** $v(t) = 6t^2 - 6t$; $a(t) = 12t - 6$; $v(2) = 12$ m/s; $a(2) = 18$ m/s^2
   b. $t = 0, 1$; retreating $0 < t < 1$; advancing $t > 1$.
   c. 4
7. **a.** $C'(5) = 5.4(\$100) = \$540$
   b. $C(6) - C(5) = 5.44(\$100) = \$544$
8. Output increases by approximately $\dfrac{75\,000}{7}$ units
9. 0.001586 m^2 per week
10. **a.** 36π cm^3 per cm
    b. $\dfrac{8}{3}\%$

CHAPTER 2   Review Exercises

1. $f'(x) = 2x - 3$
3. $f'(x) = 24x^3 - 21x^2 + 2$
5. $\dfrac{dy}{dx} = \dfrac{-14x}{(3x^2 + 1)^2}$
7. $f'(x) = 10(20x^3 - 6x + 2)(5x^4 - 3x^2 + 2x + 1)^9$
9. $\dfrac{dy}{dx} = 2\left(x + \dfrac{1}{x}\right)\left(1 - \dfrac{1}{x^2}\right) + \dfrac{5}{2\sqrt{3x^3}}$
11. $f'(x) = 3\sqrt{6x + 5} + \dfrac{9x + 3}{\sqrt{6x + 5}}$
13. $\dfrac{dy}{dx} = \dfrac{-7}{2(3x + 2)^2}\sqrt{\dfrac{3x + 2}{1 - 2x}}$
15. $y = -x - 1$
17. $y = -\dfrac{2}{3}x + \dfrac{5}{3}$
19. **a.** $f'(0) = 0$
    b. $f'(1) = -\dfrac{1}{4}$

21. a. -400%

b. -100%

23. a. $\dfrac{dy}{dx} = -2(2 - x)$

b. $\dfrac{dy}{dx} = -\dfrac{1}{(2x + 1)^{3/2}}$

25. a. 2

b. $\dfrac{3}{2}$

27. a. $f''(x) = 24x$

b. $f''(x) = 24(x + 4)(x + 2)$

c. $f''(x) = \dfrac{2(x - 5)}{(x + 1)^4}$

29. a. $\dfrac{dy}{dx} = -\dfrac{2y}{x}$

b. $\dfrac{dy}{dx} = -\left[\dfrac{1 + 10y^3(1 - 2xy^3)^4}{4 + 30xy^2(1 - 2xy^3)^4} \right]$

31. a. $m = -\dfrac{5}{9}$

b. $m = -1$

33. $\dfrac{d^2y}{dx^2} = \dfrac{6y^2 - 9x^2}{4y^3} = \dfrac{-9}{2y^3}$

35. a. 75 000 per year

b. 0 per year

37. a. $v(t) = \dfrac{-2(t + 4)(t - 3)}{(t^2 + 12)^2}$,

$a(t) = \dfrac{2(2t^3 + 3t^2 - 72t - 12)}{(t^2 + 12)^3}$.

The object is advancing for $0 < t < 3$, retreating for $3 < t < 4$. It is always decelerating for $0 < t < 4$.

b. $\dfrac{11}{42}$

39. a. Output will increase by approximately 12 000 units.

b. Output will increase by 12 050 units.

41. Output will decrease by approximately 5000 units per day.

43. Pollution will increase by approximately 10%.

45. a. 0.2837 individuals per square kilometre

b. 55 years; 60.67 animals per year

47. 1.5%

49. a. $\dfrac{dF}{dC} = \dfrac{-kD^2}{2\sqrt{A - C}}$; F decreases as C increases

b. $\dfrac{50}{A - C}\%$

51. $425.25 \le A \le 479.53$; accurate to 6%

53. $100\dfrac{\Delta Q}{Q} \approx 0.67\%$

55. $17.01 \le S \le 19.18$; accurate to 6%

57. decreasing by 2 toasters per month

59. $P(3) = 215$ fish after 3 months

$P'(3) = 3.5$ fish/year growth rate

61. 5 seconds; 240 m

63. a. $-\$30$/month

b. $-\$20$ per month per month

65. $-\$99$ per month

67. 456 cm^3

69. 1.1 m/s

71. 0.804 m/s

73. -1.26 m/s

75. $\dfrac{8}{5}$ m/s^2

77. The percentage rate of change approaches 0 since, if $y = mx + b$, $\dfrac{100y'}{y} = \dfrac{100m}{mx + b}$, which approaches 0 as x approaches ∞.

79.

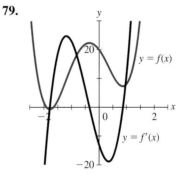

$f'(x) = 0$ when $x \approx -1.78, -0.35, 0.88$

81. a.

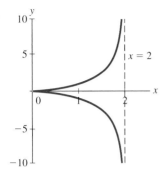

b. At (1, 1), the tangent line is $y = 2x - 1$ and at (1, −1), it is $y = -2x + 1$.

c. As $x \to 2^-$, the top branch of the curve rises indefinitely ($y \to +\infty$) while the bottom branch falls indefinitely ($y \to -\infty$).

d. The x axis ($y = 0$) is a double tangent line to the curve at the origin.

CHAPTER 3 Section 1

1. $f'(x) > 0$ for $-2 < x < 2$; $f'(x) < 0$ for $x < -2$ and $x > 2$

3. $f'(x) > 0$ for $x < -4$ and $0 < x < 2$; $f'(x) < 0$ for $-4 < x < -2$, $-2 < x < 0$, and $x > 2$

5. B

7. D

9. $f(x)$ is increasing for $x > 2$; $f(x)$ is decreasing for $x < 2$.

11. $f(x)$ is increasing for $x < -1$ and $x > 1$; $f(x)$ is decreasing for $-1 < x < 1$.

13. $g(t)$ is increasing for $t < 0$ and $t > 4$; $g(t)$ is decreasing for $0 < t < 4$.

15. $f(t)$ is increasing for $0 < t < 2$ and $t > 2$; $f(t)$ is decreasing for $t < -2$ and $-2 < t < 0$.

17. $h(u)$ is increasing for $-3 < u < 0$; $h(u)$ is decreasing for $0 < u < 3$.

19. $F(x)$ is increasing for $x < -3$ and $x > 3$; $F(x)$ is decreasing for $-3 < x < 0$ and $0 < x < 3$.

21. $f(x)$ is increasing for $x > 1$; $f(x)$ is decreasing for $0 < x < 1$.

23. $x = 0, 1$; (0, 2) relative minimum; (1, 3) neither

25. $t = -1$; (−1, 3) neither

27. $x = 1$; (1, 0) neither

29. $t = -\sqrt{3}, \sqrt{3}$; $\left(\sqrt{3}, \dfrac{\sqrt{3}}{6}\right)$ relative maximum; $\left(-\sqrt{3}, -\dfrac{\sqrt{3}}{6}\right)$ relative maximum

31. $t = -2, 0, 1, 4$; (0, 0) relative maximum; $\left(4, \dfrac{8}{9}\right)$ relative maximum

33. $t = 0, -1, 1$; (0, 1) relative maximum; (−1, 0) and (1, 0) relative minima

35.

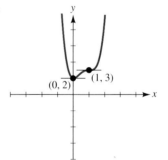

37.

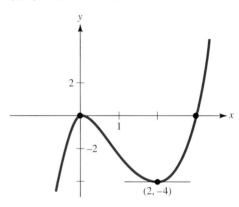

39.

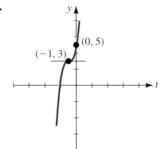

41.

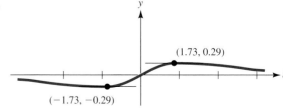

43.

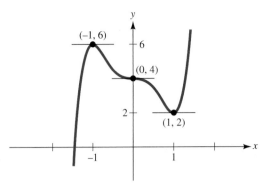

45.

Critical Numbers	Classification
-2	Relative minimum
0	Neither
2	Relative maximum

47.

Critical Numbers	Classification
-1	Neither
$\dfrac{4}{3}$	Relative maximum

49. One possibility:

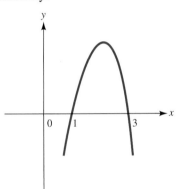

51. One possibility:

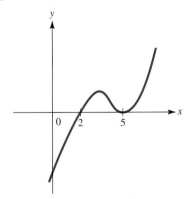

53. **a.** $A'(x) = 2x - 20 - \dfrac{242}{x^2}$

b. Increasing: $x > 11$; decreasing: $0 \le x < 11$

c. Average cost is minimized when $x = 11$ minimum average cost: \$102 000/unit.

55. $R(x) = x(10 - 3x)^2$; $\dfrac{dR}{dx} = (10 - 3x)(10 - 9x)$;

Revenue is maximized when $x = \dfrac{10}{9}$,

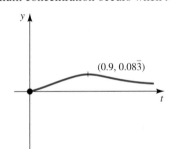

57. Maximum concentration occurs when $t = 0.9$ hours.

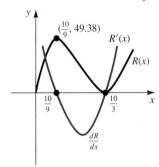

59. **a.**

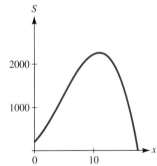

b. 207

c. \$11 000; 2264 units

61. **a.** $1 \le r \le 5495$

b. 5.495%; 1137 mortgages

63. **a.** January; there was a recession in 2009
  b. February–March 2007
  c. Approx. 3250/month
  d. August–September 2005

65. **a.** $Y(t) = \dfrac{9300}{31 + t}(3 + t - 0.05t^2)$

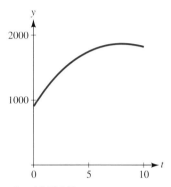

  b. 8 weeks; 1860 kilograms

67. $H(t)$

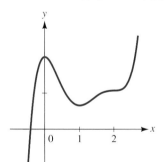

Maximum of 85.81% at 23.58°C.

69.

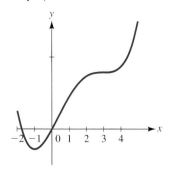

71.

73. $a = 2; b = 3; c = -12; d = -12$

75.

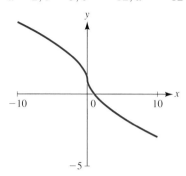

77. By the product rule, $\dfrac{dy}{dx} = (x - p)(1) + (1)(x - q)$

$= 2x - p - q$. Solving $\dfrac{dy}{dx} = 0$ yields $x = \dfrac{p + q}{2}$, the point midway between the x intercepts p and q.

79. $f'(x) = 0$ at $x = -3$; $-2.529, -1.618, -0.346, 0.618$

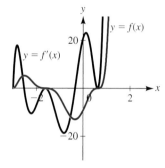

81. $f'(x)$ is never 0.

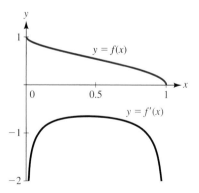

CHAPTER 3 Section 2

1. $f''(x) > 0$ for $x > 2$; $f''(x) < 0$ for $x < 2$

3. $f''(x) > 0$ for $x < -1$ and $x > 1$; $f''(x) < 0$ for $-1 < x < 1$

5. Concave upward for $x > -1$; concave downward for $x < -1$; inflection at $(-1, 2)$

7. Concave upward for $x > -\frac{1}{3}$; concave downward for $x < -\frac{1}{3}$; inflection at $\left(-\frac{1}{3}, -\frac{1}{27}\right)$

9. Concave upward for $t < 0$ and $t > 1$; concave downward for $0 < t < 1$; inflection at $(1, 0)$

11. Concave upward for $x < 0$ and $x > 3$; concave downward for $0 < x < 3$; inflection at $(0, -5)$ and $(3, -65)$

13. Increasing for $x < -3$ and $x > 3$; decreasing for $-3 < x < 3$; concave upward for $x > 0$; concave downward for $x < 0$; maximum at $(-3, 20)$; minimum at $(3, -16)$; inflection at $(0, 2)$

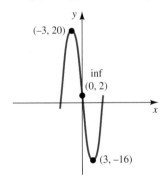

15. Increasing for $x > 3$; decreasing for $x < 3$; concave upward for $x < 0$ and $x > 2$; concave downward for $0 < x < 2$; minimum at $(3, -17)$; inflection at $(0, 10)$ and $(2, -6)$.

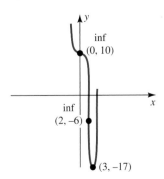

17. Increasing for all x; concave upward for $x > 2$; concave downward for $x < 2$; inflection at $(2, 0)$

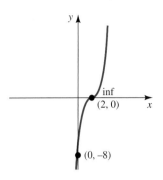

19. Increasing for $x > 0$; decreasing for $x < 0$; concave upward for $x < -\sqrt{5}$, $-1 < x < 1$, $x > \sqrt{5}$; concave downward for $-\sqrt{5} < x < -1$ and $1 < x < \sqrt{5}$; minimum at $(0, -125)$; inflection points at $(-\sqrt{5}, 0)$, $(\sqrt{5}, 0)$, $(-1, -64)$ and $(1, -64)$

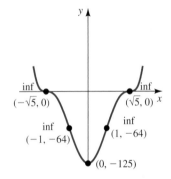

21. Increasing for $s > -1$; decreasing for $s < -1$; concave upward for $s < -4$ and $s > -2$; concave downward at $-4 < s < -2$; minimum at $(-1, -54)$; inflection at $(-4, 0)$ and $(-2, -32)$

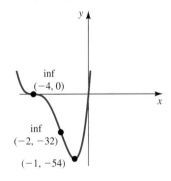

23. Increasing for $x > 0$; decreasing for $x < 0$; concave upward for all real x; minimum at $(0, 1)$

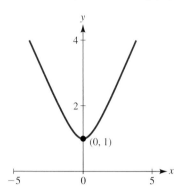

25. Increasing for $x < -\dfrac{1}{2}$; decreasing for $x > -\dfrac{1}{2}$; concave upward for $x < -1$ and $x > 0$; concave downward for $-1 < x < 0$; maximum at $\left(-\dfrac{1}{2}, \dfrac{4}{3}\right)$; inflection at $(-1, 1)$ and $(0, 1)$

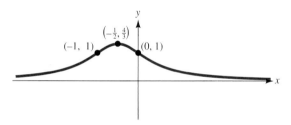

27. $f''(x) = 6(x + 1)$; maximum at $(-2, 5)$; minimum at $(0, 1)$

29. $f''(x) = 12(x^2 - 3)$; maximum at $(0, 81)$; minimum at $(3, 0)$ and $(-3, 0)$

31. $f''(x) = \dfrac{36}{x^3}$; maximum at $(-3, -11)$; minimum at $(3, 13)$

33. $f''(x) = 12x^2 - 60x + 50$; maximum at $\left(\dfrac{5}{2}, \dfrac{625}{16}\right)$; minimum at $(0, 0)$ and $(5, 0)$

35. $h''(t) = \dfrac{4(3t^2 - 1)}{(1 + t^2)^3}$; maximum at $(0, 2)$

37. $f''(x) = \dfrac{24(x - 2)}{x^4}$; maximum at $(-4, -13.5)$. Test fails for $x = 2$ [there is an inflection point at $(2, 0)$].

39. Concave upward for $x < 0$, for $0 < x < 1$, and for $x > 3$; concave downward for $1 < x < 3$; inflection points at $x = 1, 3$

41. Concave upward for $x > 1$; concave downward for $x < 1$; inflection point at $x = 1$

43. **a.** Increasing for $x < 0$ and $x > 4$; decreasing for $0 < x < 4$

   b. Concave upward for $x > 2$ and concave downward for $x < 2$

   c. Relative minimum at $x = 4$, relative maximum at $x = 0$; inflection point at $x = 2$

   d.

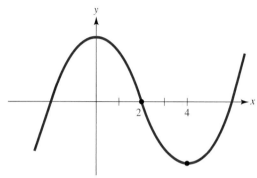

45. **a.** Increasing for $-\sqrt{5} < x < \sqrt{5}$; decreasing for $x > \sqrt{5}$ and $x < -\sqrt{5}$

   b. Concave upward for $x < 0$ and concave downward for $x > 0$

   c. Relative maximum at $x = \sqrt{5}$ and relative minimum at $x = -\sqrt{5}$; inflection point at $x = 0$

   d.

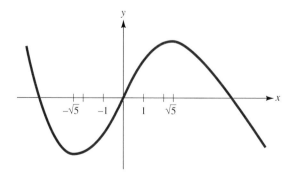

47. A typical graph is shown.

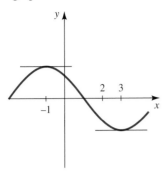

49. $f(x)$ is increasing for $x > 2$.
   $f(x)$ is decreasing for $x < 2$.
   $f(x)$ is concave upward for all real x.
   $f(x)$ has a relative minimum at $x = 2$.
   $f(x)$ has no inflection points.

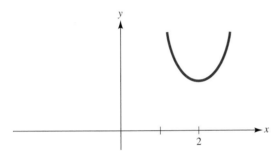

51. $f(x)$ is increasing for $x > 2$.
   $f(x)$ is decreasing for $x < -3$ and $-3 < x < 2$.
   $f(x)$ is concave upward for $x < -3$ and $x > -1$.
   $f(x)$ is concave downward for $-3 < x < -1$.
   $f(x)$ has a relative minimum at $x = 2$.
   $f(x)$ has inflection points at $x = -3$ and $x = -1$.

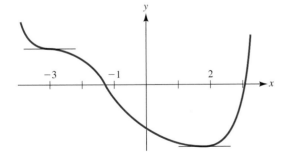

53. $C(x) = 0.3x^3 - 5x^2 + 28x + 200$
   a. $C'(x) = 0.9x^2 - 10x + 28$

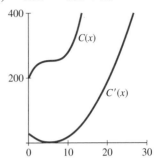

   b. Only inflection number of $C(x)$ is $x = 5.56$.
   It corresponds to a minimum on the graph
   of $C'(x)$.

55. a. 1000 units will be sold.
   b. Inflection point when $x = 11$. Sales are increasing at the largest rate when \$11 000 is spent on
   marketing.

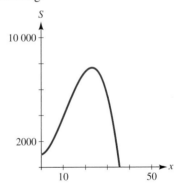

57. Output rate is $Q'(t) = -3t^2 + 9t + 15$.
   a. Rate is maximized at $t = 1.5$ (9:30 A.M.).
   b. Rate is minimized at $t = 4$ (noon).

59. Rate of growth is $P'(t) = -3t^2 + 18t + 48$.
   a. Rate is largest when $t = 3$ years.
   b. Rate is smallest when $t = 0$ years.
   c. Rate of growth of $P'(t)$ is $P''(t) = -6t + 18$,
   which is largest when $t = 0$.

61. a. $M'(r) = \dfrac{0.02 - 0.018r - 0.00018r^2}{(1 + 0.009r^2)^2}$

$M''(r) = \dfrac{-0.018 - 0.00108r + 0.000486r^2 + 0.00000324r^3}{(1 + 0.009r^2)^3}$

b. M

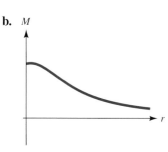

c. 7.10%

63. a. $N'(t) = \dfrac{60 - 5t^2}{(12 + t^2)^2};$

$N''(t) = \dfrac{10t^3 - 360t}{(12 + t^2)^3}$

b. 3.5 weeks after the outbreak; 722 new cases

c. 6 weeks after the outbreak; 62.5 new cases

65. $R = k\,N(P - N)$, so $R' = k(P - 2N)$ and $R' = 0$

when $N = \dfrac{P}{2}$. Since $R''\left(\dfrac{P}{2}\right) = -2k < 0$, there is a

maximum at $N = \dfrac{P}{2}$.

67. a. $R'(t) = A''(t) = \dfrac{d}{dt}[(k\sqrt{A(t)})(M - A(t))]$

$= k\left(\dfrac{1}{2}\right)\dfrac{A'(t)}{\sqrt{A(t)}}[M - A(t)] + k\sqrt{A(t)}(-A'(t))$

$= \dfrac{kA'(t)}{2\sqrt{A(t)}}[M - 3A(t)]$

$= 0$ when $A = \dfrac{M}{3}$.

b. Greatest

c. The graph of $A(t)$ has an inflection point where

$A(t) = \dfrac{M}{3}$.

69. a. Relative max at $t = 0$, relative min at $t = 1$.

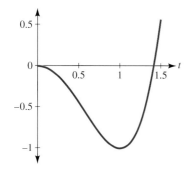

b. relative min of velocity at $t = \sqrt{\dfrac{1}{3}}$

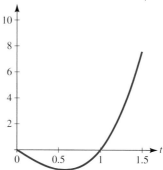

c. relative minimum acceleration as he passes his mother at $t = 0$

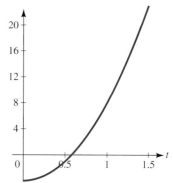

71. $y = ax^2 + bx + c$
$y' = 2ax + b$, and $y'' = 2a$
$y'' < 0$ when $a < 0$, the curve is concave downward and the graph has a maximum. Similarly, when $a > 0$, the graph has a minimum and is concave up.

73. $P(x) = f(x)g(x)$
$P'(x) = f'(x)g(x) + f(x)g'(x)$
$P''(x) = f''(x)g(x) + f'(x)g'(x) + f'(x)g'(x) + f(x)g''(x)$
$\quad\;\; = f''(x)g(x) + 2f'(x)g'(x) + f(x)g''(x)$
$P''(c) = f''(c)g(c) + 2f'(c)g'(c) + f(c)g''(c)$
$\quad\;\; = 0$ since $f'(c) = 0$

Thus, $P(x)$ has an inflection point at $x = c$.

CHAPTER 3   Section 3

1. Vertical asymptote, $x = 0$; horizontal asymptote, $y = 0$

3. No vertical asymptotes; horizontal asymptote at $y = 0$

5. Vertical asymptotes, $x = -2$, $x = 2$; horizontal asymptotes, $y = 2$ and $y = 0$ (x axis)

7. Vertical asymptote, $x = 2$; horizontal asymptote, $y = 0$

9. Vertical asymptote, $x = -2$; horizontal asymptote, $y = 3$

11. No vertical asymptotes; horizontal asymptote, $y = 1$

13. Vertical asymptotes, $t = 2$, $t = 3$; horizontal asymptote, $y = 1$

15. Vertical asymptotes, $x = 0$, $x = 1$; horizontal asymptote, $y = 0$

17.

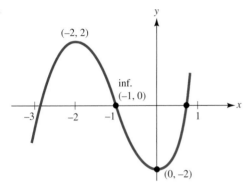

19.

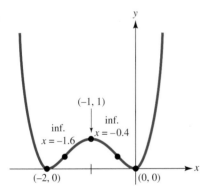

21.

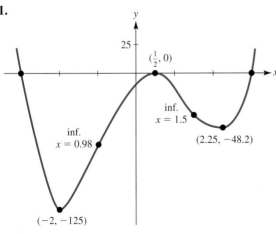

23.

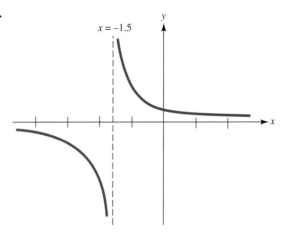

25.

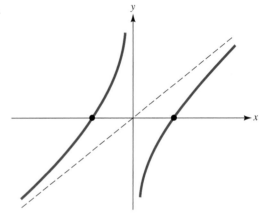

27.

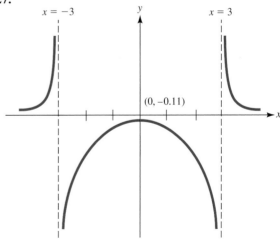

29.

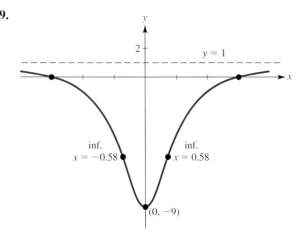

31.

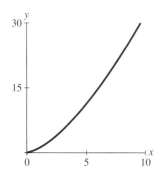

33. Answers may vary.

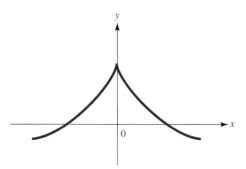

35. Answers may vary.

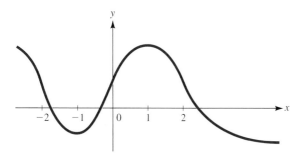

37. Answers may vary.

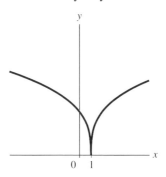

39. a. $f(x)$ is increasing ($f'(x) > 0$) for $0 < x < 2$ or $x > 2$; $f(x)$ is decreasing ($f'(x) < 0$) for $x < 0$.

b. Relative minimum at $x = 0$

c. $f''(x) = x^2(x - 2)(5x - 6)$; $f(x)$ is concave up for $x < 0, 0 < x < \dfrac{6}{5}$, and $x > 2$; concave down for $\dfrac{6}{5} < x < 2$.

d. $x = \dfrac{6}{5}$ and $x = 2$

41. a. $f(x)$ is increasing $f'(x) > 0$ for $-3 < x < 2$ or $2 < x$; $f(x)$ is decreasing ($f'(x) < 0$) for $x < -3$.

b. Relative minimum at $x = -3$

c. $f''(x) = \dfrac{-x - 8}{(x - 2)^3}$; $f(x)$ is concave up for $-8 < x < 2$; concave down for $x < -8$ and $x > 2$.

d. $x = -8$

43. $B = -\dfrac{5}{2}; A = -10$

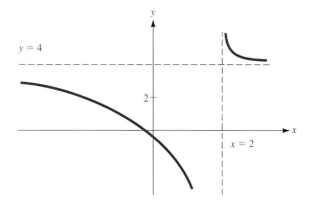

45. a. Vertical asymptote, $x = 0$; no horizontal asymptotes

   b. The average cost curve $A(x)$ approaches the line $y = 3x + 1$ as x gets larger.

   c.

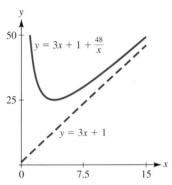

47. a.

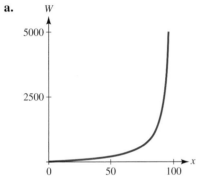

   b. 11.8%

49. a.

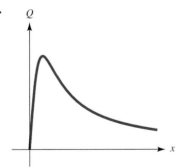

   b. $5196; 674 units

   c. $9000

51. Answers may vary.

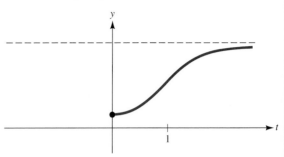

53.

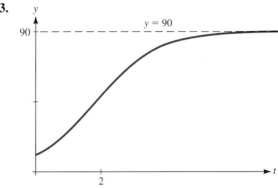

55. a.

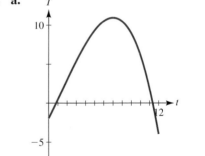

   b. The temperature is greatest at 1:00 P.M. The high temperature is 10.9°C.

57. a.

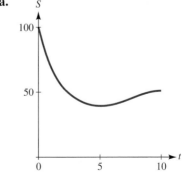

   b. $t = 5$; 41.2%

   c. Positive; decreasing

59. **a.** $C(x) = \dfrac{22\ 056}{x} + 2.58x$

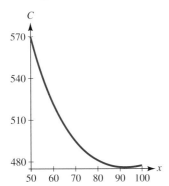

b. 92.5 km/h; $477

61. **a.** $f'(x) = \dfrac{10(x - 1)}{3x^{1/3}}$; $f(x)$ is increasing for $x < 0$ and $x > 1$; $f(x)$ is decreasing f or $0 < x < 1$; relative minimum at $(1, -3)$; relative maximum at $(0, 0)$.

b. $f''(x) = \dfrac{10(2x + 1)}{9x^{4/3}}$; $f(x)$ is concave upward for $x > -\dfrac{1}{2}$, $f(x)$ is concave downward for $x < -\dfrac{1}{2}$ inflection point at $\left(-\dfrac{1}{2}, -3\sqrt[3]{2}\right)$.

c. $(0, 0), \left(\dfrac{5}{2}, 0\right)$; no asymptotes

d.

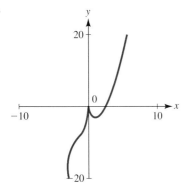

63. **a.**

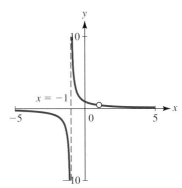

The graph has a hole at $x = 1$.

b.

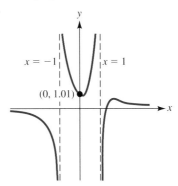

Vertical asymptotes at $x = 1$, $x = -1$; horizontal asymptote $y = 0$.

CHAPTER 3 Section 4

1. Absolute maximum at $(1, 10)$; absolute minimum at $(-2, 1)$

3. Absolute maximum at $(0, 2)$; absolute minimum at $\left(2, -\dfrac{40}{3}\right)$

5. Absolute maximum at $(-1, 2)$; absolute minimum at $(-2, -56)$

7. Absolute maximum at $(-3, 3125)$; absolute minimum at $(0, -1024)$

9. Absolute maximum $\left(3, \dfrac{10}{3}\right)$; absolute minimum at $(1, 2)$

11. Absolute minimum at $(1, 2)$; no absolute maximum

13. $f(x)$ has no absolute maximum or minimum for $x > 0$.

15. Absolute maximum at $(0, 1)$; no absolute minimum

17. **a.** $R(q) = 49q - q^2$; $R'(q) = 49 - 2q$;

$$C'(q) = \frac{1}{4}q + 4$$

$$P(q) = -\frac{9}{8}q^2 + 45q - 200;$$

maximum when $q = 20$

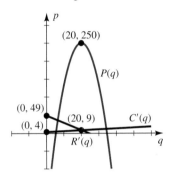

b. $A(q) = \frac{1}{8}q + 4 + \frac{200}{q}$; $A(q)$ is minimized

at $q = 40$.

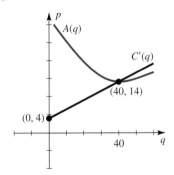

19. **a.** $R(q) = 180q - 2q^2$; $R'(q) = 180 - 4q$; $C'(q) = 3q^2 + 5$; $P(q) = -q^3 - 2q^2 + 175q - 162$; $P(q)$ is maximized at $q = 7$.

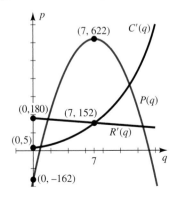

b. $A(q) = q^2 + 5 + \frac{162}{q}$; $A(q)$ is minimized at $q = 4.327$.

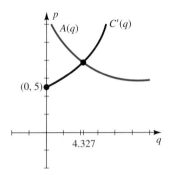

21. **a.** $R(q) = 1.0625q - 0.0025q^2$;

$R'(q) = 1.0625 - 0.005q$

$$C'(q) = \frac{q^2 + 6q - 1}{(q + 3)^2};$$

$$P(q) = \frac{-0.0025q^3 + 0.055q^2 + 3.1875q - 1}{q + 3};$$

Profit $P(q)$ is maximized when $P'(q) = 0$; when $q = 17.3$ units.

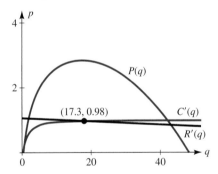

b. $A(q) = \frac{q^2 + 1}{q(q + 3)}$ is minimized when

$$q = \frac{1 + \sqrt{10}}{3} \approx 1.3874.$$

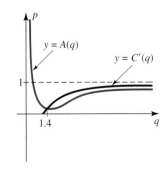

23. $E(p) = \dfrac{-1.3p}{-1.3p + 10}$; $E(4) = -\dfrac{13}{12}$, elastic

25. $E(p) = \dfrac{2p^2}{p^2 - 200}$; $E(10) = -2$, elastic

27. $E(p) = \dfrac{30}{p - 30}$; $E(10) = -\dfrac{3}{2}$, elastic

29. The slope $f'(x) = 4x - x^2$ has its largest absolute value when $x = -1$. The graph is steepest at $\left(-1, \dfrac{7}{3}\right)$. The slope of the tangent is -5.

31. **a.** $P'(q) = -4q + 68$

$A(q) = \dfrac{P(q)}{q} = -2q + 68 - \dfrac{128}{q}$

b. $P'(q) = A(q)$ when $q = 8$

c. A is increasing ($A' > 0$) if $0 < q < 8$ and decreasing ($A' < 0$) if $q > 8$

d.

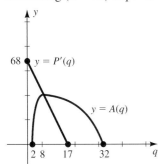

33. **a.** Largest in 2010 ($x = 15$). Smallest in 2006 ($x = 11$).

b. Largest: 58 500 in 2010. Smallest: 12 100 in 2006.

35. Maximum displacement 5 m when $t = 0$.

37. The speed of the blood is greatest when $r = 0$, that is, at the central axis.

39. **a.** $E = \dfrac{-3p}{2q + 3p}$

b. When $p = 3$, $q = 2$, and $E = -\dfrac{9}{13}$; demand is inelastic.

41. **a.** $225 \le p \le 250$

b. $E(p) = \dfrac{p}{p - 250}$

Demand is elastic when $p > 125$, inelastic when $p < 125$, and of unit elasticity when $p = 125$.

c. Total revenue is increasing for $p < 125$ and decreasing for $p > 125$.

d. If any number of prints are available, then $p = 125$ maximizes total revenue. If only 50 prints are available, then $p = 225$ maximizes total revenue.

43. 400 basic dump trucks and 700 electronic dump trucks

45. 11:00 A.M.

47. 150 boxes

49. $V(T)$

$= \left(-\dfrac{6.8}{10^8}\right)T^3 + \left(\dfrac{8.5}{10^6}\right)T^2 - \left(\dfrac{6.4}{10^5}\right)T + 1$

for $0 \le T \le 10$

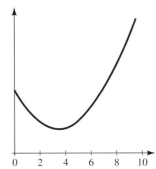

The minimum occurs at $T = 3.95$.

b. Writing exercise. Answers will vary.

51. **a.** $E(p) = \dfrac{ap}{ap - b}$

b. $E(p) = -1 \Rightarrow ap = (-1)(ap - b)$

$\Rightarrow 2\,ap = b \Rightarrow p = \dfrac{b}{2a}$

c. Elastic for $\dfrac{b}{2a} < p \le \dfrac{b}{a}$, inelastic for

$0 \le p < \dfrac{b}{2a}$

53. $E(p) = \dfrac{p}{q}\dfrac{dq}{dp} = \dfrac{p}{\left(\dfrac{a}{p^m}\right)}\left(\dfrac{-ma}{p^{m+1}}\right) =$

$\left(\dfrac{p^{m+1}}{a}\right)\left(\dfrac{-ma}{p^{m+1}}\right) = -m$

If $m = 1$, demand is of unit elasticity. If $m > 1$, demand is elastic, and if $0 < m < 1$, demand is inelastic.

55. **a.** $R(x) = \dfrac{AB + A(1 - m)x^m}{(B + x^m)^2}; x = \left(\dfrac{B}{m - 1}\right)^{1/m}$

b. $R'(x) = \dfrac{-Amx^{m-1}[(1 - m)x^m + (1 + m)B]}{(B + x^m)^3}$

$x = 0$ and $x = \left[\dfrac{B(m + 1)}{m - 1}\right]^{\frac{1}{m}}$

c. Relative maximum; use the first derivative test

CHAPTER 3   Section 5

Algebra Warm-Up

1. 4.33
2. 9
3. 5
4. 4
5. 4/3
6. 2.84
7. −5
8. 2.83

Exercises

1. $\dfrac{1}{2}$

3. $x = 25, y = 25$

5. $\$40.83 \approx \$41.00$

7. 80 trees

9. $p = \$8.12$ (or \$8.13) per ice cream carton

11. Make the playground square with side $S = 60$ m.

13. Let x be the length of the rectangle and y the width, and let P be the fixed value of the perimeter, so that

$P = 2(x + y)$ and $y = \dfrac{1}{2}(P - 2x)$. The area is

$A = xy = x\left[\dfrac{1}{2}(P - 2x)\right] = -x^2 + \dfrac{1}{2}Px$

Differentiating, we find that

$A' = -2x + \dfrac{1}{2}P = 0$

when $x = \dfrac{P}{4}$. Since $A'' = -2 < 0$, the maximum

area occurs when $x = \dfrac{P}{4}$ and

$y = \dfrac{1}{2}\left[p - 2\left(\dfrac{P}{4}\right)\right] = \dfrac{P}{4}$

that is, when the rectangle is a square.

15. 6 by 2.5

17. 2 by 2 by $\dfrac{4}{3}$ metres

19. Yes, he has 5 minutes, 17 seconds to spare.

21. Entirely under water

23. Width: 22 cm; length: 44 cm

25. $r = 3.82$ cm; $h = 7.64$ cm

27. $r = \dfrac{2}{3}h$

29. 17 floors

31. **a.** 200 bottles
    b. Every three months

33. **a.** 10 machines
    b. \$400
    c. \$200

35. 5 years from now

37. $x = 0.5$ m; $y = 1$ m; $V = 0.25$ m^3

39. $C(x) = 1200 + 1.20x + \dfrac{100}{x^2}; x = 6$

41. The point P should be $\dfrac{5\sqrt{3}}{3} \approx 2.9$ km from A.

43. $S = Kwh^3 = Kh^3\sqrt{1600 - h^2}$;

$S'(h) = \dfrac{K(4800h^2 - 4h^4)}{\sqrt{1600 - h^2}} = 0$ when $h \approx 34.6$ cm.;

$w \approx 20.1$ cm.

45. 4.5 km from plant A

47. Suppose the setup cost and operating cost are aN and $\dfrac{b}{N}$, respectively, for positive constants a and b. Total cost is then $C = aN + \dfrac{b}{N}$. The minimum cost occurs when

$C' = a - \dfrac{b}{N^2} = 0$

$aN = \dfrac{b}{N}$

that is, when setup cost aN equals operating cost $\dfrac{b}{N}$.

49. Frank is right; in Example 3.5.5 replace 3000 with any fixed distance $D \geq 1200$. The result is the same since D drops out when $C'(x)$ is computed.

51. a. Let x be the number of machines and t the number of hours required to produce Q units. The setup cost is $C_s = xs$ and the operating cost (for all x machines) is $C_o = pt$. Since n units can be produced per machine per hour, $Q = nxt$ or $t = \dfrac{Q}{nx}$. The total cost is

$$C = C_s + C_o = xs + \frac{pQ}{nx}$$

so that

$$C' = s - \frac{pQ}{nx^2} = 0$$

when $x = \sqrt{\dfrac{pQ}{ns}}$. Since $C'' > 0$, this corresponds to a minimum.

b. $C_s = xs = \sqrt{\dfrac{pQs}{n}}$, $C_o = \dfrac{pQ}{nx} = \sqrt{\dfrac{pQs}{n}}$

53. a. $P(x) = x\left(15 - \dfrac{3}{8}x\right) - \dfrac{7}{8}x^2 - 5x - 100 - tx$;

Thus, $P'(x) - \dfrac{5}{2}x + 10 - t = 0$

When $x = \dfrac{2}{5}(10 - t)$

b. $t = 5$

c. The monopolist will absorb $4.25 of the $5 tax per unit. $0.75 will be passed on to the consumer.

d. Writing exercise; responses will vary.

CHAPTER 3 Checkup

1. (a) is the graph of $f(x)$ and (b) is the graph of $f'(x)$. Answers will vary. One reason is that the x intercepts of the graph in (b) correspond to the high and low points in (a).

2. a. Increasing for $x < 0$ and $0 < x < 3$; decreasing for $x > 3$

Critical Numbers	Classification
0	Neither
3	Relative maximum

b. Increasing for $t < 1$ and $t > 2$; decreasing for $1 < t < 2$

Critical Numbers	Classification
1	Relative maximum
2	Relative maximum

c. Increasing for $-3 < t < 3$; decreasing for $t < -3$ and $t > 3$

Critical Numbers	Classification
-3	Relative maximum
3	Relative maximum

d. Increasing for $x < -1$ and $x > 9$; decreasing for $-1 < x < 9$

Critical Numbers	Classification
-1	Relative maximum
9	Relative maximum

3. a. Concave upward for $x > 2$; concave downward for $x < 0$ and $0 < x < 2$; inflection at $x = 2$

b. Concave upward for $-5 < x < 0$ and $x > 1$; concave downward for $x < -5$ and $0 < x < 1$; inflection at $x = -5$, $x = 0$, and $x = 1$

c. Concave upward for $t > 1$; concave downward for $t < 1$; no inflection points

d. Concave upward for $-1 < t < 1$; concave downward for $t < -1$ and $t > 1$; inflection at $t = -1$ and $t = 1$

4. a. Vertical asymptote, $x = -3$; horizontal asymptote, $y = 2$

b. Vertical asymptotes, $x = -1$, $x = 1$; horizontal asymptote, $y = 0$

c. Vertical asymptotes, $x = -\dfrac{3}{2}$, $x = 1$; horizontal asymptotes, $y = \dfrac{1}{2}$

d. Vertical asymptote, $x = 0$; horizontal asymptote, $y = 0$

5. a. No asymptotes; intercepts at $(0, 0)$ and $\left(\dfrac{4}{3}, 0\right)$; relative minimum at $(1, -1)$; inflection points at $(0, 0)$ and $\left(\dfrac{2}{3}, \dfrac{-16}{27}\right)$;

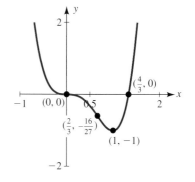

b. No asymptotes; y intercept at $(0, 1)$; relative minimum at $(0, 1)$; inflection points at $(1, 2)$ and $\left(\dfrac{1}{2}, \dfrac{23}{16}\right)$

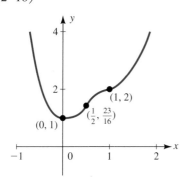

c. Vertical asymptote, $x = 0$; horizontal asymptote, $y = 1$; x intercept at $(-1, 0)$; relative minimum at $(-1, 0)$; inflection point at $\left(-\dfrac{3}{2}, \dfrac{1}{9}\right)$

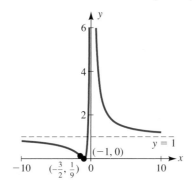

d. Vertical asymptote, $x = 1$; horizontal asymptote, $y = 0$; x intercept at $\left(\dfrac{1}{2}, 0\right)$; y intercept at $(0, 1)$; relative maximum at $(0, 1)$; inflection point at $\left(-\dfrac{1}{2}, \dfrac{8}{9}\right)$

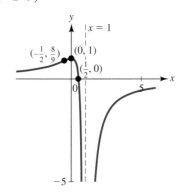

6. Answers will vary.

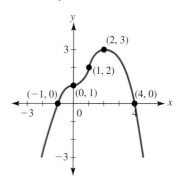

7. **a.** Absolute maximum value of 6 where $x = -1$; absolute minimum value of -26 where $x = 3$

b. Absolute maximum value of 23 where $t = 2$; absolute minimum value of -69 where $t = 4$

c. Absolute maximum value of 19 where $u = 16$; absolute minimum value of 3 where $u = 0$

8. $f''(t) = 0$ when $t = \dfrac{7}{3}$; 8: 20 A.M.

9. $135

10. **a.**

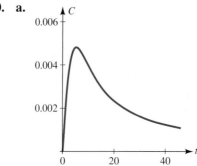

b. $t = 9$

c. The concentration tends to 0.

11. **a.** 1.667 million

b. $t = 2$ hours; 5 million

c.

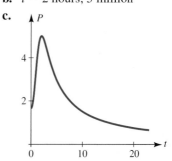

The population dies off.

CHAPTER 3   Review Exercises

1. $f(x)$ is increasing for $-1 < x < 2$; decreasing for $x < -1$ and $x > 2$; concave up for $x < \frac{1}{2}$, concave down for $x > \frac{1}{2}$, relative maximum $(2, 15)$; relative minimum $(-1, -12)$; inflection point $\left(\frac{1}{2}, \frac{3}{2}\right)$.

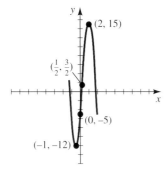

3. $f(x)$ is increasing for $x < -0.79$ and $1.68 < x$; $f(x)$ is decreasing for $-0.79 < x < 1.68$; $f(x)$ is concave downward for $x < \frac{4}{9}$; $f(x)$ is concave upward for $x > \frac{4}{9}$.

There is a relative maximum at $(-0.79, 22.51)$, and a relative minimum at $(1.68, -0.23)$.

There is one inflection point, at $(0.44, 11.14)$.

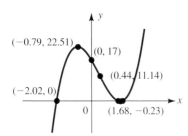

5. $f(t)$ is increasing for $t < -2$ and $t > 2$; decreasing for $-2 < t < 2$; concave up for $-\sqrt{2} < t < 0$ and for $t > \sqrt{2}$; concave down for $t < -\sqrt{2}$ and for $0 < t < \sqrt{2}$. Relative maximum at $(-2, 64)$; relative minimum at $(2, 64)$; inflection points $(-\sqrt{2}, 39.6)$ and $(\sqrt{2}, -39.6)$.

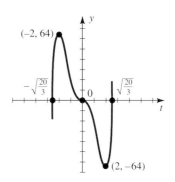

7. $g(t)$ is increasing for $t < -2$ and for $t > 0$; decreasing for $-2 < t < -1$ and for $-1 < t < 0$; concave up for $t > -1$; concave down for $t < -1$. Relative maximum $(-2, -4)$; relative minimum $(0, 0)$; no inflection points.

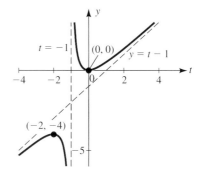

9. $F(x)$ is increasing for $x < -2$ and for $x > 2$; decreasing for $-2 < x < 0$ and for $0 < x < 2$; concave up for $x > 0$; concave down for $x < 0$. Relative maximum $(-2, -6)$; relative minimum $(2, 10)$; no inflection points.

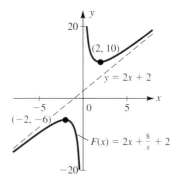

11. (b) is the graph of $f(x)$ and (a) is the graph of $f'(x)$. Answers will vary. One reason is that the graph in (b) is always increasing and the graph in (a) is always positive.

13.

Critical Numbers	Classification
−1	Relative minimum
0	Relative maximum
3/2	Neither
7	Relative minimum

15.

Critical Numbers	Classification
0	Relative minimum
2	Neither

17. Here is one possible graph.

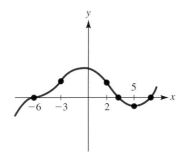

19. Here is one possible graph.

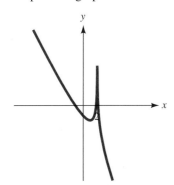

21. Relative maximum at $(2, 15)$; relative minimum at $(-1, -12)$

23. Relative maximum at $(-2, -4)$; relative minimum at $(0, 0)$

25. Absolute maximum value of 40 where $x = -3$; absolute minimum value of -12 where $x = -1$

27. Absolute maximum value of $\dfrac{1}{2}$ where $s = -\dfrac{1}{2}$ or $s = 1$; absolute minimum value of 0 where $s = 0$

29. **a.** $f(x)$ is increasing for $0 < x < 1$ and $x > 1$; it is decreasing for $x < 0$.

    b. $f(x)$ is concave upward for $x < \dfrac{1}{3}$ and $x > 1$; it is concave downward for $\dfrac{1}{3} < x < 1$.

c. $f(x)$ has a relative minimum at $x = 0$ and inflection points at $x = \dfrac{1}{3}$ and $x = 1$.

d.

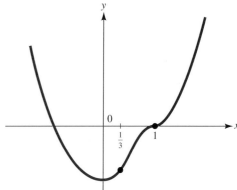

31. $12.50

33. $r = \dfrac{2}{3}h$

35. **a.** An 80 m by 80 square

    b. 160 m parallel to the wall, 80 m wide

37. Paddle all the way to the campsite.

39. 12 machines

41. **a.** $E = \dfrac{-2p^2}{100 - p^2}$

    b. At $p = 6$, $E = -\dfrac{9}{8}$, so $|E| = \left|-\dfrac{9}{8}\right| = \dfrac{9}{8}$, Since $|E| > 1$ demand is elastic (i.e., as price increases, revenue decreases).

    c. $5.77

43. **a.** $E(p) = \dfrac{1.4p^2}{0.7p^2 - 300}$

    b. $E(8) = -0.351$; raise the price

45. Rectangle: 1.56 m by 1.66 m; side of triangle: 1.56 m

47. 4000 maps per batch

49. Hint: If x units are ordered, $C = k_1 x + \dfrac{k_2}{x}$.

51. **a.** Relative minimum at $x = \dfrac{1}{c}$

    b. Maximum of $\dfrac{\pi}{3}(5 - 3\sqrt{2})$; minimum of $\dfrac{\pi}{6}$

    c. Maximum of $\dfrac{\sqrt{3}\pi}{16}$; minimum of $\dfrac{\sqrt{3}\pi}{4(2 + \sqrt{2})^2}$

    d. $\lim\limits_{x \to \infty} f(x) = Kc^2$, when r is much larger than R, the packing fraction depends only on the cell structure in the lattice.

ANSWERS

53. a.

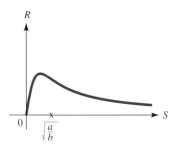

The graph appears to have a highest point $\left(\text{at } x = \sqrt{\dfrac{a}{b}}\right)$, a lowest point (at $x = 0$), and a point of inflection. The growth rate seems to level off (toward 0) as S grows larger and larger.

CHAPTER 4 Section 1

1. $e^2 \approx 7.389$, $e^{-2} \approx 0.135$, $e^{0.05} \approx 1.051$, $e^{-0.05} \approx 0.951$, $e^0 = 1$, $e \approx 2.718$, $\sqrt{e} \approx 1.649$, $\dfrac{1}{\sqrt{e}} \approx 0.607$

3.

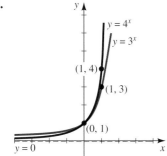

5. a. 9

   b. $\dfrac{1}{27}$

7. a. 12

   b. $\dfrac{189}{1331}\sqrt{7}$

9. a. 3

   b. 4

11. a. 243

   b. $e^{14/3}$

13. a. $9x^4$

   b. $2x^{2/3}y$

15. a. $\dfrac{1}{x^{1/3}y^{1/2}}$

   b. $x^{1.1}y^2$

17. a. $\dfrac{1}{t}$

   b. t

19. $\dfrac{3}{2}$

21. 1

23. 1

25. $-2, 2$

27. $0, \dfrac{3}{2}$

29.

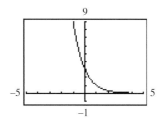

31.

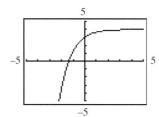

33. $b = 2, C = 3$

35. a. \$1967.15

   b. \$2001.60

   c. \$2009.66

   d. \$2013.75

37. \$3534.12

39. a. \$6361.42

   b. \$6342.19

41. a. \$40.60

   b. \$4060

   c. Revenue is less by \$1458 when 100 units are produced.

43. a. 50 000 000

   b. 91 105 940

45. a. 3 mg/mL; 1.78 mg/mL

   b. -0.72 mg/mL per hour

47. a. $A = \dfrac{10\ 000}{2^{0.01}} \approx 9931$

   b. 9931; 10 070; 10 353

   c. 439 bacteria per hour or 7.32 bacteria per minute

49. The sellers would have gotten the better deal by 2755 billion dollars

51. 13 570

53. **a.** 0.13 g/cm^3
b. 0.1044 g/cm^3; 0.0795 g/cm^3
c. -0.0016 g/cm^3 per minute
d. As $t \to \infty$, $C(t) \to 0.065$ g/cm^3
e.

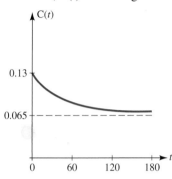

55. $r_e \approx 6.14\%$

57. $r_e \approx 5.13\%$

59. d, c, b, a

61. \$608.33

63. **a.** 0.5488
b. 0.1813
c. 0.1215

65. **a.** 12 000 people per square kilometre
b. 5959 people per square kilometre

67. $\dfrac{1}{\sqrt[3]{10}} l_0 \approx 0.46 l_0$

69. **a.** $A = 214.37$, $B = 126.35$
b. 88 million
c. 251 million

71. \$1206.93

73. **a.** No. A fair monthly payment is \$166.07.
b. Writing exercise; responses will vary.

CHAPTER 4   Section 2

Algebra Warm-Up

a. $x = -\dfrac{8}{3}$

b. $x = -\dfrac{14}{3}$

c. $x = \dfrac{16}{23}$

d. $x = \dfrac{450}{151}$

e. $p = 10\sqrt{\dfrac{2}{3}} \approx 8.16$

Exercises

1. $\ln 1 = 0$, $\ln 2 \approx 0.693$, $\ln e = 1$, $\ln 5 \approx 1.609$, $\ln \dfrac{1}{5} \approx -1.609$, $\ln e^2 = 2$; $\ln 0$ and $\ln(-2)$ are undefined; e^x cannot be negative or equal to zero.

3. 3

5. 5

7. $\dfrac{8}{25}$

9. $3 + \log_3 2 + \log_3 5$

11. $2 \log_3 2 + 2 \log_3 5$

13. $4 \log_2 x + 3 \log_2 y$

15. $\dfrac{1}{3}[\ln x + \ln(x - 1)]$

17. $2 \ln x + \dfrac{2}{3} \ln(3 - x) - \dfrac{1}{2} \ln(x^2 + x + 1)$

19. $3 \ln x - x^2$

21. $\dfrac{\ln 53}{\ln 4} \approx 2.864$

23. 5

25. $\dfrac{\ln 2}{0.06} \approx 11.552$

27. $\dfrac{\ln 5}{4} \approx 0.402$

29. $e^{-C - t/50}$

31. 4

33. $\dfrac{2}{\ln 3} \approx 1.820$

35. $10 \ln 2$

37. $5 \ln 2 \approx 3.4657$

39. $7 \ln 5 - \ln 2 \approx 10.5729$

41. -5.5

43. $\dfrac{\ln 2}{0.06} \approx 11.55$ years

45. $\dfrac{\ln 2}{13} \approx 5.33\%$

47. $\dfrac{12 \ln 3}{\ln 2} \approx 19.02$ years

49. 5.83%

51. **a.** 0.765 g/cm^3; 0.784 g/cm^3

 b. $-50 \ln \left(\dfrac{0.125}{0.13}\right) \approx 1.96$ seconds

53. 5614 years

55. $Q(t) = 6000e^{0.0203t}$; 20 283

57. $Q(t) = 500 - 200e^{-0.1331t}$; 459.5 units

59. 10 523 years

61. 24.84 years ago; 95.6%

63. $f(t) = 22 + 78e^{-0.0741t}$; temperature is 72.01°C

65. **a.** $10 - \ln 11 \approx \$7.60$

 b. $\ln 102 \approx \$4.62$

 c. $x_e = \dfrac{-3 + \sqrt{1 + 4e^{10}}}{2} \approx 147$ units

 $P_e \approx \$5$

67. Stan; Wednesday morning at 2.15 A.M.

69. **a.** 8.25

 b. 7.94×10^{14} joules

71. **a.** 45%

 b. 2.34%

73. **a.**

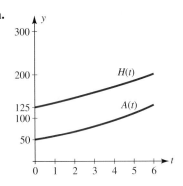

 b. $A = \dfrac{2H^2}{625}$

75. **a.** $51 + 100 \ln 3 \approx 161$ thousand

 b. $e^{271/100} - 3 \approx 12$ years

 c. $10 \ln \dfrac{13}{3} \approx 14\ 700$ people/year

77. For $y = Cx^k$, let $Y = \ln y$ and $X = \ln x$. Then $Y = mX + b$ where $m = k$ and $b = \ln C$.

79. **a.** $(\log_a b)(\log_b a) = \left(\dfrac{\ln b}{\ln a}\right)\left(\dfrac{\ln a}{\ln b}\right) = 1$

 b. $\log_a x = \dfrac{\ln x}{\ln a}$

 $= \dfrac{(\ln x)(\ln b)}{(\ln b)(\ln a)} = (\log_b x)(\log_a b)$

 $= \dfrac{\log_b x}{\log_b a}$ (using part a)

81. 10^x and $\log_{10} x$ are reflections about $y = x$.

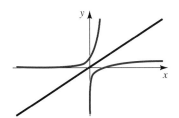

CHAPTER 4 Section 3

1. $f'(x) = 5e^{5x}$

3. $f'(x) = xe^x + e^x$

5. $f'(x) = -0.5e^{-0.05x}$

7. $f'(x) = (6x^2 + 20x + 33)e^{6x}$

9. $f'(x) = -6e^x(1 - 3e^x)$

11. $f'(x) = \dfrac{3}{2\sqrt{3x}}e^{\sqrt{3x}}$

13. $f'(x) = \dfrac{3}{x}$

15. $f'(x) = 2x \ln x + x$

17. $f'(x) = \dfrac{2}{x}e^{2x/3}$

19. $f'(x) = \dfrac{-2}{(x + 1)(x - 1)}$

21. $f'(x) = -2e^{-2x} + 3x^2$

23. $g'(s) = (e^s + 1)(2e^{-s} + 1) +$
 $(e^s + s + 1)(-2e^{-s} + 1)$
 $= 1 + 2s + e^s + se^s - 2se^{-s}$

25. $h'(t) = \dfrac{te^t \ln t + t \ln t - e^t - t}{t(\ln t)^2}$

27. $f'(x) = \dfrac{e^x - e^{-x}}{2}$

29. $f'(t) = \dfrac{t+1}{2t\sqrt{\ln t + t}}$

31. $f'(x) = \dfrac{1 - e^{-x}}{x = e^{-x}}$

33. $g'(u) = \dfrac{1}{\sqrt{u^2 + 1}}$

35. $f'(x) = \dfrac{2^x(x \ln 2 - 1)}{x^2}$

37. $\dfrac{1 + \ln x}{\ln 10}$

39. $e;\ 1$

41. $3e^{-4/3};\ -1$

43. $\dfrac{3\sqrt{3}}{8}e^{-3/2};\ 0$

45. $\dfrac{1}{e};\ 0$

47. $y = x$

49. $y = e^2$

51. $y = \dfrac{1}{2}x - \dfrac{1}{2}$

53. $f''(x) = 4e^{2x} + 2e^{-x}$

55. $f''(t) = 2 \ln t + 3$

57. $f'(x) = f(x)\left[\dfrac{4}{2x + 3} + \dfrac{1 - 10x}{2(x - 5x^2)}\right]$

59. $f'(x) = f(x)\left[\dfrac{5}{x + 2} - \dfrac{1}{2(3x - 5)}\right]$

61. $f'(x) = f(x)\left[\dfrac{3}{x + 1} - \dfrac{2}{6 - x} + \dfrac{2}{3(2x + 1)}\right]$

63. $f'(x) = (2 \ln 5)x5^{x^2}$

65. a. $E(p) = -0.04p$; elastic for $p > 25$, inelastic for $p < 25$, of unit elasticity for $p = 25$
   b. Demand will decrease by approximately 1.2%.
   c. $R(p) = 3000pe^{-0.04p};\ p = 25$

67. a. $E(p) = \dfrac{-p^2 - p}{10(p + 11)}$; elastic when $p > 15.91$, inelastic when $p < 15.91$, of unit elasticity when $p = 15.91$
   b. Demand will decrease by approximately 1.85%.
   c. $R(p) = 5000\,p(p + 11)\,e^{-0.1p};$ $p = 15.91$

69. a. $C'(x) = 0.2e^{0.2x}$
   b. 5 units

71. a. $C'(x) = \dfrac{6\,e^{x/10}}{\sqrt{x}}\left(1 + \dfrac{x}{5}\right)$
   b. 5 units

73. a. Value is decreasing at the rate of $1082.68 per year.
   b. Constant rate of -40% per year

75. a. Population is increasing at the rate of 1.22 million people per year.
   b. Constant rate of 2% per year

77. a. Approximately 406 copies
   b. 368 copies

79. a. $F'(t) = -k(1 - B)e^{-kt}$ is the rate at which you are forgetting material.
   b. $F'(t) = -k(F(t) - B)$, which says that your rate of forgetting is proportional to the fraction you have left to forget.
   c.

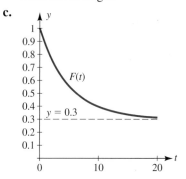

81. a. $E'(p) = 3000e^{-0.01p}(1 - 0.01p)$
   b. $p = 100$
   c. $p = 200$

83. a. $N'(t) = \dfrac{36e^{-0.02t}}{(1 + 3e^{-0.02t})^2}$ the population is increasing at all times t.
   b. Increasing for $t < 50 \ln 3$, decreasing for $t > 50 \ln 3$
   c. $N(t)$ approaches 600.

85. a. $P_1'(10) \approx 1.556$ cm/day
   $P_2'(10) = -0.257$ cm/day per day; decreasing
   b. Plants have the same height, approximately 20 cm, after 20.71 days; $P_1'(20.71) \approx 0.286$, $P_2'(20.71) \approx 0.001$, so the first plant is growing more rapidly.

ANSWERS

87. $\dfrac{R'(t_0)}{R(t_0)} = \dfrac{0.09(11) - 0.02(8)}{19} \approx 0.0437$, or 4.37%

89. a. $b^x = e^{x \ln b}$

$\dfrac{d}{dx}(b^x) = e^{x \ln b}[\ln b]$

$= (\ln b)b^x$

b. $y = b^x$

$\ln y = x \ln b$

$\dfrac{1}{y}\dfrac{dy}{dx} = \ln b$

$\dfrac{dy}{dx} = (\ln b)y = (\ln b)b^x$

CHAPTER 4 Section 4

1. $f_5(x)$

3. $f_3(x)$

5. $f(t)$ is increasing for all real t; concave upward for all real t. There is a horizontal asymptote at $y = 2$.

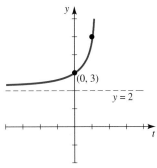

7. $g(x)$ is decreasing for all real x; concave downward for all real x; $y = 2$ is a horizontal asymptote.

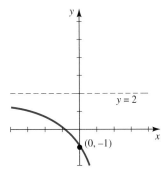

9. $f(x)$ is increasing for all real x; concave upward for $x < 0.549$; concave downward for $x > 0.549$. Inflection point is (0.549, 1), and $y = 2$ and $y = 0$ are horizontal asymptotes.

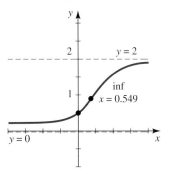

11. $f(x)$ is increasing for $x > -1$; decreasing for $x < -1$; concave upward for $x > -2$; concave downward for $x < -2$. Relative minimum is $\left(-1, -\dfrac{1}{e}\right)$ and inflection point is $\left(-2, -\dfrac{2}{e^2}\right)$. The x axis ($y = 0$) is a horizontal asymptote.

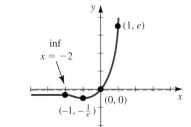

13. $f(x)$ is increasing for $x < 1$; decreasing for $x > 1$; concave upward for $x > 2$; concave downward for $x < 2$. Relative maximum is $(1, e)$, inflection point is $(2, 2)$. The x axis ($y = 0$) is a horizontal asymptote.

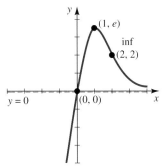

15. $f(x)$ is increasing for $0 < x < 2$; decreasing for $x < 0$ and $x > 2$; concave upward for $x < 0.6$ and $x > 3.4$; concave downward for $0.6 < x < 3.4$. Relative minimum is $(0, 0)$; relative maximum is $\left(2, \dfrac{4}{e^2}\right)$, inflection points are $(0.6, 0.2)$ and $(3.4, 0.4)$. The x axis is a horizontal asymptote.

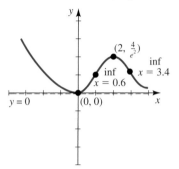

17. $f(x)$ is increasing for all real x; concave upward for $x < 0$; concave downward for $x > 0$. Inflection point is $(0, 3)$. The x axis $(y = 0)$ and $y = 6$ are horizontal asymptotes.

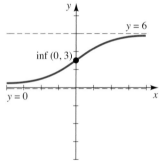

19. $f(x)$ is increasing for $x > 1$; decreasing for $x < 1$; concave upward for $x < e$; concave downward for $x > e$. Relative minimum is $(1, 0)$; inflection point is $(e, 1)$. The y axis $(x = 0)$ is a vertical asymptote.

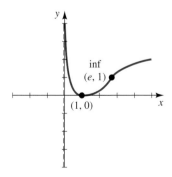

21. 36 billion hamburgers

23. 202.5 million

25. **a.** As $t \to \infty$, $f(t) \to 1$

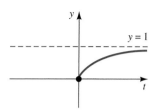

 b. 0.741

 c. 0.089

27. **a.** $A = 85$, $k = \dfrac{1}{20} \ln \dfrac{17}{6}$

 b.

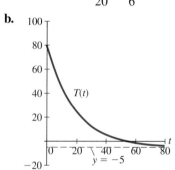

 The temperature approaches $-5°C$.

 c. $12.8°C$

 d. 54.4 minutes

29. **a.**

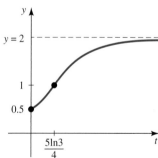

 b. 500

 c. 1572

 d. 2000

31. 37.5 shelves per day

33. **a.** Approximately 403 copies

 b. 348 copies

35. **a.** $e - 1 \approx 1.7$ years

 b. The learning rate $L'(t)$ is largest when $t = 0$ (at birth).

ANSWERS

37. **a.** minimum at $x = 0$
   b. 2.55 m

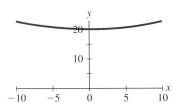

39. **a.** $P(x) = 1000e^{-0.02x}(x - 125)$

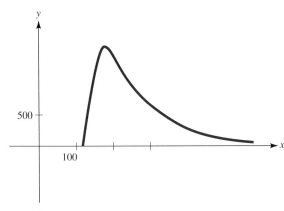

   b. \$175

41. **a.** 0.15% per year
   b. 70.24 years; 0.15% per year

43. 69.44 years from now

45. **a.** $V(5) = \$207.64$;
$$V'(t) = V_o\left(1 - \frac{2}{L}\right)^t \ln\left(1 - \frac{2}{L}\right)$$
$V'(5) \approx -\$60/\text{year}$
   b. $100 \ln\left(1 - \frac{2}{L}\right)$

47. **a.** $E(t) = 1000w(t)p(t)$
   b. $t = 82$; 3527 kg
   c.

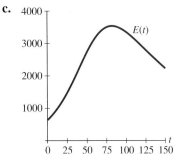

49. **a.** $C = 9$; $k = \frac{1}{2}\ln 3$
   b. 4 hours

51. **a.** $N(0) = 15$ employees; $N(5) = 482$ employees; 2.10 years; 500 employees
   b.

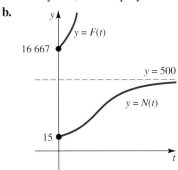

53. **a.** $C'(t) = Ae^{-kt}(1 - kt)$; $C(t)$ is increasing for $t < \frac{1}{k}$ and decreasing for $t > \frac{1}{k}$. Maximum concentration is $\frac{A}{ke}$ at $t = \frac{1}{k}$.
   b. $C''(t) = kAe^{-kt}(kt - 2)$, the graph of $C(t)$ is concave upward for $t > \frac{2}{k}$ and is concave downward for $t > \frac{2}{k}$. Then $\left(\frac{2}{k}, \frac{2A}{ke^2}\right)$ is the inflection point. The rate of change of drug concentration is a minimum at the inflection point.
   c.

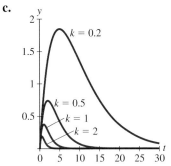

The point where the maximum occurs shifts left as k increases and the maximum height decreases.

55.

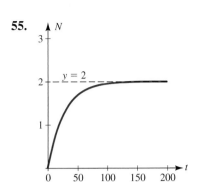

The value of N approaches the maximum of 2 million people.

57. Answers will vary.

59. a. $f'(x) = -\dfrac{1}{\sigma^3\sqrt{2\pi}}(x - \mu)e^{-(x-\mu)^2/2\sigma^2}$

$f''(x) = -\dfrac{1}{\sigma^5\sqrt{2\pi}}[-\sigma^2 + (x - \mu)^2]e^{-(x-\mu)^2/2\sigma^2}$

$f'(x) = 0$ when $x = \mu$; $f'(x) > 0$ when $x < \mu$; $f'(x) < 0$ when $x > \mu$. Thus, $f(x)$ has one absolute maximum at $x = \mu$.
$[(\mu \pm \sigma) - \mu]^2 = \sigma^2$ so $f''(x) = 0$ when $x = \mu + \sigma$ and $x = \mu - \sigma$. Thus, $f(x)$ has inflection points at $x = \mu + \sigma$ and $x = \mu - \sigma$.

b. $[(\mu + c) - \mu]^2 = c^2$ so $f(\mu + c) = f(\mu - c)$ for every number c. Thus, the graph of $f(x)$ is symmetric about the line $x = \mu$.

CHAPTER 4   Checkup

1. a. 1

b. $\dfrac{10}{3}$

c. 0

d. $\dfrac{16}{81}$

2. a. $27x^6y^3$

b. $\dfrac{1}{\sqrt{3}xy^{2/3}}$

c. $\dfrac{y^{7/6}}{x^{1/6}}$

d. $\dfrac{1}{x^{6.5}y^8}$

3. a. $x = 3, x = -1$

b. $x = \dfrac{1}{\ln 4}$

c. $x = -4, x = 4$

d. $t = 2\ln\dfrac{11}{3}$

4. a. $\dfrac{dy}{dx} = \dfrac{e^x(x^2 - 5x + 3)}{(x^2 + 3x)^2}$

b. $\dfrac{dy}{dx} = \dfrac{3x^2 + 4x - 3}{x^3 + 2x^2 - 3x}$

c. $\dfrac{dy}{dx} = x^2(1 + 3\ln x)$

d. $\dfrac{dy}{dx} = y\left(-2 + \dfrac{6}{2x - 1} + \dfrac{2x}{1 - x^2}\right)$

5. a. Relative maximum at $x = 2$, relative minimum at $x = 0$

b. Relative maximum at $\sqrt{e}$

c. Relative maximum at $x = 1/4$

d. $f(x)$ is increasing for all x, no relative max or min

6. $\$2323.67$; 8.1 years

7. a. $\$4323.25$

b. $\$4282.09$

8. a. Increasing for $0 \le t < e - 1$; decreasing for $t > e - 1$

b. $t = e^{3/2} - 1$

c. The price approaches $\$500$.

9. a. $q'(p) = -1000e^{-p}(p + 1) < 0$ for $p \ge 0$

b. $p = \sqrt{2}$ $\$141.42$; $\$117\,384.14$

10. 6601 years old

11. a. 80 000

b. 2 hours; 81 873

c. The population dies off completely.

CHAPTER 4   Review Exercises

1.

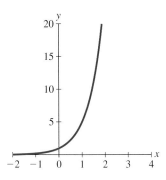

3.

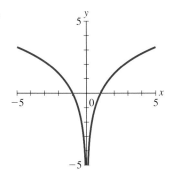

5. **a.** $f(4) = \dfrac{3125}{8}$

    b. $f(3) = \dfrac{100}{3}$

    c. $f(9) = \dfrac{65}{2}$

    d. $f(10) = \dfrac{6}{2}$

7. $x = 25 \ln 4$

9. $x = e^2$

11. $x = \dfrac{41}{2}$

13. $x = 0$

15. $\dfrac{dy}{dx} = xe^{-x}(2 - x)$

17. $\dfrac{dy}{dx} = 2 \ln x + 2$

19. $\dfrac{dy}{dx} = \dfrac{2}{x \ln 3}$

21. $\dfrac{dy}{dx} = e^x$ (Note that $y = e^x$.)

23. $\dfrac{dy}{dx} = \dfrac{-(1 + 2e^{-x})}{1 + e^{-x}}$

25. $\dfrac{dy}{dx} = \dfrac{-e^{-x}(x^2 + x + 1 + x \ln x)}{x(x + \ln x)^2}$

27. $\dfrac{dy}{dx} = \dfrac{ye^{x-x^2}(2x - 1) + 1}{e^{x-x^2} - 1}$

29. $\dfrac{dy}{dx} = 2y\left[\dfrac{3x + 3e^{2x}}{x^2 + e^{2x}} - 1 - \dfrac{1 - 2x}{3(1 + x - x^2)}\right]$

31. $f(x)$ is increasing for all x; concave upward for $x > 0$; concave downward for $x < 0$. There is an inflection point at $x = 0$.

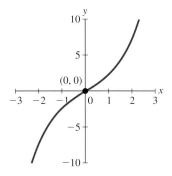

33. $f(t)$ is increasing for $t > 0$; decreasing for $t < 0$; concave upward for all t. There is a relative minimum at $t = 0$. There are no inflection points.

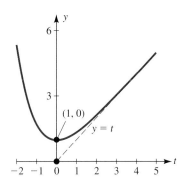

35. $F(u)$ is increasing for $-2 < u < -1$ and $u > -1$; concave upward for $u > -1$; concave downward for $-2 < u < -1$. There is an inflection point at $u = -1$, and $u = -2$ is a vertical asymptote.

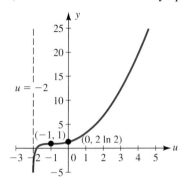

37. $G(x)$ is decreasing for all x. $G(x)$ is concave upward for all x.

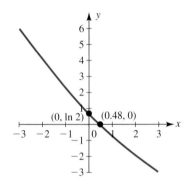

39. $\ln 4$; $\ln 3$

41. $\left(e + \dfrac{1}{e}\right)^5$; 32

43. $y = 2x - 2$

45. $y = 4x$

47. 8

49. The original investment will have quadrupled.

51. 204.8 grams

53. 20 480 bacteria

55. a.

(0, 10)

$y = 50$

b. 10 000

c. 32 027

d. 9.81 thousand dollars ($9808.29)

e. Just under 50 000 units

57. a. 11.57 years

 b. 11.45 years

59. a. $4975.96

 b. $5488.12

61. 8.20% per year compounded continuously

63. a.

$y = 30$

b. 10 million

 c. 17.28 million (17 283 507)

 d. The population will approach 30 million.

65. a. 0.13 parts per million per year

 b. Constant rate of 3%

67. After 200 years

69. a. Since $\lambda = \dfrac{\ln 2}{k}$, we have $k = \dfrac{\ln 2}{\lambda}$ and

$Q(t) = Q_0 e^{-(\ln 2/\lambda)t}$.

 b. $Q_0(0.5)^{kt} = Q_0 e^{-(\ln 2/\lambda)t}$

$kt \ln 0.5 = -\left(\dfrac{\ln 2}{\lambda}\right)t$

So, $k = \dfrac{1}{\lambda}$

71. a. 2.31×10^{-200} percent left, too little for proper measurement.

 b. Writing exercise.

73. 0.8110 minutes = 48.66 seconds; $-8.64°C$ per minute

75. a. $A = 5e, k = \dfrac{1}{2}$

b. $t = 9.78$ hours

77. $C = \dfrac{37}{3}, k = 0.021,$

$P(50) \approx 7.52$ billion

79. a. $D(10) = 0.00195; D(25) = 0.000591$

b.

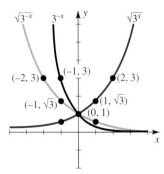

81. Let $G(t)$ denote the country's GDP in billions of dollars in t years where $t = 0$ represents 2000. $G(t)$ has the form $G(t) = Ae^{kt}$. We know
$$G(0) = Ae^0 = A = 100 \quad \text{and}$$
$$G(10) = Ae^{10k} = 100e^{10k} = 165.$$

Solving for k gives $k = \dfrac{1}{10} \ln \dfrac{165}{100}$. In the year 2010, $t = 20$, and the GDP is estimated to be
$$G(20) = 100e^{[(\ln 1.65)/10](20)}$$
$$\approx 272.25 \text{ billion dollars}$$

83.

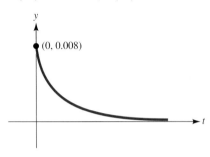

85. $y = 3^x$

$y = 4 - \ln\sqrt{x}$

The graphs intersect at (1.2373, 3.8935).

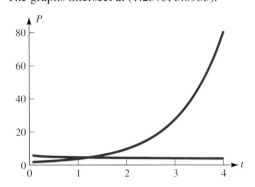

CHAPTER 5   Section 1

1. $-3x + C$

3. $\dfrac{x^6}{6} + C$

5. $-\dfrac{1}{x} + C$

7. $4\sqrt{t} + C$

9. $\dfrac{5}{3}u^{3/5} + C$

11. $t^3 - \dfrac{2\sqrt{5}}{3}t^{3/2} + 2t + C$

13. $2y^{3/2} + y^{-2} + C$

15. $\dfrac{e^x}{2} + \dfrac{2}{5}x^{5/2} + C$

17. $\dfrac{u^{1.1}}{3.3} + \dfrac{u^{2.1}}{2.1} + C$

19. $x + \ln x^2 - \dfrac{1}{x} + C$

21. $-\dfrac{5}{4}x^4 + \dfrac{11}{3}x^3 - x^2 + C$

23. $\dfrac{2}{7}t^{7/2} - \dfrac{2}{3}t^{3/2} + C$

25. $\dfrac{1}{2}e^{2t} + 2e^t + t + C$

27. $\dfrac{1}{3}\ln|y| - 10\sqrt{y} - 2e^{-y/2} + C$

29. $\dfrac{2}{5}t^{5/2} - \dfrac{2}{3}t^{3/2} + 4t^{1/2} + C$

31. $y = \dfrac{3}{2}x^2 - 2x - \dfrac{3}{2}$

33. $y = \ln x^2 + \dfrac{1}{x} - 2$

35. $f(x) = 2x^2 + x - 1$

37. $f(x) = -\dfrac{1}{3}x^3 - \dfrac{1}{2}x^2 + \dfrac{31}{6}$

39. $f(x) = \dfrac{x^4}{4} + \dfrac{2}{x} + 2x - \dfrac{5}{4}$

41. $f(x) = -e^{-x} + \dfrac{x^3}{3} + 5$

43. \$22 360

45. \$646.20

47. 3253

49. 10 128 people

51. **a.** $18\dfrac{1}{3}$ (18 items)

   b. $48\dfrac{1}{3}$ (48 items)

53. **a.** $T(t) = 16 - 20e^{-0.35t}$
   b. 6.1°C
   c. 3.44 hours

55. **a.** $P(q) = 100q - q^2 - 200$
   b. $q = 50$; \$2300

57. $c(x) = 0.9x + 0.2x^{3/2} + 10$

59. The car travels $39.2 + 1.96$ m $= 41.16$ m so does not hit the moose.

61. **a.** $f'(x)$ is maximized when $x = 10$; 7 items per minute
   b. $f(x) = x + 0.6x^2 - 0.02x^3$
   c. $f'(20.8) = 0; f(20.8) \approx 100$ items

63. $v(r) = \dfrac{1}{2}a(R^2 - r^2)$

65. 20 metres

67. $\displaystyle\int b^x\, dx = \int e^{x \ln b}\, dx = \dfrac{e^{x \ln b}}{\ln b} + C = \dfrac{b^x}{\ln b} + C$

69. **a.** $v(t) = -23t + 67; s(t) = -\dfrac{23}{2}t^2 + 67t$

   b.

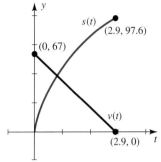

   c. $v(t) = 0$ when $t = 2.9$ min and $s(2.9) = 97.6$ cm; $s(t) = 45$ cm when $t \approx 0.77$ min or 5.05 min and $v(0.77) \approx 49.2$ cm/min while $v(5.05) \approx -49.15$.

CHAPTER 5   Section 2

1. **a.** $u = 3x + 4$
   b. $u = 3 - x$
   c. $u = 2 - t^2$
   d. $u = 2 + t^2$

3. $\dfrac{1}{12}(2x + 6)^6 + C$

5. $\dfrac{1}{6}(4x - 1)^{3/2} + C$

7. $-e^{1+x} + C$

9. $\dfrac{1}{2}ex^2 + C$

11. $\dfrac{1}{12}(t^2 + 1)^6 + C$

13. $\dfrac{4}{21}(x^3 + 1)^{7/4} + C$

15. $\dfrac{2}{5}\ln|y^5 + 1| + C$

17. $\dfrac{1}{26}(x^2 + 2x + 5)^{13} + C$

19. $\dfrac{3}{5}\ln|x^5 + 5x^4 + 10x + 12| + C$

21. $-\dfrac{3}{2}\left(\dfrac{1}{u^2 - 2u + 6}\right) + C$

23. $\dfrac{1}{2}(\ln 5x)^2 + C$

ANSWERS

25. $\dfrac{-1}{\ln x} + C$

27. $\dfrac{1}{2}[\ln (x^2 + 1)]^2 + C$

29. $\ln |e^x - e^{-x}| + C$

31. $\dfrac{1}{2}x - \dfrac{1}{4}\ln |2x + 1| + C$

33. $\dfrac{1}{10}(2x + 1)^{5/2} - \dfrac{1}{6}(2x + 1)^{3/2} + C$

35. $2 \ln (\sqrt{x} + 1) + C$

37. $y = -\dfrac{1}{6}(3 - 2x)^3 + \dfrac{9}{2}$

39. $y = \ln |x + 1| + 1$

41. $y = \dfrac{1}{2} \ln |x^2 + 4x + 5| - \dfrac{1}{2}\ln 2 + 3$

43. $f(x) = \dfrac{1}{5} - \dfrac{1}{5}(1 - 2x)^{5/2}$

45. $f(x) = \dfrac{3}{2} - \dfrac{1}{2}e^{4-x^2}$

47. **a.** $x(t) = -\dfrac{4}{9}(3t + 1)^{3/2} + \dfrac{40}{9}$
    b. $x(4) = -16.4$
    c. $t = 0.4$

49. **a.** $x(t) = \sqrt{2t + 1} - 1$
    b. $x(4) = 2$
    c. $t = \dfrac{15}{2}$

51. **a.** $C(q) = (q - 4)^3 + 64 + k$, where k is the overhead
    b. $1500

53. 2.3 metres

55. **a.** $R(x) = 50x - 175e^{-0.01x^2} + 175$
    b. $610.6

57. **a.** $C(t) = \dfrac{1}{e^{0.01t} + 1}$
    b. 0.3543 mg/cm^3; 0.1419 mg/cm^3
    c. 294 minutes

59. **a.** $L(t) = 0.03\sqrt{-t^2 + 16t + 36} + 0.07$; at $t = 8$
    (3:00 P.M.); 0.37 parts per million

b. The ozone level at 11:00 A.M. ($t = 4$) is
$L(4) = 0.345$. The same level occurs at $t = 12$
(7:00 P.M.).

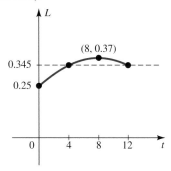

61. **a.** $p(x) = \dfrac{300}{\sqrt{x^2 + 9}} + 15$
    b. $66.45; $115
    c. 265

63. Profit declines by $93 733.

65. $\dfrac{3}{7}(x^{2/3} + 1)^{7/2} - \dfrac{3}{5}(x^{2/3} + 1)^{5/2} + C$

67. $e^x + 1 - \ln (e^x + 1) + C$

CHAPTER 5   Section 3

1. 15

3. $\dfrac{95}{2}$

5. $\dfrac{6}{5}$

7. $-\dfrac{6}{5}$

9. $3 - \dfrac{4}{e}$

11. 1.95

13. 144

15. $\dfrac{8}{3} + \ln 3 \approx 3.7653$

17. $\dfrac{2}{9}$

19. 3.2

21. $\dfrac{4}{3}$

23. $\dfrac{7}{6}$

25. e

27. $\dfrac{8}{3}$

29. $e^3 - e^2$

31. -20

33. 0

35. 5

37. 3

39. $\dfrac{33}{5}$

41. $\dfrac{112}{9}$

43. 4

45. $\dfrac{3}{2} \ln 3 \approx 1.6479$

47. $V(5) - V(0)$

49. \$480

51. 0.75 ppm

53. About 98 people

55. \$75

57. $1500\left(\dfrac{3}{2} + \dfrac{5}{4} \ln \dfrac{11}{9}\right) \approx 2626$ cell phones

59. **a.** $-\$48\,036.33$
    b. \$28 546.52

61. The concentration decreases by 0.8283 mg/cm^3.

63. A decrease of \$1870

65. $2 \ln 2 \approx 1.386$ grams

67. $8\sqrt{11} - 8\sqrt{6}$ or about 7 facts

69. 45.9 m

71. **a.** $\dfrac{\pi}{4}$

    b. $\dfrac{\pi}{4}$; part of the area under the circle
        $(x - 1)^2 + y^2 = 1$

CHAPTER 5   Section 4

1. $\dfrac{5}{12}$

3. $2 \ln 2 - \dfrac{1}{2}$

5. Area = 1

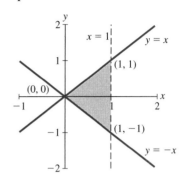

7. Area = $\dfrac{4}{3}$

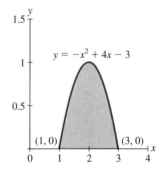

9. Area = $\dfrac{4}{3}$

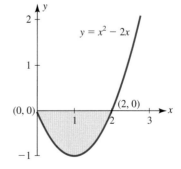

11. Area = 9

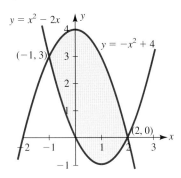

13. Area = $\dfrac{443}{6}$

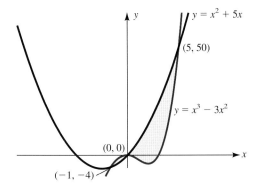

15. Area = 18

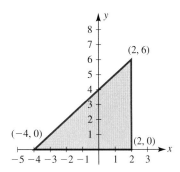

17. Area = 14

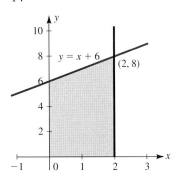

19. -2

21. $\dfrac{3}{2}\left(e - \dfrac{1}{e}\right)$

23. $\dfrac{\ln 5 - \ln 3}{\ln 3}$

25. Average value = $\dfrac{2}{3}$

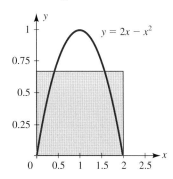

27. Average value = $\dfrac{\ln 2}{2}$

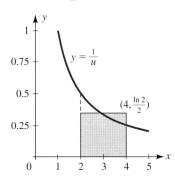

29. $\dfrac{1}{2}$

31. 0.1833

33. 0.383

35. 2400 units

37. 30 000 kg

39. 2272.2

41. a. $11 361.02

    b. Writing exercise; responses will vary.

43. a. 16 years
b. $209 067
c.

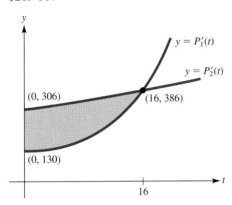

45. a. 14.7 years
b. $582 221
c.

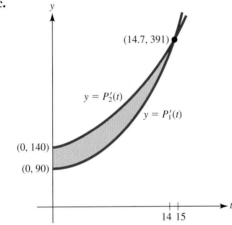

47. 0.412 million (412 000)

49. $\frac{1}{40}$ mg/cm^3

51. a. 0°C
b. 8 A.M. and 2 P.M.

53. a. 39.25 km/hour
b. 3:30 P.M.

55. a. $M_0 + 20.833$
b. $t = \sqrt{5}, M(\sqrt{5}) = M_0 + 50\sqrt{5}/e$

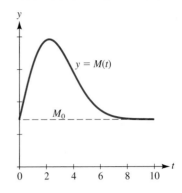

57. Baseball: $\frac{1}{3}$; football: $\frac{5}{18}$; basketball: $\frac{9}{25}$. Football is the most equitable sport, basketball the least.

59. 5710 people

61. $241 223.76

63. a. $S' = F''(M) = \frac{1}{3}(2k - 6M) = 0$ for $M = \frac{k}{3}$.
Maximum since $S'' = -2 < 0$.
b. $\frac{k^3}{108}$

65.

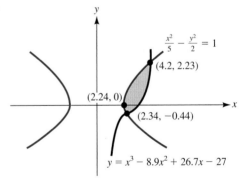

$$A = \int_{\sqrt{5}}^{2.34}\left[\sqrt{\frac{2x^2}{5} - 2} - \left(-\sqrt{\frac{2x^2}{5} - 2}\right)\right]dx$$
$$+ \int_{2.34}^{4.2}\left[\sqrt{\frac{2x^2}{5} - 2} - (x^3 - 8.9x^2 + 26.7x - 27)\right]dx$$
$$\approx 2.097$$

67. Writing exercise; responses will vary.

CHAPTER 5   Section 5

1. $17 182.82

3. **a.** 11 years
   b. $26 620
   c.

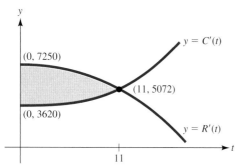

5. **a.** 8
   b. $15 069
   c. Net earnings are represented as the area between the curve $R'(t) = 6537e^{-0.3t}$ and the horizontal line $y = 593$.

7. $237 730; $319 453

9. $5308.78

11. The $50 000 plan is better, producing net income of $37 465 versus $22 479 for the $30 000 plan over 5 years.

13. **a.** $P(t) = 32.5e^{0.04t} - 32.5$; 4.14 billion barrels; 4.67 billion barrels
   b. 12 years
   c. 1646.44 billion dollars
   d. Answers will vary.

15. **a.** $P(t) = 60e^{0.02t} - 60$; 3.71 billion barrels; 3.94 billion barrels
   b. 9.12 years
   c. 1218 billion dollars
   d. Answers will vary.

17. $1 929 148

19. **a.** $137 334.29
   b. $44 585.04

21. $1 287 360

23.
$$FV = \int_0^T f(t)\, e^{r(T-t)}\, dt$$
$$= \int_0^T M e^{r(T-t)}\, dt$$
$$= M e^{rT} \int_0^T e^{-rt}\, dt$$
$$= M e^{rT}\left(\frac{1}{r} - \frac{e^{-rT}}{r}\right)$$
$$= \frac{M}{r}(e^{rT} - 1)$$

CHAPTER 5   Section 6

1. 30 484

3. 468 130

5. 451 404

7. $7\pi \approx 21.99$ cubic units

9. $\dfrac{1532}{15}\pi \approx 320.86$ cubic units

11. $\dfrac{32}{3}\pi \approx 33.51$ cubic units

13. $2\pi \approx 6.28$ cubic units

15. 61 070 138

17. About 80 members

19. 4097.62 (4098 people)

21. 515.48 billion barrels

23. 4207 members

25. **a.** The LDL level decreases 6.16 units.
   b. $L(t) = \dfrac{3}{28}(49 - t^2)^{1.4} + 120 - \dfrac{21}{4}(49)^{0.4}$
   c. 5.8 days

27. Approximately 208 128 people

29. After t days, the first population is
$$P_1(t) = \left(100\,000 - \frac{50}{0.011}\right)e^{-0.011t} + \frac{50}{0.011}$$
Then $P_1(t) < P(t)$ for $t = 50$ and $t = 100$, but $P_1(300) > P(300)$.

31. 1565.83 (1566 animals)

33. 10 125 people

35. $\int_0^{12} [W'(t) - D'(t)]\,dt = 0.363$;

about 36 people; 18.1%

37. a. 55 years
   b. 70.78 years
   c. 86.36 years; they have already exceeded their life expectancy.
   d. 71.69 years

39. a. 2.37 s
   b. 0.905 L
   c. 0.382 L/s

41. a.

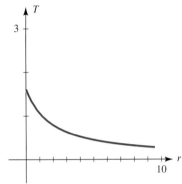

   b. $r(T) = \dfrac{3}{T} - 2$

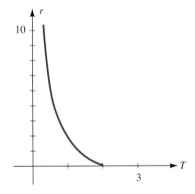

   c. $\pi\left[12\left(\ln\dfrac{1}{3} - \ln\dfrac{3}{2}\right) + \dfrac{77}{3}\right] \approx 23.93 \text{ m}^3$

43. a. $100\pi \ln\dfrac{23}{5} \approx 479.42$ units

   b. $L = \dfrac{3\sqrt{10}}{2} \approx 4.74$ kilometres;

   $100\pi \ln 10 \approx 723.38$ units.

45. The hypotenuse of the triangle has equation $y = \dfrac{r}{h}x$, and the volume is

$$V = \pi\int_0^h \left(\frac{r}{h}x\right)^2 dx = \pi\int_0^h \frac{r^2}{h^2}x^2\,dx$$

$$= \frac{\pi r^2}{h^2}\left(\frac{1}{3}x^3\right)\Big|_0^h = \frac{\pi r^2}{3h^2}(h^3 - 0)$$

$$= \frac{1}{3}\pi r^2 h$$

CHAPTER 5   Checkup

1. a. $\dfrac{x^4}{4} - \dfrac{2\sqrt{3}}{3}x^{3/2} - \dfrac{5}{2}e^{-2x} + C$

   b. $\dfrac{x^2}{2} - 2x + 4\ln|x| + C$

   c. $\dfrac{2}{7}x^{7/2} - 2x^{1/2} + C$

   d. $\dfrac{-1}{2\sqrt{3 + 2x^2}} + C$

   e. $\dfrac{1}{4}(\ln x)^2 + C$

   f. $\dfrac{1}{2}e^{1-x^2} + C$

2. a. $\dfrac{62}{5} + 4\ln 2$

   b. $e^3 - 1$

   c. $1 - \ln 2$

   d. $\sqrt{31} - 2$

3. a. $\dfrac{73}{6}$

   b. 36

4. $1 - 2\ln 2$

5. $10\,333.33

6. 71.14 billion dollars; increase

7. $16\,183.42

8. 45\,055

9. 0.1 mg/cm^3

ANSWERS

CHAPTER 5 Review Exercises

1. $\dfrac{1}{4}x^4 + \dfrac{2}{3}x^{3/2} - 9x + C$

3. $\dfrac{x^5}{5} + \dfrac{5}{2}e^{-2x} + C$

5. $\dfrac{5}{3}x^3 - 3\ln|x| + C$

7. $\dfrac{1}{6}t^6 - t^3 - \dfrac{1}{t} + C$

9. $\dfrac{2}{9}(3x + 1)^{3/2} + C$

11. $\dfrac{1}{12}(x^2 + 4x + 2)^6 + C$

13. $\dfrac{-3}{4(2x^2 + 8x + 3)} + C$

15. $\dfrac{1}{14}(y - 5)^{14} + \dfrac{5}{13}(y - 5)^{13} + C$

17. $-\dfrac{5}{2}e^{-x^2} + C$

19. $\dfrac{2}{3}(\ln x)^{3/2} + C$

21. 0

23. $\dfrac{1}{2}(e^2 + 5)$

25. 1,710

27. $1 - \dfrac{1}{e}$

29. $e - 2$

31. Area $= \dfrac{101}{6}$

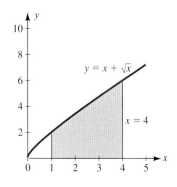

33. Area $= \ln 2 + \dfrac{7}{3}$

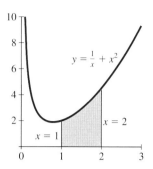

35. Area $= \dfrac{15}{2} - 8\ln 2$

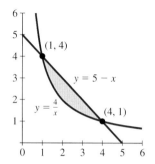

37. Area $= \dfrac{9}{2}$

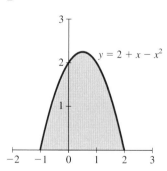

39. $\dfrac{11\,407}{84} - \dfrac{2\sqrt{2}}{21} \approx 135.7$

41. $\dfrac{1}{4}\left(1 - \dfrac{1}{e^4}\right)$

43. $GI = \dfrac{1}{5}$

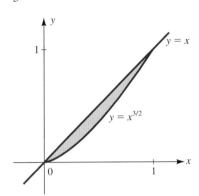

45. $GI = \dfrac{1}{10}$

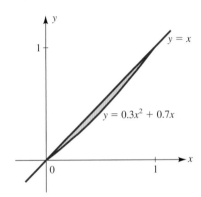

47. 43 984

49. 14 308

51. $\dfrac{78}{5}\pi \approx 49.01$ cubic units

53. $\pi \ln 3 \approx 3.45$ cubic units

55. $y = 2x + 10$

57. $x = \dfrac{9}{2} - \dfrac{1}{2}e^{-2t}$

59. $y = \dfrac{1}{2}\ln(x^2 + 1) + 5 - \dfrac{1}{2}\ln 2$

61. $87.57

63. 1220 people

65. 11 250 commuters

67. In 2006 (0.2554 billion barrels versus 0.1003 billion barrels in 2009)

69. $7377.37

71. 61.65 (about 62 homes)

73. 14 868 kilograms

75. $3 447 360

77. Temperature decreases by 2.88°C

79. **a.** $p_1(x) = 0.2x + 0.001x^3 + 250$; $p_1(10) = 2.53$ per dozen
  b. $p_2(x) = 0.3x + 0.001x^3 + 250$; $p_2(10) = 2.54$ per dozen

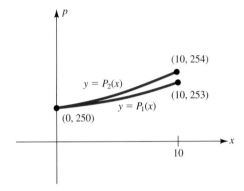

81. 30 metres

83. Physical therapists

85. 2255 trout

87. **a.** $\dfrac{1}{N}\displaystyle\int_0^N S(t)\,dt$

  b. $\displaystyle\int_0^N S(t)\,dt$

  c. The average speed is equal to the total distance divided by the total number of hours.

89. The region bounded by the curves is between $x = -4.66$ and $x = -1.82$; the curves also intersect at $x = 4.98$. The area is approximately 3.

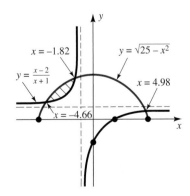

CHAPTER 6 Section 1

1. $-(x + 1)e^{-x} + C$

3. $(2 - x)e^x + C$

5. $\dfrac{t^2}{2}\left(\ln 2t - \dfrac{1}{2}\right) + C$

7. $-5(v + 5)e^{-v/5} + C$

9. $\dfrac{2}{3}x(x - 6)^{3/2} - \dfrac{4}{15}(x - 6)^{5/2} + C$

11. $\dfrac{1}{9}x(x + 1)^9 - \dfrac{1}{90}(x + 1)^{10} + C$

13. $2x\sqrt{x + 2} - \dfrac{4}{3}(x + 2)^{3/2} + C$

15. $\dfrac{8}{3}$

17. $\dfrac{1}{4}(1 - 3e^{-2})$

19. $\dfrac{1}{12}(3e^4 + 1)$

21. $\dfrac{1}{16}(e^2 + 1)$

23. $-\dfrac{1}{x}(\ln x + 1) + C$

25. $\dfrac{1}{2}e^{x^2}(x^2 - 1) + C$

27. $\dfrac{1}{25}(3 - 5x - 3\ln|3 - 5x|) + C$

29. $\dfrac{-\sqrt{4x^2 - 9}}{x} + 2\ln|2x + \sqrt{4x^2 - 9}| + C$

31. $\dfrac{1}{2}\ln\left|\dfrac{x}{2 + 3x}\right| + C$

33. $\dfrac{\sqrt{3}}{24}\ln\left|\dfrac{4 + \sqrt{3}u}{4 - \sqrt{3}u}\right| + C$

35. $x(\ln x)^3 - 3x(\ln x)^2 + 6x\ln x - 6x + C$

37. $-\dfrac{1}{25}\left[\dfrac{5 + 4x}{x(5 + 2x)} + \dfrac{4}{5}\ln\left|\dfrac{x}{5 + 2x}\right|\right] + C$

39. $-(x + 1)e^{-x} - e^{-x} + 5 + \dfrac{3}{e}$

41. Approximately \$13 212.06

43. 176 units

45. Approximately 2 008 876 people

47. 29.4 mg/mL

49. \$39 220.72

51. \$11 417.35

53. 4367

55. For lawyers, the Gini index is 0.
For engineers, the Gini index is 0.47152
Lawyers show a more equitable distribution
of income.

$1 - \dfrac{2}{e} \approx 0.2642$

57. $I = \displaystyle\int u^n e^{au} du$

Let $\quad f = u^n \quad$ and $\quad dV = e^{au} du$

then $\quad df = nu^{n-1} du \quad$ and $\quad V = \dfrac{1}{a}e^{au}$

so that

$$I = \dfrac{1}{a}u^n e^{au} - \int \dfrac{1}{a}e^{au} \cdot nu^{n-1}\, du$$

$$= \dfrac{1}{a}u^n e^{au} - \dfrac{n}{a}\int u^{n-1}e^{au} du$$

59. (0.2437, 0.3528)

61. a. (3.481, 2.402)
    b. Writing exercise; responses will vary.

63. Area ≈ 0.7583

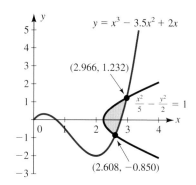

65. Area ≈ 1.95482

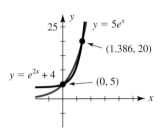

67. 4.227

69. 0.4509

CHAPTER 6 Section 2

1. $\dfrac{1}{2}$

3. Diverges

5. Diverges

7. $\dfrac{1}{10}$

9. $\dfrac{5}{2}$

11. $\dfrac{1}{9}$

13. Diverges

15. $\dfrac{2}{e}$

17. $\dfrac{2}{9}$

19. Diverges

21. Diverges

23. 2

25. $60 000

27. $600 000

29. 200

31. 50

33. **a.** Machine 1: $28 519
Machine 2: $20 222
The company should purchase machine 2.
 b. Writing exercise; responses will vary.

35. $PV = \lim\limits_{T \to +\infty} \displaystyle\int_0^T Q e^{-rt} dt$

$\quad = \lim\limits_{T \to +\infty} Q \left(\dfrac{e^{-rt}}{-r} \right) \Big|_1^T$

$\quad = \lim\limits_{T \to +\infty} Q \left(\dfrac{e^{-rt}}{-r} - \dfrac{1}{-r} \right)$

$\quad = \dfrac{Q}{r}$

37. $\displaystyle\int_0^{\infty} C(e^{-at} - e^{-bt}) dt$

$\quad = 1$ if $C = \dfrac{ab}{b-a}$.

CHAPTER 6 Section 3

1. **a.** 2.3438
 b. 2.3333

3. **a.** 0.7828
 b. 0.7854

5. **a.** 1.1515
 b. 1.1478

7. **a.** 0.7430
 b. 0.7469

9. **a.** 1.930756
 b. 1.922752

11. **a.** 1.096997
 b. 1.094800

13. **a.** 0.849195
 b. 0.836203

15. **a.** 0.5090 with $|E| \leq \dfrac{1}{32}$
 b. 0.5004 with $|E| \leq \dfrac{1}{384}$

17. **a.** 2.7968 with $|E| \leq \dfrac{1}{600}$
 b. 2.7974 with $|E| \leq \dfrac{1}{60 000}$

19. **a.** 1.4907 with $|E| \leq \dfrac{e}{32}$
 b. 1.4637 with $|E| \leq \dfrac{19e}{11 520}$

21. **a.** 164
 b. 18

23. **a.** 36
 b. 6

25. **a.** 179
 b. 8

27. **a.** 3.0898
 b. 3.1212

29. 0.1386

31. 0.358531 cubic units

33. $26 072.45

35. 51.75 km

37. $5949.70

39. 235 m^2

41. 475 197 people

43. GI $\approx$ 0.39425

CHAPTER 6   Checkup

1. **a.** $\dfrac{4\sqrt{2}x^{3/2}}{9}(-2 + 3\ln|x|) + C$

 b. $25 - 20e^{1/5}$

 c. $-\dfrac{298}{15}$

 d. $-xe^{-x} + C$

2. **a.** 10

 b. $\dfrac{3}{4}e^{-2}$

 c. Diverges

$$\lim_{N\to\infty}\int_1^N \frac{x}{(x+1)^2}dx$$

$$f_{ys} = 2x(x^2y + 1)e^{x^2y};\ f_{yy} = x^4e^{x^2y}$$

$$= \lim_{N\to+\infty}\left(\frac{1}{N+1} + \ln|N+1| - \frac{1}{2} - \ln 2\right) = \infty$$

 d. $\dfrac{1}{2}$

3. **a.** $\dfrac{x}{4}[(\ln|3x|)^2 - 2\ln|3x| + 2] + C$

 b. $-\dfrac{1}{2}\ln\left|\dfrac{\sqrt{4 + x^2} + 2}{x}\right| + C$

 c. $\dfrac{\sqrt{x^2 - 9}}{9x} + C$

 d. $-\dfrac{1}{4}\ln\left|\dfrac{x}{3x - 4}\right| + C$

4. The waste will increase without bound.

5. Approximately $1 666 666.67

6. 3.5

7. 16 000 units

8. By the trapezoidal rule, $\displaystyle\int_3^4 \frac{\sqrt{25 - x^2}}{x}dx \approx 1.0276.$

The exact answer is $-1 + 5\log\dfrac{3}{2} \approx 1.0273.$

The error of the approximation is 0.0003.

CHAPTER 6   Review Exercises

1. $-(1 + t)e^{1-t} + C$

3. $\dfrac{x}{3}(2x + 3)^{3/2} - \dfrac{1}{15}(2x + 3)^{5/2} + C$

5. $-2 + 4\ln 2$

7. $\dfrac{74}{7}$

9. $\dfrac{x^2}{9}(3x^2 + 2)^{3/2} - \dfrac{2}{135}(3x^2 + 2)^{5/2} + C$

11. $\dfrac{5}{8}\ln\left|\dfrac{2 + x}{2 - x}\right| + C$

13. $-3(18 + 6w + w^2)e^{-w/3} + C$

15. $x[-6 + 6\ln 2x - 3(\ln 2x)^2 + (\ln 2x)^3] + C$

17. Diverges

$$\lim_{N\to\infty}\int_0^N \frac{1}{\sqrt[3]{1 + 2x}}dx = \lim_{N\to\infty}\left[-\frac{3}{4} + \frac{3(1 + 2N)^{2/3}}{4}\right]$$
$$= \infty$$

19. Diverges

$$\lim_{N\to\infty}\int_0^N \frac{3t}{t^2 + 1}dt = \lim_{N\to\infty}\frac{3}{2}\ln(1 + N^2) = \infty$$

21. $\dfrac{1}{4}$

23. $\dfrac{1}{4}$

25. Diverges

$$\lim_{N \to \infty} \int_1^N \frac{\ln x}{\sqrt{x}}\, dx = \lim_{N \to \infty} 2(2 - 2\sqrt{N} + \sqrt{N}\ln N) = \infty$$

27. \$320 000

29. The population increases without bound.

31. 15 000 kg

33. a. 1.1016 with $|E| \leq \dfrac{1}{75}$

    b. 1.0987 with $|E| \leq \dfrac{4}{9,375}$

35. a. 3.0607 with $|E| \leq \dfrac{e}{1,200}$

    b. 3.0591 with $|E| \leq \dfrac{e}{200\,000}$

37. a. 58

    b. 8

39. a. $\displaystyle\int_0^8 \sqrt{q}\, e^{0.01q}\, dq$

    b. 15.6405 (\$15.64)

41. a.

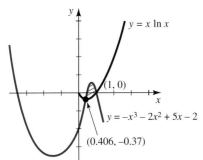

    b. 0.1692

43. $\dfrac{2}{3}\ln 2 \approx 0.4621$

45. $I(1) \approx 0.4214$

    $I(10) \approx 0.5$

    $I(50) \approx 0.5$

    $I(N)$ converges to $\dfrac{1}{2}$ as $N \to \infty$.

CHAPTER 7   Section 1

1. $f(-1, 2) = 1; f(3, 0) = 15$

3. $g(1, 1) = 0; g(-1, 4) = -5$

5. $f(2, -1) = -3, f(1, 2) = 16$

7. $g(4, 5) = 3, g(-1, 2) = \sqrt{3} \approx 1.7321$

9. $f(e^2, 3) = \dfrac{3}{2}; f(\ln 9, e^3) = 25.515$

11. $g(1, 2) = 2.5; g(2, -3) = -2.167$

13. $f(1, 2, 3) = 6; f(3, 2, 1) = 6$

15. $F(1, 1, 1) \approx 0.2310; \; F(0, e^2, 3e^2) \approx 0.1048$

17. All ordered pairs (x, y) of real numbers for which $y \neq \dfrac{-4}{3}x$

19. All ordered pairs (x, y) of real numbers for which $y \leq x^2$

21. All ordered pairs (x, y) of real numbers for which $x > 4 - y$

23.

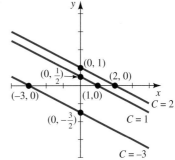

25.

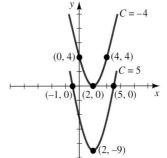

27.

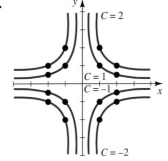

29.

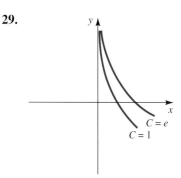

$C = e$
$C = 1$

c.

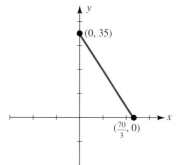

$(0, 35)$

x

$\left(\frac{70}{3}, 0\right)$

31. a. 160 000 units
  b. Production will increase by 16 400 units.
  c. Production will increase by 4000 units.
  d. Production will increase by 20 810 units.

33. a. $R(x_1, x_2) = 200x_1 - 10x_1^2 + 25x_1x_2 + 100x_2 - 10x_2^2$
  b. $7230

35. a. If $a + b > 1$, production is more than doubled.
  b. If $a + b < 1$, production is increased (but not doubled).
  c. If $a + b = 1$, production is doubled.

37. $R(x, y) = 60x - \dfrac{x^2}{5} + \dfrac{xy}{10} + 50y - \dfrac{y^2}{10}$

39. a. $S(15.83, 87.11) = 0.5938$

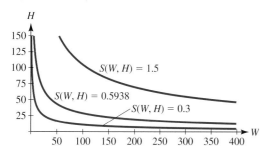

$S(W, H) = 1.5$
$S(W, H) = 0.5938$
$S(W, H) = 0.3$

The curves represent different combinations of height and weight that result in the same surface area.
  b. Height = 90.05 cm
  c. 254%

41. a. 70 units
  b. $y = -\dfrac{3}{2}x + 35$

d. Unskilled labour should be decreased by three workers.

43. 260

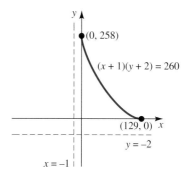

$(0, 258)$
$(x + 1)(y + 2) = 260$
$(129, 0)$ x
$y = -2$
$x = -1$

45. a. 0.866 cm/s
  b.

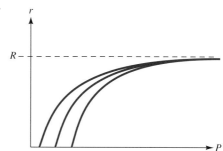

r
R
P

The curves represent different combinations of pressure and distance from the axis that result in the same speed.

47. a.

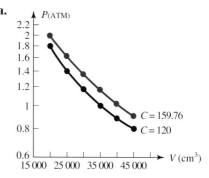

$P(\text{ATM})$
$C = 159.76$
$C = 120$
V (cm^3)

  b. 159.76°C

49. a. 2105.03 kilocalories

    b. 1428.84 kilocalories

    c. Approximately 27 years

    d. Approximately 24.4 years

51. a. \$2003.13; \$110 563.40

    b. \$1435.20; \$266 672.00

53. 23.54

55. For $Q(K, L) = A[\alpha K^{-\beta} + (1 - \alpha) L^{-\beta}]^{-1/\beta}$,

$$Q(sK, sL) = A[\alpha(sK)^{-\beta} + (1 - \alpha)(sL)^{-\beta}]^{-1/\beta}$$
$$= A[\alpha s^{-\beta} K^{-\beta} + (1 - \alpha) s^{-\beta} L^{-\beta}]^{-1/\beta}$$
$$= A[s^{-\beta}\{\alpha K^{-\beta} + (1 - \alpha) L^{-\beta}\}]^{-1/\beta}$$
$$= A[s^{-\beta}]^{-1/\beta}[\alpha K^{-\beta} + (1 - \alpha) L^{-\beta}]^{-1/\beta}$$
$$= sA[\alpha K^{-\beta} + (1 - \alpha) L^{-\beta}]^{-1/\beta}$$
$$= sQ(K, L)$$

CHAPTER 7 Section 2

1. $f_x = 7; f_y = -3$

3. $f_x = 12x^2 - 6xy + 5; f_y = -3x^2$

5. $f_x = 2y^5 + 6xy + 2x; f_y = 10xy^4 + 3x^2$

7. $\dfrac{\partial z}{\partial x} = 15(3x + 2y)^4; \dfrac{\partial z}{\partial y} = 10(3x + 2y)^4$

9. $f_s = -\dfrac{3t}{2s^2}; f_t = \dfrac{3}{2s}$

11. $\dfrac{\partial z}{\partial x} = (xy + 1)e^{xy}; \dfrac{\partial z}{\partial y} = x^2 e^{xy}$

13. $f_x = \dfrac{-e^{2-x}}{y^2}; f_y = \dfrac{-2e^{2-x}}{y^3}$

15. $f_x = \dfrac{5y}{(y - x)^2}; f_y = \dfrac{-5x}{(y - x)^2}$

17. $\dfrac{\partial z}{\partial u} = \ln v; \dfrac{\partial z}{\partial v} = \dfrac{u}{v}$

19. $f_x = \dfrac{1}{y^2(x + 2y)}; f_y = \dfrac{2[y - (x + 2y) \ln (x + 2y)]}{y^3(x + 2y)}$

21. $f_x(1, -1) = 2; f_y(1, -1) = 3$

23. $f_x(0, -1) = 2; f_y(0, -1) = 0$

25. $f_x(-2, 1) = -22; f_y(-2, 1) = 26$

27. $f_x(0, 0) = 1; f_y(0, 0) = 1$

29. $f_{xx} = 60x^2y^3; f_{xy} = 2(30x^3y^2 + 1);$

    $f_{yx} = 2(30x^3y^2 + 1); f_{yy} = 30 x^4y$

31. $f_{xx} = 2y(2x^2y + 1)e^{x^2y}; f_{xy} = 2x(x^2y + 1)e^{x^2y};$

    $f_{yx} = 2x(x^2y + 1)e^{x^2y}; f_{yy} = x^4 e^{x^2y}$

33. $f_{ss} = \dfrac{t^2}{\sqrt{(s^2 + t^2)^3}}; f_{st} = \dfrac{-st}{\sqrt{(s^2 + t^2)^3}};$

    $f_{ts} = \dfrac{-st}{\sqrt{(s^2 + t^2)^3}}; f_{tt} = \dfrac{s^2}{\sqrt{(s^2 + t^2)^3}}$

35. Substitute

37. Neither

39. Substitute

41. Yes

43. No

45. Daily output will increase by approximately 10 units.

47. a. $Q_K = 60 K^{-3/2}[0.4 K^{-1/2} + 0.6L^{-1/2}]^{-3}$

    $Q_L = 90 L^{-3/2}[0.4 K^{-1/2} + 0.6L^{-1/2}]^{-3}$

    b. If $K = 5041$ and $L = 4900$,

      $Q_K \approx 58.48$ and $Q_L \approx 91.54$.

    c. Labour

49. $F(L, r) = \dfrac{kL}{r^4}$

    a. $F(3.17, 0.085) = 60\ 727.24k;$

      $\dfrac{\partial F}{\partial L} = \dfrac{k}{r^4} = 19\ 156.86k;$

      $\dfrac{\partial F}{\partial r} = -\dfrac{4kL}{r^5} = -2\ 857\ 752.58k$

    b. $F(1.2L, 0.8r) = \dfrac{k(1.2L)}{(0.8r)^4} = 2.93F(L, r);$

      $\dfrac{\partial F}{\partial r}(1.2L, 0.8r) = 3.66\dfrac{\partial F}{\partial r}(L, r);$

      $\dfrac{\partial F}{\partial L}(1.2L, 0.8r) = 2.44\dfrac{\partial F}{\partial L}(L, r);$

51. The monthly demand for bicycles decreases by approximately 16 bicycles (actually, 16.45).

53. The volume is increased by 72π cm^3.

55. a. An increase in x will decrease the demand $D(x, y)$ for the first brand of mower. An increase in y will increase the demand $D(x, y)$ for the first brand of mower.

    b. $\dfrac{\partial D}{\partial x} < 0, \dfrac{\partial D}{\partial y} > 0$

    c. $b < 0, c > 0$

ANSWERS

57. $P(x, y, u, v) = \dfrac{100xy}{xy + uv}$

$P_x = \dfrac{(xy + uv)100y - 100xy^2}{(xy + uv)^2} = \dfrac{100uvy}{(xy + uv)^2}$

$P_y = \dfrac{(xy + uv)100x - 100x^2y}{(xy + uv)^2} = \dfrac{100uvx}{(xy + uv)^2}$

$P_u = \dfrac{-100xyv}{(xy + uv)^2}; P_v = \dfrac{-100uxy}{(xy + uv)^2}$

All of these partials measure the rate of change of percentage of total blood flow with respect to the quantities x, y, u, and v, respectively.

59. $\dfrac{\partial F}{\partial z} = \dfrac{-c\pi x^2}{8\sqrt{y - z}}$; decreasing since $F_{zz} < 0$

61. a. $\dfrac{\partial^2 Q}{\partial L^2} < 0$; for a fixed level of capital investment, the effect on output of the addition of one worker-hour is greater when the workforce is small than when it is large.

b. $\dfrac{\partial^2 Q}{\partial K^2} < 0$; for a fixed workforce, the effect on output of the addition of $1000 in capital investment is greater when the capital investment is small than when it is large.

63. a. $Q(37, 71) = 304\ 691$; $Q(38, 71) = 317\ 310$; $Q(37, 72) = 309\ 031$

b. $Q_x(37, 71) = 12\ 534$ units; $Q(38, 71) - Q(37, 71) = 12\ 619$ units

c. $Q_y(37, 71) = 4344$ units; $Q(37, 72) - Q(37, 71) = 4340$ units

65. $\dfrac{dz}{dt} = 4t + 15$

67. $\dfrac{dz}{dt} = \dfrac{3}{y} - \dfrac{6xt}{y^2}$

69. $\dfrac{dz}{dt} = 2ye^{2t} - 3xe^{-3t}$

71. The number of units produced decreases by about 55.

73. a. 424 units/month
b. 16.31

75. The number of units produced increases by about 61.6 units per day.

77. a. $P_x = -8x + 10y - 10$
$P_y = 14y + 10x + 185$

b. For $x = 70$ and $y = 73$,
$P_x = 160$ and $P_y = -137$

c. Decreases profit by 114 cents

d. Increases profit by 457 cents.

79. a. $C(R, H) = 0.001\pi(R^2 + RH + R^2H)$

b. The cost increases by about 0.08 cents per can.

81. $\dfrac{dy}{dx} = \dfrac{1}{2}$; $x - 2y = -1$

83. a. $U_x = 2(y + 5)$
$U_y = 2x + 3$

b. For $x = 27$ and $y = 12$,
$U_x = 34$ and $U_y = 57$

c. Decrease satisfaction by 12 units.

d. Increase bond units by about 0.6.

CHAPTER 7 Section 3

	Relative Maximum	Relative Minimum	Saddle Point
1.	$(0, 0)$	None	None
3.	None	None	$(0, 0)$
5.	None	None	$(2, -1)$
7.	$(-2, -1)$	$(1, 1)$	$(-2, 1); (1, -1)$
9.	None	$\left(4, \dfrac{19}{2}\right)$	$\left(2, \dfrac{7}{2}\right)$
11.	$(0, 0)$	None	$(3, 6); (3, -6)$
13.	$(0, 1); (0, -1)$	$(0, 0)$	$(1, 0); (-1, 0)$
15.	None	$\left(\dfrac{4}{3}, \dfrac{4}{3}\right)$	$(0, 0)$
17.	$(1, 1); (-1, -3)$	None	$(0, -1)$
19.	$\left(-\dfrac{3}{2}, 1\right)$	None	None
21.	$(e, 1); (e, -1)$	None	None

23. Sunset shirts $x = 2.70; Mermaid shirts $y = 2.50

25. The base of the box is a 2 m by 2 m square. The height is 8 m.

27. $x = 100/3$, $y = 100/3$ or approximately 33 of each.

29. $x = \dfrac{\sqrt{2}}{2}; y = \dfrac{\sqrt{2}}{2}$

31. $x = y = z = \sqrt[3]{V_0}$

33. $x = 200$; $y = 300$

35. $S\left(\dfrac{5}{4}, \dfrac{1}{4}\right)$

37. Maximum of $P = \dfrac{2}{3}$, when $p = q = r = \dfrac{1}{3}$.

39. a. Dan should wait at $x = 0.424$ km; Maria should wait at $y = 2.236$ km (1.6397 km from F); optimum time is 1.748 hours.

  b. Tom, Dan, and Maria will win by 0.2080 hours (12.5 minutes).

  c. Writing exercise; responses will vary.

41. Problem can be stated as:

Maximize $V = 2\left[x - 2\left(\dfrac{2}{\sqrt{3}}\right)\right]y$

subject to $2xy + 2\left(\dfrac{1}{2}\dfrac{\sqrt{3}}{2}x^2\right) = 55$

Solution is $x \approx 4.61$ m; $y \approx 3.97$ m.

43. $\dfrac{\partial f}{\partial x} = 2x - 4y; \dfrac{\partial f}{\partial y} = 2y - 4x$. Thus, $(0, 0)$ is a

critical point. Since $\dfrac{\partial^2 f}{\partial^2 x} = 2 > 0$, the second

derivative test tells us there is a minimum in the

x direction. Likewise, $\dfrac{\partial^2 f}{\partial^2 x} = 2 > 0$, implies a

minimum in the y direction. However, along the curve determined by $y = x$, we have $f = -2x^2$, which has a relative maximum at $(0, 0)$.

45. $f_x = \dfrac{x^2 - 7y^2}{x^2 \ln y};$

$f_x = \dfrac{y(x + 14y)\ln y - (x^2 + xy + 7y^2)}{xy(\ln y)^2}$

Critical points: $(\sqrt{7}e, e), (-\sqrt{7}e, e)$

47. $f_x = 8x^3 - 22xy + 36x; f_y = 4y^3 - 11x^2$
Critical point: $(0, 0)$

CHAPTER 7   Section 4

1. $y = \dfrac{1}{4}x + \dfrac{3}{2}$

3. $y = 3$

5. $y = \dfrac{7}{9}x + \dfrac{19}{18}$

7. $y = -\dfrac{1}{2}x + 4$

9. $y = 1.018x + 0.802$

11. $y = -0.915x + 1.683$

13. $y = 15.018e^{0.04x}$

15. $y = 20.03e^{-0.201x}$

17. a.

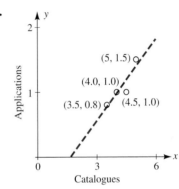

  b. $y = 0.42x - 0.71$

  c. When 4800 catalogues ($x = 4.8$) are requested, $y = 0.42(4.8) - 0.71 = 1.306$ or 1306 applications are predicted to be received.

19. a.

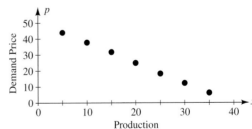

  b. $p = -1.29x + 50.71$

  c. The predicted price is negative. All 4000 units cannot be sold at any price.

21. a. Let x denote the number of hours after the polls open and y the corresponding percentage of registered voters that have already cast their ballots. Then

x	2	4	6	8	10
y	12	19	24	30	37

x	y	xy	x^2
2	12	24	4
4	19	76	16
6	24	144	36
8	30	240	64
10	37	370	100
$\sum x$ $= 30$	$\sum y$ $= 122$	$\sum xy$ $= 854$	$\sum x^2$ $= 220$

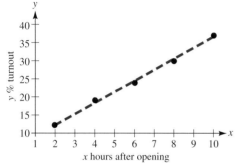

b. $y = 3.05x + 6.10$

c. When the polls close at 8:00 P.M., $x = 12$ and so $y = 3.05(12) + 6.10 = 42.7$, which means that approximately 42.7% of the registered voters can be expected to vote.

23. a. $y = 24.15 + 0.3013t - 0.000143t^2$
$y = 24.22 + 0.2404t + 0.00652t^2 - 0.000178t^3$
Quadratic predicted 35.68 million, cubic predicted 32.95 million. Written answers will vary.

b. Written answers will vary.

25. a. $V = 53.90e^{0.041t}$; 4.09%

b. $122\,380$

c. Approximately 42 years

d. Frank's formula is $V = 54.52e^{0.044t}$. It is easier to find but does not take into account the trend in account value.

27. a. $y = 666 + 4019$

b. \$8013

29. a.

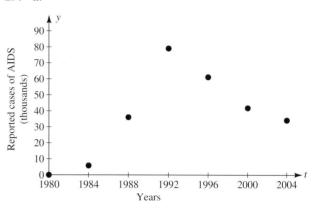

b. $y = 1871t + 15\,122$

c. $67\,510$ cases

d. No. The line has positive slope, but the number of cases appears to be decreasing.

31. a.

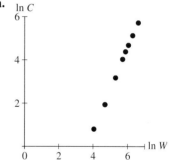

b. $y = 1.631x + 4.975$

c. $C = 0.0069W^{1.631}$

CHAPTER 7   Section 5

1. $f\left(\dfrac{1}{2}, \dfrac{1}{2}\right) = \dfrac{1}{4}$

3. $f(1, 1) = f(-1, -1) = 2$

5. $f(0, 2) = f(0, -2) = -4$

7. $f\left(\dfrac{\sqrt{3}}{2}, -\dfrac{1}{2}\right) = f\left(-\dfrac{\sqrt{3}}{2}, -\dfrac{1}{2}\right) = \dfrac{3}{2}$(max);
$f(0, 1) = -3$(min)

9. $f(8, 7) = -18$

11. $f(\sqrt{2}, \sqrt{2}) = f(-\sqrt{2}, -\sqrt{2}) = e^2$(max);
$f(\sqrt{2}, -\sqrt{2}) = f(-\sqrt{2}, \sqrt{2}) = e^{-2}$(min);

13. $f\left(8, 4, \dfrac{8}{3}\right) = \dfrac{256}{3}$(max)

15. $f\left(\dfrac{4}{\sqrt{14}}, \dfrac{8}{\sqrt{14}}, \dfrac{12}{\sqrt{14}}\right) = \dfrac{56}{\sqrt{14}}$(max);

$f\left(-\dfrac{4}{\sqrt{14}}, -\dfrac{8}{\sqrt{14}}, -\dfrac{12}{\sqrt{14}}\right) = \dfrac{-56}{\sqrt{14}}$(min);

17. 12 500 Deluxe and 17 500 Standard

19. **a.** \$36 000 on development and \$24 000 on promotion
    b. Approximately 4320 more books will be sold.

21. 40 metres by 80 metres

23. 11 664 cm³, when $x = 18$ cm, $y = 36$ cm

25. $r = 4.15$ cm; $h = 8.31$ cm

27. **a.** \$40 000 on labour and \$80 000 on equipment
    b. Approximately 31.75 more units will be produced.

29. $H = 2R$

31. $s_{max} = 4L$

33. $x = 8.93$ cm, $y = 10.04$ cm

35. $x = y = z = \sqrt[3]{V_0}$

37. Front length 11.5 m; side length 15.4 m; height 7.2 m

39. $\lambda = 306.12$, which gives the approximate change per \$1000. Since the difference is only \$100, the maximum profit is increased by approximately $0.1(\$306.12) = \$30.61$.

41. **a.** $x = 35$ units, $y = 42$ units
    b. $\lambda = 14.33$ is the approximate change in the maximum utility resulting from a one-unit increase in the budget.

43. Increases by $\lambda = \left(\dfrac{\alpha}{a}\right)^{\alpha}\left(\dfrac{\beta}{a}\right)^{\beta}$

45. Let $Q(x, y) =$ production; $C(x, y) = px + qy = k$. $C_x = p$, $C_y = q$. Therefore, $Q_x = \lambda p$; $Q_y = \lambda q$; $\dfrac{Q_x}{p} = \dfrac{Q_y}{q}$.

47. $Q(40, 14) \approx 1398$

49. The Lagrange equations are

$$A\alpha K^{-\beta-1}[\alpha K^{-\beta} + (1 - \alpha)L^{-\beta}]^{-1/\beta-1} = c_1\lambda$$
$$A(1 - \alpha)L^{-\beta-1}[\alpha K^{-\beta} + (1 - \alpha)L^{-\beta}]^{-1/\beta-1} = c_2\lambda$$
$$c_1K + c_2L = B$$

Solve the first two equations for λ and simplify to get

$$c_2\alpha K^{-\beta-1} = c_1(1 - \alpha)L^{-\beta-1}$$

51. $x = 0, y = -2$. The critical point $(0, -2)$ is an inflection point, not a relative extremum.

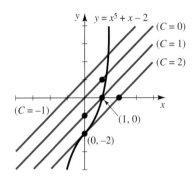

53. $\dfrac{\partial P}{\partial K} = \lambda\partial\dfrac{\partial C}{\partial K}; \dfrac{\partial P}{\partial L} = \lambda\dfrac{\partial C}{\partial L}; \lambda = \dfrac{\dfrac{\partial P}{\partial K}}{\dfrac{\partial C}{\partial K}} = \dfrac{\dfrac{\partial P}{\partial L}}{\dfrac{\partial C}{\partial L}}$

55. $\dfrac{dy}{dx} = \dfrac{\dfrac{y}{x^2} - (1 + xy^2)e^{xy^2} - \ln(x + y) - \dfrac{x}{x + y}}{2x^2ye^{xy^2} + \dfrac{1}{x} + \dfrac{x}{x + y}}$

57. $f(2.1623, 1.5811) = 1.6723$

59. $f(0.9729, -0.1635) = 2.9522$

CHAPTER 7   Section 6

1. $\dfrac{7}{6}$

3. -1

5. $4 \ln 2 = \ln 16$

7. 0

9. $\ln 3$

11. 32

13. $\dfrac{1}{3}$

15. $\dfrac{32}{9}$

17. $\dfrac{e^2 - 1}{8}$

19. Vertical cross sections: $0 \le x \le 3$
$$x^2 \le y \le 3x$$
Horizontal cross sections: $0 \le y \le 9$
$$\dfrac{y}{3} \le x \le \sqrt{y}$$

21. Vertical cross sections: $-1 \leq x \leq 2$
$\qquad\qquad\qquad\qquad\quad 1 \leq y \leq 2$

Horizontal cross sections: $1 \leq y \leq 2$
$\qquad\qquad\qquad\qquad\qquad\; -1 \leq x \leq 2$

23. Vertical cross sections: $-1 \leq x \leq e$
$\qquad\qquad\qquad\qquad\quad 0 \leq y \leq \ln x$

Horizontal cross sections: $0 \leq y \leq 1$
$\qquad\qquad\qquad\qquad\qquad\; e^y \leq x \leq e$

25. $\dfrac{3}{2}$

27. $\dfrac{1}{2}$

29. $\dfrac{44}{15}$

31. 1

33. $\dfrac{3}{2} \ln 5$

35. $2(e - 2)$

37.

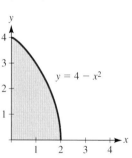

$$\int_{y=0}^{y=4} \int_{x=0}^{x=\sqrt{4-y}} f(x, y)\, dx\, dy$$

39.

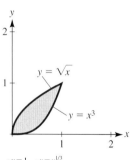

$$\int_{y=0}^{y=1} \int_{x=y^2}^{x=y^{1/3}} f(x, y)\, dx\, dy$$

41.

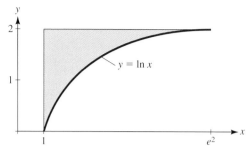

$$\int_{y=0}^{y=2} \int_{x=1}^{x=e^y} f(x, y)\, dx\, dy$$

43.

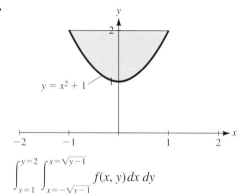

$$\int_{y=1}^{y=2} \int_{x=-\sqrt{y-1}}^{x=\sqrt{y-1}} f(x, y)\, dx\, dy$$

45. $\displaystyle\int_{x=-4}^{x=2} \int_{y=0}^{y=x+4} 1\, dy\, dx = 18$

47. $\displaystyle\int_{x=0}^{x=4} \int_{y=\frac{1}{2}x^2}^{y=2x} 1\, dy\, dx = \dfrac{16}{3}$

49. $\displaystyle\int_{x=1}^{x=3} \int_{y=x^2-4x+3}^{0} 1\, dy\, dx = \dfrac{4}{3}$

51. $\displaystyle\int_{x=1}^{x=e} \int_{y=0}^{y=\ln x} 1\, dy\, dx = 1$

53. $\displaystyle\int_{y=0}^{y=3} \int_{x=y/3}^{x=\sqrt{4-y}} 1\, dx\, dy = \dfrac{19}{6}$

55. $\displaystyle\int_{y=0}^{y=2} \int_{x=0}^{x=1} (6 - 2x - 2y)\, dx\, dy = 6$

57. $\displaystyle\int_{x=1}^{x=2} \int_{y=1}^{y=3} \dfrac{1}{xy}\, dy\, dx = (\ln 3)(\ln 2)$

59. $\displaystyle\int_{x=0}^{x=1} \int_{y=0}^{y=2} xe^{-y}\, dy\, dx = \dfrac{1}{2}\left(1 - \dfrac{1}{e^2}\right)$

61. $\displaystyle\int_{y=0}^{y=1} \int_{x=y}^{x=2-y} (2x + y)\, dx\, dy = \dfrac{7}{3}$

63. $\displaystyle\int_{x=-2}^{x=2}\int_{y=x^2}^{y=8-x^2}(x+1)\,dy\,dx=\frac{64}{3}$

65. Area $=\displaystyle\int_{x=-2}^{x=3}\int_{y=-1}^{y=2}1\,dy\,dx=15$

Average $=\dfrac{1}{15}\displaystyle\int_{x=-2}^{x=3}\int_{y=-1}^{y=2}xy(x-2y)\,dy\,dx=\dfrac{1}{6}$

67. Area $=\displaystyle\int_{x=0}^{x=1}\int_{y=0}^{y=2}1\,dy\,dx=2$

Average $=\dfrac{1}{2}\displaystyle\int_{x=0}^{x=1}\int_{y=0}^{y=2}xye^{x^2y}\,dy\,dx=\dfrac{e^2-3}{4}$

69. Area $=\displaystyle\int_{x=0}^{x=3}\int_{y=x/3}^{y=1}1\,dy\,dx=\dfrac{3}{2}$

Average $=\dfrac{2}{3}\displaystyle\int_{x=0}^{x=3}\int_{y=x/3}^{y=1}6xy\,dy\,dx=\dfrac{9}{2}$

71. Area $=\displaystyle\int_{x=-2}^{x=2}\int_{y=0}^{y=4-x^2}1\,dy\,dx=\dfrac{32}{3}$

Average $=\dfrac{3}{32}\displaystyle\int_{x=-2}^{x=2}\int_{y=0}^{y=4-x^2}x\,dy\,dx=0$

73. $\displaystyle\int_{x=1}^{x=3}\int_{y=2}^{y=5}\dfrac{\ln(xy)}{y}\,dy\,dx$

$=(3\ln 3-2)\ln\dfrac{5}{2}+(\ln 5)^2-(\ln 2)^2$

75. $\displaystyle\int_{x=0}^{x=1}\int_{y=0}^{y=1}x^3e^{x^2y}\,dy\,dx=\dfrac{e-2}{2}$

77. Area $=\displaystyle\int_{x=0}^{x=5}\int_{y=0}^{y=7}1\,dy\,dx=35$

Average $=\dfrac{1}{35}\displaystyle\int_{x=0}^{x=5}\int_{y=0}^{y=7}(2x^3+3x^2y+y^3)\,dy\,dx$

$=\dfrac{943}{4}$

79. Average $=\dfrac{1}{25.19}\displaystyle\int_{x=100}^{x=125}\int_{y=70}^{y=89}[(x-30)(70+5x-4y)$

$+(y-40)(80-6x+7y)]\,dy\,dx$

$=24\,896.5(2\,489\,650)$

81. Average $=\dfrac{1}{12}\displaystyle\int_{x=0}^{x=4}\int_{y=0}^{y=3}90(2x+y^2)\,dy\,dx$

$=630$ m

83. Value $=\displaystyle\int_{x=-1}^{x=1}\int_{y=-1}^{y=1}(300+x+y)e^{-0.01x}\,dy\,dx$

$=79\,800e^{0.01}-80\,200e^{-0.01}\approx 1\,200$

85. a. 62 949

b. 2,518 people per square unit

c. Writing exercise; responses will vary.

87. a. 0.991 square metres

b. No, it can only be considered the average surface area from birth until the time at which the person reached adulthood.

89. $\dfrac{17\,408}{105}\approx 166$ m^3

91. $\dfrac{304}{27}\approx 11.26$

93. $\dfrac{7e^{-6}}{9}+\dfrac{17}{9}\approx 1.891$ cubic units

CHAPTER 7 Checkup

1. a. Domain: all ordered pairs (x, y) of real numbers

$f_x=3x^2+2y^2$

$f_y=4xy-12y^3$

$f_{xx}=6x$

$f_{yx}=4y$

b. Domain: all ordered pairs (x, y) of real numbers for which $x\neq y$

$f_x=\dfrac{-3y}{(x-y)^2}$

$f_y=\dfrac{3x}{(x-y)^2}$

$f_{xx}=\dfrac{6y}{(x-y)^3}$

$f_{yx}=\dfrac{-3(x+y)}{(x-y)^3}$

c. Domain: all ordered pairs (x, y) of real numbers for which $y^2>2x$

$f_x=2e^{2x-y}-\dfrac{2y}{y^2-2x}$

$f_y=-e^{2x-y}+\dfrac{2y}{y^2-2x}$

$f_{xx}=4e^{2x-y}-\dfrac{4}{(y^2-2x)^2}$

$f_{yx}=-2e^{2x-y}+\dfrac{4y}{(y^2-2x)^2}$

2. **a.** Circles centred at the origin and the single point $(0, 0)$
   b. Parabolas with vertices on the x axis and opening to the left

3. **a.** Relative maximum: $(0, 0)$; relative minimum: $(1, 4)$; saddle points: $(1, 0)$, $(0, 4)$
   b. Saddle point: $(-1, 0)$
   c. Relative minimum: $(-1, -1)$

4. **a.** $\dfrac{16}{5}$ at $\left(\dfrac{4}{5}, \dfrac{8}{5}\right)$
   b. Maximum value of 4 at $(1, 2)$ or $(1, -2)$; minimum value of -4 at $(-1, 2)$ or $(-1, -2)$

5. **a.** 16
   b. $\dfrac{1}{4}(e^2 + 3e^{-2})$
   c. $2\ln 2 - \dfrac{3}{4}$
   d. $1 - \dfrac{1}{e^2}$

6. $Q_K = 180$; $Q_L = 3.75$

7. 20 DVDs and 2 video games

8. 30 units of drug A and 25 units of drug B, which results in an equivalent dosage of $E(30, 25) = 83.75$ units. Since the total number of units is 55, which is less than 60, there is no risk of side effects, and since $E(30, 25) > 70$, the combination is effective.

9. $\dfrac{5}{2}(1 + e^{-2}) \approx 2.84°C$

10. **a.**

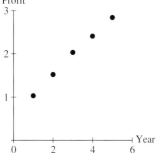

   b. $y = 0.45x + 0.61$
   c. 3.31 million dollars

CHAPTER 7   Review Exercises

1. $f_x = 6x^2y + 3y^2 - \dfrac{y}{x^2}$; $f_y = 2x^3 + 6xy + \dfrac{1}{x}$

3. $f_x = \dfrac{3x - y^2}{2\sqrt{x}}$; $f_y = -2y\sqrt{x}$

5. $f_x = \dfrac{1}{2\sqrt{xy}} - \dfrac{\sqrt{y}}{2x^{3/2}}$; $f_y = \dfrac{1}{2\sqrt{xy}} - \dfrac{\sqrt{x}}{2y^{3/2}}$

7. $f_x = \dfrac{2x^3 + 3x^2y - y^2}{(x + y)^2}$; $f_y = \dfrac{-x^2(x + 1)}{(x + y)^2}$

9. $f_x = \dfrac{2(x^2 + xy + y^2)}{(2x + y)^2}$; $f_y = \dfrac{-x^2 - 4xy - y^2}{(2x + y)^2}$

11. $f_{xx} = (4x^2 + 2)e^{x^2+y^2}$; $f_{yy} = (4y^2 + 2)e^{x^2+y^2}$;
    $f_{xy} = 4xy\, e^{x^2+y^2}$; $f_{yx} = 4xye^{x^2+y^2}$

13. $f_{xx} = 0$; $f_{yy} = -\dfrac{x}{y^2}$; $f_{xy} = \dfrac{1}{y}$; $f_{yx} = \dfrac{1}{y}$

15. **a.**

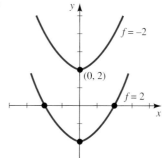

   b.

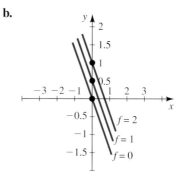

   $f_y = \dfrac{x}{y(x + 3y)} = \dfrac{1}{y} - \dfrac{3}{x + 3y}$

17. Saddle point at $(6, -6)$

19. Relative maximum at $(-2, 0)$; relative minimum at $(0, 2)$; saddle points at $(0, 0)$ and $(-2, 2)$

21. Relative minimum at $\left(-\dfrac{23}{2}, 5\right)$; saddle point at $\left(\dfrac{1}{2}, 1\right)$

23. Saddle points at $\left(\dfrac{2}{3}, -\dfrac{5}{6}\right)$ and $\left(-\dfrac{2}{3}, \dfrac{5}{6}\right)$

25. Maximum value of 12 at $(1, \pm\sqrt{3})$; minimum value of 3 at $(-2, 0)$

27. Maximum value of 17 at $(1, 8)$; minimum value of 17 at $(-1, -8)$

29. Daily output will increase by approximately 16 units.

31. The level of unskilled labour should be decreased by approximately two workers.

33. Maximize area $A = xy$ subject to fixed perimeter $P = 2x + 2y$. Lagrange conditions are $y = \lambda(x)$, $x = \lambda(y)$, and $2x + 2y = C$. We must have $\lambda > 0$ since x and y are positive, so $x = y$ and the optimum rectangle is actually a square.

35. Development $x = \$4000$; promotion $y = \$7000$

37. We have $f_x = y - \dfrac{12}{x^2}$ and $f_y = x - \dfrac{18}{y^2}$.
so $f_x = f_y = 0$ when $x = 2$ and $y = 3$. Since $f(x, y)$ is large when either x or y is large or small, a relative minimum is indicated at $(2, 3)$. To verify this claim, note that
$$D = \left(\dfrac{24}{x^3}\right)\left(\dfrac{24}{y^3}\right) - 1 \text{ and } f_{xx} = \dfrac{24}{x^3}$$
so that $D(2, 3) > 0$ and $f_{xx}(2, 3) > 0$.

39. 0.5466

41. $\dfrac{1}{4}(e^2 - e^{-2})$

43. $2e - 2$

45. $\dfrac{3}{2}(e - 1)$

47. 2

49. $\dfrac{3}{2}(e^{-2} - e^{-3})$ cubic units

51. $x = y = z = \dfrac{20}{3}$

53. $\sqrt{10}$; at $(0, \pm\sqrt{10}, 0)$

55. a.

b. $y = 11.54x + 44.45$
c. Approximately $\$102\ 150$

57. 5.94; demand is increasing at the rate of about 6 litres per month.

59. -3; demand is decreasing at the rate of 3 pies per week.

61. The amount of pollution is decreasing by about 113 units per day.

63. About 7.056 units

65. $Q(x, y) = x^a y^b$
$Q_x = ax^{a-1}y^b$; $Q_y = bx^a y^{b-1}$
$xQ_x + yQ_y = a(x^{a-1}y^b) + y(bx^a y^{b-1})$
$\qquad\qquad = (a + b)x^a y^b = (a + b)Q$
If $a + b = 1$, then $xQ_x + yQ_y = Q$.

CHAPTER 8   Section 1

1. $y = x^3 + \dfrac{5}{2}x^2 - 6x + C$

3. $y = Ce^{3x}$

5. $y = -\ln(C - x)$

7. $y = \pm\sqrt{x^2 + C}$

9. $y = \left(\dfrac{1}{3}x^{3/2} + C\right)^2$

11. $y = C|x - 1|$

13. $y = -3 + Ce^{[-(2x-5)^{-5}]/10}$

15. $x = \dfrac{Ce^{t/z}}{(2t + 1)^{1/4}}$

17. $y = \ln(xe^x - e^x + C)$

19. $y = Ce^{t(\ln t - 1)/2}$

ANSWERS

21. $y = \dfrac{1}{5}(e^{5x} + 4)$

23. $y = \left(\dfrac{3}{2}x^2 + 21\right)^{1/3}$

25. $y = \dfrac{6}{4(4 - x)^{3/2} + 3}$

27. $y - 2 \ln |y + 1| = \ln |t| + 2(1 - \ln 3)$

29. $\dfrac{dQ}{dt} = 0.07Q$ where $Q(t) =$ amount of investment at time t

31. $\dfrac{dQ}{dt} = kQ$ where $Q(t) =$ number of bacteria at time t

$k =$ constant of proportionality

33. $\dfrac{dP}{dt} = 500$ where $P(t) =$ number of people at time t

35. $\dfrac{dT}{dt} = k(T - T_m)$ where $T(t) =$ temperature of the

object at time t
$T_m =$ temperature of the medium
$k =$ constant of proportionality

37. $\dfrac{dR}{dt} = k(F - R)$ where $R(t) =$ number of facts

recalled at time t
$F =$ total number of facts
$k =$ constant of proportionality

39. $\dfrac{dP}{dt} = kP(C - P)$ where $P(t) =$ number of people

implicated at time t
$C =$ number of people involved
$k =$ constant of proportionality

41. $\dfrac{dy}{dx} = \dfrac{d}{dx}(Ce^{kx}) = kCe^{kx} = ky$

$\dfrac{dy}{dx} = C_1e^{kx} + C_2(xe^x + e^x)$

43. $\dfrac{d^2y}{dx^2} = C_1e^x + C_2(xe^x + 2e^x)$

$\dfrac{d^2y}{dx^2} - 2\dfrac{dy}{dx} + y$

$= [C_1e^x + C_2(xe^x + 2e^x)] - 2[C_1e^x + C_2(xe^x + e^x)]$
$\quad + [C_1e^x + C_2xe^x] = 0$

45. About $1\ 252\ 608$

47. **a.** $\dfrac{S(t)}{40}$ kg per minute

b. $\dfrac{dS}{dt} = -\dfrac{S}{40}$

c. $S(t) = 400e^{-t/40}$

$\dfrac{dP}{dt} = kP(C - P)$

49. $\dfrac{d^2P}{dt^2} = k\dfrac{dP}{dt}(C - P) - kP\dfrac{dP}{dt}$

$\quad = k(C - 2P)\dfrac{dP}{dt}$

$\dfrac{d^2P}{dt^2} = 0 \Leftrightarrow \dfrac{dP}{dt} = 0$ or $P = \dfrac{C}{2}$

We want the maximum of $\dfrac{dP}{dt}$, so we eliminate

$\dfrac{dP}{dt} = 0$ and check $P = \dfrac{C}{2}$.

$\dfrac{d^2P}{dt^2} = k(kP(C - P))(C - 2P)$, where $k > 0, P \geq 0$

Assume $P < \dfrac{C}{2}$; then $\dfrac{d^2P}{dt^2} > 0$.

Assume $P > \dfrac{C}{2}$; and $P < C$; then $\dfrac{d^2P}{dt^2} < 0$.

Thus, $\dfrac{dP}{dt}$ has a local maximum at $P = \dfrac{C}{2}$.

51. $\dfrac{dP}{dt} = k(1 - p),\ p(8) = 0.05$

$p(t) = 1 - (0.95)^{t/8}$

53. 45

55. **a.** $R(S) = k \ln\left(\dfrac{S}{S_0}\right)$

b.

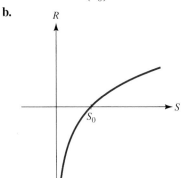

57. a. $p(t) = 8 - 3e^{-5t/2}$ **b.** $p_e = 8$

c. As $t \to \infty$, $e^{-5t/2} \to 0$ and $p(t) \to 8$

59. $\dfrac{b}{a} \ln S + \dfrac{C}{a} S + \dfrac{1}{2a} S^2 = t + K$

61. $\dfrac{dP}{dx} = \dfrac{kP}{x}$

$P(x) = Cx^k$

CHAPTER 8   Section 2

1. $y = \dfrac{x^2}{5} + \dfrac{C}{x^3}$

3. $y = \dfrac{1}{\sqrt{x}(xe^x - e^x + C)}$

5. $y = \dfrac{2}{x} \ln x + \dfrac{C}{x}$

7. $y = \dfrac{x}{2e^{2x}} + \dfrac{C}{xe^{2x}}$

9. $y = \dfrac{1}{1 + x}(Ce^x - 1)$

11. $y = \dfrac{1}{1 + t}\left(\dfrac{t^2}{2} + \dfrac{t^3}{3} + C\right)$

13. $y = (t + 1)(1 + Ce^{-t})$

15. $y = 2x^3$

17. $y = 1 + (x - 2)e^{-x^2/2}$

19. $y = \dfrac{1}{x}(\ln x - 2)$

21. $y = \dfrac{5}{3} + Ce^{-3x}$

23. $y = 2 - \dfrac{C}{x + 1}$

25. $y = -x - 1 + 2e^{x+1}$

27. About \$260 578

29. a. The change in the value of the account is due to accrued interest, rV, plus deposits, D.

b. About \$245 108.19

c. About \$11 488.68

31. About \$3860.03

33. a. $\dfrac{dS}{dt} = \dfrac{-3}{40 - 2t} S, \; S(0) = 5$

$S(t) = \dfrac{1}{8\sqrt{5}}(20 - t)^{3/2}$ for $0 \le t \le 20$

b. ≈ 2.76 min

c. 0 kg

35. a. $\dfrac{dg}{dt} = 0.2 - 0.0001g; \; g(0) = 0$

$g(t) = 2000 - 2000e^{-0.0001t}$

b. ≈ 100.5 minutes

37. About \$39 346.93

$\dfrac{dA}{dt} = 3 - kA; \; A = 0$ when $t = 0$

39. a. $A(t) = \dfrac{3}{k}(1 - e^{-kt})$

b. $A(8) \approx 5.32$ mg

$\lim_{t \to \infty} A(t) \approx 5.39$ mg

c. The rate should be about 5.013 mg/h.

41. a. $p(t) = \dfrac{49}{5} e^{5t/3} - \dfrac{9}{5}$

b. Unstable

CHAPTER 8   Section 3

1. $xy = C$, orthogonal family $y^2 - 2x^2 = C$

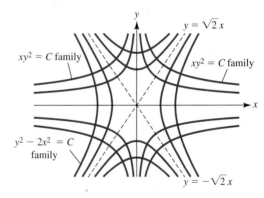

3. $y = Cx^2$, orthogonal family $2y^2 + x^2 = C$

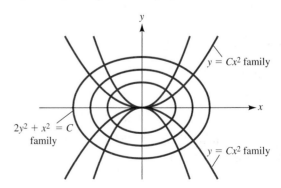

ANSWERS

5. **a.** $\dfrac{dp}{dt} = 0.02(8 - 4p)$, $p(t) = 2 - e^{-0.08t}$

   b. $p(4) = 2 - e^{-0.32} \approx \$1.27$

   c. $\lim\limits_{t\to\infty} p(t) = 2$, the price approaches $2.

7. **a.** $\dfrac{dp}{dt} = 0.02(1 + 7e^{-t} - p)$,

   $p(t) = 1 - \dfrac{1}{7}e^{-1} + \dfrac{22}{7}e^{-0.02t}$

   b. $p(4) \approx \$3.90$

   c. $\lim\limits_{t\to\infty} p(t) = 1$, the price approaches $1.

9. Money market: about $9 499.10; Stock fund: about $30 274.65

11. 6 minutes

13. About 97.27°C

15. Time of death $\approx$ 12:28 P.M.

17. $\dfrac{DA}{dt} = 6(0.16) - 6\left(\dfrac{A}{200}\right)$

   $A(t) = 32(1 - e^{-0.03t})$

19. $\dfrac{-V}{A} \ln\left(1 - \dfrac{L}{c_0}\right)$ seconds

21. **a.** $\dfrac{dN}{dt} = 0.2te^{-t/40}\dfrac{75}{3000} - N\left(\dfrac{75}{3000}\right)$

   $N(t) = \dfrac{1}{400}t^2 e^{-t/40}$

   b. $N'(t) = 0$ when $t = 80$ minutes

   $\dfrac{N(80)}{3000} = 0.000722$ g/L

23. $P(t) = \dfrac{1000}{3}t + \dfrac{100\,000}{9} + \dfrac{800\,000}{9}e^{0.03t}$

25. $\dfrac{dP}{dt} = 0.02P + 100e^{-t}$; $P(0) = 300\,000$

   $p(t) \approx -98.04e^{-t} + 300\,098.04e^{0.02t}$

27. $N(t) = 60(1 + 29e^{-0.423t})^{-1}$

   $N(t) = 20$ when $t \approx 6.32$; he does not complete his mission.

29. **a.** If $A(t)$ is the amount of drug at time t, then

   $\dfrac{dA}{dt} = 5 - \dfrac{4A}{800 + 2t}$ where $A(0) = 50$

   $A(t) = \dfrac{5}{6}(800 + 2t) - \dfrac{296\,000\,000}{3(t + 400)^2}$

   $C(t) = \dfrac{A(t)}{800 + 2t} = \dfrac{5}{6} - \dfrac{148\,000\,000}{3(t + 400)^3}$

   b. About 12.41 hours

31. $y = x^{1/3}$

33. **a.** With populations in thousands,

   $N(t) = \dfrac{2.5(250 - N_0)e^{-9.9t} + 250(N_0 - 2.5)}{(250 - N_0)e^{-9.9t} + (N_0 - 2.5)}$

   b. If $N_0 = 5(5000$ people$)$, then as $t \to \infty$

   $N(t) = \dfrac{2.5(250 - 5)e^{-9.9t} + 250(5 - 2.5)}{(250 - 5)e^{-9.9t} + (5 - 2.5)} \to 250$

   or 250 000 people

   c. If $N_0 = 1(1000$ people$)$, then $N(t) = 0$ when the numerator satisfies

   $2.5(250 - 1)e^{-9.9t} + 250(1 - 2.5) = 0$

   $t = 0.051$ days

   d. In general, if $N_0 > m$, then $N(t) \to N_s$ as $t \to \infty$. However, if $N_0 < m$, then $N(t) = 0$ when

   $t = \dfrac{1}{(N_s - m)k} \ln\left(\dfrac{m(N_s - N_0)}{N_s(m - N_0)}\right)$

35. **a.** $R_e = 26$; $S_e = 21$

   b. $26 \ln R - R = 63 \ln S - 3S - 43.3$

37. **a.** $k = \dfrac{1}{30} \ln 10$

   b. $D(t) = 50e^{-\left(\frac{\ln 10}{30}\right)t}$

   $S(t) = 5e^{\left(\frac{\ln 10}{15}\right)t}$

   c. About 23.2 units

39. **a.** $\dfrac{dP_1}{dt} = \dfrac{-15P_1}{7000}$; $P_1(0) = 2000$

   $P_1(t) = 2000e^{-3t/4400}$

   $\dfrac{dP_2}{dt} = \dfrac{15P_1}{7000} - \dfrac{15P_2}{4000}$; $P_2(0) = 0$

   b. $a = \dfrac{15}{7000}$ and $b = \dfrac{15}{4000}$

   $P_2(t) = \dfrac{8000}{3}(e^{-3t/1400} - e^{-3t/800})$

   c. $P'_2(t) = 0$ when $t = 348.2$ hours.

   $P_2(348.2) \approx 542$ pounds of pollutant, the maximum

41. **a.** $P(t) = \left(\dfrac{1}{-ckt + \dfrac{1}{p_0^c}}\right)^{1/c}$

   b. $t_a = \dfrac{1}{ck\,P_0^c}$, $\lim\limits_{t\to t_a} P(t) = \infty$

   c. $P(t) = \left(\dfrac{1}{2^{-0.02} - 0.00679t}\right)^{50}$

   d. 145 months or 12 years and 1 month

CHAPTER 8   Section 4

1.

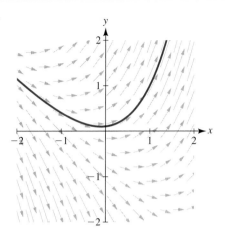

3.

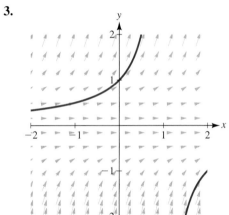

5.

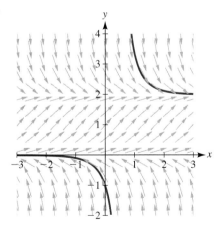

7.

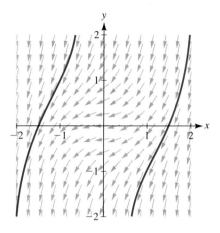

9.

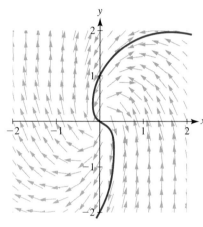

11.

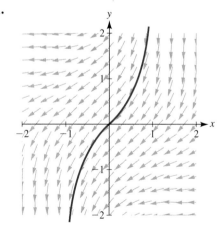

13.

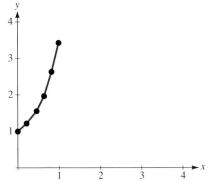

15.

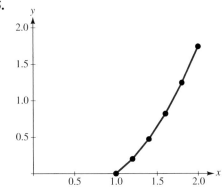

17.

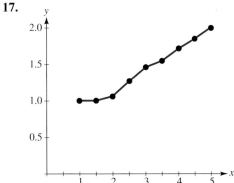

19.

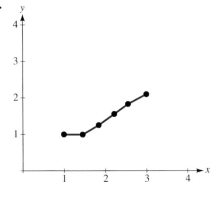

21. $y(1)$ is approximately 5.9728.

23. $y(1)$ is approximately 0.666546.

25. $y(2)$ is approximately 2.77973.

27. $L(4)$ is approximately 3.4526.
$n = 6, h = 0.5$

29. About 870 units will be produced.

31. **a.** At time $t = 2$, the fish will be about 4.9415 cm long.

b. $L(t) = 5.3 - 5.1e^{-1.03t}$
$L(2) = 4.65$ cm

CHAPTER 8   Section 5

1. $y_0 = 1$
$y_1 = 3$
$y_2 = 9$
$y_3 = 27$
$y_4 = 81$

3. $y_0 = y_1 = y_2 = y_3 = y_4 = 1$

5. $y_0 = 1$
$y_1 = 1$
$y_2 = 2$
$y_3 = 3$
$y_4 = 5$

7. $y_0 = 1$
$y_1 = \dfrac{1}{3}$
$y_2 = \dfrac{1}{7}$
$y_3 = \dfrac{1}{15}$
$y_4 = \dfrac{1}{31}$

9. $A = 1, B = -1$

11. $y_n = (-1)^n$

13. $y_n = 2 - \dfrac{1}{2^{n-1}}$

15. $y_n = 2\left(\dfrac{9}{8}\right)^n - 2$

17. $y_n = \dfrac{11^n}{10^{n-1}}$

19. $y_n = (1.05)\, y_{n-1} + 50,\ y_0 = 1000$
$y_n = 1000\,[2(1.05)^n - 1]$

21. **a.** $25\,000\,(1.08^n + 1)$
b. $h = 4000$

23. a. $S_{n+1} - S_n = k(P - S_n) - 0.2S_n$

where $P = 10\ 000$, $S_0 = 500$, $S_1 = 1100$, $k = \dfrac{7}{95}$

b. $S_n = \left(0.8 - \dfrac{7}{95}\right)^n 500$

$+ \left[\dfrac{1 - (0.8 - \frac{7}{95})^n}{1 - (0.8 - \frac{7}{95})}\right]\dfrac{7}{95}(10\ 000)$

$\lim\limits_{n \to \infty} S_n = 2692.3$

25. a. $C_n = 0.75\,C_{n-1} + 50$

$C_n = (0.75)^n\,100 + 200$

b. $205; 200$

27. $M = \dfrac{iD}{1 - (1 + i)^{-tk}}$, with $i = \dfrac{r}{k}$

29. $446.08

31. $A = \$180.63$

33. $I_n = \left(I_0 = \dfrac{h}{c}\right)(1 + c)^n + \dfrac{h}{c}$

35. a. $C_n = C_{n-1} + 18 - \left(\dfrac{C_{n-1}}{5000}\right)18$, $C_0 = 0$

$C_n = 5000\left(1 - \left(\dfrac{4982}{5000}\right)^n\right)$

b. About 638.46 years

c. $t = 639.61$ years; very similar times

d. Writing exercise; responses will vary.

37. a. $p_e = 0.825$

b. $p_n = -3\,p_{n-1} + 3.3; p_0 = 1$

$p_n = 0.175(-3)^n + 0.825$

c.

n	P_n	S_n	D_n
1	0.3	1	1
2	2.4	-1.1	-1.1
3	-3.9	5.2	5.2

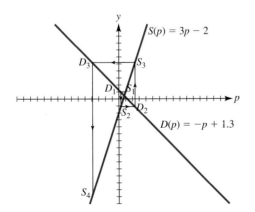

$S(p) = 3p - 2$

$D(p) = -p + 1.3$

39. a. $p_e = 2$

b. $p_n = -\dfrac{1}{2}p_{n-1} + 3$

$p_n = (-0.5)^n + 2$

c.

n	P_n	S_n	D_n
1	1.5	4	4
2	2.25	2.5	2.5
3	1.875	3.25	3.25

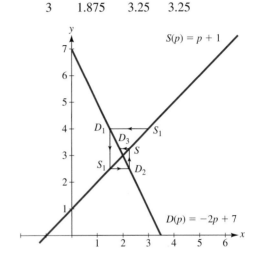

$S(p) = p + 1$

$D(p) = -2p + 7$

41. P_n in thousands

a. $P_n = \dfrac{1000}{n}\left[\dfrac{1 - 1.05^{-n}}{0.05}\right]$

b. $A = \$30\ 242.59$

43. a. $D_n = 0.94\,D_{n-1}$

$D_n = (0.94)^n\,D_0$

b. Approximately 71.4 mg

45. a. $A_n = 1.07A_{n-1} + 25\ 000$

$A_n = 25\ 000\left[1.07^n\left(1 + \dfrac{1}{0.07}\right) - \dfrac{1}{0.07}\right]$

b. At age 40 (after 19 years)

c. \$283\ 389.88

47. a. Suppose $P_m = P_{m+1} = L$ for some m.

Then $P_{m+2} = P_{m+1} + kP_{m+1}\left(1 - \dfrac{1}{K}P_{m+1}\right)$

$\qquad = P_m + kP_m\left(1 - \dfrac{1}{K}P_m\right)$

$\qquad = L$

Thus, once two consecutive values are the same, all the rest of the values will be the same.

b. $K = 105,\ k = \dfrac{21}{110},\ L = 105$

c. $K = 110,\ k = \ln\dfrac{5}{6},\ L = K = 110$

The values are similar and $L = K$ in both cases.

CHAPTER 8   Checkup

1. a. $y = C\sqrt{x^2 + 1}$

 b. $y = \dfrac{1}{2}x^4 + Cx^2$

 c. $y = -\ln(|x + 1 - Ce^x|) + x$

2. a. $y = \sqrt{\dfrac{4 + 5x}{x}}$

 b. $y = \dfrac{5}{2}\left(x - \dfrac{1}{x}\right)$

 c. $y = \dfrac{1}{3}\left[x^2 + 1 - \dfrac{1}{\sqrt{x^2 + 1}}\right]$

3. a. $y_n = -\left(-\dfrac{1}{2}\right)^n$

 b. $y_n = 2\left(\dfrac{4}{3}\right)^n$

 c. $y_n = 2\left(1 - \left(\dfrac{5}{4}\right)^n\right)$

4. a.

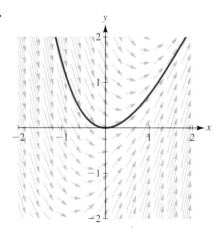

b.

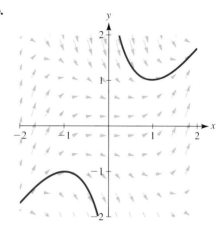

c.

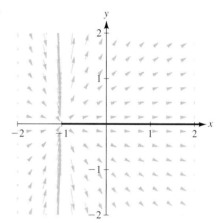

5. $\dfrac{dP}{dt} = -0.11p$

6. $\dfrac{dP}{dt} = k[S(p) - D(p)]$

7. $A_0 \approx 2{,}546$ mg

8. $t = 16$ years

9. $\dfrac{dA}{dt} = 8 - \dfrac{A(t)}{200 + 3t}$

$S(t) = 400 + 6t - 325\left[\dfrac{200}{200 + 3t}\right]^{1/3}$

There are about 760.54 g of salt when the tank is full.

10. a. $p_e = 10$

 b. $p_n = -\dfrac{1}{2}p_{n-1} + 15$

 $p_n = 10\left(1 + \left(-\dfrac{1}{2}\right)^n\right)$

c.

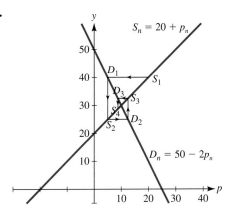

31.

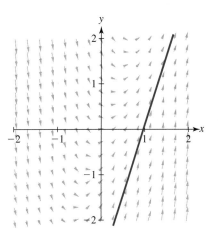

CHAPTER 8   Review Exercises

1. $y = \dfrac{1}{4}x^4 - x^3 + 5x + C$

3. $y = 80 + Ce^{-kx}$

5. $y = -\ln(-e^x - C)$

7. $y = \dfrac{C}{x^4} - e^{-x}\left[1 + \dfrac{4}{x} + \dfrac{12}{x^2} + \dfrac{24}{x^3} + \dfrac{24}{x^4}\right]$

9. $y = \dfrac{1}{(x+1)^2}[2x^2 + 4\ln(x+2) + C]$

11. $y = \pm\sqrt{(\ln x)^2 + C}$

13. $y = x^5 - x^3 - 2x + 6$

15. $y = 100\,e^{0.06x}$

17. $y = -\dfrac{1}{2}x^3 - \dfrac{7}{2}x^5$

19. $y = (4 + x)e^{x^2/2}$

21. $y = -4x^{x/2}e^{(1-x)/2}$

23. $y = 2e^{1-\sqrt{1-x^2}}$

25. $y_n = 1 + 2n$

27. $y_n = 1 + 3n$

29. $y_n = \dfrac{1}{2}(-1 + 5^{1+n})$

33.

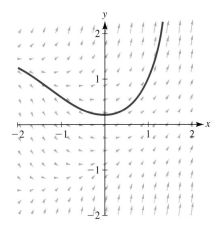

35.

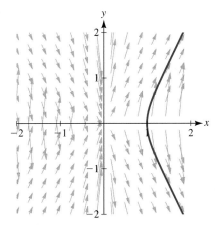

37. $x = 0.1$ $y = 0$
$x = 0.2$ $y = 0.03$
$x = 0.3$ $y = 0.087$
$x = 0.4$ $y = 0.1683$
$x = 0.5$ $y = 0.27147$

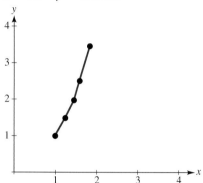

39. $x = 1.2$ $y = 1.4$
$x = 1.4$ $y = 1.91333$
$x = 1.6$ $y = 2.56933$
$x = 1.8$ $y = 3.40437$

41. $y(1)$ is approximately 19.7642.

43. $y(2)$ is approximately 678.6282.

45. $\dfrac{dN}{dt} = k(A - N), N(0) = 0$

$N(t) = A(1 - e^{-kt})$

47. About 81.20 g

49. $\dfrac{dP}{dt} = kP(10 - P), P(0) = 4, P(4) = 7.4$

$P(t) = \dfrac{20}{3e^{-kt} + 2}$

$P(10) = 9.617$ million

51. **a.** $\dfrac{dC}{dt} = R - kC, k > 0$

$C(t) = \dfrac{1}{k}(R - (R - C_0 k)e^{-kt})$

b. $\lim_{t \to \infty} C(t) = \dfrac{R}{k}$

c. $T = -\dfrac{1}{k} \ln\left(\dfrac{1}{2}\left(1 + \dfrac{R}{R - C_0 k}\right)\right)$

53. $E_n = 1.05 \, E_{n-1}, E_0 = 100$
$E_n = (105)^n \, 100$

55. **a.** $p_n = 0.3 \, p_{n-1}$
$p_n = (0.3)^n \, p_0$

b. The likelihood that a defective unit obtains its defect at some time during production equals 1.
$p_n = 0.7(0.3)^n$
$p_3 = 0.0189$

57. **a.** Since $P_0 > 0, a > 0, b > 0,$

then $P_1 = \dfrac{b P_0}{a + P_0} > 0$

likewise $P_2 > 0, P_3 > 0, \ldots, P_{n-1} > 0, \ldots$
Since $P_{n-1} > 0,$
then $a + P_{n-1} > a$

Then $P_n = \dfrac{bP_{n-1}}{a + P_{n-1}} < \dfrac{b}{a}P_{n-1}$

$< \left(\dfrac{b}{a}\right)^2 P_{n-2}$

$< \cdots < \left(\dfrac{b}{a}\right)^n P_0$

b. Assume $a > b$. Then

$\lim_{n \to \infty} P_n < \lim_{n \to \infty}\left[\left(\dfrac{b}{a}\right)^n P_0\right] = 0$

Hence, $\lim_{n \to \infty} P_n = 0$.

c. By assumption

$x = \lim_{n \to \infty} P_n = \lim_{n \to \infty} P_{n-1} < \infty$

Then x satisfies $x = \dfrac{bx}{a + x}$

or $x = 0$ (which is discarded)
or $x = b - a$. So, $\lim_{n \to \infty} P_n = b - a$.

59. **a.** Each month you have all the pairs of rabbits from the previous month, R_{n-1}, plus a pair of rabbits for each pair of rabbits that is at least two months old, so $Rn = R_{n-1} + R_{n-2}$.

b. $\dfrac{1 + \sqrt{5}}{2}, \dfrac{1 - \sqrt{5}}{2}$

c. $A = \dfrac{1}{\sqrt{5}}$ $B = -\dfrac{1}{\sqrt{5}}$

d. Writing exercise; responses will vary.

CHAPTER 9   Section 1

1. $\sum\limits_{n=1}^{\infty} \dfrac{1}{3^n}$

3. $\sum\limits_{n=2}^{\infty} \dfrac{n-1}{n}$

5. $\sum\limits_{n=1}^{\infty} (-1)^{n+1} \dfrac{n^2}{n+1}$

7. $\dfrac{15}{16}$

9. $-\dfrac{7}{12}$

11. 5

13. 3

15. Diverges

17. $\dfrac{3}{20}$

19. 45

21. $\dfrac{3}{16}$

23. 100

25. 10

27. 4

29. 1

31. $\dfrac{1}{2}$

33. $\dfrac{1}{3}$

35. $\dfrac{25}{99}$

37. \$625 billion, including government spending

39. \$25 503

41. $\dfrac{A}{1+r} + \dfrac{A}{(1+r)^2} + \cdots = \sum\limits_{n=1}^{\infty} \dfrac{A}{(1+r)^n} = \dfrac{A}{r}$

43. $\dfrac{20}{\sqrt{e}-1}$ units

45. $\dfrac{d(1-q)}{q}$

47. a. $1 + 2\sqrt{0.36} + 2(0.36) + 2(0.36)^{3/2} + \cdots = 4$

     b. $\dfrac{\sqrt{H}}{\sqrt{4.9}}(1 + 2\sqrt{r} + 2r + 2r^{3/2} + \cdots)$

$$= \dfrac{\sqrt{H}}{\sqrt{4.9}}\left(1 + 2\sum\limits_{n=1}^{\infty} r^{n/2}\right) = \dfrac{\sqrt{H}}{\sqrt{4.9}}\left(\dfrac{1+\sqrt{r}}{1-\sqrt{r}}\right)$$

49. 2037

51. 8 hours

53. 856.52125

55. a. When $r = 1$, $S_n = na$. Hence $\lim\limits_{n\to\infty} S_n$ is not finite if $a \neq 0$.

     b. When $r = -1$, S_n, $= a$ and S_n, even $= 0$. Hence $\lim\limits_{n\to\infty} S_n$ does not exist.

CHAPTER 9   Section 2

1. $\lim\limits_{n\to\infty} a_n = 2 \neq 0$

3. $\lim\limits_{n\to\infty} a_n$ does not exist

5. Diverges

7. Diverges

9. Diverges

11. Converges

13. Converges

15. Diverges

17. Diverges

19. Converges

21. Diverges

23. Diverges

25. Converges

27. Diverges

29. Converges

31. Converges

33. Converges

35. Diverges

37. $\sum\limits_{k=1}^{\infty} \dfrac{1}{2^k}, \sum\limits_{k=1}^{\infty} \dfrac{1}{k^2}$

39. If $p > 1$, then $\displaystyle\int_1^{\infty} \dfrac{dx}{(1+2x)^p} = \dfrac{3^{1-p}}{2(p-1)d}$.

If $p \leq 1$, then $\displaystyle\int_1^{\infty} \dfrac{dx}{(1+2x)^p}$ does not converge.

41. $c < 1$

43. $\displaystyle\sum_{n=0}^{\infty} \frac{1}{3n+1} > \sum_{n=0}^{\infty} \frac{1}{3n+3}$

$\displaystyle = \frac{1}{3}\sum_{n=0}^{\infty}\frac{1}{n+1} = \frac{1}{3}\sum_{m=1}^{\infty}\frac{1}{m}$

and this last series is the divergent harmonic series.

45. As in Example 9.2.2, determining the breaking strength of the first beam will set a record minimum breaking strain and also break the beam. The next beam will, with probability $\frac{1}{2}$, set a new record minimum breaking strain and also break the beam. The total number of broken beams is the number of new record minimum breaking strains, or

$$1 + \frac{1}{2} + \frac{1}{3} + \cdots + \frac{1}{100} \approx 5.19$$

47. An infinite amount of time

49. Compare to $\displaystyle\sum_{k=1}^{\infty}\frac{1}{k^{2/3}}$; therefore it diverges.

51. Compare to $\displaystyle\sum_{k=1}^{\infty}\left(\frac{2}{5}\right)^{k}$; therefore it converges.

53. Diverges, because $\displaystyle\sum_{n=3}^{\infty}\frac{1}{(\ln n)^2} > \sum_{n=3}^{\infty}\frac{1}{\ln n}$, which diverges by the integral test.

CHAPTER 9   Section 3

1. $R = \dfrac{1}{5}$; the interval of absolute convergence is $-\dfrac{1}{5} < x < \dfrac{1}{5}$.

3. $R = 1$; the interval of absolute convergence is $-1 < x < 1$.

5. $R = \dfrac{1}{4}$; the interval of absolute convergence is $-\dfrac{1}{4} < x < \dfrac{1}{4}$.

7. $R = \infty$; the interval of absolute convergence is all real x.

9. $f(x) = \displaystyle\sum_{n=1}^{\infty}(-1)^{n+1}x^{n}$; the interval of absolute convergence is $-1 < x < 1$.

11. $f(x) = \displaystyle\sum_{n=1}^{\infty}x^{2n}$; the interval of absolute convergence is $-1 < x < 1$.

13. $f(x) = \ln 2 + \displaystyle\sum_{n=1}^{\infty}\frac{(-1)^{n+1}}{n\,2^{n}}x^{n}$; the interval of absolute convergence is $-2 < x < 2$.

15. $f(x) = \displaystyle\sum_{n=0}^{\infty}\frac{3^{n}}{n!}x^{n}$

17. $f(x) = \displaystyle\sum_{n=0}^{\infty}\frac{1}{(2n)!}x^{2n}$

19. $f(x) = \displaystyle\sum_{n=0}^{\infty}\frac{e^{3}(-3)^{n}}{n}(x+1)^{n}$

21. $f(x) = \displaystyle\sum_{n=1}^{\infty}\frac{(-1)^{n+1}2^{n}}{n}\left(x-\frac{1}{2}\right)^{n}$

23. $f(x) = \dfrac{2}{3} + \displaystyle\sum_{n=1}^{\infty}\frac{(x-2)^{n}}{(-3)^{n+1}}$

25. **a.** $f(x) = 3x^{2}\displaystyle\sum_{n=0}^{\infty}\frac{x^{3n}}{n!} = \sum_{n=0}^{\infty}\frac{3}{n!}x^{3n+2}$

   b. $f(x) = \dfrac{d}{dx}\left(\displaystyle\sum_{n=0}^{\infty}\frac{x^{3n}}{n!}\right) = \sum_{n=1}^{\infty}\frac{3x^{3n-1}}{(n-1)!}$

   $= \displaystyle\sum_{n=0}^{\infty}\frac{3}{n!}x^{3n+2}$

27. $\displaystyle\sum_{n=0}^{\infty}\frac{(-1)^{n}}{(2n+3)n!}x^{2n+3}$

29. 1.0955

31. -0.3565

33. 0.6065

35. 0.4591

37. 0.2970

39. 0.4854

41.

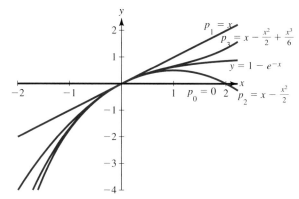

43.

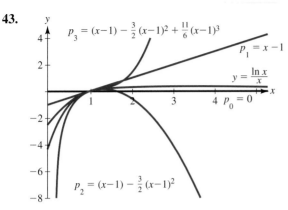

45. Approximately $1862

47. 13 548

49. a. $c = 0: P_{ave} = 2$

$c = 5: P_{ave} \approx 0.98280$

$c = 10: P_{ave} \approx 0.99981$

b. Using $c = 5$ yielded the best approximation.

CHAPTER 9   Checkup

1. a. Converges to $\dfrac{1}{8}$

   b. Diverges

2. a. p-series test with $p = \dfrac{3}{2}$

   b. Ratio test $L = \lim\limits_{n\to\infty}\left|\dfrac{a_{n+1}}{a_n}\right| = \lim\limits_{n\to\infty}\dfrac{3}{n+1} = 0$

3. a. Divergence test: $\lim\limits_{n\to\infty}\dfrac{n(n^2+1)}{100n^3+9} = \dfrac{1}{100} \neq 0$

    Therefore, diverges.

   b. $\displaystyle\sum_{n=3}^{\infty}\dfrac{\ln n}{\sqrt{n}} > \sum_{n=3}^{\infty}\dfrac{1}{\sqrt{n}}$ which diverges by p-series test. Therefore, diverges.

4. a. Diverges

   b. Converges

   c. Converges

   d. Converges

5. a. $-1 < x < 1$

   b. All real x

6. a. $\displaystyle\sum_{n=0}^{\infty}(-1)^n x^{2n+1}$

   b. $\displaystyle\sum_{n=0}^{\infty}\dfrac{5}{2}\left(\dfrac{-3}{2}\right)^n x^n$

7. a. $\displaystyle\sum_{n=0}^{\infty}\dfrac{2x^{2n}}{(2n)!}$

   b. $\displaystyle\sum_{n=1}^{\infty}\dfrac{(-1)^{n+1}x^{2n}}{n}$

8. $\dfrac{25}{e^{1/3}-1} \approx 63.2$

9. $122\ 517

10. Using $n = 10\ 000$, the estimate is 0.7854.

CHAPTER 9   Review Exercises

1. $\dfrac{1}{7}$

3. $\dfrac{65}{6}$

5. $\dfrac{\sqrt{e}}{\sqrt{e}-1}$

7. Diverges

9. Diverges

11. Diverges

13. Diverges

15. Converges

17. Converges

19. $-\dfrac{\sqrt{3}}{2} < x < \dfrac{\sqrt{3}}{2}$

21. $-\dfrac{1}{e} < x < \dfrac{1}{e}$

23. $\displaystyle\sum_{n=0}^{\infty}\dfrac{2}{(2n+1)!}x^{2n+1}$

25. $\displaystyle\sum_{n=1}^{\infty}\dfrac{(-1)^n(2^n-1)x^n}{n}$

27. $\dfrac{25\ 417}{495}$

29. 10 metres

31. $102\ 521

33. $k = \ln\dfrac{19}{14} \approx 0.305$

35. **a.** $\sum_{k=1}^{\infty} \dfrac{T}{k(k+1)}$ is the number of different words; it should sum to T.

  b. $\sum_{k=1}^{\infty} \dfrac{T}{k(k+1)} = T$

  c. Writing exercise; responses will vary.

37. $\sum_{n=0}^{\infty} (-1)^n (n+1)(x-2)^n$

39. $x - 1 + \sum_{n=2}^{\infty} \dfrac{(-1)^n}{n(n-1)}(x-1)^n$

41. 0.9487

43. **a.** The total time to collision is

$$\frac{1000}{90} + \frac{1000}{90}\left(1 - \frac{2}{3}\right) + \frac{1000}{90}\left(1 - \frac{2}{3}\right)^2 + \cdots$$
$$= \frac{50}{3} \text{ seconds.}$$

  The distance the bee travels is
  $$\frac{50}{3} \cdot 60 = 1000 \text{ metres.}$$

  b. The trains will collide in $\dfrac{1000}{90} = \dfrac{50}{3}$ seconds.

  The distance the bee travels is
  $$\frac{50}{3} \cdot 60 = 1000 \text{ metres.}$$

  c. The trains travel for $\dfrac{1000 - 180}{60} = \dfrac{820}{60}$

  seconds before starting to decelerate. They then travel for another 6 seconds before stopping with the bee cradled gently between their headlights. Thus, the total travel time for the bee is $\dfrac{820}{60} + 6 = \dfrac{1180}{60}$ seconds, and in that time, it travels $\left(\dfrac{1180}{60}\right)(60) = 1180$ metres.

45. **a.**

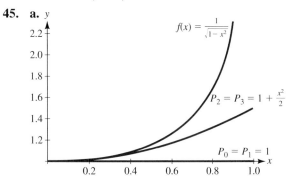

47. **b.** The integral is approximately $\dfrac{25}{48}$.

47. **a.** $t = \dfrac{\lambda}{\ln 2}\left[\sqrt{2\dfrac{S(t) - S(0)}{R(t)} + 1} - 1\right]$

  b. $t = 5.587 \times 10^8$ years

  c. Writing exercise; responses will vary.

CHAPTER 10   Section 1

1. Discrete

3. Continuous

5. Continuous

7. Discrete

9. Discrete

11. $\{0, 1, 2\}$

13. $\{0, 1, 2\}$

15. **a.** $\{11, 12, 13, 14, 15, 16, 21, 22, 23, 24, 25, 26,$ $31, 32, 33, 34, 35, 36, 41, 42, 43, 44, 45, 46\}$
  (*Note*: The number on the ticket is listed first.)

  b. $11, 13, 15, 22, 24, 26, 31, 33, 35, 42, 44, 46$

  c. $11, 22, 33, 44$

17. **a.** H H H H, $X = 0$ H T T H, $X = 2$
  H H H T, $X = 1$ T H T H, $X = 2$
  H H T H, $X = 1$ T T H H, $X = 2$
  H T H H, $X = 1$ H T T T, $X = 3$
  T H H H, $X = 1$ T H T T, $X = 3$
  H H T T, $X = 2$ T T H T, $X = 3$
  H T H T, $X = 2$ T T T H, $X = 3$
  T H H T, $X = 2$ T T T T, $X = 4$

  b. H T T T, T H T T, T T H T, T T T H

  c. All outcomes

  d. $P(X = 0) = \dfrac{1}{16}$, $P(X \le 3) = \dfrac{15}{16}$,
  $P(X > 3) = \dfrac{1}{16}$

19. $E(X) = \dfrac{21}{8}$, $\text{Var}(X) = \dfrac{143}{64}$, $\sigma(X) = \dfrac{\sqrt{143}}{8}$

21. $E(X) = \dfrac{20}{7}$, $\text{Var}(X) = \dfrac{160}{49}$, $\sigma(X) = \dfrac{4}{7}\sqrt{10}$

23.

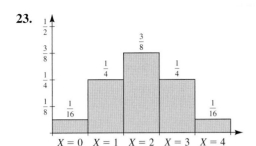

$E(X) = 2$

$\text{Var}(X) = 1$

$\sigma(X) = 1$

25. $20

27. $-$0.027 (loss of 2.7 cents)

29. a.

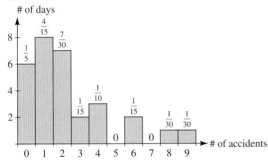

b. $E(X) = \dfrac{23}{10}$

We expect about two accidents to occur on any given day.

c. $\text{Var}(X) = \dfrac{1603}{300}$

$\sigma(X) = \dfrac{\sqrt{4809}}{30}$

31. $E(X) = 0.096$

33. a. $E(X) = pw + (1 - p)(-1)$

$\qquad\quad = pw + p - 1$

b. The game is fair when $E(X) = 0$.

$pw + p - 1 = 0$

$pw = 1 - p$

$w = \dfrac{1 - p}{p}$

Fair odds are $\dfrac{1 - p}{p}$ to 1.

35. $\dfrac{19}{18}$ to 1

37. $P(X = 1) = 0.8$, $P(X = 2) = 0.16$,
$P(X = 3) = 0.032$, $P(X = 4) = 0.0064$,
$P(X = 5) = 0.00128$

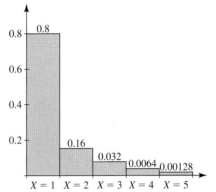

39. a. 0.00567

b. 0.99757

c. 0.00243

d. About 1.4 chips

41. a. About 0.00050

b. Page 2000

c. About 0.98757

43. a. About 0.03397

b. About 0.18463

c. About 0.61271

d. 25 students

45. About 0.13533

47. $E(X) = \displaystyle\sum_{n=1}^{\infty} np(1 - p)^{n-1}$

$\qquad\quad = p\displaystyle\sum_{n=1}^{\infty} n(1 - p)^{n-1}$

$\qquad\quad = p\,\dfrac{1}{(1 - (1 - p))^2}$

$\qquad\quad = \dfrac{1}{p}$

49. $\displaystyle\sum_{n=1}^{\infty} P(X = n) = \sum_{n=1}^{\infty} \dfrac{a}{n^b} = a\sum_{n=1}^{\infty} \dfrac{1}{n^b} = 1$

Therefore, $a = \dfrac{1}{\zeta(b)}$ where $\zeta(b) = \displaystyle\sum_{n=1}^{\infty} \dfrac{1}{n^b}$.

CHAPTER 10 Section 2

1. Yes
3. No
5. Yes
7. Yes
9. No
11. Yes
13. Yes
15. No such number exists; $\displaystyle\int_{-\infty}^{+\infty} f(x)\,dx = 1$ for $k = \dfrac{5}{2}$, but then $f(1) = -\dfrac{1}{2}$.
17. No such number exists; $\displaystyle\int_{-\infty}^{+\infty} f(x)\,dx = 1$ for $k = \dfrac{3}{4}$, but then $f(x) < 0$ for $\dfrac{4}{3} < x < 2$.
19. $k = -\dfrac{3}{2}$
21. $k = 1$
23. $k = 3$
25. a. 1
    b. $\dfrac{1}{3}$
    c. $\dfrac{1}{3}$
27. a. 1
    b. $\dfrac{3}{16}$
    c. $\dfrac{9}{16}$
29. a. 1
    b. $\dfrac{7}{8}$
    c. $\dfrac{1}{8}$
31. a. 1
    b. $\dfrac{1}{e} - \dfrac{1}{e^4}$
    c. $1 - \dfrac{1}{e^4}$

33. a. $\dfrac{9}{14}$
    b. $\dfrac{43}{70}$
    c. $\dfrac{13}{28}$
35. a. $\dfrac{3}{5}$
    b. $\dfrac{3}{20}$
37. a. About 0.05772
    b. About 0.45119
    c. About 0.54881
39. a. $\dfrac{1}{e^2}$
    b. $\dfrac{1}{e^{1/4}} - \dfrac{1}{e^{5/4}} \approx 0.49230$
41. a. $1 - \dfrac{9}{4}\left(\dfrac{1}{e^{5/4}}\right) \approx 0.35536$
    b. $\dfrac{7}{2}\left(\dfrac{1}{e^{5/2}}\right) \approx 0.28730$
43. $1 - e^{-0.96} \approx 0.61711$
45. a. $f(t) - \begin{cases} \frac{1}{60} & \text{if } 0 \le t \le 60 \\ 0 & \text{otherwise} \end{cases}$
    b. $\dfrac{1}{4}$
    c. $\dfrac{1}{6}$
47. $\dfrac{1}{e} \approx 0.36788$
49. a. About 0.11612
    b. About 0.06346
51. a. About 0.33353
    b. About 0.66647
53. About 0.45816
55. 20 seconds
57. $\dfrac{A + B}{2}$
59. a. $r(x) = \displaystyle\int_{x}^{\infty} f(x)\,dx = 1 - \int_{0}^{x} f(x)\,dx$
    b. e^{-200}

61. a.

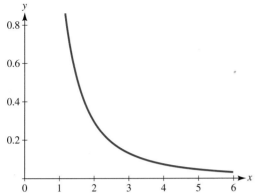

b. 0.06

c. Writing exercise; responses will vary.

CHAPTER 10 Section 3

1. $E(X) = \dfrac{7}{2}$, Var $(X) = \dfrac{3}{4}$

3. $E(X) = \dfrac{4}{3}$, Var $(X) = \dfrac{8}{9}$

5. $E(X) = \dfrac{3}{2}$, Var $(X) = \dfrac{3}{4}$

7. $E(X) = \dfrac{1}{4}$, Var $(X) = \dfrac{1}{16}$

9. $E(X) = \dfrac{2}{3}$, Var $(X) = \dfrac{2}{63}$

11. $E(X) = 2$, $E(Y) = 1$

13. $E(X) = 1$, $E(Y) = 2$

15. $E(X) \approx 4.685$ years

17. $E(X) = 10$ minutes

19. $E(X) = 20$ months

21. a. $\lambda = \dfrac{1}{5}$

 b. About 0.32968

 c. About 0.24660

23. a. $A = b = -\ln 0.9$

 b. About 35.9%

 c. About 22.9 weeks

 d. About 9.5 weeks

 e. About 90.1 (weeks)^2; the variability is large compared to the expected value.

25. 2 hours

27. $E(X) = \displaystyle\int_a^b \left(\dfrac{x}{b-a}\right) dx = \dfrac{x^2}{2(b-a)}\Big|_{x=a}^{x=b}$

$$= \dfrac{(b^2 - a^2)}{2(b-a)} = \dfrac{b+a}{2}$$

29. $E(X) = \displaystyle\int_0^\infty \lambda x e^{-\lambda x} dx = \lim_{N\to\infty}\int_0^N \lambda x e^{-\lambda x} dx$

$$= \lim_{N\to\infty}\left[-xe^{-\lambda x}\Big|_0^N + \int_0^N e^{-\lambda x} dx \right]$$

$$= \lim_{N\to\infty}\left(-xe^{-\lambda x} - \dfrac{1}{\lambda}e^{-\lambda x} \right)\Big|_0^N = \dfrac{1}{\lambda}$$

31. a. $f(x) = \begin{cases} 2e^{-2x} & \text{if } x \geq 0 \\ 0 & \text{otherwise} \end{cases}$

 b. $e^{-2} \approx 0.13534$

 c. $e^{-1/2} - e^{-1} \approx 0.23865$

33. Patient in 107A: 4 days

Patient in 107B: 3 days

35. a. $\dfrac{7}{30}$

 b. A: $\dfrac{24}{5}$ worker-hours, B: $\dfrac{16}{3}$ worker-hours

 c. About $193.07

CHAPTER 10 Section 4

1. $E(X) = 0$, Var $(X) = 4$, $\sigma = 2$

3. $E(X) = -1$, Var $(X) = 3$, $\sigma = \sqrt{3}$

5. 0.8925

7. 1.0000

9. 0.6827

11. 1.00

13. 1.10

15. a. 0.0228

 b. 0.5

 c. 0.6826

 d. 0.0000

17. a. 5

 b. -3.5

ANSWERS

19. a. 0.683 or 68.3%
    b. 0.954 or 95.4%

21. a. $3e^{-3} \approx 0.149$

b. $\dfrac{9}{2}e^{-3} \approx 0.224$

c. $\dfrac{27}{8}e^{-3} \approx 0.168$

d. $\dfrac{3^8}{8!}e^{-3} \approx 0.008$

23.

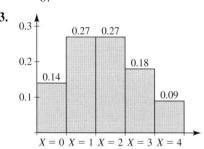

25. a. False; only true if mean = 0
    b. True; it is symmetric about 0.

27. a. Roughly 68.3% of adult basset hounds have weight w in the range $21 \leq w \leq 25$ kg.
    b. Roughly 95.4% of adult basset hounds have weight in the range $19 \leq w \leq 27$ kg.
    c. Roughly 99.7% of adult basset hounds have weight in the range $17 \leq w \leq 29$ kg.

29. 0.3829

31. a. 0.9544
    b. 0.0013

33. About 0.0388

35. 11.51%

37. 44.2 years old

39. a. 8.08%
    b. 498 g

41. a. 7.5%
    b. 2.66 mmol/L

43. a. 0.0668
    b. 0.3538
    c. 0.4554

45. 0.98177

47. a. 0.3472, 0.4599
    b. 0.2047
    c. 0.7953

49. Shop B

51. a. 0.9544
    b. Yes, 0.0062
    c. Yes, 0.00003
    d. Writing exercise; responses will vary.

53. a. $\lambda = 6.1$
    b. 0.0022
    c. 0.0848
    d. 0.4298

55. The difference in probabilities is 0.00027.

57. a. $\lambda = \ln 20 \approx 3$
    b. 0.1005
    c. ≈ 0.9165
    d. 0.95
    e. 3 hits

59. a. $\displaystyle\int_{x=\mu-2\sigma}^{x=\mu+2\sigma} \frac{1}{\sigma\sqrt{2\pi}}e^{\frac{-(x-\mu)^2}{2\sigma^2}}\,dx$

$= \displaystyle\int_{-2}^{2} \frac{1}{\sqrt{2\pi}}e^{-x^2/2}\,dx \approx 0.954$

b. $\displaystyle\int_{x=\mu-3\sigma}^{x=\mu+3\sigma} \frac{1}{\sigma\sqrt{2\pi}}e^{\frac{-(x-\mu)^2}{2\sigma^2}}\,dx$

$= \displaystyle\int_{-3}^{3} \frac{1}{\sqrt{2\pi}}e^{-x^2/2}\,dx \approx 0.997$

CHAPTER 10   Checkup

1.

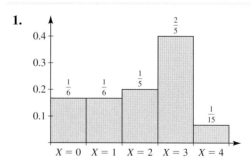

$E(x) = \dfrac{61}{30} \approx 2.0333$

$\mathrm{Var}(x) = \dfrac{1349}{900} \approx 1.4989$

$\sigma(x) = \dfrac{\sqrt{1349}}{30} \approx 1.2243$

2. a. Yes
   b. No

3. a. $e^{-2/3} - e^{-1} \approx 0.1455$
   b. $e^{-1} \approx 0.3679$

4. $E(X) = 3$, $\text{Var}(X) = \dfrac{9}{5}$

5. a. 0.1977
   b. 0.9026

6. a. 0.683
   b. 0.954

7. $\dfrac{2}{3}$

8. 0.7738

9. a. $e^{-1.5} \approx 0.2231$
   b. 2 years; the life expectancy of those on the drug

10. 170 points

11. 0.5595

12. a. 0.9698
   b. The second component; 3 years

CHAPTER 10   Review Exercises

1.

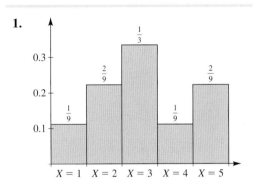

$$E(x) = \frac{28}{9} \approx 3.1111$$

$$\text{Var}(x) = \frac{134}{81} \approx 1.6543$$

$$\sigma(X) = \frac{\sqrt{134}}{9} \approx 1.2862$$

3. Finite discrete

5. Finite discrete

7. $P(2 \le X \le 7) = 0.8$
   $P(X \ge 5) = 0.6$

9. $P(X \le 1) = 0.3125$
   $P(1 \le X \le 3) = 0.6875$

11. $E(X) = 1.2$, $\text{Var}(x) = 0.16$

13. $c = \dfrac{1}{16}$

15. 0.1587

17. Up 16%

19. $11.40

21. a. 15 people
   b. 15 people
   c. 0 people

23. 0.0004

25. a. $1 - e^{-4} \approx 0.9817$
   b. $e^{-8} \approx 0.0003$
   c. 2.5 years

27. a. $1 - 6e^{-5} \approx 0.9596$
   b. 4 m

29. a. $P(X = 10\,000) = 0.98$
       $P(X = -90\,000) = 0.02$
   b. $8000
   c. $18\,000

31. a. $\lambda = \dfrac{\ln 2}{10}$
   b. $1 - \dfrac{\sqrt{2}}{2}$
   c. $\dfrac{10}{\ln 2} \approx 14.2$ weeks

33. a. 1.0000
   b. 352.5 mL

35. 0.4784; 39 students

37. a. 0.5
   b. 5

39. a. $\dfrac{32}{3} e^{-4} \approx 0.1954$
   b. $e^{-4} \approx 0.0183$
   c. $\dfrac{71}{3} e^{-4} \approx 0.4335$

41. a. $e^{-2.4} \approx 0.0907$
   b. $1 - e^{-1.2} \approx 0.6988$
   c. $e^{-0.1} \approx 0.9048$

43. **a.** $\int_{y=0}^{y=1} \int_{x=0}^{x=1} 0.4(2x + 3y)\, dx\, dy = 1$

and for $0 \le x \le 1, 0 \le y \le 1, f(x, y) \ge 0$

b. 0.275

c. 0.072

d. $\dfrac{1}{3} \approx 0.333$

45. **a.** $(e^{-5/4} - 1)(e^{-5/2} - 1) \approx 0.6549$

b. $1 + e^{-4} - 2e^{-2} \approx 0.7476$

47. **a.** $\dfrac{2}{3}$

b. $\dfrac{1}{2}$

c. $\dfrac{5}{18}$

49. **a.** $\dfrac{25}{92} \approx 0.2717$

b. $\dfrac{67}{192} \approx 0.3490$

c. This distribution does not have a finite expected value.

d. Writing exercise; responses will vary.

CHAPTER 11   Section 1

1. **a.** 30°
  b. 60°

3. **a.** −120°
  b. 390°

5. **a.** $\dfrac{3\pi}{4}$

  b. $\dfrac{\pi}{3}$

7. **a.** $-\dfrac{\pi}{3}$

  b. $\dfrac{5\pi}{6}$

9. **a.** 60°

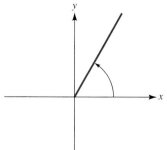

  b.

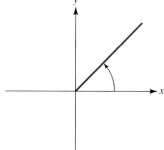

y 120°

11. **a.** 45°

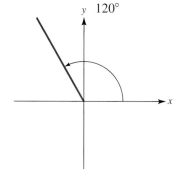

  b. −150°

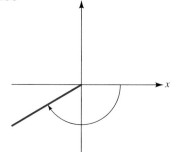

13. a. $\dfrac{3\pi}{4}$

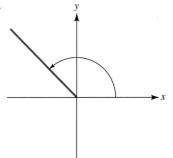

b. $-\dfrac{2\pi}{3}$

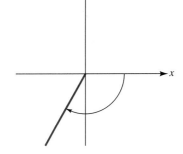

15. a. $-\dfrac{5\pi}{3}$

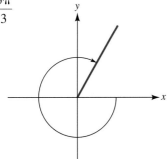

b. $\dfrac{3\pi}{2}$

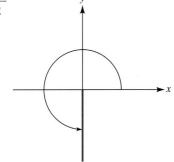

17. a. $\dfrac{\pi}{12}$

    b. $-\dfrac{4\pi}{3}$

19. a. $\dfrac{3\pi}{4}$

    b. 3π

21. a. $150°$

    b. $-15°$

23. a. $540°$

    b. $\dfrac{180°}{\pi}$

25. 0

27. 1

29. 0

31. 1

33. $-\dfrac{1}{2}$

35. $\dfrac{1}{2}$

37. $\dfrac{\sqrt{3}}{3}$

39. -1

41. $\dfrac{2\sqrt{3}}{3}$

43. $-\sqrt{2}$

45. $\cos\dfrac{\pi}{3} = \dfrac{1}{2}, \sin\dfrac{\pi}{3} = \dfrac{\sqrt{3}}{2}$

47.

θ	$\dfrac{7\pi}{6}$	$\dfrac{5\pi}{4}$	$\dfrac{4\pi}{3}$	$\dfrac{3\pi}{2}$	$\dfrac{5\pi}{3}$	$\dfrac{7\pi}{4}$	$\dfrac{11\pi}{6}$	2π
$\sin\theta$	$-\dfrac{1}{2}$	$-\dfrac{\sqrt{2}}{2}$	$-\dfrac{\sqrt{3}}{2}$	-1	$-\dfrac{\sqrt{3}}{2}$	$-\dfrac{\sqrt{2}}{2}$	$-\dfrac{1}{2}$	0
$\cos\theta$	$-\dfrac{\sqrt{3}}{2}$	$-\dfrac{\sqrt{2}}{2}$	$-\dfrac{1}{2}$	0	$\dfrac{1}{2}$	$\dfrac{\sqrt{2}}{2}$	$\dfrac{\sqrt{3}}{2}$	1

49. $\dfrac{3}{4}$

51. $\dfrac{4}{3}$

53. $\dfrac{5}{4}$

55. $\theta = \dfrac{\pi}{2}, \dfrac{3\pi}{2}$

57. $\theta = 0, \dfrac{2\pi}{3}$

59. $\theta = \dfrac{\pi}{6}, \dfrac{5\pi}{6}$

61. $\theta = \dfrac{\pi}{3}, \pi$

63. **a.** $E_0 \sin\left(\dfrac{2\pi}{28} \cdot 21 \cdot 365\right) = -E_0$

b. $P_0 \sin\left(\dfrac{2\pi}{23} t\right)$

$$P_0 \sin\left(\dfrac{2\pi}{23} \cdot 21 \cdot 365\right) = P_0 \sin\left(\dfrac{15\,330}{23}\pi\right)$$
$$\approx 0.998\, P_0$$

c. The E_0 cycle has a maximum when $t = 7(4n + 1)$ days and n is an integer. The P_0 cycle has a maximum when $t = \dfrac{23}{4}(4m + 1)$ days and m is an integer. There are no integers n and m which give the same t. Therefore Jake's 2 cycles will never both be at their peak levels simultaneously.

65. **a.** $A = 106, B = 24, k = \dfrac{79\pi}{30}$

b. 130

67. $\sin\left(\dfrac{\pi}{2} - \theta\right) = \sin\left(\dfrac{\pi}{2}\right)\cos(-\theta) + \cos\left(\dfrac{\pi}{2}\right)\sin(-\theta)$
$$= \cos(-\theta)$$
$$= \cos(\theta)$$

69. An angle θ corresponds to a point on the unit circle with coordinates $(x, y) = (\cos\theta, \sin\theta)$. The correspondence is one-to-one for $0 \le \theta \le 2\pi$. From the following graphs we see that for $0 \le \theta \le 2\pi$, if (x_0, y_0) corresponds to angle θ, then (y_0, x_0) corresponds to angle $\dfrac{\pi}{2} - \theta$. This gives us

$$(y_0, x_0) = \left(\cos\left(\dfrac{\pi}{2} - \theta\right), \sin\left(\dfrac{\pi}{2} - \theta\right)\right)$$

If $(x_0, y_0) = (\cos\theta, \sin\theta)$, then If $(y_0, x_0) = (\sin\theta, \cos\theta)$, Combining this with the earlier result, we have $(y_0, x_0) = \left(\cos\left(\dfrac{\pi}{2} - \theta\right), \sin\left(\dfrac{\pi}{2} - \theta\right)\right) = (\sin\theta, \cos\theta)$.

1st quadrant

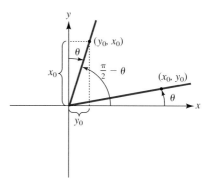

2nd quadrant

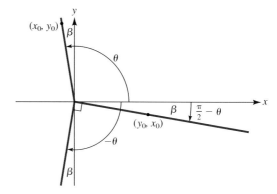

3rd quadrant

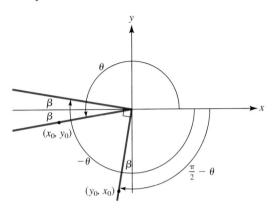

4th quadrant

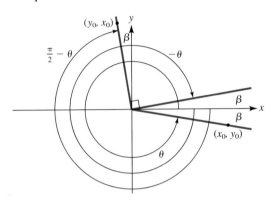

CHAPTER 11   Section 2

1. $3 \cos 3t$

3. $-2 \cos(1 - 2t)$

5. $-3t^2 \sin(t^3 + 1)$

7. $2 \cos\left(\dfrac{\pi}{2} - t\right) \sin\left(\dfrac{\pi}{2} - t\right)$

9. $-6(1 + 3x) \sin(1 + 3x)^2$

11. $e^{-u/2}\left(-\dfrac{1}{2} \cos 2\pi u - 2\pi \sin 2\pi u\right)$

13. $\dfrac{\cos t}{(1 + \sin t)^2}$

15. $-3t^2 \sec^2(1 - t^3)$

17. $-2\pi \sec\left(\dfrac{\pi}{2} - 2\pi t\right) \tan\left(\dfrac{\pi}{2} - 2\pi t\right)$

19. $2 \cot t$

21. $-2 \cos\left(\dfrac{t}{2}\right) + C$

23. $-\dfrac{1}{6} \sin(1 - 3x^2) + C$

25. $\dfrac{1}{4} \cos(2x) + \dfrac{x}{2} \sin(2x) + C$

27. $\dfrac{1}{2} \sin^2 x + C$

29. $\sin x - x \cos x + C$

31. Period 4, amplitude 1.5, phase shift 1, vertical shift 2

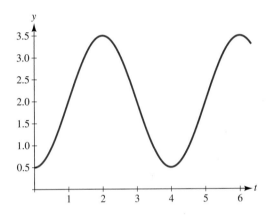

33. Period π, amplitude -1, phase shift $\dfrac{\pi}{2}$, vertical shift 3 vertical shift 3

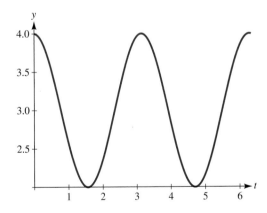

35. $\dfrac{3\sqrt{3} - 4}{2\pi}$

37. $\dfrac{2}{\pi} + \dfrac{1}{6}$

39. 2π

41. $2\sqrt{3}\pi$

43. $y(t) = -1 + \sqrt{5 - \cos 2t}$

45. $\sin y = -1 - \cos x$

47. ≈ -3.69 deg/hr, cooler

49. a.

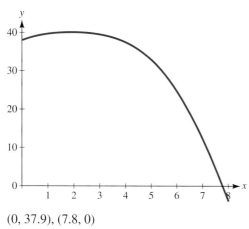

$(0, 37.9)$, $(7.8, 0)$

b. ≈ 244.73

51. $B_0 + 32\sqrt{6}$

53. a. Minimum of 36.8 at time $t = 3$ (3:00 A.M.)
Maximum of 37.2 at time $t = 15$ (3:00 P.M.)

b. $\approx 0.037°$C/hr

c. At $t = 3$ (9:00 A.M.)

55. a. $\approx -5.60 \dfrac{\text{thousand units}}{\text{month}}$, decreasing

b. $\approx 28.53 \dfrac{\text{thousand units}}{\text{month}}$

57. $\dfrac{d\theta}{dt} = \dfrac{4}{15}$ rad/sec

59. 60 rad/hr

61. $\dfrac{d}{dt}(\cos t) = \lim\limits_{h \to 0} \dfrac{\cos(t + h) - \cos(t)}{h}$

$= \lim\limits_{h \to 0} \dfrac{\cos(t)\cos(h) - \sin(t)\sin(h) - \cos(t)}{h}$

$= \lim\limits_{h \to 0} \dfrac{\cos(t)(\cos(h) - 1) - \sin(t)\sin(h)}{h}$

$= \lim\limits_{h \to 0} \dfrac{\cos(h) - 1}{h} = 0.$

and $\lim\limits_{h \to 0} \dfrac{\sin(h)}{h} = 1$, so

$\dfrac{d}{dt}(\cos t) = -\sin t.$

63. $\lim\limits_{x \to 0} \dfrac{\cos x - 1}{x} = 0$

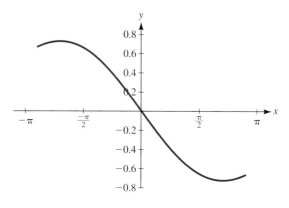

65. a. $f(t + p) = a + b\sin\left[\dfrac{2\pi}{p}(t + p - d)\right]$

$= a + b\sin\left[\dfrac{2\pi}{p}(t - d) + 2\pi\right]$

$= a + b\left[\sin\left(\dfrac{2\pi}{p}(t - d)\right)\cos(2\pi)\right.$

$\left. + \sin(2\pi)\cos\left(\dfrac{2\pi}{p}(t - d)\right)\right]$

$= a + b\sin\left[\dfrac{2\pi}{p}(t - d)\right]$

$= f(t)$

b. $f(t + p) = a + b\cos\left[\dfrac{2\pi}{p}(t + p - d)\right]$

$= a + b\cos\left[\dfrac{2\pi}{p}(t - d) + 2\pi\right]$

$= a + b\left[\cos\left(\dfrac{2\pi}{p}(t - d)\right)\cos(2\pi)\right.$

$\left. - \sin\left(\dfrac{2\pi}{p}(t - d)\right)\sin(2\pi)\right]$

$= a + b\cos\left[\dfrac{2\pi}{p}(t - d)\right]$

$= f(t)$

CHAPTER 11   Section 3

1. FV ≈ 9,401; PV ≈ 7,697

3. FV ≈ 2,000; PV ≈ 1,828

5.

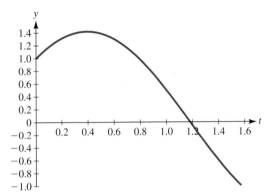

Maximum at $t = \dfrac{\pi}{8} \approx 0.39$,

minimum at $t = \dfrac{\pi}{2} \approx 1.57$

7.

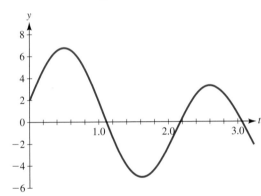

Maximum at $t \approx 0.502$,
minimum at $t \approx 1.615$

9.

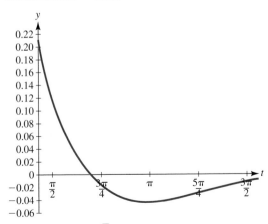

Maximum at $t = \dfrac{\pi}{2}$, minimum at $t = \pi$

11. Maximum value $= f\left(\dfrac{\pi}{4}\right) = -1$,

minimum value, $\lim_{t \to \pi} f(t) = -\infty$

13. **a.** Maximum volume = 0.9 litre at time
$t = (2n + 1)(1.75)$ where n is an integer

    b. Minimum volume = 0.1 litre
at times $t = (2n)(1.75)$ where n is an integer

    c.

(graph)

15. $\displaystyle\int \dfrac{dP}{P} = \int (b + a \cos 2\pi t)\, dt$

$\ln |P| = bt + \dfrac{a}{2\pi} \sin 2\pi t + C_1$

$P(t) = C_2 e^{bt + \frac{a}{2\pi} \sin 2\pi t}$, where $C_2 = e^{C_1}$

$P_0 = P(0) = C_2$

17. $\sin \theta = \dfrac{E_L}{E_W}$

19. Length of 4th wall = 40 m
Maximum area $= 300\sqrt{3}$ m²

21. $P(10) = 13\ 499$

23. **a.** $L = \dfrac{2\pi}{0.65} \approx 9.67$ sec

    b. 2.68 litres; yes

    c. $I = 9.67$ sec; inhale 6.47 litres; no

25. $\theta = \dfrac{\pi}{2}$

ANSWERS

27.

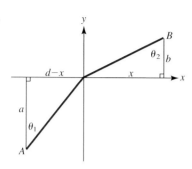

$$\text{Time} \; = \; T = \frac{\text{distance}}{\text{speed}}$$

$$T = \frac{\sqrt{a^2 + (d-x)^2}}{v_1} + \frac{\sqrt{b^2 + x^2}}{v_2}$$

$$\frac{dT}{dx} = \left(\frac{1}{9}\right)\frac{2(d-x)(-1)}{v_1\sqrt{a^2 + (d-x)^2}} + \frac{x}{v_2\sqrt{b^2 + x^2}}$$

$$= \frac{-(d-x)}{v_1\sqrt{a^2 + (d-x)^2}} + \frac{x}{v^2\sqrt{b^2 + x^2}}$$

$$= \frac{-\sin\theta_1}{v_1} + \frac{\sin\theta_2}{v_2}$$

$$\frac{DT}{dx} = \frac{\sin\theta_1}{\sin\theta_2} = \frac{v_1}{v_2}$$

29. 12.76 m

31. $x = (v_0 \cos\theta)t$

$y = -4.9t^2 + (v_0 \sin\theta)t$

a. The maximum height occurs when $y' = 0$.

$$-9.8t + v_0 \sin\theta = 0$$

$$t = \frac{v_0 \sin\theta}{9.8}$$

Substitute this result into the equation for y.

$$y_{\max} = -4.9\left(\frac{v_0^2 \sin^2\theta}{9.8^2}\right) + \frac{(v_0^2 \sin^2\theta)}{9.8} + s_0$$

$$y_{\max} = \frac{v_0^2 \sin^2\theta}{19.6} + s_0$$

b. The maximum distance (range) occurs when $y = 0$.

$$0 = -4.9t^2 + (v_0 \sin\theta)t$$

$$0 = t(-4.9t + v_0 \sin\theta)$$

So, $t = 0$ or

$$-4.9t + v_0 \sin\theta = 0$$

$$4.9t = v_0 \sin\theta$$

$$t = \frac{v_0 \sin\theta}{4.9}$$

Substitute this result into the equation for x.

$$\text{Range} = v_0 \cos\theta\left(\frac{v_0 \sin\theta}{4.9}\right) = \frac{v_0^2 \sin\theta \cos\theta}{4.9} = \frac{v_0^2 \sin 2\theta}{9.8}$$

The range is largest when $\sin 2\theta$ is largest; that is, when $2\theta = \dfrac{\pi}{2}$, or $\theta = \dfrac{\pi}{4}$ or $45°$.

33. a. $T'(t) = -\dfrac{0.46\,\pi}{12}\sin\left[\dfrac{\pi(t - 16.37)}{12}\right]$

b. Maximum at $t = 16.37$
Maximum temperature $= 37.75°C$

c. Minimum temperature $= 36.83°C$
at time $t = 4.37$

d. The average over period from 7 A.M. to 10 P.M. $= 37.48°C$.

35. $F'(\theta) = kh\cos\theta\left(\dfrac{\csc\theta}{R^4} - \dfrac{\cot\theta}{r^4}\right)$, $0 < \theta < \dfrac{\pi}{2}$

Critical point at θ_0 where $\cos\theta_0 = \dfrac{r^4}{R^4}$.

$\csc\theta > 0$ for $0 < \theta < \dfrac{\pi}{2}$ and $k > 0$, $h > 0$

$\Rightarrow$ sign of $F'(\theta) = $ sign of $\left(\dfrac{\csc\theta}{R^4} - \dfrac{\cot\theta}{r^4}\right)$

$$\frac{\csc\theta}{R^4} - \frac{\cot\theta}{r^4} = \frac{1}{R^4 \sin\theta} - \frac{\cos\theta}{r^4 \sin\theta}$$

$$= \frac{1}{\sin\theta}\left(\frac{1}{R^4} - \frac{\cos\theta}{r^4}\right)$$

$\sin\theta > 0$ for $0 < \theta < \dfrac{\pi}{2}$, so, consider

$$\left(\frac{1}{R^2} - \frac{\cos\theta}{r^4}\right)$$

$\cos\theta$ is strictly decreasing on $\left(0, \dfrac{\pi}{2}\right)$

$\Rightarrow$ if $\theta < \theta_0$ then $\cos\theta > \cos\theta_0$

$\Rightarrow \left(\dfrac{1}{R^4} - \dfrac{\cos\theta}{r^4}\right) < \left(\dfrac{1}{R^4} - \dfrac{\cos\theta_0}{r^4}\right) = 0$

$\Rightarrow F'(\theta) < 0$.

Likewise, if $\theta > \theta_0$, then $\cos\theta < \cos\theta_0$

$\Rightarrow \left(\dfrac{1}{R^4} - \dfrac{\cos\theta}{r^4}\right) > \left(\dfrac{1}{R^4} - \dfrac{\cos\theta_0}{r^4}\right) = 0$

$\Rightarrow F'(\theta) > 0$.

There is a minimum at θ_0 since
$F(\theta)$ is decreasing $(F'(\theta) < 0)$ for $\theta < \theta_0$
and increasing $(F'(\theta) > 0)$ for $\theta > \theta_0$.

37. a. 490, 2178, 9680, respectively
b.

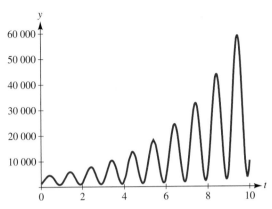

c. Writing exercise; responses will vary. Answers should mention that 145 days into the year is early summer, and this is the time when the most ticks are present.

39. a. $t = 35$ (February 5); 47
b. $t = 217.5$ (August 6); 3
c. January 1: ≈ 0.2146; increasing
April 1: ≈ -0.3110; decreasing
June 1: ≈ -0.3420; decreasing
d.

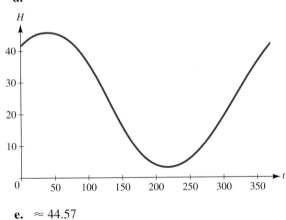

e. ≈ 44.57

CHAPTER 11 Checkup

1. a. $\dfrac{1}{2}$

b. $-\dfrac{1}{2}$

c. -1

d. $-\dfrac{2\sqrt{3}}{3}$

2. $\cos \theta = \dfrac{3\sqrt{13}}{13}$

3. a. $2 \cos (2x)$
b. $\cos x - x \sin x$
c. $\dfrac{x \sec^2 x - \tan x}{x^2}$
d. $\sin(x^2) + 2x^2 \cos(x^2)$

4. a. $-\dfrac{1}{2} \cos (2x) + C$

b. $\dfrac{1}{2} \sin (x^2) + C$

c. ∞

d. $\dfrac{1}{24}$

5. $\theta = \dfrac{\pi}{6}, \dfrac{\pi}{2}, \dfrac{5\pi}{6}$

6. $\sqrt{2} - 1$

7. $\sin^2 x + \cos^2 x = 1$
$\left(\dfrac{1}{\sin^2 x}\right)(\sin^2 x + \cos^2 x) = \left(\dfrac{1}{\sin^2 x}\right)$
$1 + \cot^2 x = \csc^2 x$

8. a. $\approx -4.92 \dfrac{\text{thousand units}}{\text{month}}$; decreasing
b. Largest sales: 25.9 thousand units; February and May
Smallest sales: 6.5 thousand units; October

9. a. $3750
b. $1500 per unit, $\approx$ $1871 per unit

10. a. $k \approx 8.38$
b.

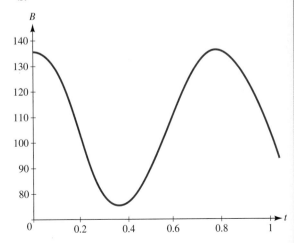

c. Systolic: 136; diastolic: 74
d. 105

CHAPTER 11   Review Exercises

1. a. $120° = \dfrac{2\pi}{3}$ radians

b. $-225° = -\dfrac{5\pi}{4}$ radians

3. a. $\dfrac{45°}{\pi} \approx 14.32°$

b. $\dfrac{180°}{\pi} \approx 57.30°$

c. $-\dfrac{270°}{\pi} \approx 85.94°$

5. $\dfrac{4}{3}$

7. $\dfrac{\pi}{2}, \dfrac{3\pi}{2}$

9. $\dfrac{\pi}{6}, \dfrac{5\pi}{6}, \dfrac{7\pi}{6}, \dfrac{11\pi}{6}$

11. $\cos(2\theta) = \cos^2\theta - \sin^2\theta$

$\Rightarrow \cos(2\theta) = \cos^2\theta - (1 - \cos^2\theta)$

$\Rightarrow \cos(2\theta) = -1 + 2\cos^2\theta$

$\Rightarrow \cos^2\theta = \dfrac{1}{2}(1 + \cos 2\theta);$

$\sin^2\theta = 1 - \cos^2\theta$

$= 1 - \dfrac{1}{2}(1 + \cos(2\theta))$

$= \dfrac{1}{2}(1 - \cos(2\theta))$

13. a. $\cos\left(\dfrac{\pi}{2} + \theta\right) = \cos\left(\dfrac{\pi}{2}\right)\cos\theta - \sin\left(\dfrac{\pi}{2}\right)\sin\theta$

$= -\sin\theta$

$\sin\left(\dfrac{\pi}{2} + \theta\right) = \sin\left(\dfrac{\pi}{2}\right)\cos\theta + \sin\theta\cos\left(\dfrac{\pi}{2}\right)$

$= \cos\theta$

b.

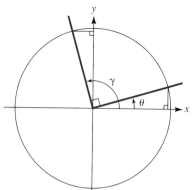

Adding 90° to θ effectively interchanges the x and y axes, so that if $\gamma = \theta + \dfrac{\pi}{2}$, then

$|\cos\gamma| = |\sin\theta|$ and
$|\sin\gamma| = |\cos\theta|.$

By quadrant,

sin > 0	sin > 0
cos < 0	cos > 0
sin < 0	sin < 0
cos < 0	cos > 0

Adding 90° to θ moves the angle to the next quadrant. We see that $\cos\left(\dfrac{\pi}{2} + \theta\right) = -\sin\theta$ and

$\sin\left(\dfrac{\pi}{2} + \theta\right) = \cos\theta.$

15. $f'(x) = -2\cos x \sin x$

17. $f'(x) = 6\tan(3x + 1)\sec^2(3x + 1)$

19. $f'(x) = -\dfrac{2\sin x}{\cos x}$

21. $-\dfrac{1}{2}\cos 2t + C$

23. $\dfrac{1}{2}\sin^2 x + C$

25. $\ln|\tan t| + C$

27. Angle $A = -52°\ 14'\ 20''$, $\theta = 232°\ 14'\ 20''$

29. $x = 0.436, 1.468$

31. $\dfrac{2}{3}\pi^2 - \dfrac{3\pi}{4}$

33. a. $p = 25$
$b = 27$
$d = 11$
$a = 33$

b.

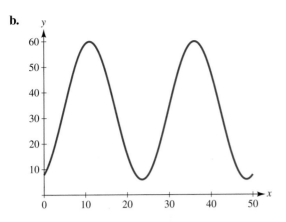

35. $C = (3000 - 900 \cot \theta)\, 4 + (900 \csc \theta)\, 5$

$C' = -900\,(-\csc^2 \theta)\, 4 + 900\,(-\csc \theta \cot \theta)\, 5$

$\quad = 900 \csc^2 \theta (4 - 5 \cos \theta)$

$C' = 0 \Rightarrow \cos \theta = \dfrac{4}{5}$ or $\csc \theta = 0$

but $\csc \theta \neq 0$

37. a.

t	0.125	0.25	0.325	0.5
$p(t)$	70 570	548 317	255 673	500

b.

t	0.125	0.25	0.325	0.5
$p'(t)$	0.2195×10^7	0	-0.5105×10^7	$-21.991.15$

c.

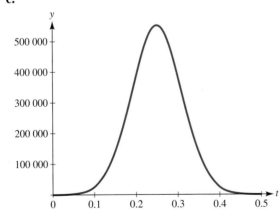

39. a. 1st year 3300
   2nd year 3300

b. Largest month is 12; smallest month is 1.

41. a.

Jan. 1	Mar. 15	June 21
9.11	11.12	15.20

b. Maximum on 186th day, July 4
   Minimum on 3rd day, January 3

c. 12.2 hours

43. -0.12 radians/s

45. a. $L = \dfrac{4.25}{\cos \theta \sin^2 \theta}$

b. L minimized $\theta = 0.9553$;
   minimum length $= 11.04$ centimetres

c. Area $= \dfrac{1}{2} L \sin \theta = \dfrac{1}{8} 8.5^2 \sec \theta (\csc \theta)^3$;
   area is minimized when $\theta = \dfrac{\pi}{3}$;
   minimum area is 27.81 square centimetres.

47. a. $A = 46$; $B = 24$; $C = 431$; $D = 204$; the cycle for both species is 24 years.

b. 431 rabbits; 431 rabbits

c. 46 foxes; 46 foxes

d.

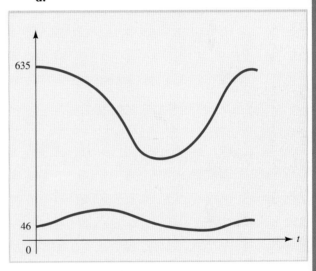

49. $y = -\cos x + \dfrac{\sin x}{x}$

51. a. 0.828155   **b.** 0.828116

53. $-\ln |\cos x| + C$

55. $\dfrac{3\sqrt{3}}{8} \approx 0.65$

57. a. $\dfrac{du}{dt} = (-c^2 k^2 e^{-c^2 k^2 t}) \sin kx = -c^2 k^2 e^{-c^2 k^2 t} \sin kx$

   $\dfrac{du}{dt} = e^{-c^2 k^2 t}(k \cos kx)$

   $\dfrac{d^2 u}{dx^2} = e^{-c^2 k^2 t}(-k^2 \sin kx) = -k^2 e^{-c^2 k^2 t} \sin kx$

   $\Rightarrow u_t = c^2 u_{xx}$

b. Writing exercise; responses will vary.

59. 1.848; at $\left(\dfrac{-\pi}{8}, \dfrac{\pi}{8}\right)$

Appendix Section A.1

1. $1 < x < 5$

3. $x > -5$

5.

7.

9. 4

11. 5

13. $-3 \le x \le 3$

15. $-6 \le x \le -2$

17. $x \le -7$ or $x \ge 3$

19. 125

21. 4

23. 4

25. $\dfrac{1}{2}$

27. $\dfrac{1}{2}$

29. $\dfrac{1}{4}$

31. 2

33. $\dfrac{1}{4}$

35. $n = 10$

37. $n = 1$

39. $n = 4$

41. $n = \dfrac{13}{5}$

43. $a^5 b^8 c^8$

45. $\dfrac{a^8 c^{12}}{b^4}$

47. $\dfrac{a^{10}}{b^2 c^{14}}$

49. $\dfrac{a^{18} b^{12}}{c^6}$

51. $\dfrac{1}{a^3 b^5 c^3} + \dfrac{1}{a^4 b c^3} + \dfrac{1}{a b c^4}$

53. $a^{-1} b^2 + a^2 b$

55. -2

57. $1{,}350 \sqrt[3]{900}$

59. $38\sqrt{2}$

61. $9\sqrt{6}$

63. $a^3 b^4 c^7$

65. $\dfrac{5b}{7a}$

67. $\dfrac{a\sqrt[3]{a^{2b^2}}}{b^3 c^3}$

69. $\dfrac{1}{a^2 b c^6}$

71. $\dfrac{a^2 c^3}{b^4}$

73. $a - \sqrt{b}$

75. $\dfrac{a^5 b^3 \sqrt[3]{a^2 c}}{c^4}$

77. $\dfrac{1}{5(\sqrt{3} + \sqrt{2})}$

79. $\dfrac{7(3 + \sqrt{3})}{6}$

81. $\dfrac{1}{3(\sqrt{5} - 2)}$

83. $\dfrac{5(\sqrt{5} - 1)}{4}$

85. $\sqrt{x+h} - \sqrt{x} = \dfrac{(\sqrt{x+h} - \sqrt{x})(\sqrt{x+h} + \sqrt{x})}{\sqrt{x+h} + \sqrt{x}}$

$$= \dfrac{x + h - x}{\sqrt{x+h} + \sqrt{x}}$$

$$= \dfrac{h}{\sqrt{x+h} + \sqrt{x}}$$

87. **a.** Surface area is approximately 5.212×10^8 km^2; mass of the atmosphere is 5.212×10^{18} kg.

     b. 127 400 years

Appendix Section A.2

1. $3x^2 - 27x$

3. $x^2 - 5x - 14$

5. $-6x^2 + 26x - 28$

7. $x^3 + x^2 - 5x + 3$

9. $x^5 - 3x^4 - x^3 + 13x^2 - 18x + 8$

11. $\dfrac{2x^2 + 3x + 9}{x^2 - 9}$

13. $\dfrac{x^2}{x^2 + 2x - 3}$

15. $\dfrac{3}{2x^2 - 7x - 15}$

17. $-\dfrac{x + 10}{x^2 + x - 2}$

19. $\dfrac{7}{x^2 + 7x + 12}$

21. $\dfrac{-x + 3}{x + 3}$

23. -2

25. $\dfrac{x}{3x - 1}$

27. $\dfrac{x^2 + x - 1}{3x - 1}$

29. $(x + 2)(x - 1)$

31. $(x + 3)(x - 4)$

33. $(x - 1)^2$

35. $(4x + 5)(4x - 5)$

37. $(x - 1)(x^2 + x + 1)$

39. $x^5(x + 1)(x - 1)$

41. $2x(x - 5)(x + 1)$

43. $(x + 4)(x - 3)$

45. $(2x + 5)(x - 3)$

47. $(x + 2)(x - 9)$

49. $2(2x + 1)(7x - 3)$

51. $x(x + 5)(x - 3)$

53. $(x + 3)(x^2 - 3x + 9)$

55. $x^2(x + 1)(x^2 - x + 1)$

57. $(3x + 1)(x + 2)^2$

59. $x = 4; x = -2$

61. $x = -5$

63. $x = 4; x = -4$

65. $x = -\dfrac{1}{2}; x = -1$

67. $x = -\dfrac{3}{2}$

69. $x = 1; x = -5$

71. $x = 1; x = -2$

73. $x = -1$

75. $x = 1; x = -3$

77. $x = \dfrac{1}{3}; x = \dfrac{3}{5}$

79. No real solutions

81. $x = \dfrac{-17 + \sqrt{385}}{12}; x = \dfrac{-17 - \sqrt{385}}{12}$

83. $x = -\dfrac{1}{2}; x = -1$

85. No real solutions

87. $x = -\dfrac{3}{2}$

89. $x = 3; y = 2$

91. $x = 4, y = 2$

93. $x = -7; y = -5$ and $x = 1, y = -1$

Appendix Section A.3

1. 0

3. $\dfrac{1}{2}$

5. $-\dfrac{1}{3}$

7. 0

9. 0

11. 0

13. 0

15. e^2

Appendix Section A.4

1. 34

3. 0

5. $\displaystyle\sum_{j=1}^{6} \dfrac{1}{j}$

7. $\displaystyle\sum_{j=1}^{6} 2x_j$

9. $\displaystyle\sum_{j=1}^{8} (-1)^{j+1} j$

Appendix Review Exercises

1. $-2 \le x < 3$

3.

5.

7. 3

9. $2 \le x \le 4$

11. 243

13. 4

15. $16\sqrt[4]{2}$

16. 73

19. $\dfrac{3}{2}$

21. $n = \dfrac{7}{18}$

23. $n = -1$

25. 21

27. 95

29. $1 + \displaystyle\sum_{k=2}^{7} \dfrac{(-1)^k}{k}$

31. $x^2(x + 3)(x - 3)$

33. $x^4(x^6 + 4)(x^3 + 2)(x^3 - 2)$

35. $x(x - 1)^2$

37. $(x + 5)(x - 3)$

39. $(2x + 3)^2$

41. $(x + 1)(x - 1)(x + 3)$

43. $x = -4; x = 1$

45. $x = -7$

47. $x = -1; x = 2$

49. $x = -\dfrac{3}{7}; x = \dfrac{1}{2}$

51. No real solutions

53. $x = -2; x = \dfrac{1}{3}$

55. $x = -2; y = 1$

57. $x = 1, y = 2$ and $x = 15, y = -26$

59. $\dfrac{2}{7}$

61. 0

63. 1

INDEX

Index of Selected Applications